W9-AKU-797

BIOLOGY
THE STUDY OF LIFE

Prentice Hall dedicates this book to the Bengal tiger. This animal was placed
on the endangered species list twenty years ago and is still listed there. The cover
image is of two Bengal tigers, one, a rare white individual.

BIOLOGY
THE STUDY OF LIFE

PRENTICE HALL

Fifth Edition

William D. Schraer
Chairperson, Science Department
Middletown High School
Middletown, NY

Herbert J. Stoltze
Professor of Biology
Northeastern Illinois University
Chicago, IL

Needham, Massachusetts

Englewood Cliffs, New Jersey

Credits

Staff Credits

Editorial Director:	Roland Boucher
Senior Editor:	Lois B. Arnold
Project Editor:	Natania Mlawer
Editors:	Thomas M. Frado, Rosemary E. Previte
Art Director and Cover Art Direction:	L. Christopher Valente
Book Design/Production Coordinator:	Jonathan Pollard, Stuart Wallace
Book Design/Photography and Illustration Coordinator:	Betty Fiora
Production/Manufacturing Coordinator:	Bill Wood, Leanne Cordischi, Valerie Rhoades
Photo Researcher:	Russell Lappa
Marketing Director:	Arthur C. Germano
Associate Product Manager:	Michael D. Buckley

Outside Credits

Editorial/Writing Services:	Barbara Branca, Lee H. Bruno, Camillo Cimis, Ann L. Collins, James A. Craig, Joseph Degnan, Mary Jo Diem, Lisa P. Easley, Patricia A. Mitchell, Gene Rogers, Dotty Burstein, Dorothy Marshall
Editorial Production Assistant:	Helen M. Webber
Art and Design Assistants:	Daniel Ashton, James Brown, Carolyn Champigny, Pamela Donahue, Hannah Fogarty, David Gerratt, Marie McAdam, Eve Melnechuk, Helen Reebenacker, Dayle Silverman, Kevin W. Spotts, Melodie Wertelet, Linda Dana Willis, Evelyn Young
Cover Design:	Martucci Studio
Book Design:	Martucci Studio
Photo Researcher:	Laurel Anderson/Photosynthesis
Illustrators:	Edmond S. Alexander/Alexander and Turner, Sally J. Bensusen/Visual Science Studio, Kent Boughton, Carmella M. Clifford, Barbara Cousins, Paul Foti/Boston Graphics, Inc., Howard Friedman, Andrew Grivas, Steven J. Harrison/Fine Line Studio, Jackie Heda, Floyd E. Hosmer, Keith Kasnot, Elizabeth McClelland, Robert Margulies/Margulies Medical Art, Martucci Studio/Jerry Malone, Fran Milner, Sanderson Associates, Lois Sloan, Cynthia Turner/Alexander and Turner

Copyright © 1993 by Prentice-Hall, Inc.

All rights reserved. No part of the material protected by this copyright notice may be reproduced or utilized in any form or by any means, electronic or mechanical, including photocopying and recording, or by any information storage or retrieval system without written permission from the copyright owner.

Printed in the United States of America

ISBN 0-13-085390-9

3 4 5 6 7 8 9 98 97 96 95 94 93

Acknowledgments

Program Consultants

Critical and Creative Thinking
Robert J. Swartz
Codirector of Critical and Creative Thinking Program
University of Massachusetts, Boston, MA

Sarah Carolyn Dolde Duff
Baltimore City Public Schools, Baltimore, MD

Problem Solving
Sarah Carolyn Dolde Duff
Baltimore City Public Schools, Baltimore, MD

Problem Finding
Eileen Jay
Harvard University School of Education, Cambridge, MA

Study Skills
Gary I. Krasnow
Educational Consultant
Pacific Grove, CA

Careers in Biology
Cara Adler
Massachusetts Medical Society, Waltham, MA

Cooperative Learning
Deborah S. Haber
Educational Consultant, Health and Science
Framingham, MA

Science Content

Donald W. Deters (Respiration)
Department of Biological Sciences
Bowling Green State University
Bowling Green, OH

Gary B. Ellis (Vertebrates)
Office of Technology Assessment for the U.S. Congress
600 Pennsylvania Avenue, S.E.
Washington, D.C.

Kenneth R. Miller (Photosynthesis)
Biology Department
Brown University, Providence, RI

M. V. Parthasarathy (Plant Structure and Function)
Biology Department
Cornell University, Ithaca, NY

Misconceptions in Biology
Mary Ann Brearton
Maryland State Department of Education
Baltimore, MD

Ann Benbow
American Chemical Society
Washington, D.C.

Themes in Biology
Bonnie A. Branch
Cajon High School
San Bernardino, CA

Concept Laboratories
Susan Offner
Milton High School, Milton, MA

Laboratory Investigations
Susan Stone Plati
Brookline High School, Brookline, MA

Testing
Walter B. MacDonald
Trenton State College, Trenton, NJ

Graphic Organization
Paula Borinsky
Science Teaching Center
University of Maryland, College Park, MD

Computers in Biology
Donald P. Kelley
Educational Software Consultant, Pedalogic
South Burlington, VT

Irwin Rubenstein (The Cell)
Department of Genetics and Cell Biology
University of Minnesota, Minneapolis, MN

Charles F. Stevens (Nervous and Endocrine Systems)
Section of Molecular Neurobiology
Yale University Medical School, New Haven, CT

Daryl Sweeney (Invertebrates)
Department of Biology
University of Illinois, Champaign-Urbana, IL

Marjorie B. Zucker (Circulatory Systems)
Pathology Department
New York University School of Medicine, New York, NY

Many people contributed their ideas and services in the preparation
of the current and past editions of *Biology: The Study of Life.*
Among them are many biology teachers whose names are listed below.
Their contributions are gratefully acknowledged.

Cathy Banks
Wheeler High School, Marietta, Georgia

Diane G. Bemis
Watertown High School, Watertown, Massachusetts

Cathy Bennett
Dunbar High School, Dunbar, West Virginia

Carol J. Bershad
formerly, Learning for Life, Boston, Massachusetts

Warren Bjork
Glenbrook South High School, Glenview, Illinois

Carole Brenkacz
West Seneca East Senior High School, West Seneca,
New York

Lornie D. Bullerwell
Dedham High School, Dedham, Massachusetts

Barbara A. Cauchon
Brookline High School, Brookline, Massachusetts

Brenda L. Dorsey
York Community High School, Elmhurst, Illinois

Michael J. Flanagan
Dedham High School, Dedham, Massachusetts

Karen L. Fout
Fenwick High School, Tiffin, Ohio

Anne R. Fraulo
Career High School, New Haven, Connecticut

Richard L. Gaume
Plain Local Schools, Canton, Ohio

Norm Grimes
Columbian High School, Tiffin, Ohio

Deborah S. Haber
formerly, Watertown High School, Watertown,
Massachusetts

Emiel Hamberlin
Du Sable High School, Chicago, Illinois

Sister M. Francis Hopcus
Pomona Catholic High School, Pomona, California

Mic Jaeger
East Union High School, Manteca, California

Nevin Longenecker
John Adams High School, South Bend, Indiana

Ernest Nichol
Newton North High School, Newton, Massachusetts

Susan Offner
Milton High School, Milton, Massachusetts

Susan Stone Plati
Brookline High School, Brookline, Massachusetts

Harold Pratt
Jefferson County Schools, Lakewood, Colorado

Deborah A. Sandall
Pennichuck Junior High School, Nashua, New Hampshire

Hazel M. Schroder
Shrewsbury High School, Shrewsbury, Massachusetts

Helen Louis Shafer
Science and Technology Magnet School, Dallas, Texas

Raymond D. Spencer
Somerville High School, Somerville, Massachusetts

Harlow B. Swartout
Woodstock High School, Woodstock, Illinois

Robert C. Wallace
formerly, Reavis High School, Burbank, Illinois

Frederick J. Watson
Silver Lake Regional High School, Pembroke, Massachusetts

Harold A. Wiper
Newton North High School, Newton, Massachusetts

Melanie Wojtulewicz
Whitney M. Young Magnet High School, Chicago, Illinois

Russell G. Wright
Montgomery County Public Schools, Rockville, Maryland

Michael A. Zunno
West Hempstead High School, West Hempstead, New York

Contents

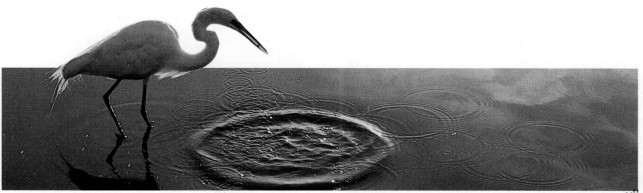

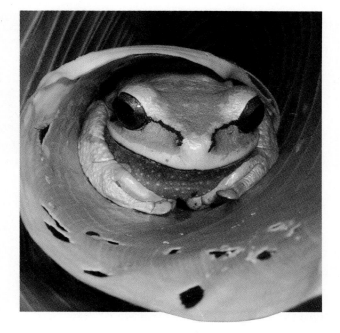

Ecology Unit **8**

Laboratory Investigations

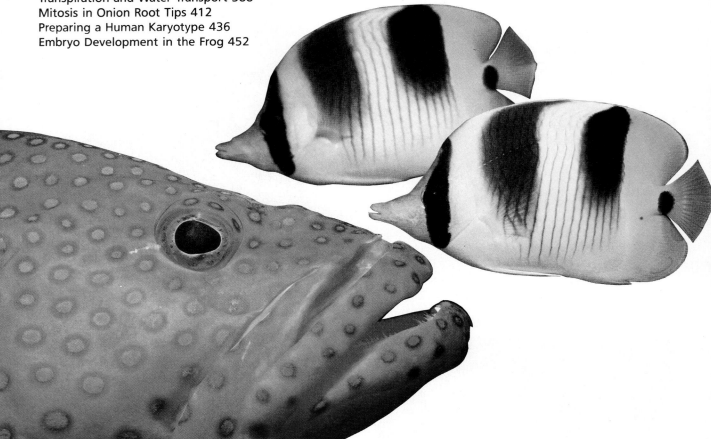

FEATURES

Critical Thinking in Biology

Concept Laboratories

Biology and Problem Solving

Can You Explain This?

Careers in Action

Computer Projects

Biology and You

Science, Technology, and Society

Introduction to Biology

Life is everywhere in the sea—fish and living coral and millions of one-celled organisms that you cannot see. Life is everywhere on land, as well. From deserts to snow-capped mountains, incredibly varied forms of life exist. Each organism is well adapted for carrying out its life processes. In this unit, you will consider the traits all organisms have in common despite their amazing diversity. You will see that at the cellular and molecular level the similarities are even more striking.

▲ The odd-looking starfish has characteristics in common with all living things.

The Nature of Life

1

A starfish moves along living coral, bathed by a salty sea. Fish flash exotic colors, and a crab crawls on the ocean floor. Somewhere far off, a lizard scurries across desert sand, and a sidewinder snake lives up to its name. In still another setting, sidewalk cracks go green with life while ants make pie-shaped mounds. Life is everywhere, in forms incredibly diverse. But what is life? In this chapter, you will learn how biologists answer that difficult question.

Guide for Reading

Key words: biology, organism, homeostasis, nutrition, transport, respiration, growth, reproduction, metabolism

Questions to think about:

📖 What are the characteristics of life?

📖 What are the functions of each of the life processes?

📖 What is homeostasis?

1-1 Understanding Life

Section Objectives:

- *Identify* some unifying themes in biology.
- *List* nine general characteristics that distinguish living from nonliving things.
- **Laboratory Investigation:** *Observe* and *compare* living and nonliving specimens (p. 9).

The word **biology** is easy to define. It is the study of living things. Although biology is unique, it is not an isolated science. You will discover, for example, that information and ideas from chemistry, physics, and geology appear throughout your biology text. Scientists think that biology is linked to other sciences by major unifying themes, or big ideas.

Unifying Themes

Energy Organisms require energy to grow and to reproduce. Just as a car stops when there is no gasoline to fuel its engine, an organism dies when no energy is available to carry out its life processes.

Evolution In biology, evolution, or changes in living things through time, explains the inherited similarities as well as the diversity in all forms of life. Evolution, which is explained in detail in Unit 6, is the major unifying theme of biology.

Patterns of Change Scientists try to make sense of change by

▲ **Figure 1–1**

An Organism. This horned puffin shares many characteristics with all other living things.

▲ **Figure 1–2**

Signs of Life in a Forest. A forest during the growing season (left) shows obvious signs of life. During winter (right), the same forest may appear to be dead.

noting cycles, trends, or random events that occur before or during change. For example, all living things, including humans, have a life cycle.

Scale and Structure Living things are described in terms of their structures and levels of organization. At one level of organization, for example, you might describe the main structures of a green plant—its roots, stems, and leaves; at another level, you might describe the cells in each of these structures.

Systems and Interactions Examples of biological systems include ecosystems, in which communities of organisms interact with the environment, and human body systems, in which organs interact to carry on life functions.

Stability Biological systems tend to achieve a stable equilibrium. A good example of this kind of stability is your body temperature, which tends to remain constant.

Environment All the external factors that make up an organism's surroundings make up its environment. Organisms interact with their environment and have an impact on it.

Unity and Diversity Although each life form, from the ameba to the great blue whale, differs in appearance and organization (diversity), it also displays certain similarities to other organisms (unity). We begin the study of biology by talking about the most basic similarities: the general characteristics of life.

Characteristics of Life

What do you mean when you say that something is alive? Look at the photo on the left in Figure 1-2. Do you see a dead tree? What makes you think it is dead? Now look at the photo on the right. Are all these trees dead? Although they look like the dead tree, they are probably not dead. In the spring, tiny buds at the ends of twigs will grow into leaves.

The growth of new leaves is a sign of life. What other signs of life can you think of? Biologists have not been able to agree on a simple definition of life. But they have agreed on what the "signs of life" are. Taken together, these signs can become a definition of life.

Each living thing is called an **organism** (OR guh nih zum). Organisms have all the following characteristics: ▪ Living things are highly organized and contain many complex chemical substances. ▪ Living things are made up of one or more cells, which are the smallest units that can be said to be alive. ▪ Living things use energy. ▪ Living things have a definite form and a limited size. ▪ Living things have a limited life span. ▪ Living things grow. ▪ Living things respond to changes in the environment. ▪ Living things are able to reproduce. ▪ Groups of living things evolve, or change, over time.

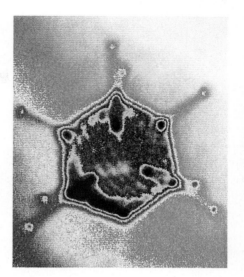

Non-living objects may show one or a few of these characteristics, but they never show all of them. In some borderline cases, however, it may not be clear whether an object is living or nonliving. One example is the virus. Viruses can be stored like chemicals in a bottle but, when inside living cells, can reproduce. Even so, viruses do not exhibit most of the other characteristics of living things.

1-1 Section Review

1. Define the term *organism*.
2. What are three unifying themes in biology?
3. Name three characteristics of life.

Critical Thinking

4. Look back over the list of the characteristics of life. Which characteristics are found *only* in living things? Which are found in both living and nonliving things? (Classifying)

1-2 Life Processes

Section Objectives:

▪ *Name* and *define* eight general processes by which the life of an organism is maintained.
▪ *Define* the term *metabolism*.

Living things carry out many different kinds of processes. Some of these processes, such as growth, reproduction, and the use of energy, already have been discussed as basic characteristics of life. These, of course, are not the only processes carried out by living things. Biologists have identified a number of other general processes that relate to the functioning of living things. Many of these life processes are necessary for maintaining a fairly constant environment within an organism in spite of its constantly changing external environment. The condition of a constant internal environment is known as **homeostasis** (hoh mee oh STAY sis).

Nutrition

Every organism takes materials from its external environment and changes them into forms it can use. This activity is called **nutrition. Nutrients** (NOO tree unts) are the substances that an organism needs for energy, growth, repair, or maintenance.

There are only two basic types of nutrition. In one type, the organism can produce complex nutrients from simple substances found in the environment. All green plants and some bacteria and other one-celled organisms are able to make their own nutrients in this way.

In the second type, organisms that cannot make their own nutrients obtain them ready-made from the environment. Animals, for example, get their nutrients by eating other organisms in their environment.

The taking in of food from the environment is called **ingestion.** Usually, the nutrients in food are not in forms that an organism can use directly. They are too complex chemically, and the organism must break them into simpler forms. The breakdown of complex food materials into simpler forms that an organism can use is called **digestion.**

▲ **Figure 1–4**
Nutrition in Animals. All animals must obtain food by ingesting other organisms.

Transport

The process by which substances enter and leave cells and become distributed within the cells is known as **transport.** In the smallest and simplest organisms, materials are exchanged directly with the external environment. Usable materials enter the cells directly from the environment; waste materials pass from cells directly into the environment.

In larger, multicellular organisms, however, most cells are not in direct contact with the external environment. In many animals, for example, a *circulatory system* transports materials to, and wastes away from, the cells of the organism. The fluid, or blood, of the circulatory system is kept in motion, distributing these materials among the cells of the organism. In plants, specialized conducting structures transport substances from the roots and leaves to all parts of the plant.

Respiration

All life processes require a constant supply of energy. Organisms obtain their energy by releasing the chemical energy stored in nutrients. The process of releasing chemical energy is called **respiration** (res puh RAY shun).

Respiration involves a complex series of chemical reactions. In one type of respiration, sugar or another food substance is broken down to produce water and carbon dioxide. This process requires oxygen from the air and is known as *aerobic respiration.* Some organisms break down food without using oxygen. This is called *anaerobic respiration.*

Synthesis and Assimilation

Organisms are able to combine simple substances chemically to form more complex substances. This process is called **synthesis** (SIN thuh sis). Usually, in animals, the substances used in synthesis are the products of digestion.

Synthesis produces materials that can become part of the structure of an organism. In this way, the organism can repair or replace worn-out parts. These materials also allow the organism to grow. The incorporation of materials into the organism's body is called **assimilation** (uh sim uh LAY shun).

Growth

The process by which living organisms increase in size is called **growth.** It is one result of the assimilation of nutrients. In one-celled organisms, growth is simply an increase in the size of the cell. In organisms made up of many cells, growth is usually the result of an increase in both the number and the size of cells. Growth in multicellular organisms is accompanied by *cellular specialization.* This process involves different cells becoming specialized for different functions. In animals, growth usually follows a particular pattern and ends after a certain period of time. Some plants, however, continue growing indefinitely.

Excretion

Every organism produces waste substances that it cannot use and that may be harmful if accumulated in the body. These wastes are the products of many of the chemical reactions that occur within cells. The removal of these wastes from the organism's body is called **excretion.**

Regulation

All the activities that help to maintain an organism's homeostasis make up the process of **regulation.** In animals, systems such as the digestive, transport, excretory, nervous, and endocrine (EN duh krin) systems are involved in some part of the process of regulation. Each of these systems contributes to maintaining homeostasis. The digestive system allows an animal to provide nutrients to its body. The transport system ensures that all areas of the body are constantly supplied with necessary materials. The excretory system removes wastes that are produced in the body.

▲ **Figure 1–5**
Growth. This tiny sequoia seedling (top) can grow into a mature tree with a height of over 100 meters (bottom).

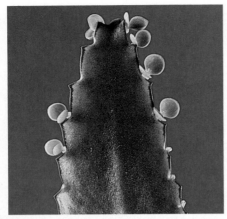

▲ **Figure 1–6**
Reproduction. In one type of asexual reproduction, the plantlets on the leaf's edge (top) fall off and grow into new plants. Sexual reproduction (bottom) involves two parents.

The nervous and endocrine systems regulate these systems. The nervous system carries messages—nerve impulses—throughout the body through a network of specialized cells. The endocrine system is made up of a number of organs that release chemicals, called *hormones,* into the bloodstream. Hormones act as chemical messengers. Both nerve impulses and hormones can bring about changes in the organism in response to changes in either the internal or the external environment.

Plants do not have nervous systems, but they do have parts that produce hormones. These hormones allow a plant to respond to various changes in its environment.

Reproduction

Reproduction is the process by which living things produce new organisms of their own kind. Reproduction is not necessary for the continued life of a single organism. However, it is necessary for the continued existence of that kind of organism.

There are two types of reproduction—**asexual** (ay SEK shuh wul), or *vegetative,* **reproduction** and **sexual reproduction** (see Figure 1–6). In asexual reproduction, a single individual produces offspring that are identical to that parent. In sexual reproduction, there are two parents, and the offspring are not identical to either parent.

Metabolism

All the chemical reactions occurring within the cells of an organism are called its **metabolism** (muh TAB uh liz um). Metabolism includes processes that build complex substances from simpler ones and processes that break down complex substances into simpler ones. Metabolism also involves the continuous release and use of energy. Many biologists consider metabolic activity to be the single most important characteristic of life.

1-2 Section Review

1. What life process involves obtaining material and changing it into useful forms?
2. Name the process by which organisms release chemical energy from nutrients.
3. Define the term *growth.*
4. What is homeostasis?

Critical Thinking

5. How is the transport system in your community (roads, sidewalks, etc.) similar to the transport system in your body? How do these systems differ? (*Comparing and Contrasting*)

Laboratory Investigation

Comparing Living and Nonliving Things

It is difficult to formulate a concise definition of life. But, all living things share certain characteristics. In this investigation, you will learn what these characteristics are. You will examine a variety of specimens and classify them as living or nonliving, based on the presence or absence of these characteristics.

Objectives

- *Recognize* the characteristics of living things.
- *Classify* things as living or nonliving.

Materials

Various materials provided by your teacher, such as:

chicken egg	fish	snail	water
millipede	elodea	earthworm	spider
flashlight	plant	ant	crystal
burning candle	moss	rock	match
fern	wood	tadpole	seeds

ameba (set up for observation on a microscope stage)

Procedure 🔥 🐁 🌸 ✋ *

A. Copy the chart shown on the bottom of the page. Along the top of this chart are listed the characteristics that all living things must have. Remember that a nonliving thing may have a few of these characteristics but not all.

B. Select a specimen from the demonstration table to examine. Study it carefully. Write the name of the specimen in the first column of your chart. Ask yourself if this specimen possesses each of the listed characteristics. Record your answers as yes, no, or CD (cannot determine). Base your answers on your immediate observations, prior knowledge, or information you can obtain from reference books.

C. Copy the chart shown below. Write the name of the specimen you are examining in the first column. In the second column, identify the specimen as living or nonliving. In the third column, explain the reasoning for your choice.

Specimen	Living or Nonliving	Explanation

D. Repeat steps A through C for each specimen on the demonstration table.

Analysis and Conclusions

1. Which specimens did you classify as living? Which did you classify as nonliving?

2. Were there any specimens that you had difficulty placing in either category? Explain your answer.

3. What characteristics of life were most difficult to evaluate? Explain your answer.

4. One purpose of space exploration is to find out if life exists anywhere else in the universe. If there were life on another planet, would it have all of the characteristics of living things identified in your chart? Why or why not?

5. Determine which of the following statements are true: (a) Living things may have some but not all of the characteristics identified in the chart; (b) Nonliving things may have some but not all of these characteristics; (c) Living things *must* have all of these characteristics.

Specimen	Highly Organized	One or More Cells	Uses Energy	Grows	Limited Form/Size	Life Span	Responds to Changes	Reproduces

*For detailed information on safety symbols, see Appendix B.

Chapter Review

Study Outline

1–1 Understanding Life

■ Biology is the study of living things.

■ Major unifying themes link biology to other sciences.

■ All living things share certain characteristics that distinguish them from nonliving things.

■ Some things, such as viruses, are difficult to classify as either living or nonliving.

1–2 Life Processes

■ The function of many life processes is to maintain homeostasis, a constant internal environment.

■ All living things carry on certain life processes that are characteristic of life. These processes include nutrition, transport, respiration, synthesis and assimilation, growth, excretion, regulation, and reproduction.

■ Metabolism includes all the chemical reactions occurring within the cells of an organism.

Vocabulary Review

biology (3)	synthesis (7)
organism (4)	assimilation (7)
homeostasis (6)	growth (7)
nutrition (6)	excretion (7)
nutrients (6)	regulation (7)
ingestion (6)	reproduction (8)
digestion (6)	asexual reproduction (8)
transport (6)	sexual reproduction (8)
respiration (6)	metabolism (8)

A. Sentence Completion—Fill in the vocabulary term that best completes each statement.

1. Organisms need _____ for energy, growth, repair, or maintenance.

2. In _____, a single organism produces offspring identical to itself.

3. Organisms release chemical energy by the process of _____.

4. Substances are incorporated into the body by the process of _____.

5. _____ is the removal of cellular wastes from the body.

6. Simple substances are combined chemically into more complex substances by _____.

7. _____ is the process by which substances enter and leave cells and are distributed among the cells of an organism.

8. Individual living things are referred to as _____.

B. Matching—Select the vocabulary term that best matches each definition.

9. The process of maintaining homeostasis

10. The study of living things

11. An increase in size

12. To take in food from the environment

13. The process by which two parents produce offspring

14. The condition of a constant internal environment

15. The chemical reactions occurring in the cells of an organism

Content Review

16. List the nine characteristics of living things.

17. Why are viruses difficult to place in the category of living things?

18. How does nutrition in green plants differ from nutrition in animals?

19. Explain the relationship between the processes of ingestion and digestion.

20. Explain how nutrients are related to respiration.

21. How does aerobic respiration differ from anaerobic respiration?

22. In animals, what materials are used in synthesis?

23. Discuss the relationship between growth and assimilation.

24. How does growth occur in multicellular organisms?

25. How does the excretory system help to maintain homeostasis in the body?

26. Describe the difference in function between the nervous system and the endocrine system.

27. Compare the appearance of a parent with its offspring produced by asexual reproduction.

28. Describe the two opposing processes that occur in metabolism.

Graphic Organizing

For information on graphic organizers, see Appendix G at the back of this text.

29. **Word Map:** Copy the incomplete word map for *metabolism* onto a piece of paper. Then complete it, using the word map for *biology* as a model.

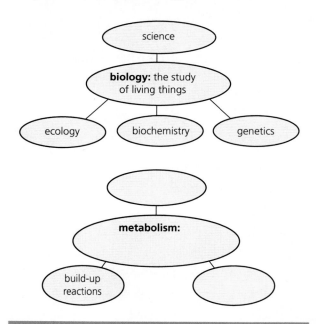

Critical Thinking

30. List four ways in which an automobile is like a living thing. (*Comparing and Contrasting*)

31. List four ways in which an automobile is **not** like a living thing. (*Comparing and Contrasting*)

32. Would you classify a virus as living or nonliving? Explain the reasons for your classification. (*Classifying*)

Creative Thinking

33. A "computer virus" is a program that can reproduce itself inside a computer. In time, it can grow and evolve. Is a computer virus alive? Why or why not?

34. The unmanned Viking space probe that was sent to Mars checked for signs of life on the planet. How could such a probe confirm the existence of life?

Problem Solving

35. Design a simple experiment to determine if seeds that have been stored for many years are viable.

36. Baby chickens require a constant source of food. As chicks grow, more energy is needed for daily activities. The following table gives the grams of food eaten by a chick over a five day period.

Number of Days	Food Eaten (grams)
0	0.0
1	1.0
2	3.2
3	6.5
4	10.6
5	15.4

On a separate piece of paper construct a line graph using this data. Based on your graph, predict the amount of grain that will be eaten by the chick on the sixth and seventh day.

Projects

37. Prepare a collection of various small organisms. Collect and place each one in a separate clean jar with several leaves of vegetation. Moisten the leaves with a few drops of water every day. Poke holes in the top of each jar for ventilation. Refer to field guides in your school library to classify the organisms. Then, label each jar with the name of the organism. Ask your teacher if you may display the collection in the classroom.

38. National park rangers have a variety of duties, including protecting inhabitants of the park and teaching visitors about the park and its many plants and animals. In a written report, explain why a national park ranger must be familiar with the characteristics of life and the life processes.

▲ With a leg band and a transmitter, scientists monitor the movements of this sandhill crane.

Biology as a Science

2

he sandhill crane folds its wings and lands gently in the marsh grass. It drinks from the placid water before returning to its nest to tend to its young. The marsh is undisturbed by humans, but the crane is being observed nonetheless. A transmitter on one of the crane's legs allows biologists to track its every move. Transmitters are only one tool that biologists use to study life. In this chapter, you will learn about the methods of science and some of the instruments scientists use in their work.

Guide for Reading

Key words: scientific method, hypothesis, controlled experiment, variable, theory, magnification, resolution

Questions to think about:

📖 What is the scientific method?

📖 How is a hypothesis useful to scientists?

📖 How do instruments aid biologists?

2-1 The Nature of Science

Section Objectives:

- *State* the essential steps of a scientific investigation.
- *Explain* what is meant by a controlled experiment.
- *Define* the terms *hypothesis, theory,* and *scientific law.*
- *List* the basic units of measurement in the metric system.
- **Laboratory Investigation:** *Perform* and *analyze* the results of a controlled experiment (p. 28).

Broadly speaking, science is an attempt to understand the world we live in. By this, we mean that science goes beyond the simple observation and description of objects and events. It tries to find general principles to explain why things are as they are and why things happen the way they do.

The Scientific Method

There are so many different kinds of phenomena to be explained that scientists have had to become specialists—physicists, chemists, astronomers, earth scientists, biologists, and so on. Within each of these fields, there are numerous subdivisions. Even so, scientists in all fields approach their problems in the same way. When a scientist announces a finding or proposes a new idea, other scientists may repeat the work or test its conclusions. This universal approach to scientific problems is called the **scientific method.** Its main features are the same in all areas of science.

▲ **Figure 2–1**

Observation in Science. Biologists often observe organisms in controlled environments and in nature as part of their research.

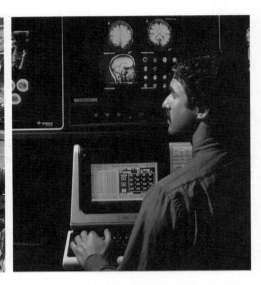

▲ **Figure 2–2**

Scientific Investigation. Careful observations of phenomena in nature and in the laboratory help scientists define a specific problem for investigation.

Defining the Problem   The scientific method begins when a person asks a question about a particular phenomenon or set of facts he or she has observed. See Figure 2–2. For example, a scientist interested in seed germination might ask, "Do the seeds of a particular plant species need light to germinate?" The subject may be one about which little is known, or it may be a well-understood one. In either case, by asking a question, the scientist consciously defines a specific problem for investigation.

The question is usually followed by a thorough search for information about the topic. Most of the information is derived from the data of experiments performed by other scientists and reported in scientific journals. By becoming familiar with existing knowledge, the scientist can avoid duplicating work already done and can plan the best approach to the problem.

Formulating a Hypothesis   In analyzing a problem, the scientist may find a specific pattern of events or some definite relationship between certain factors. However, by themselves, these observations do not explain anything. What a scientist really wants to know is what causes the pattern. It is here that reasoning, guesswork, and inspiration enter. At this stage in the scientific method, the scientist usually formulates a **hypothesis,** a possible explanation for an observed set of facts. This is a critical step in the scientific method.

Testing the Hypothesis—Experimentation   A hypothesis may offer a possible explanation for everything that is known about a problem. Until it is tested, however, it remains only a hypothesis—a logical guess. A hypothesis cannot be tested by carrying out the same types of experiments that established the pattern or relationship. That would only verify the known pattern rather than provide the explanation. A good hypothesis will predict other kinds of patterns or interactions that have not yet been observed. Thus, the scientist must test a hypothesis by designing experiments that will either verify or disprove the predictions of the hypothesis. A hypothesis is accepted as probably correct if all its predicted effects are observed and if these effects are repeatable. A hypothesis can never be completely proved. However, at any time it may be disproved by a single experiment! (Of course, that experiment, like all experiments, must be repeated and checked to make sure its results are correct.)

The design of an experiment is critical to its success. In an experiment, the scientist sets up a situation in which a particular observation can be made. The scientist makes changes in the situation and observes the results or response. In biology, research often involves the use of **controlled experiments.** In a controlled experiment, a situation is set up in duplicate. A single factor, called a **variable,** is changed in one setup. The variable may be any factor, such as temperature or light intensity. In the other setup, no factor is changed. If any difference occurs in the results or response of the two setups, it can be assumed to be caused by the changed factor. The setup in which no change was made, serves as a reference and is called the **control.** See Figure 2–3.

Observing and Measuring Since the goal of science is to explain what is observed, every investigation must include observations. In the early history of science, observations were often imprecise. One thing might be described as larger than another, or an event might be described as more likely to happen at warm temperatures than at cold temperatures. Today, we know that generalizations are not very useful. The heart of modern science is

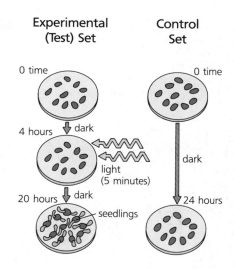

Experimental (Test) Set   Control Set

Seeds on moist filter paper in the dark for 4 hours, followed by 5 minutes light, followed by 20 hours dark at 25° C.

Seeds on moist filter paper in the dark for 24 hours at 25° C.

▲ **Figure 2–3**
A Controlled Experiment.

Critical Thinking in Biology

Evaluating Firsthand Observations

Tanya and Joel measured the temperature of the water in the classroom aquarium.

Tanya, the classroom lab assistant, used a thermometer calibrated to tenths of degrees. She bent down to read the thermometer at eye level and reported a temperature of 20.2°C.

Joel, who will be the new lab assistant, used a thermometer calibrated to degrees. He stood upright and looked down as he read the thermometer. Joel reported a water temperature of 21°C.

1. Whose temperature reading do you consider more reliable, Tanya's or Joel's? Why?

2. Suppose that before Tanya had measured the temperature, her teacher had told her it should be 20.2°C. How might this have affected Tanya's reading?

3. What are some factors that should be considered in evaluating the reliability of someone's observations?

4. What other steps could Tanya and Joel have taken to ensure accurate readings?

5. Think About It Describe the thinking process you followed to evaluate the reliability of the temperature readings.

accurate measurement and the statement of results in numerical, or quantitative, form. To obtain precise, quantitative results, scientists use many special tools and instruments.

Analyzing and Drawing Conclusions Once obtained, the data from an experiment need to be analyzed. Analysis can reveal patterns or relationships that are not apparent from unanalyzed, or raw, data. Analysis allows a scientist to interpret results and to draw conclusions. Ultimately, experimental data can provide evidence to support, modify, or reject a hypothesis or to formulate a new hypothesis.

Reporting Observations For progress to occur in any field of science, there must be no secrecy. The materials and procedures used in all investigations, as well as all observations and results, must be recorded accurately and reported in detail. If an experiment cannot be repeated by other investigators, the results of the original investigation cannot be considered valid.

Careers *in Action*

Medical Writer

Genetic engineering, AIDS, cures for cancer—these are a few of the topics that medical writers report on. Their job is to keep the public informed about current medical research and how it may affect society.

Medical writers must write clearly and accurately about technical subjects. They begin by gathering information from libraries or personal interviews with scientists, physicians, and editors. The finished piece, whether a book, magazine article, or television script, puts a biological issue in terms that people can easily understand. Readers or viewers can then discuss the issue intelligently and make informed decisions.

Problem Solving Suppose you are a medical writer writing an article on cigarette smoking.

You must translate the following summary into everyday language:

We investigated the effects of cessation of smoking on 1893 men and women who were 55 years of age or older and had angiographically documented coronary artery disease. The six-year mortality rate was greater among continuing smokers than among those who quit smoking and abstained throughout the study. There was no diminution of the beneficial effect with increasing age. Smoking cessation was seen to lessen the risk of death or myocardial infarction in older persons with coronary heart disease.

1. Look up and write a simple definition for any unfamiliar vocabulary terms.

2. What is the main idea of the paragraph?

3. Because older smokers have smoked for so long, many claim their health cannot be improved by quitting. Do you believe this? What evidence can you cite to support your belief?

4. Write a short article about the information given above. Your article should be aimed at older Americans and interesting.

■ **Help Wanted:**
Medical writer. Bachelor's degree required, preferably in science or journalism. Contact: American Medical Writers Association, 9650 Rockville Pike, Bethesda, MD 20814.

When a research project has been completed, the investigator may publish a paper describing the project in a scientific journal. These journals are usually publications of scientific societies that specialize in a particular branch of science. The journals serve as sources of information on recent developments in various scientific fields. Before a research paper is accepted for publication, it is reviewed by several scientists. The reviewers are usually scientists working in an area the same as, or related to, that of the research paper. Reviewers look to see that correct scientific methodology has been used, that results are reported clearly, and that conclusions are supported by the experimental data. If they find deficiencies, their criticisms and suggestions are passed on to the original scientist. The scientist can then perform additional experiments, analysis, and/or interpretation of data to correct and improve the paper before it is resubmitted. In this way, the quality of science that is reported is maintained at a high level.

Theories and Laws

As hypotheses are tested through experimentation, new and better hypotheses are proposed. In fact, scientists are constantly trying to refine hypotheses, or to have them describe nature more accurately. While hypotheses are important, each one is usually an idea limited to observations in a particular investigation. Explanations that apply to a broad range of phenomena and that are supported by experimental evidence are called **theories.** Theories are harder to establish than hypotheses. The germ theory of disease, developed from the work of Louis Pasteur, is an example of a well-tested theory. According to this theory, diseases are the effects of microscopic organisms living and reproducing inside the body of the diseased individual. The germ theory led to methods of treating and preventing many human diseases. The theory does have limits, however, because many diseases, such as arthritis, diabetes, and heart disease, are not caused by germs.

A theory attempts to explain everything about a phenomenon, including its cause. A **scientific law,** on the other hand, is a statement that describes some aspect of a phenomenon that is always true. A law does not explain how or why something occurs as it does, only that it occurs. Scientists can state a law only after they have observed that a particular event or relationship always exists under a given set of circumstances. The law of gravity, for example, states that any two objects attract each other. This law is based on thousands of observations. There are also scientific laws in biology. In Unit 5, you will learn about the laws of genetics.

Scientific Measurement

In scientific investigations, measurements need to be expressed in units of a standardized system that everyone understands. The International System of Units, abbreviated as **SI,** is the system used by scientists. Many SI units are units of the *metric system.* In SI, the

---

Can You Explain This?

"I have this **theory** that red uniforms will result in more wins."

"The cell **theory** states that the cell is the basic unit of all living things."

Note the use of the term *theory* in the statements above.

■ *Is there any difference between the two uses of the term? Explain. Which is the meaning used by scientists? Explain.*

Figure 2–4
Common Metric System Prefixes. ▶

Common SI Prefixes		
Prefix	Meaning	
mega-	1 000 000	(1 million)
kilo-	1 000	(1 thousand)
deci-	0.10	(one-tenth)
centi-	0.01	(one-hundredth)
milli-	0.001	(one-thousandth)
micro-	0.000 001	(one-millionth)
nano-	0.000 000 001	(one-billionth)

basic unit of length is the meter (m); the unit of mass is the gram (g); the unit of volume is the liter (L); the unit of time is the second (s); and the unit of temperature is either the Celsius degree or the Kelvin. In the system of measurement commonly used in the United States, units of different size have different names. For example, inches, feet, yards, and miles are all units of length. For a particular measurement, we choose the unit that is most convenient. Thus, we express the size of a sheet of paper in inches; the dimensions of a room in feet; and the distance between cities in miles. In the SI system, a simpler method is used to make units of convenient size. A prefix is attached to the basic unit name to make it larger or smaller. The most commonly used prefixes are defined in Figure 2–4. A complete listing of SI units appears in Appendix E at the back of this text. Some familiar examples of the use of SI units are shown in Figure 2–5.

The biologist needs to use small units of measurements because of the small size of the structures in living cells. Many of these structures are no more than a few millionths of a meter (micrometers) in length or diameter. Since so many measurements in biology are in this range, biologists often use the term *micron* in place of micrometer. One micron (symbol, μ) is the same as one micrometer, or one-millionth of a meter. It is also equal to one-thousandth of a millimeter (0.001 mm).

Figure 2–5
SI Units. SI units are used worldwide for all types of measurements. ▶

In this text, when a number has more than four digits, a space rather than a comma will be used to mark off groups of three digits. For example, 5976 remains as 5976 but 85,463 becomes 85 463. Also, 0.1523 remains as 0.1523 but 0.52186 becomes 0.52 186.

2-1 Section Review

1. Name the principal steps in the scientific method.
2. Define the term *hypothesis.*
3. List the basic units of measurement in the metric system.

Critical Thinking

4. Suppose you were given seed packages for two varieties of radishes. How would you determine experimentally the variety that had the better rate of germination? (*Problem Solving*)

2-2 Tools of the Biologist

Section Objectives:

- *State* why instruments are necessary for scientific research.
- *Name* and *state* the functions of the parts of a compound microscope.
- *Distinguish* between magnification and resolution in a microscope.
- *Describe* the steps in preparing a specimen for examination with a compound microscope.
- **Concept Laboratory:** *Develop* an understanding of the importance of instruments in making observations (p. 21).

Observation and measurement are the backbone of scientific investigation. The observations that can be made by the unaided senses are limited. Therefore, every branch of science makes use of instruments that increase the range and accuracy of the human senses. Even in daily life we use instruments to help our senses. Eyeglasses are an obvious example. But we also use thermometers, measuring cups, scales, and rulers. We seldom think of these things as scientific instruments, but that is what they are. In this section of the text, we will describe a few of the important instruments that are used in biological research.

The Light Microscope

A **light microscope,** or optical microscope, is any device that uses light to produce an enlarged view of an object. What we see when we use a microscope to examine an object is called an *image.* The ratio of the image size to the object size is the **magnification,** or magnifying power, of the instrument. Light microscopes depend on the fact that light rays change direction when they pass from one transparent medium to another. Optical microscopes contain lenses, pieces of glass with curved surfaces. The lenses cause light rays from an object to bend in such a way as to produce an enlarged image.

The Simple Microscope   The **simple microscope** is what we know as a magnifying glass, such as the one shown in Figure 2–6. It consists of a single lens. Lenses of this type were used as early as the tenth century. They are still used by biologists to identify specimens in the field and to make quick observations that do not require the high magnifications of a laboratory instrument.

The Compound Microscope   A **compound microscope,** like the one in Figure 2–7, uses two lenses. One lens produces an enlarged image that is further magnified by the second lens. A compound microscope has an optical system, a mechanical system, and a light system. See Figure 2–8. Its use has led to dramatic advances in nearly all fields of science.

Lenses make up the **optical system** of the compound microscope. The two lenses of the optical system are the *objective* and the *ocular* (AHK yuh ler), or eyepiece. In modern microscopes, the objective and the ocular each consist of several lenses combined to give the desired optical properties. As far as the operation of the instrument is concerned, each set of lenses acts as a single lens.

A compound microscope usually has two or more objectives of different magnifying powers. A *low-power objective* is used to locate the region of the specimen to be examined. A *high-power objective* is moved into position if further magnification is wanted. The ocular can be removed and replaced by a lens of different power.

The **mechanical system** is made up of the structural parts that hold the specimen and lenses and permit focusing of the image.

The *base* is the structure on which the microscope stands. Most of the other mechanical parts are attached to the *arm*. The *stage,* which is a platform coming out from the arm, has a round opening over which the specimen is placed. The specimen is usually mounted on a glass or plastic slide for observation. Two *clips* attached to the stage hold the slide in place. Attached to the top of the arm is the cylindrical *body tube,* which holds the lenses. The eyepiece is at the top of the body tube. At the bottom of the body tube is a revolving *nosepiece,* which holds the objective lenses. The objectives are changed by turning the nosepiece.

To focus the microscope, two adjustment knobs are used. The large knob is the *coarse adjustment,* which is used for approximate focusing of the low-power objective. The smaller knob is the *fine adjustment,* which is used for final focusing of the low-power objective and for all focusing of the high-power objective. Both knobs vary the distance between the objective and the specimen by moving either the body tube or the stage. The specimen is in focus when the image appears sharp. When the high-power objective is in position, it lies close to the slide on which the specimen is mounted. For this reason, only the fine adjustment knob should be used in focusing the high-power objective.

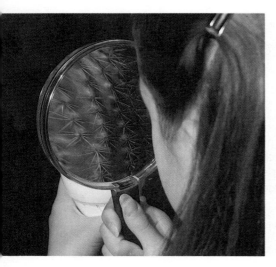

▲   **Figure 2–6**
A Simple Microscope. A magnifying glass is also known as a simple microscope.

Figure 2–7
The Modern Compound Light Microscope. Work in numerous fields of science and technology is dependent on the use of the light microscope. ▼

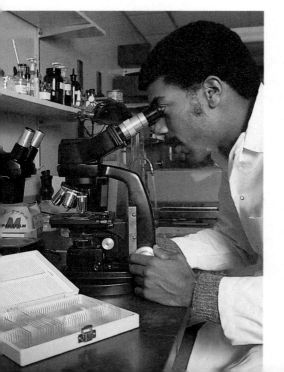

Concept Laboratory

Observation Using the Compound Microscope

Goal

To gain an understanding of the value of instrumentation and careful observation in science.

Scenario

Imagine that you work in a police crime lab. Your job is to compare strands of hair found at the scene of a crime with hair samples from several suspects. Your results will be used in court to establish the guilt or innocence of the suspects.

As you work, think about the need for careful observation in scientific investigations.

Materials

scissors
mirror
clear tape
envelopes
compound
    microscope

Procedure  *

A. Using scissors, cut off two short strands of your hair. Separately sandwich each strand between two pieces of clear tape. Place your two samples in an envelope marked with your name and give it to your teacher.

B. Your teacher will number one sample from each student. He or she also will select an unnumbered sample from one student as "evidence from the scene of the crime." The numbered samples and the sample serving as "evidence" will be returned.

C. Without using a microscope, examine each numbered hair sample and the evidence sample. On a chart like the one below, record the characteristics, such as color, thickness, and shape of each sample.

D. Using a microscope under low, then under high, power, reexamine each hair sample. Make a detailed drawing of each at the magnification that most clearly shows the hair's features. Based on the magnified view, revise the descriptions you recorded in your chart.

Organizing Your Data

1. What differences and similarities did you observe?

2. How do your data compare with those of the other members of your group?

Drawing Conclusions

1. Based on your observations in step C, which of the numbered samples matches the hair found at the scene of the crime?

2. Based on your observations from step D, which of the numbered samples matches with the hair from the crime scene? Are you more, or less, confident of this conclusion than of your conclusion in question 1? Explain.

3. Based on your experience, write a statement on the importance of observation and instrumentation.

Thinking Further

1. What if each hair sample had been labeled with the name of the suspect from whom it came? How might this have affected your conclusions? Why is it important for researchers to remain unbiased?

2. Suppose that the prosecuting attorney asks you to do further testing on the hair samples to confirm your conclusions. What other kinds of scientific analyses might you use? How would this help increase your confidence level?

3. For each biologist in Figure 2–2, describe the role that careful observation plays in his or her research. Suggest some ways that each of the scientists is aided by instrumentation.

Sample	Color	Relative Thickness	Shape (e.g., curly, straight)	Texture (e.g., rough, smooth)	Other
1					

*For detailed information on safety symbols, see Appendix B.

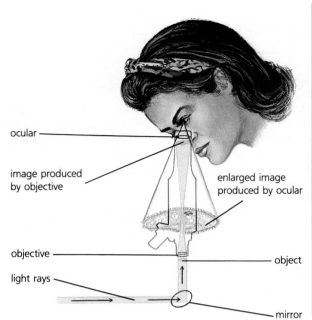

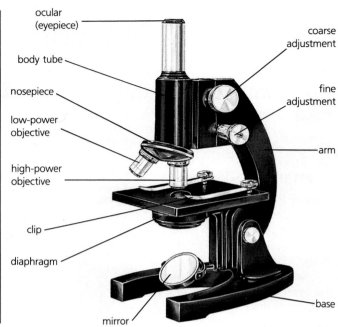

▲ **Figure 2–8**

A Compound Microscope. The objective produces an enlarged image of the specimen. When this image is viewed through the ocular, a still larger image of the specimen is seen.

In its simplest form, the **light system** consists of a mirror and a diaphragm. The *mirror,* which is under the opening in the stage, can be adjusted to direct light up through the specimen into the objective. In some microscopes, light is supplied directly by a *substage illuminator,* which is a small electric light. The amount of light reaching the objective is regulated by the *diaphragm* (DY uh fram), which is mounted below the stage. Some microscopes have *condensers* located in or below the stage. Condensers are lenses that concentrate the light on the specimen.

Magnification   **Magnification** refers to the enlargement of an image. The extent to which a microscope magnifies images is known as its magnifying power. It is expressed as a number followed by a multiplication sign, such as 100 ×. The magnifying power refers to enlargement in one direction. For example, the image of a line 1 millimeter long will appear to be 100 millimeters long when viewed through a microscope with a magnifying power of 100 ×.

In a compound microscope, the total magnification can be found by multiplying the magnifying power of the objective by the magnifying power of the ocular. In most student microscopes, the power of the high-power objective is 43 × and the power of the ocular is 10 ×. With the high-power objective in use, the total magnifying power of the microscope is 43 × 10, or 430 ×. This means that the distance between two points in the image is 430 times greater than it is in the actual object.

Resolution   Although a microscope enlarges images, it does not add detail to objects. The details are always there. A microscope spreads the details apart so the human eye can make them out. To the unaided eye, two tiny spots close together appear as one. They are not seen as separate spots. Under the microscope, these two spots appear farther apart, which allows us to see them separately.

The ability of a microscope to show two points that are close together as separate images is called **resolution** (rez uh LOO shun), or resolving power. Resolution is another term for the sharpness of an image. It does no good to increase the magnifying power of a microscope if its resolving power is not also increased. If only the magnification is increased, the image becomes larger, but you cannot make out any more detail. Small blurred spots simply become large blurred spots.

Up to a point, the resolving power of a microscope depends on the precision and quality of the lenses. However, there is a limit to the resolving power of any optical lens system. Because of the properties of light, a light microscope cannot distinguish two points that are less than 0.2 micrometers apart. This property of light limits what can be discovered about the structure of cells through the optical microscope. This limit remained until the development of the electron microscope, which does not depend on light. The electron microscope will be described later in this chapter.

Preparation of Specimens   To observe a specimen under the compound microscope, the specimen must be thin enough for light to pass through it. This is the case for all *microorganisms*— organisms too small to be seen clearly with the naked eye. Most other organisms or biological materials, however, are too thick. For this reason, they must be fixed, embedded, and sliced into thin sections. *Fixation* is done by cutting the material into small pieces and allowing it to soak in a fixative, such as formalin. The fixed material is then *embedded* in liquid wax or plastic, which is allowed to harden. The wax or plastic holds the material in place so that it can be sliced or *sectioned.* The instrument used for slicing thin sections is called a *microtome,* which is shown in Figure 2–9.

The thin sections are then attached to a glass slide and stained. Without staining, few structural details can be observed with the compound microscope since cells are transparent. However, as Figure 2–10 illustrates, by using one or more colored stains, which

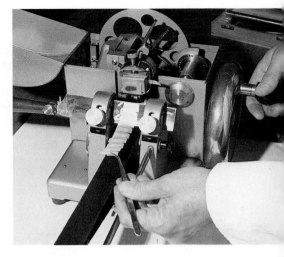

▲ **Figure 2–9**
A Microtome. Thin sections are being collected off the knife of this microtome.

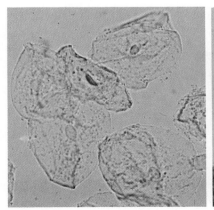 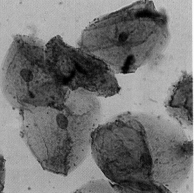

◀ **Figure 2–10**
Staining. Unstained (left) and stained (right) preparations of human cheek cells are seen through a light microscope.

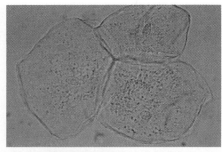

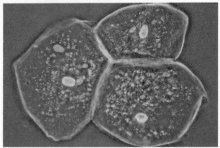

▲   **Figure 2–11**

Phase-Contrast Microscopy. Human cheek cells as seen through an ordinary light microscope (top) and through a phase-contrast microscope (bottom). (Magnification 396 ×)

Figure 2–12

A Transmission Electron Microscope and Micrograph. With the transmission electron microscope, scientists can study cell structure at extremely high magnifications. Nerve cell mitochondria magnified 124 800 times are shown in the electron micrograph on the right. ▼

are absorbed only by certain structures in the section, the details can be seen. There are some stains, called *vital stains,* that can be used with living tissues. Although they are taken in by the tissue, they do not kill it, and the structural details can be seen with the microscope.

The Phase-Contrast Microscope   A **phase-contrast micro-scope** is a special type of compound microscope that allows the details within living specimens to be seen without staining. This microscope enhances the differences that occur in light as it passes through different regions of a cell. As a result, structures in living cells that cannot be seen with the ordinary compound microscope are made visible. See Figure 2–11.

The Stereomicroscope   The type of light microscope used in studying the external, or surface, structure of specimens is the **stereomicroscope.** A stereomicroscope has an ocular and objective for each eye and a low magnifying power, usually from 6 × to 50 ×. These characteristics provide a more three-dimensional image of the specimen. In addition, the image produced by a stereomicroscope is not reversed as it is with a compound microscope. For this reason, procedures, such as dissections that require magnification, are performed using stereomicroscopes.

The Electron Microscope

Magnification and resolution beyond the limits of the light microscope were not available until the development of electron microscopes in the 1930s. The **transmission electron microscope** can magnify images more than 250 000 times. It uses electron beams, rather than light, and electromagnetic lenses, rather than glass lenses. The electron beam is directed through a vacuum chamber that contains a series of electromagnets. The electromagnets serve as lenses to focus the electron beam. When the electrons hit the specimen, some pass through, some are absorbed, and some are scattered. Those that are transmitted through the specimen are focused on a viewing screen similar to a television screen. Denser portions of the specimen absorb more electrons than less dense portions and thus appear darker on the viewing screen. Electron microscopes also contain cameras to photograph the image of the specimen. See Figure 2–12.

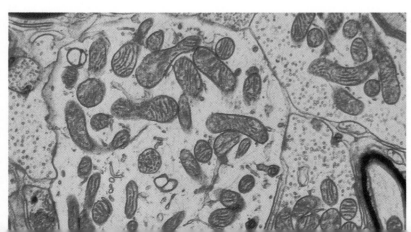

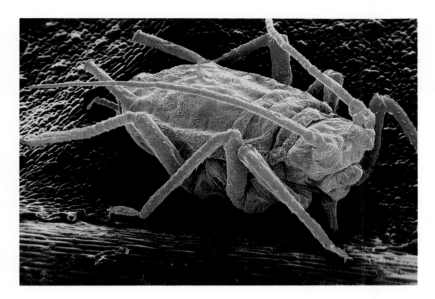

◀ **Figure 2–13**
Scanning Electron Micrograph. The scanning electron microscope allows the viewing of whole organisms. It produces a three-dimensional image, such as this one of a peach aphid on a leaf. (Magnification 68 ×)

Specimens that are going to be viewed in an electron microscope first must be dried, embedded in plastic, and sliced into thin sections no more than one micron in thickness. After this has been done, the sections are mounted on fine grids and stained with a metal to increase contrast.

The **scanning electron microscope** operates in a somewhat different way. It uses an electron beam that has been focused to a fine point. The beam is passed back and forth over the surface of the specimen. Electrons reflected or ejected from the specimen's surface are collected and used to produce an image of great depth. See Figure 2–13. The magnification of the scanning electron microscope is not as great as that of the transmission electron microscope. However, it can reveal fine details of the surface structure of whole specimens. There are many scanning electron microscope photographs in this text.

Laboratory Techniques

A number of important techniques are used routinely in day-to-day biological research. Many of these techniques are improved or refined continually as new and better instruments and other materials become available.

Centrifugation   Materials of different densities suspended in a liquid can be separated from each other by the process of **centrifugation** (sen truh fyoo GAY shun). The material that is to be separated is suspended in liquid in a test tube that is put into a centrifuge. The centrifuge spins the tube around. The heaviest particles in the liquid settle to the bottom the fastest. The next heaviest form a layer on top of the heaviest, and so on. The lightest layer is left on top. Each layer, or fraction, then can be removed from the tube.

The *ultracentrifuge* is much more powerful than a regular centrifuge. It spins at rates from 40 000 to 100 000 revolutions per minute. It can be used to separate very light particles, including various parts of the cell, from one another.

Microdissection In **microdissection** (my kroh dis EK shun), tiny instruments are used to perform various operations on living cells. This work must be done under a microscope. First, a *micromanipulator* (my croh muh NIP yuh lay ter) is attached to the microscope stage. This apparatus controls the tools used in micro-dissection. Among the tools that can be used with a micromanipu-lator are *microelectrodes,* which are used to measure or produce electrical currents in the cell; *microknives,* or *microneedles,* which are used to remove cell structures; and *micropipettes* (my croh py PETS), which are used to introduce materials into, or remove materials from, the cell.

Tissue Culture The technique of maintaining living cells or tissues in a culture medium outside the body is called **tissue culture.** Cells from living organisms are placed in culture tubes and bathed in fluid containing all necessary nutrients, oxygen, and other factors. Cells grown in tissue cultures are used in many types of biological and medical research.

Chromatography Any technique that separates different substances from each other on the basis of their chemical or physical properties is known as **chromatography** (kroh muh TAHG ruh fee). In chromatography, the mixture to be separated is placed on a solid material to which it adheres. A solvent then is intro-duced. Those substances that adhere loosely to the material will be carried away first in the solvent. Those substances that adhere more tightly will be carried away last. In this way, the different substances in the mixture separate. If the test substances are colored, they will form colored bands or spots, as shown in Figure 2–14. If they are colorless, they can react with chemicals that give them a color. The rate at which a given substance moves in a given solvent is a characteristic of that substance. By comparing the distance that the test substances have moved with the pattern of known substances, the test substances can be identified.

Electrophoresis The technique for separating substances made up of particles that have an electrical charge is called **electropho-resis** (ih lek truh fuh REE sis). An electric current is run through a solution containing a variety of dissolved substances. Since differ-ent substances move at different rates in the electrical field, the substances that make up the mixture are separated. Again, the rate at which each substance moves is a characteristic of that substance.

Spectrophotometry Sometimes, a scientist can determine what a substance is and how much is present in a sample by knowing the kind and amount of light absorbed by the sample. This method of using light to analyze samples is called **spectrophotom-etry** (spek truh fuh TAH muh tree). Measurements of this type are made with an instrument called a *spectrophotometer.*

Figure 2–14

Chromatography. Each band of color in this paper chromatogram indicates the presence of a separate substance. ▼

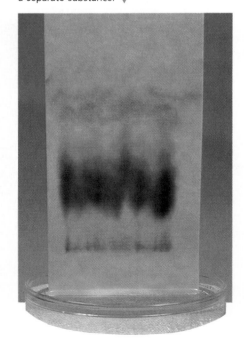

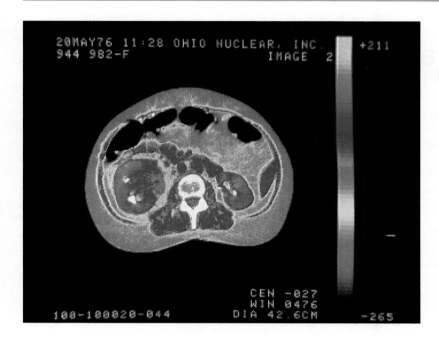

Figure 2–15
Computerized Axial Tomography. CAT scans, such as this one showing a cross-sectional view through the trunk of a body, are important in medical diagnosis and treatment.

Computers Computers have become increasingly important in all areas of biological study. Computers are used to collect, store, and analyze data. In these uses alone, computers have greatly increased the progress of science. To gain an understanding of complex biological processes, scientists are using computers to simulate processes. Our understanding of the complex relationships between organisms and their environments has been advanced as a result of computer analysis and simulation. Images produced by various types of microscopes can be enhanced by computers to show details otherwise not visible. In medicine, sophisticated diagnostic methods depend upon computers. Magnetic resonance imaging (MRI), computerized axial tomography (CAT), and sonography, for example, are three different methods that use computers to generate images of body tissues. See Figure 2–15. In these and countless other ways, computers are increasing our knowledge and improving our lives.

2-2 Section Review

1. How do instruments improve observations?
2. List the systems of a compound microscope.
3. Explain the difference between magnification and resolution.

Critical Thinking

4. Suppose you wanted to study a living microorganism and record its feeding behavior. Which type of microscope, light or electron, would you use to carry out this study? Explain your answer. (*Judging Usefulness*)

Laboratory Investigation

A Controlled Experiment

A control is an important part of scientific experimentation. In this investigation, you will perform a controlled experiment to test the effect of sugar on the metabolic activity of yeast.

Objectives

- *Perform* a controlled experiment, and analyze the results.
- *Test* the effects of sugar on the metabolic activity of yeast.

Materials

2 test tubes
test-tube rack
glass-marking pencil
graduated cylinder
yeast culture
beaker
water
water bath, 30°C
table sugar, ½ packet
metric ruler

Procedure  *

A. Work with a partner. Obtain two test tubes, and mark one with a large "C" for control, the other with a large "T" for test. Place them in a test-tube rack.

B. Using a graduated cylinder, obtain 25 mL of yeast suspension. Pour 10 mL of the suspension into each of the test tubes.

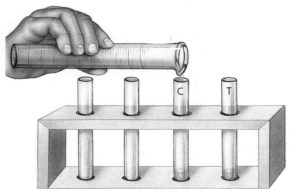

*For detailed information on safety symbols, see Appendix B.

C. Add 10 mL of water at 30°C to both test tubes. **Caution:** exercise care when using the hot water bath. To test tube T, also add a half-packet of sugar.

D. Mark the level of the liquid in each tube. Mix each tube's contents with a gentle swirling motion, being careful to avoid spilling any liquid. Keep the tubes warm. Note the time.

E. When 15 minutes have passed, mark the levels of the liquids in each test tube again. The mark should include the volume resulting from any bubbles. With a centimeter ruler, measure the distance, if any, between the two marks on each tube. Record the data on a chart like the one shown below.

Tube	Distance Between Marks
C	
T	

Analysis and Conclusions

1. In your experiment, which variables were the same in both test tubes? Which were different?

2. At the end of the experiment, how does the height of the liquid in test tube T compare with the height of the liquid in test tube C?

3. What accounts for any difference between the two tubes in the heights of the liquid?

4. If you had not used test tube C as a control in your experiment, what might you have concluded about the water alone as the cause of the bubbling in test tube T?

5. What differences, if any, do you think would have resulted if the whole packet of sugar had been added to tube T? Explain.

6. Can a valid control have more than one variable different from the test (experimental) set? Why or why not?

7. Design a controlled experiment that tests whether or not temperature has an effect on the bubbling, that is, the metabolic activity, of yeast.

Chapter Review

Study Outline

2–1 The Nature of Science

■ Scientists use the scientific method to find answers to questions and to solve problems. The method consists of a series of logical steps aimed at establishing facts about some object, event, or process.

■ Through observation and review of the literature, a scientist may find a problem of interest. He or she then formulates a hypothesis that may solve or explain the problem. Then, the scientist tests the hypothesis, often by means of a controlled experiment.

■ An accepted hypothesis is subject to verification and further refinement based on the results of new experiments. When a hypothesis is shown by experimentation to be incorrect, a new hypothesis is formulated and tested.

■ Theories are explanations that apply to a broad range of events. A theory is based on well-established evidence, usually from many sources.

■ Experimental observation frequently includes measurement using the International System of Units (SI). Some SI units are the meter, gram, liter, second, and Celsius degree.

2–2 Tools of the Biologist

■ Instruments increase the range and accuracy of human senses. Instruments commonly used by biologists include the simple and compound microscopes, the stereomicroscope, and the electron microscope.

■ To be useful, a microscope must provide both magnification and resolution. Magnification refers to enlargement of an image. Resolution refers to sharpness of an image.

■ Special techniques used in biological research include centrifugation, microdissection, tissue culture, chromatography, electrophoresis, and spectrophotometry.

■ Computers are used to collect, analyze, and store data, to carry out simulations of various processes, and to generate and enhance images from a variety of instruments.

Vocabulary Review

scientific method (13)
hypothesis (14)
controlled experiment (15)
variable (15)
control (15)
theories (17)
scientific law (17)
SI (17)
light microscope (19)
magnification (19)
simple microscope (20)
compound microscope (20)
optical system (20)
mechanical system (20)
light system (22)
resolution (23)
phase-contrast microscope (24)
stereomicroscope (24)
transmission electron microscope (24)
scanning electron microscope (25)
centrifugation (25)
microdissection (26)
tissue culture (26)
chromatography (26)
electrophoresis (26)
spectrophotometry (26)

A. Multiple Choice— Choose the letter of the answer that best completes each statement.

1. A possible explanation for an observed set of facts is called a(n) (a) variable (b) control (c) hypothesis (d) theory.

2. The ability of a microscope to show two points close together as separate images is called (a) magnification (b) mechanical system (c) resolution (d) microdissection.

3. An explanation that applies to a broad range of phenomena and that is based on well-established evidence is known as a(n) (a) control (b) variable (c) hypothesis (d) theory.

4. A type of microscope that does not use light is the (a) stereomicroscope (b) phase-contrast microscope (c) scanning electron microscope (d) compound microscope.

5. In a controlled experiment, the factor that changes is known as the (a) variable (b) hypothesis (c) control (d) theory.

Chapter Review

B. Matching—Select the vocabulary term that best matches each definition.

6. A microscope used to observe the details of living cells without staining

7. Separation of materials by density

8. Measurements involving the amount and kind of light

9. Performing operations on living cells

10. Separation from each other of the chemical substances of a mixture.

11. Growing living cells for study

12. Holds the lenses and permits focusing of the image

13. A microscope able to magnify more than 250 000 times

14. Consists of a mirror and diaphragm

Content Review

15. What is the main purpose of a scientific experiment?

16. Explain the function of a control in a scientific experiment.

17. Why are scientific papers reviewed by other scientists before they are published?

18. Explain the difference between a theory and a scientific law.

19. Why would the size of an elephant be expressed in meters but that of a bacterium in microns?

20. What parts of a compound microscope make up the optical system?

21. How is the total magnification of a compound microscope calculated?

22. What are the magnifications of the following combinations of objective and ocular: (a) 43X, 10X; (b) 20X, 5X; (c) 43X, 8X?

23. In microscopy, what is meant by resolution?

24. Explain the steps necessary in preparing materials to be viewed with the light microscope.

25. What are vital stains, and why are they used?

26. What advantage does a phase-contrast microscope have over an ordinary compound microscope?

27. Why is a compound microscope unsuitable for microscopic dissections?

28. Describe how a transmission electron microscope produces the image of a specimen.

29. Explain the technique of centrifugation.

30. How is a micromanipulator used in biology?

31. Explain the principles behind chromatography.

32. How is spectrophotometry useful to a scientist?

33. In what ways do computers aid scientists?

Graphic Organizing

For information on graphic organizers, see Appendix G at the back of this text.

34. **Flow Chart:** Copy the flow chart below onto a separate sheet of paper. Complete the chart by filling in the missing steps of the scientific method.

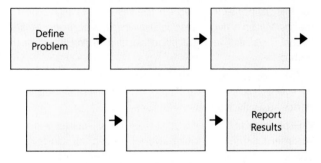

Critical Thinking

35. Describe the situation in the *Biology and You* feature on page 108 using only the clues in the picture. Decide who the person is, what the person is doing, where the picture was taken, and so on. Then, make a list of the assumptions you made in order to describe the situation. For example, perhaps you said the picture showed an athlete. That would be an assumption, since the person may not be an athlete. How many of your assumptions are you able to identify? (*Identifying Assumptions*)

36. Can hypotheses and theories change? Explain your answer. (*Generalizing*)

37. List some reasons why scientists make careful observations and records of experimental data. (*Identifying Reasons*)

38. How are the compound microscope and the electron microscope alike? How are they different? (*Comparing and Contrasting*)

39. In Figure 2–2, biologists are making observations. What information about the scientists and their work would help you decide whether their observations are reliable? (*Evaluating Firsthand Observations*)

40. Suppose you centrifuged a test tube containing particles of five different densities. What would you expect to see in the test tube? (*Predicting*)

Creative Thinking

41. Think of a person you know, or know about, whom you think of as creative. What is it about this person that makes him or her creative? How would these traits be useful for a scientist?

42. Louis Pasteur said, "Chance favors only the mind that is prepared." Explain this statement by relating it to the qualities of a good scientist.

Problem Solving

43. A water plant placed in a bright light gives off bubbles of oxygen. In the lab, it was noticed that if the light were placed at different distances from the plant in the aquarium, the rate of bubbling varied. The data are shown in the table below.

Distance From Light (cm)	O_2 Bubbles/Minute
10	40
20	20
30	10
40	5

On a separate sheet of paper, plot the data on a line graph. What is the variable in this investigation? Draw a conclusion from the data. Predict the amount of oxygen released at 50 and at 60 centimeters.

44. A biologist studying life in a lake noticed that from day to day the depth at which single-celled, green organisms lived varied greatly. These organisms were as much as five feet below the surface of the water on sunny days but were only six inches below on cloudy days. Propose a hypothesis to explain these findings. Design an experiment to test the hypothesis.

45. In the experiment illustrated in Figure 2–3, what conclusions can you draw from the results? What other questions could you ask about how seed germination is controlled?

46. In the same experiment, suppose none of the seeds in the test set germinated. What could you conclude? Suppose the control seeds germinated but the test seeds did not. What could you conclude?

Projects

47. Think of a scientific problem you would like to pursue. Develop a hypothesis. Then, design a controlled experiment to test your hypothesis. If possible, perform your experiment.

48. Prepare a report for the class on one of the following career opportunities: (a) electron microscopist, (b) laboratory technician, (c) microbiologist. If possible, interview someone working in that field. Be sure to prepare your questions in advance. You may wish to ask about training, future opportunities, and daily routine.

49. Obtain seeds from a variety of plants. Check their light requirements for germination by using the procedure described in Figure 2–3. Be sure to soak the seeds in water before testing light requirements.

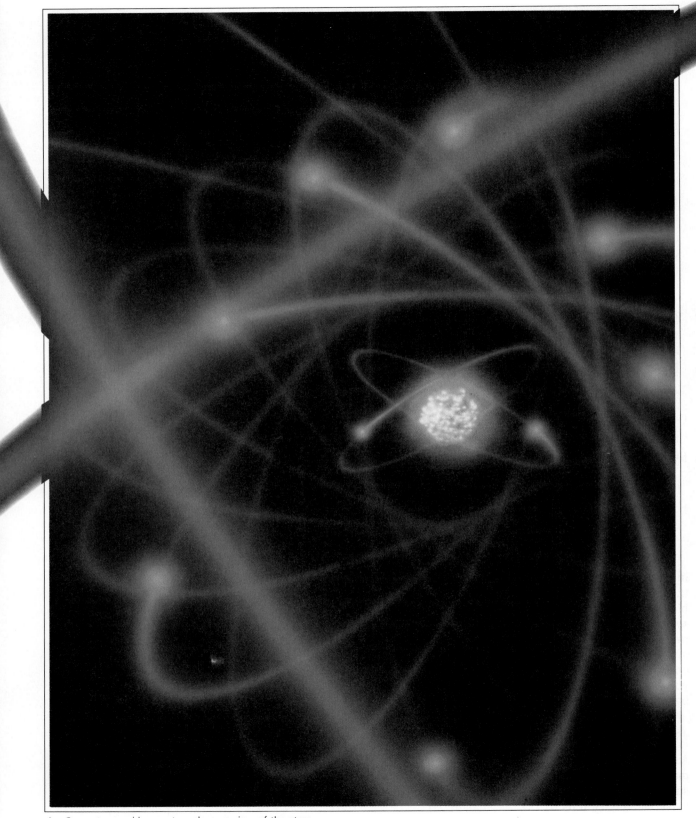

▲ Computer graphics create a close-up view of the atom.

Basic Chemistry

An atom is a miniature world—a small universe in which tiny particles move about. Despite the tiny scale, distances seem vast. Nothing but empty space separates orbiting particles and a compact inner core. The entire atom pulses energetically. One atom binds to another in a network that contains countless more. They join to form blades of grass, rocks, and the hairs on a cat—everyting the world contains. This chapter will introduce you to the structure of the atom, and explain how atoms interact and arrange themselves into groups. Welcome to the fascinating world of chemistry!

Guide for Reading

Key Words: atom, element, compound, covalent bond, ionic bond, solution, acid, base

Questions to think about:

📖 How does the structure of an atom determine its chemical properties?

📖 How do substances combine to form different compounds? What holds a compound together?

📖 What are the properties of acids and bases?

3-1 Atomic Theory of Matter

Section Objectives:

- *Define* the terms *element, compound, atomic number, mass number,* and *isotope.*
- *Describe* the structure of the atom and the arrangement of electrons around the nucleus.
- *Explain* how radioisotopes and other isotopes are used in biological and chemical research.

In this century, people have made great progress in understanding the processes of life. Because these processes are chemical, the biology student needs to know some basic chemistry. Living systems, from the smallest organism to a forest filled with living things, are made of the same basic building blocks as are nonliving systems. These substances react according to the same laws of chemistry that other substances obey. In this chapter and the next, you will study the chemistry that will enable you to understand the basic processes of life.

Atoms, Elements, and Compounds

As you can see just by looking around you, the world is made of many different substances. Hundreds of thousands of different substances are known. Hundreds of thousands of others probably exist. Chemistry tells us that all of these different kinds of matter

▲ **Figure 3–1**
Chemistry and Matter. All things, whether living or nonliving, are made up of the same basic chemical building blocks.

▲ **Figure 3–2**
Elements and compounds. Glass (top) can look very much like diamond (bottom). However, diamond is a crystalline form of the element carbon. Glass is a mixture of compounds.

are made of atoms combined in various ways. **Atoms** are the basic building blocks of matter. They cannot be subdivided any further by any ordinary chemical means. You will learn more about the structure of atoms later in this chapter.

In spite of the large number of different substances, there are only about 100 kinds of atoms. Some substances are made of only one kind of atom. These substances are called **elements.** Iron, for example, is an element. All iron consists entirely of iron atoms. Oxygen is an element made entirely of oxygen atoms. Since there are about 100 different kinds of atoms, there are about 100 different elements.

Most substances are **compounds.** In a compound, two or more kinds of atoms are combined in definite proportions. For example, water is a compound made of hydrogen atoms and oxygen atoms in the proportion of 2 to 1. This means that there are always 2 hydrogen atoms for each oxygen atom.

Just looking at substances like the ones in Figure 3–2 will not tell you whether they are elements (made of a single kind of atom) or compounds (made of two or more kinds of atoms). For example, both oxygen and carbon dioxide are gases, but oxygen is an element and carbon dioxide is a compound. For this reason, scientists use chemical means to decide whether a substance is an element or a compound. Compounds can be separated into the elements that make them up. Carbon dioxide can be separated into carbon and oxygen. Carbon and oxygen cannot be separated into anything else. Elements, unlike compounds, cannot be broken down into simpler substances by ordinary means.

Chemists have given names and symbols to all the elements. The symbols are a kind of shorthand for showing the makeup of compounds and for showing what happens during chemical reactions. Figure 3–3 lists the names and symbols of some elements that are important in living things. You will find it useful to learn these symbols. Most of the symbols are abbreviations or the initials of the name of the element in English. The other symbols come from the Latin names of the elements. These are elements that were known to scientists hundreds of years ago, when Latin was used for most scholarly writing. Although the name of an element is not capitalized, the first letter of the symbol is capitalized.

Structure of Atoms

The idea that matter is made of atoms is very old. It goes back at least 2500 years to the philosophers of ancient Greece. Early scientists believed that an atom was an extremely small particle that could not be changed in any way. Research in the twentieth century has shown that atoms are not hard, solid balls. Instead, atoms consist of still smaller particles. Each atom has a small central part called the **nucleus.** The nucleus of every atom contains particles called **protons** (PROH tahnz) and **neutrons** (NOO trahnz). Protons and neutrons have about the same mass, or quantity of matter. Protons have one unit of positive electric charge, while

Elements of Importance in Biology		
Name of Element	**Symbol**	**Atomic Number**
hydrogen	H	1
carbon	C	6
nitrogen	N	7
oxygen	O	8
sodium (*natrium*)	Na	11
magnesium	Mg	12
phosphorus	P	15
sulfur	S	16
chlorine	Cl	17
potassium (*kalium*)	K	19
calcium	Ca	20
iron (*ferrum*)	Fe	26

◀ **Figure 3–3**
Elements of Importance in Biology. The Latin name is given when the symbol is derived from it.

neutrons are neutral. That is, they have no electric charge. The nucleus of the atom is stable. That is, it does not change under normal circumstances. In the space outside the nucleus, there are other particles, called **electrons** (ih LEK trahnz). Electrons have one unit of negative electric charge. They also have much less mass than protons or neutrons.

The Nucleus   The number of protons in the nucleus determines the atomic number of an element. The **atomic number** also tells you how many units of positive charge are in the nucleus. Each element has a different atomic number which is used to identify it. For example, a single atom of hydrogen has 1 proton in its nucleus. Therefore, the atomic number of hydrogen is 1. The nucleus of an oxygen atom has 8 protons. Therefore, the atomic number of oxygen is 8. See Figure 3–4.

Figure 3–4
Diagrams of some atoms. The atomic number of an atom equals the number of protons in the nucleus. It also equals the number of electrons surrounding the nucleus. ▼

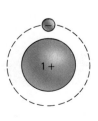

Hydrogen (H)
atomic number: 1

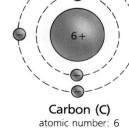

Carbon (C)
atomic number: 6

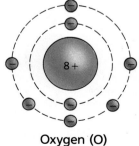

Oxygen (O)
atomic number: 8

 = electron   + = proton   (Neutrons in nucleus are not shown.)

Most of an atom's mass lies in its nucleus. This is because protons and neutrons have about 2000 times the mass of electrons. If you think of each proton and each neutron as having one unit of mass, then the mass of the entire atom is equal, roughly, to the sum of its protons and neutrons. This sum is called the **mass number.** For example, hydrogen has 1 proton and 0 neutrons in its nucleus. Therefore, the mass number of hydrogen is 1, the sum of the proton and neutron. Oxygen has a mass number of 16 because it has 8 protons and 8 neutrons in its nucleus.

Electrons Normally, atoms have the same number of electrons as protons. For example, the hydrogen atom has 1 electron and 1 proton. The oxygen atom has 8 electrons and 8 protons. In each atom, the total positive charge of the protons is balanced by the equal number of negative charges of its electrons. Therefore, under normal circumstances, an atom is electrically neutral. The electrons in an atom determine the atom's chemical properties. The electrons are arranged in levels, known as energy levels. Each energy level is at a different distance from the nucleus. Based on modern atomic theory, there are rules for determining how electrons are arranged in an atom. While you do not need to know all of this theory, you do need to know that the first energy level can hold only 2 electrons. Once it has 2 electrons, it is said to be filled. An atom with more than 2 electrons has some of its electrons occupying other energy levels. In all of these atoms, the outside level can hold only 8 electrons.

A filled outer energy level is a stable electron arrangement. Elements that have filled outer levels are chemically inactive. Except for a few special cases, they do not form compounds with other elements. These elements are all gases under ordinary conditions. Examples are helium, neon, and argon. The structures of these atoms are shown in Figure 3–5.

If the outside level of an atom has fewer than 8 electrons, it is unfilled. Atoms with unfilled outer energy levels can form compounds with other elements. When the atoms combine to form

Figure 3–5
Elements with Filled Outer Energy Levels. ▼

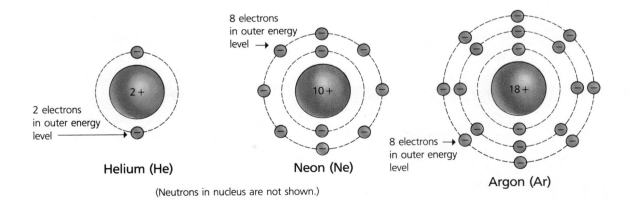

2 electrons in outer energy level

Helium (He)

8 electrons in outer energy level →

Neon (Ne)

8 electrons → in outer energy level

Argon (Ar)

(Neutrons in nucleus are not shown.)

Atomic number: 1   Atomic number: 1   Atomic number: 1
mass number: 1   mass number: 2   mass number: 3

⊖ = electron   + = proton   N = neutron

▲   **Figure 3–6**

Isotopes of Hydrogen. The number of protons in the nucleus is the same for all isotopes of the same element. Only the number of neutrons in the nucleus is different for each isotope.

compounds, their outside electrons are rearranged to give each atom a filled outer level. The ways in which this can happen are described in Section 3–2.

Isotopes

The atoms of an element may have different numbers of neutrons. For example, most hydrogen atoms have no neutrons. The nucleus is simply a single proton. However, as shown in Figure 3–6, there are hydrogen atoms with 1 neutron in the nucleus and others with 2 neutrons. Although these atoms are not exactly alike, they behave the same chemically because they all have only 1 electron. Therefore, these three kinds of atoms are considered atoms of the element hydrogen. Varieties of an element that differ only in the number of neutrons in their atomic nuclei are called **isotopes** (I suh tohps).

All elements have isotopes. Isotopes of the same element have the same atomic number but different mass numbers since the mass is the total of protons and neutrons. See Figure 3–7. For example, the three common isotopes of oxygen have mass numbers of 16, 17, and 18.

Figure 3–7

Isotopes of Oxygen. The mass number of each isotope of an element is different. The atomic number remains the same. ▼

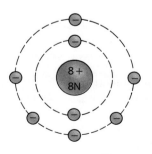

Oxygen – 16(^{16}O)
atomic number: 8
mass number: 16

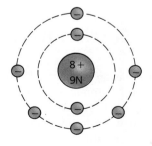

Oxygen – 17(^{17}O)
atomic number: 8
mass number: 17

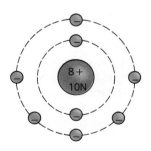

Oxygen – 18(^{18}O)
atomic number: 8
mass number: 18

Biology and You

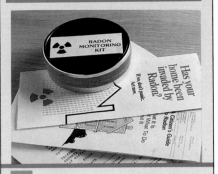

Q: My parents bought a radon testing kit for our house. What is radon, and how dangerous is it?

A: Recently, radon, an odorless, invisible gas, has become a household word. Some reports estimate that one out of every eight American homes has radon problems. Reports linking exposure to high levels of radon with lung cancer have prompted many people to test for the presence of this gas.

Radon is produced when uranium undergoes radioactive decay. As the uranium breaks down, radon, which is also radioactive, is released. Many homes have been built on or near rock formations containing uranium. Radon can seep into these homes through the basement, foundation, or drains. Trapped inside, radon builds up and may be harmful.

Test kits allow you to test for radon in your home. If there is a dangerous level, you can take steps to prevent the gas from leaking in. One option is to seal the locations where the radon enters. Another is to increase the amount of ventilation in your home.

■ *Survey five to ten neighbors. Are they aware of the radon problem? How many have tested for radon? What were the results?*

To distinguish one isotope from another, a number is placed next to the chemical symbol to show the mass number of the isotope. The number is written as a superscript, a small number usually written to the left of the chemical symbol. For example, ^{18}O stands for the oxygen isotope with a mass number of 18. The isotope ^{18}O also could be spelled out as oxygen-18.

Radioactive Isotopes

The nuclei of many isotopes are unstable. An unstable nucleus emits, or gives off, charged particles and radiation at times that cannot be predicted. This causes the number of protons or neutrons in the nucleus to change. During emission, the atom changes to another isotope, usually an isotope of a different element. The process is called **radioactivity** (ray dee oh ak TIV uh tee). It was discovered in 1896 during experiments with minerals containing the element uranium.

Radioactivity is a nuclear process, not a chemical process. That is, it changes the nuclear structure, not the electron structure of the atom. It is mentioned here because radioactivity is an important means of studying biological processes. Instruments sensitive to radiation can detect and measure the radioactivity given off by radioactive isotopes, or **radioisotopes** (ray dee oh I suh tohps). Thus, radioisotopes can be used to detect abnormalities in the size, shape, or function of organs. See Figure 3–8. Sometimes, radioisotopes are used in the treatment of certain cancers. They can also be used to study chemical reactions in living things. The atoms of the radioisotope act as *tracers,* or tagged atoms. As they move from one compound to another, they can be detected and followed. Thus, the detailed chemical steps of a process can be determined.

Isotopes do not need to be radioactive to act as tracers. An instrument called a *mass spectrometer* (spek TRAHM uh ter) can detect the different mass numbers of isotopes. Oxygen-18, a stable isotope, has been used in this way to study the process of photosynthesis.

3-1 Section Review

1. How is an element different from a compound?
2. Which atomic particles have electric charges? What are these charges?
3. Define the terms *atomic number* and *mass number.*
4. How are the electrons arranged in an atom?

Critical Thinking

5. Explain the relationship between the nucleus of an atom and its electrons, protons, and neutrons. (*Relating Parts and Wholes*)

3-2 Chemical Bonding and Chemical Reactions

Section Objectives:

- *Explain* the formation of covalent and ionic bonds.
- *Define* the following terms and give examples of each: *molecule, diatomic molecule, ion, chemical formula,* and *structural formula*.
- *Describe* the changes that can occur when a chemical reaction takes place.
- *Determine* whether or not a chemical equation is balanced and label the reactants and the products.

How do chemists know that two substances react with each other to form a compound? Only observing the substances in the laboratory will answer this question. A change in temperature, a change in color, and the formation of a gas are some of the signs that a reaction has taken place. When a reaction does take place, it can take a great deal of work to determine just what substance or substances have been formed.

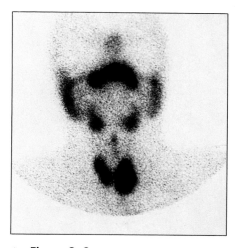

▲ **Figure 3–8**

Use of Radioactive Iodine to Diagnose Thyroid Disorders. Iodine is absorbed by the cells of the thyroid gland. This nuclear scan shows iodine concentrated in an enlarged thyroid gland (below the neck).

Covalent Bonds

Water is a compound of the elements hydrogen and oxygen. Each particle of water has 2 atoms of hydrogen and 1 atom of oxygen. A particle of this kind, in which two or more atoms are combined and act as a single particle, is called a **molecule** (MAHL ih kyool). Let us see how a molecule of water is formed.

The structures of a hydrogen atom and an oxygen atom are shown in Figure 3–9. The hydrogen atom has only 1 electron. This electron occupies the first energy level. As you have read, this level can hold 2 electrons. One more electron can be added to the level to fill it. The oxygen atom has 8 electrons. Two of its electrons are in

Figure 3–9

Covalent Bonding in Water. The hydrogen atom needs one electron to fill its outer energy level. The oxygen atom needs two electrons to fill its outer energy level. By sharing two pairs of electrons, the outer levels of the three atoms in the water molecule (H_2O) are filled. Each pair of shared electrons forms one single covalent bond. ▼

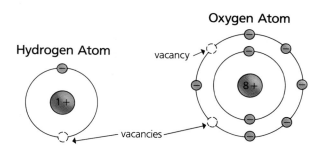

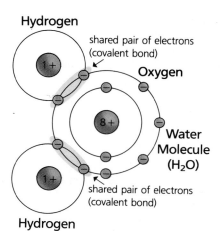

the first energy level, which, therefore, is filled. The other 6 electrons are in the second level. This level, which is the outer energy level, can hold 8 electrons. Two more electrons can be added to fill it.

In Figure 3–9, there is also a diagram of a water molecule, which shows 2 hydrogen atoms combined with 1 oxygen atom. Each hydrogen atom is sharing its electron with the oxygen atom. At the same time, the oxygen atom is sharing one of its electrons with each hydrogen atom. In this arrangement, the outer levels of all three atoms are filled. That is, the first energy level of each hydrogen atom is now filled with 2 electrons. At the same time, the second energy level of the oxygen atom is filled with 8 electrons.

The sharing of a pair of electrons by two atoms creates a force of attraction that holds the atoms together. This force of attraction is called a **chemical bond.** When a chemical bond is formed by the sharing of electrons, it is called a **covalent** (koh VAY lent) **bond.** In 1 molecule of water, there are two covalent bonds holding the molecule together.

Figure 3–10 shows the electron structure of chlorine. The outer energy level of a chlorine atom has 7 electrons. The chlorine atom needs only 1 more electron to fill its outer level. It can fill this level by sharing a pair of electrons with a hydrogen atom. When this happens, a molecule of the compound hydrogen chloride is formed. This molecule is held together by a covalent bond between the chlorine atom and the hydrogen atom.

A few elements have atoms that react with atoms of the same element to make two-atom molecules. For example, a hydrogen atom will react with another hydrogen atom to form a molecule made up of 2 hydrogen atoms. This type of two-atom molecule is called a **diatomic** (dy uh TAHM ik) **molecule.** Most elements that form diatomic molecules are gases under ordinary conditions. These elements are hydrogen, oxygen, nitrogen, chlorine, and fluorine. See Figure 3–11.

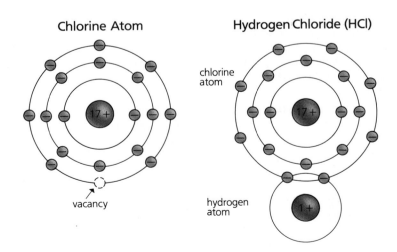

Figure 3–10

Covalent Bonding in Hydrogen Chloride. Chlorine needs one electron to fill its outer energy level. In hydrogen chloride, a covalent bond is formed by the sharing of one pair of electrons. ▶

Chlorine Molecule (Cl₂)

single bond

Diatomic Molecules of Some Gases. In a diatomic molecule of hydrogen or chlorine, two atoms form one covalent bond. In diatomic molecules of oxygen, atoms form two covalent bonds (a double bond).

Oxygen Molecule (O₂)   ## Hydrogen Molecule (H₂)

single bond →

double bond →

Ionic Bonds

Figure 3–12 shows the electron structures of a sodium atom and a chlorine atom before bonding. The outer energy level of a sodium atom has only 1 electron. To fill this level, 7 more electrons would be needed. Could a sodium atom fill its outside energy level by combining with a chlorine atom? No, there is no way for this to happen. Electrons are usually shared in pairs, 1 electron from each atom. The sodium atom has only 1 outside electron to share and, therefore, could share in only one pair.

However, the sodium atom can transfer its outer electron to a chlorine atom. This transfer will fill the outer energy level of the chlorine atom. At the same time, the sodium atom, having lost its outer electron, will be left with a new outer energy level of 8 electrons. Thus, the new outer level of the sodium atom will also be filled.

After the chlorine atom has received an electron from the sodium atom, the chlorine atom has 17 protons and 18 electrons. Thus, it has an excess negative charge of 1 unit. An atom that has an excess charge is called an **ion** (ɪ ahn). When the chlorine atom has

Figure 3–12
Ionic Bonding in Sodium Chloride. After sodium gives one outer electron to chlorine, each atom becomes an ion with a filled outer energy level. The sodium ion has a single positive charge. The chloride ion has a single negative charge. An ionic bond is formed between them. ▶

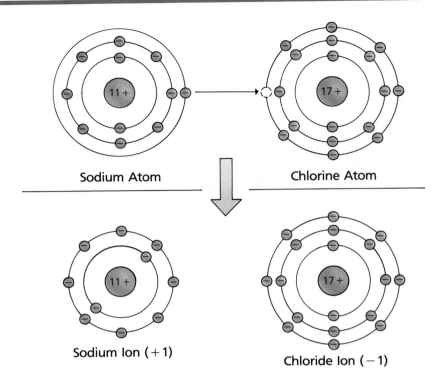

Sodium Atom   Chlorine Atom

Sodium Ion (+1)   Chloride Ion (−1)

become an ion, it has a negative charge of 1 unit. It is then called a chloride ion and is represented by the symbol Cl^-. Look again at Figure 3–12.

When a sodium atom gives up an electron, it is left with 11 protons and only 10 electrons. Therefore, it has an excess positive charge of 1 unit. The sodium atom has become a sodium ion, written as Na^+.

According to the laws of electric charge, particles with opposite charges attract each other. When a sodium atom loses an electron to a chlorine atom, the two ions that form are attracted to each other. The force of attraction between two ions is called an **ionic** (i AHN ik) **bond.** The compound of sodium and chlorine, called sodium chloride, consists of sodium and chloride ions held together by ionic bonds. Notice that the sodium and chloride ions do not form molecules. See Figure 3–13. Molecules are formed only when atoms share electrons and form covalent bonds. In sodium chloride, the ions remain separate. Each sodium ion is attracted to several chloride ions around it. Each chloride ion is attracted to several sodium ions around it. Materials with this type of structure are called crystals. No distinct molecules are present in crystals.

Chemical Formulas

Every compound consists of atoms combined in definite proportions. In water, for example, the proportion is 2 hydrogen atoms to 1 oxygen atom. In sodium chloride, the proportion is 1 sodium atom to 1 chlorine atom. This information about a compound can be

▲ **Figure 3–13**

Structure of Sodium Chloride. Ionic compounds do not form molecules. Each ion in the crystal structure is attracted to several oppositely charged ions around it. This results in the cubic shape of sodium chloride crystals.

given in a **chemical formula.** In a chemical formula, each element is represented by its chemical symbol. The proportions in which the atoms combine are shown by subscripts. Subscripts are small numbers written after the symbol and slightly below the line. The subscript 1 is not written in the formula, but it is understood to be there when no other subscript is shown. For example, the chemical formula for water is H_2O, and the formula for sodium chloride is NaCl.

A **structural formula** is a kind of chemical formula. It shows not only the number and kind of atoms in a molecule but also how the atoms are bonded to one another. In a structural formula, each pair of shared electrons—that is, each covalent bond—is shown by a short line joining the atoms that are connected by the bond. The structural formulas of benzene and glucose are shown in Figure 3–14. Glucose is one of the sugars made by plants. It is the chief source of energy for living things.

Note that the formula of glucose has several OH symbols. The covalent bond between the O and the H has been omitted. This is often done to simplify structural formulas that contain certain common groups of atoms.

Water (H_2O)

Acetylene (C_2H_2)

Benzene (C_6H_6)

Glucose ($C_6H_{12}O_6$)

◄ **Figure 3–14**

Structural Formulas. Structural formulas can be written only for molecular compounds. Note that every carbon atom has four bonds, every oxygen atom has two bonds, and every hydrogen atom has one bond.

▲ **Figure 3–15**

A Familiar Chemical Reaction. Whenever chemical bonds are broken or formed, energy is involved. Some chemical reactions absorb energy. Others, like the burning of wood, release energy.

Chemical Reactions

Every compound is a combination of the atoms of certain elements bonded to one another in definite proportions and patterns. Chemical bonds can be broken, and atoms can form new bonds in new combinations. When this happens, different substances are formed. Whenever different substances are formed, we say that a chemical change, or **chemical reaction,** has taken place. See Figure 3–15. The substances that were present before the reaction started are called **reactants** (ree AK tunts). The new substances produced by the reaction are called **products.**

A chemical equation can be written to represent the changes in a chemical reaction. A chemical equation is a short way of explaining a chemical reaction. For example, consider the reaction that occurs when water is formed from hydrogen and oxygen. The equation for this reaction could be written as:

$$\text{hydrogen} + \text{oxygen} \longrightarrow \text{water}$$

or

$$H_2 + O_2 \longrightarrow H_2O$$

The reactants appear to the left of the arrow. The products always appear on the right.

There is, however, something wrong with this equation. Look carefully at Figure 3–16a. There are 2 oxygen atoms on the left side of the equation but only 1 oxygen atom on the right side of the equation. This violates the **law of conservation of mass,** which states that mass can be neither created nor destroyed. Because atoms have mass, they are not created or destroyed in chemical reactions. They are merely rearranged. Thus, there should be equal numbers of each kind of atom on the left and right sides of a chemical equation. An equation written this way is called a *balanced equation.* The balanced equation for the reaction of hydrogen with oxygen is:

$$2H_2 + O_2 \longrightarrow 2H_2O$$

This equation states that 2 molecules of hydrogen combine with 1 molecule of oxygen to form 2 molecules of water. You can confirm that this equation is balanced if you count the atoms in Figure 3–16b. There are 4 hydrogen atoms and 2 oxygen atoms on each side of the equation.

a. Unbalanced Equation

$$H_2 + O_2 \longrightarrow H_2O$$

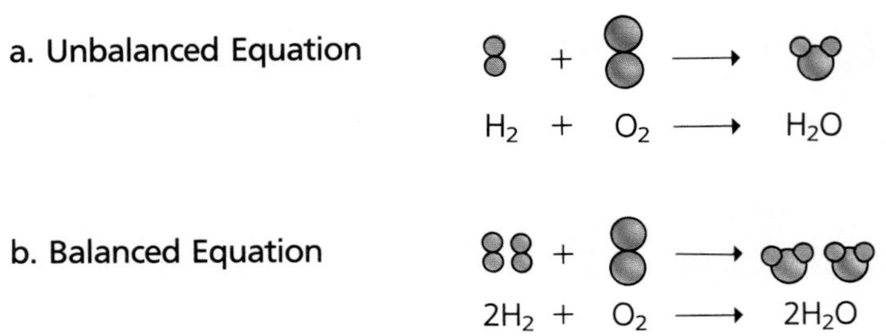

Figure 3–16

Unbalanced and Balanced Equations. Note that in the balanced equation there are equal numbers of hydrogen and oxygen atoms on each side of the equation. ▶

b. Balanced Equation

$$2H_2 + O_2 \longrightarrow 2H_2O$$

Chemical equations are almost always written in balanced form. Here is the balanced equation for the breakdown of glucose.

$$\underset{\text{glucose}}{C_6H_{12}O_6} + \underset{\text{oxygen}}{6O_2} \longrightarrow \underset{\substack{\text{carbon} \\ \text{dioxide}}}{6CO_2} + \underset{\text{water}}{6H_2O}$$

This reaction provides most of the energy for living organisms.

3-2 Section Review

1. Name two types of chemical bonds. Which type of chemical bond forms molecules?
2. Describe the two ways that an ion can be formed.
3. On which side of a chemical equation do the products appear?
4. Explain why the equation for the breakdown of glucose is a balanced equation.

Critical Thinking

5. As you know, there are no molecules in ionic compounds. What, then, does the formula for an ionic compound such as $MgCl_2$ represent? (*Relating Parts and Wholes*)

3-3 Mixtures

Section Objectives

- *Define* the following terms and give an example of each: *mixture, solvent,* and *solute.*
- *Explain* the difference between solutions, suspensions, and colloidal dispersions.

In every compound, the atoms or ions are joined by chemical bonds. The atoms or ions are present in fixed proportions and in a definite arrangement in space. It is possible, however, for many substances to be physically mixed without forming new chemical bonds. The result is called a **mixture.** The substances in a mixture may be present in any proportions. In fact, the proportions can change as one substance is added to or removed from the mixture. No matter what their proportions, however, the different substances in the mixture retain their usual properties.

For example, consider a mixture of table salt (sodium chloride) and iron filings. If this mixture is placed in water, the salt will dissolve, as it normally does. The iron filings will remain undissolved. On the other hand, a magnet will attract the iron filings in the mixture and leave the salt behind. See Figure 3–17.

The substances in a mixture may be spread evenly throughout the mixture. Such a mixture is said to be *homogeneous* (hoh muh JEE nee us). Air, for example, is a homogeneous mixture of several different gases. The gases in air include nitrogen, oxygen, carbon dioxide, water vapor, and a few others.

Figure 3–17

A Mixture. In a mixture of salt and iron filings, each substance retains its own properties. A magnet can remove the iron filings from the dry mixture, leaving the salt behind. ▼

Solutions

Although any homogeneous mixture can be called a **solution,** the term is usually used for mixtures that are liquid. The liquid substance that makes up the bulk of the solution is called the **solvent.** The other substances, which are dissolved in the solvent, are called **solutes** (SAHL yoots). Solutes may be solids, liquids, or gases before they are dissolved in the solvent. The most common solutions have water as the solvent.

When a molecular substance dissolves in a liquid, the substance separates into its individual molecules. The solute is spread through the solvent in the form of separate molecules. For example, when sugar dissolves in water, the molecules of sugar spread throughout the water. When an ionic substance dissolves, the compound breaks into its ions. Thus, when sodium chloride dissolves in water, it breaks into sodium and chloride ions. See Figure 3–18. This process is called *dissociation* (dis soh see AY shun). It can be shown by the following equation:

$$NaCl \longrightarrow Na^+ + Cl^-$$

You will learn in later chapters that many important processes in living cells and tissues depend upon the presence of dissolved ions.

Suspensions

There are many substances that do not dissolve noticeably in water. For example, sand is insoluble in water. If you put sand in a pail of water and shake the water, the sand will form a cloudy mixture with the water. If you let this mixture stand, tiny particles of sand will slowly settle to the bottom of the pail. A mixture that separates on standing is called a **suspension** (suh SPEN shun).

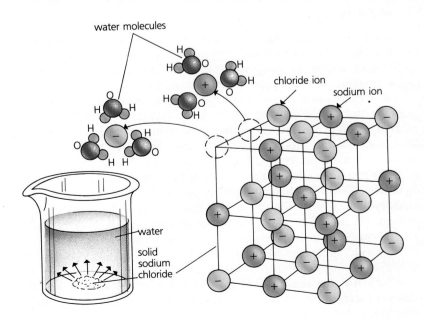

Figure 3–18

Dissociation of an Ionic Compound. When sodium chloride is dissolved in water, the structure of the salt breaks down. The compound separates into its individual ions, which become surrounded by water molecules. A solution of sodium chloride is formed. ▶

◄ **Figure 3–19**
Colloidal Dispersions and Solutions. Note the cloudiness of the colloidal dispersion of clay particles in water on the left. The solution on the right is completely transparent.

Colloidal Dispersions

In a true solution, the particles of the solute are either molecules or ions. They remain spread throughout the solvent indefinitely. In a suspension, the particles are large enough to give the liquid a cloudy appearance, and the force of gravity gradually causes them to settle out of the solvent. There is still another type of mixture known as a **colloidal dispersion** (kuh LOYD ul dis PER zhun). In a colloidal dispersion, the particles are larger than molecules or ions but too small to settle out.

The medium in which a colloidal dispersion forms does not need to be a liquid. It can be a gas or even a solid. The dispersed substance also may be a solid, liquid, or gas. Smoke, for example, is a colloidal dispersion of carbon particles in air. Milk and mayonnaise are colloidal dispersions of several liquids. Whipped cream is a colloidal dispersion of a gas (air) in a liquid.

Solutions, suspensions, and colloidal dispersions are all present in living cells and tissues. All of the activities of life depend upon the special properties that each one of these three different types of mixtures possesses.

3-3 Section Review

1. What is a mixture?
2. What are substances that dissolve in a solvent called?
3. What happens to particles in a suspension?

Critical Thinking

4. How large are the particles in a colloidal dispersion compared to the particles found in a solution and to the particles found in a suspension? (*Ordering*)

Science, Technology and Society

Issue: Acid Rain

Many scientists blame acid rain for the destruction of pine forests and lakes in the eastern United States. Studies also link acid rain to the corrosion of buildings and statues and to possible harmful effects on human health. These scientists believe that industry and automobile emissions are responsible for the acid rain problem. They fear that the environmental damage soon may be irreversible. To control acid rain, they claim, strict regulations must be imposed, especially on industrial emissions.

Opponents of regulation argue that the causes of acid rain are not yet understood. They point to studies showing that natural processes, not acid rain, are to blame for the damage. Other studies have concluded that motor vehicle emissions are the major source of acid rain. Thus, this group argues, industry controls will not cure the problem. Instead, controls will reduce the industries' efficiency and waste the public's money.

■ *Should we regulate industries to control acid rain? Why or why not?*

3-4 Acids, Bases, and Salts

Section Objectives:

■ *Define* and *compare* acids and bases and give examples of each.
■ *Describe* what happens in a neutralization reaction.
■ *Explain* the meaning of the pH scale and what is indicated by pH values of 1, 7, and 14.
■ *Explain* how pH indicators are used.
  Laboratory Investigation: *Measure* the pH of various substances and perform a neutralization reaction (p. 52).

There are many compounds that form ions when they are dissolved in water. Two important groups of these compounds are the acids and the bases. How are these produced?

Acids and Bases

An **acid** is any compound that produces hydrogen ions in solution. All acids consist of molecules that contain hydrogen covalently bonded to another atom or group of atoms. When these molecules are dissolved in water, the hydrogen breaks loose as a hydrogen ion, H^+. The rest of the molecule forms a negative ion.

Hydrochloric acid, HCl, is one example of an acid that is important to life activities. When not mixed with water, hydrochloric acid is a gas made of HCl molecules. However, when it is dissolved in water, it separates into H^+ and Cl^- ions:

$$HCl \longrightarrow H^+ + Cl^-$$

Another example of an important acid is acid rain. Acid rain occurs when fossil fuels are burned, which releases oxides of sulfur and nitrogen into the air. These compounds then form acids when they dissolve in rain.

A compound that produces *hydroxide* (hy DRAHK syd) *ions* (OH^-) when dissolved in water is called a **base.** In the dry state, many bases are ionic compounds. Sodium hydroxide (NaOH) is one example. This is a solid compound made of sodium and hydroxide ions. When it is dissolved in water, it separates into its ions:

$$NaOH \longrightarrow Na^+ + OH^-$$

Neutralization

When solutions of an acid and a base are mixed, a reaction takes place. The hydrogen ions from the acid combine with the hydroxide ions from the base to form molecules of water:

$$H^+ + OH^- \longrightarrow HOH \text{ (or } H_2O)$$

If the quantities of acid and base are right, all the H^+ ions and the OH^- ions will combine, and there will be no excess of either one in the solution. The solution will be neither an acid nor a base. When

◀ **Figure 3–20**
Effects of Acid Rain. The damage to this leaf was caused by acid rain.

this occurs, the solution is said to be neutral. The process of reacting an acid and a base to produce a neutral solution is called **neutralization** (noo truh luh ZAY shun).

Acids and bases are both caustic. That is, in concentrated solutions, they can damage both living tissue and nonliving material. The best first-aid treatment for an acid or base spill is to flush the spill with water and consult a doctor. When an acid or base is accidentally spilled on nonliving matter, it can be neutralized. An acid can be neutralized with a base. A base can be neutralized with an acid.

Salts

When an acid and a base react, their hydrogen and hydroxide combine to form water molecules. This reaction removes these two ions from the solution. That is, they are no longer present as separate ions. However, the negative ions of the acid and the positive ions of the base are still present. For example, when hydrochloric acid reacts with sodium hydroxide, sodium and chloride ions remain in the solution. In other words, the neutralization reaction produces a solution of sodium and chloride ions.

$$HCl + NaOH \longrightarrow Na^+ + Cl^- + HOH$$

Solid sodium chloride can be obtained from the solution by evaporating the water.

The ionic compound produced by the neutralization reaction between an acid and a base is called a **salt.** Sodium chloride, or table salt, is actually only one of many different salts that can be formed by a neutralization reaction. Most of the substances in food that are called minerals are salts. Salts provide many essential ions for body processes.

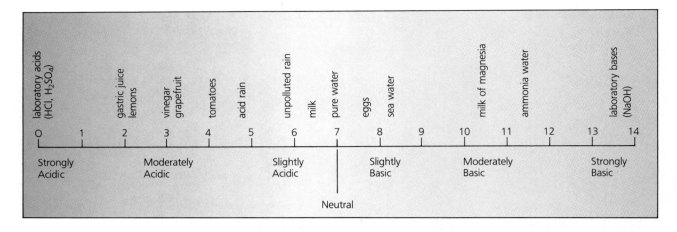

▲ **Figure 3–21**

The pH Scale. Each change of 1 unit on this scale is a change of 10 times in the hydrogen ion concentration. For example, a solution with a pH of 4 has 10 times as many hydrogen ions as a solution with a pH of 5.

Figure 3–22

Indicators. Phenolphthalein is an indicator that changes from colorless to red in moderately basic solutions. ▼

The pH Scale

We have said that water is the molecular compound, H_2O. However, at any given moment a small fraction of the molecules in water are broken into hydrogen and hydroxide ions:

$$H_2O \longrightarrow H^+ + OH^-$$

Since each water molecule produces 1 hydrogen ion and 1 hydroxide ion, there will be equal numbers of these ions in pure water. Water is therefore neutral because it has no excess of either H^+ or OH^- ions.

When an acid is dissolved in water, the concentration of H^+ ions increases. This means that in acid solutions, the H^+ concentration is greater than it is in pure water.

When a base is dissolved in water, the concentration of OH^- ions increases. These excess OH^- ions will react with the H^+ ions already present in the water. In doing so, they form additional molecules of water. This, in turn, reduces the concentration of H^+ ions. As a result, the H^+ concentration in basic solutions is less than it is in water.

The H^+ concentration is indicated by a unit of measure called **pH.** The pH scale has been set up in such a way that high concentrations of H^+ (acid solutions) have low values of pH. See Figure 3–21. Low concentrations of H^+ (basic solutions) have high pH values. The pH scale runs from 0 (highly acidic) to 14 (highly basic). A neutral solution has a pH of 7. This is the pH of pure water.

Whether a solution is acidic or basic can be shown by indicators. An **indicator** is a substance that changes color when the pH goes above or below a certain value. *Litmus* (LIT mus) *paper,* for example, turns red when the pH is moderately acidic (below 5); it turns blue when the pH is slightly basic (above 8). *Methyl orange* changes from yellow to red in moderately acidic solutions (pH below 3). *Phenolphthalein* (feen ul THAL leen) changes from colorless to red in moderately basic solutions (pH above 10). See Figure 3–22.

Soil Chemistry. Soil can be acidic, neutral, or basic. Most plants grow best in soil that has a pH between 6 and 7. However, some plants, such as blueberries, flourish in acidic soil that has a pH of about 5.

Indicators can show whether a solution is acid or basic. They do not, however, tell the actual value of the pH. There are special indicator papers that can be used to find pH more closely. This is done by wetting the paper with the solution and comparing the color of the indicator paper with a chart. These papers will determine pH within a few tenths of a unit. More accurate measurements can be made with pH meters. These work by measuring the electrical properties of the solution.

If you are a gardener, you probably know that the pH of the soil must be right for the plants you want to grow in your garden. See Figure 3–23. The pH levels of body tissues are also important for the body's activities. For example, the contents of the stomach must be slightly acid for digestion to take place normally. Keeping the pH at the correct levels in different parts of the body is a part of homeostasis.

3-4 **Section Review**

1. When acids are dissolved in water, what positive ion is produced in excess?
2. When bases are dissolved in water, an excess of which negative ion is produced?
3. What substances are formed from the reaction of an acid and a base?
4. What would be the pH of a neutral solution? An acid solution? A basic solution?

Critical Thinking

5. You are given a solution with a pH of 9. Predict the color produced after adding some of this solution to litmus paper. To phenolphthalein? To methyl orange? (*Predicting*)

Laboratory Investigation

An Investigation of pH

In this investigation, you will use pH indicators to measure the pH of various substances and to explore the effect of an acid on a base. You will also observe how a biological molecule is affected by a change in pH.

Objectives

- *Determine* the pH of a variety of common substances.
- *Perform* a neutralization reaction.
- *Observe* the effect of pH on the protein in milk.

Materials

white vinegar	bleach
fresh milk	shampoo
boiled water	litmus paper
lemon juice	wide-range pH paper
apple juice	test tubes
cola	droppers
ammonia	graduated cylinder

Procedure 🧍 🥽 🧪 🧤 *

A. Tear off eight 1-cm strips of litmus paper. Space the strips evenly along the length of a paper towel.

B. Place a single drop of each substance on a strip of litmus paper. Copy the chart below and record whether the substance is acidic, basic, or neutral. **Caution:** Do not mix ammonia and bleach.

C. Test the same substances again using wide-range pH paper. Determine the pH of each substance by matching the color of the test paper with the reference color chart.

Substance	Litmus Test (acidic basic, or neutral)	pH Paper (pH)
vinegar		
milk		

D. Copy the chart above. Pour 1 mL of ammonia into a test tube. Using wide-range pH paper, find the pH. Record it on the chart. Add vinegar, one drop at a time.

*For detailed information on safety symbols, see Appendix B.

After each drop, mix well, test the pH, and record it. Stop when the pH reaches 7.

Drops of Vinegar	pH
0	
1	

E. To explore the effect of low pH on milk protein, label two clean test tubes A and B. Pour 2 mL of milk into each tube. Add 6 drops of distilled water, one drop at a time, to tube A. After each drop, mix well and note any change. Using vinegar, repeat the procedure with test tube B.

Analysis and Conclusions

1. Which of the substances are acidic? Which are basic? Which are neutral?

2. Copy the scale below on a separate sheet. Write the name of each substance you tested at the point along the scale that corresponds to its pH.

3. Did the results using wide-range pH paper agree with the results using litmus paper? What additional information did the wide-range pH paper provide?

4. What happened to the pH of the ammonia as acid was added to it? Explain the results in terms of the chemical reaction that occurred.

5. Predict what would happen to the pH of the ammonia-vinegar solution if you continued to add acid to the tube.

6. In the controlled experiment in step E, what problem was being investigated? What was the variable? What served as the control? What can you conclude from the results?

7. In biological systems, pH is maintained within a narrow range. Based on your results in step E, why is this an important aspect of homeostasis?

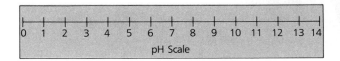

Chapter Review

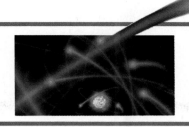

3

Study Outline

3–1 Atomic Theory of Matter

■ Matter is divided into elements, which are made of one kind of atom, and compounds, which are made of two or more kinds of atoms combined in definite proportions.

■ The nucleus of the atom contains neutrons, which have no charge, and protons, which are positively charged. Electrons, found outside the nucleus, are negatively charged.

■ The atomic number is the number of protons in the nucleus.

■ The mass number equals the sum of an atom's protons and neutrons.

■ Electrons are found in energy levels. Each energy level is a different distance from the nucleus.

■ Isotopes are varieties of the same element that have the same number of protons but different numbers of neutrons.

■ Radioisotopes have unstable nuclei that emit charged particles and radiation.

3–2 Chemical Bonding and Chemical Reactions

■ An atom can form a chemical bond if its outermost energy level is unfilled. There are two types of chemical bonds: ionic bonds, formed by exchanging electrons, and covalent bonds, formed by sharing electrons.

■ Diatomic molecules form when a covalent bond joins two atoms of the same element.

■ In a chemical formula for a compound, each element is represented by its chemical symbol. The subscripts in the formula give the proportions of each element.

■ A chemical equation represents the results of a chemical reaction. In a chemical reaction, matter is neither created nor destroyed.

3–3 Mixtures

■ Mixtures contain substances that are physically mixed but are not chemically bonded to each other.

■ The substances in a mixture can be present in any proportions.

■ There are three types of mixtures: solutions, suspensions, and colloidal dispersions.

3–4 Acids, Bases, and Salts

■ Acids release excess hydrogen ions (H^+) in solution. Bases release excess hydroxide ions (OH^-) when dissolved in water. A neutralization reaction results when an acid and a base react together to form a salt and water.

■ The hydrogen ion concentration of a solution is measured in units on the pH scale. Acids have a pH less than 7, bases have a pH greater than 7. The pH of a solution can be determined by indicators.

Vocabulary Review

atoms (34)	structural formula (43)
elements (34)	chemical reaction (44)
compounds (34)	reactants (44)
nucleus (34)	products (44)
protons (34)	law of conservation of mass (44)
neutrons (34)	
electrons (35)	mixture (45)
atomic number (35)	solution (46)
mass number (36)	solvent (46)
isotopes (37)	solutes (46)
radioactivity (38)	suspension (46)
radioisotopes (38)	colloidal dispersion (47)
molecule (39)	acid (48)
chemical bond (40)	base (48)
covalent bond (40)	neutralization (49)
diatomic molecule (40)	salt (49)
ion (41)	pH (50)
ionic bond (42)	indicator (50)
chemical formula (43)	

Chapter Review

A. Definitions—Replace the italicized definition with the correct vocabulary term.

1. *Negatively charged particles* are located in energy levels around the atomic nucleus.

2. Isotopes of an element have the same atomic number but a different *sum of protons and neutrons.*

3. Chemical bonds do not form in *substances that are physically mixed.*

4. The *number of protons in the nucleus* is different for each element.

5. Some substances in our surroundings are *composed entirely of one kind of atom.*

6. Atoms of the same element sometimes form *molecules consisting of two atoms.*

7. *A reaction of a base with an acid* produces a neutral solution.

8. *Two or more atoms combined in definite proportions* form most of the substances that are found in nature.

B. Multiple Choice—Choose the letter of the answer that best completes each statement.

9. Unstable isotopes that emit charged particles are known as: (a) compounds (b) diatomic molecules (c) ions (d) radioactive

10. A pH greater than 7 represents: (a) an element (b) a base (c) an acid (d) a salt

11. Varieties of an element that differ only in the number of neutrons are: (a) neutrons (b) ions (c) molecules (d) isotopes

12. A chlorine atom that gets an electron from a sodium atom becomes: (a) an ion (b) diatomic (c) a molecule (d) an isotope

13. A mixture in which particles settle to the bottom is called a: (a) colloidal dispersion (b) suspension (c) solution (d) solute

14. Excess hydrogen ions are produced in solution by: (a) acids (b) bases (c) salts (d) mixtures

15. In a neutral atom, the number of protons in the nucleus equals the number of: (a) electrons (b) the mass number (c) neutrons (d) molecules

Content Review

16. How can you determine chemically whether a substance is an element or a compound?

17. About how many elements exist?

18. Explain why electrons determine an atom's chemical properties.

19. For the following neutral atoms, give the number of electrons in the outer level: carbon (atomic number: 6), oxygen (8), chlorine (17), sulfur (16), argon (18), beryllium (4), magnesium (12), and phosphorus (15).

20. What is the pH scale?

21. What changes occur to a radioactive isotope?

22. Describe a covalent bond.

23. Explain what happens when an ionic bond forms between magnesium (atomic number: 12) and oxygen (atomic number: 8).

24. What is the difference between a chemical formula and a structural formula?

25. What is a balanced equation?

26. What happens when a molecular substance dissolves in a liquid?

27. What happens when an ionic compound dissolves in a liquid?

28. How does a colloidal dispersion differ from a suspension?

29. Why is water neutral on the pH scale?

30. Write the symbols for the three isotopes of oxygen having mass numbers of 16, 17, and 18.

31. What color is litmus paper if the pH is above 8? What color will it turn if the pH is below 5?

Graphic Organizing

For information on graphic organizers, see Appendix G at the back of this text.

32. **Scale:** Copy the scale below onto a separate piece of paper. Add the following elements to the scale (the mass number is indicated by the superscript): carbon (^{12}C), nitrogen (^{14}N), sodium (^{23}Na), phosphorus (^{31}P), sulfur (^{32}S), chlorine (^{35}Cl), and potassium (^{39}K).

^1H, neutron, proton ^{17}O

^2H ^3H ^{16}O ^{18}O ^{40}Ca

Low Mass Number   High Mass Number

Critical Thinking

33. How are the atoms of different elements alike? How do they differ? (*Comparing and Contrasting*)

34. Lithium fluoride (LiF) is an ionic compound like sodium chloride. What kind of three-dimensional structure does LiF have? Explain your answer. (*Reasoning by Analogy*)

35. If you put a small amount of lithium fluoride into water, what would happen? Explain why. (*Reasoning by Analogy*)

36. Acids produce excess H^+ ions in water. Bases, on the other hand, produce excess OH^- ions in solution. Which of the following are likely to be acids? Which are probably bases? HBr, LiOH, HCl, HNO_3, and H_2SO_4, NH_4OH, NaOH, $Zn(OH)_2$. (*Classifying*)

37. Suppose a person's stomach wall were irritated by an oversecretion of stomach acid. What would happen to the pH in the stomach if the person took milk of magnesia (a basic solution). Explain. (*Predicting*)

Creative Thinking

38. You are given two clear, colorless liquids and told that one is a compound and the other is a mixture. How could you determine which is the compound and which is the mixture?

39. What steps could be taken to reduce the increased acidity of rivers and lakes that have been damaged by acid rain?

Problem Solving

40. For each of the following mixtures, design an experiment that will separate the individual components: (a) sugar and water, (b) NaCl (salt) and sand, (c) sand and pebbles.

41. Using radioisotopes to monitor the flow of water, design a controlled experiment to test the following hypothesis: The flow of water in plants is faster in thinner branches than in thicker branches. What should be the variable in your experiment? What response should be measured? What factors should be held constant?

42. To study the structure of atoms, Ernest Rutherford did the experiment shown in the diagram below. He bombarded a thin sheet of metal foil with tiny, subatomic particles. He found that most of the particles passed through the metal foil without seeming to have any effect on it. However, a few of the particles bounced back. What did his results tell him about the structure of the atom?

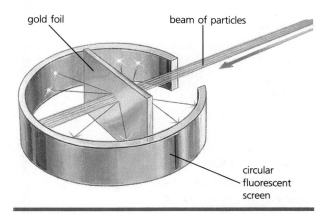

gold foil   beam of particles

circular fluorescent screen

Projects

43. Unpolluted rain water has a pH of 5.6, not 7, as you might expect. Acid rain has a pH below 5.6. Collect samples of rain water, tap water, and pond water (if any is available) from your community. Measure the pH of each sample with pH paper supplied by your teacher. From the results of your example tests, is acid rain a problem in your area?

44. Radioactive isotopes play an important role in the diagnosis of disease. Research three isotopes, and find out which diseases they are used to detect. Determine how physicians use them for diagnosis and how they are traced in the body. Present your findings in an oral report to the class.

45. Astronomers believe that elements found in the human body, such as carbon, calcium, and iron, were once a part of the stars. Use the library to gather information on the origin of the elements. You may wish to read the article by Robert Kirschner entitled "Super Nova—Death of a Star," which appeared in *National Geographic*, May 1988. Write a science news article about this topic.

▲ Chemicals known as pigments give both the salamander and the leaf their color.

Chemical Compounds of Life

4

A salamander rests for a moment on a glossy, red leaf. Chemicals in both the salamander and the leaf produce pigments that give each its color. In this case, the colors blend, which may save the salamander from a predator. In this chapter, you will learn about the chemicals inside living things. Some chemicals form pigments, some form cell structures, others store and release energy, and still others contain genetic information. Each chemical has unique characteristics and plays an essential role in the processes of life.

Guide for Reading

Key words: organic compound, carbohydrate, lipid, nucleic acid, DNA, RNA, protein, enzyme

Questions to think about:

📖 Why is water so important for living things?

📖 What are the compounds that make up living things?

📖 Why are enzymes needed for most chemical reactions that take place in cells?

4-1 Biologically Important Compounds

Section Objectives:

- *Compare* and *contrast* organic and inorganic compounds.
- *Define* each of the following terms: *polar molecule, cohesion,* and *adhesion.*
- *Explain* why organic compounds are usually larger and more complex than inorganic compounds.
- *Name* the four major types of organic compounds found in living cells.

Organic and Inorganic Compounds

Every chemical compound is either an organic compound or an inorganic compound. All **organic compounds** contain the element carbon. Most organic compounds occur naturally only in living organisms or in their products, although many of them can be produced in chemistry laboratories. In addition to carbon, most organic compounds contain hydrogen, and many contain either oxygen or nitrogen, or both. Less frequently, the elements phosphorus, sulfur, iron, calcium, sodium, chlorine, magnesium, and potassium are found in organic compounds. Thus, of the more than 90 elements to be found in nature, only a few appear in organic compounds.

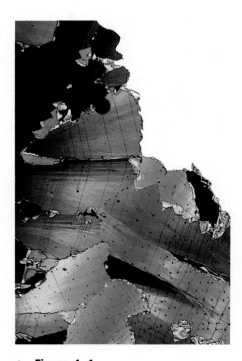

▲ **Figure 4–1**

Crystals of Sulfur. Of the more than 90 elements found in nature, only a few, such as sulfur, appear in organic compounds.

Every compound that is not an organic compound is an **inorganic compound.** Usually, inorganic compounds do not contain carbon. Carbon dioxide (CO_2) and other carbonate compounds like calcium carbonate ($CaCO_3$) are exceptions. Living organisms contain inorganic compounds as well as organic compounds. Water, carbon dioxide, various salts, and inorganic bases and acids are some of the inorganic compounds that are commonly found in living things.

At one time, people believed that there was something special about the chemistry of life. Many chemists were sure that the substances found in living things could not be made in a laboratory. Today, we know that all the chemical changes in living cells can be explained by the same principles that apply to chemistry in a test tube. The only difference is that most of the compounds in living cells are enormously complex. The reactions among these compounds are also complicated.

Water—An Important Inorganic Compound

Of all the inorganic compounds found in living things, water is the most important. All living organisms need water to survive, and most organisms contain water. About 65 percent of your body weight is water. People have been known to survive for weeks without food but only a few days without water. Water is essential because many biological processes can take place only in water solutions. The properties of water, which are discussed below, will help you to understand the way living things function.

Cohesion   As you read in Chapter 3, the oxygen atom in a water molecule shares a pair of electrons with each of the hydrogen atoms. But, as you can see from Figure 4-2, the electrons are not shared equally. Because oxygen has a stronger attraction for electrons than hydrogen does, the electrons are held closer to the oxygen atom. As a result, the oxygen end of a water molecule has a partial negative charge. The hydrogen end has a partial positive charge.

A molecule with regions of partial negative and partial positive charges is called a **polar molecule.** Because opposite charges attract each other, the negative end of one polar molecule will be attracted to the positive end of another polar molecule. This force of attraction between molecules of the same substance is called **cohesion** (koh HEE zhun).

It is the cohesion between water molecules that holds a drop of water together. Cohesion also explains why water can store heat better than most other liquids. When water is heated, the water molecules move faster and faster as the temperature rises. But, the molecules can move faster only if the cohesion that holds the molecules together is overcome. Therefore, some of the heat that is applied to the water will raise the temperature, but much of the heat must be used to overcome the cohesion. As a result, water is able to

▲ **Figure 4–2**

Water—A Polar Molecule. The oxygen end of the water molecule has a partial negative charge. The hydrogen end of the molecule has a positive charge.

absorb a great deal of heat without undergoing an abrupt change in temperature. This ability to store heat protects organisms from damaging changes in temperature. For example, if you become overheated, your body responds to this situation by producing sweat. As the sweat evaporates, the water in it absorbs heat from your body. Thus, the evaporation of sweat cools you by removing heat from your body.

Adhesion   The attraction between the molecules of one substance and the molecules of another substance is called **adhesion** (ad HEE zhun). Water adheres well to many substances because of its polar molecules. Adhesion makes water one of the best solvents. Water can dissolve most polar substances because the adhesion between the water molecules and the polar solute molecules is greater than the cohesion among the solute molecules. For example, HCl (a polar molecule) dissolves easily in water, but O_2 (a nonpolar molecule) does not. Water also can dissolve most ionic compounds.

The adhesion between water molecules and glass molecules is particularly strong. If you were to put two dry, glass slides together and then dip their ends into water, the water would be drawn up between the two slides. See Figure 4–3. This action is called **capillary action,** or **capillarity.**

Adhesion and cohesion both play important roles in the transport of water in plants. These forces help water to rise up through the roots into the plant. You will be reading more about this process in Chapter 19.

▲ **Figure 4–3**
Capillary Action. Water rises between the glass slides because of the adhesion between water and glass.

Structure and Types of Organic Compounds

The big difference between organic compounds and inorganic compounds lies in the greater size and complexity of many organic molecules. Organic molecules are often larger and more complex than inorganic molecules. This is because of the electron structure of the carbon atoms they contain.

A carbon atom has 6 electrons. As you can see in Figure 4–4, 2 of these electrons occupy the first energy level, which leaves 4 electrons for the second level. This means that the carbon atom can fill its outer energy level by forming 4 covalent bonds with other atoms. One electron in each bond comes from the carbon atom. The other electron comes from another atom bonded to the carbon. The atoms bonded to a carbon atom can be other carbon atoms. These atoms may be bonded into long chains. A long chain may have other groups of chains of atoms branching from it. Carbon atoms may also be bonded into rings with side branches or bonded with other rings. The possible size and variety of these arrangements is unlimited.

In many cases, the bond that is formed by the carbon atoms present in organic compounds is a double bond. That is, there are two bonds between the same pair of atoms. In a few cases, a pair of

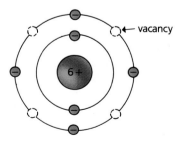

▲ **Figure 4–4**
Structure of the Carbon Atom. The four vacancies in the outer energy level allow a carbon atom to form four covalent bonds. This special characteristic accounts for the great variety of organic compounds.

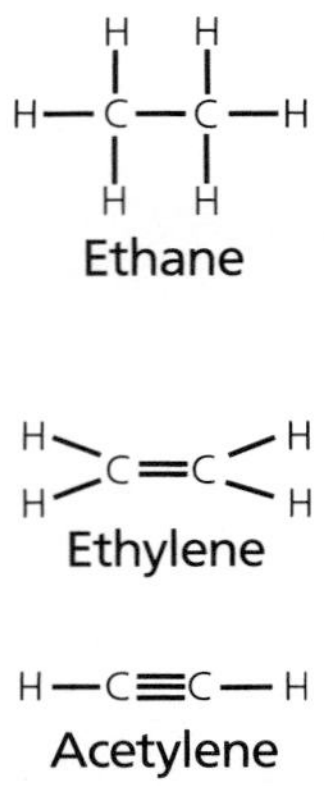

H—C—C—H
Ethane

C==C
Ethylene

H—C≡C—H
Acetylene

▲ **Figure 4–5**
Bonds Between Carbon Atoms. Carbon atoms can form single bonds (top), double bonds (middle), or triple bonds (bottom).

carbon atoms may be joined by a triple bond. Figure 4–5 provides several simple examples of possible single, double, and triple carbon-to-carbon bonds.

Although there are many organic compounds, they can be classified into a fairly small number of types. Four types of organic compounds will be discussed in this chapter—carbohydrates, lipids, nucleic acids, and proteins.

4-1 Section Review

1. In what ways is an organic compound different from an inorganic compound?
2. What four elements are most often found in organic compounds?
3. Which end of the water molecule is positively charged? Which end is negatively charged?
4. How many covalent bonds can a carbon atom form?

Critical Thinking

5. Compare the following compounds and classify them into three groups: organic, inorganic, not sure. Explain all "not sure" answers. (*Classifying*)

C_2H_6 NaOH C_2H_5OH Fe_2O_3 HCl   CO $C_{18}H_{36}O_2$

4-2 Carbohydrates and Lipids

Section Objectives:

- *Describe* the basic chemical makeup of carbohydrates.
- *Explain* the relationship between dehydration synthesis and hydrolysis.
- *Describe* how a fat is formed by dehydration synthesis.
- *Explain* the relationship between saturated fats, cholesterol, and circulatory disorders.

Carbohydrates

Carbohydrates (kar boh HY drayts) are compounds of carbon, hydrogen, and oxygen. Carbohydrates have the same ratio of hydrogen to oxygen as water has (2 atoms of hydrogen for every 1 atom of oxygen). The simplest carbohydrates are the simple sugars, or **monosaccharides** (mahn uh SAK uh ryds). The most common monosaccharides have the chemical formula, $C_6H_{12}O_6$. However, these atoms can be arranged in several different structures, each corresponding to a different sugar. Figure 4–6 shows the structural formulas of three simple sugars. Simple sugars with 5 carbons ($C_5H_{10}O_5$) and 4 carbons ($C_4H_8O_4$) also exist. The chemical names of the sugars always end in *-ose*.

The sugars are important because they contain large amounts of energy. This energy can be released in the presence of oxygen by breaking the sugars down to carbon dioxide and water. Nearly all organisms use glucose as a source of energy.

Dehydration Synthesis   Sugar molecules can be bonded together by a process called **dehydration synthesis** (dee hy DRAY shun SIN thuh sis). *Synthesis* means "putting together" and *dehydration* means "removing water." Thus, dehydration synthesis means "putting together by removing water." When sugar molecules are bonded together, the bond forms where an OH group is present in each molecule. At this point, 1 OH combines with the H from the other OH, forming a molecule of water. The 2 molecules are then joined through the remaining O. See Figure 4–7. In living cells, dehydration synthesis is brought about by the action of *enzymes,* which are discussed on page 68. It is an important process in making the many complex organic compounds that an organism needs.

The molecule formed by joining two simple sugars is called a double sugar, or **disaccharide** (dy SAK uh ryd). The disaccharide formed by the dehydration synthesis shown in Figure 4–7 is maltose. When several simple sugars are joined by dehydration synthesis, they form **polysaccharides** (pahl ee SAK uh ryds), or long chains of repeating sugar units. Large molecules consisting of chains of repeating units are called **polymers** (PAHL uh merz). The polysaccharides are sugar polymers. The largest polysaccharides may contain hundreds or thousands of sugar units.

Organisms store excess sugar in the form of polysaccharides. In plants, this form of stored sugar is called **starch.** Starch is found in seeds and in roots and stems specialized for food storage. In humans, surplus sugar is stored in the liver as the polysaccharide *glycogen* (GLY kuh jen), sometimes called "animal starch." Other types of polysaccharides form the tough structural parts of organisms. For example, *cellulose* (SEL yuh lohs) is a polysaccharide found in plants. *Chitin* (KYT un) is a polysaccharide that makes up the shells of insects.

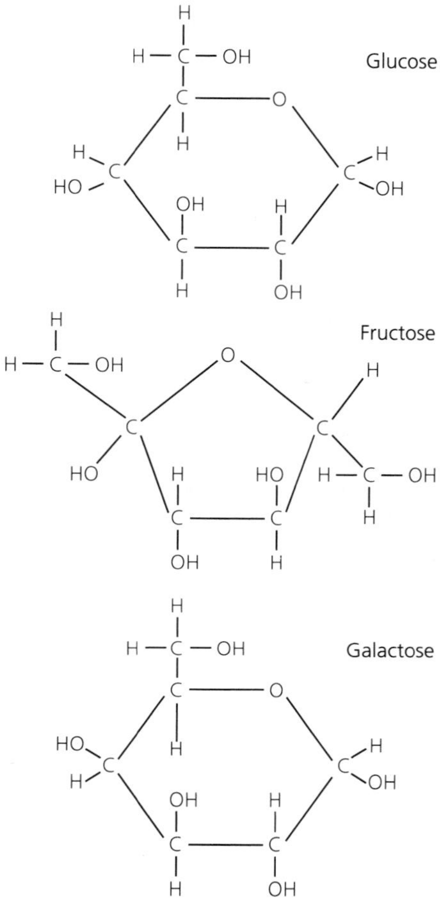

▲ **Figure 4–6**

Structure of Three Simple Sugars. Glucose, fructose, and galactose have the same chemical formula ($C_6H_{12}O_6$), but their atoms are arranged differently.

Figure 4–7

Dehydration Synthesis. In dehydration synthesis, two simple molecules bond together to form a more complex molecule. In this process, one or more molecules of water are released. ▼

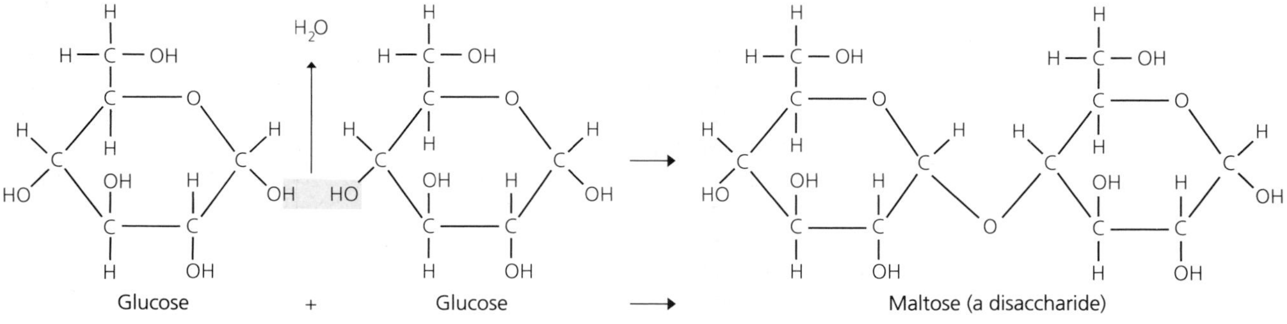

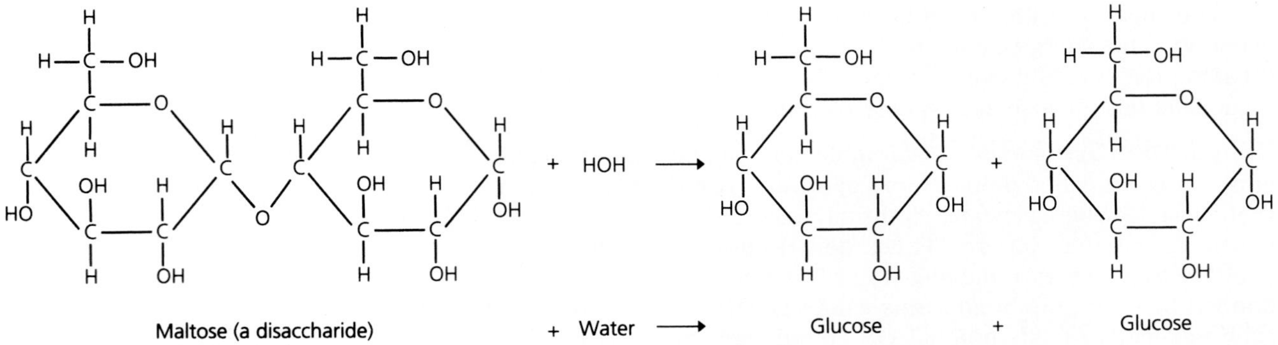

Maltose (a disaccharide)   + Water ⟶   Glucose   +   Glucose

▲ **Figure 4–8**
Hydrolysis. A complex molecule can be broken down into simpler molecules by the addition of a water molecule. Hydrolysis is the most common process used by organisms to change organic compounds into usable forms.

Hydrolysis Disaccharides and polysaccharides may be broken apart by a process called **hydrolysis** (hy DRAHL uh sis). As Figure 4–8 shows, in this type of reaction, a water molecule reacts with a chain of sugar molecules to produce two simpler sugars. When hydrolysis occurs, a bond between two simple sugars is broken, and the original OH groups are restored. Hydrolysis can be repeated on long chains until an entire polysaccharide has been split into its simple sugars. In living organisms, the process of hydrolysis, like dehydration synthesis, is brought about by the action of different enzymes.

Lipids

Lipids (LIP idz) include the substances commonly called fats, oils, and waxes. Like carbohydrates, lipids are made of carbon, hydrogen, and oxygen. However, there is less oxygen in lipids than in carbohydrates. Lipids are a part of cell structures and serve as a reserve energy supply in an organism. Lipids furnish about twice as much energy as the same amount of carbohydrates.

Fats and oils are chemically similar. Unlike fats, however, oils remain liquid at room temperature. Plants store oils in seeds. Some familiar oils are peanut oil, corn oil, and castor oil. Mammals store fat under the skin. There, it cushions the body and helps to stop heat loss. Although fats in animals are storage products, they are not stored for long. Instead, they are constantly broken down and replaced. Investigations have shown that mice, for example, replace about one-half of their stored fat each week.

The Formation of Lipids Waxes are lipids formed by the combination of fatty acids with compounds that are similar to glycerol. Fats and oils, on the other hand, are formed when fatty acids combine with glycerol. Glycerol is a simple 3-carbon chain with an OH group bonded to each carbon. A **fatty acid** molecule has two parts: a chain of carbon atoms to which hydrogen atoms are bonded and a **carboxyl** (kar BAHK sul) **group.** The carboxyl group consists of 1 carbon atom that is bonded to 1 oxygen atom by a double bond and to an OH group. Figure 4–9 shows a typical fatty acid.

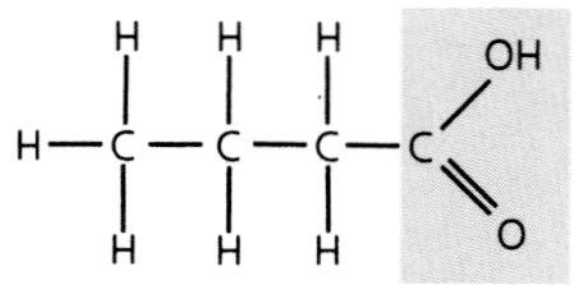

▲ **Figure 4–9**
Structure of a Fatty Acid. A fatty acid consists of a chain of carbon and hydrogen atoms with a carboxyl group at one end.

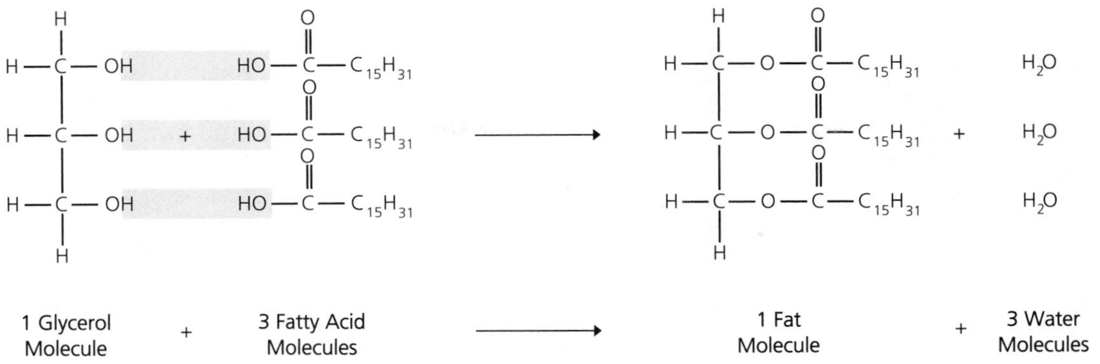

| 1 Glycerol Molecule | + | 3 Fatty Acid Molecules | ⟶ | 1 Fat Molecule | + | 3 Water Molecules |

As you can see in Figure 4–10, the dehydration synthesis of 3 fatty acid molecules and 1 glycerol molecule produces 1 molecule of fat or oil. When glycerol reacts with the 3 fatty acids, each fatty acid becomes attached to the glycerol molecule at one of the OH groups. In this dehydration synthesis, 3 molecules of water are released for each molecule of fat or oil that is formed.

Saturated and Unsaturated Fats　When all the carbon-to-carbon bonds in a fatty acid are single bonds, the acid is said to be saturated. Fats that are formed from fatty acids with single carbon-to-carbon bonds are called **saturated** (SATCH uh rayt ed) **fats.** In **unsaturated fats,** one or more pairs of carbon atoms in the fatty acid molecules are joined together by a double bond or even a triple bond. A fat that has chains with more than one double or triple bond is called *polyunsaturated.* A typical unsaturated fatty acid is shown in Figure 4–11.

In a saturated fatty acid, each carbon atom along the chain is bonded to 2 hydrogen atoms. (The carbon at the start of the chain is bonded to 3 hydrogens.) An unsaturated acid of the same length will have fewer hydrogen atoms. Unsaturated fats can be changed to saturated fats by adding hydrogen to them. This process is called *hydrogenation* (hy drahj uh NAY shun). Many processed foods contain partially hydrogenated fats.

There is an easy way to tell what kind of fats you are eating. Unsaturated fats tend to be oils; that is, they are liquid at room temperature. Saturated fats usually are solid.

There is evidence that saturated fats, such as those found in animal products like butter and meat, tend to increase the amount of cholesterol produced in the body. **Cholesterol** (kuh LES tuh rohl) is an essential compound, found in most animal tissues. However, it also plays an important part in the buildup of deposits that harden

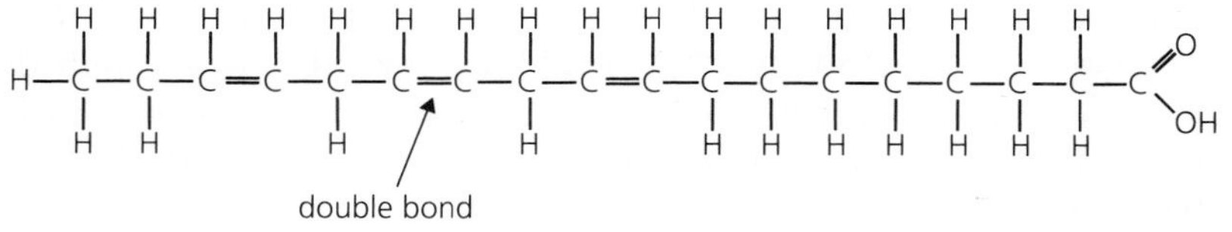

▲ **Figure 4–10**
Synthesis of a Fat. A molecule of fat is formed by the dehydration synthesis of three fatty acid molecules and one glycerol molecule.

double bond

◄ **Figure 4–11**
Structure of an Unsaturated Fatty Acid. An unsaturated fatty acid contains at least one double or triple carbon-carbon bond.

Biology and You

Q: I am a healthy, active teenager who is not overweight. Several times a week, my friends and I stop by a fast-food restaurant for hamburgers, fries, and shakes. My parents say that this is not good for me. Is it doing me any harm?

A: It may be. Fast-food meals are usually low in certain vitamins and minerals and high in fat, salt, and calories.

Diets high in fat and salt have been linked to heart disease and certain cancers in later years. Fried foods are especially high in cholesterol. Extra cholesterol can collect in your bloodstream and can eventually block the arteries to your heart, causing major heart problems. Because cholesterol starts to build up early in life, you need to limit your cholesterol intake even as a teenager.

Regarding calories, a typical fast-food meal (cheeseburger, fries, and a shake) provides as many as 1000 calories. Eating these meals can result in a gradual weight gain. Once growth has stopped and a taste for these foods has formed, the extra pounds are difficult to lose.

■ *List some ways that people can eat more nutritiously when eating at fast-food restaurants.*

and narrow the arteries. This condition can lead to heart attacks and strokes. Unsaturated fats, like those found in plant products, tend to decrease blood cholesterol levels. For these reasons, many medical authorities recommend a reduced intake of saturated fats and an increased intake of unsaturated fats in the diet.

4-2 Section Review

1. What is a monosaccharide, and how can two monosaccharides combine to form a larger molecule?
2. How are the complex sugar molecules broken apart?
3. What substances are commonly called lipids?
4. How is a fat formed from glycerol and a fatty acid?

Critical Thinking

5. Use what you know about carbohydrates and lipids to group the following compounds: (*Classifying*)

$$C_{12}H_{24}O_2 \quad C_6H_{12}O_6 \quad C_{10}H_{18}O_2 \quad C_{12}H_{22}O_{11}$$

4-3 Nucleic Acids and Proteins

Section Objectives:

■ *Describe* where the two types of nucleic acids are found and give the functions of these acids.
■ *Compare* the structures of DNA and RNA.
■ *Illustrate* the general molecular structure of an amino acid.
■ *Describe* a peptide bond and explain the difference between a polypeptide and a protein.

Nucleic Acids

Nucleic (noo KLAY ik) **acids** are compounds that contain phosphorous and nitrogen in addition to carbon, hydrogen, and oxygen. There are two kinds of nucleic acids. One is called **DNA**—*deoxyribonucleic* (dee AHK see ry boh noo KLAY ik) *acid.* The other is called **RNA**—*ribonucleic* (ry boh noo KLAY ik) *acid.* These substances were first found in the part of the cell called the nucleus. DNA is the hereditary material that is passed on from one generation to the next during reproduction. As you will learn in Chapter 26, working with RNA, DNA directs and controls the development and activities of all the cells in an organism.

The Structure of DNA   The DNA molecule is a long chain of repeating units, called **nucleotides** (NOO klee uh tyds). Each nucleotide consists of a 5-carbon sugar (*ribose* or *deoxyribose*) bonded to a *phosphate* group (PO_4) and a *nitrogenous base*. A nitrogenous base is an organic base that contains nitrogen. There

are only four nitrogenous bases in a DNA molecule. They are
adenine (AD uh neen), *thymine* (THY meen), *cytosine* (SYT uh seen),
and *guanine* (GWAH neen). In any particular pair of bases, an
adenine is always bonded to a thymine. A cytosine is always bonded
to a guanine. These four bases can be attached in any sequence
along the length of the molecule.

As you can see in Figure 4–12a, the shape of the DNA molecule
resembles the shape of a ladder—two sides connected to each
other by rungs. Each side of the molecule is a chain of nucleotides.
The bases act like the rungs of the ladder, bonding together the
nucleotides on each side. Notice that each "rung" consists of two,
and only two, bases. In human cells, a single DNA molecule may
have as many as 3 billion pairs of these bases. As you will learn in
Chapter 26, the sequence of bases acts as a code that determines
what proteins will be made in the cell. In turn, the proteins
determine the nature and activities of the cell.

The entire DNA molecule is coiled into the form called a
double helix (HEE liks). See Figure 4–12b. The double helix is
coiled upon itself many times. A human DNA molecule would be
about four centimeters long if it were stretched out in a straight line.
Repeated coilings enable it to fit into tiny structures in the cell.

The Structure of RNA   Although an RNA molecule is similar in
chemical composition to DNA, it has certain differences. The RNA
molecule consists of only one chain, or strand, of bases. The sugar
in RNA is ribose, not deoxyribose. The base thymine is replaced by
uracil (YOOR uh sil). RNA is involved in protein synthesis.

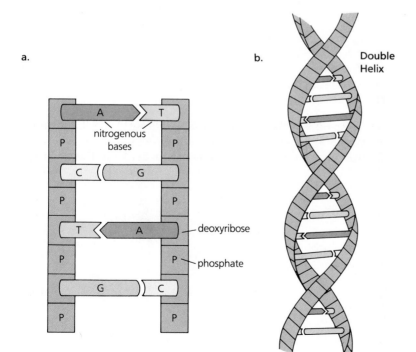

a.

b.   Double
Helix

nitrogenous
bases

deoxyribose

phosphate

◀   **Figure 4–12**

DNA Molecule. The nitrogenous bases in
DNA bond together, adenine (A) with thymine
(T) and cytosine (C) with guanine (G). The side
chains are formed of alternating deoxyribose
and phosphate groups (P). The entire molecule
is coiled in a shape called a double helix.

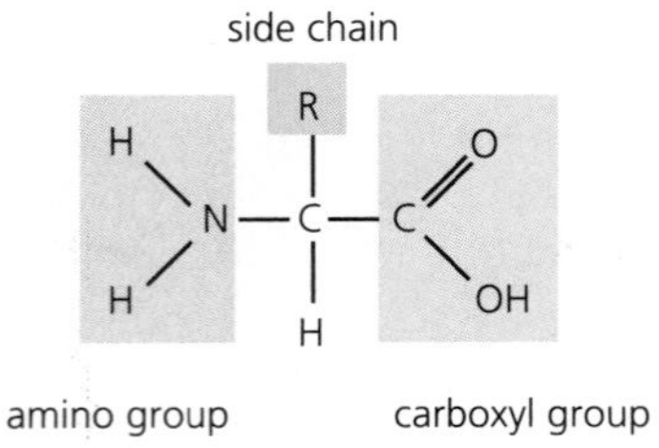

▲ **Figure 4–13**
Structure of an Amino Acid. The side chains of the 20 amino acids give each its special chemical properties.

Proteins

Proteins (PROH teenz) are compounds that contain nitrogen as well as carbon, hydrogen, and oxygen. Some proteins also contain sulfur and phosphorus. The number of possible proteins is virtually unlimited. Proteins have an astonishing range of properties. The reasons for this variety will become clear when you look at the way in which proteins are formed.

Proteins make life possible in its present degree of complexity. You can begin to understand the importance of proteins when you realize that they are found throughout living organisms. There are proteins in the structural parts of cells and body tissues, such as cartilage, bones, and muscles. There are proteins in hormones, the chemical messengers that regulate body functions in plants and animals. There also are proteins in antibodies, the substances that protect animals against disease organisms. Even enzymes, which allow complex chemical reactions to take place are proteins.

To understand how proteins do all this and much more, you need to understand their chemical makeup and structure.

Amino Acids   **Amino** (uh MEE noh) **acids** are the structural units of proteins. As you can see in Figure 4–13, an amino acid is a simple compound. It consists of a central carbon atom to which are bonded

- 1 carboxyl group (COOH)
- 1 **amino group** (NH_2)
- 1 hydrogen atom
- 1 side chain, symbolized by the letter R, which is different in each amino acid.

In the simplest amino acid, glycine, the side chain is just another H. In alanine, it is a CH_3 group. While other side chains may be more complex, none of the amino acid molecules is especially large. There are 20 different amino acids that are commonly found as parts of proteins.

The Peptide Bond   As you can see in Figure 4–14, two amino acids may be bonded together by dehydration synthesis. The bond forms between the amino group of one amino acid and the carboxyl

Figure 4–14

Formation of a Peptide Bond. In this dehydration synthesis reaction, a peptide bond forms between two amino acids. The resulting molecule is a dipeptide. ▼

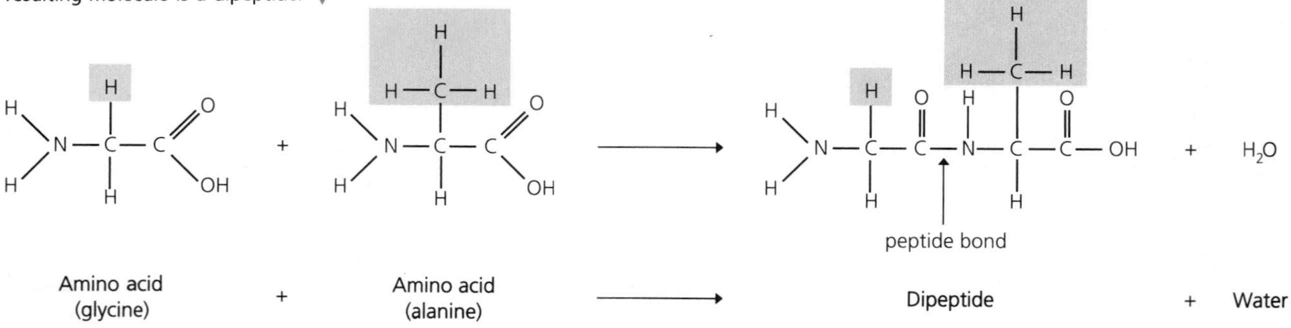

group of the other, with the resulting loss of 1 water molecule. The bond between two amino acids is called a **peptide** (PEP tyd) **bond.** The resulting molecule is called a **dipeptide.** (dy PEP tyd).

Amino acids can be added to either end of a dipeptide by dehydration synthesis. In this way, a long chain of amino acids can be formed. This type of a chain is called a **polypeptide.** All proteins are made of one or more polypeptides bonded together.

The Structure of Proteins   Amino acids can be linked together in any sequence and in chains of varying length. Each different sequence makes a different protein. Furthermore, the chains can fold and twist in space. Typical shapes of protein molecules are coils or helixes, pleated sheets, and globules. Often, neighboring sections of a folded chain become bonded to each other by what are called *cross-links.* It is the variations in the shapes and formations of cross-links that make possible the enormous variety of proteins.

All protein molecules contain many atoms and amino acid units. The smallest protein molecules have about 50 amino acids, or about 1000 atoms. The largest have over 100 000 amino acids and millions of atoms.

Determining the order of the amino acids in a particular protein is obviously a difficult task. The first protein structure to be determined was that of *insulin,* a hormone that controls blood-glucose levels. This was done by Frederick Sanger at Cambridge University, England, in 1954. Sanger later received the Nobel Prize for his work.

Today, the molecular structures of several hundred proteins have been worked out by painstaking methods. These methods include breaking the molecule into smaller and smaller pieces and identifying the amino acid at the end of each broken section. Machines are now being built that will analyze a protein and print out its amino acid sequence.

4-3 Section Review

1. Name the two types of nucleic acids.
2. Which nucleic acid occurs as a double helix?
3. List the elements found in both proteins and nucleic acids. What additional element is found in proteins?
4. What groups are found in each amino acid?

Critical Thinking

5. A molecule has been isolated from an animal's muscle cells. The molecule has over 1000 atoms and contains the elements carbon, hydrogen, nitrogen, oxygen, and phosphorus. What kind of molecule is it likely to be? Explain. (*Reasoning Categorically*)

Science, Technology and Society

Technology: Scanning Tunneling Microscope

A new device, called a scanning tunneling microscope (STM) can make high-magnification images of objects too small to be examined by an electron microscope. The objects an STM examines are molecules. The pictures it creates appear on a computer screen.

The key to the STM is a tiny probe with a slight electrical charge. The probe is positioned about .001 microns above the surface of a molecule. As the probe moves above the molecule's surface, electrons on the surface cross the gap between the surface and the probe tip. The number of electrons that "tunnel" through the gap depends on the gap's size. Because the molecule's surface is irregular, the probe moves up and down to keep the size of the gap constant. A computer translates the probe's movements into a map of the molecule's surface profile.

The resolution of the STM is so good that scientists can now view DNA (see photo), look at enzymes, or watch polymers form.

■ *Why would the ability to view an enzyme's structure be useful?*

4-4 Enzymes

Section Objectives:

- *Explain* the functions of enzymes in living cells.
- *Describe* the lock-and-key model and the induced-fit model of enzyme action.
- *Explain* the effects of temperature, pH, and enzyme and substrate concentrations on enzyme action.
- *Define* the term *coenzyme.*

Laboratory Investigation: *Test* for enzyme activity in animal tissues (p. 72).

The Importance of Enzymes

Enzymes (EN zymz) are protein substances that are necessary for most of the chemical reactions that occur in living cells. There is a major difference between chemical reactions in nonliving things and the reactions in living things. Consider, for example, the burning of gasoline in an automobile engine. The gasoline vapor is admitted to the engine cylinder. It is ignited by a spark. The vapor burns in a fraction of a second. In fact, the burning is so rapid that it produces a small explosion, which helps to drive the engine.

The chemical reactions of life are different. Although we sometimes say that glucose is "burned" to release energy in the cell, the "burning" occurs in dozens of small steps. In some of these steps, a small part of a molecule is removed. In others, a small group of atoms is added. In still others, atoms are rearranged within the molecule. These steps must occur with great precision and in the right order. They also must occur at ordinary temperatures inside the cell and must not give off large amounts of heat. Otherwise, the cell will be destroyed.

Enzymes make this possible in the living cell. For each step of a reaction, there is a particular enzyme at work. Enzymes enter into a chemical reaction only temporarily. Enzymes are not changed by the reaction. They are used again and again for the same chemical step with other molecules. A substance that brings about a reaction without being changed itself is called a **catalyst** (KAT uh list). Enzymes are organic catalysts.

The substance that an enzyme acts upon is called its **substrate** (SUB strayt). The names of enzymes usually end with the suffix -*ase*. The rest of the name is often derived from the name of the substrate. For example, the enzyme that splits maltose into 2 glucose molecules is called *maltase.* Enzymes that break down proteins are called *proteases* (PROH tee ay zez). Enzymes that break down lipids are called *lipases* (LY pay zez).

All enzymes in a living organism are made by the cells of the organism. Most of these enzymes are used within the cell in which they are made. However, some enzymes are passed out of the cell to catalyze reactions outside the cell. All the digestive enzymes that are produced in the human digestive tract are of this type. For example, pepsin is an enzyme made inside the cells of glands in the stomach wall. It leaves the cells and mixes with food in the stomach. Here, pepsin breaks down proteins in the food into simpler molecules.

How Enzymes Work

The ability of enzymes to act as catalysts depends on their shape. Somewhere on the surface of each enzyme there is a region called the **active site.** The substrate molecules fit the shape of the active site. See Figure 4–15a. When the substrate molecule comes in contact with the active site of the enzyme, it forms a temporary union with the enzyme. This is called an *enzyme-substrate complex.* During this time, the enzyme may break bonds within the substrate molecule. The substrate is thus separated into 2 smaller molecules.

An enzyme may also cause 2 molecules to join. In this case, there are two substrates. Each fits into the active site in such a way that they are brought into close contact. This enables bonds to form between the two substrate molecules.

There is a theory to explain how the enzyme and substrate fit together at an active site. It is called the *lock-and-key model.* Just as the notched surface of a key can open only one lock, the shape of the active site of an enzyme only fits certain substrates. Thus, each enzyme can catalyze a reaction only of those substrates.

Recently, another theory has been proposed to explain how some enzymes work. It is called the *induced-fit model.* In this model, the enzyme is not a rigid shape. Instead, the enzyme changes shape slightly as the substrate enters the active site. Once it has changed shape, the enzyme fits snugly around the substrate, as shown in Figure 4–15b. This induced fit is similar to the way a hand grasps a baseball. As the enzyme embraces the substrate, it can weaken the chemical bonds in the substrate, helping the reaction to proceed. In the induced-fit model, as in the lock-and-key model, each enzyme catalyzes a reaction only with certain substrates.

Figure 4–15

Two Models of Enzyme Action. a. In the lock-and-key model, the substrate fits exactly into the active site on the enzyme. **b.** In the induced fit model, the enzyme changes shape slightly to "grasp" the substrate at the active site. ▼

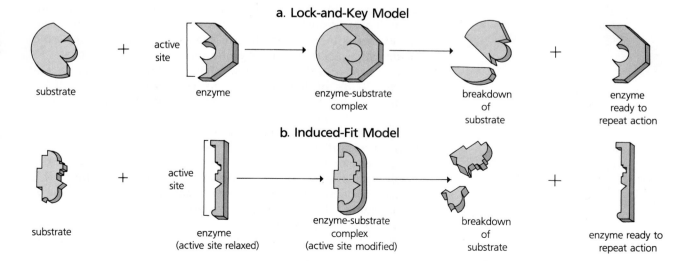

a. Lock-and-Key Model

substrate   +   active site   enzyme → enzyme-substrate complex → breakdown of substrate   +   enzyme ready to repeat action

b. Induced-Fit Model

substrate   +   active site   enzyme (active site relaxed) → enzyme-substrate complex (active site modified) → breakdown of substrate   +   enzyme ready to repeat action

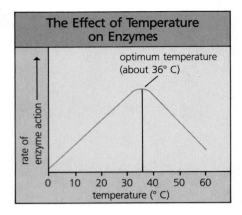

The Effect of Temperature on Enzymes

optimum temperature
(about 36° C)

rate of enzyme action

0   10   20   30   40   50   60
temperature (° C)

▲ **Figure 4–16**

The Effect of Temperature on the Rate of Enzyme Action. Most enzymes work best at normal cell temperatures.

Factors Affecting Enzyme Action

Experiments have shown that many factors can affect the action of enzymes in living cells. These factors, outlined below, affect only the rate of a catalyzed reaction. The products formed by the reaction do not change.

Small amounts of an enzyme can cause the reaction of large quantities of substrate. The time needed for an enzyme-substrate complex to form and a reaction to occur is very short. A single enzyme molecule can catalyze thousands of substrate reactions each second. Thus, only small amounts of any enzyme need to be present in a cell at any given time.

Enzymes enable cell reactions to take place at normal temperatures. Many chemical reactions that take place slowly at ordinary temperatures can be speeded up by raising temperatures. However, high temperatures can kill living cells. Enzymes speed up reactions in the cell without requiring high temperatures.

Enzymes work best at certain temperatures. Enzyme action depends on the random motion of molecules because this motion brings the substrates into contact with the enzymes. The motion increases as the temperature rises. If the temperature is low, the rate at which enzyme-substrate complexes form will be low, and the effect of the enzyme will be reduced. See Figure 4–16. At higher temperatures, the enzyme becomes more effective, because complexes are forming at a faster rate. At still higher temperatures, however, the enzyme itself starts to break down. This process is called *denaturation.* When the shape of the enzyme molecule changes, its active site no longer fits the substrate molecule, and it loses its effectiveness. There is a particular temperature—the optimum temperature—at which an enzyme is most effective. Optimum temperatures for enzymes in living cells are usually close to the normal cell temperature.

Each enzyme works best at a certain pH. The effectiveness of an enzyme depends on the pH of the surrounding medium. See Figure 4–17. The pH of the contents of the human stomach, for example, is slightly acidic. The enzyme *pepsin* starts the digestion of proteins in the stomach. Pepsin is most effective at this pH level. The pH in the intestine is slightly basic. At this pH, the enzyme *trypsin,* which continues the digestion of proteins, works best.

The rate of an enzyme-controlled reaction depends on the concentrations of the enzyme and substrate. The rate of an enzyme-controlled reaction depends on how often enzyme and substrate molecules bump into each other. When there is little enzyme but a great deal of substrate, the concentration of enzyme limits the rate of the reaction. In this case, the total number of enzyme molecules are acting on only a small fraction of the available substrate molecules. Adding more enzyme, therefore, increases the number of substrate molecules that can be reacting at any moment. Consequently, the reaction rate increases until a maximum rate is reached, as shown in Figure 4–18a. The reaction rate reaches a maximum when all substrate molecules are occupied

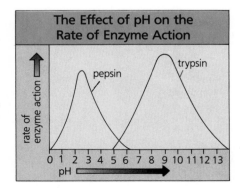

The Effect of pH on the Rate of Enzyme Action

rate of enzyme action

trypsin

pepsin

0 1 2 3 4 5 6 7 8 9 10 11 12 13
pH

▲ **Figure 4–17**

The Effect of pH on the Rate of Enzyme Action. Different enzymes work best at different ranges of pH.

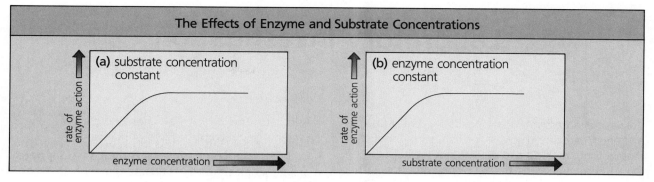

The Effects of Enzyme and Substrate Concentrations

continuously. After this point, adding more enzyme will not increase the rate of reaction.

Similarly, when there is less substrate than enzyme, the amount of substrate limits the reaction rate. That is, there are not enough substrate molecules to occupy all the enzyme molecules all the time. Thus, adding more substrate allows all the enzyme molecules to function simultaneously. Therefore, the reaction rate increases until a maximum rate is reached, as shown in Figure 4–18b. This rate is reached because all the enzyme molecules are occupied. Adding more substrate will have no effect.

The concentrations of enzyme or substrate determine reaction rates as long as both are freely accessible to each other. However, enzyme and substrate may not be accessible to each other when, for example, a membrane or other structure separates them. In this case, the rate at which the substrate crosses the membrane or other barrier limits the rate of the reaction.

Some enzymes need substances called coenzymes in order to function. **Coenzymes** (koh EN zymz) are organic substances but not proteins. A coenzyme allows an enzyme to perform its catalytic function. Some coenzymes are built into the structure of an enzyme. Others are separate molecules. During the formation of the enzyme-substrate complex, the coenzyme is changed in a way that helps the reaction. After the reaction, the coenzyme is restored to its original form. It is now known that some vitamins are coenzymes or are made into coenzymes in the cell.

▲ **Figure 4–18**
The Effects of Enzyme and Substrate Concentration on the Rate of Enzyme Action.

4-4 Section Review

1. What are enzymes?
2. Describe two models of enzyme action.
3. What factors affect the action of enzymes?

Critical Thinking

4. As you know, enzymes are not changed in chemical reactions. If enzymes were changed or were used up by reactions, how would this affect the rate of catalyzed reactions? (*Predicting*)

Laboratory Investigation

Enzyme Function

Catalase is an enzyme that speeds the breakdown of hydrogen peroxide (H_2O_2), a poisonous byproduct of metabolism, to H_2O and O_2. Without catalase, cells would be damaged by the H_2O_2 they produce. In this activity, you will use catalase in liver cells to see how various factors affect enzyme activity.

Objectives

■ *Observe* the effects of organic and inorganic catalysts on chemical reactions.

■ *Determine* how temperature, pH, and enzyme concentration affect enzyme activity.

Materials

fresh liver	test-tube rack	stirring rod
fine sand	test tubes	pH paper
distilled water	test-tube	boiling water
hydrogen peroxide	holder	bath
manganese dioxide	graduated	boiling chips
powder (MnO_2)	cylinder	salt-ice bath
sodium hydroxide	spatula	
hydrochloric	mortar and	
acid	pestle	

Procedure  *

A. Set up a test-tube rack with four test tubes, labeled 1 through 4. Pour 2 mL of H_2O_2 into each tube.

B. With a spatula, add a pinch of sand to tube 1 and a pinch of MnO_2 powder to tube 2. Watch for bubbling, which indicates the release of oxygen. Estimate the reaction rates using this scale: 0 = no bubbling; 1 = slow; 2 = moderate; 3 = fast. Record the rates on a chart like the one to the right.

C. To see how grinding tissue affects enzyme activity, grind one pea-sized piece of liver in a mortar with a little sand.

*For detailed information on safety symbols, see Appendix B.

Add it to tube 4. Place an unground piece in tube 3. Estimate and record the rates of bubbling.

D. Label four more test tubes, 5 through 8. Separately grind two pieces of liver, adding one each to tubes 6 and 8. Add a pinch of MnO_2 to tubes 5 and 7.

E. Place tubes 5 and 6 in a boiling water bath for 4 minutes. **Caution:** Exercise care when using the hot water bath. Place tubes 7 and 8 in a freezer or salt-ice bath for 4 minutes. Using a test-tube holder, remove the tubes and allow them to come to room temperature.

F. Add 2 mL of hydrogen peroxide to each tube and record the reaction rates on your chart.

G. Label three new test tubes 9, 10, and 11 and to each add ground tissue from one pea-sized piece of liver. Add 2 mL of water to tube 9, 2 mL of hydrochloric acid to tube 10, and 2 mL of sodium hydroxide to tube 11. **Caution:** HCl and NaOH are corrosives.

H. Test the pH of each tube, using a stirring rod and pH paper. Record the pH values.

I. Add 2 mL of hydrogen peroxide to each tube. Record the rates of bubbling in each tube on your chart.

Test Tube	Substance	Treatment	Reaction Rate
1	sand	———	
2	MnO_2	———	
3	liver	whole	

Analysis and Conclusions

1. Why did you test sand for its catalytic activity?

2. What effect did grinding the tissue have on the reaction rate? Propose a hypothesis that will explain this effect.

3. Using your data, draw a bar graph showing the relative reaction rates for all the substances.

4. How does pH affect the activity of catalase?

5. Predict the effect of high pH or low pH on the catalytic activity of manganese dioxide. Explain.

Chapter Review

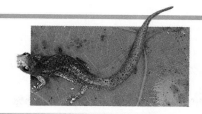

4

Study Outline

4–1 Biologically Important Compounds

■ All organic compounds contain carbon, whereas inorganic compounds do not.

■ Water's unique properties help account for the functioning of living things.

■ Organic compounds are often large and complex because of the chemical nature of the carbon atom.

4–2 Carbohydrates and Lipids

■ Carbohydrates are compounds containing carbon, hydrogen, and oxygen. The simplest carbohydrates are the simple sugars or monosaccharides.

■ Two monosaccharides can be bonded together by the process of dehydration synthesis to form a disaccharide.

■ Polysaccharides are polymers consisting of monosaccharides bonded together by dehydration synthesis. Polysaccharides can be broken into simple sugars by hydrolysis.

■ Lipids are composed of carbon, hydrogen, and oxygen and include fats, oils, and waxes. Fats and oils are formed by combination of fatty acids and glycerol.

■ The carbon-to-carbon bonds in saturated fats are all single. One or more of the carbon-to-carbon bonds in unsaturated fats are double or triple. Saturated fats lead to increased production of cholesterol, which can contribute to heart disease.

4–3 Nucleic Acids and Proteins

■ Nucleic acids contain carbon, hydrogen, oxygen, phosphorus, and nitrogen. The two kinds of nucleic acids are DNA and RNA.

■ DNA contains hereditary information. Each DNA molecule contains two strands made up of many repeating nucleotide units. Each nucleotide consists of a nitrogenous base, a 5-carbon sugar, and a phosphate. The molecule has the shape of a long double helix.

■ RNA molecules are similar to DNA but have only a single strand. RNA plays a role in protein synthesis.

■ Proteins are composed of amino acid molecules containing the elements carbon, hydrogen, oxygen, and nitrogen. Amino acids are linked by peptide bonds.

■ Proteins differ in their sequence, number of amino acids, and cross-links.

4–4 Enzymes

■ Enzymes are proteins that act as catalysts for biochemical reactions that occur in the living cell.

■ Each enzyme reacts with a specific substance called a substrate. The enzyme may break bonds within substrate molecules.

■ The two theories of enzyme action are the lock-and-key model and the induced-fit model.

■ Reaction rates depend on the temperatures, the pH, and the concentration of enzyme and substrate. Enzymes sometimes work in conjunction with coenzymes.

Vocabulary Review

organic compounds (57)

inorganic compounds (58)

polar molecule (58)

cohesion (58)

adhesion (59)

capillary action (59)

capillarity (59)

carbohydrates (60)

monosaccharides (60)

dehydration synthesis (61)

disaccharide (61)

polysaccharides (61)

polymers (61)

starch (61)

hydrolysis (62)

lipids (62)

fatty acid (62)

carboxyl group (62)

saturated fats (63)

unsaturated fats (63)

cholesterol (63)

nucleic acids (64)

DNA (64)

RNA (64)

nucleotides (64)

proteins (66)

amino acids (66)

amino group (66)

peptide bond (67)

dipeptide (67)

polypeptide (67)

enzymes (68)

catalyst (68)

substrate (68)

active site (69)

coenzymes (71)

Chapter Review

A. Word Relationships—In each of the following sets of terms, three of the terms have a common characteristic. One term does not belong. Identify the term that does not belong.

1. simple sugar, disaccharide, polysaccharide, amino acid

2. fatty acid, unsaturated fat, cholesterol, polypeptide

3. protein, disaccharide, amino acid, amino group

4. DNA, fatty acid, RNA, nucleic acid

5. enzyme, catalyst, carboxyl group, substrate

6. inorganic compound, lipid, carbohydrate, protein

B. Multiple Choice—Choose the letter of the answer that best completes each statement.

7. A type of reaction in which a bond is formed between sugar molecules is called (a) hydrolysis (b) dehydration synthesis (c) cohesion (d) adhesion.

8. The active site of an enzyme will fit only a specific (a) catalyst (b) lipid (c) substrate (d) cholesterol.

9. One type of carbohydrate is a(n) (a) dipeptide (b) polypeptide (c) amino acid (d) polysaccharide.

10. A reaction between a fat and water producing fatty acids and glycerol is (a) hydrolysis (b) dehydration synthesis (c) adhesion (d) cohesion.

11. A chemical group found in all fatty acids is a(n) (a) amino group (b) carboxyl group (c) peptide (d) starch.

12. Proteins are composed of (a) monosaccharides (b) fatty acids (c) lipids (d) amino acids.

13. DNA and RNA are examples of (a) nucleotides (b) nucleic acids (c) carbohydrates (d) proteins.

14. The force of attraction between molecules of the same substance is called (a) dehydration synthesis (b) hydrolysis (c) adhesion (d) cohesion.

Content Review

15. What element is present in all organic compounds?

16. Explain why water is a polar molecule.

17. How do cohesion and adhesion differ?

18. Describe capillary action.

19. What is a disaccharide?

20. In what form does the human body store excess sugar? In what form do plants store excess sugar?

21. List several functions of lipids in living organisms.

22. What structural feature of glycerol is important in the formation of fats?

23. What kind of organic compound involved in fat formation contains carboxyl groups?

24. Explain the difference between saturated and unsaturated fats.

25. Describe the general structures of DNA and RNA.

26. List several functions of proteins in living cells.

27. What are amino acids?

28. Describe the formation of a peptide bond.

29. Explain why the possible number of proteins is virtually unlimited.

30. How are most enzymes named?

31. Name the enzymes that react with the following substrates: (a) lipids, (b) maltose, (c) sucrose, (d) protein, (e) lactose.

32. Explain the lock-and-key model of enzyme action.

33. Describe how enzyme action is affected by temperature, pH, enzyme concentration, and substrate concentration.

34. What is a coenzyme, and what is its function?

Graphic Organizing

For information on graphic organizers, see Appendix G at the back of this text.

35. **Compare/Contrast Matrix:** Use the matrix below to compare the structure and functions of carbohydrates and proteins. Copy the matrix onto a separate sheet and complete it.

Characteristic	Carbohydrates	Proteins
building blocks	?	?
functions in organisms	?	?
examples	?	?

Critical Thinking

36. Classify the following substances as carbohydrates, lipids, proteins, or nucleic acids: maltose, chlorophyll,

deoxyribonucleic acid, vegetable oil, fructose, RNA, wax, glycogen, insulin, albumin. (*Classifying*)

37. Compare and contrast the dehydration synthesis of starch with the hydrolysis of glycogen. (*Comparing and Contrasting*)

38. Which of the functions of proteins would you rank as the most important? Why? (*Ranking*)

39. What might happen to a cell whose DNA is destroyed? What might happen to a cell whose DNA is changed? (*Predicting*)

40. In what respects is the fitting of a lock and key like the fitting of an enzyme and substrate? In what respects are the two processes different? (*Comparing and Contrasting*)

Creative Thinking

41. Organic molecules—and therefore life on this planet—are based on carbon. What might organic molecules be like if they were based on another element such as hydrogen?

42. "You are what you eat!" Explain the ways in which that statement is correct and the ways in which it is incorrect.

Problem Solving

43. Amylase is an enzyme that converts starch, by hydrolysis, into sugars. An experiment was performed to determine how rapidly amylase works at different temperatures. The data from this experiment are listed in the table below.

Temperature (°C)	Rate of Starch Conversion (grams/minute)
0	0.0
10	0.4
20	0.6
30	0.8
40	1.0
50	0.4
60	0.2
70	0.0

A. Identify the variable in this experiment. What response is being measured?
B. Graph the data in the table.

C. Draw a conclusion about the effect of temperature on the rate of activity of amylase.
D. According to the data, at what temperature does amylase function most efficiently? (This value is called the optimum temperature of the enzyme.)

44. If you were to study the enzyme amylase more carefully, you would find that its actual optimum temperature is 37°C. Design an experiment that would reveal this.

45. Amylase is present in saliva. If you were to chew a cracker for a long time without swallowing it, the cracker would start to taste sweet. Explain this change in terms of the enzyme action of amylase.

46. Egg white is made almost entirely of albumin, a protein. Raw egg white is translucent. When egg white is cooked, its properties change. Develop a hypothesis to explain these changes.

47. Two substances are found in the leaves of a desert plant. A biochemist analyzed the substances and determined that molecules in Substance 1 were made up of 12 atoms of carbon, 22 atoms of hydrogen, and 11 atoms of oxygen. Molecules in Substance 2 were made up of 2 atoms of carbon, 5 atoms of hydrogen, 1 atom of oxygen, and 1 atom of nitrogen. Identify the class to which each compound belongs.

Projects

48. Make a list of foods that you typically eat in a day. Use the labels to identify ingredients in prepared foods. Using the library as a resource, identify the principal organic nutrients in each food or food ingredient as carbohydrates, lipids, or proteins.

49. Heart disease and some kinds of cancer have been linked to diets that are high in fat and cholesterol. List ways in which you can lower the fat and cholesterol content of your diet. You may wish to consult *The American Heart Association Cookbook* (Ballantine Books, New York, 1987) as a reference.

50. Write a report on the life and work of Frederick Sanger. In your report, include a discussion of Sanger's investigation of the chemical structure of the hormone insulin.

51. Mix together some meat tenderizer and tap water. Place a small amount of hamburger that contains both red meat and fat into the mixture you have prepared. Describe your observations over a 24-hour period. Attempt to account for any changes you observe.

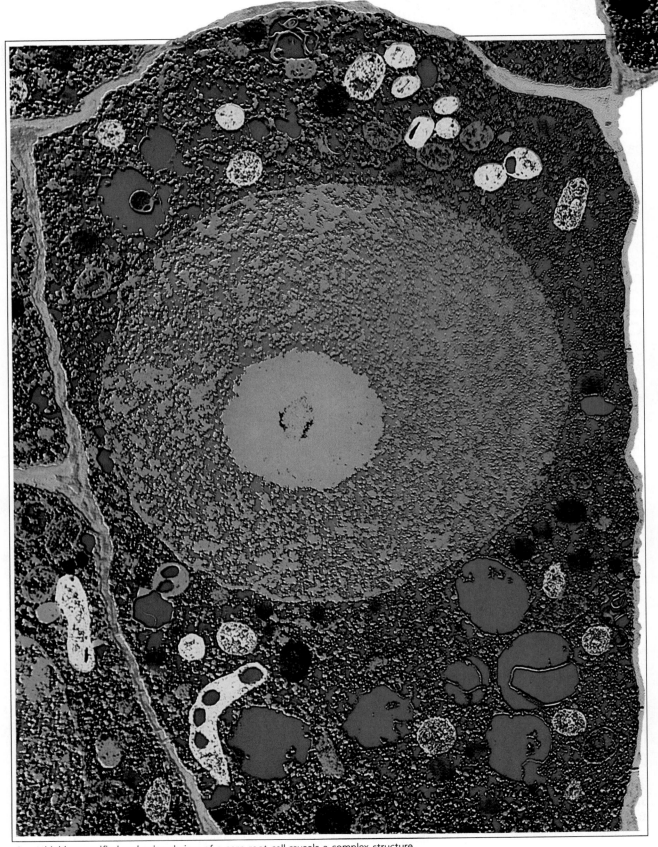

▲ A highly magnified and colored view of a corn root cell reveals a complex structure.

The Cell

5

The cell biologist carefully adjusts the microscope to focus on a slice of a corn root cell. The cell had been chemically fixed and then thinly sliced to form the specimen under study. Thus, the cell is frozen in time, capturing in exquisite detail the structure that underlies its functions. In life, the cell was only one of millions. Their collective and cooperative activities formed the whole plant. Each of these millions was the product of cell divisions that began with a single cell. In this chapter, you will discover some of what we know about the structure and functions of cells.

Guide for Reading

Key words: cell, prokaryotic, eukaryotic, organelles, diffusion, osmosis, tissue, organ
Questions to think about:
What are the basic features of a cell?
How is selective permeability involved in cellular homeostasis?
What are the different levels of organization in organisms?

5-1 What Is a Cell?

Section Objectives:

- *Describe* the contributions of the following scientists to the development of the cell theory: Robert Hooke, Anton van Leeuwenhoek, Robert Brown, Matthias Schleiden, Theodor Schwann, and Rudolf Virchow.
- *State* the cell theory.
- *Explain* the differences between the two basic types of cells, prokaryotic and eukaryotic.
- **Concept Laboratory:** *Gain* an understanding of how cell size is limited by the surface area-to-volume ratio (p. 81).

As you recall from Chapter 1, all living things are made up of small, individual units that usually cannot be seen with the naked eye. These units are called **cells.** Some organisms consist of one cell; others are made up of many cells. Whether one-celled or many-celled, however, the life processes of the organism are carried on by its cells.

Development of the Cell Theory

It was not until the mid-1600s that microscopes were used to study biological materials. In England, a scientist named Robert Hooke examined thin slices of cork and other plant tissues with a compound microscope. He found that these substances were made

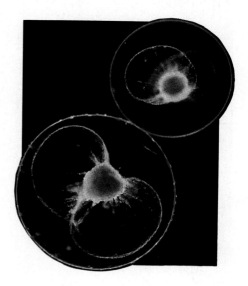

▲ **Figure 5–1**
Cellular Reproduction. This single-celled organism is dividing to produce two new cells.

of boxlike structures that he called *cells*. What Hooke saw were the walls of dead cells. He never studied living materials in which the contents of cells could be seen.

At the time that Hooke was making and using compound microscopes, Anton van Leeuwenhoek (LAY ven huke) in Holland was making single-lens microscopes of amazing power. Looking at drops of pond water with these microscopes, Leeuwenhoek saw living things that no one had ever seen before. We know now that many of these living things were one-celled organisms. Leeuwenhoek also observed and described human blood cells. In 1683, he described what must have been bacteria—the smallest kinds of cells. Leeuwenhoek, however, did not know that he was seeing single cells, and he drew no conclusions about the cellular nature of organisms.

It was not until the early 1800s that the cellular nature of biological materials began to receive attention. In 1824, Henry Dutrochet (doo troh SHAY) of France proposed that all living things were made of cells. In 1831, Robert Brown noticed that the small, dense, round body that had been observed in cells by other users of microscopes also appeared in all plant cells. He called this structure the *nucleus* (NOO klee us). However, the role of the nucleus in cell function was not known at this time.

In 1838, Matthias Schleiden (SHLY den) developed the theory that all plants were made of cells. In the following year, Theodor Schwann (shvahn) proposed that all animals were made of cells. In that same year, Johannes Purkinje (per KIN jee) used the term *protoplasm* to refer to the jellylike material that fills the cell. The last part of the cell theory was expressed by Rudolph Virchow (VIHR koh) in 1855, when he stated that all cells arise only from preexisting cells.

In 1861, Max Schultze (shults) defined protoplasm as "the physical basis of life." He proposed that this material was found in the cells of all types of organisms. At about the same time, Felix Dujardin (doo zhar DAHN) recognized the existence of one-celled organisms.

By the end of the 1800s, biologists had discovered many of the structures that lie within the cell. They were also able to describe the events of cell division in which one cell divides to form two cells.

The ideas that make up the cell theory are:

■ All organisms are made of one or more cells and the products of those cells. An organism may be a single cell, such as a bacterium, or many cells organized to function together, as in an animal or plant. In many-celled organisms, there may be intercellular material made by the cells.

■ All cells carry on life activities. The life activities of a many-celled organism are the combined result of the activities of its individual cells.

■ New cells arise only from other living cells by the process of cell division.

Figure 5–2

A Single-Lens Microscope. Van Leeuwenhoek's single-lens microscope (shown approximately actual size at top) could magnify single-celled organisms like those below about 200 times. ▼

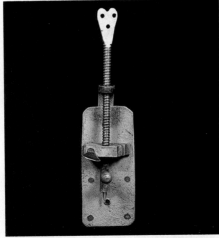

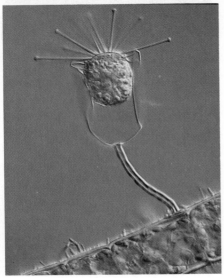

The Two Basic Cell Types

New and better instruments, such as the electron microscope, have allowed scientists to probe into the structure of living things in increasing detail. In doing so, biologists have discovered that there are two basic kinds of cells: prokaryotic (proh kar ee AHT ik) and eukaryotic (yoo kar ee AHT ik). These two types of cells have marked structural differences.

Prokaryotic cells, which are shown in Figure 5–3a, lack any internal membrane-bound structures. In other words, within a prokaryotic cell, membranes do not separate different areas of the cell from each other. Prokaryotic cells make up the smallest single-celled organisms, bacteria. Although membranes are present in some bacteria, these membranes are in contact with the rest of the cell's contents.

In contrast to prokaryotic cells, **eukaryotic cells** are present in all living things except bacteria. Eukaryotic cells have many kinds of internal membrane-bound structures. See Figure 5–3b. The most important of these is the **nucleus,** the structure in which the cell's hereditary material (DNA) is located. In fact, the term *eukaryotic* means "true nucleus." *Prokaryotic* means "without nucleus." Compared to prokaryotic cells, eukaryotic cells are much more compartmentalized. You will learn more about the nucleus and the other membrane-bound structures that are present in eukaryotic cells in the following sections.

Apart from their structural differences, prokaryotic and eukaryotic cells are fundamentally alike. Both types of cells, for example, are surrounded by a membrane that helps to keep their internal environment constant and different from their external environment. Both kinds of cells carry out the same life processes, using the same kinds of organic compounds—carbohydrates, fats, proteins, and nucleic acids—and the same kind of metabolic machinery.

Cell Size

The diameters of prokaryotic cells can range between 1 and 10 micrometers. Eukaryotic cells are, on the average, about 10 times larger, with diameters that can range between 10 and 100 micrometers. There are, however, exceptions to these various sizes. A chicken egg cell, for example, may be as large as 6 centimeters across, and some nerve cells, although thin, reach a total length of close to 1 meter.

The small size of cells has to do with the necessity of getting materials into and out of the cell at rates that will meet the cell's needs. Nutrients must be able to get into a cell at rates that will meet the cell's needs for nutrients. Similarly, wastes must be able to move out of a cell rapidly, so that they do not build up to harmful levels. What limits the rate of exchange of materials between the contents of a cell and its surroundings is the cell's surface area-to-volume ratio.

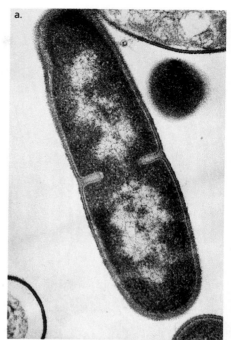

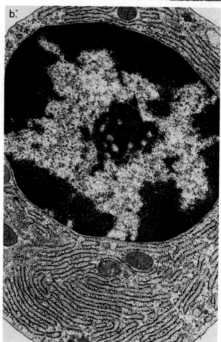

▲ **Figure 5–3**

Prokaryotic and Eukaryotic Cells. A prokaryotic cell (top) lacks a nucleus or any kind of membrane-bound organelles. In contrast, the eukaryotic cell (bottom) contains both a nucleus and many membrane-bound organelles.

Surface Area and Volume Relationships			
	1 cm 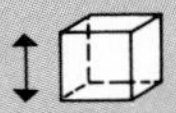	2 cm 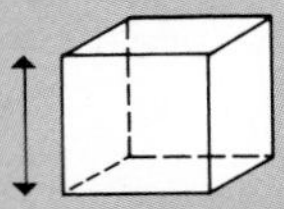	3 cm 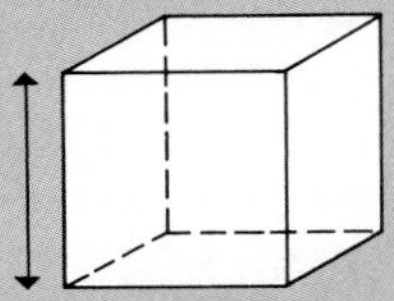
Length of Side (cm)	1	2	3
Total Surface Area (cm²)	(1 x 1 x 6) = 6	(2 x 2 x 6) = 24	(3 x 3 x 6) = 54
Total Volume (cm³)	(1 x 1 x 1) = 1	(2 x 2 x 2) = 8	(3 x 3 x 3) = 27
Surface Area-to-Volume Ratio	6/1 = 6	24/8 = 3	54/27 = 2

▲ **Figure 5–4**

Surface Area and Volume Relationships. The ratio of surface area to volume decreases as cell size increases.

To understand why this is so, consider that a cell is the approximate shape of a cube or a sphere. Substances move into and out of the cell by passing through the cell membrane that covers the entire surface of the cell. The more surface area there is for a given amount of cell volume (that is, the larger the surface area-to-volume ratio), the more materials a cell can exchange with its environment in a given amount of time. For small cells, the surface area-to-volume ratio is high. This means all parts of the cell are close to the external environment. As cell size increases, however, the ratio of surface area to volume becomes smaller, and many parts of the cell are farther from the environment. This is because volume increases proportionately more than surface area for any increases in the size of an object like a cube or sphere. As you can see in Figure 5–4, if the sides of a cube are doubled, the surface area increases fourfold, but the volume increases eightfold. Similarly, if the cube dimensions are increased by a factor of 3, the surface area increases 9 times, while the volume increases 27 times. Thus, as the surface area-to-volume ratios indicate, there is proportionately less surface area per cell volume in a large cell than in a smaller cell of the same shape. Surface area-to-volume relationships, therefore, place a limit on cell size.

5-1 Section Review

1. What instrument led to the discovery of cells?
2. How many cells make up an organism?
3. How do new cells arise?

Critical Thinking

4. List three similarities and three differences between eukaryotic and prokaryotic cells. (*Comparing and Contrasting*)

Concept Laboratory

Cell Size and Surface Area-to-Volume Ratios

Goal

To gain an understanding of how a cell's surface area-to-volume ratio limits its size.

Scenario

Imagine you are a cell biologist. You want to show that, as a cell gets larger, some areas within the cell become too distant from the external environment for adequate rates of exchange of materials.

As you perform this experiment, think about how the surface area-to-volume ratio of a cell affects how rapidly it can exchange materials with its environment.

Materials

raw potatoes	tweezers
paper cups	paper towels
potassium	beaker
permanganate	millimeter ruler
knife	calculator

Procedure  *

A. Using a knife, cut a slice about 5 cm thick from a potato. From the slice, cut 10 mm, 20 mm, and 40 mm cubes.

B. Place the cubes in a paper cup and cover them with potassium permanganate. Every 5 minutes for 20 minutes turn the cubes over with tweezers.

C. While the cubes are in solution, calculate the surface area (length × width × 6), volume (length × width × height), and surface area-to-volume ratio (surface area ÷ volume) for each cube. Record these numbers in a chart like the one below.

D. After 20 minutes, remove the cubes from the cup with tweezers and blot them dry on a paper towel. Pour the solution into a waste beaker.

E. Slice each cube in half. Measure and record the distance the permanganate diffused into each cube. See the figure below.

distance penetrated by permanganate

distance not penetrated

Organizing Your Data

1. For each cube, calculate the volume that was not penetrated by the permanganate. First, measure the distance not penetrated in each cube. Then, cube this number for the total, unpenetrated volume.

2. For each cube, calculate the percent of the total volume not penetrated: divide the unpenetrated volume by the total volume. Record these values.

Drawing Conclusions

1. What do the cubes of potato tissue represent? What does the permanganate represent?

2. How does the surface area-to-volume ratio and the percent of the total volume not penetrated vary as cube size increases?

3. How small must a potato cube be for the permanganate to completely penetrate it in 20 minutes?

Thinking Further

1. What would happen to a single-celled organism the size and shape of a 40 mm cube? Explain.

2. Suppose the organism in question 1 were made up of 64 000 cube-shaped cells, each 1 mm on a side. Why would this organism need a circulatory system?

3. Based on your results, explain why you think nearly all cells are smaller than 1 mm on a side.

Side of Cube (mm)	Surface Area (mm²)	Volume (mm³)	Surface Area-to-Volume	Distance Penetrated (mm)	Distance not Penetrated (mm)	Vol. not Penetrated (mm³)	% total vol. Not Penetrated
10	600	1000	0.6	2	6	216	21.6

*For detailed information on safety symbols, see Appendix B.

5-2 Cell Structure

Section Objectives:

- *Describe* the structure and functions of the following cell parts: *cell wall, cell membrane,* and *cytoplasm.*
- *Describe* the structure and functions of the following cell organelles: *nucleus, endoplasmic reticulum, ribosomes, Golgi bodies, lysosomes, mitochondria, microtubules, microfilaments, centrioles, cilia, flagella, vacuoles,* and *plastids.*
- *Compare* and *contrast* the structures of an animal cell and a plant cell.

Eukaryotic cells have many specialized internal structures, called **organelles.** Some organelles are enclosed in their own membrane while other organelles are not. Figure 5–5 shows the organelles of animal and plant cells. In reality, most plant and animal cells are specialized for specific functions and, therefore, do not appear the same as these idealized drawings. Referring to these two illustrations, however, will help as you read descriptions of the various organelles.

Cell Walls

The cells of most bacteria, various other microorganisms, and all plants are enclosed by a rigid **cell wall,** which lies just outside the cell membrane. The cell wall gives the cell its shape and provides protection for the cell. In plants, this wall is composed largely of *cellulose* (SEL yuh lohs). In other organisms, it may contain other compounds. The cell wall has many small openings that allow materials to pass to and from the cell membrane. Thin strands of cytoplasm sometimes extend through the walls of neighboring cells, which allow materials to pass directly from one cell to another. Animal cells do not have a cell wall.

The Cell Membrane

The **cell membrane,** or *plasma* (PLAZ muh) *membrane,* separates the cell from its surrounding environment. The membrane controls the movement of materials into and out of the cell, which makes it possible for the cell contents to be chemically different from the environment. The membrane keeps the internal conditions of the cell constant. It maintains homeostasis.

Structure of the Cell Membrane The cell membrane is a two-layered structure that is composed of lipids, proteins, and carbohydrates. The two layers are made of lipids with proteins embedded in them like mosaic tiles. Some of the proteins are on the outer surface of the membrane, some are on the inner surface, and some are thought to extend through the membrane. Carbohydrates, which are linked chemically to some membrane proteins or lipids, branch from the external surface of the membrane. Evidence

Animal Cell

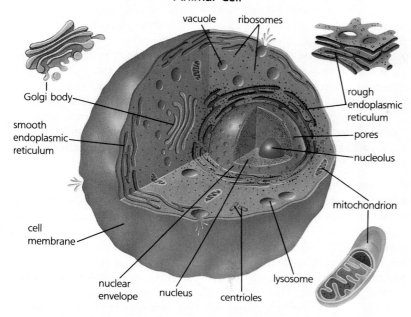

vacuole   ribosomes

Golgi body

smooth
endoplasmic
reticulum

rough
endoplasmic
reticulum

pores

nucleolus

mitochondrion

cell
membrane

nuclear
envelope   nucleus   centrioles   lysosome

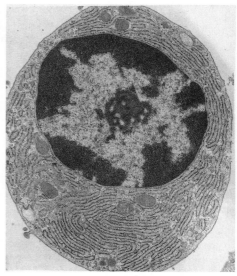

Plant Cell

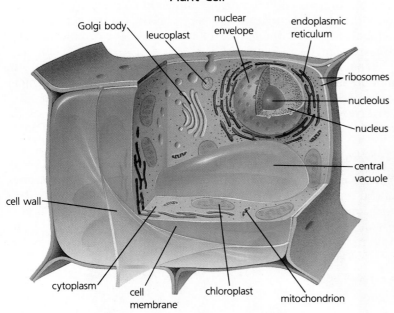

Golgi body   leucoplast   nuclear
envelope   endoplasmic
reticulum

ribosomes

nucleolus

nucleus

central
vacuole

cell wall

cytoplasm   cell
membrane   chloroplast   mitochondrion

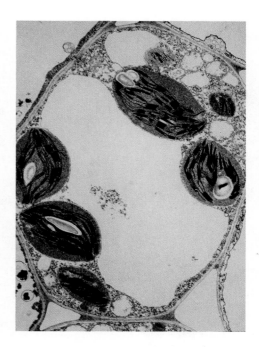

▲ **Figure 5–5**
Generalized Structure of Animal and Plant Cells. Animal and plant cells are
eukaryotic in structure with many membrane-bound organelles. Plant cells differ from
animal cells in having a cell wall, a large central vacuole, and plastids.

Outside Surface

carbohydrates

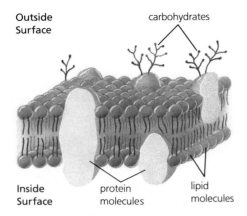

Inside Surface

protein molecules

lipid molecules

▲ **Figure 5–6**

Fluid-Mosaic Model of Cell Membrane. According to the fluid-mosaic model of the cell membrane, proteins are embedded in, but still able to move along, the lipid layer, which is in a fluid state.

indicates that the lipid and protein molecules of the membrane actually move along the membrane. This "fluid" property of the cell membrane has resulted in the concept that the membrane behaves like a fluid mosaic structure. See Figure 5–6.

The proteins of the cell membrane serve a number of functions. Some are *transport proteins.* They control the movement of substances through the membrane. Some act as *receptors,* that is, binding sites for specific messenger molecules that signal the cell to begin or to stop some metabolic activity. Other membrane proteins act as enzymes. Still others help bind the membrane to neighboring cells or to structural elements in the *cytoplasm* (SYT uh plaz um) of the cell.

Permeability of the Cell Membrane   The cell membrane is **selectively permeable.** That is, some substances can pass through it freely. Other substances can pass through only to some slight extent or only at certain times. Still other substances cannot pass through it at all. Through its selective permeability, the cell membrane regulates the chemical composition of the cell. The selectively permeable nature of the membrane is the result of the chemical and electrical properties of the membrane's molecules. The passage of materials through cell membranes is discussed in detail later in this chapter.

The Nucleus

The cell **nucleus** (plural, nuclei), shown in Figure 5–7, is a round, membrane-bound structure that serves as the control center for cell metabolism and reproduction. If it is removed, the cell dies. It is the largest organelle.

The membrane that surrounds the nucleus is called the **nuclear envelope.** It is actually a double membrane—two membranes lying close to each other. Like the cell membrane, it is selectively permeable. The inner and outer membranes of the nuclear envelope fuse at certain points, forming well-defined pores.

Figure 5–7

The Cell Nucleus. A double-membrane nuclear envelope with pores surrounds the nucleus, seen in this electron micrograph. The round, darkly stained structure within the nucleus is the nucleolus. ▼

cytoplasm

nuclear envelope

nucleoplasm

nucleolus

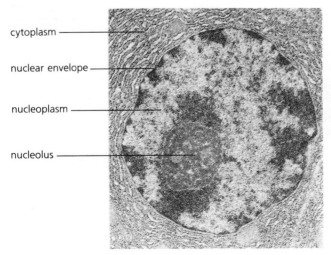

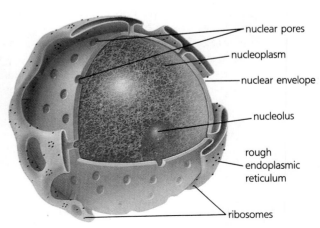

nuclear pores

nucleoplasm

nuclear envelope

nucleolus

rough endoplasmic reticulum

ribosomes

The pores control the passage of certain substances into and out of the nucleus. The selective permeability of the nuclear membrane allows the contents of the nucleus, the *nucleoplasm,* to remain chemically different from the rest of the cell.

Within the nucleus are one or more **nucleoli** (noo KLEE uh ly) (singular, nucleolus). These are dense, granular bodies that disappear at the beginning of cell division and reappear at the end. They are made up of DNA, RNA, and protein. Nucleoli are the sites of production of ribosomes.

Much of the nucleoplasm consists of chromatin. *Chromatin* (KROH muh tin) is DNA bound to various proteins. Chromatin in the form of long, thin threads makes up the structures called *chromosomes* (KROH muh sohmz). During cell division, the chromosomes shorten by coiling and become thick enough to be clearly visible when they are stained. The DNA in the chromosomes is the hereditary material of the cell.

The Cytoplasm

The watery material lying within the cell between the cell membrane and the nucleus is the **cytoplasm.** Many of the substances involved in cell metabolism are dissolved in the cytoplasm. In fact, many of the chemical reactions of cell metabolism take place in the cytoplasm. The cytoplasm also contains a variety of organelles that have specific functions in cell metabolism.

Endoplasmic Reticulum   The **endoplasmic reticulum** (en duh PLAZ mik rih TIK yuh lum) is a system of fluid-filled canals, or channels, enclosed by membranes. As shown in Figure 5–8, these canals usually form a continuous network throughout the cytoplasm. The canals of the endoplasmic reticulum serve as paths for the transport of materials through the cell. In addition, the membranes of the network provide a large surface area on which many biochemical reactions may occur. The endoplasmic reticulum also divides the cell into compartments, making it possible for a number of different reactions to go on at the same time.

The membranes of the endoplasmic reticulum are similar in structure to the cell membrane. In places, the membranes of the endoplasmic reticulum are joined to the outer membrane of the nuclear envelope. In electron micrographs, endoplasmic reticulum has either a rough or smooth appearance. In *rough endoplasmic reticulum,* the outer surfaces of the membranes are lined with tiny particles called *ribosomes* (RY buh sohmz). The ribosomes give the membrane a granular appearance. On *smooth endoplasmic reticulum,* there are no ribosomes.

Ribosomes   Small particles, called **ribosomes,** are sites of protein synthesis in the cell. They are found lining the membranes of the endoplasmic reticulum and in the cytoplasm. In cells that synthesize proteins to be released from the cell, the ribosomes are usually attached to the outer membrane surface of the endoplasmic reticulum. The proteins pass through these membranes into the

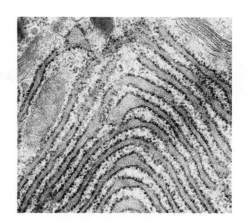

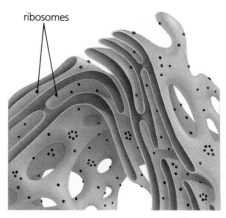

ribosomes

▲   **Figure 5–8**

Rough Endoplasmic Reticulum. The outer surfaces of the endoplasmic reticulum, shown in this electron micrograph (top), are covered with ribosomes, giving them a rough appearance. The endoplasmic reticulum serves as a system of channels through which proteins may be transported within the cell. It also divides the cell into many compartments.

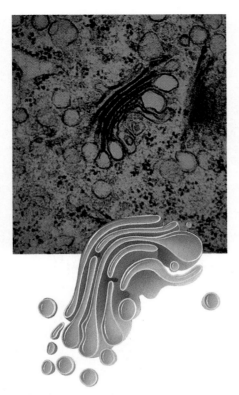

▲ **Figure 5–9**
Golgi Bodies. A Golgi body, shown here in cross section, is made up of stacks of flattened membrane sacs. Golgi bodies aid in the preparation and storage of molecules to be secreted by the cell.

canals, which carry the proteins to the cell membrane and out of the cell. Proteins that are to be used within the cell are synthesized on ribosomes that are free in the cytoplasm. These proteins are usually enzymes that function in the cell's cytoplasm.

Golgi Bodies Stacks of flattened membrane sacs make up the organelles called **Golgi** (GOHL jee) **bodies.** See Figure 5–9. Golgi bodies serve as processing, packaging, and storage centers for the products released from the cell. Animal cells usually have only one Golgi body, which frequently is located near the nucleus. Plant cells may have several hundred Golgi bodies.

In some studies, scientists have found connections between the Golgi body and the endoplasmic reticulum. Proteins synthesized on the ribosomes attached to the endoplasmic reticulum pass through the canals of the endoplasmic reticulum into the Golgi bodies. Here, they are packaged in vesicles. The vesicles migrate to the cell surface, where their membranes fuse with the cell membrane. The materials in the vesicle are then released outside the cell. The process in which materials made by the cell are released to the outside of the cell is called **secretion**. Cell products other than proteins also may be packaged in the Golgi body. In plant cells, the Golgi bodies are thought to be involved in assembling materials for the cell wall.

Lysosomes Small, saclike structures surrounded by a single membrane and containing strong digestive, or hydrolytic, enzymes are **lysosomes** (LY suh sohmz). Lysosomes are thought to be produced by the Golgi bodies. Lysosomes are found in most animal cells and in some plant cells. In one-celled organisms, lysosomes are involved in the digestion of food within the cell. In multicellular organisms, lysosomes serve several different functions. They break down worn-out cell organelles. In some animals, they are part of the body's defense against disease. White blood cells, which ingest disease-causing bacteria, contain lysosomes that break down the bacteria. Lysosomes are also involved in certain developmental processes. For example, as a frog develops from a tadpole to a mature frog, it loses its tail. Lysosomes are involved in the digestion and absorption of the tail.

Mitochondria Round or slipper-shaped organelles that release the energy in food molecules for use by the cell are called **mitochondria** (myt uh KAHN dree uh) (singular, mitochondrion). As shown in Figure 5–10, mitochondria are surrounded by a double membrane. The inner membrane is highly folded, forming *cristae* (KRIS tee) that extend into the middle of mitochondrion. The cristae of the mitochondria provide a large surface area on which many biochemical reactions occur. Cells that require large amounts of energy, such as muscle cells, contain large numbers of mitochondria. Because most of the energy needed by cells is released in the mitochondria, this organelle is often called "the powerhouse of the cell." The process by which the energy of food is released in the mitochondria and elsewhere in the cell is called *cellular respiration.*

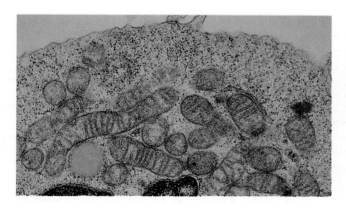

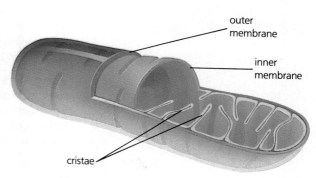

outer
membrane

inner
membrane

cristae

A cell may contain from 300 to 800 mitochondria, depending on its activity. Within the cell, the mitochondria are usually in motion, moving individually or in groups. They may also be found at specific locations within the cell. For example, in muscle cells, the mitochondria are found along the protein fibers that cause the muscle cell to contract. Mitochondria contain their own DNA and are capable of duplicating themselves.

Microtubules   Long, hollow, cylindrical structures found in the cytoplasm are **microtubules** (my kroh TOOB yoolz). As shown in Figure 5–11, they serve as a sort of "skeleton" for the cell, giving it shape. Microtubules are found in *centrioles* (SEN tree ohlz), *cilia* (SIL ee uh), and *flagella* (fluh JEL uh)—organelles discussed later in this section. They may also be involved in movement of the chromosomes during cell division. Microtubules are composed of a protein called *tubulin* (TOOB yuh lin). The molecules of this protein consist of two subunits that stack alternately in a helix. This gives the microtubule its form.

Microfilaments   Long, solid, threadlike strands found in some cells are called **microfilaments** (my kroh FIL uh ments). Most are composed of the protein *actin* (AK tin) and are associated with cell movement. Microfilaments are thought to have the capacity to contract and to be involved in the movement of cytoplasm within

▲ **Figure 5–10**

Mitochondria. Mitochondria, seen here in longitudinal and cross sections, carry out many of the reactions of cellular respiration. (Magnification 144 000 X)

Figure 5–11

Microtubules. The system of microtubules present throughout the cytoplasm is evident in this photomicrograph (left) of a specially prepared cell. (Magnification 2695 X) Cross sectional views of the microtubules within a flagellum can be seen in the electron micrograph (right). ▼

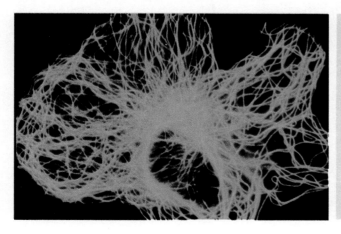

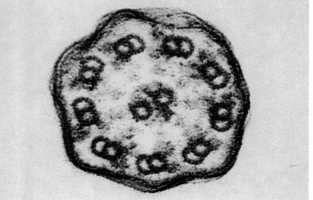

▲ **Figure 5–12**
Centrioles. An animal cell has a pair of centrioles near the nucleus. Each centriole consists of a ring of nine sets of three microtubules.

the cell, a phenomenon known as *cyclosis* (sy KLOH sis), or cytoplasmic streaming. Actin microfilaments are also found in muscle cells and are involved in muscle contraction. Some microfilaments are not made of actin and may serve as supporting structures for the cell.

Centrioles   Near the nucleus in animal cells is a pair of cylindrical **centrioles** that lie at right angles to each other. See Figure 5–12. Each centriole consists of a ring of nine groups of three microtubules. Centrioles are involved in cell division in animal cells. They are also found in the moving cells of algae, fungi, and plants.

Cilia and Flagella   Hairlike organelles with the capacity for movement are called **cilia** and **flagella.** See Figure 5–13. They extend from the surface of many different types of cells. Their structure is identical, except that flagella are longer than cilia. There are usually only a few flagella on a cell, but cilia may cover the entire cell surface. In one-celled organisms, cilia and flagella are involved in cell movement. In larger, many-celled animals, cilia serve to move substances over the surface of the cells.

Cilia and flagella arise from structures called *basal bodies.* The structure of a basal body is similar to that of a centriole. The cilia and flagella are slightly different in structure from the basal body. They have a ring of nine pairs of microtubules and, in the center of the ring, another pair of microtubules.

Vacuoles   **Vacuoles** (VAK yoo wohlz) are fluid-filled organelles enclosed by a membrane. The vacuoles found in plant cells are filled with a fluid called *cell sap.* In most mature plant cells, a single, large vacuole occupies most of the interior of the cell. In various microorganisms and simple animals, food is digested in special *food vacuoles* within the cells. Many of these organisms also have *contractile vacuoles* in which excess water from the cell collects. The water is periodically pumped out from the cell directly into the environment. Vacuoles may also serve as storage sites for certain cell products.

Figure 5–13
Cilia. The respiratory cilia (left) line the human trachea. An electron micrograph shows a row of cilia in cross section (right). Each cilium is made up of one central pair of microtubules surrounded by nine other pairs. (Magnification 192 500 X) ▼

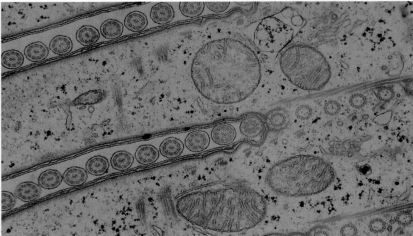

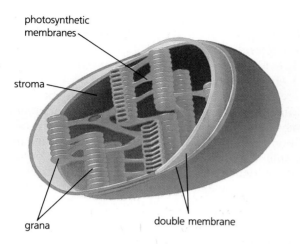

photosynthetic membranes

stroma

grana

double membrane

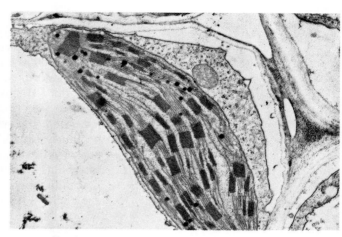

▲ **Figure 5–14**
The Chloroplast. The internal membranes of chloroplasts contain the photosynthetic pigments. These stacks of photosynthetic membranes are called grana. (Magnification 25 450 X)

Plastids   Membrane-enclosed organelles that are found only in the cells of photosynthetic, eukaryotic organisms are called **plastids.** Plastids are not present in the cells of animals or fungi. Like mitochondria, plastids are bounded by a double membrane and have systems of membranes within the organelle. There are two types of plastids. **Leucoplasts** (LOO kuh plasts) are colorless plastids in which starch or other plant nutrients are stored. **Chromoplasts** (KROH muh plasts) contain the pigments that give certain colors to fruits, flowers, and leaves. The most important type of chromoplasts are **chloroplasts** (KLOR uh plasts), which contain the green pigment *chlorophyll* (KLOR uh fil). The chloroplasts are the site of *photosynthesis,* the food-making process that uses light energy.

As shown in Figure 5–14, the inside of the chloroplast contains a system of photosynthetic membranes. These membranes are often arranged in the form of stacks, called *grana* (GRAH nuh). The pigments involved in photosynthesis are located in these membranes. The protein-containing material that fills the rest of the chloroplast is called the *stroma* (STROH muh). Chloroplasts, like mitochondria, contain their own DNA and have the ability to duplicate themselves.

Origins of the Eukaryotic Cell

The structural differences between eukaryotic and prokaryotic cells are so great that biologists have wondered how these two kinds of cells are related. The answer is not certain, but based on the evidence they have collected, biologists have developed a theory to explain the connection. According to this theory, known as the *endosymbiotic theory,* eukaryotic cells are the result of **endosymbiosis**—the condition in which one organism lives inside the cell of another organism to the benefit of both. Mitochondria and chloroplasts, for example, are thought to have evolved from bacteria that were engulfed by, and then lived within, other, larger cells. The bacteria that evolved into mitochondria were probably oxygen-consuming bacteria that supplied their host cells with energy, as mitochondria now do. The bacteria that evolved into chloroplasts may have been photosynthetic bacteria similar to today's blue-green bacteria.

There is strong evidence to support the endosymbiotic theory. Mitochondria and chloroplasts contain their own DNA and can reproduce themselves as can bacteria. They are about the same size as bacteria, and they have the same metabolic machinery, including ribosomes, on which they make some of their own proteins. In addition, endosymbiosis involving eukaryotic and prokaryotic organisms is known to occur in many modern organisms. This and other evidence has led many biologists to conclude that mitochondria and chloroplasts, and perhaps other organelles, are bacteria or other prokaryotic forms in disguise.

5-2 **Section Review**

1. What is the function of the cell membrane?
2. Where in the cell is the hereditary material located?
3. Name the function of mitochondria.
4. What organelle is the site of protein synthesis?

Critical Thinking

5. Suppose that a great improvement is made in the magnifying power and resolution of microscopes. What effect might this have on our understanding of the cell? (*Predicting*)

5-3 Maintaining a Constant Cell Environment

Section Objectives:

- *Relate* the structure of the cell membrane to the role of the cell membrane in maintaining homeostasis.
- *Describe* the roles of diffusion, facilitated diffusion, and osmosis in the passage of materials into and out of cells.
- *Explain* what is meant by *selective permeability, concentration gradient, turgor pressure,* and *plasmolysis.*
- *Compare* passive transport with active transport.

As you know, the internal environment of a cell remains relatively constant at all times. This condition, known as homeostasis, means that factors such as pH and the concentrations of all cell substances are held constant over the life of a cell. Cellular homeostasis occurs even though the external environment may be different from the conditions in the cell. If homeostasis is disrupted, a cell will die. The ability of cells to establish and maintain homeostasis is the result of the special properties of the cell membrane. These special properties are the subject of this section.

Selective Permeability of the Cell Membrane

Certain types of substances pass through cell membranes more easily than others. For example, lipid molecules and molecules that dissolve in lipids, such as alcohol, ether, and chloroform, pass through cell membranes easily. Small molecules, such as water, glucose, amino acids, carbon dioxide, and oxygen, also can pass through cell membranes easily. Large molecules, such as starch and proteins, cannot. Electrically neutral molecules enter and leave cells more easily than electrically charged ions. In addition, the permeability of cell membranes to certain substances varies from one type of cell to another. Even in the same cell, the permeability may vary from one moment to another. Thus, a given substance may pass freely through the cell membranes of one type of cell but not another, or it may pass through a cell membrane at one time but be held back at another.

Some of the mechanisms by which substances move through cell membranes will be explained later in the chapter. Before reading about these mechanisms, however, it is necessary to understand how molecules move from place to place and what determines the direction of their movement.

Diffusion

Molecules of gases and liquids are in constant motion. They move in straight lines in all directions until they run into other molecules or into the walls of their container. Collisions send them off in new directions so that their paths zigzag. As a result of this motion, the molecules of a substance tend to spread from a region in which they are more concentrated to regions in which they are less concentrated. For example, if instant coffee granules are placed at the bottom of a glass of water, a concentrated solution of coffee will first form near the bottom. See Figure 5–15. However, gradually, the coffee molecules will spread upward through the liquid. Similarly, if a perfume bottle is opened in one corner of a room, molecules of perfume will evaporate into the air near the bottle. At first, the odor will be noticeable only in the vicinity of the bottle. Eventually, the

Figure 5–15
Diffusion. (A) Instant coffee will, at first, be concentrated at the bottom of a glass of water. (B) With time, the coffee diffuses into other regions of the glass. (C) Eventually, the coffee becomes evenly distributed throughout the water. ▼

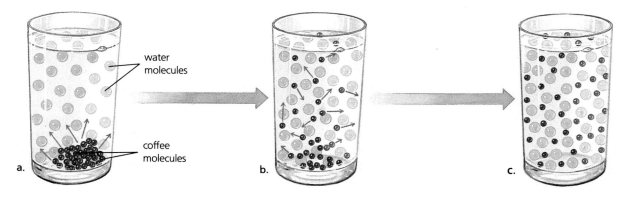

water
molecules

coffee
molecules

a.   b.   c.

odor will spread to all parts of the room. Both of these examples illustrate the process of diffusion. **Diffusion** is the movement of molecules or particles from an area of high concentration to an area of lower concentration. Diffusion occurs because molecules are in constant motion.

The difference in concentration between a region of high concentration and a region of lower concentration is called the **concentration gradient.** Diffusion occurs only if there is a concentration gradient. As a result of diffusion, the molecules become evenly distributed throughout the available space. Once they are evenly distributed, no further change in concentration occurs. The molecules are still in motion, but over any time period, as many molecules move out of a given area as move into the area. Such a situation is called an *equilibrium* (ee kwuh LIB ree um). In the example with instant coffee in water, equilibrium is reached when every drop of water in the glass contains the same amount of coffee.

Diffusion is important in the movement of molecules into and out of cells. Depending on the concentration gradient, certain materials will either enter or leave cells by diffusion. You can understand the importance of diffusion when you consider the role of diffusion in a cell using oxygen and producing carbon dioxide during gas exchange.

Oxygen and carbon dioxide will be in solution inside the cell and in the liquid medium surrounding the cell membrane. As the cell uses the oxygen dissolved in its cytoplasm, the concentration of oxygen inside the cell will decrease. At first, the concentration outside the cell will not change. Therefore, a concentration gradient toward the inside of the cell will develop across the cell membrane. As a result, oxygen will diffuse into the cell.

The opposite situation will develop over carbon dioxide. As the cell produces carbon dioxide, its concentration inside the cell increases, while its concentration outside remains the same. Therefore, a concentration gradient for carbon dioxide develops across the cell membrane toward the outside of the cell. Carbon dioxide diffuses out of the cell. If the concentration gradients of oxygen and carbon dioxide were reversed, then oxygen would leave the cell and carbon dioxide would enter. Thus, the process of diffusion plays an important role in the entry and exit of molecules in living cells. It should be noted that many different substances may be diffusing through the cell membrane at the same time without affecting each other's rate of diffusion.

Facilitated Diffusion

Some substances diffuse across cell membranes more rapidly than you might expect from the chemical properties of these substances. This type of transport, called **facilitated diffusion,** occurs because of specialized transport proteins in the cell membrane. Each type of transport protein is specific to the substance that it carries, just as an enzyme is specific for its substrate. Unlike enzymes, however, a transport protein does not cause a chemical reaction. Rather, it

---

Can You Explain This?

A   B

molecules of iodine   molecules of water

Suppose an apparatus is set up as shown above. Two chambers of a container are separated by a membrane permeable to both water and iodine. Chamber A contains a solution of water and iodine. Chamber B contains pure water.

■ *Describe the movement of molecules through the membrane. What will happen to the molecules of iodine in Chamber A? What will happen to the molecules of water in Chamber A? What will happen to the molecules of water in Chamber B?*

Shown below are the two liquids in a state of equilibrium.

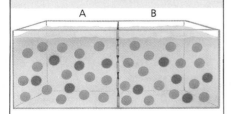

A   B

■ *What will happen to the molecules of water and iodine after equilibrium has been reached?*

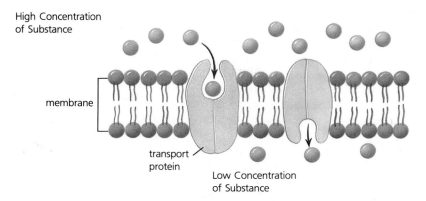

High Concentration
of Substance

membrane

transport
protein

Low Concentration
of Substance

◀ **Figure 5–16**
Facilitated Diffusion. In facilitated diffusion, specialized transport molecules speed the movement of molecules across a membrane.

binds the substance on one side of the membrane and then releases it on the other side of the membrane. See Figure 5–16. Facilitated diffusion works only in the direction of the concentration gradient. It speeds up the movement across a membrane from a region of high concentration to one of a lower concentration. An example of facilitated diffusion in your body is the transport of glucose from your blood into the cells of your body.

Osmosis—The Diffusion of Water

Thus far, diffusion has been discussed only for substances dissolved in water. A water solution is simply a mixture of water molecules and molecules of the dissolved substance, or the solute. Water molecules are small. They can pass through a selectively permeable membrane easily, just as other small molecules do. Therefore, water itself will diffuse through cell membranes. The direction of this diffusion will depend only on the difference in concentration of water on opposite sides of the membrane. The diffusion of water across a selectively permeable membrane from a region of high water concentration to a region of low water concentration is called **osmosis** (os MOH sis).

What do we mean by "concentration" of water? The meaning is the same as it is for any other substance—it is the amount of water present in a given volume. Since two things cannot occupy the same space at the same time, the number of particles dissolved in a given amount of water determines the concentration of water molecules present. The concentration of water molecules is highest in pure water—water with nothing else in it. The more particles dissolved in water, the fewer the water molecules present. For example, the concentration of water is higher in 100 milliliters of pure water than it is in 100 milliliters of a water and sugar solution.

Figure 5–17 shows an experiment that demonstrates osmosis. The glass bulb of a thistle tube is filled with a concentrated sugar solution. The bulb is then covered with a membrane of cellophane and placed in a jar of pure water. The membrane has pores through which the water molecules, but not the sugar molecules, can pass. The water concentration in the jar is 100 percent. The water concentration in the thistle tube is less because the sugar solution

Figure 5–17

Demonstration of Osmosis. A tube containing a concentrated sugar solution and covered with a semipermeable membrane is placed in a beaker of pure water (left). Water molecules pass through the membrane into the sugar solution by osmosis, causing the solution to rise in the tube (right). ▼

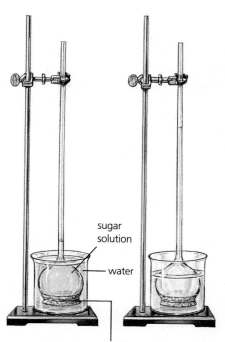

sugar
solution

water

selectively permeable membrane

contains fewer water molecules than the same volume of pure water. Since the water molecules can pass through the membrane, they diffuse from the area of high concentration (the jar of pure water) to the area of lower concentration (the sugar solution in the thistle tube). As osmotic diffusion of water occurs, the level of liquid in the thistle tube rises.

Osmotic Pressure

In the osmosis experiments in Figure 5–17, the water eventually stops rising in the thistle tube, even though the concentration of water below the membrane is higher than the concentration above the membrane. This happens because that diffusion depends on pressure as well as concentration. As the water-sugar solution rises in the thistle tube, its weight results in increased pressure on the upper side of the membrane. This increasing pressure increases the rate of water diffusion from the water-sugar solution across the membrane into the pure water. Eventually, equilibrium will be reached and the rate of water diffusion across the membrane in both directions will be the same. Water molecules are still passing into the thistle tube because of the concentration gradient across the membrane. But, they are passing out at the same rate because of the pressure gradient in the opposite direction. Thus, there is no net change in the amount of water inside the thistle tube, and the water level remains constant. There is, however, a steady excess of pressure inside the thistle tube. The increased pressure resulting from osmosis is equal to the tendency of water to enter the solution. This value is called **osmotic** (os MAH tik) **pressure.**

Critical Thinking in Biology

Relating Parts and Wholes

To understand how a complex object, such as a car, works, it is helpful to examine how its parts function in relation to the whole object. Examine the apparatus in Figure 5–17.

1. Identify its parts.
2. What is the function of the apparatus?
3. How are each of the following parts related to the function of the apparatus: (a) the thin part of the thistle tube; (b) the semipermeable membrane; (c) the beaker; (d) and the clamp?
4. Could an impermeable material be substituted for the semipermeable membrane? Why or why not?
5. If the beaker broke, what could you use in its place?
6. What other changes could be made to the apparatus without changing its function?
7. Think About It Describe your thought process as you answered question 3.

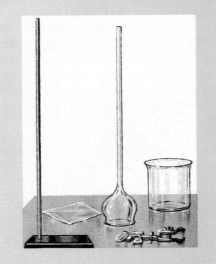

Effects of Osmosis

The cytoplasm of cells is made mainly of water that contains a variety of dissolved substances. Water can pass freely through the cell membrane in both directions. If the concentration of water inside and outside the cell is the same, no net water diffusion occurs. If the water concentration is higher outside the cell than inside, water will tend to move into the cell. If the concentration of water is higher inside the cell than outside, water will move out of the cell. The effects of placing cells in solutions containing different concentrations of water can be seen in Figure 5–18.

An **isotonic** (i suh TAHN ik) **solution** is one that has the same concentration of dissolved substances as the living cell placed in it. The concentration of water molecules in the cell and in an isotonic solution is the same. Since the concentration gradient is zero, there is no net gain or loss of water by the cell.

A **hypotonic** (hy puh TAHN ik) **solution** contains a lower concentration of dissolved substances than the cell. Therefore, the concentration of water molecules is higher in the hypotonic solution than it is in the cell. Since the concentration of water is higher outside the cell than inside, there is a net movement of water into

Figure 5–18
Effects of Osmosis on Living Cells. Cells placed in an isotonic solution neither gain nor lose water. In a hypotonic solution, animal cells swell and burst; the vacuoles of plant cells swell, pushing cell contents out against the cell wall. In a hypertonic solution, animal cells shrink; plant cell vacuoles collapse. ▼

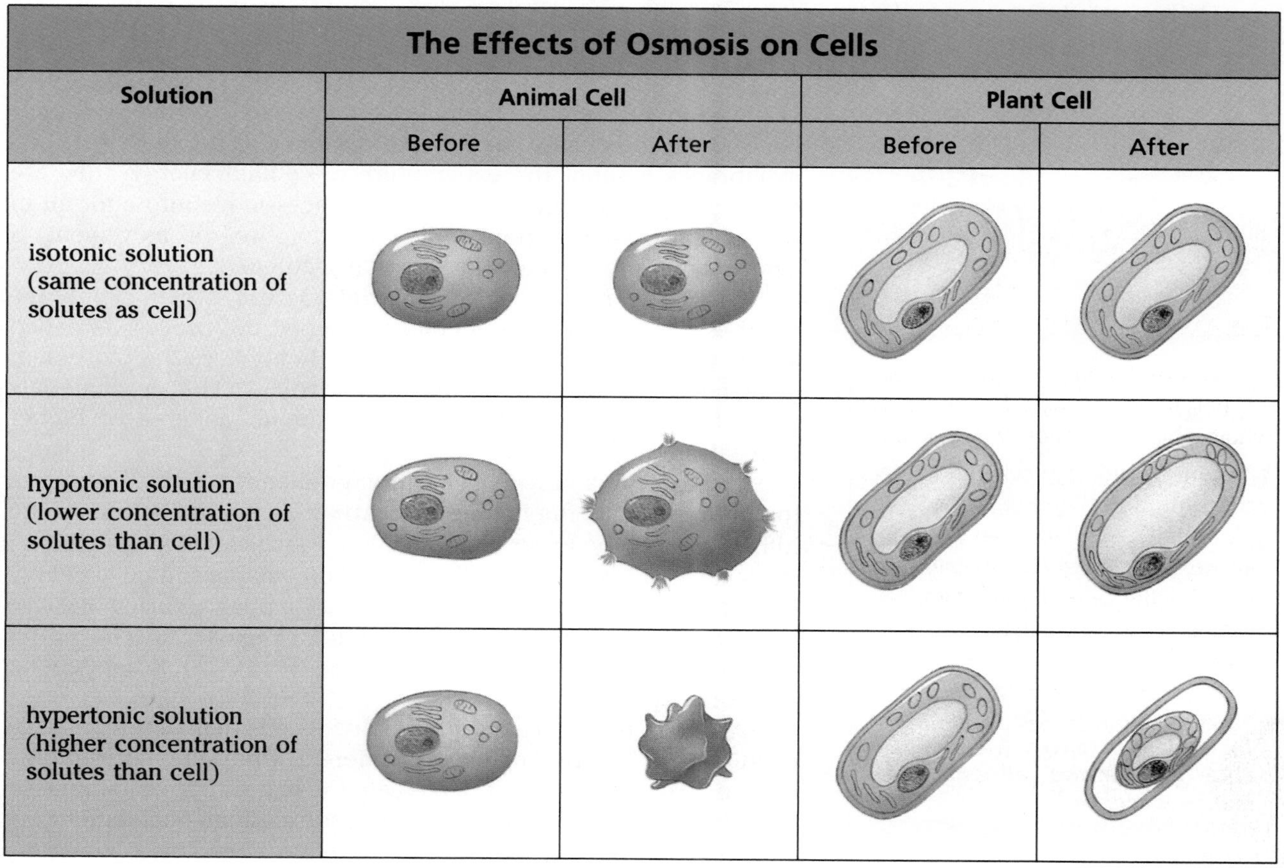

The Effects of Osmosis on Cells				
Solution	**Animal Cell**		**Plant Cell**	
	Before	After	Before	After
isotonic solution (same concentration of solutes as cell)				
hypotonic solution (lower concentration of solutes than cell)				
hypertonic solution (higher concentration of solutes than cell)				

the cell by osmosis. Because of the osmotic pressure produced by the water that enters the cells, animal cells swell and burst when placed in a hypotonic solution. When a plant cell is placed in a hypotonic solution, excess water collects in the large, central vacuole, causing the cell to swell and push against the cell wall. Plant cell walls, although flexible, resist being stretched. Consequently, if water moves into a plant cell, pressure builds as the increasing cell volume pushes against the cell wall. The pressure that develops inside a plant cell due to osmosis is called *turgor pressure*. It is similar to the pressure of the water on the cellophane membrane in the thistle tube. Eventually, the rate of diffusion out of the plant cell (due to pressure) balances the rate of diffusion into the cell (due to the concentration gradient). At this point of equilibrium, net diffusion of water into the plant cell stops.

A **hypertonic** (hy per TAHN ik) **solution** contains a higher concentration of dissolved substances than the cell. The concentration of water molecules in a hypertonic solution, therefore, is lower than that in the cell. The concentration gradient results in a net movement of water out of the cell. Animal cells will shrink when they are placed in a hypertonic solution; in plant cells, the vacuole will collapse, which will cause the cytoplasm to shrink within the cell wall. The shrinking of cytoplasm by osmosis is called **plasmolysis** (plaz MAH luh sis).

Passive and Active Transport

In diffusion, no cellular energy is used to move substances into or out of the cell. Because these processes move materials across cell membranes without the expenditure of cellular energy, they are called **passive transport.** Many substances move into and out of the cell by passive transport. The direction of movement is determined only by the concentration gradient.

When the movement of materials across a cell membrane requires the expenditure of cellular energy, the process is called **active transport.** Active transport usually involves the movement of materials against a concentration gradient. That is, materials move from an area of low concentration to an area of higher concentration.

You can think of a concentration gradient as a hill. The top of the hill represents the higher concentration, and the bottom of the hill represents the lower concentration. A bicycle moving downhill along the concentration gradient needs no outside source of energy in order to keep rolling. This is what happens during passive transport. To move the bicycle up the hill, or against the concentration gradient, however, requires energy. This is what happens in active transport.

Active transport makes it possible for cells to maintain internal conditions that are chemically different from the surrounding medium. For example, in a nerve cell, the concentration of potassium is higher inside the cell than it is in the medium outside the cell.

Can You Explain This?

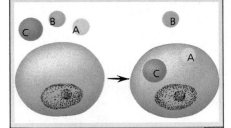

A, B, and C represent three different substances found outside a cell's membrane. Each substance is present outside the cell in a higher concentration than inside the cell.

As shown in the diagram above, substances A and C are able to enter the cell without using cellular energy, but substance B cannot.

■ *Propose a possible explanation for why substances A and C are able to enter the cell while substance B cannot.*

The concentration of sodium, on the other hand, is lower inside the cell than it is outside. The cell uses active transport to maintain these differences in concentration. In certain seaweeds, minerals such as potassium and iodine build up in concentrations that are 1000 times the concentrations found in ocean water. In the human kidney, active transport is involved in concentrating wastes in urine.

There are two processes by which active transport can occur. In one process, a substance is moved across a membrane molecule by molecule. In this case, each molecule that is transported must first bind to a transport protein on one side of a membrane and then be released by the same protein on the other side of the membrane. As with facilitated diffusion, the transport protein is specific for each kind of transported substance. Unlike facilitated transport, however, the process requires the expenditure of cellular energy in the form of *ATP*. You will learn about ATP in the discussion found in Chapter 6.

The other kind of active transport involves membrane vesicles. Materials that enter a cell by this method become enclosed within an inpocketing of the cell membrane, shown in Figure 5–19. The outer surface of the membrane closes over, pinching off the pouch, which becomes a sac, or vesicle, within the cell. Inside the cell, the contents of the vesicle are released. This process of transporting material into a cell by means of a vesicle is known as **endocytosis.** When small amounts of liquid are taken into the cell in this way, the process is called **pinocytosis.** Another form of endocytosis, called **phagocytosis,** occurs when solid particles are ingested into a cell. In this case, part of the cell surrounds a particle. The cell membrane fuses, sealing the particle within its own vesicle, or food vacuole. Enzymes then digest the particle.

The movement of materials out of the cell by the reverse of endocytosis is called **exocytosis.** In this case, a vacuole carrying materials within the cell fuses to the inside surface of the cell membrane. The contents of the vacuole then are released to the outside of the cell membrane.

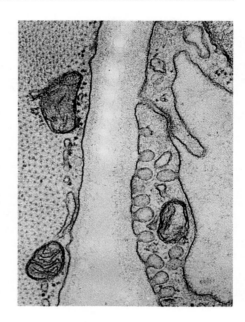

▲ **Figure 5–19**

Pinocytosis. In pinocytosis, inpocketings of the cell membrane close over small particles or liquid droplets, forming small vacuoles in the cytoplasm.

5-3 Section Review

1. Define the terms *diffusion* and *osmosis.*
2. What is meant by the term *facilitated diffusion?*
3. What is a hypertonic solution?
4. Name two kinds of active transport.

Critical Thinking

5. A person is more likely to recover from a near drowning in sea water than in fresh water. How would osmosis account for this? (*Identifying Causes*)

5-4 Organization of Cells in Living Things

Section Objectives:

- *Describe* and *compare* the different levels of organization and specialization that exist in unicellular, colonial, and multicellular organisms.
- **Laboratory Investigation:** *Observe* cells from a variety of plant and animal cells (p. 102).

A cell may exist alone, or it may be part of a larger organism made up of many cells. A cell that exists independently is regarded as a one-celled, or unicellular, *organism.* Many-celled, or multicellular, organisms may be made up of hundreds, thousands, millions, or billions of cells.

Unicellular and Colonial Organisms

Unicellular organisms are able to carry on all the life processes. They synthesize and obtain nutrients, break them down for energy, synthesize new materials, reproduce, and perform all the other activities that living things do. Unicellular organisms include bacteria, protozoa, many algae, and some fungi. These organisms vary in size and in the complexity of their structure.

The simplest level of multicellular organization occurs in *colonial organisms.* A colonial organism, also called a *colony,* is an organism of a few to many cells that are loosely attached to each other and that show little or no specialization among themselves. In some colonies, the cells are all alike, and each cell carries on all the life processes. These colonies are like a group of unicellular organisms that are stuck together. Any one of the cells has the capacity to reproduce and form a new colony.

In more complex colonies, the cells show some specialization. That is, the cells forming the colony vary in their structure and function. For example, *Volvox,* an alga, shown in Figure 5–20, forms spherical colonies that can include many thousands of cells. However, only about 20 of these cells in one part of the colony are capable of reproducing and forming new colonies. In another part of the colony are smaller cells containing large, light-sensitive organelles. These cells control the positioning and movement of the colony in the water. The cells of a *Volvox* colony are connected by thin strands of cytoplasm.

Multicellular Organisms

True multicellular organisms can consist of hundreds to billions of cells of many different types. In multicellular organisms, the cells are specialized. They cannot function as independent, single-celled organisms, each performing all life functions. Instead, in multicellular organisms, each cell carries out only some of its own life

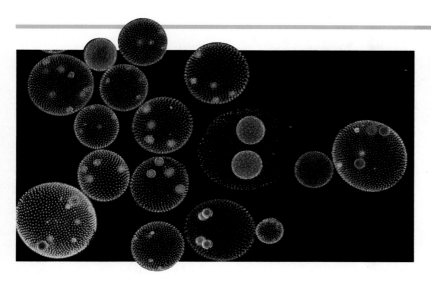

***Volvox* Colonies.** Each sphere in this light micrograph is a colony of *Volvox* made up of thousands of cells. The flagella of all the cells beat in sequence, causing the colony to rotate through the water.

processes, while accomplishing the specific whole-organism functions for which it is specialized. Each type of cell depends on all the other types of cells. These complex relationships require many levels of multicellular organization and interaction.

Tissues   In multicellular organisms, a group of cells that are structurally similar and perform the same function forms a **tissue.** Each cell in a tissue carries on many life processes at the same time. Each cell also carries on some special processes that are related to the specific function of the tissue. In plants, for example, there are tissues that transport water and nutrients throughout the plant, tissues that cover and protect the parts of the plant, tissues that carry on photosynthesis, and so on. The structures and functions of plant tissues are discussed in detail in Chapter 18. Complex multicellular animals contain a greater variety of tissues than plants, and many of these tissues are more highly specialized than plant tissues.

Figure 5–21 lists the main types of tissues found in animals. Most animal tissues are a form of either *epithelial tissue* or *connective tissue.* Muscle, nerve, and blood are highly specialized tissues that do not belong to either of these groups.

Tissues that cover body surfaces and line body cavities and organs are **epithelial tissues.** They also form glands, structures made of a few to many cells that produce and release one or more substances. As you can see in Figure 5–22, epithelial tissues are generally in the form of sheets of closely packed cells. The simplest epithelial tissues consist of sheets only one cell layer thick. More complex forms consist of several cell layers.

Tissues that support other body tissues and bind tissues and organs together are called **connective tissues.** They give the body form. Unlike the cells of epithelial tissues, the cells of connective tissues are widely separated. The space between them is filled with various types of substances. In bone, for example, the bone-producing cells secrete the hard, bony material that fills the spaces between the cells. In tendons, which connect muscles to bones, dense bundles of tough elastic fibers lie between cells.

Tissues that are specialized for contraction, or shortening, are **muscle tissues.** There are three types of muscle tissues. *Cardiac muscle* tissue is found only in the heart. *Skeletal muscle* tissue makes up the muscles that are attached to the bones of the skeleton. Movement of these muscles is under voluntary control by the animal. *Smooth muscle* tissue is found in various organs of the body, such as the stomach, intestine, and blood vessels. Contraction of smooth muscles is involuntary and automatic. Skeletal and smooth muscles are discussed in Chapter 13, blood is discussed in Chapter 10, and nervous tissue is discussed in Chapter 14.

Organs and Organ Systems A group of tissues that work together to perform a specific function forms an **organ.** The stomach, which has nerves, muscles, and blood vessels, is an organ.

Figure 5–21
Types of Animal Tissues. ▼

Types of Animal Tissues			
Type of Tissue	**Structure**	**Functions**	**Location**
Epithelium (epithelial tissue)	cells arranged in sheets one or more cell layers in thickness	protection (outer layer of skin); absorption (inner lining of intestine); secretion (glands)	lines cavities; covers surfaces; forms glands
Connective tissue	cells and fibers embedded in an extra-cellular "ground substance," or matrix	support of other tissues and binding of organs; connects or binds tissues and organs together	throughout body
Adipose tissue	specialized ovoid fat cells in connective tissue fibers	stores fat	throughout body; found in large numbers in some areas
Bone and cartilage	cells and fibers embedded in an extracellular formless "ground substance," or matrix	make up the skeleton giving the body support and form, and, with muscles, allowing for movement	*bone:* in skeleton; *cartilage:* in skeleton, trachea, outer ear, nose, and discs of spinal column
Blood	specialized liquid tissue	transport of nutrients, wastes, oxygen, and carbon dioxide throughout body	within vessels of the circulatory system
Nerve tissue	specialized cells, called neurons, that are bound together by connective tissue to form nerves	conduction of impulses	brain and spinal cord; nerves and sense receptors throughout body
Muscle tissue	individual cells or fused cells bound together by connective tissue to form bundles or sheets	*skeletal muscle:* voluntary movement of body parts; *smooth muscle:* involuntary movement of internal organs; *cardiac muscle:* makes up heart and generates heart beat stimulus	skeletal muscles; internal organs; heart

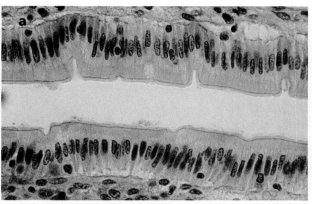

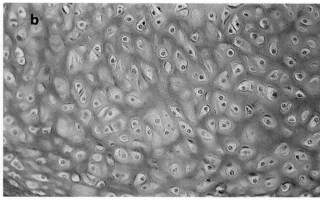

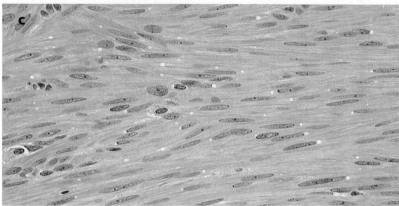

▲ **Figure 5–22**

Animal Tissues. (a) Epithelial cells cover surfaces and line cavities. (b) Connective tissues support and bind together other body tissues. Cartilage, one type of connective tissue, is firm and flexible. (c) Muscle tissue is made up of cells that are able to contract. Smooth muscle cells are long and spindle shaped.

A group of organs that work together to perform a specific function forms an **organ system.** An example of an organ system is the digestive system, which includes the mouth, esophagus, stomach, intestines, pancreas, liver, and so forth.

Although the parts of a multicellular organism can be described in terms of separate cells, tissues, organs, and organ systems, all these parts must function together for the organism to carry on its life processes.

5-4 Section Review

1. What is a colonial organism?
2. List the levels of organization, simple to complex, in multicellular organisms.
3. Describe the locations of epithelial tissue.
4. Define an organ system and give an example.

Critical Thinking

5. Why do you think the cells of all multicellular organisms are specialized? (*Reasoning Conditionally*)

Laboratory Investigation

Observing Plant and Animal Cells

A cell is the smallest unit that can carry out life processes. Although all cells have certain characteristics in common, cells can vary in size, shape, and structure. In this investigation, you will use a microscope to examine a variety of plant and animal cells.

Objectives

- *Examine* the structure of cells.
- *Identify* similarities and differences in animal and plant cells.

Materials

dropper	coverslips	tweezers
slides	microscope	paper towel
flat toothpicks	onion	elodea
iodine solution	scalpel	

Procedure *

A. Place a drop of water on a clean slide. Gently scrape the inside lining of your cheek with the flat end of a toothpick to remove cheek cells. Swirl the end of the toothpick in the drop of water on the slide. Add a drop of iodine to the water drop, and cover it with a coverslip. Blot off excess liquid with a paper towel.

B. Under low power, locate single cheek cells. Switch to high power to observe them more closely. Draw a single cheek cell, and label the nucleus, nuclear envelope, cytoplasm, and cell membrane.

C. Use tweezers to peel the epidermis from the inner surface of an onion bulb leaf as shown below. Remove a small piece of the epidermis, about 5 mm square, and place it on a slide. Add a drop of water and cover with a

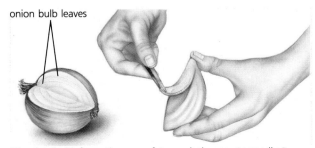

onion bulb leaves

*For detailed information on safety symbols, see Appendix B.

coverslip. Examine the onion skin under low, then high, power. Identify the cell wall, cytoplasm, nucleus, and vacuole of a cell. Draw and label an epidermal cell.

D. Remove the slide from the stage. Place a drop of iodine at one end of the coverslip. Hold a piece of paper towel near the opposite edge to draw the iodine underneath the coverslip, as shown below. Reexamine the epidermis. Revise your drawing to include structures made visible by the iodine.

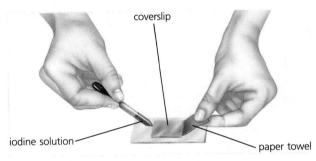

coverslip

iodine solution

paper towel

E. Remove a leaf from the tip of an elodea plant. Place it on a slide, add a drop of water, and cover it with a coverslip.

F. Examine the elodea cells under low power. Notice that the elodea leaf is only a few layers thick. Adjust the focus to observe just one layer. Switch to high power. Draw and label an elodea cell.

Analysis and Conclusions

1. How does the structure of cheek cells compare with the structure of onion epidermal cells?

2. How does the iodine stain change the appearance of the onion cells?

3. Describe the shape and location of the chloroplasts in the elodea leaf.

4. In which cells did you see movement of cytoplasm?

5. Draw and label a generalized structure of a plant cell and an animal cell.

6. A detective, called to the scene of a crime, collected samples of tissues found nearby. How could he tell whether these tissues were from a plant or an animal?

Chapter Review

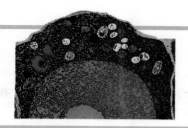

Study Outline

5–1 What Is a Cell?

■ All living things are made up of one or more cells that carry on life-sustaining metabolic processes. New cells arise only from other living cells.

■ Cells are of two basic types: prokaryotic, without membrane-bound organelles, and eukaryotic, with a membrane-bound nucleus and other distinct organelles.

■ A living cell exchanges materials with its environment across its surface. As a cell grows, its surface enlarges less than its volume, thus limiting cell size.

5–2 Cell Structure

■ Every cell has a selectively permeable cell membrane. A cell wall surrounds the cell membrane in plants and in some microorganisms.

■ Plant and animal cells have specialized structures, called organelles, in the cytoplasm. One organelle, the nucleus, contains the cell's hereditary material and controls cell metabolism and reproduction.

■ Other cellular organelles are specialized for different functions, such as protein synthesis or release of energy. Unlike animal cells, plant cells have organelles, called plastids, that have several functions, including the use of light energy for making food.

5–3 Maintaining a Constant Cell Environment

■ Cellular homeostasis is maintained by the cell membrane, which controls movement of substances into and out of the cell.

■ Some substances pass through the cell membrane by diffusion, moving from higher to lower concentration. Water enters or leaves the cell by osmosis, the diffusion of water through a selectively permeable membrane.

■ Diffusion, facilitated diffusion, and osmosis are types of passive transport, requiring no cell energy. Active transport requires energy to move substances into or out of the cell. Larger particles are actively moved into or out of the cell by endocytosis and exocytosis.

5–4 Organization of Cells in Living Things

■ A one-celled organism carries on all the life processes within itself. A colonial organism is a group of loosely attached unicellular organisms, some of which may be slightly specialized.

■ In multicellular organisms, all cells are specialized to perform specific functions for the entire organism. They are, therefore, dependent on the rest of the cells in carrying out their life processes.

■ Cells of similar structure and function form a tissue. Tissues working together form an organ; organs that function together, in turn, form an organ system.

Vocabulary Review

cells (77)

prokaryotic cells (79)

eukaryotic cells (79)

nucleus (79)

organelles (82)

cell wall (82)

cell membrane (82)

selectively permeable (84)

nucleoli (85)

cytoplasm (85)

endoplasmic reticulum (85)

ribosomes (85)

Golgi bodies (86)

lysosomes (86)

mitochondria (86)

microtubules (87)

microfilaments (87)

centrioles (88)

cilia (88)

flagella (88)

vacuoles (88)

plastids (89)

leucoplasts (87)

chromoplasts (89)

chloroplasts (89)

endosymbiosis (89)

diffusion (92)

concentration gradient (92)

facilitated diffusion (92)

osmosis (93)

osmotic pressure (94)

isotonic solution (95)

hypotonic solution (95)

hypertonic solution (96)

plasmolysis (96)

passive transport (96)

active transport (96)

endocytosis (97)

phagocytosis (97)

exocytosis (97)

tissue (99)

epithelial tissues (99)

Chapter Review

connective tissues (99)   organ (100)

muscle tissues (100)   organ system (101)

A. Matching—Select the vocabulary term that best matches each definition.

1. Plastids that use light energy for making food

2. Similar cells that are grouped to perform the same function

3. The network of membrane-enclosed canals along which materials move through the cell

4. Stacked sacs that package and store cell products for release by the cell

5. Organelles containing digestive enzymes

6. Fluid-filled organelles that are usually much larger in plant than in animal cells

B. Multiple Choice—Choose the letter of the answer that best completes each statement.

7. Cell energy is needed in (a) osmosis (b) active transport (c) passive transport (d) diffusion.

8. Materials that cannot diffuse through the cell membrane may be brought into the cell by (a) exocytosis (b) plasmolysis (c) endosymbiosis (d) endocytosis.

9. The cell material lying between the nucleus and cell membrane is called the (a) cytoplasm (b) protoplasm (c) centriole (d) chromoplast.

10. Because bacteria lack a nuclear membrane, they are classified as (a) prokaryotic cells (b) eukaryotic cells (c) leucoplasts (d) nucleoli.

11. Tissue that lines body cavities and covers body surfaces is (a) connective tissue (b) muscle tissue (c) epithelial tissue (d) endothelial tissue.

Content Review

12. State three main points of the cell theory.

13. Explain what each scientist contributed to the cell theory: Hooke, Schleiden, Schwann, Virchow.

14. To check whether a one-celled organism is prokaryotic or eukaryotic, what single difference would you look for with the aid of a microscope?

15. List four functions of the proteins that are present in the cell membrane.

16. Explain the difference between the nucleus and the nucleolus.

17. Is the structure of the endoplasmic reticulum uniform throughout the cell? Explain your answer.

18. Where are ribosomes found in cells, and what is their function?

19. Describe both the structure and the function of the microtubules.

20. How are cilia and flagella alike? How do they differ?

21. Which cell structures are found in plant cells but not animal cells?

22. How does diffusion differ from facilitated diffusion?

23. Explain what happens to an animal cell in each kind of solution: hypertonic, hypotonic, isotonic.

24. Describe how turgor pressure develops and how it affects plant cells.

25. Why is active transport important to a cell?

26. What is the difference between a tissue, an organ, and an organ system?

Graphic Organizing

For information on graphic organizers, see Appendix G at the back of this text.

27. **Line Graph:** Copy the incomplete graph below onto a separate sheet. Refer to Figure 5–4 and add data points for cubes with sides of 2 and 3 centimeters. Then, compute the surface area-to-volume ratios for cubes with sides of 4 and 5 centimeters. Complete the graph. What conclusion can you draw from the graph?

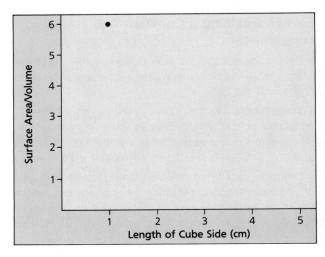

Critical Thinking

28. Leeuwenhoek's microscope revealed one-celled living things he called "animalcules." Yet, he never identified cells as the smallest units of living things. Why not? (*Identifying Reasons*)

29. If the cell membrane were completely permeable, how would this affect the cell? (*Predicting*)

30. How does a one-celled organism differ from the single cell of a multicellular organism? (*Comparing and Contrasting*)

31. Why are the cells of a huge elephant not larger than the cells of a cat? What limits the cell's growth as the elephant grows? (*Relating Parts and Wholes*)

Creative Thinking

32. A living cell has been compared to a factory. Explain which cell part matches each of the following: factory manager, power plant, assembly line, storage and shipping center, liquid storage tank, security guard.

33. Suppose you were the size of a bacterium. What type of clothing would help you enter an animal cell? How would you enter the cell?

Problem Solving

34. The graph below shows the speed of movement of a substance's molecules at different temperatures. Develop a hypothesis about diffusion based on the data in the graph. How could you test your hypothesis?

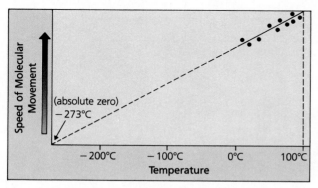

35. When living yeast cells were placed in congo red dye and examined under the microscope, the yeast cells remained colorless. However, when placed in methylene blue, they became blue. Later, dead yeast cells were placed in congo red dye. These cells turned red. Explain these three observations.

Computer Project

36. As you read in Section 5–1, a cell's surface area-to-volume ratio is important in determining how efficiently the cell can exchange materials with its environment.

In this activity, you will use a spreadsheet to compare the surface area-to-volume ratios of cells of different shapes—spherical (liver cells, for example), cube-shaped (epithelial cells), cylindrical (nerve cells), and disk-shaped (red blood cells). The table below provides formulas for computing the surface area and volume for each shape. (Note: $\pi = 3.14$)

Shape	Surface Area	Volume
cube (x = side)	$6x^2$	x^3
sphere (r = radius)	$4\pi r^2$	$4/3\pi r^3$
cylinder or disk (r = radius, h = height)	$2\pi r (h + r)$	$\pi r^2 h$

A. Compute the surface area, volume, and surface area-to-volume ratio for a cube with a side of 3. Increase the sides in increments of 0.5 up to 6 and compute the same information for each cube. What happens as the size increases?

B. Compute the surface area-to-volume ratios for various spheres. Start with a radius of 2, then increase in increments of 0.25 up to a radius of 3.5. What happens as the size increases?

C. To compute the surface area-to-volume ratios of various cylinders, keep the radius constant at 1 and vary the height from 10 to 70, increasing by 10 each time. What happens as a cylindrical cell gets longer?

D. A disk-shaped cell is similar to a low, flat cylinder. To compute surface area-to-volume ratios, keep the height constant at 1 and increase the radius from 3 to 9 by 1 unit each time. What happens as a disk-shaped cell becomes wider?

E. To compare the various cell shapes, construct a line graph with volume on the horizontal axis (range 20 to 260) and surface area-to-volume on the vertical axis. Plot data points for each shape and connect the points. Which shape has the greatest surface area-to-volume ratio at any given volume? How do the shapes compare in terms of cell efficiency?

F. Problem Finding: What other questions could you answer using this spreadsheet?

▲ Sockeye salmon leap up over a waterfall in Katmai National Park, Alaska.

Cellular Respiration

A sockeye salmon darts through a swift stream almost too quickly to be seen and then suddenly shoots out of the water. Using its powerful muscles, it leaps into the air, over the rushing waterfall, and continues its upstream journey from the ocean back to the waters in which it was spawned. How does the salmon get the energy it needs for its difficult migration? How do all living things—animals, plants, single cells—obtain the energy they need to live? In this chapter, you will learn how energy for life is created from the chemical energy stored in food.

Guide for Reading

Key words: ATP, aerobic respiration, anaerobic respiration, glycolysis, fermentation, Krebs cycle, electron transport chain

Questions to think about:

How do living things obtain the energy to carry out their life functions?

In what ways do aerobic respiration and anaerobic respiration differ? In what ways are they similar?

6-1 Energy for Life

Section Objectives:

- *Explain* why cellular respiration is important for living things.
- *Describe* the role of ATP in energy transfer.
- *Contrast* oxidation and reduction and explain why one cannot take place without the other.
- *Discuss* the function of hydrogen acceptors in cellular respiration.

The Uses of Energy

Energy is the ability to do work. Just open any newspaper and you are sure to find articles about energy—energy needs, energy policies, and sources of energy. Some of the energy that fuels the world is gotten from falling water used by hydroelectric power plants. Some comes from nuclear energy. Still more comes directly from solar radiation.

Most of our energy, however, comes from the burning of fuels such as oil, gas, and coal. Burning a fuel releases energy in the form of heat and light. The heat can be used to run engines and electric generators, which turn the heat energy into other forms of energy. The burning of fuel is a chemical process. Carbon and hydrogen in the fuel combine with oxygen from the air to form carbon dioxide and water. The fuels contain stored chemical energy. The stored chemical energy is released mostly as heat during the chemical changes of burning.

▲ **Figure 6–1**

Energy for Life. Like all organisms, this hummingbird must use energy to carry out its daily activities.

Biology and You

Q: Is "carbo-loading" a good way for athletes to improve their performance in endurance events?

A: Athletes are always looking for ways to improve their game. One technique is carbohydrate (carbo) loading, eating foods high in carbohydrates before athletic events. Some athletes feast on spaghetti and other complex carbohydrates the night before a race. They hope to store extra carbohydrates in their muscles and liver. The stored carbohydrates may act as reserve energy, thus delaying exhaustion during the event.

Reports show that carboloading may be effective only for the highly conditioned athlete who has trained for a long event, such as a marathon. Although it may increase the amount of stored carbohydrates, carbo-loading does not increase speed or power.

For the casual athlete, carboloading probably will not increase energy and may cause water retention and discomfort. These effects may make an athlete feel stiff and bloated. The best plan is to eat a balanced diet, rich in complex carbohydrates, on a regular basis.

■ *Write a sample menu that you would suggest to the head of an athletic training camp.*

Just as machines need energy to run, all living things need energy to carry on their life activities. Some of this energy is needed for physical or mechanical work. A flying bird needs energy just as an airplane does. A beaver building a dam or a worm burrowing in the soil needs energy just as earth-moving equipment does. Even tree frogs singing on a spring evening need energy just as a radio does. Most of the movements of living things require energy. Energy is also needed for less obvious purposes. Making complex compounds from simpler ones and transferring some materials across cell membranes require energy. In fact, cells need a continuous supply of energy to stay alive.

Energy from Food

Living things rely on the chemical energy stored in their food. Carbohydrates are the foods most commonly broken down for energy. In most cases, this energy is released by chemical changes that resemble burning. However, when organisms break down food, only part of the energy is released as heat. This heat maintains your body temperature. The rest of the energy is stored in chemical form. Living things can only use chemical energy to carry out their life functions. They cannot use heat energy to do work. It is not surprising, therefore, that the breakdown of food, and the resulting release of energy, takes place without a direct reaction between carbohydrate and oxygen. Instead, the breakdown takes place in many small chemical steps that are linked to the formation of new, high-energy compounds.

The release of the energy stored in food goes on inside the cells of both autotrophs and heterotrophs. Organisms as diverse as plants, animals, and microorganisms show remarkable similarities in the way they obtain energy from food. This energy-releasing process is called **cellular respiration.** In this chapter, you will learn how cells obtain from food the energy they need to carry out life functions.

ATP and ADP

The energy released during cellular respiration is not used directly. It is first "packaged" in the molecules of certain compounds. One such compound is called *adenosine triphosphate* (uh DEN uh seen try FAHS fayt), which is abbreviated as **ATP.** Figure 6–2 shows the structure of an ATP molecule. The main part of the molecule is made up of 1 molecule of adenine joined to 1 molecule of ribose. As you learned in Chapter 4, adenine is one of the nitrogenous bases found in DNA and RNA. Ribose is the 5-carbon sugar found in RNA. The combination of these two molecules is called *adenosine*. There are three phosphate groups bonded end-to-end to the adenosine in ATP. Phosphate groups are also part of the structure of DNA and RNA. Notice how the cell uses the same molecular units for different purposes. There are many examples of such multiple uses of chemical groups in the chemistry of life.

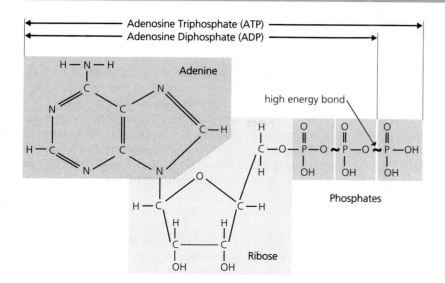

Structure of ATP and ADP. When the third phosphate group is removed from ATP, chemical energy is released. The remaining molecule, ADP, has only two phosphate groups. ADP has less energy than ATP.

For energy storage, the bond linking the last phosphate group to the molecule is the important part of the ATP molecule. The bond is shown as a wavy line. This symbol means that the bond contains a large amount of energy. It is called a *high-energy bond.* When the third phosphate in ATP is removed and bonded to another compound, it transfers energy to the other compound. This transfer of energy is called *phosphorylation* (fahs for uh LAY shun). Phosphorylation is a common way for chemical energy to be transferred in living cells. See Figure 6–3.

When the third phosphate is removed from ATP, the remaining molecule is called *adenosine diphosphate,* or **ADP.** ADP has less energy than ATP. Although ADP's second phosphate is attached by a high-energy bond, this bond is used less often as a source of energy. Look again at Figure 6–2 to compare ATP with ADP.

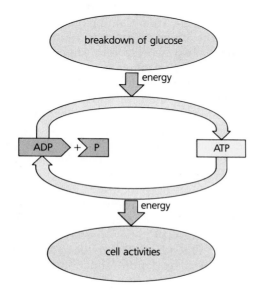

◀ **Figure 6–3**

The Energy Cycle in the Cell. ATP supplies the cell with the energy needed for cell activities. As glucose is broken down, the energy released is used to attach a third phosphate group to ADP, forming ATP. When the third phosphate group is detached from the ATP, the energy released is used for cell activities. The low-energy ADP is returned for reuse. The energy from a single glucose molecule can form 36 molecules of ATP.

The Source of Energy for ATP

During cellular respiration, the energy released by the gradual breakdown of food molecules is used to attach a third phosphate to ADP. This changes the ADP molecule to a molecule of ATP. The ATP then can be used in any part of the cell that needs its energy.

The sugar glucose is the most common food substance from which cells obtain energy. The food of most organisms, however, does not contain glucose in its simple form. Instead, it is found in the form of complex carbohydrates. The simple form of glucose is obtained from the breakdown, or digestion, of these more complex carbohydrates.

From the energy in 1 molecule of glucose, a cell can make a number of molecules of ATP from ADP. That is, the energy produced from breaking down 1 molecule of glucose is divided into many small units. If all this energy were released in a single burst, it would be too much for the cell to handle. A cell cannot use that much energy all at once. However, the amount of energy held in a single molecule of ATP is just about right for the average cellular reaction that requires energy. Thus, the packaging of energy in ATP is convenient and efficient for the needs of the cell.

Oxidation-Reduction Reactions

There are several steps by which the energy in glucose is used to produce ATP. The idea of chemical oxidation and reduction can help you understand these steps. The term **oxidation** (ox sih DAY shun) refers to any chemical change in which an atom or a molecule loses electrons. For example, when sodium combines with chlorine, the sodium atom loses an electron (see Chapter 3, section 3–2). This is an example of oxidation. We say that the sodium atom is oxidized.

At the same time that the sodium atom loses an electron, the chlorine atom gains an electron. Gaining electrons is called **reduction.** We say that the chlorine atom is reduced. Oxidation and reduction are simply two aspects of a single chemical reaction. When one substance in a reaction is oxidized, another must be reduced. That is, the electrons given up by the substance that is oxidized are taken up by another substance that is reduced. A reaction of this kind is called an **oxidation-reduction reaction.**

In some oxidation-reduction reactions, an electron is transferred from one substance to another as part of a hydrogen atom. That is, one compound may transfer hydrogen atoms to another. The loss of hydrogen atoms is a form of oxidation. Gaining hydrogen atoms is a form of reduction.

Oxidation-reduction reactions involve a transfer of energy. The substance that is oxidized (loses electrons or hydrogen) usually loses energy. The electrons or the hydrogen atoms carry the energy to the substance that is reduced. The reduced substance thus gains energy. As you will see, oxidation-reduction reactions play a key role in cellular respiration.

Hydrogen Acceptors

In cellular respiration, after complex carbohydrates have been broken down to obtain glucose, the glucose itself is broken down in a series of chemical steps. A sequence of chemical reactions that leads to a particular result in a living cell is called a *biochemical pathway*. At several points along the biochemical pathway of cellular respiration, a compound is oxidized by giving up hydrogen atoms. As you have read, for oxidation to occur, some other compound must accept the hydrogen and thus be reduced. Each of these oxidation-reduction steps requires the action of a specific enzyme. In turn, each enzyme requires a coenzyme (see Chapter 4, section 4–4) to act as the hydrogen acceptor in the reaction that the enzyme catalyzes.

One of the coenzymes that act as a hydrogen acceptor in cellular respiration is *NAD* (*nicotinamide adenine dinucleotide*). Another is *FAD* (*flavin adenine dinucleotide*). Each of these molecules can accept two hydrogen atoms, thus undergoing reduction:

$$NAD + 2H \longrightarrow NADH_2{}^*$$
$$FAD + 2H \longrightarrow FADH_2$$

As the hydrogen atoms are transferred to NAD or FAD, these coenzyme molecules gain energy because they are carrying extra hydrogen. This energy gain is temporary. In another series of reactions, the coenzymes give up the hydrogen and return to their oxidized form. At the same time, the extra energy the coenzymes were carrying can be used to form ATP from ADP. Hydrogen is passed along in this manner until the last step in the pathway. At that point, either oxygen or another substance acts as the final acceptor of the hydrogen. In later sections of this chapter, we will examine some of the details of this process.

6-1 Section Review

1. How many phosphate groups are there in a molecule of ATP? In a molecule of ADP?
2. What is the main source of the energy that is released during cellular respiration?
3. What is an oxidation-reduction reaction?
4. What is the function of the coenzymes NAD and FAD?

Critical Thinking

5. ATP is often termed the "currency" of the cell. In what ways is the function of ATP in the cell similar to the function of money in society? (*Reasoning by Analogy*)

---

*This is a simplified way of showing the reduction of NAD. The oxidized form of NAD actually carries a positive charge. A more accurate equation for its reduction is:

$$NAD^+ + 2H \longrightarrow NADH + H^+$$

6-2 Anaerobic Respiration

Section Objectives:

- *Distinguish* between aerobic respiration and anaerobic respiration in organisms.
- *Describe* the overall scheme of glycolysis.
- *Explain* the process of fermentation.

Types of Respiration

In most organisms, respiration is carried on in the presence of free oxygen, or O_2. Free oxygen is oxygen that is not combined with another element. Free oxygen is obtained from the air or from the water in which it is dissolved. Respiration requiring free oxygen is known as **aerobic respiration.** In aerobic respiration, glucose is completely oxidized to carbon dioxide and water. Complete oxidation allows the maximum amount of energy to be removed from the glucose.

A number of organisms, including yeast and many forms of bacteria, carry on cellular respiration without oxygen. This is called **anaerobic respiration.** In anaerobic respiration, the cell receives little of the chemical energy in the glucose.

The first steps of both aerobic and anaerobic respiration are the same. For this reason, you will first study the chemical pathway that includes the steps common to both forms of respiration.

Splitting of Glucose (Glycolysis)

The first steps in respiration take place in the cytoplasm of the cell. The process begins with phosphorylation reactions. In these reactions, two phosphate groups are bonded to a glucose molecule. These steps need energy. The energy and the phosphate groups are gotten from the breakdown of 2 ATP molecules to ADP. The energized glucose molecule then goes through a series of chemical reactions. These reactions split the glucose molecule into 2 molecules of a 3-carbon compound, *PGAL (phosphoglyceraldehyde)*. PGAL is then oxidized by the loss of 2 hydrogen atoms and changes to another 3-carbon compound called *pyruvic* (py ROO vik) *acid.* The oxidation of PGAL gives off energy. Some of this energy is used directly to form 2 ATP molecules. At the same time, the hydrogen removed from PGAL is accepted by the coenzyme NAD, forming $NADH_2$. The process of breaking down the glucose molecule into 2 3-carbon pyruvic acid molecules is called **glycolysis** (gly KAHL uh sis). Figure 6–4 shows this process.

For each pyruvic acid molecule produced by glycolysis, 2 ATP molecules are formed. Since the splitting of 1 glucose molecule produces 2 pyruvic acid molecules, a total of 4 ATP molecules are formed per glucose molecule. Remember, however, that 2 ATP molecules were used to energize the glucose molecule. The net

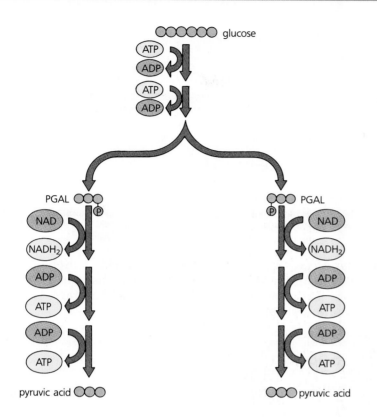

KEY
○ carbon atom

ADP, NAD (low energy)

ATP, NADH₂ (high energy)

◀ **Figure 6–4**

Glycolysis. The first stage of both aerobic and anaerobic respiration is glycolysis, which takes place in the cytoplasm of the cell. In glycolysis, 1 molecule of glucose is split into 2 molecules of pyruvic acid, a 3-carbon compound. Two ATP molecules are used up in the reaction, and 4 ATP molecules are made. Thus, glycolysis produces 2 molecules of ATP for every 1 molecule of glucose. In addition, 2 molecules of $NADH_2$ are formed.

energy output of glycolysis, then, is 2 ATP molecules for each molecule of glucose, plus the energy in the coenzyme $NADH_2$. The energy carried by $NADH_2$ may be used to form more ATP at a later stage but only if oxygen is present.

Fermentation

In anaerobic respiration, all of the energy obtained by the cell comes from the process of glycolysis. Several different chemical changes may follow glycolysis, but no additional ATP is produced. The kind of change that follows glycolysis depends on the particular organism. In all cases, however, the pyruvic acid molecules act as hydrogen acceptors. They accept the hydrogens from $NADH_2$ and oxidize them to NAD. The NAD molecules are then available to be used again in the next round of glycolysis.

In anaerobic respiration, after glycolysis has occurred, the pyruvic acid is changed to other compounds. In yeast cells, it is converted to ethyl alcohol and carbon dioxide. See Figure 6–5a. In certain bacteria, the end product is lactic acid. See Figure 6–5b.

When glycolysis is followed by the conversion of pyruvic acid to some end product with no further release of energy, it is called **fermentation** (fer men TAY shun). There are several industrial processes that make use of natural fermentation. One of the most

Figure 6–5

Fermentation. In the absence of oxygen, pyruvic acid generated by glycolysis accepts hydrogen from $NADH_2$. Lactic acid or ethyl alcohol and carbon dioxide form, and NAD is regenerated. ▶

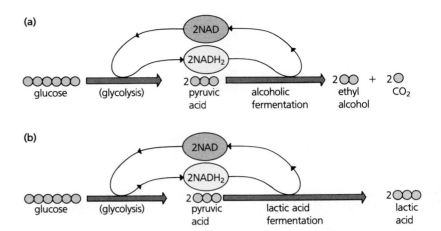

familiar processes is yeast fermentation, which is used in making bread. The carbon dioxide produced by the yeast causes the bread to "rise." The manufacture of ethyl alcohol for beverages is another well-known example.

6-2 Section Review

1. What does aerobic respiration require that is not required by anaerobic respiration?
2. How many ATP molecules must be used to activate one glucose molecule for glycolysis?
3. Name the product of fermentation that is important in making bread.

Critical Thinking

4. Why might winemakers try to minimize the amount of air available to yeast during fermentation? (*Identifying Reasons*)

6-3 Aerobic Respiration

Section Objectives:

- *Describe* the function of the Krebs cycle.
- *Explain* where and how the electron transport chain operates in an organism's body.
- *Compare* the efficiency of aerobic and anaerobic respiration.
- *Relate* muscle fatigue to oxygen debt.
- **Laboratory Investigation:** *Test* for respiration in germinating seeds (p. 120).

The Importance of Oxygen

In anaerobic respiration, the only energy-yielding process is the formation of pyruvic acid. The end products of fermentation, ethyl alcohol or lactic acid, have almost as much energy as the glucose from which they are made. In contrast, a cell that uses oxygen for respiration can remove much more energy from glucose. It can do this because, after a series of reactions, oxygen will accept the hydrogen atoms removed by oxidation during aerobic respiration.

Pyruvic Acid Breakdown

Aerobic respiration begins with glycolysis—the same step as in anaerobic respiration. In anaerobic respiration, you will recall, pyruvic acid accepts the hydrogen from $NADH_2$. No further energy is released. In aerobic respiration, however, the pyruvic acid undergoes a further breakdown and energy release. Some energy is also obtained from the $NADH_2$ formed during glycolysis.

The remaining steps of aerobic respiration take place inside the mitochondria of the cell. The pyruvic acid produced by glycolysis enters the mitochondrion. As you read in Chapter 5, a mitochondrion has a double membrane. The inner membrane is deeply folded. This gives it a large surface area. Most of the enzymes, coenzymes, and other special molecules needed for aerobic respiration are located inside or on the surface of this inner membrane. It is the presence of these molecules within or on the membrane that makes aerobic respiration possible.

Inside the mitochondrion, the pyruvic acid breaks down into carbon dioxide, $NADH_2$, and a 2-carbon compound known as an *acetyl group*. The acetyl group then combines with a coenzyme, called coenzyme A (CoA), to form **acetyl CoA.** The acetyl CoA molecule enters into the next stage of aerobic respiration.

Krebs Cycle

The series of chemical reactions that begin with the acetyl CoA formed from pyruvic acid is called the **Krebs cycle.** See Figure 6–7. Its details were discovered by Sir Hans Krebs of Oxford University in England. He received a Nobel Prize in 1953 for this accomplishment. Krebs found that the series of reactions has the form of a repeating cycle. Certain organic acid molecules that are part of the cycle are used over and over again. During the cycle, they are changed to other compounds. Then, before the cycle begins again, they change back to their original form.

Each "turn" of the Krebs cycle requires 1 molecule of acetyl CoA. Each turn yields 2 molecules of carbon dioxide and 4 pairs of hydrogen atoms. The hydrogen atoms are picked up by NAD and FAD. Almost all the chemical energy removed from the pyruvic acid is carried by these hydrogens. Only 1 ATP molecule is produced by each turn of the Krebs cycle.

Electron Transport Chain

So far you have learned that in aerobic respiration, each time 1 molecule of glucose is split into 2 molecules of pyruvic acid, 2 ATP molecules are produced. An additional ATP molecule is produced by each turn of the Krebs cycle (2 ATP molecules for each glucose

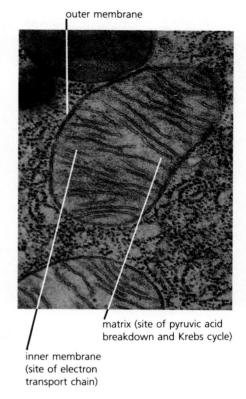

outer membrane

matrix (site of pyruvic acid breakdown and Krebs cycle)

inner membrane (site of electron transport chain)

▲ **Figure 6–6**

A Mitochondrion. The mitochondrion has a double membrane. The inner membrane contains the enzymes and coenzymes that make up the electron transport chain. The area enclosed by the inner membrane is the matrix. Within the matrix, pyruvic acid breakdown and the Krebs cycle take place. (Magnification 52 400 X)

Figure 6–7

Pyruvic Acid Breakdown and the Krebs Cycle. Pyruvic acid from glycolysis reacts to form acetyl CoA, which then enters the Krebs cycle. For every turn of the cycle, 2 molecules of CO_2, 4 pairs of hydrogen molecules, and 1 ATP are produced. Note that 3 water molecules are used during the Krebs cycle. ▶

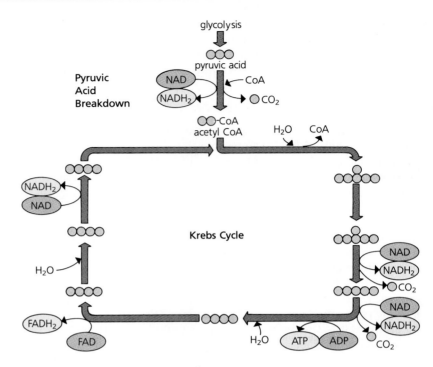

molecule). This is a total of 4 ATP molecules per glucose molecule. All the remaining energy released by the breakdown of glucose is carried by the hydrogen in $NADH_2$ and $FADH_2$.

There are 12 pairs of these hydrogens—2 from glycolysis, 2 from pyruvic acid breakdown, and 8 from the Krebs cycle. The energy still contained in these hydrogen atoms is used to form additional ATP. These energy-releasing reactions are carried out by the **electron transport chain,** a highly organized system of enzymes and coenzymes located in the inner mitochondrial membrane. In the electron transport chain, a series of oxidation-reduction reactions take place. Hydrogen atoms are carried into the chain by $NADH_2$ and $FADH_2$. The electrons from the hydrogen atoms are then passed along from one compound to another. See Figure 6–8. At various places along the chain, the electrons give up some energy, and molecules of ATP are formed.

The final step in this process involves free oxygen, which acts as the final hydrogen acceptor. One oxygen atom combines with each pair of hydrogen atoms and a pair of electrons to form water.

Water produced by cellular respiration is called the water of metabolism. It may be used by the cell, or it may be excreted as a waste product. For desert animals, such as the kangaroo rat, the water of metabolism is an important source of water for survival. Altogether, in most cells 32 ATP molecules are produced by the electron transport chain for each molecule of glucose. This is in addition to the 2 ATP molecules that come directly from glycolysis and the 2 ATP molecules that come from the Krebs cycle. Therefore, aerobic respiration can produce a total of 36 ATP molecules from each molecule of glucose.

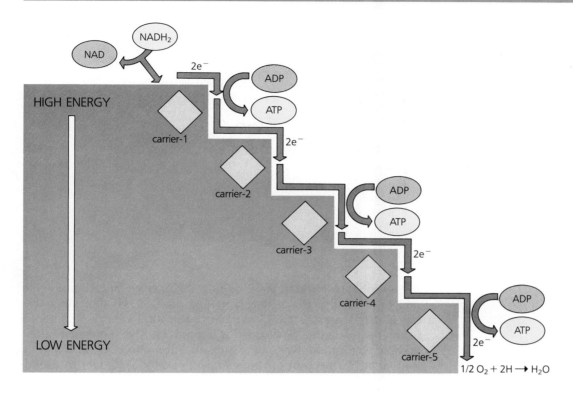

Net Reactions of Aerobic Respiration

The net result of all the steps of aerobic respiration is usually summarized in the following chemical equation:

$$C_6H_{12}O_6 + 6\ O_2 \longrightarrow 6\ CO_2 + 6\ H_2O + \text{Energy (36 ATP)}$$

This equation is somewhat oversimplified. Water is needed as a raw material for the Krebs cycle. Looking back at Figure 6–7, you can see three places where a molecule of water enters the cycle. Since the Krebs cycle runs twice for each glucose molecule, 6 molecules of water are needed for each glucose molecule that is broken down. This water should be shown as a raw material in the equation. Therefore, the equation should be written as follows:

$$C_6H_{12}O_6 + 6\ H_2O + 6\ O_2 \longrightarrow 6\ CO_2 + 12\ H_2O + \text{Energy (36 ATP)}$$

Efficiency of Cellular Respiration

Figure 6–9 summarizes all of the biochemical pathways involved in anaerobic and aerobic respiration. In anaerobic respiration, you may recall, the glycolysis pathway produces a net yield of 2 ATP molecules per molecule of glucose. This type of respiration is fairly inefficient. It leaves most of the potential energy of the glucose in the end products of fermentation. Anaerobic respiration, however, does meet the energy needs of many simple organisms, such as yeast and bacteria.

▲ **Figure 6–8**
The Electron Transport Chain. The electron transport chain is a series of electron carriers on the inner membrane of the mitochondrion. NADH$_2$ and FADH$_2$ deliver hydrogen to the electron transport chain. The hydrogen is split into hydrogen ions (H$^+$) and electrons (e$^-$). As the electrons pass from one electron carrier to the next, they release energy, and ATP is formed. At the end, the electrons, hydrogen ions, and free oxygen combine to form water molecules.

Figure 6–9

Cellular Respiration—A Summary. During glycolysis, glucose is broken down to pyruvic acid, producing 2 molecules of ATP. In anaerobic respiration, pyruvic acid is further broken down, but no more ATP is produced. In aerobic respiration, the Krebs cycle and the electron transport chain yield 34 ATP (for a total of 36 molecules of ATP). ▶

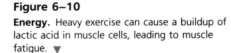

Aerobic respiration yields almost 20 times as much energy per molecule of glucose as fermentation does. It is, moreover, an efficient process. About 45 percent of the total energy obtainable from the oxidation of glucose is stored as ATP molecules after aerobic respiration. By comparison, an automobile engine converts only about 25 percent of the chemical energy of its fuel to useful work.

Muscle Fatigue and Oxygen Debt

Some organisms that have the capacity for aerobic respiration can function by anaerobic respiration alone when free oxygen is not available. Yeast cells, for example, use aerobic respiration when oxygen is present, but they can live and grow by anaerobic respiration in the absence of oxygen. Muscle cells in humans and other animals normally rely on aerobic respiration to meet their energy needs. They can, however, function for a short time without oxygen. They do this by using the energy obtained from glycolysis alone.

During periods of hard or prolonged physical activity, the muscle cells may use oxygen faster than the respiratory and circulatory systems can supply it. When the oxygen supply becomes too low, the electron transport chain cannot work. This means that $NADH_2$ and $FADH_2$ build up in the mitochondria and are not recycled. This forces the Krebs cycle to stop.

When the Krebs cycle stops, the muscle cells continue to release energy by glycolysis. Pyruvic acid becomes the acceptor for hydrogen and is converted to lactic acid. The buildup of lactic acid in the muscle cells produces feelings of fatigue and gradually reduces the ability of the cells to do their normal work.

When lactic acid builds up, the cells need a period of rest or reduced activity to recover to a normal condition. During this time, fresh supplies of oxygen allow the lactic acid to be oxidized back to pyruvic acid, and the accumulated hydrogen is passed down

Figure 6–10

Energy. Heavy exercise can cause a buildup of lactic acid in muscle cells, leading to muscle fatigue. ▼

through the electron transport chain. The amount of oxygen needed to get rid of the lactic acid is called **oxygen debt.** You know that during periods of strenuous activity, your breathing and heart rates increase in order to deliver more oxygen to your muscles. When the intense activity stops, your breathing and heart rates remain high for a time. During this time, extra supplies of oxygen are delivered to the muscles to pay back the oxygen debt.

Respiration of Fats and Proteins

This discussion of aerobic respiration has focused on the breakdown of glucose to supply energy for the cell. Cells that carry on aerobic respiration can also extract energy from other types of food substances, such as fats and proteins. These substances are broken down and converted into compounds that can enter the respiratory pathway at some intermediate point in the glucose breakdown pathway. Figure 6–11 shows where these compounds enter into the pathway.

When fats are used as a source of energy, twice as much ATP is produced as when glucose is used. This is why a diet high in fat can lead to weight gain. You must be twice as active to burn a gram of fat as you must be to burn a gram of carbohydrate. A gram of protein, on the other hand, yields about the same amount of energy as a gram of carbohydrate. However, proteins are not a preferred source of energy for the cell.

The pathways shown in Figure 6–11 help to explain the importance of a well-balanced diet. By supplying your body with enough carbohydrates, fats, and proteins, your body cells can balance their energy needs with their needs for building materials. In a balanced diet, for example, many of the proteins you consume are broken down into amino acids, which are then used to build enzymes and other proteins needed by your cells. If, however, your diet is low in carbohydrates and fats, the amino acids are used as an energy source instead.

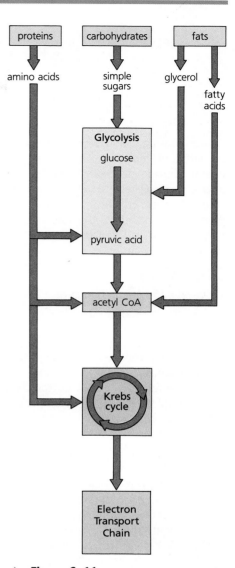

▲ **Figure 6–11**

The Role of Proteins and Fats in Respiration. Proteins, fats, and carbohydrates can be used as fuel for respiration. Once the molecules are broken down, they enter the respiratory pathway at various points.

6-3 Section Review

1. What is the function of oxygen in aerobic respiration?
2. Which system of enzymes and coenzymes uses the energy from $NADH_2$ and $FADH_2$ to form ATP?
3. What are the three end products of aerobic respiration?
4. During strenuous exercise, what substance becomes an end product of respiration in muscle cells?

Critical Thinking

5. Imagine a hot potato being passed down a line, from person to person. In what ways can this be thought of as similar to aerobic respiration? How does it differ? (*Comparing and Contrasting*)

Laboratory Investigation

Respiration in Plant Tissue

Carbon dioxide produced by respiration can be detected with a pH indicator. The indicator changes color as carbon dioxide dissolves in water, forming carbonic acid ($CO_2 + H_2O \longrightarrow H_2CO_3$). The rate at which the color changes indicates the rate of carbon dioxide production and, thus, the rate of respiration. In this investigation, you will use a pH indicator to detect respiration in plant cells.

Objectives

- *Observe* the color change of a pH indicator.
- *Detect* respiration in germinating seeds.

Materials

6 germinating seeds	3 large test tubes
6 boiled seeds	forceps
carbonated water	cotton balls
bromthymol blue	wax pencils
10 mL graduated cylinder	test tube stoppers
dropper	

Procedure   *

A. Use a wax pencil to label three test tubes, A, B, and C. Using a graduated cylinder, add 10 mL of bromthymol blue solution to each test tube. To Tube A, add carbonated water one drop at a time, mixing the contents after each drop. Note and record the color change, then set the test tube aside.

B. Using forceps, place a loose ball of cotton in tubes B and C, about 3 cm above the solution, as shown below.

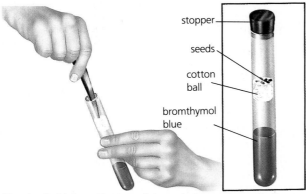

stopper
seeds
cotton ball
bromthymol blue

*For detailed information on safety symbols, see Appendix B.

C. Place 6 germinating seeds on top of the cotton in tube B and seal the tube with a stopper. Note the time on a chart like the one shown below.

Test Tube	Time Tube Sealed	Time Color Change Complete	Elapsed Time
B			
C			

D. In tube C, place 6 boiled seeds on top of the cotton, seal the tube with a stopper, and record the time on your chart.

E. Observe test tubes B and C for 20 minutes. For each tube, record the time at which the color change, if any, occurs.

Analysis and Conclusions

1. What color change does bromthymol blue undergo in the presence of dissolved carbon dioxide?

2. How does the pH of the water change as carbon dioxide dissolves in it?

3. What was the purpose of using boiled seeds?

4. From your data, what evidence is there that the germinating seeds are respiring?

5. Why was cotton and not a tight-fitting piece of plastic wrap or rubber used to support the seeds in the test tubes?

6. Based on your results, what would you expect to happen to the concentration of oxygen in each tube?

7. What would happen to the respiration rate in test tube B if the seeds were kept sealed in the test tube for many days?

8. How would your results have been affected if you had used more seeds? Fewer seeds?

9. Suppose you repeated this experiment using an equivalent amount of tissue from another organism. You found that the indicator changed color more quickly than it did with the seeds. How would you compare its respiration rate with the respiration rate of the seeds?

Chapter Review

Study Outline

6–1 Energy for Life

■ All living things obtain energy for their life processes through cellular respiration.

■ Cellular respiration consists of the step-by-step breakdown of a nutrient, most commonly glucose, to release its energy in order to produce ATP.

■ Energy for cell function is stored in the high-energy compound ATP. When the third phosphate group of an ATP molecule is removed—leaving ADP—and bonded to another compound, it transfers energy to the other compound.

■ In the breakdown of glucose, energy is transferred from substance to substance in a series of oxidation-reduction reactions that are controlled by enzymes. Coenzymes such as NAD and FAD act as hydrogen acceptors.

6–2 Anaerobic Respiration

■ Both aerobic and anaerobic respiration begin with glycolysis. Each glucose molecule is broken apart into two pyruvic acid molecules, with a net gain of two ATP molecules.

■ In anaerobic respiration, glycolysis is followed by further chemical reactions in which pyruvic acid is converted to lactic acid or to ethyl alcohol and carbon dioxide. No additional ATP is produced.

■ Many industrial processes make use of anaerobic respiration, also called fermentation. These processes include bread making and alcohol production.

6–3 Aerobic Respiration

■ In aerobic respiration, pyruvic acid is broken down after glycolysis to yield additional energy. These steps take place within the mitochondria and in the presence of oxygen.

■ Acetyl CoA, formed from pyruvic acid, enters the Krebs cycle, which yields 2 molecules of carbon dioxide, 4 pairs of hydrogen atoms, and 1 ATP per "turn." There are two "turns" of the cycle per glucose molecule.

■ The hydrogen atoms enter the electron transport chain, where they are carried along by hydrogen acceptors. As their electrons are transferred, energy is released. Thirty-two ATP molecules are formed.

■ Aerobic respiration is an efficient process. It yields 20 times more energy per molecule of glucose than does fermentation.

■ In human muscle cells, during vigorous activity, the pyruvic acid produced in glycolysis may be converted to lactic acid until enough oxygen is available again for aerobic respiration to continue.

■ Fats and proteins may also be used as a source of energy, entering the respiratory pathway at intermediate points.

Vocabulary Review

cellular respiration (108)	anaerobic respiration (112)
ATP (108)	glycolysis (112)
ADP (109)	fermentation (113)
oxidation (110)	acetyl CoA (115)
reduction (110)	Krebs cycle (115)
oxidation-reduction reaction (110)	electron transport chain (116)
aerobic respiration (112)	oxygen debt (119)

A. Sentence Completion—Fill in the vocabulary term that best completes each statement.

1. When nutrients are broken down in respiration, their energy is then transferred to molecules of _____.

2. Both anaerobic and aerobic respiration begin with the process of _____.

3. After glycolysis in some organisms, ethyl alcohol and carbon dioxide may be the end products of _____.

4. The loss of electrons or hydrogen atoms by an atom or molecule is called _____.

5. During heavy exercise, the build-up of lactic acid in muscle cells results in _____.

Chapter Review

B. Definition—Replace the italicized definition with the correct vocabulary term.

6. *The breakdown of food molecules to release energy for cell activities* takes place in all living things.

7. *Reactions in which electrons released by one substance are gained by another* are essential to aerobic respiration.

8. The *repeating series of reactions that begins with acetyl CoA* takes place within the mitochondria.

9. *The complete breakdown of glucose in the presence of oxygen* results in the most efficient use of the energy stored in glucose.

10. In aerobic respiration, most of the ATP is produced by the *series of oxidation-reduction reactions that releases energy contained in hydrogen atoms.*

Content Review

11. Why must all living things carry on cellular respiration?

12. Which bond of the ATP molecule is most often broken? Why?

13. Explain what happens during a phosphorylation reaction. Why is this reaction important to cells?

14. Why are oxidation-reduction reactions always coupled?

15. Does a substance acting as a hydrogen acceptor gain energy or lose energy? Explain your answer.

16. State two differences between anaerobic respiration and aerobic respiration.

17. For each glucose molecule entering glycolysis, 4 ATP molecules are produced. Why is there an actual gain of only 2 ATP molecules per glucose molecule?

18. After glycolysis, what happens to pyruvic acid if no oxygen is present?

19. How is yeast useful in the baking industry?

20. In anaerobic respiration, what is the net gain in ATP from one glucose molecule?

21. Where do the following processes take place in the cell: glycolysis, Krebs cycle, electron transport chain?

22. What is the chief function of the Krebs cycle?

23. What happens to the hydrogen atoms carried into the electron transport chain?

24. Which element serves as the final hydrogen acceptor in aerobic respiration? What compound is formed?

25. In anaerobic respiration, what happens to most of the energy stored in glucose?

26. What happens to most of the energy stored in glucose in aerobic respiration?

27. Explain what happens in muscle cells to form lactic acid.

28. In addition to carbohydrates, what other nutrients may be used in respiration?

Graphic Organizing

For information on graphic organizers, see Appendix G at the back of this text.

29. **Compare/Contrast Matrix:** Use the matrix below to compare aerobic respiration and anaerobic respiration. Copy the matrix onto a separate sheet of paper and complete it.

Characteristic	Aerobic Respiration	Anaerobic Respiration
starting material	?	?
pathways involved	?	?
final hydrogen acceptor	?	?
end products	?	?
energy produced	?	?

Critical Thinking

30. What would happen if the energy in glucose were released all at once instead of in a series of small steps? (*Predicting*)

31. Do you think yeast cells would grow more rapidly when carrying out fermentation or aerobic respiration? Support your answer. (*Reasoning Conditionally*)

32. Why are NAD and FAD classified as coenzymes rather than as enzymes? (*Reasoning Categorically*)

33. Why are oxidation-reduction reactions important to the cell? (*Generalizing*)

Creative Thinking

34. Victims of a heart attack often have small amounts of lactic acid in the blood leaving the heart. What does this suggest about a possible factor in heart attacks?

35. Compounds that participate in energy changes in living cells have been compared to "charged batteries" and "discharged batteries." Which cell compound would the charged battery represent? Which compound would the discharged battery represent? How does the discharged battery become charged again?

Problem Solving

36. A student had a packet of seeds several years old. To avoid planting seeds that might not grow, he first tested them as shown below. In jar A he placed 10 bean seeds that had been soaked in water overnight. In jar B he placed the indicator bromthymol blue, which turns yellow-green when mixed with carbon dioxide. By the next day, the bromthymol blue was a strong yellow-green color. The student decided to plant the seeds.

  Why did the student conclude that the seeds were alive? What control should he have set up to verify the cause of the color change?

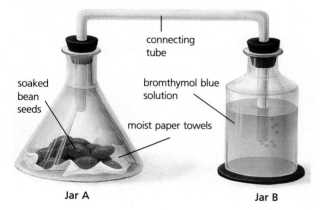

connecting tube

soaked bean seeds

bromthymol blue solution

moist paper towels

Jar A   Jar B

37. How many molecules of ATP could be formed from the breakdown of 500 molecules of glucose if oxygen were present during cellular respiration? How many molecules of ATP could be formed if oxygen were absent?

38. A person who carries out aerobic exercise regularly (at least 3 times a week) is less likely to suffer muscle discomfort than one who does so only occasionally. Develop a hypothesis to explain this difference. Then, do some library research to determine whether your hypothesis is correct.

Projects

39. Before going to sleep, place a body thermometer near your bed. When you awaken the next morning, without getting out of bed, take your temperature and record it. Then, in the late afternoon, take your temperature again and record it. For at least one week, keep a record of your waking and afternoon temperatures. Do the morning and afternoon temperatures differ? Do you find a pattern in the differences? If so, how would you explain the pattern? Design an experiment to test your hypothesis.

40. At least three of the B vitamins are essential to cellular respiration: *niacin* to form NAD, *riboflavin* to form FAD, and *thiamine* to convert pyruvic acid to acetyl CoA. Do library research and write a report on one of these vitamins. Describe the symptoms and treatment of conditions arising from its deficiency. How can the deficiency be prevented?

41. When food is scarce, some animals, such as bears, bats, and woodchucks, hibernate or enter some other inactive state. Choose one of these animals. Research and prepare an oral report on this animal's adaptations. Explain how the animal enters and leaves the inactive state and how it obtains energy while inactive.

▲ The classification of living things takes into account their similarities and differences.

Classification of Living Things

A lone giraffe strolls past grazing zebras on a lazy afternoon in Kenya. Trees, surrounded by a carpet of green grass, stretch for the sky. Somewhere nearby, an ant zigzags across the dirt, a bird watches from a branch, and insects buzz. Suppose you had to organize all of this life into groups. Would trees and giraffes go together, since both are tall? Would insects and birds go together since both have wings? In this chapter, you will learn how biologists classify living things—from those on the grasslands of Kenya to those in a city park.

Guide for Reading

Key words: taxonomy, genus, kingdom, binomial nomenclature, species
Questions to think about:
📖 How are biologists able to classify organisms?
📖 How is the theory of evolution the basis for modern taxonomy?
📖 What are the major characteristics of each of the five kingdoms?

7-1 Classification

Section Objectives:

- *Explain* the function of classification systems.
- *Describe* the naming system used in modern biology.
- *Explain* how the theory of evolution has affected taxonomy.
- *Describe* the types of evidence now used to determine relationships between groups of organisms.

About 1.5 million kinds of living organisms are known today, and each year thousands more are identified. Some experts believe there are as many as 10 million different kinds. They vary in form and size from microscopic bacteria to giant redwood trees.

To deal with this huge number of organisms, biologists name and group, or classify, the organisms according to an established international system. This makes it easier for scientists to discuss the types and characteristics of living things. The branch of biology that deals with the classification and naming of living things is called **taxonomy** (tak SAHN uh mee).

Classification Systems

You face many everyday situations in which you need to find objects or information. If the objects or information have been organized or classified into groups, you should be able to find what you are looking for without checking every item in a large group. To

▲ **Figure 7–1**

Diversity of Life. A great diversity of plant and animal species is found in the rain forests.

Figure 7–2

Organization in a Supermarket. Think of how long you might search for a single item if products in a supermarket were not arranged according to some system. ▶

help you understand what is involved in classification, consider the arrangement of goods in a supermarket.

A large supermarket, like the one shown in Figure 7–2, carries 7000 to 10 000 items. If these items were placed on the shelves randomly, shopping for the week's groceries might take an entire day or even longer. However, the basis for any classification system is the grouping of things according to similarities. You can find the items you need fairly quickly because related items are arranged in groups. First, items are grouped into broad categories—frozen foods, meat, produce, cleaning supplies, paper goods, and dairy products. Each of these departments is subdivided into a series of smaller, related categories. For example, the frozen food department has separate sections for vegetables, juices, cakes, fish, TV dinners, and ice cream. Each of these sections is further subdivided. The ice cream section is divided into half-gallons, quarts, pints, and cups. Within each size range, the ice cream may be grouped by flavor or brand name.

Once you are familiar with the organization of the supermarket, it is easy to find an item. If the market manager gets a new kind of frozen cake, it is a simple matter to place it with the other frozen cakes. In a similar manner, the classification system used in modern biology allows biologists to identify an organism and place it in the correct group with related organisms.

Early Classification Schemes

In early attempts at classification, living things were separated into two major groups—the plant kingdom and the animal kingdom. These two groups were then subdivided in various ways. In early historical documents, for example, plants were divided into grasses, herbs, and trees, while animals were classified as fish, creeping creatures, fowl, beasts, and cattle.

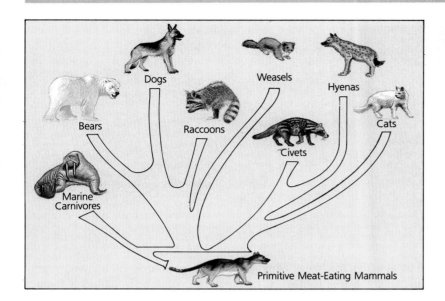

◀ **Figure 7–7**
Phylogenetic Tree. The evolutionary relationship among the different types of meat-eating animals is shown in this phylogenetic tree.

Bears

Dogs

Weasels

Hyenas

Raccoons

Cats

Civets

Marine Carnivores

Primitive Meat-Eating Mammals

Structural Information   Structural similarities, such as those in skeletal structure or leaf anatomy, are the primary basis for grouping organisms. There are instances, however, in which additional information may be needed to make a classification judgment. Furthermore, it is always desirable to pair structural information with other kinds of available information.

Biochemical Information   Today, taxonomists frequently use biochemical data when classifying organisms. Information about the DNA, RNA, and proteins of different species can be compared to establish relationships. This kind of information was used recently to show that the giant panda of Asia probably is closely related to bears.

Cytological Information   Details about the cellular structure of organisms can also offer clues to evolutionary relationships. Many single-celled organisms, for example, have been classified on the basis of their similarities in cellular structure, which can be seen with the electron microscope. Another source of cytological information that taxonomists have used is information on numbers of chromosomes in a species. Many plant species have been classified, in part, according to their chromosome numbers.

Embryological Information   The term *embryo* refers to any multicellular organism in its early stages of development. While the adult forms of different animal species may look very different, the embryos can look very similar. Taxonomists use embryological data to establish taxonomic relationships that otherwise might be overlooked.

Behavioral Information   Sometimes, behavior may be the only basis for distinguishing one species from another. For example, some species of crickets can be distinguished only on the basis of their mating calls.

▲ **Figure 7–8**

Modern and Fossil Fish. A modern-day fish (left) shows striking similarity to the fossil of an ancient fish (right).

Fossil Information Any preserved evidence of an organism, such as a bone, footprint, or body impression, is called a **fossil.** Fossils are useful in establishing likely relationships between modern-day species and species that lived thousands or even millions of years ago.

Taxonomy in Perspective

Although taxonomy has come a long way since Aristotle, not all classification problems have been solved. In fact, taxonomists constantly reexamine and refine previous classifications. A good part of taxonomy is subjective. While taxonomists follow general, established principles, their decisions are judgment calls that often vary from one taxonomist to another. Taxonomists may disagree over whether a species belongs in one genus or another or whether several species should be considered one species. Add to this process the task of classifying and naming newly discovered organisms and you realize taxonomy is a dynamic and important field of biology.

7-1 Section Review

1. Describe the system of binomial nomenclature used by biologists.
2. How has the theory of evolution affected the science of taxonomy?
3. What is the modern definition of a species?
4. What kinds of information are used by taxonomists to determine relationships between groups of organisms?

Critical Thinking

5. Explain the relationship between a class and an order. (*Relating Parts and Wholes*)

7-2 Major Taxonomic Groups

Section Objectives:

■ *Name* the five kingdoms and describe the characteristics of each.
■ *Explain* the advantages of the five-kingdom system.
Laboratory Investigation: *Classify* some organisms by means of a taxonomic key (p. 140).

In all early classification schemes, living things were divided into two kingdoms—plants and animals. This system works with large organisms. Trees, grass, and flowers are plants, while frogs, fish, insects, birds, and cats are animals. However, some organisms have both plant and animal characteristics. An organism called euglena (yoo GLEEN uh), for example, is a unicellular organism that carries on photosynthesis like a plant yet moves like an animal.

To solve the problem of classifying organisms that are not obviously animals or plants, taxonomists have added new kingdoms to the classification system. However, there is no universal agreement about how many kingdoms are needed and which organisms should be placed in them. Each possible arrangement has some advantages and some disadvantages.

Most taxonomists use a five-kingdom system of classification. This is the system used in this text. The five kingdoms are Monera, Protista, Fungi, Plantae, and Animalia. The five-kingdom system emphasizes certain basic differences among large groups of organisms. The general characteristics of the five kingdoms are described briefly in the following sections and in Figure 7–9.

Figure 7–9
Characteristics of the Five Kingdoms. ▼

Characteristics of the Five Kingdoms					
CHARACTERISTIC	**KINGDOM**				
	Monera	**Protista**	**Fungi**	**Plantae**	**Animalia**
cell type	prokaryotic	eukaryotic	eukaryotic	eukaryotic	eukaryotic
body form	most unicellular; some colonial	most unicellular; some simple multicellular	most multicellular	multicellular	multicellular; organs, and organ systems
cell wall composition	polysaccharides and amino acids	present in some composition varies	usually chitin	cellulose	no cell wall
mode of nutrition	photosynthesis, chemosynthesis, absorption	photosynthesis, ingestion, or absorption	absorption	photosynthesis	ingestion
nervous system	absent	absent	absent	absent	present
locomotion	present in some	present in some	absent	absent	present

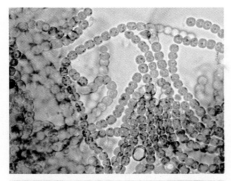

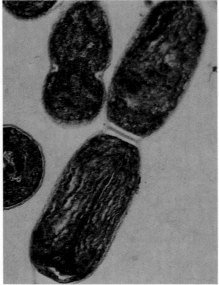

▲ **Figure 7–10**

Examples from the Kingdom Monera. A colonial (filamentous) blue-green bacterium (top) and a blue-green bacterium at higher magnification are examples of monerans. (Magnification 99 900X)

Kingdom Monera

The kingdom **Monera** (muh NER uh) includes all prokaryotic organisms. These organisms, known as *monerans,* or bacteria, are mostly unicellular, although some form chains, clusters, or colonies of connected cells. As you know from Chapter 5, prokaryotic cells do not have an organized nucleus with a nuclear membrane. They also lack most of the organelles, such as mitochondria, lysosomes, and Golgi bodies, found in eukaryotic cells. Although prokaryotic cells have cell walls, the cell walls are chemically different from the cell walls of other organisms.

Most bacteria do not carry on photosynthesis. They must absorb nutrients from the environment. However, blue-green bacteria (once called blue-green algae) and a few other types of bacteria contain chlorophyll and carry on photosynthesis. See Figure 7–10. Chloroplasts are not present in these bacteria. Rather, chlorophyll is contained in layers of membranes that lie within the bacterial cell.

Kingdom Protista

Members of the kingdom **Protista** (proh TIST ah) are mostly unicellular, although there are colonial and simple, multicellular forms. The cells of *protists* are eukaryotic. That is, like the cells of multicellular organisms, each contains a membrane-bound nucleus and other organelles. Some protists are motile and feed upon bacteria and bits of organic matter. These animal-like protists are called *protozoa.* Other protists carry out photosynthesis. These plantlike protists are members of a group of organisms called algae.

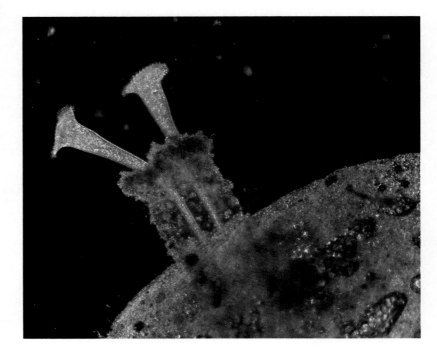

Figure 7–11

An Example from the Kingdom Protista. This colonial species of *Stentor* is a type of protozoan. (Magnification 52X) ▶

Algae is the term used for all eukaryotic, unicellular and simple multicellular, organisms that carry out photosynthesis and live in water. The different groups of algae are not thought to be directly related to each other.

Kingdom Fungi

The kingdom **Fungi** (FUN jy) includes molds, yeasts, mushrooms, rusts, and smuts. *Fungi* (singular, fungus) live either as parasites on other living things or as decomposers of dead matter. Some fungi are unicellular, but most are multicellular. Fungi are eukaryotic. That is, they have cell organelles and distinct nuclei surrounded by nuclear membranes. Although fungi have cell walls, the walls are chemically different from those of other organisms. In the past, members of the kingdom Fungi were placed in the plant kingdom because they resemble plants more than animals. See Figure 7–12. The differences between fungi and plants are so great, however, that biologists have placed fungi in their own kingdom. Unlike plants, fungi contain no chlorophyll and cannot synthesize food. Instead, they secrete enzymes that digest food material outside the organism. The fungi then absorb the nutrients.

▲ **Figure 7–12**
An Example from the Kingdom Fungi. Mushrooms, like the poisonous one shown here, are just one of many types of fungi.

Kingdom Plantae

The *plants,* members of the kingdom **Plantae** (PLAN tee), include mosses, liverworts, ferns, and seed plants. See Figure 7–13. The cells of plants have cell walls. Members of the plant kingdom show a true tissue and organ organization. Plants cannot move from place to place on their own. Nearly all plants carry on photosynthesis and most live on land. Within plant cells, chlorophyll is found in chloroplasts.

Kingdom Animalia

There are more species in kingdom **Animalia** (an uh MAL ee ah) than in any other kingdom. Members of the animal kingdom are multicellular and usually show an organ and organ system level of organization. During some part of their life cycle, most animals can move from place to place on their own. Since animals cannot carry on photosynthesis, they must obtain food from their environment. Most animals actively search for food, relying upon highly specialized sensory systems, brains, and nerve-muscle systems to do so. These same systems permit other complex types of behavior. Animals such as sponges, worms, and insects have no backbone and are called *invertebrates.* Animals with backbones, such as fish, snakes, and humans, are called *vertebrates.* Sexual reproduction is more common in animals than asexual reproduction. In some species, there is specialized courtship behavior, and there may be extensive parental care of the young.

A more complete list of all the taxonomic groups is found in Appendix F at the back of this book.

▲ **Figure 7–13**
Examples from the Plant and Animal Kingdoms. The leaf is part of a plant, while the lizard is a member of the animal kingdom.

The Taxonomic Key

What questions would you ask if you were playing a game in which you had to identify a famous person of whom another player was thinking? You might begin by asking: Is the person male or female? If female, is she American, or non-American? If non-American, is she European or non-European? By continuing with questions that became more and more specific, you could identify the person.

Biologists use a similar procedure to identify an organism. The procedure they use is a series of instructions called a taxonomic key. A **taxonomic key** is a tool used to identify organisms already classified by taxonomists. Most keys are dichotomous. That is, they consist of a series of paired statements that describe alternative possible characteristics of the organism. These paired statements usually describe the presence or absence of certain characteristics or structures that are easily seen. For example, an animal may or may not have a spinal column. If it has a spinal column, it may or may not have gills. If gills are absent, its body may or may not be covered with scales, and so on. Each set of choices must be arranged in such a way that each step produces a smaller grouping. Figure 7–14 shows a sample taxonomic key that is used for identifying vertebrates.

Critical Thinking in Biology

Reasoning Categorically

Something that is true about a group or a *category* will be true for every member of the group or category. For example, you know that all fish can swim. You also know that a perch is a fish. Therefore, you could reason that a perch can swim. This type of reasoning is called categorical reasoning. The argument can be set up formally:

Premise: All fish can swim.
Given: A perch is a fish.
Conclusion: A perch can swim.

1. Using the categorical argument above as a model, construct a categorical argument to show that a bird is a vertebrate.
2. The cells shown at right are from a multicellular organism. Construct a categorical argument to determine whether they are plant or animal cells.
3. What is wrong with this categorical argument?
Premise: All cats are meat-eaters.
Given: Organism B is a meat-eater.
Conclusion: Therefore, Organism B is a cat.
4. Think About It Explain your thought processes as you answered Question 1.

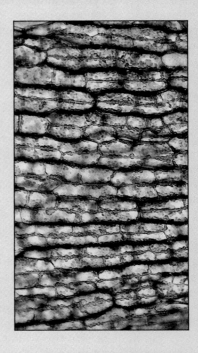

Key for Identifying Vertebrates	
1	1A. spinal column present . . . go to 2.
	1B. spinal column absent . . . Invertebrate.
2	2A. fins and gills present . . . Fish.
	2B. fins and gills absent . . . go to 3.
3	3A. scales present . . . Reptile.
	3B. scales absent . . . go to 4.
4	4A. feathers present . . . Bird.
	4B. feathers absent . . . go to 5.
5	5A. hair or fur present . . . Mammal.
	5B. hair or fur absent . . . Amphibian.

◀ **Figure 7–14**
A Sample Taxonomic Key For Identifying Vertebrates.

Representative Organisms

As you learned in Chapter 1, all living things carry on certain processes such as nutrition, respiration, transport, excretion, and regulation. Size greatly influences how an organism performs these processes. The cells of single-celled and small multicellular organisms are in close contact with the external environment. Because they are small, they can carry on their life processes in a simple fashion. However, as animals become larger, most of their cells are not in contact with the environment. In these animals, the life functions are carried out by groups of organs arranged in systems.

In Unit 2, you will study how certain organisms solve the problems of life. Since you cannot consider every organism, you will study six representative organisms in order of increasing complexity. Two of these organisms, the ameba and the paramecium, are unicellular and are classified as protists. The others are animals: the hydra, the earthworm, the grasshopper, and the human. All these organisms are **heterotrophs** (HET uh ruh trohfs) —organisms that obtain their food from the environment. In contrast, plants and certain other organisms make their own food. These organisms are called **autotrophs.**

Ameba and Paramecium   The ameba and paramecium are common inhabitants of ponds and streams. As protists, they carry out their life processes within a single cell. Although they are small, they can be seen without a microscope. Under the microscope, the ameba, shown in Figure 7–15, appears as a transparent mass that constantly changes shape. It has cytoplasm with a cell membrane and a nucleus. It creeps along through the flowing of cytoplasm into temporary structures called *pseudopods,* or false feet.

The paramecium, shown in Figure 7–16, is easy to recognize because of its slipperlike shape. It contains two nuclei. The larger macronucleus controls general cell activities. The smaller micronucleus is involved in reproduction. A stiffened cell membrane with

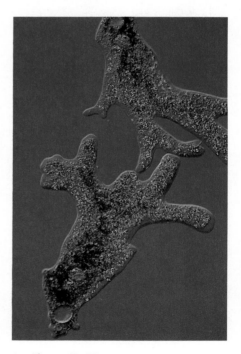

▲ **Figure 7–15**
Ameba. The ameba is just one of many types of protozoan. Organelles and other particles are evident in the cytoplasm of these ameba. (Magnification 108X)

Figure 7–16

Paramecium. The paramecium is another type of protozoan. The cilia that cover the surface of a paramecium can be seen along the edge of this specimen. (Magnification 325X) ▶

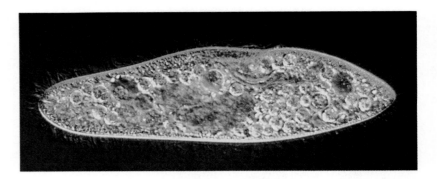

numerous, hairlike cilia surrounds the paramecium. The cilia allow the paramecium to swim. On one side of the paramecium is a depression, known as the oral groove, that leads into a tubular gullet. Both the oral groove and the gullet are involved in nutrition.

Hydra   The hydra, shown in Figure 7–17, belongs to the phylum **Coelenterata.** In its structure and function, the hydra is a simple animal. It is about five millimeters long and lives in fresh water. Usually it attaches itself to an underwater plant or some other solid object. The hydra has a tubelike body with only one opening, a mouth. The mouth, which leads into an internal cavity, is surrounded by tentacles. The body wall is made of only two cell layers.

Earthworm   The earthworm, shown in Figure 7–18, belongs to the phylum **Annelida.** Its long, round body is composed of many segments. The earthworm has a well-developed digestive system, a circulatory system, excretory organs, and a well-defined nervous system.

Grasshopper   The grasshopper, shown in Figure 7–19, belongs to the class *Insecta* of the phylum **Arthropoda.** The grasshopper has a well-developed digestive system, a circulatory system, a respiratory system, excretory organs, and a nervous system.

Human   Humans belong to the class *Mammalia* of the subphylum *Vertebrata* of the phylum **Chordata.** Mammals nourish their young with milk. Their bodies are covered with hair or fur. Many have well-developed brains.

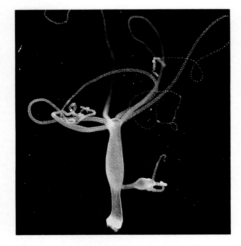

▲   **Figure 7–17**

Hydra. The hydra represents a simple, multicellular organism. This hydra is reproducing by budding a new hydra.

Figure 7–18

Earthworm. Earthworms have a more complex body organization than hydra. ▶

◀ **Figure 7–19**
Grasshopper. The grasshopper shows a level of complexity in its body structure that is greater than that in the earthworm.

Evolution: A Unifying Theme

As you have read, the modern taxonomy system attempts to classify organisms into groups that reflect their hereditary and, therefore, evolutionary relationships. To classify an organism, taxonomists must gather whatever is known about the organism's structure, chemistry, embryological development, and behavior and try to reconstruct, based on the modern theory of evolution, a reasonable picture of the organism and its relationship to other organisms.

When scientists observe striking similarities between some species, they are led to conclude that similar species have evolved from a common ancestor. Scientists may note, for example, that while a bird's wing and a seal's flippers look very different and have very different functions, their internal structure as well as their embryological development is quite similar. Using these similarities as clues to a common ancestor, scientists have some basis for classifying the bird and the seal.

As a unifying theme in biology, evolution explains why there is such a diversity of life on earth today as well as how various groups of organisms are related. It also explains how the organisms of today are related to organisms of the past.

7-2 Section Review

1. Name the five kingdoms.
2. In what fundamental way are members of the Monera different from organisms in the other kingdoms?
3. What are protozoa?
4. To what phylum and class do humans belong?

Critical Thinking

5. Why do scientists prefer the five-kingdom system of classification over a three- or four-kingdom system? (*Identifying Reasons*)

Laboratory Investigation

Classifying Insects

Biologists look at many features before deciding how to classify organisms. In this investigation, you will observe the features of insects from several orders. Based on your observations, you will develop and use your own classification scheme as well as a formal taxonomic key to classify the insects into their orders.

Objectives

- *Develop* and practice observation skills.
- *Classify* specimens according to your own classification scheme.
- *Use* a dichotomous key for classifying organisms.

Materials

insect kit
stereomicroscope
dissecting needle
petri dishes

Procedure 👉 *

A. With a stereomicroscope, study the specimens in your kit. Keep track of the specimens by keeping each one in a Petri dish labeled with a letter assigned by your teacher.

B. Based on similarities and differences, divide the specimens into four groups. The number of specimens in each group may vary. Draw a chart like the one below and list the specimens in each group.

Group	Specimen
1	B, D
2	
3	
4	

C. Using the taxonomic key shown here, and the insect classification chart on pages 722 and 723, compare your classification with that of taxonomists. Select one specimen and go to step 1 of the key. Choose the statement from step 1 that describes your specimen. Continue through the key until you identify the order to which the specimen belongs.

*For detailed information on safety symbols, see Appendix B.

Key to the Eight Major Insect Orders	
1	1a. Wings entirely membranous . . . go to 2 1b. Wings *not* membranous; front wings horny or leathery . . . go to 6
2	2a. Two wings . . . **Diptera** 2b. Four wings . . . go to 3
3	3a. Wings covered with scales . . . **Lepidoptera** 3b. Wings *not* covered with scales . . . go to 4
4	4a. Hind wings as long as front wings . . . **Odonata** 4b. Hind wings shorter than front wings . . . go to 5
5	5a. Wings held stiff over back, tentlike . . . **Homoptera** 5b. Wings *not* held over back, tentlike; waist constricted . . . **Hymenoptera**
6	6a. Front wings leathery or partially leathery . . . go to 7 6b. Front wings horny . . . **Coleoptera**
7	7a. Front wings leathery, back wings fan-like . . . **Orthoptera** 7b. Front wings half-leathery; triangle on back just behind head . . . **Hemiptera**

D. Repeat this procedure for each specimen.

E. Examine all the specimens in one order. Note the similarities and differences. Assume that each specimen is from a different family within that order. List the differences you would use to separate these specimens into families.

Analysis and Conclusions

1. What characteristics do all the specimens have in common?

2. How did your classification compare with the classification based on the taxonomic key?

3. In your classification scheme, did you use characteristics that were the same as, or different from, those in the taxonomic key?

4. Which characteristics did you use in step E?

5. Why are most classification systems based on structural, rather than behavioral, characteristics?

6. How do you think a taxonomist would begin to make a taxonomic key for a group of organisms?

Chapter Review

Study Outline

7–1 Classification

■ All living things are classified according to an established, international classification system.

■ Linnaeus classified organisms with a genus name followed by a specific name. This system of binomial nomenclature is still in use.

■ Each species belongs to one genus, one family, one order, one class, one phylum, and one kingdom. Related species are grouped in a genus, related genera in a family, and so forth.

■ The theory of evolution is the basis of the modern taxonomy system in which species are defined in terms of interbreeding populations.

■ In addition to structural similarities, taxonomists base their classifications on similarities in biochemistry, cell structure, embryological development, behavior, and fossil evidence.

7–2 Major Taxonomic Groups

■ The kingdoms of the five-kingdom system of classification are **Monera, Protista, Fungi, Plantae,** and **Animalia.**

■ The monerans include all prokaryotic organisms, the bacteria.

■ The protists are simple eukaryotic organisms such as the protozoa and algae.

■ The fungi include molds, yeasts, and mushrooms.

■ The plant kingdom includes the mosses, ferns, and seed plants.

■ The animal kingdom is divided into vertebrates and invertebrates. Vertebrates are animals with backbones. Invertebrates are animals without backbones.

■ A taxonomic key is a tool used to identify and classify organisms.

■ The six representative organisms used to study the life processes in Unit 2 are the ameba, paramecium, hydra, earthworm, grasshopper, and human.

Vocabulary Review

taxonomy (125)	species (130)
genus (127)	phylogeny (130)
kingdom (128)	fossil (132)
phylum (128)	Monera (134)
class (128)	Protista (134)
order (128)	Fungi (135)
family (128)	Plantae (135)
nomenclature (129)	Animalia (135)
binomial nomenclature (129)	taxonomic key (136)
	heterotrophs (137)
theory of evolution (130)	autotrophs (137)

A. Sentence Completion—Fill in the vocabulary term that best completes each statement.

1. Most members of the kingdom _____ carry out photosynthesis and are not motile.

2. A(n) _____ is a group of related phyla.

3. A(n) _____ is an organism that must obtain its food from the environment.

4. The branch of biology called _____ deals with the classification and naming of organisms.

5. A(n) _____ is a group of similar organisms that can interbreed in nature.

6. _____ is a two-word system of naming each type of organism.

7. A(n) _____ is a group of closely related species.

8. A(n) _____ is a tool used to identify previously classified organisms.

B. Matching—Select the vocabulary term that best matches each definition.

9. The largest group within a phylum

10. A system for naming organisms

11. A group of related families

Chapter Review

12. Organisms that make their own food

13. The kingdom that includes mushrooms, yeasts, and molds

14. The evolutionary history of a group of organisms

15. A group of related classes

16. Gradual change in a species over long periods of time

Content Review

17. On what basis did Aristotle classify animals?

18. How did Theophrastus classify plants?

19. How did John Ray define a species?

20. What was the basis that Carolus Linnaeus used to classify organisms?

21. To what order and family do humans belong?

22. Why is binomial nomenclature useful to biologists?

23. To what genus does *Canis familiaris* belong?

24. Name one organism with a common name that is misleading or inexact.

25. Give an example of cytological evidence used by modern taxonomists in classifying organisms.

26. Why have modern taxonomists added new kingdoms to their classification schemes?

27. What are the distinguishing basic characteristics of members of the kingdom Monera?

28. Describe the basic characteristics of the kingdom Protista.

29. How do fungi digest their food?

30. Describe the basic characteristics of plants.

31. Describe the basic characteristics of animals.

32. What is the major difference between vertebrates and invertebrates?

33. How do plants and animals differ?

34. Describe what a taxonomic key is, and explain how it is used.

35. Briefly describe each one of the six representative organisms.

36. Which of the six representative organisms are animals?

37. How does evolution act as a unifying theme in biology?

Graphic Organizing

For information on graphic organizers, see Appendix G at the back of this text.

38. **Concept Map:** Copy the incomplete concept map that is shown below onto a separate sheet of paper. Then, fill in the missing terms. You may add additional concepts and relationships.

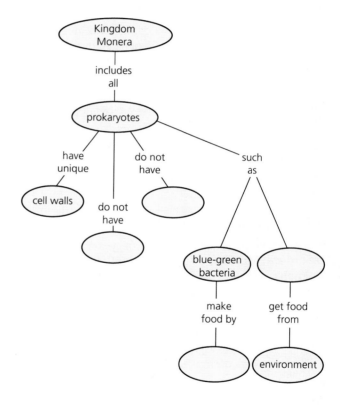

Critical Thinking

39. Place the following groups of classification in order, from the largest to the smallest: class, family, genus, kingdom, phylum, and species. (*Ordering*)

40. Why is it useful for a biologist to be able to place an unidentified organism in its proper group with related organisms? (*Judging Usefulness*)

41. *Homo erectus,* a species known only from fossils, is in the same genus as humans, *Homo sapiens.* What are the kingdom, class, and phylum of *Homo erectus?* Explain your answer. (*Reasoning Categorically*)

42. Two organisms can interbreed if they belong to the same species. Saint Bernards and Great Danes belong to the species, which is known as *Canis familiaris.* What can you conclude about Saint Bernards and Great Danes? What factors could make this conclusion incorrect? (*Reasoning Conditionally*)

43. The modern taxonomy's two-word system of identifying organisms is the same as Linnaeus's two-word system. Why do you think taxonomists have not changed this? (*Identifying Reasons*)

44. Suppose that you are given a jar containing hundreds of various organisms. After examining a dozen randomly picked specimens, you discover that each of the dozen belongs to the class Insecta. Make a generalization about the entire group of organisms in the jar. What could make this generalization invalid? What steps could you take to increase your confidence in your generalization? (*Generalizing*)

Creative Thinking

45. List 10 objects that are in the room you are now in. List some of the things that all these objects have in common. Name one thing for each object that makes it different from all of the others.

46. Based on any system or organizing principle you choose, divide the list of objects from question 45 into 3 groups. Give a title to each group and explain the value of your system. Then, classify the objects again, according to a different system or principle. How are your classifications alike? How are they different?

Problem Solving

47. Construct a taxonomic key that would help you to determine whether or not an organism is one of the following: monkey, elephant, earthworm, clam, swordfish, or robin.

48. You are a taxonomist interested in estimating how many new species of orchids exist in a tropical rain forest of 22 square kilometers. You randomly select five sites, each 0.1 square kilometer in area, and count the number of new species at each site. Once you identify a new species, it is no longer considered new. Based on the data in the table below, calculate the average number of new species of orchids present in a 0.1 square kilometer area. Then, when you have found the average, estimate how many new species of orchids are present in the forest.

Rain Forest Samples	
Site	**Number of New Species of Orchids**
A	3
B	2
C	0
D	8
E	1

Projects

49. Collect a variety of leaves, insects, or fungi. Use a local field guide to identify them. Prepare a display of your collection, labeling each item as accurately as possible.

50. Use a field guide and a microscope to identify protists found in a pond or fish tank. Write a report to the class on your findings.

51. Prepare a report for the class on one of the career opportunities listed below. If possible, interview someone who is working in the field. Be sure to prepare your questions in advance. You may wish to inquire about training, future opportunities, and the daily routine. A tape recorder will be helpful.
a. taxonomist   c. librarian
b. taxidermist

Unit 1

Biology and Problem Solving

How do homing pigeons find their way home?

Did you ever receive an important message delivered by a bird? Probably not. But for over 2000 years, homing pigeons have provided dependable message-carrying services. These pigeons carry messages across many miles and deliver them to their home lofts. People have long been puzzled about how homing pigeons, released at distant, unfamiliar places, find their way back home.

1. Before reading further, develop some hypotheses to explain how you think homing pigeons might find their way home.

Some biologists have hypothesized that homing pigeons use familiar landmarks to find their way. To test this hypothesis, the following experiments were conducted:

■ Homing pigeons were taken in covered cages to places they had never seen. When released, these pigeons found their way home.

■ Pigeons were taken under deep anesthesia to a release site. These pigeons also found their way home.

■ Contact lenses were placed over the eyes of a group of homing pigeons. (The lenses would dissolve after the experiment.) Half the pigeons were given clear lenses. The other half were given frosted lenses through which they could not see objects more than a few yards away. When released 80 miles from home, both groups of pigeons flew home.

2. Which of the three experiments support the familiar landmark hypothesis? Which of them refute the hypothesis? Explain.

3. Why was it necessary to conduct the experiment with the contact lenses even after the two other experiments had been done?

A second hypothesis was proposed: homing pigeons use the sun to find their way. However, observers knew that pigeons find their way on cloudy, as well as sunny, days. Therefore, biologists decided to test yet another hypothesis: homing pigeons use the earth's magnetic field to help them find their way. An experiment was done:

■ Scientists attached bar magnets to one group of pigeons. They attached brass bars to a second group. (A bar magnet distorts the earth's natural magnetic field around the bird, while a brass bar has no effect.) The two groups of birds were released at unfamiliar places on both sunny and overcast days. On sunny days, both sets of birds found their way home. On overcast days, birds with brass bars headed homeward, while birds with bar magnets flew away in different directions.

4. How would you interpret the results of the experiment done on sunny days? The experiment done on cloudy days?

5. Why are the birds with brass bars an essential part of this experiment?

6. Based on all of the described experiments, propose a theory to explain the pigeons' homing ability.

Problem Finding

In science, answers to questions often give rise to new questions. List some questions that you still have about the homing ability of pigeons.

Animal Maintenance

In the shallows of an Alaskan lake, a moose and her calf wade peacefully. Soon, however, winter will dramatically transform the landscape. How do animals survive such changes? Their bodies must maintain a stable internal state while external conditions vary. In this unit, you will see how the highly regulated systems of animals and animal-like protists carry out the activities essential to life.

▲ Like hummingbirds and bees, this bat has a taste for nectar.

Nutrition

A bat flaps through the darkness in search of food. Lured to a flower by the smell of nectar, the bat lands and clasps the flower with its wings. Most bats eat insects, plucking them in midflight. But, nectar contains the nutrients this bat needs. Some animals eat only meat, others only plants. Whatever the source of food, its role is the same. It provides the energy to carry out life processes. In this chapter, you will learn about the nutrients in food and how they are processed by the body.

Guide for Reading

Key words: nutrients, kilocalorie, absorption, digestion, alimentary canal, intestine, stomach, rectum
Questions to think about:
📖 How does nutrition in simple organisms compare with nutrition in more complex organisms?
📖 How is food converted into usable nutrients by the human digestive system?

8-1 The Process of Nutrition

Section Objectives:

- *Contrast* autotrophs and heterotrophs.
- *Define* the terms *calorie* and *kilocalorie* and explain how the energy content of food is measured.
- *Describe* the functions of the six basic types of nutrients found in the human diet.
- *Describe* a healthy human diet in terms of the Four Basic Food Groups.

Nutrients

Living organisms need food. All food contains **nutrients.** Nutrients are substances that provide the energy and materials needed for metabolic activities—growth, repair and maintenance of cells, and regulation. **Nutrition** is the process by which organisms get food and break it down so it can be used for metabolism.

Nutrients include proteins, carbohydrates, fats, vitamins, minerals, and water. Inorganic nutrients, such as minerals and water, are simple inorganic compounds that must be obtained from the environment. As you read in Chapter 4, water is necessary for most biological processes. **Minerals** are chemical elements that organisms need for normal functioning. Plants absorb minerals such as iron, calcium, phosphorus, and iodine from the soil in the form of salts or ions dissolved in water. Animals obtain minerals by eating plants or by eating other animals that have eaten plants.

▲ **Figure 8–1**

Nutrition. This hungry meadowlark obtains its nutrients by feeding on caterpillars.

Organic nutrients include essential organic compounds such as proteins, carbohydrates, fats, and vitamins. You read about the structure and function of these organic compounds in Chapter 4. Recall that **vitamins** are coenzymes or are converted into coenzymes in cells. Many important biological reactions need vitamins.

Organisms get the organic nutrients they need in two basic ways. Some organisms are capable of making, or synthesizing, organic nutrients from simple inorganic substances. Such organisms are **autotrophs** (AWT uh trohfs). Green plants, algae, and various other types of microorganisms are autotrophs. Most autotrophs are photosynthetic—that is, they use energy from sunlight, and carbon dioxide and water from the environment to make their own food. These organisms are called *phototrophs*. However, certain types of bacteria that are autotrophs do not use light as a source of energy. They are chemosynthetic, i.e., they make their own food using the energy from special types of chemical reactions. Such organisms are called *chemotrophs* (KEE muh trohfs). Photosynthesis and chemosynthesis are discussed in Chapter 17.

Organisms that cannot synthesize their own organic nutrients from inorganic compounds are called **heterotrophs** (HET ur ruh trohfs). All animals and certain types of microorganisms are heterotrophs. Such organisms must take in, or ingest, food containing "ready-made" organic nutrients from other plants or animals.

Energy Content of Food

Living organisms need energy to carry on their life processes. This energy is provided in most cases by the chemical breakdown of carbohydrates, fats, and proteins. As explained in the discussion of cellular respiration in Chapter 6, the energy is released in a series of small steps and stored in molecules of ATP for later use.

For any given quantity of food, the total energy released by cellular respiration is the same as would be released by burning the food. The energy content of a food sample is determined by completely burning a sample of the food and measuring the amount of heat given off during burning. The instrument used to measure the energy content of a food sample is a *calorimeter* (kal uh RIM uh ter). See Figure 8–2.

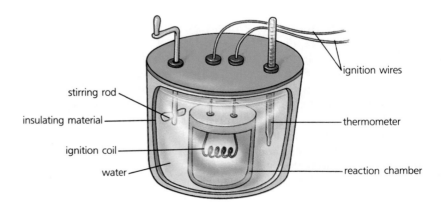

Figure 8–2
Calorimeter. In a calorimeter, a food sample is burned, and the heat produced is measured. The amount of heat produced is equal to the energy content of the food. ▶

Although the *joule* (jool) is the unit of energy in the International System of Units, the **calorie** (KAL uh ree) is the unit commonly used in measuring the energy content of food. One calorie equals 4.2 joules. It also is defined as the amount of heat that is needed to raise the temperature of 1 gram of water 1°C. Because the calorie is a very small unit, the preferred unit is the **kilocalorie** (KIL uh kal uh ree). A kilocalorie is 1000 calories. In tables giving the "calorie" content of foods, the unit of measurement used is actually a kilocalorie.

With the use of a calorimeter, it has been determined that the amount of heat given off by 1 gram of carbohydrate or 1 gram of protein is about 4 kilocalories. One gram of fat on the other hand, releases 9 kilocalories. Fat contains more than twice as many calories as an equal weight of carbohydrate or protein.

Human Nutritional Needs

Energy from Food   Age, sex, life style, weight, and body condition are factors that affect daily calorie needs. Usually, younger people need more calories than older people, males need more calories than females, and active people need more calories than inactive people. A person whose diet contains more calories than are needed gains weight. A person whose diet includes fewer calories than are needed loses weight. The energy contents of some common foods are shown in Figure 8–3.

Energy Content of Some Common Foods		
Food	**Portion**	**Calories**
apple	1 medium (150g)	70
bacon	2 slices (16g)	100
banana	1 (150g)	85
bread, white	1 slice (23g)	70
candy bar	1 plain (57g)	300
carrot	1 cup (145g)	45
cheese, American	1 oz. (28g)	105
corn	1 cup (256g)	170
cupcake	1 (50g)	185
egg	1 large (50g)	80
frankfurter	1 (51g)	155
ham	3 oz. (85g)	245
hamburger	3 oz. (85g)	245
ice cream	½ cup (74g)	175
milk	1 cup (244g)	150
orange	1 (180g)	60
peas	1 cup (160g)	115
potato	1 medium (130g)	105
tomato	1 medium (150g)	35

◄ **Figure 8–3**
Energy Content of Some Common Foods. The calorie content of different foods varies greatly.

Healthy Diets   Humans, like any other living organisms, need six basic nutrients. These are proteins, carbohydrates, fats, vitamins, minerals, and water. Sources and functions of these nutrients are given in Figure 8–4.

The foods we eat can be classified into five groups. Of the five, four are good sources of nutrients. These are called the Four Basic Food Groups. They include the Milk-Cheese Group, the Meat-Poultry-Fish-Bean Group, the Fruit-Vegetable Group, and the Bread-Cereal Group. See Figure 8–5. The fifth food group is the Fat-Sweet Group. These foods are high in fat, oil, or sugar content but low in nutrient value.

Figure 8–4
Nutrients Important for Human Metabolism. Each of the six basic nutrients is essential for good health. Too much or too little of any one nutrient can result in poor health. ▼

Nutrients Important for Human Metabolism		
Nutrient	**Dietary Sources**	**Function**
Proteins	Meat, fish, poultry, milk, eggs, nuts, beans	Supply building materials to form new cells for growth; repair and maintain tissue.
Carbohydrates: Sugar	Molasses, jelly & jam, candy, cake, brown and white sugar	Supply energy for body functions. Help the body to use fat, spare protein.
Starch	Pasta, bread, grains, cereal, corn, potatoes, rice, beans	
Fiber	Fruits, vegetables, whole grains	Gives bulk to digestive materials.
Fats	Butter, margarine, bacon, meat, egg yolk, cream, cooking oils	Supply energy. May be stored as fuel for the body. Carry vitamins and flavors.
Vitamins: A	Liver, carrots, spinach, sweet potatoes, whole milk products	Maintains healthy skin, bones, and eyes
B Complex (thiamine)	Pork, liver, legumes, whole grain products, fresh vegetables	Aids in carbohydrate use. Necessary for heart, nervous system and appetite.
C (ascorbic acid)	Citrus fruits, melons, green vegetables, potatoes	Maintains healthy cells and tissues. Helps to maintain healthy blood vessels, heal wounds, and resist infection. Promotes iron absorption.
D	Fortified milk, cod liver oil, eggs	Aids in calcium and phosphorus use.
K	Dark-green leafy vegetables, liver	Aids in blood clotting
Minerals: Calcium	Milk and dairy products, dark-green vegetables, sardines, canned salmon	Aids in bone/tooth formation and blood clotting.
Chlorine	Table salt, meat, milk, eggs	Helps digestion and cellular water balance.
Iodine	Seafood, added to salt	Essential for normal metabolism. Part of the thyroid hormone.
Iron	Liver, red meat, eggs, green leafy vegetables	Prevents anemia. Part of hemoglobin.
Potassium	Orange juice, citrus fruits, bananas, green leafy vegetables	Helps to maintain heartbeat, water balance, and nerve transmission; aids in carbohydrate and protein metabolism.
Water	Milk, fruit, vegetables, beverages	Serves as building material in all cells. Aids digestion, carries nutrients through the body and transports wastes.

▲ **Figure 8–5**
The Four Basic Food Groups.

Nutritionists recommend eating a variety of foods from the Four Basic Food Groups. Note that it is possible for a vegetarian, a person who does not eat meat, to have a healthy diet because the Meat-Poultry-Fish-Bean Group includes legumes. *Legumes* are plants that bear seeds in pods, such as beans, peas, and peanuts. Legumes are good sources of proteins.

In addition to nutrients, a healthy human diet must contain bulky, indigestible materials called **fiber.** Fiber is made of cellulose and other indigestible materials found in the cell walls of fruits, vegetables, and grains. Fiber stimulates the muscles of the digestive system to keep food moving through it. Eating adequate amounts of fiber also provides other healthful benefits, such as reducing the risk of colon and rectal cancers.

Many foods have been linked to health problems, such as obesity, heart disease, and cancer. The average American diet is high in sugar, saturated fat, and sodium, which contribute to these health problems. Most people can improve their diets by recognizing their nutritional needs, developing a plan for meeting these needs, and making wise food choices. Nutritionists have developed the following guidelines for a healthy diet.

- Eat a variety of foods. Vary the foods you choose to eat from the Four Basic Food Groups. No one food group provides all the nutrients your body needs.

- Maintain your ideal weight.

- Reduce your intake of fats, especially saturated fats and cholesterol. Eat lean meats, fish, poultry, and legumes instead of fatty meats. Avoid fried foods.

- Eat foods with adequate starch and fiber. Choose whole-grain breads and cereals, fruits, vegetables, and legumes.

- Don't eat too much sugar. Limit your intake of desserts, candy, and soft drinks. Eat fresh fruit instead of sweets for snacks.

- Avoid too much sodium. Use herbs and spices instead of salt to season your food. Do not add salt to food and limit the salty foods you eat.

8-1 Section Review

1. What is nutrition?
2. Name six nutrients required by living organisms.
3. Define the terms *heterotroph* and *autotroph.*
4. Identify the Four Basic Food Groups.

Critical Thinking

5. Would you say humans are heterotrophs? Why or why not? (*Reasoning Categorically*)

Critical Thinking in Biology

Ordering

Arranging your record albums alphabetically is one way to order them. When you order items or information, you organize things in a sequence, based on some characteristic they have in common.

1. Using the information given in Figure 8–3, order all the desserts (fruits and sweets) from the one with the fewest calories per serving to the one with the most calories per serving.
2. How would a list of desserts ordered by calorie content be useful?
3. Name two characteristics besides calorie content by which someone might order a list of foods. Explain how it would be useful to have a list ordered according to those characteristics.
4. Think About It Describe your thinking process as you answered question 1.

8-2 Adaptations for Nutrition and Digestion

▲ **Figure 8–6**

Mechanical Breakdown of Food. Giant pandas feed primarily on bamboo. When a giant panda bites and gnaws a bamboo shoot, the mechanical breakdown of food results.

Section Objectives:

- *Distinguish* between mechanical breakdown and chemical digestion of food.
- *Contrast* intracellular digestion and extracellular digestion.
- *Compare* digestive processes in protozoa, the hydra, the earthworm, and the grasshopper.

Digestion and Absorption

For a nutrient to be used by the cells of an organism, it must pass through the cell membranes. This process is called **absorption.** The nutrient molecules in food are usually too large to pass through cell membranes. Thus, to be absorbed by the cells, most food molecules must be broken down into smaller, simpler forms. The process by which food molecules are broken down is called **digestion** (dy JES chun).

Digestion is part of the process of nutrition. The term *digestion* usually refers to the chemical breakdown of food substances into simpler compounds. In many organisms, pieces of food are first cut, crushed, or broken into smaller particles without being changed chemically. This results in the mechanical breakdown of the food. Mechanical breakdown increases the surface area of the food particles. Chemical digestion is carried out by digestive enzymes, which act only on the surface of food particles. Thus, mechanical breakdown prepares the food for faster chemical digestion by exposing more food surface to the action of the digestive enzymes. Chemical digestion, like mechanical breakdown, takes place in stages. Large molecules are broken down into smaller molecules. In turn, these are broken down into still simpler forms. The usable, simplest products of digestion are the end products of the process of nutrition.

Nutrition in Protists

Among the protists digestion is *intracellular* (in truh SEL yuh ler)—that is, it happens inside the cell. However, members of this group have different ways of getting food. The ameba and paramecium are protists that live in fresh water and feed on small organisms. Both can move in response to various stimuli. They appear to be attracted to food by chemical stimuli.

Amebas crawl along solid surfaces using projections of the cell called *pseudopods,* or "false feet." The pseudopods move when cytoplasm flows into or out of them. When an ameba comes in contact with a food particle, pseudopods surround the particle. The cell membranes of the pseudopods join so that the particle is taken into the cell but is enclosed within a membrane. Although the food

Figure 8–7

Food-Getting in the Ameba. As the ameba senses its food (left), its pseudopods reach out to surround it (center). The engulfed food is enclosed in a food vacuole inside the ameba (right). (Magnification 310 X) ▶

is inside the cell, it is separated from the other cell contents by a membrane. A food vacuole then forms, free to move about within the cell cytoplasm. The food vacuole fuses with a lysosome, and digestive enzymes from the lysosome break down the food in the vacuole into forms usable by the cell. See Figure 8–7. These products of digestion, because they are small particles, can diffuse across the vacuole membrane into the cytoplasm. Indigestible materials remain in the food vacuole. The food vacuole eventually fuses with the cell membrane, and its contents are expelled from the cell.

The paramecium moves by the beating of hairlike cilia that cover the outside of the organism. The movement of the cilia also sweeps food particles down the **oral groove** into the **gullet** (GUHL et). See Figure 8–8. As food collects at the end of the gullet, the cell membrane bulges inward and pinches off, forming a food vacuole. The food vacuole travels through the cytoplasm and fuses with a lysosome, which contains digestive enzymes. Digestion occurs within the vacuole, and the usable products diffuse into the cytoplasm. Indigestible material is discharged from the cell through an opening called the **anal** (AYN ul) **pore.**

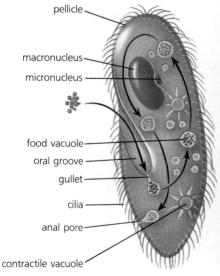

pellicle

macronucleus

micronucleus

food vacuole

oral groove

gullet

cilia

anal pore

contractile vacuole

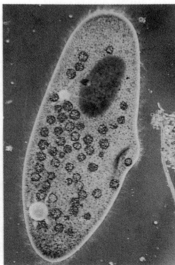

Figure 8–8

Food-Getting in the Paramecium. Food particles are swept down the oral groove into the gullet by the beating of the cilia. (Magnification 416 X) ▶

Nutrition in the Hydra

The hydra is a relatively simple multicellular animal about five millimeters long from the tip of its tentacles to its base. The body of the hydra is a hollow cylinder made up of two layers of cells. See Figure 8–9. The outer layer is the *ectoderm* (EK tuh derm), and the inner layer is the *endoderm* (EN duh derm). The tentacles, which surround the mouth, contain stinging cells called *cnidoblasts* (NYD uh blasts). Inside each cnidoblast is a capsule called a *nematocyst* (neh MAT uh sist), which contains a coiled, hollow thread.

The hydra captures its food with its tentacles. When a water flea or some other small animal comes in contact with a tentacle, the nematocysts release their long threads. Some of the threads wind around the prey, while others inject a poison that paralyzes the animal. The movement of the tentacles pushes the food through the mouth and into the **gastrovascular cavity,** where digestion begins. See Figure 8–10.

Digestion in hydra is both intracellular and extracellular. *Extracellular digestion* takes place outside the cells. Specialized cells in the endoderm secrete digestive enzymes into the gastrovascular cavity. These enzymes partially break down the food. Nutrients are then absorbed into the cells.

Some endoderm cells have flagella, and the waving of these organelles circulates the food particles through the gastrovascular cavity. *Intracellular digestion* takes place when endoderm cells form pseudopods and engulf the small food particles, thus forming food vacuoles. Digestion is completed by enzymes secreted into the food vacuoles.

Because the hydra is only two cell layers thick, the end products of digestion pass easily from the cells of the endoderm into the cells of the ectoderm by diffusion. Wastes from the ectoderm cells diffuse directly into the surrounding water. Wastes from the endoderm diffuse back into the gastrovascular cavity and are carried out through the mouth by water currents. Because hydras have only one body opening, the mouth is both an entrance for food and an exit for wastes.

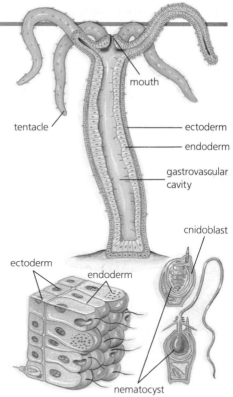

▲ **Figure 8–9**

Structure of Hydra. The hydra's body has two cell layers, the ectoderm and the endoderm. Tentacles containing stinging cells called cnidoblasts surround the mouth.

Figure 8–10

Food-Getting in Hydra. The hydra uses its tentacles to capture a water flea (daphnia) and stuff it into its gastrovascular cavity, where digestion will occur. ▼

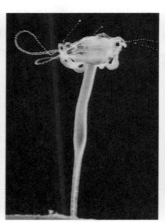
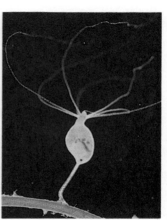

Nutrition in the Earthworm

The earthworm is a complex multicellular animal with a "tube-within-a-tube" body plan. The inner tube is the digestive system, while the outer tube is the body wall. See Figure 8–11. The digestive tube, or **alimentary** (al uh MENT uh ree) **canal,** has two openings—the mouth, through which food enters the body, and the **anus** (AYN us) through which waste matter leaves. Food travels through the digestive system in one direction—from the mouth to the anus. The food is broken down both mechanically and chemically in the digestive tract. Usable nutrients are then absorbed into the body cells.

As earthworms burrow through the ground, they ingest large quantities of soil. They also come to the surface to eat leaf litter and other decaying plant matter. Food is pulled into the mouth by the sucking action of the muscular **pharynx** (FA rinks). The food is then pushed through the digestive tube by waves of muscular contraction. From the pharynx, food passes through the **esophagus** (eh SAHF uh gus) into a round, thin-walled organ called the **crop.** The crop, which functions as a storage chamber, gradually releases food into the **gizzard** (GIZ urd). The gizzard is a thick-walled grinding organ that crushes the food. Mechanical breakdown is accomplished by the muscular movements of the gizzard, which grind the organic material against sand grains from the soil.

From the gizzard, the pastelike food mass passes into the long intestine. The **intestine** is where chemical digestion and absorption take place. The surface area of the intestine is increased by a fold in the wall called the **typhlosole** (TIF luh sohl). Cells lining the intestine secrete enzymes that break down large food molecules into smaller molecules. The products of digestion are absorbed by cells of the intestine and are picked up by the blood. The food molecules are transported in the blood to all parts of the body. Undigested materials and soil from which the food has been removed pass out of the worm through the anus.

Figure 8–11
The Digestive System of the Earthworm. ▼

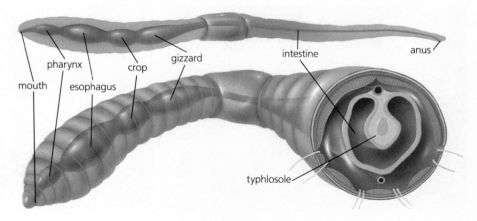

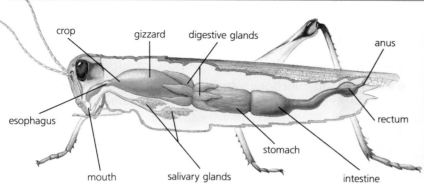

▲ **Figure 8–12**

The Grasshopper. As the grasshopper chews on leaves, digestive enzymes begin the chemical breakdown of food, a process that is completed in the stomach.

Nutrition in the Grasshopper

The grasshopper, like the earthworm, has a tubular digestive system, shown in Figure 8–12. Food is broken down mechanically by the mouthparts, which are well-adapted for chewing leafy vegetation. In the mouth the food is mixed with **saliva** (suh LY vuh) secreted by the **salivary** (SAL uh ver ee) **glands.** Saliva contains enzymes that begin the chemical breakdown of food. The food then passes through the esophagus into the crop, where it is stored temporarily. From the crop the food passes into the muscular gizzard, where it is ground into smaller particles by the action of teethlike plates made of *chitin* (KYT un). From the gizzard, food passes into the **stomach,** where chemical digestion and absorption take place. Digestive enzymes produced by glands just outside the stomach pass into the stomach, where they act on food particles. The products of digestion are absorbed into the bloodstream through the stomach walls and are transported to all the cells of the body. Undigested material passes through the intestine and is stored temporarily in the **rectum** where water is absorbed. The dried wastes are eliminated through the anus.

8-2 Section Review

1. Where does digestion happen in the ameba and paramecium?
2. Name an organism that uses both intracellular and extracellular digestion.
3. In the earthworm, what is the function of the typhlosole?
4. Describe the path food takes in the digestive system of the grasshopper.

Critical Thinking

5. Compare and contrast intracellular and extracellular digestion. (*Comparing and Contrasting*)

8-3 The Human Digestive System

Section Objectives:

- *Describe* the functions of the different parts of the human digestive system—the mouth, esophagus, stomach, small intestine, liver, gallbladder, pancreas, large intestine, rectum, and anus.
- *List* the principle digestive enzymes, where they are produced, the type of food they act upon, and the end products of enzymatic breakdown.

Laboratory Investigation: *Experiment* with the digestion of starch by the enzyme amylase (p. 168).

Parts of the Human Digestive System

The structure and function of the human digestive system are basically similar to those of the earthworm and the grasshopper. The digestive tube is made up of a series of specialized organs, with different phases of digestion taking place in each organ. See Figure 8–13. Food passes through the digestive tube in the following order: oral cavity (mouth), pharynx (throat), esophagus (gullet),

Figure 8–13

The Human Digestive System. Each organ in the digestive system carries out specific functions. Digestion is aided by accessory digestive glands. These glands secrete digestive enzymes and juices into the digestive tube. ▼

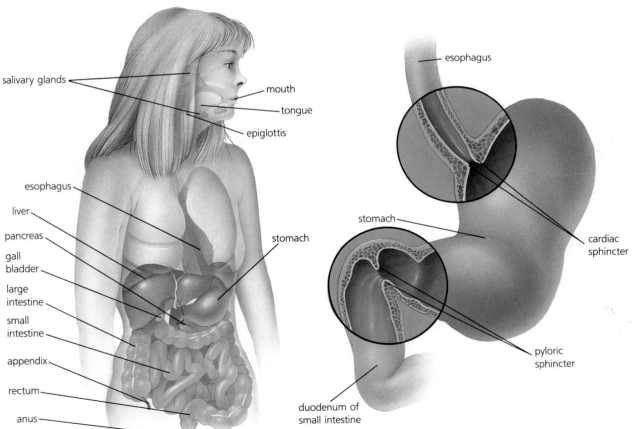

stomach, small intestine, large intestine, rectum, and anus. Several glands secrete digestive enzymes and juices into the digestive tube, where extracellular digestion occurs.

The *digestive glands* are groups of specialized secretory cells that are found in the lining of the alimentary canal or in separate accessory organs. The accessory glands lie outside the digestive tract. Their secretions pass into the digestive tract by way of a tube or duct. Food is never found within the accessory glands, only within the alimentary canal itself. The accessory glands include the salivary glands, the **liver,** and the **pancreas** (PAN kree us). The liver and pancreas have many functions. These organs aid digestion by the secretion of digestive fluids.

Cells in the lining of the walls of the alimentary canal also secrete *mucus* (MYOO kus), which acts as a lubricant for the food mass. It also provides a coating that protects the delicate cells of the digestive tube from the action of acid, digestive enzymes, and abrasive substances in food.

The Mouth and Pharynx

Food enters the body through the mouth, where both mechanical breakdown and chemical digestion occur. Chunks of food are bitten off with the teeth and ground into pieces small enough to swallow. The tongue moves and shapes the food mass in the mouth.

As food is chewed, it is mixed with saliva, which is secreted into the mouth by three pairs of salivary glands. There are actually two types of saliva. One is a thin, watery secretion that wets the food. The other is a thicker, mucous secretion that acts as a lubricant and causes the food particles to stick together to form a food mass, or *bolus* (BOH lus). Saliva also contains a digestive enzyme called **salivary amylase** (AM uh layz). This enzyme breaks down starch, which is a polysaccharide, into maltose, which is a disaccharide.

When the food has been chewed sufficiently, it is pushed by the tongue to the back of the throat, or pharynx. See Figure 8–14. This starts the automatic swallowing reflex, which forces food into the esophagus, the tube leading to the stomach. However, air as

Figure 8–14

Swallowing. The epiglottis prevents food or liquid from entering the air passages during swallowing. ▼

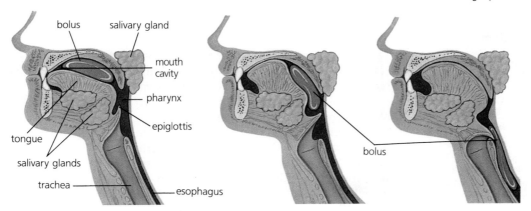

Biology and You

Q: My best friend always thinks about losing weight. She makes herself vomit after eating. Should I be worried about her?

A: Your friend has bulimia, an eating disorder, which is common among young, high-achieving women. Although there are also cases of men with bulimia, about 90 percent are female. Bulimics fear becoming fat, even though most are of average weight or only slightly overweight. They go on eating binges and then purge themselves by vomiting or with laxatives.

Many doctors think that society's overconcern with thinness is a major cause of bulimia. Bulimics often have distorted views of their bodies. They are perfectionists who have trouble feeling good about themselves. No matter how thin they are, it is never thin enough for them.

Bulimia is a psychological and a physical problem. Constant vomiting may damage teeth, gums, stomach, kidneys, and heart. Purging may result in dehydration and vitamin and mineral deficiencies. Your friend should get counseling and medical help immediately.

Look through a teenage magazine for articles and advertisements that emphasize thinness. Write a letter to the editor expressing your views on your findings.

well as food passes through the pharynx. The air must pass through the voice box, or *larynx* (LA rinks), and down the *trachea* (TRAY kee uh) to the lungs. To prevent food and liquids from entering the larynx, it is automatically closed off during swallowing by a flap of tissue called the **epiglottis** (ep uh GLAHT is). At the same time, breathing stops momentarily, and the passageways to the nose, ears, and mouth are blocked. When a person "swallows the wrong way" and food enters the trachea, it is brought back up into the throat by coughing.

The Esophagus

The esophagus is a tube through which food passes from the pharynx to the stomach. Beginning in the esophagus, the movement of food down the digestive tube is aided by alternate waves of relaxation and contraction in the muscular walls of the alimentary canal. This is called **peristalsis** (pehr uh STAHL sis). The muscles in front of the food mass relax, while those behind the food mass contract, pushing the food forward.

Aided by peristaltic contractions, food passes quickly down the esophagus. Where the esophagus opens into the stomach, there is a ring of muscle called a **sphincter** (SFINK ter). The sphincter acts as a valve and controls the passage of food from the esophagus into the stomach. When the wave of peristalsis reaches the sphincter, it relaxes and opens, and the food enters the stomach. The sphincter between the esophagus and the stomach is called the *cardiac sphincter*. During vomiting, a wave of peristalsis passes upward— reverse peristalsis—causing the cardiac sphincter to open, and the contents of the stomach to be "thrown up."

Sometimes, when the cardiac sphincter relaxes, hydrochloric acid (HCl) from the stomach backs up into the esophagus. When the acid comes into contact with the sensitive lining of the lower esophagus, the result is discomfort, commonly known as acid indigestion or "heartburn."

The Stomach

The stomach is a thick-walled, muscular sac that can expand to hold more than two liters of food or liquid. Food is stored temporarily in the stomach. The mechanical breakdown of food and the partial digestion of proteins also occur there. Food is broken down mechanically by contractions of the muscular stomach walls. The food mass is churned and mixed with acidic **gastric juice** secreted by glands in the stomach walls.

The lining of the stomach contains two types of glands. The first type, *pyloric* (py LOR ik) *glands,* secrete mucus, which covers the stomach lining and protects it from being digested. The second type, *gastric* (GAS trik) *glands,* secrete *gastric juice,* which has a pH of 1.5 to 2.5. The acidic pH of gastric juice is due to its high concentration of hydrochloric acid. Hydrochloric acid kills most of the bacteria that are swallowed in food. Gastric juice also contains the digestive enzyme **pepsin** (PEP sin). Pepsin is secreted in an inactive form called *pepsinogen* (pep SIN uh jen), which is activated after it is mixed with the hydrochloric acid. Pepsin breaks down large protein molecules into shorter chains of amino acids called *polypeptides* (pol ee PEP tydz).

The breakdown of starch by salivary amylase, which begins in the mouth, continues for some time after the food mass reaches the stomach. Gradually, however, the low pH of the acid in the stomach inactivates this enzyme, and starch breakdown stops.

When the stomach is empty, only small amounts of gastric juice are present. When food is eaten, the flow of gastric juice increases. There are three mechanisms involved in stimulating the flow of gastric juice.

1. The thought, sight, smell, or taste of food stimulates the brain to send messages to the gastric glands, causing them to secrete moderate amounts of gastric juice.

2. Food touching the lining of the stomach stimulates the secretion of moderate amounts of gastric juice.

3. When a food mass enters the stomach, it stretches the stomach walls. The stretching of the stomach wall, as well as the presence of proteins, caffeine, alcohol, and certain other sub-stances, stimulates the lining of the stomach to secrete a hormone called *gastrin* (GAS trin) directly into the blood. (A *hormone* is a substance that is secreted directly into the bloodstream and that produces a specific effect on a particular tissue.) Gastrin further stimulates the gastric glands in the stomach to secrete large amounts of gastric juice.

Liquids pass through the stomach in 20 minutes or less. Solids, on the other hand, must first be reduced to a thin, soupy liquid called **chyme** (kyme). The chyme passes in small amounts at a time through the *pyloric sphincter,* the muscle that controls the passage of food from the stomach into the small intestine. The stomach empties from 2 to 6 hours after a meal. Hunger is felt when an empty stomach is churning.

If the thick mucous layer that protects the stomach wall breaks down, a part of the stomach wall may be digested. When this happens, a painful ulcer develops. See Figure 8–15. Ulcers are

Figure 8–15
Stomach Ulcer. A mucous layer protects the sensitive cells of the stomach wall. If this layer is damaged, digestive juices eat away at the cells creating an ulcer. ▼

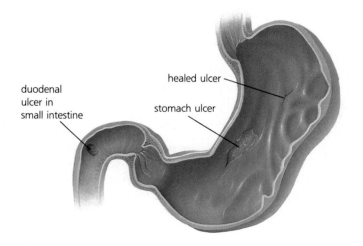

duodenal ulcer in small intestine

healed ulcer

stomach ulcer

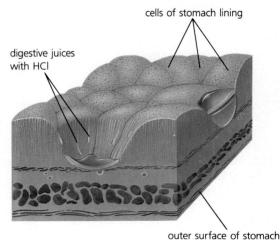

cells of stomach lining

digestive juices with HCl

outer surface of stomach

Figure 8–16

Structure of a Villus. The small intestine wall is covered by fingerlike projections called villi. Within each villus are a network of blood vessels and a lacteal. An electron micrograph shows how the intestinal epithelial cell membrane folds inward, creating microvilli. (Magnification 16 200 X) ▼

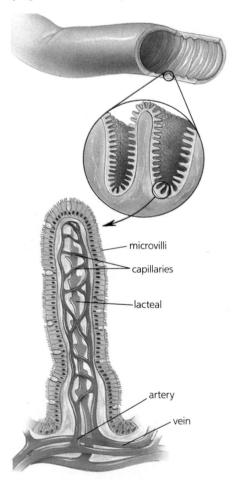

microvilli

capillaries

lacteal

artery

vein

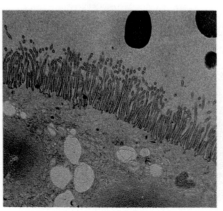

painful because hydrochloric acid comes into contact with the exposed stomach wall. It is thought that some ulcers are caused by the oversecretion of gastric juice, which may be brought about about by nervousness or stress. Ulcers are treated by diet, various medications, or in severe cases, by surgery.

The Small Intestine

The **small intestine** is a coiled tube about 6.5 meters long and about 2.5 centimeters in diameter. The small intestine has three parts. Food leaving the stomach through the pyloric sphincter enters the *duodenum* (doo uh DEE num). This is the shortest section of the small intestine, about 25 centimeters long. The middle section is called the *jejunum* (jeh JOO num). The last section is the *ileum* (ILL ee um).

Most chemical digestion takes place in the small intestine. This also is the site of absorption. After digestion is complete, simple sugars, amino acids, vitamins, minerals, and other substances are absorbed through the wall of the small intestine into the blood vessels of the circulatory system. At the same time, fatty acids and glycerol are absorbed into tiny vessels of the lymphatic system called **lacteals** (LAK tee uls). The lymphatic system is discussed in greater detail in Chapter 9.

The small intestine has a number of structural features that increase its surface area and make it ideally suited for absorption. These features are illustrated in Figure 8–16. First, the small intestine is very long. Second, its lining has many folds. Third, the lining is covered with millions of fingerlike projections, which are called **villi** (VIL ly). Fourth, the epithelial cells that make up the intestinal lining have *brush borders*. In the brush borders, the membranes of cells that face into the intestinal opening have tiny projections called *microvilli* that further increase the surface area of the cells.

Within each villus there is a network of blood capillaries, and in the center is a lacteal. The outer covering of each villus is a layer of epithelial cells with microvilli. During absorption, digested nutrients pass through the epithelial cells and enter either the capillaries or the lacteal. Absorption involves both diffusion and active transport.

When food is present, the small intestine is in constant motion. These peristaltic movements have four main effects: (1) they squeeze chyme through the intestine; (2) they mix the chyme with the digestive enzymes present in the small intestine; (3) they break down food particles mechanically; and (4) they speed up absorption of digestive end products by bringing the intestinal contents into contact with the intestinal wall.

Unlike the stomach with its acid secretions, fluids in the small intestine are generally alkaline. Chyme is mixed with **pancreatic** (pan kree AT ik) **juice** from the pancreas, **bile** from the liver, and **intestinal juice** from glands in the wall of the intestine. These three secretions contain the enzymes and other substances necessary to complete digestion.

Pancreatic Juice   When the acid chyme from the stomach enters the small intestine, it stimulates cells in the intestinal lining to secrete two hormones into the blood. These hormones are *secretin* (sih KREET in) and *cholecystokinin* (koh luh sis tuh KY nin). These hormones stimulate the pancreas to secrete pancreatic juice and pancreatic enzymes, which pass through the *pancreatic duct* into the upper part of the small intestine. Pancreatic juice contains sodium bicarbonate, which neutralizes the acid in the chyme and makes the pH of the contents of the small intestine slightly alkaline (pH 8). The enzymes secreted by the pancreas act on proteins, carbohydrates, fats, and nucleic acids.

The pancreatic enzymes include *amylase,* which hydrolyzes any remaining starch to maltose; *proteases* (PRO tee ay zez) (protein-splitting enzymes), including *trypsin* (TRIP sin) and *chymotrypsin* (ky muh TRIP sin), which continue the breakdown of large protein molecules begun in the stomach; and *lipase,* which breaks down fats.

Bile   The cells of the liver produce bile, which passes through ducts into the **gallbladder,** where it is stored. Bile passes from the gallbladder to the upper part of the small intestine through the *bile duct.* The release of bile from the gallbladder is stimulated by the hormone cholecystokinin, which also acts on the pancreas. Bile contains no enzymes, but it aids in the digestion of fats and oils by breaking them up into tiny droplets. This process, called *emulsification* (ih mul suh fuh KAY shun), increases the surface area for enzyme action. Since bile is alkaline, it aids in neutralizing the acid chyme from the stomach.

Intestinal Juice   The walls of the small intestine contain millions of intestinal glands, which secrete intestinal juice. Intestinal juice contains the enzymes peptidase and maltase that complete the digestion of carbohydrates, fats, and proteins.

In the small intestine, molecules of proteins, carbohydrates, and fats are broken down into the end products of digestion. Proteins are broken down into amino acids, carbohydrates into simple sugars, and fats into fatty acids and glycerol. A summary of the secretions of the human digestive system and their functions is given in Figure 8–17.

The Large Intestine

Undigested and unabsorbed materials pass from the small intestine through a sphincter into the **large intestine.** The large intestine is about 1.5 meters long and 6 centimeters in diameter. No digestion occurs in this portion of the digestive system.

On the lower right side of the abdomen, where the small intestine joins the large intestine, is a small pouch, the **appendix** (uh PEN diks). The appendix plays no part in the functioning of the human digestive system. Occasionally, however, the appendix becomes infected or inflamed, a condition known as **appendicitis** (uh pen duh SY tus). If the condition is not treated, the appendix may burst, spreading the infection.

Science, Technology and Society

Issue: Food Irradiation

Imagine storing food for weeks without spoiling. Food irradiation may make this possible. Food is exposed to the radioactive elements cesium or cobalt. The food absorbs some of the energy without becoming radioactive. Bacteria and insects are destroyed and the ripening process is halted. The Food and Drug Administration (FDA) allows irradiation of pork, wheat, spices, and fruits and vegetables.

People who favor irradiation point to its potential for fighting world hunger. Food might be stored indefinitely. Many diseases caused by bacteria and insects could be eliminated. Chemical pesticides, which are used to kill microorganisms, may not be needed.

People opposed to food irradiation believe that research is incomplete. They are unsure of the dangers associated with it. Some fear that the chemical byproducts may cause cancer. Others note that the process changes the texture and taste of some foods and may decrease nutritive values. These people believe that other safe alternatives should be pursued.

■ *Do you think the FDA should allow food irradiation? Why or why not?*

One of the principal functions of the large intestine is the reabsorption of water from the food mass. During digestion, water is mixed with the food as it moves through the digestive system. Under normal conditions, about three-fourths of the water is reabsorbed. This reabsorption into the capillaries of the large intestine helps the body conserve water. If too little water is absorbed, diarrhea results; if too much water is absorbed, constipation results.

A second function of the large intestine is the absorption of the vitamins that are produced by bacteria that normally live in the large intestine. These vitamins are absorbed with the water from the food mass. Intestinal bacteria live on undigested food material. They produce vitamin K, which is essential for blood clotting, and some of the B vitamins. When large doses of antibiotics are given to overcome an infection, they can destroy the intestinal bacteria, and a vitamin K deficiency may result.

The third function of the large intestine is *elimination*—removal of undigested and indigestible material from the digestive tract. This material consists of cellulose from plant cell walls; large quantities of bacteria, bile, and mucus; and worn-out cells from the digestive tract. As this material travels through the intestine, it

Figure 8–17
Secretions of the Human Digestive System and Their Functions. ▼

Secretions of the Human Digestive System		
Digestive Secretions and Enzymes	**Origin**	**Function**
Saliva Salivary amylase	Salivary glands	Wets food. Helps to form food into bolus. Breaks down starch into maltose.
Gastric juice: Pepsin Hydrochloric acid	Stomach (gastric glands)	Breaks down proteins into smaller molecules (peptones and proteoses). Activates pepsin.
Bile	Liver	Breaks down fat mechanically into small droplets (emulsification).
Pancreatic juice: Amylase Trypsin Lipase	Pancreas	Continues the digestion of starch to disaccharides. Digests peptones and proteoses into peptides. Digests fat droplets into fatty acids and glycerol.
Intestinal juice: Peptidase Maltase	Small intestine (intestinal glands)	Breaks down peptides into amino acids. Breaks down maltose (a disaccharide) into glucose (a monosaccharide).

becomes **feces** (FEE seez), or stool. Fecal matter is stored in the last part of the large intestine, the rectum, and periodically eliminated, or defecated, through the anus.

8-3 Section Review

1. What are the functions of saliva and its enzyme, salivary amylase?
2. Name the process that causes food to move through the digestive tube.
3. List three fluids that mix with food in the small intestine.
4. What are the functions of the large intestine?

Critical Thinking

5. Which of these nutrients need to be digested in order to be absorbed? starch, maltose, amino acids, proteins, fat, glucose, glycerol, peptides (*Classifying*)

Careers *in Action*

Dietitian

The meals in your school cafeteria are not selected at random. They are carefully planned by a dietitian to promote healthy eating habits and meet the nutritional needs of teenagers. Dietitians use the principles of nutrition to develop diets and menus for groups of people, such as students, patients, or the elderly. They work in many different settings, including hotels, hospitals, and nursing homes.

In planning menus, dietitians must consider nutrition, cost, taste, and appeal. Administrative and managerial skills are important because dietitians often supervise food buying and meal preparation. They also manage budgets and enforce health and safety regulations. Dietitians may do nutritional research and run education programs to teach proper nutrition.

Problem Solving Suppose you are the dietitian at your school. You know that some of the foods students like are not nutritionally sound. You would like to come up with meals that are nutritionally balanced and well liked. You poll the students to discover their favorite foods.

1. Survey 10 students to learn what their three favorite lunch menus are. Which four foods are mentioned most often?
2. What nutrients do each of the four foods contain?
3. Propose two meals that are both nutritionally balanced and include some of the students' favorite foods.
4. Suppose that some of the students in the school are vegetarians. Propose a vegetarian meal that is appealing and nutritionally balanced.
5. You decide to hang a poster in the cafeteria to teach the students about good nutrition. Design a poster that is motivational and informative.

■ **Help Wanted:**
Dietitian. Bachelor's degree required; postgraduate internship and certification recommended. Contact: The American Dietetic Association, 430 N. Michigan Ave., Chicago, IL 60611.

Laboratory Investigation

Digestion of Starch

The digestion of food allows nutrients to cross the membranes of cells lining the digestive tract and to enter the bloodstream. In this activity, you will see how digestion and membrane permeability account for the way in which some organisms obtain nutrition.

Objectives

- *Observe* how molecular size can affect diffusion through a membrane.
- *Observe* the effect of a digestive enzyme on its substrate.
- *Perform* chemical tests for starch and sugar.

Materials

starch solution	test tubes
amylase solution	test-tube rack
iodine solution	boiling water bath
Benedict's solution	boiling chips
cellophane tubing	string
graduated cylinders	scissors
droppers	tags
large beaker	

Procedure 🖐 🥽 🔬 ☢ 🧤 🗑 *

A. Using scissors, cut two 15-cm-long sections of cellophane tubing and four 10-cm-long pieces of string. Tie one end of each length of tubing with a string and attach a tag. Label one tube A and the other B.

starch solution

B. Pour the starch solution into tube A until it is about two-thirds full, as shown here. Squeeze out as much air as you can from the tube and tie the end with a string. Place the tube in an empty beaker.

C. Pour starch solution into tube B until it is about two-thirds full. Then, add 10 drops of amylase. Squeeze out the air and tie the end with a string. Invert tube B several times to mix the contents.

*For detailed information on safety symbols, see Appendix B.

D. Rinse off both cellophane tubes with water. Place each tube in a graduated cylinder labeled A or B. Add just enough water to cover each tube.

E. Wait 10 minutes. During this time, set up the boiling water bath, label four large test tubes A1, A2, B1, and B2, and review the upcoming procedures. **Caution:** Exercise care when using the hot water bath.

F. After 10 minutes, carefully raise and lower the cellophane tube in each cylinder several times to mix the surrounding liquid. Set the tubes aside. Into tubes A1 and A2, pour 2 mL of liquid from cylinder A. Into tubes B1 and B2, pour 2 mL of liquid from cylinder B.

G. Add 10 drops of Benedict's solution to tubes A1 and B1. Place each tube in the boiling water bath for 5 to 10 minutes. If sugar is present, the solution will turn green or red. Record your results in a chart like the one below.

Test	Test Tube			
	A1	**A2**	**B1**	**B2**
sugar				
starch				

H. Add iodine drops to tubes A2 and B2. A blue-black color indicates starch. Record your results.

Analysis and Conclusions

1. Why was the outside of each cellophane tube rinsed before it was placed in the cylinder?

2. Why was it necessary to wait 10 minutes before testing for sugar?

3. Which graduated cylinder, A or B, held liquid that contained sugar? Explain your answer.

4. Which graduated cylinder, A or B, held liquid that contained starch? How would you explain this?

5. How does a digestive enzyme affect what diffuses across a membrane?

6. Why do hospitals give sugar solutions instead of starch solutions to patients who are fed intravenously?

Chapter Review

Study Outline

8–1 The Process of Nutrition

■ Nutrition is the process by which organisms obtain nutrients and use them to carry on their life activities. These nutrients include proteins, carbohydrates, fats, vitamins, minerals, and water.

■ Autotrophs are capable of making nutrients from simple inorganic substances. Heterotrophs must depend on other living things for food.

■ Energy to carry on life processes comes from the breakdown of fats, carbohydrates, and proteins. The kilocalorie (Calorie) is commonly used to measure the energy content of foods.

■ A healthy human diet includes foods from the Four Basic Food Groups.

8–2 Adaptations for Nutrition and Digestion

■ Digestion is the process by which foods are broken down into simpler compounds. Absorption is the movement of digested food molecules across cell membranes and into cells.

■ In protists, digestion is intracellular, and food particles are digested within food vacuoles. In the hydra, digestion is both extracellular and intracellular. It begins in the gastrovascular cavity and is completed by endoderm cells that engulf food particles.

■ The digestive systems of the earthworm and grasshopper are tubular. The inner tube, the alimentary canal, contains the organs of digestion, and the outer tube is the body wall.

8–3 The Human Digestive System

■ Humans, like earthworms and grasshoppers, have a "tube-within-a-tube" body plan, in which the inner tube is the digestive tract.

■ The human digestive tract includes the mouth, pharynx, esophagus, stomach, small intestine, large intestine, rectum, and anus. Accessory glands include the salivary glands, liver, and pancreas.

Vocabulary Review

nutrients (149)	salivary glands (159)
nutrition (149)	stomach (159)
minerals (149)	rectum (159)
vitamins (150)	liver (161)
autotrophs (150)	pancreas (161)
heterotrophs (150)	salivary amylase (161)
calorie (151)	epiglottis (162)
kilocalorie (151)	peristalsis (162)
fiber (153)	sphincter (162)
absorption (155)	gastric juice (162)
digestion (155)	pepsin (162)
oral groove (156)	chyme (163)
gullet (156)	small intestine (164)
anal pore (156)	lacteals (164)
gastrovascular cavity (157)	villi (164)
alimentary canal (158)	pancreatic juice (164)
anus (158)	bile (164)
pharynx (158)	intestinal juice (164)
esophagus (158)	gallbladder (165)
crop (158)	large intestine (165)
gizzard (158)	appendix (165)
intestine (158)	appendicitis (165)
typhlosole (158)	feces (167)
saliva (159)	

A. Definitions—Replace the italicized definition with the correct vocabulary term.

1. *Organisms capable of making organic nutrients from simple inorganic substances* include green plants, algae, and various other microorganisms.

2. The *digestive tube* of the earthworm has two openings—the mouth and the anus.

3. The saliva produced by the salivary glands in humans contains *a digestive enzyme that breaks down starch*.

Chapter Review

4. In humans, a *ring of muscle* at the point at which the esophagus opens into the stomach acts as a valve controlling the passage of food into the stomach.

5. The *thick-walled grinding organ* of the earthworm crushes food before the food passes into the intestine.

6. Food and liquids are prevented from entering the larynx in humans by the *flap of tissue that covers the trachea.*

7. The *amount of heat needed to raise the temperature of 1 gram of water 1°C* is a unit commonly used in measuring the energy content of food.

8. A healthy diet includes *bulky indigestible materials.*

9. The lining of the small intestine is covered with millions of *fingerlike projections.*

B. Diagram Labeling—Draw a diagram of the human digestive system.

10. Label the following parts: (a) salivary glands; (b) esophagus; (c) stomach; (d) small intestine; (e) liver; (f) pancreas; (g) large intestine; (h) rectum; and (i) anus. Briefly describe the function of each organ.

Content Review

11. Compare phototrophs and chemotrophs.

12. How is the energy content of food determined, and in what unit of energy is it expressed?

13. Why is fiber important in the human diet?

14. How does the average American diet contribute to health problems?

15. Why does food have to be digested?

16. What is the difference between mechanical and chemical digestion?

17. How is food digested in protists?

18. Describe digestion in the hydra.

19. Trace the path of food through the digestive system of the earthworm.

20. How does digestion in the grasshopper differ from digestion in the earthworm?

21. Explain why food and air are not usually mixed during swallowing.

22. How is the passage of food from the esophagus into the stomach controlled?

23. Describe the functions of hydrochloric acid and pepsin in the stomach.

24. What mechanisms stimulate the flow of gastric juice?

25. How do stomach ulcers develop?

26. Describe the structure and function of villi.

27. What are the effects of peristalsis in the small intestine?

28. Describe the process and function of emulsification.

29. What is the function of intestinal juice?

Graphic Organizing

For information on graphic organizers, see Appendix G at the back of this text.

30. **Circle Graph:** The completed circle graph below shows the percentages of major nutrients in a typical current American diet. To the right is an incomplete graph for a diet recommended by nutritionists. The percentages of these nutrients that nutritionists recommend are: refined sugars: 10%; complex carbohydrates and naturally occurring sugars: 48%; proteins: 12%; saturated fats: 10%; unsaturated fats: 20%. Copy the incomplete graph on the right onto a sheet of paper and complete it.

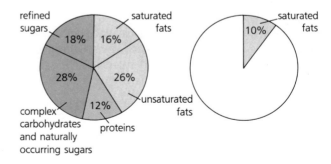

Current Diet   Recommended Diet

31. **Flow Chart:** Construct a flow chart tracing the path of food through the human digestive system.

Critical Thinking

32. Referring to Figure 8–3, compare the energy content of: a medium apple and a slice of white bread; three ounces of hamburger and three ounces of Ameri-

can cheese; a candy bar and two cups of milk. Should calorie comparisons alone be used to make nutritional choices? Explain. (*Comparing and Contrasting; Judging Usefulness*)

33. Why do starchy foods taste sweet if they have been chewed for a while? (*Identifying Causes*)

34. Why are the salivary glands, liver, and pancreas usually considered a part of the digestive system? Why might they not be considered a part of the digestive system in the same way that the esophagus, stomach, or intestines are? (*Reasoning Categorically*)

Creative Thinking

35. Individuals who have had much (or even all) of their stomachs removed can survive very well. Suggest an explanation for this. Do you think the same individuals could survive without a small intestine? Explain.

36. Why might two individuals have different energy needs in performing the same tasks?

Problem Solving

37. Pancreatic juice contains both sodium bicarbonate and digestive enzymes. It is secreted by the pancreas in response to the specific composition of chyme in the upper portion of the small intestine. The graph below shows data from a study to determine the secretion of the pancreas in response to the presence of three different substances in the small intestine. What does the graph show about the secretion of bicarbonate and enzymes in response to hydrochloric acid, fat, and protein? How would you interpret these findings?

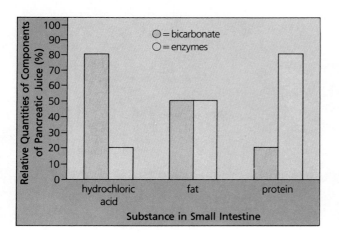

38. Four newly weaned rats are fed identical diets except for the amounts of vitamin A. Rat 1 receives no vitamin A. Rat 2 receives 10 units daily; rat 3,100 units; rat 4,1000 units. The rats are weighed every week for two months. What is the variable? What response is being measured? What factors are held constant? Do you see any problems with the setup?

Computer Project

39. High fat intake, especially saturated fats, is linked to obesity and cardiovascular disease. Nutritionists recommend limiting daily fat intake to no more than 30 percent of the total calories. No more than 10 percent of the total calories should be from saturated fats.

In this activity, you will use a spreadsheet to compute the fat content of some typical lunches. Your teacher will give you a table that provides nutritional information for common lunch foods. (Note: To convert grams of fat to calories, multiply the number of grams by 9.)

A. Suppose you ate a 3 oz. cheeseburger on a roll, 20 french fries, a 12 oz. cola, a piece of cake, and 3 oz. of ice cream. Calculate the percentage of total calories that come from fat and from saturated fat. Is this meal within the recommended guidelines?

B. How would substituting a double cheeseburger and 30 french fries affect the percentage of fat and saturated fat in the meal?

C. Suppose your lunch was a 3 oz. tuna sandwich on 2 slices of white bread with 1 tablespoon of mayonnaise, 1 cup of tomato soup, a mixed salad (no dressing), 8 oz. of skim milk, and an apple. How does this meal compare with the recommendations? Which item contributes most to the percentage of fat in this meal?

D. By how much would the percentage of fat change if you substituted whole milk for the skim milk?

E. Choose your favorite foods from the list to make a typical meal. Compute the percentage of fat and saturated fat and compare it to the recommendations.

F. Nutritionists recommend that a person's intake of carbohydrates equal about 60 percent of total calories. This should be made primarily of starchy carbohydrates (breads and grains) rather than sugary carbohydrates (sweets and soft drinks). Use the spreadsheet to design a meal that meets the recommendations for fat, saturated fat, and carbohydrate intake. (Note: To convert grams of carbohydrates to calories, multiply by 4.)

G. Problem Finding   What other questions could you answer using this spreadsheet?

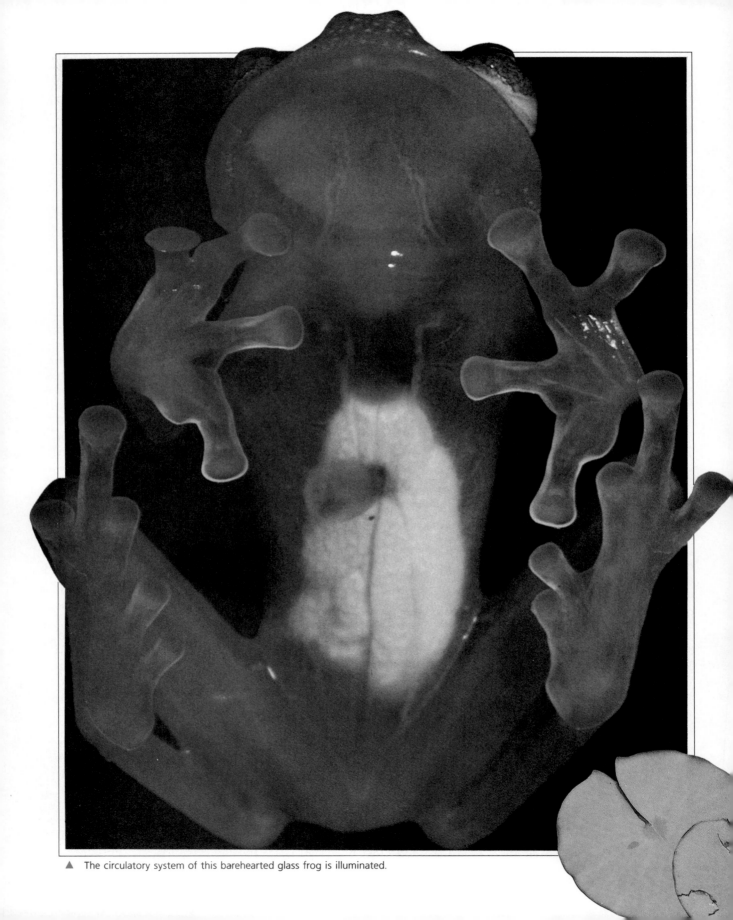

▲ The circulatory system of this barehearted glass frog is illuminated.

Transport

9

A cool, dark pond rests silent and still, not a whisper of wind or sign of motion. Beneath the placid surface of the pond, a complex world pulses with the energy and activity of life. Within every organism, a similar, unseen world exists. Cells are bathed in a fluid that carries needed nutrients to them. Waste products wash away in this same fluid. A network of passage-ways is used to transport nutrients to every single cell in the organism and to carry away wastes. In this chapter, you will explore the mechanisms through which living things keep their individual cells nour-ished and healthy.

Guide for Reading

Key words: transport, circulatory system, heart, capillaries, arteries, veins, lymphatic system
Questions to think about:
📖 How is transport accomplished in protists, hydra, earthworms, and grasshoppers?
📖 How does the heart pump blood to all parts of the human body?
📖 What are the major circulatory pathways in the body?

9-1 Adaptations for Transport

Section Objectives:

- *Explain* the importance of the transport process, for both simple and complex organisms.
- *Describe* transport in the ameba, paramecium, and hydra.
- *Compare* the circulatory system of the earthworm with that of the grasshopper.

As you may recall from Chapter 1, **transport** is the process by which substances move into or out of cells or are distributed within cells. Every cell needs substances from the environment to carry on its life processes. In order to enter the cell, these substances must move across cell membranes. Once they are inside the cell, the substances must be moved to places where they can be used or stored. In simple organisms, no special system is needed to move materials within cells and between cells. This is because the cells of these organisms are in close contact with their outside environ-ment. However, in large or complex organisms, many cells are far from the outside environment. These organisms need a special system, called the **circulatory system,** to transport materials to and from all parts of the organism. The circulatory system acts as a link between the cells of a complex organism and its environment.

In the following sections, you will learn how transport occurs in four representative organisms.

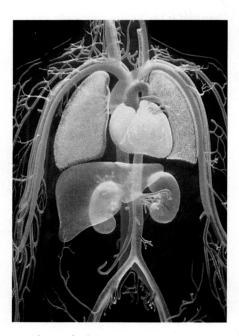

▲ **Figure 9–1**

The Human Circulatory System. The heart, lungs, liver, and kidneys are shown in this model of the human circulatory system.

Figure 9–2

Transport in Ameba and Paramecium. Because one-celled animals such as the ameba (left) and paramecium (right) have no circulatory system, the exchange of materials between the cells and the environment occurs by diffusion through the cell membrane. Within the cell, cytoplasmic streaming helps movement of materials. ▶

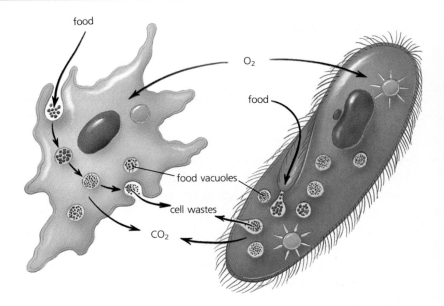

A circulatory system is made up of three parts: (1) a fluid in which materials are transported; (2) a network of tubes or body spaces through which the fluid flows; and (3) a means of driving the fluid through the tubes or spaces. For example, in humans, the most familiar circulatory fluid is the *blood.* The organ that pumps blood through the circulatory system is the **heart.**

Transport in Protists

Protists are usually one-celled animals that have no circulatory systems. Even in colonies, most protist cells are in direct contact with their surroundings. Diffusion and active transport alone are enough to move materials into and out of the cell. Within the cell, the movement of the material is aided by *cyclosis,* the streaming of the cytoplasm.

In the ameba and paramecium, food vacuoles move around in the cytoplasm by cyclosis. See Figure 9–2. As the food is digested, the absorption of nutrients takes place by diffusion or active transport across the food vacuole membrane.

Figure 9–3

Transport in the Hydra. In the hydra, cells of both the ectoderm and the endoderm are in direct contact with the environment. The exchange of materials between the cells and the environment takes place by diffusion through the cell membranes. ▼

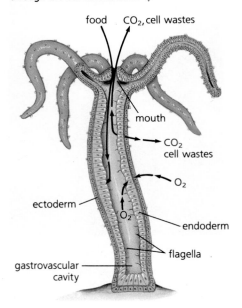

Transport in Hydra

Simple multicellular animals, such as the hydra, can also get along without a circulatory system. The hydra lives in fresh water. As you can see in Figure 9–3, its body is shaped like a hollow sac. The body wall of the hydra is composed of two layers of cells. The outer layer, the ectoderm, is in direct contact with the surrounding water. The inner layer, the endoderm, lines the gastrovascular cavity. The endoderm is also in contact with water because water enters and leaves the gastrovascular cavity through the mouth. Therefore, through diffusion, both cell layers can exchange dissolved oxygen, carbon dioxide, and wastes with their surroundings.

Nutrients from the gastrovascular cavity pass into the cells of the endoderm by both active transport and diffusion. The outer layer of ectodermal cells absorbs nutrients from the endoderm cells by diffusion. Within all cells, nutrients and other substances move around by cyclosis.

The muscular movements of the hydra as it stretches and contracts help to distribute materials within the gastrovascular cavity. This movement carries materials to all the inner cells of the endoderm. It also stops wastes from collecting near the surface of the endoderm. The flagella of endoderm cells also help to move materials. Thus, the gastrovascular cavity serves both to transport and to digest materials in the hydra.

Transport in the Earthworm

The earthworm is much more complex than the hydra. It contains true organs and organ systems. Most of its cells are not in direct contact with its surroundings. It is the circulatory system of the earthworm that makes possible the exchange of materials between its outside environment and its body cells.

Figure 9–4 illustrates the main features of the earthworm's circulatory system. The blood carries dissolved nutrients, gases, wastes, water, and other substances. It is red because it contains the red, iron-containing pigment **hemoglobin** (HEE muh gloh bin). Hemoglobin increases the amount of oxygen the blood can carry. The circulatory system of the earthworm is an example of a **closed circulatory system.** In a closed circulatory system, the blood is always contained within tubes or vessels in the body.

There are two major blood vessels in the earthworm. One is the *dorsal* (DOR sul) *vessel,* which runs along the top of the digestive tract. The other is the *ventral* (VEN trul) *vessel,* which runs below the digestive tract. These two vessels are connected near the head end of the worm by five pairs of blood vessels known as *aortic* (ay ORT ik) *arches,* or "hearts." The beating of these heartlike blood vessels pumps the blood from the dorsal to the ventral vessel.

The ventral vessel divides into many smaller vessels that go to all parts of the body. These small blood vessels gradually branch into still smaller and smaller vessels. The smallest blood vessels of

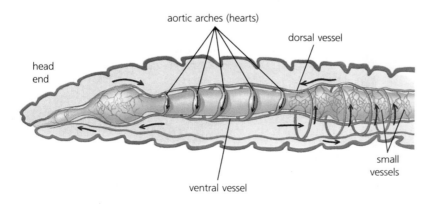

aortic arches (hearts)

dorsal vessel

head end

small vessels

ventral vessel

◀ **Figure 9–4**

Closed Circulatory System of the Earthworm. In the closed circulatory system of the earthworm, five pairs of contracting "hearts" pump blood through a system of vessels.

all are the microscopic **capillaries** (KAP uh ler eez). There are so many capillaries that every cell in the body of an earthworm is near one. The exchange of materials between the blood and the body cells takes place through the walls of the capillaries. Dissolved materials are able to diffuse across the thin walls of capillaries. The capillaries then join to form larger vessels. These larger vessels carry the blood back to the dorsal vessel. The dorsal blood vessel contracts rhythmically. By doing so, it forces the blood back into the aortic arches.

Transport in the Grasshopper

The grasshopper has an **open circulatory system.** In an open circulatory system, the blood is not always enclosed in blood vessels. Instead, it flows directly into body spaces where it bathes the tissues.

The colorless blood of the grasshopper does not contain hemoglobin, and it does not transport oxygen or carbon dioxide. Instead, these respiratory gases are transported through a series of tubes that are separate from the circulatory system. In the grasshopper, the blood serves mainly to transport nutrients and nitrogen-containing wastes.

As you can see in Figure 9–5, the open circulatory system of the grasshopper is quite different from the closed system of the earthworm. Along the back, above the digestive and reproductive systems, is a single vessel, the *aorta* (ay OR tah), and a tubular heart. Contraction of the heart, which is near the rear of the animal, forces the blood forward through the aorta toward the head. In the head, the blood flows out of the aorta. It then trickles through the body spaces and over the body tissues. The exchange of materials between the blood and the body cells takes place while the blood is in the body spaces. The blood is kept moving through the spaces by breathing and other movements of the body. Eventually, the blood circulates back into the heart through valvelike openings in the heart wall.

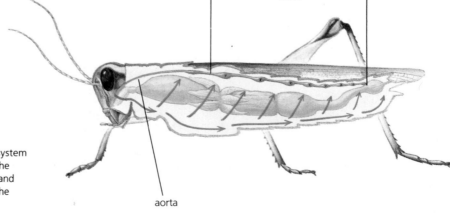

Figure 9–5

Open Circulatory System of the Grasshopper. In the open circulatory system of the grasshopper, blood pumped by the tubular heart passes through the aorta and into the body spaces, where it bathes the body tissues. ▶

There is an important difference between a closed circulatory system and an open circulatory system. In a closed circulatory system, the blood is under pressure. This pressure, which does not exist in an open circulatory system, causes the blood of an organism to move faster. However, an open circulatory system does move blood fast enough to meet the needs of those organisms that have this type of system.

9-1 Section Review

1. What are the three main parts of a circulatory system?
2. Name the processes that are involved in the transport of materials in protists.
3. How does a closed circulatory system differ from an open circulatory system?
4. What does the blood of a grasshopper transport?

Critical Thinking

5. What type of circulatory system would you expect to find in a bee? Explain your reasoning. (*Reasoning by Analogy*)

9-2 The Human Circulatory System

Section Objectives:

- *Compare* the structure and function of an artery, a vein, and a capillary.
- *Trace* the path of blood through the heart.
- *Explain* the heartbeat cycle and the mechanisms that control the rate and strength of the heartbeat.
- *List* the factors that cause variations in blood pressure.
- **Laboratory Investigation:** *Examine* the structure of a mammalian heart (p. 190).

Blood Vessels

Humans, like other vertebrates, have a closed circulatory system. The system is similar to that of an earthworm but more complex. It includes a single heart and a network of blood vessels. The heart pumps the blood, and the vessels carry the blood to and from all the cells of the body. There are three kinds of blood vessels— arteries, veins, and capillaries. These blood vessels are shown in Figure 9-6.

Arteries   The **arteries** (AR tuh reez) are the blood vessels that carry blood away from the heart to the organs and tissues of the body. The walls of arteries are thick and elastic. They contain

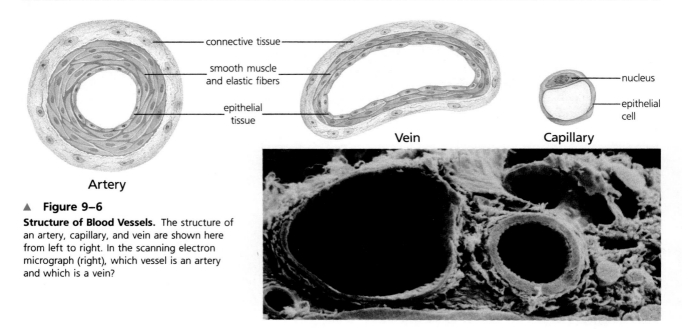

▲ **Figure 9–6**

Structure of Blood Vessels. The structure of an artery, capillary, and vein are shown here from left to right. In the scanning electron micrograph (right), which vessel is an artery and which is a vein?

Figure 9–7

Valves in a Vein. Pressure on the veins opens valves and allows blood to flow toward the heart (left). The valves close when the pressure decreases (right), thus preventing a backflow of blood. ▼

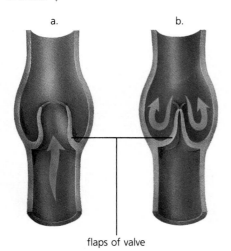

layers of connective tissue, muscle tissue, and epithelial tissue. As an artery enters a tissue or organ, it divides and subdivides many times to form smaller and smaller arteries. The smallest arteries are called *arterioles* (ar TEER ee olz).

Veins The **veins** (vaynz) are the blood vessels that return blood from the body tissues to the heart. The smallest veins are called *venules* (VEEN yoolz). The venules join together to form veins, which also merge, forming larger and larger veins. Unlike the artery walls, the walls of veins are thin and only slightly elastic. Inside the veins are flaplike **valves** that allow the blood to flow in only one direction—toward the heart. See Figure 9-7. When the valves do not work properly, blood tends to build up within the vein. The walls of the vein become stretched and lose their elasticity. This condition, called *varicose* (VAR uh kohs) *veins,* sometimes occurs in the veins of the leg.

Capillaries Arterioles and venules are connected by networks of microscopic capillaries. The walls of the capillaries consist of a single layer of epithelial cells. These vessels are so narrow that red blood cells pass through them in single file. The thin walls of the capillaries allow the exchange of dissolved nutrients, wastes, oxygen, and other substances between the blood and the body cells.

The Heart

The heart acts as a pump. Its regular contractions force the blood through the vessels. This muscular organ is somewhat larger than your fist and can be found slightly to the left of the middle of the chest cavity. It is made mostly of cardiac muscle, which consists of

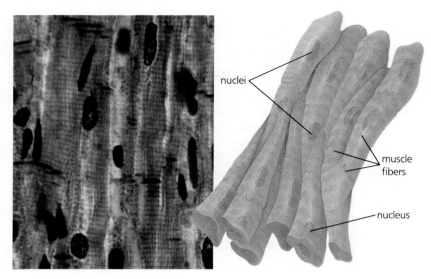

◀ **Figure 9–8**
Structure of Cardiac Muscle. The nuclei of muscle cells are visible in this scanning electron micrograph of cardiac muscle tissue (left). (Magnification 648 X) Fibers in cardiac muscle are interconnected (right), which helps to regulate the heartbeat.

nuclei

muscle fibers

nucleus

individual cells, each with a single nucleus. See Figure 9–8. The cardiac muscle cells form a branching, interlocking network. This allows them to contract with greater force.

A tough membrane, the **pericardium** (per uh KARD ee um), covers the heart and protects it. Inside, the heart is divided into four chambers, as you can see in Figure 9–9. The two upper, thin-walled chambers are the **atria** (AY tree uh), or *auricles* (OR ih kulz). The two lower, thick-walled chambers are the **ventricles** (VEN trih kulz). The right and left sides of the heart are separated by

Figure 9–9

Structure of the Human Heart. The cross section of the heart shows the four chambers, the valves, and the blood vessels that connect with the heart chambers. In the drawing below arrows trace the path of blood through the heart. ▼

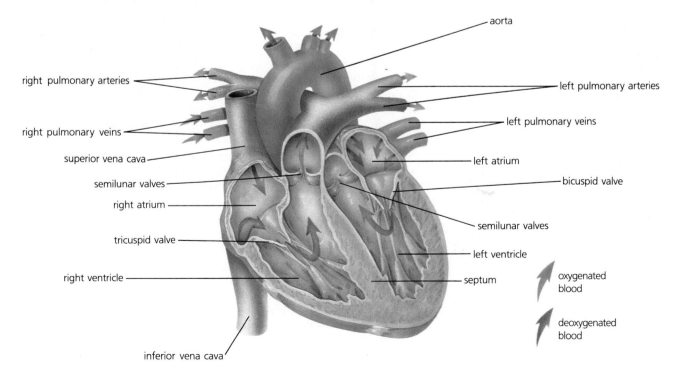

aorta

right pulmonary arteries

left pulmonary arteries

right pulmonary veins

left pulmonary veins

superior vena cava

left atrium

semilunar valves

bicuspid valve

right atrium

semilunar valves

tricuspid valve

left ventricle

right ventricle

septum

oxygenated blood

deoxygenated blood

inferior vena cava

Q: Why is it that some people die from using crack?

A: Crack is a strongly addictive drug that dramatically alters the delicate chemistry of the brain. The drug enters the bloodstream almost instantly, causing blood pressure to skyrocket. It is transported immediately to the brain and within seconds starts to affect the functioning of brain cells. The "high" feeling is just one effect of the drug's action on metabolism in the brain.

Other immediate effects of the drug can be deadly. Some people are extremely sensitive to crack. These people can die within minutes from *cardiac arrest*—the stoppage of the heart. This occurs because crack dramatically affects the region of the brain that controls the heartbeat. Many people who have died after using crack for the first time had no way of knowing that their bodies would react so violently to the drug.

Crack is a threat to everyone who uses it. It can also cause lung damage, severe weight loss, strokes, and birth defects.

■ *Design a poster that you feel would be effective in discouraging people from using crack.*

a wall, called the *septum* (SEP tum). The septum prevents the oxygen-poor blood found in the right side of the heart from mixing with the oxygen-rich blood of the left side. Entering from the back are the arteries and veins.

Four flaplike valves control the direction of the blood flow inside the heart. Two of these valves are called the *atrioventricular* (ay tree oh ven TRIK yoo ler), or *A-V, valves*. They allow blood to flow only from the atria into the ventricles. In the right side of the heart, the A-V valve is called the *tricuspid* (try KUS pid) *valve* because it has three flaps. In the left side, it is called the *bicuspid* (by KUS pid), or *mitral* (MY trul), *valve*. The other two valves are called the *semilunar* (sem ee LOON er) *valves*. When open, these valves allow blood to move from the ventricles into the arteries that carry blood away from the heart. When closed, the valves stop blood from flowing back into the ventricles.

Actually, the heart is a double pump. The right side of the heart sends oxygen-poor blood to the lungs. The left side sends oxygen-rich blood to the rest of the body.

The Heartbeat Cycle   The pumping action of the heart has two main periods. During one of these periods, the heart muscle is relaxed. This period of relaxation is called **diastole** (dy AS tuh lee). During the other period, the heart muscle contracts. The period of contraction is called **systole** (SIS tuh lee).

During diastole—the period of relaxation—the A-V valves are open. Blood flows from the atria into the ventricles. By the end of diastole, the ventricles are about 70 percent filled. Systole—the period of contraction—begins with contraction of the atria. The contraction of the atria forces more blood into the ventricles, filling them. The ventricles then contract. The pressure of this contraction closes the A-V valves and opens the semilunar valves. Blood flows out of the right ventricle into the **pulmonary** (PUL muh nair ee) **artery.** This artery divides into two branches, each of which goes to one of the lungs. Blood flows out of the left ventricle into the **aorta,** the largest artery of the body. The aorta branches and divides into many smaller arteries. These carry blood to all the body tissues.

While the ventricles are contracting, the atria relax. This permits blood to flow into the atria from the veins. Blood returning from all the body tissues except the lungs enters the right atrium. Blood returning from the lungs enters the left atrium. When the ventricles relax, a new period of diastole begins, and the cycle repeats.

As the heart valves open and close, they make a "lub-dup" sound. This sound can be heard clearly through a stethoscope. The "lub" sound is produced by the closing of the A-V valves. The "dup" sound is made by the closing of the semilunar valves. If the septum or any of the heart valves are damaged, there will be a leak, or backflow, of blood at certain times during the heartbeat cycle. This produces an abnormal heart sound, commonly known as a "heart murmur."

Control of the Heartbeat   The cardiac muscle that makes up the heart is different from the other muscle tissues of the body. As you have read, cardiac muscle fibers form a network, or lattice. The interlocking arrangement of muscle fibers in the two atria cause the atria to function together as one unit. Similarly, the two ventricles function together as another unit.

Although the nervous system controls the contraction of other types of muscle, cardiac muscle has a built-in ability to contract. Even when it is removed from the body, the heart will keep beating for a while if it is kept in a special solution. Each heart-muscle fiber has its own rate of contraction. However, the heart as a whole must work as a unit. This is made possible by a structure in the heart called the *sinoatrial* (sy no AY tree ul) *node,* or **S-A node,** also known as the pacemaker. The S-A node is a small group of specialized muscle cells in the wall of the right atrium. Contraction of the heart begins when the heart receives electrical impulses from the S-A node. This causes the atria to contract. Almost immediately the impulse reaches the *atrioventricular node,* or **A-V node.** This small bundle of muscle cells is located at the base of the right atrium. The A-V node triggers an impulse that causes the ventricles to contract. See Figure 9–10.

The tiny electrical current produced each time the heart contracts can be recorded on a machine that produces an *electrocardiogram* (eh lek troh KARD ee oh gram), or ECG. Physicians use electrocardiograms to check the health of the heart.

Figure 9–10

Controlling and Recording the Heartbeat. (Left). Electrical impulses from the S-A node cause the atria to contract. These contractions trigger an A-V node impulse, which causes the ventricles to contract. (Right). An electrocardiogram, or ECG, machine records the changing currents produced by contractions of the heart and helps to detect any abnormalities. ▼

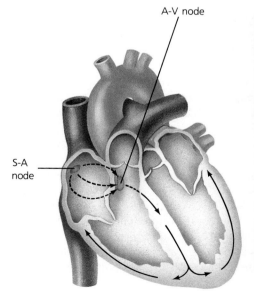
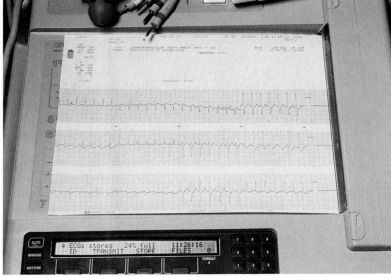

The rate of the heartbeat is regulated by certain nerves that enter the pacemaker. Impulses from the *vagus* (VAY gus) *nerves* slow down the pacemaker. Those from the *cardioaccelerator* (KARD ee oh ak sel uh ray tur) *nerves* speed up the pacemaker. The built-in rhythm of the heart is also affected by changes in body temperature and by certain chemicals circulating in the blood.

Sometimes, the natural pacemaker of the heart fails to work properly, and the rhythm of the heart is disturbed. When this happens, a battery-powered electronic pacemaker can be surgically placed inside a person's body. This artificial pacemaker regulates the heartbeat by delivering an electric shock to the heart at regular intervals.

Blood Pressure and the Flow of Blood

The thick, muscular walls of the arteries are elastic. When the ventricles contract, blood is forced out of them under great pressure into the arteries. Because they are elastic, the arteries are able to expand and to absorb this great pressure. As the ventricles relax, the pressure decreases. The elasticity of the artery walls helps to maintain the pressure between heartbeats. In this way, blood is kept flowing throughout the body continuously. The **pulse** is the expansion (high pressure) and relaxation (lower pressure) that can be felt in an artery each time the left ventricle contracts and relaxes. Both the rate and the force of the heartbeat can be measured by the pulse.

Physicians measure the pressure on the blood in the artery of the upper arm with an instrument called a *sphygmomanometer* (sfig moh muh NAHM uh ter). Pressure is measured in terms of the height of a column of mercury in a tube in this instrument. In an average adult at rest, the pressure during systole is enough to support a column of mercury about 120 millimeters high. During diastole, the pressure drops. The maximum height of the mercury is only about 80 millimeters. Usually, blood pressure is stated in the form of systolic pressure/diastolic pressure. Thus, the normal blood pressure in a resting adult is 120/80. During exercise and times of stress, blood pressure increases.

High blood pressure, or *hypertension* (HY per ten shun), is a medical condition. The blood pressure in people with this condition remains above normal throughout the heartbeat cycle. It is a serious and fairly common health problem. One frequent cause of high blood pressure is *atherosclerosis* (ath uh roh skluh ROH sis), a disease commonly called "hardening of the arteries." In this disease, deposits of cholesterol and other fatty materials collect on the inner walls of the arteries. See Figure 9–11. The arteries become narrower and the walls become more rigid, which causes blood pressure to increase. The condition puts a strain on both the heart and the blood vessels. If untreated, atherosclerosis can lead to heart attacks and strokes. Studies show that the chance of developing atherosclerosis increases with the amount of cholesterol

Figure 9–11

Atherosclerosis. The large deposit of fatty material in this artery has narrowed the space through which blood can flow. ▼

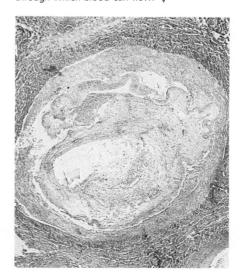

Preventing Heart Disease
■ Do not smoke tobacco. The more you smoke, the greater the risk of heart disease.
■ If you do smoke, quit. The risks of heart disease are greatly reduced within two years of quitting.
■ Monitor your blood pressure. It should be checked regularly by a doctor or nurse.
■ Reduce the amount of high-cholesterol foods you eat. Eat less butter, mayonnaise, and fatty meat. Eat more chicken, fish, fresh fruits, and vegetables.
■ Cut down on salt. Reducing the amount of salt you eat will help prevent high blood pressure.
■ Exercise regularly. Twenty minutes of exercise at least three times a week strengthens the heart and improves blood circulation.
■ Avoid obesity. If you are overweight, your heart has to work harder than it should.
■ Learn to manage stress. Feelings of stress and anxiety can contribute to heart disease.

◀ **Figure 9–12**

Preventing Heart Disease. Following these tips promotes good health and may prevent serious illness.

in the blood. In turn, the amount of cholesterol in the blood seems to be related to the amount of fat, particularly animal fat, in the diet. Many physicians suggest that the amount of animal and certain other fats in the diet should be kept low for this reason. Figure 9–12 lists other ways to reduce your risk of atherosclerosis and other heart diseases.

As blood flows through the arteries, there is little drop in pressure. However, there is a large drop in pressure when the blood reaches the arterioles. At the capillary ends of the arterioles there are rings of muscle that control the blood flow through the capillaries. The capillaries are the most numerous blood vessels in the body. If all the capillaries were open at the same time, there would not be enough blood in the body to fill them. The opening and closing of the rings of muscle at the capillary ends of the arterioles direct the flow of blood to the parts of the body where it is needed. For example, when an individual runs, the flow of blood into the capillaries of the skeletal muscles is increased. At the same time, the supply of blood to the capillaries of the digestive tract is decreased.

By the time blood reaches the veins, pressure is low. It is too low to return the blood to the heart, especially from the lower parts of the body. Blood flow in the veins is helped by the squeezing action of the skeletal muscles as the body moves. As a contracting muscle presses against a vein, the blood in the vein is forced to move. The blood moves toward the heart since the valves in the veins prevent flow in the opposite direction.

9-2 Section Review

1. Name the three kinds of blood vessels.
2. Describe the structure of the human heart.
3. What structure makes the heart-muscle fibers function as a unit?
4. What is the pulse?

Critical Thinking

5. Where would you expect blood pressure to be higher—in an artery in the arm or in an artery in the leg? Explain. (*Ordering*)

9-3 Pathways of Human Circulation

Section Objectives:

- *Trace* the path of the blood through the pulmonary circulation.
- *Discuss* the exchange of gases that occurs in the lungs.
- *Describe* the path of the blood through the systemic circulation, including the coronary, hepatic-portal, and renal circulations.
- *Identify* the structures of the lymphatic system and explain their functions.

Circulation of Blood

In the second century A.D., a Greek physician proposed that blood flowed back and forth from the heart to the rest of the body through the veins. It was not until 1628 that the correct pathway for blood circulation was identified. William Harvey, an English physician, showed that the heart pumps blood to the organs through arteries and that veins carry blood back to the heart. Harvey thought that connections between the ends of tiny arteries and the ends of tiny veins must exist. However, he could not find these connecting vessels. In 1660, an Italian anatomist showed that capillaries connect arteries to veins. Thus, Harvey's theory of the circulation of the blood was proven correct.

The major arteries and veins of the human circulatory system are shown in Figure 9–13. The circulatory system consists of two major pathways. **Pulmonary circulation**—the first pathway—carries blood between the heart and the lungs. **Systemic** (sys TEM ik) **circulation**—the second pathway—carries blood between the heart and the rest of the body. These two pathways are illustrated in Figure 9–14.

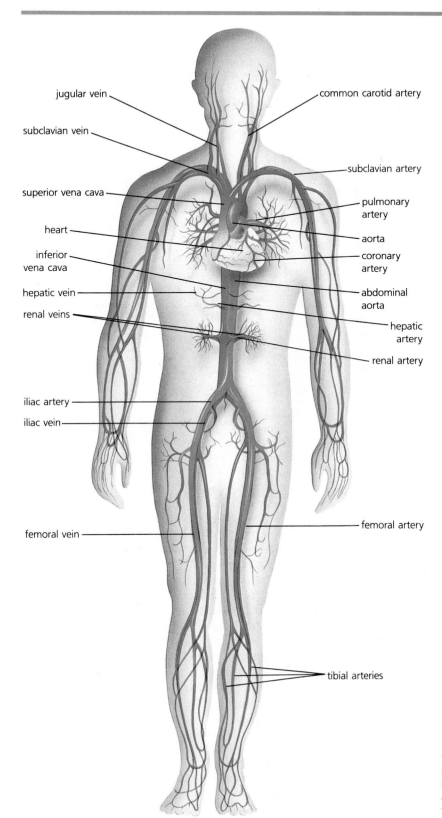

jugular vein

common carotid artery

subclavian vein

subclavian artery

superior vena cava

pulmonary artery

heart

aorta

inferior vena cava

coronary artery

hepatic vein

abdominal aorta

renal veins

hepatic artery

renal artery

iliac artery

iliac vein

femoral artery

femoral vein

tibial arteries

◄ **Figure 9–13**

Major Arteries and Veins of the Human Body. Notice that the names of many veins and arteries provide clues to their location in the body.

Figure 9–14
Pulmonary and Systemic Circulatory Systems. Pulmonary circulation carries blood between the heart and lungs. Systemic circulation carries blood from the heart to the rest of the body. ▶

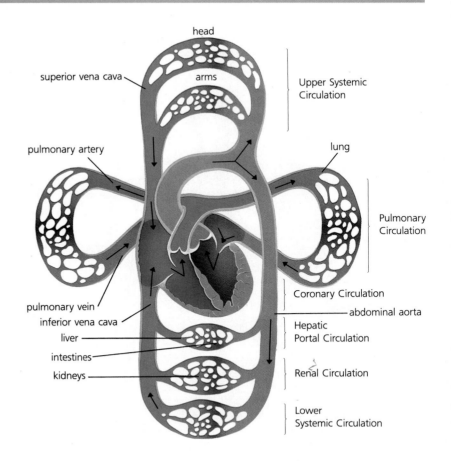

Pulmonary Circulation

Pulmonary circulation adds oxygen and removes carbon dioxide from the blood. Blood returning to the heart from the body tissues is low in oxygen and high in carbon dioxide. This blood enters the right atrium and flows into the right ventricle. The right ventricle pumps it through the pulmonary arteries to the lungs. The pulmonary arteries are the only arteries that carry oxygen-poor blood. All other arteries carry oxygen-rich blood. As the blood travels through the capillaries in the lungs, it gains oxygen and gets rid of carbon dioxide. The pulmonary capillaries merge into pulmonary veins. These veins carry the oxygen-rich blood to the left atrium of the heart. The pulmonary veins are the only veins that carry oxygen-rich blood. All other veins carry oxygen-poor blood.

Systemic Circulation

From the left atrium, the blood enters the left ventricle. Systemic circulation begins in the left ventricle of the heart. The powerful left ventricle has thicker walls than the other chambers of the heart because it pumps blood throughout the body. From the left ventricle, the blood is pumped into the aorta. The aorta branches, forming arteries that serve all parts of the body. The arteries divide and subdivide, forming smaller and smaller vessels. The smaller vessels finally form capillaries. Every cell in the body is near a capillary. The exchange of materials between the blood and the body tissues takes place through the walls of the capillaries. Capillaries merge to form veins. These veins finally return the blood to the heart. The largest veins of the body are the **superior vena cava** (VEE nuh KAY vuh) and the **inferior vena cava.** These large veins empty into the right atrium of the heart. The superior vena cava returns blood from the head, arms, and chest to the heart. The inferior vena cava returns blood from the lower body regions to the heart.

The systemic circulation includes three branches of special importance—the *coronary* (KOR uh ner ee) *circulation,* the *hepatic-portal* (heh PAT ik PORT ul) *circulation,* and the *renal* (REEN ul) *circulation.*

Coronary Circulation   **Coronary circulation** is the branch of the systemic circulation that supplies blood to the muscle of the heart. The right and left coronary arteries branch off the aorta just after the aorta leaves the heart. The coronary arteries run down either side of the heart, with branches entering the heart muscle. Within the heart, the arteries divide, eventually forming capillaries. The veins in the heart muscle drain the blood directly into the chambers of the heart, mostly the right atrium.

The cells of the heart need a constant supply of nutrients and oxygen. When a coronary artery is blocked by a blood clot or fat deposit, a heart attack can occur. Today, surgeons routinely perform coronary bypass surgery in these cases. A vein, usually from the person's leg, or an artificial vessel is used to construct a detour around the blocked artery. See Figure 9–15.

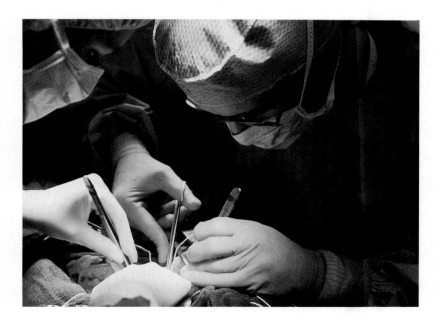

◄ **Figure 9–15**
Coronary Bypass Surgery. To improve the flow of blood to the heart, doctors use a vein (often from the leg) or an artificial vessel to build a detour around a blocked artery.

Hepatic-Portal Circulation Generally, blood travels through only one set of capillaries before it returns to the heart. An exception is the hepatic-portal circulation. The **hepatic-portal circulation** is the branch of systemic circulation that carries blood from the digestive tract to the liver. Blood passing through the capillaries of the digestive tract picks up nutrients. The veins draining these capillaries do not lead directly back to the heart. Instead, they form the *portal vein,* which goes to the liver. Within the liver, the vein divides into smaller veins. These veins divide into vessels similar to capillaries—the *hepatic sinuses.* Fluids, nutrients, and even blood proteins diffuse easily out of the blood and into the spaces between the cells of the liver. Blood in the sinuses of the liver is collected by a number of hepatic veins, which empty into the inferior vena cava.

Hepatic-portal circulation helps to maintain the balance of glucose in the blood. As blood passes through the liver, excess glucose is absorbed by the liver cells. In the liver cells, the glucose is converted to glycogen, which is then stored. If no food has been eaten for a long time, the blood reaching the liver from the digestive tract will be low in glucose. The liver then converts some of its stored glycogen to glucose, which diffuses out of the liver cells and into the blood. Thus, the liver helps to keep the amount of glucose in the blood at a constant level.

Renal Circulation One of the functions of the blood is to carry off some of the wastes of the body tissues. These wastes must be disposed of, or excreted. One waste product is the gas carbon dioxide. It is carried by the pulmonary circulation to the lungs, where it is excreted. Other wastes are removed from the blood and excreted by the kidneys. **Renal circulation** is the branch of the systemic circulation that carries blood to and from the kidneys. Renal circulation will be discussed in more detail in Chapter 12.

Circulation of Lymph

All the cells of the body are bathed in a colorless, watery fluid called the **intercellular** (in ter SEL yoo ler) **fluid,** or *interstitial* (in ter STISH ul) *fluid.* This fluid helps move materials between the capillaries and the body cells. All substances exchanged between blood and body cells diffuse through the intercellular fluid.

The intercellular fluid is formed from the parts of the blood that diffuse out of the capillaries. Intercellular fluid consists of water and salts and proteins and nutrients. Diffusion of intercellular fluid into the body tissues occurs when capillaries merge with arterioles. At the opposite end of the capillaries, close to the venules, most of the intercellular fluid and some of the substances it contains diffuse into the capillaries. However, all the proteins and some fluid remain outside the capillaries.

The excess fluid and proteins from the intercellular spaces are returned to the blood by a system of vessels called the **lymphatic** (lim FAT ik) **system.** See Figure 9–16. Without the lymphatic system,

Figure 9–16

Human Lymphatic System. The lymphatic system collects the excess fluid and proteins from intercellular spaces in the body and returns them to the blood. ▼

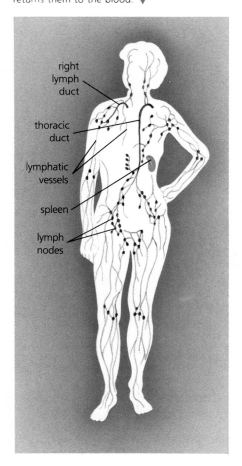

right lymph duct

thoracic duct

lymphatic vessels

spleen

lymph nodes

the constant loss of fluid from the blood eventually would drain the circulatory system. At the same time, body tissues would become flooded and swell. The lymphatic system begins in the body tissues with *lymph capillaries.* Lymph capillaries are microscopic tubes that are closed at one end. The walls of these tubes are only one cell thick. Intercellular fluid and proteins pass readily into the lymphatic capillaries. Once inside the lymphatic system, the fluid is called **lymph.**

The lymphatic capillaries merge to form larger and larger vessels. Like veins, lymphatic vessels have flaplike valves that allow the lymph to flow only in one direction. Muscular activity squeezes the lymph vessels and pushes the lymph along. Eventually, all the lymph from the lower part of the body, the left side of the head and chest, and the left arm flows into the *thoracic* (thuh RAS ik) *duct.* This is the largest lymphatic vessel in the body. Lymph from the thoracic duct is emptied into a large vein at the left side of the neck. All lymph from the right side of the head, the right arm, and the right side of the chest enters the *right lymph duct,* which drains into a large vein on the right side of the neck. In this way, the fluid and proteins lost from the blood in the capillaries are returned to the blood.

At various places along the lymphatic vessels, there are **lymph nodes,** or *lymph glands.* These glands play an important role in the body's defense against disease. See Figure 9–17. They filter foreign matter from the lymph. This prevents cancer cells, bacteria, and other disease-causing organisms from entering the bloodstream. Lymph nodes also produce some types of white blood cells that contain products able to destroy bacteria and other foreign substances. In the area of an infection, the lymph node may become enlarged and sore. These "swollen glands" show that the body is fighting an infection.

Lymphoid tissue like that found in lymph nodes is also found in the *spleen,* an organ near the stomach. In the spleen, the lymphoid tissues filter out bacteria and worn-out red cells from the blood.

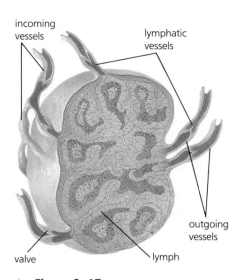

▲ **Figure 9–17**

Structure of a Lymph Node. Foreign matter, including disease-causing organisms, is filtered out of the lymph in the lymph nodes.

9-3 Section Review

1. How do the pulmonary arteries differ from all other arteries in the body?
2. What parts of the body are served by the systemic circulation?
3. What are the three major branches of the systemic circulation?
4. What is the role of the lymphatic system?

Critical Thinking

5. Filaria are parasitic worms that live in and block the lymphatic vessels of the legs. What do you think would be the result of filarial infection? (*Predicting*)

Laboratory Investigation

The Mammalian Heart

The mammalian heart has an elegant design. In this investigation, you will use a sheep heart to study the design of a mammalian heart. Since a sheep heart is similar to a human heart, use Figure 9–9 in your text to help identify its parts.

Objectives
- *Identify* the parts of a mammalian heart and explain their functions.
- *Trace* the pathway of blood through the heart.

Materials

sheep heart	scalpel	dissecting pan
latex gloves	dissecting scissors	glass rod

Alternate Approach: A model or a diagram may be used as the basis for an alternative procedure to the one below.

Procedure *

A. Wearing latex gloves, place a sheep heart in your dissecting pan. The pointed (ventricle) end should be close to you and tilting slightly to your right. The heart should be on its back side with the left ventricle on your right. To be sure, squeeze one ventricle, then the other. The one that feels more firm is the left ventricle.

B. Draw and label the outside of the heart. The line of fat on the upper part of the heart marks the boundary between the atria and the ventricles. In a preserved heart, the atria look like two small caps at the top of the heart. Fat and blood vessels run diagonally across the lower part of the heart, marking the boundary between the right and left ventricles.

C. Draw and label the blood vessels of the heart. The thick-walled pulmonary artery leaves the right ventricle and branches a short distance above the heart. Behind the pulmonary artery is the thick-walled aorta. It leaves the left ventricle, rises up, and bends above the left side of the heart. On the back side of the heart, the thin-walled inferior vena cava leads into the right atrium. The superior vena cava at the top of the heart leads into the right atrium. The coronary vessels on the outer wall of the heart mark the boundary between the

*For detailed information on safety symbols, see Appendix B.

right and left ventricles. The remaining four vessels (the pulmonary veins) enter the left atrium.

D. Using your scalpel and scissors, make a slit down the middle of the right side of the heart. Cut from the top to the bottom through the right atrium and right ventricle, as shown below. Pull back the sides of the opening and note the tricuspid valve. Cut upward into the pulmonary artery to locate the pulmonary semilunar valve. Draw and label the valves.

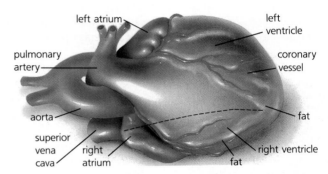

E. Cut another slit down the middle of the left side of the heart. Draw the bicuspid valve located between the left atrium and left ventricle. Cut into the aorta. Draw the aortic semilunar valve. Push a glass rod against the semilunar valve to see how it allows blood to flow out of the left ventricle.

F. Starting with the right atrium, use the glass rod to trace the pathway of blood through the heart.

Analysis and Conclusions

1. How do the atria and ventricles compare with respect to the thickness of their walls?

2. Describe how the tricuspid and bicuspid valves differ structurally from the semilunar valves.

3. Why should the walls of the left ventricle be firmer than the walls of the right ventricle?

4. Why do you think the heart needs coronary arteries to supply it with blood when blood is always flowing through the heart's chambers?

5. Describe the pathway of blood through the heart from the right atrium to the aorta.

Chapter Review

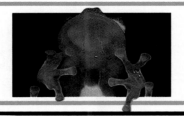

Study Outline

9–1 Adaptations for Transport

- All cells need materials from their environment to sustain their life processes. A circulatory system helps provide complex organisms with these substances.

- Protists have no circulatory system. Cells are in direct contact with the environment and obtain materials through diffusion and active transport. Streaming of the cytoplasm moves materials within the cell.

- Simple multicellular organisms, like the hydra, are also able to obtain materials through diffusion and active transport, because their body cells are in direct contact with their environment.

- The earthworm has a closed circulatory system in which blood is always confined in vessels. Its circulatory system consists of two major blood vessels, aortic arches, and a network of capillaries to provide cells with dissolved materials.

- The grasshopper has an open circulatory system in which blood flows into open spaces and bathes the tissues. Unlike in the closed circulatory system, there is no blood pressure in the open system.

9–2 The Human Circulatory System

- Humans have a closed circulatory system. It includes a single heart and a network of arteries, veins, and capillaries.

- Arteries transport blood away from the heart to the organs and tissues of the body. Veins return blood from the body tissues to the heart. Capillaries connect the smallest arteries to the smallest veins. The thin walls of the capillaries allow exchange of materials between the blood and the body cells.

- The heart acts as a double pump to transport blood to the lungs and the rest of the body. It is divided into four chambers with valves between the atria and ventricles and the ventricles and arteries.

- The heartbeat cycle has a period of relaxation, diastole, and a period of contraction, systole. The sinoatrial, or S-A, node controls the contraction of the heart.

- The pulse is created by the contraction and relaxation of the heart. Blood is forced out of the heart and into the arteries under great pressure.

- High blood pressure is a serious and fairly common medical condition. It is often caused by atherosclerosis.

9–3 Pathways of Human Circulation

- Pulmonary circulation oxygenates the blood by pumping it to and from the heart and lungs.

- Systemic circulation carries blood to and from the heart and all parts of the body except for the lungs. Coronary circulation supplies blood to the muscle of the heart. Hepatic-portal circulation transports blood from the digestive tract to the liver. Renal circulation carries blood to and from the kidneys.

- The lymphatic system returns excess fluid and proteins from the intercellular spaces to the blood.

Vocabulary Review

transport (173)

circulatory system (173)

heart (174)

hemoglobin (175)

closed circulatory system (175)

capillaries (176)

open circulatory system (176)

arteries (177)

veins (178)

valves (178)

pericardium (179)

atria (179)

ventricles (179)

diastole (180)

systole (180)

pulmonary artery (180)

aorta (180)

S-A node (181)

A-V node (181)

pulse (182)

pulmonary circulation (184)

systemic circulation (184)

superior vena cava (187)

inferior vena cava (187)

coronary circulation (187)

hepatic-portal circulation (188)

renal circulation (188)

intercellular fluid (188)

lymphatic system (188)

lymph (189)

lymph nodes (189)

Chapter Review

A. Matching

A. Matching—Select the vocabulary term that best matches each definition.

1. The tough, protective membrane surrounding the outside of the heart

2. A circulatory system in which blood is always confined in vessels

3. The period of relaxation in the heartbeat cycle

4. Blood vessels that carry blood from the body tissues to the heart

5. The fluid inside the lymph vessels

6. The lower, thick-walled chambers of the heart

7. Structure that triggers an impulse causing ventricles to contract

8. A circulatory system that carries blood to and from the kidneys

B. Sentence Completion

B. Sentence Completion—Fill in the vocabulary term that best completes each statement.

9. The _____ returns blood to the heart from the head, arms, and chest.

10. _____ connect the smallest arteries to the smallest veins.

11. The _____ returns excess fluid and proteins from the intercellular spaces to the blood.

12. The major artery carrying oxygenated blood away from the heart is called the _____.

13. _____ supplies blood to the tissues of the heart.

14. _____ is a pigment in red blood cells that increases the oxygen-carrying capacity of the blood.

15. The period of contraction during the heartbeat cycle is called _____.

Content Review

16. How is the body form of the hydra limited by its lack of a circulatory system?

17. Compare transport in protists with transport in hydra.

18. What is the function of hemoglobin?

19. Describe blood circulation in the grasshopper.

20. Why is it impossible for a grasshopper to have high blood pressure?

21. Why does it not make sense for the grasshopper to have a lymphatic system?

22. How do valves function in a circulatory system?

23. Trace the path of blood through the human heart.

24. Why is the heart referred to as a double pump?

25. How are the aortic arches of the earthworm similar to the human heart?

26. How is the rate of the heartbeat controlled?

27. How is pulse related to systole and diastole?

28. What happens when arteries lose elasticity?

29. How is the blood moved through the veins?

30. Explain the function of pulmonary circulation.

31. What is the difference between the superior vena cava and the inferior vena cava?

32. Describe hepatic-portal circulation.

33. What is the difference between coronary circulation and renal circulation?

34. Trace the path of lymph from a lymph capillary until it is returned to the blood.

35. What does the swelling of the lymph nodes mean?

Graphic Organizing

For information on graphic organizers, see Appendix G at the back of this text.

36. **Bar Graph:** The bar graph on page 193 has been constructed using data from the table below for death rates from heart disease among American females in 1987. Copy the graph onto graph paper. Then, construct a similar bar graph using the data given for males.

Ages	Deaths per 100 000	
	Females	**Males**
15–24	6	10
25–34	19	39
35–44	64	175
45–54	233	574
55–64	700	1575
65–74	1802	3114
75–84	4804	6813
85 and over	12 720	14 407

Death Rate from Heart Disease in Females, 1987

Chart showing Number of Deaths, 1987 (per 100 000 females) by Age Group (15–24, 25–34, 35–44, 45–54, 55–64, 65–74, 75–84, 85+), with the y-axis ranging from 0 to 13 000.

Critical Thinking

37. Large animals require a "pump" as part of their circulatory system. Why do you think there is a need for such a pump? (*Judging Usefulness*)

38. How are the human circulatory system and lymphatic system alike? How are they different? (*Comparing and Contrasting*)

39. How might a large, complex animal, such as a mammal, be affected if it had an open circulatory system? (*Relating Parts and Wholes*)

40. List some arguments for and against development of an artificial heart. Do you favor development? Why or why not? (*Making Ethical Judgments*)

Creative Thinking

41. Design a program that can be implemented in your school to help students lower their risk of heart disease.

42. If you were a scientist trying to construct artificial arteries for use in surgery, what qualities would you look for in the material to be used?

Problem Solving

43. An investigation was undertaken to determine the effect of exercise on pulse rate. Nonsmoking, healthy students of the same age, sex, and weight did different numbers of jumping jacks. One group did 10, a second group did 20, a third did 30, and a fourth did 40 jumping jacks. Pulse rates were taken for 15 seconds immediately before and after exercise. Identify the variable in this experiment. What factors are kept constant? What response is being measured? Design a control for this experiment.

44. Most fitness experts suggest that to get the maximum conditioning of the heart and circulatory system, individuals should exercise at their *target heart rate* for 30 minutes at least three times a week. To calculate your target heart rate (in beats per minute), subtract your age from 220 and multiply by 0.7. What is your target heart rate? What is the target heart rate for a 45 year old?

45. Using the data in the table below, calculate the number of times the heart beats in a day and in a year. Also, calculate the volume of blood pumped in a day and in a year. Then, calculate how long (in minutes) the heart relaxes in an hour.

Characteristics of an Average Human Heart	
Normal beat rate	72 per minute
Length of contraction	0.39 second
Length of relaxation	0.445 second
Volume of blood pumped per beat	70 mL
Volume of blood pumped per minute	5 liters

46. Do you think that changes in atmospheric temperature have an effect on the heart rate of a human being? Design an experiment to test whether or not your answer is correct.

Projects

47. Obtain prepared slides of artery and vein cross sections from your teacher. Examine them under the microscope and draw labeled diagrams of what you see.

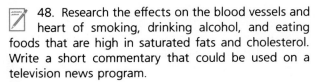

48. Research the effects on the blood vessels and heart of smoking, drinking alcohol, and eating foods that are high in saturated fats and cholesterol. Write a short commentary that could be used on a television news program.

▲ Red and white blood cells float in liquid plasma as blood is pumped through the body with each heartbeat.

The Blood and Immunity

Through the vast, intricate passageways of the circulatory system, red and white blood cells make their appointed rounds, driven by the forceful contractions of the heart. Like robotic maintenance workers in some complex factory, red blood cells dutifully pick up life-sustaining oxygen and nutrients, deliver these to the other cells of the body, and carry off their wastes. Meanwhile, the armada of ever-vigilant white blood cells searches for and destroys any foreign invaders. In this chapter, you will explore these and other amazing functions of the blood.

Guide for Reading

Key words: red blood cells, white blood cells, platelets, immune response, antibodies, active immunity, ABO blood groups, AIDS

Questions to think about:

📖 How is blood able to carry out so many different functions in the body?

📖 What are the components of the immune system, and how does each function?

10-1 Blood—A Multipurpose Fluid

Section Objectives:

- *Name* the different parts of the blood and explain the function of each part.
- *Trace* the sequence of events that results in blood clotting.
- **Laboratory Investigation:** *Observe* the blood cells of frogs and humans (p. 214).

Blood is a remarkable tissue. It contains both dissolved and suspended materials that travel through the blood vessels to every part of the body. Each one of the many components of blood plays an essential role in the maintenance of homeostasis throughout your body.

Functions of Blood

Blood is a liquid tissue that has three major functions: transportation, regulation, and protection. In humans and other vertebrates, blood transports materials to and from all the cells of the body. The nutrients and oxygen in blood are supplied to cells at levels that meet the needs of the cells. Wastes, produced by the cells, are carried away in blood to organs where the wastes are removed. Chemical messengers produced and released in one part of the

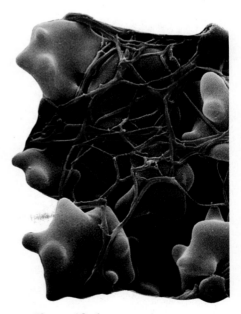

▲ **Figure 10–1**

Blood Clotting. Clotting will stop the flow of blood from a wound. Here, protein strands trap red blood cells, forming a clot.

body are carried in the blood to other areas where they regulate cell activity. In this way, the activities of different tissues throughout the body are coordinated.

Blood itself acts as a regulator. For example, blood absorbs heat from warm areas of the body and releases heat in cooler areas. Blood also tends to maintain a constant pH and water balance. Some substances dissolved in blood resist changes in pH. Others prevent too much water from leaving the blood and entering body tissues.

Blood also protects the body. It carries specialized cells and chemicals that defend the body against disease-causing organisms. Blood also has the ability to clot, thus protecting the body against blood loss from an injury.

The Components of Blood

As you might expect from reading about its functions, blood is made up of many different parts. The liquid part of blood is called **plasma** (PLAZ muh). Plasma makes up about 55 percent of the total volume of blood. The other 45 percent of the blood is made up of *red blood cells, white blood cells,* and *platelets.* See Figure 10–2. An adult human has between four and six liters of blood in his or her body.

Plasma   Plasma is the clear, straw-colored liquid portion of blood. Ninety percent of plasma is water. The remaining 10 percent is made of a variety of substances that are dissolved in the water. These dissolved substances include salts, glucose, amino acids, fatty acids, vitamins, enzymes, hormones, cellular wastes, and proteins.

Three major types of proteins are present in plasma: *albumin* (al BYOO min), *fibrinogen* (fy BRIN uh jen), and *globulins* (GLAHB yoo linz). Each of these proteins has its own function. Albumin, the most abundant of the plasma proteins, keeps water from leaving the blood and entering the surrounding cells by osmosis. See Chapter 5. It does this by helping to maintain the concentration of water in the blood at the same concentration as the water in the body tissues.

Fibrinogen is involved in the clotting of blood. Its role in this process will be discussed later in this chapter. The globulins have several different functions. Some globulins are involved in the transport of proteins and other substances from one part of the body to another. Other globulins, called antibodies (AN tee bod eez), help to fight infection. **Antibodies** are proteins that bind to, and thus help to destroy, foreign substances in the body. These substances include disease-causing organisms.

Red Blood Cells   Cells that are red in color and carry oxygen and carbon dioxide are called **red blood cells,** or *erythrocytes* (eh RITH ruh syts). They are present in enormous numbers in the blood. The human body contains 30 trillion red blood cells, or about 5 million

Figure 10–2

Plasma. When whole blood is centrifuged, the red blood cells, white blood cells, and platelets collect at the bottom of the test tube, leaving the clear, straw-colored plasma at the top. ▼

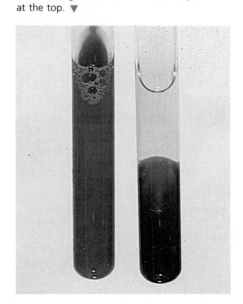

cells per cubic millimeter of blood. Red blood cells transport oxygen from the lungs to the body tissues. They also carry carbon dioxide from the body tissues to the lungs. Red blood cells are disk-shaped but are thinner in the center than around the rim. See Figure 10–3. In humans, mature red blood cells contain no nuclei. Their cytoplasm is filled with an iron-containing protein known as hemoglobin. **Hemoglobin** is the substance that gives blood its red color. In Chapter 11, you will learn how hemoglobin functions in the transport of oxygen and carbon dioxide.

Human red blood cells are made by bone marrow and have an average life span of 120 days. New red blood cells are formed at the same rate that old red cells are destroyed—about 2 million every second. Worn-out red cells are removed from the blood by the liver and spleen and then broken down. The iron from hemoglobin is held by the body and reused.

A condition in which a person has too few red blood cells or an insufficient amount of hemoglobin is called **anemia** (uh NEE mee uh). Either of these conditions lowers the amount of oxygen that can be carried in the blood. Anemia, therefore, results in the cells of the body not receiving enough oxygen. Some forms of anemia can be treated by eating iron-rich foods or by injections of vitamin B12. Sickle-cell anemia, a hereditary disorder, is caused by an abnormal form of hemoglobin rather than a vitamin deficiency.

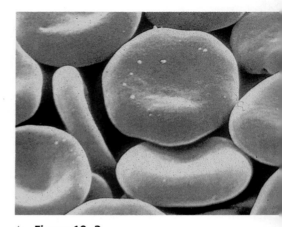

▲   **Figure 10–3**

Red Blood Cells. Red blood cells affect the survival of all the cells in the body because the red blood cells transport oxygen.

White Blood Cells   A variety of colorless blood cells make up the **white blood cells,** or *leukocytes* (LOO kuh syts). White blood cells are the defenders of the body. They protect the body from disease-causing organisms, such as bacteria and viruses. Mature white cells have a nucleus and are larger than red cells. They are less numerous than red blood cells, but their numbers are still impressive—about 60 billion in the adult human body. Bone marrow and lymphatic tissue produce about 1 million white blood cells every second. White blood cells are carried throughout the body in the circulatory system. Moreover, they can move on their own in the same way that ameba move. Using ameboid movement, they can squeeze between the cells of capillary walls and move through the body tissues. When there is an infection in the body, white blood cells collect in the infected area and attack the invading organisms.

As you can see in Figure 10–4, there are five different kinds of white blood cells. Most function in some way to protect the body. Some white cells—*neutrophils* and *monocytes*—are phagocytic. They protect the body by engulfing bacterial invaders, foreign substances, and cancer cells. **Lymphocytes** (LIM fuh syts) are responsible for the production of antibodies and cells that destroy foreign cells and substances.

Normally, there are 7000 to 10 000 white blood cells per cubic millimeter of blood. When there is an infection in the blood, however, the number may increase to 30 000 or more per cubic millimeter. As shown in Figure 10–5, the phagocytic white blood

Figure 10–4

White Blood Cells and Platelets. Note how the five types of white blood cells and the platelets differ in shape, size, and function. ▶

White Blood Cells and Platelets

Type	Function	Appearance
neutrophils	phagocytosis of small particles	
monocytes	phagocytosis of large particles	
eosinophils	release clot-digesting enzyme; combat allergy-causing substances	
basophils	release heparin, an anticoagulant, and histamine, a substance causing inflammation	
lymphocytes	involved in immune response	
platelets	involved in blood clotting	

Figure 10–5

A Phagocytic White Blood Cell. At the center of this micrograph is a white blood cell ingesting a chain of bacteria. ▼

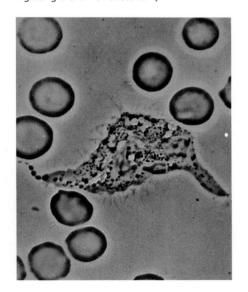

cells eat bacteria. The pus that forms at the site of an infected wound consists, in part, of white cells that have died after eating bacteria.

Cancer of the cells that produce white blood cells is called *leukemia*. Persons with leukemia have abnormally high levels of some types of white blood cells. Fortunately, some forms of leukemia can be controlled by drugs.

Platelets   Cell fragments that are involved in blood clotting are called **platelets** (PLAYT lets). Platelets are formed by the pinching off of bits of cytoplasm from large cells within the bone marrow. Although these bits of cytoplasm contain no nuclei, they are surrounded by a membrane. About 300 000 platelets are present in a cubic millimeter of blood. That is a total of about 1.5 trillion platelets in the blood of an adult human. They live for about seven days and are produced at a rate of 200 billion a day.

Blood Clotting

Although cuts and scrapes are common occurrences, bleeding from these injuries seldom becomes life-threatening. The torn blood vessels are quickly patched. The solid mass that plugs the hole in the torn vessel is called a blood clot. This solidification of blood at the site of an injured blood vessel is called **clotting.**

The Clotting Process   When a blood vessel is injured, the platelets in the blood stick to the wall of the damaged vessel and rupture. If the vessel damage is minor, the material from the ruptured platelets seals the leak. If the break is more serious, then the clotting process is triggered. More than 30 different substances are known to be involved in clotting. The major steps in this process are summarized here and in Figure 10–6.

- The ruptured platelets and the wall of the injured blood vessel release an enzyme, *thromboplastin* (throm boh PLAS tin).

- Thromboplastin initiates a series of enzyme-controlled reactions. The result of these reactions is the conversion of *prothrombin* (proh THROM bin), a plasma protein, into *thrombin*.

- Thrombin, an enzyme, converts soluble plasma fibrinogen into insoluble strands of *fibrin*.

- Fibrin forms a network of strands that traps red blood cells and platelets to form a clot.

The clot stops the bleeding, contracts, and hardens. In time, the wound is repaired by the growth of cells that replace the cells damaged by the injury. When healing is completed, a plasma enzyme called *plasmin* is activated and dissolves the fibrin clot.

Two factors prevent clots from forming inside uninjured blood vessels. First, the smoothness of the inner wall of the vessels prevents platelets from becoming activated. Second, substances in the blood act as *anticoagulants* (an tee koh AG yuh lunts) and prevent clot formation. One of these anticoagulants, *heparin,* is used as a drug after surgery to prevent clotting.

Figure 10–6
The Process of Blood Clotting. An injured blood vessel triggers a complex chain of reactions known as the clotting process. ▼

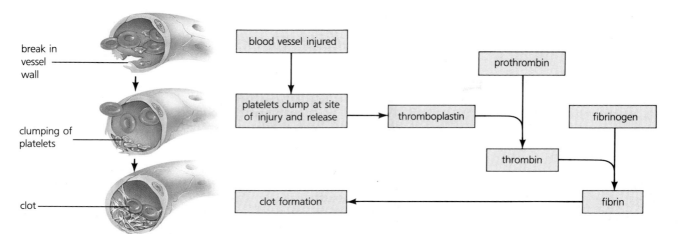

Figure 10–7

Use of Anticoagulants. The patient shown here is receiving an intravenous injection of heparin to prevent postsurgical clotting. ▶

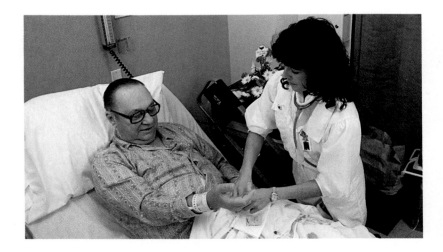

Clotting Problems Various conditions can cause the clotting system to malfunction. Not enough platelets in the blood or lack of vitamin K in a diet reduces the ability of blood to clot. Vitamin K is needed for the synthesis of prothrombin. Persons with **hemophilia** (hee muh FIL ee uh), a hereditary disease, lack one of the clotting factors. Now, they can receive injections of the missing factor, which is obtained from genetically engineered bacteria.

Not all clotting problems involve a failure to clot. Sometimes, a clot forms within a blood vessel when there is no injury. Clots can even form in one part of the circulatory system and travel through the body. If a clot cuts off or reduces the flow of blood to a whole organ, the effects can be disastrous. A blocked coronary artery, for example, can cause a heart attack. A block in an artery to the brain can cause a stroke. An obstructed artery in the lung can drastically reduce the oxygen supply of the body. Recently, scientists have produced bacteria that can manufacture large amounts of clot-digesting enzyme. If the enzyme is injected within a few hours of a heart attack, it can prevent, or limit, damage to the heart or other vital organs.

10-1 Section Review

1. Name the liquid part of blood.
2. What is the function of red blood cells?
3. What is the function of white blood cells?
4. What plasma protein forms the strands in a blood clot?

Critical Thinking

5. What would you expect to be the condition of a person whose blood contained twice the normal levels of white blood cells? (*Predicting Consequences*)

10-2 The Immune System

Section Objectives:

- *Explain* the three lines of defense existing against disease-causing organisms.
- *Describe* what occurs during an immune response.
- *Compare* a primary immune response to a secondary immune response.
- *Explain* the ABO and Rh blood groups.

Concept Laboratory: *Develop* an understanding of the regulation of the immune system (p. 207).

The immune system carries out a major part of the protective function of blood. Cells of the immune system are on constant patrol, ready to attack foreign invaders that get past other body defenses. While the immune system is not the only infection-fighting system in the body, it is the most complex.

Defenses Against Infection

Viruses, bacteria, and other microorganisms that cause disease are called **pathogens** (PATH uh jenz). These microorganisms are present everywhere in the environment. In fact, you are exposed to various pathogens every day in the food you eat, the water you drink, and the air you breathe. Fortunately, your body has several effective defenses against pathogens. These defenses are what keep you healthy most of the time.

First-Line Defenses   The body's first line of defense against pathogens involves several kinds of physical and chemical barriers. These include skin, sweat, tears, saliva, membranes lining body passages, mucus, stomach acid, and urine. Unbroken skin and the membranes lining body passages are effective barriers to most pathogens. Sweat, tears, and saliva contain chemicals that kill or inhibit some bacteria. Mucus that covers internal membranes entraps pathogens that are then washed away or destroyed by chemicals. Stomach acid destroys many pathogens that may be present in food.

Second-Line Defenses   If a pathogen gets past the first line of defense and starts an infection, parts of the second line of defense become activated. This results in the **inflammatory response,** a reaction of the body that causes swelling, redness, warmth, and pain in the area of an infection. Cells that are damaged from the infection release certain chemicals. These chemicals increase the flow of blood to the area. The increased blood flow causes puffiness and warmth and attracts phagocytes—neutrophils and macrophages. **Macrophages** (MAC ruh fay jez) are giant white blood cells that can ingest large numbers of bacteria. They develop from monocytes.

As the inflammatory response proceeds, the phagocytes ingest the pathogen and any damaged tissue. Eventually pus, a mixture of phagocytes, dead cells, bacteria, and body fluid, collects in the wound. The pus either drains or is absorbed by the body. Most of the time, the pathogen is destroyed, the inflammation dies down, and the wound heals.

When the pathogen is a virus, the infected cells produce a protein called **interferon** (in tuh FIR ahn). This substance causes nearby uninfected cells to produce enzymes that block the reproduction of the virus. In this way, healthy cells are protected from attack by the virus. As you will read in Chapter 27, through genetic engineering, scientists have produced bacteria that can make large amounts of interferon.

Third-Line Defenses   When the inflammatory response defense is insufficient, the pathogen is targeted for destruction by the body's last line of defense—the immune system. The immune system recognizes, attacks, destroys, and "remembers" each kind of pathogen or foreign substance that enters the body. It does this by producing antibodies and specialized cells that bind to and inactivate pathogens. Unlike the first two lines of defense, the immune system discriminates between different kinds of pathogens. For each kind of pathogen, the immune system produces antibodies or cells that are specific to that pathogen.

The Immune Response

The immune system includes all the parts of the body that are involved in the recognition and destruction of foreign materials. Bone marrow, white blood cells, especially phagocytes and lymphocytes, and various tissues of the lymphatic system, such as the lymph nodes, tonsils, thymus, and spleen, make up the immune system. The immune system provides **immunity** (ih MYOO nuh tee), the ability of the body to fight infection through the production of antibodies or cells that inactivate foreign substances or cells.

The basis of immunity lies in the body's ability to distinguish between its own substances or "itself," and foreign substances, or "nonself." This recognition is based on differences in certain large molecules, such as proteins, between one organism and another. When the body recognizes foreign cells or molecules, it produces antibodies or special cells that bind to the foreign substance and inactivate it. The production of antibodies and specialized cells that bind to and inactivate foreign substances is called the **immune response.**

Antigens   Any substance that can cause an immune response is called an **antigen** (ANT uh jin). Viruses and microorganisms have substances on their outer surfaces that are antigens. Most antigens are proteins, but carbohydrates and nucleic acids also may be antigens. The cells of each human contain a unique combination of proteins that no other human has. As a result, tissue from one person transplanted into another will act as an antigen. An immune response to an antigen acts to destroy the antigen.

Lymphocytes

Lymphocytes are the cells of the immune system that recognize specific antigens and either produce antibodies or kill foreign cells directly. There are two types of lymphocytes: B lymphocytes, or **B cells,** and T lymphocytes, or **T cells.** See Figure 10–8. B cells and T cells are produced in the bone marrow. B cells remain in the marrow and mature there, while T cells mature in the thymus gland. Millions of B and T cells are produced. At maturity, these cells are released and move into the circulatory and lymphatic systems.

B and T cells are capable of recognizing different antigens. In fact, B and T cells develop this ability while they are maturing—before they are ever exposed to any of the millions of antigens in the environment. As a result, each of the millions of individual B and T cells in a person will respond to a different antigen from among the millions of possible antigens.

When an antigen enters your body for the first time, your immune system undergoes a **primary immune response.** During the first 5 days following exposure to the antigen, no measurable amounts of antibodies or specialized immune cells are present. Then, over the next 10 to 15 days, there is a gradual rise in the levels of these products. If the same antigen enters the body another time, a more rapid **secondary immune response** happens. Within 1 to 2 days after infection, high levels of antibodies or specialized immune cells are present in the blood.

Whether primary or secondary, an immune response can involve two categories of reactions. One involves specialized B cells that produce antibodies. The other involves specialized T cells that attack foreign cells directly. The activation of both B and T cells, however, depends on the activities of regular phagocytic white blood cells as they engulf all foreign materials.

B Cells and Antibodies   When a B cell comes in contact with the antigen that it recognizes for the first time, it does not immediately produce antibodies. Instead, the B cell must be stimulated by a helper T cell that recognizes the same antigen. The helper T cell recognizes the antigen only after the antigen has been ingested by a phagocyte and has been "displayed" on the phagocyte's cell membrane. Following this stimulation, the B cell undergoes cell division, producing plasma cells and memory B cells. Plasma cells make antibody molecules that are released into the blood or lymph and bind to the antigen. The binding of antibodies to an antigen attracts other phagocytes that engulf and then destroy the antigen-antibody complex. When the antigen is a bacterium, an additional system, which is known as the complement system, helps to

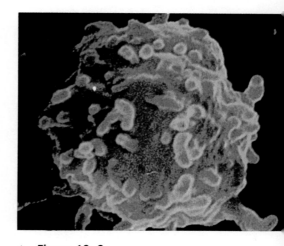

▲   **Figure 10–8**

Lymphocytes. B cells and T cells have specific surface receptors that recognize specific antigens.

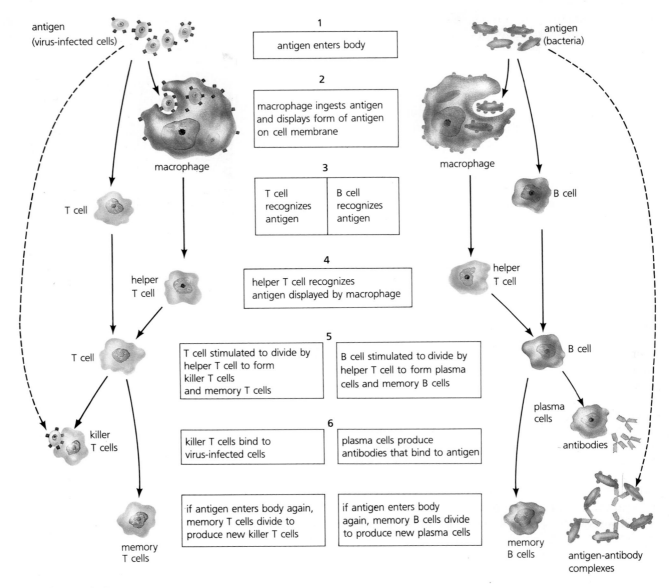

antigen
(virus-infected cells)

macrophage

T cell

helper
T cell

T cell

killer
T cells

memory
T cells

1
| antigen enters body |

2
| macrophage ingests antigen and displays form of antigen on cell membrane |

3
| T cell recognizes antigen | B cell recognizes antigen |

4
| helper T cell recognizes antigen displayed by macrophage |

5
| T cell stimulated to divide by helper T cell to form killer T cells and memory T cells | B cell stimulated to divide by helper T cell to form plasma cells and memory B cells |

6
| killer T cells bind to virus-infected cells | plasma cells produce antibodies that bind to antigen |

| if antigen enters body again, memory T cells divide to produce new killer T cells | if antigen enters body again, memory B cells divide to produce new plasma cells |

antigen
(bacteria)

macrophage

B cell

helper
T cell

B cell

plasma
cells

antibodies

memory
B cells

antigen-antibody
complexes

▲ **Figure 10–9**

The Immune Response. Trace the sequence of actions and interactions involved in the immune response.

destroy the bacteria. The **complement system** is a series of enzymes in blood that catalyze reactions that result in the bursting of the bacterial cell.

As the antibodies and phagocytes overcome the infection, other T cells, called suppressor T cells, release substances that slow down and eventually stop the plasma cells from producing antibodies. Thus, the immune response is kept under control. See Figure 10–9.

While plasma cells live for only a few days, the memory cells produced by the stimulated B cell can live for a lifetime. These cells produce a secondary immune response if the same antigen enters the body again. If this happens, the memory cells divide rapidly, forming new plasma cells that produce large quantities of antibodies, which act to destroy the invader. This rapid secondary response explains why the same disease usually does not strike the same person twice.

T Cells and Immunity   When an antigen is a virus-infected cell or a tumor cell, a single T cell, rather than a B cell, recognizes the antigen on the cell surface. In addition, a helper T cell recognizes the antigen when it is displayed by a phagocyte and stimulates the T cell into dividing. See Figure 10–9, which shows the sequences of the immune response. Cell division in the T cell results in killer T cells and memory T cells. Receptors on the surface of the killer T cells are like antibodies and cause the cell to bind to the antigens on the infected cell. Eventually, through enzyme-controlled reactions, the killer T cell causes the infected cell to burst.

As the foreign cells are brought under control, suppressor T cells release substances that shut down the killer T cells. The memory T cells, like the memory B cells, will cause a secondary response if the same antigen appears again.

Types of Immunity

There are two types of immunity: active and passive. In **active immunity,** the body produces its own antibodies or killer T cells to attack a particular antigen. In **passive immunity,** a person is given antibodies obtained from the blood of either another person or an animal. Passive immunity is "borrowed" immunity.

Active immunity may develop as the result of having had a disease. For example, a person who has had chicken pox rarely gets the disease a second time. Memory cells remaining in the body tissues quickly produce antibodies or killer T cells if the chicken pox virus invades the body again.

Thanks to the British physician Edward Jenner, the father of vaccination, active immunity may also develop through the use of a vaccine. See Figure 10–10. A vaccine consists of dead or weakened bacteria or viruses, or modified bacterial poisons. In each case, the organism or poison can still act as an antigen, but because it is weakened or modified, it can no longer cause disease. When the vaccine is injected into the body, the immune system responds to it as it would to the pathogen and produces antibodies or killer T cells. In this way, a person develops an immunity to a disease without actually suffering through it.

In contrast, passive immunity is only temporary. It usually does not last for more than a month because the body destroys the borrowed antibodies. Although it is short-lived, it is fast-acting. One type of passive immunity, called maternal immunity, occurs in infants. Antibodies from the mother enter the baby's blood before birth and provide the infant with passive immunity. Antibodies are also present in the mother's milk. Maternal immunity protects a child against most infectious diseases for the first few months of the child's life.

Science, Technology and Society

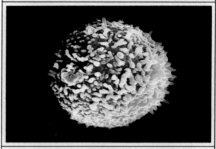

Technology: Monoclonal Antibodies

An antibody sorts through all the antigens in the body and attaches itself to one kind. This property makes antibodies perfect for locating specific cells. But how can large amounts of a single antibody be produced? Monoclonal antibodies—identical antibodies produced by clones (descendants) of a single cell—offer the solution.

Scientists have formed monoclonal antibodies by first fusing an antibody-producing cell (B-lymphocyte) with a tumor cell. The new cell, called a hybridoma, has the properties of both parent cells. It divides forever (as does a tumor cell), and it has one kind of antibody (as does a B-lymphocyte).

Researchers are experimenting with monoclonal antibodies to treat cancer. They have created monoclonal antibodies that recognize antigens on the surfaces of cancer cells. When these antibodies are attached to cell-killing drugs, they can deliver the drugs selectively to cancer cells, leaving normal cells unharmed.

■ *How does the development of monoclonal antibody technology illustrate the creative use of scientific knowledge?*

Figure 10–10
The Father of Vaccination. Edward Jenner (1749–1823) developed a vaccine for smallpox. ▶

Blood Groups and Transplants

Antibodies do more than fight infections. On the surface of every human cell are antigens that cause an immune response when recognized as foreign. In this section, you will see how the immune system is involved in blood types, transfusions, and transplants.

ABO Blood Group   In the early 1900s, Dr. Karl Landsteiner discovered that there are four major human blood types, or groups. These types, named A, B, AB, and O, make up the **ABO blood group.** A person's blood type depends on the presence or absence of two antigens, called A and B, on the surface of red blood cells. Figure 10–11 shows that individuals with type A blood have A antigens on their red cells. Individuals with B blood have B

Figure 10–11
Major Blood Groups. Human blood types depend on the presence of particular antigens on the surface of red blood cells. ▶

ABO Blood Groups		
Blood Type	**Antigens (on red cells)**	**Antibodies (in plasma)**
A	A	anti-b
B	B	anti-a
AB	A, B	none
O	none	anti-a, anti-b

Concept Laboratory

The Immune System

Goal

To gain an understanding of the immune response.

Scenario

Imagine you are a troubleshooter for Immunosystems, Inc. The company installs systems, patterned after the human immune system, that protect customers' homes by recognizing and destroying microbial invaders. Your job is to analyze customer complaints to identify the parts of the system that are malfunctioning.

Materials

Immunosystems
  master plan
customer
  problem reports

Procedure

A. Refer to the Immunosystems master plan below and to Figure 10–9 to identify and understand the function of each part of the system.

Immunosystems Master Plan

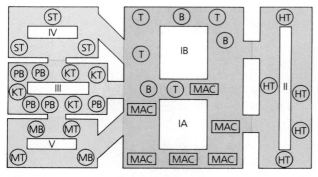

Circuit IA: a squad of vigilant, free-roaming, hungry macrophages (*MACs*). MACs recognize, engulf, and digest any invader and alert other circuits to the presence and nature of the invader.

Circuit IB: a platoon of highly trained and specialized T units (*Ts*) and B units (*Bs*) that are stimulated by contact with invaders. After stimulation, Ts and Bs wait for attack messages from HT units (*HTs*) in circuit II. Then Ts and Bs produce units that fill circuit III.

Circuit II: consists of helper T units (*HTs*). HTs receive information about invaders from MACs; then they signal Ts and Bs to launch an attack.

Circuit III: consists of plasma B units (*PBs*) and killer T units (*KTs*) produced by the Ts and Bs in circuit IB after attack orders. PBs release chemicals to immobilize the intruder. KTs bind to and kill other invaders.

Circuit IV: populated by suppressor T units (*STs*), which shut down PBs and KTs when invader is defeated.

Circuit V: consists of memory T and memory B units (*MTs* and *MBs*). As soon as former intruder attempts a new invasion, MTs and MBs reactivate circuit III by producing PBs and KTs, thus bypassing the need for circuits I and II. With each new invader, a new invader-specific MB or MT is added to the circuit.

B. Read each customer problem report and determine where the problem exists.

Customer Problem Report 1: System continues at full speed even after invader is destroyed.

Customer Problem Report 2: Invaders enter but are never detected. No alarms are sounded.

Customer Problem Report 3: Previous invaders reestablish themselves before being driven back.

Organizing Your Data

1. For each customer problem report, list the possible source or sources of the malfunction.

2. Discuss your findings with another troubleshooter and modify your report as appropriate.

Drawing Conclusions

Based on the master plan diagram, how many different invaders has this system battled?

Thinking Further

1. Write a hypothetical customer problem report that indicates one of the following conditions: (a) an immune deficiency disorder; (b) an autoimmune disorder.

2. Which condition described in the preceding question would be similar to that of a person with AIDS? Explain.

3. What strategy might an invader use to avoid detection by the system?

antigens on their red cells, and people with type AB blood have A and B antigens on their red cells. People with type O blood have neither A nor B antigens on their red cells.

The ABO system has an odd feature. An individual is born with antibodies against red blood cell antigens that his or her blood does not have. In other words, the plasma of a person contains antibodies that will bind to "foreign" red blood cells. Thus, as shown in Figure 10–11, type A blood contains anti-b antibodies. Type B blood contains anti-a antibodies. Type AB blood does not have either of these antibodies. Type O blood has both anti-a and anti-b antibodies. This kind of information is necessary to give safe blood transfusions.

Rh Factors   The **Rh factors** are another group of antigens found on the surface of red blood cells. They are called Rh factors because they were first found in rhesus monkeys. About 85 percent of the human population have the Rh factors on their red cells and are said to be Rh positive, or Rh+. The remaining 15 percent lack the Rh antigens and are said to be Rh negative, or Rh−. Unlike the ABO system, antibodies to the Rh factors are not produced until the individual is exposed to Rh factors.

The Rh factors may present a problem during pregnancy when the mother is Rh− but the baby has inherited Rh+ factors from its father. During birth, there may be some leak between the circulatory systems of the baby and the mother. Some of the baby's Rh+ red cells enter the mother's blood. Her immune system detects the Rh antigens as foreign and begins to form anti-Rh antibodies. In later pregnancies, anti-Rh antibodies from the mother's blood can enter the new baby's blood. If the baby is Rh+, the anti-Rh antibodies destroy the baby's red blood cells. The Rh problem in pregnancies can be eliminated if the Rh− mother is given an injection of anti-Rh antibodies, called RhoGAM, shortly after the birth of each Rh+ child. See Figure 10–12. These antibodies

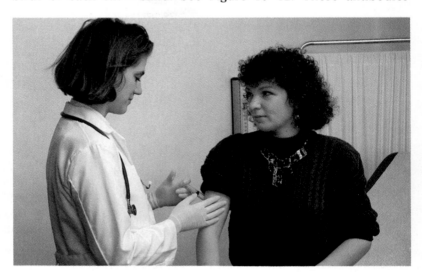

Figure 10–12

Rh Factors. The problems of Rh incompatibility between mother and child can be eliminated if an Rh-negative mother is injected with anti-Rh antibodies within 72 hours of the birth of each Rh-positive child. ▶

destroy any Rh antigens on the baby's blood cells that have entered the mother's circulatory system. In this way, the mother's immune system does not produce anti-Rh antibodies, and there is no problem with the next Rh+ baby.

Transfusions   For a blood transfusion to be safe, the person receiving blood, the recipient, must *not* have antibodies that will react with any A, B, or Rh antigens in the donor's blood. Knowing this, it is a simple matter to determine ABO and Rh blood types and match the blood types of the recipient and donor before a transfusion.

Figure 10–13 shows which blood types a recipient can safely receive. People with the same type of blood can donate blood to each other. People who have type O blood are called **universal donors.** Their blood can be given to anyone because type O blood does not contain A or B antigens. A person with type AB blood can receive a transfusion of any type of blood because type AB blood does not have anti-a or anti-b antibodies. People with AB blood are called **universal recipients.** In a transfusion, if the wrong blood groups are mixed, the red blood cells of the donor may clump together and break open because of the antigen-antibody reactions. This antigen-antibody reaction clogs the blood vessels and can cause kidney failure.

In matching blood types for transfusions, the Rh factor also must be determined. An Rh+ individual can receive both Rh+ and Rh− blood, whereas an Rh− individual can receive only Rh− blood. Although there are no ill effects the first time an Rh− person receives Rh+ blood, such a transfusion stimulates the formation of anti-Rh antibodies. If a second transfusion of Rh+ blood is given, an antigen-antibody reaction will cause the donated red blood cells to clump.

In emergency situations, plasma, instead of whole blood, is used for transfusions. The plasma restores blood volume and maintains blood pressure. When plasma is used, since no red blood cells are present, the donor and recipient need not be matched and no blood typing is necessary.

You might wonder why only the antigens and not the antibodies in the donor's blood are considered when matching blood types for transfusions. For example, when type A blood is given to a person with type AB blood, why don't the anti-b antibodies in the donor's blood bind to and clump the recipient's red blood cells? The reason is that the anti-b antibodies are greatly diluted in the much greater volume of the recipient's blood. At such low concentrations, there are too few antibodies to cause clumping.

Transplants   When an organ or tissue, such as a heart, kidney, or skin, is transplanted from one person (the donor) to another (the recipient), the transplant is recognized by the recipient's immune system as foreign. This activates the immune response, and the transplant is destroyed in a process called rejection. The rejection

Matching ABO Blood Types for Transfusions	
Recipient	Donor
A	A, O
B	B, O
AB	A, B, O
O	O

▲   **Figure 10–13**
Matching Blood Types for Transfusions.
When transfusions are given, the blood types of recipient and donor must be matched for ABO antigens to prevent clumping of red blood cells.

response to a transplant can be lessened if the donor and recipient are closely related. Rejection also may be controlled by suppressing the immune system with drugs, such as cyclosporine, an antibiotic. Unfortunately, when immunity is suppressed, the individual is more susceptible to infection.

10-2 Section Review

1. What are the two types of lymphocytes?
2. Define the term *antigen.*
3. What happens to an antigen-antibody complex?
4. What antibodies are in the plasma of a person with type B blood?

Critical Thinking

5. Compare and contrast the primary and secondary immune responses. (*Comparing and Contrasting*)

10-3 AIDS and Immune System Disorders

Section Objectives:

- *Describe* how AIDS affects the immune system.
- *Explain* how the spread of AIDS can be prevented.
- *Describe* some immune system disorders.

A healthy immune system is central to the health of the whole body. As you would expect, serious conditions result when parts or all of the immune system do not work correctly. The causes of many immune system disorders are not known. One immune system disease—AIDS—has been studied only since the 1980s. Its spread, and the efforts to cure and control it, make one of the most dramatic episodes in modern public health history.

What Is AIDS?

The disease **AIDS—Acquired Immune Deficiency Syndrome—** affects the immune system. *Acquired* means people pick up the disease from other people. *Immune deficiency* means a breakdown of the body's immune system. *Syndrome* means a group of symptoms that indicate disease.

The cause of AIDS is a virus called the **human immunodeficiency virus,** or **HIV.** HIV attacks the helper T cells of the immune system. How viruses reproduce and cause infection will be discussed in Chapter 30.

When HIV enters the body, the immune system recognizes it as an antigen and responds to it as it would to any antigen. However, for reasons that are not understood, the body's defenses are

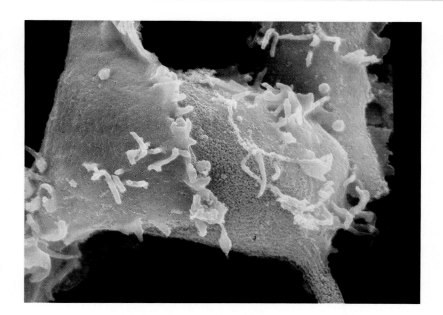

◀   **Figure 10–14**
HIV Attacking a T Cell. HIV, shown here as small blue particles, invades helper T cells, destroying them and weakening the immune system.

unsuccessful. HIV infects the helper T cells and remains within these cells for months or even years without producing any symptoms. When HIV suddenly becomes active, it reproduces, spreads, and destroys the helper T cells. See Figure 10–14. The decrease in helper T cells weakens the immune system. The body cannot fight infections, and the person develops a full-blown case of AIDS.

The first symptoms of AIDS are similar to those of a cold. Swollen lymph glands, fever, weakness, and unexplained weight loss are common. Usually, the AIDS patient goes on to develop other infections that the weakened immune system cannot overcome. One is a type of pneumonia caused by the microorganism *pneumocystis carinii* (noo moh SIS tis kar RIN ee ee). Another is a blood vessel cancer called **Kaposi's sarcoma** (KAH poh sheez sar KOH muh). These two diseases are the most common causes of death in AIDS patients.

HIV also attacks the nervous system, including the brain. This attack may show up as memory loss, loss of coordination, partial paralysis, or mental disorder.

How Is AIDS Spread?

AIDS is primarily a sexually transmitted disease. This means that HIV is passed from an infected person to an uninfected person during intimate sexual contact that involves the exchange of body fluids, namely semen and vaginal secretions. AIDS also can be transmitted by blood-to-blood contact. In the United States, this mode of transmission occurs most frequently among intravenous drug users who sometimes share needles and syringes. Prior to 1985, some people became infected with HIV when they received HIV-contaminated blood in a transfusion. However, donated blood

Biology and You

Q: Can you get AIDS from kissing a person on the mouth?

A: To date, there have been no reported cases of AIDS transmission from kissing a person on the mouth. Though small quantities of the AIDS virus have been found in the saliva of infected individuals, research suggests that saliva may contain a component that inactivates the virus.

Some people are also afraid that they can get AIDS by casual contact with an infected person. This fear is unfounded. You cannot get AIDS from being sneezed on, coughed on, or cried on by an infected person. Compared to other viruses, the AIDS virus is easily destroyed outside the body.

Another myth is that people can become infected with AIDS by donating blood. Some people believe that the needles used to draw blood may be contaminated with the virus. In fact, there is *no* risk of infection because hospitals and blood collection centers use sterile equipment as well as disposable needles.

■ *Find two current newspaper articles about AIDS. Prepare a brief oral report in the form of a television news story and present it to the class.*

in the United States now is tested for the presence of HIV. Finally, HIV can be transmitted from an infected pregnant woman to her unborn child before birth.

How Can AIDS Be Prevented?

It is estimated that about 2 million people in the United States are infected with AIDS. All of these individuals are capable of spreading the virus. However, the spread of AIDS is preventable. Preventing it is a matter of avoiding those behaviors that put an individual at risk. Since AIDS is sexually transmitted, the only no-risk sexual behavior is sexual abstinence—avoiding intimate sexual contact with another person. According to the Surgeon General, the next safest course of action is the use of latex condoms and a spermicide called nonoxynol 9, which kills HIV on contact. These products will reduce, but unfortunately not eliminate, the risk of infection. Of course, avoiding intravenous drug use prevents transmission of AIDS by blood-to-blood contact.

There is a blood test that detects antibodies to HIV. The test tells whether or not a person has been exposed to the virus. Anyone who thinks that he or she may have been exposed to HIV should be tested. Testing is available at hospitals and clinics. Information about AIDS and AIDS testing can be obtained by calling the AIDS hotline number. This number is available by calling the toll-free information operator at 1-800-555-1212.

Immune Disorders

Because the immune system is so complex, it is not surprising that it sometimes fails. A number of common and not-so-common human ailments are the result of immune system disorders.

Allergies   A rapid overreaction to an antigen that is not normally harmful is known as an **allergy** (AL ur gee). The wind-carried pollen of some plants causes a common allergy known as hay fever. Dust mites, insect stings, certain foods, and animal "hair" are a few other examples of allergy-causing antigens, or *allergens*. The typical symptoms of an allergy include a runny nose, swollen eyes, sneezing, coughing, and a rash. These symptoms are caused by the release of the substance *histamine* from the body cells at the site of the immune reaction. Histamine induces an inflammatory response, as it does whenever there is an injury or infection. Antihistamines are drugs that are commonly used to counteract the effects of histamine.

Autoimmune Diseases   In an **autoimmune disease,** the immune system of an individual fails to recognize some of the person's body cells as "self" and, therefore, produces antibodies against them. In juvenile diabetes, antibodies destroy the insulin-producing cells in the pancreas. When a person has rheumatoid arthritis, an immune reaction causes inflammation and crippling of the joints of the body. In multiple sclerosis, antibodies attack the

◀ **Figure 10–15**
Killer T Cell Attacking a Tumor Cell. The smooth-surfaced tumor cell (on the right) is being attacked by a killer T cell (on the left).

fatty covering of nerve cells. A person with *lupus erythematosus* (uh rith ma TOH sis) forms antibodies to different parts of the body, such as kidneys. Some viral or bacterial infections may trigger autoimmune diseases. For example, rheumatic fever is caused by a bacterium that triggers an immune response against heart and joint tissues.

Cancer is a variety of diseases in which cells of the body multiply without control. Cancer cells probably form in the body continuously. Fortunately, the immune system recognizes abnormal proteins on the surface of cancer cells as antigens. In healthy people, some T cells search the body and destroy the cancer cells. See Figure 10–15. Sometimes, the immune response does not recognize cancer cells as "nonself." Ignored by the immune system, the cancer grows and spreads. That is why the suppression of the immune system, whether by drugs or a pathogen, such as the AIDS virus, often results in cancer.

10-3 Section Review

1. What is the pathogen that causes AIDS?
2. How can a person infected with AIDS not show any symptoms of the disease?
3. Name three ways AIDS can be transmitted.
4. What is histamine?

Critical Thinking

5. If a two-year-old child were found to have AIDS, what would be the most likely cause of the infection? (*Identifying Probable Causes*)

Laboratory Investigation

Observing Blood Cells

Red blood cells, white blood cells, and platelets carry out different but critically important functions. In this investigation, you will use prepared slides to examine the structure of the cells and the platelets of human blood.

Objectives

- *Observe* and *determine* the relative amounts of platelets and types of blood cells.
- *Compare* the blood cells of frogs and humans.

Materials

prepared slides of frog and human blood
microscope

Procedure *

A. Using a microscope under low power, examine a prepared slide of human blood. Find platelets and red and white blood cells. After you have identified them, compare their sizes.

B. Count the number of red and white blood cells that appear in a single field of view. Repeat this in two other areas of the slide, and record the numbers in a chart like the one below.

	Number of Red Blood Cells	Number of White Blood Cells
human	1. 2. 3. Ave.	1. 2. 3. Ave.
frog	1. 2. 3. Ave.	1. 2. 3. Ave.

C. Switch to high power and examine a red blood cell. Note if it has a nucleus. Make a drawing of the red blood cell.

*For detailed information on safety symbols, see Appendix B.

D. Using Figure 10–4 in your text as a reference, try to identify the different types of white blood cells on your slide. The shapes of the nuclei and the staining of the cytoplasm will help. Basophils, neutrophils, and eosinophils have numerous, evenly distributed, stained granules in their cytoplasm. Lymphocytes and monocytes do not. Locate, draw, and name as many types of white blood cells as you can.

E. In a stained preparation, platelets appear as small blue specks. Observe the platelets under high power. Note if they have nuclei. Compare the size, appearance, and frequency of the platelets with those of the red and white blood cells.

F. Examine a prepared slide of frog blood under low power. Locate platelets and red and white blood cells. Compare the size, appearance, and frequency of each type of cell with the size, appearance, and frequency of the same type of cell in human blood.

Analysis and Conclusions

1. Based on your observations, which type of blood cell—red or white—is smaller?

2. According to your data for human blood, what is the average number of red blood cells for each white blood cell? How could you obtain a more accurate number? Does your number agree with the information presented in your text?

3. What is the most obvious structural difference between human and frog red blood cells?

4. What kinds of human white blood cells did you identify?

5. Based on your observations, which component— red blood cells, white blood cells, or platelets—is most abundant in human blood? In frog blood?

6. How does the shape of red blood cells contribute to their function?

7. How might the absence of a nucleus in mammalian red blood cells be an adaptation for the efficient transport of oxygen?

8. Why might you expect frogs to have a lower concentration of red blood cells than mammals?

Chapter Review

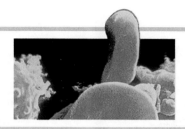

Study Outline

10–1 Blood—A Multipurpose Fluid

■ Blood is a liquid tissue that transports nutrients and oxygen to all the cells in the body and carries away wastes. It regulates body functions and cell activities by carrying chemical messengers. It protects the body against disease-causing organisms.

■ Plasma, the liquid portion of blood, accounts for about 55 percent of the total volume of blood in the body. The remaining 45 percent is made up of red blood cells, white blood cells, and platelets.

■ Red blood cells carry oxygen and carbon dioxide. Hemoglobin gives red blood cells their color. Anemia is a condition in which there is too little hemoglobin or too few red blood cells.

■ White blood cells play key roles in the body's natural defense mechanisms against disease-causing bacteria and viruses.

■ Platelets are cell fragments involved in blood clotting.

10–2 The Immune System

■ The body has three lines of defense against infection: physical barriers, such as skin and membranes lining body passages; the inflammatory response; and the immune system.

■ The immune system protects the body by destroying foreign substances that can cause disease. It includes bone marrow, white blood cells, especially phagocytes and lymphocytes, and certain lymphoid tissues, such as the lymph nodes and the spleen.

■ Antigens are substances that cause an immune response. Lymphocytes, which include T cells and B cells, respond to the presence of antigens either by attacking them directly or by producing antibodies that bind specifically to the antigens.

■ Human blood types are determined by the presence or absence of certain antigens on the surface of red blood cells. Each human blood type belongs to two groups: the ABO group includes types A, B, AB, and O; the Rh factor group includes Rh+ and Rh−.

10–3 AIDS and Immune System Disorders

■ Acquired Immune Deficiency Syndrome (AIDS) is caused by the human immunodeficiency virus (HIV), which attacks helper T cells.

■ AIDS can be transmitted by intimate sexual contact involving the exchange of body fluids, by blood-to-blood contact, and by an infected pregnant woman to her unborn child.

■ The spread of AIDS is preventable. AIDS can be prevented by avoiding behaviors that place a person at risk for infection by HIV.

■ Allergies, autoimmune diseases such as juvenile diabetes and multiple sclerosis, and cancer represent functional disorders of the immune system.

Vocabulary Review

plasma (196)

antibodies (196)

red blood cells (196)

hemoglobin (197)

anemia (197)

white blood cells (197)

lymphocytes (197)

platelets (198)

clotting (199)

hemophilia (200)

pathogens (201)

inflammatory response (201)

macrophages (201)

interferon (202)

immunity (202)

immune response (202)

antigen (202)

B cells (203)

T cells (203)

primary immune response (203)

secondary immune response (203)

complement system (204)

active immunity (205)

passive immunity (205)

ABO blood group (206)

Rh factors (208)

universal donors (209)

universal recipients (209)

AIDS—Acquired Immune Deficiency Syndrome (210)

HIV—human immunodeficiency virus (210)

Kaposi's sarcoma (211)

allergy (212)

autoimmune disease (212)

Chapter Review

A. Definition—Replace the italicized definition with the correct vocabulary term.

1. *A rapid overreaction to an antigen, such as those in pollen, pet fur, or certain foods,* is a common disorder of the immune system.

2. The immune system protects the body from *disease-causing viruses, bacteria, and other microorganisms.*

3. *The body's ability to produce antibodies to attack a particular antigen* may result from an actual infection or a vaccination.

4. Blood from *people with type O blood* can be given to anyone because such blood does not contain A or B antigens.

5. AIDS is caused by *a virus that attacks the helper T cells of the immune system.*

6. Ninety percent of *the liquid part of the blood* is water.

7. *The substance that gives blood its red color* is an iron-containing protein that functions in the transport of oxygen and carbon dioxide.

8. Hemophiliacs lack a factor in their blood that allows *the solidification of blood at the site of an injured blood vessel.*

B. Multiple Choice—Choose the letter of the answer that best completes each statement.

9. Cell fragments necessary in clotting are (a) pathogens (b) platelets (c) B cells (d) T cells.

10. A disease spread primarily through sexual contact is (a) anemia (b) Kaposi's sarcoma (c) AIDS (d) hemophilia

11. Cells that pick up and release oxygen and carbon dioxide as they pass throughout the body are called (a) red blood cells (b) lymphocytes (c) white blood cells (d) macrophages

12. People with blood type AB are known as (a) antibodies (b) universal recipients (c) universal donors (d) pathogens

13. A condition in which a person has too few red blood cells or an insufficient amount of hemoglobin is called (a) AIDS (b) hemophilia (c) anemia (d) Kaposi's sarcoma

14. A substance that is made by cells when they are attacked by viruses and that protects nearby uninfected cells from the virus is (a) plasma (b) interferon (c) hemoglobin (d) Rh factor

15. Antibodies from the mother's body entering a baby's blood before birth is an example of (a) inflammatory response (b) red blood cells (c) active immunity (d) passive immunity

Content Review

16. Describe the three functions of blood.

17. How is the composition of red blood cells related to their function?

18. Explain how white blood cells protect the body from infection.

19. Explain the process of blood clotting.

20. Describe the body's three lines of defense against disease.

21. Explain how body cells respond when they are attacked by viruses.

22. What is the basis of immunity?

23. In what two ways do lymphocytes destroy antigens and provide immunity?

24. Describe how passive immunity develops.

25. What is the ABO blood group?

26. Under what circumstances is the Rh factor a problem during pregnancy?

27. Describe the donor-recipient relationship necessary for a safe blood transfusion.

28. What causes the rejection of organs transplanted from one person to another?

29. Name two ways to avoid or reduce the risk of HIV infection.

30. Explain why antihistamines are used by people who have allergies.

31. Why might the suppression of the immune system result in cancer?

Graphic Organizing

For information on graphic organizers, see Appendix G at the back of this text.

32. **Concept Map:** Use the concept map on the following page to help you organize the information on blood. Copy the map onto a separate sheet of paper and fill in the missing concepts and linking words.

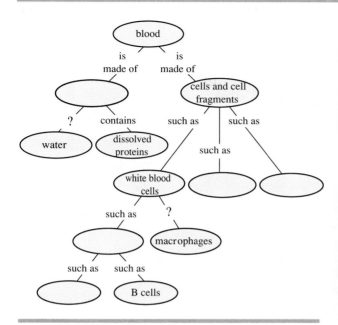

Critical Thinking

33. Compare and contrast the action of phagocytes responding to a bacterial infection with the action of T lymphocytes in the presence of an antigen. (*Comparing and Contrasting*)

34. Drugs must be given to recipients of organ transplants to lessen the immune system's response to the foreign tissue. Predict how this necessity could affect the way the recipients live their lives. (*Predicting*)

35. Person X donated blood to a friend, person Y, who was undergoing surgery. One year later, X needed a transfusion, but the doctor would not allow Y's blood to be used. Suggest some probable reasons for the doctor's decision. (*Identifying Reasons*)

36. Suppose a person's immune response is operating at a below-normal level. Suggest several possible general causes for this deficiency. (*Identifying Causes*)

37. Despite screening, there is a small chance that blood used for a transfusion could contain HIV. Should a person who has just had an emergency transfusion be told of the resulting small chance of contracting AIDS? (*Making Ethical Judgments*)

Creative Thinking

38. Why do you think a person can come down with the common cold over and over again without developing immunity to it?

39. Imagine that you could create an ideal artificial blood. Describe some of the characteristics that you would want it to have.

Problem Solving

40. A laboratory technician carried out a differential white-blood-cell count on a patient. Of the 200 white blood cells counted, 122 were neutrophils, 61 were lymphocytes, 5 were eosinophils, 1 was a basophil, and 11 were monocytes. Calculate the percentage of white blood cells of each type.

41. Design an experiment to determine the distribution of the ABO blood types among the students in your school. Indicate how you would present your data.

42. Normally, in humans, there are about 7000 white blood cells and 5 million red blood cells per cubic millimeter of blood. What is the ratio of white cells to red cells?

Projects

43. Write a dramatic sketch or draw cartoons that describe how the human body reacts when invaded by an antigen.

44. Conduct library research to find out how bone marrow transplants can be used in the treatment of disease.

45. Write a report on why an effective vaccine against AIDS has not yet been developed.

46. Antibodies have been divided into five classes based on function. Use references to identify these classes of antibodies and make a chart describing their functions.

47. Write a report describing how donated blood is stored, how it can be separated into its components, and how these components are used.

▲ A diver explores a colorful underwater world, breathing with the aid of an oxygen tank.

Gas Exchange

At the approach of a bubbling intruder, fish dart easily away. Weightless, the diver moves among the wonders of the Red Sea. Wonderful perhaps, yet for the diver the sea is hostile, a place without air. Fish and diver alike require oxygen. The fish get their oxygen from the sea itself, the diver from the tank she carries. To survive, animals must take in oxygen and expel carbon dioxide from their bodies. In this chapter, you will learn about the mechanisms by which various organisms accomplish this process of gas exchange, including the mechanisms of the human respiratory system.

Guide for Reading

Key words: gas exchange, respiratory pigments, gills, lungs, bronchi, alveoli

Questions to think about:

📖 How is gas exchange accomplished in simple organisms, earthworms, grasshoppers, and fish?

📖 Which organs in humans function in gas exchange?

📖 Why does smoking cigarettes lead to respiratory problems?

11-1 Adaptations for Gas Exchange

Section Objectives:

- *Define* the term *gas exchange* and list the characteristics of a respiratory surface.
- *Describe* the physical methods of gas exchange in protists and in hydra.
- *Explain* why specialized gas exchange systems are necessary in large, multicellular animals.
- *Compare* and *contrast* gas exchange in earthworms, grasshoppers, and animals with gills.

As you may recall from previous chapters, most protists and animals need to obtain oxygen from their surroundings and to remove carbon dioxide from their bodies. The oxygen is needed for *aerobic cellular respiration*. As you may remember from Chapter 6, cellular respiration is a chemical process that allows organisms to release energy from substances such as glucose. During aerobic cellular respiration, carbon dioxide is produced. Protist and animal cells must get rid of the excess carbon dioxide that has been produced. **Gas exchange** refers to the physical methods that organisms have for obtaining oxygen from their surroundings and removing excess carbon dioxide.

▲ **Figure 11–1**

Gas Exchange. Like all organisms, this wood bison must exchange oxygen and carbon dioxide with the environment.

Often, biologists use the term *respiration* to describe gas exchange. In this book, however, the term *gas exchange* will be used wherever possible. This will avoid confusion between the process of gas exchange and the energy-releasing process of cellular respiration.

The Respiratory Surface

When oxygen and carbon dioxide are exchanged between an organism and its environment, the gases pass through a boundary surface. The surface through which gas exchange takes place is called the **respiratory surface.** A respiratory surface must have the following characteristics.

- The surface must be thin-walled so that diffusion across it can occur rapidly.

- It must be moist because the oxygen and carbon dioxide must be in solution.

- It must be in contact with a source of oxygen that exists in the surroundings.

- In most multicellular organisms, it must be in contact with the transport system that carries dissolved materials to and from the cells of the organism.

Gas exchange through the respiratory surface takes place by diffusion. The direction of the gas exchange depends on the amounts of the gases on each side of the respiratory surface. For example, when oxygen is used up inside an organism's tissues, more oxygen diffuses into the tissues. When carbon dioxide builds up within tissues, the excess diffuses out of the tissues. The larger the area of the respiratory surface, the greater the amount of gas exchange that can occur during a given time period.

In protists and small multicellular animals, the exchange of gases takes place directly between the cells and the environment. In larger animals, however, the majority of the body cells are not near the outside environment. Therefore, gas exchange cannot take place directly between all the cells and the environment. Furthermore, larger animals often have an outer protective layer, such as scales, feathers, or dry skin, that prevents direct gas exchange through the skin. Therefore, most large, multicellular animals have respiratory surfaces in specialized organs or organ systems.

Gas Exchange in Protists

The exchange of gases with the surroundings is fairly simple in protists. It takes place directly through the body surface—the cell membrane. In the ameba and paramecium, oxygen, which is dissolved in the surrounding water, passes through the cell membrane into the cytoplasm by diffusion. See Figure 11–2. The carbon dioxide formed by cellular respiration diffuses out of the cytoplasm into the surrounding water.

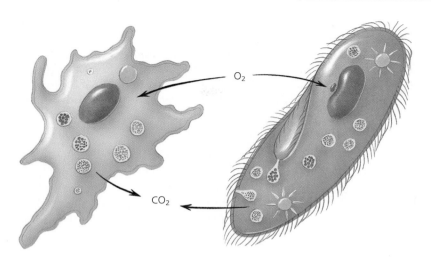

◀ **Figure 11-2**
Gas Exchange in Protists. The exchange of respiratory gases in the ameba (left) and paramecium (right) takes place directly through the cell membrane.

Gas Exchange in Hydra

Hydra are small and simple in structure. As you can see in Figure 11-3, the cells of the two layers that make up a hydra's body are in direct contact with water. Because they are, the exchange of respiratory gases can take place by direct diffusion between the body cells and the environment. There are no special structures for gas exchange.

Gas Exchange in Large, Multicellular Animals

A large, multicellular animal must exchange large amounts of gases across a respiratory surface. Animals that depend on gas dissolved in water have different exchange problems than animals that breathe air. First, the amount of oxygen dissolved in water is usually less than 1 percent. In contrast, oxygen makes up about 21 percent of the air. (The oxygen that is chemically part of the water molecules is, of course, not available for gas exchange. Only the free oxygen dissolved in the water can be used.) Second, oxygen diffuses more slowly in water than in air. Therefore, to obtain enough oxygen, an animal living underwater must constantly move a large amount of water over its respiratory surface.

Since gases must be in solution before they can diffuse across living membranes, air-breathing animals must keep their respiratory surfaces moist. Most air-breathing animals have respiratory systems that extend inside the organism. This protects the respiratory surface and lowers the amount of water lost by evaporation.

Respiratory Pigments

Many multicellular animals have colored substances in their blood. These substances, called **respiratory pigments,** carry oxygen and carbon dioxide between the respiratory surface and the body cells. Respiratory pigments allow the blood to carry more oxygen and

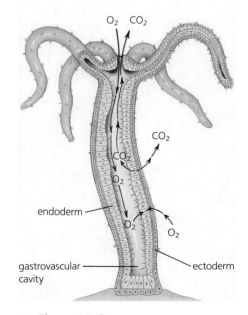

▲ **Figure 11-3**
Gas Exchange in Hydra. The exchange of respiratory gases in the hydra takes place by direct diffusion between the body cells and the environment.

▲ **Figure 11–4**
Gas Exchange in the Earthworm. The moist skin of the earthworm is its respiratory surface. Gases are exchanged with the environment through the skin and are carried to and from the body cells by the blood.

carbon dioxide than plain water can. For example, 100 milliliters of water can carry about 0.2 milliliters of oxygen and 0.3 milliliters of carbon dioxide. Hemoglobin, the most common respiratory pigment in the blood, can carry large amounts of respiratory gases. It allows 100 milliliters of human blood to carry about 20 milliliters of oxygen and 30 to 60 milliliters of carbon dioxide. (These are not the volumes of the gases when in solution but their equivalent volumes as gases in the air.)

Gas Exchange in the Earthworm

In earthworms, which live in moist soil, the skin is the respiratory surface. See Figure 11–4. The skin is thin, and mucus secreted by special cells helps to keep it moist. Just below the skin is a large number of capillaries. Oxygen diffuses from the air in the soil through the moist skin into the capillaries. Blood in the capillaries picks up the oxygen and carries it to the cells of the body. The blood contains hemoglobin, which aids in the transport of oxygen and carbon dioxide. At the body cells, the blood gives up oxygen and picks up carbon dioxide. The blood carries the carbon dioxide to the capillaries in the skin. The carbon dioxide diffuses through the skin into the air.

Damp soil keeps the earthworm's skin moist and helps its respiratory system to work well. If earthworms are exposed to air, their skin soon dries out, and they suffocate. When the weather is dry, they burrow deeper into the soil until they reach a moist area. Rain causes other problems for earthworms. Rain can flood their burrows. Because the water contains too little dissolved oxygen, earthworms have to leave their burrows to avoid drowning.

Gas Exchange in the Grasshopper

In grasshoppers, gas exchange does not depend on the circulatory system. A grasshopper's blood does not carry oxygen or carbon dioxide. Instead, a system of branching air tubes carries air directly to all the cells of the body. See Figure 11–5. These tubes are called **tracheal** (TRAY kee ul) **tubes.** Air enters and leaves the grasshop-

Figure 11–5
Gas Exchange in the Grasshopper. In the grasshopper, air enters and leaves through the spiracles. A branching system of tracheal tubes carries the air to and from the body tissues. ▶

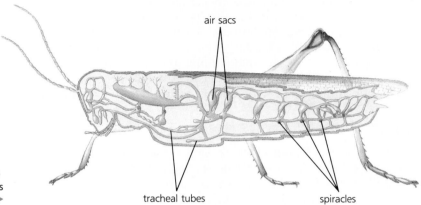

air sacs

tracheal tubes

spiracles

per's body through 10 pairs of openings called **spiracles** (SPEER uh kulz). From each spiracle, the tracheal tubes branch into smaller and smaller tubes. The fluid-filled ends of these microscopic air tubes act as a respiratory surface. They are in direct contact with the body cells. Oxygen in the air diffuses from the tracheal tubes to the body cells. Carbon dioxide diffuses from the body cells into the tracheal tubes.

Air is pumped into and out of the tracheal system by contraction of the grasshopper's muscles. When the area around the tracheal tubes expands, air flows in through the front four pairs of spiracles. Several large, collapsible, balloonlike chambers, called **air sacs,** are connected to the tubes. The air sacs help to pump air in and out of the tracheal system. When the area around the tracheal tubes contracts, the four pairs of spiracles close, and air is pumped out of the tracheal tubes through the six rear pairs of spiracles.

The system of tracheal tubes works well for gas exchange in small animals, such as grasshoppers and other insects. However, in a large animal, it would be impossible to move the needed volume of gases through this type of system.

Gas Exchange Through Gills

The gas-exchange organs of many animals that live in water, including fish, clams, oysters, and lobsters, are called **gills.** Gills are thin layers of tissue that are richly supplied with blood vessels. See Figure 11–6. Gills provide a large surface area for gas exchange. As water passes over them, dissolved oxygen diffuses from the water across the gill tissue and into the blood. The blood carries the oxygen to all parts of the body. Carbon dioxide from the blood diffuses out of the gills and into the water. There must be a constant flow of water over the gills. If the water flow is stopped, the animal will die from too little oxygen. Without water, the gills dry and stick to each other, and gas exchange cannot take place.

▲ **Figure 11–6**
Gills of a Fish. The thin filaments of a gill are richly supplied with blood vessels and provide a large surface area for the exchange of respiratory gases.

11-1 **Section Review**

1. Which two gases do organisms exchange with their surroundings?
2. What is a respiratory surface?
3. How does gas exchange occur in protists? How does it occur in the earthworm?
4. Name the gas-exchange organs of such water-dwellers as fish, clams, and lobsters.

Critical Thinking

5. Suppose that earthworms were short and thick rather than long and thin. Would their current method of gas exchange meet their needs? Why or why not? (*Predicting*)

11-2 The Human Respiratory System

Section Objectives:

- *Identify* the structures of the human respiratory system and state their functions.
- *Describe* the four phases of gas exchange in humans.
- *Explain* how body cells obtain oxygen and get rid of carbon dioxide.
- *Name* five diseases of the human respiratory system.
- **Laboratory Investigation:** *Discover* how the rate of breathing is controlled in humans (p. 232).

Structure of the Human Respiratory System

The human respiratory system is made up of the *lungs* and the system of air tubes that carry air to and from the lungs. See Figure 11-7. **Lungs** are the gas exchange organ in air-breathing vertebrates and some other animals. They are made up of many small

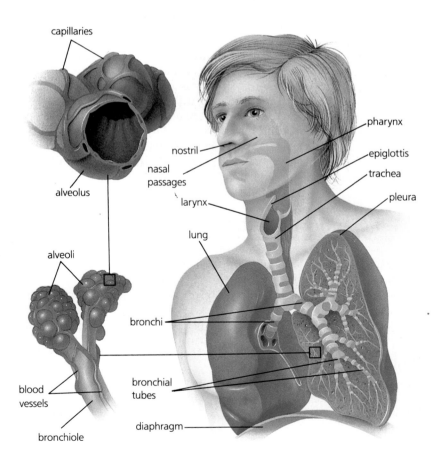

Figure 11-7

The Human Respiratory System. Air enters the body through the nose and travels down the trachea and bronchi to the bronchioles and alveoli in the lungs. ▶

chambers. Each chamber is surrounded by capillaries. Inside these small chambers is a huge respiratory surface for the diffusion of oxygen into the blood and the diffusion of carbon dioxide out of the blood.

The lungs fill a large part of the chest cavity in humans. They are separated from the abdominal cavity by the **diaphragm** (DY uh fram). The diaphragm is a muscle that forms the floor of the chest cavity. Each lung is completely enclosed by a two-layered membrane, which is called the **pleura** (PLUR uh). One layer of the pleura covers each lung, while the other layer is in contact with the diaphragm and the other organs of the chest cavity. A lubricating fluid between the two layers allows the lungs to move freely in the chest during breathing.

Air passes from the environment to the respiratory surface in the lungs. It goes through the nose, pharynx, larynx, trachea, bronchi, bronchial tubes, bronchioles, and alveoli. These structures are also shown in Figure 11–7.

The Nose   Air normally enters the respiratory system through the *nostrils*. These lead into hollow spaces in the nose called the **nasal passages.** Hairs at the openings of the nostrils stop various foreign particles from entering. The walls of the nasal passages and other air passageways are lined with a mucous membrane. Many of the cells that make up this membrane have cilia. Others secrete mucus, which is a sticky fluid. The mucus and the cilia trap bacteria, dust, and other particles in the air. The mucus also moistens the air. Just below the mucous membrane are a large number of capillaries. As air passes through the nose, it is warmed by the blood in these capillaries. Thus, the nasal passages serve to filter, moisten, and warm the air before it reaches the delicate lining of the lungs. When you breathe through your mouth, you lose these advantages.

Pharynx and Larynx   From the nasal passages, air travels through the **pharynx** (FAR inks), or throat. After leaving the pharynx, air passes into the **larynx** (LAR inks), or voice box. The voice box is made mainly of *cartilage,* which is a flexible connective tissue. The **vocal cords** are two pairs of membranes that are stretched across the inside of the larynx. As air is breathed out, the vocal cords vibrate. By controlling the vibrations of the vocal cords, humans are able to make sounds. To prevent choking during swallowing, food and liquids are blocked from entering the opening of the larynx by the *epiglottis* (ep ih GLOT is).

Trachea   The larynx runs directly into the **trachea** (TRAY kee uh), or windpipe. The trachea is a tube about 12 centimeters long and 2.5 centimeters wide. The trachea is kept open by horseshoe-shaped rings of cartilage embedded in its walls. Like the nasal passages, the trachea is lined with a ciliated mucous membrane. See Figure 11–8. Normally, the cilia move mucus and trapped foreign matter to the pharynx. There, they leave the air passages and are usually swallowed.

▲   **Figure 11–8**
The Lining of the Trachea. The cilia of the cells lining the trachea beat rhythmically, moving mucus and foreign particles toward the pharynx. (Magnification 10 500 X)

Science, Technology and Society

Issue: Passive Smoking

Recent research shows that nonsmokers may suffer negative health effects from passive smoking, inhaling the cigarette smoke of others. Some people are calling for a "smoke-free society," in which smoking is prohibited in public places.

Many nonsmokers find it impossible to avoid inhaling other people's cigarette smoke. This passive smoking has been linked to a high rate of lung cancer, bronchitis, and pneumonia. It can also irritate the eyes, nose, and throat. Some people feel that nonsmokers have the right to a healthy environment. Thus, they say, smoking should be banned in public.

Other people claim that research linking passive smoking to cancer and other illnesses is incomplete. They feel that a ban on public smoking would make smokers a persecuted minority. They contend, for example, that smokers might be discriminated against when they apply for jobs. Smoking, they say, is a personal freedom that should be protected.

■ *Do you think smoking should be banned in public places? Why or why not?*

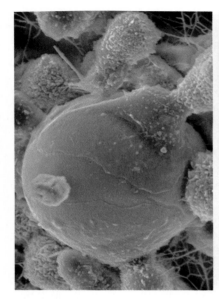

▲ **Figure 11–9**

The Consequences of Smoking. The phagocytic macrophages ingest foreign particles (left). However, when an excess of particles from smoking accumulates, the macrophages become overwhelmed and can no longer prevent damage to the lungs. Compare the nonsmoker's lung (bottom right) with the smoker's lung (top right).

As many people have discovered, the respiratory system cannot handle tobacco smoke. Smoking stops the cilia from moving. Just one cigarette stops their motion for about 20 minutes. Furthermore, tobacco smoke increases the amount of mucus in the air passages. See Figure 11–9. When smokers cough, their bodies are trying to remove the extra mucus.

Bronchi In the middle of the chest, the trachea divides into two cartilage-ringed tubes called **bronchi** (BRAHN kee). Like the trachea, this part of the respiratory system is lined with ciliated cells. The bronchi enter the lungs and branch in a treelike fashion into smaller tubes called **bronchial** (BRAHN kee ul) **tubes.**

Bronchioles The bronchial tubes divide and subdivide. As they do so, their walls become thinner with less and less cartilage. Finally, they become a group of tiny tubes called **bronchioles** (BRAHN kee ohlz).

Alveoli Each bronchiole ends in a tiny air chamber that looks like a cluster of grapes. Each chamber contains several cup-shaped cavities called **alveoli** (al VEE uh ly). The walls of the alveoli, which are only one cell thick, are the respiratory surface. They are thin, moist, and surrounded by a large number of capillaries. It is through these walls that the exchange of oxygen and carbon dioxide between blood and air takes place. It has been estimated that the lungs contain about 300 million alveoli. Their total surface area would be about 70 square meters. This would be 40 times the surface area of the skin.

Smoking makes it hard for oxygen to be taken through the alveoli. When cigarette smoke is inhaled, about one-third of the particles remain in the alveoli. Phagocytic cells called *macrophages* (MAK ruh fay jez) can slowly remove many of the particles. Look again at Figure 11–9. Too many particles from smoking or from other sources of air pollution damage the walls of the alveoli. This causes inelastic, scarlike tissue to form. This reduces the working area of the respiratory surface and leads to a disease called *emphysema* (em fuh ZEE muh).

Gas Exchange in Humans

Gas exchange in humans can be divided into four stages.

- Breathing is the movement of air into and out of the lungs.
- External respiration is the exchange of oxygen and carbon dioxide between the air and the blood in the lungs.
- Internal respiration is the exchange of oxygen and carbon dioxide between the blood in the capillaries and the body cells.
- Oxygen and carbon dioxide transport is the movement between the lungs and other body parts.

These stages of gas exchange are *physical* processes. They should not be confused with the chemical processes that take place within cells during cellular respiration. As you read in Chapter 6, during cellular respiration, nutrients are broken down, and energy is released.

Breathing   Breathing moves air into and out of the lungs. There are two phases of breathing. **Inhalation** (in huh LAY shun) draws air into the lungs. **Exhalation** (eks huh LAY shun) forces air out of the lungs. Since the lungs contain no muscle tissue, they cannot move by themselves. However, they are elastic. During breathing, they are forced to expand or contract as a result of pressure changes caused by the movement of the diaphragm, ribs, and rib muscles, as well as the force of air pressure.

Inhalation is the active phase of breathing. As the ribs are pulled up and out and the diaphragm is pulled downward, the chest cavity becomes larger. See Figure 11–10. As a result, the pressure within the chest cavity is reduced. Air from outside the body rushes down the air passageways into the lungs. This forces the lungs to expand.

Exhalation is the passive phase of breathing. The diaphragm relaxes and moves upward. The rib muscles relax, causing the ribs to drop. This causes the chest cavity to become smaller, and the pressure of the lungs to become greater. Thus, air is squeezed out of the lungs. Normal rates of breathing vary from about 12 to 25 times per minute.

Although people can control their breathing to some extent, it is, for the most part, an involuntary process. It is controlled by the *respiratory center* in the brain. There are also special structures in the aorta and several other larger arteries that can sense the

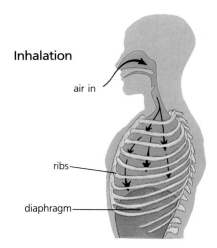

Inhalation

air in

ribs

diaphragm

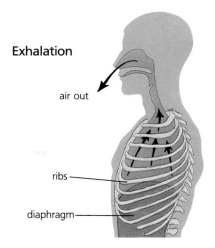

Exhalation

air out

ribs

diaphragm

▲ **Figure 11–10**

Breathing. Pressure changes caused by the movements of the diaphragm, ribs, and rib muscles force air into and out of the lungs.

Figure 11–11

External and Internal Respiration. In human body tissues, all cells exchange gases with the internal environment by diffusion across the moist cell membranes. The gases are transported to and from the lungs by the blood and the circulatory system. Exchange with the external environment occurs by diffusion across moist cell membranes in the alveoli of the lungs. ▶

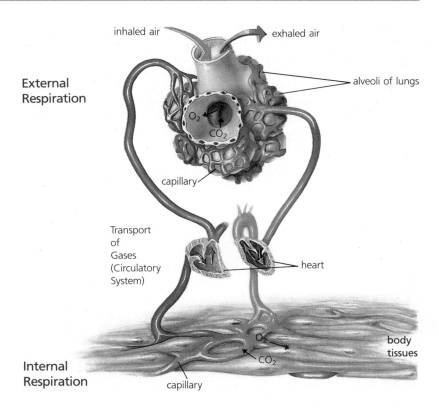

amount of oxygen and carbon dioxide in the blood. These *chemoreceptors* send messages to the respiratory center. When the amount of carbon dioxide in the blood increases, the respiratory center of the brain is stimulated. Nerves from the respiratory center carry impulses to the diaphragm and chest muscles that raise the rate and depth of breathing. This lowers the amount of carbon dioxide and raises the amount of oxygen in the blood.

During heavy exercise, lactic acid is produced by muscle cells. This increases the acidity of the blood and stimulates the respiratory center of the brain. Once the respiratory center is stimulated, the rate of breathing increases.

External and Internal Respiration   External respiration is the exchange of oxygen and carbon dioxide between the air and the blood in the lungs. See Figure 11–11. After inhalation, the amount of oxygen in the alveoli is higher than the amount of oxygen in the blood. Oxygen dissolves into the moist lining of the alveoli and diffuses from the region of higher concentration (the alveoli) to the region of lower concentration (the blood).

As the blood is pumped through the vessels of the body by the heart, oxygen-rich blood from the lungs is carried to the body tissues. Blood that is rich in carbon dioxide from the body tissues is returned to the lungs. In the lungs, then, the amount of carbon dioxide in the blood is higher than it is in the alveoli. Carbon dioxide diffuses out of the blood and into the alveoli—in the opposite direction from oxygen.

Internal respiration is the exchange of oxygen and carbon dioxide between the blood in the capillaries and the body cells. In the capillaries of the body tissues, oxygen diffuses from the blood through the intercellular fluid in the body cells. Carbon dioxide diffuses from the cells through the intercellular fluid into the blood. Each gas diffuses down a concentration gradient. That is, each gas diffuses from a region of higher concentration to a region of lower concentration.

Oxygen Transport   Most oxygen is carried from the lungs to the body tissues by the hemoglobin in the red blood cells. Little of it is dissolved in the plasma. Hemoglobin (abbreviated Hb) is a red, iron-containing protein that combines easily with oxygen. Hemoglobin holds the oxygen loosely.

The amount of oxygen in the surrounding tissues determines whether hemoglobin will combine with oxygen or will release oxygen. In the lungs, where there is a large amount of oxygen, hemoglobin combines with oxygen to form *oxyhemoglobin* (HbO_2). Oxygen-rich blood is a bright red color because of the oxyhemoglobin. When the blood reaches the capillaries of the body tissues, where the amount of oxygen in the surrounding tissues is low, the oxyhemoglobin breaks down into oxygen and hemoglobin. The

Critical Thinking in Biology

Identifying Reasons

When you identify reasons, you uncover the ideas people use to justify their thoughts, actions, or statements. At a city council meeting, townspeople were debating a ban on smoking in public places.

"We know exposure to cigarette smoke can cause respiratory problems," Mario said. "Therefore, I move that no smoking be permitted in any public areas in this town."

"I second the motion," called Evelyn. "I'd like a smoke-free environment because I always cough when I breathe cigarette smoke."

1. Identify each speaker's reasons for a smoke-free environment.
2. Cue words such as *since* and *thus* are sometimes used to link reasons with conclusions. What cue words helped you identify the speakers' reasons?
3. Suppose you attended the same meeting as Mario and Evelyn. Write a statement expressing your views on smoking in public places. Make sure you offer a reason for your view.
4. Study the poster shown at right carefully. What reason is offered for not smoking?
5. Think About It Describe your thought process as you answered question 1.

oxygen diffuses from the blood into the body cells. Once in the cells, the oxygen is used in aerobic respiration. The blood, now low in oxygen, is a dark red or dull purple color.

People who smoke cigarettes have a significantly lower level of oxygen in their blood. This is because cigarette smoke contains the gas *carbon monoxide*. Carbon monoxide has a greater attraction for hemoglobin than does oxygen. This means that, when carbon monoxide is present, it prevents oxygen from joining with hemoglobin. Because their blood contains too little oxygen, smokers often experience a shortness of breath when they are active.

Carbon Dioxide Transport   As you may recall from Chapter 6, cellular respiration produces carbon dioxide. Thus, the amount of carbon dioxide tends to be greater in the body cells than in the capillary blood. Therefore, the carbon dioxide diffuses out of the cells and into the blood. Carbon dioxide is carried by the blood to the lungs in three ways.

Most (about 70 percent) of the carbon dioxide that diffuses into the blood combines with water, forming carbonic acid, H_2CO_3.

$$CO_2 + H_2O \longrightarrow H_2CO_3$$

The H_2CO_3 quickly breaks down into hydrogen ions, H^+, and bicarbonate ions, HCO_3^-.

$$H_2CO_3 \longrightarrow H^+ + HCO_3^-$$

These two reactions take place quickly because of an enzyme in the red blood cells. Therefore, most of the carbon dioxide from the body cells is carried away in the plasma in the form of bicarbonate ions.

Some (about 20 percent) of the carbon dioxide that diffuses into the blood combines with hemoglobin. This carbon dioxide is carried in the red blood cells as carboxyhemoglobin, $HbCO_2$.

$$CO_2 + Hb \longrightarrow HbCO_2$$

A small amount (about 10 percent) of the carbon dioxide that diffuses into the blood is dissolved in the plasma. This carbon dioxide is carried away from the body cells to the lungs in the plasma.

All these reactions are easily reversed. In the lungs, carbon dioxide is released from the blood.

Diseases of the Respiratory System

The following list gives some of the common disorders of the respiratory system.

- *Asthma* (AZ muh) is a severe allergic reaction that causes wheezing, coughing, and breathing difficulties. During an asthma attack, the bronchioles go into spasms, squeezing the air passages.

- *Bronchitis* is a condition in which the linings of the bronchial tubes become irritated and swollen. The passageways to the alveoli may swell and clog with mucus. This often causes severe coughing

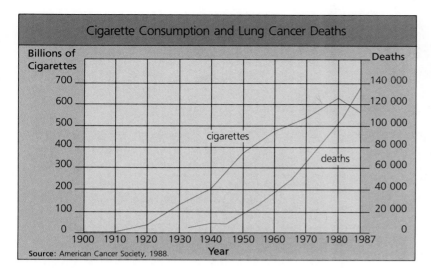

◀ **Figure 11–12**
Cigarette Consumption and Lung Cancer Deaths. As cigarette consumption has increased, so have the number of deaths caused by lung cancer. The time lag between the two curves reflects the 20 years it takes for cancer to develop. Because the rate of cigarette smoking has declined recently, a decrease in lung cancer deaths may soon follow.

and makes it hard to breathe. Bronchitis is more common in smokers than in nonsmokers.

■ *Emphysema* is a condition in which the lungs lose their elasticity. The walls of the alveoli become damaged, making the respiratory surface smaller. Emphysema causes shortness of breath. Some people with emphysema have difficulty blowing out a match. As you already know, smoking greatly increases a person's chances of getting emphysema. The damage done to the lungs by emphysema cannot be undone. Even if an emphysema victim were to quit smoking, the condition would not improve.

■ *Pneumonia* (noo MOH nyuh) is a condition in which the alveoli become filled with fluid. This prevents the exchange of gases in the lungs.

■ *Lung cancer* is a disease in which tumors (masses of tissue) form in the lungs as a result of irregular and uncontrolled cell growth. Many studies have shown a relationship between lung cancer and smoking. See Figure 11–12.

11-2 Section Review

1. What role do the nasal passages and the diaphragm play in gas exchange?
2. What is the name of the respiratory surface in humans?
3. What are the four phases of gas exchange in humans?
4. Name three respiratory diseases that are more common in cigarette smokers than in nonsmokers.

Critical Thinking

5. People with emphysema often suffer from heart problems as well. Offer an explanation for this finding. (*Identifying Causes*)

Laboratory Investigation

Control of Human Breathing Rate

Regulation of the breathing rate ensures that the rates at which oxygen (O_2) is supplied to, and carbon dioxide (CO_2) is removed from, the body increase or decrease as the body's needs increase or decrease. In this investigation, you will demonstrate the effect of carbon dioxide concentration on your breathing rate.

Objectives

- *Demonstrate* that exhaled air has a higher concentration of CO_2 than inhaled air.
- *Measure* the effects of increased concentration of CO_2 on breathing rate.

Materials

2 beakers	solid stopper
limewater	flexible plastic straws
tap water	3 Erlenmeyer flasks
carbonated water	tape
graduated cylinder	paper bag
two 2-hole stoppers	watch with second hand

Procedure *

A. Label two beakers A and B. Using a graduated cylinder, add 100 mL of limewater to each beaker. Add 30 mL of carbonated water to beaker A and add 30 mL of tap water to beaker B, the control. Note how carbonated water, which contains dissolved CO_2, changes the appearance of the limewater.

lime water

For detailed information on safety symbols, see Appendix B.

B. As shown in the first column, prepare a two-hole stopper with one straw (labeled A) below the level of the liquid and the other (labeled B) above the liquid.

C. Label one flask E for exhaled air and the other I for inhaled air. Add 100 mL limewater to each. Attach the stopper with the two straws to flask E. Place a solid stopper on flask I and set it aside.

D. Blow slowly (exhale) into straw A, which is below the surface of the limewater in flask E. Count the number of bubbles needed to turn the limewater cloudy. Record the number in a data table.

E. Remove the stopper from flask E and place it on flask I. Prepare a second 2-hole stopper with two straws, one (labeled C) inserted 6 cm below the stopper, the other (labeled D) inserted 2 cm below the stopper, as shown in the first column.

F. Using tape, connect straw B to straw C and place the second stopper on an empty flask. Inhale slowly on straw D. Count and record the number of bubbles needed to turn the limewater cloudy.

G. Relax in a chair for several minutes. Count the number of breaths you take in one minute. Repeat this process twice. Record your results in a data table.

H. Firmly hold a paper bag over your nose and mouth, and breathe normally. **Caution**: stop if you feel faint. While your partner keeps time, count the number of breaths you take in one minute. Repeat this process twice. Record your findings in your data table.

Analysis and Conclusions

1. How does limewater indicate the presence of carbon dioxide?

2. From your results, how does the concentration of CO_2 in the air we exhale compare with the concentration in the air that surrounds us? Explain.

3. How does your normal breathing rate compare with the rate while breathing into the bag?

4. What effect does breathing into a paper bag have on CO_2 concentration in the blood?

5. Do your results suggest any relationship between CO_2 levels in the blood and the breathing rate?

Chapter Review

Study Outline

11–1 Adaptations for Gas Exchange

- Gas exchange takes place by diffusion of gases across a thin, moist respiratory surface. The direction of exchange depends on the amount of the gas on each side of the respiratory surface.

- In simple organisms, such as protists and hydra, cells exchange gases directly with the environment through the cell membrane.

- In complex, multicellular organisms, where most cells are not near the surrounding environment, complex respiratory and transport systems provide for gas exchange.

- In earthworms, gas exchange takes place through the moist skin, which is well supplied with capillaries. The blood contains the respiratory pigment hemoglobin.

- In grasshoppers, gas exchange is not dependent on the circulatory system. Air enters a system of tracheal tubes which carries oxygen to and carbon dioxide from the body cells.

- In fish and many other aquatic animals, the respiratory organs are gills. Respiratory gases are exchanged as water flows over gills.

11–2 The Human Respiratory System

- In humans, the respiratory system consists of the lungs and a system of tubes that carry gases to and from the lungs.

- Air enters the human respiratory system through the nostrils, which lead into the nasal passages. The nasal passages filter, moisten, and warm the air before it reaches the lungs.

- From the nasal passages, the air travels through the pharynx and larynx and into the trachea, where ciliated mucous membranes trap foreign matter. Air continues through the bronchial tubes and into the bronchioles within the lungs.

- Within the lungs are tiny, thin-walled alveoli surrounded by capillaries. Gas exchange takes place through the walls of the alveoli and capillaries.

- Respiration in humans occurs in four phases: breathing, external respiration, internal respiration, and oxygen and carbon dioxide transport.

- The hemoglobin that is found in the red blood cells carries most oxygen from the lungs to the body tissues. Carbon dioxide diffuses out of the body cells and into the blood. From the blood, carbon dioxide is carried to the lungs.

- Cigarette smoking has been linked to respiratory disorders such as bronchitis, emphysema, and lung cancer. Other respiratory disorders include pneumonia and asthma.

Vocabulary Review

gas exchange (219)	pharynx (225)
respiratory surface (220)	larynx (225)
respiratory pigments (221)	vocal cords (225)
tracheal tubes (222)	trachea (225)
spiracles (223)	bronchi (226)
air sacs (223)	bronchial tubes (226)
gills (223)	bronchioles (226)
lungs (224)	alveoli (226)
diaphragm (225)	inhalation (227)
pleura (225)	exhalation (227)
nasal passages (225)	

A. Matching—Select the vocabulary term that best matches each definition.

1. Two-layered membrane that encloses the human lung

2. Moist surface through which the exchange of respiratory gases takes place

3. Tube through which air passes from the larynx to the bronchi

4. Phase of breathing in which air is forced out of the lungs

5. Muscle that forms the floor of the chest cavity

6. Respiratory organ of many aquatic animals

Chapter Review

7. Throat

8. Openings through which air enters and leaves the body of the grasshopper

B. Multiple Choice—Choose the letter of the answer that best completes each statement.

9. Inhaled air is filtered, moistened, and warmed in the (a) nasal passages (b) air sacs (c) larynx (d) tracheal tubes.

10. The two pairs of membranes that are stretched across the inside of the larynx in humans are called (a) nasal passages (b) tracheal tubes (c) vocal cords (d) bronchioles.

11. In humans, the exchange of respiratory gases between the air and the blood occurs through the walls of the (a) spiracles (b) larynx (c) alveoli (d) trachea.

12. The trachea divides into two cartilage-ringed tubes called (a) pleura (b) bronchi (c) bronchioles (d) alveoli.

13. The gas exchange organs in humans and some other air-breathing animals are called (a) tracheal tubes (b) pleura (c) air sacs (d) lungs.

14. The phase of breathing during which air is drawn into the lungs is called (a) inhalation (b) gas exchange (c) exhalation (d) diaphragm.

15. In the blood, hemoglobin is one of the (a) air sacs (b) respiratory pigments (c) bronchioles (d) alveoli.

Content Review

16. What are the primary characteristics of a respiratory surface?

17. How is the direction of diffusion across a respiratory surface determined?

18. In protists, how is gas exchange with the environment accomplished?

19. How do the cells of hydra obtain oxygen and get rid of carbon dioxide?

20. Why must aquatic animals move large volumes of water over their respiratory surfaces?

21. Why are respiratory pigments, such as hemoglobin, so important to respiration?

22. Why is it important for earthworms to maintain a moist skin surface?

23. Why does flooding drown earthworms?

24. Describe gas exchange in the grasshopper.

25. How do gills function in gas exchange?

26. Where is the diaphragm located in relation to the lungs?

27. How is foreign matter normally removed from the human respiratory system?

28. What advantages are lost when a person breathes through the mouth instead of the nasal passages?

29. What is the function of the epiglottis?

30. How does smoking affect the air passages?

31. Trace the path of air from the nasal passages to the alveoli.

32. Describe the relationship between the bronchioles and the alveoli.

33. How does cigarette smoking cause emphysema?

34. What causes the lungs to contract and expand?

35. Describe the transport of oxygen in the blood from the lungs to the body cells.

36. Describe the transport of carbon dioxide in the blood from the body cells to the lungs.

Graphic Organizing

For information on graphic organizers, see Appendix G at the back of this text.

37. **Compare/Contrast Matrix:** Use the matrix below to compare gas exchange in earthworms, grasshoppers, and humans. Copy the matrix onto a separate sheet and complete it.

Characteristic	Earth-worms	Grass-hoppers	Humans
respiratory surface	?	?	?
respiratory pigment	?	?	?
open or closed system?	?	?	?
gas exchange organs	?	?	?

Critical Thinking

38. List three ways in which gas exchange in protists is similar to gas exchange in humans. (*Comparing and Contrasting*)

39. Place the following organs in the order in which air passes through them on the way to the lungs: alveoli, bronchi, bronchioles, larynx, nose, pharynx, and trachea. On what other basis could the same organs be ordered? (*Ordering*)

40. How is the blood of humans like the blood of grasshoppers? How is it different? (*Comparing and Contrasting*)

41. Do you think that a store, such as a pharmacy, that sells medicines for people who are sick should also sell cigarettes and other tobacco products? Why or why not? (*Making Ethical Judgments*)

Creative Thinking

42. Design a program that might help a person stop smoking cigarettes.

43. In terms of gas exchange, explain why it is not possible to have the giant insects, such as 50 foot high grasshoppers or 40 foot long ants, that are seen in old science fiction movies.

Problem Solving

44. An investigation was carried out to determine what factors cause an increase in the rate of breathing during and after exercise. Individuals' blood and breathing rates were analyzed before and immediately after exercise, with the results shown in the table below. Make a bar graph of the data, and use it to answer the following questions: (a) What is the relationship between exercise and oxygen concentration? Between exercise and carbon dioxide concentration? (b) Why does the lactic acid concentration increase during exercise? (See page 118.)

45. The statements that follow are based upon the data in the table in question 44. For each statement, indicate whether: (a) it is a logical hypothesis supported by the data; (b) it is a hypothesis not supported by the data; or (c) it is not a hypothesis.
A. Decrease of oxygen in the blood is the stimulus for increased breathing rate.
B. An excess of carbon dioxide in the blood is the stimulus for increased breathing rate.
C. The concentration of oxygen in the blood is decreased and that of carbon dioxide increased during exercise.
D. A decrease of carbon dioxide is the stimulus for increased breathing rate.
E. A change in the concentration of some chemical substance(s) in the blood is the stimulus for increased breathing rate.
F. An increase in lactic acid is the stimulus for increased breathing rate.
G. An increase of oxygen in the blood is the stimulus for increased breathing rate.
H. An increase in the acidity of the blood is the stimulus for increased breathing rate.
I. The concentration of some substances in the blood changes during exercise.

46. On the basis of the data presented in the table *only,* which of the hypotheses suggested in question 45 is best? Why is it the best?

Projects

47. Research and write a report in the form of a feature article for a newspaper on one of the following topics: (a) scuba diving equipment; (b) artificial respiration; (c) emphysema; (d) asthma.

48. Build a balloon-lung model of the respiratory system. The following materials should be used: gallon-size plastic mayonnaise or pickle jar, one-hole rubber stopper, Y tube, two balloons, thin rubber sheeting, and string. Demonstrate how the model works.

49. Write to the local chapter of the American Lung Association for information about the effects of smoking and chewing tobacco on health. Be sure to ask for information on respiratory diseases, various types of cancer, and damage to fetuses. Prepare a bulletin board display showing the results of your study.

Analysis	Before Exercise	After Exercise
oxygen concentration	15 units/mL	10 units/mL
carbon dioxide concentration	50 units/mL	55 units/mL
lactic acid concentration	10 units/mL	35 units/mL
breathing rate	12 breaths/min	28 breaths/min

▲ A runner, sweating profusely, rounds the corner of a race track.

Excretion

12

A runner pounds out step after demanding step. Flushed and driven, his body is soaked in sweat. With the energy expended on every step, more heat is generated, causing sweat to form and evaporate from his skin. As the sweat evaporates, it cools his body. All complex animals produce excess body heat and various metabolic wastes that must be removed. The skin, lungs, kidneys, and liver help rid the body of these substances. In this chapter, you will learn about the processes by which excess heat and body wastes are removed.

Guide for Reading

Key words: excretion, urine, uric acid, urinary system, kidneys, nephrons, epidermis, dermis

Questions to think about:

📖 What is excretion?

📖 How does excretion in earthworms and grasshoppers differ?

📖 What role do the liver, the kidneys, the lungs, and the skin play in the process of human excretion?

12-1 Adaptations for Excretion

Section Objectives:

- *Define* the term *excretion* and describe how this process helps maintain homeostasis.
- *Name* the major metabolic wastes and the processes by which they are formed.
- *List* the metabolic wastes of protists and hydra and explain how each type of waste is removed from the cell.
- *Describe* and *compare* the excretory structures and products of the earthworm and grasshopper.

As an organism carries out its life processes, waste products build up in the body fluids. If these metabolic wastes were not removed from the body, the organism would die. Therefore, the organism must be able to remove metabolic wastes and other excess substances that build up over time. **Excretion** (ek SKREE shun) is the process by which these wastes and excess substances are removed from the organism. The process of excretion also removes excess heat from the body, thus helping to keep the temperature of the body constant.

In humans and other complex animals, the organs of excretion are the lungs, kidneys, liver, and skin. These organs work with the circulatory, nervous, and endocrine systems to keep the body's internal environment constant. In other words, these organ systems maintain homeostasis.

▲ **Figure 12–1**

Excretion. The skin is just one of the excretory organs of the body.

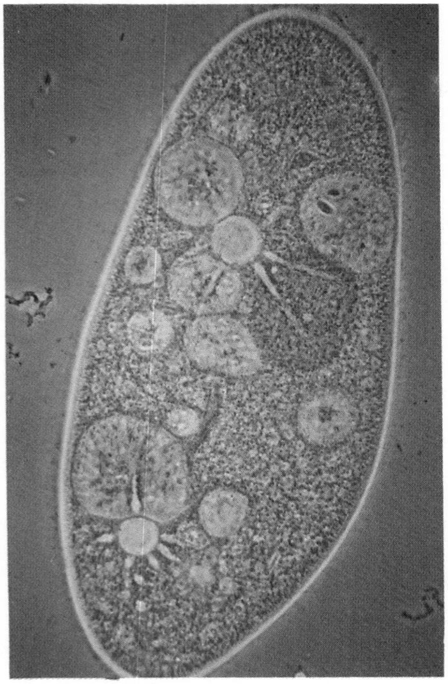

▲ **Figure 12–2**
Expanded Contractile Vacuole in Paramecium. Water balance in the paramecium is maintained by contractile vacuoles. When excess water collects in the vacuoles, the vacuoles expand, as shown here. When the expanded vacuoles are full, they contract, ejecting water from the organism.

Major Metabolic Wastes

The most important of the metabolic wastes are carbon dioxide, water, certain nitrogen compounds, and mineral salts. Carbon dioxide and water are formed during cellular respiration. Dehydration synthesis, described in Chapter 4, also produces water. Nitrogen compounds, such as ammonia, urea, and uric acid, are produced by the breakdown of amino acids. Mineral salts, such as sodium chloride and potassium sulfate, build up during metabolism. All of these wastes are poisonous in high concentrations.

Many people confuse excretion with elimination. Elimination, or *defecation,* is the removal from the digestive tract of unabsorbed and undigested food in the form of *feces.* Since these materials have never entered the body cells, they are not metabolic wastes.

Excretion in Protists

Excretion in protists is a simple process. Wastes diffuse out of the cell through the cell membrane into the surrounding water. Metabolic wastes include carbon dioxide, mineral salts, and ammonia. Ammonia (NH_3) is the chief nitrogenous waste of all microorganisms and many aquatic multicellular animals. Although ammonia is poisonous to cells, it is soluble in water. Thus, ammonia can be excreted as a waste product if there is water to wash it away.

Freshwater protists, such as ameba and paramecium, must use active transport to maintain homeostasis. Water constantly diffuses into the cell by osmosis. Because water is also produced as a byproduct of cellular respiration, excess water must be "pumped out" of the cell against a concentration gradient. The excess water collects in *contractile vacuoles.* From time to time, when it is full, the vacuole contracts, ejecting water from the cell. See Figure 12–2.

Excretion in Hydra

Hydra are small, freshwater organisms. Most of their cells are in contact with their aquatic environment. This allows the metabolic wastes, which include carbon dioxide, ammonia, and mineral salts, to diffuse directly through the cell membrane of each cell into the surrounding water.

Water tends to enter a hydra's cells by osmosis. However, no contractile vacuole has been seen in the cells of the hydra. For this reason, scientists believe that the excess water may be pumped out of the organism through the cell membrane by some other means of active transport.

Excretion in Earthworms

When most of the cells of an animal are not in contact with its surroundings, special excretory organs must remove the metabolic wastes. The excretory organs of the earthworm are the **nephridia**

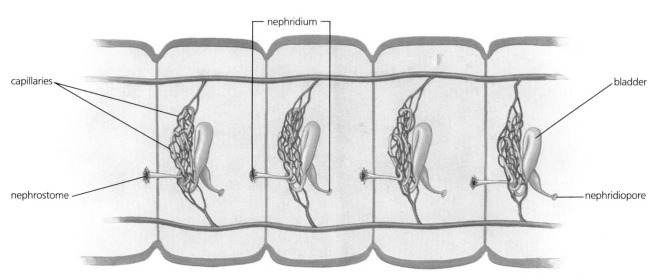

▲ **Figure 12–3**

Excretory System of the Earthworm. A pair of nephridia surrounded by capillaries is found in almost every segment of the earthworm. Fluid from the body cavity enters the nephridium, and useful substances are reabsorbed into the bloodstream. Wastes, in the form of urine, pass through the bladder and leave the body through the nephridiopore.

(nih FRID ee uh). These structures are found in pairs—one on each side—in most segments of the earthworm's body. Each nephridium extends slightly into the neighboring segment. See Figure 12–3.

Some cellular wastes diffuse directly into the fluid in the body cavity of the earthworm. This body fluid enters the nephridium at the *nephrostome* (NIH fruh stom), the funnel-shaped opening of each nephridium. The beating of cilia then moves the fluid through a tubule to the major part of the nephridium in the next segment. Here, the tubule loops several times and widens into a large bladder. The bladder drains to the outside of the body through an external opening called the *nephridiopore* (nih FRID ee oh por).

The coiled loops of the nephridium are surrounded by capillaries. As you read in Chapter 10, the blood carries metabolic wastes. Wastes from the bloodstream pass from the capillaries into the nephridium. At the same time, useful substances, such as glucose and water, pass from the body fluid in the nephridium into the blood. This exchange of wastes and useful substances is found in the excretory systems of most complex animals.

The wastes remaining in the nephridium leave the body through the nephridiopore as a dilute solution called **urine** (YUR en). The urine is made up of water, mineral salts, ammonia, and **urea** (yuh REE uh). Urea is formed from ammonia and carbon dioxide. Like ammonia, urea is soluble in water. However, it is less poisonous to cells than ammonia.

In the earthworm, carbon dioxide is excreted through the moist skin. See Chapter 11.

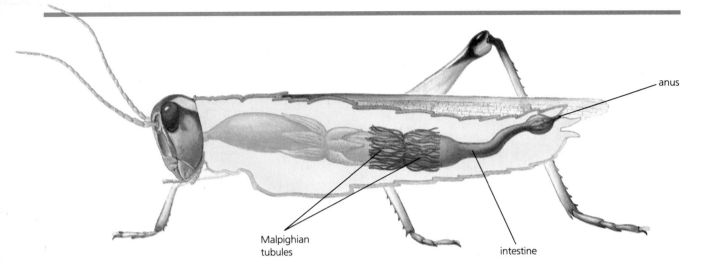

anus

Malpighian tubules

intestine

▲ **Figure 12–4**

Excretory System of the Grasshopper. The Malpighian tubules of the grasshopper remove wastes from the blood by diffusion and active transport. A dry nitrogenous waste product, uric acid, leaves the body with the feces through the anus.

Excretion in Grasshoppers

The excretory organs of grasshoppers and other insects are the **Malpighian** (mal PIG ee en) **tubules** shown in Figure 12–4. As you read in Chapter 9, insects have open circulatory systems. Thus, the slender excretory tubules are bathed directly by the blood, which moves freely within the body spaces. Wastes and other substances from the blood enter the tubules by diffusion and active transport. From the tubules, these materials pass into the intestine. Water, nutrients, and other useful substances are reabsorbed from the tubules and the digestive tract and are returned to the body fluids. The dry nitrogenous waste product, **uric** (YUR ik) **acid,** passes out of the body with the feces through the anus.

Of all wastes containing nitrogen, uric acid is the least poisonous. In fact, because it does not dissolve much in water, it is almost completely harmless. It is excreted as a solid or semisolid by birds and reptiles, as well as insects. Because so little water leaves the body in the excretion of uric acid, this type of excretion helps to save water in land animals that have a limited water supply.

Carbon dioxide diffuses from the body tissues into the tracheal tubes. From there, it diffuses out of the grasshopper through the spiracles. See Chapter 11.

12-1 Section Review

1. What is excretion?
2. List the major metabolic wastes.
3. Why does excretion in the earthworm require specialized organs?
4. What animals excrete uric acid, and how does it help them?

Critical Thinking

5. Order the following nitrogenous waste products from least to most poisonous: urea, ammonia, uric acid. (*Ordering*)

12-2 The Human Excretory System

Section Objectives:

- *Identify* the principal metabolic wastes of the human body.
- *Describe* the excretory functions of the liver.
- *Draw* and *label* the parts of the human urinary system, and describe the process of urine formation.
- *Explain* the excretory functions of the lungs and skin.
- **Laboratory Investigation:** *Examine* prepared slides of kidney cells through the microscope (p. 248).

The complex and highly developed excretory system of humans plays a major role in the maintenance of homeostasis. Carbon dioxide, urea, water, and mineral salts are the metabolic wastes of humans. The organs of excretion are the liver, kidneys, lungs, and skin. See Figure 12–5.

The Role of the Liver in Excretion

The role of the liver in digestion was discussed in Chapter 8. As an excretory organ, the liver works in several ways to regulate the makeup of body fluids.

Detoxification   The liver removes harmful substances, such as bacteria, certain drugs, and hormones, from the blood. Within the liver, these substances are changed into inactive or less poisonous forms. Thus, the liver purifies, or *detoxifies,* the blood. The inactive substances formed in the liver are returned to the bloodstream and are finally excreted from the body by the kidneys.

Overloading the liver with harmful materials, such as alcohol, can lead to a disease called *cirrhosis.* In this disease, the liver becomes overgrown with excess tissue. The excess tissue cuts down the blood flow through the liver and limits the amount of purification that it can perform. Eventually, the liver may cease to function altogether, resulting in death. Cirrhosis causes the death of about 13 000 Americans yearly.

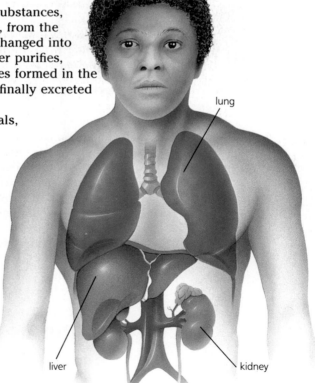

Figure 12–5

Organs of the Human Excretory System. The excretory system prevents the buildup of wastes produced by body cells. The organs of excretion in humans are the liver, kidneys, lungs, and skin. ▶

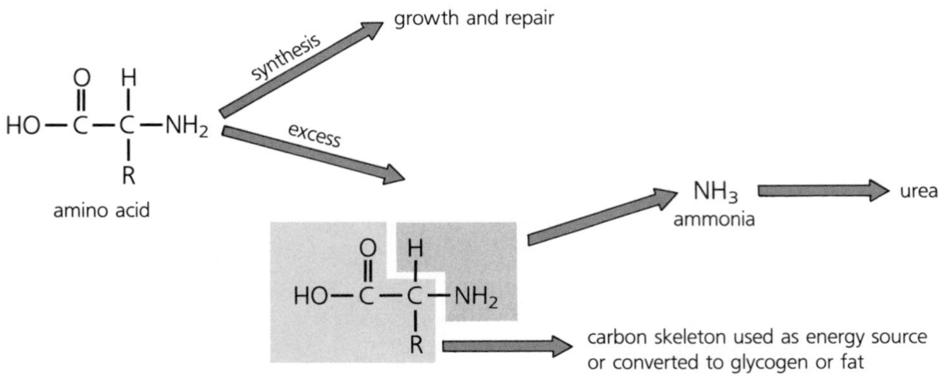

▲ **Figure 12–6**

Excess Amino Acids in the Body. Excess amino acids are broken down in the liver. The amino group is removed and converted to ammonia. The ammonia is quickly converted to urea, which is excreted from the body in urine. The carbon skeleton of the amino acid can be converted to pyruvic acid and used in cellular respiration, or it can be converted to glycogen or fat and stored.

Excretion of Bile   Bile, which is made by the cells of the liver, contains bile salts, cholesterol, and part of the hemoglobin molecule from worn-out red blood cells. Because some of the ingredients of bile are metabolic wastes, bile is considered an excretory product. Bile collects in the gall bladder and passes through the bile duct to the small intestine. There, it helps in the digestion and absorption of fats. In the last part of the small intestine, most of the bile salts are reabsorbed into the blood and returned to the liver. From the liver, they again pass to the small intestine. Thus, bile salts are reused. The rest of the bile passes into the large intestine and leaves the body in the feces. When the bile is not excreted properly, its metabolic wastes are reabsorbed into the blood, resulting in a condition called *jaundice.* Reabsorbed hemoglobin fragments in the bloodstream cause the skin to look yellow.

Formation of Urea   Amino acids are the breakdown products of proteins. Because excess amino acids cannot be stored in the body, they are broken down in the liver. The parts of the amino acids are changed into other substances. From each amino acid, the amino group (NH_2) is changed into ammonia (NH_3). The remainder of the amino acid molecule either is changed into pyruvic acid and used as an energy source in cellular respiration or is changed into glycogen or fat for storage. See Figure 12–6.

Because the ammonia produced from the amino group is very poisonous, it is changed into the less harmful substance urea by a series of enzyme-catalyzed reactions. The urea diffuses from the liver into the bloodstream. The bloodstream then carries the urea to the kidneys. The kidneys filter the urea from the blood, and it is finally excreted from the body in the urine.

The Urinary System

The **urinary** (YUR uh ner ee) **system** is made up of the kidneys, the ureters, the bladder, and the urethra. The two **kidneys** are the organs that produce urine. Urine passes from each kidney through a tube called a **ureter** (YUR et ur) to the **urinary bladder,** where it

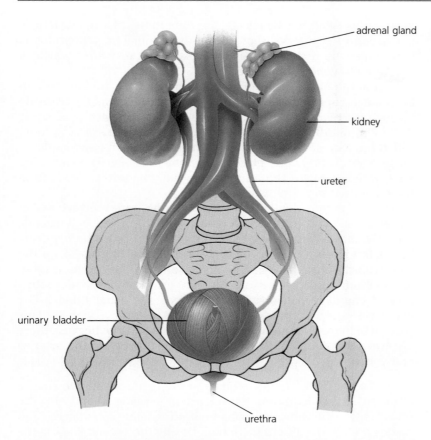

adrenal gland

kidney

ureter

urinary bladder

urethra

◀ **Figure 12–7**
The Human Urinary System. The urinary
system consists of the kidneys, ureters,
bladder, urethra, and associated blood vessels.
The adrenal glands rest on the kidneys but are
not part of the excretory system.

is stored. During urination, the stored urine travels from the
bladder to the outside of the body through the **urethra** (yuh REE
thruh). See Figure 12–7.

The kidneys are bean-shaped organs that are about 10 centimeters long. They lie against the muscles of the back in the
abdomen just below the diaphragm. The kidneys are important for
two reasons. First, they remove the wastes of cellular metabolism
from the blood. Second, they regulate the concentrations of the
substances found in the body fluids. If the kidneys cannot perform
these two functions, a person will die.

Structure of the Kidneys   The kidney has three parts. The
outer part is the *cortex*. The middle part is the *medulla* (meh DUHL
uh). The inner region is the *pelvis*. Blood is filtered in the cortex.
The medulla is made up of tubes, called *collecting ducts*. The
collecting ducts carry the filtered substances, called the *filtrate,* to
the pelvis. The pelvis is a cavity connected to the ureter. Urine
formed from the filtrate drains from the pelvis into the ureter. The
three parts of the kidney are shown in Figure 12–8.

The most important work of the kidneys—the filtering of
wastes from the blood—takes place in the **nephrons** (NEF rahnz).
Each kidney has about 1.25 million nephrons. A part of each
nephron is in the cortex. The remainder of the nephron lies in the
medulla. At one end of a nephron, shown in Figure 12–8, is the

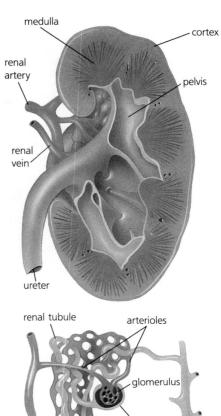

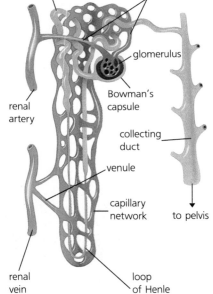

▲ Figure 12–8
The Kidney and the Nephron. The cross section of the kidney (top) shows three distinct regions. Blood is filtered in the outermost region, the *cortex*. The filtrate passes through the tubes in the middle region, the *medulla,* and drains into the ureter from the innermost region, the *pelvis.* The nephron (bottom) is the functional unit of the kidney. Each nephron is made of a glomerulus, Bowman's capsule, and renal tubule.

glomerulus (glah MER yuh lus). The glomerulus (plural, *glomeruli*), is a group of capillaries that form a tight ball. The glomerulus is surrounded by a double-walled, cup-shaped structure, which is known as **Bowman's capsule.**

From the blood in the glomerulus, substances are filtered into Bowman's capsule. The filtrate then exits Bowman's capsule through the *renal tubule,* a long tubule that empties into a collecting duct. The middle section of the renal tubule forms a long loop, called the *loop of Henle* (HEN lee), which extends into the medulla. Some reabsorption takes place in the loop of Henle. The collecting duct, which receives the filtrate from many nephrons, leads from the medulla into the pelvis.

Blood enters the kidneys through the **renal arteries** and leaves the kidneys through the **renal veins.** Each nephron has one arteriole that carries blood from the renal artery to the nephron. The arteriole branches to form the capillaries that make up the glomerulus. Before leaving Bowman's capsule, the capillaries join together into a single arteriole. The new arteriole subdivides into a second capillary network that surrounds the renal tubule. The capillaries then merge again, this time to form a venule. This pattern of blood flow around the nephrons is unusual because there are two sets of capillaries, rather than one, between the arteries and veins.

Urine Formation   Urine is made in the nephrons in two stages: filtration and reabsorption. It is important to realize that during the first stage—the filtration stage—both useful substances and wastes are removed from the blood. During the second stage—the reabsorption stage—some of the useful substances reenter the blood to be used by the body.

Filtration   To understand reabsorption, we need to look at filtration in more detail. Filtration takes place in the glomeruli and Bowman's capsules. The blood that enters a glomerulus is under pressure. The pressure forces the filtrate, which includes water, urea, glucose, amino acids, and various salts, through the thin walls of the glomerulus into Bowman's capsule. Blood cells and blood proteins, however, are too large to pass through the walls of the glomerulus. These substances remain in the blood. The filtrate that enters Bowman's capsule is like blood plasma, but it does not contain proteins.

The kidneys form about 180 liters of filtrate in a 24-hour period. However, only about 1 liter to 1.5 liters of urine (only a fraction of the filtrate) are actually produced by the kidneys in 24 hours. If all of the filtrate that is formed were excreted, the body would lose too much water along with important nutrients and salts dissolved in that water.

Reabsorption   After the filtrate has left Bowman's capsule, reabsorption occurs in the renal tubule. It is the process of *reabsorption* that reduces the volume of filtrate and returns various important substances to the blood. Normally, as the filtrate passes through the renal tubules of the nephrons, about 99 percent of the water, all of the glucose and amino acids, and many of the salts are reabsorbed. These substances are reabsorbed into the blood by the capillaries that surround the tubules. The reabsorption of water from the renal tubules is an important means of water conservation in mammals. Since most of the water is reabsorbed, the substances left in the filtrate are highly concentrated.

While water is reabsorbed by osmosis, glucose, amino acids, and salts need active transport to be reabsorbed. ATP, the energy source for active transport, is supplied by the many mitochondria found in the cells of the renal tubule. The tubules are lined with microvilli that greatly increase the surface area through which reabsorption can occur. The large area allows the reabsorption of huge amounts of water and other substances.

Most substances have what is called a *kidney threshold level.* If the concentration of a substance in the blood is greater than a certain level, the excess substance is not reabsorbed. The excess remains in the urine and is excreted from the body. For example, the blood sugar level of a person who has diabetes is so high that not all the glucose in the filtrate can be returned to the blood. As a result, glucose appears in the urine.

After reabsorption, the fluid remaining in the tubules is urine. The urine is made up of water, urea, and various salts. Urine flows from the tubules into the collecting ducts. It passes out of the kidneys through the ureters to the bladder, which is emptied from time to time through the urethra.

Sometimes, substances crystallize out of the urine in the urinary tract or kidney. These crystallized substances are called kidney stones. If the stones are too big to be passed with the urine, they must be surgically removed or shattered into small pieces with sound waves or lasers.

If one kidney stops working, the second kidney can take over its work. If both kidneys fail, however, excess fluid and wastes build up rapidly in the body, which can lead to death. In this case, an artificial kidney machine may be used to filter the blood. This process is known as *dialysis.* See Figure 12–10. The patient is connected to the machine with tubes leading into blood vessels. Blood flows from the body into the dialysis machine, where wastes are filtered out. The filtered blood is then returned to the body. By undergoing several hours of dialysis each week, a person can live for years after kidney failure.

The Lungs

The lungs are considered a part of the excretory system because they rid the body of carbon dioxide and water (in the form of water vapor). Both of these substances are the end products of aerobic cellular respiration. The lungs are discussed in greater detail in Chapter 11.

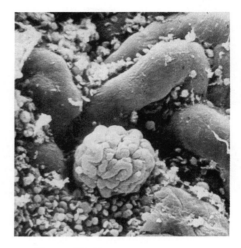

▲ **Figure 12–9**
A Glomerulus. At the center of this scanning electron micrograph is a glomerulus surrounded by renal tubules. The Bowman's capsule of the glomerulus has been removed. (Magnification 921 X)

Figure 12–10
The Kidney Dialysis Machine. This patient is undergoing kidney dialysis. When a person's kidneys do not function properly, a dialysis machine can be used to filter excess fluids and wastes from the blood. ▼

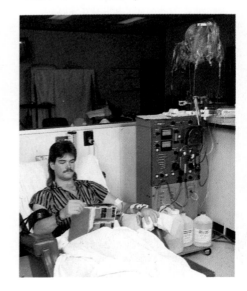

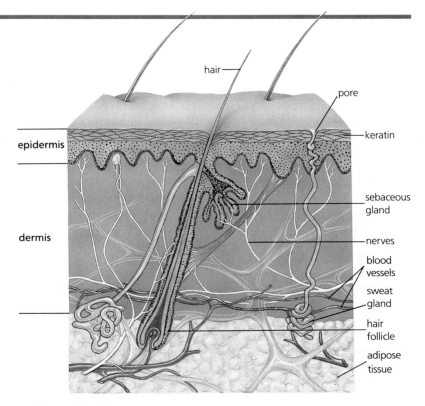

▲ **Figure 12–11**

Structure of the Skin. The epidermis acts as protection for the dermis, which contains blood and lymph vessels, sebaceous glands, sweat glands, nerves, sense receptors, and many hair follicles.

The Skin

The skin, which is made up of many different kinds of tissues, performs a number of functions. One of these functions is the excretion of wastes.

Structure of the Skin   As shown in Figure 12–11, the skin has two layers. The outer layer is the *epidermis* (ep uh DER mis) and the inner layer is the dermis.

The **epidermis** is formed of layers of tightly packed epithelial cells. The deepest portion of the epidermis is made up of rapidly dividing cells. As these cells are pushed farther and farther away from the dermis by the new cells that are forming, they receive less nourishment and die. Before dying, they produce large amounts of a tough, waterproofing protein, which is called *keratin* (KER uh tin). The outer part of the epidermis consists of these hardened, dead epithelial cells. This part of the epidermis is always wearing away. As it wears away, it is replaced by new cells from the dividing layer underneath it. The tough, waterproof epidermis acts as protection for the dermis.

The **dermis** lies below the epidermis. It is made up of elastic connective tissue. The dermis is a thick layer that supports the skin and binds it to the muscle and bone lying beneath it. Within the dermis are blood vessels, lymph vessels, nerves, sense receptors, sebaceous glands, sweat glands, and hair follicles. Beneath the dermis is a layer of fat, or *adipose,* tissue. Thin people have little fat in this layer. On the other hand, in overweight people, this adipose layer of tissue is very thick.

The **sebaceous** (sih BAY shus) **glands** produce oily secretions that provide a protective coating to the skin and hair and keep them soft and pliable. **Sweat glands** are made of tiny coiled tubes that open to the surface of the skin through holes called *pores.* Sweat is released through these holes.

Functions of the Skin   The skin keeps microorganisms and other foreign materials from entering the body. Since the outermost layer is waterproof, the skin also keeps the body from drying out. The skin excretes a small amount of urea and salts in sweat, which is 99 percent water. However, the skin's major role in excretion is the removal of excess heat.

When the body becomes too warm, extra heat is lost in two ways. First, the blood vessels in the skin open wider. This increases the blood flow through the skin's capillaries (causing a flushed appearance) and allows more heat to be given off to the air. Second, sweat begins to evaporate. As you read in Chapter 4, energy for the evaporation of sweat comes from body heat. Therefore, the evaporation of sweat cools the body by using, and thus removing, some of its heat.

The skin also helps keep heat in when the body is too cool. Blood vessels in the skin narrow slightly, reducing the supply of blood to the skin capillaries. The body also sweats less. In these ways, less heat is lost from the body. If the body is too cold, it can even produce heat by muscle tension and shivering. These ways by which the body holds a constant temperature are good examples of homeostatic control.

12-2 **Section Review**

1. Name the organs of excretion in humans.
2. Name the parts of the urinary system.
3. Describe the two stages involved in the formation of urine by the nephron.
4. In what way is the nephron's filtrate different from urine?

Critical Thinking

5. Explain the relationship between a renal tubule and Bowman's capsule. (*Relating Parts and Wholes*)

Biology and You

Q: My friends and I enjoy sunbathing during the summer. Is suntanning bad for your skin?

A: Too much sun can permanently damage your skin and lead to skin cancer. The sun emits short, intense waves called ultraviolet (UV) light. It is these waves, and not the sun's heat, that can injure your skin. Ultraviolet light can damage the DNA in your skin cells. This is why suntanning can cause aging skin and may lead to skin cancer.

Some individuals are more prone to skin cancer than others. The lighter your skin, the greater the danger. Yet, even if you are at high risk, you can protect your skin. Use a sunscreen when in the sun, especially in the middle of the day when the sun is most intense. You can also shield your skin from ultraviolet rays by covering up with clothing, a hat, or an umbrella.

Contrary to what some people think, tanning machines are *not* a safe alternative. These machines emit ultraviolet rays. You may be surprised to learn that most of the sun's harmful rays also can reach you through haze and clouds, but they cannot penetrate glass.

■ *Create a public service announcement for television warning teenagers about sunbathing.*

Laboratory Investigation

The Kidneys

The kidneys remove metabolic wastes from the blood while preventing significant losses of other blood components. This vital and remarkable feat is the work of the nephrons, the functional units of the kidneys. In this investigation, you will examine the structure of the whole kidney and its microscopic nephrons.

Objectives

- *Study* the organs that make up the urinary system.
- *Identify* the cortex, medulla, and pelvis regions of the kidney.
- *Examine* the structure of nephrons and relate it to their function.

Materials

chart of human urinary system
kidney model
prepared slides of nephrons, c.s.
microscope

Procedure *

A. From models and charts provided by your teacher, and/or from figures in your text, study the parts of the human urinary system. Make a drawing showing the kidneys, ureters, urinary bladder, and urethra.

B. Examine a model of a human kidney that shows an internal view of the organ. Using Figure 12–8 in your text as a reference, locate the cortex, medulla, and pelvis regions. Also locate the renal artery and the renal vein. Draw a longitudinal section of a kidney and label cortex, medulla, pelvis, renal artery, and renal vein.

C. Under low power, examine a prepared slide from the cortex of a kidney. Refer to the photo above. Look for circular clusters of thin, red-stained cells. These areas are the highly compact capillary networks called glomeruli. Surrounding each glomerulus is a layer of cells that forms Bowman's capsule. See Figure 12–8. Filtrate forms when pressure from the blood in the glomeruli forces plasma into Bowman's capsule.

D. Examine the well-defined circular cavities that are the most prominent feature of the tissue. These are

*For detailed information on safety symbols, see Appendix B.

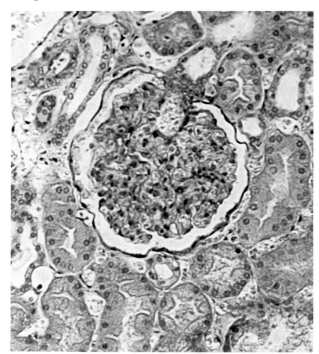

cross-sections of the tubules that make up the nephrons. See Figure 12–8 in your text. Note how the tubules are made of a single layer of cells. Through these tubules flows the filtrate produced at the site of Bowman's capsule. As the filtrate moves through the tubules, useful substances, including water, are reabsorbed into the blood of the surrounding blood vessels. As you examine the slide, look for blood vessels—arterioles, venules, and capillaries—that intertwine among the tubules.

Analysis and Conclusions

1. Trace the pathway through the body of a molecule of urea from its production in the liver to its elimination from the body in urine.

2. What is the functional relationship between the tubules of a nephron and the blood vessels that surround the tubules?

3. How does the tubular nature of a nephron contribute to its efficiency as a urine-producing unit?

Chapter Review

Study Outline

12–1 Adaptations for Excretion

■ Waste products of the metabolic processes of cells build up in the body fluids. Metabolic wastes include carbon dioxide, water, nitrogen compounds, and mineral salts.

■ Organisms must maintain homeostasis—a constant internal environment. These cellular wastes are removed from the organism by the process of excretion.

■ In simple organisms, wastes diffuse directly from the cells into the environment. Ammonia is the primary nitrogen-containing waste of microorganisms and many aquatic animals.

■ Complex multicellular organisms have specialized excretory organs, such as the nephridia, which are found in earthworms, and the Malpighian tubules, which are found in grasshoppers.

■ Urine is composed of water, mineral salts, ammonia, and urea. Urea is soluble in water and is formed from ammonia and carbon dioxide.

■ Insects, birds, and reptiles excrete the dry, nitrogenous waste product uric acid.

12–2 The Human Excretory System

■ In humans, excretion involves the liver, kidneys, lungs, and skin. The functioning of these organs is important in the maintenance of homeostasis.

■ The liver functions to detoxify harmful substances in the blood and to synthesize and excrete bile. Urea is formed in the liver by the breakdown of amino acids, and it is transported through the bloodstream to the kidneys for excretion.

■ The urinary system consists of the kidneys, ureters, urinary bladder, and urethra. The kidneys remove metabolic wastes from the blood and regulate the concentrations of substances in the body fluids.

■ In the kidneys, the nephrons produce urine by the processes of filtration and reabsorption. Urine passes from the kidneys and through the ureters to the bladder and leaves the bladder through the urethra.

■ During respiration, the lungs remove carbon dioxide and water vapor from the body. The skin excretes small amounts of urea and salt in perspiration, protects the internal tissues of the body, and helps maintain a constant body temperature.

Vocabulary Review

excretion (237)	urethra (243)
nephridia (238)	nephrons (243)
urine (239)	glomerulus (244)
urea (239)	Bowman's capsule (244)
Malpighian tubules (240)	renal arteries (244)
uric acid (240)	renal veins (244)
urinary system (242)	epidermis (246)
kidneys (242)	dermis (247)
ureter (242)	sebaceous glands (247)
urinary bladder (242)	sweat glands (247)

A. Sentence Completion—Fill in the vocabulary term that best completes each statement.

1. The excretory organs of the earthworm are known as _____.

2. The glomerulus of the nephron is surrounded by a cup-shaped structure called _____.

3. In the process of _____, metabolic wastes are removed from an organism.

4. _____ is a dry, nitrogen-containing waste product.

5. The _____ is a tight ball of capillaries found at one end of the nephron.

6. Blood enters into the kidneys by way of the _____.

7. Oily secretions are produced by the _____ in the skin.

8. Urine is stored in the _____ prior to excretion.

9. In the kidneys, the filtering of the blood takes place in the _____.

Chapter Review

B. Matching

B. Matching—Select the vocabulary term that best matches each definition.

10. Glands composed of tiny coiled tubes that open to pores on the surface of the skin

11. The excretory organs of grasshoppers and other insects

12. The tubes that connect the kidneys to the urinary bladder

13. A waste product formed from ammonia and carbon dioxide

14. Human organs that remove wastes from the blood and regulate the concentrations of substances in the body fluids

15. An excretory fluid composed of water, urea, and salts

Content Review

16. Why must metabolic wastes be removed from an organism?

17. How do protists excrete wastes?

18. Explain how water balance is maintained in freshwater protists.

19. How is excretion in hydra similar to excretion in protists?

20. How is carbon dioxide excreted by the earthworm?

21. What is the relationship between the nephridium and the capillaries of the earthworm?

22. Explain the function of the Malpighian tubules of the grasshopper.

23. Why might you expect to find excretion of uric acid by animals living in a very dry environment?

24. What is the role of the human liver in regulating the composition of body fluids?

25. What is the function of bile?

26. How do the liver and the kidneys work together within the human excretory system?

27. What are the two main functions of the kidneys?

28. Draw a diagram of the nephron and label the following parts: (a) glomerulus; (b) Bowman's capsule; (c) renal tubule; (d) loop of Henle; (e) collecting duct.

29. Compare the process of filtration in the kidneys with the process of reabsorption.

30. What is meant by the kidney threshold level?

31. Draw a diagram of the skin and label the following parts: (a) epidermis, (b) dermis, (c) keratin, (d) sebaceous gland, (e) hair follicle, (f) hair, (g) sweat gland, (h) pore.

32. What is the primary role of the skin in excretion?

33. How does the human body eliminate excess heat? How does it conserve heat?

Graphic Organizing

For information on graphic organizers, see Appendix G at the back of this text.

34. **Word Map:** Below are two incomplete word maps for the terms *kidney* and *nephron*. Copy the word maps onto a separate sheet and complete them.

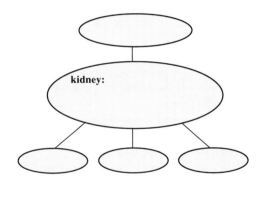

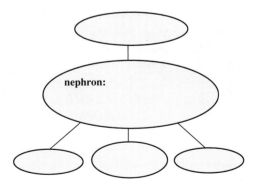

Critical Thinking

35. Contrast the process of excretion in organisms with the process of elimination or defecation. (*Comparing and Contrasting*)

36. What substances might you find in a person's urine if the walls of the glomeruli were damaged or broken? (*Relating Parts and Wholes*)

37. Why are the lungs and skin considered excretory organs? (*Identifying Reasons*)

38. If a person were in a blazing hot desert environment, how would the kidneys function to maintain the quality and quantity of body fluids? (*Predicting*)

39. Compare and contrast the major characteristics of uric acid with the major characteristics of urea. (*Comparing and Contrasting*)

Creative Thinking

40. Suggest a list of foods that a person hiking through the desert might pack in order to minimize the kidneys' job of maintaining homeostasis.

41. An organism with an increased metabolic rate could move faster, obtain more food, and defend its territory more vigorously than one without. Suggest some possible reasons why the organism's excretory requirements would limit how much its metabolic rate could increase.

Problem Solving

42. Design a controlled experiment to determine the effect of exercise on body temperature. What hypothesis will be tested in your experiment? What variable will you change? What factors will be kept constant? What response will you measure?

43. An investigation was carried out to determine the effect of drinking an excessive amount of water on urine flow. A subject drank one liter of water, and then urine was collected through a tube into a collection vessel. Urine output was measured every 10 minutes and the results were recorded on the following graph. Analyze the data and explain how the subject's urine output changed over the course of the experiment.

How does the body compensate when too much water is consumed? How might the body compensate when too little water is consumed?

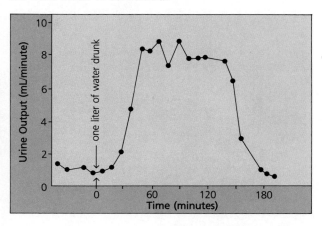

Projects

44. When a person's kidneys fail, dialysis is needed several times a week to remove wastes from the blood. For some individuals, a kidney transplant is an option. Research the advantages and disadvantages of kidney transplants, and write an article that could be used in a magazine. You might also interview someone who has had such a transplant.

45. Do library research on the harmful effects of the sun on the skin and the effectiveness of sun screens in providing protection. Report your findings to the class in an oral presentation.

46. In the course of a physical examination, the physician usually has a sample of the patient's urine analyzed. Write a report explaining what information about the patient's condition can be obtained from urinalysis.

47. Write a report on one of the following career opportunities: (a) urologist; (b) dialysis technician; (c) dermatologist.

▲ A brightly colored lorikeet maneuvers effortlessly through the air.

Support and Locomotion

13

A rainbow lorikeet glides gracefully past a cluster of yellow flowers. She lowers her feet and fans out her feathers. Motion slows, and she lands, clasping a branch. A bird is built to fly. Its body is light, its wings are strong, and its feathers catch the wind. Every animal travels in its own way. An earthworm tunnels through earth. A dolphin swims the open sea. A kangaroo bounds across a grassy meadow. In this chapter, you will find out how the design of an organism—the organization of muscles, skeleton, and certain other structures—makes motion possible.

Guide for Reading

Key words: exoskeleton, endoskeleton, bone, cartilage, skeletal muscle, ligament

Questions to think about:

📖 Why is locomotion essential to many organisms?

📖 How do simple organisms, earthworms, and grasshoppers move from place to place?

📖 How do bones and muscles work together to allow movement in humans?

13-1 Adaptations for Locomotion

Section Objectives:

- *List* the advantages of locomotion.
- *Compare* and *contrast* the two types of skeletons: exoskeletons and endoskeletons.
- *Describe* methods of locomotion in the protist, hydra, earthworm, and grasshopper.

The Advantages of Locomotion

Many types of living things are able to move on their own from one place to another. Being able to move oneself from place to place is called *locomotion*. Organisms that are capable of locomotion are said to be *motile* (MOH til). Although not all animals are motile, most animals and many protists are motile. On the other hand, plants are not motile.

Some animals that live in water, such as corals and adult sponges, cannot move from place to place on their own. They live fastened to the ocean floor or to some object. These animals are stationary, or *sessile* (SES il). By moving parts of their bodies, however, they create water currents that allow them to get food and oxygen and to carry out their life processes.

▲ **Figure 13–1**

Locomotion. This bright orange tropical crab scuttles over rocks and sand with its five pairs of walking legs.

▲ **Figure 13–2**
Exoskeleton of a Mollusk. The exoskeleton of a clam is a hard double shell composed of calcium compounds.

Being able to move from place to place offers a number of advantages for an organism.

- Locomotion makes it easier for organisms to get food. (A cougar may hunt for food over a territory of more than 160 square kilometers.)
- Locomotion allows organisms to find suitable places to live and to move away from harmful conditions in the environment. (Some kinds of fish swim away from warm, oxygen-poor water toward cooler, oxygen-rich water.)
- Locomotion allows organisms to escape enemies or to seek shelter. (Rabbits and deer escape from danger by moving quickly.)
- Locomotion allows organisms to find mates and reproduce. (Male and female salmon swim thousands of kilometers to reach their spawning grounds.)

Living things have many different methods of locomotion. In single-celled organisms, such as protists, *pseudopods* or various cell structures may be used for locomotion. In multicellular animals, locomotion always involves specialized muscle tissue. Whatever the method, the basis for nearly all protist and animal movement is *contractile proteins*—proteins that can change in length.

Muscles and Skeletons

In all but the simplest animals, locomotion uses both muscles and a skeleton to which the muscles are fastened. Muscles can exert force when they *contract,* or shorten. When they contract, they move the parts of. the skeleton to which they are fastened.

Most skeletons are made up of hard materials. If the skeleton is outside the body, enclosing the soft parts, it is called an **exoskeleton** (eks oh SKEL uh tun). Some protists and many invertebrates have exoskeletons. For example, clams, oysters, and other mollusks have hard shells made of calcium compounds. See Figure 13–2. The animal lives inside the shell, and its movement is limited. Crabs, spiders, insects, and other arthropods have exoskeletons made of **chitin** (KYT in), which is a tough, but lightweight, carbohydrate material. Exoskeletons serve as the site of attachment for muscles. In arthropods, the exoskeleton is jointed, so that it is flexible and can move in various ways. Exoskeletons protect the soft parts of the body. However, because they are not made of living cells, they cannot grow. For this reason, from time to time, arthropods shed, or *molt,* their exoskeletons and replace them with new, larger ones. See Figure 13–3. During the time between molting and growth of the new exoskeleton, the animal's soft body is unprotected.

In vertebrates, the skeleton is made of *bone* and *cartilage,* two types of connective tissue, and is located within the body walls. This type of skeleton, found inside the body, is called an **endoskeleton** (en doh SKEL uh tun). An endoskeleton does not protect the animal as well as an exoskeleton. However, because bones and cartilage contain living cells and can grow, the skeleton grows

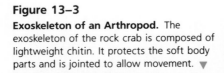

Figure 13–3

Exoskeleton of an Arthropod. The exoskeleton of the rock crab is composed of lightweight chitin. It protects the soft body parts and is jointed to allow movement. ▼

larger along with the rest of the animal. Skeletal muscles are fastened to the endoskeleton, making the movement of body parts possible. Endoskeletons are found in fish, amphibians, reptiles, birds, and mammals.

Locomotion in Protists

Among the protists, some forms have no means of locomotion, while others are highly motile. Motile protists usually move by pseudopods, cilia, or flagella.

Pseudopods   **Pseudopods** are temporary projections of the cell surfaces. The organism moves when cytoplasm flows into or out of the pseudopods. This type of locomotion is best known in amebas. It is also found in other organisms as well as in some white blood cells. Locomotion by means of pseudopods is also known as *ameboid movement.*

Studies of the ameba show that the cytoplasm within the organism is in two states. In the central part of the cell, the cytoplasm is more fluid and is called the *endoplasm.* As the ameba moves, the endoplasm flows forward through the center of the pseudopod. See Figure 13–4. At the tip, it spreads out in all directions and changes into a firmer, less fluid state. In this state, the cytoplasm is known as *ectoplasm.* The ectoplasm travels backward along the sides of the cell. Near the "back," it changes back into endoplasm and joins the forward-flowing stream. Energy for movement comes from the breakdown of ATP to ADP.

Cilia and Flagella   Protists having cilia, such as the paramecium, move quickly compared to the ameba. The paramecium is covered with thousands of short, hairlike **cilia,** whose rhythmic, oarlike

Figure 13–4
Ameboid Movement. In ameboid movement, the cell cytoplasm alternates between a more fluid endoplasm and a less fluid ectoplasm. Cytoplasm in the endoplasm flows forward into newly forming pseudopods and then changes to ectoplasm. At the rear of the cell, cytoplasm flows forward as it changes from ectoplasm to endoplasm. In this way, the cell contents move in the direction of the new pseudopods. ▼

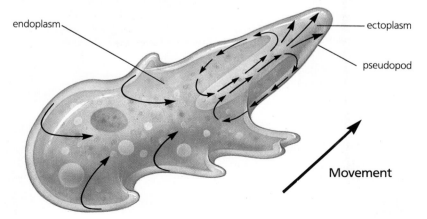

endoplasm

ectoplasm

pseudopod

Movement

Figure 13–5

Locomotion in the Paramecium. The entire outer surface of a paramecium is covered with cilia that beat rhythmically to propel it through the water. The beating of the cilia is coordinated by a network of fibrils. ▶

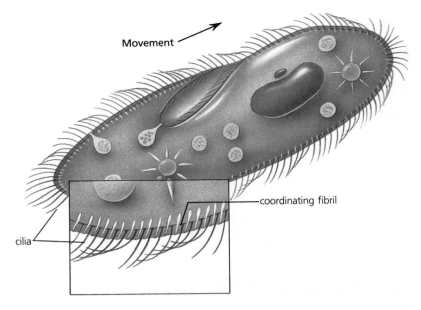

Movement

coordinating fibril

cilia

beating moves the organism through the water. See Figure 13–5. A system of tiny fibers connects the cilia at their bases. These tiny fibers, or fibrils, cause the cilia to beat in the right order.

Flagella are like cilia except that they are longer, and there are usually only one or two per cell. Euglena is a protist that moves by means of one long, thin, flagellum. See Figure 13–6. The whiplike movements of the flagellum pull the euglena through the water.

Critical Thinking in Biology

Comparing and Contrasting

When you compare and contrast, you examine two or more items to see how they are alike and how they differ. By doing so, you may notice characteristics you would otherwise have overlooked. You may also see new relationships between the items.

1. Pictured at right are an ameba (top) and a macrophage (bottom). List some similarities between the two.

2. List some differences between an ameba and a macrophage.

3. Suppose you were looking at an unknown cell under a microscope. Which characteristics from your lists could help you determine whether the cell was an ameba or a macrophage?

4. Imagine that you see an unfamiliar bird while on a hike. You pull out your field guide to help you identify the bird. How would comparing and contrasting help you to identify the bird?

5. Think About It Explain your thought processes as you answered questions 1 and 2.

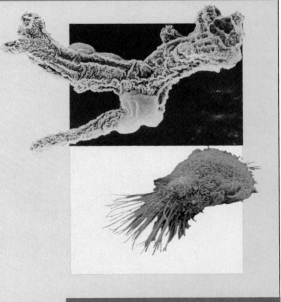

Locomotion in the Hydra

The hydra has cells specialized for contraction. Although the hydra tends to stay in one place, contractile fibers allow it to move about in several ways. The presence of mucus-secreting cells and ameboid cells allow it to "glide" along on its base. It can move quickly by somersaulting its base completely over its tentacles. It can also inch along by bending over and fastening its tentacles to an object and then pulling its base closer. The hydra can float upside down in the water by making an air bubble on its base.

Locomotion in the Earthworm

The earthworm uses muscles to burrow through the soil. Within its body wall are two layers of muscles. See Figure 13–7. An outer layer of circular muscles goes around the worm. An inner layer of longitudinal muscles goes along the full length of the body. When the circular muscles contract, the worm becomes longer and thinner. When the longitudinal muscles contract, the body becomes shorter and thicker. Within the earthworm, the body cavity is filled with fluid. When the surrounding muscle layers contract, the fluid stiffens the body of the worm and allows it to push through the soil.

On almost all of the earthworm's segments, there are four pairs of tiny bristles called **setae** (SEE tee). In locomotion, the setae in the rear of the earthworm hook into the ground while the circular muscles contract. This lengthens the body and pushes the worm forward. Then, the setae near the front of the worm anchor into the ground, and the setae in the rear relax. The longitudinal muscles contract, shortening the body and pulling forward the hind end of the worm. The earthworm moves by repeating these movements over and over again.

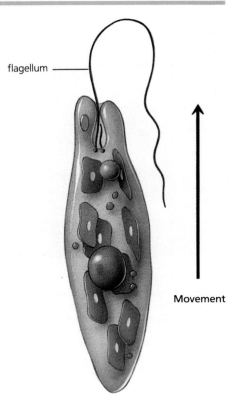

flagellum

Movement

▲ **Figure 13–6**
Locomotion in the Euglena. Movement of the long, whiplike flagellum pulls the organism through the water.

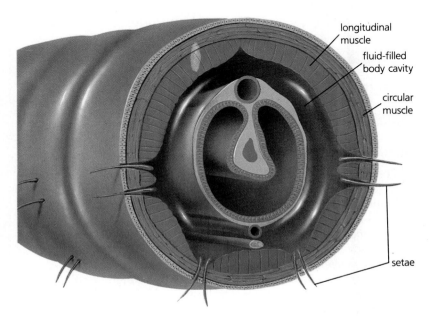

longitudinal muscle

fluid-filled body cavity

circular muscle

setae

◄ **Figure 13–7**
Locomotion in the Earthworm. Both circular and longitudinal muscle layers can be seen in this cross section of an earthworm. When the outer layer of circular muscles contracts, the worm lengthens and becomes thinner. When the inner layer of longitudinal muscles contracts, the worm becomes shorter and thicker.

▲ **Figure 13–8**

A Short-Horned Grasshopper Jumping. The first two pairs of legs of the grasshopper are used for walking. The powerful hind legs are used in jumping. The tough outer wings protect the delicate inner wings that are used for flying.

Locomotion in the Grasshopper

The body of the grasshopper is covered by an exoskeleton of chitin. The exoskeleton is divided into plates separated from each other by flexible joints. This arrangement allows the grasshopper to move freely. Grasshoppers can walk, jump, and fly. See page 721, Figure 33–15.

Like other insects, the body of the grasshopper has three major divisions—the *head, thorax* (THOR aks), and *abdomen* (AB duh men). Fastened to the thorax are three pairs of jointed legs. The first two pairs are used for walking, while the powerful hind pair is used for jumping. A grasshopper can jump more than 20 times its body length. Also fastened to the thorax are two pairs of wings. The outer pair is hard and protects the delicate inner pair, which is used in flying. The powerful muscles used in flight are fastened to the exoskeleton of the thorax. The muscles have no direct connection with the wings but move the wings by changing the shape of the body wall of the thorax.

The muscles of a grasshopper work in pairs. When one muscle of a pair contracts and bends a joint, the other muscle relaxes. When the second muscle contracts, extending the joint, the first muscle relaxes.

13-1 Section Review

1. What is locomotion?
2. What is an exoskeleton? An endoskeleton?
3. Name some means used by protists for locomotion.
4. Name the two layers of muscles used in locomotion in the earthworm.

Critical Thinking

5. An exoskeleton is sometimes referred to as a "coat of armor." In what ways is this a good analogy? (*Reasoning by Analogy*)

13-2 The Human Musculoskeletal System

Section Objectives:

- *Describe* the functions and structure of the bones and cartilage found in the human musculoskeletal system.
- *Name* the major parts of the human skeleton and the types of joints found in it.
- *Describe* the structure of skeletal muscle and explain how voluntary movement is accomplished.
- *Compare* skeletal muscle with smooth muscle.
- **Laboratory Investigation:** *Examine* the skeletal and muscular structure of a vertebrate forelimb (p. 266).

Bones and Cartilage

Bone is a type of connective tissue that is hard and inflexible. The bones of the human skeletal system serve a number of different purposes.

■ They serve as sites of attachment for skeletal muscles, and they serve as levers that make body parts move when these muscles contract.

■ They give the body its general shape and support body structures.

■ They protect delicate structures, such as the brain, spinal cord, heart, and lungs.

■ They serve as storage places for minerals, such as calcium and phosphorus.

■ They serve as the places where red blood cells and some white blood cells are produced.

Bone is made up of living bone cells, connective tissue fibers, and inorganic compounds. It is an active tissue. There is a constant destruction of old tissue and laying down of new tissue. A basic part of the structure is **collagen** (KAHL uh jen), a protein material with great strength. When bones are being formed, living cells called *osteoblasts* (AHS tee uh blasts) secrete collagen and certain polysaccharides. The collagen forms fibers that are bound together by the polysaccharides, which act as cement. Bone is formed when calcium and phosphate ions from the body fluids combine, forming calcium phosphate. The calcium phosphate precipitates as crystals within the mass of collagen fibers and cement. The hardness and heaviness of bone are due to the presence of the calcium phosphate. The osteoblasts are trapped in small cavities inside the bone substance to form bone cells called **osteocytes** (AHS tee uh syts).

In the bone, the osteocytes are arranged in a series of smaller and smaller circles with a common center. See Figure 13–9. In the center of each series of circles, there is a cavity called the **Haversian** (huh VER zhun) **canal,** which contains blood vessels and nerves. Tiny canals connect the osteocytes to each other and to the Haversian canal. The blood vessels within the Haversian canals carry oxygen and nutrients to the bone cells and remove wastes. If a bone is broken, the osteocytes become active, producing new bone tissue to heal the wound.

The outside of a bone, except at its ends where it connects to other bones, is covered by a tough membrane called the **periosteum** (pehr ee AHS tee um). The main purpose of the periosteum is to make new bone for growth and repair. The periosteum is also the point at which muscles are fastened to bones. This membrane contains blood vessels and nerves that enter the bone.

There are two types of bony tissue—*compact bone* and *spongy bone*. They are made of the same material, but compact bone is very dense and strong, while spongy bone is more porous. Most bones contain both types of tissue, as shown in Figure 13–9.

Biology and You

Q: Do people shrink as they grow older?

A: Older adults become noticeably shorter as they age, although the whole body does not actually shrink. As you age, your bones lose minerals and living materials. As calcium strengthens and hardens bones, losing calcium can soften bones in your neck and back. If these bones compress or fracture, height decreases.

Osteoporosis, thinning of the bones, is a common disorder. Not only can it shorten your height, but it can also cause your bones to break easily. Many older adults break wrists and hips that are weakened by osteoporosis. Although the disease is most common among women over fifty, it occurs among men also.

You probably have been told, "If you want to grow strong, drink your milk." Milk and other dairy products are rich in calcium. If you provide your body with lots of calcium when young, your bones will grow strong. Strengthening your bones now will help prevent problems like osteoporosis as you age. Regular exercise may also help strengthen your bones.

■ *Research and list foods that are high in calcium. Then create a calcium-rich menu for a day.*

Figure 13-9

Internal Structure of Bone. The photo (left) shows a cross section of a human thigh bone. Surrounding the central Haversian canals are concentric circles of cavities in which osteocytes are found. The Haversian canals contain blood vessels and nerve cells and are connected to the osteocytes by tiny canals. The drawing (right) shows the structure of a long bone. ▶

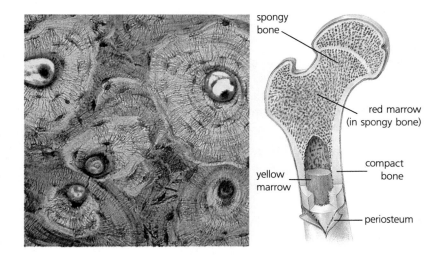

Some of the bones of the body are hollow. There is also a great deal of space in spongy bone. These spaces are filled with a soft tissue called **marrow** (MAR oh). There are two types of marrow—*red marrow* and *yellow marrow.* Red marrow makes red blood cells, platelets, and some types of white blood cells. In adults, red marrow is found in the spongy bone of the vertebrae, ribs, breastbone, cranium, and long bones. Yellow marrow is made of fat cells. In adults, it is found in the hollow center of long bones.

Cartilage (KART il idj), like bone, is a type of connective tissue. While bone is rigid, cartilage bends easily. In the embryo, most of the skeleton is cartilage. As the embryo develops, minerals are laid down, and much of the cartilage slowly changes into bone. This process, called **ossification** (ahs if fih KAY shun), goes on into adulthood. The bones of small children contain more cartilage than the bones of adults. Children's bones are therefore more elastic and not as easily broken. In adults, cartilage is found at the ends of ribs, at joints, and in the nose and outer ear. Cartilage gives support while still allowing some bending motion. It allows the bones to bend more easily at joints and cushions against impact or pressure.

The Human Skeleton

Parts of the Skeleton   The adult human skeleton contains 206 bones. See Figure 13-10. The skeleton has two main parts: the *axial* (AK see ul) *skeleton* and the *appendicular* (ap en DIK yuh ler) *skeleton.*

The **axial skeleton** is made up of the skull, vertebrae, ribs, and breastbone. The upper part of the skull, the **cranium** (KRAY nee um), houses and protects the brain. The rest of the skull is made up of the facial and jaw bones. The **spinal column,** or *backbone,* has 33 bones called **vertebrae** (VERT uh bree). The vertebrae are separated from each other by disks of cartilage. The disks act as shock absorbers and allow the spine to bend. The ribs are fastened at the back to the upper vertebrae and at the front of the breastbone, or

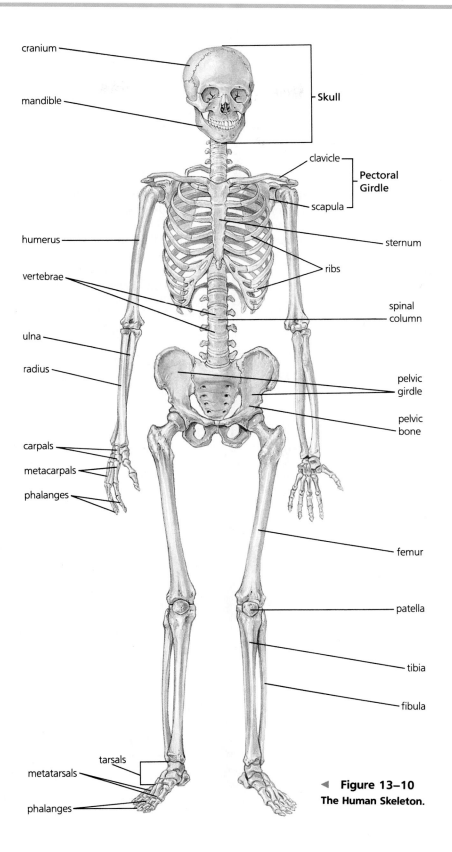

cranium

mandible

Skull

clavicle

Pectoral
Girdle

scapula

humerus

sternum

vertebrae

ribs

ulna

spinal
column

radius

pelvic
girdle

pelvic
bone

carpals

metacarpals

phalanges

femur

patella

tibia

fibula

tarsals

metatarsals

phalanges

◄ **Figure 13–10**
The Human Skeleton.

Science, Technology and Society

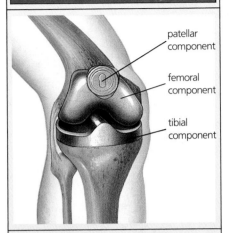

patellar component

femoral component

tibial component

Technology: Artificial Joints

For millions of Americans, everyday movement is difficult. These people suffer from arthritis, a painful inflammation of the joints. Today, the surgical replacement of diseased joints can mean a new freedom from pain for many of these arthritis sufferers.

Over half a million joints are replaced with artificial joints each year. Knees and hips are the most frequently replaced joints. The knee replacement joint is made up of three components (see above). Two curved metal plates are arranged side by side. They rotate in a plastic dish attached to the bone and allow the leg to move. Many artificial joints used today can last as long as 15 years. With the development of new materials, researchers are hopeful that the lifetime of these replacement joints will be extended significantly.

■ *Aside from arthritis sufferers, who else might stand to benefit from developments in artificial joint technology?*

sternum. The space enclosed by the sternum, ribs, and backbone is the *chest cavity.* Within the chest cavity, the heart and lungs are supported and protected by the ribs and sternum.

The **appendicular skeleton** is made up of arm and leg bones and the *pectoral* (PEK tuh rul) *girdle* and the *pelvic girdle.* The shoulder blades (scapula) and collar bones (clavicle) make up the pectoral girdle. It connects the arms to the spine. The pelvic girdle is made up of the hip bones and connects the legs to the spine.

Joints A point in the skeleton where bones meet is called a **joint.** There are several different types of joints in the body. See Figure 13–11. Joints in which bones are tightly fitted together, as in the cranium, are **immovable joints.** Most joints, however, are movable. **Hinge joints,** such as those at the elbow and knee, permit back-and-forth motion. **Ball-and-socket joints,** such as those at the shoulder and hip, allow movement in all directions. In this type of joint, the ball-shaped end of one bone fits into the cuplike hollow, or socket, of another bone. **Pivot** (PIV it) **joints,** such as those at the base of the skull, allow side-to-side as well as up-and-down movement. **Gliding joints,** such as the wrists or the joints between the vertebrae, allow some bending and twisting movements.

At movable joints, bones are held together by tough, fibrous bands of connective tissue called **ligaments** (LIG uh ments). A fluid, called *synovial* (sih NOH vee ul) *fluid,* is secreted into movable joints. This fluid acts as a lubricant and reduces friction at the joint.

Figure 13–11

Types of Joints. Hinge joints permit back-and-forth motion. Ball-and-socket joints permit a wide range of movement. Gliding joints offer limited flexibility in all directions. The bones of the cranium are joined by immovable joints. Pivot joints allow side-to-side and up-and down movement. ▼

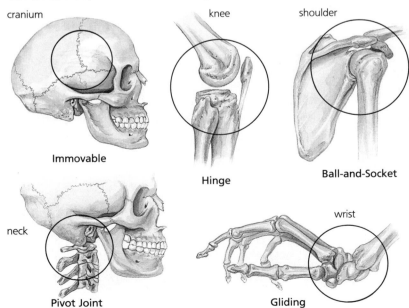

cranium

knee

shoulder

Immovable

Hinge

Ball-and-Socket

neck

wrist

Pivot Joint

Gliding

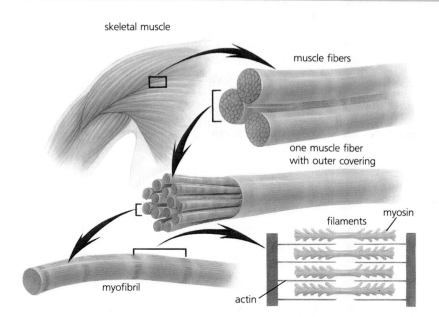

skeletal muscle

muscle fibers

one muscle fiber
with outer covering

filaments

myosin

myofibril

actin

◀ **Figure 13–12**

Structure of Skeletal Muscle. A skeletal muscle is composed of a bundle of muscle fibers. Each muscle fiber is made up of a bundle of myofibrils. A myofibril consists of protein filaments arranged in an overlapping pattern.

Skeletal Muscle

Skeletal, or *striated* (STRY ay ted), **muscles** are used in locomotion and all other voluntary movement. They are fastened to the bones of the skeleton. Skeletal muscle tissue is not made up of clearly defined, separate cells. Instead, during development, cells fuse together, forming individual *muscle fibers*. A skeletal muscle is made up of bundles of muscle fibers bound together by connective tissue.

Under a light microscope, the muscle fibers appear striped, or striated. That is, they show alternating bands of light and dark. See Figure 13–12. Electron microscope studies have shown that each fiber is actually a bundle of smaller fibers, called **myofibrils** (my oh FY brilz). Each myofibril is made up of still finer protein filaments, one thick and one thin. The thick filaments are *myosin*. The thin ones are *actin*. The two types of filaments are arranged in an overlapping pattern that makes the whole muscle fiber look striped.

According to the *sliding filament theory* of muscle contraction, the muscle fibers shorten when the actin slides over the myosin. As the overlap increases, the fiber shortens. See Figure 13–13. Cross bridges between actin and myosin allow the fibers to exert a pull. Energy for the sliding of the filaments is supplied by ATP. ATP is produced in the mitochondria of many of the muscle fibers.

Voluntary Movement

All voluntary movement is started and coordinated by impulses from the brain and spinal cord. Voluntary movement includes any action that is under conscious control. For example, picking up a pencil and throwing a baseball are voluntary movements. Voluntary movements require the action of skeletal muscles.

Figure 13–13

Contraction of Striated Muscle. According to the sliding filament theory of muscle contraction, the muscle fibers shorten when the actin and myosin filaments that make up the myofibrils slide over one another, increasing the amount of overlap. Muscle relaxed (top); muscle contracted (bottom). ▼

Relaxed Muscle

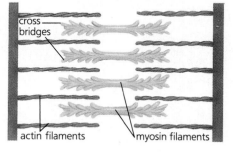

cross bridges

actin filaments

myosin filaments

Contracted Muscle

Skeletal muscles are fastened to bones by strong fibers of connective tissue called **tendons.** Muscles can pull when they contract, but they cannot push when they relax. Therefore, they must always work in *antagonistic* pairs. The two muscles in an antagonistic pair move a bone in opposite directions. The bending and extending of the arm at the elbow show how muscles and bones work together to produce movement. See Figure 13–14. When the *biceps* (BY seps) muscle on the front of the upper arm contracts, the arm bends. Because the biceps muscle bends, or flexes, the joint, it is called a **flexor** (FLEK ser). Whenever the biceps contracts, the *triceps* (TRY seps) muscle in the back of the arm relaxes, making it possible for the arm to bend. When the triceps muscle contracts, the biceps muscle relaxes, and the arm is extended, or straightened. Because the triceps muscle extends the joint, it is called an **extensor** (ek STEN ser). Throughout the body, antagonistic pairs of muscles cause the bones of the skeleton to move.

■ Careers *in Action*

Athletic Trainer

Athletes, young and old, amateur and professional, are injured at times. Working with athletes, athletic trainers try to prevent injuries and to treat those that do occur.

Athletic trainers work in high schools and colleges, for professional sports teams, or as private consultants. To help prevent injuries, trainers design exercise programs that strengthen muscles and increase endurance. Some trainers are responsible for selecting and maintaining sports equipment and creating special protective devices such as braces and pads.

For athletes who are prone to injury, trainers may use tapes, braces, or bandages to help support a weak joint. When injuries happen, the trainer works with the doctor to set up treatment programs that may include ice, bandages, heat, whirlpool baths, and rest. Next, the trainer develops and monitors a program of exercises and massages to improve the flexibility and strength of the injured area.

Problem Solving Suppose that your school has hired you as an athletic trainer. You are asked to figure out why there have been many injuries during sports practice. You notice students do not warm up before starting to play. You also note that more than the usual number of students suffer sprains and muscle cramps.

1. From your observations, propose a hypothesis to explain why so many students are being injured.
2. What changes would you make to prevent sprains and cramps during practice? Why?
3. A baseball pitcher has read that it is important to warm up by "stretching both muscles in a pair." Explain what this means.
4. List the protective gear that each of these players should wear: (a) softball catcher, (b) ice-hockey player.

■ **Help Wanted:**
Athletic trainer. Bachelor's degree, postgraduate internship, and national certification required. Contact: National Athletic Trainers Association, P.O. Drawer 1865, Greenville, NC 27834.

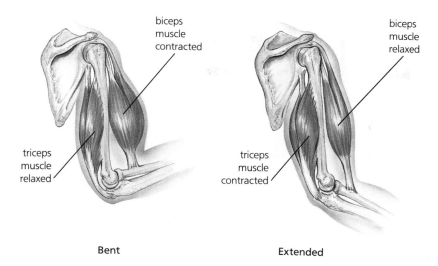

biceps
muscle
contracted

triceps
muscle
relaxed

biceps
muscle
relaxed

triceps
muscle
contracted

Bent   Extended

◀ **Figure 13–14**

Muscles of the Upper Arm. The biceps and triceps muscles of the upper arm work as an antagonistic pair. When the biceps muscle in the front of the arm contracts (left), the triceps muscle in the back of the arm relaxes and the arm bends. When the triceps muscle contracts (right), the biceps muscle relaxes and the arm straightens.

As long as you are conscious, your skeletal muscles are never completely relaxed. Instead, your brain keeps all your muscles partly contracted. This is called **muscle tone.** Muscle tone keeps the muscles ready for the powerful contractions of movement. It also maintains posture by keeping the muscles of the back and neck partly contracted.

Smooth Muscle

In addition to skeletal muscle, which is under voluntary control, the body also contains muscle that works involuntarily, or without conscious control. This involuntary muscle tissue is called **smooth muscle.** It is found in the walls of the digestive organs, in the walls of arteries and veins, in the diaphragm, and in other internal organs. The cells of smooth muscle are long and overlap to form sheets of muscle rather than bundles of fibers. Smooth muscle does not look striated. See Figure 13–15.

Cardiac muscle is a type of muscle that is found only in the heart. Cardiac muscle was discussed in Chapter 9.

Figure 13–15

Smooth Muscle. This micrograph shows the smooth muscle fibers of the urinary bladder. (Magnification 630 X). ▼

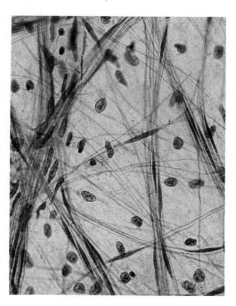

13-2 Section Review

1. What is the difference between bone and cartilage?
2. Name the two main parts of the skeleton.
3. What is a joint?
4. What are the three types of muscles in the human body?

Critical Thinking

5. Explain the central role played by actin-myosin cross bridges in muscle contraction. (*Relating Parts and Wholes*)

Laboratory Investigation

Anatomy of the Chicken Wing

The limbs of vertebrates are complex structures, adapted for a variety of movements. In this investigation, you will examine the muscles, bones, and joints of a chicken wing to relate these structures to their functions.

Objectives

- *Observe* the muscles, bones, and joints of a chicken wing.
- *Compare* the structure of the chicken wing with that of the human arm.

Materials

raw chicken wing	scissors
latex gloves	scalpel
dissecting pan	blunt probe
dissecting pins	forceps
paper towels	

Alternate Approach: A model or a diagram of a bird wing may be used as the basis for an alternative procedure to the one below.

Procedure 🔒 ☞ *

A. Put on the latex gloves and examine the external characteristics of the wing. Carefully bend the wing to see how many major sections it has. Draw a diagram of the external structure. Label the upper arm, elbow, lower arm, wrist, and hand.

B. Place the wing in the dissecting pan. Use your dissecting tools to remove as much of the skin as possible. Be careful not to damage the muscles and other structures. Draw a diagram of the chicken wing with the skin removed. Label the upper arm, elbow, lower arm, wrist, hand, thumb, fingers, and muscles.

C. Examine the muscles. In the upper arm, there are two groups of muscles. Pull on each of these muscles, and observe what happens to the rest of the wing. Repeat this step for the muscles in the lower arm.

D. Locate the tough, shiny, ribbonlike structures at the ends of muscles. These are called tendons. Notice where the tendons attach the upper and lower arm muscles to the bones. Sketch the tendons in your drawing from step B.

*For detailed information on safety symbols, see Appendix B.

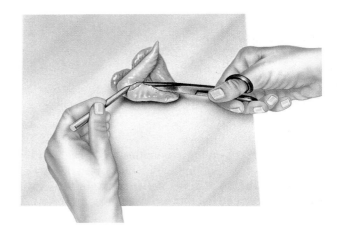

E. Remove the muscles and tendons to expose the bones of the upper and lower arm. Locate the shiny, ribbonlike structures connecting two bones. These are ligaments. Also, locate the shiny, white substance covering the ends of the bones. This is cartilage. Feel its texture. Draw a diagram of the chicken wing with the muscles removed. Label the humerus (upper arm bone), radius (smaller lower arm bone), ulna (larger lower arm bone), ligaments, and cartilage.

F. Examine the elbow joint. Move the arm bones back and forth to observe the motion allowed by this joint.

G. Study a model of the human skeleton or refer to Figures 13–10 and 13–11. Compare the structure of the human arm with that of the chicken wing.

Analysis and Conclusions

1. What happens to the wing when you pull on each of the upper and lower arm muscle groups?

2. Draw the muscles of the wing and indicate those that would need to contract in order to stretch out the wing.

3. Describe the freedom of motion allowed by the elbow joint. How many ligaments hold it in place?

4. What similarities can you see between the chicken wing and the human arm? How are the chicken wing and the human arm adapted to serve different functions?

Chapter Review

Study Outline

13-1 Adaptations for Locomotion

■ Locomotion is the ability to move oneself from place to place. In all but the simplest animals, locomotion involves muscles and a skeleton to which the muscles are attached.

■ The two types of skeletons are exoskeletons, which are found in some protists and many invertebrates, and endoskeletons, which are found in all vertebrates.

■ Among protists, locomotion is usually carried out by pseudopods (as in the ameba), cilia (as in the paramecium), or flagella (as in the euglena).

■ The hydra, although often sessile, has specialized contractile cells that make movement possible. Mucus-secreting cells allow it to glide along on its base.

■ The earthworm uses its longitudinal and circular muscles, its fluid-filled body cavity, and its setae for locomotion. Setae are tiny bristles that hook into the ground as the worm's muscles contract or relax.

■ The grasshopper has a flexible, jointed exoskeleton that enables it to move freely. Three pairs of legs and two pairs of wings allow it to walk, jump, and fly.

13-2 The Human Musculoskeletal System

■ The human skeleton is made up of bones and cartilage. Bones allow movement, give the body shape and support, protect internal organs, store minerals, and serve as a site of blood cell production.

■ Bone is living tissue made up of bone-forming cells called osteocytes, connective tissue fibers, and inorganic compounds. Within some hollow bones is a soft tissue called marrow.

■ Cartilage is a flexible connective tissue. In children, some cartilage slowly changes into bone in a process called ossification.

■ The axial skeleton includes the skull, vertebrae, ribs, and breastbone. The appendicular skeleton includes arm and leg bones, the pectoral and pelvic girdles.

■ A joint is a point in the skeleton where bones meet. Most joints are movable to some extent.

■ Skeletal muscle is made up of bundles of muscle fibers. Muscle contraction occurs when the fibers shorten. Muscles act in antagonistic pairs, with one muscle contracting and the other relaxing.

■ Smooth muscle works without conscious control. It is found in the walls of arteries and veins and in some internal organs. It is made up of long, overlapping cells that form sheets of muscle.

Vocabulary Review

exoskeleton (254)

chitin (254)

endoskeleton (254)

pseudopods (255)

cilia (255)

flagella (256)

setae (257)

bone (259)

collagen (259)

osteocytes (259)

Haversian canal (259)

periosteum (259)

marrow (260)

cartilage (260)

ossification (260)

axial skeleton (260)

cranium (260)

spinal column (260)

vertebrae (260)

appendicular skeleton (262)

joint (262)

immovable joints (262)

hinge joints (262)

ball-and-socket joints (262)

pivot joints (262)

gliding joints (262)

ligaments (262)

skeletal muscles (263)

myofibrils (263)

tendons (264)

flexor (264)

extensor (264)

muscle tone (265)

smooth muscle (265)

A. Analogies—Select the vocabulary term that relates to the single term in the same way that the paired terms are related.

1. pelvic girdle *is to* appendicular skeleton as skull *is to*

_____

2. euglena:_____ as paramecium:cilia

Chapter Review

3. shoulder:ball-and-socket joint as elbow: _____

4. involuntary movement:smooth muscle as voluntary movement:_____

5. clam:_____ as human:endoskeleton

6. rigid:bone as flexible:_____

7. biceps:flexor as triceps:_____

B. Definitions—Replace the italicized definition with the correct vocabulary term.

8. *Bone cells* are arranged in a series of smaller and smaller circles with a common center.

9. *A tough, lightweight carbohydrate material* makes up the exoskeleton of arthropods.

10. The *bone cavity* contains the blood vessels and nerves that serve the osteocytes.

11. In certain protists, a *temporary projection of the cell surface* allows the organism to move.

12. The *soft tissue that fills the hollow spaces in bone* is of two kinds: red and yellow.

13. *Strong fibers of connective tissue* attach skeletal muscles to bone.

14. *Muscle tissue that is not under voluntary control* is made up of sheets of individual cells that are not striated.

15. *The process by which cartilage is replaced by bone* goes on into adulthood in humans.

Content Review

16. How does the capacity for locomotion benefit an organism?

17. Why do organisms with exoskeletons usually molt from time to time?

18. How does the ameba move?

19. Compare locomotion in paramecia with locomotion in euglena.

20. What are the four methods by which hydra can move?

21. Describe how setae function in helping an earthworm move.

22. Identify the parts of the grasshopper that are used in locomotion, and give the function of each.

23. What are the functions of bones?

24. Describe the role of osteoblasts in human bone formation.

25. How do broken bones heal?

26. What three functions does the periosteum serve?

27. How does the skeleton of a child differ from that of an adult?

28. Of what parts do the human axial and appendicular skeletons consist?

29. Name and compare the motion of the four types of movable joints. Give an example of each.

30. What produces the striped appearance of striated muscle?

31. Describe the sliding filament theory of muscle contraction. Draw a labeled diagram to support your description.

32. Why must muscles work in antagonistic pairs?

33. Briefly describe the muscle movements involved in bending and extending an arm at the elbow.

34. How does the structure of smooth muscle differ from that of striated muscle?

35. In which parts of the body is smooth muscle found?

Graphic Organizing

For information on graphic organizers, see Appendix G at the back of this text.

36. **Concept Map:** Copy the incomplete concept map below onto a separate sheet of paper. Fill in the missing concepts and the missing linking words.

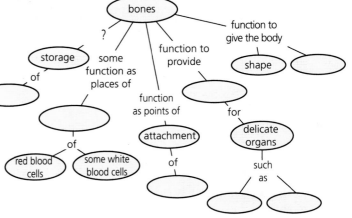

Critical Thinking

37. Compare and contrast exoskeletons and endoskeletons. (*Comparing and Contrasting*)

38. Classify the following bones as either part of the axial or the appendicular skeleton: carpals, cranium, femur, humerus, mandible, metatarsals, phalanges, ribs, sternum, vertebrae. (*Classifying*)

39. If the biceps muscle were paralyzed, what arm motion would not be possible? Explain. (*Relating Parts and Wholes*)

40. Suppose that after adding a chemical compound to a culture of normally fast-swimming paramecia you found the paramecia unable to move. They were in no other way affected. What is the most probable cause of their inability to swim? (*Identifying Causes*)

41. Based on the examples provided in the chapter, make a general statement about how *all* muscles work. (*Generalizing*)

Creative Thinking

42. If smooth muscle had to be consciously controlled, as skeletal muscle does, what types of activities would you have to "think" about?

43. Aquatic animals need less skeletal support than land animals of similar size. Why do you think this is so?

Problem Solving

44. Design a controlled experiment to determine whether calcium phosphate plays a role in controlling bone strength. (Hint: Calcium phosphate can be dissolved in dilute acid.) What is the variable in your experiment? What factors are kept constant? What response is measured?

45. A scientist wanted to find out if the distance that teenagers can hike is affected by environmental temperature. The results of an investigation are presented in the graph below. Summarize the relationship shown in the graph. Can you draw any other conclusions from the data? What factors should the scientist have kept constant for the experiment to be valid?

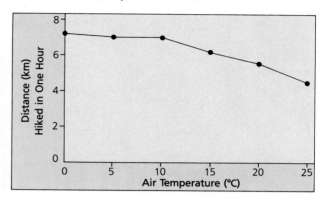

Projects

46. Prepare a skeleton of a chicken or a fish. First cook the chicken or fish; remove the skin and meat; allow the bones to dry; then glue and wire them together. Mount the skeleton on a board. **Caution:** Do not touch any hot objects.

47. Choose any sport, and research injuries common to athletes in that sport. You may wish to speak with a sports medicine specialist or physical therapist in gathering your information. Find out about the types of injuries, their causes, prevention, and treatment. Present your findings in an oral report to the class. Include illustrations or three-dimensional models in your presentation.

48. Place a chicken bone in vinegar for four to five days. Observe the bone and describe any changes that occur. Write up your observations in a report. Include diagrams if necessary.

▲ Signals carrying information and controlling movements travel along nerve cells in the fox's body.

Nervous Regulation

14

A red fox stops, listens, sniffs, and lightly paws the grass. A frightened field mouse darts off. The fox gives chase, zig-zagging in pursuit. Such immediate reactions to stimuli are the function of nerve cells in the body. These nerve cells allow an organism to sense and react to its environment. In this chapter, you will learn about nerve cells and how they transmit signals throughout the body.

Guide for Reading

Key words: stimulus, neuron, dendrites, axon, synapse, threshold, neurotransmitters
Questions to think about:
📖 How does an organism respond to changes in its environment?
📖 What changes take place as an impulse moves along a nerve?

14-1 The Regulatory Process

Section Objectives:

- *Explain* the functions of a nervous system and describe its basic types of structures.
- *Define* the term *stimulus*.
- *Describe* the neuron's structure and the function of its parts.
- *Name* the different types of nerves and describe their functions.

Functions of Regulation

An organism's environment is always changing. Some of the changes take place outside the body. These changes may include a change in temperature, the appearance of food, and the appearance of an enemy. Changes also take place within the organism. For example, the amount of a waste product may increase; a disease-causing organism may enter the body, and the supply of a needed substance may decrease.

To stay alive, an organism must respond to these changes. It must maintain homeostasis. To do so, the organism must keep all the factors of its internal environment within certain limits.

Responses usually are not single, independent events. An organism is always responding to a wide variety of changes that take place both inside and outside its body. In fact, the various life activities of an organism are in themselves examples of complex responses. These responses must be *regulated*. That is, they must be controlled in amount and directed to the right place. They must also be *coordinated*. That is, they must be made to take place in the right order or relationship.

▲ **Figure 14–1**

A Nervous Response. A well-developed nervous system allows this owl chick to respond to a perceived threat by assuming a complex defensive posture.

In one-celled organisms and some simple multicellular organisms, regulation and coordination of responses are functions of each cell as a whole. The ability of a cell to respond to its environment is called **irritability.** In more complex multicellular animals, a *nervous system* and an *endocrine system* control the regulation and coordination of responses. In this chapter and in Chapter 15, you will learn how nervous systems work. In Chapter 16, you will read about the human endocrine system.

Mechanisms of Nervous Regulation

A nervous system is a network of specialized cells, known as *nerve cells,* that carry messages, or **impulses,** throughout the organism. Along with the nerve cells, two other types of body structures—receptors and effectors—interact with this network. **Receptors,** or *sense organs,* are specialized structures that are sensitive to certain changes, physical forces, or chemicals both inside and outside the organism. Stimulation of a receptor causes impulses to be carried, or transmitted, over a pathway of nerve cells. These impulses finally reach an **effector,** which is a specialized structure that responds to the commands of the nervous system. For example, in humans, an effector is either a gland or a muscle. If the effector is a gland, it will respond to the impulse by either decreasing or increasing its activity, depending on the nerve pathways the impulse has followed. If the effector is a muscle, a nerve impulse will cause it to contract.

Anything that causes a receptor to start impulses in a nerve pathway is called a **stimulus** (STIM yuh lus). The stimulus causes electrical and chemical changes in the receptor. These changes start the nerve impulses. Thus, there are three basic events in nervous regulation. First, a stimulus activates a receptor. Second, impulses are started in associated nerve pathways. Third, an effector responds to the impulse.

Although these three steps sound simple, a nerve pathway is not a straightforward connection from a particular receptor to a particular effector. In most animals, each nerve pathway crosses and interconnects with many other pathways. Impulses from a single receptor are usually carried to a number of different nerve pathways. Impulses reaching an effector are the result of the combination and interaction of many impulses from many different pathways.

Multicellular animals have several different types of receptors. In general, each type of receptor is sensitive to a different type of stimulus. For example, a particular receptor may respond to a certain stimulus, such as heat. Other receptors may respond to cold, light, sound, pressure, or chemicals. The way in which each receptor works is described in detail in the next chapter.

All but the simplest animals have a **brain,** a specialized group of nerve cells that controls and coordinates the activities of the nervous system. The more complex the organism is, the more complex are the structure and functions of the brain.

Structure of Neurons

The *nerve cell,* or **neuron** (NOO rahn), is the basic structure in the nervous systems of all multicellular animals. Neurons can send both electrical and chemical (electrochemical) impulses. Being able to send impulses is a property of the nerve cell membrane. Nerve impulses do not pass through the cytoplasm of neurons. They are transmitted only along the cell membrane.

A nerve cell usually is made of three basic parts: a cell body, dendrites, and an axon. These are illustrated in Figure 14–2. The **cell body** contains the nucleus and the cell organelles. The metabolic activities that take place in all cells are carried out in the cell body, which also controls the growth of the nerve cell. Materials that are needed for the maintenance of the nerve cell are made in the cell body. They are then moved to other parts of the cell where they are needed.

The **dendrites** (DEN dryts) are short, highly branched fibers that receive impulses. Dendrites generally conduct impulses toward the cell body. In some neurons, the dendrites branch out around the cell body, giving the cell a bushy appearance.

The **axon** (AK sahn) is usually a long, thin fiber that extends from the cell body. Axons usually carry impulses away from the cell body and send them either to other neurons or to effectors. Axons range in length from a part of one centimeter to more than one meter. Sometimes, either the axon or the dendrite of a neuron is called a *nerve fiber.*

Many vertebrate axons are surrounded by cells known as **Schwann** (shwahn) **cells.** On some axons, the Schwann cells produce layers of a white, fatty substance called **myelin** (MY uh lin). The myelin forms a covering around the axon, and axons having such a covering are said to be *myelinated.* At places along a myelinated axon, there are gaps in the myelin that expose the axon membrane to the surrounding medium. These gaps, which are between neighboring Schwann cells, are called the *nodes of Ranvier* (RAHN vee ay). See Figure 14–3.

Figure 14–2
Structure of a Neuron. ▼

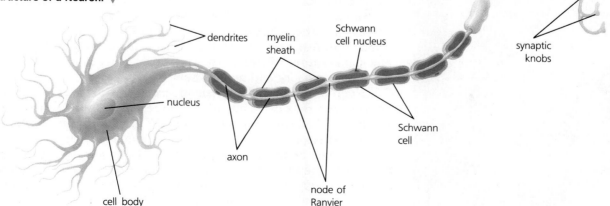

Figure 14–3
Cross Section of a Myelinated Axon. The myelin, which covers the axon, is produced by Schwann cells. ▶

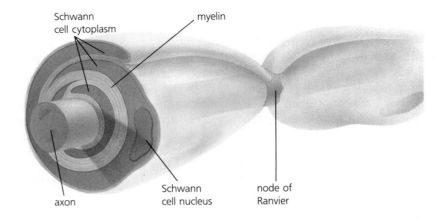

Schwann cell cytoplasm

myelin

axon

Schwann cell nucleus

node of Ranvier

Unlike other cells in the body, the nerve cells of mature animals cannot divide, so neurons cannot be replaced. If, however, the cell body of the neuron is unhurt, damaged axons and dendrites outside the brain and spinal cord can grow back.

The Synapse

The axon of a neuron usually has no branches along its length, but it may have a great many branches at its end. Each of these *terminal branches* almost touches another cell. The place between the terminal branch of a neuron and the membrane of another cell is called a **synapse** (SIN aps). The synapse has a microscopic gap between the end of the terminal branch and the neighboring cell. Impulses are carried across this gap from the axon to the neighboring cell. The structure of a synapse and the way in which impulses are carried across it are described on page 279. Each axon may have one or more synapses with as many as 1000 other neurons. Axons from other neurons may also make contact with these cells. All of these synapses make the interconnections and impulse pathways of a typical nervous system very complex.

Types of Nerves and Neurons

Figure 14–4
Structure of a Nerve. Nerves are made up of bundles of neurons bound together by connective tissue. Note the blood vessels that provide nutrients for the nerve fibers. ▼

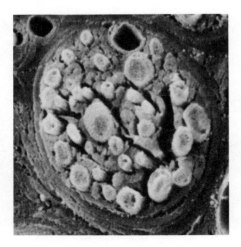

Nerves are bundles of axons or dendrites that are bound together by connective tissues. See Figure 14–4. Nerves are called *sensory nerves* if they carry impulses from receptors to the spinal cord and brain. They are *motor nerves* if they carry impulses from the brain and spinal cord to effectors. They are *mixed nerves* if they are made up of both sensory and motor fibers.

Neurons are usually grouped according to what they do. **Sensory neurons** carry impulses from receptors to the spinal cord and brain. **Motor neurons** carry impulses from the brain and spinal cord to effectors, usually muscles. **Interneurons,** or *associative* (uh SOH shee ay tiv) *neurons,* relay impulses from one neuron to another in the brain and spinal cord. Most of the neurons in the human nervous system are interneurons. Figure 14–5 shows how the structure of each kind of neuron is slightly different.

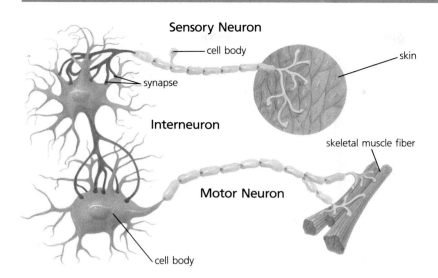

Sensory Neuron
cell body
synapse
skin
Interneuron
skeletal muscle fiber
Motor Neuron
cell body

◀ **Figure 14–5**
Pathway of Nerve Impulses. Sensory neurons receive stimuli and trigger impulses in other neurons. Motor neurons carry impulses toward effectors, such as skeletal muscle. Interneurons, which are found in the brain and spinal cord, relay impulses from one neuron to another.

14-1 Section Review

1. What are the three types of structures found in a true nervous system?
2. What happens when a receptor is stimulated?
3. List the three parts of a nerve cell.
4. What is a synapse?

Critical Thinking

5. What would happen if the motor neurons in one of your legs stopped functioning? *(Predicting)*

14-2 The Nerve Impulse

Section Objectives:

- *Describe* the electrical state of a resting neuron and the function of the sodium-potassium pump.
- *List* the changes that occur as an impulse travels along an axon.
- *Explain* how the nervous system distinguishes between stimuli of different types and strengths.
- *Identify* the structures of a synapse and describe the transmission of an impulse across a synapse and at a neuromuscular junction.

The Resting Neuron

It is the difference in electrical charge between the outer and inner surfaces of the nerve cell membrane that makes the transmission of a nerve impulse possible. When the neuron is resting (not transmitting an impulse), the outside of the membrane has a net positive charge, and the inside has a net negative charge. See Figure 14-6.

Figure 14–6

Electrical State of a Resting Neuron. As a result of the action of the sodium pump, the outside of the membrane of a resting neuron has a net positive charge, and the inside of the membrane has a net negative charge. ▶

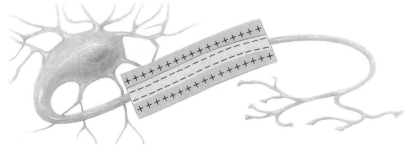

The cell membrane is said to be electrically *polarized* because there is a difference in electrical charge between its outer and inner surfaces. The polarization is caused by different concentrations of certain ions outside and inside the cell. While some of the concentration differences result from the selective permeability of the membrane, most of the differences are the result of the active transport of ions across the membrane. The ions that take part in the polarization of the nerve cell membrane are mainly sodium ions and potassium ions, both of which have a positive electrical charge.

The nerve cell membrane pumps sodium ions out of the cell and potassium ions into the cell by means of active transport. The active transport mechanism that performs this pumping action is called the **sodium-potassium pump.** In its resting state, the nerve cell membrane is freely permeable to potassium ions but not to sodium ions. As a result, the potassium ions pumped into the cell tend to diffuse back out. Because the sodium ions pumped out of the cell cannot diffuse freely through the membrane, the number of them increases outside the cell. As a result, an excess of positive charge (due to sodium ions) builds up outside the membrane. An excess of negative charge is left behind inside the membrane.

The Nerve Impulse

Membrane Changes in the Area of the Impulse   The arrival of an impulse from a neuron or a stimulus from a receptor starts a nerve impulse in the membrane of a neuron. At the place on the neuron where the impulse is started, the permeability of the membrane to sodium ions suddenly increases. Because there are more sodium ions outside the membrane than inside, sodium ions diffuse rapidly to the inside of the membrane. This flow of positive sodium ions reverses the polarization of the membrane. In the area of the impulse, the inside of the nerve cell membrane becomes positively charged; the outside becomes negatively charged. See Figure 14–7.

This reversal of polarity takes place in only a small area of the membrane. However, it results in a flow of electrical current that affects the permeability of neighboring areas of the membrane. Sodium ions rush through these new regions of increased permeability, causing the polarization in these regions to become reversed. In this way, the reversal of polarization travels over the entire length of the nerve cell. The nerve impulse is the reversal of

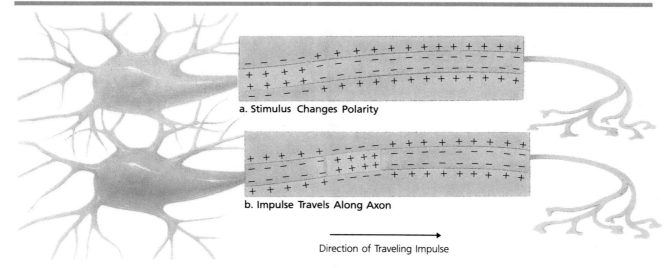

a. Stimulus Changes Polarity

b. Impulse Travels Along Axon

Direction of Traveling Impulse

polarization. The passage of the impulse along an axon is similar to a relay race. Once the impulse moves to the next section of the neuron, the first section returns to its original polarized state.

In the area of the nerve impulse, the high permeability of the cell membrane to sodium ions lasts for only a part of a second. It then returns to normal, which stops the sodium ions from diffusing across the membrane. The diffusion of potassium to the outside of the membrane, together with the action of the sodium pump, restores the normal distribution of ions. Once this happens, the polarity of the membrane is returned to normal, with a positive charge outside and a negative charge inside.

Following the passage of an impulse, there is a brief period during which the nerve cell membrane cannot be stimulated to carry impulses. This time, which lasts only a few thousandths of a second, is called the **refractory period.** When it is over, the membrane is again ready to carry impulses.

Rate of Impulse Conduction   The rate at which impulses travel depends on two factors: the size of the nerve fiber, and whether or not it has a myelin covering. In small fibers without myelin, the nerve impulse travels at a rather slow two meters per second. In large myelinated fibers, it may travel more than 100 meters per second.

Myelinated fibers carry impulses more quickly because the impulse travels in "jumps" from one node of Ranvier, where the axon is bare, to the next. This is called *saltatory* (SAL tuh tor ee) *conduction.* Because myelin prevents the flow of ions, depolarization occurs only at the nodes that are highly sensitive. Less active transport is needed for restoring the normal distribution of ions after the impulse has passed. Thus, saltatory conduction is faster and uses less energy.

Nerve Cell Thresholds   For an impulse to be started in a nerve cell, the stimulus must have a certain minimum strength. Each nerve cell has a minimum level of sensitivity, or **threshold.** If the

▲ **Figure 14–7**

A Nerve Impulse. In the area of an impulse, the nerve cell membrane becomes permeable to sodium ions, which enter the cell. This causes a reversal of the polarity of the membrane. The area of reversed polarity is the nerve impulse, and it travels quickly down the axon membrane.

strength of the stimulus is below that threshold, the stimulus cannot start impulses in the neuron. Any stimulus above the threshold will start impulses in the neuron. All the impulses transmitted by a given neuron are alike. That is, they are all the same "size," and they pass along the neuron at the same rate. Thus, a neuron works on an "all-or-none" basis. Either an impulse is started or it is not started, depending only on whether the stimulus is above or below the threshold level. The situation is like the firing of a gun. The gun does not fire until enough force is exerted on the trigger.

Distinguishing Strength and Type of Stimulus   If all nerve impulses are basically alike, how does an organism know what type of stimulus caused the impulses or how strong the stimulus was? For example, why does touching a hot stove feel different from touching a warm surface? How do you tell the difference between a bright light and a loud sound?

The strength of a stimulus is measured by two effects. First, a stronger stimulus causes more impulses to be transmitted each second. See Figure 14–8. That is, the impulses follow each other more closely. Second, different neurons have different thresholds. Some need a stronger stimulus than others to transmit an impulse. When a stimulus is stronger, both low-threshold and high-threshold neurons will transmit impulses.

Recognition of the *type* of stimulus depends on the particular pathways that carry the nerve impulse. Each type of receptor is sensitive to a certain type of stimulus. For example, light-sensitive receptors in the retina of the eye transmit nerve impulses only when light strikes them. Impulses from the retina travel along the optic nerve to a part of the brain that interprets them as sight. Artificial stimulation of the optic nerve causes a person to "see"

Figure 14–8

Strength of a Stimulus. A strong stimulus (hot object, bottom) triggers more impulses in more neurons than a weak stimulus (warm object, top). These differences enable the brain to determine the strength of the stimulus. ▼

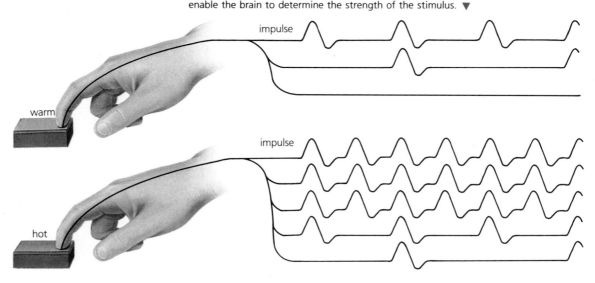

flashes of light. Sound waves, on the other hand, have no effect on the receptors in the eye. Instead, if the sounds are within the range of human hearing, they trigger receptors inside the ear. The impulses then travel to the brain by way of the auditory nerve. When these impulses reach the brain, they are interpreted as sound.

Transmission at the Synapse

As you read earlier, there is a tiny gap, called a synapse, between the end of one neuron and the dendrite or cell body of another cell. The structure of a synapse is shown in Figure 14–9. At the synapse, the axon ends in a *synaptic knob.* When an impulse arrives at the synaptic knob, it must be carried across a narrow space to the membrane of the neighboring cell. This narrow space is sometimes known as a synaptic gap.

The transmission of impulses across the synaptic gap is a chemical process. Within the synaptic knob are many small sacs, *synaptic vesicles,* that contain special chemicals called **neurotransmitters.** Among the most common of these are *acetylcholine* (uh seet uh KOH leen) and *norepinephrine* (NOR ep uh NEF rin). When an impulse reaches the synaptic knob, some of the neurotransmitters are released into the synaptic gap. The neurotransmitter diffuses across the synaptic gap and starts impulses in the neighboring nerve cell by changing the permeability of its membrane. To do this, the neurotransmitters react with special receptor proteins in the membrane of the dendrites.

Note that it is not the nerve impulse that crosses the synaptic gap. Instead, it is a chemical compound—the neurotransmitter—that is sent across the gap. Each impulse that reaches a synapse causes the release of a certain amount of neurotransmitter. When the impulses are arriving quickly (representing a stronger initial stimulus), more neurotransmitter is released into the synaptic gap. This greater amount of neurotransmitter acts as a stronger stimulus on the neighboring neuron, and the neuron then carries more impulses per second. In this way, information about the strength of the original stimulus is passed across the synapse and down the nerve pathway. As soon as the neurotransmitter has done its work, it must be removed from the synaptic gap to clear the way for new signals. This is usually done by enzymes present in the synaptic gap. These enzymes quickly break down the molecules of neurotransmitter after the neuron has responded to them.

Usually, neurotransmitters are released only by the ends of axons, and they exert their effects only at specialized receptor sites. This means that information usually travels in only one direction across synapses. This direction is from axons of one neuron to the dendrites or cell bodies of another. Thus, synapses control the direction in which information flows over nerve pathways.

Different types of neurons release different neurotransmitters. Some neurons release *excitatory neurotransmitters.* These chemicals start impulses in their neighboring neurons. Acetylcholine,

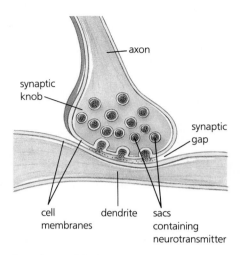

cell membranes   dendrite   sacs containing neurotransmitter

synaptic knob   axon   synaptic gap

▲ **Figure 14–9**

Structure of the Synapse. When a nerve impulse reaches the synaptic knob, some neurotransmitters are released into the synaptic gap. The neurotransmitters diffuse across the synaptic gap and initiate nerve impulses in the neighboring neuron.

norepinephrine, and the amino acids histamine and glutamic acid are excitatory neurotransmitters. Still other neurons release neurotransmitters that block, or *inhibit,* the start of impulses in neighboring neurons. *Inhibitory neurotransmitters* include serotonin, epinephrine, and the amino acid glycine. Thus, while some synapses transmit impulses from one neuron to the next, other synapses block the transmission of impulses.

As you have learned, the axon of a single neuron may form 1000 or more synapses. These may include many synapses on the same neuron. The dendrites of one neuron may also have synaptic connections with 1000 or more other neurons. Thus, the dendrites of a single neuron receive impulses from many neurons. Some of these impulses may be excitatory, while others may be inhibitory. The cell body totals or averages these impulses. If the overall results are excitatory, impulses are sent down the axon to the next set of synapses. If the results are inhibitory, no impulses are sent. Thus, in a nerve pathway, stimulation of certain neurons results in the inhibition of certain other neurons. A great deal of the complex behavior of an organism results from the great number and variety of synaptic circuits that are formed when neurons are "switched" on and off.

Neuromuscular Junctions

Impulses pass from motor neurons to muscles at special points of contact called **neuromuscular** (noor oh MUS kyoo ler) **junctions.** See Figure 14–10. The axons of motor neurons end in structures called *motor end plates.* Like synaptic knobs, motor end plates contain neurotransmitters. When impulses reach the motor end plates, they cause the release of the chemical transmitter acetylcholine. The acetylcholine diffuses across the gap between the end of the axon and the membrane of the muscle cell. Then, it combines with receptor molecules on the muscle cell membrane. The acetylcholine increases the permeability of the muscle cell membrane to sodium, causing impulses to travel along the muscle cell membrane. These impulses cause the muscle cell to contract. As in the synapses between neurons, enzymes quickly destroy the acetylcholine at the neuromuscular junction.

Drugs and the Synapse

Many poisons and drugs affect the activity of chemical transmitters at synapses. Nerve gas, *curare* (kyoo RAH ree), *botulin* (BAHCH uh lin) *toxin* (a bacterial poison), and some insecticides are poisons that interfere with the functioning of acetylcholine at neuromuscular junctions. These poisons cause muscle paralysis. If the muscles of the respiratory system become paralyzed, death follows.

Drugs that affect the mind and the emotions or alter the activity of body systems also act on synapses. *Stimulants* are drugs that speed up the body's activity. Overuse can cause heart damage and other problems. Among the stimulants, *amphetamines* (am FET uh meenz), or "uppers," produce their effects by binding to certain receptors. This causes short-lived feelings of well-being and excitement, followed by depression. *Caffeine* (kah FEEN), which is found in coffee, tea, and some cola drinks, increases synaptic transmission. This can result in sleeplessness and nervousness.

Depressants are drugs that slow down body activities. *Barbiturates* (bar BICH uh ritz), or "downers," produce a depressant effect by blocking the formation of norepinephrine.

Some of the mind-altering or hallucinatory drugs, such as *LSD* ("acid") and *mescaline* (MES kuh lin), interfere with the effect of serotonin, an inhibitory transmitter.

14-2 **Section Review**

1. What name is given to a neuron that is not sending a nerve impulse?
2. What is the sodium-potassium pump?
3. What crosses the synapse when a nerve impulse is traveling over a pathway of nerve cells?
4. What are neuromuscular junctions?

Critical Thinking

5. What would happen if the enzymes that break down neurotransmitters were not present at a given synapse? *(Predicting)*

14-3 Adaptations for Nervous Regulation

Section Objectives:

- *Describe* the responses of protists to stimuli.
- *Compare* and *contrast* the nervous systems of the hydra, earthworm, and grasshopper.
- **Laboratory Investigation:** *Observe* how planaria respond to various stimuli (p. 284).

Regulation in Protists

Although protists do not have true nervous systems, they are able to respond to certain stimuli in a coordinated way. Amebas have no specialized sense receptors, but they can tell the difference between food and nonfood and move away from such things as strong light and harmful chemicals. How these responses are brought about is not yet understood.

Some protists have specialized filaments that work in a manner similar to the neurons of more complex animals. In the paramecium, a system of interconnected fibers, found at the bases of the cilia, controls the cilia's beating. The paramecium can respond to

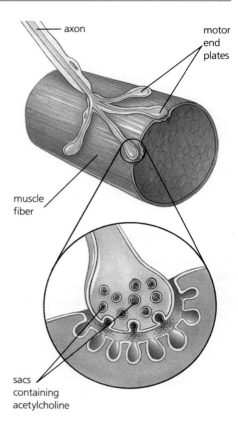

▲ **Figure 14–10**

Structure of the Neuromuscular Junction. When a nerve impulse reaches the motor end plates of a motor neuron, acetylcholine is released. The acetylcholine diffuses across the synaptic gap and initiates impulses that cause the muscle cell to contract.

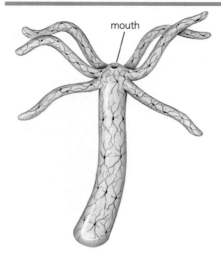

mouth

▲ **Figure 14–11**
Nervous System of the Hydra. The nerve net of the hydra allows the muscles of the organism to react to stimuli in a coordinated manner.

various stimuli. It can move toward food or away from strong acids and can change direction to stay away from solid matter in its path. Some protists have organelles that are sensitive to certain stimuli and initiate responses in the organism.

Regulation in Hydra

The nervous system of the hydra is in the form of a **nerve net.** See Figure 14–11. In this system, the nerve cells form an irregular network between the two layers of the body wall. This network connects special receptor cells in the body wall with muscle and gland cells. There is no organized center, such as a brain or nerve cord, to control and coordinate the nerve impulses. Instead, when a stimulus is received by any part of the body, impulses spread slowly from the stimulated area throughout the nerve net. Thus, all the muscle fibers in the organism respond, but the response shows coordination. For example, when a tentacle touches food, the impulses travel slowly through the entire organism. In response, the animal stretches toward the food, and the tentacles work together to capture the food and stuff it into the mouth.

Regulation in the Earthworm

The nervous system of the earthworm is more complex. It includes a **central nervous system** and a **peripheral nervous system.** See Figure 14–12. The central nervous system is made up of a "brain" connected to a pair of solid, ventral nerve cords. The nerve cords enlarge into *ganglia* (GANG lee uh) in each segment. A **ganglion** (GANG lee un) is a group of cell bodies and interneurons that switch, relay, and coordinate nerve impulses. The so-called "brain" is actually a pair of ganglia joined together.

The peripheral nervous system is made of the nerves branching from the central nervous system and passing to all parts of the

Figure 14–12
Nervous System of the Earthworm. The nervous system is made up of a brain and two ventral nerve cords. Sensory and motor nerves branch from the nerve cords. ▼

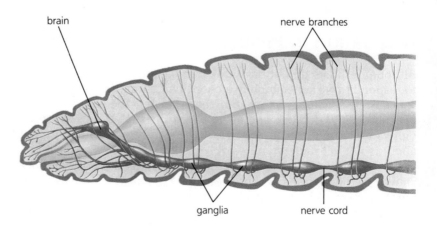

brain

nerve branches

ganglia

nerve cord

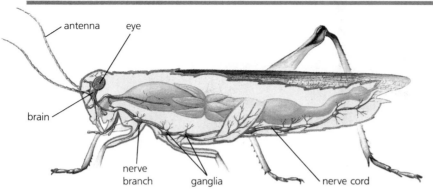

◀ **Figure 14–13**
Nervous System of the Grasshopper. The nervous system of the grasshopper includes a brain, a pair of ventral nerve cords, and branching nerves. In addition, the grasshopper has well-developed sense organs.

body. These nerves contain sensory neurons and motor neurons. The sensory neurons carry impulses from receptors in the skin to the nerve cords. The motor neurons carry impulses from the nerve cords to muscles and glands (effectors). The specialized receptors in the skin are sensitive to light, vibrations, chemicals, and heat.

In the earthworm, the nerves of the peripheral nervous system connect receptors and effectors to the central nervous system. Impulses travel over definite pathways in only one direction. The nervous systems of more complex animals are similar to the nervous system of the earthworm.

Regulation of the Grasshopper

The nervous system of the grasshopper is similar to that of the earthworm. See Figure 14–13. It is made up of a brain in the head region; a pair of solid, ventral nerve cords that run the length of the body; and ganglia. Nerves branch from the ganglia to all parts of the body. The sense organs of the grasshopper are more highly developed than those of the earthworm. The grasshopper has eyes, *antennae* (an TEN ee), or "feelers," and taste organs that respond to a variety of stimuli. Grasshoppers are also sensitive to sound. Because the grasshopper has a more highly developed nervous system than the earthworm, it is able to behave in more complex ways.

14-3 Section Review

1. Name the types of stimuli to which some protists respond.
2. What type of nervous system does the hydra have?
3. Name the two parts of the nervous system of the earthworm.
4. What specialized sense organs does the grasshopper possess?

Critical Thinking

5. Based on their capacity for nervous regulation, place each of the following organisms in order from most complex to least complex: ameba, earthworm, grasshopper, human, hydra, and paramecium. *(Ordering)*

Laboratory Investigation

Observing the Behavior of Planarians

A nervous system detects stimuli and generates responses to the stimuli that help the organism to survive. In this activity, you will observe the behavior of a planarian to learn about its nervous system. The planarian has a simple brain and two nerve cords that run the length of its body. Two eyespots appear at its head end, and receptor cells are present in its skin.

Objective

- *Observe* the behavior of a planarian in response to various stimuli.

Materials

planarian	pond water	stereomicroscope
raw liver	dropper	lamp
culture dish	blunt metal probe	tweezers

Procedure

A. With a dropper, transfer a planarian to a culture dish half-filled with pond water. Be sure the planarian is covered with water.

B. Observe the planarian with a stereomicroscope. In your lab book, describe how the planarian moves.

C. Sketch your planarian. Label its front (anterior) end, eyespots, rear (posterior) end, and back (dorsal) side. *Gently* touch the rear of the planarian with a probe. Record its response. Then, gently touch the head and record its response.

D. Gently turn the planarian over with a blunt probe. The opening in the middle of the bottom (ventral) side

pond water
planarian

* For detailed information on safety symbols, see Appedix B.

284

is the mouth. You may see the feeding tube, or pharynx, sticking out of the mouth. Draw and label the ventral side.

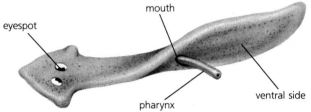

eyespot
mouth
pharynx
ventral side

E. Carefully turn the planarian over so that its dorsal side is up. Record the response of the planarian.

F. To test the planarian's response to light and dark, cover the dish with a thick piece of colored paper, leaving only a small area uncovered. With the lamp illuminating the dish, record the planarian's response.

G. Using tweezers, place a small piece of liver close to the planarian. Record the planarian's response. *Gently* turn the planarian over and record what you see.

H. When you have finished observing your planarian, return it to its original container of pond water.

Analysis and Conclusions

1. How does the planarian respond to touch?

2. Judging from its response, is the planarian more sensitive to touch in its head end or in its tail end? What would explain how one area is more sensitive than another area?

3. What behavior indicates that planarians are sensitive to gravity?

4. Based on your observations, does a planarian prefer bright or dim light? Explain.

5. How did the planarian respond to the liver? From this response, what can you infer about the kind and location of receptors that detect food?

6. How would you test to see if planarians are sensitive to temperature?

7. Based on your observations, where in a pond would you expect planarians to live? What kind of food do you think they normally eat?

Chapter Review

14

Study Outline

14–1 The Regulatory Process

■ The nervous system and the endocrine system regulate and coordinate responses to external and internal change. The functioning of a true nervous system involves receptors, nerve cells, and effectors.

■ Various physical forces and chemical substances stimulate receptors, which in turn trigger impulses in nerve pathways. Impulses cause effectors to respond.

■ Neurons are cells specialized for the rapid conduction of impulses. A neuron is usually made up of dendrites, a cell body, and an axon.

■ Impulses usually travel along the neuron from dendrites to cell body to axon. Impulses are carried across synapses to neighboring cells by chemicals called neurotransmitters.

■ The three basic types of neurons are sensory neurons, interneurons, and motor neurons. Nerves are bundles of axons or dendrites that are bound together by connective tissue.

14–2 The Nerve Impulse

■ In a resting neuron, the outside of the cell membrane is electrically positive, and the inside is negative. The membrane is polarized because of an unequal distribution of sodium and potassium ions inside and outside the membrane.

■ When a neuron is stimulated, cell membrane permeability changes at the point of stimulation, reversing its polarity. This change alters the permeability of an adjacent area of the membrane and reverses its polarity. In this way, an impulse is transmitted along the length of the neuron.

■ The rate at which an impulse travels depends on the size of the nerve fiber and on the presence or absence of a myelin covering.

■ An impulse is an "all or nothing" response. Any stimulus above a minimum strength or threshold will start an impulse in the neuron. Any stimulus below the threshold will not start an impulse.

■ When impulses reach the synaptic knob at the end of the axon, neurotransmitters are released into the synaptic gap. These chemical compounds stimulate or inhibit the firing of impulses in the neighboring neuron.

■ At neuromuscular junctions, the motor end plates of axons release the chemical transmitter acetylcholine, which stimulates the contraction of neighboring muscle fibers.

■ Some poisons interfere with the functioning of acetylcholine at neuromuscular junctions and can cause muscle paralysis. Drugs, such as stimulants, depressants, LSD, and mescaline, act on synapses and alter body activity.

14–3 Adaptations for Nervous Regulation

■ Protists do not have true nervous systems, but they can respond to stimuli. In the hydra, a nerve net transmits impulses throughout the organism, producing coordinated responses.

■ The earthworm has a primitive "brain" and two ventral nerve cords that enlarge into ganglia in each segment. Sensory and motor nerves branch from the "brain" and nerve cords.

■ The nervous system of the grasshopper is similar to that of the earthworm, but the sense organs are more highly developed in the grasshopper. The grasshopper has eyes, antennae, and taste organs, and is sensitive to sound.

Vocabulary Review

irritability (272)	axon (273)
impulses (272)	Schwann cells (273)
receptors (272)	myelin (273)
effector (272)	synapse (274)
stimulus (272)	nerves (274)
brain (272)	sensory neurons (274)
neuron (273)	motor neurons (274)
cell body (273)	interneurons (274)
dendrites (273)	

Chapter Review

sodium-potassium pump (276)

refractory period (277)

threshold (277)

neurotransmitters (279)

neuromuscular junctions (280)

nerve net (282)

central nervous system (282)

peripheral nervous system (282)

ganglion (282)

A. Definition—Replace the italicized definition with the correct vocabulary term.

1. The *short, highly branched fibers of the neuron* receive impulses.

2. The *irregular network of nerve cells* in the hydra is located between the two layers of the body wall.

3. Impulses are transmitted to the *specialized structure that responds to the commands of the nervous system.*

4. Impulses travel across the *gap between the terminal branch of a neuron and the membrane of another cell.*

5. In the earthworm, the nerve cords enlarge in each segment into a *group of cell bodies and interneurons that control nerve impulses.*

6. Following the passage of an impulse, there is a *brief period during which the nerve cell membrane cannot be stimulated to carry impulses.*

7. The *part of the earthworm's nervous system that includes the "brain" and ganglia* regulates nerve impulses.

8. In complex animals, *bundles of axons or dendrites bound together* carry impulses from receptors to the spinal cord and brain and from the brain and spinal cord to effectors.

B. Multiple Choice—Choose the letter of the answer that best completes each statement.

9. Neurons that carry impulses from the brain and spinal cord to effectors are the (a) interneurons (b) sensory neurons (c) nerves (d) motor neurons.

10. Impulses pass from motor neurons to muscles at special points of contact called (a) synapses (b) neuromuscular junctions (c) thresholds (d) interneurons.

11. All but the simplest animals have an (a) brain (b) nerve net (c) cell body (d) sodium-potassium pump.

12. Each nerve cell has a minimum level of sensitivity, or (a) stimulus (b) threshold (c) irritability (d) refractory period.

13. Many vertebrate axons are surrounded by (a) nerves (b) motor neurons (c) dendrites (d) Schwann cells.

14. Throughout the organism, a network of nerve cells carries (a) sensory neurons (b) effectors (c) impulses (d) receptors.

15. The long, thin fiber that extends from the cell body of a neuron is the (a) ganglion (b) myelin (c) axon (d) synapse.

Content Review

16. Describe the relationship between a receptor and an effector.

17. Explain the basic sequence of events in nervous regulation.

18. What is the brain's function in the nervous system?

19. Draw a labeled diagram of a neuron. Then, compare the structure and function of the three basic parts of the neuron.

20. What are the functions of the three types of neurons?

21. Describe the electrical state of the membrane of a resting neuron.

22. How is the difference in electrical charge across the nerve-cell membrane maintained?

23. Describe the electrical state of the nerve-cell membrane in the area of an impulse.

24. What changes in ion distribution occur in the area of an impulse?

25. How does an impulse travel along a nerve-cell membrane?

26. When is the nerve-cell membrane unable to carry impulses?

27. Why do myelinated fibers carry impulses faster than those without a myelin covering?

28. Explain the following statement, "A neuron fires on an all-or-nothing basis."

29. How are varied strengths and types of stimuli recognized by the nervous system?

30. Explain how a nerve impulse crosses a synapse.

31. How do impulses pass from a motor neuron to a muscle cell?

32. Describe the effects of nerve gas on the nervous system.

33. How does caffeine affect the nervous system?

34. How does the hydra respond when a part of its body receives a stimulus?

35. Describe the structures of the central and peripheral nervous systems in the earthworm.

36. How do the nervous systems of the grasshopper and the earthworm differ? How are they alike?

Graphic Organizing

For information on graphic organizers, see Appendix G at the back of this text.

37. **Flow Chart:** Construct a flow chart that shows the path of a nerve impulse from a stimulated sensory organ to a responding muscle.

Critical Thinking

38. Compare and contrast a nerve and a neuron. (*Comparing and Contrasting*)

39. Suggest some possible causes for a person's inability to move his or her muscles. (*Identifying Causes*)

40. Imagine that the nerve fibers from pain receptors in your thumb could be exchanged with those in your index finger. If you were to prick your thumb with a needle, where would you feel the pain? (*Predicting*)

41. How would a rabbit's behavior be different if it had no myelinated nerve fibers? (*Relating Parts and Wholes*)

42. Compare the nervous systems of the hydra and the earthworm. (*Comparing and Contrasting*)

43. How do you think a painkiller works to stop the sensation of pain? (*Identifying Causes*)

Creative Thinking

44. A rare genetic disease in which babies are born with no pain receptors results in their feeling no pain. How would this affect the babies as they grow up? How might it affect the way their parents care for them?

45. Sometimes people who have had an arm or leg amputated feel pain in the limb that is no longer there. This is called "phantom limb syndrome." How can you explain it?

Problem Solving

46. In a myelinated nerve fiber, the nerve impulse travels at about 100 meters per second, and in an unmyelinated fiber, at about 2 meters per second. How long would it take for an impulse to travel: (a) 88 centimeters through a myelinated fiber running from the spinal cord to the tip of the toe? (b) 45 centimeters through an unmyelinated fiber running from the base of the brain to the end of the spinal cord?

47. A scientist removed the hearts from two living frogs. Two nerves leading to the heart were left attached to heart A but were cut from heart B. Both hearts were kept moist with salt solution and continued to beat at the same rate. As shown below, heart A was connected to heart B by a tube that allowed the salt solution from heart A to be pumped into heart B.

When the scientist stimulated one nerve leading to heart A, its beat decreased. When the salt solution from heart A flowed into heart B, heart B's beat also decreased. When the other nerve leading to heart A was stimulated, both heartbeats increased.

What conclusion(s) can you draw from this experiment? What additional experiments would be useful?

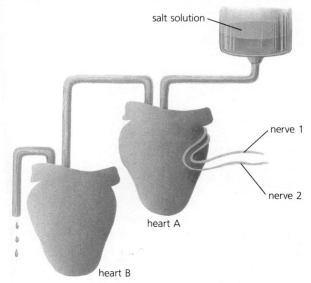

salt solution

nerve 1

nerve 2

heart A

heart B

Projects

48. Design and conduct a controlled experiment on how paramecia, hydras, or earthworms respond to stimuli such as light and dark, acidic and basic substances, or warmth and cold.

 49. Research the effects of drugs on the synapse. Write up your findings as a magazine article.

▲ The body's nervous system allows this skydiver to experience and respond to an unaccustomed environment.

The Human Nervous System

A skydiver freefalls through crisp California air. As he drops, the roar of the engine quickly fades. He is alone, listening only to the whoosh of air. Every instant, inside the skydiver's body, a different sort of action takes place. Nerves flash information to the brain. The brain orders subtle adjustments in the muscles. Thoughts come and go. This chapter will introduce you to the structure of the human brain and the elaborate system of senses and nerves that connect the brain to the body and to the outside world.

Guide for Reading

Key Words: cerebrum, cerebellum, spinal cord, reflex arc, retina, cochlea, olfactory cells

Questions to think about:

📖 How does the human nervous system compare with that of other vertebrates?

📖 What important difference is there between the somatic and autonomic nervous systems?

15-1 The Central Nervous System

Section Objectives:

- *Describe* the functions of the skull, spinal column, meninges, and cerebrospinal fluid.
- *Name* the major parts of the brain.
- *Explain* how momentary memory, short-term memory, and long-term memory differ from one another.
- *Describe* the structure and functions of the spinal cord.

The human nervous system, like nervous systems of other vertebrates, can be divided into two subsystems. One of these subsystems is the *central nervous system*. The central nervous system is made up of the brain and the spinal cord. The interneurons and the cell bodies of most motor neurons are also a part of the central nervous system. The other subsystem is called the *peripheral nervous system*. This system is a vast network of nerves that conducts impulses between the central nervous system and the receptors and effectors of the body. The peripheral nervous system is made up of sensory neurons, including their cell bodies, and the axons of all the motor neurons.

The central nervous system—the brain and the spinal cord—controls most of the activities of the body. Impulses from sense receptors throughout the body bring a constant flow of information about conditions inside and outside the body. The information is

▲ **Figure Figure 15–1**
Body Control. The nervous system enables fine coordination of body movements, as shown with these ballet dancers.

Figure 15–2

The Brain and the Spinal Column. The spinal cord is protected by the vertebrae of the spinal column (below), and the brain (right) is protected by the cranium. Added protection is provided by the meninges and by the cerebrospinal fluid, which cushions the tissues against shock. ▼

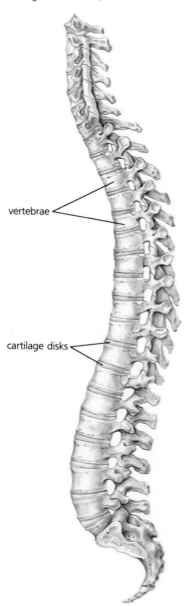

vertebrae

cartilage disks

interpreted, or responded to, by the brain or the spinal cord. Then, impulses that cause appropriate responses are sent out to muscles and glands.

The Skull and Spinal Column

The brain and the spinal cord are protected by bone. See Figure 15–2. The brain is enclosed by the cranium, which is the upper part of the skull. The spinal cord is surrounded by the vertebrae of the spinal column, or backbone. Disks of cartilage between the vertebrae absorb the shocks of running, walking, or jumping. The brain and the spinal cord are also covered and protected by three tough membranes known as the *meninges* (muh NIN jeez). A liquid, the *cerebrospinal* (suh ree broh SPYN ul) *fluid,* fills the space between the inner and middle membranes. This fluid cushions the delicate nervous tissues against shock. A *concussion* occurs when the brain is severely shaken, causing it to bump against the cranium. Inside the brain, four spaces, called *ventricles,* are filled with the cerebrospinal fluid. The ventricles connect with the fluid-filled space between the meninges and with the central canal of the spinal cord, which is also filled with cerebrospinal fluid.

The Brain

The brain is one of the most active organs in the human body. It receives 20 percent of the blood that is pumped from the heart, and it replaces most of its protein every three weeks. The brain is the major user of glucose in the body. Unlike the cells of other tissues, the cells of the brain usually metabolize only glucose for the release of energy.

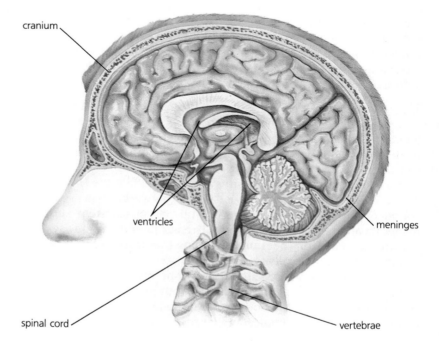

cranium

ventricles

meninges

spinal cord

vertebrae

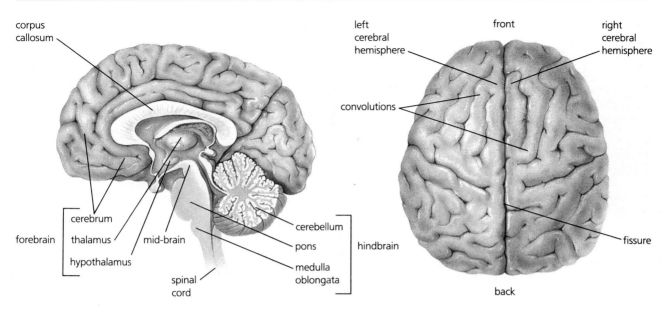

▲ **Figure 15–3**
Structure of the Brain. The major regions of the brain can be seen in the lengthwise section of the brain (left). A top view of the brain (right) shows the division of the cerebrum into two hemispheres.

The major parts of the brain are the *cerebrum* (suh REE brum), *cerebellum* (sehr uh BEL um), and the *medulla oblongata* (muh DUHL uh ahb lon GHAT uh). See Figure 15–3. Other parts of the brain are the thalamus, hypothalamus, and pons. The **thalamus** serves as a relay center between various parts of the brain and the spinal cord. It also receives and changes all sensory impulses, except those involved in smell, before they travel to the cerebral cortex. It may also be involved in feeling pain and keeping a person conscious. The **hypothalamus** helps control body temperature, blood pressure, sleep, and emotions. It also plays a role in the functioning of the endocrine system. See Chapter 16. The **pons** serves as a relay system, linking the spinal cord, medulla oblongata, cerebellum, and cerebrum. The pons, the medulla oblongata, and the midbrain are often called the **brainstem.**

The cerebrum, thalamus, and the hypothalamus make up the *forebrain.* The small *midbrain* connects the forebrain with the hindbrain and is a center for some visual and reflex functions. The *hindbrain* is another name for the brainstem, the cerebellum, the pons, and the medulla oblongata.

The Cerebrum   The **cerebrum** is the largest part of the human brain. It makes up about two-thirds of the entire organ. The greatest difference between the human brain and the brains of other vertebrates is the larger size and greater development of the human cerebrum. The cerebrum is divided in half from front to back by a deep groove, or fissure. This groove separates the cerebrum into the right and left **cerebral hemispheres.** See Figure 15–3. Nerve fibers from each hemisphere pass to the other hemisphere and to

Science, Technology and Society

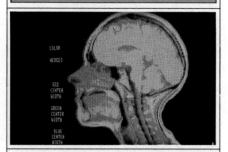

Technology: Magnetic Resonance Imaging

A new tool called Magnetic Resonance Imaging (MRI) allows doctors to "see" inside their patients. It does this by displaying the inside of the body on a computer screen (see photo).

When MRI is used, a patient is placed in the hollow cylinder of a giant electromagnet. The strong magnetic field acts upon the hydrogen atoms in the body, causing them to enter a high energy state. In this state, the atoms are unstable. As the atoms return to a lower energy state, they emit radio waves.

The intensity of the radio waves is in proportion to the concentration of hydrogen atoms present in the area scanned. This concentration differs in different organs and in tumors. A computer maps the intensity differences that correspond to the various organs and to any abnormalities.

MRI enables doctors to "see" vital organs, identify tumors, and spot early signs of disease. Although MRI scans are expensive, they often eliminate the need for exploratory surgery and harmful X rays.

■ *List some potential benefits and drawbacks of MRI.*

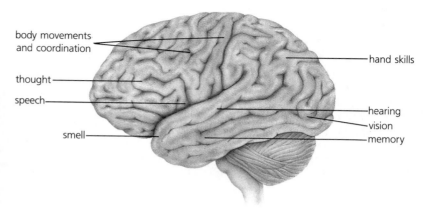

▲ **Figure 15–4**
Motor and Sensory Regions of the Cerebral Cortex.

other parts of the nervous system. These nerve fibers make a bridgelike connection between hemispheres that is called the **corpus callosum** (KOR puhs kuh LOH suhm).

The outermost layer of the cerebrum is the **cerebral cortex,** or *gray matter.* The gray matter is made up of the cell bodies of motor neurons and a huge number of interneurons. These are interconnected by unmyelinated fibers. The outer surface of the cortex has many folds. These folds, or *convolutions* (kahn vuh LOO shunz), greatly increase the surface area of the gray matter.

The cerebral cortex has three major functions—sensory, motor, and associative. Each part of the cortex carries out a particular function. Some functions, however, may be carried out in two or more areas of the cortex as well as in other parts of the brain. Some functions of the various parts are shown in Figure 15–4.

The sensory areas of the cortex receive and interpret impulses from the sense receptors. These receptors include the eyes, ears, taste buds, and nose, as well as the touch, pain, pressure, heat, and cold receptors in the skin and other organs. The motor areas of the cortex start impulses that are responsible for all voluntary movement and for the position of the movable parts of the body. Impulses from the motor cortex may be slightly changed by other parts of the brain. The associative areas of the brain are responsible for memory, learning, and thought.

The two cerebral hemispheres do not usually function in exactly the same way. Instead, some functions are carried out by the left hemisphere and others by the right hemisphere. For example, in many people, the left hemisphere is the center for mathematical thinking. The right hemisphere is often the center for artistic and musical ability.

Beneath the gray matter of the cerebrum is an inner area called the *white matter.* This area is made up of myelinated nerve fibers. One of the bundles, or tracts, of fibers in the white matter is the *corpus callosum.* This connects the right and left hemispheres, so that information can pass between the two halves of the cerebrum.

Other tracts from the white matter connect the cortex to other parts of the nervous system.

Nerve fibers leaving the cerebral hemispheres pass through the brain and spinal cord. At some point along their pathway, these fibers cross over to the opposite side of the brain or spinal cord before going to various parts of the body. Thus, the left cerebral hemisphere controls the right side of the body, and the right hemisphere controls the left side. For this reason, an injury to one side of the cerebrum will affect the other side of the body.

The Cerebellum   The **cerebellum** is found below the rear part of the cerebrum. See Figure 15–5. The cerebellum, like the cerebral cortex, is divided into two hemispheres. It also has gray matter and white matter. The highly folded outer layer of the cerebellum consists of gray matter, while the inner portion is white matter.

The cerebellum controls all voluntary movements and some involuntary movements. Motor impulses from the cerebral cortex are carried by nerve pathways that send some branches directly to the muscles involved and other branches to the cerebellum. The muscles also send impulses over sensory nerve pathways to the cerebellum. These impulses provide information about the muscles' position, rate of contraction, and so on. The cerebellum then sends impulses to the cerebral cortex to correct and coordinate the movement of the muscles. Thus, the cerebral cortex and the cerebellum work together to produce smooth, orderly voluntary movement. With certain involuntary movements, the cerebellum works in the same manner with other parts of the brain. For example, the cerebellum, using information from receptors in the inner ear, maintains balance, or *equilibrium*. It also plays a role in the maintenance of *muscle tone* (keeping the muscles slightly tensed). Damage to the cerebellum results in jerky movements, tremor, or loss of equilibrium. Staggering and other signs of loss of coordination—seen when someone has had too much to drink—are caused by a temporary loss of cerebellar function.

The Medulla Oblongata   As you saw in Figure 15–5, beneath the cerebellum and connected to the spinal cord is the **medulla oblongata.** In this lowest part of the brain, the white matter makes up the outer layer, while the gray matter is the inner layer. The medulla oblongata is made mainly of nerve fibers that connect the spinal cord to other parts of the brain. Nerve centers in the medulla oblongata control many involuntary activities. These include breathing, heartbeat, blood flow, and coughing.

Memory

Most learning depends on being able to store and recall memories of past experiences. Memory is thought to be a function of the cerebral cortex. Although there are several theories about how memory is stored, little is really known about the process. However, scientists now believe that there are three kinds of memory—momentary, short-term, and long-term memory.

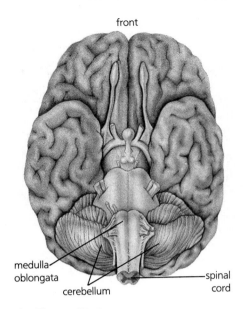

front

medulla oblongata

cerebellum

spinal cord

▲ **Figure 15–5**

The Underside of the Brain. The cerebellum coordinates and controls voluntary movements and some involuntary movements. The medulla controls many involuntary activities, such as breathing, heart beat, and blood flow.

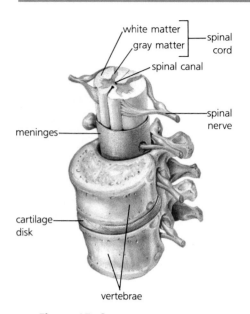

white matter ⎤
gray matter ⎦ spinal cord

spinal canal

spinal nerve

meninges

cartilage disk

vertebrae

▲ **Figure 15–6**

Structure of the Spinal Cord. Note that the spinal cord is completely surrounded by the bone tissue of the vertebrae. Cartilage disks between most of the vertebrae serve as shock absorbers.

Momentary memory lasts for a few minutes at most. Remembering a phone number only long enough to dial it is an example of momentary memory. *Short-term memory* can be recalled for as long as several hours. Memory lasting weeks or years is *long-term memory.* By some process not yet known, a short-term memory is sometimes changed into a long-term memory. A person can lose momentary and short-term memories and still keep long-term memory. This is sometimes found in older people, who may be unable to remember something that just happened but can recall in detail things that took place many years earlier.

At present no one knows how memories are stored, transferred, or recalled in the human brain. There is some evidence that short-term memory is associated with patterns of impulses circulating repeatedly through particular pathways of neurons. Long-term memory may be the result of permanent changes in particular synapses, or changes within certain neurons.

The Spinal Cord

The **spinal cord,** which is about 45 centimeters long, goes from the base of the brain down through the vertebrae of the spinal column. A cross section of the spinal cord shows an inner H-shaped region of gray matter surrounded by an outer layer of white matter. See Figure 15–6. The gray matter contains many interneurons, as well as the cell bodies of motor neurons. The white matter contains myelinated fibers that carry impulses between all parts of the body and the spinal cord and brain. In the center of the cord is the *spinal canal,* which is filled with cerebrospinal fluid.

The spinal cord is important for two reasons. First, it connects the nerves of the peripheral nervous system with the brain. Impulses reaching the spinal cord from sensory neurons travel up the cord through interneurons to the brain. Impulses from the brain are sent down the spinal cord by interneurons to motor neurons. These impulses travel through peripheral nerves to muscles and glands. Second, the spinal cord controls certain reflexes, which are automatic responses. See page 296.

15-1 **Section Review**

1. What are the two main subdivisions of the human nervous system?
2. List the three major parts of the brain.
3. Identify the three kinds of memory.
4. Give two functions of the spinal cord.

Critical Thinking

5. If someone had an injury that damaged the cerebellum, how would that affect the person's ability to drive a car?   *(Relating Parts and Wholes)*

15-2 The Peripheral Nervous System

Section Objectives:

- *Identify* the nerves that make up the peripheral nervous system.
- *Compare* and *contrast* the structures and functions of the somatic and autonomic nervous systems.
- *Name* the two divisions of the autonomic nervous system and state their functions.
- *Distinguish* between a reflex and a voluntary behavior.

Structure of the Peripheral Nervous System

The peripheral nervous system is made up of all the neurons and nerve fibers outside the brain and spinal cord. The neurons of the peripheral nervous system are connected to either the brain or the spinal cord. The neurons are in bundles that form nerves. The nerves connected to the spinal cord are called the **spinal nerves,** while those connected to the brain are called the **cranial** (KRAY nee ul) **nerves.**

There are 31 pairs of spinal nerves. Each pair serves a particular part of the body. Each nerve contains both sensory and motor fibers. See Figure 15–7. The cell bodies of the sensory neurons are found in ganglia outside the spinal cord. The cell bodies of the motor neurons and all of the interneurons are found in the gray matter of the spinal cord. As a spinal nerve gets close to the spinal cord, the sensory and motor fibers separate. The sensory fibers enter the *dorsal root* (toward the back) of the spinal cord, while the motor fibers leave through the *ventral root* (toward the front) of the cord.

There are 12 pairs of cranial nerves. Those serving the eyes, ears, and nose are made up mostly of sensory fibers. The other cranial nerves contain more equal numbers of sensory and motor fibers. Most of the cranial nerves serve the sense organs and other structures of the head.

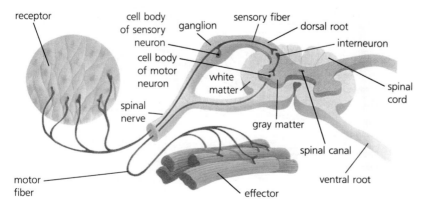

◀ **Figure 15–7**
Structure of a Spinal Nerve. A spinal nerve contains both motor and sensory fibers. Sensory fibers transmit impulses into the dorsal root of the spinal cord, and motor fibers transmit impulses from the ventral root to effectors. Note that the cell bodies of sensory neurons are outside the spinal cord.

The peripheral nervous system is divided into two parts. One part, called the *somatic* (soh MAT ik) *nervous system,* is under voluntary control. The other part, the *autonomic* (awt uh NAHM ik) *nervous system,* usually is not under voluntary control.

The Somatic Nervous System   The **somatic nervous system** contains both sensory and motor neurons that connect the central nervous system to skeletal muscles, the skin, and the sense organs. This system is responsible for body movements over which the individual has some conscious awareness or voluntary control.

The Autonomic Nervous System   The **autonomic nervous system** is made up of certain motor fibers from the brain and spinal cord that serve the internal organs of the body. Usually, there is no voluntary control over the activities of the autonomic system, which is made up of two divisions, the **parasympathetic** (par uh sim puh THET ik) **nervous system** and the **sympathetic nervous system.** See Figure 15–8. The entire system controls many important functions of the body. These include the rate of heartbeat, blood flow through arteries, breathing movements, movements of the digestive system, and secretions of certain glands.

The autonomic system is made entirely of motor neurons. The same sensory nerves that serve the somatic system provide sensory information for this system. Impulses in the autonomic system

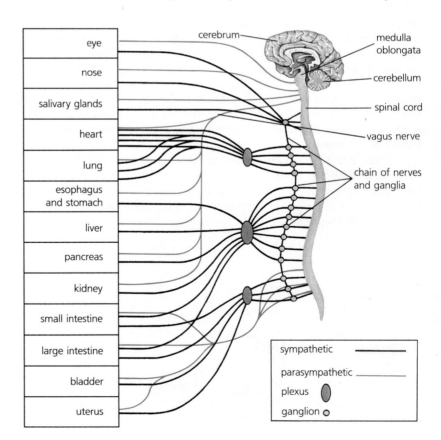

Figure 15–8
Nerve Pathways to the Sympathetic and Parasympathetic Nervous Systems. ▶

start in motor neurons in the brain or spinal cord. However, the axons of these neurons do not go to the organ involved. Instead, each axon synapses with a second motor neuron, which then carries the impulses to the muscle or gland. Some of the cell bodies of these second motor neurons are found in ganglia just outside the brain and spinal cord. The ganglia, which are interconnected by nerves, form two chains alongside the spinal column. Other ganglia are found elsewhere in the body. Some of them form large clusters called *plexuses* (PLEKS us sez).

Organs served by the autonomic nervous system generally contain nerve endings from both the sympathetic and parasympathetic divisions. The effects of these two types of nerve endings are antagonistic, or opposite, because they release different neurotransmitters in the organs. Nerves of the sympathetic system release *norepinephrine,* while those of the parasympathetic system release *acetylcholine.* When the sympathetic system speeds up an activity, the parasympathetic system slows down the same activity. For example, the beating of the heart is speeded up by the accelerator nerves of the sympathetic system and slowed down by the vagus nerves of the parasympathetic system. The antagonistic relationship that exists between the two divisions of the autonomic nervous system helps to control the organs and to maintain the homeostatic balance of the body.

Figure 15–9 lists the actions of the parasympathetic and sympathetic systems on various structures. Generally, the sympathetic system helps the body deal with emergency situations. In many cases, it does so by accelerating body activities. The parasympathetic system promotes normal, relaxed body functioning.

Reflexes

A **reflex** is an involuntary, automatic response to a given stimulus. It makes use of a relatively simple pathway between a receptor, the spinal cord or brain, and an effector. Many normal body functions

Figure 15–9
Functions of the Autonomic Nervous System. ▼

Effects of the Autonomic Nervous System		
Organ	**Sympathetic Division**	**Parasympathetic Division**
heart	speeds up and strengthens beat (stimulates)	slows and weakens beat (inhibits)
digestive tract	slows peristalsis, slows activity (inhibits)	speeds peristalsis increases activity (stimulates)
blood vessels	mostly constricts	mostly dilates
bladder	relaxes	constricts
bronchi	widens passages	constricts passages
eye	makes pupil larger	makes pupil smaller

are controlled by reflexes. These include blinking, sneezing, coughing, breathing movements, heartbeat, and peristalsis. The knee-jerk reflex and the reflex that changes the size of the pupil of the eye in response to light are used by doctors to check the condition of the nervous system. When there is no reflex response or when the response is slow, a disorder of the nervous system may be present.

Reflex Arcs The pathway over which the nerve impulses travel in a reflex is called a **reflex arc.** The simplest reflex arcs use only two neurons—one sensory and one motor. The pathway of the knee-jerk reflex is of this type. Most reflexes, however, use three or more neurons. Withdrawal reflexes, for example, use a three-neuron reflex arc. See Figure 15–10. As Figure 15–10 shows, pulling your hand back from a hot object is an example of a withdrawal reflex involving three neurons: a sensory neuron, an interneuron, and a motor neuron.

When your hand touches a hot stove, it is pulled back before you feel the sensation of heat or pain. Removing your hand is brought about by a withdrawal reflex. The parts of this reflex arc are as follow.

1. A receptor in the skin is stimulated by the heat.

2. The receptor starts impulses in a sensory neuron, which carries impulses to the spinal cord.

3. Within the spinal cord, the sensory neuron synapses with an interneuron, which synapses with a motor neuron.

4. The motor neuron sends impulses to the effector. In this example, impulses are carried to certain muscles of the arm.

5. The muscles receiving impulses from the motor neuron contract, moving the hand and arm.

The withdrawal reflex takes place without the use of the brain. However, shortly after the hand is withdrawn from the hot object, there may be sensations of heat and pain. These result from impulses that pass up the spinal cord to the brain.

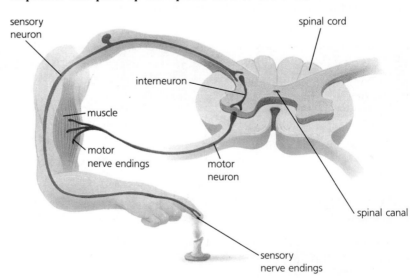

Figure 15–10

A Reflex Arc. The withdrawal reflex involves a sensory neuron, an interneuron, and a motor neuron. ▶

Voluntary Behavior

Unlike reflexes, voluntary behavior is under conscious control. It includes all the physical and mental activities that a person chooses to do, such as writing a story, running a race, or cooking. All voluntary behavior is controlled by the cerebrum and makes use of a combination of memory of past experiences, associations, reasoning, and judgments. This most complex type of behavior is more developed in humans than in other animals.

15-2 Section Review

1. What are the names of the nerves of the peripheral nervous system that are connected to the spinal cord and to the brain?
2. Name the two divisions of the peripheral nervous system.
3. Name the two divisions of the autonomic nervous system.
4. Give an example of a reflex and a voluntary behavior.

Critical Thinking

5. Compare and contrast the somatic and autonomic nervous systems. *(Comparing and Contrasting)*

15-3 Sense Receptors

Section Objectives:

- *Name* the parts of the eye and explain how vision works.
- *Indicate* the parts of the ear and explain how the ear functions in hearing and balance.
- *Identify* the sense receptors of the skin.
- *Describe* the structures and functions of taste buds and olfactory cells.

  Laboratory Investigation: *Explore* how human sense receptors respond to stimuli (p. 308).

In animals, sense receptors provide information about what is taking place both inside and outside the body. The receptors of the human nervous system range from those in the skin, which are relatively simple structures, to the eye and ear, which are very complex organs.

The Eye

Sight is the most important sense in humans. It gives us more than 80 percent of the information we receive about the outside world.

Structure of the Eye   The walls of the human eye are made of three layers. The tough outer layer of the eye is the **sclera** (SKLEHR uh), or the "white" of the eye. It helps to give the eye shape and

Figure 15–11
Structure of the Human Eye. ▶

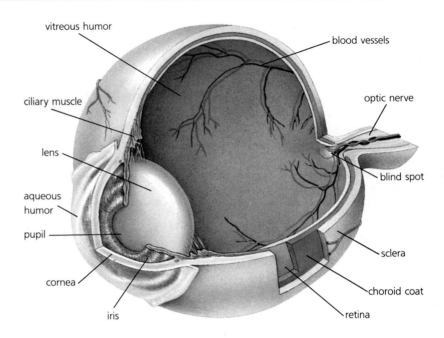

vitreous humor

blood vessels

ciliary muscle

optic nerve

lens

blind spot

aqueous humor

pupil

sclera

cornea

choroid coat

iris

retina

protects the inner parts of the eye. In the front of the eye, the outer layer bulges and becomes the transparent **cornea** (KOR nee uh). See Figure 15–11. Light enters the eye through the cornea.

Just inside the sclera is a dark-brown middle layer, the **choroid** (KOR oyd) **coat.** This layer, which contains many blood vessels, stops the reflection of light within the eye. At the front of the eye, the choroid layer forms the **iris** (I ris), which is the colored part of the eye. In the center of the iris is an opening called the **pupil.** The iris works like the diaphragm of a camera. It has muscles that can make the size of the pupil larger or smaller. In dim light, the pupil becomes larger, or dilates, which allows more light to enter the eye. In bright light, the pupil becomes smaller, or constricts, which lets less light into the eye. The size of the pupil is controlled by the autonomic nervous system.

Behind the iris is the **lens.** The lens focuses the light on the back of the innermost layer, the **retina** (RET in uh), producing an image similar to the image on the film in a camera. *Ciliary* (SIL ee ehr ee) *muscles* attached to the choroid layer hold the lens in place. These muscles also change the shape of the lens. This allows the eye to focus on objects.

The retina contains light receptors. At the rear of the eye, the retina is attached to the **optic** (OP tik) **nerve.** This nerve carries impulses from the light-sensitive cells to the brain.

The eyeball is a hollow sphere that is divided into two cavities. The cavity between the cornea and the lens is filled with a transparent watery fluid called the *aqueous* (AHK wee us) *humor.* The large cavity behind the lens is filled with a colorless, jellylike liquid called the *vitreous* (VIH tree us) *humor.* The vitreous humor gives the eye a firm shape.

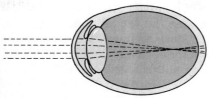

Nearsightedness

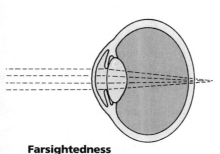

Farsightedness

◄ **Figure 15-12**
Nearsightedness and Farsightedness. In nearsightedness (top), the eyeball is too long. Light rays are brought into focus in front of the retina, causing a person to see near objects clearly and distant objects out of focus. In farsightedness (bottom), the eyeball is too short. Light rays are focused behind the retina, causing a person to see distant objects clearly and near objects out of focus.

Vision   Light entering the eye passes through the cornea, aqueous humor, pupil, lens, and vitreous humor, and forms an image on the retina. People who are **nearsighted** can see objects near to them more clearly than objects far from them. This happens because they have a long-shaped eye. Light rays from far away form an image that is blurred. See Figure 15–12. In **farsighted** people, the eyeball is too short. This causes the lens to focus light from nearby objects at a point beyond the retina. This results in a blurred image of close objects. Both conditions are easily corrected by prescribed eyeglasses or contact lenses.

The retina is made up of several different layers of cells. One inner layer contains light-sensitive cells—the **rods** and the **cones.** See Figure 15–13. Rods are sensitive to weak light but not to color. They allow a person to see in dim light, which is black-and-white vision. Cones are sensitive to color but must have bright light to function. There are three types of cones in the retina. One type is sensitive to red light, one to green light, and one to blue light. The retina contains about 125 million rods and 6.5 million cones.

Both black-and-white and color vision make use of the light-sensitive pigment *retinal* (RET in al), which is made from vitamin A. Retinal combines with proteins within the rods and cones. The

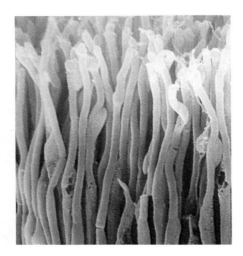

▲ **Figure 15–13**
Rods and Cones. One of the inner layers of the retina is made up of rods and cones, seen in this scanning electron micrograph.

proteins in the rods and in the three types of cones are all different, and each binds with retinal differently. It is the effect of each type of protein on the retinal that allows this pigment to respond to different colors and intensities of light.

When light strikes a rod or cone, it breaks the chemical bond between the retinal and the protein with which it was combined. This starts impulses traveling from that rod or cone. Nerve fibers from the rods and cones join to form the optic nerve, which carries the impulses to the brain. The brain interprets them as vision. The point where the optic nerve leaves the eye contains no rods or cones and is called the *blind spot.*

Too little vitamin A leads to a condition called *night blindness.* Night-blind people have trouble seeing in dim light. In this condition, the amount of retinal in both the rods and cones is decreased, and therefore both the rods and the cones become less sensitive to light. Thus, vision in dim light is affected, but there is enough pigment for vision in bright light.

Color blindness, which is an inability to see certain colors, is a hereditary condition in which the proteins of one or more of the three types of cones do not work properly.

The Ear

The human ear has two sensory functions. One is hearing. The other is helping to keep balance, or equilibrium.

Structure of the Ear   The three parts of the ear are the *outer ear,* the *middle ear,* and the *inner ear.* See Figure 15–14. The outer ear is the part that can be seen. It consists of the *pinna* (PIN uh), a flap of skin supported by cartilage, and a short **auditory** (AW dih tor ee) **canal.** The pinna functions mainly as a collecting funnel for sound

Figure 15–14
Structure of the Human Ear. ▼

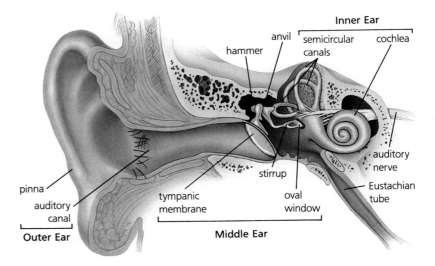

waves. Along the auditory canal are special glands that secrete a waxy material. This wax prevents foreign objects from entering the ear. Stretched across the inner end of the auditory canal is the delicate **tympanic** (tim PAN ik) **membrane,** or *eardrum.*

The middle ear is an air-filled chamber that begins at the eardrum. It contains three tiny bones—the *hammer, anvil,* and *stirrup.* These bones form a chain across the middle ear that links the eardrum to another membrane, which is known as the **oval window.** The hammer is attached to the eardrum, the anvil connects the hammer to the stirrup, and the stirrup is connected to the oval window. Going from the middle ear to the throat is the **Eustachian** (yoo STAY shun) **tube.** The function of the Eustachian tube is to make the pressure in the middle ear equal to the pressure of the atmosphere outside the body.

Critical Thinking in Biology

Judging Usefulness

When you judge usefulness, you decide whether or not an item would help to achieve a desired result. If you had to write a number of reports for school, for example, you might decide that a word processor would be useful.

Doctors often must judge the usefulness of some medical procedures for their patients. For example, the cochlear implant is an experimental device that has been developed for profoundly deaf people. As shown in the diagram, sound waves are picked up by a microphone worn behind the ear. They are sent to the speech processor, usually worn in a pocket. Here, the sound waves are changed to electronic signals. These are sent to the transmitter and then to a receiver surgically implanted under the skin behind the ear. Wires carry the signals to electrodes that have been implanted in the cochlea. Nerve endings in the inner ear receive the signals and transmit them to the brain.

1. Ben has been totally deaf since he suffered severe middle ear damage as an infant. Do you think a cochlear implant would be useful for Ben? Explain.
2. Rachel has been totally deaf since childhood because of damage to the auditory nerve. Would a cochlear implant be useful? Why or why not?
3. Harry, now in his 70s, has trouble distinguishing among some sounds and sometimes must ask people to speak a bit louder. Would the cochlear implant be a good choice for him? Why or why not?
4. Think About It. Explain your thought process as you answered question 1.

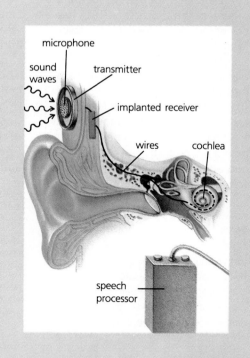

Science, Technology and Society

Issue: Noise Pollution

Airplanes, cars, construction machinery, and stereos are just a few of the factors that contribute to daily noise. Studies have linked excessive noise with hearing loss and other health effects. This has prompted some people to favor laws that regulate public noise.

People favoring such laws argue that they have the right to a healthful environment. They feel that communities should regulate car and truck noise, radios, and loud parties. They also believe that manufacturers of cars, airplanes, and appliances should be required to produce quieter equipment.

Other people argue that anti-noise laws would be impossible to enforce. They claim that the regulation of noise should be left to individuals. People who work in noisy occupations, for example, can wear protective ear equipment. Homes can be made less noisy by using sound-dampening materials and by lowering the volume of televisions, radios, and stereos. In addition, people can make an effort to avoid loud parties, rock concerts, and other noisy settings.

■ ***Do you think public noise should be regulated? Why or why not?***

Infections can travel from the nose and throat through the Eustachian tube to the middle ear causing an ear infection. The symptoms of an ear infection can include an ear ache, the ear feeling warm, and a sensation of fluid in the ear. Doctors usually prescribe antibiotics to help fight an ear infection. Severe ear infections can cause the eardrum to rupture.

The inner ear is made up of the **cochlea** (KAHK lee uh) and the **semicircular canals.** The cochlea is the organ of hearing. It is made of coiled, liquid-filled tubes that are separated from one another by membranes. Lining one of the membranes are specialized hair cells that are sensitive to vibration.

The semicircular canals allow the body to maintain balance. They are made of three, interconnected, loop-shaped tubes at right angles to one another. These canals contain fluid and hairlike projections that detect changes in body position.

Hearing Sound waves are vibrations in air or some other medium, such as water. Hearing takes place when these vibrations travel to the inner ear, where they start impulses that are carried to the brain by the **auditory nerve.**

Sound waves collected by the outer ear pass down the auditory canal to the eardrum. They cause the eardrum to vibrate. The vibrations are carried across the middle ear by the hammer, anvil, and stirrup. Vibrations of the stirrup cause vibrations in the oval window, which in turn cause the fluid within the cochlea to vibrate. The movement of the fluid causes vibrations in specialized hair cells lining one of the membranes within the cochlea. The vibrations start impulses in nerve endings around the cells. These impulses are carried to the cerebral cortex, where they are interpreted. The number of cells in the cochlea stimulated, and therefore the number of impulses sent to the brain, determines the sound's loudness. The more cells stimulated, the louder the sound.

Balance Balance, or equilibrium, is a function of both the inner ear and the cerebellum. In the inner ear, the fluid-filled semicircular canals lie at right angles to one another. As the head changes position, the fluid in the canals also changes position, which causes the hairlike projections to move. This in turn stimulates nerve endings, which start impulses traveling through a branch of the auditory nerve to the cerebellum. The cerebellum interprets the direction of movement and sends impulses to the cerebrum. Impulses started by the cerebrum correct the position of the body.

If you spin around for a time, the fluid in the semicircular canals also moves. When you stop suddenly, you feel as though you are still moving. You become dizzy because the fluid in the canals keeps moving and stimulating the nerve endings. In some people, the rhythmic motions of a ship, plane, or car overstimulate the semicircular canals, resulting in motion sickness.

Also in the inner ear are two sacs containing crystals, or stones, that rest on sensory hairs. The pressure of the stones on the sensory hairs provides information on the body's position in relation to the direction of gravitational pull.

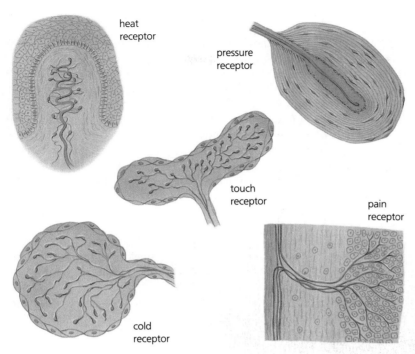

▲ **Figure 15–15**

Sense Receptors of the Human Skin. Much of what we know about our external environment comes from information that is detected by the sense receptors found in the skin.

Receptors of the Skin

There are sense receptors in the skin for touch, pressure, heat, cold, and pain. See Figure 15–15. Each type of receptor differs in structure from the others and is sensitive to only one kind of stimulus. When a receptor is stimulated, it produces impulses that travel over sensory nerve pathways to the brain, where they are interpreted.

Although all types of receptors are present all over the human body, the different types are not evenly distributed. For example, receptors sensitive to touch are farthest apart on the back, much closer together on the fingertips, and closest together on the tip of the tongue. The ability to judge the size of an object by the way it feels is partly determined by the number of touch receptors stimulated.

Pressure receptors lie deep in the skin. They are stimulated only when firm pressure is applied to the skin. Heat receptors and cold receptors respond to the direction of heat flow. The sensation of warmth is the result of heat flowing into the skin and stimulating the heat receptors. The sensation of coolness is the result of heat flowing out of the skin and stimulating the cold receptors. Pain receptors warn against injury. They respond to all types of massive stimulation. The sensation of pain is the same no matter what is causing it.

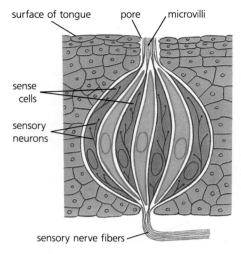

surface of tongue   pore   microvilli

sense cells

sensory neurons

sensory nerve fibers

▲ **Figure 15–16**
Structure of a Human Taste Bud.

Taste

The surface of the tongue is covered with small projections called *papillae* (puh PIL ee). Within the papillae are the taste receptors, or **taste buds.** Each taste bud is made of a number of sense cells and opens to the surface of the tongue through a pore. See Figure 15–16. Microvilli from the sense cells extend through the pore. Nerve fibers branch among the cells of the taste bud, and each cell is in contact with one or more neurons.

Only substances in solutions can stimulate the taste buds. Many substances dissolve in the saliva in the mouth. Taste buds are sensitive to only four basic tastes—sour, bitter, sweet, and salty. Each taste bud is particularly sensitive to one of these tastes and responds only slightly to the others. The taste buds for each taste tend to be found on specific areas of the tongue. Taste buds for sourness, for example, are found along the sides of the tongue. Taste buds for bitterness are located at the back of the tongue. Taste buds for sweetness and saltiness are on the tip of the tongue. See Figure 15–17.

When taste buds are stimulated, impulses are started by the sensory cells of the structure and carried by sensory pathways to the brain, where they are interpreted. Actually, most of the flavor of food comes from its smell. This is why most food has little taste to a person with a stuffy nose.

Smell

The receptors for smell, the **olfactory** (ol FAK tuh ree) **cells,** are found in the mucous membrane lining the upper nasal cavity. Odor is detected when molecules of a gaseous substance enter the nose, dissolve in the mucus, and stimulate the olfactory receptors. The olfactory cells are specialized nerve cells. See Figure 15–18. When they are stimulated, impulses are carried by the *olfactory nerves* to the brain, where they are interpreted.

Figure 15–17
Location of Taste Buds on a Human Tongue. Licking an ice cream cone makes sense because the taste buds for sweetness are found on the tip of the tongue. ▼

sour   bitter   sweet   salty

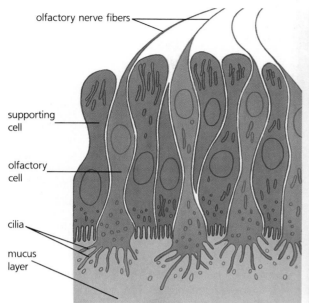

olfactory nerve fibers

supporting cell

olfactory cell

cilia

mucus layer

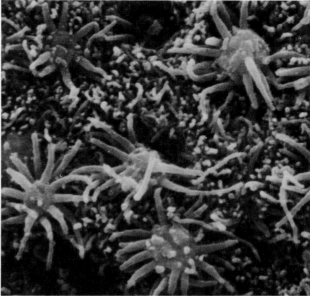

▲ **Figure 15–18**

Structure of Human Olfactory Receptors. Humans smell by using olfactory receptors located in the mucous membrane of the nasal cavity. The receptor cells are neurons. Sensory cilia (right) project from the neuron into the mucous layer.

Unlike taste buds, which respond to only four basic tastes, olfactory cells appear to respond to more than 50 different basic odors. Like the taste buds, each olfactory cell appears to be more sensitive to one basic odor than to all the others. Continuous exposure to a specific odor quickly leads to an inability to detect that odor but does not interfere with the detection of other odors. This is called *adaptation* and is thought to be partly a response of the central nervous system.

Both taste and smell result from the chemical stimulation of receptors. However, olfactory receptors are much more sensitive than the cells of the taste buds. Olfactory receptors are stimulated by much smaller amounts of chemicals, and they are sensitive to many more types of chemicals.

15-3 Section Review

1. Which part of the eye regulates the amount of light entering the eye?
2. What are the two kinds of light-sensitive cells found in the retina?
3. Name the three main divisions of the ear and the structure in the ear that is involved in balance or equilibrium.
4. List the five kinds of sense receptors of the skin and the four basic taste receptors found on the human tongue.

Critical Thinking

5. If you had to find an object in a darkened room, which senses would be most useful? Which would be least useful?   (*Ranking*)

Laboratory Investigation

Investigation of Human Senses: Sight

The various sense organs provide information about what is happening both inside and outside your body. In this investigation, you will explore some aspects of the sense of sight.

Objective

- *Explore,* and draw conclusions about, several characteristics of vision in humans.

Materials

small object (cup)
meter ruler
centimeter ruler
white, unlined paper

Procedure

A. Humans have binocular vision, meaning that both eyes cover the same field of vision. Stand about 5 m from a small object, such as a cup. While pointing your index finger at the object, look at it and close one eye. Then, open the eye and close the other eye. Do this several times. How does the field of vision of each eye compare? Record your observations in a data table like the one below.

Characteristic	Observations
binocular vision	
focusing	
blind spot	
retinal fatigue	

B. The lens of the eye changes shape, allowing you to focus on nearby or distant objects. Hold your hand 15 cm from your face, and spread your fingers apart. Look through your fingers at a distant object. Then, look back at your fingers. Repeat this several times. Record whether the background and fingers are in focus simultaneously.

C. The blind spot is the area of the retina where the optic nerve enters the eye. There are no rods or cones in this area, so no vision occurs at the blind spot. Cover

your left eye. Then, while holding your book about 10 cm away, focus on the star shown below. Continue looking at the star while slowly moving the book away from you. Note what happens to the dot as you move the book away. Record the approximate distance at which this change occurs. Repeat this process for your other eye.

D. Receptors in the retina that are stimulated by light for an extended period can become fatigued. As a result, an afterimage can be produced. Stare for 30 seconds at the dot in the center of the four colored squares below. Then stare at a sheet of white, unlined paper for 5 to 10 seconds. Draw the square and colors that you see. Record your observations.

Analysis and Conclusions

1. How do your observations from step A indicate that the fields of vision for each eye overlap but are not identical?

2. Based upon your observations regarding binocular vision, how do you think the ability to perceive distance would be affected if each eye had an entirely separate field of vision?

3. Do humans focus through a continuous depth of field or at one particular distance at a time? How do your results from step B support your answer?

4. Why are you not usually aware of the blind spot in each eye?

5. Propose a hypothesis to explain why the colors in the afterimage are different from the original colors.

Chapter Review

Study Outline

15–1 The Central Nervous System

■ The human nervous system consists of the brain and the spinal cord of the central nervous system and the nerve network of the peripheral nervous system.

■ The major parts of the brain are the cerebrum, cerebellum, and medulla oblongata. The largest part of the brain is the cerebrum, which consists of an outer layer of gray matter, called the cerebral cortex, and an inner layer of white matter.

■ The cerebral cortex performs sensory, motor, and associative functions (including memory). The white matter consists of bundles of nerve fibers connecting the cerebral cortex with various parts of the nervous system.

■ The cerebellum coordinates all voluntary movements of the body and maintains physical balance and muscle tone. The medulla oblongata controls many involuntary activities of the body.

■ The spinal cord connects the nerve network of the peripheral nervous system with the brain. It controls certain reflexes that do not involve the brain.

15–2 The Peripheral Nervous System

■ The peripheral nervous system contains all the neurons and nerve fibers outside the brain and spinal cord. It is divided into the somatic nervous system and the autonomic nervous system.

■ The somatic nervous system is made up of both sensory and motor neurons that connect the central nervous system to skeletal muscles, the skin, and the sense organs. It controls voluntary body movements.

■ The autonomic nervous system controls involuntary body movements. It is made up entirely of motor neurons. It is divided into two systems—the sympathetic and parasympathetic nervous systems—whose effects are antagonistic.

■ A reflex is an involuntary, automatic response to a stimulus. The nerve pathway followed by impulses in a reflex is called a reflex arc.

15–3 Sense Receptors

■ The eye is made up of three layers—the sclera, the choroid, and the retina. The retina contains light-sensitive cells called rods and cones. Rods are sensitive to weak light; cones are sensitive to color.

■ Hearing takes place when sound vibrations travel to the inner ear, where they start impulses that are carried to the brain by the auditory nerve. Balance is a function of the inner ear and the cerebellum.

■ Sense receptors in the skin are sensitive to touch, pressure, heat, cold, and pain. Each type of receptor has a different structure from the others and is sensitive to only one kind of stimulus.

■ Taste buds are receptors that are sensitive to taste. They are found within papillae on the surface of the tongue.

■ Olfactory cells are receptors for smell and are found in the mucous membrane of the nasal cavity.

Vocabulary Review

thalamus (291)

hypothalamus (291)

pons (291)

brainstem (291)

cerebrum (291)

cerebral hemispheres (291)

corpus callosum (292)

cerebral cortex (292)

cerebellum (293)

medulla oblongata (293)

spinal cord (294)

spinal nerves (295)

cranial nerves (295)

somatic nervous system (296)

autonomic nervous system (296)

parasympathetic nervous system (296)

sympathetic nervous system (296)

reflex (297)

reflex arc (298)

sclera (299)

cornea (300)

choroid coat (300)

iris (300)

pupil (300)

lens (300)

retina (300)

optic nerve (300)

nearsighted (301)

farsighted (301)

rods (301)

cones (301)

Chapter Review

auditory canal (302)

tympanic membrane (303)

oval window (303)

Eustachian tube (303)

cochlea (304)

semicircular canals (304)

auditory nerve (304)

taste buds (306)

olfactory cells (306)

A. Sentence Completion—Fill in the vocabulary term that best completes each statement.

1. The _____ of the eye focuses light on the retina, producing an image.

2. A _____ is an involuntary, automatic response to a given stimulus.

3. The function of the _____ is to make the pressure in the middle ear equal to that of the atmosphere outside.

4. The pons, the medulla oblongata, and the midbrain are often called the _____.

5. In _____ people, the eyeball is too short.

6. The _____ is the largest part of the human brain.

7. The _____ is made up mainly of nerve fibers that connect the spinal cord to other parts of the brain.

8. Toward the front of the eye is an opening called the _____.

9. The hindbrain consists of the brainstem, the _____, the pons, and the medulla oblongata.

B. Analogies—Select the vocabulary term that relates to the single term in the same way that the paired terms are related.

10. spinal cord **is to** spinal nerves as brain **is to** _____

11. rods: black and white as _____: color

12. cochlea: hearing as _____: balance

13. thalamus: relay center as _____: relay system

14. norepinephrine: sympathetic nervous system as acetylcholine: _____

15. taste: taste buds as smell: _____

Content Review

16. What is the function of the central nervous system?

17. How is the central nervous system protected?

18. What are the functions of the hypothalamus?

19. Describe the structure of the cerebrum.

20. Compare the gray matter of the cerebrum with the white matter as to location and composition.

21. Explain the three main functions of the cerebral cortex.

22. What is the function of the white matter of the cerebrum?

23. Describe the structure and functions of the cerebellum.

24. What involuntary activities does the medulla oblongata control?

25. What are the three types of memory? How does each type differ?

26. Describe the structure of the spinal cord.

27. Contrast the spinal nerves and cranial nerves of the peripheral nervous system.

28. What body functions does the autonomic nervous system control?

29. How do the parasympathetic and sympathetic nervous systems work together?

30. What is the end result of the reflex arc that is stimulated by touching a hot object?

31. Identify the structures that light would pass through as it travels through the eye.

32. What are the functions of rods and cones, and in what kinds of light do they function?

33. Briefly explain how your sense of hearing works.

34. How is the ear involved in the maintenance of equilibrium?

35. How does the tongue distinguish different tastes?

36. What happens when the receptors for smell are stimulated?

Graphic Organizing

For information on graphic organizers, see Appendix G at the back of this text.

37. Use the incomplete concept map on the next page to organize information about the central nervous system. Copy the concept map onto a sheet of paper, and fill in the missing concepts and linking words.

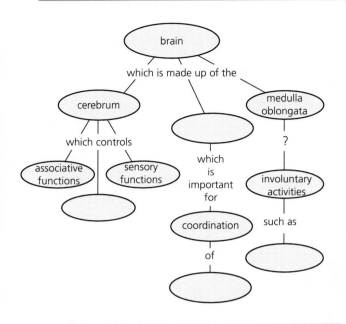

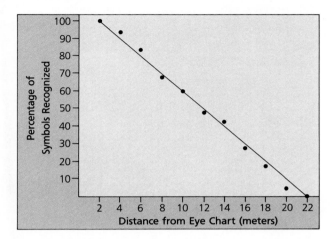

at different distances from the chart. The results of the study are shown in the line graph below. Identify the variable in this study and the response that was measured. What is the relationship between the variable and the response?

Critical Thinking

38. How is the arrangement of the semicircular canals in the inner ear related to their function? (*Relating Parts and Wholes*)

39. Suggest some possible reasons why reflex arcs would be advantageous to a species. (*Identifying Reasons*)

40. A physician observes that the pupils in a patient's eyes are unable to change size. What effect would this have on light reaching the retina? (*Predicting*)

41. People who produce abnormally low quantities of saliva have difficulty distinguishing the taste of certain foods. How could this be explained? (*Identifying Causes*)

Creative Thinking

42. "One sees with the brain rather than with the eyes." Do you agree with this statement? Why or why not?

43. Devise a plan that would help you commit some of the information in this chapter to long-term memory.

Problem Solving

44. A scientist was asked to develop an eye chart that used symbols of different sizes instead of letters. To standardize the eye chart, she determined the percentage of all the symbols that could be correctly identified

45. Design an experiment to determine the role of color in food selection by aquatic turtles. (Hint: Turtles can be fed raw fish.) What hypothesis is being tested in your experiment? Identify the variable being manipulated and the response being measured.

46. An investigation was carried out to study memory in rats. Seven rats were trained to run through a simple maze. Immediately after learning the maze, two rats were anesthetized with ether. Half an hour after learning the task, four other rats were anesthetized with chloroform. One rat was not anesthetized at all. After 24 hours, all of the rats were tested to see if they could recall how to go through the maze. The two rats that had been anesthetized immediately after learning the maze could not remember the task, while the other five rats went through the maze perfectly. What conclusion can you draw from this experiment? How can the experimental setup be improved? What further experiments should be done?

Projects

47. Construct a three-dimensional model of the human brain showing the major regions. Label the parts.

48. The brain produces its own painkillers called endorphins. Do some library research on endorphins, and prepare a short oral or written report.

49. Prepare a chart showing the effects of alcohol and other drugs on the human nervous system.

▲ Chemical substances in the bloodstream affect the behavior of this male ringneck pheasant.

Chemical Regulation

16

A pheasant's wings blur with a rapid motion that produces a whirring sound. With his chest puffed, the young male crows as loudly as he can in an attempt to attract a mate. Every spring, young male pheasants go through the same routine. Their actions are triggered by chemicals that flow through their bloodstreams. Secreted by various organs within the body, these chemicals regulate everything from behavior to metabolism to daily cycles. This chapter will introduce the mechanisms of chemical regulation and then the functions of the human endocrine system.

Guide for Reading

Key words: endocrine system, glands, hormones, pituitary gland, hypothalamus, thyroid gland, adrenal glands, islets of Langerhans

Questions to think about:

📖 How do hormones control the activity of particular target tissues?

📖 What are the functions of the human endocrine glands and their hormones?

16-1 Glands and Hormones

Section Objectives:

- *Compare* the operations of the nervous system and the endocrine system.
- *Define* the following terms: *exocrine gland, endocrine gland,* and *hormone.*
- *Explain* the regulation of hormone secretion through negative feedback.
- *Identify* the two basic mechanisms of hormone action and describe each.
- **Laboratory Investigation:** *Explore* the effects of hormones on a daphnia's heartbeat (p. 330).

Chemical vs. Nervous Regulation

The body systems of animals are never at rest. To maintain *homeostasis,* they constantly make adjustments to changing conditions outside and inside the body. You have already read how the nervous system takes part in this process in complex multicellular animals. In these animals, the nervous system sends electrochemical impulses through nerve fibers. Neurotransmitters bridge the tiny gaps between adjacent neurons. To help maintain homeostasis, the nervous system acts quickly and directs its messages to specific parts of the body.

▲ **Figure 16–1**

Courtship Behavior of Masai Giraffe. Mating behavior is just one of many functions controlled by both chemical and nervous signals.

313

Animals also use another system to maintain homeostasis. Through chemicals released into the bloodstream, the **endocrine system** regulates overall metabolism, homeostasis, growth, and reproduction. The bloodstream carries these chemicals to all the tissues of the body. When the chemicals reach the correct organ, a reaction occurs.

It takes time for chemicals to reach the correct organ through the bloodstream. For this reason, the endocrine system is slower than the nervous system in producing an effect. Its effects also tend to last longer. For the most part, the nervous system allows the body to make rapid responses for short periods of time. The endocrine system produces effects that last for hours, days, or even years. However, these two systems also work together. For example, when you run from danger, the nervous system not only controls your muscles but also stimulates parts of the endocrine system. The endocrine system then stimulates your heartbeat and respiration rate. Together, these two systems provide the energy and oxygen needed to maintain the body's activity.

Glands

Glands are organs made up of *epithelial* cells that specialize in the secretion of substances needed by the organism. Some glands, such as the digestive glands, discharge their secretions into ducts, which then carry the secretions to where they are used. These glands are called **exocrine** (EK suh krin) **glands.** Other glands release their secretions directly into the bloodstream. These glands are called **endocrine glands,** or *ductless glands.* They make up the endocrine system. See Figure 16–2.

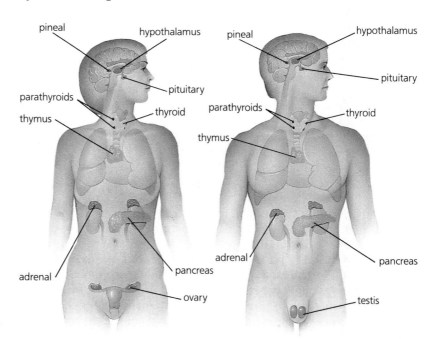

Figure 16–2
Glands of the Human Endocrine System. ▶

The secretions produced by the endocrine glands are called **hormones.** Traveling through the bloodstream, hormones regulate overall metabolism, maintenance of homeostasis, growth, and reproduction as well as other body processes.

Hormones

Although hormones are released into the bloodstream in one part of the body, they do not affect that area. Instead, they exert their effects elsewhere in the body. Because of this, hormones are sometimes called "chemical messengers."

Usually, there are low hormone concentrations in the bloodstream. Each type of hormone is recognized only by specific tissues. The tissue regulated by a given hormone is called the *target tissue.* The hormone may stimulate the target tissue and increase its activities, or it may inhibit the target tissue and decrease its activities. This is done by changing the rates of certain biochemical reactions in the target tissue. A hormone may cause a reaction to start, to speed up, to slow down, or to stop. Unlike enzymes, however, hormones do not act directly on the reacting substances. They appear to act always through some intermediate cellular process. The mechanisms of hormone action will be discussed in detail later in this chapter.

Most hormones fall into two classes. *Protein-type hormones* consist of chains of amino acids or related compounds. These hormones cannot pass through the cell membrane because they cannot dissolve in the lipids that are present in the cell membrane. **Insulin** (IN suh lin) is an example of this type of hormone. *Steroid* (STIHR oyd) *hormones* are lipidlike, carbon-ring compounds that are able to pass through the cell membrane. Steroid hormones are produced in the outer layer of the adrenal gland and in the gonads. Estrogen is an example of a steroid hormone.

Prostaglandins

Prostaglandins (prahs tuh GLAN dinz) are "local hormones." That is, they produce their effects without entering the bloodstream. They also may modify the effects of other hormones. They are thought to influence a wide variety of metabolic activities, including heartbeat, blood pressure, excretion of urine, and contraction of the uterus at childbirth. Prostaglandins are being studied for use in the treatment of high blood pressure, stroke, asthma, and ulcers.

The Regulation of Hormone Secretion

Usually, endocrine glands do not secrete their hormones at a constant rate. The rate varies with the needs of the body. A nerve impulse may cause a gland to speed up, slow down, or stop its production of a hormone. However, in most cases, chemical stimuli, including other hormones, regulate the secretions of the endocrine system.

Biology and You

Q: Some athletes in my school use steroids. Is steroid use common among high-school students?

A: Many people were shocked to learn that more than 1 in 15 males in high school admitted to using steroids in a recent survey. When asked why they used steroids, some students mentioned pressure from parents and coaches to excel. Others said they wanted to improve their appearance.

Anabolic steroids, synthetic versions of the male hormone testosterone, increase muscle bulk and strength in both males and females. Most people who take steroids do so without medical supervision and are unaware of their negative side effects.

Side effects of steroid use may range from acne, headaches, and fatigue to more serious problems. These may include sterility, stunted growth, heart problems, aggressive behavior, cancer, and even death. Teenagers risk effects because their growing bodies can be permanently damaged.

■ *Create a poster warning teenagers about steroid use. You may want to use photos from the sports section of the newspaper. Be sure to explain why steroid use is dangerous.*

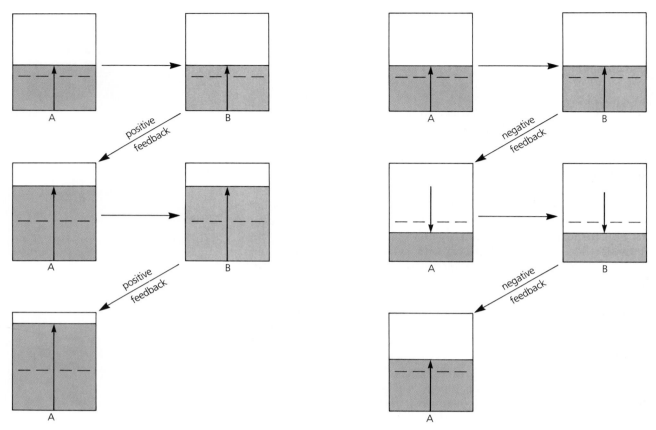

▲ **Figure 16–3**

Positive Feedback and Negative Feedback. A and B represent two related metabolic processes. An increase in A causes an increase in B. The change in B causes a change in A, which is called feedback. Positive feedback occurs when the change in A, which was caused by B, is the same as the change in B, which is caused by A. If the change caused by B is the opposite of the change in A, negative feedback has occurred. Negative feedback prevents A from changing greatly in either direction. It is an important factor in maintaining homeostasis.

The chemical regulation of glandular secretions is accomplished by *feedback*. Feedback occurs when one change causes another change, which, in turn, affects the original change. In cases of **positive feedback,** the feedback reinforces the original change. As you can observe in Figure 16–3, an increase in the level of A produces an increase in the level of B. Because the increase in B causes a further increase in A, there is said to be positive feedback from B to A. This type of feedback tends to amplify the original change in A.

In **negative feedback,** the feedback to A opposes the original change in A. In the example shown in Figure 16–3, an increase in the level of A produces an increase in the level of B. However, the increase in B then causes a decrease in A. Thus, when A increases above its normal level, negative feedback from B causes A to decrease. When A falls below its normal level, then negative feedback from B will cause A to increase. Negative feedback tends to ward the maintenance of homeostasis by restoring A to its original level.

The thermostat that keeps an oven at a constant temperature is a good example of negative feedback. When the temperature rises above the set value, the thermostat turns the oven off, allowing it to cool down. When the temperature drops below the set value, the thermostat turns the oven on again.

Most glands are regulated by negative feedback because it maintains homeostasis. The secretion of a hormone usually is controlled by the concentration of another substance in the blood, often another hormone. For example, the secretion of the hormone **thyroxine** (thy RAHK sin) by the thyroid gland is regulated by *thyroid stimulating hormone* (TSH). TSH is produced by the pituitary gland. When the thyroxine level is low, the pituitary is stimulated to secrete TSH. In turn, TSH stimulates the thyroid to produce thyroxine. When the thyroxine level reaches a certain point, the secretion of TSH by the pituitary is reduced. The pituitary then stops secreting TSH, and the thyroid stops secreting thyroxine. This is an example of the secretions of one gland being regulated by the secretions of another. In other cases, endocrine glands are controlled by the levels of simple substances, such as calcium and glucose, in the blood.

The Mechanisms of Hormone Action

Each hormone controls the activity of a particular target tissue. Since hormones are carried in the bloodstream, and since they reach all body tissues, each target tissue must have a way of recognizing the hormone that is intended for it. There also must be a mechanism through which the hormone can produce its effect within the target cells.

Recent research indicates that there are two basic mechanisms governing hormone action. One of these mechanisms, called the *one-messenger model,* is mainly characteristic of steroid hormones. The other, called the *two-messenger model,* is characteristic of protein-type hormones.

The One-Messenger Model

Steroid hormones are small, lipid-soluble molecules that are able to pass through cell membranes. These hormones enter most of the cells of the body but produce their effects only in the target cells. In these cells, there are receptor proteins that recognize a particular steroid hormone and react with it, forming an active factor. See Figure 16-4. It is the active factor that changes the rate of some chemical reaction in the cell, producing the hormonal effect. There is evidence that the active factors produce their effect by entering the cell nucleus and acting on the genetic material (DNA) that controls the cell's activities.

Figure 16–4

One-Messenger Model of Hormone Action. Steroid hormones produced by endocrine glands pass through the cell membranes of target cells and combine with special receptors. The resulting active factors trigger the response of the target cell. Nontarget cells do not have the appropriate protein receptors and are not affected by the hormones. ▶

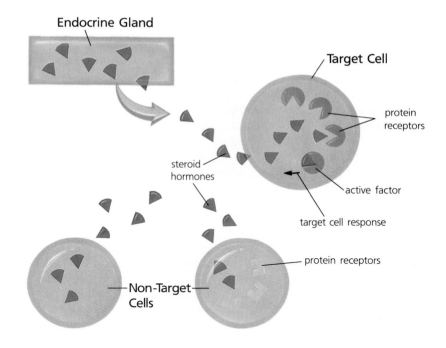

The Two-Messenger Model

Protein-type hormones usually cannot pass through cell membranes. However, specific receptors on the outer surface of target cell membranes recognize them. See Figure 16–5. When the hormone combines with the receptor on the membrane surface, the combination activates enzymes within the membrane. These enzymes produce a compound that acts as a second messenger. The second messenger diffuses throughout the

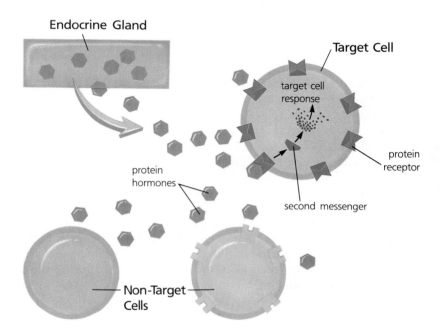

Figure 16–5

Two-Messenger Model of Hormone Action. Protein hormones produced by endocrine glands cannot pass through cell membranes. Instead, they react with receptors on target cell membranes, which then produce a second messenger. The second messenger enters the cell and produces the hormonal effect. ▶

cell interior and produces the hormonal effect. In this model, the hormone does not enter the cell. Instead, a second messenger does its work.

Although the same compound may be a second messenger in different types of target cells, it produces different effects in different cells. For example, the same compound can cause thyroid cells to produce thyroxine, adrenal cortex cells to produce cortisol, and kidney tubule cells to reabsorb more water.

16-1 Section Review

1. What two systems regulate and coordinate body functions?
2. Name the two types of glands found in the human body.
3. What are hormones?
4. Name the two basic models of hormone action.

Critical Thinking

5. What kind of feedback would occur if an increase in hormone A caused hormone B to increase which then caused hormone A to decrease? Explain. (*Reasoning Conditionally*)

16-2 The Human Endocrine System

Section Objectives:

- *Name* the major endocrine glands of the human body.
- *Describe* the role of the hypothalamus within the endocrine system.
- *List* the hormones released by each endocrine gland and briefly describe their functions.
- *Identify* disorders caused by hypersecretion or hyposecretion of specific hormones.

    Concept Laboratory: *Gain* an understanding of how negative feedback maintains homeostasis (p. 329).

The Functioning of Endocrine Glands

The human endocrine system consists of a number of endocrine glands that regulate a wide range of activities. The improper functioning of an endocrine gland may result in a disease or disorder of the body. An excess, or **hypersecretion** (hy per suh KREE shun), of a hormone may cause one type of disorder. A deficiency, or **hyposecretion** (hy poh suh KREE shun), of a hormone may cause another type of disorder. See Figure 16–6 for an outline of these disorders. In the following sections, you will read about the human endocrine system in more detail.

Glandular Disorders			
Gland	Hormone	Effects of Oversecretion (Hypersecretion)	Effects of Undersecretion (Hyposecretion)
anterior pituitary	growth hormone	*in childhood* effects include giantism. Individual grows tall but is normally proportioned. Mental development is not affected. *In adulthood* effects include acromegaly. Individual has abnormally large hands and feet and enlarged facial structures. Does not affect mental processes.	*in childhood* effects include dwarfism. Individual is small but is normally proportioned. Adult sexual development often does not occur.
adrenal cortex	aldosterone, cortisol	Cushing's disease. Individual has excess fat deposits in the upper body, a puffy face, excess growth of facial hair, and a high blood glucose level. Decreased immunity to disease also occurs.	Addison's disease. Normal blood glucose level cannot be maintained. Individual becomes sluggish, weak, loses weight, and develops increased skin pigmentation. Tolerance to stress is reduced. Without medication, the disease causes death.
thyroid	thyroxine	hyperthyroidism. Individual is nervous, irritable, loses weight, and cannot sleep. Often, eyes protrude, a condition called exophthalmos. Hyperthyroidism is often accompanied by a goiter, or enlarged thyroid.	*in infancy,* effects include cretinism. Individual is a dwarf whose body parts are out of proportion. Mental retardation occurs. *in adulthood* results include hypothyroidism. Individual is sluggish, gains weight.
pancreas (β cells, islets of Langerhans)	insulin	diabetic shock. The blood glucose level falls dangerously, and convulsions, unconsciousness, and death may occur if untreated.	diabetes. Individual has an abnormally high blood glucose level, becomes dehydrated, loses weight, and cannot resist infections. If untreated, can cause death.
pancreas (α cells, islets of Langerhans)	glucagon	abnormally high blood glucose level. Results are similar to diabetes.	abnormally low blood glucose level also known as hypoglycemia.

▲ **Figure 16–6**   Effects of Oversecretion and Undersecretion of Hormones.

The Pituitary Gland and the Hypothalamus

The **pituitary** (pih TOO uh tehr ee) **gland** is a small gland attached to the brain. See Figure 16–7. Often called the "master gland" of the body, the pituitary controls the activity of a number of other endocrine glands. It consists of an anterior, or front, lobe and a

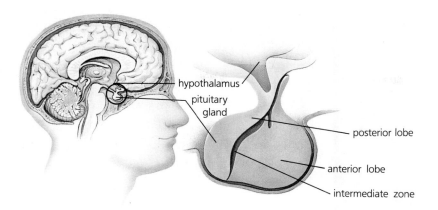

hypothalamus

pituitary
gland

posterior lobe

anterior lobe

intermediate zone

◀ **Figure 16–7**
Structure of the Pituitary Gland. The
pituitary gland, located beneath the brain,
consists of an anterior lobe and a posterior
lobe. The release of pituitary hormones is
regulated by the hypothalamus.

posterior, or back, lobe. Between these two lobes is a small
intermediate zone that does not work in humans. In other animals,
this area is larger and is an active part of the endocrine system.

The part of the brain that is connected to the pituitary is called
the **hypothalamus** (hy poh THAL uh mus). It acts like an endocrine
gland and controls the release of hormones from the pituitary. The
hypothalamus receives information from the nervous system. This
information helps to determine when the hypothalamus should
stimulate the pituitary to release a hormone. The hypothalamus is
the major link between the body's two regulating systems. It is also
influenced by the concentration of various hormones in the blood.
Thus, the hypothalamus directs the work of the pituitary.

The Anterior Pituitary   The anterior lobe of the pituitary secretes several different hormones. Many of them help to control the
body's metabolism. Hormones produced by the hypothalamus,
called *releasing hormones,* or **releasing factors,** control the release
of hormones from the anterior lobe. The releasing factors are
produced in the ends of specific neurons within the hypothalamus.
When they are released by the neurons, the factors are absorbed
directly into capillaries that carry them to the anterior pituitary. A
specific factor controls the release of each type of hormone from
the anterior pituitary. The hormones released by the pituitary
stimulate other endocrine glands to secrete their hormones.

The major hormones of the anterior pituitary are as follows:

▪ *Thyroid-stimulating hormone,* or TSH, stimulates the production
and release of thyroxine by the thyroid gland.

▪ *Adrenocorticotropic* (uh DREE noh kort ih koh troh pik) *hormone,*
or ACTH, stimulates the production and release of hormones from
the cortex layer of the adrenal glands. It is used in the treatment of
arthritis, asthma, and allergies.

▪ *Growth hormone,* or GH, controls growth. It indirectly affects the
growth of bone and cartilage by regulating production of another
factor that acts directly on these tissues. GH directly affects protein,
carbohydrate, and fat metabolism at a cellular level. See Figure
16–6 for the effects of oversecretion and undersecretion of GH.

Figure 16–8

Human Egg Cell Being Released from Ovary. The release of this egg cell from the ovary is caused by luteinizing hormone. ▶

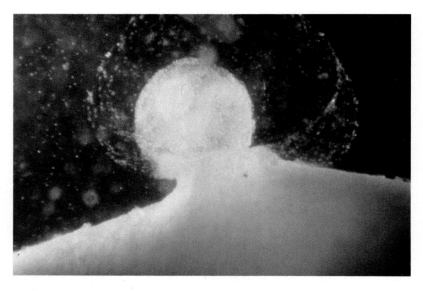

■ *Follicle-stimulating hormone,* or FSH, stimulates the development of egg cells in the ovaries of females. In males, it controls the production of sperm cells in the testes.

■ *Luteinizing* (LOOT ee in iz ing) *hormone,* or LH, causes the release of egg cells from the ovaries in females. It controls the production of sex hormones in both males and females.

■ *Prolactin* (proh LAK tin) stimulates the secretion of milk by the mammary glands of the female after she gives birth. Otherwise, it is secreted only in small amounts. It is thought that the production of prolactin is normally inhibited by a factor secreted by the hypothalamus. Following childbirth, the secretion of this inhibitory factor is blocked, and prolactin is produced.

The Posterior Pituitary   The posterior lobe of the pituitary is directly connected to the hypothalamus. Two tracts of nerve fibers beginning in the hypothalamus have their endings in the posterior pituitary. Two hormones, *oxytocin* (ahk sih TOH sin) and *vasopressin* (vay zoh PRES in), are produced by these nerve cells in the hypothalamus. The hormones then pass down the axons to the posterior lobe of the pituitary for storage and eventual release.

Oxytocin stimulates contractions of the smooth muscles of the uterus during childbirth. Vasopressin, also known as *antidiuretic* (an tee dy uh RET ik) *hormone,* or ADH, controls the reabsorption of water by the nephrons of the kidneys. ADH increases the permeability of the tubules to water, so that water is reabsorbed by osmosis.

Figure 16–9

The Thyroid Gland. The hormones of the thyroid regulate the blood calcium level and the rate of metabolism in the body. ▼

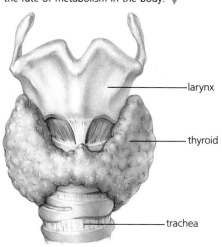

larynx

thyroid

trachea

The Thyroid Gland

As seen in Figure 16–9, the **thyroid** (THY royd) **gland** is located in the neck just below the larynx and in front of the trachea. This gland secretes the iodine-containing hormone thyroxine, which you read about earlier in this chapter. Thyroxine regulates the rate of metabolism in the body. It increases the rate of protein, carbohy-

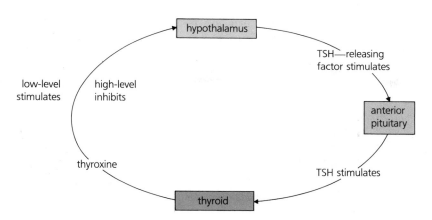

◀   **Figure 16–10**
Regulation of Thyroxine Secretion.

drate, and fat metabolism, and the rate of cellular respiration. This hormone is needed for normal mental and physical development. The thyroid also secretes another hormone, *calcitonin* (kal suh TOH nin), which is involved in the regulation of the blood calcium level.

The interaction of several hormones regulates the secretion of thyroxine. See Figure 16–10. If the concentration of thyroxine in the blood falls below a certain level, the hypothalamus is stimulated to produce TSH-releasing factor. The releasing factor stimulates the anterior pituitary to secrete thyroid-stimulating hormone (TSH). TSH, in turn, stimulates the release of thyroxine by the thyroid. Increasing levels of thyroxine in the blood inhibit the production of releasing factor by the hypothalamus. This inhibits the production of TSH and decreases the stimulation of the thyroid. Thus, by negative feedback, the concentration of thyroxine in the blood controls the system that produces it. Refer to Figure 16–6 for the effects of oversecretion and undersecretion of thyroxine.

The Parathyroid Glands

Four tiny, oval glands called the **parathyroid** (par uh THY royd) **glands** are embedded in the back of the thyroid. See Figure 16–11. They secrete parathyroid hormone, or *parathormone* (par uh THOR mohn). This hormone regulates calcium and phosphate metabolism.

Calcium is necessary for proper growth, the health of bones and teeth, blood clotting, nerve function, and muscle contraction. Phosphate is found in bones and in many important compounds in the body, including ATP, DNA, and RNA.

For the nerves and muscles to function normally, the concentration of calcium ions in the blood must be kept within fairly narrow limits. While some calcium is stored within cells, most of it is stored in bones in the form of calcium phosphate compounds. When the blood calcium level drops even slightly, the parathyroids are stimulated to secrete parathormone. This hormone causes the release of calcium from bone into the plasma. When the blood calcium concentration rises above a certain level, calcium is stored in bones. Excess calcium can be excreted by the kidneys and intestines.

Figure 16–11

The Parathyroid Glands. The four small parathyroid glands embedded in the back of the thyroid produce hormones that regulate calcium and phosphate metabolism in the body. ▼

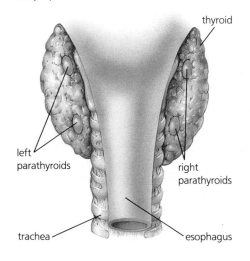

A deficiency of parathormone results in low blood calcium levels. If the level is low enough, the skeletal muscles become hypersensitive and contract violently. This condition is called *tetany* (TET uh nee). Oversecretion of parathormone results in the removal of calcium from bones to the point where they become brittle and break easily.

The Adrenal Glands

Capping the two kidneys are the **adrenal** (uh DREEN ul) **glands.** See Figure 16–12. Each gland consists of an inner layer, called the *medulla,* and an outer layer, called the *cortex.* The hormones of the adrenal gland help the body to deal with stress. The medulla releases hormones that handle sudden stress. Hormones of the cortex help the body deal with long-term stress.

The Adrenal Medulla   The tissue of the adrenal medulla is related to nerve tissue. **Epinephrine,** or *adrenalin* (uh DREN uh lin), and **norepinephrine,** or *noradrenalin* (nor uh DREN uh lin), are the two hormones secreted by the adrenal medulla. About 80 percent of the secretion is epinephrine and 20 percent norepinephrine. The nerves in the sympathetic nervous system regulate the secretion of these hormones by the adrenal medulla. The effects of these hormones are the same as those produced by stimulation of the sympathetic nervous system. However, the effects of these hormones last much longer.

Epinephrine and norepinephrine produce what is called the "emergency response," or "fight-or-flight" reaction. They are secreted in response to sudden stresses, such as fear, anger, pain, or physical exertion. Both hormones constrict the blood vessels of the body. Epinephrine increases the rate of metabolism and the release of glucose by the liver. It also increases the rate and strength of the heartbeat, blood pressure, breathing rate, blood clotting rate, and sweating.

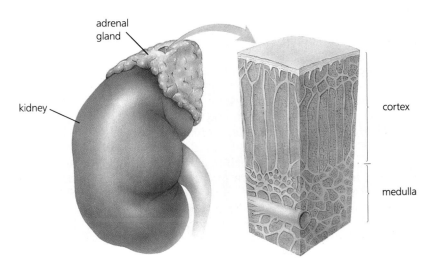

Figure 16–12

The Adrenal Glands. The adrenal glands are located on top of the kidneys. Hormones of the adrenal medulla deal with sudden stress, while hormones of the adrenal cortex deal with long-term stress. ▶

The Adrenal Cortex   The hormones of the adrenal cortex are compounds called **corticosteroids** (kort ih koh STIHR oyds). They are all synthesized from cholesterol. The major hormones of the adrenal cortex are cortisol and aldosterone, but more than 30 others are also known.

Cortisol (KORT uh sahl), or *hydrocortisone* (hy druh KORT uh sohn), affects the metabolism of carbohydrates, proteins, and fats. Its major action involves the synthesis of glucose in the liver and other tissues. It is important in regulating the glucose level in the blood.

Cortisone is a compound that is closely related to cortisol. Produced synthetically, cortisone is used as a drug for the treatment of arthritis. It is also used to counteract the symptoms of allergies.

Aldosterone and related hormones maintain the normal mineral balance in the blood. Aldosterone increases both the reabsorption of sodium by the kidney tubules and the excretion of potassium by the kidney tubules. By controlling the concentrations of these ions, aldosterone also controls the volume of the intercellular fluid and the blood.

The adrenal cortex also secretes male and female sex hormones, but the amount of female hormones produced is slight. The male sex hormones may play some role in regulating sexual development in males. See Figure 16–6 for the effects of oversecretion and undersecretion of the hormones of the adrenal cortex.

The Pancreas — Islets of Langerhans

You read about the function of the pancreas in the digestive system in Chapter 8. The pancreas is both an exocrine gland and an endocrine gland. The exocrine portion secretes digestive juices into the pancreatic duct. The endocrine portion consists of small clusters, or islands, of hormone-secreting cells. These cells, called the **islets of Langerhans** (LAHNG er hahnz), are scattered throughout the pancreas. There are two types of cells in the islets — *alpha* (*α*) *cells* and *beta* (*β*) *cells*. Alpha cells secrete the hormone **glucagon** (GLOO kuh gahn). Beta cells secrete the hormone insulin. Both of these hormones function in the control of carbohydrate metabolism.

Insulin   Insulin affects glucose metabolism in several ways. It increases the rate at which glucose is moved through cell membranes in most of the tissues of the body. Insulin secretion is controlled by the concentration of glucose in the blood. When the level of glucose in the blood is high after the ingestion of glucose, the beta cells of the pancreas are stimulated to secrete insulin. The insulin promotes the passage of the glucose into the body cells, which lowers the blood glucose level. Within the cells of the liver and skeletal muscle, insulin promotes the change of glucose to glycogen. In fatty tissues, it promotes the change of glucose to fat. It also increases the rate of oxidation of glucose within cells.

Biology and You

Q: When I'm nervous, my hands feel icy. Is this normal?

A: Icy hands, a "knotted" stomach, and a racing heart are all common when you are nervous, angry, or frightened. These emotions cause the release of the hormone epinephrine, which prepares your body for "fight or flight"—that is, to fight against or run away from the threat or challenge.

Extremities, such as hands and feet, feel cold because there is less blood flowing to them. Epinephrine constricts certain blood vessels, pushing blood away from your skin and toward vital organs—your heart, lungs, brain, and muscles. To increase blood flow, your heart beats faster and harder, perhaps making your chest feel tight. Breathing quickens to send more oxygen to your cells.

There are other common feelings associated with fright. Your stomach may feel "knotted" or gassy as epinephrine slows the flow of digestive enzymes. You may feel a "lump" in your throat as your throat muscles contract to open the airway to your lungs.

Write a short story of one page about a person who is nervous or frightened. Include details on how the body reacts and why.

Glucagon Usually, the effects of glucagon on glucose metabolism are opposite, or antagonistic, to those of insulin. While insulin lowers the blood glucose level, glucagon raises it. When the glucose concentration in the blood falls below a certain level, the alpha cells of the pancreas are stimulated to secrete glucagon. Glucagon promotes the conversion of glycogen to glucose in the liver. This glucose quickly diffuses out of the liver into the bloodstream.

When the supply of liver glycogen is exhausted, glucagon causes the conversion of amino acids and fatty acids to glucose. Thus, when there are not enough carbohydrates for the body, body fat and proteins are broken down to provide glucose to meet energy requirements.

Diabetes When the islets of Langerhans fail to produce enough insulin, the amount of glucose that can enter the body cells is decreased. As a result, the concentration of glucose in the blood increases, and the excess sugar is excreted in the urine. This condition, called **diabetes** (dy uh BEET eez) **mellitus,** can be inherited. Symptoms of diabetes include loss of weight despite increased appetite, thirst, and general weakness. If untreated, diabetes causes death. Proper diet, the use of oral medication, and/or daily injections of insulin can control the disease. See Figure 16–6 for the effects of oversecretion and undersecretion of the hormones of the islets of Langerhans.

The Gonads

The **gonads** (GOH nadz), or sex glands, are the ovaries of the female and the testes of the male. The *ovaries* (OH vuh reez) produce egg cells and the *testes* (TES teez) produce sperm cells. The gonads also secrete *sex hormones,* which control all aspects of sexual development and reproduction. The role of sex hormones in reproduction is discussed in Chapter 23.

Figure 16–13

A Woman Buying Insulin. Daily injections of insulin allow many people with diabetes to lead normal lives. The amount of insulin administered must be adjusted according to the kinds of foods eaten and the amount of exercise performed. ▶

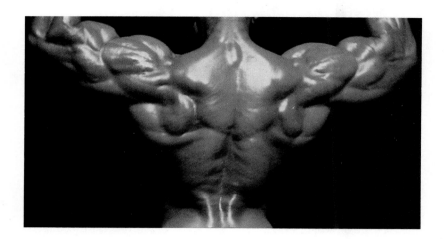

Anabolic Steroids. Anabolic steroids, which some athletes take to build body mass and strength, also pose serious health risks.

The Ovaries   The ovaries produce two hormones, **estrogen** (ES truh jen) and **progesterone** (proh JES tuh rohn). During development, estrogen stimulates the development of the female reproductive system. Estrogen also promotes the development of female *secondary sex characteristics,* such as broadening of the hips and development of breasts. Estrogen acts with progesterone to regulate the menstrual cycle.

The Testes   The testes secrete male sex hormones called *androgens.* The most important androgen is **testosterone** (tes TAHS tuh rohn). During fetal development, testosterone stimulates development of the male reproductive system. This hormone also promotes development of male *secondary sex characteristics,* such as a deep voice, beard, body hair, and the male body form.

Anabolic steroids are derived from this male hormone. Unfortunately, many athletes use these steroids to build body mass and strength. See Figure 16–14. Steroid use, however, poses serious health risks. Liver and kidney disorders, high blood pressure, and addiction can occur. Women may experience menstrual irregularities and masculinization. In men, aggressive behavior, a decreased sperm count, and impotence often develop. Little is known about long-range side effects. Medical experts believe some of these effects may be irreversible. As a result, anabolic steroids have been banned in many sports to protect the health of athletes.

Other Glands and Hormones

The Thymus   The **thymus** (THY mus) is a gland located in the upper chest near the heart. It is large in infants and children but shrinks after the start of adolescence. Early in life, the thymus helps in the processing of lymphocytes, which are part of the body's defense against infection. See Chapter 10. Current research indicates that the thymus produces a hormone called *thymosin* (THY muh sin) throughout childhood. Thymosin is thought to stimulate development of T lymphocytes, which are important in immunity. The thymus appears to serve no function in adults.

Can You Explain This?

To perform better in competitions, an athlete took anabolic steroids in order to build up muscle mass. Although he developed significant muscle growth, a medical examination revealed that his body was producing less than normal amounts of testosterone.

■ *Propose an explanation for why the athlete's testosterone level was so low.*

▲ **Figure 16–15**

A White Flounder. Flounders are famous for their ability to match the color of their background. This effect is produced by melatonin, which acts on pigment cells of the flounder.

The Pineal Gland   The **pineal** (PY nee uhl) **gland** is a pea-sized structure that is attached to the base of the brain. It produces a hormone called *melatonin* (mel uh TOH nin). In flounder and other ectotherms, melatonin acts on the pigment cells. See Figure 16–15. In rats, it inhibits the functioning of the ovaries and testes. Some recent research indicates that melatonin may inhibit sexual development in human males. At puberty, when sexual development begins, the secretion of melatonin decreases. Melatonin may also inhibit sexual development in human females.

The Stomach and Small Intestine   Hormones also function in the digestive system. Special cells in the lining of the stomach secrete the hormone *gastrin,* which stimulates the flow of gastric juice. In the lining of the small intestine, there are cells that secrete the hormone *secretin,* which stimulates the flow of pancreatic juice. Secretin was the first hormone to be discovered.

16-2 Section Review

1. Define the terms *hypersecretion* and *hyposecretion.*
2. What are releasing factors?
3. What two hormones are secreted by the thyroid?
4. What causes diabetes?

Critical Thinking

5. What ethical values do you think are violated by the use of anabolic steroids by athletes? (*Making Ethical Judgments*)

Concept Laboratory

Negative Feedback

Goal

To gain an understanding of how negative feedback maintains homeostasis.

Scenario

Imagine you are a teacher. A friend who is an engineer has given you plans for a simple device that she claims will illustrate negative feedback control. You and your class must construct and use the device.

As you perform this investigation, think about how negative feedback works.

Materials

plastic soft drink bottle	eye screws
scissors	string
rubber ball, 2.5 to 4.0 cm. diameter	fishing float
	glass rod
	marker

Procedure *

 A. Using scissors, cut off the bottom of a clear, two-liter soft drink bottle.

B. Insert an eye screw into a rubber ball. Cut a piece of string 20 cm long. Attach one end to the eye screw and the other end to a float.

C. Insert a second eye screw into the rubber ball opposite the first eye screw. Cut a piece of string 10 cm long. Attach one end to the eye screw and the other end to a glass rod.

D. With the bottle held upside down, lower the float, the rubber ball, and the attached glass rod into the bottle. Have the glass rod pass through and hang below the mouth of the bottle, as shown in the picture at the top of the column.

E. Hold the bottle over a sink or bucket. Pour water into the bottle until the string is stretched to its full length but does not lift the ball from the bottle opening. Indicate the level of water with a mark on the outside of the bottle.

F. Add more water to the bottle slowly and note what happens. Then, add more water rapidly and note any change.

*For detailed information on safety symbols, see Appendix B.

Organizing Your Data

1. Once the string has been stretched to its full length, what happens when more water is added?

2. How does adding water cause the device to lose water?

3. How does the device adjust for different rates of water flow?

4. How would a longer or a shorter string between the float and the ball affect the water level?

Drawing Conclusions

1. Suppose the water represents blood glucose. What does the height of the water represent?

2. In the system controlling blood glucose, how does an increase in the level of blood glucose lead to a decrease in the blood glucose level?

3. Relate the operation of the device to the control of glucose levels by insulin.

4. Based on this activity, explain how negative feedback maintains homeostasis in the body.

Thinking Further

1. Design a device that would add water to the bottle only when the water level starts to fall below the minimum.

2. How does the operation of the device you designed for question 1 relate to the control of blood glucose levels by glucagon?

329

Laboratory Investigation

Effect of Chemicals on Heart Rate in Daphnia

Hormones regulate the activity of nearly all tissues and organs. Drugs also can stimulate or inhibit the body's activities. In this investigation, you will determine how the heart rate in the water flea, daphnia, is affected by chemicals.

Objective

- *Observe* the effects of a hormone and several drugs on the heart rate in daphnia, the water flea.

Materials

daphnia culture	toothpicks	aquarium water
droppers	petroleum jelly	epinephrine, 0.1%
microscope	paper towels	ethyl alcohol, 0.5%
slides	watch	cola, regular
coverslips	beaker	cola, caffeine-free

Procedure *

A. Using a dropper, transfer a single daphnia in a drop of culture water onto a microscope slide. Dab petroleum jelly on the corners of a coverslip with a toothpick. This will allow the daphnia to fit underneath the coverslip.

B. Gently lower the coverslip onto the drop. Under low power, observe the beating heart of the daphnia. The figure to the right shows the location of the heart.

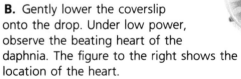

C. While one partner keeps time with a watch, the other should count the number of heartbeats in 15 seconds. Multiply by 4 to find the beats per minute. Repeat this procedure twice. Calculate and record the average beats per minute in a chart like the one above.

D. Place two drops of epinephrine solution at one edge of the coverslip, and draw out water from the opposite side with a piece of paper towel. Wait one minute, then, following the procedure in step C, calculate and record the average heartbeats per minute.

E. Lift the coverslip, and rinse the daphnia into a beaker of aquarium water labeled "treated daphnia."

*For detailed information on safety symbols, see Appendix B.

Treatment	Trial	Average Beats/Min.	Class Average Beats/Min.
water	1.		
	2.		
	3.		
adrenalin	1.		

F. Using a fresh daphnia each time, repeat the above procedures, separately testing alcohol, regular cola, and caffeine-free cola. As you finish testing each substance, rinse the daphnia that you used into the treated-daphnia beaker.

G. Following your teacher's instructions, combine your data with the data from the rest of the class, and calculate the class average values.

Analysis and Conclusions

1. How did epinephrine affect the heart rate in daphnia? How did alcohol affect the heart rate? What was the control?

2. How did caffeine affect the heart rate? What was the control?

3. Do the effects of epinephrine on heart rate in daphnia agree with what you know about the effects of epinephrine in humans?

4. Using the class average values, draw a bar graph that shows the heart rates resulting from treatment with each of the substances.

5. What was the purpose of recording three trials for each treatment? What was the purpose of combining data for a class average?

6. Why was each substance tested on a different daphnia instead of being tested on the same daphnia?

7. Based on the results of this experiment, what effects might you expect alcohol and caffeine to have on the heart rate of humans?

8. Predict what would happen to the heart rate if norepinephrine were added to a daphnia that had just been exposed to epinephrine.

Chapter Review

16

Study Outline

16–1 Glands and Hormones

■ The endocrine system regulates overall metabolism, homeostasis, growth, and reproduction. It is composed of specialized glands and tissues that secrete hormones.

■ The hormones, which act as chemical messengers, are released directly into the bloodstream and travel through the body to specific target tissues.

■ Protein-type hormones consist of chains of amino acids or related compounds. Steroid hormones are lipidlike, carbon-ring compounds that can pass through cell membranes.

■ Prostaglandins produce effects without entering the bloodstream.

■ The activity of most endocrine glands is controlled by a negative-feedback mechanism, whereby the concentration of a certain substance in the blood stimulates or inhibits gland function.

■ There are two mechanisms by which a hormone produces its effect within target cells. The one-messenger model applies to steroid hormones; the two-messenger model applies to protein-type hormones.

16–2 The Human Endocrine System

■ The pituitary gland, often called the body's "master gland," controls a number of other endocrine glands. The hypothalamus connects the pituitary to the brain and controls the release of hormones by the pituitary.

■ The major hormones of the anterior pituitary include: thyroid-stimulating hormone (TSH), adrenocorticotropic hormone (ACTH), growth hormone (GH), follicle-stimulating hormone (FSH), luteinizing hormone (LH), and prolactin. The hypothalamus produces the hormones oxytocin and vasopressin, which are transported to the posterior pituitary for storage and release.

■ The thyroid gland secretes the hormone thyroxine, which regulates the rate of metabolism in the body. The parathyroid glands secrete the hormone parathormone, which regulates calcium and phosphate metabolism.

■ The adrenal glands are composed of the medulla and the cortex. The medulla releases the hormones epinephrine and norepinephrine; the cortex releases corticosteroids.

■ The islets of Langerhans, located in the pancreas, secrete glucagon and insulin. Insulin lowers the blood glucose level, and glucagon raises it. Diabetes mellitus is caused by a deficiency of insulin.

■ Estrogen and progesterone, which are produced by the ovaries, control female sexual development and reproduction. Testosterone, which is produced by the testes, stimulates the development of the male reproductive system.

■ Research indicates that the thymus produces the hormone thymosin, which is thought to stimulate the development of T lymphocytes. The pineal gland produces the hormone melatonin.

■ The stomach secretes the hormone gastrin, which stimulates the flow of gastric juice. The small intestine secretes the hormone secretin, which stimulates the flow of pancreatic juice.

Vocabulary Review

endocrine system (314)	thyroid gland (322)
glands (314)	parathyroid glands (323)
exocrine glands (314)	adrenal glands (324)
endocrine glands (314)	epinephrine (324)
hormones (315)	norepinephrine (324)
insulin (315)	corticosteroids (325)
prostaglandins (315)	islets of Langerhans (325)
positive feedback (316)	glucagon (325)
negative feedback (316)	diabetes mellitus (326)
thyroxine (317)	gonads (326)
hypersecretion (319)	estrogen (327)
hyposecretion (319)	progesterone (327)
pituitary gland (320)	testosterone (327)
hypothalamus (321)	thymus (327)
releasing factors (321)	pineal gland (328)

Chapter Review

A. Multiple Choice— Choose the letter of the answer that best completes each statement.

1. The adrenal medulla secretes the hormone (a) thyroxine (b) estrogen (c) epinephrine (d) testosterone.

2. Glands that discharge their secretions into ducts are called (a) endocrine glands (b) exocrine glands (c) prostaglandins (d) gonads.

3. The most important androgen secreted by the testes is (a) testosterone (b) estrogen (c) progesterone (d) prostaglandin.

4. The thyroid secretes the iodine-containing hormone (a) glucagon (b) thyroxine (c) norepinephrine (d) progesterone.

5. Diabetes mellitus results when insulin is not produced in sufficient quantities by the (a) pituitary gland (b) parathyroid glands (c) gonads (d) islets of Langerhans.

6. The regulatory mechanism that tends to return a hormone level to its normal value is called (a) positive feedback (b) hypersecretion (c) hyposecretion (d) negative feedback.

7. Local hormones that produce their effects without entering the bloodstream are called (a) releasing factors (b) islets of Langerhans (c) prostaglandins (d) corticosteroids.

8. Hormones of the adrenal cortex that include cortisol and aldosterone are called (a) corticosteroids (b) releasing factors (c) gonads (d) prostaglandins.

9. The development of the female reproductive system is stimulated by a hormone produced by the ovaries called (a) glucagon (b) estrogen (c) epinephrine (d) thyroxine.

B. Diagram Labeling— Draw a diagram of the human endocrine system.

10. In your diagram, label the following parts: (a) adrenal glands, (b) thyroid gland, (c) hypothalamus, (d) pineal gland, (e) pancreas, (f) thymus, (g) pituitary gland, (h) parathyroid glands, and (i) gonads. Identify the hormone(s) secreted by each.

Content Review

11. How do exocrine and endocrine glands differ?

12. How do hormones affect the metabolism of target tissues?

13. Describe the chemical makeup of the two major classes of hormones.

14. Why are prostaglandins called local hormones?

15. Compare positive feedback with negative feedback.

16. Describe the two-messenger model of hormonal action.

17. Why is the pituitary called the master gland?

18. How is the release of hormones by the anterior pituitary controlled?

19. What are the target tissues of each of the major hormones secreted by the anterior pituitary?

20. What are the functions of the two hormones secreted by the hypothalamus and stored in the posterior pituitary?

21. Why must blood calcium levels be kept fairly constant?

22. What are the effects of the hormones epinephrine and norepinephrine?

23. Explain the functions of the adrenal cortex hormones cortisol and aldosterone.

24. Compare the functions of alpha and beta cells in the islets of Langerhans.

25. What is the function of insulin?

26. How does glucagon raise the blood glucose level?

27. What is diabetes? How is it caused, and what are its symptoms?

28. List the functions of the principal sex hormones.

29. What are the health risks of anabolic steroid misuse?

30. What are the functions of the hormones thymosin and melatonin?

31. How do the hormones gastrin and secretin function in the digestive system?

Graphic Organizing

For information on graphic organizers, see Appendix G at the back of this text.

32. **Word Map:** Construct a word map for the term *endocrine gland.* Include at least five examples.

33. **Line Graph:** Construct a line graph using the data in the chart. The data are from an experiment on two rats in which only rat B received daily injections of growth hormone. Plot body weight (grams) on the vertical axis and time (days) on the horizontal axis. Plot the data for both rats on the same graph, using a different color for each line. What can you conclude about the effect of growth hormone on body weight?

Time	Body Weight (grams)	
(days)	Rat A	Rat B
0	20	20
100	140	230
200	200	310
300	240	370
400	245	440
500	250	460
600	250	500

Critical Thinking

34. Compare and contrast the structure, secretion, and action of hormones with those of enzymes. (*Comparing and Contrasting*)

35. Why do you think iodine is usually added to table salt? (*Identifying Reasons*)

36. Some athletes take anabolic steroids to improve their performance. Some nonathletes take steroids simply to improve their appearance. Do you think the use of anabolic steroids is justified in either circumstance? (*Making Ethical Judgments*)

37. Predict the effects of permanent damage to the beta cells of the islets of Langerhans. (*Predicting*)

Creative Thinking

38. Explain why a patient whose thyroid gland has been removed and one with an excessively large thyroid gland may both show symptoms of hypothyroidism.

39. Use an everyday example that is not described in the text to explain the process of negative feedback.

Problem Solving

40. Design an experiment to show the effects of the removal of the thyroid gland in rats. Identify the physiological characteristics that should be observed, and explain how the data should be recorded.

41. A diabetic has a blood glucose concentration of 220 mg/100 mL of blood. If 1 "unit" of insulin will reduce the person's glucose by 3 mg/100 mL, how many units would be needed to reduce it to the more normal level of 100 mg/100 mL?

42. In normal individuals, the blood glucose concentration before breakfast is between 80 and 90 mg/100 mL of blood. A blood-sugar level above 110 mg/100 mL of blood is indicative of diabetes.

Each of two patients suspected of having diabetes is given a glucose tolerance test, in which ingestion of a glucose solution is followed by the periodic measurement of blood samples for glucose levels. The results of the test are recorded in the table below. Prepare a line graph of the data; identify the normal individual (A) and the diabetic individual (B), and describe what happens to the glucose concentration in each. Using the data in your graph, make a hypothesis regarding insulin secretion in the diabetic person.

Time	Blood Glucose (mg/100 mL)	
(hrs.)	Individual A	Individual B
0	90	150
$\frac{1}{2}$	120	180
1	140	220
$1\frac{1}{2}$	110	250
2	90	240
$2\frac{1}{2}$	85	230
3	90	210
$3\frac{1}{2}$	85	190
4	90	170

Projects

43. Prepare a chart on hormonal diseases. Include the names of the gland and hormone involved, the symptoms, and treatment, and state whether the disease is due to oversecretion or undersecretion.

44. Research and write a brief television news commentary on the use of hormones in livestock breeding.

45. Construct a simple flow chart to illustrate the stages in the human body's response to sudden stress. Include major chemical pathways, involved organs and body systems, and resulting behaviors.

46. Do library research on the discovery by Bayliss and Starling of the hormone secretin, and write up your findings in the form of a television script.

Biology and Problem Solving

What caused the outbreak of disease?

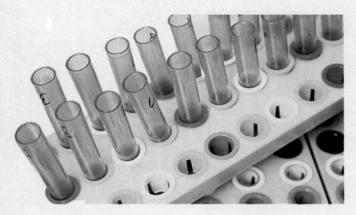

The patient's face was flushed. His heart was beating fast, and he was short of breath. He complained of feeling restless and irritable. When the doctor performed a blood test, results showed a high level of thyroxine hormone in the bloodstream. This result was odd. The patient's thyroid gland was not enlarged, as would have been expected.

Soon, other people in the area showed similar symptoms. In all, over 120 people in a nine-county area developed the same illness. What was causing such an outbreak?

To solve the mystery, doctors kept careful records of those who became ill. They reported the following information:

- The patients ranged in age from 2 to 73.

- An equal number of males and females were affected. (Ordinarily, females develop hormonal diseases four times as often as males.)

- In most instances, all or nearly all members of a household were affected.

- Friends and coworkers of the sick people did not develop the illness.

- In one family, a mother and son who did not live in the same household both became ill.

- About half the patients lived on farms. Only two cases were found in the largest city in the area.

1. Based on what you know about the disease at this point, develop a hypothesis about its cause.

2. One person wondered, "Can you catch the disease from a person who has it?" What would you tell this person? Explain your reasoning.

MEDICAL QUESTIONNAIRE

Name

Address

City _____ State

Date of Birth _____ Sex

Height _____ Weight

1. Have you recently been diagnosed as suffering from thyroid disease?

If so, when was the diagnosis made?

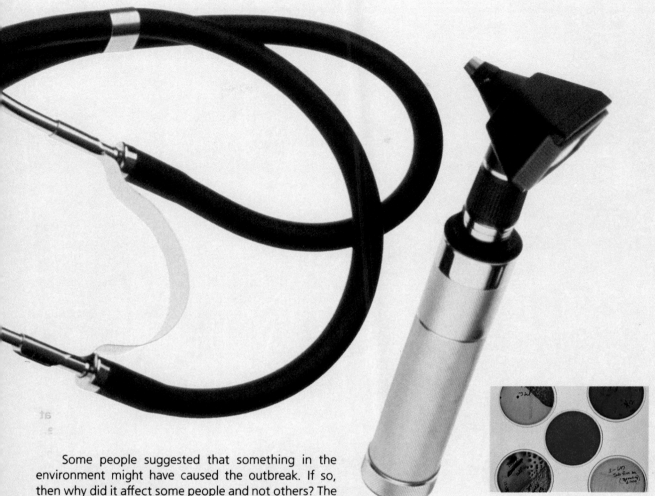

Some people suggested that something in the environment might have caused the outbreak. If so, then why did it affect some people and not others? The doctors needed to identify some experience that was common to the sick people but not to the well people. They decided to conduct a study of 50 of the ill people. They also identified 50 well people who were similar in age, sex, and place of residence.

3. Why is it necessary to include well people in the study?

4. What types of questions would you ask the participants in the study? Remember that the goal is to pinpoint differences in the experiences of the two groups.

After completing the study and analyzing the data collected, the doctors discovered that the sick people all had one thing in common—they all had eaten lean ground beef from the same meat-packing house. None of the well people had eaten the meat.

Although the doctors had pinpointed the source of the problem, one puzzling question remained: How could the beef have caused the high thyroxine levels?

5. Propose a hypothesis to explain how the beef might have caused this disease.

6. How could you find out whether your hypothesis is correct? What observations or experiment(s) would you perform? What would the results indicate?

Problem Finding

Suppose you were one of the doctors who unraveled the mystery of the thyroid disease outbreak. What additional questions might you still have about the disease?

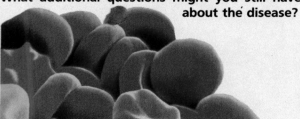

Plant Maintenance

Unit 3

Flowers flourish along the margins of an alpine range. Shoots sprout everywhere—showered by rays of morning sun. In another setting far away, towering pine trees stretch toward that same sun. Still elsewhere, lush canopies of jungle growth catch sunbeams and blanket a darker world beneath. Everywhere, plants in all their variety spread their greenery to capture the energy of the sun, and so begins the earth's food chain. In this unit, you will learn how different plants are adapted to perform the energy conversion process known as photosynthesis. You will also learn how plants use the energy they produce to carry out their own life processes.

▲ Soil, sun, and air are essential to a sapling's growth.

Plant Nutrition

Bright sunlight filters through a forest canopy. Below, a sapling holds eight green leaves toward the sky, capturing some of the sun's energy. The sapling's roots soak up water and minerals while the leaves take in carbon dioxide through tiny pores. The sun's energy will transform these raw materials into food for the sapling. Someday its slender stem will be a massive trunk, and its small twigs will be sturdy branches. In this chapter, you will study the processes plants use to nourish themselves.

Guide for Reading

Key words: photosynthesis, chlorophyll, grana, stroma, light reactions, dark reactions, carbon fixation, chemosynthesis

Questions to think about:

📖 How is light used to drive the reactions of photosynthesis?

📖 How are the light reactions and the dark reactions of photosynthesis dependent on each other?

17-1 Plants and Light

Section Objectives:

- *Describe* the early experiments that provided the basic facts about the process of photosynthesis.
- *Explain* what happens when light is absorbed by a pigment.
- *List* some of the characteristics of chlorophyll.
- *Draw* a chloroplast and label its parts.

🔬 **Laboratory Investigation:** *Identify* the different photosynthetic pigments in green leaves (p. 350).

"Tall oaks from little acorns grow." That simple fact was a puzzling scientific mystery for hundreds of years. It was well known that animals depend on food (that is, other organisms) for growth. But, what was the source of food for plants?

Historical Background

In the early 1600s, the Flemish physician Jan van Helmont grew a small willow tree in a pot for five years, adding only water to the pot. At the end of the five years, he found that the tree had gained 75 kilograms, but there was no change in the weight of the soil. Van Helmont concluded that the new plant material came directly from water.

Over 100 years later, in the 1770s, Joseph Priestley, an English chemist, wanted to know what would happen to a plant placed in air "damaged" by a burning candle. It was known that a burning

▲ **Figure 17-1**

Capturing Light Energy. Plants capture energy from the sun and use it to make the high-energy compounds essential for life.

▲ **Figure 17–2**
Antoine Lavoisier. Through his experiments, Lavoisier showed that burning removes oxygen from the air.

Figure 17–3
The Meaning of Wavelength. Light is energy that travels in waves. The distance between the crests of two consecutive waves is the wavelength. ▼

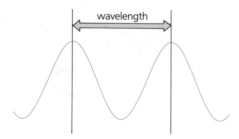

wavelength

candle placed in a closed container went out and that animals could not live in air in which an object could no longer burn. Priestley found that plants were able to grow well in this "damaged" air. In fact, the plant could restore the ability of the air to support a flame and an animal. This was the first real evidence that plants interact with air in some way.

Within a few years, the French chemist Antoine Lavoisier (lah vwahz ee AY) showed that oxygen is removed from the air during burning. See Figure 17–2. Scientists then realized that animals need oxygen from the air, just as a flame does. When the oxygen in air has been used up, air can no longer support animal life or burning. On the other hand, plants give off oxygen to the air.

Other discoveries about plant growth followed throughout the 1700s. The Dutch physician Jan Ingenhousz (ING en howz) found that plants give oxygen to the air only in sunlight. Jean Senebier, a Swiss clergyman, found that plants take in carbon dioxide during growth in sunlight. By the beginning of the 1800s, scientists had identified the basic requirements for plant growth: carbon dioxide, water, and light.

Photosynthesis

Some organisms, such as green plants, capture the energy of sunlight and transform it into chemical energy. This process of capturing and transforming the energy of sunlight into chemical energy is called **photosynthesis** (foh tuh SIN thuh sis). When green plants carry out photosynthesis, they use carbon dioxide and water to make glucose, and they release oxygen. Most of the oxygen in the atmosphere is thought to be the result of photosynthesis.

As you read in Chapter 8, organisms that are capable of making food from simple inorganic substances are called *autotrophs* (AWT uh trohfs). All green plants, many protists, and some forms of bacteria are autotrophic. *Heterotrophs* (HET uh ruh trohfs) are organisms that cannot make their own food and must depend on other plants and animals as their source of food.

There are two major types of autotrophs. Both types use carbon dioxide as a source of carbon to make food. Those that use light energy to drive the reactions needed to make food are called **photoautotrophs** (foh toh AWT uh trohfs). Normally, these organisms are referred to as photosynthetic organisms. The other, less familiar, types of autotrophs are certain kinds of bacteria. These bacteria oxidize inorganic chemicals for the energy to drive their food-making reactions. These bacteria are called **chemoautotrophs** (keem oh AWT uh trohfs).

Light Energy Sunlight is a form of energy that is known as *radiation*. Radiation travels in waves. The distance between the crest of one wave and the crest of the next wave is the *wavelength* of the light, as shown in Figure 17–3. Sunlight is a mixture of all visible wavelengths. If all wavelengths of light are reflected equally by an object, the object appears white to the human eye. Thus,

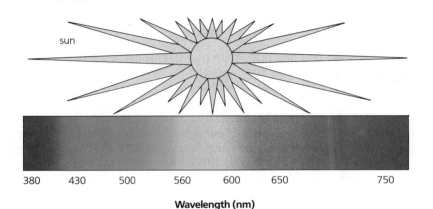

The Visible Light Spectrum. We perceive different wavelengths of light as different colors (top). Shorter wavelengths of light have higher energy than longer wavelengths. White light is made up of all visible wavelengths. When white light is passed through a prism (bottom), the light rays of different wavelengths spread out to form a spectrum. The colors in a spectrum appear in order of wavelength.

sunlight is called "white light." When a beam of white light passes through a prism, the rays of different wavelengths are bent by different amounts. This causes the light to spread out, forming a *spectrum* (SPEK trum) similar to the one in Figure 17-4 (bottom). The colors appear in the order of their wavelengths with the shortest wavelength (violet) at one end and the longest (red) at the opposite end. Figure 17-4 (top) shows the visible light spectrum.

Although light travels in waves, it acts as if it were made up of particles. Each particle of light, which is called a **photon** (FOH tahn), has a fixed amount of energy. The shorter the wavelength of light, the more energy its photons carry.

When light strikes matter, some of the matter's atoms may absorb photons. When the energy of a single photon is transferred to one of the electrons in the atom, it raises the atom's energy. In most cases, the absorbed energy is changed to heat. In photosynthetic organisms, however, the absorbed energy is used to make chemical bond energy.

A substance that absorbs light is called a **pigment.** Wavelengths of light that are not absorbed either pass through the material or are reflected off it. What your eye sees as the color of an object is the color of the reflected light. For example, a red object absorbs all the visible colors of the spectrum except red, which is reflected. The different colors, or wavelengths, of light absorbed by a particular pigment make up its **absorption spectrum.**

Photosynthetic Pigments

The most abundant and important photosynthetic pigments are the **chlorophylls** (KLOR uh fils). In plants, there are two types of chlorophyll—chlorophyll *a* and chlorophyll *b*. Both forms absorb red and blue light and reflect green light. Chlorophyll *a* is the primary photosynthetic pigment. It is involved directly in converting light energy to chemical energy. Chlorophyll *b* and other pigments, known as *carotenes* and *xanthophylls,* absorb light and transfer the energy to chlorophyll *a.* Carotenes, which are orange, and xanthophylls, which are yellow, absorb light in regions of the spectrum different from chlorophyll *a.* Because of this, plants can absorb and use light from a wider region of the spectrum than would be possible if chlorophyll *a* were the only photosynthetic pigment. See Figure 17–5.

Normally, the presence of chlorophyll *a* hides the carotenes and xanthophylls in leaves. Carotenes, xanthophylls, and other pigments can be seen in autumn when chlorophyll starts to break down and the leaves turn color.

Critical Thinking in Biology

Generalizing

Suppose you had eaten at one restaurant five times in the past few months. Each time, you ordered a different dish and enjoyed it. When a friend asks for a restaurant recommendation, you might recommend the restaurant saying, "*All* the food there is delicious."

Your recommendation is based on a generalization. When you generalize, you draw a conclusion about an entire group, based on information about a sample from that group. Although generalizations are not necessarily correct, they become more accurate when the sample is large enough, is representative of the whole group, and is obtained with care and accuracy.

1. From records kept over a five-year period, a naturalist in a state park noted that the leaves on 10 sugar maple trees began to change color between October 10 and October 15 of each year. The naturalist also noted that 14 white oak trees began to change color between October 18 and October 25. The observed trees were located in a variety of places in the park. Formulate a generalization based on the naturalist's observations.
2. How accurate do you think your generalization is? Explain.
3. What additional data would support your generalization?
4. Can you include any information about the elm trees in the state park in your generalization? Why or why not?
5. Think About It Explain your thought process as you answered question 1.

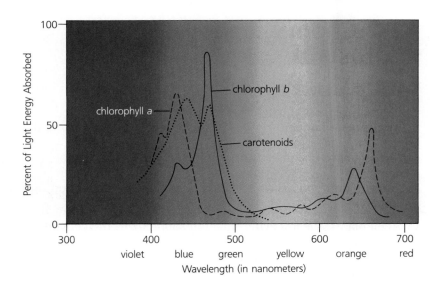

Figure 17–5
Absorption Spectrum for Chlorophylls and Carotenes. The peaks on this graph represent wavelengths of high absorption. The chlorophylls absorb light in the red and blue ranges of the spectrum and reflect light in the green-yellow range. Carotenes absorb only in the blue and blue-green ranges.

The Chloroplasts

In green plants, photosynthesis occurs within **chloroplasts** (KLOR uh plasts). As you may recall from Chapter 5, section 5–2, chloroplasts are organelles containing *photosynthetic membranes* in which the photosynthetic pigments are found. The photosynthetic membranes are arranged in the form of flattened sacs called **thylakoids** (THY luh koidz). Stacks of thylakoids are called **grana** (GRAH nuh). The regions between the grana make up the part of the chloroplast known as the **stroma** (STROH muh).

The combination of chlorophyll and photosynthetic membranes is vital for converting light energy to chemical energy. If chlorophyll is removed from photosynthetic membranes and exposed to light, it absorbs the energy in light but immediately loses it as heat and light of a longer wavelength. Only when chlorophyll is combined with the specialized proteins and other substances in photosynthetic membranes can the light energy be captured *and* stored as chemical energy.

17-1 Section Review

1. Name the process used by plants to make food.
2. What substance is restored to air by photosynthesis?
3. Name the pigments found in chloroplasts.
4. What are stacks of thylakoids called?

Critical Thinking

5. Classify the following organisms as autotrophic or heterotrophic: oak tree, mushroom, seaweed, wolf, human, ameba, tomato plant, fish. (*Classifying*)

17-2 Chemistry of Photosynthesis

Section Objectives:

- *Write* the general equation for photosynthesis.
- *Explain* what happens during the light reactions and during the dark reactions.
- *List* the environmental conditions that affect the rate of photosynthesis.
- *Compare* and *contrast* the functions of cellular respiration and photosynthesis.

In green plants, the following equation summarizes the conversion of light energy into chemical energy:

$$6CO_2 + 12H_2O \xrightarrow{\text{light}} C_6H_{12}O_6 + 6O_2 + 6H_2O$$

This equation for photosynthesis represents many separate chemical reactions that occur in chloroplasts in the light. These reactions are classified into two types. Reactions in the first type are called the **light reactions.** As their name suggests, light reactions take place only in the presence of light. Light supplies the energy for these reactions. Reactions in the second type are called the **dark reactions.** Although these reactions can occur without light, they depend upon the high-energy chemical products made in the light reactions. Both sets of reactions are part of the process of photosynthesis. In green plants, light reactions produce high-energy compounds that the dark reactions use to make glucose. Figure 17–6 illustrates how the light reactions and the dark reactions depend on each other.

Figure 17–6

Photosynthesis. The chemical reactions of photosynthesis may be divided into two types—the light reactions and the dark reactions. The light reactions require sunlight; they take place in the grana of the chloroplast. The dark reactions do not require light; they take place in the stroma of the chloroplast. The two sets of reactions work together to convert light energy into the chemical energy of glucose. ▶

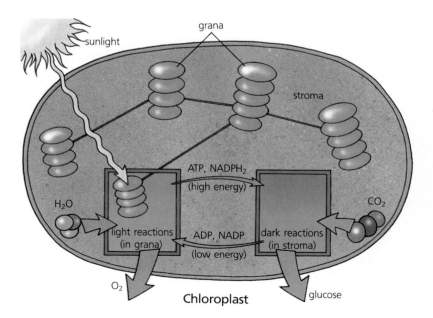

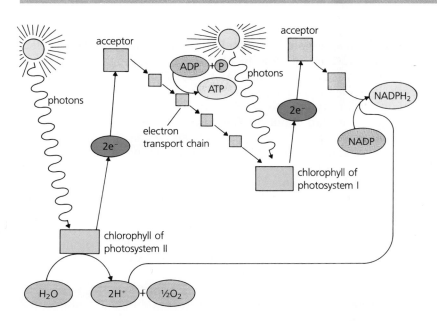

◀ **Figure 17–7**

The Light Reactions. Light absorbed by photosystem II raises the electrons to a higher state of energy. The electrons are passed from the electron acceptor along an electron transport chain to a lower energy state. In the process, some of their energy is packaged in the form of ATP. Light absorbed by photosystem I boosts the electrons to another electron acceptor. These electrons are passed to NADP, forming NADPH$_2$. The electrons removed from photosystem I are replaced by those from photosystem II. The final products of the light reactions are ATP and NADPH$_2$.

Light Reactions

Light reactions begin when the pigments in the photosynthetic membranes of the chloroplasts absorb light. The chlorophyll molecules in these membranes are packaged into two light-absorbing forms, called photosystem I and photosystem II. Each of these *photosystems* is made up of several hundred chlorophyll molecules. In the photosynthetic membrane of a single chloroplast, there are millions of these photosystems.

Photosystems I and II are linked together structurally and functionally. As you can see in Figure 17–7, when photosystem II absorbs light, electrons are passed to an electron transport chain. Like the electron transport chain used in cellular respiration (see Chapter 5), this chain uses the energy of the electrons to make ATP. At the end of the chain, the electrons are passed to photosystem I. When photosystem I absorbs light, the electrons in it are passed to NADP to form NADPH$_2$. NADP is similar to the hydrogen acceptor NAD, which is used in cellular respiration. Photosystem II provides a continuous supply of electrons for the reactions. When it absorbs light and loses electrons, it replaces the lost electrons by removing electrons from water. As a result, oxygen is produced. In light, both photosystems simultaneously absorb light. One photosystem generates ATP, the other, NADPH$_2$. These two high-energy products are used to power the remaining reactions of photosynthesis—the dark reactions.

Dark Reactions

While the light reactions are happening in the membranes of the chloroplasts, the dark reactions are taking place in the stroma. There, carbon dioxide, which diffuses into stroma from the external

Can You Explain This?

Erica knew that plants release oxygen in the process of photosynthesis. She assumed she could determine the amount of photosynthesis a plant performed by measuring the amount of oxygen released.

On a sunny morning, Erica placed a plant inside a bell jar and measured the amount of oxygen. At the end of the day she measured the oxygen a second time to determine the amount of increase. Erica's teacher pointed out that the increase was not a true measure of the amount of photosynthesis. Her teacher said there was *more* oxygen involved in photosynthesis than was released in the bell jar.

■ ***Propose an explanation for what happened to the "missing" oxygen.***

environment, is used to form glucose. The incorporation of carbon dioxide into an organic compound during photosynthesis is called **carbon fixation.** Carbon fixation occurs through a series of enzyme-controlled reactions called the *Calvin cycle.* The Calvin cycle is named after its discoverer, an American scientist, Melvin Calvin. Calvin won the Nobel Prize in 1961 for his discovery of this cycle. As you read about the Calvin cycle, refer to Figure 17–8.

The starting and ending compound in the Calvin cycle is a 5-carbon sugar called *ribulose bisphosphate,* or *RuBP.* The cycle begins when carbon dioxide reacts with RuBP. The products of this reaction are two molecules of a 3-carbon compound, called *phosphoglyceric acid,* or *PGA.* Because the first compound has 3 carbon atoms, the Calvin cycle is also called the C_3 photosynthetic pathway. The PGA molecules formed in the first reaction are converted to *phosphoglyceraldehyde,* or *PGAL.* $NADPH_2$ and ATP, which are produced in the light reactions, provide the hydrogen and energy for this reaction.

As you can see in Figure 17–8, most of the PGAL is used to make more RuBP, so that the cycle can continue. For example, for every 6 molecules of CO_2 that react in the cycle, 12 molecules of PGAL are formed. Ten of these molecules are used to form 6 molecules of RuBP, while 2 PGAL molecules react to form 1 glucose molecule.

Factors Affecting Photosynthesis

Light intensity, temperature, and water and mineral availability are only a few of the factors that affect the rate of photosynthesis. As the intensity of sunlight increases, the rate of photosynthesis increases, but only up to a point. Usually, photosynthesis takes

Figure 17–8

The Dark Reactions (Calvin cycle). In the Calvin cycle, carbon dioxide from the environment reacts with ribulose bisphosphate (RuBP) to produce the 3-carbon compound PGA. ATP and $NADPH_2$ are used to convert PGA to PGAL. Most of the PGAL reacts with ATP to make more RuBP, allowing the cycle to begin again. A small portion of the PGAL is used to produce glucose. ▶

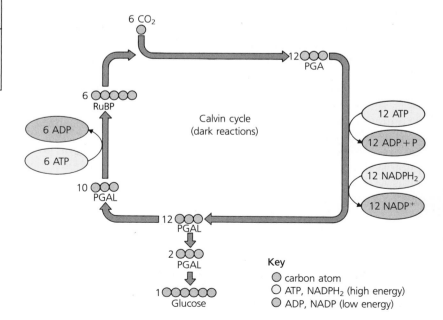

Photosynthesis and Respiration		
	Photosynthesis	**Respiration**
Function	energy storage	energy release
Location	chloroplasts	mitochondria
Reactants	CO_2, H_2O	$C_6H_{12}O_6$, O_2
Products	$C_6H_{12}O_6$, O_2	CO_2, H_2O
Equation	$6CO_2 + 12H_2O +$ light energy $\rightarrow$ $C_6H_{12}O_6 + 6O_2 + 6H_2O$	$C_6H_{12}O_6 + 6O_2 + 6H_2O \rightarrow$ $6CO_2 + 12H_2O +$ chemical energy

▲ **Figure 17–9**
Comparison of Photosynthesis and Respiration.

place most rapidly at a specific temperature. At extremes of temperature—below 0°C or above 35°C—the enzymes are damaged, and the rate of photosynthesis is slowed. A shortage of water tends to slow photosynthesis. If the shortage is severe, photosynthesis may stop. Of course, many minerals play a role in photosynthesis. If these minerals are in short supply, photosynthesis, as well as other metabolic processes, is affected.

Photosynthesis and Cellular Respiration

In its effect, photosynthesis is the reverse of cellular respiration. Both processes occur simultaneously in light. Respiration, however, takes place in the cytoplasm and mitochondria. Photosynthesis occurs in the chloroplasts. In respiration, glucose and oxygen are used to produce carbon dioxide and water and to release energy. (See Chapter 6.) In photosynthesis, carbon dioxide, water, and energy are used to produce glucose and to release oxygen. Thus, photosynthesis captures light energy, storing it as chemical energy, and cellular respiration releases chemical energy. Figure 17–9 compares photosynthesis and cellular respiration.

17-2 Section Review

1. Name the two sets of reactions in photosynthesis.
2. What are the end products of the light reactions?
3. What is another name for the dark reactions?
4. Name two factors that affect the rate of photosynthesis.

Critical Thinking

5. Compare and contrast photosynthesis with cellular respiration. (*Comparing and Contrasting*)

▲   **Figure 17–10**
Sugar Cane. The C₄ pathway allows such plants as sugar cane and corn to photosynthesize rapidly.

Figure 17–11
A Cyanobacterium. Photosynthetic bacteria do not contain chloroplasts. Instead, photosynthetic membranes are present throughout the cytoplasm. (Magnification 6500 X). ▼

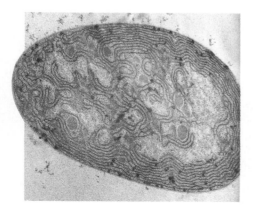

17-3 Special Cases

Section Objectives:

■ *Explain* the adaptive advantages of C_4 photosynthesis.
■ *Compare* photosynthesis in bacteria and plants.
■ *Summarize* the main events in chemosynthesis.
■ *Describe* the two types of heterotrophic nutrition found in plants.

Although the process of photosynthesis is the same in all photosynthetic organisms, there are some interesting variations on the process. In addition to the Calvin cycle, some plants have another pathway that improves the efficiency of the Calvin cycle. Moreover, plants are not the only photosynthetic organisms. Many protists and bacteria carry out photosynthesis, and in some forms of bacterial photosynthesis, as you will read, oxygen is not produced.

C₄ Plants

Some flowering plants use more than the Calvin cycle to fix CO_2. In these plants, called C_4 plants, CO_2 is first fixed in some of the leaf cells into a 4-carbon compound. This compound is transported into other nearby leaf cells where the Calvin cycle operates. In these cells, the C_4 compound breaks down, releasing the CO_2 previously incorporated into the compound. The released CO_2 enters the Calvin cycle and is made into glucose.

The extra photosynthetic pathway, called the C_4 pathway, acts as a CO_2 "pump." That is, it increases the concentration of CO_2 in the cells in which the Calvin cycle operates. This allows C_4 plants to fix CO_2 more rapidly than plants without a C_4 pathway. However, with this extra pathway, C_4 plants need more energy to fix CO_2 than plants with only the Calvin cycle. This is not a problem for C_4 plants because they grow best under conditions of high light intensity where there is more than enough light to generate the extra ATP needed. Sugar cane and corn are examples of C_4 plants.

Photosynthetic and Chemosynthetic Bacteria

All of the photosynthetic bacteria have photosynthetic membranes throughout their cytoplasm. See Figure 17–11. The most familiar of these bacteria are the blue-green bacteria, or cyanobacteria. These were once called blue–green algae. Blue–green bacteria and another similar group carry out photosynthesis in the same way as plants. Chlorophyll *a* is their primary photosynthetic pigment. Like plants, blue–green bacteria use water in the light reactions as the source of hydrogen. Therefore, they release oxygen.

Other photosynthetic bacteria contain bacteriochlorophyll instead of chlorophyll *a*. These bacteria do not release oxygen because they use hydrogen sulfide (H_2S) instead of water as a hydrogen source.

Some types of bacteria carry out **chemosynthesis**—a process in which food is made from CO_2 by using the energy of inorganic substances. Like photosynthetic organisms, chemosynthetic bacteria fix CO_2 through the reactions of the Calvin cycle. However, the energy to make ATP and $NADPH_2$ comes from the oxidation of inorganic substances, not from the absorption of light.

▲ **Figure 17–12**

Carnivorous Plants. Each leaf of a pitcher plant (left) is made up of a funnel with a pool of water at its base. If an insect falls into the funnel, it drowns in the pool. The insect is digested by enzymes secreted into the water. When an insect lands on the Venus flytrap (right), it is quickly trapped as the modified leaves snap shut. Glands within the trap secrete digestive enzymes, and nutrients are later absorbed by the leaves.

Heterotrophic Plants

Some plants have developed heterotrophic methods of nutrition in addition to, or instead of, photosynthesis. Some of these plants are parasitic. The mistletoe, for example, is parasitic on oaks and other trees. Although mistletoe is photosynthetic, it adds to its nutrition by siphoning sap from the vascular tissue of the host tree. The dodder plant is also parasitic, but it cannot photosynthesize. The roots of the dodder plant grow into the tissues of its host, from which it draws nutrients and water. The roots of the Indian pipe form a combination with a fungus. The Indian pipe depends on the fungus for nutrients. The fungus acts as a bridge, carrying nutrients to the Indian pipe from the roots of nearby photosynthetic plants.

Some plants, like those in Figure 17–12, digest insects to supplement their mineral nutrition. In bogs and marshes, the acidic water slows the rate of decay of dead organisms and limits the amount of nitrogen in the soil. In nitrogen-poor areas like these, insects provide a source of nitrogen for making plant proteins.

17-3 **Section Review**

1. What is the C_4 pathway named after?
2. What substance is used in place of water by photosynthetic bacteria?
3. Name the autotrophic process that does not use light energy.
4. Name two parasitic plants.

Critical Thinking

5. What would happen to a C_4 plant growing in a shaded area among C_3 plants? Explain. (*Predicting*)

Laboratory Investigation

Chromatography of Plant Pigments

A biological process like photosynthesis involves hundreds of substances. Separating and identifying these substances is a way of analyzing such a process. Chromatography is a method of separating substances from each other. In this activity, you will use paper chromatography to discover the photosynthetic pigments in spinach leaves.

Objectives

- *Use* paper chromatography to separate leaf pigments.
- *Identify* the photosynthetic pigments in leaves.

Materials

isopropyl alcohol, 70%   pencil
spinach leaf extract   metric ruler
#3 filter paper   large test tube
scissors   test tube stopper
micropipette   test tube rack
plastic eyedropper   beaker

Procedure 🔳 🥽 🧪 ☠️ 🧤 *

A. With scissors, cut a strip of filter paper narrow enough to fit inside the test tube without touching its sides. Cut it to be about 2 cm shorter than the test tube. Cut an arrow-shaped point 1.5 cm long, at one end of the strip. This end will be at the bottom of the test tube, as shown in the figure above.

B. Measure 2.5 cm up from the bottom tip of the strip and mark a point with a pencil at the midline of the strip. Do not use ink. Also place a pencil line across the strip about 1 cm below the top edge.

C. Place the paper strip on a paper towel. Draw some leaf extract into the micropipette. Briefly touch the tip of the pipette to the pencil mark on the strip so a small spot of extract seeps onto the paper. Let the spot dry completely. Repeat until the spot is dark green.

☠️ **D.** Make a mark 1 cm from the bottom of the test tube and add alcohol with an eyedropper up to the mark. Place the paper strip into the test tube, being sure that the pigment spot is above the solvent. Seal the tube with a stopper. See the figure above.

*For detailed information on safety symbols, see Appendix B.

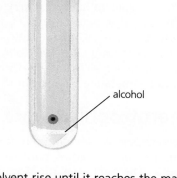

alcohol

📚 **E.** Let the solvent rise until it reaches the mark at the top of the paper. Remove the paper strip and set it on a paper towel to dry. Pour the solvent remaining in the test tube into a beaker labeled "waste solvent."

F. Observe the spots of color on the chromatogram. Draw a circle around each spot and label its color. The pigments, from top to bottom, are carotenes, xanthophylls, chlorophyll *a*, and chlorophyll *b*.

Analysis and Conclusions

1. How many different colors were separated on your chromatogram? Describe the colors.

2. Why do you think pencil, but not ink, was used to make marks on the chromatogram?

3. Based on your results and the discussion of paper chromatography in Chapter 2, which pigment would you conclude is most soluble in the solvent?

4. In addition to pigments, what other substances may have been separated, but are not visible, on your chromatogram?

5. What would you predict about the photosynthesis of a plant whose leaf extract showed 3, instead of 4, pigments?

6. Using paper chromatography, how could you determine if the leaves of two different plant species contained the same or different pigments?

Chapter Review

17

Study Outline

17–1 Plants and Light

- Our understanding of how plants are able to grow has developed through the efforts of numerous scientists, including van Helmont, Priestley, Lavoisier, Ingenhousz, and Senebier.

- In plant photosynthesis, light energy, carbon dioxide, and water are used to make glucose. Oxygen is released into the atmosphere as a byproduct.

- Autotrophs make food from simple inorganic compounds. To carry out this process, photoautotrophs use light energy, whereas chemoautotrophs use the energy of certain inorganic chemicals.

- Photosynthetic organisms contain pigments that absorb light in some regions of the spectrum. Some of the light energy is converted to chemical bond energy in certain chemical compounds.

- Chlorophylls *a* and *b* are pigments that absorb blue and red light and reflect green light. Other pigments, such as xanthophylls and carotenes, absorb light of other colors and transfer the energy to chlorophyll.

- Chloroplasts are organelles that contain photosynthetic membranes arranged in sacs called thylakoids. Stacks of thylakoids, called grana, are surrounded by material called stroma.

17–2 Chemistry of Photosynthesis

- In the light reactions of photosynthesis, light energy is absorbed and used to remove electrons from water molecules and produce ATP, $NADPH_2$, and oxygen.

- ATP and $NADPH_2$ from the light reactions drive the dark reactions, in which carbon dioxide combines with a 5-carbon sugar, forming two molecules of a 3-carbon compound. This process, called carbon fixation, repeats in a cyclic series of reactions called the Calvin cycle.

- The rate of photosynthesis depends on a number of environmental factors, including temperature, light intensity, and availability of water and minerals.

- Photosynthesis is essentially the reverse of cellular respiration.

17–3 Special Cases

- Some flowering plants use a C_4 pathway in addition to the Calvin cycle to speed up the fixation of carbon dioxide.

- Blue-green bacteria have chlorophyll *a* and carry out photosynthesis in the same way as plants. Other photosynthetic bacteria contain a different form of chlorophyll and use hydrogen from H_2S instead of H_2O.

- Chemosynthetic bacteria make ATP and $NADPH_2$ from the energy produced by the oxidation of inorganic substances.

- Some species of plants have become parasitic on green plants. Others that live in a nitrogen-poor environment capture insects to supplement their nutrition.

Vocabulary Review

photosynthesis (340)   thylakoids (343)

photoautotrophs (340)   grana (343)

chemoautotrophs (340)   stroma (343)

photon (341)   light reactions (344)

pigment (341)   dark reactions (344)

absorption spectrum (341) carbon fixation (346)

chlorophylls (342)   chemosynthesis (349)

chloroplasts (343)

A. Sentence Completion—Fill in the vocabulary term that best completes each statement.

1. A(n) _____ is a substance that absorbs light.

2. _____ are the flattened sacs of photosynthetic membranes in chloroplasts.

3. In the presence of sunlight, food is synthesized from inorganic substances in a process called _____.

4. The dark reactions of photosynthesis occur in the _____ of the chloroplasts.

5. The _____ are powered by ATP and $NADPH_2$ made in the light reactions.

Chapter Review

6. In photosynthesis, the _____ begin when light is absorbed by chlorophyll.

7. A particle or packet of light energy is known as a(n) _____.

B. Word Relationships—In each of the following sets of terms, three of the terms have a common characteristic. One term does not belong. Identify the term that does not belong.

8. photoautotrophs, pigment, photon, chemoautotrophs

9. chemosynthesis, light reactions, chlorophyll, photosynthesis

10. chlorophyll, carbon fixation, grana, thylakoids

Content Review

11. Briefly describe the findings of Ingenhousz and Senebier.

12. Explain what is meant by the term photoautotroph.

13. Explain what happens when a substance absorbs light.

14. How would you explain why a pigment has a particular color?

15. When a nonphotosynthetic pigment absorbs light, what normally happens to the absorbed energy?

16. What function do photosynthetic pigments other than chlorophyll serve?

17. Explain the importance of photosynthetic membranes in photosynthesis.

18. Write the balanced chemical equation for the overall reaction of photosynthesis.

19. Describe the light reactions of photosynthesis, including the end products.

20. Describe how photosystems I and II work together.

21. Describe the dark reactions of photosynthesis, including the end products.

22. Discuss the role of PGAL in glucose formation in the dark reactions.

23. Explain why the minimum number of carbon dioxide molecules needed to make a molecule of glucose is six.

24. How does the location of the stroma with respect to the thylakoids make for efficient photosynthesis?

25. Explain how temperature, light intensity, and the availability of water and minerals affect the rate of photosynthesis.

26. How do the reactants and end products of photosynthesis differ from those of cellular respiration?

27. What are C_4 plants?

28. Explain why photosynthetic bacteria that have bacteriochlorophyll instead of chlorophyll *a* do not release oxygen.

29. What is chemosynthesis?

30. By what pathway do chemosynthetic organisms fix CO_2?

31. Why are some plants that carry out photosynthesis adapted for capturing and digesting insects as well?

Graphic Organizing

For information on graphic organizers, see Appendix G at the back of this text.

32. **Compare/Contrast Matrix:** Construct a compare/contrast matrix to compare photosynthesis and aerobic respiration (see Chapter 6). Use the following characteristics in your comparison: starting reactants; location of process within cell; whether energy-requiring or energy-releasing; end products; and organisms (autotrophs, heterotrophs, or both) that carry out the process.

Critical Thinking

33. What would happen to a plant placed in an airtight jar by a window? Explain. (*Predicting*)

34. Seeds of Indian pipe plants were planted on the floor of a dense forest. Once they grew, would they be likely to have a competitive advantage over purely autotrophic plants? Why or why not? (*Predicting*)

35. Suggest some physical characteristics of an environment in which chemosynthetic bacteria would thrive but photosynthetic bacteria would not. (*Comparing and Contrasting*)

Creative Thinking

36. Suggest some conditions that would maximize the likelihood that plants and an animal could help to keep each other alive in a sealed environment.

37. What do you think was the order of development on earth of the following types of organisms: oxygen-consuming heterotrophs, nonoxygen-consuming heterotrophs, oxygen-producing autotrophs? Support your answer.

Problem Solving

38. A student is told that sunlight causes a higher rate of photosynthesis than would an equally bright electric light. Design a controlled experiment that the student could use to test this statement.

39. In 1882, an experiment was carried out by T. W. Engelmann in which oxygen-using bacteria and a filamentous strand of a green alga were exposed to light of different colors. As you can see in the graph below, the bacteria clustered in the areas of the alga exposed to violet light and red light. Propose an explanation for these results.

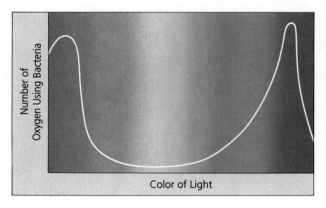

40. A green water plant was placed in a test tube filled with water. A light source was positioned at varying distances from the plant. Bubbles of oxygen given off by the plant were counted. The data collected during this investigation are shown in the table below. Graph the data, and draw a conclusion about how the rate of photosynthesis varies according to the distance from the light source.

Distance from Light (cm)	Bubbles per Minute
10	60
20	25
30	10
40	5

41. Technicians often find masses of gray growth covering submerged pumps and pipes removed from water wells. The organisms that make up this growth live with very little light and no nutrients except for the substances dissolved in the water. Develop a hypothesis about how these organisms are able to grow in this environment.

Projects

42. Grow a number of seedlings of the same type of plant. Make four groupings of the plants, growing each under a tent made of cellophane of a different color. Keep all other growing conditions the same. Which colors produce the best growth? Explain your results.

 43. Write a brief report on the life and scientific contributions of Melvin Calvin.

44. Prepare an oral report for the class on one of the following career opportunities: (a) plant physiologist, (b) forest ranger, (c) plant pathologist. If possible, interview a person working in the field. Be sure to prepare your questions in advance. You may wish to inquire about training, future opportunities, and daily routine. A tape recorder will be helpful.

▲ In the Sonoran Desert, a barrel cactus thrives.

Plant Structure

A barrel cactus basks in the hot desert sun. While its flowers lure pollinators, its spines, the true leaves of the plant, discourage predators. Unlike in most other plants, a thickened, water-storing stem carries out photosynthesis. As in other plants, tubelike tissues carry food and water throughout the plant. At the tips of roots and stem, areas of rapidly dividing cells provide new cells for growth. In this chapter, you will learn about the various special features of plant structure.

Guide for Reading

Key words: roots, stems, leaves, meristem, xylem, phloem, guard cells, mesophyll.
Questions to think about:
📖 How do roots and stems develop from apical meristems?
📖 In what way do herbaceous and woody stems differ?
📖 How is the structure of a leaf suited to the function of the leaf?

18-1 Plant Tissues

Section Objectives:

- *Name* the organs of a plant and describe their functions.
- *Explain* the functions of meristematic, protective, vascular, and fundamental tissues.
- *Describe* in detail the structure and function of xylem and phloem.
- *Name* and *describe* the three types of cells that make up fundamental tissues.

Organization of Tissues

Plants, like animals, are made of tissues that form organs. The organs of a plant are its roots, stems, leaves, and reproductive structures. See Figure 18–2. Unlike animals, however, plants do not have organ systems.

A plant is held in the soil and takes up water and minerals from the soil through its **roots. Stems** hold the leaves and allow them to receive sunlight. They also hold flowers, fruits, and seeds. **Leaves** are where photosynthesis, the process by which plants make food, takes place. *Flowers* and *cones* are reproductive structures. You will read about plant reproduction in Chapter 24.

Plants have fewer types of tissues than animals. Some plant tissues are made of only one type of cell. Others are made of two or more types of cells that work together. Some tissues are found

▲ **Figure 18–1**

A Flowering Plant. These sunflowers, and all other plants, are made of various cell types organized into tissues.

Figure 18–2

Organs of a Plant. A longitudinal section through the stem tip shows the location of meristematic tissue. The cells of the apical meristem undergo rapid cell division, producing the cells that form new leaves and new stem tissue. The cells of the lateral buds are inactive but, at some future time, may start to divide and become the apical meristems of branches. ▶

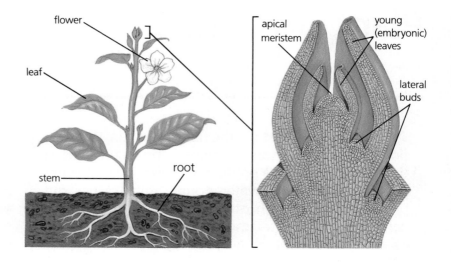

throughout the plant, while others are found only in specific structures. The main types of tissues are meristematic (mur ah stuh MAT ik), protective, vascular (VAS kyuh luhr), and fundamental.

Meristematic Tissues

In most animals, cell division takes place throughout the animal's body during periods of growth. In plants, however, cell division takes place in certain regions called **meristems** (MUR ah stehms). **Meristematic tissues** are made of cells that undergo mitosis and cell division frequently. The cells are thin-walled and lack vacuoles and are usually much smaller than mature plant cells. As you can see in Figures 18–2 and 18–10, meristems are present in the growing tips of stems and roots. These places are called apical meristems because they are found at the *apex* of the root or the stem. The cells made by apical meristems become the mature tissues of the plant body. They cause roots and stems to grow longer. However, many woody plants also grow wider. In woody plants, another type of meristem, called a **cambium,** adds tissues that increase the thickness of stems and roots. The **vascular cambium** produces layers of tissues that transport water and nutrients. Another type of cambium, the **cork cambium,** produces a layer of protective tissue called cork.

Protective Tissues

The **epidermis** is the *protective tissue* that forms the outer layer on leaves, green stems, and roots. The epidermal layer is usually one cell thick and its cells fit tightly together. The cells of the epidermis, which covers above-ground parts of a plant, secrete a waxy substance called *cutin* (KYOOT in). Cutin forms a layer over the outer surface of the epidermis. This layer, which is called the **cuticle** (KYOOT ih kul), cuts down on water loss and protects against infection by microorganisms.

Cork is a protective tissue that covers the surface of woody stems and roots. See Figure 18–3. It protects the more delicate inner tissues from mechanical injury. It also waterproofs the outer surface and prevents infection. Cork is produced by the cells of the *cork cambium.* Cork cells live for only a short time. Fully grown cork cells are dead. It is these dead cells with their waxy cell walls that protect the underlying living tissues.

Vascular Tissues

Xylem (ZY lum) and **phloem** (FLOH em) are the **vascular,** or conducting, **tissues** of the plant. Xylem, which is the material you think of as wood, conducts water and minerals from the roots upward through the stems and into the leaves of the plant. It also helps to support the plant and to hold it upright. Phloem conducts food and other dissolved materials in both directions along the length of the plant.

Most of the cells that form mature xylem are dead. They do not have cytoplasm. They form tubes that go from the roots up through the stems and leaves. Xylem is made mainly of two types of cells—*tracheids* (TRAY kee idz) and *vessel elements.* See Figure 18–4. Tracheids contain pits, or depressions, in their cell walls. The pits of neighboring tracheids are lined up, which permits the passage of water and minerals. Vessel elements are cells that form conducting tubes. These cells, which do not have end walls, are placed end to end, forming long, thin, hollow tubes called *vessels.* Xylem also has some living cells that serve as storage cells.

Unlike the tracheids and vessel elements of xylem, most of the cells that make up phloem are alive and contain cytoplasm. The substances transported by phloem are mainly organic compounds dissolved in water. Among these compounds are amino acids, sugars, and other carbohydrates. Food made in the leaves moves through the phloem to other parts of the plant. Unneeded food is

▲   **Figure 18–3**

Cork. The nonliving cork cells protect the internal tissues of woody stems and roots.

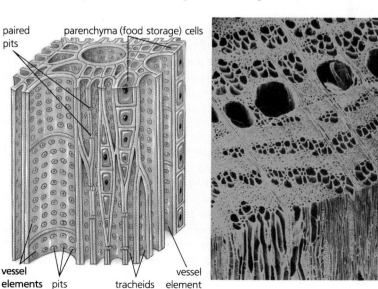

paired pits

parenchyma (food storage) cells

vessel elements   pits   tracheids   vessel element

◄   **Figure 18–4**

Structure of Xylem. The water-conducting cells of the xylem are the elongated tracheids and open-ended vessel elements. Vessel elements form conducting tubes, vessels, which are seen in this scanning electron micrograph.

Figure 18–5

Structure of Phloem. Dissolved nutrients are transported through the phloem of the plant. Phloem is made up of sieve cells and companion cells. The dark-staining end-walls (sieve plates) of sieve cells are seen in the photograph on the right. ▶

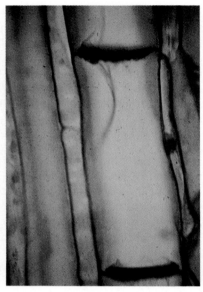

companion cells

sieve plate structure

sieve cells

food storage cells

often carried to the roots where it is stored. In the spring, sap, containing dissolved materials, is moved upward through the phloem.

Phloem is made of two types of cells, *sieve cells* and *companion cells.* See Figure 18–5. A sieve cell contains cytoplasm but does not have a nucleus when it is fully grown. The end walls of sieve cells have many small openings. Sieve cells line up end to end to form tubes, called *sieve tubes,* through which dissolved nutrients are transported. Companion cells are connected to neighboring sieve tubes by thin strands of cytoplasm. These cells, which have nuclei and cytoplasm, are thought to control the transport activities of sieve cells.

Fundamental Tissues

Tissues used in the production and storage of food and in the support of the plant are called **fundamental tissues.** The three types of fundamental tissues are parenchyma (puh REN kuh muh), collenchyma (kuh LEN kuh muh), and sclerenchyma (skluh REN kuh muh). These are shown in Figure 18–6.

Parenchyma is a tissue made of unspecialized cells with thin cell walls. These cells are found in roots, stems, leaves, and fruits. The parenchyma cells in leaves and young stems have chloroplasts and make food by photosynthesis. In roots, fruits, and portions of stems, parenchyma cells are used for food storage.

Collenchyma cells are similar to parenchyma cells, but they are longer and have thick, but flexible, cell walls. They support stems and leaves and other parts of the plant.

Sclerenchyma tissue is made of cells with greatly thickened cell walls that are stiffened with a substance called *lignin*. Lignin is what makes wood rigid. Sclerenchyma tissue is found where support is needed. When fully grown, the cells usually do not have cytoplasm.

Figure 18–6

Fundamental Tissues. The three types of cells that make up fundamental tissues are (a) parenchyma cells, (b) collenchyma cells, (c) sclerenchyma cells. ▼

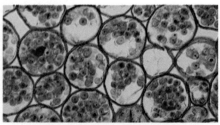

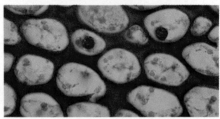

In fact, the cell walls are so thick that the space inside the cell is nearly eliminated. *Fibers* are a type of sclerenchyma cell. They are long cells that have tapered ends. They are often found in xylem and phloem. Fibers are used to make twine, rope, and thread. The term *fiber,* meaning a type of schlerenchyma cell, should not be confused with the term *fiber* when it is used in the dietary sense. Dietary fiber is the indigestible material that is present in all plant cell walls.

18-1 **Section Review**

1. Name the organs of a plant.
2. What type of tissue produces new plant cells?
3. What two conducting tissues are found in plants?
4. List the fundamental tissues found in plants.

Critical Thinking

5. If, for some reason, the phloem in the branches of a tree was not able to conduct food, how would the tree be affected? (*Relating Parts and Wholes*)

18-2 Roots

Section Objectives:

- *Explain* the functions of the root.
- *Describe* each of the following: primary root, secondary root, taproot, fibrous root, and adventitious root.
- *Name* the different zones of the root tip, and describe what happens to the cells in each zone.
- *Name* and *describe* the tissues of the root, and explain their arrangement and functions.

Types of Roots

The roots of a plant are usually found underground. As you have read, they hold the plant in the soil and take in water and minerals from the soil. They carry the water and minerals upward to the stem and transport dissolved food downward from the stem. In addition, the roots of some plants are specialized for food storage. Usually, the root system underground is as large as the system of stems and branches above ground. The roots spread out, covering a large area. In many plants, they grow no deeper than one meter into the soil.

The first structure to emerge from a sprouting seed is the **primary root.** See Figure 18–7. As the plant grows and matures, new roots form from within the tissues of the primary root. These new branches of the primary root are called **secondary roots.** As roots grow, the direction of their growth is affected by obstructions,

Figure 18–7

Primary and Secondary Roots. Secondary roots branch from the tissues of the primary root. ▼

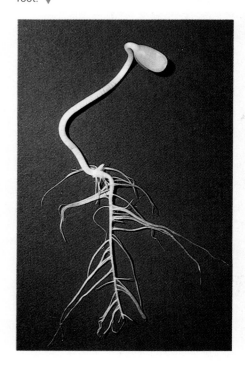

▲ **Figure 18-8**

Types of Root Systems. The taproot of the dandelion (left) grows rapidly and deeply into the soil. In the fibrous root system of grass (right), the numerous roots are all about the same size.

such as rocks, or other roots, in the soil. Other factors, such as moisture and the chemical makeup of the soil, also influence growth. These factors cause roots to grow in irregular ways with frequent bends and kinks.

Figure 18-8 shows the two common types of root systems: taproots and fibrous roots. A *taproot* system develops when the primary root grows rapidly and remains the largest root in the root system. Taproot systems grow deep into the soil and become thick and fleshy. Oak trees, carrots, turnips, and dandelions have taproots. A *fibrous root* system is made up of numerous roots, many of which are nearly equal in size. This type of system develops when branching secondary roots are as large as or larger than the primary root. Corn and grasses have fibrous root systems. In some plants, taproots and fibrous roots are modified for food storage. The carrot, radish, and beet are storage taproots, while the sweet potato and tapioca are fibrous storage roots.

Less common than taproot and fibrous root systems are various types of *adventitious* root systems. See Figure 18-9. These roots do not come from the primary root or from one of its

Figure 18-9

Adventitious Roots. The prop roots of corn (left) help to brace the root. The climbing roots of ivy (right) grow from the stem and attach the plant to a solid support. ▶

branches. Instead, they grow from stems or leaves. *Prop roots* grow from the stem down into the soil. These roots help to brace the plant. *Climbing roots,* for example, grow out from the stem and fasten the growing plant to a solid support. The climbing roots of ivy allow the plant to grow on building walls. Corn plants are an example of plants that have prop roots. Some plants, such as Spanish moss, live attached to trees and develop *aerial roots* that absorb moisture directly from the air.

Root Growth

Although the branches of a root system may be many meters long, only a small region at the tip of a root grows. There may be thousands of these root tips gradually extending into the soil. Other parts of the roots may become thicker, but they do not become longer. A mark made on the surface of a root will be found in the same spot year after year.

If you look closely at a root tip through a microscope, you will see that it is made up of a number of different zones. As you can see in Figure 18–10, each zone contains cells at different stages of development.

Root Cap   The **root cap** is a thimble-shaped group of cells that form a protective covering for the delicate meristematic tissues of the root tip behind it. As the addition of cells behind the root tip pushes it through the soil, the outer cells of the root cap are crushed. The crushed cells release a fluid that helps the passage of the root tip through the soil. To replace the crushed cells, new root cap cells are continuously formed by the meristematic tissue.

Meristematic Zone   The **meristematic zone** is a region of actively dividing cells just behind the root cap. The cells of this region are small and thin-walled. All the other cells of the root are formed from these cells.

Elongation Zone   Behind the meristematic zone is the **elongation zone.** In this zone, the cells, which were produced earlier in the meristematic zone, enlarge, pushing the root tip forward.

Maturation Zone   Behind the elongation zone is the **maturation zone,** or root hair zone, where the cells differentiate. **Differentiation** (dif uh RHEN she ay shen) is the process during which unspecialized cells develop into specialized cells. In the root, as in the stem, cells develop into fully grown, functioning cells of various types, such as xylem, phloem, and parenchyma.

Root Structure and Function

A cross section through the maturation zone will show that the root is made up of several tissue layers. See Figure 18–11. The outermost layer—the epidermis—is only one cell thick. The taking-in of water and minerals from the soil is the primary function of the epidermis. Many epidermal cells have hairlike extensions that are

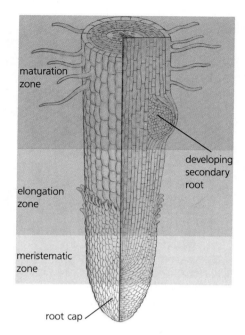

▲   **Figure 18–10**
Zones of the Root Tip. The meristematic zone is made of rapidly dividing, undifferentiated cells. In the elongation zone, the newly formed cells grow in length, forcing the root tip through the soil. In the maturation zone, the cells develop into specialized tissues.

Figure 18–11
Cross Section of Root Tip. The mature tissues of a root are seen in a cross section through the maturation zone. ▼

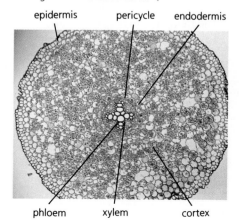

Science, Technology and Society

Technology: Nitrogen-Fixing Corn

Plants need nitrogen to grow and carry out life processes. Although nitrogen gas is abundant in the atmosphere, plants cannot use it in that form. The nitrogen must be converted to ammonia or nitrate.

Plants in the pea family benefit from a relationship with nitrogen-fixing bacteria living in nodules on their roots. But, other crops, such as corn, have no such beneficial association. Farmers must apply fertilizers, which are expensive and environmentally damaging.

Now scientists are attempting to "engineer" nitrogen-fixing corn plants. First, they isolated genes from bacteria that control nitrogen-fixation. Then, they transferred the genes into strains of bacteria that they hope will live in corn roots. The next, and most difficult, step is to isolate genes that control nodule-formation in pea plants and transfer them into corn plants. The result, scientists hope, will be a symbiotic relationship between bacteria and corn plants—and increased crop yields.

■ *List some potential benefits and dangers of this nitrogen-fixing technology.*

called **root hairs.** These root hairs greatly increase the surface area for the absorption of water. They are present only in the small zone of maturation that lies behind the growing root tips. As the root tips grow, new root hairs are formed, and older ones die and fall off. It is in the region at the ends of the root branches that practically all water absorption occurs.

Just beneath the epidermis is the **cortex.** The parenchyma cells of the cortex store the plant's food, which is mainly starch. These cells also transport the water taken in by the root hairs to the conducting tissues in the center of the root. The innermost layer of the cortex is called the **endodermis** (en duh DER mis). The cells of this layer control the movement of water into the central cylinder.

The **vascular cylinder,** or central cylinder, is the core of the root. It is surrounded by a layer of parenchyma cells called the *pericycle* (PER uh sy kul), which is just inside the endodermis. All secondary roots grow from the pericycle layer. These roots push their way through the cortex and epidermis into the soil. Look again at Figure 18–11.

At the center of the vascular cylinder are the conducting tissues, xylem and phloem. The xylem carries water and minerals up the root to the stem and leaves. The phloem carries dissolved food, made in the leaves, throughout the plant. In the roots of woody plants, a vascular cambium develops between the xylem and phloem. The vascular cambium adds new xylem to its inside and new phloem to its outside.

Roots and Microorganisms

Plants often have bacteria or fungi living with their roots. Because this relationship helps at least one of the organisms, it is called a *symbiotic* relationship. In many cases, the symbiotic relationship is beneficial to both the plant and the organism.

Some plants, such as those in the pea family, have bacteria living in nodules on their roots. These bacteria convert nitrogen gas from the air in the soil into forms of nitrogen that the plant can use. In this symbiotic relationship, the plant benefits from the constant supply of usable nitrogen, and the bacteria benefit from an environment that offers plenty of food and water. Farmers may plow plants with nitrogen-fixing nodules into the ground in order to increase the nitrogen in their soil.

As you can see in Figure 18–12, many plants have fungi that grow on their roots. When a fungus grows in a symbiotic relationship with the roots of a plant, the resulting fungus root is called *mycorrhiza* (my kuh RY zuh) (plural, mycorrhizae). Some fungi grow meshlike coverings over the entire root surface, while others grow mostly within the root cortex. All mycorrhizal fungi have hairlike filaments that grow into the soil. These filaments act like a vast network of root hairs and provide the plant with water and nutrients that are absorbed from the soil. The fungi, in turn, seem to gain nutrients that are necessary for growth and development from the plant. Some plants with few or no root hairs at all are so dependent on the fungi living on their roots, that they cannot survive without them.

18-2 Section Review

1. What types of root systems are found in plants?
2. What happens in the maturation zone?
3. Where in the root are the xylem and phloem found?
4. What function do root hairs and mycorrhizae share?

Critical Thinking

5. If a permanent dye was injected into a cell in the apical meristem of a root, in what zone of the root would you expect to find the marker one week later? Six weeks later? (*Predicting*)

18-3 Stems and Leaves

Section Objectives:

- *Compare* the internal structures of herbaceous and woody stems.
- *Describe* the external structure of a woody dicot stem.
- *Make* and then *label* a drawing that shows the external structure of a leaf.
- *Describe* the internal structure of a typical leaf and relate it to its function.

  Laboratory Investigation: *Observe* the internal structure of various stems and roots (p. 370).

Types of Stems

Like the roots of a plant, the stems of a plant also have several functions. Vascular tissue runs through the stem, transporting water, food, and minerals between the roots and the leaves. The stem also displays the plant's leaves to sunlight. Taller stems may hold leaves above other plants, thus increasing the leaves' exposure to sunlight. In addition, stems are adapted to different environments. Some underground stems, such as the white potato tuber, are specialized for food storage. The stem of the cactus is modified for water storage and photosynthesis. The stems of the strawberry plant run along the surface of the ground and sprout independent plants. This allows the plant to reproduce quickly.

Based on their stem structure, plants are grouped as herbaceous (her BAY shus) or woody. Herbaceous plants have soft, green, juicy stems that are called **herbaceous stems.** These plants usually live for one or two years. Corn and tomatoes are typical herbaceous

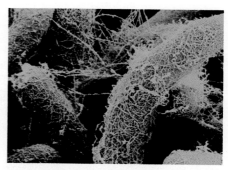

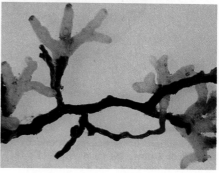

▲   **Figure 18–12**

Mycorrhizae. A scanning electron micrograph (top) reveals the meshlike structure of mycorrhizae. Mycorrhizae appear as swollen regions at the tips of the roots of an aspen tree (bottom).

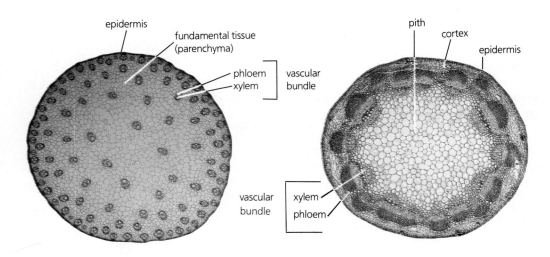

epidermis

fundamental tissue
(parenchyma)

phloem ⎱ vascular
xylem ⎰ bundle

vascular
bundle

pith

cortex

epidermis

xylem
phloem

vascular
bundle

Monocot Stem

Dicot Stem

▲ **Figure 18–13**

Cross Sections of Herbaceous Monocot and Dicot Stems. The stem tissues of herbaceous monocots and dicots differ in organization. In cross section, the vascular bundles of a monocot are randomly scattered (left). In a dicot (right), the vascular bundles form a ring around the pith.

plants. Woody plants have **woody stems** that are made up of the thick, tough tissue that you know as wood. Plants with woody stems normally live for more than two years. Trees, such as oaks and maples, and shrubs, such as lilac and forsythia, have woody stems.

Herbaceous Stems—Internal Structure All the tissues of herbaceous plants develop from the cells produced by the apical meristems. Although the source of these tissues is the same for every herbaceous plant, the organization of the tissues depends on the type of plant. For example, the two groups of flowering plants, the *monocots* and *dicots,* each have a different type of stem structure. Refer to Figure 18–13 as you read about the stem structure of herbaceous monocots and dicots.

Corn is a typical herbaceous monocot. A protective epidermis encloses its soft, green stem. The epidermis is dotted with small openings, called **stomates** (STOH mayts). These openings allow an exchange of gases between the tissues inside the stem and the atmosphere.

Under the epidermis is a layer of chloroplast-containing cells, which are involved in photosynthesis. These cells include fiber cells that stiffen and support the stem. The interior of the corn stem is made of parenchyma cells. Bundles of vascular tissue are scattered throughout these cells. Each bundle, called a **vascular bundle,** contains xylem and phloem that are surrounded by supporting thick-walled cells. Because the stems of all herbaceous monocots have no cambium, they show little growth in diameter.

Typical herbaceous dicots include sunflowers, geraniums, buttercups, and alfalfa. In the herbaceous dicots, the stem is enclosed by a protective layer of epidermis. Inside the epidermis is the cortex, which is made up of collenchyma and parenchyma cells. These tissues provide support for the stem and serve for food storage. Inside the cortex is a ring of *vascular bundles*. Each bundle is made of an outer group of phloem cells, an inner group of xylem cells, and the vascular cambium, which is between the xylem and the phloem. The cambium may undergo a short period of cell division, adding a small amount of new xylem to its inside and a small amount of phloem to its outside. The central region of the stem is called the **pith.** Pith is made of parenchyma cells that store food.

Woody Stems—Internal Structure As you read earlier, the stems and roots of woody plants grow in thickness. This is the result of new tissues that are produced by vascular cambium throughout the plant's life. Almost all woody plants are dicots. Their stems are tough because of the large amounts of xylem that are added to the thickness of the stem. As shown in Figure 18–14, round layers of wood increase the thickness of the stem as xylem builds up on the inside of the vascular cambium. New phloem is also produced in round layers, but these are on the outside of the cambium. Unlike the xylem, the phloem produced by the vascular cambium does not build up. Instead, its older, outer layers break off as new phloem is produced.

Figure 18–14
Cross Section of a Woody Dicot Stem. The cell division activity of the vascular cambium produces xylem to its inside and phloem to its outside. The xylem accumulates as annual growth rings. The phloem, and the cork that develops within it, continuously peels off as it is being replaced. ▼

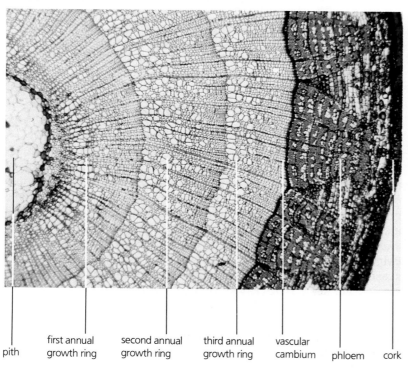

| pith | first annual growth ring | second annual growth ring | third annual growth ring | vascular cambium | phloem | cork |

Biology and You

Q: What are the health benefits of a high-fiber diet?

A: Recently, a great deal of attention has been paid to the health benefits of fiber, the nondigestible material found in plant cell walls. Fiber helps your digestive system function by providing bulk needed to move wastes through the colon. Vegetables, fruits, whole-grain breads, and cereals are rich in fiber.

Some studies show that a fiber-rich diet plays an important role in preventing colon cancer, gallstones, appendicitis, hemorrhoids, and colon infections. By strengthening your intestinal muscles and absorbing water, fiber may allow solid wastes to pass more quickly through the colon, giving only minimal exposure to infectious bacteria and cancer-causing agents.

For adults, 25 to 35 grams of fiber a day is recommended. Too much fiber may aggravate digestive conditions and prevent your body from absorbing essential minerals. A sensible approach to daily intake of fiber is to eat a normal diet that includes fresh fruits and vegetables, whole-grain breads, and cereals.

■ *Plan a day's menu that includes fiber-rich foods at each meal. Then, ask some friends if they would choose to eat the foods on your list.*

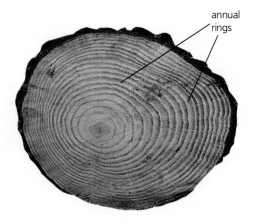

▲ **Figure 18–15**
Annual Rings. Each year's growth is visible as an annual ring in the cross section because of differences in spring and summer wood.

Figure 18–16

External Structure of Dormant Woody Dicot Stem. Woody stems produce a terminal bud at the end of each growing season. A longitudinal section through a terminal bud shows the bud scales surrounding the apical meristem and embryonic leaves. In spring, the bud scales fall off and the bud may develop into a new stem, leaves or flowers. ▼

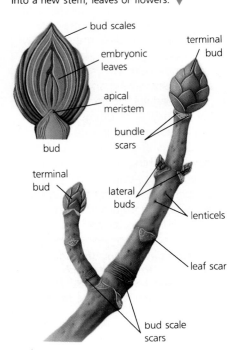

The growth of new xylem during each growing season results in the formation of *annual rings,* shown in Figure 18–15. The age of a woody dicot stem may be found by counting these rings. Each annual ring shows one year of growth. In some woody dicot stems, the cells of the xylem formed in the spring (known as *spring wood*) are larger and lighter in color than those formed in the summer (known as *summer wood*).

The width of the annual rings may vary according to climate that existed during the growing season. Favorable conditions produce wide growth rings. Scientists know by counting the annual rings that some trees, like the giant sequoias of California, live for thousands of years. These trees provide a historical record of changes in climate.

In young woody dicots, the center of the stem is filled with pith, and there is a cortex layer inside the epidermis. In older woody stems, the cells of the pith die, and the cortex is replaced by phloem from the vascular cambium. The living xylem cells, which conduct water, lie next to the cambium. This light-colored xylem is called *sapwood.* The thickness of the sapwood remains fairly constant from year to year. This is because old sapwood is converted to *heartwood* as new sapwood is produced by the cambium. Heartwood is the older, inner, dark-colored region of xylem. Since this layer is always being added to, the area occupied by heartwood increases over time.

The outermost layer of a woody stem is the bark, a protective tissue. On young stems, the bark may be thin, but older stems and trunks have bark that is much thicker. Bark is made of phloem, cork cambium, and cork cells. The cork cells are made by the cork cambium. The inner, younger part of the bark is alive, while the outer, older part is dead tissue. Bark is made as the stem grows in diameter. As the stem size gets larger, and the older outer bark cracks and peels off, new bark takes its place.

Woody Stems—External Structure

Figure 18–16 shows the external features of a dormant twig. A dormant twig is one that has lost its leaves for the winter. At the tip of the twig is the **terminal bud.** The terminal bud is made up of apical meristem, enclosed by overlapping protective scales, called *bud scales.* The bud scales are produced at the end of the previous growing season. When growth begins again in the spring, the apical meristem begins active cell division and forms new stem tissues and leaves. At this time, the bud scales fall off, leaving scars on the twig, called *bud scale scars.* These scars mark the point at which the season's growth began. The length of stem between two successive sets of bud scale scars is one year's growth.

Another feature of the dormant twig is the *leaf scar. Leaf scars* are formed when leaves drop off the stem in autumn. The scars mark the points where leaves from other growing seasons were fastened to the stem. A layer of protective tissue forms at the scar to protect the tissues inside the stem. Within the leaf scars are small dots called *vascular bundle scars*. These are the points at which vascular bundles containing xylem and phloem passed from the stem into the leaf.

Above each leaf scar is a **lateral,** or *axillary,* **bud.** Lateral buds are found just above the point where a leaf is or was attached to the stem. These buds may develop into new branches, or they may remain small and dormant. The points along the stem where leaves and lateral buds form are called *nodes*. The space between two nodes is called an *internode*. Along the surface of the twig, there are small raised openings called lenticels. **Lenticels** (LENT uh sels) are holes that pass through the cork tissue. They allow the exchange of oxygen and carbon dioxide between the atmosphere and the internal tissues.

Types of Leaves

Leaves are specialized to capture light for photosynthesis. The broad, flat structure of most leaves exposes a large surface area to the sun. In addition, the arrangement of leaves around the stem of the plant maximizes their exposure to whatever sunlight is available. The shape of the blade, the arrangement of the leaves, and the pattern of the veins make up some of the characteristics of each plant species.

Often, the environment in which a plant lives determines the type of leaf it has. For example, some plants are adapted to life in dry climates. The cactus plant is so highly specialized for life in the desert that its leaves have been reduced to spines. The stem is used for both photosynthesis and water storage, while the spines protect the stem.

A typical leaf has a thin, flat blade, and a stalk, or **petiole** (PET ee ohl). The petiole joins the leaf to the stem, as shown in Figure 18–17. A **simple leaf** has only one blade and one petiole. In a **compound leaf,** the blade is divided into several parts, or *leaflets,* that are attached to a petiole. The leaves of some plants, such as corn, lilies, and irises, do not have petioles. Instead, the leaf blades are fastened directly to the stem. A network of veins runs through the leaf. The veins contain the vascular tissues—xylem and phloem—of the leaf.

Figure 18–17

Simple and Compound Leaves. Leaves with one undivided blade are called simple leaves (left). Leaves with blades made of several divisions are called compound leaves (middle, and right). ▼

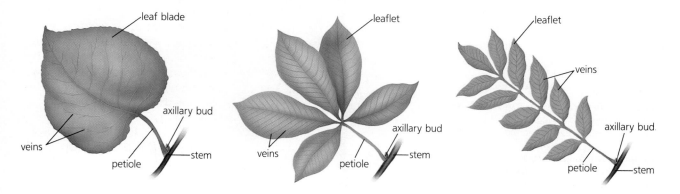

Internal Structure of the Leaf

No matter how they differ in their external appearance, all leaves are made of three types of tissues. These are protective, fundamental, and vascular tissue. Figure 18–18 shows a cross section of a typical leaf.

Cuticle and Epidermis The outermost layer of both the upper and lower leaf surfaces is the clear, waxy cuticle. This layer protects the inner tissues and slows down water loss from the leaf. Beneath the cuticle is the epidermis, which also protects the inner tissues. The epidermal layer, like other epidermal layers, is only one cell thick. The cells are flattened and fit together like the pieces of a jigsaw puzzle. Most of the cells of the epidermis are clear because they have little or no pigment. This allows light to reach the photosynthetic tissues below.

Like the epidermis of the stem, the leaf epidermis has many stomates. Usually, there are many more of these openings on the lower surface of the leaf than on the upper surface. The stomates allow the exchange of carbon dioxide and oxygen between the tissues inside the leaf and the environment. Water vapor also passes out of the leaf through the stomates. The stomates are not open all the time. Instead, they open and close according to the needs of the leaf. Each stomate is surrounded by a pair of specialized epidermal cells called **guard cells.** The kidney-shaped guard cells regulate the opening and closing of the stomates. The way this is done is described in Chapter 19.

Mesophyll Between the upper and lower layers of epidermis is a layer of photosynthetic tissue called **mesophyll** (MEZ uh fil). In some plants, the mesophyll contains two types of thin-walled cells. The upper portion of the mesophyll is called **palisade** (pal uh SAYD)

Figure 18–18

Internal Structure of a Leaf. A leaf is made up of different tissue types. Most of the photosynthesis takes place in the mesophyll. Water is carried to the mesophyll and food is carried away through vascular bundles, which are contained within veins. The epidermis protects the inner tissues. ▶

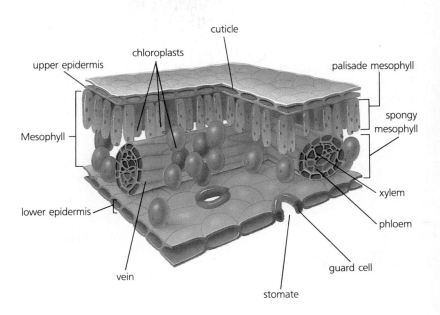

▲ **Figure 18–19**

Venation. Monocot plants have parallel venation (left). Dicot plants have netted venation (right).

mesophyll. It is one or two cells thick. This layer is made of tall, tightly packed cells filled with chloroplasts. Below the palisade layer is the **spongy mesophyll.** This layer is made of irregularly shaped cells with large air spaces between them. The stomate openings of the lower epidermis are next to the intercellular air spaces of the spongy mesophyll. The cells of the spongy mesophyll have fewer chloroplasts than the cells of the palisade layer.

Veins Within the mesophyll layer is a network of veins. The veins contain the vascular tissues. The vein network is so fine that no mesophyll cell is far from a vein. The xylem and phloem of the leaf veins are continuous with the xylem and phloem of the stem and roots. The arrangement of veins in a leaf is called its **venation** (ven AY shun). There are distinct differences between the vein patterns of monocot and dicot leaves. In the leaves of monocots, the main veins usually run parallel to one another along the length of the leaf. In dicots, the veins form a network of branches, as seen in Figure 18–19.

18-3 **Section Review**

1. What tissue produces growth in stem and root thickness?
2. What layer slows down water loss from a leaf?
3. Where does photosynthesis occur in the leaf?

Critical Thinking

4. Compare and contrast a simple and a compound leaf. (*Comparing and Contrasting*)

Laboratory Investigation

The Structure of Stems and Roots

The structure of stems and roots can be readily observed in thin cross sections. In this investigation, you will make and stain cross sections of roots and stems from herbaceous plants.

Objectives

- *Observe* and *compare* the root and stem structure of a herbaceous plant.
- *Compare* tissue development between young and mature regions of stems and roots.

Materials

herbaceous plant	slides	toluidine blue
scalpel	coverslips	culture dishes
dropper	microscope	paper towel
forceps		

Procedure  *

A. Label four small culture dishes *MS, YS, MR, YR*. Cover the bottoms of the dishes with toluidine blue stain.

B. Obtain a main or branch stem from a house or garden plant. With a scalpel, cut a thin cross section from a point about 3 cm behind the tip of the stem. See the figure below. The slice should be as thin as possible—paper thin is best. Place the slice in the staining dish labeled MS (mature stem).

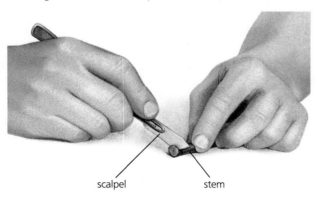

scalpel   stem

C. Cut another cross section about 4 mm behind the tip of the stem. Place this section in the staining dish labeled YS (young stem).

*For detailed information on safety symbols, see Appendix B.

D. Obtain a root from the plant used in step B. Slice a cross section 3 cm from the root tip. Place it in the staining dish labeled MR (mature root).

E. Cut another cross section from the root about 2 to 4 mm behind the root tip. Place this in the staining dish labeled YR (young root).

F. Make a wet mount of the mature stem cross section. With forceps, remove the cross section from the stain, blot it lightly on a paper towel, and place it on a slide. Add a drop of water, then a coverslip. Using low power and referring to Figure 18–15 in your text, identify the vascular bundles, xylem, phloem, cortex, pith, and epidermis. Sketch your specimen, and label each of these tissues. Switch to high power and examine each of these tissues separately.

G. Make and examine a wet mount of the young stem cross section. Draw and label the specimen and compare it with the section in step F.

H. Prepare a wet mount of the mature root cross section. Under low power, find the vascular cylinder, and identify the xylem, phloem, pericycle, endodermis, cortex, and epidermis. Refer to Figure 18–11 in your text. Examine each of these tissues under high power. Sketch your specimen, labeling each tissue.

I. Prepare a wet mount of the young root section and compare it to the section in step H. Draw and label the specimen.

Analysis and Conclusions

1. How do the tissues close to, and farther from, the tip of the stem compare? How do tissues close to and far from the tip of the root compare?

2. How many vascular bundles were present in your stem cross section? How were they arranged?

3. How is the arrangement of vascular tissue in the root different from that in the stem?

4. How can you tell that the cortex cells in the stem and root are parenchyma cells?

5. Suppose your stem cross section revealed vascular bundles scattered throughout the stem. In which group of flowering plants would your plant belong?

Chapter Review

Study Outline

18–1 Plant Tissues

- Plants are made of tissues that form plant organs, which are the roots, stems, leaves, and reproductive structures.

- Meristematic tissues consist of rapidly dividing cells. Protective tissues form the outer layers on certain parts of plants. Vascular tissues carry nutrients and water. Fundamental tissues produce and store food and support the plant.

18–2 Roots

- Roots anchor the plant and absorb water and minerals from the soil. The two common root systems are taproots and fibrous roots.

- A root tip consists of a root cap, a meristematic zone, an elongation zone, and a maturation zone.

- The mature tissues of the root are the epidermis, the cortex, including the endodermis, and the vascular cylinder (pericycle, xylem, and phloem).

- Symbiotic relationships often exist between plant roots and bacteria or fungi.

18–3 Stems and Leaves

- Stems transport and store water and nutrients and support leaves and reproductive structures. Plants are herbaceous or woody, depending on stem structure.

- Herbaceous stems are soft and green. Monocots, which are nearly all herbaceous, have vascular bundles scattered throughout the stem. Herbaceous dicot stems have vascular bundles arranged in a ring. Woody dicot stems are thick and hard, containing successive layers of xylem (wood) produced by the vascular cambium.

- A typical leaf consists of a stalk, or petiole, and a thin, flat blade with a network of veins. Leaves are the major sites of photosynthesis in the plant.

- The leaf epidermis is covered by a clear, waxy layer, the cuticle. In the epidermis, openings called stomates allow exchange of gases between leaf tissues and the environment.

Vocabulary Review

roots (355)
stems (355)
leaves (355)
meristems (356)
meristematic tissues (356)
cambium (356)
vascular cambium (356)
cork cambium (356)
epidermis (356)
cuticle (356)
cork (357)
xylem (357)
phloem (357)
vascular tissues (357)
fundamental tissues (358)
primary root (359)
secondary roots (359)
root cap (361)
meristematic zone (361)
elongation zone (361)
maturation zone (361)

differentiation (361)
root hairs (362)
cortex (362)
endodermis (362)
vascular cylinder (362)
herbaceous stems (363)
woody stems (364)
stomates (364)
vascular bundle (364)
pith (365)
terminal bud (366)
lateral bud (367)
lenticels (367)
petiole (367)
simple leaf (367)
compound leaf (367)
guard cells (368)
mesophyll (368)
palisade mesophyll (368)
spongy mesophyll (369)
venation (369)

A. Matching—Select the vocabulary term that best matches each definition.

1. Made up of thin-walled cells that undergo frequent cell division

2. Layers of photosynthetic tissue between the lower and the upper leaf epidermis

3. Process by which unspecialized cells develop into specialized cells

4. Structure located at the tip of a dormant twig

5. Stems that are soft, green, and juicy

6. Anchor the plant in the soil and absorb water

7. Openings that allow for the exchange of gases in leaves

Chapter Review

8. Conducts water and minerals from the roots to the stems

9. The first structure to emerge from a sprouting seed

10. Central core of the root

B. Sentence Completion— Fill in the vocabulary term that best completes each statement.

11. Food is conducted throughout a plant by vascular tissue called _____.

12. A branch is formed by the growth of a(n) _____.

13. The _____ regulate the opening and closing of stomates.

14. The _____ covers and protects the root tip.

15. In woody stems, exchange of gases occurs through openings called _____.

16. Each _____ in a herbaceous stem is made up of xylem and phloem.

Content Review

17. Identify the functions of the four organs of plants.

18. Describe the epidermis and state its main function.

19. What is cork? What is its function, and how is it produced?

20. Describe the kinds of cells that form vascular tissue.

21. Describe the three types of fundamental tissues.

22. List the different zones of the root tip, and describe the cellular activity in each zone.

23. What occurs when root cells differentiate?

24. What are the primary functions of the root? How is its structure suited to its functions?

25. What structure produces the cells that account for growth in the length of a stem?

26. Compare the internal structure of herbaceous monocot stems with that of herbaceous dicot stems.

27. How are annual rings produced?

28. How do sapwood and heartwood differ?

29. How do woody dicot stems grow in width?

30. What are lenticels? What is their function?

31. Describe the two main types of leaves.

32. Describe the structure of a leaf viewed in cross section.

33. Explain how leaves exchange gases with the environment.

Graphic Organizing

For information on graphic organizers, see Appendix G at the back of this text.

34. **Concept Map:** Copy the incomplete concept map below onto a separate sheet of paper. Add the following concepts: *epidermis, cork, cambium, phloem, apical meristems, xylem.* Include other appropriate concepts of your own choosing. Be sure to add linking words between concepts.

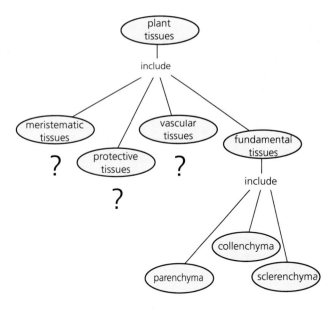

Critical Thinking

35. What would happen to a plant if its leaves were coated with petroleum jelly? What would happen if just the stem were coated? (*Predicting*)

36. What is the significance to the functioning of the xylem that the tracheids and vessel elements are dead at maturity? (*Judging Usefulness*)

37. Compare the arrangement of mature tissues in the youngest region of a tree stem, just behind the growing

tip, with that in a region where the stem has increased in width. (*Comparing and Contrasting*)

38. Why is gas exchange with the environment essential for plants? (*Identifying Reasons*)

Creative Thinking

39. Root cells absorb water by osmosis. In view of this, how would a high salt concentration in the soil affect a plant's ability to grow?

40. You are given a leaf and a cross section of the stem from the same plant. What features of these parts would you use to classify and identify the plant?

Problem Solving

41. Botanists claim that roots grow in a crooked, erratic pattern because of obstacles in the soil, such as rocks or other roots. Design a controlled experiment to test this hypothesis.

42. A student notices that leaves at the very top of a shrub are relatively small and thick. Leaves lower down on the shrub are broader and thinner. Develop a hypothesis to account for the two types of leaves.

43. A botanist wanted to determine the average total surface area of the root hairs of mature bean plants. She estimated the surface area of a typical root hair to be 0.50 cm^2. The number of root hairs on the roots of five bean plants she recorded as shown in the table above. Calculate the total surface area of the root hairs for each plant and the average total surface area for the five plants.

Plant Number	Number of Root Hairs
1	1900
2	1700
3	1800
4	2000
5	1400

Projects

44. Collect dormant twigs from ten different trees or shrubs in your area. Use a field guide to find the common names, scientific names, and basic characteristics of the plants from which they came. Write a brief description of each plant on an index card.

45. Make a collection of different types of leaves. Examine them individually, and try to classify them as belonging to monocots or dicots. Consult a reference to check your classification.

46. Prepare an oral report on one of the following career opportunities: (a) botanist, (b) florist, (c) tree surgeon. If possible, interview a person who is working in the field. Be sure to prepare your questions in advance. You may wish to inquire about training, future opportunities, and daily routine. A tape recorder will be helpful.

47. Prepare a brief written report on the life and scientific contributions of either Matthias Schleiden or Asa Gray.

▲ Golden leaves reach toward the sun's rays, seeking their life-giving energy.

Plant Function

To all eyes, the tree appears immobile. It is firmly rooted in the soil where the seedling fell. The roots draw water and nourishment into the young tree and anchor it securely to the earth. The seedling grows; the branches reach skyward. The leaves unfurl, opening to the air and sunlight. Buds form flowers, giving birth to seeds that form still more new seedlings. This rhythm of living reflects forces in the tree itself as well as forces in the external world. In this chapter, the various factors that bring about growth and regulate the interaction of plants with their environment will be explored.

Guide for Reading

Key words: transpiration, root pressure, transpiration pull, auxins, gibberellins, tropism, nastic movement, photoperiodism

Questions to think about:

📖 How is water in the soil transported to leaves?

📖 How is food made in the leaves transported to the rest of the plant?

📖 What determines when a plant flowers?

19-1 Transport

Section Objectives:

- *Explain* how loss of water from a plant is regulated.
- *Explain* how water travels upward throughout a plant.
- *Summarize* the mechanism of food transport in plants.
- **Laboratory Investigation:** *Observe* and *measure* water transport in the xylem (p. 388).
- **Concept Laboratory:** *Gain* an understanding of the relationship between leaf surface area and transpiration (p. 387).

Plants, like all other organisms, carry out the basic life processes. This chapter focuses on some of these processes. It describes how plants transport water and food and discusses the mechanisms that regulate their development.

Transpiration

Plants that live on land must be adapted to supply water to all their tissues. Much of the water that plants take up, however, is lost to the atmosphere by evaporation. The evaporation of water vapor from plant surfaces is called **transpiration** (tranz puh RAY shun). Most transpiration takes place through the stomates. As you know from Chapter 18, stomates are the pores in the epidermis of leaves. Through the stomates, leaves exchange carbon dioxide and oxygen

▲ **Figure 19–1**

Plants and Water. Plants need water and other raw materials to carry out their life processes.

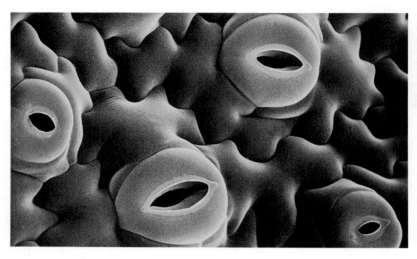

▲ **Figure 19–2**
Stomates. The stomates of a tobacco leaf are open in the light when the plant has enough water. (Magnification 660 X)

with the atmosphere during photosynthesis. See Figure 19–2. For these gases to diffuse through the cell membranes, plant cell surfaces must be kept moist by water supplied from the roots. This water continually evaporates into the spaces between the leaf cells. When the stomates are open, water vapor passes out through them and into the atmosphere. Rates of transpiration can be enormous. For example, a large tree may lose as much as 720 liters of water in a 12-hour day.

Regulation of Transpiration Rate The rate of transpiration, that is, the amount of water lost from leaves in a given amount of time, is regulated by the size of the opening of the stomates. Stomates are usually closed when there is too little water available to a plant, when the temperature is low, or when there is little light. Stomates generally open in light if a plant has enough water.

The opening and closing of each stomate are controlled by the pair of guard cells that surrounds it. The guard cells are sausage-shaped, with thick walls along their inside edge and thin walls along their outside edge. The stomates open or close as the guard cells gain or lose water. When the guard cells become swollen with water, or *turgid* (TER jid), the uneven thicknesses of their walls cause them to bow outward, as shown in Figure 19–3. The opening created between the two cells is the stomate. When the guard cells lose water, they become less turgid, and the stomate closes. The stomates of most plants open during the day and close at night.

Guard cells gain and lose water by osmosis. For a review of osmosis, see Chapter 5. When guard cells build up their concentration of solutes by active transport, water from surrounding cells diffuses into the guard cells. This causes the guard cells to swell and, thus, the stomates to open. When guard cells release the solutes, water diffuses out from the guard cells and back into the neighboring tissues. As a result, the stomates close.

Figure 19–3

Stomates and Guard Cells. A stomate opens when the walls of the guard cells swell with water and bow outward. ▼

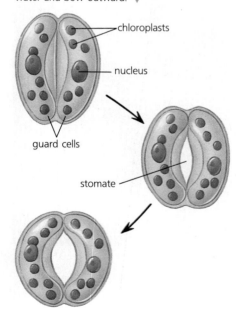

chloroplasts

nucleus

guard cells

stomate

Guard Cell Osmotic Control   The solute most important in changing the osmotic conditions of the guard cells is the potassium ion, K^+. The concentration of potassium ions in the guard cells of open stomates is many times higher than the concentration in the guard cells of closed stomates. ATP is used to power an active transport system that pumps potassium ions into guard cells. Thus, ATP is required to keep the stomates open. During photosynthesis, the chloroplasts of the guard cells supply at least part of the ATP needed for this process. No ATP is needed for stomates to close. The active transport system simply stops, allowing potassium and other solutes to diffuse freely from the guard cells.

Although light is an important factor affecting stomate opening, carbon dioxide concentration *in the leaf* appears to be even more important. A low concentration of carbon dioxide causes stomates to open, even in complete darkness. A high concentration causes stomates to close, even in light. However, scientists do not know how these factors act to turn on or off the active transport system that controls guard cell solute concentrations.

Water Transport

The large amounts of water lost through transpiration must be replaced, or the leaves of the plant will wilt and die. Water in the soil is taken up by the roots and moved upward through the xylem of the roots, stems, and into the leaves.

In tall trees, water is supplied to leaves that may be more than 100 meters above the ground. See Figure 19–4. How water moves so high against the pull of gravity has long puzzled botanists. It cannot be due to active transport within the xylem, because the cells of the xylem that conduct water are not alive. Other processes, such as capillary action and root pressure, offer only a partial answer.

Capillary Action   The tendency of liquid to rise inside a narrow tube is called *capillary action.* The details of this process are discussed in Chapter 4. In plant roots and stems, the water-conducting cells of the xylem form a system of narrow, capillary-type tubes through which water moves. Because of the strong attraction between water and the cell walls that form these tubes, water rises in the xylem. However, capillary action can raise water in the xylem no more than several centimeters. Nevertheless, the forces of attraction between water and the walls of the water-conducting cells help to keep the columns of water intact.

Root Pressure   When the stem of a well-watered plant is cut off close to the soil, sap flows from the cut stem. If a glass tube is attached to the cut end of the stem, sap rises in the tube to a height of about one meter. The pressure that holds up the column of water is called **root pressure.** Root pressure is actually osmotic pressure caused by a buildup of solutes in the xylem of roots.

As you read in Chapter 5, osmosis is the diffusion of water across membranes from areas of high water concentration to areas of lower water concentration. In a root, the endodermis—the layer

Figure 19–4

Water Transport. Water must travel from the roots to the leaves of tall trees, such as this Jeffrey Pine in Yosemite National Park. ▼

▲ **Figure 19–5**
Guttation. In guttation, plants exude water as droplets along the edges of their leaves.

Figure 19–6
Transpiration Pull. Water is pulled up the xylem as water evaporates from leaves. ▼

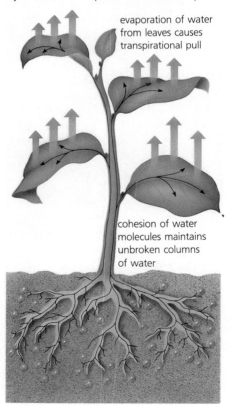

evaporation of water from leaves causes transpirational pull

cohesion of water molecules maintains unbroken columns of water

of cells that surrounds the root vascular cylinder—pumps solutes into the vascular cylinder by active transport. As a result, solute concentration increases in the root xylem, lowering the concentration of water there. Thus, water in the soil diffuses into the root xylem, causing a buildup of pressure. The pressure in the root xylem forces water upward through the xylem of the stem.

Root pressure does happen occasionally in some plants. When it does, a process known as guttation occurs. **Guttation** (gyoo TAY shun) is the formation of water droplets at the edges, or tips, of leaves as a result of root pressure. See Figure 19–5. Guttation usually occurs during the night and only in small plants growing under moist conditions. Because of the pressure buildup, water in the xylem of the leaf is forced out onto the surface of leaves in the form of droplets. These droplets usually can be seen early in the morning. They should not be confused with dew drops, which appear randomly on a leaf. Dew drops are the result of water vapor in the air condensing on cool leaf surfaces.

Even when root pressure occurs, it can cause water to rise no more than about one meter. Therefore, root pressure alone cannot account for the rise of water to the tops of trees. Nor can it cause water to move through a plant when it is growing with little water or when transpiration occurs. During transpiration, large volumes of water move through the roots, preventing solutes from building up in concentration.

Although root pressure is not the mechanism for water transport in plants, it does explain how water first enters the roots of plants when they are seedlings.

Transpiration Pull   Water in the xylem exists as thin, unbroken columns, stretching from the roots, through the stem, and into the leaves. Each water column is held together by strong attractive forces between the water molecules themselves (cohesion) and between the water molecules and the walls of the conducting cells (adhesion).

In roots, the water columns in the xylem are continuous with water that is present throughout the root tissue and in contact with films of water surrounding soil particles. Similarly, water in the leaf xylem is continuous with films of water surrounding all the leaf cells. This water also is continuous with water vapor in the leaf air spaces and atmosphere. See Figure 19–6.

During transpiration, water molecules evaporate from the surface of leaf cells into leaf air spaces and diffuse out through stomates. The lost water is replaced by water in contact with water at the end of a water column in the leaf xylem. When this happens, other water molecules in the column are drawn into the leaf tissue by the strong attractive forces between water molecules. Consequently, the loss of water by evaporation creates a pull or tension on the columns of water in the xylem. Water, therefore, is pulled up, not pushed up, through the xylem. This process, which is called **transpiration pull,** can account for the movement of water to the tops of the tallest trees.

Food Transport

Sugars produced in the leaves during photosynthesis are distributed to other cells in the plant where they are stored or used for energy. This movement of dissolved food through a plant is called **translocation** (tranz loh KAY shun). Translocation occurs only in the phloem.

While materials flow through the xylem in one direction—from the roots to the leaves—flow of materials through the phloem can occur in any direction, depending upon where food is being used.

As you may recall from reading Chapter 18, the conducting cells of the phloem are the sieve cells. They are stacked end-to-end. This formation results in the creation of sieve tubes. Sieve cells are alive. If they are killed, materials in the phloem stop moving. This is in contrast to the xylem where the conducting cells are dead. Scientists have found that sieve cells in the leaves use active transport to build up high concentrations of sugars as they are made during photosynthesis. This causes water to diffuse into these cells, which results in a buildup of pressure within the sieve tubes. In other areas of the plant, sugars move from the sieve cells to the cells in which the sugars are used. As a result, water diffuses out of the sieve cells in these areas, causing a drop in pressure. Thus, there is a pressure gradient—high pressure in the leaf phloem and lower pressure in the phloem in other parts of the plant. This gradient, which is illustrated in Figure 19–7, is what causes the liquid in the phloem to move. This explanation of translocation is known as the **pressure-flow theory.**

Figure 19–7

Pressure Flow Theory. Fluid in the phloem moves from areas of high pressure (where sugar concentration is highest) to areas of lower pressure (where sugar concentration is lower). ▼

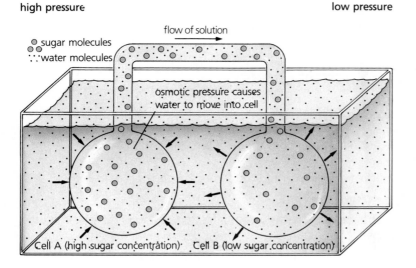

high pressure   low pressure

flow of solution

sugar molecules
water molecules

osmotic pressure causes water to move into cell

Cell A (high sugar concentration)   Cell B (low sugar concentration)

Science, Technology and Society

Issue: Vanishing Habitats

Tropical rain forests grow in hot, wet regions near the equator in countries such as Brazil. They are the natural habitat for over half of the plant and animal species on this planet. Yet, these forests are disappearing at an alarming rate.

Some people contend that the destruction of the forests is unavoidable. Preservation of the environment should not be a priority when basic needs must be met. The people of tropical countries clear forests for timber, farmland, and cattle ranches. Timber, crops, and beef are major exports for poor, developing nations.

Other people argue that deforestation contributes to drought and mud slides. Some scientists warn that changes in the ecology of the rain forests may change temperature patterns and water cycles around the world. Those who oppose deforestation point out that many of the species being destroyed may be valuable to humankind. They argue that the rain forests must be saved.

■ *How can nations work together to save the tropical rain forests?*

Can You Explain This?

Pictured above are plants that have been carefully taken from soil without damage to their roots. Now their roots are placed in running water laced with minerals. Sunlight is also provided. The plants will never be in soil again.

■ *Do you think the plants will continue to thrive? Explain your reasoning.*

▲ **Figure 19–8**

Nitrogen Deficiency. Compare the squash plants grown in nitrogen-deficient soil (left) with those grown in soil containing nitrogen (right).

Plant Mineral Nutrients

Plants are able to synthesize all the carbohydrates, proteins, and other organic compounds they need. To do so, however, they need inorganic raw materials. To make carbohydrates and fats, plants need water and carbon dioxide, which provide hydrogen, oxygen, and carbon. To make amino acids and proteins, they also need a source of nitrogen. Some plants get usable nitrogen compounds from bacteria that live in their roots. Most, however, must get nitrogen compounds from the soil. Besides nitrogen, plants must obtain a number of other inorganic substances, called minerals, from the soil. The minerals needed by plants include compounds of phosphorus, potassium, magnesium, iron, calcium, and sulfur. See Figure 19–8. Many of these are required for enzymes to function. Others form parts of essential plant substances. Magnesium, for example, forms part of the chlorophyll molecule. Iron is part of many electron carrier molecules. Plant growth can be seriously affected by a lack of any of these. Plant fertilizers supply varying amounts of the minerals that plants require.

19-1 Section Review

1. What is transpiration?
2. Why it is impossible for the water-conducting cells of xylem to carry on active transport?
3. What is guttation?
4. What is the transport of food through a plant called?

Critical Thinking

5. What beneficial effect do you think transpiration might have on the temperature of leaves? (*Predicting*)

19-2 Regulation of Plant Growth

Section Objectives:

- *Describe* the effects of auxins on different parts of the plant.
- *Explain* the difference between positive and negative tropisms.
- *List* the various types of plant hormones and describe the function of each.
- *Define* the terms *nastic movement* and *photoperiodism*.

A plant on a windowsill grows toward the sunlight. Plant roots grow toward gravity. What causes these specific growth patterns? Changes in the rate and direction of growth in plants are caused by changes in the levels of certain chemicals in response to environmental factors.

Plant Hormones

Chemical messengers, called hormones, regulate many plant functions. Most hormones are made in minute quantities by actively dividing tissues at the tips of roots and stems. Once produced, they are transported to various parts of the plant where they affect cell metabolism, cell division, and plant growth.

There are a number of different types of hormones in plants, but three of the most important kinds are auxins, gibberellins, and cytokinins. These hormones affect the growth of various plant tissues.

Auxins   Hormones that affect the growth of plant tissues are called **auxins** (AWK sinz). Auxins may stimulate or slow growth, depending on the type of tissue and the amount of hormone. These hormones increase plant growth by stimulating cells to lengthen. In addition, they cause cells to differentiate. See Figure 19–9.

◄ **Figure 19–9**

Apical Dominance. The apical meristem on the left produces an auxin that inhibits the growth of lateral branches below it. If the apical meristem is snipped off, as with the plant on the right, the lateral branches grow, creating a bushy appearance.

▲ **Figure 19–10**
Effect of Gibberellins on Plant Growth.
Compare the height of these two-day old kidney bean seedlings. The taller seedlings on the right have been sprayed with gibberellins.

The most common auxin in nature is *indoleacetic* (in dohl uh SEE tik) *acid,* or IAA. IAA is produced in high concentrations in the terminal bud of plants. As the auxin diffuses down the stem, it prevents the growth of lateral buds. When a lateral bud on a stem is far enough away from the terminal bud on the same stem, the lateral bud will begin to grow. When this happens, it becomes the terminal bud of a branch. This explains why cutting off the terminal buds of house plants causes lateral buds to grow, and the plant to become bushier.

Auxins also affect the process of *abscission* (ab SIZH in)—the dropping off of leaves, flowers, or fruits from a plant. Synthetic auxins are sprayed on apple trees to prevent apples from dropping before they are fully ripe. Auxins are also used commercially to stimulate the production of roots on stem or leaf cuttings and to control weeds. Spraying weeds with synthetic auxins causes them to overgrow so much that they die.

Gibberellins   Hormones that affect plant growth and the development of fruits and seeds are called **gibberellins** (jib uh REL inz). Unlike auxins, gibberellins are distributed evenly throughout the plant tissues. Gibberellins have important effects on stem growth in plants. In fact, plant stems that are typically short can be made to grow by spraying them with gibberellins. See Figure 19–10. Commercially, these hormones are used to stimulate flowering and also to increase fruit size. Many commercial grapes are sprayed with gibberellins to increase cluster size.

Other Hormones   Another group of naturally occurring plant hormones is the **cytokinins** (syt uh KY ninz). Cytokinins stimulate cell division and growth during seed germination. They are thought to work together with auxins in stimulating cell differentiation.

A gas, **ethylene** (ETH uh leen), is another plant hormone. Ethylene, along with auxins, plays a role in abscission. It also stimulates the ripening of many fruits. As bananas ripen, the dark specks you see on their "skin" are concentrated areas of ethylene. In fact, clusters of green bananas are ripened at the supermarket by exposing them to low concentrations of ethylene.

The hormone **abscisic** (ab SIZ ik) **acid** increases in concentration as the days become shorter in the fall, and temperatures decrease. Abscisic acid is associated with the shedding of leaves and the seasonal slowing down of plant activities.

Auxins and Tropisms

The growth of a plant in a specific direction in response to a stimulus is called a **tropism** (TROH piz um). Plant growth or movement toward a stimulus is called a *positive tropism,* while movement away from a stimulus is called a *negative tropism.* The kinds of stimuli that plants grow toward or away from include light (phototropism), gravity (geotropism), touch (thigmotropism), and water (hydrotropism). The stem of a plant that is growing toward

◀ **Figure 19–11**
Phototropism. No matter which direction this plant is turned, it will always grow toward the sunlight.

the light is an example of positive **phototropism** (foh tuh TROH piz um). See Figure 19–11. Roots, unlike stems, show negative phototropism—they grow away from the light source.

Roots generally show positive **geotropism** (jee uh TROH piz um). They grow down into the ground in the direction of the force of gravity. Stems, on the other hand, show negative geotropism. They grow up away from the force of gravity.

When the tendrils of a grapevine wind themselves around the stem of another plant, they are showing **thigmotropism** (thig muh TROH piz um), growth in response to touch. **Hydrotropism** (hy droh TROH piz um) is observed in plants such as willow trees in which roots grow toward water.

The growth responses seen in tropisms are thought to be caused by uneven distribution of auxins in the affected plant parts. In the positive phototropism of stems, for example, the concentration of auxins becomes higher on the shaded side of the stem than on the lighted side. Thus, the cells on the shaded side grow faster than the cells on the lighted side. The uneven rates of growth on opposite sides of the stem result in a bending toward the side of less rapid growth. In this case, the stem bends toward the light.

Nastic Movements

A plant movement that is in response to a stimulus but independent of the direction of the stimulus is called a **nastic** (NAS tik) **movement.** Some nastic movements are reversible and do not involve growth. For instance, the leaves of the prayer plant are spread out flat during the day but become vertical at night. Many other kinds of plants show similar movements of their leaves and flower parts over a 24-hour period.

▲ **Figure 19–12**

Nastic Movement. The leaves of the oxalis plant are open during the day (left) but closed at night (right).

Most nastic movements involve changes in the internal pressure, or turgor pressure, of specific cells. For instance, if you touch the leaflets of the sensitive plant, they will instantly collapse. They collapse because of changes in osmotic pressure. When the leaflets are touched, the concentration of ions in the cells at the base of each leaflet drops rapidly, causing water to flow out of the cells by osmosis. When this happens, the cells lose their turgor, and the leaflets drop.

The rapid movement of the leaves of the Venus flytrap is another example of a nastic movement. When the trigger hairs on a leaf are touched, cells on the upper surface of the leaf along the midline rapidly lose solute. As a result, water moves out of the cells, the cells collapse, and pressure from the other leaf cells causes the leaves to snap shut. In this way, the plant traps insects. After a while, the water pressure in the cells is restored and the leaves reopen.

Photoperiodism

Light affects plant growth and development in ways that are unrelated to its role in photosynthesis. For instance, the flowering of many plants is in response to changes in the length of day over the course of the year. The response of a plant to changes in the length of day or night is called **photoperiodism** (foh toh PIR ee uh diz em).

In many types of plants, flowering and other processes, such as leaf abscission, are controlled photoperiodically. At first, it was thought that the length of the light period was what caused flowering. Plants that flowered during short days were called *short-day plants.* Those that flowered during long days were called *long-day plants.* However, scientists later discovered that it is the length of uninterrupted darkness that causes flowering. Short-day

▲ **Figure 19–13**

Short-Day and Long-Day Plants. The poppy (left) is a long-day plant, and the morning glory (right) is a short-day plant.

plants, therefore, are better described as *long-night plants*. They require long periods of darkness in order to flower. Similarly, long-day plants are more appropriately called *short-night plants*. They flower when there are short periods of darkness. Nevertheless, the terms short-day plant and long-day plant continue to be used.

Some short-day plants are morning glory, forsythia, tulip, chrysanthemum, aster, and goldenrod. These plants flower in the early spring, late summer, or fall when the days are short and the nights are long. Long-day (short-night) plants include clover, potato, beet, poppy, and gladiolus. Long-day plants usually bloom in the summer. Plants whose flowering is unaffected by the lengths of light and dark are called *day-neutral plants*. Tomato, cucumber, dandelion, string bean, and corn are examples of day-neutral plants. Day-neutral plants have a long flowering season.

The effects of photoperiodism on flowering and on other processes are the result of changes that take place in a pigment called *phytochrome* (FY tuh krohm). This pigment, which is found in low concentration in plant cells, has a dramatic effect on plant growth and development. Although the details are not known, changes that take place in the phytochrome molecule during the dark period bring about metabolic changes in certain plant tissues. These changes may affect the production or release of certain hormones or may affect plant cells at other levels of metabolism.

19-2 **Section Review**

1. Define the term *positive phototropism*.
2. List three plant hormones.
3. What causes the leaflets of the sensitive plant to collapse when they are touched?
4. What is the term for the response of a plant to the changing duration of light and darkness?

Critical Thinking

5. How is a nastic movement different from a tropism? (*Comparing and Contrasting*)

■ Careers *in Action*

Nursery Manager

Plant nurseries grow and sell trees, shrubs, flowers, and other plants. A nursery manager's job is to supervise and also coordinate these and many other activities.

A nursery manager's tasks are varied and depend on the size of the nursery. Nursery managers may propagate plants from seeds, root cuttings, or grafts and tend the plants while they grow to marketable size. Managers may also prepare soil, breed plants, and control insects, weeds, and disease. On the business side, they plan what to buy from wholesalers; they prepare budgets, oversee staff, and make long-range development plans.

While many nursery managers work for private nurseries, others may work for botanical gardens, parks, institutions with outdoor areas, or government agencies.

Problem Solving Suppose a customer comes to you for advice. The plants she bought from you flowered in her garden late in the fall. She wants the plants to grow and flower indoors during the winter. She is keeping some transplanted in her basement under artificial lights. She turns the lights on every morning and then turns them off just before going to bed. The plants, however, are not flowering.

1. From what the customer has told you, does she have short-day, long-day, or day-neutral plants? Explain.
2. What is keeping the plants from flowering? What can be done to induce flowering in the plants?
3. How can the customer determine what would be the best light conditions for inducing flowering?
4. Suppose the customer creates light conditions under which the plants will flower, but she then finds that if the lights are on, even briefly, in the middle of the night, the plants do not flower. How would you explain this finding?

■ **Help Wanted:**
Nursery manager. High-school diploma required. Contact: American Association of Nurserymen, Inc., 1250 I St., N.W., Suite 500, Washington, D.C. 20005.

Concept Laboratory

Leaf Surface Area and Transpiration

Goal

To gain an understanding of the size of a plant's leaf surface area and its relationship to transpiration.

Scenario

Imagine you are designing the landscaped area within the entranceway of a large building. One requirement is that the daily amount of water transpired by the plants cannot exceed 50 kg. The plants you plan to use have an average transpiration rate of 2400 grams water/day/m² of leaf surface area. To know how many plants can be present, you must calculate the approximate total leaf surface area of the plants that you want to use.

Materials

pencil
graph paper ruled
   in centimeters
plant with medium-to-large leaves

Procedure

A. Remove three leaves of average size from a plant, and trace their outlines on a sheet of graph paper.

B. Count the number of squares covered by each leaf. Include all squares that are more than half-covered but none that are less than half-covered. Count the squares by row from left to right. Draw a line through each row after you have counted it.

C. Each square on the graph paper equals 1 cm². Thus, the number of squares covered by a leaf equals the size of the leaf in cm². In a chart like the one below, record the size of each leaf. Multiply this number by 2 to obtain the area of both sides of each leaf.

Sizes of Leaves (cm²)	Surface Areas of Leaves	Total Number Leaves	Total Leaf Surface Area (m²)
1. 2.			

D. Calculate the average surface area of the three leaves. Add the three leaf areas and divide by 3.

E. Estimate and record the number of leaves on a branch and the number of branches on the plant. Multiply these numbers for the total leaf number.

Organizing Your Data

1. Calculate the total surface area of the leaves. Multiply the number of leaves on the plant by the average surface area of a leaf. Divide your answer, in cm², by 10 000 to convert to m².

2. How does the leaf surface area of your plant compare with the surface areas of the following objects: a twin bed—about 2m²; a classroom—about 100m²; and a football field—about 5000 m²?

Drawing Conclusions

1. Calculate the amount of water transpired per day by the plant used in this exercise.

2. Calculate the maximum number of plants (of the size and type of the one used in this activity) that could be planted in the entranceway.

Thinking Further

1. In what way is the large surface area of leaves useful to a plant?

2. In what way can the large surface area of leaves be harmful to a plant? What features of leaves help to reduce this effect?

3. If a tree has 300 000 leaves and an average leaf surface area of 10 cm², what is the total leaf surface area of the tree? Assume the average number of stomates per cm² of leaf area is 8 000. Calculate the total number of stomates in the tree.

4. How much water is transpired in a day by the tree in question 3? Assume the same transpiration rate of 2 400 g/day/m².

5. Imagine a park of 100 trees, a woodland of 1 000 trees, and a forest of a million trees like the tree in question 3. How much water would be transpired by the trees in each area in a day? In a week?

Laboratory Investigation

Transpiration and Water Transport

Water is supplied to plant tissues through continuous columns of water in the xylem. In this investigation, you will demonstrate and measure transport of water in the xylem.

Objectives

- *Demonstrate* water transport in the xylem.
- *Measure* the rate of transpiration.

Materials

celery stalk with leaflets	electric fan	transparent millimeter ruler
small beaker	scalpel	microscope
water	slide	calculator
food coloring	coverslip	
	dropper	

Procedure

A. Half fill a small beaker with water, add a few drops of food coloring, and mix the contents.

B. Select a celery stalk with leaflets at its end. Use a scalpel to cut crosswise through the stalk about 10 cm from the leaflet end. To eliminate air from the xylem, make a second cut to remove 1 cm from the cut end.

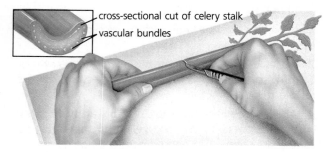
cross-sectional cut of celery stalk
vascular bundles

C. Place the cut end of the stalk into the beaker, and set it in front of a fan so that air blows across the leaflets. Record the time in a chart.

D. When the colored water has moved at least 5 cm up the stalk, remove the stalk from the beaker and blot it dry with a paper towel. Record the time.

E. Measure and record in a chart the length of the colored column of water in three of the vascular bundles. Calculate and record the average length.

*For detailed information on safety symbols, see Appendix B.

F. With a scalpel, make a cross-sectional cut through the region of the stalk containing colored water. Slice a paper-thin cross section from one of the cut surfaces, and place it on a slide. Add a drop of water, then a coverslip, and set it aside.

G. Under low power, position a transparent millimeter ruler so that its scale edge passes through the center of the field of view. The number of millimeters that you can see is the actual diameter of the field of view. Record this number.

H. Now examine the thin section of the celery stalk under low power. Locate one colored area—the xylem of a single vascular bundle—and determine its diameter. (Estimate what fraction of the width of the whole field of view it occupies. Then multiply this by the width of the field of view.) Record this in your chart. Repeat this procedure for two other vascular bundles. Calculate and record the average xylem diameter.

I. Calculate the volume of colored water in the stalk. (This is equal to the volume of water transpired during the period.) Assume the colored water in each vascular bundle forms a long, thin cylinder. The volume of a cylinder is its length multiplied by its cross-sectional area (area = $3.1416 \times \text{radius}^2$). Using the average radius and length of the colored water columns, calculate and record the average volume of colored water in one vascular bundle. Multiply this number by the number of vascular bundles in the stalk. Record the total volume.

Analysis and Conclusions

1. Why was the celery placed in front of a fan?

2. Where in the celery stalk would you expect the food coloring to end up eventually?

3. What was the rate of transpiration (cm^3 water per minute) in the celery stalk? To convert mm^3 to cm^3, divide by 100.

4. How would the transpiration rate differ if the celery stalk had larger leaflets? Explain.

5. If all stalks have the same transpiration rate, what is the transpiration rate of the whole plant?

Chapter Review

Study Outline

19–1 Transport

■ Transpiration is the loss of water by evaporation from plant surfaces. It takes place mainly through the stomates.

■ The opening and closing of stomates by guard cells controls the rate of transpiration. Guard cells change shape as they gain or lose water due to changes in their solute concentration.

■ The loss of water from leaves results in a "pull" on the continuous columns of water that run through the xylem of the roots, stems, and leaves. This transpiration pull accounts for the movement of water from roots to leaves.

■ The movement of dissolved food through a plant is called translocation. It takes place in the phloem and is thought to occur by a gradient of turgor pressure in the sieve tubes of the phloem.

■ Plants need nitrogen and other minerals from the soil to carry out their normal metabolic activities.

19–2 Regulation of Plant Growth

■ Growth responses toward or away from stimuli are known as tropisms. Tropisms, which can be stimulated by light, gravity, touch, and water, are thought to result from an uneven distribution of auxins in the responding plant part.

■ Auxins, gibberellins, cytokinins, ethylene, and abscisic acid are some of the known plant hormones. They regulate many aspects of plant growth and development.

■ Nastic movements are usually reversible movements of plant parts that always occur in the same direction, regardless of the direction of the stimulus. They result from changes in the turgor pressure of certain cells.

■ In many plants, flowering, leaf abscission, and other developmental events occur in response to the changing lengths of day and night. Such responses, which involve the plant pigment phytochrome, are examples of photoperiodism.

Vocabulary Review

transpiration (375)	ethylene (382)
root pressure (377)	abscisic acid (382)
guttation (378)	tropism (382)
transpiration pull (378)	phototropism (383)
translocation (379)	geotropism (383)
pressure-flow theory (379)	thigmotropism (383)
auxins (381)	hydrotropism (383)
gibberellins (382)	nastic movement (383)
cytokinins (382)	photoperiodism (384)

A. Multiple Choice—Choose the letter of the answer that best completes each statement.

1. The movement of dissolved food through a plant is called (a) translocation (b) transpiration (c) geotropism

2. The process that can account for the movement of water to the tops of tall trees is (a) tropism (b) transpiration pull (c) pressure-flow theory (d) translocation.

3. The loss of leaves in the fall is influenced by (a) geotropism (b) nastic movements (c) abscisic acid

4. Plant growth toward or away from a stimulus is caused by uneven distribution of (a) ethylene (b) cytokinins (c) gibberellins (d) auxins

5. The loss of water by evaporation from plant surfaces is known as (a) photoperiodism (b) transpiration (c) tropism (d) root pressure

6. Pressure caused by a buildup of solutes in root xylem is called (a) root pressure (b) hydrotropism (c) translocation (d) guttation

7. Hormones that affect stem growth and fruit and seed development are called (a) abscisic acids (b) gibberellins (c) auxins (d) cytokinins.

8. The bending of a stem toward light is a growth response called (a) thigmotropism (b) photoperiodism (c) phototropism (d) transpiration

9. The formation of water droplets at the edges and tips of leaves is called (a) geotropism (b) guttation (c) thigmotropism (d) transpiration.

Chapter Review

B. Definitions—Replace the italicized definition with the correct vocabulary term.

10. Prayer plant leaves illustrate *movement in response to a stimulus but independent of its direction*.

11. Grapevine tendrils illustrate *growth in response to touch*.

12. *The response of a plant to changes in duration of light* results in flowering and leaf fall.

13. *A gaseous plant hormone* hastens the ripening of fruit.

Content Review

14. Why is transpiration unavoidable in a plant?

15. How does the structure of their cell walls affect how guard cells control the opening and closing of stomates?

16. What role does potassium play in guard cell movements?

17. Describe transpiration pull.

18. Explain why the columns of water in the xylem do not break apart during transpiration.

19. Explain how the pressure-flow theory accounts for translocation.

20. What are the raw materials that plants require?

21. How do auxins affect plant tissues?

22. What processes are affected by gibberellins?

23. What is the effect of cytokinins and auxins working together within a plant?

24. Explain the difference between a negative and a positive tropism.

25. What causes most nastic movements?

26. What is photoperiodism?

Graphic Organizing

For information on graphic organizers, see Appendix G at the back of this text.

27. **Line Graph:** Construct a line graph based on the information in the table in the next column, which shows the hourly rate of transpiration in a plant over a 16-hour period. Plot time on the x-axis and transpiration rate on the y-axis. At what times are the rates highest? Lowest? How do you account for this?

Time	Transportation Rate (g/h)	Time	Transportation Rate (g/h)
8 AM	190	5 PM	220
9	200	6	213
10	209	7	208
11	215	8	190
12 Noon	221	9	120
1 PM	227	10	100
2	233	11	98
3	230	12	90
4	227		

Critical Thinking

28. Explain how each of the following features of many desert plants minimizes transpirational water loss: (a) thick, leathery leaves, (b) thickened cuticle, (c) leaves that drop off during dry season, (d) few stomates per unit of leaf surface area. (*Identifying Reasons*)

29. How might it benefit a plant to have its terminal bud grow for a period of time before any lateral buds grow? (*Identifying Reasons*)

30. A substance that binds potassium ions (K^+) was dissolved in the water supplied to a plant. What effect, if any, do you think this substance would have on the plant's leaves and ultimately on the whole plant? (*Predicting*)

Creative Thinking

31. What developmental problems do you think would arise if gardens were grown in large space stations orbiting the earth?

32. Imagine that you must produce flowers at times of the year that do not coincide with the plants' natural flowering periods. How would you go about this task?

Problem Solving

33. How would you measure the rate of transpiration of a potted plant?

34. An investigator tested the effect of light on the direction of growth of grass seedlings. The series of diagrams on the next page illustrate the effects of three kinds of treatments on the seedlings. Develop a hypothesis to account for the experimental results.

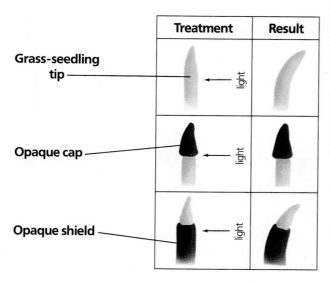

	Treatment	Result
Grass-seedling tip	light	
Opaque cap	light	
Opaque shield	light	

Long-day lighting conditions

Long-day lighting conditions

Leaf in light box under short-day conditions

Short-day plant

Short-day plant

35. An experiment to determine how plants are induced to flower photoperiodically is shown in the illustration on the right. Identify the control and propose a hypothesis to explain the results.

Projects

36. Write to the United States Department of Agriculture in Washington, D.C. to inquire about recent research projects in genetic engineering carried out to improve food products.

37. Write to an industrial producer of plant hormones such as auxins. Ask for materials describing how the hormones are produced and how they are used commercially. After receiving the information, prepare a short presentation for the class.

38. Start six bean plants from seeds. Remove both cotyledons from two seedlings, one cotyledon from two others, and leave the remaining two seedlings intact. Measure the growth over a period of time and record the data. Develop a hypothesis to explain differences in the growth of the plants.

39. Prepare a report on one of the following career opportunities: farmer; agricultural engineer; plant breeder. If possible, interview someone who is working in the field. Be sure to prepare your questions in advance. You may wish to inquire about training, future opportunities, and daily routine. A tape recorder will be helpful.

Biology and Problem Solving

How do plants defend themselves?

Throughout their lives, plants are attacked by insect predators. Attacks can be dramatic, as when hordes of insects invade a wide area, destroying virtually all the plant life. Other attacks are less severe—plants are damaged but not destroyed.

How do plants defend themselves from insect attacks? They cannot run away. They cannot fight back physically. Yet most survive. In fact, some trees even live for hundreds of years. The following case studies have suggested some answers to this question.

■ Case Study A: Some insects, including locusts, go through several stages of development: larva, pupa, and adult. Two types of hormones that control insect development are the juvenile hormones and the ecdysteroids.

After swarms of locusts destroyed an extensive area of vegetation in Kenya, one observer noted that the only remaining plants in that area were bugleweed. Extracts of bugleweed were fed to a number of insects, including armyworm caterpillars. The armyworm caterpillars did not develop normally from the larval to the pupal state. They developed extra heads, which blocked their mouthparts and caused them to starve. Analysis of the bugleweed showed the presence of ecdysteroidlike chemicals.

1. Develop a hypothesis to explain how bugle-weed plants defend themselves against insect attacks.

■ Case Study B: Black swallowtail butterflies are attracted to celery but avoid cabbage. Cabbage is known to contain a chemical called sinigrin, which celery plants do not have. Researchers induced celery plants to take up sinigrin in varying amounts. Then they fed leaves of these celery plants to butterfly larvae. Celery leaves that contained small amounts of the chemical were seen to inhibit the growth of the larvae. Celery leaves that contained as much sinigrin as cabbage caused the larvae to die.

2. What conclusion(s) can be drawn from this experiment? Explain your reasoning.

■ Case Study C: Tent caterpillars eat the leaves of willow trees. A scientist placed some tent caterpillars in one group of willows but none in a nearby group. After two weeks, he fed leaves from both infested and uninfested trees to another group of caterpillars. Leaves from the infested trees caused the caterpillars to grow much more slowly than normal. To his surprise, the scientist found that leaves from the uninfested trees also caused the caterpillars to grow more slowly.

It was found that, in response to the insect attack, the leaves of both the infested and uninfested trees produced a chemical that slowed the growth of caterpillars. The scientist hypothesized that the presence of the caterpillars in the infested trees stimulated the production of this chemical. But how were the uninfested trees stimulated to produce the chemical?

3. List some possible explanations for how the uninfested trees could have been stimulated to produce the chemical.

■ Case Study D: A group of scientists placed 30 poplar plants, separately potted, in one room and 15 such plants in a second room far away. They damaged 15 of the plants in the first room by tearing their leaves. The other 15 plants in the first room and the 15 plants in the second room were not damaged. They found that both the damaged and undamaged plants in the first room produced an increased amount of phenolics, chemicals that are offensive to insects. The 15 plants in the distant room showed no such change.

4. What can you conclude from this experiment?

5. For each explanation you offered in question 3, do the results of Case Study D support it or rule it out? Explain.

6. Based on all of the case studies described, formulate a general statement explaining how plants defend themselves against insect predators.

Problem Finding

Review each of the case studies you have just read. What questions do you still have about plant defenses against insects?

Reproduction and Development

Unit 4

Penguins thrive in a harsh and forbidding wintry habitat. It is in this frigid environment that vast numbers of penguins pair, nest, reproduce, and live together in large communities. Everywhere in the world, all living things, from the smallest one-celled organisms to the largest mammals, reproduce, ensuring the survival of their species. In this unit, you will study the process of reproduction in all its diversity. You will also learn about different adaptations that can increase a species' reproductive efficiency.

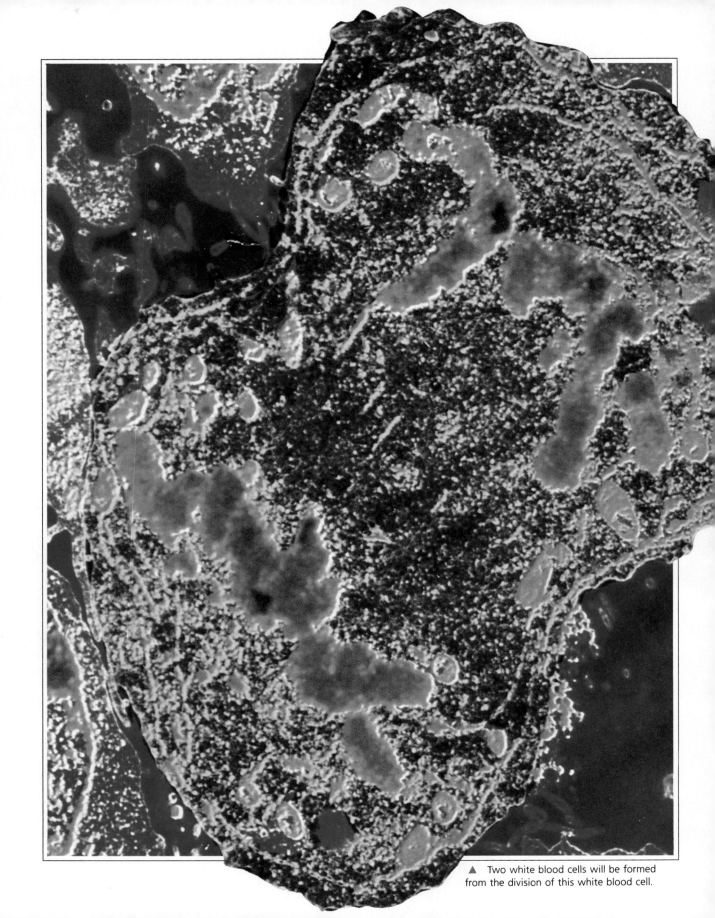

▲ Two white blood cells will be formed from the division of this white blood cell.

Mitosis and Asexual Reproduction

20

A white blood cell undergoes cell division, repeating the process by which it was itself produced. This remarkable event, in which a single cell becomes two cells, occurs in all living things. Think of it—every living creature, from the smallest bacterium to the largest whale or tallest tree, is the product of cell division. Whether it results in new single-celled organisms or, in the case of multicelled organisms, in growth and in replacement of old, worn-out cells, cell division is a basic fact of life. In this chapter, you will explore the processes of cellular reproduction and various modes of asexual reproduction.

Guide for Reading

Key words: mitosis, cytokinesis, asexual reproduction, sexual reproduction, chromosomes, binary fission, budding, spores

Questions to think about:

How do the processes of mitosis and cytokinesis produce two daughter cells from a single cell?

What is the role of mitosis in asexual reproduction?

How do the various types of asexual reproduction differ?

20-1 Mitosis

Section Objectives:

- *Name* and briefly describe the two types of reproduction and the two basic processes involved in cell division.
- *Explain* what happens in interphase and in the stages of mitosis.
- *Compare* and *contrast* mitotic cell division and cytokinesis in plant and animal cells.
- *Discuss* how mitotic cell division in unicellular and multicellular organisms may be controlled.

Laboratory Investigation: *Observe* the stages of mitosis in onion root tip cells (p. 412).

Asexual and Sexual Reproduction

All cells arise from other cells. New cells are formed when one cell divides into two cells. In a one-celled organism, cell division results in the creation of another individual of that species. In contrast, cell division in a multicellular organism usually does not result in the creation of a new organism. Instead, the organism grows larger and replaces some of its tissue. In some multicellular organisms, however, division of body cells may be a method of reproduction if the new cells separate from the parent and form a complete, independent individual.

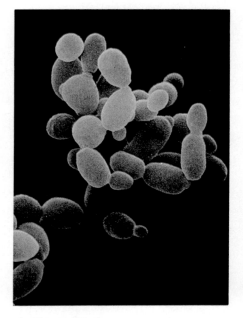

▲ **Figure 20–1**
Budding. Yeast is a one-celled organism that reproduces either by cell fission or by budding.

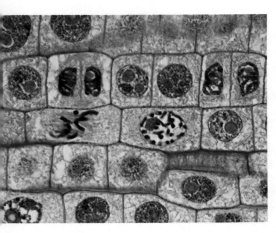

▲ **Figure 20–2**
Onion Cells Undergoing Mitosis. Stained onion root tip cells in various stages of mitosis as seen under a light microscope. (Magnification 700 X)

Figure 20–3

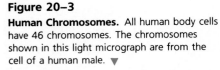

Human Chromosomes. All human body cells have 46 chromosomes. The chromosomes shown in this light micrograph are from the cell of a human male. ▼

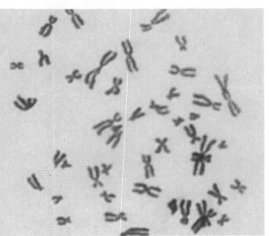

When a cell with a distinct nucleus divides, two processes take place. In one process, the nucleus divides to form two nuclei. This process is called **mitosis** (my TOH sis). During mitosis, the hereditary material in the parent cell duplicates and then divides into two identical sets. In the second process, called **cytokinesis** (sy toh kih NEE sis), the cytoplasm divides into two parts. Each part contains one of the newly formed nuclei, with one complete set of hereditary material, and about half of the other contents of the parent cell. Cytokinesis may take place at the same time as mitosis or after mitosis.

The first section of this chapter discusses mitosis and cytokinesis. The second section discusses *asexual reproduction.* **Asexual reproduction,** which is one of two forms of reproduction, involves only one parent. No special reproductive cells or organs are used to produce the new organism. Instead, asexual reproduction is accomplished by the processes of mitosis and cytokinesis. The new organism is simply a separated part of the parent organism. **Sexual reproduction**—the other form of reproduction—involves special reproductive cells. Usually, these reproductive cells are produced by two separate parent organisms. The details of sexual reproduction are discussed in Chapter 21. Some organisms reproduce only asexually, others reproduce only sexually, and still others reproduce by both methods.

Changes in the Nucleus

As you learned in Chapter 5, the nucleus is the control center of almost all cells. If the nucleus and the hereditary material in it are removed, the cell dies. The nucleus also plays a major role in cell division.

Before mitosis begins, a series of changes takes place in the nucleus and results in the duplication of the hereditary material. When the nucleus divides, each new nucleus, called a daughter nucleus, receives a complete copy of this material.

DNA (deoxyribonucleic acid), the hereditary material of the cell, is found in the cell's nucleus. The information needed to make the parts of each cell is stored in the DNA. The DNA also has information that determines how the organism is made up and works as a whole. This information must be passed on to all cells produced.

In nondividing cells, nuclear DNA exists in a mass of thin, twisted threads called **chromatin** (KROH muh tin). Chromatin is made up of DNA wound around small groups of proteins called **histones** (HIS tones). In cells undergoing mitosis, the threadlike chromatin shortens and thickens into the rodlike structures called **chromosomes** (KROH muh sohmz). See Figure 20–3.

Each type of organism has a specific number of chromosomes in its body cells. For example, humans have 46 chromosomes, wheat has 42, potatoes have 48, crayfish have 20, and fruit flies have 8. Look again at Figure 20–3. The number of chromosomes in the

Interphase and Mitotic Cell Division in Animal Cells

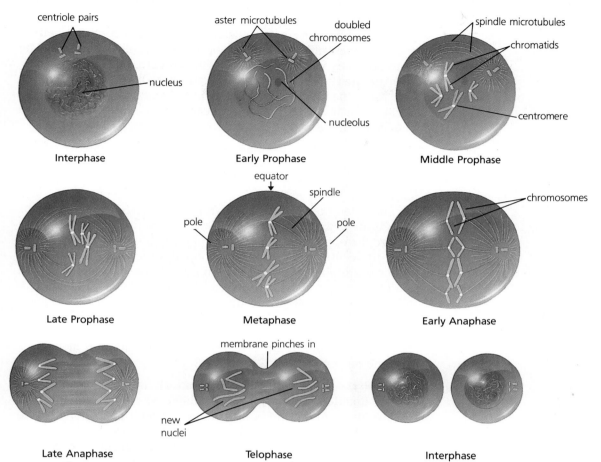

▲ **Figure 20–4**
Interphase and Mitotic Cell Division in Animal Cells.

body cells of an organism is constant. Because each chromosome has only part of the total hereditary information, each cell must have a complete set of chromosomes to work correctly.

Interphase and Mitosis in Animal Cells

Once begun, mitosis is a continuous process. However, it is easier to understand if it is divided into stages, or phases. These stages are known as prophase, metaphase, anaphase, and telophase. There is no sharp break between these stages. Each merges into the next. When a cell is between mitotic cycles, it is in the **interphase** (IN ter fayz) stage. Figure 20–4 shows interphase and the major events of each stage of mitosis in animal cells.

Interphase Although interphase is also called the resting stage, the cell is never really at rest. Interphase lasts from the end of one cell division to the beginning of the next. During interphase, the cell grows in size, and more nucleic acids, proteins, and cellular organelles are produced. At some point before mitosis begins, each chromosome makes a copy of itself, or *replicates* (REP luh kayts), and becomes a doubled chromosome.

During interphase, the nucleus of the cell is contained within the nuclear membrane, and one or more nucleoli are present. The chromosomes cannot be seen with a microscope at this time. Instead, the DNA appears as a tangled, threadlike mass of chromatin. Near the nucleus are the **centrioles,** two tiny, cylindrical bodies that lie at right angles to each other. The centrioles also replicate during interphase and form two pairs.

Prophase   During **prophase** (PROH fayz), the doubled chromosomes become visible as long threads that coil and contract into thick rods. Each strand of doubled chromosomes is called a **chromatid** (KROH muh tid). The chromatids are connected at a region called the **centromere** (SEN truh meer). See Figure 20–5b.

At the beginning of prophase, the two pairs of centrioles move toward the opposite ends, or *poles,* of the cell. Microtubules extend from the centrioles to form star-shaped structures called **asters.** Other microtubules go from pole to pole. These microtubules form a football-shaped structure called the **spindle.** Some of the microtubules become fastened to the centromeres of the chromosomes. As prophase goes on, the doubled chromosomes begin to move toward the *equator,* which is the place midway between the poles. See Figure 20–5b. By the end of prophase, the nuclear membrane and the nucleolus have disappeared.

Metaphase   During **metaphase** (MET uh fayz), the centromeres of the doubled chromosomes are lined up on the equator. See Figure 20–5c. At the end of metaphase, the centromeres divide, and the two chromatids of each chromosome become separate chromosomes. In other words, each doubled chromosome gives rise to two single-stranded, identical chromosomes.

Anaphase   In **anaphase** (AN uh fayz), the duplicate chromosomes move to opposite poles. See Figure 20–5d. The microtubules of the spindle help in this movement. As a result, one complete set of chromosomes goes to one pole while the other identical set goes to the other pole.

Telophase   When the chromosomes reach the poles, **telophase** (TEL uh fayz) begins. During telophase, the chromosomes uncoil, get longer, and slowly take on the threadlike appearance of chromatin. See Figure 20–5e. The spindle and asters disappear. A nuclear membrane forms around each daughter nucleus, and the nucleoli reappear. This ends the nuclear division of an animal cell.

Cytokinesis in Animal Cells

Cytokinesis often begins during late anaphase and finishes during telophase. In animal cells, the division of the cytoplasm comes about by a pinching-in of the cell membrane. The pinching-in takes place in the middle of the cell and results in the formation of two daughter cells of about the same size. Each daughter cell receives one of the newly-formed nuclei and about half of the organelles from the parent cell.

Mitosis and Cytokinesis in Plant Cells

Cell division in plants can be seen fairly easily in developing seeds and in the growing regions of roots and stems. The main events of nuclear division are the same in plants as in animals. See Figure 20–6. However, plant cell division differs from animal cell division in two ways. First, because plant cells do not have centrioles, asters do not form. However, a spindle does form, and the chromosomes

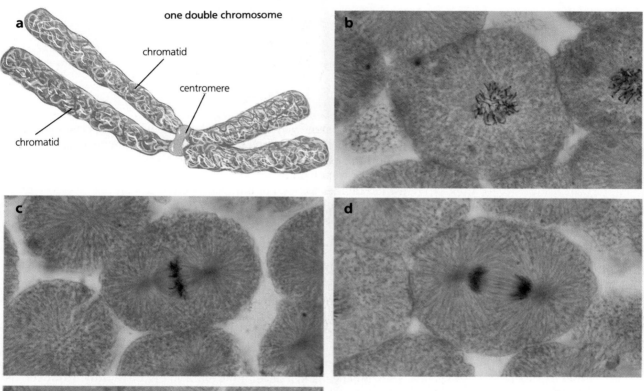

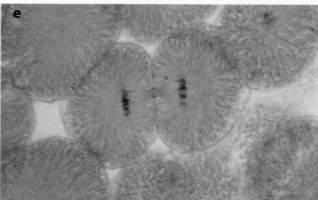

◄ **Figure 20–5**

The Stages of Mitosis. Following the duplication of chromosomes, which occurs during interphase, mitosis takes place. (a) Early in prophase, the double chromosomes become visible. Each chromosome is made up of two chromatids that are connected at the centromere. (b) Later in prophase, the double chromosomes become more distinct and begin to move toward the equator of the cell. In addition, the nuclear membrane slowly disintegrates. (c) In metaphase, the double chromosomes line up at the equator of the cell. At the end of metaphase, the centromeres divide. (d) During anaphase, the chromatids of each double chromosome move to opposite poles of the cell. (e) In the last stage of mitosis, telophase, the nuclear membrane reappears. At this point, the cell body divides into two daughter cells. (Magnification 625 X)

Figure 20–6

Interphase and Mitotic Cell Division in Plant Cells. ▶

Interphase and Mitotic Cell Division in Plant Cells

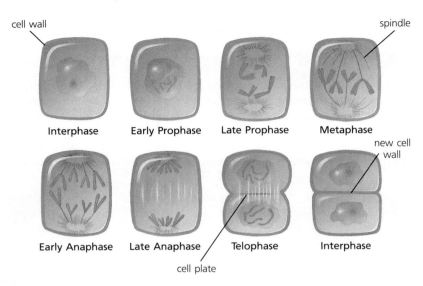

cell wall

spindle

Interphase   Early Prophase   Late Prophase   Metaphase

new cell wall

Early Anaphase   Late Anaphase   Telophase   Interphase

cell plate

move in the same way as in animal cells. Second, the rigid cell walls of plant cells do not pinch in during telophase. Instead, a structure called the **cell plate** forms across the middle of a cell. See Figure 20–7. The cell plate grows outward and joins the old cell wall, which divides the cell in half. New cell wall material is secreted on each side of the cell plate.

Time Span of Mitotic Cell Division

The time needed for a cell to pass through all the stages of mitosis varies from one type of organism to another and from one type of tissue to another. In general, interphase is long compared with the stages of mitotic cell division. For example, a human cell in tissue culture takes about 1 hour to divide and then remains in interphase for 16 to 20 hours.

Mitosis takes place most often in cells that are least specialized. Most cells in a developing embryo divide rapidly. However, as the embryo matures, the cells become specialized, and the rate of mitosis decreases. In adults, cells divide frequently only in certain tissues. Some cells, such as cambium and root tip cells in plants and bone marrow and skin epithelial cells in animals, divide rapidly. Specialized cells, such as xylem, nerve, and muscle cells, seldom or never divide once they are formed.

Control of Mitotic Cell Division

What causes cell division to start and what controls it are not known. In unicellular organisms, it is thought that an increase in cell size triggers mitotic cell division. As a cell gets larger, its volume increases faster than its surface area—in this case, the area

Figure 20–7

Cytokinesis in Plant Cells. In plant cells, such as this onion root tip, the cytoplasm is divided in half by the formation of a cell plate. ▼

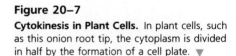

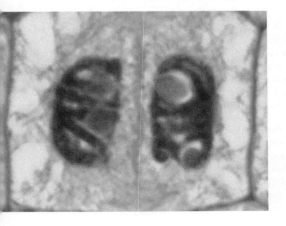

of the cell membrane. When cell size increases above a certain critical point, the surface area of the cell membrane is no longer large enough to permit the necessary exchange of materials into and out of the cell. Thus, in order for a cell to work properly, there must be an upper limit to its size. When a cell reaches the largest size that is characteristic of its type, then either it stops growing or else it divides.

In complex multicellular organisms, some cells divide regularly. Others divide rarely or not at all once they are formed. Still other cells divide only under certain circumstances. Normally, cell division takes place only as needed for the repair or growth of tissue. It is possible that the cells themselves regulate their own division by secreting a control substance that acts through a negative feedback mechanism. See Chapter 16, section 16–1. According to this hypothesis, when the number of cells is normal, the concentration of the control substance that they secrete is high enough to stop cell division. If the number of cells decreases, then

■ Careers *in Action*

Biology Teacher

Teachers of high-school biology have challenging jobs. They educate students about biology and help them to learn more about themselves and their world.

Biology teachers prepare lessons, grade tests and papers, plan demonstrations and laboratory exercises, and organize field trips. Using a variety of teaching aids, such as films, computers, and laboratory equipment, teachers must devise ways to teach and inspire students. Although high-school teachers teach as many as four or five classes a day, they must be aware of the students' individual needs.

In a rapidly changing field like biology, teachers must keep up with new developments. To do this, teachers attend conferences, take courses, and read books and periodicals.

Problem Solving Suppose you are a biology teacher teaching your class about mitosis and asexual reproduction.

1. After you have finished the lesson, you find that students do not understand how, at the end of mitosis, each daughter cell has the same number of chromosomes as the parent cell. They think the daughter cells should have half as many chromosomes. How would you clarify this concept?
2. One of the laboratory exercises asks students to look at cells from the root tip of an onion through a microscope. Why are these cells a good place to see mitosis?
3. At the end of the class, you set up the onion root tip exercise. Most of the cells the students see are in interphase. Is something wrong? Explain.
4. You ask the class whether offspring produced asexually are

different from offspring produced sexually. A student answers that the offspring are the same. Is the student's answer correct? Explain.

■ Help Wanted:
Biology Teacher. Bachelor's degree and state certification required. Contact: National Association of Biology Teachers, 11250 Roger Bacon Dr., Reston, VA 22090.

the concentration of control substance also decreases. The decrease in the concentration of control substance then allows cell division to take place until there are the normal number of cells present.

Once in a while a group of cells begins to divide in an uncontrolled fashion. These cells then may go into neighboring tissue and interfere with normal organ functions. Such uncontrolled dividing of cells is called cancer. Understanding the factors that normally start and control cell division would be of help in controlling cancer.

20-1 **Section Review**

1. Name the two basic types of reproduction.
2. Identify the two processes involved in cell division.
3. List the stages of mitosis.
4. In what two ways does division in plant cells differ from division in animal cells?

Critical Thinking

5. Compare and contrast asexual and sexual reproduction. (*Comparing and Contrasting*)

20-2 Asexual Reproduction

Section Objectives:

- *Describe* the processes of binary fission, budding, and spore formation.
- *Explain* regeneration, and name three types of animals in which regeneration can be a form of asexual reproduction.
- *Describe* the various types of natural and artificial vegetative reproduction.

Producing Identical Offspring

Unicellular organisms, many simple animals, and many plants reproduce asexually, at least during one part of their life cycles. In asexual reproduction in multicellular organisms, the offspring develop from undifferentiated, unspecialized cells of the parent organism.

Because asexual reproduction takes place only by mitotic cell division, each offspring has exactly the same hereditary information as its parent. The offspring show little variation. That is, they are all nearly identical to each other and to the parent. Thus, asexual reproduction results in the same characteristics within a species from one generation to the next.

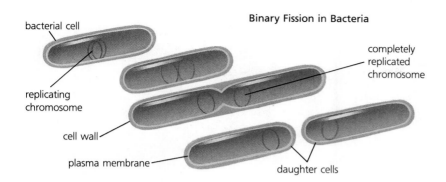

Binary Fission in Bacteria

bacterial cell

completely replicated chromosome

replicating chromosome

cell wall

plasma membrane

daughter cells

◄ **Figure 20–8**
Binary Fission in Bacteria. When bacteria reproduce by binary fission, the chromosome replicates, and a wall divides the cell into two approximately equal parts.

Asexual reproduction is usually rapid and often results in the production of large numbers of offspring. There are several methods of asexual reproduction, including binary fission, budding, spore formation, regeneration, and vegetative reproduction.

Binary Fission

In **binary fission** (BY nehr ee FISH un), the simplest form of asexual reproduction, the parent organism divides into two parts that are about equal. Each of the daughter cells becomes a separate individual and grows to normal size. No parent is left in this method of reproduction because the parent has become two individuals. Binary fission is the usual method of reproduction among one-celled organisms, including bacteria, protozoa, and many algae. When binary fission occurs in cells that have a distinct nucleus, the nucleus divides by mitosis.

Fission in Bacteria   Bacteria lack an organized nucleus. The hereditary material is in the form of a single circular chromosome. Before cell division, the chromosome attaches to the plasma membrane and then replicates. A cell wall forms between the chromosome and its copy. The wall divides the cell into two daughter cells, each containing one chromosome. See Figure 20–8. Each daughter cell grows to normal size before it too divides. Sometimes, the daughter cells do not separate from each other and thus form chains of bacteria. Under favorable conditions, some bacteria can divide every 20 minutes.

Fission in Protozoa   When an ameba reaches full size, it becomes round, and the nucleus undergoes mitosis. After nuclear division, the cytoplasm in the middle of the cell pinches in, or constricts, producing two daughter cells. See Figure 20–9. The two resulting cells are both smaller than the original parent cell, but they soon grow to full size.

The paramecium has two nuclei, a *micronucleus* (my kroh NOO klee us) and a *macronucleus* (mak roh NOO klee us). The small micronucleus controls the reproductive functions of the cell. During binary fission, the micronucleus divides by mitosis. The macronucleus divides by a modified form of mitosis. One of each kind of

Binary Fission in the Ameba

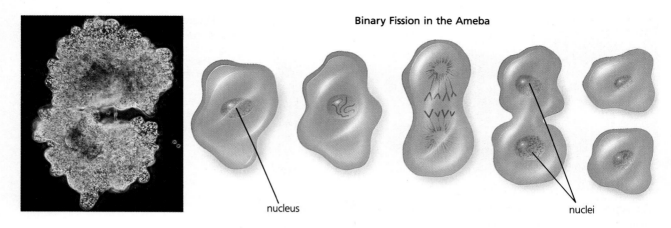

nucleus

nuclei

▲ **Figure 20–9**

Binary Fission in Ameba. Binary fission in ameba involves division of the nucleus by mitosis, followed by division of the cytoplasm into two approximately equal parts.

nucleus goes to each daughter cell. The oral groove and gullet also replicate, and two new contractile vacuoles appear. Thus, before separation takes place, the parts needed for two complete organisms are present. Division of the cytoplasm takes place when the middle of the cell pinches in. The paramecium can also reproduce sexually.

Budding

Budding is a type of asexual reproduction in which the parent organism divides into two unequal parts. New individuals develop as small outgrowths, or buds, on the outer surface of the parent organism. The buds may break off and live independently, or they may remain attached, forming a colony. Budding differs from binary fission in that the parent and offspring are not the same size. Budding takes place in yeast and hydra, as well as in sponges and some worms.

Budding in Yeast   When a yeast cell reaches a certain size, the nucleus moves toward the side of the cell. An enzyme softens the cell wall near the nucleus so that it bulges outward, forming a small knoblike structure called a *bud*. See Figure 20–10. The nucleus then undergoes mitosis, producing two daughter nuclei. One daughter nucleus moves into the bud, while the other remains in the parent cell. A cell wall forms between the parent cell and the bud. The bud may remain fastened to the parent, or they may separate. In either case, the bud is an independent cell that can increase in size and finally produce its own buds. Yeast can also reproduce sexually.

Budding in Yeast

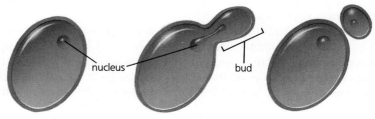

nucleus

bud

Figure 20–10

Budding in Yeast. In budding, the cytoplasm divides into unequal parts. ▶

◄ **Figure 20–11**
Budding in Hydra. The bud begins as a small mound of cells on the side of the parent. The cells divide, producing a complete hydra, which eventually separates from the parent.

Budding in Hydra Budding in hydra is different from budding in yeast. Hydras are made of several kinds of cells. As budding begins, undifferentiated cells on the side of the parent undergo several mitotic divisions, producing a small mound of cells. These cells go on dividing, and in a few days, a small, complete hydra with a mouth and tentacles is formed. See Figure 20–11. The bud finally separates from the parent. Hydras can also reproduce sexually.

Spore Formation

Spores are single, specialized cells that are produced by certain organisms. When released from the parent organism, spores germinate and grow to form new individuals. The term *spore* can refer to a large number of different single-celled structures. Although spores differ greatly in appearance, structure, and origin, all work as single units of reproduction. Each spore has the usual parts of a cell. Often, a spore is surrounded by a special thick, hard outer wall. Sometimes, however, spores lack these walls and have flagella.

Spores can be formed sexually or asexually. (Sexually formed spores are discussed in Chapters 24 and 31.) Asexually formed spores are a common method of reproduction in many simple organisms, such as fungi, algae, and protozoa. These spores are the products of mitotic cell division. Large numbers of them are generally produced during division. They are formed within, and released from, a single cell structure that is the remains of the original parent cell from which the spores came.

Spore Formation in Bread Mold Bread mold, which is a fungus, can often be seen growing as a dark, cottony mass on bread and other foods. The spores are produced by mitotic cell division in spore cases on specialized stalks that grow upward from the surface. Thousands of black spores develop within each spore case. When fully grown, the walls of the spore case break down, and the tiny, light spores are carried away by air currents. When a spore lands in an environment where there is warmth, food, and moisture, it germinates and grows to form a new mass of mold. Bread molds also reproduce sexually, as described in Chapter 31.

Figure 20–12
Regeneration. The planarian, like many
simple animals, can regenerate lost parts. ▶

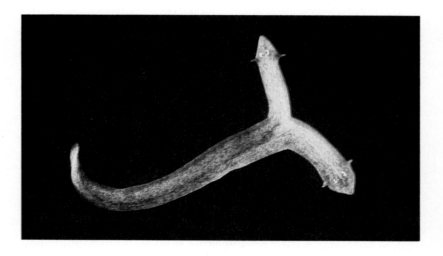

Regeneration

Regeneration (rih jen uh RAY shun) is the ability of an organism to
regrow lost body parts. See Figure 20–12. Relatively simple animals, such as the hydra, planarian, starfish, and earthworm, can
regenerate lost parts. If a hydra is cut in half, each half will
regenerate into a new individual. A planarian can be cut into
several pieces, each of which will grow into a complete worm.

Starfish feed on oysters. Workers gathering oysters used to try
to destroy the starfish they caught by chopping them into pieces
and tossing the pieces back into the water. However, each part of a
starfish can regenerate into a whole new organism as long as it
contains a piece of the central disk. Thus, the workers were really
helping the starfish to multiply rather than destroying them.

The power of regeneration decreases as animals become more
complex. A crab can regrow a lost claw but cannot regenerate a
whole animal from small pieces. Mammals can repair damaged
tissue but cannot regenerate a leg or even a toe. Although simple
organisms have great powers of regeneration, they do not usually
reproduce in this manner.

Vegetative Reproduction

Although most plants reproduce sexually by means of seeds,
asexual reproduction involving roots, stems, and leaves is also
common. Roots, stems, and leaves are called *vegetative* (VEJ uh tay
tiv) structures. They normally play a part in the nutrition and
growth of plants. When they give rise to a new plant, the process is
called **vegetative reproduction,** or *vegetative propagation*.

In vegetative reproduction, undifferentiated cells, such as
cambium and epidermal cells, divide mitotically and then differentiate to produce an independent plant. The new plant has the same
hereditary characteristics as its parent. Vegetative reproduction
takes place naturally, but it can be brought about artificially.

Natural Vegetative Reproduction   Vegetative propagation takes place naturally in several different ways. See Figure 20–13.

Some plants, such as tulips, onions, and lilies reproduce by bulbs. A **bulb** is a short underground stem surrounded by thick, fleshy leaves that contain stored food. As the plant grows, small new bulbs sprout from the old one. Each of the new bulbs can give rise to a new plant. Other plants, such as gladioli, crocuses, and water chestnuts grow from corms (kormz). A **corm** is like a bulb, but it does not contain fleshy leaves. Rather, corms are short, stout underground stems that contain stored food.

A tuber (TOO ber) is another means of natural vegetative propagation. A **tuber** is an enlarged part of an underground stem that contains stored food. White potatoes are tubers. Along the surface of a tuber are indentations called "eyes." These eyes are tiny buds. When a farmer plants white potatoes, the tuber is cut into pieces, each piece having at least one "eye." Each eye grows into a

◀ **Figure 20–13**
Natural Vegetative Propagation. When a plant reproduces asexually, complete plants can develop from specialized structures other than the seeds, such as bulbs, corms, tubers, runners, or rhizomes. Shown here are: (a) silverweed runners; (b) an iris rhizone and a potato tuber; and (c) daffadil and saffron crocus corms.

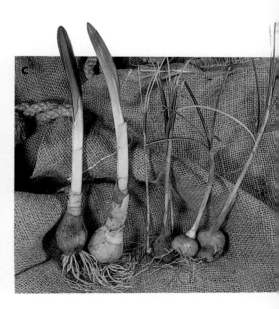

shoot that grows upward through the soil surface and also produces roots. The young shoot uses the stored food of the tuber until it can carry on photosynthesis.

Strawberry plants and many kinds of grasses that reproduce quickly use runners. A **runner,** or *stolon* (STOH lun), is a stem that grows sideways and has buds. It usually grows along the surface of the ground. Where buds from a runner touch the soil, new independent plants develop.

Finally, a few plants, such as ferns, irises, cattails, and water lilies, reproduce by rhizomes (RY zohms). A **rhizome** is a stem that grows sideways underground. It is usually thick and fleshy and contains stored food. Along the rhizome are enlarged portions called *nodes.* Buds produced at nodes on the upper surface of the rhizome give rise to leaf-bearing branches. The lower surface of the rhizome produces roots.

Artificial Vegetative Reproduction Farmers and gardeners have developed several methods of artificial vegetative reproduction. See Figure 20–14. These techniques allow them to grow plants with desirable traits.

A **cutting** is any vegetative part of a plant—stem, leaf, or root—used to produce a new individual. In a *stem cutting,* a branch, or slip, is cut from a plant and placed in water or moist sand. Usually the bottom of the cutting is dipped into hormones to stimulate root growth. When roots develop, the cutting becomes an independent plant and is transplanted to soil. Geraniums, roses, ivy, and grapevines are propagated in this manner.

In a *leaf cutting,* a leaf or part of a leaf is placed in water or moist soil. After a while, a new plant develops from certain cells in the leaf. African violets, snake plants, and begonias are often propagated by leaf cuttings.

Under natural conditions, the leaves of kalanchoe give rise to tiny plants along their edges. These plantlets have tiny leaves,

Figure 20–14

Artificial Vegatative Propagation. (A) Many important ornamental plant species are propagated by means of leaf or stem cuttings. Here, a begonia leaf has been placed in soil to root. (B) Grafting makes it possible to combine the best qualities of different plant species or varieties into a single plant. ▼

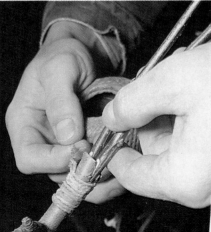

stems, and sometimes roots. When they fall from the parent plant, they take root and go on growing in the soil. They can be separated and used to produce new plants.

The sweet potato is an enlarged root containing stored food. Farmers place it in moist sand or soil until it sprouts several new plants. Then the sprouts are removed and planted.

In **layering,** a stem is bent over so that part of it is covered with soil. After the covered part forms roots, the new plant may be cut from the parent plant. Layering is used to reproduce such plants as raspberries, roses, and honeysuckle. It also takes place naturally.

In **grafting,** a stem or bud is removed from one plant and joined permanently to the stem of a closely related plant. The part of this combination providing the roots is called the *stock;* the added piece is called the *scion* (SY en). The cambium layers (growing regions) of the scion and stock must be in close contact. Usually they are held together by tape and coated with wax to protect the growing tissue from water loss and disease. After a time the cambiums of the two pieces form new xylem and phloem, which grow together and connect the scion and stock. The stock supports and nourishes the scion. However, the scion keeps its own characteristics. For example, although scions of McIntosh apples can be grafted onto stock of any kind of apple tree, the scions will produce only McIntosh apples. Grafting is used to propagate roses, peach trees, plum trees, grapevines, and various seedless fruits, including navel oranges, grapes, and grapefruits.

Advantages of Artificial Vegetative Propagation   Plants grown from seeds do not always show the same characteristics as the parent plant. Vegetative propagation, however, produces new plants exactly like the parent. There is little variation because all offspring have the same hereditary makeup as the parent. Furthermore, the development of a plant by vegetative propagation is often faster than development from seed. In the development of an improved plant variety, stem cuttings or grafts made onto mature plants will produce fruit in much less time than it takes for small plants to bear fruit. Plants bearing *seedless fruit* can be grown only by vegetative propagation. See Figure 20–15. Grafting also can be used to obtain higher yields of fruits or nuts.

▲   **Figure 20–15**
A Seedless Navel Orange.

20-2 **Section Review**

1. How is binary fission different from budding?
2. Name an organism that reproduces by means of spores.
3. What is regeneration?
4. Give examples of natural and artificial vegetative reproduction.

Critical Thinking

5. Why would a farmer use artificial vegetative propagation instead of sexual reproduction by means of seeds? (*Identifying Reasons*)

Laboratory Investigation

Mitosis in Onion Root Tips

At the tip of any stem or root is a small area, the apical meristem, where most of the cells undergo mitosis and cell division. In this investigation, you will look at prepared slides of root tips. The tissues have been fixed and stained so that chromosomes are visible in the cells undergoing mitosis.

Objectives

- *Observe* the stages of mitosis.
- *Calculate* the relative and actual length of time for some of the stages of mitosis.

Materials

prepared slide of a longitudinal
   section of onion root tip
compound microscope

Procedure   *

A. Using low power, examine the prepared slide. Position the slide with the rounded, root cap end pointing toward you as you look through the microscope. Directly behind the root cap is the apical meristem. The nuclei of the cells in this region are large in comparison to the size of the cells.

B. Using Figure 20–7 in your text as a guide, look for a cell in prophase. Its nucleus should appear darkly stained and granular. Center the cell in your field of view and switch to high power. Draw and label the cell.

C. Find, with low power, and then examine under high power, cells in metaphase, anaphase, and telophase. Draw and label each stage as you see it.

D. Using low power, select an area that has at least 50 cells in mitosis. Determine the stage of mitosis for each cell. Keep a running tally, then record the total count on a chart like the one shown below. Examine the cells row by row to avoid counting any twice.

E. Total the number of cells at each stage and enter the numbers in the Total Count column of your chart.

F. Calculate the percent of cells in each stage. Divide the number of cells in a stage by the total number of cells counted and multiply by 100. Record these numbers, then graph them in a circle chart. Compare your chart with those of your classmates.

G. The percent of the total time of mitosis that any given stage takes is equal to the percent of cells in that stage at any moment. For example, if 20 percent of the cells are seen to be in prophase, then prophase takes 20 percent of the total time. Mitosis in onion takes 80 minutes. Using this information and the percentages you have just determined, calculate the actual time for each stage: (percent/100) × 80. Record these values in your chart and plot them on a bar graph.

Analysis and Conclusions

1. Why is the root tip a good area to study cells undergoing mitosis? What other regions of a plant would be suitable?

2. In which stage of mitosis were the greatest number of cells? In which stage were the fewest?

3. Based on your findings, list, from the longest to the shortest, the stages of mitosis.

4. How did your data on percentage of cells in each stage of mitosis compare with the data from other groups? Why were there differences?

5. Do you think the relative length of time spent in the various stages of mitosis in onion would be similar to, or different from, those in an animal? Explain your answer. How could you verify your answer?

6. Cancerous tissue is made up of cells undergoing rapid, uncontrolled cell division. Do you think the procedure of counting cells in mitosis could be used to identify cancerous tissue? Explain your answer.

Stage	Total Count	Percent in Stage (No. in Stage/Total No. Cells) × 100	Actual Time in Stage (Percent/100) × 80 Min
Prophase			

*For detailed information on safety symbols, see Appendix B.

Chapter Review

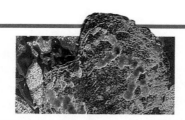

20

Study Outline

20–1 Mitosis

■ New cells arise from existing cells by division of the original cell into two cells. Cell division consists of two processes: division of the nucleus, or mitosis, and division of the cytoplasm, or cytokinesis.

■ The nucleus contains the hereditary material. In cells undergoing mitosis, the hereditary material is found in the chromosomes. Every organism has a specific and characteristic number of chromosomes in its body cells.

■ Mitotic division is a continuous process during which the chromosomes are duplicated, and each new cell receives an entire set. The process is divided into four stages: prophase, metaphase, anaphase, and telophase.

■ The period between the end of one cell division and the beginning of the next is called interphase. During interphase, each chromosome (in the form of chromatin) replicates and becomes a doubled chromosome.

■ The basic stages of mitotic division are the same in plant and animal cells. However, cytoplasmic division in plant cells differs from that in animal cells, since the dividing plant cells form cell plates and do not form asters.

20–2 Asexual Reproduction

■ Asexual reproduction involves mitotic cell division of only one parent. The simplest form of asexual reproduction is binary fission, in which the parent divides into two daughter cells of approximately equal size.

■ In budding, the parent organism divides into two individuals of unequal size. The new individual is smaller than the parent and grows to full size after separation from the parent.

■ Spores are specialized reproductive cells that can be formed sexually or asexually. The asexual formation of spores involves mitotic cell division. When environmental conditions become favorable, the spore develops into a new organism.

■ Regeneration is the ability to regrow lost body parts; it occurs in relatively simple animals, such as the starfish

and earthworm. Regeneration is not usually a method of reproduction.

■ Most plants reproduce sexually, but asexual reproduction involving roots, stems, and leaves is also common. In vegetative reproduction, undifferentiated cells divide mitotically, differentiate, and give rise to an independent plant.

■ Vegetative reproduction occurs naturally with bulbs, corms, tubers, runners, and rhizomes. It can also be accomplished artificially with cuttings, layering, and grafting.

Vocabulary Review

mitosis (398)	telophase (400)
cytokinesis (398)	cell plate (402)
asexual reproduction (398)	binary fission (405)
	budding (406)
sexual reproduction (398)	spores (407)
chromatin (398)	regeneration (408)
histones (398)	vegetative reproduction (408)
chromosomes (398)	
interphase (399)	bulb (409)
centrioles (400)	corm (409)
prophase (400)	tuber (409)
chromatid (400)	runner (410)
centromere (400)	rhizome (410)
asters (400)	cutting (410)
spindle (400)	layering (411)
metaphase (400)	grafting (411)
anaphase (400)	

A. Word Relationships—In each of the following sets of terms, three of the terms have a common characteristic. One term does not belong. Identify the term that does not belong.

1. telophase, interphase, prophase, metaphase

2. spore, rhizome, bulb, tuber

3. chromosome, histone, chromatin, cell plate

Chapter Review

4. cutting, layering, budding, grafting

5. vegetative reproduction, sexual reproduction, binary fission, regeneration

6. centromere, spindle, aster, cytokinesis

7. binary fission, layering, runner, bulb

B. Definition—Replace the italicized definition with the correct vocabulary term.

8. During prophase, each *strand of doubled chromosomes* is connected at the centromere.

9. *The stage of mitosis during which the chromosomes uncoil* is the final step of nuclear division.

10. *A type of vegetative propagation accomplished by permanently joining a part of one plant to another plant* is used to propagate roses and peach trees.

11. *A mass of thin, twisted threads of nuclear DNA* is found in nondividing cells.

12. In some plants, a *horizontal stem with buds* forms independent plants by vegetative propagation.

13. During interphase, *two tiny cylindrical bodies lying at right angles to each other* are found near the nucleus of animal cells.

14. Although most plants reproduce sexually, some may reproduce asexually by means of *mitotic division and differentiation of root, stem, or leaf cells.*

Content Review

15. Compare the processes of mitosis and cytokinesis.

16. What is the hereditary material in the cell, and where is it usually located?

17. What is chromatin made up of? Describe the structure of it.

18. What events occur during interphase before mitosis begins?

19. Using an animal cell with four chromosomes as an example, draw the four stages of mitosis and briefly describe each stage.

20. How does cytokinesis occur in animal cells?

21. Contrast cytokinesis in plant cells with that in animal cells.

22. Why might an increase in size trigger mitotic cell division in unicellular organisms?

23. How does cancer affect mitotic division?

24. Describe binary fission in bacteria.

25. What is unusual about binary fission in paramecia?

26. Compare bud formation in yeast and in hydra.

27. How and where are asexually formed spores produced?

28. Why does cutting individual starfish into pieces increase the population of starfish?

29. Briefly describe in your own words the process of vegetative reproduction.

30. Explain how vegetative reproduction occurs in a tuber.

31. How do bulbs, corms, tubers, runners, and rhizomes differ?

32. Briefly explain three methods of artificial vegetative reproduction.

Graphic Organizing

For information on graphic organizers, see Appendix G at the back of this text.

33. **Circle Graph:** Using the data in the table below, construct a circle graph of a cell cycle for a cell that completes a cycle in 24 hours. The cell cycle is made up of interphase (which is divided into three phases: G_1—the presynthesis period; S—the period when DNA replicates; and G_2—the postsynthesis period) and mitosis. Use the graph to explain why most of the cells in a population are in interphase at any given time.

Stage	Time Length
Interphase:	
G_1	10 hours
S	9 hours
G_2	4 hours
Mitosis (M)	1 hour

34. **Word Map:** Construct word maps for the terms *asexual reproduction* and *vegetative reproduction.*

Critical Thinking

35. What might be the uses for a chemical that can control the rate of mitotic division of human cells? (*Judging Usefulness*)

36. Cancer, which is the uncontrolled division of cells, is sometimes treated with radiation or drugs that attack rapidly growing cells. Explain why this type of treatment may harm the patient. (*Identifying Reasons*)

37. In some animal cells, certain cell structures divide prior to anaphase. Among these structures are the chromosomes, the centrioles, and the centromeres. What would be the effect on the daughter cells if each of these structures did not divide? (*Predicting*)

38. Nerve cells and muscle cells seldom divide after they are formed. What effect does this have on the human body in the event of injury to these cells? (*Predicting*)

Creative Thinking

39. In what ways might the process of regeneration be useful to humans?

40. Suggest a possible explanation for the fact that the ability to regenerate decreases as animals become more complex.

Problem Solving

41. An experiment was conducted to determine if plant hormones (see Chapter 19) will stimulate mitosis at the site of a plant graft. One of two different hormones—indoleacetic acid and kinetin—were applied to several grafts of rose and orange plants. As a control, some of the grafts were not exposed to either hormone. After seven days, a tissue sample was taken within 0.5 mm of the grafts, and the cells were examined to determine the number undergoing mitosis. Using the data in the table below, calculate the percentage of mitotic cells and nonmitotic cells present in each sample. Interpret the data and draw some conclusions regarding the effect of each hormone on mitosis.

42. An investigation was carried out to determine if the movement of the spindle microtubules during mitosis depends on adenosine triphosphate (ATP). Cells that were about to divide were placed in a solution containing radioactive ATP, in which the last phosphate group was radioactively labeled. After 24 hours at 25° C, the cells were examined, and radioactive phosphate was found on the spindle microtubules. What conclusion could be drawn from this experiment? Is there a control? How might you improve the design of the experiment? What further steps might be performed?

43. Design an experiment to investigate the effect of caffeine on mitosis. Hint: Mitosis can be easily observed in the onion root tip.

44. Some bacteria can divide every 20 minutes. Starting with one such bacterium, how many would be present after 4, 8, and 12 hours?

Projects

45. A tissue culture of human cancer cells, called HeLa cells, was started in 1952 and is still growing today. Do library research on the tissue culture technique and write up your findings as a magazine article.

46. Interview a physician or a nurse who specializes in cancer (an oncologist or an oncology nurse) about new treatments for cancer. Write up your findings as a feature for a television health program.

47. Prepare a poster or a display on the use of grafting in plant propagation.

48. Grow plants by means of vegetative reproduction. Begonias and African violets can be raised by planting stems with attached leaves in moist soil. Geraniums and coleuses can be grown from stem cuttings kept in water; when they develop sturdy roots, plant them in soil. Experiment with other plants, taking cuttings below stem nodes. Raise potato plants in soil, using potato pieces with eyes. To grow flowering tulips, hyacinths, and narcissi, plant the bulbs to half their depth in a bowl filled with pebbles; keep the pebbles moist.

Plant	Indoleacetic Acid		Kinetin		No Hormone	
	Mitotic Cells	Nonmitotic Cells	Mitotic Cells	Nonmitotic Cells	Mitotic Cells	Nonmitotic Cells
rose	38	112	50	75	33	92
orange	15	68	23	71	17	84

▲ The diversity of a litter of guinea pigs illustrates the varied results of sexual reproduction.

Meiosis and Sexual Reproduction

21

A multicolored litter of guinea pigs rests comfortably. They vary in fur color from white to black to brown-and-white. Although all have the same parents, each guinea pig has its own characteristics distinct from those of its parents and siblings. In this chapter, you will learn about the process of sexual reproduction as it takes place in various organisms, including humans. You will also learn about meiosis, the special kind of cell division that affects only the reproductive cells.

Guide for Reading

Key words: gametes, zygote, meiosis, diploid, haploid, conjugation, ovaries, testes
Questions to think about:
📖 How does meiosis produce four haploid cells from one diploid cell?
📖 Why does conjugation increase genetic variety?
📖 How do gametes function in sexual reproduction in animals?

21-1 Meiosis

Section Objectives:

- *Explain* the importance of meiosis in sexual reproduction.
- *Define* the following terms: *gamete, zygote, diploid,* and *haploid.*
- *Describe* the stages of meiosis.
- **Laboratory Investigation:** *Identify* pairs of homologous chromosomes in a human cell(p. 436).
- **Concept Laboratory:** *Gain* an understanding of the process of meiosis by creating a model(p. 423).

Another Kind of Cell Division

As you may recall from Chapter 20, asexual reproduction involves only one parent cell. When reproduction occurs asexually, the parent cell divides to produce offspring with the same genetic makeup as the parent cell. Sexual reproduction, on the other hand, requires two different parent cells from two separate organisms or from two sexually different parts of a single organism. Sexual reproduction produces offspring that are genetically different from either parent.

In simple organisms, sexual reproduction involves a transfer of genetic material from one organism to another. In more complex organisms, two special sex cells, called **gametes** (GAM eets), are needed. Usually, there are two kinds of gametes, one male and one female. A new offspring results when the male and the female

▲ **Figure 21–1**
Sexual Reproduction. The sperm and the egg each contain half the hereditary information needed to create a new organism.

417

gametes fuse, or join together. The fusion of the nuclei of the male and the female gametes is called **fertilization.** The single cell formed from this fusion is known as a **zygote** (ZY goht).

Gametes are formed by **meiosis** (my OH sis), a kind of cell division that results in gametes with half the number of chromosomes as the parent cell. Because of meiosis, each gamete has half of the regular number of chromosomes. The process of fertilization restores the regular number of chromosomes to the zygote.

Diploid and Haploid Chromosome Numbers

Somatic (soh MAH tic) **cells,** or body cells, are all the cells of an organism except for the specialized cells that play a part in sexual reproduction. Every species of organism has a certain number of chromosomes in its body cells. For example, human somatic cells typically have 46 chromosomes. The body cells of bull frogs have 26 chromosomes. Fruit fly body cells have 8 chromosomes.

The chromosomes of most organisms can be grouped into pairs of similar chromosomes. For example, in humans, the 46 chromosomes in body cells can be arranged into 23 pairs. The chromosomes that make up each pair are called **homologous** (hoh MAL uh guhs) **chromosomes.** See Figure 21–2. Homologous chro-

Figure 21–2

Human Chromosomes. The chromosomes of cells undergoing mitosis were stained and photographed. The photographs were then cut apart and assembled into pairs of homologous chromosomes. The set of chromosomes to the left were from a male. The pair labeled X and Y are the sex chromosomes. The set of chromosomes to the right were from a female cell and have a pair of X chromosomes. Chromosomes in the male are the same size as those in the female; they look smaller here because they are magnified less than the female's chromosomes. ▼

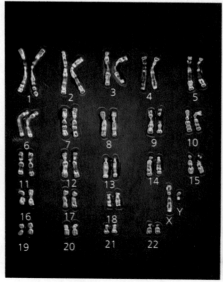
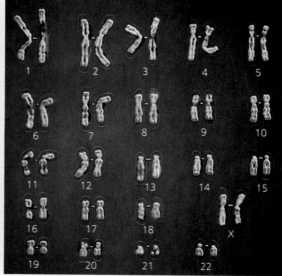

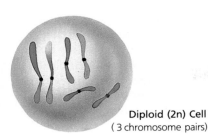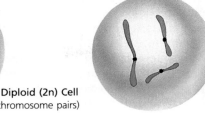

Diploid (2n) Cell
(3 chromosome pairs)

Haploid (n) Cell
(3 chromosomes)

◀ **Figure 21–3**
Diploid and Haploid Cells. The haploid cell contains only one chromosome from each pair of homologous chromosomes. The diploid cell contains both chromosomes from each pair of homologous chromosomes.

mosomes are similar in size and shape, and they have similar genetic content. (In some organisms, including humans, one pair of chromosomes, the sex chromosomes, are not homologous in one of the sexes.) Those cells that have all the homologous chromosomes that are characteristic of the species are referred to as **diploid** (DIP loyd), or *2n*.

Unlike the body cells, gametes do not contain pairs of homologous chromosomes. Instead, they have only one chromosome from each pair. As a result, they have only half the diploid number of chromosomes. For example, human gametes contain 23 chromosomes—one from each of the 23 pairs of chromosomes in body cells. Cells that have only one chromosome from each pair are said to be **haploid** (HAP loyd), or *monoploid* (MAHN uh ployd). Haploid cells contain *n* chromosomes, rather than the *2n* of diploid cells. See Figure 21–3.

If gametes were diploid cells, the number of chromosomes per cell would double with each generation. The doubling would take place when two gametes were united at the time of fertilization. Because of the process of meiosis, the doubling of the chromosome number does not take place. Meiosis, which produces gametes in animals and spores in plants and in some fungi, ensures that the gametes receive only half the number of chromosomes that are present in the parent cells.

Stages of Meiosis

Meiosis, which is also known as *reduction division,* takes place in special cells. At the start of meiosis, these cells have the diploid number of chromosomes. In meiosis, each cell divides twice. The chromosomes, however, replicate only once. This replication takes place before the first division. In the second division, no replication of the chromosomes takes place. As a result of the two meiotic divisions, each original cell produces four cells, which are known as *daughter cells.* Each of these cells contains the haploid number of chromosomes.

Both the first and second meiotic divisions can be divided into stages similar to the stages of mitosis. Thus, both divisions show a prophase, metaphase, anaphase, and telophase stage. The events of the first and second meiotic divisions are shown in Figure 21–4 using an animal cell as an example.

Prophase I At the beginning of prophase in the first meiotic division, each chromosome has already replicated, producing two chromatids, as in mitosis. However, the pairs of chromatids do not move independently to the equator. Instead, each pair of chromatids lines up with its homologous pair, and they become fastened at their centromeres. This pairing process is called **synapsis** (suh NAP sis), and each group of four chromatids is called a **tetrad.** The strands of the tetrad sometimes twist about each other, and at this point, they may exchange segments. The exchange of segments between chromatids during synapsis is called *crossing-over.*

While these chromosomal changes are taking place, the nuclear membrane disappears and the spindle fibers form. As prophase I ends, the homologous chromosome pairs, each made up of four chromatids, move toward the equator of the cell.

Metaphase I In metaphase in the first meiotic division, the centromeres of the tetrads line up on the equator. The tetrads are fastened to the spindle microtubules at their centromeres.

Anaphase I During anaphase in the first meiotic division, the homologous chromosomes of each tetrad separate from each other and move to opposite ends of the cell. This process of separation is called **disjunction** (dis JUNK shun). The cluster of chromosomes around each pole is haploid—there are half as many chromosomes as in the original cell. However, each chromosome is double-stranded.

Telophase I Telophase marks the end of the first meiotic division. The cytoplasm divides, forming two daughter cells. Each of the newly formed daughter cells has half the number of the parent cell's chromosomes, but each chromosome is already in replicated form.

Sometimes at the end of telophase I, nuclear membranes form and a short interphase follows. However, in most cases, the cells immediately begin the second division. No further replication of the chromosomes takes place, but the remainder of the division is exactly like mitosis.

Prophase II During prophase II, each of the daughter cells forms a spindle, and the double-stranded chromosomes move toward the middle of the spindle.

Metaphase II During metaphase II, the chromosomes become fastened to spindle microtubules at their centromeres, and the centromeres of the chromosomes line up on the equator. Each chromosome still consists of two strands, or chromatids.

Anaphase II During anaphase II, the centromeres divide, and the two chromatids separate, each becoming a single-stranded chromosome. The two chromosomes then move toward the opposite ends of the spindle.

Telophase II During telophase II, both daughter cells divide, forming four haploid cells. In each cell, chromosomes return to their interphase state, and the nuclear membrane forms again.

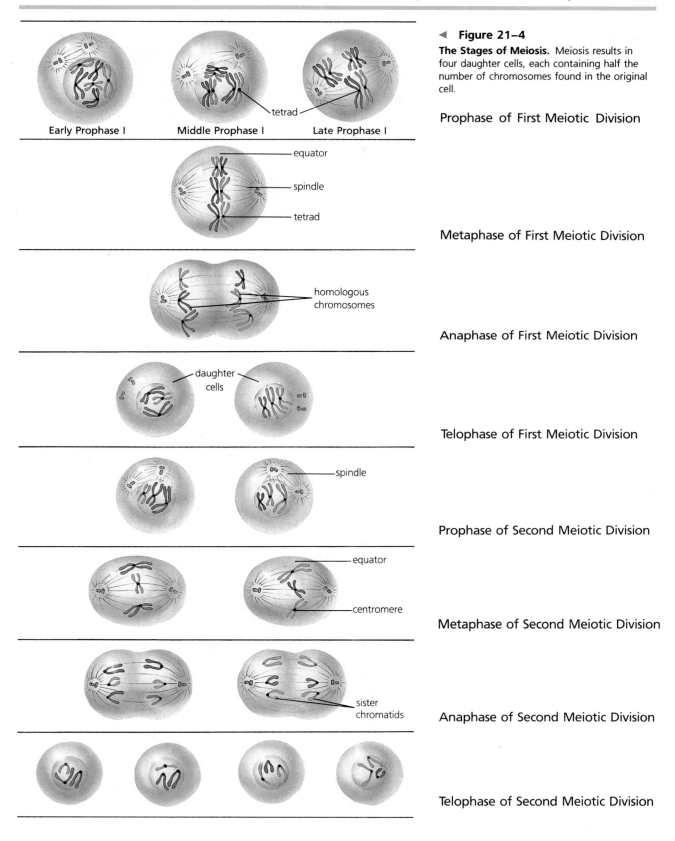

◀   **Figure 21–4**

The Stages of Meiosis. Meiosis results in four daughter cells, each containing half the number of chromosomes found in the original cell.

Prophase of First Meiotic Division

Metaphase of First Meiotic Division

Anaphase of First Meiotic Division

Telophase of First Meiotic Division

Prophase of Second Meiotic Division

Metaphase of Second Meiotic Division

Anaphase of Second Meiotic Division

Telophase of Second Meiotic Division

Figure 21–5

Comparison of Mitosis and Meiosis. Both mitosis and meiosis result in the production of new cells. Mitosis produces two diploid cells from one diploid cell. Meiosis produces four haploid cells from one diploid cell. ▶

Comparison of Mitosis and Meiosis	
Mitosis	**Meiosis**
Occurs in growth and asexual reproduction	Occurs in production of gametes in animals, and spores in plants and in some simple organisms
Homologous chromosomes not paired up during prophase. There is no exchange of parts between homologous chromosomes.	Homologous chromosomes paired up during prophase of first division. While paired, there may be an exchange of parts between homologous chromosomes.
Involves one cell division. In the course of division, the double-stranded chromosomes line up at cell equator, centromeres divide, and one chromatid of each chromosome goes to each daughter cell.	Involves two cell divisions. During first division, pairs of homologous two-stranded chromosomes line up at equator. The members of each pair separate, and one two-stranded chromosome of each pair goes to each daughter cell. During second division, centromeres of two-stranded chromosomes divide, and chromatids separate, one going to each daughter cell.
As a result of mitosis, each daughter cell receives the same number of chromosomes as the original cell. Mitosis maintains the chromosome number.	As a result of meiosis, each daughter cell receives only one member of each pair of homologous chromosomes. It therefore has only one-half the number of chromosomes in the original cell. Meiosis reduces the chromosome number by one-half.

The chart in Figure 21–5 outlines the similarities and differences between the two different kinds of cell division, mitosis and meiosis.

21-1 Section Review

1. What is meiosis?
2. Which cells of an animal are diploid and which are haploid?
3. How many pairs of chromosomes are found in human body cells?
4. What are the products of meiosis in animals? In plants?

Critical Thinking

5. Suppose that disjunction did not occur in one homologous pair of chromosomes during anaphase I. How would this affect the chromosome number in the daughter cells? (*Predicting*)

Concept Laboratory

Chromosome Movements During Meiosis

Goal

To gain an understanding of meiosis by using a model.

Scenario

Imagine a basketball-sized cell with four chromosomes. Your task is to move the chromosomes through meiosis. As you do so, think about the orderly movements of the chromosomes in meiosis.

Materials

6 pipe cleaners, 3 of one
    color, 3 of another
colored pencils to match
    pipe cleaners
4 sheets of unlined paper
8 Cheerios
transparent tape
scissors

Procedure 👉 *

A. Insert two full-size pipe cleaners of different colors into separate Cheerios. Do the same with two half-size pipe cleaners. The pipe cleaners represent chromosomes, and the Cheerios are their centromeres. Randomly place the pipe cleaners on a sheet of paper representing the cell. Using colored pencils, sketch them.

B. During interphase, each chromosome becomes a double chromosome made of two chromatids. Insert into each Cheerio a second pipe cleaner of the same length and color as the one already present. Sketch the chromosomes at this stage.

C. Begin prophase I of meiosis by pairing homologous chromosomes to form tetrads. Put the two long chromosomes (one of each color) side by side with centromeres touching. Do the same with the two short ones.

D. During metaphase I, line up the chromosomes so their centromeres are in the same plane. Same-colored chromosomes may be on one side of the plane or on opposite sides. Draw the chromosomes again.

E. Move the pipe cleaners through the remaining stages of meiosis I. As you do, sketch each step. After telophase I, divide the nucleus in half with a line. Each

*For detailed information on safety symbols, see Appendix B.

nucleus should contain one of each kind of chromosome, still in the two-chromatid form.

F. Continue with the steps of meiosis II. Again, draw each stage. At the end of telophase II, divide each nucleus in half again.

Organizing Your Data

1. Using your model, complete a table like the one below.

2. Summarize your results, showing chromosome size and "color" in: (a) each of the two nuclei after meiosis I; (b) each of the four nuclei after meiosis II.

3. How did the arrangement of chromosomes at metaphase I affect the combination of chromosomes in the four nuclei at telophase II?

Stage	Number of Cells	Chromosomes per Cell	Double Chromosomes (yes/no)
Interphase			
Telophase I			
Telophase II			

Drawing Conclusions

1. What do same-colored pipe cleaners represent?

2. At what stage of meiosis is the number of chromosomes halved?

3. Summarize what happens to homologous chromosomes during meiosis.

4. How do the events of metaphase I determine if chromosomes of different "colors" end up in the same nucleus or in different nuclei after meiosis?

Thinking Further

1. When the chromosome number is halved during meiosis, why are not some characteristics lost?

2. If homologous chromosomes failed to separate at anaphase I, how would it affect the final products?

21-2 Sexual Reproduction in Simple Organisms

Section Objectives:

- *State* the advantages of sexual reproduction.
- *Define* the term *conjugation.*
- *Explain* the importance of mating types.
- *Describe* the process of conjugation in bacteria, spirogyra, and paramecium.

Advantages of Sexual Reproduction

All living things give rise to new members of their own kind by means of either asexual or sexual reproduction or both. As you read in Chapter 20, in asexual reproduction, each new individual receives a set of chromosomes that is exactly like its parent's chromosomes. This means that no inherited differences, or variations, are likely to occur. As long as the environment of the organisms remains the same, variations are not necessary for the survival of the species. If, however, the environment changes and the organisms are unable to adapt to the change, they are likely to die out.

In sexual reproduction, the offspring are not identical to either parent. Instead, they show new combinations of characteristics. Thus, in any species in which sexual reproduction takes place, the members of the species will show differences in structure and/or function. Increasing the amount of variation in members of a species increases the possibility that some individuals of that species will be better able than others to survive both short-term and long-term changes in the environment. The better-adapted individuals are more likely to survive environmental changes and to pass on the helpful variations to their offspring. Variations may also allow certain individuals in a population to move into new environments. Sexual reproduction helps to ensure the survival of the species by making a population more varied.

Conjugation and Mating Types

The simplest type of sexual reproduction takes place in protists and other simple organisms. Although these organisms usually reproduce asexually, some also reproduce sexually. In these organisms, sexual reproduction restores the organism's ability to grow and reproduce. If sexual reproduction is prevented in some species, the species may die out. Sexual reproduction also allows a recombination of hereditary material, which introduces variation within the species.

Among simple organisms that reproduce sexually, there are no distinct sexes. That is, all members of a single species look almost exactly the same. Although no male or female cells can be

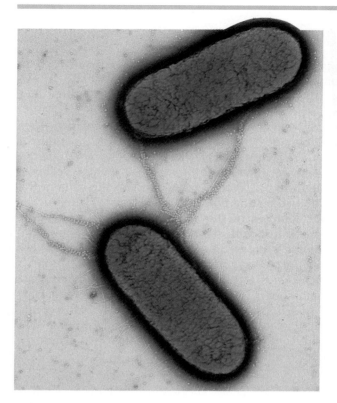

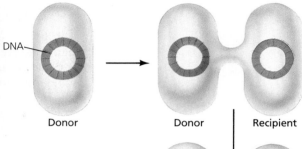

DNA

Donor

Donor   Recipient

Donor   Recipient

Donor is   Recipient
the same

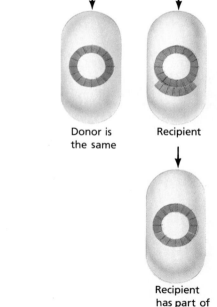

Recipient
has part of
donor's DNA

distinguished, there are usually two different *mating types* or *strains*. These are commonly called plus (+) and minus (−). It appears that there are biochemical and chromosomal differences between different mating types.

The type of sexual process most commonly found among simple organisms is called **conjugation** (kahn juh GAY shun). In conjugation, a bridge of cytoplasm forms between two cells, and an exchange or transfer of nuclear material takes place through the bridge. Conjugation takes place only between two cells of different mating types.

Conjugation in Bacteria

Among certain species of bacteria, one mating type, called a *donor,* is able to give a copy of all or part of its DNA to another mating type, called a *recipient.* This is done by means of the process of conjugation.

During conjugation in bacteria, the donor bacterium extends a long, thin tube toward one or more of the recipients. See Figure 21–6. The tube has a sticky end that fastens onto the recipient. A copy of the donor's circle-shaped DNA slowly pushes through the tube toward the recipient. The amount of DNA that enters the recipient depends on the length of time that conjugation lasts. Usually, conjugation stops before a copy of all the donor's DNA has entered the recipient.

▲   **Figure 21–6**

Conjugation in Bacteria. This transmission electron micrograph (left) shows conjugation between two bacteria. (Magnification 23 220 X) During conjugation (right), DNA is transferred from the donor to the recipient through a fine tube.

Conjugation in Spirogyra

Spirogyra is a type of green alga that has threadlike filaments. The filaments are made of haploid cells fastened end to end. These organisms usually reproduce asexually by binary fission, but sometimes they reproduce sexually by conjugation.

During conjugation, two filaments of opposite mating types come to lie side by side. Projections are formed on the sides of neighboring cells. See Figure 21–7. These projections meet, and the walls where the projections meet break down. This creates a passageway, called the *conjugation tube,* between the cells of the two filaments. The two mating types of spirogyra are called *active* and *passive.* The contents of the active cells flow through the conjugation tube and fuse with the nucleus and cytoplasm of the passive cells, forming zygotes. Since the cells of the filaments are haploid, the newly formed zygotes are diploid.

Each zygote secretes a thick, protective wall and becomes a **zygospore** (ZY guh spor). The original filaments decay, releasing the zygospores. After a period of rest, and when favorable conditions return, each zygospore undergoes meiosis forming four haploid cells. Only one cell survives, and it divides by mitosis, giving rise to a new haploid filament.

Figure 21–7

Conjugation in Spirogyra. A photomicrograph of two spirogyra filaments (right) shows the connecting conjugation tubes. (Magnification 156 X) During conjugation in spirogyra (bottom), genetic material flows from the active cell into the passive cell. When the nuclei of the two cells fuse, a diploid zygospore is formed. ▶

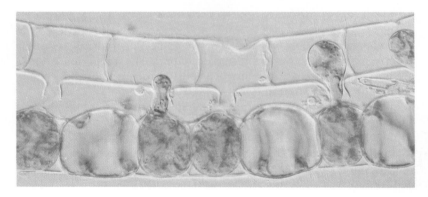

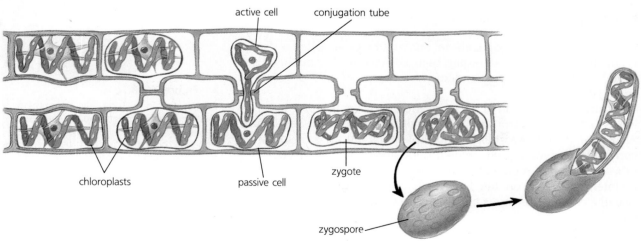

active cell

conjugation tube

chloroplasts

passive cell

zygote

zygospore

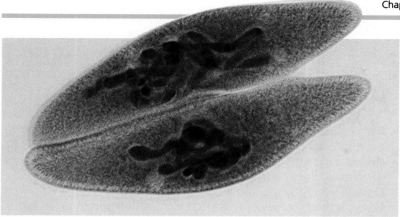

Conjugation in Paramecia. Although paramecia reproduce asexually, they also reproduce by conjugation.

Conjugation in Paramecia

Paramecia usually reproduce asexually by binary fission. From time to time, however, they reproduce by conjugation. Conjugation takes place between two different mating types—plus and minus. In some species, the exchange of hereditary material must take place from time to time by conjugation in order for the cells to reproduce asexually. If it does not take place, the cell stops dividing and dies.

During conjugation, two paramecia—one plus, the other minus—stick together at their oral grooves. See Figure 21–8. A protoplasmic bridge forms between them. A complex series of changes takes place in the nuclei of each. Paramecia and other ciliates have two types of nuclei—macronuclei and micronuclei. Each cell may contain more than one of each type of nucleus. During conjugation, the macronucleus disappears. The micronucleus divides by meiosis. Only two of the resulting haploid micronuclei do not disappear. One of the two haploid micronuclei from each paramecium moves across the protoplasmic bridge into the opposite cell where it fuses with the remaining haploid micronucleus of that cell. Both cells now contain a diploid micronucleus. The conjugating paramecia now separate and the new micronuclei undergo several mitotic divisions. These divisions result in the formation of new macronuclei and micronuclei. Both organisms then divide twice without nuclear division. Eight new organisms are produced.

21-2 Section Review

1. Why is sexual reproduction important to the survival of a species?

2. What are mating types?

3. What type of sexual reproduction is found among protists?

Critical Thinking

4. How is conjugation in bacteria and spirogyra similar? How does it differ? (*Comparing and Contrasting*)

21-3 Sexual Reproduction in Animals

Section Objectives:

- *Define* the following terms: *gonad, ovary, ovum, testis, sperm, hermaphrodite, gametogenesis,* and *parthenogenesis.*
- *Describe* the processes of oogenesis and spermatogenesis and the structures of a sperm and an egg.
- *Explain* what takes place during fertilization.
- *Contrast* the processes of external fertilization and internal fertilization in animals.

Reproductive Systems

Sexual reproduction in animals usually involves two sexes—male and female. In many animals, the sex of an individual can be identified by physical appearance. Even in animals in which there is little or no difference in appearance between the sexes, there are internal differences. The gametes of animals develop in specialized organs called **gonads** (GOH nadz). The female gonads are called **ovaries** (OH vuh reez). The ovaries produce the female gametes, which are called *egg cells,* or **ova** (OH vuh) (singular, *ovum*). The male gonads are called **testes** (TES teez). The testes (singular, *testis*) produce male gametes, which are called **sperm cells.** Egg cells are commonly referred to as eggs, and sperm cells are referred to as sperm (either singular or plural). In addition to the gonads, most animals have other organs that are needed for reproduction. These organs together with the gonads form the reproductive system.

Figure 21–9

Earthworms Mating. Earthworms are hermaphroditic. Each earthworm has both testes and ovaries. ▼

Separation of Sexes and Hermaphroditism

In most animals, the sexes are separate. That is, each individual has either testes or ovaries and is either male or female. In some animals, however, the sexes are not separate. Instead, each individual has both testes and ovaries. These organisms are called **hermaphrodites** (her MAF ruh dyts). *Hermaphroditism* (her MAF ruh dit iz um) is usually found among animals that move slowly or those that are attached to a surface. Examples include earthworms, snails, and hydras.

Even though hermaphroditic organisms can produce both eggs and sperm, self-fertilization is rare. Instead, these organisms exchange sperm with another individual of the same species. For example, during the mating of earthworms, two individuals lie parallel to each other. See Figure 21–9. Each earthworm transfers sperm to the *sperm receptacle* of its partner. After they have separated, each worm uses the stored sperm from the partner to fertilize its own eggs.

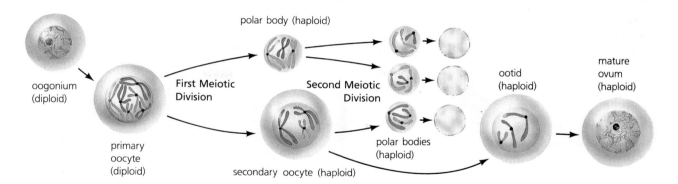

Oogenesis. A single functional egg cell is produced by meiotic division of a primary oocyte. The oocyte is diploid; the egg cell is haploid.

Gametogenesis: Meiosis in Females and Males

The process by which gametes develop in the gonads is called **gametogenesis** (guh meet uh JEN uh sis). More specifically, the formation of eggs in the ovaries is called **oogenesis** (oh uh JEN uh sis), while the formation of sperm in the testes is called **spermatogenesis** (sper mat uh JEN uh sis). Although the same basic processes are involved in the production of both eggs and sperm, there are some differences.

Oogenesis   Oogenesis is the production of eggs in the ovary. The major steps in oogenesis are shown in Figure 21–10. Eggs develop in the ovary from immature cells called *oogonia* (oh uh GOH nee uh) (singular, *oogonium*). In many animals, the oogonium is surrounded by a *follicle,* a small spherical sac of cells within which the mature egg develops. Oogonia contain the diploid number of chromosomes. During the early development of the female organism, the oogonia divide many times by mitosis to form a supply of oogonia. In human females, the production of oogonia stops at birth. Thus, each human female is born with all the oogonia she will ever have. Before birth and by the third month of development of a human female, oogonia within the baby's ovaries begin to develop into cells called *primary oocytes* (OH uh syts). By birth, the primary oocytes are in prophase in the first meiotic division. At this point, meiosis stops until the female reaches sexual maturity. Then, about once a month in most women, one of these primary oocytes finishes meiosis and develops into a functional egg.

When the first meiotic division takes place in the primary oocyte, the cytoplasm of the cell divides unequally. One of the daughter cells is large and receives most of the cytoplasm. This cell is called the *secondary oocyte.* The other daughter cell is small and is called the first *polar body.* Each of these daughter cells has the haploid number of chromosomes.

During the second meiotic division, the secondary oocyte divides unequally into a large cell called an **ootid** (OH uh tid) and another polar body. The first polar body may also divide into two polar bodies. The ootid grows into a mature egg, having the haploid chromosome number. The polar bodies break apart and die.

Figure 21–11

Spermatogenesis. Four functional sperm cells are produced by meiotic division of a primary spermatocyte. ▶

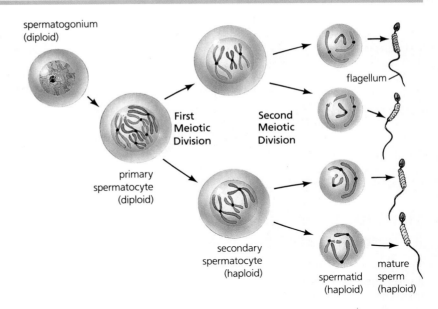

spermatogonium (diploid)

First Meiotic Division

Second Meiotic Division

flagellum

primary spermatocyte (diploid)

secondary spermatocyte (haploid)

spermatid (haploid)

mature sperm (haploid)

Spermatogenesis Spermatogenesis is the production of sperm in the testes. The major steps in spermatogenesis are shown in Figure 21–11.

Within the testes, the sperm develop from immature sex cells called **spermatogonia** (sper mat uh GOH nee uh). Throughout childhood, these spermatogonia (singular, *spermatogonium*) divide mitotically many times to produce additional spermatogonia. The spermatogonia contain the diploid number of chromosomes. In humans, after a male matures sexually, there is a continual development of some spermatogonia into functional sperm. Other spermatogonia go on dividing mitotically, producing more spermatogonia. Thus, while the number of human eggs is limited in each female, the number of sperm is not limited in males.

In the course of development, a spermatogonium increases in size to become a *primary spermatocyte* (sper MAT uh syt). The primary spermatocyte undergoes the first meiotic division, forming two cells of equal size. These are *secondary spermatocytes*. Each secondary spermatocyte then undergoes the second meiotic division, forming four *spermatids* (SPER muh tids), all of equal size. The spermatids contain the haploid number of chromosomes. Without any further division, each one of the spermatids develops into a mature sperm with a flagellum. The flagellum is a long, thin tail that helps sperm swim. Thus, each primary spermatocyte gives rise to four haploid sperm.

Comparison of Egg and Sperm

Female gametes of animal species are different in structure and appearance from male gametes. Usually eggs are round and unable to move by themselves. They contain a nucleus and often have stored food in the form of yolk. The egg is usually larger than the sperm of the same species. The size of the egg is different from one

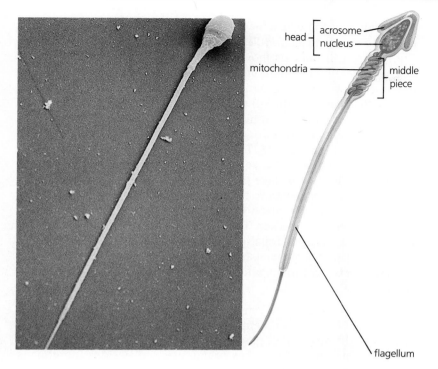

head { acrosome
nucleus

mitochondria

middle piece

flagellum

◄ **Figure 21–12**

Structure of a Human Sperm Cell. The sperm is specialized for swimming toward and penetrating the egg. The overall length of the sperm is about 0.05 millimeters, half the length of a human egg. The human sperm shown in this scanning electron micrograph is magnified about 2000 X.

species to another and depends on the amount of yolk stored in it. Yolk is used as food for the developing animal. For example, the yellow center of a chicken egg is the egg cell. It contains a great deal of yolk because the developing chicken receives no other nourishment while in the egg. The eggs of humans and other mammals are usually microscopic and contain no yolk because the developing animal gets nourishment from the mother. The human egg is about 0.1 millimeter in diameter.

Most sperm cells are microscopic. A typical sperm is made up of a head, a middle piece, and a flagellum. See Figure 21–12. The head is made up of the nucleus, which contains the chromosomes, and an *acrosome* (AK ruh sohm). The acrosome contains enzymes that help the sperm penetrate the egg. The middle piece is packed with mitochondria, which give the sperm energy so that it can move. The long, whiplike flagellum enables the sperm to swim through liquids.

Fertilization and Zygote Formation

The joining of the haploid sperm cell with the haploid egg cell produces a diploid zygote. Fertilization restores the species number of chromosomes.

Eggs cannot move by themselves. Sperm, on the other hand, are specialized for fast movement. When sperm are released by the male, their flagella beat rapidly, pushing them along in all directions. When a sperm comes in contact with an egg, the acrosome releases enzymes that dissolve an opening through the protective membranes of the egg. This allows the nucleus in the head of the sperm to enter the egg, while the rest of the sperm remains outside. The sperm cell nucleus moves through the cytoplasm toward the egg cell nucleus. The haploid sperm nucleus joins with the haploid egg nucleus to form a diploid zygote, which thus has $2n$ chromosomes. Because only the sperm nucleus enters the egg, all cytoplasmic DNA, such as DNA in mitochondria, comes from the egg.

A *fertilization membrane* forms around the egg after a sperm enters it. See Figure 21–13. This membrane stops other sperm from entering the egg and also serves as a protective covering.

For fertilization to take place, there must be a fluid medium so that the sperm can swim to the egg. Also, because sperm and eggs live for only a short period of time, the male and female gametes must be released together. There are two basic ways in which the gametes are brought together. One is **external fertilization,** in which the gametes fuse outside the body of the female. The other is **internal fertilization,** in which the gametes fuse inside the body of the female.

External Fertilization In external fertilization, the eggs are fertilized in the environment outside the body of the female. This type of fertilization takes place only in animals that breed in water. In these animals, the only sex organs needed besides the gonads are the ducts that carry the gametes from the gonads to the water. Fertilization takes place directly in the water after each parent releases its gametes. The sperm swim through the water to the eggs. Although there is no problem about moist surroundings for fertilization, there are many hazards in the environment. The sperm

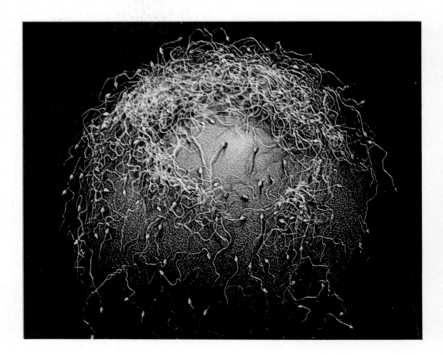

Figure 21–13

Fertilized Human Egg Surrounded by Sperm. Once a sperm has penetrated the egg, other sperm are prevented from entering by formation of a fertilization membrane around the egg. ▶

Amplexus. Amplexus ensures that male and female frogs release their gametes into the water at the same time. This increases the probability that some of the eggs will be fertilized.

and eggs may not meet, the eggs or developing offspring may be eaten by other animals, they may die because of changes in the temperature and/or the amount of oxygen in the water, and so on. To overcome the hazards of external fertilization, large numbers of eggs and sperm are released. External fertilization takes place in almost all aquatic invertebrates, most fish (but not sharks), and many amphibians.

To improve the chances of eggs and sperm meeting, gametes are not released at random during external fertilization. There are many hormonally controlled behavior patterns that make certain that sperm and eggs are released at about the same time and place. In some fish, the female lays thousands of eggs, and the male swims over them releasing sperm. This process is known as *spawning*.

Spawning takes place with salmon, for example. Salmon hatch in freshwater streams. The young fish then travel downstream to the ocean where they mature. When the salmon are ready to spawn, they return to the freshwater stream where they were hatched. This behavior ensures that the males and females are in the same place and in the right environment for spawning.

In frogs, when a female is full of eggs and ready to mate, she goes to a male. The male embraces the female with his front legs in a process called *amplexus*. See Figure 21–14. This stimulates the female to release her eggs, and at the same time, the male releases his sperm. Because they are in close contact and the gametes are released at the same time, the sperm reach many of the eggs. Amplexus brings about the release of gametes at the same time and place.

Internal Fertilization   Fertilization within the body of the female is called internal fertilization. This kind of fertilization is found most often in animals that reproduce on land. It is also found in some aquatic animals, such as sharks and lobsters. Internal fertilization requires a specialized sex organ to carry the sperm from the body of the male into the body of the female. After the sperm are placed within the female's body, they travel to the eggs and fertilize them. The moist tissues of the female provide the watery environment needed for the sperm to swim to the egg. After fertilization, either the zygote is enclosed in a protective shell and released by the female, or it remains and develops within a special part of the female's body.

Internal fertilization does away with the scattering of gametes and the dangers of the outside environment. Fewer eggs are needed because they are well protected, and the chances of fertilization are much greater than they are when outside the body in water. However, even with internal fertilization, large numbers (often in the millions) of sperm are released by the male into the body of the female. Because the sperm store little food, they live only a short time. Therefore, even within the female's body, the sperm can fertilize the egg for only a brief period of time. Also, the egg can be penetrated by the sperm for only a brief time. In humans, the egg can be fertilized for only about 24 hours.

In animals with internal fertilization, many of the specialized adaptations involved in reproduction are concerned with the timing of the release of sperm and eggs. Because the gametes live only for a short time, mating must take place within certain time periods for fertilization to take place. Many of the reproductive adaptations are controlled by hormones. Reproductive adaptations include such things as singing, the display of special feathers, color patches on the skin, and the release of chemicals called *pheromones* (FER uh mohnz), which have distinctive odors. These adaptations stimulate the mating response and trigger the release of eggs and sperm.

In many insects and in bats, the problem of timing is solved in an interesting way. After mating, the sperm are stored in specialized structures in the female and then used to fertilize the eggs at some later time. In the queen honeybee, for example, enough sperm are stored from one mating to fertilize the hundreds of thousands of eggs she lays during her lifetime. In bats, mating takes place in the fall, and the sperm are stored until the following spring, when fertilization takes place. It is not known how the sperm remain alive for such long periods.

Parthenogenesis

The development of an unfertilized egg into an adult animal without fusion with sperm is called **parthenogenesis** (par thuh noh JEN uh sis). In nature, it takes place in many insects, including bees, wasps, aphids (plant lice), and certain ants, and in rotifers and other microscopic animals. For example, in bees, the queen bee mates

◀ **Figure 21–15**

Beehive with Drones, Workers, and Queen. The queen honeybee, seen at the center of this photograph, lays both fertilized and unfertilized eggs. Fertilized eggs produce females, and unfertilized eggs produce males.

only once. She can then produce either unfertilized eggs or fertilized eggs. The unfertilized eggs become male drones while the fertilized eggs become female workers or queens. See Figure 21–15. Female aphids reproduce by parthenogenesis during the spring and summer. In the fall, the eggs produce both males and females. These insects mate, and the females produce fertilized eggs that hatch in the spring.

21-3 Section Review

1. Where do the gametes of animals develop?
2. How many mature sperm are produced from each primary spermatocyte? How many mature egg cells are produced from each primary oocyte?
3. Name one animal in which external fertilization occurs and one animal in which internal fertilization occurs.
4. What is parthenogenesis?

Critical Thinking

5. List three differences between oogenesis and spermatogenesis. (*Comparing and Contrasting*)

Laboratory Investigation

Preparing a Human Karyotype

A karyotype is a picture of a cell's chromosomes arranged in homologous pairs. Refer to Figure 21–2 in your text. In this investigation, you will use photographic images of the chromosomes from a human cell to prepare a human karyotype.

Objectives

- *Identify* the pairs of homologous chromosomes in a human cell.
- *Determine* the sex of an individual from a karyotype.

Materials

photograph of a human cell nucleus in
  metaphase of mitosis
scissors   paper
glue or scotch tape   envelope

Procedure *

A. Obtain from your teacher a copy of a photograph of the nucleus of a human cell in metaphase of mitosis. The chromosomes, which are stained, consist of two chromatids joined together at a single point, the centromere. Chromosomes that have a central centromere appear roughly in the shape of the letter X, with the centromere at the center of the X.

B. Cut out each chromosome from the photograph. Work carefully and be sure you do not lose any chromosomes. If you have to stop, store all the chromosomes in an envelope.

C. Arrange homologous chromosomes in pairs. First, arrange the chromosomes according to size. Then, try

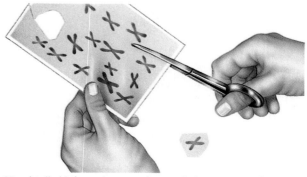

to match them according to position of the centromere. If banding is evident on the chromosomes, use this feature as another means of matching homologous chromosomes.

D. Determine the sex of the person whose chromosomes are pictured. If the person is female, each chromosome will have a matching partner. If the person is male, a pair will consist of a large chromosome, the X chromosome, and a much smaller chromosome, the Y chromosome. See Figure 21–2.

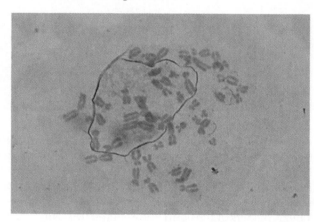

E. Line up the pairs of homologous chromosomes in order of decreasing size; then glue or tape them to a piece of paper. Use the karyotype in Figure 21–2 as a model.

Analysis and Conclusions

1. For each of the following, list the number that are present in your karyotype: (a) chromosomes; (b) centromeres; (c) pairs of homologous chromosomes; (d) chromatids.

2. How many chromosomes have a centromere at one end? How many have a centromere at the center?

3. Are the chromosomes of the karyotype from the cell of a female or a male?

4. For any pair of homologous chromosomes, what is the origin of each member of the pair?

5. What is the relationship between sexual reproduction and having two chromosomes of each type?

*For detailed information on safety symbols, see Appendix B.

Chapter Review

Study Outline

21–1 Meiosis

■ Sexual reproduction involves hereditary material from two parents. The offspring are genetically different from each parent.

■ Complex organisms have sex cells, or gametes, that join during fertilization to form a zygote. A new offspring results from the fusion of gametes.

■ Every species has a characteristic number of chromosomes in its somatic cells. In humans, the characteristic chromosome number is 46, which is known as the diploid number ($2n$).

■ Gametes, which are formed by meiosis, are haploid cells containing n chromosomes, or half the diploid number. In meiosis, the chromosomes replicate once, and each cell divides twice. Meiosis results in the formation of four haploid cells.

21–2 Sexual Reproduction in Simple Organisms

■ Sexual reproduction increases the amount of variation in a species, thus increasing the chance that some individuals will be better adapted to survive changes in the environment.

■ Protists and other simple organisms may reproduce asexually or sexually. During conjugation, two different mating types exchange nuclear material through a bridge of cytoplasm.

■ In bacteria, there are two mating types—the donor and the recipient. During conjugation, the donor bacterium passes a copy of some of its DNA to the recipient through a thin tube.

■ In spirogyra, haploid-cell filaments of two mating types lie side by side during conjugation. The contents of active and passive cells join, and walls form around the resulting zygotes, forming zygospores.

■ When paramecia reproduce by conjugation, plus and minus mating types join. Meiotic division of micronuclei results in haploid micronuclei. Micronuclei of opposite types fuse to form diploid micronuclei.

21–3 Sexual Reproduction in Animals

■ In animals, ovaries of the female produce eggs. Because eggs often have stored food, they tend to be larger than sperm. Testes of the male produce sperm, which are specialized for movement.

■ Eggs develop by the process of oogenesis. Diploid oogonia in the ovary undergo meiosis to form mature haploid eggs. In humans, oogonia production stops at birth, limiting the number of eggs.

■ Sperm develop by the process of spermatogenesis. Diploid spermatogonia in the testes undergo meiosis to form mature haploid sperm. Sperm development in males is continuous, and therefore the number of sperm is not limited.

■ External fertilization occurs only in animals that breed in water. Behavior patterns help ensure that sperm and eggs are released at the same time and place.

■ Internal fertilization usually occurs in animals that breed on land. It also occurs in cartilagenous fishes and some bony fishes. A sex organ transports the sperm from the male's body into the female's body.

Vocabulary Review

gametes (417)

fertilization (418)

zygote (418)

meiosis (418)

somatic cells (418)

homologous chromosomes (418)

diploid (419)

haploid (419)

synapsis (420)

tetrad (420)

disjunction (420)

conjugation (425)

zygospore (426)

gonads (428)

ovaries (428)

ova (428)

testes (428)

sperm cells (428)

hermaphrodites (428)

gametogenesis (429)

oogenesis (429)

spermatogenesis (429)

ootid (429)

spermatogonia (430)

external fertilization (432)

internal fertilization (432)

parthenogenesis (434)

Chapter Review

A. Analogies—Select the vocabulary term that relates to the single term in the same way that the paired terms are related.

1. testes **is to** spermatogenesis as ovary **is to** _____

2. somatic cells : diploid as _____ : haploid

3. _____ : sperm as oogonia : eggs

4. external fertilization : water as _____ : land

5. bees : parthenogenesis as paramecia : _____

6. prophase I : synapsis as anaphase I : _____

B. Matching—Select the vocabulary term that best matches each definition.

7. A group of four chromatids

8. A zygote covered by a thick protective wall

9. Development of an unfertilized egg into an adult organism

10. Female gametes

11. Fusion of an egg cell nucleus and a sperm cell nucleus to form a zygote

12. Having half the diploid number of chromosomes

13. Formation of sperm in the testes

14. Organisms that have both testes and ovaries in each individual

15. Organs in animals in which gametes develop

Content Review

16. Explain how meiosis differs from mitosis.

17. How do diploid cells differ from haploid cells?

18. How is the chromosome content of the cell changed by the first meiotic division?

19. How is the chromosome content of the cell changed by the second meiotic division?

20. Compare the processes of synapsis and disjunction.

21. Describe the process of crossing-over. When may it occur?

22. In sexual reproduction, why are the offspring different from the parents?

23. Explain the benefit of sexual reproduction in species that normally reproduce asexually.

24. Describe conjugation in simple organisms.

25. How is the zygospore produced in spirogyra?

26. Describe conjugation in paramecia.

27. What are the functions of ovaries and testes?

28. Explain how mating and fertilization take place in the earthworm.

29. Why is the number of eggs in human females limited?

30. How many mature eggs and polar bodies are produced from each primary oocyte?

31. Describe the process of spermatogenesis.

32. Explain how an egg is fertilized by a sperm.

33. Why do organisms in which external fertilization occurs produce large numbers of gametes?

34. What adaptations in animals make internal fertilization possible?

35. Describe parthenogenesis in bees.

Graphic Organizing

For information on graphic organizers, see Appendix G at the back of this text.

36. **Flow Chart:** Construct a flow chart showing the stages of meiosis. In each box, indicate the number of chromosomes present at that stage in an organism with a diploid number of 8. Also indicate whether the chromosomes would be single-stranded or double-stranded at each stage.

Critical Thinking

37. Explain the similarities and differences between the first meiotic division and mitosis and between the second meiotic division and mitosis. (*Comparing and Contrasting*)

38. What might happen if one member of a pair of homologous chromosomes did not separate during anaphase I? (*Predicting*)

39. Hermaphroditic organisms produce both sperm and eggs, but self-fertilization is rare. What is the advantage of mating for these organisms? (*Judging Usefulness*)

40. How does the structure of a sperm cell differ from that of an egg cell? How are their structures suited to their respective functions? (*Comparing and Contrasting*)

Creative Thinking

41. Develop a hypothesis to explain why during oogenesis the meiotic divisions produce cells of unequal size—the secondary oocytes and the polar bodies.

42. Females of several different species—the bat, for example—store sperm for long periods (up to six months) after mating. If we could discover how this is done, how could the knowledge benefit humans?

Problem Solving

43. An investigation was conducted to determine the effect of temperature on the number of eggs produced by goldfish and the number of eggs that hatched. Using the data in the following table, calculate the percentage of eggs that hatched at each temperature. Interpret the data and draw some conclusions regarding the optimum temperature range for egg production and hatching. At what temperature do egg production and hatching decrease most drastically?

Temp. (°C)	Number of Eggs Produced	Number of Eggs Hatched
10°C	234	79
15°C	411	158
20°C	549	221
25°C	374	119
30°C	80	14

44. Several beekeepers observe that more male drone bees hatch on some days than on others. To investigate which factors might influence the hatching rate of the bees, they select the following variables to be tested: (a) atmospheric temperature, (b) relative humidity, (c) amount of food (pollen) available in the surrounding area, and (d) number of bees living in the hive. Each of the variables is to be studied for its effect on the hatching rate. For each variable, suggest a controlled experiment in which it would be tested.

Projects

45. The aphid, an insect that damages crops and ornamental plants, has unusual reproductive abilities. Do library research on the aphid's life cycle and eating habits. Calculate approximately how many aphids will be produced from two adults in one year. Write up your report in the form of a feature article for a magazine.

46. Using modeling clay, make models of the stages of meiosis.

47. With a microscope, study the stages of conjugation in paramecia. Prepare a written report with diagrams of your observations.

48. Write a report on the life and contributions of one of the following scientists: (a) Elizabeth Blackwell, (b) Oskar Hertwig, (c) Jacques Loeb, (d) Alice Eastwood.

▲ Rain frogs of Costa Rica pass through the tadpole stage in the egg.

Animal Development

One cell splits, leaving two. Two cells split to make four. Four become eight and eight become sixteen. A new creature is on its way. Will it be a beagle, an ostrich, a goldfish? It is too early to tell. The dividing continues, over and over again. Gradually, the lump of cells shows some form. Eyes appear. A simple heart beats. Small bumps bud into limbs—later to become legs perhaps, or wings, or fins. Finally, a new organism emerges, equipped for the outside world. In this chapter, you will learn about the process that transforms a fertilized egg into a new being.

Guide for Reading

Key words: embryo, cleavage, blastula, gastrula, differentiation, chorion, amnion, placenta

Questions to think about:

📖 Why are growth and differentiation necessary during embryonic development?

📖 How do external and internal development differ?

📖 What structural adaptations allow bird and reptile eggs to develop externally on land?

22–1 Embryonic Development

Section Objectives:

- *Explain* the following terms in this chapter: *development, embryo,* and *differentiation.*
- *Describe,* in general terms, the events of cleavage and embryonic development through the gastrula stage.
- *Explain* what roles the nucleus and cytoplasm play in controlling development.
- *Name* the three germ layers and list a few of the tissues and organs formed from each.
- *Describe* the process of embryonic induction and explain how this process controls development.

In animals, as in plants, fertilization of an egg to form a zygote is only the first step in a complex series of events. It is this series of events that finally gives rise to a full-grown organism. These events, which are the subject of this chapter, are called **development.** In the following sections, you will learn about the changes the zygote undergoes as it develops.

In the early stages of development, the organism is called an **embryo** (EM bree oh). The study of the development of embryos is called *embryology* (em bree AHL uh jee). Although embryos develop in many different ways, the basic processes are always the same in animals. These processes of development include cleavage, growth, and differentiation.

▲ **Figure 22–1**

A Red-Eyed Leaf Frog. This red-eyed leaf frog is fully developed.

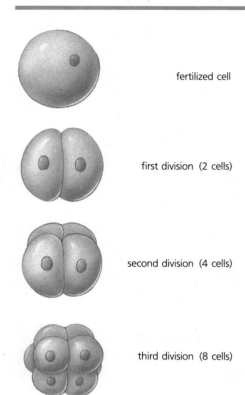

fertilized cell

first division (2 cells)

second division (4 cells)

third division (8 cells)

▲ **Figure 22–2**
Early Stages of Cleavage. Although the number of cells doubles at each division during cleavage, no growth occurs.

Cleavage

After fertilization, the zygote begins a series of cell divisions known as **cleavage** (KLEE vidj). During cleavage, the fertilized egg divides by mitosis into two cells. Each of these cells divides again, producing four cells. These four produce eight, and so on. See Figure 22–2.

During cleavage, the cells do not grow. Each division decreases cell size. Since the eggs of a species are usually larger than the average cell of the adult organism, cleavage changes a single, large fertilized egg into many small cells. Cleavage goes on until the cells of the developing embryo have been reduced to the size of the cells of the adult organism.

The early divisions of cleavage result in a solid ball of cells. At this stage, the embryo is called a **morula** (MOR yuh luh). As the cells continue to divide, they are rearranged to form a hollow sphere. Usually, the layer of cells in the sphere is only one cell thick. The inside of the sphere is filled with fluid. At this stage, the embryo is called a **blastula** (BLAS chuh luh), and the fluid-filled inside of the sphere is called the **blastocoel** (BLAS tuh seel).

Although these are the usual steps in cleavage, the arrangement of cells in the developing embryo depends on the amount and distribution of yolk in the egg. Some eggs, such as those of humans, have little yolk. In these eggs, cleavage results in a blastula in which all the cells are nearly equal in size. However, other eggs, such as those of amphibians, bony fish, birds, and reptiles, have a large amount of yolk at one end of the egg cell. Because yolk tends to slow down cell division, cleavage takes place mostly or only at the pole without yolk.

In frog eggs, the large amount of yolk found at one pole slows down cleavage at that pole. This slowdown results in a blastula that has larger cells at the yolk-filled, or *vegetal,* pole. More numerous, smaller cells form at the yolkless, or *animal,* pole, as shown in Figure 22–3. The blastocoel is formed only within the region of the animal pole.

In the chicken egg, the nucleus and cytoplasm are concentrated in the *germinal disk,* a platelike area on the surface of the ball of yolk. Only the cells of the germinal disk undergo cleavage. The blastocoel is formed when these cells separate from the yolk, leaving a space between the yolk and the cells. The developing chick uses the yolk for food and finally fills completely the space within the egg.

Gastrulation

As the blastula develops, it reaches a point at which the cells begin to grow before dividing. At this point, mitotic division continues, but it is also accompanied by growth. At the same time, various movements of the cells take place. These movements will set the shape of the embryo.

Role of the Nucleus and Cytoplasm

Within the nucleus of the fertilized egg is the hereditary material. This contains all the information needed for development. The information is encoded within the chemical structure of DNA in the chromosomes. DNA controls the chemical processes of the cell and determines which proteins are made by the cell. Because the cells of the embryo divide by mitosis, each cell has in it the same chromosomes and DNA as the original fertilized egg cell.

If DNA controls cellular activities and all cells of an organism have the same DNA, how are the many kinds of cells in the organism made? Scientists have discovered that different sections of the DNA in a cell can be turned off or on. This results in the formation of different types of cells. Thus, in muscle cells, for example, the part of the DNA that controls the making of muscle cell proteins is turned on, while the part of the DNA that controls the making of nerve cell proteins is not turned on. If this were not the case, differentiation would not take place, and all the cells of an embryo would be the same. However, little is known at this time about how the activity of DNA in different cells is controlled.

Experiments with frogs have shown that a nucleus from a fully differentiated cell contains a complete copy of the hereditary information. In 1962, the English biologist J. G. Gurden surgically replaced the haploid nuclei of unfertilized frog's eggs with diploid nuclei from the intestinal cells of tadpoles. Some of the eggs that received the intestinal nuclei became normal frogs. This means that some of the DNA that was switched off in the intestinal cell nucleus was turned on when the nucleus was placed in the egg cell. From this and other research, it appears that an interaction between the DNA and certain things in the cytoplasm controls development. As a result of the interaction, parts of the hereditary material are switched on and off, and this in turn determines the direction of cellular differentiation.

Other experimental evidence shows that differentiation begins early in development. For example, if the cells of a four-celled frog embryo are carefully separated, each cell will develop into a normal tadpole. If, however, the cells of an older embryo are separated, the cells do not develop correctly and die. In the older embryo, the cells have begun to differentiate and can no longer produce a whole, normal organism.

Role of Neighboring Cells

As an embryo grows, there must be coordination and communication between its tissues. By the late blastula or early gastrula stage, the way in which groups of cells will develop has been determined. Cells in certain regions develop along certain lines. For example, there is a particular place in the frog gastrula that normally develops into an eye. If this tissue is removed from the embryo and placed in a special nutrient solution, it develops into an irregular mass of cells. On the other hand, if it is transplanted into any other

Science, Technology and Society

Technology: Surrogate Parenting for Endangered Species

To keep endangered species from becoming extinct, scientists look for ways to increase the "baby yield"—the number of offspring per parent.

One technique now being used for mammalian species is surrogate parenting. In this technique, a fertilized egg from an endangered species is implanted into a related species. First, the scientists treat the endangered female with hormones that cause the release of many eggs. The eggs are fertilized and then flushed from the animal's uterus. At this point, the embryos are implanted into waiting surrogate mothers.

So far, the eland, a common antelope, has been used as a surrogate for the rare bongo antelope. A domestic cat has given birth to an endangered Indian desert cat. Lions will soon be used as surrogates for rare tigers. Although scientists are hopeful about this technique, one major question remains: Will the offspring be able to survive in the wild?

 Write a short essay stating your views on the surrogate parenting technology.

part of another frog embryo, it will develop into a recognizable eye, even though it does not work like an eye. Not all the tissues in the extra, nonfunctioning eye develop from the transplanted tissue. Instead, the transplanted tissue causes some of the surrounding tissue, which would not usually make up part of the eye, to develop into eye structures. A tissue, then, can affect the differentiation of neighboring tissues.

Certain parts of the developing embryo act as *organizers* and influence the development of neighboring cells. The process by which the organizers cause, or induce, other structures to differentiate is called **embryonic induction.** Scientists do not yet know what brings about embryonic induction. Some chemical substances can cause induction. Cell contact probably is also important. It may be that as development goes on, organizers, in some way and at certain times, influence which parts of a cell's hereditary material become active. In doing this, the organizers determine the course of differentiation of those cells.

22-1 Section Review

1. What is cleavage?
2. Name the three germ layers in the gastrula.
3. Define the term *differentiation*.
4. What are embryo organizers?

Critical Thinking

5. Which germ layer do you think gives rise to the human heart? Explain your conclusion. *(Reasoning by Analogy)*

22-2 External and Internal Development

Section Objectives:

- *Compare* and *contrast* external development in water and internal development on land.
- *Describe* the adaptations of reptiles, birds, nonplacental animals, and placental mammals for reproduction on land.
- **Laboratory Investigation:** *Observe* embryo development in the frog as an example of animal development (p. 452).

Embryonic development may take place either outside or inside the mother's body, depending on the type of organism that is developing. Regardless of where the development takes place, the embryo has certain needs that must be met for survival. These include nourishment, proper temperature, oxygen, protection, and a means of getting rid of wastes.

External Development in Water

In most aquatic animals, fertilization and development take place outside the mother's body in the water. Nourishment for these embryos comes from the yolk stored in the egg. The young usually develop under proper environmental conditions because mating and fertilization take place at the right times of the year. Oxygen from the surrounding water diffuses into the embryo, and wastes diffuse from the embryo into the water.

In most aquatic animals, there is little or no care of the young by the parents. Some fish, however, do provide a certain amount of care for the developing young. For example, the male stickleback guards the nest and fans the embryos with water currents to provide oxygen. Some fish are "mouthbreeders." That is, the fertilized eggs are held in one parent's mouth until they hatch. In general, the survival of species that develop in water depends on the production and fertilization of large numbers of eggs. Many of the eggs are eaten or destroyed in other ways, and only a small part of the original number survive from each mating.

External Development on Land

Fertilization inside the mother's body followed by development outside her body takes place in birds and most reptiles, as well as in a few mammals. In these animals, the fertilized egg, which contains a large amount of yolk, is enclosed in a protective shell. The shelled egg is moist inside, which provides the embryo with a self-contained watery environment. The shell is almost waterproof, but is porous enough to allow oxygen from the air to diffuse into the egg and carbon dioxide from the embryo to diffuse out. The number of embryos that survive to hatching is greater for animals whose eggs have a shell than for those whose eggs lack a shell. In fact, animals that lay eggs with a shell produce fewer eggs than animals that lay eggs without a shell. Both the hard shell of the bird egg and the tough, leathery shell of the reptile egg provide protection for the

▲ **Figure 22–7**
Male Stickleback. This male stickleback fans a nest containing eggs in order to provide the embryos with oxygen.

◄ **Figure 22–8**
Eggs. The hard shells of the black redstart bird eggs (left) and the leathery shell of the green snake eggs (right) protect the embryos from harm.

Figure 22–9
Internal Structure of the Chicken Egg. The shell protects the embryo from drying out, nourishes it, and allows it to develop outside of water. ▶

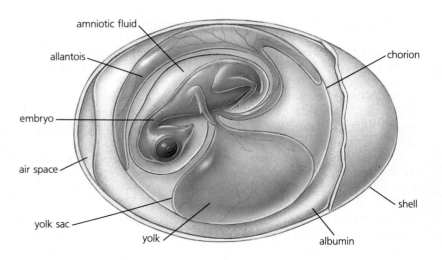

amniotic fluid

allantois

embryo

air space

yolk sac

yolk

chorion

shell

albumin

embryo. Reptiles, however, usually leave their eggs whereas bird eggs and young birds are carefully tended by the parents. Thus, the percentage of reptile eggs that survive is less than the percentage of bird eggs that survive. Knowing this, it is not surprising that reptiles lay many more eggs than birds.

Internal Structure of the Chicken Egg   As the chicken embryo develops, it forms four membranes that lie outside the embryo itself but inside the shell. These are the **extraembryonic membranes** shown in Figure 22–9. The extraembryonic membranes perform a number of important functions.

The outermost membrane, or the **chorion** (KOR ee ahn), lines the inside of the shell and surrounds the embryo and the other three membranes. The chorion aids in the exchange of gases between the embryo and the environment.

A second membrane, the **allantois** (uh LAN tuh wis), is a saclike structure that grows out of the digestive tract of the embryo. It is through the blood vessels of the allantois that the exchange of oxygen and carbon dioxide takes place. The metabolic wastes of the embryo also collect in the allantois.

The third membrane is a fluid-filled sac, known as the **amnion** (AM nee ahn), that surrounds the embryo. The amniotic fluid within the sac provides a watery environment for the embryo and acts as a cushion to protect it from shocks.

The fourth, and final, extraembryonic membrane is the **yolk sac,** which surrounds the yolk. The yolk sac is the source of food for the embryo. Blood vessels in the yolk sac carry the food to the developing embryo.

The shell-covered egg with its extraembryonic membranes is an important adaptation that allows bird embryos to develop on land. When the young bird hatches, the extraembryonic membranes are gotten rid of along with the shell. The eggs of reptiles have similar features. Figure 22–10 shows various stages of development in the chicken.

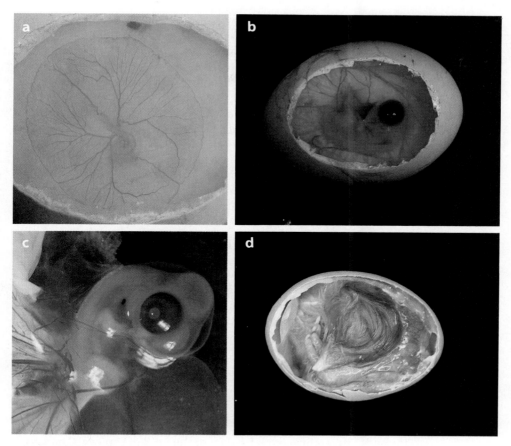

▲ **Figure 22–10**
Stages of Development of a Chicken. (a) Three days. (b) Seven days. (c) Fourteen days. (d) Nineteen days.

Internal Development

In some sharks and in some reptiles, such as garter snakes, fertilization and development take place inside the body. However, in these animals, the young do not receive food directly from the mother. Instead, the source of their food is yolk, which is stored in the egg. By the time the yolk is used up, the embryos have reached a stage of development at which they can take care of themselves, and they are born.

Among mammals, fertilization takes place inside the mother's body, and the embryos typically develop within a structure called the *womb* (woom), or **uterus** (YOO tuh rus), inside the body. They are born in an undeveloped condition and, for a period, feed on milk produced by the mother's mammary glands. The young are well protected during development and after birth, and a high percentage of them survive to adulthood. Thus, as might be expected in animals in which the young develop internally, relatively few eggs are produced.

Placental Mammals

Placental Mammals   Most mammals, including humans, are *placental* (pluh SENT ul) *mammals.* In these mammals, the blood vessels of the embryo's circulatory system are in close contact with the mother's circulatory system. See Figure 23–6. This contact takes place in a specialized structure called the **placenta** (pluh SENT uh), which is in the wall of the uterus.

In the placenta, nutrients and oxygen diffuse from the mother's blood into the embryo's blood. In turn, carbon dioxide and other wastes diffuse from the embryo's blood into the mother's blood. However, there is no direct connection between the two circulatory systems. The embryo is attached to the placenta by a structure called the **umbilical** (um BIL ih kul) **cord.** This structure contains blood vessels that connect the embryo's circulatory system to capillaries in the placenta.

Nonplacental Mammals   There are two types of *nonplacental mammals*—mammals in which no placenta forms during development of the embryo. These types are the *egg-laying mammals,* of which there are only two living species, and the *pouched mammals.*

The spiny anteater and duckbill platypus are egg-laying mammals. The embryo of the platypus is inside a leathery egg, which looks like the egg of a reptile. Unlike the eggs of other mammals, the eggs of the platypus contain a large amount of yolk. The female lays the eggs in a nest. When the young hatch out of the eggs, the female gathers them against her body. The young animals then feed on milk from the mother's mammary glands.

In the pouched mammals, or *marsupials,* some internal development of the embryo takes place in the uterus, but no placenta is formed. The embryo gets food from the yolk of the egg. The young animal is born in a very immature condition. It crawls into a pouch

Figure 22–11

A Duckbill Platypus. The duckbill platypus lays several eggs at a time. After the young hatch, they feed by licking milk from the mother's fur around the mammary glands. ▶

▲   **Figure 22–12**
A Western Grey Kangaroo. Young kangaroos complete development
within the mother's pouch.

on the outside of the mother's body and attaches itself to a
mammary gland. Development is completed in the pouch. Most
marsupials are found in Australia. While the kangaroo is the most
familiar example, the opossum, which is found in the Western
Hemisphere, is also a marsupial.

22-2 **Section Review**

1. Where may embryonic development occur?
2. Name two groups of animals that have internal fertilization and
external development.
3. What are the two types of mammals?

Critical Thinking

4. What would happen to a chicken embryo if the chorion were
impermeable? *(Relating Parts and Wholes)*

Laboratory Investigation

Embryo Development in the Frog

Fertilized frog eggs are useful for observing embryo development. The eggs incubate in water, and the transparent egg case permits direct observation of the embryo inside. In this investigation, you will observe development in the embryos of the frog.

Objectives

- *Observe* embryo development in the frog as an example of animal development.
- *Recognize* the processes of cleavage, growth, and differentiation during development.

Materials

fertilized frog eggs	plastic wrap
culture dish	thermometer
spoon	stereomicroscope

Procedure  *

A. Pour about 1 ml of water that has stood overnight at room temperature into a culture dish. With a spoon, transfer a few frog eggs from the main culture dish to a small culture dish.

B. Using your microscope, study the eggs. A fertilized egg will show a dark area on the uppermost part of the egg. Refer to the figure below to identify the embryos' approximate stages of development. Since the eggs were fertilized before shipping, the earliest stages of development have already occurred.

C. In a chart like the one above, sketch the embryo and record the time and date. Measure and record the temperature of the culture water.

*For detailed information on safety symbols, see Appendix B.

Date	Time	Stage	Temp. (° C)	Notes	Sketch

D. When you have completed your observations, store the eggs in an area not exposed to direct sunlight or to extremes of temperature.

E. Observe the embryos every day for at least one week. Each day make a sketch, and record date, time, temperature, and observations in your table. After about one week, tadpoles will have developed. If you continue your observations, feed the tadpoles according to your teacher's instructions.

Analysis and Conclusions

1. What is the major change that occurs in the first stages of embryonic development?

2. After how many days did gastrulation begin?

3. At what stage does differentiation become evident?

4. How does the embryo obtain oxygen before its gills develop?

5. How do you think development would be affected if the eggs were incubated at a lower temperature?

6. How does the embryo develop to the tadpole stage without ingesting food?

7. How do you think frog embryos could be used to help determine if a drug taken by a pregnant woman will harm her developing child?

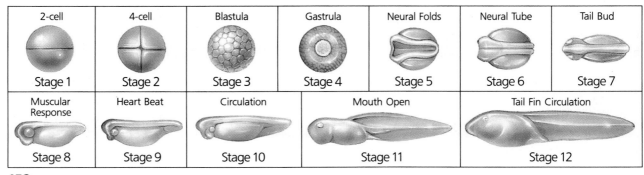

Chapter Review

Study Outline

22–1 Embryonic Development

■ Development is a complex series of events by which a single cell becomes an adult, multicellular organism. Development of all animal embryos includes cleavage, growth, and differentiation.

■ Cleavage results in the formation of a blastula—a single layer of cells surrounding a fluid-filled cavity called the blastocoel. The arrangement of cells in the blastula depends on the amount and distribution of yolk in the egg.

■ At the end of the blastula stage, gastrulation occurs, leading to the formation of a two-layered embryo called the gastrula. The blastopore (opening) of the gastrula later becomes one of the openings to the adult digestive system.

■ The three cell layers of the gastrula—ectoderm, mesoderm, and endoderm—make up the germ layers. Unspecialized embryonic cells grow and differentiate into specialized cells, tissues, and organs.

■ Differentiation is controlled by the DNA of the chromosomes, parts of which are switched on and off by substances in the cytoplasm. Interactions with neighboring cells also affect differentiation of cells.

22–2 External and Internal Development

■ Development in animals may be external or internal. All embryos need nourishment, proper temperature, oxygen, protection, and a means of getting rid of wastes.

■ In most aquatic animals, fertilization and development are external. The eggs do not have shells. Conditions in the water are usually adequate for proper development.

■ Birds, reptiles, and a few mammals produce eggs with shells that develop externally. The egg with its shell provides protection and nourishment for the externally developing embryo.

■ Development and fertilization are internal in mammals. In placental mammals, exchange of substances between the mother and developing young occurs in the placenta. The mother provides nutrients and oxygen, and the embryo releases carbon dioxide and other wastes. The umbilical cord connects the embryo to the placenta.

■ After the young of egg-laying mammals have hatched, they feed on milk produced by the mother's mammary glands. The young of pouched mammals undergo a period of internal development. Once this period has ended, the young finish their development in the mother's external pouch.

Vocabulary Review

development (441)	germ layers (444)
embryo (441)	differentiation (444)
cleavage (442)	embryonic induction (446)
morula (442)	
blastula (442)	extraembryonic membranes (448)
blastocoel (442)	chorion (448)
gastrulation (443)	allantois (448)
gastrula (443)	amnion (448)
blastopore (443)	yolk sac (448)
ectoderm (443)	uterus (449)
endoderm (443)	placenta (450)
primitive gut (443)	umbilical cord (450)
mesoderm (443)	

A. Multiple Choice— Choose the letter of the answer that best completes each statement.

1. The human embryo receives food and oxygen from the mother through the (a) extraembryonic membranes (b) placenta (c) primitive gut (d) yolk sac.

2. After fertilization, the zygote begins a series of cell divisions known as (a) gastrulation (b) embryonic induction (c) cleavage (d) differentiation.

3. The inner layer of cells in the gastrula is called the (a) ectoderm (b) endoderm (c) blastopore (d) mesoderm.

Chapter Review

4. The fluid-filled sac surrounding the chicken embryo is called the (a) yolk sac (b) allantois (c) germ layers (d) amnion.

5. The cavity within the gastrula that eventually becomes the digestive system is called the (a) uterus (b) mesoderm (c) primitive gut (d) blastocoel.

6. Unspecialized embryonic cells become specialized cells, tissues, and organs through the process of (a) cleavage (b) embryonic induction (c) allantois (d) differentiation.

7. In placental mammals, the embryo is attached to the placenta by the (a) umbilical cord (b) uterus (c) allantois (d) amnion.

B. Definitions—Replace the italicized definition with the correct vocabulary term.

8. The *opening in the gastrula created by the gastrulation process* later becomes one of the openings of the digestive system in the adult organism.

9. The *outermost membrane surrounding the chicken embryo* aids in the exchange of gases.

10. The *early stage of development in which the embryo consists of a solid ball of cells* results from cleavage of the fertilized egg.

11. In the eggs of reptiles and birds, the *four membranes that lie outside the embryo but inside the shell* perform a number of important functions.

12. During *the process in which the cells on one side of a blastula move in to form a two-layered gastrula*, a blastopore is formed.

13. In placental mammals, the *multicellular organism in the early stages of development* obtains nutrients and oxygen through the placenta.

14. All the tissues and organs of multicellular animals arise from the *three embryonic cell layers*, ectoderm, mesoderm, and endoderm.

Content Review

15. Explain how cleavage reduces the size of the cells of an embryo.

16. Briefly describe the steps leading to the formation of a blastula. Describe the blastula's structure, including its interior.

17. Describe how the arrangement of cells in the early development of human embryos differs from that in frog embryos. Explain why this occurs.

18. Describe the process of gastrulation.

19. Why does embryo size increase during gastrulation but not during cleavage?

20. How are the two openings of the digestive system formed?

21. Describe the location of each of the three germ layers in the gastrula.

22. How do the many different kinds of cells in an organism arise from the same DNA?

23. Describe the eye tissue experiment conducted to determine that embryonic cells are influenced by neighboring tissue.

24. Explain the role of embryonic induction in cell differentiation.

25. How do eggs that develop in water obtain nourishment and oxygen?

26. What are the advantages of a egg with a shell for external development on land?

27. It is known that fish produce more eggs than birds. Why do fish produce more?

28. Diagram a bird's egg containing a developing embryo. Label and describe the function of each part.

29. Explain how exchange of gases within the bird egg takes place.

30. What is the function of the amnion that surrounds the bird embryo?

31. How does a bird embryo obtain nutrition?

32. Compare fertilization and development in mammals with that in birds.

33. What are the functions of the placenta?

34. Describe embryonic development in egg-laying mammals.

35. Describe embryonic development in pouched mammals.

Graphic Organizing

For information on graphic organizers, see Appendix G at the back of this text.

36. **Scale:** Construct a scale for the gestation periods of various mammals using the information in the table on page 455. Note that some gestation periods are given in days, some in months.

Mammal	Gestation Period
antelope	9 months
beaver	3 months
deer	7 months
dog	60 days
elephant	21 months
hog	114 days
horse	11 months
lion	108 days
rat	22 days
sheep	5 months
sperm whale	16 months
zebra	11 months

Critical Thinking

37. Discuss the similarities and differences in fertilization, egg structure, and egg-laying patterns of frogs, reptiles, and birds. (*Comparing and Contrasting*)

38. What conclusion would J. G. Gurden have drawn if his frog eggs had not developed into normal frogs but into intestinal cells? (*Identifying Reasons*)

39. Arrange the factors that affect differentiation in order from most to least important. (*Ordering*)

40. What might happen to the embryo in a chicken egg if the amnion were not present? (*Predicting*)

Creative Thinking

41. Design a simple experiment to locate the part of a developing frog embryo that acts as an organizer for the neural tube.

42. Imagine that you have found two birds' nests—one on the ground, with two large speckled eggs in it, and one in a tree, with five small white eggs in it. Draw four conclusions about the two bird species and support each conclusion.

Problem Solving

43. In experiment A, a laser was used to randomly destroy half the cytoplasm in 100 fertilized frog eggs. The frog eggs were then observed. Some eggs survived and developed normally. Others survived but did not develop normally. In experiment B, a similar technique was used on another 100 fertilized frog eggs.

However, a special region of the cytoplasm was not destroyed in any of the embryos. The results of both experiments are given in the table below. For each experiment, calculate the percentage of eggs that survived, the percentage of surviving embryos that developed normally, and the percentage of surviving embryos that did not develop normally. What best accounts for the difference in the number of embryos that did not develop normally? Design a control for the experiments.

	Experiment A	Experiment B
number of eggs that survived	63	54
number of embryos that developed normally	15	54
number of embryos that did not develop normally	48	0

44. An embryologist proposed the hypothesis that in the sea urchin embryo, cells of the intestinal lining have all the same genes as cells of the brain. Design an experiment to test this hypothesis.

45. A scientist destroyed one of the cells of a two-celled frog embryo with a probe. The remaining living cell produced an abnormal organism. Another scientist separated the two cells of a two-celled sea urchin embryo with a very fine silk thread. Each cell produced a small but complete organism. What conclusion do you think each scientist reached? Explain why the scientists got different results. What further experiments should be done to determine which conclusion is correct?

Projects

46. Do library research on the effects of drugs on the embryonic development of animals such as sheep and cattle that are grown for human consumption. Write up your findings in the form of a feature article for a magazine or newspaper.

47. *Salmonella,* a tiny microorganism that can infect chicken eggs, causes food poisoning in humans. Interview a local poultry farmer to find out how eggs become infected and what measures farmers take to prevent *Salmonella* infection. Present your findings to the class in an oral report.

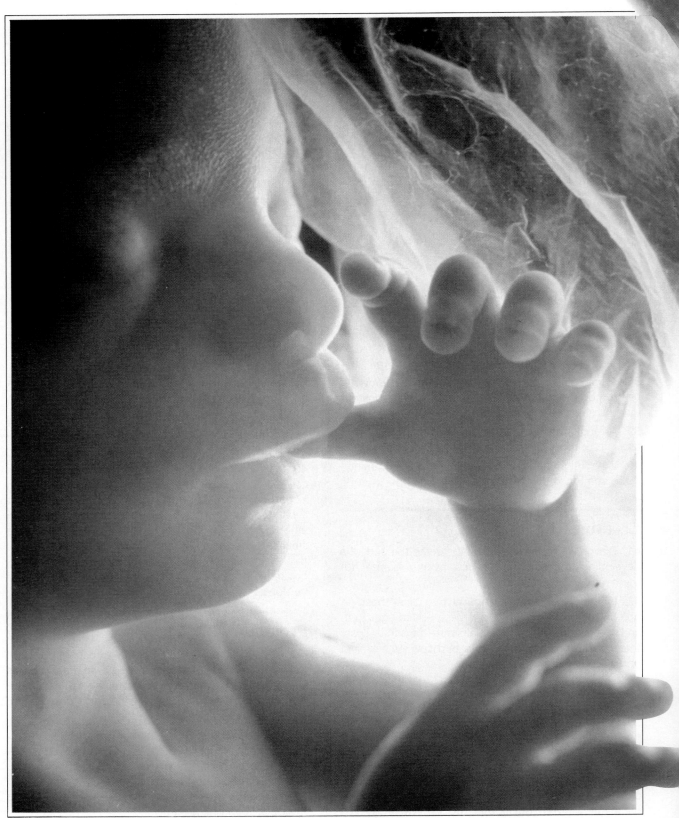

▲ A human fetus, only four and a half months old, develops within its mother.

Human Reproduction

23

Protected from the bustle of the outside world, a human fetus develops inside its mother, preparing for the day it will emerge as a squirming, crying baby. Before the fetus reached this point, a sperm and an egg were produced and united. Changes took place in the mother's body that allowed it to nourish and accommodate the growing fetus. And, in several more months, a baby will be born. In this chapter, you will learn about the body systems and processes involved in human reproduction—the process that brings a new person into the world.

Guide for Reading

Key words: uterus, menstrual cycle, implantation, gestation, labor, fraternal and identical twins

Questions to think about:

How and where do human sperm cells develop?

What are the stages of the menstrual cycle in the human female?

How is the human fetus nourished before birth?

23-1 Human Reproductive Systems

Section Objectives:

- *List* the major structures of the male and female reproductive systems and explain the functions of each structure.
- *Describe* the male and female secondary sex characteristics and name the hormones involved in their development.
- *List* the stages of the menstrual cycle and explain the role of hormones in the cycle.

Laboratory Investigation: *Observe* the microscopic structure of mammalian ovaries and testes (p. 468).

You have inherited many of your physical characteristics and your personality traits from your parents and, through time, from your grandparents and great-grandparents. In order to understand human reproduction, you first need to know how the male and female reproductive systems work.

The Male Reproductive System

The male gonads are the **testes.** The testes make sperm cells, which are the male gametes, and the male sex hormone *testosterone*. Testosterone causes the development of the **secondary sex characteristics** in the male. These are physical traits that usually

▲ **Figure 23–1**

Inherited Traits. Hair and eye color, as well as many other traits, are inherited from parents.

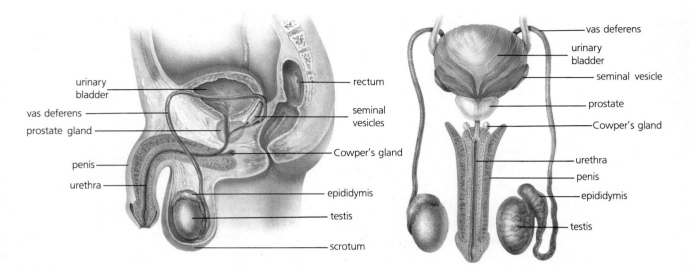

Figure 23–2 ▲

Reproductive System of Human Male. In the male, the testes produce sperm cells and the sex hormone testosterone. Fluid produced by the seminal vesicles, prostate gland, and Cowper's glands mix with sperm to form semen.

appear during adolescence but are not directly involved in reproduction. The secondary sex characteristics of men include body hair, muscle development, and a deep voice.

The male reproductive system is shown in Figure 23–2. A pair of testes hangs outside the body wall in a sac of skin called the **scrotum.** The scrotum keeps the testes at a temperature slightly lower than the rest of the body. This is necessary for the production and storage of sperm. If the testes become too warm, muscles in the scrotum relax, allowing the testes to fall away from the body. If the testes become too cool, the scrotal muscles pull them closer to the body.

The testes actually form inside the body during development and move down into the scrotum only in the last month before birth. Sometimes, one or both testes do not pass into the scrotum. Testes that stay within the body cannot make sperm because the temperature is too high. In some cases, however, they can be moved surgically or with hormone treatment.

Each testis is made up of small, coiled tubes called the *seminiferous* (sem uh NIF uh rus) *tubules.* There are 300 to 600 tubules in each testis. Immature sperm are made in the seminiferous tubules. From there, the immature sperm pass to the **epididymis** (ep uh DID uh mis), a storage area on the upper rear part of each testis. In the epididymis, the sperm mature, which takes about 18 hours. They leave the epididymis through the **vas deferens** (vas DEF uh renz), a tube that leads upward from each testis into the lower part of the abdomen.

The two vas deferens empty into the **urethra,** the passageway for the excretion of urine. In mammals, it is also the passageway through which sperm leave the body. In the human male, the urethra passes through the penis to the outside of the body. As sperm enter the urethra, the *seminal vesicles, Cowper's glands,* and the *prostate* (PRAHS tayt) *gland* all secrete fluids into the urethra. These fluids nourish the sperm and protect them from the acidity of

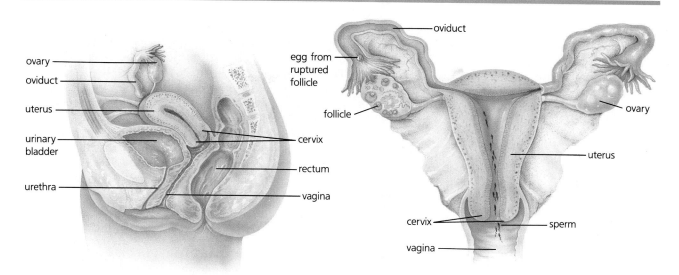

ovary
oviduct
uterus
urinary bladder
urethra

cervix
rectum
vagina

oviduct
egg from ruptured follicle
follicle
ovary
uterus
cervix
sperm
vagina

the female reproductive tract. The mixture of sperm and fluids is called **semen** (SEE men). Involuntary muscular contractions force the semen through the urethra and out of the body in a process called **ejaculation** (ih jak yuh LAY shun). For a short time before, during, and after ejaculation, reflex actions keep the outlet of the urinary bladder closed. This prevents urine from entering the urethra and mixing with the semen.

▲ **Figure 23–3**

Reproductive System of Human Female. In the female, the ovaries produce the female sex hormones estrogen and progesterone. Eggs mature in the follicles in the ovaries. When a follicle breaks, the egg is released and drawn into the oviduct, where fertilization may occur.

The Female Reproductive System

The female gonads are the **ovaries.** The ovaries make eggs, which are the female gametes. The ovaries also secrete the female sex hormone *estrogen.* Estrogen causes the development of female secondary sex characteristics and plays a large role in the menstrual cycle, which you will study later in this section. Female secondary sex characteristics include breasts, a broadened pelvis, and the distribution of body fat.

The female reproductive system is shown in Figure 23–3. There are two ovaries. These are found in the lower part of the abdomen. They are about four centimeters long and two centimeters wide. Each ovary contains about 200 000 tiny egg sacs called **follicles.** In each follicle, there is an immature egg. These immature eggs are present at the time of birth. During the life of a female, no more than 500 eggs mature.

When an egg matures, its follicle moves to the surface of the ovary. The follicle then breaks, releasing the egg as shown in Figure 23–3. This process is called **ovulation** (ahv yuh LAY shun). An egg can be fertilized for about 24 hours after ovulation.

Near each ovary, but not connected to it, is an **oviduct** (OH vuh duhkt), or *Fallopian* (fuh LOH pee un) *tube.* The oviduct is a tube with a funnel-like opening. Cilia lining the oviduct create a current that draws the released egg into the tube. The oviduct is where the egg may be fertilized if any sperm are present.

From the oviduct, the egg passes into the **uterus** (YOOT uh rus), a thick-walled, muscular, pear-shaped organ. If the egg has been fertilized, it finishes its development in the uterus, attached to the uterine wall. The narrow neck of the uterus is called the **cervix** (SER viks). The cervix opens into the **vagina** (vuh JY nuh), or *birth canal,* which leads to the outside of the body. At birth, the baby leaves the mother's body through this passageway.

In the early stages of the human female embryo, the vagina joins the urethra, as does the vas deferens in the male. During later development, however, a second opening is formed for the vagina. Thus, unlike the male, in the mature human female, the urinary and reproductive tracts are completely separate.

The Menstrual Cycle

Characteristics of the Menstrual Cycle

In the human female, a mature egg develops and leaves one of the ovaries about every 28 days. At this time, the wall of the uterus has thickened with a rich supply of blood vessels and is prepared to accept a fertilized egg for development. If the egg is not fertilized, the built-up portion of the uterine wall breaks down and along with the unfertilized egg, passes from the body. Then another egg matures, and the buildup of the uterine wall begins again. This cycle is known as the **menstrual** (MEN struhl) **cycle,** shown in Figure 23–4. The changes that take place involve the interaction of hormones made by the hypothalamus, pituitary gland, and ovary.

The menstrual cycle begins at puberty, which usually takes place in human females sometime between the ages of 10 and 14. The cycle stops during the time when a woman is pregnant. It stops permanently sometime in middle age, usually between the ages of 45 and 50. The permanent stopping of the menstrual cycle is called *menopause* (MEN uh pawz).

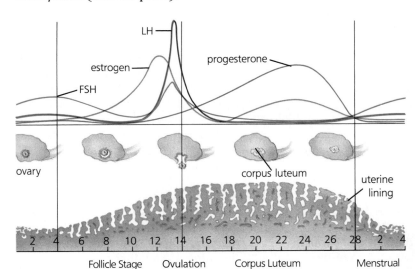

Figure 23–4

The Menstrual Cycle. During the menstrual cycle, the lining of the uterus thickens in preparation for a fertilized egg. The cycle is controlled by four interacting hormones and repeats itself about every 28 days. ▶

Stages of the Menstrual Cycle   The female human menstrual cycle can be divided into four distinct stages, of which menstruation is the final stage.

The first stage is the follicle stage. The pituitary gland secretes follicle-stimulating hormone (FSH), which you read about in Chapter 16. FSH causes several follicles in the ovary to begin developing. Usually, only one follicle matures. As the follicle develops, it secretes estrogen. The estrogen stimulates the uterine lining to thicken with mucus and a rich supply of blood vessels. These changes in the lining prepare the uterus for a possible pregnancy. This stage lasts 10 to 14 days.

The second stage is called ovulation. A high level of estrogen in the blood causes the pituitary to decrease the secretion of FSH and begin the secretion of luteinizing hormone (LH). When the concentration of LH in the blood reaches a certain level, ovulation takes

Critical Thinking in Biology

Evaluating Secondhand Information

How do you know whether or not to believe something you read? Consider the following situation.

A magazine is lying open on a library table. The title of an article attracts your attention. You read:

Leanness and Infertility

The loss of fat because of exercise or dieting can result in infertility in women. This infertility can be reversed when fat is gained. Fat tissue may help determine whether human females can reproduce.

In the article, the author describes scientific research conducted over 15 years. Scientists found that a certain ratio of fat body tissue to lean body tissue is required for menstruating. When fat tissue decreased below that level, menstruation stopped. When fat tissue was gained, menstruation resumed.

Although you do not recognize the name of the magazine, you see on the contents page that it has been published for over 100 years. The magazine contains articles about science topics for general readers. Notes on the author of the article state that she reported medical news for more than 20 years.

1. How reliable would you judge the information in the article to be? Explain.

2. What additional information might help you to evaluate the reliability of the information in the article?

3. Suppose the article appeared in a tabloid that carried sensational news. How might this affect your judgment?

4. Think About It Describe your thought process as you answered question 1.

place. That is, at that point, the follicle breaks, releasing a mature egg. Ovulation usually takes place in about the middle of the menstrual cycle.

The third stage is the corpus luteum stage. After ovulation, LH causes the broken follicle to fill with cells, forming a yellow body called the **corpus luteum** (KOR pus LOOT ee um). The corpus luteum begins to secrete the hormone progesterone, which brings about the continued growth of the uterine lining. Because of its part in maintaining the uterine wall, progesterone is often called the hormone of pregnancy. It also stops the development of new follicles in the ovary by inhibiting the release of FSH. The corpus luteum stage lasts 10 to 14 days.

Menstruation is the fourth stage. If fertilization does not happen, secretion of LH decreases, and the corpus luteum breaks down. This causes a decrease in the level of progesterone. With a drop in the progesterone level, the thickened lining of the uterus is no longer maintained, and it breaks down. The extra layers of the lining, the unfertilized egg, and a small amount of blood pass out of the body through the vagina. This is called **menstruation** (men strou WA shun). It lasts from about three to five days. While menstruation is taking place, the amount of estrogen in the blood is dropping. The pituitary increases its output of FSH, and a new follicle starts to mature.

Humans and other primates are the only mammals that have a menstrual cycle. Other mammals have an *estrous cycle.* This cycle is marked by periodic changes in the female's sex organs and in the desire to mate, but there is little or no bleeding at the end of the cycle. Mammals such as foxes, coyotes, and wolves have one estrous cycle each year, whereas cats and dogs have two.

For most of the estrous cycle, the female is not fertile and will not mate. Mating takes place only when the female is fertile—when the uterus and associated structures increase in size in preparation for pregnancy. Ovulation takes place spontaneously during the fertile period, or after mating, depending on the species. During fertile periods, the female is said to be "in heat." If fertilization does not happen, the sex organs go back to their normal state.

23-1 Section Review

1. Name the male and the female gonads.
2. What glands add secretions to the sperm?
3. Define ovulation.
4. Name the four stages of the menstrual cycle.

Critical Thinking

5. If a woman's fallopian tubes are blocked, how is her body affected? *(Relating Parts and Wholes)*

23-2 Fertilization, Implantation, and Development

Section Objectives:

- *Summarize* the processes of fertilization and implantation in humans.
- *Explain* the roles of the placenta and umbilical cord in pregnancy.
- *Define* the following terms: *amniotic fluid, gestation period, labor, fraternal twins,* and *identical twins.*

The primary function of human reproductive systems is to create new individuals. This occurs in a process that begins with fertilization and ends with the birth of a baby.

Fertilization

In human mating, or sexual intercourse, hundreds of millions of sperm are ejaculated into the vagina. The sperm then pass through the cervix, up through the uterus, and into the oviducts. If an egg is passing down one of the oviducts at this time, fertilization—the fusion of a sperm and an egg nucleus—may take place. The egg secretes a chemical that attracts the sperm. One of the sperm breaks through the membranes surrounding the egg, and the sperm cell nucleus enters the cytoplasm of the egg cell. When this happens, the membranes around the egg change, stopping other sperm from entering the egg. The sperm nucleus fuses with the egg nucleus, resulting in the diploid cell called the zygote.

If a woman's Fallopian tubes are blocked or she cannot ovulate, a technique called **in vitro** (VEE troh) **fertilization** may be tried. In vitro fertilization is fertilization in a glass laboratory dish. An egg taken from the woman or from an anonymous donor is fertilized by sperm contributed by the woman's husband or, if the husband is infertile, an anonymous donor. Two days after fertilization, the zygote is placed in the woman's uterus, where it may become implanted and continue to develop.

Implantation and Development

After fertilization, the zygote undergoes cleavage and develops into a blastula as it moves down the oviduct toward the uterus. About 5 to 10 days after fertilization, the embryo enters the uterus. Within the uterus, the outer layer of cells of the embryo secretes enzymes that digest part of the thick lining of the uterus, and the embryo attaches itself at this spot.

The fastening of the embryo to the wall of the uterus is called **implantation** (im plan TAY shun). Implantation marks the beginning of **pregnancy,** the period during which the baby develops in the uterus. Sometimes, the embryo implants somewhere other than the uterus, such as in the oviduct or out in the abdomen. This is

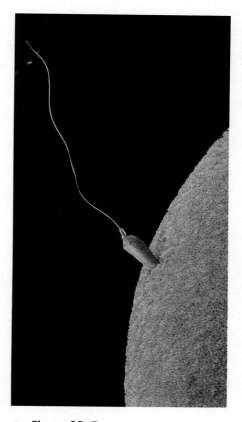

▲ **Figure 23–5**

Human Sperm Fertilizing an Egg. Fertilization occurs when a sperm unites with an egg.

Can You Explain This?

A young wife gave birth to her firstborn, a baby boy. At first, the baby appeared normal. Soon, however, a nurse noticed the newborn was showing signs of drug withdrawal. The baby had not been given any drugs since its birth.

The distraught mother was confused. She knew that her blood did not mix with the blood of the developing baby during her pregnancy. She had therefore assumed that the presence of drugs in her system would not affect her baby.

■ *Suggest an explanation for how this baby developed an addiction to drugs.*

called an *ectopic* (ek TAHP ik) *pregnancy*. Ectopic pregnancies usually end in the death of the embryo and sometimes cause severe bleeding inside the mother.

After implantation, the embryo undergoes gastrulation. The three germ layers are formed, and all the tissues and organs of the body develop from these layers by growth and differentiation. The developing human is called an *embryo* from the time of fertilization up until about eight weeks. After this time, the embryo is usually called a **fetus** (FEET us).

Extraembryonic Membranes

The human embryo develops the same extraembryonic membranes as birds and reptiles. However, these membranes play different parts in the development of humans.

In humans, the outer cell layer of the blastula becomes the **chorion**—the outermost of the extraembryonic membranes. As seen in Figure 23–6, the chorion completely surrounds the embryo and the other membranes. Small fingerlike projections called *chorionic* (kor ee AHN ik) *villi* form on the outer surface of the

Figure 23–6

Nourishment of the Fetus. Nutrients and oxygen diffuse from the mother's blood into fetal capillaries in the chorionic villi. Wastes from fetal blood diffuse into the mother's blood. Materials are transported to and from the fetus through blood vessels in the umbilical cord. There is no direct connection between the circulatory systems of the fetus and the mother. ▼

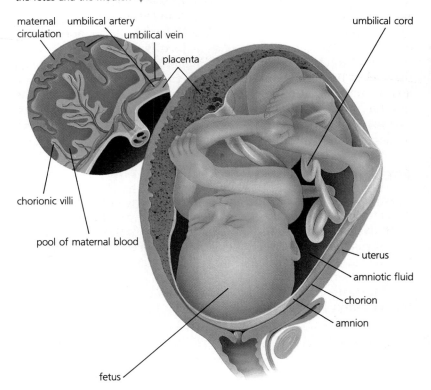

chorion and extend into the uterine lining. The chorionic villi and the uterine lining together form a temporary organ called the **placenta.**

The placenta allows the exchange of nutrients and wastes between the embryo and the mother. It also secretes a hormone that stops the menstrual cycle during pregnancy by preventing the breakdown of the corpus luteum. The corpus luteum goes on secreting high levels of progesterone, which, in turn, maintains the thickened wall of the uterus. The high progesterone level also prevents the development of new follicles in the ovaries. Thus, neither ovulation nor menstruation takes place during pregnancy.

In the human, the yolk sac and the allantois develop into the **umbilical cord.** This ropelike structure connects the developing fetus to the placenta.

The amnion is the innermost of the membranes. It completely surrounds the fetus within the chorion, like a balloon within a balloon. The amnion is filled with **amniotic fluid,** which protects the fetus, giving it a stable environment and absorbing shocks.

Nourishment of the Embryo   From the time of fertilization until implantation in the uterus, food stored in the egg nourishes the embryo. Soon after implantation, the embryo begins receiving food and oxygen from the mother's body through the placenta.

Fetal blood carrying wastes flows to the placenta through arteries in the umbilical cord. These arteries branch again and again, finally forming capillaries in the chorionic villi. The villi are surrounded by the mother's blood, which has been forced by arteriole blood pressure out of her blood vessels into spaces around the villi. Nutrients and oxygen in the blood of the mother diffuse into the fetus's blood in the chorionic villi. Wastes diffuse from the fetal blood into the mother's blood. However, the blood of the fetus and that of the mother do not mix. The enriched fetal blood then travels back to the fetus through veins in the umbilical cord. The waste materials are excreted by the mother. Thus, the placenta allows the fetus to make use of the mother's organ systems while its own are developing.

The placenta acts as a barrier that protects the fetus from some harmful substances in the mother's blood. However, many dangerous substances can pass through the placenta from the mother's blood to the fetus's blood. Viruses, such as German measles and AIDS, as well as nicotine, alcohol, and many drugs, can pass through the placenta and harm the fetus. It is the responsibility of the mother to avoid such harmful substances.

Birth

The length of pregnancy is called the **gestation** (jes TAY shun) period. The human gestation period is a little over 9 months. The gestation periods for other placental mammals are different. For example, the gestation period for a mouse is only 20 days. For an elephant, it is about 21 months.

Biology and You

Q: My aunt is pregnant and refuses to drink any alcohol. How would alcohol affect an unborn child?

A: Your aunt is taking every precaution to ensure a healthy baby. Women who drink while pregnant risk giving birth to children with irreversible physical, mental, and behavioral problems. When a pregnant woman drinks, the alcohol passes from her bloodstream into the baby's. If a mother drinks enough to get drunk, the baby also gets drunk. The unborn infant has a difficult time ridding its undeveloped body of the alcohol.

Alcohol in a baby's bloodstream can lead to Fetal Alcohol Syndrome, or FAS. Infants with FAS may suffer from such birth defects as low birth weight, malformed facial features, and heart defects. Later, they may experience delayed growth, mental retardation, or coordination difficulties.

There is only one sure way to prevent FAS and that is to refuse all alcoholic beverages while pregnant. Nobody knows exactly how much alcohol causes birth defects, so pregnant women should avoid alcohol completely.

Research and write a report on how either tobacco or drug use during pregnancy affects unborn babies.

Science, Technology and Society

Issue: Surrogate Motherhood

A surrogate mother bears a baby for others, usually for a fee. A couple may employ a surrogate if the wife cannot become pregnant or if pregnancy would endanger her health. Typically, the surrogate mother is artificially inseminated with sperm from the husband. She signs a contract to turn the baby over to the couple.

Those who favor surrogate motherhood say it allows infertile couples to have a child who is genetically related to the husband. They favor laws that make surrogate contracts legally binding. They also feel that surrogates should be carefully screened and counseled.

Other people believe that contracts should allow the mother time to change her mind after the baby is born. Some believe that a woman's body and baby should not be for sale. They fear that wealthy people may take advantage of poor women and that the children involved may suffer.

■ *Do you believe surrogate motherhood should be legal? Why or why not?*

When the human fetus is ready to be born, the uterine muscles begin slow, rhythmic contractions. This is called **labor.** At this time, the opening of the cervix begins to get larger. Its diameter must go from 1 or 2 centimeters to 11 or 12 centimeters before the baby can pass out of the uterus and into the birth canal. When the opening of the cervix has gotten large enough, the contractions force the baby, head first, from the uterus into the vagina and out of the mother's body. During labor, the amniotic membrane bursts, releasing the amniotic fluid. This fluid eases the passage of the baby through the birth canal.

When the baby passes out of the mother's body, the umbilical cord still connects it to the placenta. The umbilical cord is then tied and cut, leaving a scar on the baby called the *navel.* Shortly after the birth of the baby, additional uterine contractions expel the placenta and amnion, which are known as the *afterbirth.*

During pregnancy, progesterone and estrogen prepare the breasts for nursing. After birth, the pituitary hormone prolactin causes the mammary glands in the breasts to secrete milk.

While most pregnancies take place without difficulty, problems can occur. Delivery of a fetus before it is ready to be born is *premature birth.* Because premature babies are not fully developed, they are placed in an *incubator,* a special chamber designed to protect the baby until it is more developed. See Figure 23–7. Sometimes delivery through the cervix and vagina is not possible or is not safe for mother and baby. In these circumstances, the doctor will make an incision into the abdomen and uterus and deliver the baby by *cesarian section.*

Figure 23–7

A Premature Baby. Premature births most often occur during the last three months of pregnancy. This premature baby rests safely within an incubator. ▼

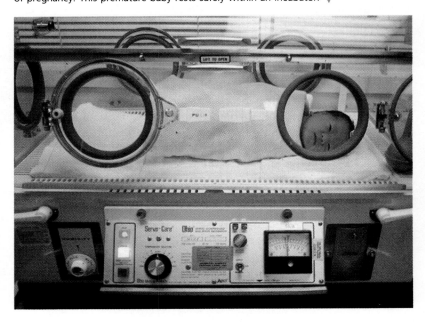

◀   **Figure 23–8**
Identical Twins. Identical twins develop from one fertilized egg and inherit identical traits.

Multiple Births

Once in a while, two eggs mature at the same time and pass into the oviduct during the same cycle. Each egg then may be fertilized by a different sperm. Both embryos may become implanted in the uterus and develop separately. Because the two embryos have received different hereditary material, they develop into people with different characteristics. Two people born from the same pregnancy are called *twins*. When they have developed from different eggs, they are **fraternal twins.** Fraternal twins may be of opposite sex and are no more alike than any two children of the same parents.

In other cases, a single fertilized egg divides into two embryos at a very early stage of development. The two individuals that then develop have the same hereditary makeup and are physically very much alike. Two individuals that develop from the same egg are called **identical twins.** Identical twins are always of the same sex.

Twins are the most common type of multiple birth. However, in rare cases, three or more embryos may form during the same pregnancy. Such multiple births may be fraternal, identical, or a combination of both types.

23-2 Section Review

1. In which part of the female reproductive system does fertilization generally occur?
2. What organ provides food and oxygen for the developing embryo?
3. Name the extraembryonic membranes that surround a fetus.
4. Define the term gestation.

Critical Thinking

5. Compare fraternal and identical twins. How are they alike? How are they different?   *(Comparing and Contrasting)*

Laboratory Investigation

Mammalian Gonads and Gametes

Gonads—testes and ovaries—are the organs of animals in which meiosis and the production of gametes take place. In this investigation, you will examine prepared slides of human gonads and gametes.

Objectives

- *Observe* the structure of human gonads and gametes.
- *Examine* the stages of human gamete development.

Materials

compound microscope
prepared slides of human testis, ovary, and sperm

Procedure *

A. Under low power, examine a cross section of a human testis. Locate an area showing well-defined circular structures. These are cross sections of seminiferous tubules. Look for cross sections of the small blood vessels lying outside the tubules.

B. Switch to high power. Using the figure below and Figure 21–11 in your text as guides, examine the internal structure of the tubules. The cells within each tubule are in various stages of development leading to sperm formation. Cells just inside the walls of the tubules are spermatogonia—the cells in which meiosis begins. Cells closer to the center of the tubules are at later stages of meiosis. The cells closest to the center of the tubules are differentiating into sperm cells. Look for sperm cells with flagella in some of the tubules.

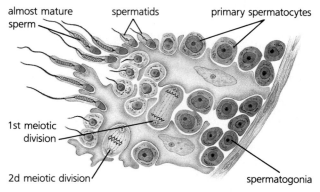

almost mature sperm
spermatids
primary spermatocytes
1st meiotic division
2d meiotic division
spermatogonia

* For detailed information on safety symbols, see Appendix B

C. Under low power, examine a prepared slide of human sperm cells. Switch to high power to observe individual sperm cells. Compare what you observe with Figure 21–12 in your text.

D. Obtain a prepared slide of a human ovary. Under low power, look for larger-than-average cells. Use the diagram below to help you. These cells, called oocytes, have a distinct nucleus surrounded by a lightly stained area of cytoplasm. Each one is at some stage of oogenesis. Refer to Figure 21–10 in your text.

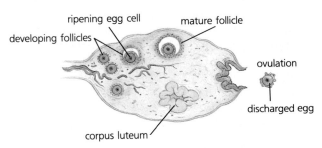

ripening egg cell
mature follicle
developing follicles
ovulation
discharged egg
corpus luteum

E. Each oocyte is surrounded by several layers of follicle cells that become more prominent as the egg matures. At later stages of oogenesis, a follicle enlarges, forming a fluid-filled cavity. Find the largest follicle. Using high power, examine the follicle and the egg in it. Try to see the layers of follicle cells and the nucleus of the egg. Draw and label a follicle and its oocyte.

Analysis and Conclusions

1. Where does meiosis I occur in the seminiferous tubules?

2. Based on your observations of the structure of the seminiferous tubules, where would you predict that new cells are produced by mitosis?

3. How does the structure of the testis allow for large numbers of sperm to be produced?

4. According to what you observed, how does the volume of egg cytoplasm compare to the volume of sperm cytoplasm?

5. By looking at cross sections of an ovary, what can you infer about the number of offspring produced by the organism at one time?

Chapter Review

Study Outline

23–1 Human Reproductive Systems

■ The male gonads, the testes, produce sperm and secrete testosterone. Testosterone causes the development of male secondary sex characteristics.

■ The female gonads, the ovaries, produce eggs and secrete estrogen. Estrogen causes the development of female secondary sex characteristics.

■ The menstrual cycle is a hormone-controlled monthly cycle in the human female. During the menstrual cycle, an egg is released from an ovary, and the uterine lining thickens in preparation for possible pregnancy.

■ Only humans and other primates have a menstrual cycle. Other mammals have an estrous cycle, which is marked by periodic changes in the female's sex organs and in the desire to mate.

■ The menstrual cycle is divided into four stages—the follicle stage, ovulation, the corpus luteum stage, and menstruation. The egg is released from the ovary at midcycle during ovulation.

■ If the egg is unfertilized, menstruation occurs, and the uterine lining is shed. If the released egg is fertilized, menstruation does not occur.

23–2 Fertilization, Implantation, and Development

■ After the egg is fertilized, it undergoes cleavage and forms a blastula as it passes down the oviduct into the uterus. In the uterus, the embryo attaches itself to the uterine wall in a process called implantation.

■ Following implantation, the embryo forms three germ layers. All of the tissues and organs of the body develop from these layers. The embryo also forms extraembryonic membranes, parts of which extend into the uterine lining to form the placenta.

■ Nutrients and oxygen from the mother's blood diffuse across the placenta into the fetus's blood. Wastes diffuse from the fetus's blood into the mother's blood.

■ The human gestation period lasts about nine months. At the end of this period, rhythmic contractions of the uterus push the baby into the birth canal and then out of the mother's body.

■ When two eggs are fertilized, the result is fraternal twins. Fraternal twins have different hereditary material. Identical twins result when one fertilized egg divides into two embryos. They have the same genetic makeup.

Vocabulary Review

testes (457)

secondary sex characteristics (457)

scrotum (458)

epididymis (458)

vas deferens (458)

urethra (458)

semen (459)

ejaculation (459)

ovaries (459)

follicles (459)

ovulation (459)

oviduct (459)

uterus (460)

cervix (460)

vagina (460)

menstrual cycle (460)

corpus luteum (462)

menstruation (462)

in vitro fertilization (463)

implantation (463)

pregnancy (463)

fetus (464)

chorion (464)

placenta (465)

umbilical cord (465)

amniotic fluid (465)

gestation (465)

labor (466)

fraternal twins (467)

identical twins (467)

A. Word Relationships—In each of the following sets of terms, three of the terms share a common characteristic. One term does not belong. Identify the term that does not belong.

1. oviduct, cervix, scrotum, ovaries

2. in vitro fertilization, corpus luteum, menstrual cycle, menstruation

3. epididymis, testes, semen, follicles

4. chorion, vas deferens, umbilical cord, placenta

5. uterus, cervix, placenta, vagina

6. menstruation, labor, gestation, pregnancy

7. fetus, chorion, urethra, amniotic fluid

Chapter Review

B. Definition—Replace the italicized definition with the correct vocabulary term.

8. *Two individuals who develop from the same fertilized egg* are always of the same sex.

9. The beginning of pregnancy is marked by *the fastening of the embryo to the wall of the uterus.*

10. Both sexes have *physical traits controlled by a sex hormone but not directly involved in reproduction.*

11. After fertilization, the zygote develops into a blastula as it moves down the *tube that carries the egg away from the ovary.*

12. In humans, the *ropelike structure that connects the fetus to the placenta* develops from the yolk sac and the allantois.

13. The *female gonads* produce eggs, which are the female gametes.

14. The *progesterone-secreting yellow body in the ovary* forms during the third stage of the menstrual cycle.

15. Sperm cells, the male gametes, are produced in the *male gonads.*

Content Review

16. What are the functions of the testes?

17. What are the secondary sex characteristics in males? What causes their development?

18. Describe the structure of a testis.

19. What are the functions of the seminal vesicles, Cowper's glands, and the prostate gland?

20. Trace the path of sperm from the testes until it leaves the body.

21. What are the functions of the ovaries?

22. What are the functions of estrogen?

23. Trace the path of an unfertilized egg from a follicle until it leaves the body.

24. In what ways does the estrous cycle differ from the menstrual cycle?

25. Describe what happens to the ovaries and uterus during each stage of the menstrual cycle.

26. How do hormones regulate the menstrual cycle?

27. What prevents more than one sperm cell from fertilizing an egg cell?

28. Describe the process of in vitro fertilization.

29. Trace the development of a zygote from fertilization through implantation.

30. What is an ectopic pregnancy?

31. Describe the structure of the chorion.

32. What are the amnion and amniotic fluid? What are their functions?

33. How does the embryo obtain nutrients and get rid of wastes?

34. Briefly describe what happens during the birth of a baby.

35. What is the afterbirth?

36. How and why do fraternal twins differ from identical twins?

Graphic Organizing

For information on graphic organizers, see Appendix G at the back of this text.

37. **Bar Graph:** Mortality rates are reliable indicators of a nation's health. The infant mortality rate (IMR) is the number of deaths of children under one year of age per 1000 live births. Using the data table below, construct a bar graph of infant mortality rates for the countries listed. Graph the data for 1979 and 1989 in two separate colors. According to your graph, which three countries have experienced the greatest drop in infant mortality rates over the last ten years? The 1989 infant mortality rate for Brazil is 63.0. What factors might contribute to such a high rate?

Country	Infant Mortality Rate	
	1979	**1989**
Canada	14.0	7.9
Finland	12.0	5.8
France	11.0	7.6
Ireland	16.0	7.4
United Kingdom	14.0	9.0
United States	14.0	9.9
West Germany	16.0	9.3

Critical Thinking

38. Some scientists believe semen and amniotic fluid are adaptations by organisms to life on land. What reasons can you give to support this? (*Identifying Reasons*)

39. How would failure of the corpus luteum to develop affect the rest of a menstrual cycle? (*Predicting*)

40. Sperm can survive for about 24 hours in the female body, yet they survive much longer in the male testes. Explain why. (*Identifying Causes*)

41. The women of a South African tribe often nurse their children until they are four or five years old, because their diet provides no soft food. As in all primates, ovulation is usually less likely during nursing. What effect would this prolonged absence of ovulation have on the tribal population? (*Predicting*)

Creative Thinking

42. Some men and women are unable to have children. Suggest some possible explanations for their infertility.

43. What could prevent the embryo from implanting in the uterus? Could it survive and develop anyway?

Problem Solving

44. In order to investigate the effect of temperature on sperm production, a scientist conducts the following experiment: A male dog is placed in a room at a temperature of 40° C. After two weeks, tests show that the dog is sterile. The scientist concludes that high temperatures stop sperm production, thereby causing sterility. Is this a valid conclusion? How would you improve this experiment?

45. Researchers conducted an investigation to determine the effect of secondhand cigarette smoke on pregnant women. A group of 1366 nonsmoking women participated in the study. The researchers divided the group into women whose husbands smoked and those whose husbands did not smoke. Data were gathered on the average weight of their babies at birth and on the number of miscarriages. Using the data presented in the table below, calculate the percentage of miscarriages for each group. Next, calculate the percentage difference in average birth weight between the two groups. What conclusions can you draw about the effects of secondhand smoke on pregnant women?

Effects of Secondhand Smoke on Pregnant Women		
	Wife Nonsmoker/ Husband Nonsmoker	Wife Nonsmoker/ Husband Smoker
Number of couples	837	529
Number of miscarriages	92	93
Average weight of baby at birth	3.2 kg	2.9 kg

Projects

46. Conduct library research on in vitro fertilization. Include information on what procedures are involved and how successful the technique is. Present your findings orally to the class.

47. Abnormal development of a fetus may be caused by smoking, by certain drugs such as alcohol, and by some diseases such as the German measles. Conduct library research on various types of fetal abnormalities and how they are caused. Include information on medical treatment for these abnormalities. Write up your findings in the form of a feature article for a magazine or a newspaper.

48. Prenatal care is essential to maintaining the health of the mother and the developing baby. Interview a nurse-midwife or obstetrician to determine exactly what is involved in prenatal care. Include specific information on what may be beneficial or harmful to the mother and the fetus. Summarize your findings on a poster for your class.

▲ While gathering nectar in a hibiscus flower, this honey bee becomes dusted with pollen grains.

Sexual Reproduction in Plants

24

A bee hovers above a flower. Drawn by the scent and color of the blossom, it lands to gather nectar. Unavoidably, however, as the bee flits from flower to flower, pollen grains cling to its body. And so, the bee becomes an instrument of flower fertilization, carrying the fertilizing pollen from one plant to another. In this chapter, you will learn how sexual reproduction takes place in flowering plants and in other plants.

Guide for Reading

Key words: angiosperm, gymnosperm, alternation of generations, seed, pollen grain, ovule, double fertilization

Questions to think about:

📖 How do plant and animal life cycles differ?

📖 How does a seed differ from a spore?

24-1 Life Cycles of Nonseed Plants

Section Objectives:

- *Define* the terms *sporophyte* and *gametophyte*.
- *Explain* alternation of generations.
- *List* the reasons why green algae are considered to be relatives of land plants.
- *Describe* the life cycle of green algae, mosses, and ferns.

Sexual reproduction in plants, as in other organisms, involves two processes: meiosis and fertilization. Plant groups differ in when, where, and how in their life cycles these events occur.

The major land plant groups are the mosses, ferns, and seed plants. The seed plants, which evolved more recently than the mosses and ferns, include the **gymnosperms** (JIM noh spermz) and the **angiosperms** (AN jee oh spermz). Most gymnosperms are cone-bearing plants, such as the pines. Their seeds develop on the surface of the cone scales. In contrast, the angiosperms—the flowering plants—have seeds that develop within protective structures. At maturity, the protective structure is called a fruit.

In mosses and ferns, moisture, in the form of dew or rain, is necessary for sexual reproduction. In these plants, the sperm have flagella, and water is needed for the sperm to swim to the egg. Gymnosperms and angiosperms, on the other hand, have adaptations for fertilization that do not require water. Thus, seed plants can thrive and reproduce sexually on land under conditions that make sexual reproduction in mosses and ferns impossible.

▲ **Figure 24–1**

A Seed Plant. The major land plant groups include the mosses, ferns, and seed plants. This common milkweed is a seed plant.

Alternation of Generations

The life cycle of a plant switches back and forth, or alternates, between two different plant forms. One form is always haploid and produces gametes. The other form is diploid and produces spores. This alternating between haploid and diploid plant forms is called **alternation of generations.** Figure 24–2 shows alternation between the haploid and diploid generations.

The gamete-producing plant is called the **gametophyte** (guh MEET uh fyt). All the cells of the gametophyte are haploid (*n*). The spore-producing plant is called the **sporophyte** (SPOR uh fyt). All the cells of the sporophyte are diploid (*2n*). During the growth of the sporophyte, some of the cells undergo *meiosis* and form haploid spores. When the spores are released from the sporophyte, they grow by mitosis into haploid, multicellular gametophytes. Some cells differentiate to form haploid gametes. When two gametes combine during fertilization, a diploid zygote is created. The zygote develops into a young sporophyte—the embryo. Through mitosis, the young sporophyte grows into a multicellular, mature sporophyte, and the life cycle begins again.

Usually one generation is more obvious than the other. This generation is said to be the *dominant generation.* The plant of the dominant generation is larger and lives longer than the plant of the other generation. In all plants, except the mosses, the diploid sporophyte is the obvious, dominant generation.

Life Cycle of Algae

Most botanists believe that plants are related to green algae. They believe this because green algae and plants share certain features, such as similar chlorophylls, the storage of food as starch, and cell walls made of cellulose. In fact, in some species of green algae there even is alternation of generations.

Figure 24–2
Alternation of Generations in Plants. In plants, there is a haploid, gamete-producing generation (the gametophyte) and a diploid spore-producing generation (the sporophyte). The gametophyte and sporophyte are not two stages in the life of an individual plant but rather two individual plants of the same species. ▶

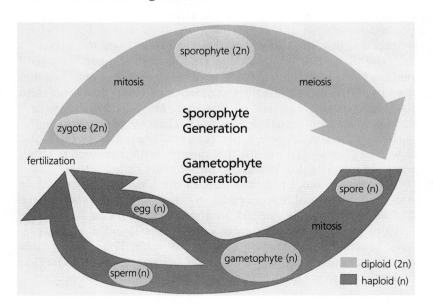

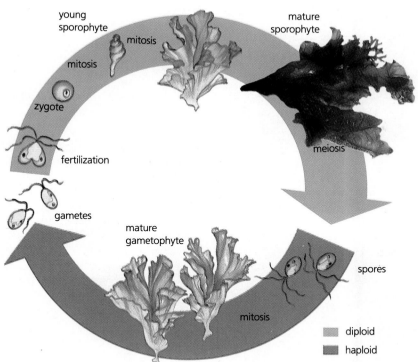

young
sporophyte

mitosis

mitosis

zygote

fertilization

gametes

mature
gametophyte

mitosis

mature
sporophyte

meiosis

spores

diploid

haploid

◀ **Figure 24–3**
Life Cycle of *Ulva*. In sea lettuce, or *Ulva*, the
haploid gametophyte and diploid sporophyte
generations resemble each other closely.

Figure 24–3 illustrates the life cycle of a green alga, the sea
lettuce *Ulva*. In this alga, alternation of generations happens. The
diploid sporophyte of *Ulva* is a multicellular structure that looks
similar to a lettuce leaf. Some of the cells of the sporophyte undergo
meiosis to produce haploid spores. These spores use their flagella
to swim away from the sporophyte. Eventually, the spores lose their
flagella and become attached to rocks. They then develop into
multicellular, leaflike gametophytes that are identical in form to the
sporophytes. Certain cells of the gametophyte produce haploid
gametes by mitosis. These gametes fuse to form a zygote, which
develops into a multicellular, leaflike sporophyte, thus completing
the life cycle.

Life Cycle of Mosses

The **mosses** are among the most primitive of plants. Mosses have
structures similar in function to the roots, stems, and leaves of
higher plants, but they are much simpler. The stemlike structure is
usually only several centimeters high. It is surrounded by small,
leaflike structures that are only one cell thick. Anchoring the plant
are rootlike structures called **rhizoids.** Unlike roots, the rhizoids
do not conduct water. Mosses do not have the specialized conducting tissues of higher plants, which makes the transport of materials
somewhat inefficient. Because transport is inefficient, mosses are
small and need a moist environment for growth and reproduction.
Therefore, they are found on the damp floors of forests, on shaded
rocks, and in swamps.

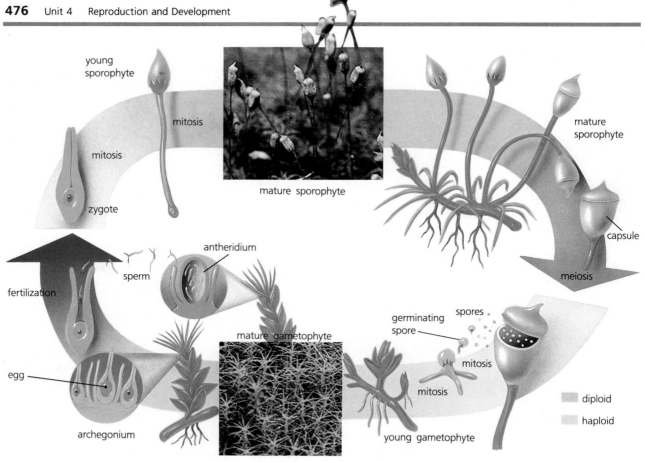

young
sporophyte

mitosis

mitosis

zygote

mature sporophyte

mature
sporophyte

capsule

meiosis

antheridium

sperm

fertilization

spores

germinating
spore

mitosis

egg

mature gametophyte

mitosis

archegonium

young gametophyte

diploid

haploid

▲ **Figure 24–4**
Life Cycle of Mosses. In mosses, the haploid,
or gametophyte, generation is dominant.

The gametophyte generation is dominant. The haploid gameto-
phyte generation is what you would recognize as the moss plant.
The sporophyte grows on the gametophyte and cannot live by itself.
Figure 24–4 illustrates the life cycle of mosses.

In some mosses, there are separate male and female gametophytes. In the male gametophyte, the reproductive organ is called
the **antheridium** (an thuh RID ee um). Within the antheridium,
sperm are produced. In the female, the reproductive structure is
called the **archegonium** (ar kuh GOH nee um). A single egg
develops within each archegonium. When sperm are released from
an antheridium, they swim through films of water to an egg in an
archegonium of the female plant.

Fertilization produces a diploid zygote that grows into a sporophyte. The sporophyte grows out of the archegonium. It is a single
leafless stalk that remains attached to the gametophyte and depends
on it for nourishment. At the tip of the mature sporophyte, a *capsule*
develops. Within the capsule, haploid spores are produced.
These spores are released into the air. When they germinate,
they form new haploid gametophytes, completing the life cycle.

Life Cycle of Ferns

As you can see in Figure 24–5, in **ferns,** the dominant generation
is the sporophyte. The diploid sporophyte has an underground
stem, called a *rhizome*, which grows just beneath the surface of

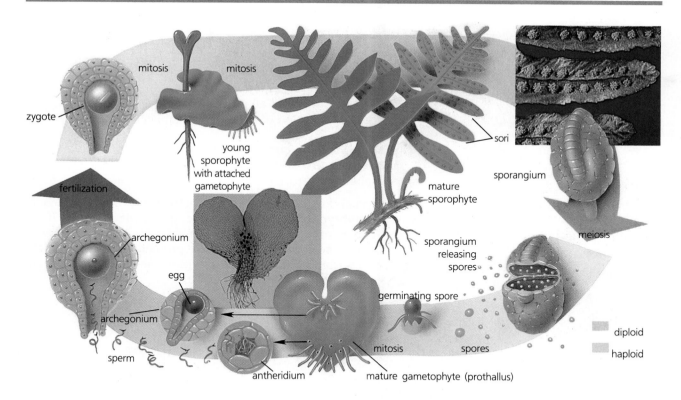

zygote
mitosis
mitosis
young sporophyte with attached gametophyte
fertilization
archegonium
egg
archegonium
sperm
antheridium
mature gametophyte (prothallus)
mitosis
spores
germinating spore
sporangium releasing spores
mature sporophyte
sori
sporangium
meiosis
diploid
haploid

▲ **Figure 24–5**
Life Cycle of Ferns. In ferns, the diploid, or sporophyte, generation is dominant.

the soil. On the lower surface of the rhizome, true roots anchor the plant and absorb water and minerals. Large leaves, called *fronds*, grow from the upper surface of the rhizome.

On the underside of some fronds are rows of small dots called *sori* (SOR eye). Within the sori, meiosis results in haploid spores, which are eventually released into the air and scattered by the wind. A germinating spore forms a small, heart-shaped gametophyte, called a *prothallus* (proh THAL us). A prothallus has both antheridia and archegonia in which gametes develop. The sperm released from the antheridia can fertilize an egg in an archegonium on the same gametophyte, or they can swim to and fertilize eggs in a nearby gametophyte. The fertilized egg then grows into a mature sporophyte, the fern plant.

24-1 **Section Review**

1. In plants, which generations produce gametes?
2. Name the female reproductive structure of the moss.
3. Which is the dominant generation in ferns?
4. What is the gametophyte of a fern called?

Critical Thinking

5. Compare and contrast the antheridium and the archegonium of a moss. (*Comparing and Contrasting*)

24-2 Life Cycles of Seed Plants

Section Objectives:

- *Describe* the life cycle of a gymnosperm.
- *Draw* a flower and label all of its parts.
- *Describe* the formation of male and female gametes in flowering plants.
- *Describe* pollination and fertilization in flowering plants.
  Laboratory Investigation: *Examine* and *identify* the structure of a plant (p. 488).

The *seed plants* are the most abundant of the plants. A **seed** is made up of the embryo, or young sporophyte, of the plant and its food supply. The embryo and its food are enclosed within a protective layer called the **seed coat.** The seed coat protects the embryo when it is released from the parent plant. Within the seed, the embryo has a ready supply of food for its growth.

The Life Cycle of Gymnosperms

The gymnosperms are seed plants that do not form flowers. The name gymnosperm, meaning "naked seed," refers to the development of seeds at the surface of, rather than enclosed within, the tissues of a reproductive structure. In most gymnosperms, called *conifers,* such as pine and spruce, seeds are borne on the scales of the reproductive structures known as cones.

Development of Gametes Figure 24–7 shows the life cycle of a gymnosperm. In these plants, the sporophyte generation is dominant. In many of the gymnosperms, the leaves are in the form of needles, and the reproductive organs are located on cones. Pine trees and other gymnosperms produce two types of cones. The male cone is called the *pollen cone.* The larger female cone is the *seed cone.* A single tree usually produces both pollen and seed cones. Spore-producing structures are found on the *scales* of cones. These scales are actually modified leaves or branches.

In a pollen cone, there are two spore cases, or *sporangia* (spaw RAN jee uh), on the underside of each scale. In each sporangium, many haploid spores are formed by meiosis. Within its spore wall, each haploid spore undergoes mitosis twice. The result is an immature, male gametophyte, called a **pollen grain.** Two of its four cells die, leaving a *tube cell* and a *generative cell.*

In the seed cone, two sporangia are found on the upper surface of each scale. Meiosis within each sporangium produces four haploid spores, three of which die. The remaining spore divides many times by mitosis forming a female gametophyte. The female gametophyte, sporangium, and associated structures make up an **ovule** (OHV yool). Each female gametophyte forms two or three archegonia, and within each of these, an egg cell develops.

Pollination The transfer of pollen grains to the vicinity of the female gametophyte is called **pollination** (pahl uh NAY shun). All

Figure 24–6

A Gymnosperm. Conifers dominate many temperate forests and may grow in particularly harsh environments. This lone pine grows at 7000 feet above sea level in Yosemite National Park. ▼

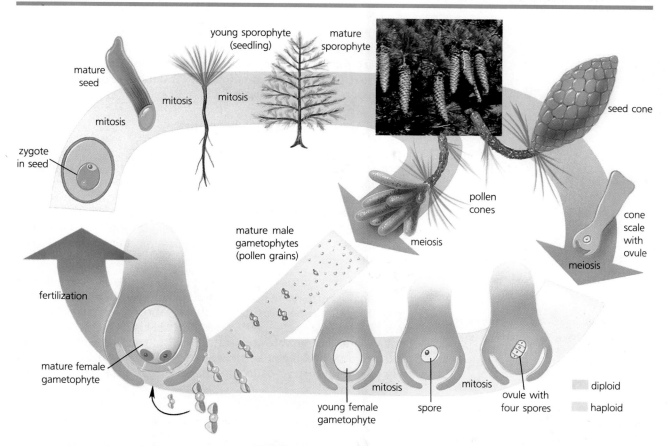

mature seed

young sporophyte (seedling)

mature sporophyte

mitosis

mitosis

mitosis

mitosis

seed cone

zygote in seed

pollen cones

cone scale with ovule

mature male gametophytes (pollen grains)

meiosis

meiosis

fertilization

mature female gametophyte

mitosis

mitosis

diploid

haploid

young female gametophyte

spore

ovule with four spores

seed plants have pollination as part of their reproduction cycle. In the gymnosperms, pollination occurs when the sporangia of mature pollen cones burst, releasing millions of pollen grains into the wind. Many of the pollen grains land on the cone scales of seed cones. Some land near a small opening on an ovule, which is known as the **micropyle** (MY kruh pyl). Although the ovule is pollinated, fertilization does not take place for more than a year. During this time, the pollen grain and the female gametophyte develop into mature male and female gametophytes.

As a pollen grain matures, its tube cell grows through the micropyle and into the ovule, forming the **pollen tube.** The pollen tube serves as a bridge between the pollen grain and the egg. At the same time, the generative cell of the pollen grain divides by mitosis to form two sperm cells. Meanwhile, in the female gametophyte, a single egg cell matures within each archegonium.

Fertilization When the pollen tube finally reaches the female gametophyte, growth stops. At this point, the tube cell nucleus degenerates, and two sperm pass from the tube into an archegonium. One of the sperm dies, and the other fuses with the egg. The resulting zygote develops into the plant embryo. The ovule, which now contains the plant embryo, develops into a seed. Eventually, the seed is released from the cone. Under favorable conditions, the seed will germinate, growing into a new sporophyte.

▲ **Figure 24–7**

Life Cycle of Gymnosperms. In most gymnosperms, the spore-bearing organs are in the form of cones and the leaves are in the form of narrow leaves called needles.

In the gymnosperms, the small gametophytes are totally dependent on the parent sporophyte for nutrition. Water is not necessary for fertilization because wind carries the male gametophyte close to the female gametophyte, and the sperm travels to the egg through the pollen tube.

Life Cycle of Angiosperms

The second group of seed plants is the angiosperms. The angiosperms, or flowering plants, are the most successful and abundant of the modern-day plants. In contrast to gymnosperms, the reproductive structures of the angiosperms are found within structures called flowers, and the seeds, while developing, are enclosed within a fruit. Like the gymnosperms, the sporophyte generation is dominant, and water is not needed for fertilization. Figure 24–8 shows the angiosperm life cycle.

Structure of a Flower   As you can see in Figure 24–9, the flower of an angiosperm is made up of rings of modified leaves on a specialized stem. Supporting the flower and connecting it to the stem of the plant is the **pedicel** (PED uh sel). The large end of the pedicel is the **receptacle** (ruh SEP tuh kul) to which the other flower parts are attached. Leaflike structures that form a ring around the base of the flower are called the **sepals** (SEEP ulz). They enclose

Figure 24–8

Life Cycle of Angiosperms. In angiosperms, the reproductive structures are found within the flowers. Seeds are enclosed within a fruit. ▼

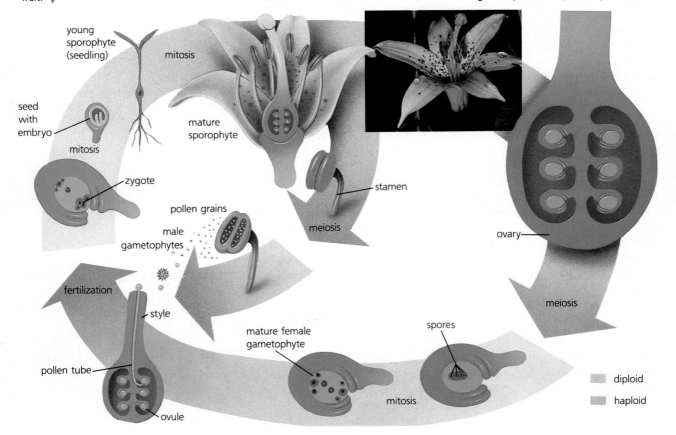

and protect the flower bud before it blossoms. The sepals may be small and green, or they may be large and brightly colored. The complete circle of sepals is called the **calyx** (KAY liks). Above the sepals are the **petals.** In some flowers, the petals are white; in others they are brightly colored. The complete circle of petals forms the **corolla** (kuh ROHL uh). The petals surround the reproductive organs of the flower.

The reproductive organs are in the center of the flower. These include the stamens and the pistil. The **stamens** (STAY menz) are usually called the male reproductive organs. They are located inside the corolla. A stamen is often made up of two parts—a stalklike **filament** and a saclike structure called the **anther** (AN ther) at the tip of the filament. Pollen grains are produced within the anthers. A flower may contain one or more stamens.

The **pistil** (PIS tul) is usually called the female reproductive organ of the angiosperms. It is located in the center of the flower. The pistil is made up of three parts. The top of the pistil is the **stigma** (STIG muh), which is an enlarged area that receives the pollen. Supporting the stigma is the **style.** At the base of the pistil is the expanded **ovary,** which contains the ovules. The ovary develops into the **fruit.** The ovules develop into seeds. A flower may have one or many pistils.

Stamens and pistils are the *essential organs* of the flower. The corolla and calyx are *accessory organs.* Some flowers contain stamens, pistil, corolla, and calyx, while others are missing one or more of these structures. *Pistillate* (PIS tuh layt) *flowers* contain pistils but not stamens, and *staminate* (STAM uh nayt) *flowers* contain stamens but no pistils. Plants that have only pistillate flowers are considered female plants, while those that have only staminate flowers are considered male plants. Some plants contain both pistillate and staminate flowers.

Development of Gametes   Haploid spores are produced by meiosis within the anthers of the stamens. See Figure 24–10. The spores undergo mitosis once, developing into pollen grains. These

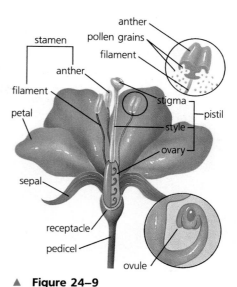

▲ **Figure 24–9**

Structure of a Flower. In a flower, the pistil is the female reproductive organ, and the stamen is the male reproductive organ.

Figure 24–10

The Stamen. The highly sculptured walls of two pollen grains are shown in the left photograph (Magnification 1400 X). In the right photo, the internal structure of a pollen grain is revealed. Note the two nuclei. (Magnification 4860 X).

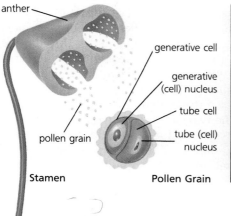

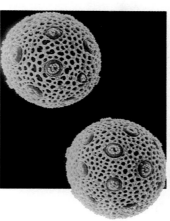

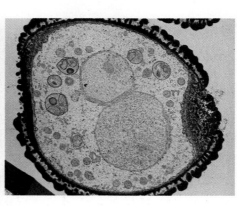

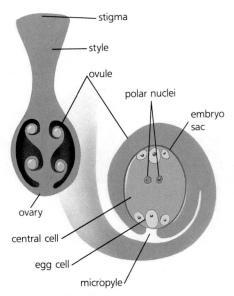

▲ **Figure 24–11**
The Pistil. After fertilization, the ovary develops into a fruit and each ovule develops into a seed.

Figure 24–12
Honeybee Gathering Nectar. As this honeybee gathers nectar at a crocus, pollen grains stick to its back. The next flower the bee visits will receive pollen from this flower. ▼

pollen grains are the young, male gametophytes. Two haploid cells are found within the thick, protective wall of each pollen grain. Like the pollen grain cells of the gymnosperms, one of the cells in the pollen grain is the tube cell, while the other is the generative cell. Once the pollen grains mature, the anther bursts, exposing the pollen to the air.

Every ovule within the ovary of the pistil has a micropyle. Every ovule also is attached to the wall of the ovary by a short stalk. In each ovule, meiosis of a single cell results in four haploid spores. Three of these spores die, and the remaining spore undergoes mitosis three times. The resulting female gametophyte, called an **embryo sac,** has only seven cells but eight haploid nuclei. See Figure 24–11. Two of the nuclei are found within a large central cell in the embryo sac. These are called the **polar nuclei.** The egg cell, which is near the micropyle, is surrounded by two cells. Three other cells are at the other end of the embryo sac.

Pollination In angiosperms, pollination is the transfer of pollen from an anther to a stigma. In some plants, pollen grains either fall or are transferred from an anther to a stigma on the same plant. This is known as *self-pollination.* When the pollen grains fall onto the stigma, the anthers usually are located above the stigmas. The transfer of pollen from the anthers of one plant to the stigma of another is called *cross-pollination. Artificial pollination* occurs when pollen is intentionally transferred by humans from one plant to another. Artificial pollination is used in plant breeding to produce plants with specific characteristics.

Cross-pollination may occur by wind, by animals or by water. This type of pollination is not as random as it seems. Flowers and flower parts are adapted for specific types of pollination. Flowers pollinated by animals are usually showy and/or give off an aroma to attract pollinators. These flowers often produce a sugary liquid, known as **nectar,** that pollinators use as food. As you can see in Figure 24–12, when pollinators gather nectar, the heavy pollen sticks easily to their bodies. Wind-pollinated flowers are not showy and produce no nectar or aroma. They do produce large amounts of light, loose pollen that is easily carried off by the wind. The stigmas of wind-pollinated flowers are expanded and feathery to help catch the wind-borne pollen.

Fertilization When a pollen grain reaches the stigma of a flower, it germinates. The protective coat of the pollen grain breaks open. As shown in Figure 24–13, a pollen tube grows down through the stigma and style and into the ovary. It then enters the ovule through the micropyle. The tube cell nucleus and the generative cell nucleus pass from the pollen grain down the pollen tube. As the generative nucleus moves down the pollen tube, it divides to form two haploid sperm nuclei. The two sperm nuclei enter the embryo sac. One fertilizes the egg cell to form a diploid zygote that develops into the sporophyte embryo. The other fuses with the two polar nuclei of the central cell to form a triploid (*3n*) *endosperm*

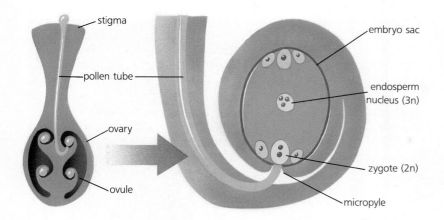

stigma

pollen tube

ovary

ovule

embryo sac

endosperm nucleus (3n)

zygote (2n)

micropyle

◀ **Figure 24–13**

Fertilization in a Flower. When a pollen grain lands on a stigma and germinates, a pollen tube grows from the pollen grain down through the style of the pistil to an ovule. The generative nucleus divides into two sperm nuclei. The tube nucleus directs the growth of the pollen tube. The sperm nuclei enter the ovule through the micropyle. One sperm nucleus fuses with the egg cell nucleus to form a diploid zygote. The other sperm nucleus fuses with the two polar nuclei, forming a triploid (3n) cell that develops into endosperm tissue.

nucleus. Because one sperm fertilizes the egg and the other fertilizes the two polar nuclei, the process is called **double fertilization.** Double fertilization is a unique characteristic of flowering plants. Following fertilization, the endosperm nucleus divides by mitosis to form the **endosperm.** Endosperm is the tissue that stores food for the developing plant embryo.

24-2 **Section Review**

1. List three examples of gymnosperms.
2. What is a pollen grain?
3. Name the female reproductive organ of the angiosperm.
4. What serves as a bridge between the pollen grain and the egg?

Critical Thinking

5. Classify the following as either wind-pollinated or animal-pollinated plants: pine tree, carnation, rose, ragweed, African violet, orchid, grass, lily, cactus. (*Classifying*)

24-3 Fruits and Seeds

Section Objectives:

- *Explain* the differences between monocots and dicots.
- *Describe* the formation of fruits and seeds in flowering plants.
- *Draw* the structure of a seed and explain the functions of the cotyledon, epicotyl, and hypocotyl.
- *Describe* several mechanisms for seed dispersal.

After fertilization, each ovule develops into a seed, and the ovary develops into a fruit. In angiosperms, the seeds are always found within the fruit, which protects them. The parts of the flower not involved in the formation of the fruit wither and die. The ovary grows larger, and its wall thickens. The wall of the ripened ovary may be hard or soft, dry or fleshy, and it may be made up of several

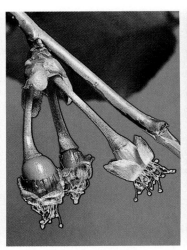

▲ **Figure 24–14**
Stages in the Development of a Cherry Fruit. In angiosperms, the fruit develops from the ovary.

Figure 24–15

Types of Fruits. A strawberry (top) is an aggregate fruit; a pineapple (center) is a multiple fruit; a grapefruit (bottom) is a simple fruit. ▼

separate layers. Figure 24–14 shows the development of a fruit from a fertilized ovary. If a flower has not been pollinated, a fruit usually does not form, and the flower withers and falls away.

Types of Fruits

There are thousands of different types of fruits. Many are rich in sugars, minerals, and vitamins. In fact, some of the foods we call vegetables are actually fruits. For example, many people think of the tomato as a vegetable, but it is, in fact, a fruit. Walnuts, pea pods, corn, and squash are all fruits.

Fruits are classified according to their origin. A fruit that develops from a single ovary is called a *simple fruit*. See Figure 24–15. Cherries and tomatoes are examples of simple fruits. When several ovaries are found within one flower, an *aggregate fruit* forms. Raspberries and strawberries are aggregate fruits. In some plants, such as the pineapple, the simple fruits of many separate flowers fuse together to form a *multiple fruit*.

Structure of the Seed

The seed, or ripened ovule, is made up of the seed coat, the embryo, and endosperm. The tough, protective seed coat develops from the wall of the ovule. On the outside of the seed coat is a scar called the *hilum* (HY lum), which marks where the ovule was attached to the ovary. The embryo develops by mitosis from the fertilized egg. The endosperm, which is a food storage tissue, develops by mitosis from the endosperm nucleus. The nutrients stored in the endosperm cells come from the parent plant.

All angiosperm embryos have at least one seed leaf, or **cotyledon** (kaht uh LEED un). In some plants, the endosperm is the only source of nourishment for the developing seedling. In other plants, however, the nutrients are stored in the cotyledons. The seeds of one group of angiosperms have only one cotyledon, and the plants that grow from these seeds are called **monocots.** Other angiosperms have two cotyledons and are called **dicots**. You can also recognize monocots and dicots by a number of other features, which are summarized in Figure 24–16. For example, in monocots, flower petals occur in groups of three, while in dicots, petals occur in groups of four or five.

In addition to one or two cotyledons, the plant embryo has three parts: the epicotyl, hypocotyl, and radicle. The part of the embryo above the point of attachment of the cotyledons is called the **epicotyl** (EP uh kaht ul). It usually gives rise to the terminal bud, leaves, and upper part of the stem of the young plant. The **hypocotyl** (HY puh kaht ul) is the part of the embryo below the point of attachment of the cotyledons but above the radicle. The **radicle** (RAD uh kul) is the lowermost part of the embryo, the embryonic root. In some plants, the stem forms entirely from the epicotyl, and the roots are formed from the hypocotyl and radicle. In other plants, the hypocotyl gives rise to the lower part of the stem, while the radicle gives rise to the roots.

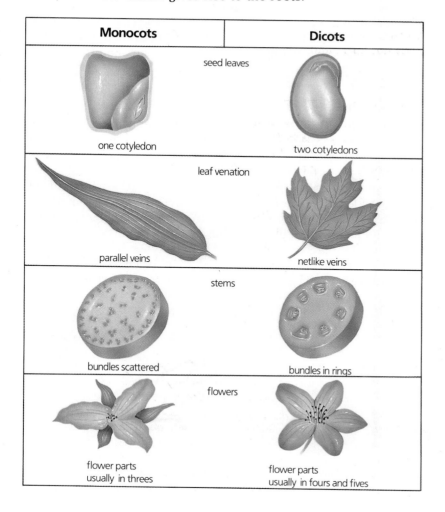

Monocots	Dicots
seed leaves	
one cotyledon	two cotyledons
leaf venation	
parallel veins	netlike veins
stems	
bundles scattered	bundles in rings
flowers	
flower parts usually in threes	flower parts usually in fours and fives

◀ **Figure 24–16**
Comparison of Monocots and Dicots.

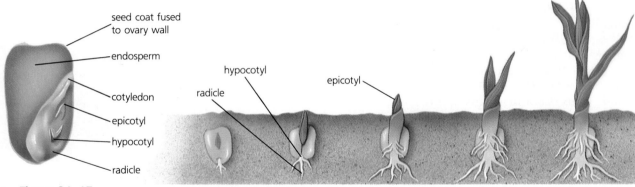

▲ **Figure 24–17**
Development of a Corn Seedling.

In a developing corn seedling, for example, the epicotyl gives rise to the stem and leaves, while the hypocotyl and radicle give rise to the roots. Figure 24–17 shows the development of a corn plant. Each kernel of corn is a single-seeded fruit. The embryo is partially surrounded by endosperm. The cotyledon stores food.

In the bean seed, which is a dicot, two large cotyledons make up most of the embryo. See Figure 24–18. In the mature bean seed, there is no endosperm, and nutrients are stored only in the cotyledons. In the developing bean seedling, the epicotyl gives rise to the terminal bud, the leaves, and the upper part of the stem. The hypocotyl gives rise to the lower part of the stem, and the radicle gives rise to the roots.

Seed Dispersal

The scattering, or dispersal, of seeds from the parent plant is important to the survival of the species. Plants that grow too close together must compete for water, minerals, and sunlight. Therefore, adaptations for seed dispersal help plants to survive. In some plants, pressure develops within the drying fruit. When the fruit bursts, seeds are released with enough force to scatter them over a large area. The snapdragon shows this type of dispersal. Many seeds and single-seeded fruits, such as those of milkweed, maple, and dandelion, are extremely light. These types of seeds can be carried great distances by the wind. Others, such as the coconut, float and are carried by water. Some seeds or fruits, such as those of

Figure 24–18
Development of a Bean Seedling. ▼

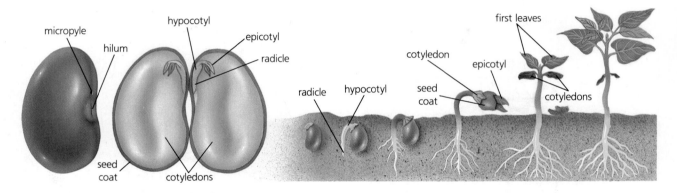

sandbur and wild carrot, have burs or hooks. When an animal brushes against them, they become attached to the fur of the animal, and they are carried away from the parent plant. Sweet, fleshy fruits are often eaten by birds and mammals. The seeds, which are usually indigestible, are later deposited elsewhere along with other digestive wastes.

Seed Dormancy

For a seed to begin to sprout, or germinate, it needs water, oxygen, and the proper temperature. Some seeds also require light. Many seeds go through a resting period before they begin to grow. During this time, growth is slowed, or stops altogether, and the seed is said to be in a state of **dormancy** (DOR mun see). The dormant seed will not sprout even if conditions are favorable. The length of the dormant period varies with the type of plant. Even within a single species, individual seeds show different lengths of dormancy. Some seeds may begin to grow after one year, some after two or three years, and so on. This characteristic is useful for the survival of the species, because not all the seeds will be killed by unusually harsh conditions during a given year.

There are several ways in which dormancy is brought about. In some species, dormancy occurs because the seed coat does not allow water and/or oxygen to reach the embryo. In others, the seed coat is so strong that the embryo cannot break through it. In still others, the embryo must undergo further development before growth can occur. Sometimes, chemical inhibitors that are present prevent germination. When dormancy is caused by the toughness of the seed coat, it lasts until the seed coat decays or is broken down enough for germination to occur. With immature embryos and chemical inhibitors, a certain amount of time must pass either for the embryo to mature or for the chemicals to break down.

In many species native to areas that have cold winters, seed dormancy is broken by a combination of exposure to low temperatures and moisture. Such a system makes sure that germination will happen only after the harsh conditions of winter have passed. Thus, the seeds will germinate in the spring, and the seedlings will have the whole growing season in which to complete their development.

24-3 Section Review

1. What kind of fruit develops from a single ovary?
2. What is a plant that has seeds with two cotyledons called?
3. What three parts of the embryo form the stem and roots?
4. How are the seeds of a dandelion dispersed?

Critical Thinking

5. What general characteristic would you expect the seed coat of a seed from a sweet, fleshy fruit to have? (*Predicting*)

Biology and You

Q: Every spring, I have sneezing fits because of hay fever. What causes hay fever?

A: Sneezing fits, itching eyes, runny nose, wheezing, all are symptoms of hay fever. Hay fever is an allergic response to pollen grains. During certain seasons, trees, flowers, shrubs, and grasses release pollen. If you are allergic to pollen, your body responds with the symptoms of hay fever.

People develop allergies because their bodies react to some substance. In the case of hay fever, your immune system releases antibodies to fight the invading pollen grains. As part of this response, certain immune cells release histamine, which causes sneezing, wheezing, and itching.

Having hay fever is like being allergic to the air you breathe. It is almost impossible to avoid breathing in pollen. Staying indoors can help. Also, you can avoid woods and fields where the pollen count is highest. A doctor may recommend antihistamines or decongestants to relieve symptoms. Serious cases of hay fever may require allergy shots.

■ *Interview some people who have hay fever. Questions should include: "When do you get it?" "What are the symptoms?" and "What do you do for relief?"*

Laboratory Investigation

Flower Structure

Different flowers vary in the number and type of their parts, as well as in color, shape, size, orientation, and other features. In this activity, you will dissect a variety of flowers to observe and compare their structure.

Objectives

- *Dissect* and *examine* the flowers of several species.
- *Observe* the structure of the parts of a flower.

Materials

tulip/lily flower	coverslips
geranium/rose flower	sucrose solution
dissecting needles	dropper
stereomicroscope	plastic bag
compound microscope	scalpel
slides	

Procedure *

A. Refer to Figure 24–9 in your text. Look at the center of one of your flowers. Note that the flower parts radiate out from the base of the flower in ring arrangements called whorls.

B. The bottom whorl is made up of sepals. In geraniums and roses, sepals look like small leaves. In tulips and lilies, they are large and look like petals. Record the number of the sepals for each flower in a chart like the one below. Break off a sepal, observe it under the stereomicroscope, and record its characteristics.

C. Look at the next whorl of flower parts—the petals. Record the number of petals. Break off a petal and examine it under the stereomicroscope. Record the color, fragrance, and other features of the petals.

D. Examine the stamens, which make up the next whorl. Record their number, arrangement, and other features.

E. Remove a stamen, and examine it with a stereomicroscope. Look at the pollen sacs and compare their appearance with Figure 24–10 in your text.

F. With a dissecting needle, tease some pollen grains from the anther onto a slide. Add a drop of sucrose solution, then put on a coverslip. With the compound microscope examine the pollen grains.

G. Place the slide in a plastic bag to prevent drying out and set it aside for a later step.

H. Locate the pistil. Some flowers have a single pistil, but others have several pistils. With a scalpel, cut the pistil from its attachment. Draw the pistil and record its features in your chart.

I. Using a scalpel, cut the pistil crosswise at its widest region, the ovary. Then cut a thin cross section from one side of the cut surface and examine it with the stereomicroscope. Draw and label the cross section of the ovary.

J. With a compound microscope, reexamine the pollen-grain slide. Look for grains that have formed pollen tubes. Try to find the nucleus of the tube cell near the end of a pollen tube.

Analysis and Conclusions

1. How many sepals, petals, stamens, and pistils do each of the flowers have?

2. What features of the petals suggest that they are modified leaves?

3. What parts of the flower, if any, produce an aroma? What is the function of the aroma?

4. Where in the flower does meiosis occur?

5. How does the number of pollen grains in a flower compare with the number of its ovules?

Flower	Characteristics of			
	sepals	petals	stamens	pistil(s)
tulip				

*For detailed information on safety symbols, see Appendix B.

Chapter Review

Study Outline

24–1 Life Cycles of Nonseed Plants

■ The life cycle of plants alternates between a multicellular diploid generation, the sporophyte, and a multicellular haploid generation, the gametophyte.

■ Sporophytes produce haploid spores by meiosis. The spores develop by mitosis into haploid gametophytes. Gametophytes produce haploid gametes by mitosis. Gametes that fuse develop into diploid sporophytes.

■ Alternation of generations occurs in some green algae. Because of this and other common characteristics, the green algae are thought to be related to plants.

■ In mosses, the gametophyte is the dominant generation, and the sporophyte is dependent on the gametophyte. In ferns, the sporophyte generation is dominant. In mosses and ferns, water is required for fertilization.

24–2 Life Cycles of Seed Plants

■ In seed plants, the gymnosperms and angiosperms, the sporophyte is dominant, and water is not required for fertilization. In gymnosperms, pollen cones produce spores that develop into male gametophytes called pollen grains. Female gametophytes develop from spores produced inside ovules on the scales of seed cones.

■ Pollination in gymnosperms occurs by wind and air currents. Fertilization in gymnosperms occurs when the pollen tube grows into the female gametophyte within an ovule and releases a sperm that fuses with an egg to form a zygote. The ovule develops into a seed.

■ In the flowers of angiosperms, stamens produce spores that develop into male gametophytes, or pollen grains. Pistils produce spores that develop into female gametophytes inside structures called ovules.

■ Pollination in angiosperms may be by wind, animals, or water.

■ Fertilization in angiosperms occurs when the pollen tube reaches the female gametophyte and releases two sperm nuclei. One fuses with the egg, forming the zygote, and the other fuses with the two polar nuclei,

forming the endosperm nucleus. The endosperm nucleus develops into the endosperm.

24–3 Fruits and Seeds

■ The ovules of a flower develop into seeds, and the ovary, which encloses the seeds, develops into the fruit.

■ An angiosperm seed has a seed coat, embryo, and endosperm. The embryo consists of an epicotyl, cotyledon, hypocotyl, and radicle. Plants whose embryos have one cotyledon are monocots. Those whose embryos have two cotyledons are dicots.

■ After seeds are released, they may undergo a period of dormancy before germinating.

Vocabulary Review

gymnosperms (473)	corolla (481)
angiosperms (473)	stamens (481)
alternation of generations (474)	filament (481)
gametophyte (474)	anther (481)
sporophyte (474)	pistil (481)
mosses (475)	stigma (481)
rhizoids (475)	style (481)
antheridium (476)	ovary (481)
archegonium (476)	fruit (481)
ferns (476)	embryo sac (482)
seed (478)	polar nuclei (482)
seed coat (478)	nectar (482)
pollen grain (478)	double fertilization (483)
ovule (478)	
pollination (478)	endosperm (483)
micropyle (479)	cotyledon (484)
pollen tube (479)	monocots (485)
pedicel (480)	dicots (485)
receptacle (480)	epicotyl (485)
sepals (480)	hypocotyl (485)
calyx (481)	radicle (485)
petals (481)	dormancy (487)

Chapter Review

A. Sentence Completion—Fill in the vocabulary term that best completes each statement.

1. The male gametophytes of angiosperms and gymnosperms are _____.

2. A fruit develops from the _____.

3. The spore-producing generation is called the _____ generation.

4. Many seeds pass through a period of _____ before they germinate.

5. A life cycle that switches between multicellular haploid and multicellular diploid forms is called _____.

6. The transfer of pollen to a female structure is called _____.

7. Pollen grains are produced in the _____ of the stamen of a flower.

8. Ovules develop into _____ after fertilization.

9. The sperm of plants such as _____ and _____ need water to swim to the egg.

10. In a dicot, each seed has two _____.

11. The young ovary of a flower contains _____.

12. In mosses, the _____ generation is dominant.

13. The seeds of _____ are formed on the scales of cones.

14. The petals of a flower collectively make up the _____.

15. _____ results in a zygote and an endosperm nucleus.

16. A ripened ovary is known as a _____.

B. Diagram Labeling

17. Draw and label a diagram of a flower that includes the following parts: (a) pedicel, (b) receptacle, (c) sepal, (d) pistil, (e) stigma, (f) style, (g) ovary, (h) stamen, (i) filament, (j) anther, (k) petal.

18. Draw diagrams of a bean seed and a corn seed. Label the following parts: (a) seed coat, (b) endosperm, (c) cotyledon, (d) epicotyl, (e) hypocotyl, (f) radicle.

Content Review

19. Explain alternation of generations.

20. For each major group of plants, state which generation is dominant.

21. What is unique about the life cycle of *Ulva*?

22. Compare the gametophytes of mosses and ferns.

23. Compare the sporophytes of mosses and ferns.

24. What are antheridia and archegonia?

25. Describe the life cycle of gymnosperms.

26. What role does pollination play in seed plant reproduction?

27. Briefly explain the function of each part of a flower.

28. What is the difference between a pistillate and a staminate flower?

29. Describe the development of the male and female gametophytes in angiosperms.

30. Explain what is meant by self-pollination and cross-pollination.

31. Describe double fertilization in flowering plants.

32. How do the gametophytes of seed plants differ from the gametophytes of mosses and ferns?

33. How is reproduction in seed plants adapted for life on land?

34. Name and define the three kinds of fruits, and give an example of each.

35. Name and explain the function of the three basic parts of an angiosperm seed.

36. Draw and label the parts of a plant embryo. Into which mature plant part(s) does each develop?

37. Why is seed dispersal important in plant reproduction?

38. What factors affect seed germination?

39. What is seed dormancy? What is the advantage to a seed in having a dormant period?

Graphic Organizing

For information on graphic organizers, see Appendix G at the back of this text.

40. **Concept Map:** Copy the unfinished concept map above onto a separate sheet of paper. Add the following concepts where they best fit: ovules, anther, style, calyx, filament, pollen, corolla, seeds, ovary, and stigma. Be sure to add linking words between concepts. Add any other terms you think are important.

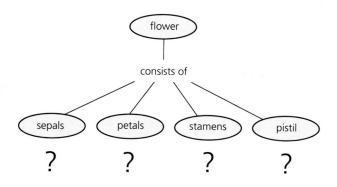

?   ?   ?   ?

Type of Tree and Fruit	Time to fall 1 meter
Silver maple	0.64 sec
Norway maple	0.98 sec
White ash	0.3 sec
Red oak	0.16 sec
Shagbark hickory	0.16 sec

Critical Thinking

41. What part of the angiosperm female gametophyte (embryo sac) probably represents an archegonium? (*Reasoning by Analogy*)

42. How are the gametes produced by a gametophyte different from the gametes produced by an animal? (*Comparing and Contrasting*)

43. Explain why the cells of the endosperm could not undergo normal meiosis. (*Identifying Reasons*)

Creative Thinking

44. Either with a drawing or a written description, design a seed that could be dispersed by wind.

45. Suppose you were given a plot of land the size of your classroom on which you could grow plants of your own choosing. What plants would you grow and how would you arrange them?

Problem Solving

46. Suppose you found a small, multicellular, green organism growing on the forest floor. How could you determine whether it is a gametophyte or a sporophyte?

47. A scientist measured the average time it took for the fruits of certain trees to fall a distance of 1 meter. The data are shown in the table above. For every second that a fruit falls, it is carried about 1.5 meters away from the parent tree. Using this value and the fall times in the data table, calculate the distance from a parent tree that each fruit would be carried during a vertical fall of 8 meters. Based on this data, develop a hypothesis regarding the relationship between seed form and fall time. How would you test your hypothesis? What benefit is there for a seed to have a long fall time?

Projects

48. Prepare a display comparing wind-pollinated flowers with animal-pollinated flowers or a display of fruits organized according to the ways in which they are dispersed.

49. Write a report on one of the following career opportunities: (a) nursery operator, (b) horticulturist, (c) herbarium worker, (d) plant collector. If possible, interview an individual who is working in that career. Be sure to prepare questions in advance. You may wish to inquire about training, future opportunities, and daily routine.

50. Write a report that describes examples of flowering plants with unusual or highly specialized mechanisms of pollination, such as the fig.

Biology and Problem Solving

How does the stress of birth prepare babies for life?

After months of comfortable, sheltered growth in its mother's uterus, a baby is born. As contractions force the fetus through the birth canal, pressure is exerted on its head, and the fetus is sometimes deprived of oxygen for short periods. During birth, a fetus produces "stress hormones" at levels higher than in a person having a heart attack. Yet, after several hours of this, a baby emerges, usually in good health and ready to survive on its own.

How do babies survive the stresses of birth so well? Scientists have begun to piece together the answer to this question.

When fetal sheep, cows, and horses are deprived of oxygen, the fetus's adrenal glands produce the stress hormones epinephrine and norepinephrine. Other tissues in the fetus also produce norepinephrine. Researchers began to suspect that these hormones contribute to infant survival.

More information was gained from an experiment with day-old rats. The rats were divided into two groups. In group 1, the rats were left alone. In group 2, some of the rats had their adrenal glands removed, while others were given drugs that prevented the action of the stress hormones. Both groups were exposed to very low levels of oxygen for 90 minutes. The rats in group 1 produced high levels of the stress hormones, and 70 percent survived. All of the rats in group 2, however, died.

1. What can you conclude from these results? Explain.

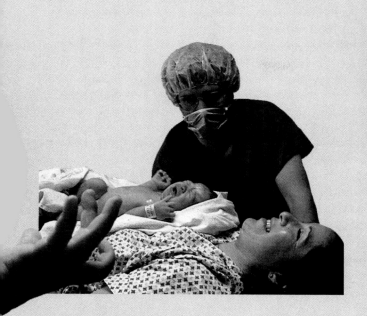

5. Explain why, unlike an adult, a fetus in stress would *not* benefit from having extra blood delivered to skeletal muscles.

Babies delivered surgically by cesarian section, rather than vaginally, do not show a surge in epinephrine and norepinephrine at birth. Surgically delivered babies often have breathing difficulties. The fluid in their lungs may not be adequately absorbed, and they may not produce enough surfactant, a soaplike substance that helps the alveoli of the lungs remain open. Scientists have hypothesized that the hormones enhance the absorption of lung fluid and the production of adequate surfactant in newborns.

6. Devise an experiment to test this hypothesis.

It has also been found that stress hormones in newborns cause nutrients to break down into a readily usable form.

7. Why would it be advantageous for a newborn to have readily available nutrients?

8. Write a brief summary of the role that hormones play in helping babies survive the stresses of birth.

Problem Finding

List some questions you still have about the stress of birth and its effects on newborns.

2. Instead of removing adrenal glands from all the rats in group 2, researchers treated some with hormone-suppressing drugs instead. Why was this step necessary?

3. Can the conclusions drawn from this experiment be applied to human fetuses? Why or why not?

The next question that scientists asked was, "How do these hormones help a fetus survive oxygen deprivation?" They knew that in adults responding to stress these hormones cause the "fight or flight" response. Heartbeat increases, and blood vessels either constrict or dilate. Blood is directed toward essential organs—the heart, brain, and skeletal muscles—and away from nonvital organs such as the skin, kidneys, and intestines. In fetuses, the adrenal glands are proportionately larger than in adults and produce more norepinephrine than epinephrine. These two hormones act in much the same way, except that norepinephrine slows the heartbeat rather than increasing it and does not direct blood to skeletal muscles.

When scientists injected fetal sheep with as much norepinephrine as produced during oxygen deprivation, the fetus's blood was directed away from nonvital organs and toward the brain, heart, placenta, and adrenal glands. This caused its blood pressure to increase and its heartbeat to decrease.

4. Why would it be helpful to a fetus experiencing oxygen deprivation for its heartbeat to decrease?

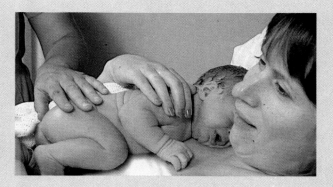

Genetics

What an impressive display! Long rows of tulips stretch to the hills behind— each row parallel to the others and straight as an arrow. Even more striking is the constancy of color—every tulip a clone of the one behind. How can there be such similarity? The answer is found in a remarkable and complex molecule known as DNA. In this unit, you will learn how DNA controls the transfer of hereditary information from parents to offspring. You will examine the structure and function of DNA and see how it ensures that both individual and species characteristics are passed from one generation to the next. You will also learn how it has recently become possible for scientists to alter the DNA of some organisms, controlling the way they function.

▲ Genetics explains the similarities between two generations of snow leopard.

Mendelian Genetics

A young snow leopard relaxes with his mother on a quiet afternoon. Look closely and family resemblances become clear. Compare the shape of their mouths and the pattern of black markings. The young leopard watches the landscape with the same confident eyes as his mother. You might say that he even has her nose. Family members often look alike, but why? How are traits passed from parent to offspring? This chapter will introduce you to the laws of heredity—the laws that explain how traits are passed from generation to generation.

Guide for Reading

Key words: genetics, dominant, recessive, gene, alleles, homozygous, heterozygous, genotype, phenotype

Questions to think about:

📖 What conclusions did Mendel reach from his experiments on pea plants?

📖 What laws govern how traits are inherited?

📖 How is the phenotype of an organism related to its genotype?

25-1 Mendel's Principles of Heredity

Section Objectives:

- *Describe* the experimental procedures that were used by Gregor Mendel.
- *State* and *give* an example of Mendel's law of dominance and his law of segregation.
- *Explain* Mendel's law of segregation in terms of chromosomes and meiosis.

The Study of Heredity

In sexual reproduction, the new individual develops from a single cell—the zygote—which was formed by the union of two gametes, one contributed by each parent. The chromosomes of each gamete bring hereditary material to the new cell. This hereditary material controls the development and characteristics of the embryo, as well as determining the features of the adult organism. Because the hereditary material comes from two different parents, the offspring is similar to both parents in some ways but differs from both parents in other ways. The offspring has all the common characteristics of its species, but it also has its own distinct, individual characteristics that make it different from all other members of the species.

▲ **Figure 25–1**

Parent and Offspring. The white coloring of polar bears is hereditary. All offspring inherit traits from their parents.

Figure 25–2
Sugar Pea Flowers. The pea plant was the experimental subject in the first scientific study of heredity. ▶

Figure 25–3

Seven Traits Studied by Mendel. Mendel used pea plants to study seven pairs of contrasting traits. ▼

1. Seed Shape
   round   wrinkled

2. Seed Color
   yellow   green

3. Seed Coat Color
   grayish brown   white

4. Pod Color
   green   yellow

5. Pod Shape
   inflated   wrinkled

6. Stem Length
   long   short

7. Flower Position
   lateral   terminal

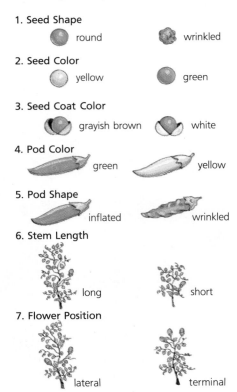

Genetics (juh NET iks) is the branch of biology that studies the ways in which hereditary information is passed on from parents to offspring. The first scientific study of heredity was carried out by Gregor Mendel (MEN dul) in the 1800s, before much was known about either chromosomes or cell division.

Mendel's Experiments

Gregor Mendel was a monk who was interested in mathematics and science. Mendel lived in a monastery in the town of Brünn in what is now Czechoslovakia. For a while, he taught science at the local high school. From 1857 to 1865, Mendel investigated the inheritance of certain traits in pea plants grown in the monastery garden. After many experiments, Mendel arrived at some basic principles of heredity that are still accepted today.

Pea plants were a good choice for Mendel's investigations. They are easy to grow, and they mature quickly. Different plants show sharply contrasting traits. For example, some are tall; some are short. Some have green pods; some have yellow pods. Each pair of traits is easily seen. In addition, the structure of the pea flower and its natural method of pollination make it easy to use in controlled experiments. Pea flowers normally self-pollinate because the stigma and anthers are enclosed by the petals, as you can see in Figure 25–2. This reduces cross-pollination in nature. By removing the stamens before they ripened, Mendel could prevent self-pollination and cross-pollinate the flower by dusting pollen from another plant onto the stigma. If he wanted certain plants to self-pollinate in the normal way, he left them alone.

Mendel kept careful records of what he did to each generation of plants. He collected the seeds from each experimental cross, and then he planted them in a definite place so that he could see the results. Mendel made a careful count of each type of offspring, and he used mathematics to understand the results. It was this use of mathematics that allowed him to draw the important conclusions that he did.

Mendel wrote a paper about his discoveries. It was published in the journal of his local scientific society and sent to other scientific organizations and libraries. Other scientists, however, do not seem to have understood its importance at the time. Mendel died in 1884, without receiving recognition for his discoveries. In 1900, however, three European scientists, all working separately, reached the same conclusions about heredity that Mendel had. Before they published their works, they read through the past scientific literature and found Mendel's papers. They gave him credit for his discoveries, and Mendel finally received the recognition he deserved.

The Law of Dominance

Mendel noticed that pea plants have certain traits that come in two forms. For example, plants are either tall or short, seeds are either yellow or green, and so on. In his experiments, Mendel studied seven pairs of contrasting traits. See Figure 25-3.

Mendel discovered that some plants "bred true" for a certain trait. For example, when short plants were allowed to self-pollinate through several generations, the offspring were always short. Mendel considered these plants to be pure for shortness. In his experiments, Mendel always started with plants that he knew were pure for the trait in which he was interested.

Mendel then wanted to find out what would happen if he cross-pollinated pure plants with contrasting traits. To do this, he stopped self-pollination by removing the stamens from a plant that was pure for one trait. He pollinated that plant with pollen from a plant that was pure for the contrasting trait. For example, he pollinated short plants with pollen from tall ones, and he pollinated tall plants with pollen from short ones. In these experiments, the pure plants made up the **parent, or P, generation.** Mendel collected the seeds produced by this cross-pollination, planted them, and allowed them to grow. Mendel found that all the offspring of this cross were tall. See Figure 25-4. That is, the short trait seemed to have disappeared in the **first filial, or F$_1$, generation.** The same kinds of results were obtained for all seven of the pairs of contrasting traits that Mendel investigated. The offspring of crosses between pure parents showing contrasting traits are called **hybrids.** In Mendel's experiments, the hybrids showed only one of the contrasting traits and not the other.

Mendel wanted to know if the trait of shortness had been lost forever as a result of the cross. To answer this question, he allowed the hybrid plants of the F$_1$ generation to self-pollinate. See Figure 25-5. When their seeds were planted and grown, about three-fourths of the offspring were tall and about one-fourth were short. The offspring of the self-pollinated hybrids made up the **F$_2$ (second**

Figure 25-4

A Cross of Pure Tall and Pure Short Pea Plants. All offspring in the first filial (F$_1$) generation are tall. ▼

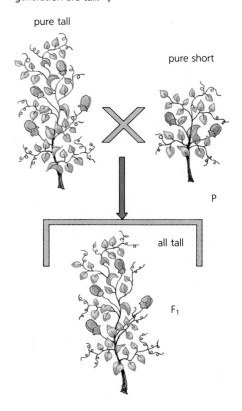

pure tall

pure short

P

all tall

F$_1$

Figure 25–5

A Cross of the Hybrid Plants of the F₁ Generation. In a cross of the F₁ generation, about three-fourths of the second filial (F₂) generation are tall, and one-fourth are short. ▶

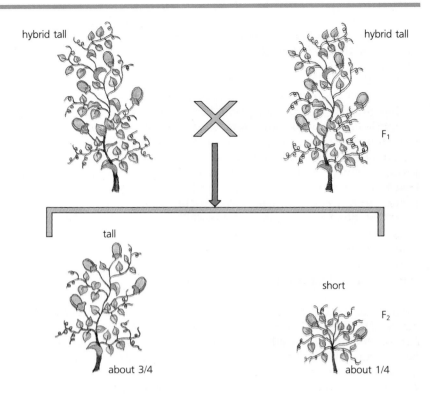

filial) **generation.** The appearance of short plants in the F_2 generation showed that the factor that determined shortness was still present in the F_1 generation.

Mendel described the traits that were expressed in the F_1 generation as **dominant** and the traits that were hidden in the F_1 generation as **recessive.** He concluded that *when an organism is hybrid for a pair of contrasting traits, only the dominant trait can be seen in the hybrid.* This is called the **law of dominance.**

The Law of Segregation

Mendel tried to explain why the recessive trait disappeared in one generation and appeared again in the next generation. He hypothesized that each trait in an individual was controlled by a pair of "factors." (Remember that in Mendel's time, the role of chromosomes and genes in heredity was not known.) Mendel also hypothesized that a factor could be one of two kinds. There was, for example, a factor for tallness and another factor for shortness. The factors in a pair could be alike or different. In a cross, the offspring received one factor from each parent. Thus, in a cross between a tall plant and a short plant, the offspring received both kinds of factors. However, only the dominant factor was expressed. The recessive factor was hidden.

Because the factor for shortness was still present in these tall plants, it was possible for the factor to show itself in the later generations. This would happen when fertilization brought two shortness factors together in the same seed. The idea that *factors that occur in pairs are separated from each other during gamete formation and recombined at fertilization is called Mendel's* **law of segregation.**

The Gene-Chromosome Theory

The importance of Mendel's work may have been overlooked in the mid-1800s because little was known about chromosomes, mitosis, and meiosis. When Mendel's research was rediscovered in 1900, however, much more had been learned about cells. Chromosomes had been stained and observed in cells, and the processes of mitosis and meiosis had been described in detail.

The idea that Mendel's "factors" might be carried by homologous chromosomes was suggested first in 1903 by an American graduate student, W. S. Sutton. Sutton was studying the formation of sperm in the grasshopper. He observed the pairs of homologous chromosomes in diploid cells and the separation of the homologous chromosomes during spermatogenesis. He realized that the chromosomes that separated during meiosis were the same as the chromosomes that had united during the fertilization process that had originally produced the animal. After reviewing Mendel's work, Sutton began to think that the factors of Mendel's theory were carried on the chromosomes. Figure 25–6 shows how Sutton's chromosome theory would apply to a Mendelian cross of two tall hybrid plants.

As you can see in Figure 25–6, the separation of homologous chromosome pairs during meiosis and their recombination during fertilization would account for the separation and recombination of the Mendelian factors.

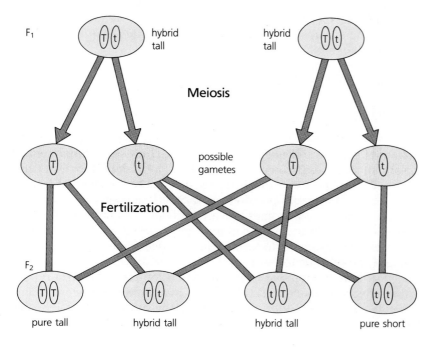

◄ **Figure 25–6**
The Gene-Chromosome Theory. T represents the factor for tallness and t the factor for shortness. Each factor is on a homologous chromosome. When the pairs of homologous chromosomes separate during gamete formation, they form two kinds of gametes: one with T, and the other with t. During fertilization, the chromosomes recombine.

Following the publication of Sutton's paper, titled "The Chromosomes in Heredity," in 1903, many experiments showed that this hypothesis was correct. At that time, the term **gene** was used in place of Mendel's "factor." Research showed not only that chromosomes carry genes but also that the genes are in a definite order along each chromosome. This work led to the modern gene-chromosome theory of heredity, which is discussed in detail in Chapter 26.

25-1 Section Review

1. Why are pea plants a good choice for genetic experiments?
2. What is a hybrid, and what is a dominant trait?
3. State Mendel's law of segregation.
4. What idea did Sutton propose?

Critical Thinking

5. What assumptions did Mendel make when he simply left plants alone that he wanted to self-pollinate? (*Identifying Assumptions*)

25-2 Fundamentals of Genetics

Section Objectives:

- *Define* the terms *alleles, homozygous, heterozygous, genotype,* and *phenotype.*
- *State* the law of probability, and explain how it applies to Mendel's experimental results.
- *Use* Punnett squares to work out the possible results of various types of genetic crosses.
- *Describe* the procedure for a test cross, and explain the significance of the results.

Alleles

To agree with Mendel's findings, each body cell of an organism should have two copies of the gene for each trait. For example, a pea plant should have two copies of the gene for height. From modern genetics, we know that this is true. One copy of the gene for height is found at the same position on each chromosome of a pair of homologous chromosomes. In an individual organism, the two copies of the gene for a certain trait may be alike or may be different. For example, in a pea plant, the two copies of the gene for height can both be for tallness, both be for shortness, or be one of each. Different copies or forms of a gene controlling a certain trait are called **alleles** (uh LEELZ). In pea plants, the gene controlling height exists as either an allele for tallness or an allele for shortness.

If the alleles for a certain trait in an organism are the same, the organism is said to be **homozygous** (hoh muh ZY gus) for that trait. For example, a pea plant that is homozygous for tallness would have two copies of the gene for tallness. If the alleles are different, the organism is said to be **heterozygous** (het uh roh ZY gous). Therefore, a pea plant that is heterozygous for tallness has one gene for tallness and one for shortness. Homozygous means *pure*. Heterozygous means *hybrid*.

Genotypes and Phenotypes

When writing about alleles, a capital letter is used for the allele for a dominant trait. For example, the allele for tallness is represented by the symbol T. A lowercase letter is used for the contrasting recessive allele. Therefore, the allele for shortness is represented by the symbol t.

A pure tall pea plant has two alleles for tallness. Its genetic makeup is represented as TT. The genetic makeup of a pure short plant is tt, while that of a hybrid is Tt. The genetic makeup of an organism is called its **genotype** (JEE nuh typ). The physical trait that an organism develops as the result of its genotype is called its **phenotype** (FEE nuh typ). It is possible for two different individuals to have the same phenotype but different genotypes. A pure tall plant and a hybrid tall plant have the same phenotype (both are tall), but they have different genotypes (TT for the pure plant and Tt for the hybrid). See Figure 25–7. An organism that shows a recessive trait is always homozygous for that trait. Thus, a short pea plant will have two copies of the gene for shortness (tt).

Probability in Genetics

Mendel's experiments involved hundreds of plants. He arrived at the laws of dominance and segregation by counting many, many offspring. To explain the *numerical* results of Mendel's experiments, you must understand the laws of chance. For example, if you toss a penny, you know that the chance of its turning up heads is 1 out of 2, or ½. If you toss the coin 100 times, you expect to get about 50 heads and 50 tails. That is, you expect to get about 1 head for every 1 tail. This can be expressed as a ratio of 1:1, or 1 head: 1 tail. In any real trial, the ratio of heads to tails is rarely exactly 1:1. In a short trial, say, 4 tosses, you might even get all heads or all tails. However, if you made a large number of tosses, 1000 or more for example, you would expect the ratio of heads to tails to be quite close to 1:1. Experiments have shown that the larger the number of trials, the closer the ratio comes to the expected value. This assumes, of course, that there is nothing special about the coin or the way in which it is tossed that would make one side *more likely* to turn up than the other.

Consider another example—the rolling of dice (singular, die). When a die is rolled, each face is as likely as any other to turn up. If you roll the die 600 times, you would expect to get about 100 of each face: 100 1's, 100 2's, 100 3's, and so on.

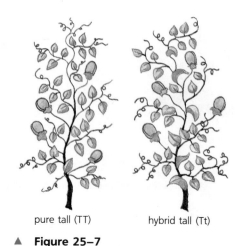

▲ **Figure 25–7**
Phenotype and Genotype. Although both of these pea plants have the same phenotype (tall), they have different genotypes (TT or Tt).

pure tall (TT)   hybrid tall (Tt)

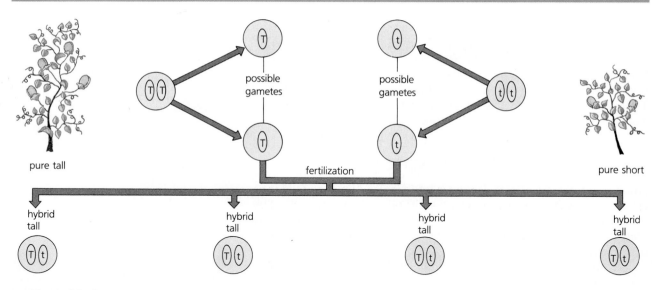

▲ **Figure 25–8**
Genetic Explanation of the Formation of Hybrids.

These are examples of the basic **law of probability,** or chance: If there are several possible events that might happen, and no one of them is more likely to happen than any other, then they will all happen in equal numbers over a large number of trials. This law allows you to predict the results of breeding experiments like those of Mendel. However, these predictions apply only when large numbers of individuals are involved.

The Punnett Square

Consider how alleles separate during meiosis and then recombine during fertilization, when a pure tall pea plant is crossed with a pure short plant. The body cells of the tall plant have two alleles for tallness. Their genotype is TT. When gametes form in this plant, each gamete receives one T allele. The genetic makeup of each gamete can be shown by the single letter T. See Figure 25–8. The cells of the short plant have two alleles for the recessive trait of shortness. Their genotype is tt, and the genotype of the gametes is t. Since each parent plant is homozygous, its gametes all contain the same allele.

Suppose that we transfer pollen from the tall plant to the pistil of a short plant. Each sperm cell nucleus will be carrying one T allele. The egg cell in each ovule of the short plant will contain one t allele. When fertilization takes place, the zygotes will receive one T allele and one t allele, and their genotype will be Tt. The plants that grow from these zygotes will be hybrid tall.

A diagram called a **Punnett square** is a helpful way to show the results of any cross. The Punnett square for the cross we have just discussed is shown in Figure 25–9. In this diagram, the alleles of the possible male gametes are written at the heads of the columns of boxes. The alleles of the possible female gametes are written at the sides of the rows of boxes. (The positions of the male and

Figure 25–9

Punnett Square for Cross of Pure Dominant (Tall) with Pure Recessive (Short). ▼

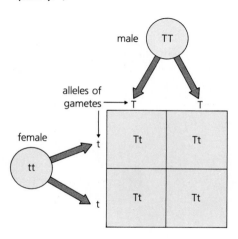

female gametes can be interchanged.) Each box contains the genotype of the zygote that forms when the allele at the top of the column and the allele at the left of the row are brought together at fertilization.

In this case, there is only one combination of alleles. All the zygotes are alike. The results are 100 percent hybrid tall (Tt).

The Punnett Square for a Hybrid Cross

A Punnett square is even more useful for a more complicated case in which hybrid tall plants either self-pollinate or are cross-pollinated. The genotype of the tall hybrids is Tt. Because they contain two different alleles for plant height, they produce two types of gametes — one type with T and the other with t. Since the T and t alleles are present in equal numbers, the two types of gametes

Critical Thinking in Biology

Reasoning Conditionally

Suppose that a friend invites you to a picnic but tells you that if it is raining, the picnic will be cancelled. On the day of the picnic, you look out the window and notice that it is raining. Therefore, you reason that the picnic will be cancelled.

This type of reasoning is known as reasoning conditionally. When you reason conditionally, you draw a conclusion based on a conditional (if-then) statement and some factual information. The conditional argument above can be set up formally:

Premise: If it is raining, then the picnic will be cancelled.
Given: It is raining.
Conclusion: Therefore, the picnic will be cancelled.

1. In pea plants, purple flowers are dominant over white flowers. Suppose you mated a homozygous, purple-flowered pea plant (PP) with a white-flowered pea plant (pp). Using the conditional argument above as a model, construct a conditional argument to show what proportion of the offspring would have purple flowers.

2. During a meeting with your biology teacher, he tells you that if you receive a grade of 70 or more on your genetics exam, you will pass the course. Suppose you receive a grade of 80. Construct a conditional argument to show whether you will pass the course. Then, suppose you receive a grade of 60. Can you

construct a conditional argument in this case? Why or why not?

3. What is wrong with the following conditional argument?

Premise: If I cross two homozygous tall pea plants (TT × TT), then I will have all tall pea plants.
Given: I have all tall pea plants.
Conclusion: Therefore, I crossed two homozygous tall pea plants.

4. Think About It Explain your thought processes as you answered question 1.

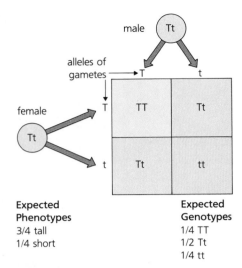

Expected Phenotypes
3/4 tall
1/4 short

Expected Genotypes
1/4 TT
1/2 Tt
1/4 tt

▲ **Figure 25–10**

Punnett Square for Crossing Two Hybrids. Of the offspring produced, about one-fourth should be pure tall (TT), one-half hybrid tall (Tt), and one-fourth short (tt).

Can You Explain This?

Father
Bb

sperm (B) (b)

eggs
(B)

Mother
Bb

	BB	Bb
(b)	Bb	bb

A brown-eyed couple, both heterozygous for eye color (Bb), have three brown-eyed children and are expecting a fourth child. The mother insists that the child she is carrying will have blue eyes. She draws a Punnett square (see above) to back up her belief.

■ *Do you agree with the mother? What do you think are the chances of the fourth child having blue eyes? Explain.*

are produced in equal numbers. This is true for both male and female gametes, and it is an important fact for the discussion that follows.

The Punnett square for the fertilizations that take place between these gametes is shown in Figure 25–10. Each letter at the head of a column stands for one type of male gamete that is formed. Each letter at the left of a row stands for one type of female gamete. Remember that these types of gametes are produced in equal numbers.

Each box in the diagram stands for a possible union of a male gamete with a female gamete. Since the types of gametes are present in equal numbers, each combination is just as likely to happen as any other. The law of probability tells us that each of the four zygotes is equally likely to appear if there is a large number of pollinations and fertilizations.

The Punnett square shows the four possible combinations. Given a large number of fertilizations, all four combinations should happen in about equal numbers. Among a large number of offspring, you would expect about ¼ to be TT (pure tall), about ½ (¼ + ¼) to be Tt (hybrid tall), and ¼ to be tt (pure short). Therefore, the genotype offspring expected ratio would be 1:2:1. In terms of the way they appear, or phenotype, about ¾ would be tall and ¼ would be short. Thus, the expected phenotype ratio of the offspring of this cross would be about 3:1 (3 tall to 1 short).

You can see that the law of probability, combined with a Punnett square, allows you to explain the results that Mendel got with his experimental crosses.

The Test Cross

As you now know, it is not possible to tell from appearance alone whether an individual showing a dominant trait is pure for the trait (homozygous) or hybrid (heterozygous). Breeders often need to know the genotypes of plants and animals. A test cross can be used to find out.

In a **test cross,** an individual of unknown genotype is mated with an individual showing the contrasting recessive trait. The genotype of the individual showing the recessive trait must be homozygous. The genotype of the individual with the unknown genotype may be homozygous or heterozygous. The test cross will show which is the case.

To understand how the test cross works, suppose a breeder wants to know whether a tall pea plant is homozygous (TT) or heterozygous (Tt). The plant with the unknown genotype is crossed, by artificial pollination, with a short plant, which must be homozygous (tt). The Punnett squares in Figure 25–11 show the results of the two possible cases.

You can see that if the test plant is pure tall (TT), all offspring of the cross will be tall. If the test plant is heterozygous tall, half the offspring, on the average, will be short. That is, the test cross shows that the recessive allele is present in the tall parent being tested. The advantage of this method is that you do not need to test and count large numbers of phenotypes. One short offspring shows that the test plant carries one recessive allele. Thus, by crossing an individual of unknown genotype with a homozygous recessive individual and looking at the offspring, it is possible to determine whether the test individual is homozygous or heterozygous.

25-2 Section Review

1. Define the terms *allele, genotype,* and *phenotype.*
2. Why is it important to use a large number of fertilizations and offspring when studying Mendelian genetics?
3. What is a Punnett square used for?
4. How can you tell whether an organism showing a dominant trait is pure or hybrid?

Critical Thinking

5. According to the law of probability, if a coin is tossed and comes up heads 10 times in a row, what are the chances of heads coming up on toss number 11? (*Predicting*)

25-3 Other Concepts in Genetics

Section Objectives:

- *Explain* Mendel's law of independent assortment in terms of genes and meiosis.
- *Use* Punnett squares to predict the phenotype ratios in a dihybrid cross.
- *Describe* incomplete dominance, codominance, and multiple alleles and give examples of each.
- **Laboratory Investigation:** *Examine* the monohybrid and dihybrid inheritance patterns in corn (p. 512).

The Law of Independent Assortment

The hybrid cross discussed in the last section is called a **monohybrid cross** because only one pair of contrasting traits is being studied. Mendel first experimented only with monohybrid crosses. In experiments on tallness and shortness, for example, he did not record the other traits of the plants. After a time, however, Mendel decided to follow two pairs of contrasting traits at the same time. From the experiments he did before, he knew that yellow color (Y) was dominant over green color (y) in pea seeds and that round seed shape (R) was dominant over wrinkled seeds (r). Using this information, Mendel made crosses in which he kept track of both seed color and seed shape.

As he had done before, he started with plants that were pure, or homozygous, for these traits. For one parent, he used plants that were pure for both dominant traits. They produced yellow, round

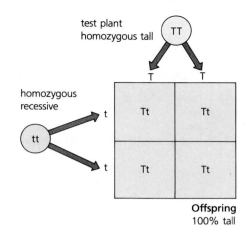

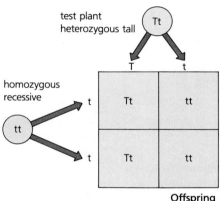

▲ **Figure 25–11**

A Test Cross. An individual showing a dominant trait is crossed with an individual showing the contrasting recessive trait. If any offspring show the recessive trait, the test individual must be hybrid.

yellow-round seeds = 315

yellow-wrinkled seeds = 101

green-round seeds = 108

green-wrinkled seeds = 32

▲ **Figure 25–12**
The Results of One of Mendel's Dihybrid Crosses.

Figure 25–13

Independent Assortment. The alignment of homologous chromosomes during metaphase I of meiosis determines the combination of chromosomes in gametes. ▼

seeds. The other parent plants were pure for both recessive traits. They produced green, wrinkled seeds. He artificially pollinated one type of plant with pollen from the other type and then observed the seeds that were produced. The results were as expected. All the seeds of the F_1 generation showed only the two dominant traits. That is, they were yellow and smooth. No green or wrinkled seeds appeared.

The next step was to plant the hybrid seeds and let the plants that grew from them self-pollinate. This would produce the F_2 generation of seeds. As expected, recessive traits appeared again in some of these seeds. Many seeds still showed both dominant traits. Some, however, were yellow and wrinkled (dominant-recessive), some were green and smooth (recessive-dominant), and a few were green and wrinkled (recessive-recessive). A breeding experiment like this one, involving two different traits, is called a **dihybrid cross.**

The data from one of Mendel's dihybrid crosses are given in Figure 25–12. Each trait considered by itself is close to the 3:1 ratio expected in a monohybrid cross. There are 416 yellow seeds and 140 green seeds. The ratio of 416 to 140 is 2.97:1, which is close to the expected 3:1 ratio. There are 423 round seeds and 133 wrinkled seeds. This ratio is 3.18:1, which is also close to 3:1. Note that there is about the same number of yellow, wrinkled seeds (dominant of one trait, recessive of the other) as green, smooth seeds (recessive of one trait, dominant of the other).

From data of this kind, Mendel concluded that different traits were inherited independently of one another. This principle is known as the **law of independent assortment.** In modern terms, this means that during meiosis, *genes for different traits are separated and distributed to gametes independently of one another*. See Figure 25–13. Today, we know that this is not always true. The reasons why are discussed in Chapter 26.

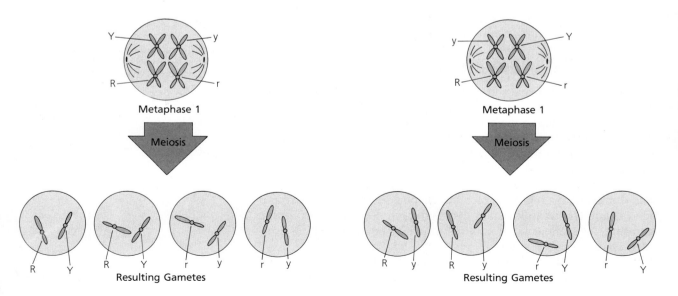

Metaphase 1

Meiosis

Resulting Gametes

Metaphase 1

Meiosis

Resulting Gametes

Phenotype Ratios in a Dihybrid Cross

With a Punnett square, you can predict the phenotype and genotype ratios expected in a dihybrid cross. First, make the diagram for the cross between the pure dominant for both traits and the pure recessive for both traits. See Figure 25–14. Note that spaces are provided for four gametes from each parent. According to the law of independent assortment, four different but equally probable combinations of two alleles, one from each of the two genes involved, can end up in the same gamete. In this case, all four possible allele combinations in the gametes have the same genotype, but they are the result of four different pairings. In the second cross, this point will be important.

As you might expect, the phenotypes of the offspring in the F$_1$ generation are 100 percent dominant for both traits (yellow and smooth). The genotype is 100 percent hybrid for both traits (YyRr).

Now consider the Punnett square for a cross between the F$_1$ dihybrids. This time, the four possible gametes will be different: YR, Yr, yR, and yr. By the law of probability, there should be equal numbers of the four types of gametes. The zygotes produced by this cross are shown in Figure 25–15.

Again, by the law of probability, all 16 of the possible zygotes will be present in equal numbers. The Punnett square shows ratios of the types of offspring produced when large numbers are involved. The phenotypes are as follows:

- 9 yellow-round (dominant-dominant)
- 3 yellow-wrinkled (dominant-recessive)
- 3 green-round (recessive-dominant)
- 1 green-wrinkled (recessive-recessive)

This phenotype ratio of 9:3:3:1 is the ratio that is seen in dihybrid crosses when the numbers of offspring are large enough. Note that each trait considered by itself has the expected 3:1 phenotype ratio. There are 12 yellow seeds to 4 green. There are 12 round seeds to 4 wrinkled.

Incomplete Dominance

While many genes follow the patterns outlined by Mendel's laws, many do not. For example, in some organisms, both alleles contribute to the phenotype of a heterozygous individual to produce a trait that is not exactly like either parent. This is known as **incomplete dominance.** For example, the inheritance of flower color in the Japanese four-o'clock plant does not follow the pattern of dominance. A cross between a plant with red flowers and one with white flowers produces offspring with pink flowers. See Figure 25–16. Note that genotypes for incomplete dominance can be written using the capital initial letter of each allele, since both alleles influence phenotype. In this case, red is represented by R and white by W. Individuals with red or white flowers are always homozygous (RR or WW). Individuals with a heterozygous genotype (RW) have an

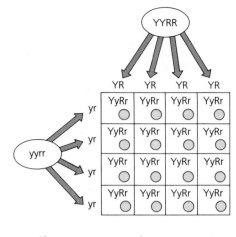

Phenotype
100%
yellow and round

Genotype
100%
hybrid for both traits

▲ **Figure 25–14**
Cross of Parents Pure for Two Contrasting Traits. All offspring are hybrid dominant for both traits.

Figure 25–15
Predicting the Results of a Dihybrid Cross. The phenotype ratios agree fairly well with Mendel's experimental results. ▼

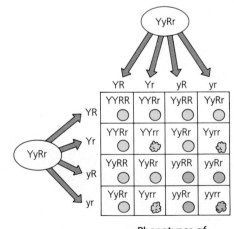

Phenotypes of Offspring
9 yellow-round
3 yellow-wrinkled
3 green-round
1 green-wrinkled

Figure 25–16
Incomplete Dominance. The hybrids of the F₁ generation (left) show a trait different from both pure traits. When these hybrids are crossed (right), one-fourth of the offspring are pure dominant, one-fourth pure recessive, and one-half hybrid intermediate. ▶

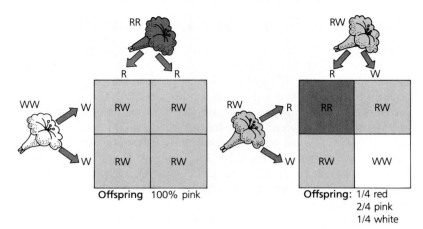

	W	R
W	RW	RW
W	RW	RW

Offspring 100% pink

	R	W
R	RR	RW
W	RW	WW

Offspring: 1/4 red
2/4 pink
1/4 white

intermediate color. When two pink hybrid four-o'clocks are crossed, a 1:2:1 ratio of red to pink to white flowers is produced in the F₂ generation.

There is also a variety of chicken, called Andalusian, in which a cross between pure black and pure white chickens produces offspring that appear blue. When the blue chickens are crossed, the F₂ generation has a 1:2:1 ratio of black to blue to white chickens. When there is incomplete dominance, the F₁ generation has a phenotype different from that of either of the parents. Also, when there is incomplete dominance, the F₂ generation shows a phenotype ratio of 1:2:1 rather than the 3:1 ratio seen in normal Mendelian inheritance.

Codominance

In **codominance,** two dominant alleles are expressed at the same time. This is different from incomplete dominance, in which neither allele is completely dominant or completely hidden. One example of codominance is the roan coat in some cattle. A cross between homozygous red shorthorn cattle and homozygous white shorthorn cattle results in heterozygous offspring with a roan coat. The roan coat consists of a mixture of all red hairs and all white hairs. See Figure 25–17. Because each hair is either all red or all white, the condition shows codominance.

Capital letters with superscripts are often used to represent genotypes in codominance. For example, the symbol

Figure 25–17

Roan Coat. White hairs and red hairs in the coat of this strawberry roan horse show the full expression of each dominant allele in different hairs. ▼

C^R can represent the allele for red coat in shorthorn cattle, and the symbol C^W can represent the allele for white coat. The genotype for homozygous red coat is then symbolized as C^RC^R, and the genotype for homozygous white coat is C^WC^W. The heterozygous animal with a roan coat has a genotype of C^RC^W.

Codominance also occurs in human heredity. The inheritance of AB blood type is an example of codominance found in humans. Blood type inheritance is discussed in the next section.

Multiple Alleles

For some traits, there are more than two alleles in the species. They are referred to as **multiple alleles.** Although a single individual cannot have more than two alleles for each trait, different individuals can have different pairs of alleles when multiple alleles exist in a population.

The alleles for human blood type are an example of multiple alleles for a trait. The ABO blood group system is described in Chapter 10. The existence of multiple alleles explains why there are four different blood types. There are three alleles that control blood type. These alleles are called A, B, and O. O is recessive. A and B are both dominant over O, but neither one is dominant over the other. When A and B are both present in the genotype of an individual, they are codominant; that is, both alleles are expressed in the individual.

The usual way to write alleles in a multiple allele system is to use the capital letter I to show a dominant allele and the lowercase i to show a recessive allele. A superscript letter then stands for each particular dominant allele. Thus, I^A stands for the dominant allele A. I^B stands for the dominant allele B. Finally, i is understood to stand for the recessive allele O.

Since there are three alleles, there are six possible genotypes: I^AI^A, I^AI^B, I^Ai, I^BI^B, I^Bi, and ii. Figure 25–18 shows the blood types and their associated genotypes.

Rh blood factors are another example of multiple alleles in human genetics. See Chapter 10.

ABO Blood Group System	
Genotype	Blood Type
I^AI^A or I^Ai	A
I^BI^B or I^Bi	B
I^AI^B	AB
ii	O

▲ **Figure 25–18**

Multiple Alleles in the ABO Blood Group System. I^A and I^B are each dominant over i but not over each other. When both dominant alleles are present, the blood type is AB. Type O blood is produced only when neither dominant allele is present (genotype ii).

25-3 Section Review

1. State the law of independent assortment in modern terms.
2. What phenotype ratios would you expect as the result of a dihybrid cross?
3. Give one example each of incomplete dominance and codominance.
4. Name a trait that is controlled by multiple alleles.

Critical Thinking

5. Compare and contrast incomplete dominance and codominance. (*Comparing and Contrasting*)

Laboratory Investigation

Genetic Corn

The kernels on an artificially pollinated ear of corn represent the offspring from a single cross. In this activity, you will count the number of different kinds of kernels on several ears of corn to discover some general patterns of inheritance.

Objectives

- *Calculate* the phenotype ratios for some monohybrid and dihybrid crosses.
- *Determine* the genotypes of parents from the phenotype ratios of their offspring.

Materials

genetic corn ears	pencil and paper
tacks	calculator

Procedure *

A. Working in groups of two, use the procedure outlined in this step for each ear of corn. Place a tack at one end of a row of kernels to mark your starting row. Using a pencil to point to the kernels, one partner should count aloud the phenotype of each kernel while the other partner keeps tally. Use a second tack to keep track of the row you are counting. Record your data in a chart like the one below.

B. Each ear of corn is labeled with a number, but you may study the ears in any order.

Cross 1: This ear has purple and yellow kernels. Purple (P) is dominant to yellow (p). Each parent of this ear developed from a purple kernel. Count the number of purple and yellow kernels. Record the data in your chart.

Cross 2: This ear has purple and yellow kernels. One parent developed from a purple kernel, the other from a yellow kernel. Count the kernels and record the data in your chart.

Cross 3: This ear has purple and yellow, as well as starchy (plump) and sweet (wrinkled) kernels. One parent developed from a purple, starchy kernel, the other from a yellow, sweet kernel. Starchy (S) is dominant to sweet (s). Count each kernel for both traits: starchy purple, starchy yellow, sweet purple, or sweet yellow. Record your data.

Analysis and Conclusions

1. For each cross, calculate and record the ratio of the phenotypes. Divide the number of purple by the number of yellow. Also, divide the number of starchy by the number of sweet.

2. From your data, determine the genotypes of the parents in each cross. Record these in your chart.

3. From the genotypes of the parents, calculate, then record, the expected ratios of the offspring phenotypes. Use a Punnett square to show your work.

4. How do the actual offspring phenotype ratios compare with the expected ratios?

5. What would the parent genotypes be if all the offspring of a cross between a purple parent and a yellow parent were purple?

Cross Number	Parental Phenotypes	Number of Each Offspring Phenotype	Ratio of Offspring Phenotypes	Parental Genotypes	Expected Ratio of Offspring Phenotypes
I	purple × purple				

For detailed information on safety symbols, see Appendix B.

Chapter Review

Study Outline

25–1 Mendel's Principles of Heredity

■ Genetics is the branch of biology that studies the ways in which hereditary information is passed on from parents to offspring. The first scientific study of heredity was carried out by Gregor Mendel in the 1800s.

■ From studies of heredity in pea plants, Mendel developed the laws of dominance, segregation, and independent assortment.

■ According to Mendel's law of dominance, when an organism is hybrid for a pair of contrasting traits, only the dominant trait can be seen in the hybrid.

■ Mendel's law of segregation states that factors (genes) occur in pairs, are separated from each other during gamete formation, and are recombined at fertilization.

■ In the early 1900s, it was shown that Mendel's "factors" were genes carried on homologous chromosomes. This discovery led to the modern gene-chromosome theory of heredity.

25–2 Fundamentals of Genetics

■ Alleles are the different forms of a gene controlling a certain trait. An individual may be homozygous with two identical alleles or heterozygous with two different alleles for the same trait.

■ The genetic makeup of an individual is called the genotype. The physical trait that an organism develops as a result of its genotype is called its phenotype.

■ To explain the numerical results of Mendel's experiments, you must apply the law of chance or probability.

■ In a simple hybrid cross involving one pair of contrasting traits, the expected genotype ratio would be 1 homozygous dominant, 2 heterozygous, 1 homozygous recessive; the expected phenotype ratio would be 3 expressing the dominant trait and 1 expressing the recessive trait.

■ In a test cross, an individual of unknown genotype showing a dominant trait is mated with an individual showing the contrasting recessive trait.

25–3 Other Concepts in Genetics

■ The law of independent assortment states that, during meiosis, genes for different traits are separated and distributed to gametes independently of one another.

■ The expected phenotype ratio of a dihybrid cross is 9 dominant-dominant, 3 dominant-recessive, 3 recessive-dominant, and 1 recessive-recessive.

■ Incomplete dominance occurs when both alleles contribute to the phenotype of a heterozygous individual to produce a trait that is not exactly like that of either parent.

■ In codominance, two dominant alleles are expressed at the same time.

■ Some traits are controlled by multiple alleles, where there are more than two alleles in a species.

Vocabulary Review

genetics (498)

parent (P) generation (499)

first filial (F_1) generation (499)

hybrids (499)

F_2 (second filial) generation (499)

dominant (500)

recessive (500)

law of dominance (500)

law of segregation (501)

gene (502)

alleles (502)

homozygous (503)

heterozygous (503)

genotype (503)

phenotype (503)

law of probability (504)

Punnett square (504)

test cross (506)

monohybrid cross (507)

dihybrid cross (508)

law of independent assortment (508)

incomplete dominance (509)

codominance (510)

multiple alleles (511)

A. Matching—Select the vocabulary term that best matches each definition.

1. Having two different alleles for a certain trait

2. A cross in which two pairs of contrasting traits are studied

Chapter Review

3. Traits that are hidden in the F₁ hybrid generation

4. Physical traits that an organism develops as the result of its genetic makeup

5. Genes for different traits are separated and distributed independently of one another during gamete formation

6. A type of diagram used to show the possible results of a genetic cross

7. Two dominant alleles of a contrasting pair fully expressed at the same time in the heterozygous individual

8. A different form of a gene that controls a certain trait

9. A method for determining if an individual showing the dominant trait is homozygous or heterozygous

10. A cross in which one pair of contrasting traits is studied

B. Sentence Completion— Fill in the vocabulary term that best completes each statement.

11. The branch of biology that studies the ways in which hereditary information is passed on from parents to offspring is called _____.

12. The genotype TT represents a _____ individual.

13. A cross between red and white flowers that produces pink flowers illustrates _____.

14. For some traits, there are more than two different alleles that exist in the species, and these are referred to as _____.

15. According to the _____, alleles are separated from each other during gamete formation and are recombined at fertilization.

Content Review

16. Why was Mendel not given recognition for his discoveries during his lifetime? When did he finally receive credit for his work?

17. How did Mendel go about crossing pure parent pea plants that showed contrasting traits? What did the offspring look like?

18. What is meant by the F₁ generation?

19. Give an example of Mendel's law of dominance.

20. How did Sutton's chromosome theory help explain Mendel's law of segregation?

21. What is the difference between a homozygous and a heterozygous individual?

22. Explain how two organisms can have the same phenotypes but different genotypes.

23. State the law of probability, and tell why it is important in the study of genetics.

24. Explain the genetic makeup and the genotypic ratio of the F₂ generation of a monohybrid cross.

25. Describe the appearance and the phenotypic ratio of the F₂ generation of a monohybrid cross.

26. How is a test cross carried out?

27. Describe the experiments that led to Mendel's law of independent assortment.

28. How can Mendel's law of independent assortment be explained in terms of genes and meiosis?

29. What is the expected phenotypic ratio of each trait in a dihybrid cross when each trait is considered by itself? When both traits are considered together?

30. When there is incomplete dominance in an organism, what phenotypic ratio would you expect to find in the F₁ generation? What ratio would you expect in the F₂ generation (when two hybrids are crossed)?

31. Explain how incomplete dominance and codominance differ.

32. When multiple alleles for a trait exist in a species, how many alleles can be present in the species as a whole? How many alleles for the trait can be present in one individual member of the species?

Graphic Organizing

For information on graphic organizers, see Appendix G at the back of this text.

33. **Circle Graph:** Construct two circle graphs: (a) one showing the relative number of different phenotypes expected from a dihybrid cross, and (b) one showing Mendel's actual results (see Figure 25–12). How do Mendel's results compare with the expected results?

Critical Thinking

34. Based on your knowledge of Mendel's law of segregation, explain why the shortness trait in pea

plants disappeared and reappeared from generation to generation. (*Identifying Causes*)

35. A mother and father are each heterozygous for curly hair, with a genotype of Cc. (C is the dominant gene for curly hair.) The couple has four children, all of whom have the recessive straight hair (genotype cc). You might expect that three of the children would have the dominant curly hair and only one the recessive straight hair. Why did the actual phenotype ratios in this family differ from the expected ratios? (*Identifying Reasons*)

36. Why do scientists require that plants and animals used in research be "genetically pure," that is, that they breed true for as many characteristics as possible? (*Identifying Reasons*)

Creative Thinking

37. After going into the florist business, you realize that a particular flower is more valuable if it is pink. If you were to grow the pink flowers for profit, would you prefer that they result from a codominant trait or an incomplete dominant trait? Explain.

38. In a hospital's maternity ward, blood typing might be used to determine the identity of a newborn's parents. It might also be used as a source of evidence in paternity cases in court. Do you think this method of identification is completely reliable? Explain.

Problem Solving

39. Assume that 50 percent of 10 000 pea-plant offspring are short. Use a Punnett square to show the probable genotypes of the parents and the offspring. Let T stand for the dominant allele, and t for the recessive allele.

40. If two pea plants hybrid for a single trait produce 60 pea plants, about how many of the 60 F_2 pea plants would you expect to show the recessive trait?

41. If a child with blood type O has a mother with blood type A and a father with blood type B, what are the parent's genotypes for blood type?

42. Use a Punnett square to show the possible genotypes and phenotypes for blood type of the offspring of two parents, one with blood type O and one with blood type AB.

43. What are the blood types that could appear in the children of parents who have the genotypes $I^A i$ and $I^B i$?

44. In a certain species of meadow mouse, a dark coat is dominant over a cream coat. If you cross a heterozygous dark-coated male with a cream-coated female, what are the expected phenotypes of their offspring? Use a Punnett square to display your answer, letting C stand for the gene for dark coat, c for cream coat.

45. Yellow fruit and dwarf vines are recessive traits in tomatoes. Red fruit and tall vines are dominant traits. Illustrate the F_2 generation in a cross between hybrid plants. Copy the Punnett square shown below to display your answer. R stands for the gene for red fruit, T for tall, r for yellow fruit, and t for dwarf. What is the F_2 phenotypic ratio?

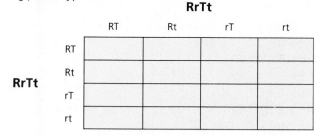

46. Use Punnett squares to illustrate test crosses to determine whether a black male guinea pig is homozygous or heterozygous for black. The black coat is produced by a dominant gene (B).

47. An albino male rat is crossed with a normally pigmented female who had an albino mother. The gene for pigmentation (A) is dominant over albinism (a). Using a Punnett square, show the types of offspring this mating would produce. Express the outcome as phenotype and genotype ratios.

48. Two women gave birth to girls in the same hospital at the same time. The nurses think they may have accidentally switched the babies' name tags and given the babies to the wrong parents. One baby, Jane, is blood type O. The other baby, Mary, is blood type A. The father in one set of parents, the Reds, is blood type A, and the mother is type B. The father in the other set of parents, the Greens, is blood type AB, and the mother is type O. Figure out which baby belongs to which parents.

Projects

49. Find out how commercial seed growers produce the various types of seeds that they sell. Present your findings orally to the class.

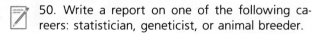 50. Write a report on one of the following careers: statistician, geneticist, or animal breeder.

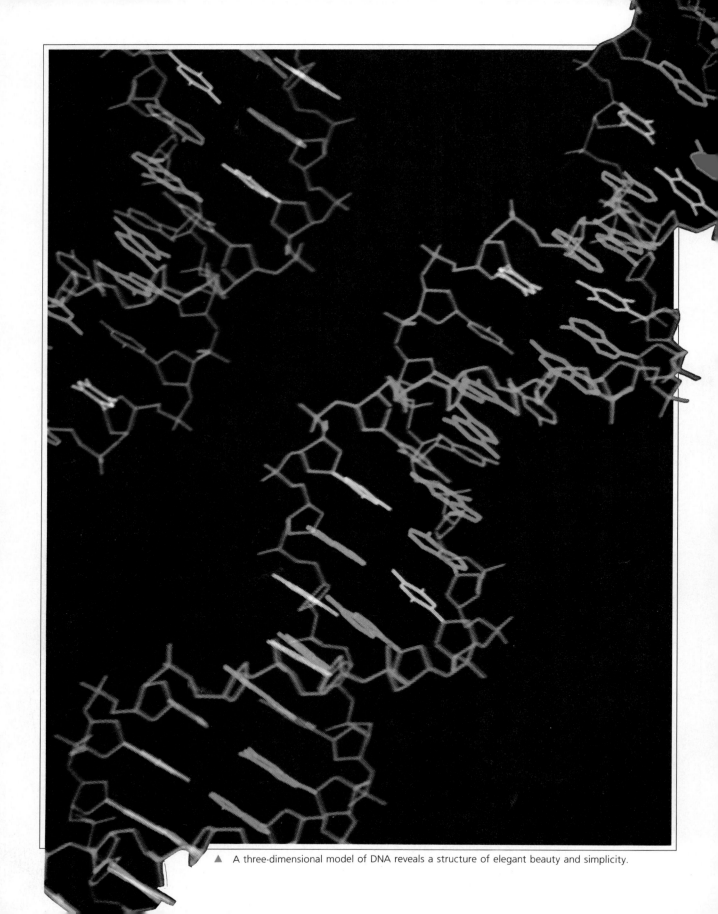

▲ A three-dimensional model of DNA reveals a structure of elegant beauty and simplicity.

Modern Genetics

26

Like a twisted ladder spiraling from end to end, each DNA molecule has the same basic shape. They are so small that scientists use computers to picture and study them. Yet, these molecules command the very look, makeup, and traits of an organism and what it will pass on to its offspring. The rungs of the ladder form a kind of genetic code. It is this code that translates into traits. In this chapter, you will learn about DNA and the ways in which it determines the traits of all living things. You will also learn about the mechanisms for passing on traits to new generations.

Guide for Reading

Key words: sex-linked trait, linkage group, nucleotide, transcription, translation, operon

Questions to think about:

📖 What evidence helped confirm that genes are found on chromosomes?

📖 How was the composition, structure, and function of DNA determined?

📖 What roles do DNA and RNA play in protein synthesis?

26-1 Chromosomal Inheritance

Section Objectives:

- *Explain* why *Drosophila* is a good experimental animal for genetics experiments.
- *State* the role of the X and Y chromosomes in determining sex.
- *Define* the terms *linkage group* and *crossing-over*.
- *Describe* multiple-gene inheritance.
- **Laboratory Investigation:** *Explore* how gene linkage affects the inheritance of traits (p. 540).

T. H. Morgan and *Drosophila*

In the early 1900s, Thomas Hunt Morgan, an American geneticist, offered the first evidence that genes are parts of chromosomes. For his work, Morgan won a Nobel Prize in 1933. One reason for Morgan's successful research was his choice of the fruit fly, *Drosophila* (droh SAHF uh luh), as his experimental animal.

The fruit fly, which is often found around ripening fruits, is a useful organism for genetic experiments. It is so tiny that large numbers can be kept in a small space. It is easy to raise, and it produces hundreds of offspring. It also has a reproductive cycle of about 14 days. This allows a geneticist to study many generations of flies in a short time. Another advantage to studying *Drosophila* is that it has only four pairs of chromosomes.

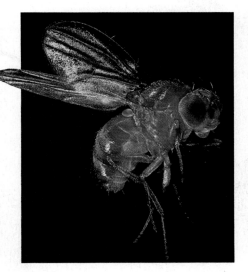

▲ **Figure 26–1**

Drosophila. Morgan used the fruit fly, *Drosophila,* as his subject for genetic experiments.

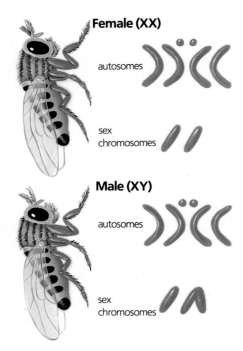

▲ Figure 26–2

***Drosophila* Sex Chromosomes.** A normal female *Drosophila* has two X chromosomes. A normal male has one X chromosome and one Y chromosome.

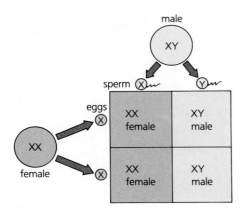

▲ Figure 26–3

Sex Determination in *Drosophila*. In *Drosophila*, as in humans, the male gamete can carry an X or a Y chromosome. For this reason, the male gamete determines the sex of the offspring.

Sex Determination and Chromosomes

Around 1890, scientists observed that chromosomes in cells from males and females were identical except for one pair. Scientists suspected that these different chromosomes determined the sex of the organism. This hypothesis now has been well confirmed. The two unmatched chromosomes are known as the **sex chromosomes;** the other homologous chromosomes are called **autosomes** (AW tuh sohmz). The discovery of sex chromosomes was important in the study of genetics because it linked an inherited trait (male or female) to a particular pair of chromosomes.

In the female *Drosophila,* the two sex chromosomes look the same. See Figure 26–2. Both are rod-shaped chromosomes, known as the **X chromosomes.** Male *Drosophila*, on the other hand, have one X chromosome and one hook-shaped chromosome, called the **Y chromosome.**

The sex of *Drosophila* is determined at fertilization. All the female gametes, or eggs, contain one X chromosome. They do because each egg cell receives one X chromosome during meiosis. The male gametes, or sperm, contain either one X chromosome or one Y chromosome. When a sperm containing a Y chromosome joins with a female gamete, the zygote will have one X and one Y chromosome. This zygote will develop into a male (XY). When a sperm containing an X chromosome joins with a female gamete, the zygote will be female (XX). Thus, the sex of the fruit fly is determined by the kind of sperm that fertilizes an egg cell. See Figure 26–3.

The sex of an organism is determined in a similar way in humans and other mammals. Not all animals, however, have the same system of sex chromosomes. In birds, butterflies, and some fish, the male has the two identical sex chromosomes, and the female has two different sex chromosomes. In these animals, the female produces two different types of gametes. The egg of the female determines the sex of the offspring for these animals.

Sex-Linked Traits

Morgan looked at thousands of fruit flies to find interesting traits to study. As you saw in Figure 26–1, the normal eye color of *Drosophila* is bright red. One day, Morgan discovered a white-eyed male fly. Since he had never seen this trait before, he decided to study it. His first step was to mate the white-eyed male with a normal red-eyed female. All offspring of this mating showed red eyes. Morgan concluded that the allele for white eyes is recessive.

If his cross obeyed the rules of Mendelian genetics, all the red-eyed flies of the F_1 generation would be heterozygous for eye color. If R represents the dominant allele for red eyes, and r the recessive allele for white eyes, the genotype of the F_1 generation should be Rr. To test this, Morgan mated males and females of the F_1 generation. The F_2 generation had the expected ratio of red eyes to white. About three-fourths of the flies had red eyes, and about one-fourth had white eyes. However, there was one peculiarity in

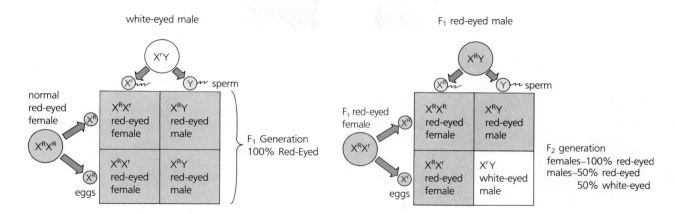

▲ **Figure 26–4**

Inheritance of the White-Eye Trait in *Drosophila*. The allele for white eyes is recessive and is carried only on the X chromosome. A white-eyed male (left) is crossed with a homozygous red-eyed female. Since all the offspring inherit the dominant red-eye allele from the female parent, all have red eyes. However, the females are heterozygous for eye color. A red-eyed male (right) is crossed with a heterozygous red-eyed female. Since all the female offspring receive the dominant allele from the male parent, they are all red eyed. However, the male offspring do not receive a gene for eye color from the male parent, only from the female parent. Half the males receive the recessive allele and are white eyed. While all the females are red eyed, half are carriers of the recessive allele. What are the genotypes of a male and female *Drosophila* that could result in a white-eyed female?

the results—all the white-eyed flies were male. All the females were red-eyed. For this reason, the inheritance of eye color seemed to be related to the sex of the offspring.

To learn more, Morgan performed a test cross. He mated the original white-eyed male with a red-eyed female from the F_1 generation. This time one-half the females had white eyes, and one-half had red. The males were also divided half white and half red.

Morgan knew that the Y chromosome is shorter than the X chromosome. He reasoned that some of the genes found on the X chromosome might be missing from the Y chromosome. It was possible, he reasoned, that the allele for eye color was carried on the X chromosome of the fruit fly, but the Y chromosome did not have a corresponding allele. If this were true, a male fly would show the recessive allele. To have white eyes, a female fly would have to have the recessive allele on both of her X chromosomes.

By means of Punnett squares, you can see that Morgan's hypothesis explains the results of the crosses he made. See Figure 26–4. In these diagrams, X^R represents an X chromosome carrying the dominant allele for red eyes. X^r indicates an X chromosome with the recessive allele for white eyes. Y stands for a Y chromosome with no gene for eye color.

Once a white-eyed female had been obtained, it was possible to make another test cross. Morgan crossed a white-eyed female with a red-eyed male. All the female offspring were red-eyed; all the males were white-eyed. You may want to draw a Punnett square that will explain this result.

Biology and You

Q: Are there any mental illnesses that are inherited?

A: Although research is incomplete, many scientists believe that some mental illnesses are passed genetically from one generation to the next. The current theory is that some mental illnesses are inherited but activated by environmental factors such as stress.

Experts generally agree that genes play a role in manic depression, a condition marked by extreme mood swings. Scientists have found a connection between manic depression and two distinct defects at the tip of the X chromosome. Another study, however, links this illness to a defect at the tip of chromosome 11. It may be that manic depression is controlled by more than one gene.

Scientists agree that genetics may also play a role in schizophrenia, a disorder in which the sufferer is out of touch with reality. Limited studies connect this illness to a gene defect in chromosome 5. Identifying the genes that cause mental illnesses may help us develop more effective treatments.

■ *List some reasons why the inheritance pattern of mental illnesses is more difficult to trace than that of hemophilia or other physical illnesses.*

A trait that is controlled by a gene found on the sex chromosome is called a **sex-linked trait.** The chance of showing the trait is affected by the sex of the individual. Most sex-linked traits are determined by genes found on the X chromosome but not on the Y chromosome. Morgan's discovery of sex-linked traits supported Sutton's hypothesis that genes are located on the chromosomes. (See Chapter 25). It was also an important discovery in its own right, since it explains the inheritance of several human diseases and disorders.

Sex-Linked Traits in Humans

Many human conditions and diseases are caused by abnormal recessive alleles of certain genes. The normal allele lets the body perform some function that the abnormal allele does not. The term *defective allele* is often used to refer to the abnormal alleles that cause genetic diseases. Several defective alleles in human genetics are sex-linked. Among the human diseases caused by defective sex-linked alleles are hemophilia, a disorder of the blood-clotting system, and muscular dystrophy, which results in the gradual destruction of muscle cells. A form of night blindness and color blindness are less serious sex-linked hereditary disorders.

Color blindness is a condition in which the individual cannot perceive certain colors, usually red and green. This condition is more common in males than in females. Few females suffer from red-green color blindness, although they may be *carriers* for it. Carriers have the allele for color blindness on one X chromosome. Females are not affected by the defective recessive allele because they have a normal dominant allele on the other X chromosome.

Every male receives an X chromosome from his mother and a Y chromosome from his father. If the mother is a carrier for color blindness, there is a 50 percent chance that any son she has will receive the defective allele. See Figure 26–5. Since the Y chromosome has no gene for color vision, when a son inherits the defective allele from his mother, the allele is expressed, and the son will be color blind. A female, on the other hand, receives an X from her mother and an X from her father. A daughter is a carrier only if she has a defective allele from her mother and a normal allele from her father or a normal allele from her mother and a defective allele from her father.

Since a father contributes only a Y chromosome to his sons, a color-blind father cannot transmit the allele to his sons. He will, however, transmit this defective allele to all his daughters. If the mother is a carrier of the defective allele, there is a 50 percent chance that a daughter will inherit the defective allele from her mother as well as from her father. Thus, on the average, half the daughters will be color blind. The other half will be carriers. Half the sons will be color blind also, but this result has nothing to do with the father's genotype. If both parents are color blind, all their offspring will be color blind because neither parent is carrying a normal dominant allele.

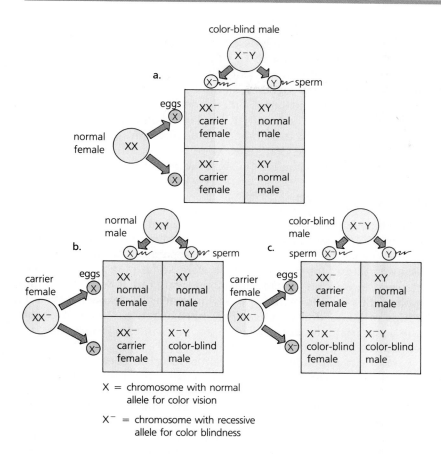

◀ **Figure 26–5**
Inheritance of Color Blindness in Humans. (A and B) The allele for color blindness cannot be transmitted from father to son. It can be transmitted only through the female to later generations. (C) For a female to be color blind, she must inherit defective alleles from both parents, a relatively rare event.

X = chromosome with normal
allele for color vision

X⁻ = chromosome with recessive
allele for color blindness

Gene Linkage

Every organism has thousands of genes. Every organism also has a certain small number of chromosomes in each body cell. Therefore, many genes must be present on each chromosome. Genes on the same chromosome are said to be *linked.* All of the genes that are on the same chromosome make up a **linkage group.** *Drosophila,* with four pairs of chromosomes, has four linkage groups. Humans, with 23 pairs of chromosomes, have 23 linkage groups.

If genes are linked on the same chromosome, they cannot be distributed independently during meiosis. Therefore, linked genes do not obey Mendel's law of independent assortment, which was discussed in Chapter 25. Mendel arrived at the law of independent assortment only because the dihybrid traits he studied happened to be controlled by genes on different pairs of chromosomes.

One of the first examples of *gene linkage* was found by R. C. Punnett and William Bateson at Cambridge University in England. They were studying the inheritance of flower color in pea plants. Purple flowers were dominant; red flowers were recessive. Long pollen grains were dominant. Round pollen grains were recessive. Plants pure for both dominant traits were crossed with plants pure for both recessive traits. The expected phenotype of 100 percent dominant for both traits in the F₁ generation did occur. However, when the dihybrids were crossed, they did not show the expected 9:3:3:1 phenotype ratios in the F₂ generation. The results were closer to the 3:1 ratio from a single hybrid cross. The two dominant traits seemed to stay together, as did the two recessive traits. That is, they were not distributed independently.

T. H. Morgan had gotten similar results from *Drosophila*. Certain traits seemed to be inherited together. Morgan thought this was further evidence that genes were parts of chromosomes and that genes on the same chromosome were inherited together.

Crossing-Over

One difficulty with the hypothesis of linked genes was that the linkage did not seem to be perfect. In a small number of offspring in the F₂ generation, the linked genes separated. In the Punnett-Bateson study, for example, there were some plants with purple flowers and round pollen, and some with red flowers and long pollen. But the numbers were nowhere near the expected 9:3:3:1 ratios of Mendelian genetics. The unusual ratios remained about the same from one experiment to another, and they were hard to explain.

Morgan concluded that the ratios occurred because pieces of homologous chromosomes were exchanged sometimes during meiosis. The exchange happened before the chromosomes separated to go to different gametes. He called this exchange process **crossing-over**. See Figure 26–6. Scientists now know that crossing-over occurs during synapsis of the first meiotic division, when the four chromatids of each homologous chromosome pair are in close contact.

Because of crossing-over, the chromosomes that go into the gametes have new gene linkages. They are not identical to the chromosomes in the parent cells. Thus, crossing-over is an important source of variations, or genetic differences, in offspring.

Morgan reasoned that genes that are far apart on the same chromosome would become separated by crossing-over more often than genes that are close together. By studying the offspring ratios of dihybrid crosses for many different pairs of linked genes, Morgan was able to figure out how close or how far apart each particular pair was. In this way, Morgan was able to make gene maps of the chromosomes in *Drosophila*. Each gene map showed the order of genes on the chromosome. The order was based on how often the genes became separated by crossing-over.

Figure 26–6

Crossing-Over. During synapsis in meiosis, segments of homologous chromatids may be interchanged. If the exchanged segments carry different alleles for certain traits, new gene combinations result. ▼

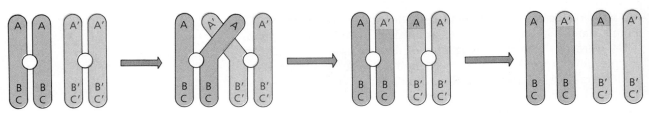

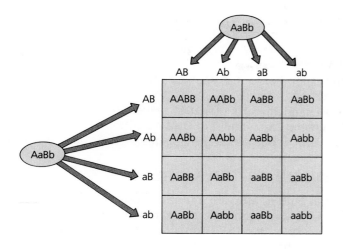

◄ **Figure 26–7**
Multiple-Gene Inheritance. In multiple-gene inheritance, when there are two genes controlling a trait, there are nine different possible genotypes. There is a range of phenotypes between the pure dominant and the pure recessive extremes.

Multiple-Gene Inheritance

Many traits in both plants and animals do not appear in two contrasting forms. For example, humans are not just either tall or short. Instead, human height varies from very short to very tall. Traits that vary between two extremes are not controlled by the alleles of a single gene, but by the alleles of two or more different genes. When two or more independent genes affect one characteristic, it is called **multiple-gene,** or *polygenic,* **inheritance.**

The simplest example of multiple-gene inheritance would involve two genes, each with its own pair of alleles. For example, the length of the ears in corn is controlled by two genes. Suppose the genes for length are Aa and Bb. With these two genes, there can be four possible gametes and nine different genotypes. See Figure 26–7. Suppose that the greater the number of capital letters in the genotype, the longer the corn ear. Then, the longest corn ears would have the genotype AABB. The shortest corn ears would have the genotype aabb. All other genotypes would show ear sizes between these two extremes.

26-1 Section Review

1. Which sex chromosomes characterize female and male *Drosophila?*
2. Name a sex-linked trait in humans.
3. In what process are pieces of homologous chromosomes exchanged during meiosis?
4. Give an example of a human trait controlled by multiple genes.

Critical Thinking

5. Why might a geneticist use fruit flies to study how genes function? Why might a scientist choose *not* to use fruit flies? (*Identifying Reasons*)

26-2 The Genetic Material

Section Objectives:

- *Describe* the experiments that showed that DNA is the genetic material.
- *List* the three chemical parts of a DNA nucleotide.
- *Explain* the Watson-Crick model of DNA.
- *Describe* how DNA replicates in a living cell.

The Chemistry of the Gene

From the work of Sutton, Morgan, and many other researchers, it was known by 1950 that chromosomes carry hereditary information. It was also certain that the information is present in distinct units, called genes, arranged along the chromosomes like beads on a string. Still, no one knew what a gene was or how it worked. Without that knowledge, heredity and genetics could not be truly understood. This understanding came in the 1950s, when the chemical nature of the gene was discovered. The first clues to the chemical nature of the hereditary material were uncovered much earlier, however.

In 1869, Friedrich Miescher, a Swiss biochemist, isolated a material from the nuclei of fish sperm. He called the material *nuclein* (NOO klee un). Other scientists showed that nuclein was made of carbon, hydrogen, oxygen, and nitrogen. It was also rich in phosphorus. When nuclein was shown to be acidic, its name was changed to *nucleic acid*. Later research found two kinds of nucleic acid—deoxyribonucleic acid, or DNA, and ribonucleic acid, or RNA. DNA occurs mainly in the nuclei of cells. RNA is found mainly in the cytoplasm.

In the 1920s, scientists found that chromosomes contained DNA. It was already known that chromosomes contained proteins. While the chemical structure of proteins was well understood, the structure of DNA was completely unknown. Although a few scientists suggested that DNA was the hereditary material, most scientists believed that only proteins were complex enough to carry genetic information.

Protein vs. Nucleic Acids

It did not become clear until the 1950s that the hereditary material of the chromosomes was DNA. To understand how this came about, you need to understand the experiments performed by Frederick Griffith and several other researchers.

Griffith's Experiments In 1928, Frederick Griffith, an English bacteriologist, was trying to find a vaccine against pneumonia. Pneumonia is a disease caused by a kind of bacteria called *pneumococcus* (noo muh KAHK us). Griffith knew that there are two types of pneumococcus. See Figure 26–8. One type, called Type S, is surrounded by an outer covering called a *capsule*. Type S bacteria

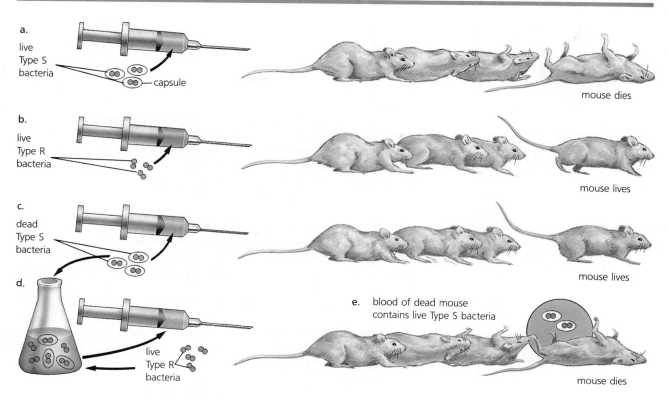

a. live Type S bacteria — capsule
mouse dies

b. live Type R bacteria
mouse lives

c. dead Type S bacteria
mouse lives

d.
e. blood of dead mouse contains live Type S bacteria
live Type R bacteria
mouse dies

▲ **Figure 26–8**

Griffith's Experiment. (A) Live Type S bacteria will kill the mouse. (B) Live Type R bacteria are harmless. (C) Dead Type S bacteria are harmless. (D) Dead Type S bacteria are mixed with live Type R bacteria. (E) The mixture kills the mouse, and live Type S are present in the mouse's tissues. Griffith concluded that the Type R bacteria had been transformed into Type S bacteria.

cause a severe case of pneumonia. The other type, called Type R, is not surrounded by a capsule. Type R bacteria do not cause pneumonia. If mice are injected with Type S bacteria, they develop pneumonia and die. Mice injected with Type R bacteria show no ill effects.

Dead Type S bacteria do not cause pneumonia when injected into mice. In Griffith's key experiment, he mixed dead Type S bacteria with live Type R. When he injected the mixture into mice, the mice developed pneumonia and died. Furthermore, the tissues of the dead mice showed living Type S bacteria.

Remember that neither dead Type S nor live Type R bacteria alone cause pneumonia. When brought together, however, they do cause pneumonia and living Type S bacteria appear. Griffith concluded that some factor from dead Type S bacteria could change, or transform, Type R bacteria into Type S. The changed bacteria were able to make capsules and to cause pneumonia in mice.

Avery, MacLeod, and McCarty   In 1944, Oswald Avery, Colin MacLeod, and Maclyn McCarty of the Rockefeller Institute in New York identified the transforming material in Griffith's experiment as DNA. In other words, DNA produced the new inherited traits in Type R bacteria. Although this was strong evidence that DNA is the genetic substance, many scientists remained unconvinced. They still thought that protein must carry the hereditary information. The conclusive evidence supporting DNA was obtained by Alfred Hershey and Martha Chase in 1952.

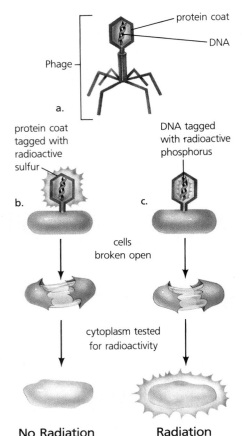

a.

protein coat
tagged with
radioactive
sulfur

b.

DNA tagged
with radioactive
phosphorus

c.

cells
broken open

cytoplasm tested
for radioactivity

No Radiation   Radiation

▲ **Figure 26–9**

The Hershey-Chase Experiment. (A)
Structure of one type of bacteriophage. (B)
Bacteria infected by phages with protein coats
are tagged with radioactive sulfur. The cell
contents do not become radioactive. (C)
Bacteria infected by phages with DNA are
tagged with radioactive phosphorus. The cell
contents become radioactive. Hershey and
Chase concluded that a phage infects a
bacterial cell by injecting its DNA into the
bacterium. The protein coat remains outside.

Hershey and Chase   Alfred Hershey and Martha Chase made
use of viruses called bacteriophages to resolve the DNA *vs.* protein
argument. A *bacteriophage,* or *phage* (FAYJ) for short, is a virus that
infects bacteria. This kind of virus is made of a DNA core surrounded by a protein coat. See Figure 26–9. A phage invades a
bacterium and makes hundreds of new phage particles once inside
the bacterial cell. The bacterial cell then breaks open, and the new
phage particles are let go. These can attack other bacterial cells.
Hershey and Chase wanted to discover whether the whole phage
entered the bacterium or whether just the DNA or the protein coat
entered. In hopes of answering this question, they tagged the
protein and the DNA of the phage particle with different radioactive
elements.

DNA contains phosphorus but no sulfur. Virus protein contains
sulfur but no phosphorus. Hershey and Chase tagged the phage
DNA with radioactive phosphorus. They tagged the protein coat
with radioactive sulfur. One group of bacteria was then exposed to
phages with radioactive DNA. Another group was exposed to
phages with radioactive protein. After large numbers of bacteria
had become infected with phages, the cytoplasm of the bacteria
was tested for radioactivity. The cells that had been infected by
phages with radioactive DNA showed a great deal of radioactivity.
The cells that had been infected by phages with radioactive protein
showed almost no radioactivity. This experiment proved that the
phage DNA enters the cells, while the phage protein stays outside
when phages infect bacteria.

If phage DNA alone can cause bacteria to make more phages, it
must be the DNA that carries the genetic instructions for making
phages. This experiment established DNA as the genetic material.
The problem then became that of finding what DNA is made of and
how it works.

Composition of DNA

The first step in analyzing an unknown organic compound is to find
out the chemical groups that form it. In the 1920s, P. A. Levene, a
biochemist, carried out a chemical analysis of DNA. Levene found
that the DNA molecule is made up of the following chemical groups:
the 5-carbon sugar **deoxyribose** (dee ahk see RY bohs); a phosphate group; and four kinds of nitrogen-containing (nitrogenous)
bases. Two of the four bases, known as **adenine** and **guanine,** are a
kind of compound called a *purine* (PYOOR een). The other two,
cytosine and **thymine,** are compounds called *pyrimidines* (pih RIM
uh deenz).

Levene found that there was one phosphate group and one
nitrogen-containing base for each sugar unit. He therefore concluded that the basic unit of DNA is a sugar, a phosphate, and one of the
four nitrogen-containing bases. He called this unit a **nucleotide**
(NOO klee uh tyd). Since there are four different bases, there are
four different kinds of nucleotides. Many, many nucleotides make
up a single DNA molecule.

Structure of DNA

After the chemical makeup of DNA was known, the second step was to work out the structure of the DNA molecule. This was done in 1953 by James Watson, an American biochemist, and Francis Crick, an English physicist, working together in Cambridge, England. To develop their model of DNA, Watson and Crick used everything that was known about DNA. An important piece of information transmitted by Maurice Wilkins came from Rosalind Franklin of Oxford University in England. She had made X-ray studies of DNA crystals. These X-ray photographs showed that the repeating units in the crystal are arranged in the form of a **helix** (HEE liks). A helix is the shape of a coiled spring.

After trying many different arrangements, Watson and Crick arrived at a model DNA molecule in which there are two chains of sugar-phosphate groups running parallel to each other. Pairs of bases link the chains together like the rungs of a ladder. See Figure 26–10. Twisting or coiling the ladder forms the helix of the molecule. Thus, the DNA molecule is a *double helix*.

Watson and Crick found that their model could work only if the pairs of bases that made each rung of the ladder were an adenine unit connected to a thymine or a guanine connected to a cytosine. This model agreed with all the data for the DNA molecule. For example, it explained why the amount of adenine in DNA is always the same as the amount of thymine. It explained why the amount of guanine is always the same as the amount of cytosine. It also explained how, since the order of bases along the chain could vary, the order could be a code for genetic information.

In the double-helix model, the order of bases along one strand determines the matching bases on the other strand. That is, every adenine (A) must be joined to thymine (T), and every guanine (G) must be joined to cytosine (C). No other pairings are possible. Suppose, for example, that the order of bases along one strand is AGGTTAC. The matching order along the second strand must be TCCAATG. The two strands are said to be *complementary*. Each strand is the complement of the other according to the A-T and G-C base pairing rule. The double-helix model of DNA was a great breakthrough in the science of genetics. Watson, Crick, and Wilkins received the Nobel Prize for this work in 1962. Had she lived, Franklin would also have been a recipient.

Replication of DNA

The double-helix model also explains how an exact copy of each chromosome is made during cell division. The base pairs that form each rung of the model are held together by a weak *hydrogen bond*. Before copying begins, these bonds break, and the two strands of the DNA molecule come apart. This exposes the bases along each strand. The bases of free nucleotides in the nucleus of the cell can then fasten onto the complementary bases on each exposed strand. When the nucleotides join together, they make a complete comple-

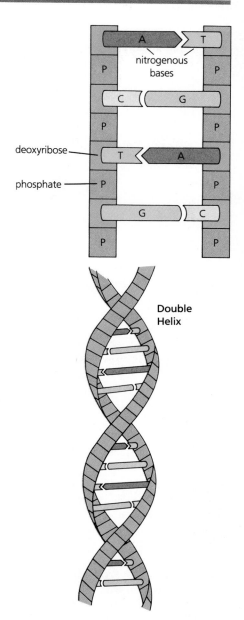

▲ **Figure 26–10**
Watson-Crick Model of DNA.

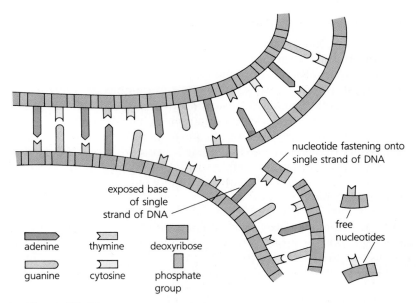

adenine thymine deoxyribose

guanine cytosine phosphate group

exposed base of single strand of DNA

nucleotide fastening onto single strand of DNA

free nucleotides

▲ **Figure 26–11**

DNA Replication. DNA replication results in the formation of two double-stranded molecules exactly like the original DNA molecule.

mentary strand exactly like the old one. In this way, two double-stranded molecules of DNA exactly like the original molecule are made. See Figure 26–11. Each double-stranded molecule contains one old strand and one new strand of DNA.

Where does replication start and end along the DNA molecule? Through experiments, scientists determined that replication does not begin at one end and continue to the other. Instead, replication begins at the same time at many points along the molecule. Enzymes then link the small segments of DNA into one long strand. In this way, a DNA molecule replicates much more rapidly than if there were only one starting point.

26-2 Section Review

1. What type of particles did Hershey and Chase use to show that DNA is the genetic material?
2. Which three chemical groups make up a nucleotide?
3. What is the shape of a DNA molecule?
4. How are the base pairs in a DNA molecule held together?

Critical Thinking

5. Suppose that a strand of DNA were to have the following base sequence: TGGCAATCTG. What would be the base sequence along the complementary strand? (*Ordering*)

26-3 Gene Expression

Section Objectives:

- *Describe* the one gene-one polypeptide hypothesis.
- *Explain* how the order of nucleotides in DNA codes for different amino acids and how this code is transcribed into RNA.
- *Compare* the structures and functions of mRNA, tRNA, and rRNA.
- *Describe* how a polypeptide is assembled.
- **Concept Laboratory:** *Gain* an understanding of DNA structure by creating a model (p. 533).

Genes and Enzymes

The idea that hereditary material controls the synthesis of enzymes was suggested in the early 1900s. Sir Archibald Garrod, an English physician, studied certain diseases that he called "inborn errors of metabolism." He hypothesized that these diseases are caused by the body's inability to make a certain enzyme. He also hypothesized that this inability was inherited. Garrod published his ideas in 1909, but they were ignored at the time. It was not until the 1930s and 1940s that scientists realized their importance.

The best evidence that genes control the production of enzymes came from the experiments of George Beadle and Edward Tatum, two American scientists, in 1941. In their experiments, Beadle and Tatum used the red bread mold *Neurospora crassa*. From the results of their experiments, Beadle and Tatum were able to conclude that each gene produces its effects by controlling the synthesis of a single enzyme. This is known as the *one gene-one enzyme hypothesis*.

One Gene-One Polypeptide Hypothesis

It is now known that genes control the synthesis of all proteins. Some proteins are enzymes. Some are hormones. Some form the structures of the cell.

As you may recall, proteins are made of polypeptides—long chains of amino acids (see Chapter 4). Some proteins consist of two or more polypeptides linked and twisted around each other. Hemoglobin, for example, is a protein made of two different polypeptide chains.

It was found that the synthesis of each polypeptide is controlled by a different gene. Because of this fact, the one gene-one enzyme hypothesis was changed to the **one gene-one polypeptide hypothesis.** According to this hypothesis, each gene directs the synthesis of a particular polypeptide chain.

The DNA Code

To direct the synthesis of a polypeptide, a gene must be able to direct the order in which amino acids are put together. It was not clear from the Watson-Crick model of DNA how this could be done.

Science, Technology and Society

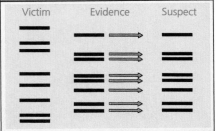

Victim   Evidence   Suspect

Technology: DNA Fingerprinting

Science has recently added a weapon to the crime lab's arsenal. From a drop of blood, strands of hair, or other biological material found at a crime scene, scientists can produce a "fingerprint" of a person's DNA. As with regular fingerprints, everyone, except identical twins, has a unique set of DNA.

To produce a DNA fingerprint, scientists isolate DNA from the blood or other organic evidence found at the scene of a crime. They also isolate DNA from a blood sample taken from a suspect. Both samples are treated with enzymes that cut the DNA into small fragments. The length of the fragments depends on the sequence of nitrogeneous bases in the DNA—which differs from person to person.

The DNA fragments in both samples are subjected to electrophoresis (see page 26) and treated to produce a black-and-white picture. The fragments in each sample create a pattern of black bands (see above). If the two patterns match, it is almost certain that the evidence came from the suspect.

■ *Imagine you are on a jury and DNA fingerprinting evidence is introduced. How would you regard such evidence? Explain.*

Among the ideas put forward was the idea that the order of bases along the DNA strands is a code that specifies the order of the amino acids.

There are 20 amino acids in the proteins of humans and most other organisms. Therefore, there must be at least 20 different code "words" to specify these amino acids. As you have read, there are 4 different bases in DNA: adenine (A), guanine (G), cytosine (C), and thymine (T). Using a 2-letter code, only 16 different 2-letter code sequences can be made from 4 bases: AA, AT, AG, AC, TT, TA, TG, TC, and so forth. This is not enough, so the code "words" must be at least 3 bases long. From 4 bases, 64 different 3-base sequences can be made—more than are needed. Research has shown that the code for specifying an amino acid is made of 3-base "words." Most amino acids are specified by more than one code "word."

Today, scientists know that a gene is made of many nucleotides. The order of the bases of three adjacent nucleotides is the code that specifies a particular amino acid to be added to a polypeptide chain. The code also contains "punctuation"—code "words" that tell where a polypeptide begins and ends. In most cases, these points are also the beginning and end of a gene.

RNA and Protein Synthesis

In cells with nuclei, the genes are found within the nucleus. Yet, protein synthesis takes place outside the nucleus. How, then, does DNA direct the synthesis of proteins? DNA does this with the help of RNA, or ribonucleic acid. The chemical makeup of RNA is similar to the chemical makeup of DNA, with two differences. In RNA, the 5-carbon sugar is ribose, and the nitrogen-containing base **uracil** (U) takes the place of thymine. Thus, the four bases found in RNA are adenine, guanine, cytosine, and uracil. The structure of RNA is also a little different. Unlike DNA, which is double-stranded, RNA is made of only a single strand of nucleotides.

Messenger RNA

The first step in directing protein synthesis is to copy the DNA code for a polypeptide into a molecule of RNA. To copy the code, the DNA strands separate for a short time and serve as a pattern, or *template,* for RNA. See Figure 26–12. Complementary RNA nucleotides take their places along the exposed strands by matching up complementary bases. When the assembled RNA sequence reaches the DNA "stop" code, it leaves the DNA strand. The RNA strand is now a separate molecule that carries the complete message for a single polypeptide in complementary form. That is, each A of the DNA is represented by a U in the RNA, each T by an A, each G by a C, and each C by a G. A strand of RNA that copies a genetic message from DNA in this way is called **messenger RNA,** or mRNA. The copying of a genetic message into a molecule of mRNA is called **transcription.** Each group of three bases on the mRNA that specifies an amino acid is called a **codon** (KOH dahn).

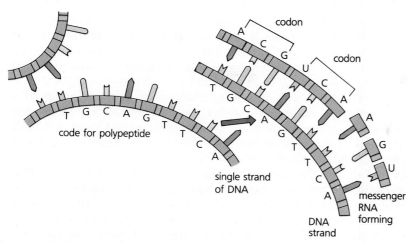

▲ **Figure 26–12**

Transcription of a Gene. The code for each polypeptide is copied from one of the DNA strands into a strand of messenger RNA. The copying process is similar to DNA replication, except that uracil replaces thymine as a complement for adenine.

Transfer RNA

Messenger RNA is only one of three kinds of RNA that are found in the cell. A second kind is called **transfer RNA,** or tRNA. While mRNA may have thousands of nucleotides along its length, tRNA has only about 80. The molecule of tRNA has an odd shape, as shown in Figure 26–13. At one end there is a short tail. A particular amino acid can become attached to this tail.

Each tRNA molecule will pick up only one kind of amino acid. There are 20 different forms of tRNA, one for each of the 20 different amino acids. At the other end of the tRNA molecule, there is a loop of exposed nucleotides. In this loop, there is a sequence of 3 bases, called an **anticodon,** that are complements of an mRNA codon. The codon that this anticodon matches is one that specifies the amino acid that each tRNA carries. Thus, tRNA is a device for bringing a certain amino acid to a certain place specified by mRNA.

Ribosomal RNA

Ribosomal RNA, or rRNA, is formed in the nucleoli of the cell. A ribosome consists of protein and rRNA. The ribosomal protein is made in the cytoplasm and then travels into the nucleus. In the nucleoli, the protein and the rRNA join together to form complete ribosomes. The ribosome is where a polypeptide is assembled during protein synthesis.

Assembly of a Polypeptide

The synthesis of the three kinds of RNA, as well as the assembly of ribosomes, occurs in the cell nucleus. The RNA and complete ribosomes migrate separately through the nuclear pores to the

Figure 26–13

Transfer RNA. The mRNA codon that matches the anticodon of the tRNA calls for the particular amino acid that is attached to the tail end of the tRNA. ▼

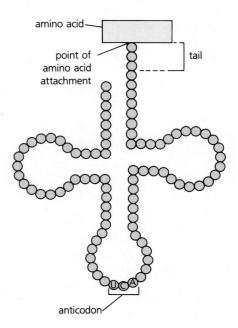

cytoplasm. Within the cytoplasm, there is contained a supply of all the amino acids that are needed to make the cell's proteins. Figure 26-14 lists all codons and the amino acids each codes for. This is called the genetic code. Within the cytoplasm polypeptides are assembled according to the instructions carried by mRNA.

In the cytoplasm, amino acid molecules become attached to their specific varieties of tRNA. See Figure 26-15. Ribosomes become attached to different places along each strand of mRNA. Where a ribosome is attached to mRNA, a molecule of tRNA with the right anticodon temporarily joins with the corresponding codon on the mRNA. The amino acid brought into position by the tRNA joins the last amino acid in the chain and separates from tRNA. The ribosome then moves along to the next codon. A new tRNA takes its place on the mRNA strand, and its amino acid joins the polypeptide chain. As the ribosome moves along, the tRNA that has delivered its amino acid is released. It is now free to pick up another amino acid and deliver it to the right place for assembly into the chain. Amino

Figure 26–14

The Genetic Code. Most of the amino acids are specified by more than one codon. For example, GCU, GCC, GCA, and GCG all code for the amino acid alanine. ▼

The Genetic Code					
First Base in Codon	**Second Base in Codon**				**Third Base in Codon**
	U	C	A	G	
U	phenylalanine	serine	tyrosine	cysteine	U
	phenylalanine	serine	tyrosine	cysteine	C
	leucine	serine	stop	stop	A
	leucine	serine	stop	tryptophan	G
C	leucine	proline	histidine	arginine	U
	leucine	proline	histidine	arginine	C
	leucine	proline	glutamine	arginine	A
	leucine	proline	glutamine	arginine	G
A	isoleucine	threonine	asparagine	serine	U
	isoleucine	threonine	asparagine	serine	C
	isoleucine (start);	threonine	lysine	arginine	A
	methionine	threonine	lysine	arginine	G
G	valine	alanine	aspartate	glycine	U
	valine	alanine	aspartate	glycine	C
	valine	alanine	glutamate	glycine	A
	valine	alanine	glutamate	glycine	G

An mRNA codon consists of three nucleotides. For example ACU codes threonine. The first letter, A, is read in the first column; the second letter C, from the second letter column; and the third letter, U, from the third letter column. Most amino acids are specified by more than one codon.

Concept Laboratory

A Model of DNA

Goal

To gain an understanding of the structure and function of DNA.

Scenario

Imagine that you are working in the laboratory of James Watson and Francis Crick in 1953. Your task is to take all the known information about DNA and come up with a three-dimensional structure that is consistent with that information.

As you do this laboratory, think about how understanding the three-dimensional structure of DNA helps you to understand the way DNA functions.

Materials

ring stand
two iron rings
string
scissors

white straws
Cheerios
round toothpicks
four different colored straws

Procedure ☞ *

A. The information you must take into account while constructing a model of a DNA molecule is as follows:
1. The repeating unit of DNA is a nucleotide, consisting of deoxyribose (a 5-carbon sugar), a phosphate group, and any one of four bases: adenine (A), guanine (G), cytosine (C), and thymine (T).
2. The nucleotides are joined one to another, forming a long-stranded molecule.
3. The backbone of a strand is made up of alternating deoxyribose and phosphate groups.
4. Two strands bond together to form a double-stranded molecule that is twisted, forming a helix.
5. The bases in each strand bond with a complementary base strand. Adenine in one strand always bonds to thymine in the other strand; similarly, cytosine in one strand always bonds to guanine in the other strand.

B. To support your model, attach one iron ring near the bottom of the stand, the other near the top. As shown below, tie two 0.5 m lengths of string to the bottom ring, about a toothpick length apart.

C. Cut all straws into 2 cm lengths. Thread white straws and Cheerios alternately on each string, and tie

*For detailed information on safety symbols, Appendix B.

the ends of completed strings to the top ring.

D. To represent the bases, use the sections of colored straws: adenine (red), guanine (blue), cytosine (yellow), thymine (green). Insert onto each toothpick complementary pairs of bases—adenine with thymine and guanine with cytosine.

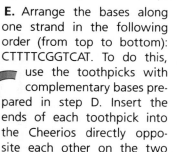

E. Arrange the bases along one strand in the following order (from top to bottom): CTTTTCGGTCAT. To do this, use the toothpicks with complementary bases prepared in step D. Insert the ends of each toothpick into the Cheerios directly opposite each other on the two strands.

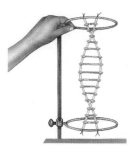

F. To make a helix, untie the strings on the top ring, and switch their positions.

Organizing Your Data

Draw a labeled diagram of your DNA model.

Drawing Conclusions

1. What do the Cheerios and white straws in your model represent?

2. Based on your model, describe the general structure of DNA.

Thinking Further

1. How is the structure of DNA related to its ability to replicate?

2. What mRNA sequence would be transcribed from the nucleotide sequence specified in step E? What amino acid sequence would result? (See Figure 26–14.)

3. Suppose that an error resulted in changing the first base of the DNA sequence from C to T. How would this affect the amino acid sequence?

Figure 26–15

Protein Synthesis. As a ribosome moves into position at a codon of a messenger RNA, a transfer RNA with the complementary anticodon temporarily bonds to the codon. The tRNA adds its amino acid to the polypeptide chain, then detaches and moves away. ▶

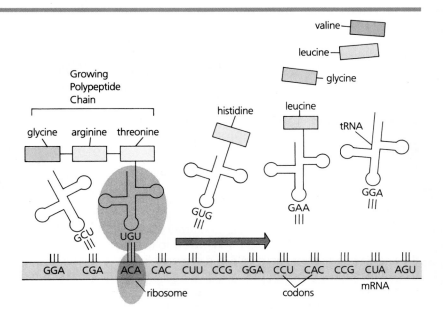

acids are added to the growing polypeptide chain until the ribosome reaches a "stop" codon. The polypeptide is then let go, and it forms itself into a complete protein molecule.

Every step in the process of translating the genetic message into a polypeptide chain is helped by a specific enzyme. There are enzymes that attach amino acids to tRNA. There are enzymes that attach tRNA to mRNA. There are enzymes that join the amino acids to the polypeptide chain. There are also enzymes in the nucleus that open the DNA molecule for transcription, and others that help in the assembly of RNA. All of these enzymes must be specified by genes.

The process by which the information coded in RNA is used for the assembly of a particular amino acid sequence is known as **translation.** Figure 26–16 shows how a genetic message is first

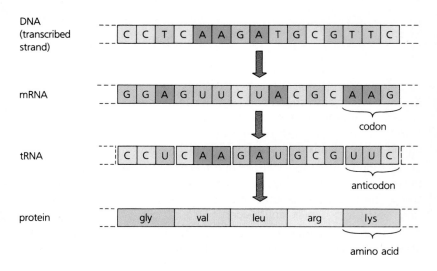

Figure 26–16

From DNA to Proteins. In transcription, the genetic code of DNA is copied into a molecule of mRNA. In translation, the information coded in mRNA is used to assemble a specific amino acid sequence, forming a polypeptide. This is done with the help of tRNA. ▶

transcribed from DNA to RNA and then translated into a polypeptide. By this remarkable system, all the cell's proteins are synthesized in the cytoplasm, while the chromosomes carrying the hereditary instructions for this synthesis remain in the nucleus.

26-3 **Section Review**

1. Which hypothesis states that each gene directs the synthesis of a particular chain of a protein?
2. How is information encoded in a gene?
3. What is the process by which genetic messages are copied into an RNA molecule?
4. Where in the cell are proteins made?

Critical Thinking

5. Suppose that, instead of the correct sequence AGC, an error occurred that changed the sequence to ATC. What would happen upon transcription? What would happen upon translation? (*Predicting*)

26-4 Control of Gene Expression

Section Objectives:

- *Explain* how gene expression is regulated in prokaryotes and in eukaryotes.
- *Compare* the operon of prokaryotes with the genes and control sections of DNA in eukaryotes.
- *Describe* homeotic genes and oncogenes.

Every cell in an organism has the complete set of genes characteristic of that organism. But, even though all the cells of an organism have the same genes, different cells perform different functions and produce different proteins. Why is a particular set of genes activated in one cell, while a different set is activated in another cell of the same organism? One of the major questions that biologists are trying to answer is what determines which proteins are produced in a given cell.

Gene Expression in Bacteria

In the early 1960s, three French biologists, François Jacob, Jacques Monod, and André Lwoff, discovered how the transcription of certain genes is controlled in the bacterium *E. coli*. They were awarded a Nobel Prize in 1965 for their work. Jacob, Monod, and Lwoff studied the production of the three enzymes the bacteria use to digest lactose, a sugar. They found that the enzymes are produced by the bacteria only when they are needed. That is, the bacteria produce the lactose-digesting enzymes when lactose is

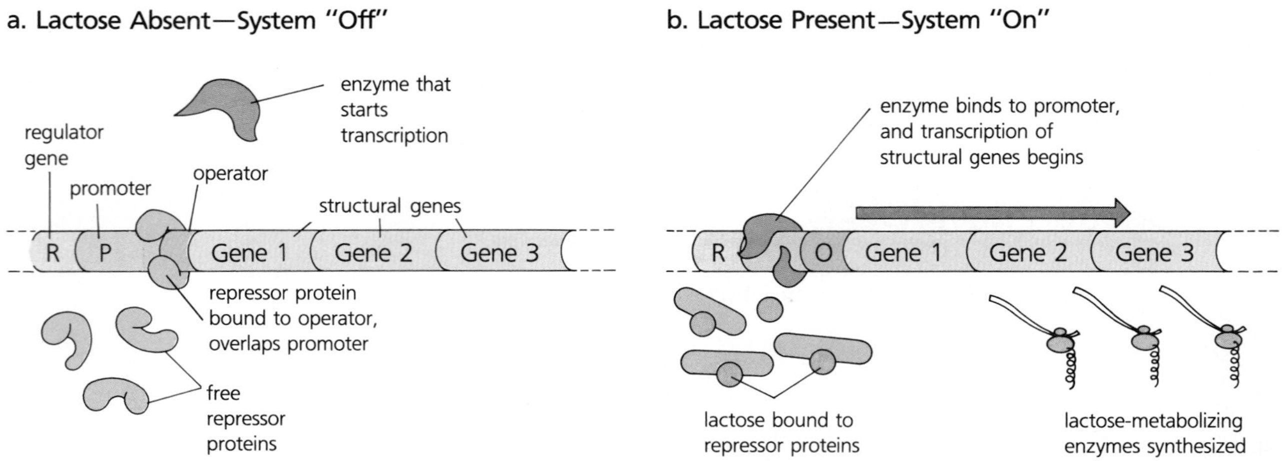

a. Lactose Absent—System "Off"

enzyme that
starts
transcription

regulator
gene

promoter

operator

structural genes

R | P | Gene 1 | Gene 2 | Gene 3

repressor protein
bound to operator,
overlaps promoter

free
repressor
proteins

b. Lactose Present—System "On"

enzyme binds to promoter,
and transcription of
structural genes begins

R | O | Gene 1 | Gene 2 | Gene 3

lactose bound to
repressor proteins

lactose-metabolizing
enzymes synthesized

▲ **Figure 26–17**

Synthesis of Lactose-Digesting Enzymes in Bacteria. (A) When lactose is not present, repressor proteins bind to the operator of the operon. Because the repressor protein is much larger than the operator gene, it overlaps the promoter gene as well. The enzyme that starts transcription of the structural genes cannot bind to the promoter. No lactose-digesting enzymes are produced.

(B) When lactose is present, the lactose binds to the repressor protein. This means that the lactose-repressor complex cannot bind to the operator. Therefore, the enzyme that starts transcription is able to bind to the promoter, which means that lactose-digesting enzymes are produced.

present and other sources of energy are not available. Thus, enzyme production is turned on and off, depending on the needs of the cell.

Jacob, Monod, and Lwoff determined that production of the lactose-digesting enzymes is controlled by a cluster of genes. The amino acid sequences of the enzymes are determined by three structural genes. A *structural gene* is a DNA segment that codes for the production of a particular polypeptide.

The investigators found that the activity of the structural genes is controlled by an *operator gene,* a sequence of nucleotides found next to the structural genes. The structural genes cannot be transcribed unless the operator gene is in an active state. We can think of the operator as being switched "on" to cause transcription or switched "off" to prevent transcription. See Figure 26–17.

The activity of the operator is controlled by a protein called a *repressor protein.* The repressor protein is the product of another gene, called a *regulator gene.* When the repressor protein binds to the operator gene, transcription of the structural genes cannot start. The repressor protein is always present in the cell, and it is normally bound to the operator gene so that the operator is off. When lactose binds to the repressor protein, the protein changes shape. This makes it unable to bind to the operator gene. Therefore, when lactose is present, the operator gene is turned on. Then, transcription of the structural genes proceeds, and lactose-digesting enzymes are produced.

There is another control gene, called a *promoter,* that also plays a role in gene expression in bacteria. The promoter attracts and binds the enzyme that starts transcription. If the enzyme cannot bind to the promoter, transcription of the structural genes will not start. The promoter and the operator control the copying of a cluster of structural genes. In prokaryotes, such as bacteria, the promoter, the operator, and their associated structural genes are called an **operon** (OP er on).

Gene Expression in Higher Organisms

Scientists first thought that the control of gene expression in higher organisms (eukaryotes) would be similar to that in prokaryotes. However, this is not so for several reasons. In the first place, eukaryote genes relating to a certain function are not clustered together. Related genes are far apart from one another on a chromosome. Sometimes, they are on different chromosomes entirely. To further complicate matters, each eukaryote gene is split into parts, called *exons,* that are not next to each other. An **exon** is a segment of DNA that codes for amino acids that will become part of a protein. In between the exons, there are sections of DNA, called *introns.* An **intron** is a segment of DNA that does not code for amino acids of a protein.

When a gene is transcribed, both the introns and exons are made into RNA. The introns are then "edited out." To do this, enzymes cut out the intron RNA and then join, or splice, together the exon RNA pieces. Before this happens, two other strings of RNA nucleotides, one called a cap and the other a tail, are added to the ends of the RNA. See Figure 26–18. Because the enzymes can cut and splice the RNA in different ways, different proteins can be made from the same DNA segment.

Changes in the amount of coiling or in the shape of the DNA also affect the expression of genes in eukaryotes. A change in

Figure 26–18

mRNA Synthesis in Eukaryotes. Both the exons and introns are transcribed, producing one long RNA molecule. More nucleotides are added to the ends of the RNA. Enzymes then cut out the RNA introns and splice together the exons to form the completed messenger RNA. ▼

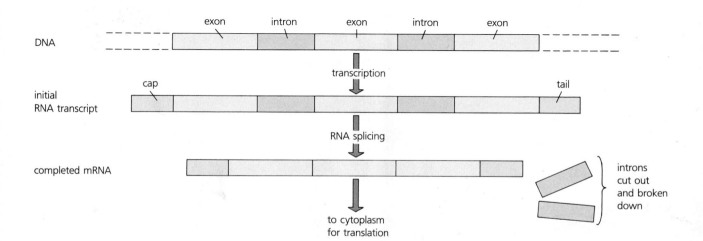

coiling or shape determines which parts of the DNA are accessible to enzymes and other proteins and to RNA. In turn, this accessibility influences the process of transcription. As you may recall from the discussion in Chapter 20, chromatin is made up of DNA wound around groups of proteins called histones. It is now thought that histones control the coiling and uncoiling of DNA in chromatin. Tightly packed DNA is not transcribed. Loosely packed DNA is transcribed. Finally, in some eukaryotes, certain chemical groups can attach to sections of DNA, changing its shape and reducing transcription.

Like the DNA of prokaryotes, the DNA of eukaryotes has sections that control the copying of genes. The control sections are not part of the gene itself. Instead, they are found elsewhere in the DNA. For example, eukaryotes have promoters that attract and bind the enzyme that starts the copying process. Another section of DNA, called the **enhancer,** controls the access of the enzyme to the promoter. Unlike the case in prokaryotes, in eukaryotes, control sections, such as enhancers, are usually far away from the genes they affect.

Gene Expression and the Environment

Some environmental factors also can switch genes on and off. You have already read about bacteria producing an enzyme when a certain sugar is present. In green plants, some development genes are switched on and off by light. Changes in temperature cause changes in the expression of the genes governing fur color in the Himalayan rabbit.

The Himalayan rabbit has white fur over most of its body, with black fur on the ear, nose, feet, and tail. See Figure 26–19. This pattern is produced by differences in temperature in certain parts of the body. A gene controls the production of black pigment. Black pigment is deposited in the fur over parts of the body in which the temperature falls below 33°C. This can be shown by placing an ice pack on a shaved area on the back of a Himalayan rabbit. Where the ice pack has lowered the temperature, the new growth of fur will be

Figure 26–19

Effect of Body Temperature on Fur Color of the Himalayan Rabbit. Cold temperatures turn on the gene that controls the production of black pigment in the Himalayan rabbit. ▼

black. Genes carry the basic information for all traits, but the phenotypes of organisms often can be changed by environmental factors that switch genes on and off.

In some reptiles, the incubation temperature of the eggs determines the sex of the offspring. In painted turtles, for example, high incubation temperatures tend to produce females, while low incubation temperatures produce mostly males.

Gene Expression in Development

During the development of an organism, different genes must be active at different times in its life cycle. The question of how genes are switched on and off during development has been studied thoroughly in the fruit fly *Drosophila.* Within the embryo of the fruit fly, scientists have discovered a group of genes, called **homeotic** (ho mee OH tik) **genes.** These genes control the key events in the development of a fruit fly. Homeotic genes switch other genes on and off. They do this by coding for homeotic proteins. It is thought that the homeotic proteins bind certain parts of the DNA and thus control the process of transcription. Similar sequences of DNA, known as *homeoboxes,* have also been found in many other animals, including humans.

Oncogenes and Cancer

Understanding the expression of genes may someday lead to a cure for cancer. **Oncogenes** (ON koh genes) are genes that cause some kinds of cancer. Oncogenes are present in most human cells, but usually they are switched off, or they are expressed in a way that does not cause cancer. When oncogenes are switched on, or when they begin to operate in an abnormal way, they lead to the uncontrolled growth of cells that we call cancer. If scientists can find a way to switch the oncogenes off, they may be able to develop a cure for many cancers.

26-4 **Section Review**

1. What is the term for the promoter, the operator, and the associated structural genes in prokaryotes?
2. What are the pieces of a split gene called in eukaryotes?
3. In what organism were homeotic genes first identified?
4. What environmental factor determines whether or not a Himalayan rabbit has black fur?

Critical Thinking

5. List some enviromental hazards that might cause oncogenes to be switched on. (*Identifying Causes*)

Laboratory Investigation

Inheritance of Human Traits

Sexual reproduction ensures that offspring are different from their parents and from each other. In this investigation, you will trace the possible path of alleles—as they are shuffled and recombined during meiosis and gamete fusion—from a wife and husband to their potential offspring.

Objective

- *Visualize* how meiosis and gamete fusion account for the possible combinations of alleles in offspring.

Materials

red and blue construction paper   scissors
pen or marker

Procedure

* **A.** Cut out two number 7 chromosomes and two number 19 chromosomes of each color, as shown below. Use red paper for the wife's chromosomes and blue paper for the husband's, and represent their alleles with the indicated letters.

Human Chromosome	Genotype			
	Wife		**Husband**	
No. 7	C / F	C / f	C / f	c / F
No. 19	G	g	g	g

B. Assume that the wife's and husband's chromosomes are each in separate cells that undergo meiosis. To treat nonhomologous chromosomes randomly with respect to each other during metaphase I, flip them over so you cannot tell one from another.

C. Pair up homologous chromosomes in each cell as they would pair during metaphase I. (For simplicity, keep each chromosome as a single rather than a double-stranded chromosome.)

*For detailed information on safety symbols, see Appendix B.

D. Segregate the homologous chromosomes, sorting the wife's chromosomes into two eggs and the husband's into two sperm. One red chromosome of each kind should be in each egg; one blue chromosome of each kind should be in each sperm. Flip over the chromosomes; record the genotypes of resulting gametes under trial 1 in a chart like the one below.

Trial	Genotypes		
	Egg	**Sperm**	**Offspring**
I	1. 2.	1. 2.	1. 2. 3. 4.

E. Combine one egg and one sperm, and record the genotype of the offspring. Record the other equally likely offspring genotypes that would result from random combinations of these eggs and sperm.

F. To again simulate independent assortment, flip the chromosomes over so you cannot identify them, and repeat steps B through E. Record the genotypes of the eggs, sperm, and potential offspring. If your results are the same as in trial 1, repeat this step a third time.

Analysis and Conclusions

1. Which two genes are linked?

2. Draw out all the possible arrangements of chromosomes at metaphase I.

3. List all the possible egg and sperm genotypes that resulted from independent assortment. List all the possible offspring genotypes.

4. Why does using single, instead of double-stranded, chromosomes in step C not affect the results?

5. Based on your results, explain how the arrangement of nonhomologous chromosomes at metaphase I affects allele combinations in the gametes.

6. How would it be possible for an offspring to have the genotype CCFFgg?

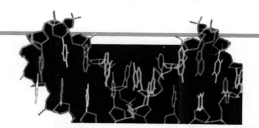

Chapter Review

Study Outline

26-1 Chromosomal Inheritance

■ T. H. Morgan provided the first evidence that genes are parts of chromosomes, using *Drosophila* as his experimental animal.

■ A pair of unmatched chromosomes, the sex chromosomes, determines the sex of an individual. In humans and many other organisms, females have XX sex chromosomes and males XY.

■ A sex-linked trait is a trait controlled by a gene located on a sex chromosome.

■ Some human diseases, such as hemophilia, color blindness, and muscular dystrophy, are caused by abnormal recessive alleles of certain genes on the X chromosome.

■ Genes located on the same chromosome cannot be distributed independently during meiosis because the genes are linked together. This condition is called gene linkage. Linked genes may be separated by crossing-over of chromosomes during synapsis.

■ Some traits, such as human height, are controlled by the alleles of two or more different genes, and not by the alleles of a single gene. When two or more genes affect an inherited characteristic, it is called multiple-gene inheritance.

26-2 The Genetic Material

■ Experiments performed by Griffith, by Avery, McLeod, and McCarty, and by Hershey and Chase showed that DNA is the hereditary material.

■ DNA is composed of deoxyribose, a phosphate group, and four nitrogen-containing bases: adenine and guanine, which are purines; and cytosine and thymine, which are pyrimidines.

■ Watson and Crick discovered the structure of the DNA molecule. Their model of DNA shows the molecule as a double helix.

■ DNA replicates by separating at the nitrogen bases and then attaching new complementary bases to the exposed bases to form a new strand of DNA.

26-3 Gene Expression

■ Proteins, which form enzymes, hormones, and cellular structures, are composed of polypeptides. According to the one gene-one polypeptide hypothesis, each gene directs the synthesis of a particular polypeptide chain.

■ The DNA code for each of the 20 amino acids consists of three nitrogen bases. The genetic message from DNA is transcribed into messenger RNA. Each group of three nitrogen bases on mRNA, called a codon, specifies an amino acid.

■ In the cytoplasm, each transfer RNA carries a specific amino acid and becomes temporarily joined by its anticodon to the codon of mRNA. A series of tRNAs carry amino acids that are assembled into a chain to form the polypeptide.

■ The process by which the information coded in RNA is used for the assembly of a certain amino acid sequence is known as translation. Each step in the process of translating the genetic message into a polypeptide chain is helped by a specific enzyme.

26-4 Control of Gene Expression

■ The expression of certain genes in prokaryotes, such as bacteria, involves an operon, which consists of a structural gene, an operator gene, a regulator gene, and a promoter gene.

■ Gene expression in higher organisms (eukaryotes) involves genes that are far apart on a chromosome, or else are on entirely different chromosomes, and are split into introns and exons. During transcription, the introns are edited out and the exons are translated into polypeptides.

■ Gene expression can be modified by environmental factors.

■ During the development of an organism, different genes are active at different times in the life cycle.

■ Oncogenes, or genes that cause some types of cancer, are present in most human cells, but they are usually switched off or expressed in a way that does not cause cancer.

Chapter Review

Vocabulary Review

sex chromosomes (518)

autosomes (518)

X chromosomes (518)

Y chromosome (518)

sex-linked trait (520)

color blindness (520)

linkage group (521)

crossing-over (522)

multiple-gene inheritance (523)

deoxyribose (526)

adenine (526)

guanine (526)

cytosine (526)

thymine (526)

nucleotide (526)

helix (527)

one gene-one polypeptide hypothesis (529)

messenger RNA (531)

transcription (531)

codon (531)

transfer RNA (532)

anticodon (532)

ribosomal RNA (532)

translation (534)

operon (537)

exon (537)

intron (537)

enhancer (538)

homeotic genes (539)

oncogenes (539)

A. Sentence Completion—Fill in the vocabulary term that best completes each statement.

1. Genes located on the same chromosome are said to be a _____.

2. Most sex-linked traits are determined by genes found on the _____.

3. Genes that control the key events in the development of a fruit fly are called _____.

4. The chromosomes in a cell that are not sex chromosomes are _____.

5. Homologous chromosomes sometimes exchange pieces in a process called _____.

6. Copying a genetic message from DNA into messenger RNA is called _____.

7. During protein synthesis, amino acids are carried by _____.

8. Some kinds of cancer are caused by _____.

9. During the process of _____, the information that is coded in RNA is used to synthesize a polypeptide.

10. The _____ is a group of three nitrogen bases that specifies an amino acid.

B. Word Relationships—In each of the following sets of terms, three of the terms have a common characteristic. One term does not belong. Identify the term that does not belong.

11. sex-linked trait, X-chromosomes, color blindness, anticodon

12. deoxyribose, adenine, thymine, cytosine

13. messenger RNA, transcription, crossing-over, codon

14. operon, intron, exon, Y chromosome

Content Review

15. Why are *Drosophila* useful for genetic experiments?

16. How do the sex chromosomes determine the sex of a human offspring?

17. Explain why hemophilia and color blindness occur much more often in men than in women.

18. How does crossing-over influence gene linkage?

19. What is a gene map and how is it constructed?

20. Describe Griffith's experiments with pneumococcus bacteria. What conclusions did he draw from them?

21. What information did the experiments of Avery, McLeod, and McCarty yield?

22. How did Hershey and Chase demonstrate that DNA carries hereditary information?

23. Draw the Watson-Crick model of DNA and describe the components.

24. How does DNA replicate?

25. Compare the chemical composition and structure of RNA and DNA.

26. Describe the experiments of Beadle and Tatum. What did they conclude from their experiments? What is the currently accepted form of their hypothesis?

27. In what form is the hereditary information encoded in the DNA molecule?

28. How is the genetic message transcribed into mRNA?

29. What are the roles of mRNA, tRNA, and ribosomes in synthesizing a polypeptide?

30. Describe how the operator and the repressor function in controlling the synthesis of lactose-digesting enzymes in the bacterium *E. coli*.

31. What are the roles of introns and exons in gene expression of higher organisms?

32. How are genes switched on and off during the development of a *Drosophila* embryo?

Graphic Organizing

For information on graphic organizers, see Appendix G at the back of this text.

33. **Bar Graph:** In multiple-gene inheritance, phenotypes vary from one extreme to another, with the phenotypes of most individuals falling between the two extremes. The chart below shows the number of plants of various heights in a sample of 64 plants of a certain species. Use the data to construct a bar graph showing the frequency of each phenotype. Plot the frequency of the phenotypes on the vertical axis and the height of the plants on the horizontal axis.

Height in cm	Number of Plants
10	1
11	6
12	15
13	20
14	15
15	6
16	1

34. **Compare/Contrast Matrix:** Construct a matrix that compares and contrasts the characteristics of DNA and RNA.

Critical Thinking

35. List some inheritance factors that may cause deviations from the Mendelian ratios that are expected. (*Reasoning Conditionally*)

36. Compare sex determination in humans and in butterflies. (*Comparing and Contrasting*)

37. How did Morgan come to the conclusion that chromosomes in *Drosophila* could be mapped? (*Identifying Reasons*)

Creative Thinking

38. In humans, ear lobe appearance—free versus attached—is an inherited characteristic. What method could you use to determine whether the free-ear-lobe trait is dominant, recessive, or sex-linked?

39. Propose an explanation for how the structure of tRNA is suited to its functions.

Problem Solving

40. What is the probability that two parents with normal color vision will have color blind sons and daughters if the mother's father is color blind?

41. In Holstein cattle, the amount of spotting depends upon multiple genes. If two pairs of multiple genes are involved, how many different degrees of spotting could be produced by parents that are heterozygous for the two genes (*SsTt* × *SsTt*)? Assume that the genotype *SSTT* produces the most numerous spots, *sstt* the fewest spots, and *SsTt* intermediate spotting.

42. The sequence of bases of a strand of a DNA molecule is AAATTTGGG. What sequence of bases of mRNA would be transcribed from this? Which amino acids would be produced during translation?

43. How would protein synthesis and the organism be affected if (a) the codon GAA were changed to GAG? (b) If the codon GAA were changed to CAA? (c) What would happen if the G were deleted from the codon?

44. As shown below, when DNA is transferred from capsulated bacteria (Type I) to noncapsulated bacteria (Type II), the Type II bacteria produce capsulated offspring (Type III) similar to Type I. As they reproduce, Type III bacteria produce capsulated cells. What conclusions can be drawn from this experiment?

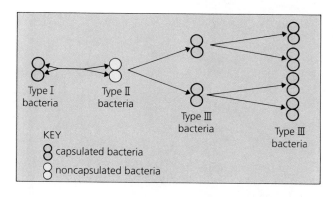

Projects

45. Write a report on recent developments in the field of oncogenes. Explain how oncogenes are expressed and what their effects are on the cell.

46. Make a model of DNA or RNA using simple, everyday materials.

▲ A small change in the gene that controls pigment creates an albino koala.

Applied Genetics

A female koala climbs through the foliage of an Australian forest. Like all koalas, this one craves eucalyptus leaves, rarely sets foot on the ground, and naps most of the day. She is a normal koala in every way—except one. She is an albino. Her fur is white and her skin is pink, not beige and chocolate brown like other koalas. The reason for this lies within the nucleus of each of her cells. Here, the directions to make pigment have been altered. This chapter looks at the mechanisms of heredity—how traits are passed from generation to generation—and how changes sometimes come about.

Guide for Reading

Key words: mutation, karyotyping, amniocentesis, clone, recombinant DNA, gene therapy

Questions to think about:

What is the original source of all variations in a species?

What are some genetic disorders that affect humans?

How is genetic engineering used to produce desired traits in organisms?

27-1 Mutations

Section Objectives:

- *List* some sources of genetic variation.
- *Distinguish* between gene mutations, chromosomal mutations, and jumping genes.
- *Describe* the kinds of chromosome and gene mutations.
- *Explain* the causes of mutations.
- **Laboratory Investigation:** *Explore* the genetic variation that exists in humans (p. 562).

The Sources of Variation

As you have read, sexual reproduction brings about variation, or inherited differences, among offspring. The offspring of sexually reproducing organisms are genetically different from either parent. It is genetic variation that allows species to adapt to changing environments. It also allows breeders to develop new *strains* of plants and animals.

Most of the variation among individuals is the result of segregation and crossing over during meiosis and recombination during fertilization.

The differences in the genetic material of the members of a population can be traced back to mutations. A **mutation** (myoo TAY shun) is a sudden change in the structure or the amount of genetic

▲ **Figure 27–1**

Genetic Variation. Variation in genetic makeup is what makes people appear different from one another.

545

material. While most mutations are harmful to an organism, some have no effect, and others are beneficial. Beneficial mutations are the source of the variations that allow species to meet the needs of their environment.

Kinds of Mutations

Plant and animal breeders have known for a long time that new inherited traits, or mutations, may suddenly appear in a strain of plant or animal. The first individuals showing the new trait are called *mutants* (MYOOT unts). The Dutch botanist Hugo De Vries, one of the scientists who rediscovered Mendel's work, developed the concept of mutations. De Vries first observed mutations in a plant known as the evening primrose.

It is now known that there are two kinds of mutations. A **chromosomal mutation** is an abnormal change in the structure of all or part of a chromosome or in the number of chromosomes an organism has. The mutations De Vries saw in the evening primrose were chromosomal mutations. The other type of mutation is a gene mutation. A **gene mutation** is a change that affects a gene on a chromosome. The white-eyed male fruit fly that T. H. Morgan discovered was the result of a mutation of the gene for eye color.

For a mutation to be inherited in a sexually reproducing organism, it must be present in the DNA of a gamete. Thus, the mutation must occur in a gamete or in any cell from which a gamete develops. Mutations that occur in body cells normally cannot be inherited in sexual organisms, since body cells are not transmitted to offspring.

Causes of Mutations

Scientists have studied mutations that produce observable changes in traits in *Drosophila* and in other organisms. Each kind of mutation seems to occur naturally at a certain low rate in large populations. Although all the causes of natural mutation are not known, many may be the result of random errors in replication of the DNA.

When factors in the environment cause mutations, the factors are called **mutagens.** Hermann Muller, a student of T. H. Morgan, found that he could greatly increase the number of mutations in fruit flies by exposing them to X rays. Scientists now know that radiation, such as X rays and ultraviolet light, and chemicals, such as chloroform and mustard gas, are mutagens. Many mutations may be partly the result of mutagens that occur in the environment.

Chromosomal Mutations

Changes in Chromosome Structure Permanent changes in chromosome structure sometimes take place during meiosis. The chromatids can become entangled and then their parts may be rearranged in several ways. These types of changes should not be

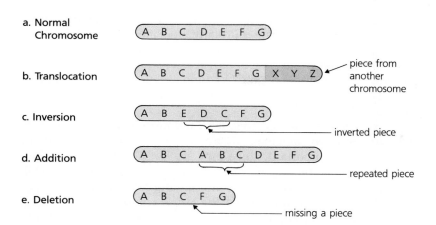

◄ Figure 27–2
Changes in Chromosome Structure.

a. Normal Chromosome

b. Translocation — piece from another chromosome

c. Inversion — inverted piece

d. Addition — repeated piece

e. Deletion — missing a piece

confused with crossing-over. **Translocation** is the transfer of a part of a chromosome to a nonhomologous chromosome. **Inversion** occurs when a piece of chromosome is rotated, which reverses the order of genes in the segment. **Addition** happens when a piece of a chromosome breaks off and attaches to a homologous chromosome. The homologous chromosome then has some gene repeated. **Deletion** occurs when a piece of a chromosome breaks off, resulting in the loss of some genes. See Figure 27–2.

Nondisjunction   The addition or loss of a whole chromosome is called **nondisjunction** (non dis JUNK shun). Nondisjunction takes place when chromosomes that normally separate during meiosis remain together. Nondisjunction causes several genetic disorders in humans.

Polyploidy   A condition in which cells have some multiple of the normal chromosome number is called **polyploidy** (PAHL ee ployd ee). Polyploidy occurs when chromosomes do not separate normally during mitosis or meiosis. This condition is commonly found in plant cells. For example, the number of chromosomes in some plant cells may be 3n, 4n, or even 5n. Polyploid plants and their fruits are often larger than normal. Plant breeders sometimes use chemicals to develop polyploid plants. See Figure 27–3.

Gene Mutations

Genes specify the order of amino acids in a polypeptide chain for a specific protein. This information is coded into the sequence of bases along a DNA strand. Any change in this sequence is likely to change the message transcribed into mRNA. This is likely to change the structure of the protein that the cell makes. These types of changes are called *gene mutations*. Once a mutation occurs in a DNA molecule, it is copied in all the later replications of the DNA.

A **point mutation** is a type of gene mutation in which only a single nucleotide in a gene has been changed. When a single nucleotide is added or removed at some point, the change is quite drastic. All the triplet codons beyond that point are changed. The

Figure 27–3
Polyploidy. The plump strawberries in the top row are from polyploid plants. The smaller ones in the bottom row are from normal plants. ▼

▲ **Figure 27–4**

Mutation. The crossed bill on this robin is the result of a gene mutation.

mutation usually makes the gene useless, and the organism lacks the protein that the gene normally specifies. Many inherited diseases are the result of this type of gene mutation.

In other cases, one base in a DNA nucleotide is substituted for another. This changes one mRNA codon and one amino acid in the protein. Changing one amino acid may result in a protein that does not function normally.

Most scientists believe that gene mutations occur from time to time randomly in all cells. Mutations in individual body cells usually are not important, since they are not likely to affect other cells or the functions of the organism as a whole. If a single cell loses the ability to make a certain protein, the cell may die, or it may obtain the protein from outside the cell. In either case, nothing noticeable happens. Mutations in sex cells, however, are important. If a mutation is present in a gamete at the time of fertilization, all the cells of the embryo and the adult organism will have the mutation.

Inherited gene mutations are usually recessive. Only about 1 in 100 gene mutations is dominant. Most gene mutations are harmful to the organism because the genes of a normal individual already meet the needs of the organism. See Figure 27–4. Any change is likely to result in a useless protein, or no protein at all, in place of one that is needed.

Jumping Genes

Genes that move or jump from chromosome to chromosome cause another kind of mutation. Although most genes stay in one place, a few are able to move to new locations on the chromosomes during replication. These mobile elements, or "jumping genes," were discovered by Barbara McClintock, an American geneticist. When a gene takes a new position in or near another gene during replication, it often causes an inactivation of that gene. Jumping genes are an important source of variation and are thought to exist in all species. McClintock was awarded the Nobel prize in 1983 for her discovery.

27-1 **Section Review**

1. What is the ultimate source of genetic differences in a population?
2. Give an example of a mutagen.
3. Name six kinds of chromosomal mutations.
4. What kind of mutation involves a change in one DNA nucleotide in a chromosome?

Critical Thinking

5. Which type of gene mutation would you expect to be more harmful to a protein—one where one nucleotide is removed or one where three nucleotides are removed? Explain. (*Ranking*)

27-2 Human Genetic Disorders

Section Objectives:

- *Explain* some of the difficulties that arise in studying human genetics.
- *List* some sex-linked disorders, recessive disorders, dominant disorders, and chromosomal disorders.
- *Describe* some techniques that are used in the diagnosis of some human genetic disorders.

Studying Human Heredity

Although a great deal is known about human genetics, heredity in humans cannot be studied in the same way it is studied in plants and other animals. The time between generations is too long, the number of offspring produced is too small, and no controlled experiments are possible.

Scientists have learned about certain genetic traits in humans by tracing the appearance of these traits in families over several generations. A **pedigree chart** is a diagram that shows the presence or absence of a particular trait in each member of each generation. Figure 27–5 shows a pedigree chart tracing the inheritance of color blindness in five generations of a family. The existence of female carriers of color blindness can be shown definitely only when the condition is found in a female's descendants.

Sex-Linked Disorders

As you may recall from Chapter 26, color blindness is a sex-linked condition in which an individual cannot perceive certain colors. Other more serious sex-linked disorders include *hemophilia* (hee muh FIL ee uh) and *Duchenne muscular dystrophy* (DIS truh fee).

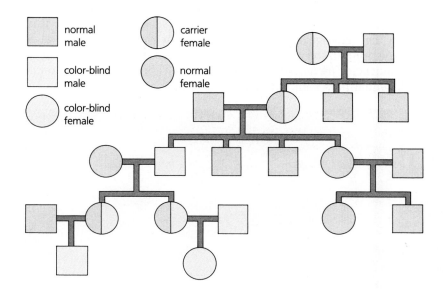

normal male

color-blind male

color-blind female

carrier female

normal female

◀ **Figure 27–5**
A Pedigree Chart Tracing Color Blindness in a Family.

Hemophilia   Hemophilia is a sex-linked disorder in which the blood is unable to clot because it lacks a certain blood-clotting protein. The recessive gene for hemophilia is carried on the X chromosome. Thus, most affected individuals are males. Females with one recessive gene are carriers but show no signs of illness.

Hemophilia is dangerous because the smallest cut or bruise can cause a person to bleed severely. Blood plasma transfusions and injections of the missing blood-clotting substance are used to treat hemophilia. Many hemophiliacs today live into adulthood.

Duchenne Muscular Dystrophy   Another sex-linked inherited disorder is Duchenne muscular dystrophy. In this disorder, the muscle tissue of affected individuals begins to break down during childhood. A recessive gene carried on the X chromosome causes the disorder. Affected individuals have an inactive form of a protein that is required for normal muscle function. These individuals usually do not live beyond their teens.

Autosomal Genetic Disorders

There are a number of other human disorders that are caused by recessive defective alleles on autosomes. These alleles rarely cause signs of illness in people who are carriers because a dominant normal allele also is present. A few human disorders, however, are caused by dominant alleles.

Sickle-Cell Disease   **Sickle-cell disease,** or sickle-cell anemia, is a recessive inherited disorder in which the red blood cells have an abnormal sickle shape. See Figure 27–6. The sickle shape causes the red blood cells to clump and to block small blood vessels. The oxygen-carrying capacity of these cells is decreased. A person with sickle-cell disease suffers from lack of oxygen in the blood and experiences pain and weakness.

The disease is caused by the change of one base in the gene that controls the production of one polypeptide chain in the hemoglobin molecule. There are about 300 amino acids in this chain. The change changes the codon for one amino acid at a specific point in the chain. The normal codon is GAA, which places glutamic acid (an amino acid) in the chain. The abnormal codon is GUA, which puts valine in place of glutamic acid.

Figure 27–6
Sickle and Normal Red Blood Cells. The normal red blood cells (left) are round and thinner in the center than near the edge. The sickle cells (right) are elongated and crescent-shaped. ▼

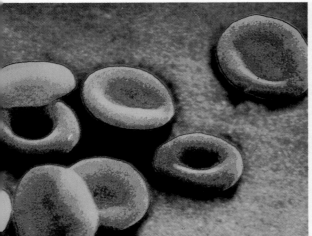

Sickle-cell disease exists mainly in people of African descent. Carriers of the trait—those with one normal and one sickle-cell allele—are more resistant to malaria than people without the defective allele. Because it offers some protection from malaria, this allele is more common than expected in African populations. Carriers of the trait usually are not troubled by symptoms of sickle-cell disease.

Screening for sickle-cell disease is done through examination of the red blood cells. Most people with sickle-cell disease do not live past childhood. With medical treatment, some individuals are able to live to early adulthood. As yet, there is no cure for the disease.

Phenylketonuria   PKU, which is short for *phenylketonuria* (fen ul kee tah NYOOR ee uh), is a recessive inherited disorder in which the enzyme that breaks down phenylalanine (an amino acid) is missing. Because of the missing enzyme, phenylalanine breaks down into chemicals that can damage the brain and cause mental retardation. In the past, it was not possible to diagnose PKU until brain damage had occurred. Now, however, PKU can be diagnosed at birth by a simple test of the infant's urine. This is done routinely in most hospitals. Brain damage can be avoided by a special diet low in phenylalanine.

Tay-Sachs Disease   Like PKU, *Tay-Sachs disease* is an incurable inherited disorder that damages the brain. The disease results from the lack of a specific enzyme for the breakdown of lipids in the brain. Without the enzyme, the lipids build up in the brain cells and destroy them. Tay-Sachs is a rare disease, found most often among Jews of Central European descent. The allele for the disease is recessive, and signs of illness occur only in the homozygous condition. Tay-Sachs disease appears before the age of one, and death occurs within several years. There is no treatment for Tay-Sachs disease at this time.

Cystic Fibrosis   The most common fatal inherited disorder among whites in the United States is *cystic fibrosis* (SIS tik fy BROH sis). This disorder is caused by a recessive allele on chromosome 7. In children with two recessive alleles, some glands produce a thick mucus that clogs and damages the lungs. As the disease progresses, it becomes increasingly difficult for the individual to breathe. With medical treatment, people with cystic fibrosis can live to early adulthood. However, there is no cure for the disease.

Huntington's Disease   *Huntington's disease* is an example of an inherited disorder that is caused by a dominant allele. Every individual who has the allele develops Huntington's disease. Any child born to a parent with Huntington's disease has a 50 percent chance of inheriting the allele and thus developing the disease.

Huntington's disease is fatal, although symptoms usually do not appear until a person is over 30 years old. The disease causes a progressive breakdown of the brain cells, leading to death.

Science, Technology and Society

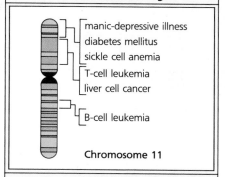

- manic-depressive illness
- diabetes mellitus
- sickle cell anemia
- T-cell leukemia
- liver cell cancer
- B-cell leukemia

Chromosome 11

Issue: Genetic Screening

Thousands of diseases can be linked to genetic factors. This means that people with a certain genetic makeup have a greater probability of contracting such illnesses. New developments in genetic engineering, however, make it possible to test people to determine their susceptibility.

Critics of genetic screening tests warn about the danger of genetic discrimination. Employers might use the data in evaluating candidates for jobs. Insurance companies might base decisions about insurance coverage on the tests. Critics argue that, like race or creed, genetic makeup should not be a factor in such decisions.

People who favor testing argue that determining susceptibility is valuable to both employer and job candidate. Individuals may be susceptible to job-related disorders. For example, a person who may develop genetic emphysema of the lung might not be an appropriate candidate for a job in which fumes are inhaled.

■ *Should genetic screening tests be allowed? Why or why not?*

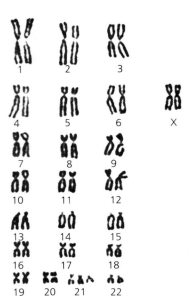

▲ **Figure 27–7**

Down Syndrome. This young girl (right) with Down syndrome has a karyotype (left) that shows three number 21 chromosomes.

Chromosomal Disorders

Nondisjunction causes several inherited disorders in humans. **Down syndrome** is a disorder that results from an extra copy of chromosome number 21. See Figure 27–7. Thus, a person with Down syndrome has three number 21 chromosomes in each cell. This causes mental retardation and physical abnormalities.

Nondisjunction of the sex chromosomes can affect sexual development in humans. One condition, called *Turner's syndrome,* is caused by the presence of only one sex chromosome, an X, in the cells. It results in a female with underdeveloped sexual characteristics. In *Klinefelter's syndrome,* a person has two X's and a Y in each cell, which results in a male with underdeveloped sex organs.

Detecting Genetic Disorders

No treatment or cure is known for many inherited disorders. It is possible, however, for people who may be carriers of a genetic disorder to be tested and counseled before having children. Genetic counselors are people who give prospective parents an accurate idea about the risks of having a child with an inherited disorder. For some inherited disorders, there are tests that show whether or not the parents are carriers.

A person's genetic makeup may be examined by a process known as **karyotyping** (kar ee oh TYP ing). In karyotyping, a cell undergoing mitosis is photographed. The photograph is enlarged, and the photos of chromosomes are cut out and arranged in pairs. This is possible because chromosome pairs differ from each other in length, shape, and position of the centromere. The karyotype of the person then can be compared with a normal human karyotype (see Figure 21–2).

Karyotyping and a number of other techniques may be used by physicians to detect genetic disorders in a fetus. In a process called **amniocentesis** (am nee oh sen TEE sis), a long needle is inserted into the amniotic sac of a pregnant woman at about the sixteenth week of pregnancy. A small sample of amniotic fluid, which contains some cells shed by the fetus, is withdrawn. After four weeks, the fluid and the cells are examined for abnormalities. For example, they may be checked for the presence or lack of a certain enzyme. Tay-Sachs disease can be detected this way. The fetal cells may also be karyotyped to determine whether they contain any abnormal, missing, or extra chromosomes. The presence of an extra chromosome number 21, which is shown in Figure 27–7, indicates that the fetus has Down syndrome. The karyotype also reveals the sex of the fetus.

Much of the same information obtained from amniocentesis may be obtained using a newer method, called **chorionic** (kor ee AHN ik) **villus sampling.** In this technique, a sample of the chorion,

Careers *in Action*

Genetic Counselor

Individuals who have questions or worries about a genetic disorder may seek the advice of a genetic counselor. Working with a medical team, a genetic counselor assesses whether a person may have an inherited disorder or may carry a gene for one. To do this, the genetic counselor assembles a detailed personal and medical history, constructs a family pedigree, and, if necessary, has blood drawn for chemical tests.

Couples with a family history of a genetic disorder may seek genetic counseling before deciding to have children. The genetic counselor will explain the couple's chances of passing on the disorder and will help the couple cope with any fears or anxieties. Sometimes, parents of children with genetic disorders may seek a genetic counselor's help to understand how

the disorder came about. Some may want to know what the chances are of their future children having the disorder.

Problem Solving Suppose you are a genetic counselor meeting with Mr. and Mrs. X, parents of a Down syndrome child. "We know that Down syndrome results from an extra chromosome 21," said Mr. X, "but we don't understand the events that led to this."

You begin by listing the facts known about the X's case: (1) Three copies of chromosome 21 are present in *all* of the child's body cells, and (2) only two copies of chromosome 21 are present in Mr. and Mrs. X's cells.

1. What information does each of these facts give you?
2. Given these facts, how could the fertilized egg receive three copies of chromosome 21?

3. Describe a mechanism that would result in a gamete with two copies of chromosome 21.
4. At which stage of meiosis would this event have occurred? Draw a diagram of this event to explain it to Mr. and Mrs. X.

■ **Help Wanted:**
Genetic counselor. Master's degree in human genetics and national certification required. Contact: Genetics Society of America, 9650 Rockville Pike, Bethesda, MD 20814.

Figure 27–8

Ultrasound. Ultrasound may reveal genetic defects or problems in the fetus, some of which can be treated during or after birth. ▶

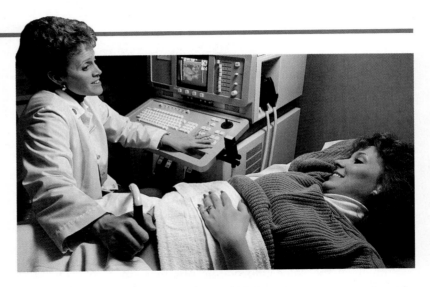

a part of the placenta, is removed for examination. The cells of the chorion are genetically identical to those of the fetus. Chorionic villus sampling may be done much earlier than amniocentesis.

To determine the size and position of a developing fetus, a procedure using high-frequency sound waves, known as **ultrasound,** may be used. The sound waves are reflected off the fetus to produce an image of the developing fetus. See Figure 27–8. By studying this image, doctors can detect abnormalities in bone, muscle, and heart formation. They may also confirm the presence of more than one fetus.

Another technique, called **fetoscopy** (fee TAH skuh pee), allows direct observation of the fetus and tissues. A needle-thin tube containing a viewing scope and a special light is inserted into the uterus. The fetus can be observed through this instrument, known as an endoscope. By inserting special tools through the endoscope, blood and cell samples may be taken for analysis.

27-2 Section Review

1. What type of mutation causes sickle-cell disease?
2. Give one example of a recessive disorder, a dominant disorder, and a chromosomal disorder.
3. Name the process in which a photograph of chromosomes is cut apart and the chromosomes are arranged into pairs.
4. What is the technique by which a small sample of amniotic fluid is removed from a pregnant woman?

Critical Thinking

5. To diagnose some genetic disorders, such as Huntington's disease, family members must submit blood samples. Suppose that one member of a family wanted to know if he or she had the disease. Do you think that other family members should be required to provide blood samples against their will? (*Making Ethical Judgments*)

27-3 Genetic Engineering

Section Objectives:

- *Describe* the methods used by plant and animal breeders to improve their crops and animals.
- *Explain* how organisms can be cloned.
- *List* some ways in which genetic engineering has been used to benefit people.

Breeding Methods

People have always tried to improve their crops and domestic animals. They have tried to increase yields, upgrade quality, and expand growing and breeding areas. Today, a knowledge of genetics is used to produce organisms with desirable traits. A breeder can choose from several methods.

Selection   The process of choosing organisms with the most desirable traits for mating is called **selection.** A dairy farmer, for example, might select and mate only the cows that are the most hardy and that give the most milk. The breeder hopes to change the population by building up desired characteristics. After mating occurs, the breeder selects only the offspring with the desired traits for further mating.

Inbreeding   The mating of closely related individuals to obtain desired characteristics is called **inbreeding.** The degree of closeness can vary. The closest possible hereditary relationship is self-pollination by plants. In organisms that require cross-fertilization, the closest relationship would be brother-sister, mother-son, and father-daughter. Inbreeding is used to produce domestic animals, such as fowl, sheep, cattle, and swine, and pets, such as purebred cats and dogs. See Figure 27-9.

◄ **Figure 27-9**

Cross-Eyed Siamese Cat. Inbreeding may result in undesirable effects, such as the crossed eyes in this Siamese cat.

Inbreeding decreases variation in a population and thus tends to increase the number of homozygous genes. Continued inbreeding and selection eventually produce a line of animals that breeds nearly pure. At the same time, however, inbreeding can result in unwanted effects. Harmful recessive alleles may be brought together by inbreeding and be expressed. This is why close relatives are prohibited from marrying in many societies.

Outbreeding   In outbreeding, individuals not closely related are mated to introduce new, beneficial alleles into the population. Special traits are often found in hybrid crosses of two close species. These superior characteristics are called **hybrid vigor,** or *heterosis* (het uh ROH sis). The mule, the offspring of a male donkey and a female horse, is one example of hybrid vigor. The mule is superior to its parents in physical endurance, strength, and resistance to disease. Mules, however, are usually sterile.

Another type of outbreeding is the mating of pure breeding lines within a species. White short-horned cattle and black Angus cattle have been crossed to produce offspring with superior beef and rapid growth.

Successful outbreeding followed by inbreeding may produce valuable new pure lines of plants and animals.

Mutations   Naturally occurring mutations are used by plant and animal breeders to improve their stock. Many fruits, such as the navel orange and seedless grape, started as natural mutations. Once discovered, a plant mutation may be reproduced by vegetative propagation. This avoids the segregation of traits that would occur in sexual reproduction. Mutations are also valuable in animal breeding. In mink ranching, for example, the most valuable colors, platinum and black cross, are the result of mutations.

Cloning

A **clone** (klohn) is a group of organisms that have exactly the same genes. Organisms that reproduce asexually produce clones, since each offspring receives an exact copy of the parent's genes. New plants produced by vegetative propagation are clones unless a mutation occurs. A great deal of research is now directed toward *cloning,* the production of clones of organisms that normally reproduce sexually.

One purpose of cloning experiments is to produce large numbers of genetically identical laboratory organisms for research. Animals with a known and controlled heredity, for example, could be useful in the study of cancer, aging, birth defects, and the regeneration of damaged body parts.

Cloning Animals   Every body cell of an animal has a complete set of genes. In theory, any of these cells should be able to develop into individuals that are identical to each other and to the parent.

Most of the work on cloning animals has been done with frogs. The method involves replacing the haploid nucleus of an egg cell with a diploid nucleus from a body cell. The egg cell is then allowed to divide and develop into a complete individual. The animal that results has the same genes as the transplanted nucleus. If the nucleus came from a cell of an adult, the new animal would be its clone. However, cloning is difficult with the nucleus of an adult cell. Apparently, too many of the genes are "turned off" and cannot be turned on again inside the egg cell. In frogs, the method works best using nuclei from the cells of an embryo at the blastula stage. By transplanting nuclei taken from one blastula, a number of genetically identical individuals can be obtained.

A similar method has been used successfully in a mammal. Nuclei from cells of the blastula of a mouse were transplanted to mouse egg cells from which the original nuclei had been removed. The altered egg cells were allowed to reach the blastula stage. Then, they were placed in the uteruses of female mice to complete their development. See Figure 27–10.

Cloning Plants   Using tissue culture techniques, scientists have successfully cloned many different kinds of plants. As you may recall from reading Chapter 2, tissue culture involves placing a cell from a living organism in a special culture medium. When a single, undifferentiated cell from a plant, such as carrot or potato, is placed in tissue culture, it can develop into a new plant. The new plant has the same genetic makeup as the plant from which the original cell was taken.

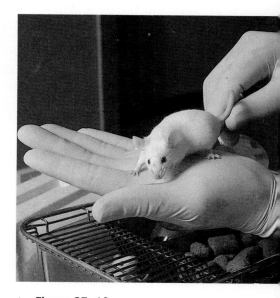

▲   **Figure 27–10**

A Mouse Clone. Cloning has been used with limited success by scientists to produce genetically identical animals. This mouse was produced by cloning.

Polyploidy

As you read earlier in this chapter, polyploidy is a condition in which an organism has some multiple of the normal chromosome number. This type of mutation is fairly common among plants. More than half of the known species of flowering plants are polyploid. See Figure 27–11. A mutant polyploid plant cannot produce offspring by mating with a plant with only one set of

◄   **Figure 27–11**
Polyploid Daylilies.

chromosomes. Thus, polyploid plants usually form separate species or produce offspring by mating with plants with similar numbers of chromosomes.

Plant breeders can produce polyploid plants by exposing plant cells to various mutagens. These mutagens prevent duplicated chromosomes from separating during meiosis. This creates diploid gametes. One mutagen that is commonly used to produce polyploidy is a chemical called *colchicine* (KAHL cheh seen). Polyploidy can produce plants that are hardier, bigger, and more productive. Look back at Figure 27–3 to see the difference between polyploid and normal strawberries.

Gene Splicing

Understanding the gene has led to the remarkable development of methods for changing a cell's DNA. When DNA from two different species is joined together, it is called **recombinant DNA.** The methods for producing recombinant DNA are referred to as **genetic engineering,** or, more popularly, as *gene splicing.* Gene splicing describes quite well the methods used at this time. These methods involve breaking a DNA molecule and inserting or attaching a new gene by means of a chemical "splice."

Genetic engineering provides a way of producing large amounts of previously rare substances. For example, *interferon* (int ur FIR on) is a rare protein that helps humans fight off viruses. Genetic engineers have inserted the human gene for interferon into bacteria. Bacteria with the human gene for interferon are then able to make the protein. When the bacteria are cloned, large amounts of interferon are produced and can be collected. Other valuable substances made in this way include human insulin, which is used in the treatment of diabetes, and the clotting substance that is needed to treat people with hemophilia.

Farmers also have been helped by genetic engineering. For example, genetic engineers have inserted genes into plants to make the plants resistant to disease, to insects, and to weed-killing substances. Growth-promoting hormones, produced by genetic engineering, are used to increase the amount of milk produced by cows and to fatten farm animals. See Figure 27–12. One of the long-range goals of genetic engineering is to correct genetic defects by transferring normal genes to cells that lack them. This is called **gene therapy.** It is possible that hereditary diseases could be treated in this way.

Gene splicing usually involves four steps. The first step is cutting DNA into small pieces that contain a certain gene. The second step is inserting the different pieces of DNA into another species, such as bacteria. The third step is cloning the organisms with the pieces of foreign DNA. The fourth step is screening the clones to see which ones contain the desirable foreign gene.

Isolating a Gene The first step in any genetic engineering experiment is to isolate pieces of DNA with the desirable gene. This is accomplished by using proteins, called **restriction enzymes,**

Figure 27–12

The Effect of Growth Hormones on Pigs. These pigs are genetically identical. The larger one has been fattened by the addition of growth hormones to its diet. ▼

that cut the DNA molecules into small pieces. Restriction enzymes cut DNA at specific nucleotide sequences. This action produces many small pieces of DNA.

Making Recombinant DNA   The second step in genetic engineering involves placing the small pieces of DNA into a host cell, such as bacteria or a virus. This foreign DNA can be inserted in several ways.

One important method of inserting foreign DNA involves small, ring-shaped DNA called **plasmids.** Many bacteria either contain plasmids or will take plasmids into their cytoplasm surroundings. Although inside the bacteria the plasmids remain separate from the bacterial chromosome, they include genes that affect the bacteria. For example, some plasmids have a gene for the production of an enzyme that destroys penicillin. Bacteria that contain these plasmids are not killed by penicillin. Plasmids replicate during every cell division. Therefore, a single bacterium with a certain plasmid can easily produce large colonies of cells with that same plasmid.

To transfer foreign DNA into bacteria, plasmids are first obtained from bacterial cells. This is done by crushing the bacteria and sorting out the plasmids. The plasmids then are placed in a solution containing the foreign DNA. See Figure 27–13. The foreign DNA usually comes from animals. In some cases, however, the DNA has actually been made by chemically assembling nucleotides in the proper sequence. A restriction enzyme is used to break open the plasmid. The foreign DNA attaches itself to the open end of the

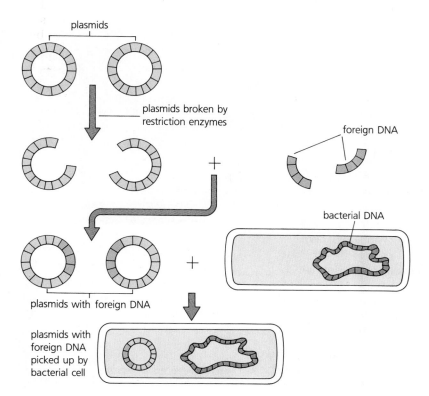

◀ **Figure 27–13**

Using Plasmids to Transfer Foreign DNA to Bacterial Cells. By introducing foreign genes into a plasmid, genetic engineers can change the genetic makeup of a cell, so that the transplanted genes are expressed.

plasmid and closes the ring. Once this is done, the plasmid includes the foreign gene.

Cloning and Screening   Bacteria are exposed to the altered plasmids, and many bacteria take them in. Once inside the bacteria, the new genes begin to be expressed. That is, the bacteria produce the protein specified by the foreign gene. At this point, the bacteria are cloned and screened for colonies that are producing the desired protein. This protein may be any one of a number of proteins that are difficult or expensive to obtain by other means.

Transformation and Transduction

Some of the methods of transferring genes to bacteria use transformation and transduction. Transformation and transduction are processes by which DNA is transferred from one bacterial cell to another. In **transformation,** living bacteria take in DNA from dead bacteria. The DNA causes traits of the dead bacteria to show in the

Critical Thinking in Biology

Making Ethical Judgments

When you make an ethical judgment, you evaluate how well something agrees with your personal values and ideals, such as truth, justice, and respect for life.

Many scientific advances bring with them difficult ethical questions. Consider the following example.

Genetic engineers are exploring ways to produce useful changes in farm animals. Some scientists have successfully transplanted a gene from a cow into the fertilized egg of a pig. The pig that developed grew faster and had leaner meat than other pigs—both desirable traits in animals raised for food. The new pig also had abnormally short legs that became crippled with arthritis. (The scientists hope to be able to eliminate this condition in future transplants.) The genetically altered pig could pass its new traits, along with all its other traits, to its offspring.

1. Based on the case described above, list some reasons for and against using genetic engineering to produce changes in farm animals.

2. What values do you hold that would be important to consider in evaluating this issue?

3. Based on your values, do you think that genetic engineering should be used to produce changes in farm animals? Why or why not?

4. Think About It   Describe your thought processes as you answered question 3.

living bacterial cells. In Griffith's experiments, described in Chapter 26, type R bacteria were transformed by DNA from dead Type S bacteria. Transformation occurs in nature as well as in laboratory experiments.

Sometimes, viruses transfer pieces of DNA from one bacterial cell to another. This process is called **transduction.** Transduction occurs when a virus infects a bacterial cell, and a piece of bacterial DNA becomes a part of the virus particle. When the virus infects a new cell, it carries with it the piece of the DNA from its previous host cell. In this way, viruses can be used to carry foreign DNA to other organisms.

The Safety of Recombinant DNA

Many scientists, including those working in the field, are aware of the possible danger of research with recombinant DNA. Experiments with recombinant DNA produce new forms of microorganisms. It is possible that an organism will be accidentally produced with the ability to cause an entirely new kind of human disease, against which the body will have no natural defenses. The escape of such an organism into the environment could lead to a massive, uncontrollable outbreak of disease.

To prevent this type of accident, the federal government and the scientists involved in recombinant DNA research have established rules and safeguards that must be observed. Since this is a new field and information is still lacking, the actual risks and the necessary safeguards are not known. Compliance with the rules ultimately depends on the efforts of individual workers. Because risks are involved, it will become ever more important to monitor safety and to update standards as recombinant DNA research expands.

Recombinant DNA is a growing field that holds promise for the future. It is important for the public to remain informed about recent developments and their potential benefits and drawbacks. In this way, citizens, scientists, and legislators can work together to assure the responsible use of this new technology.

27-3 Section Review

1. List three methods used by breeders to improve yield and quality.
2. What is a group of organisms with exactly the same genes called?
3. What happens when colchicine is applied to plants?
4. Name the small circle-shaped DNA that is used to insert foreign pieces of DNA into bacteria.

Critical Thinking

5. What do breeding methods and recombinant DNA methods have in common? How do they differ? (*Comparing and Contrasting*)

Laboratory Investigation

Genetic Variation in Humans

Genetic variation among individuals is due largely to the recombination of alleles that results from meiosis and fertilization. In this investigation, you will use a few genetically determined traits to observe genetic variation among your classmates.

Objectives

- *Observe* some genetically determined human traits.
- *Observe* variation in humans.

Materials

tracing paper
colored pencil
paper clips
mirror

Procedure

A. Place a piece of tracing paper over the chart shown at right, holding it in place with paper clips.

B. Consider the trait listed in the center of the chart. If you have the trait, color in the sector of the circle with the *yes*. If you do not have the trait, color in the sector with the *no*.

C. From the sector that you colored in, move outward to the next circle and determine whether you have the trait listed in that circle. Again, color in either the *yes* or *no* sector.

D. Continue with this procedure, working your way outward through each circle and coloring in the sectors corresponding to your traits.

E. When you complete the outermost circle, note the number that is opposite your sector. Write your name and number on the chalkboard.

Analysis and Conclusions

1. Do any classmates have the same number you have?

2. How many different combinations of the five traits are possible? How many combinations would be possible with six traits? Seven traits?

3. In terms of what happens during meiosis and fertilization, explain how all the various combinations of alleles are possible.

4. Each trait listed in the chart is for the dominant allele of the trait. If you answered *no* to any trait, you have the recessive allele. List your five traits and the possible genotypes for each. Use different letters for each trait.

5. Why is it virtually certain that except for identical twins, even offspring of the same parents will not be genetically identical?

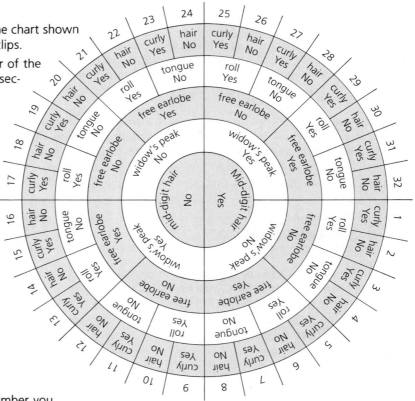

Chapter Review

27

Study Outline

27-1 Mutations

■ In sexually reproducing species, most genetic variation results from segregation and crossing-over during meiosis and from recombination during fertilization. The differences in genetic material among members of a population can be traced to new inherited traits called mutations, a concept developed by the botanist Hugo De Vries.

■ The two main types of mutations are chromosomal and gene mutations. The causes of natural mutations include random errors in DNA replication and mutagens, such as X rays, ultraviolet light, and certain chemicals.

■ Chromosomal mutations involve permanent changes in chromosome structure, such as occur in translocation, inversion, addition, and deletion; or changes in chromosome number, such as occur in nondisjunction and polyploidy.

■ Gene mutations result from a change in the sequence of bases in a DNA molecule, which ultimately alters the structure of a protein the cell synthesizes.

■ Genes that can move to new locations on the chromosomes during replication cause another kind of mutation.

27-2 Human Genetic Disorders

■ Scientists have learned about human genetics by tracing the appearance of traits in families over several generations. A pedigree chart shows the presence or absence of a particular trait in each member of each generation.

■ Sex-linked disorders, such as color blindness, hemophilia, and Duchenne muscular dystrophy, are caused by recessive genes located on the X chromosome.

■ Autosomal genetic disorders include sickle-cell disease, phenylketonuria, Tay-Sachs disease, cystic fibrosis, and Huntington's disease. Most of these disorders are caused by defective recessive alleles found on the autosomes.

■ Chromosomal disorders resulting from nondisjunction include Down syndrome, Turner's syndrome, and Klinefelter's syndrome.

■ Genetic disorders can be detected by karyotyping, amniocentesis, or chorionic villus sampling.

27-3 Genetic Engineering

■ People can improve crops and domestic animal stock by using breeding methods that include selection, inbreeding, outbreeding, naturally occurring mutations, cloning, and polyploidy.

■ Gene splicing joins the DNA from two different species. This method, referred to as recombinant DNA or genetic engineering, provides a way of producing large amounts of previously rare substances, such as interferon or human insulin.

■ Gene splicing usually entails four steps: cutting the DNA containing the desired gene into small pieces; inserting the pieces into another species, such as a bacterium; cloning the organism with the pieces of foreign DNA; and selecting the clones with the desired gene.

■ Genes can be transferred to bacteria by transformation or transduction.

■ To work with recombinant DNA safely, scientists and the federal government have established rules and safeguards that must be observed to ensure the safe use of genetically engineered organisms.

Vocabulary Review

mutation (545)	polyploidy (547)
chromosomal mutation (546)	point mutation (547)
gene mutation (546)	pedigree chart (549)
mutagens (546)	sickle-cell disease (550)
translocation (547)	Down syndrome (552)
inversion (547)	karyotyping (552)
addition (547)	amniocentesis (553)
deletion (547)	chorionic villus sampling (553)
nondisjunction (547)	

Chapter Review

ultrasound (554)

fetoscopy (554)

selection (555)

inbreeding (555)

hybrid vigor (556)

clone (556)

recombinant DNA (558)

genetic engineering (558)

gene therapy (558)

restriction enzymes (558)

plasmids (559)

transformation (560)

transduction (561)

A. Matching—Select the vocabulary term that best matches each definition.

1. An abnormal change in the structure or number of chromosomes

2. The transfer of a part of a chromosome to a nonhomologous chromosome

3. Factors in the environment that cause mutations

4. Cutting out photographs of chromosomes and arranging them in pairs

5. A diagram that shows the presence or absence of a trait in each member of each generation

6. The transfer of normal genes to cells that do not contain them

7. A change in a single nucleotide in a gene

8. The process of joining DNA from two different species

9. Process by which a virus transfers a piece of DNA from one bacterial cell to another

10. A ring-shaped strand of DNA

B. Analogies—Select the vocabulary term that relates to the single term in the same way that the paired terms are related.

11. Down syndrome **is to** extra number 21 chromosome as _____ **is to** extra set of chromosomes.

12. addition : piece of chromosome attaches as _____ : piece of chromosome breaks off.

13. _____ : red blood cells as phenylketonuria : phenylalanine

14. chorionic villus sampling : placenta as _____ : amniotic fluid

Content Review

15. What is the difference between a gene mutation and a chromosomal mutation?

16. What environmental factors can cause mutations?

17. Explain what happens in nondisjunction.

18. Describe polyploidy and its effect on plants.

19. What types of changes in DNA can result in a point mutation?

20. Explain the effects of "jumping genes."

21. State some of the difficulties encountered in scientific studies of human heredity.

22. How do sex-linked genetic disorders differ from autosomal disorders? Give an example of a human sex-linked disorder.

23. Name three autosomal genetic disorders of humans and describe the effects of each.

24. Describe the techniques of karyotyping, amniocentesis, and chorionic villus sampling.

25. Distinguish between selection, inbreeding, and outbreeding.

26. List some of the harmful effects of inbreeding.

27. How are naturally occurring mutations used by plant and animal breeders?

28. What is a clone? How does the cloning of animals differ from the cloning of plants?

29. How do genetic engineers use plasmids?

30. Describe the four steps that usually take place in genetic engineering.

31. How does transformation differ from transduction?

32. What steps have been taken by the federal government and recombinant DNA researchers to prevent accidents?

Graphic Organizing

For information on graphic organizers, see Appendix G at the back of this text.

33. **Word Map:** Construct a word map for each of the following terms: chromosomal mutation, sex-linked disorders, autosomal disorders.

Critical Thinking

34. Why do the majority of scientists assume that mutations are the ultimate sources of variation? (*Identifying Assumptions*)

35. Why are mutations that occur in body cells usually not passed to offspring? (*Identifying Reasons*)

36. Predict what would happen during meiosis if a plant had three sets of chromosomes (3n) rather than the normal diploid number. (*Predicting*)

37. Why do researchers think they must learn more before cloning can produce genetically pure laboratory animals? (*Identifying Reasons*)

Creative Thinking

38. How do gene splicing and cloning experiments pose ethical and procedural problems for society?

39. Suppose a scientist creates a bacterium that digests plastics. What precautions would you advise?

Problem Solving

40. Populations usually remain stable or change very slowly over a long period. The chart below shows data from a population genetics survey of a certain species of insect. Develop a hypothesis to account for the changes in the frequencies of the alleles. Note that the environment has remained stable for the past 50 years.

Year	Frequency of Allele B	Frequency of Allele b
1940	0.99	0.01
1950	0.95	0.05
1960	0.98	0.02
1970	0.96	0.04
1980	0.10	0.90
1990	0.08	0.92

41. The occurrence of Down syndrome in the population is recorded in the table below. Graph the data. Draw a conclusion about the relationship between the incidence of Down syndrome and the mother's age.

Age of Mother	Occurrence of Down Syndrome per 1000 Births
25	.8
30	1.0
35	3.0
40	10.0
45	30.0
50	80.0

42. A virus isolated from monkeys contains a circular double strand of DNA. The virus, called Simian Virus 40, interests scientists because it causes cancer in laboratory animals. Using a restriction enzyme, the strand is separated into six unequal segments, as shown in the following figure. A scientist hypothesizes that the segment of DNA causing cancer can contain no fewer than 600 base pairs. Divide each segment by 600 to calculate which segments have the highest potential for containing the cancer-causing sequence. Place your answers in descending order, from highest to lowest.

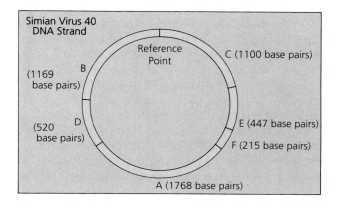

Simian Virus 40 DNA Strand
Reference Point
C (1100 base pairs)
B (1169 base pairs)
D (520 base pairs)
E (447 base pairs)
F (215 base pairs)
A (1768 base pairs)

Projects

43. Prepare a written report outlining the guidelines and rules for safe operation of genetic engineering research. You may wish to obtain information from the National Institute of Health or the National Academy of Sciences.

44. Investigate one of the following careers: genetic engineer, medical social worker, genetic counselor. Find out about aspects such as educational requirements, salaries, and demand in terms of the current job market. Present your findings to the class in an oral report.

45. Write a brief report on the life and scientific contributions of one of the following scientists: Luther Burbank, Wendell Stanley, Barbara McClintock.

46. The early inhabitants of Central America used corn, a native grass, as food. The seeds were edible but small. Over the next 5000 years, they selected larger varieties from their crops to serve as the parents of the next generation. Contact a local supplier of seeds for agriculture to get information about modern methods for genetically improving corn varieties. Prepare a poster illustrating these methods.

Biology and Problem Solving

Why is the cheetah population dwindling?

Cheetahs chasing their prey are often clocked at 70 miles per hour. These agile members of the cat family are superbly adapted for hunting. In fact, they are even more successful than lions in capturing prey.

Once cheetahs ranged around the world. Today, however, there may be as few as 20 000 cheetahs and these are found only in some regions of Africa.

Why should the population of an animal with such capabilities dwindle in this way? A group of biologists addressed this question by studying cheetahs at a breeding and research center in Africa. At this center, as at others, there was little success in breeding cheetahs in captivity—only a few reproduced, and 37 percent of the infant cheetahs died.

The scientists first examined semen from 18 male cheetahs and from a similar number of male domestic cats. They found that the concentration of sperm in cheetah semen was only one-tenth as great as the concentration of sperm in the semen of domestic cats. Furthermore, 71 percent of the cheetah sperm had abnormal shapes—the flagella were either bent or coiled, or the sperm heads were either too large or too small. In comparison, only 29 percent of the sperm from domestic cats was abnormally shaped.

1. What conclusion can be drawn from this study?

2. Why were domestic cats included in the study?

3. The formation of sperm is directed by genes. What can you conclude about the genes that control sperm formation in the cheetahs studied?

The biologists knew that high rates of sperm abnormalities in other animals are associated with infertility and with inbreeding. This led them to hypothesize that the cheetahs at the research center were genetically very similar to each other.

When inbreeding occurs naturally or is carried out deliberately in a population, the number of homozygous alleles in the population increases. The lack of genetic variability that results can be detrimental to a population. If, for example, an environmental change occurs, a lack of genetic diversity can reduce the population's chances of adapting to the change.

4. Suppose that a group of cheetahs all carried an allele that made them susceptible to the fatal disease, *feline infectious peritonitis* (FIP). What would happen if the cheetahs were exposed to this disease?

To test how genetically similar the cheetahs were, the scientists took blood samples from 19 cheetahs and from 19 domestic cats and isolated various proteins. They used a procedure called electrophoresis to determine whether each protein was the product of homozygous or heterozygous alleles. Of 52 proteins tested, *all* of the cheetah proteins were found to be the products of homozygous alleles. In domestic cats, in contrast, proteins from both homozygous and heterozygous alleles were detected.

5. Did the electrophoresis evidence support the scientists' hypothesis? Explain.

The scientists also examined physical features of the cheetahs for signs of genetic similarity. Features that are normally mirror images of each other, such as the right and left sides of a skull, are known to show less symmetry when inbreeding has occurred. In a study of cheetah skulls collected from museums, the skulls showed less symmetry than the skulls of leopards, margays, and ocelots, three other cat species.

6. How does the skull study contribute to our understanding about genetic variation in cheetahs?

Next, the scientists studied the rejection rates of skin grafts on cheetahs. Usually, an animal rejects a skin graft unless it is from a close relative with a similar genotype. Domestic cats usually reject skin grafts from unrelated animals in 10 to 12 days. Scientists grafted three patches of skin onto a group of the cheetahs: one patch from the cheetah itself, one from an unrelated cheetah, and one from a domestic cat.

7. Why were the cheetahs given grafts of their own skin?

8. If there were considerable genetic variation among the cheetahs, what would you predict would be the outcome of this experiment?

The cheetahs rejected the grafts from domestic cats by the 12th day. After 28 days both the grafts of their own skin and the grafts of skin from unrelated cheetahs had healed nicely and were growing hair and developing cheetahlike spots.

9. What can you conclude from the skin graft experiment?

10. Propose a plausible hypothesis to explain how the cheetahs came to be so genetically similar.

Problem Finding

What additional questions do you have about why the cheetah population is dwindling?

Evolution

A giant tortoise labors along the lush green growth of the Galapagos Islands. Tortoises have so labored over the face of the earth for millions of years. In fact, tortoises and many other species of today's reptiles shared the earth with the dinosaurs. How do we know? Dated fossilized bones provide evidence of their ancient existence. In this unit, you will learn how the survival of a species through the generations is linked to its ability to adapt to changing environmental pressures. You will see that this process of gradual change, called evolution, is responsible for the diverse life forms we know today. You will also learn that species that do not successfully adapt to environmental demands disappear from the earth.

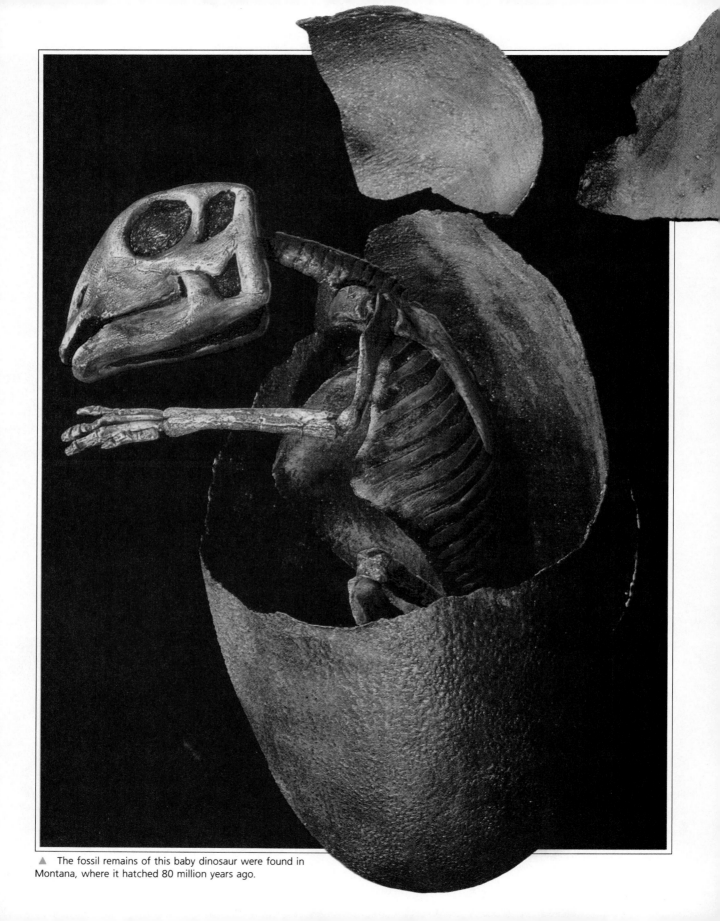

▲ The fossil remains of this baby dinosaur were found in Montana, where it hatched 80 million years ago.

Evidence of Evolution

The female dinosaur digs and prods the muddy earth, making a circular nest of mud. After she has laid a single egg, she hovers nearby, guarding the egg from predators. The egg hatches, but something is wrong, and the baby dies. The tiny body and its broken eggshell sink into the watery earth. Eighty million years later, paleontologists discover the nesting site and the skeleton. In this chapter, you will find out how clues in the earth provide evidence of the evolution of living things.

Guide for Reading

Key words: geologic evolution, organic evolution, fossil, spontaneous generation, heterotroph hypothesis
Questions to think about:
📖 What information do fossils provide about geologic and organic evolution?

📖 What is the most current hypothesis of the origin of life?

28–1 Evidence from the Past

Section Objectives:

- *Describe* at least five processes by which fossils may be formed.
- *Explain* how sedimentary rocks are formed and why fossils are often found in these rocks.
- *Define* the terms *relative dating* and *absolute dating*.
- *Explain* how radioactive dating can determine the age of rocks.
- **Laboratory Investigation:** *Examine* and *compare* a variety of fossil samples (p. 594).

Evolution is the central and unifying theme of biology. In its most general sense, the term *evolution* means a gradual change over time. Since its formation about 4.5 billion years ago, the earth itself has changed continuously. This slow change is known as **geologic evolution.** Many species also have changed since they first appeared on the earth. This process is known as **organic evolution.** How did life begin and how has it evolved into the species living today? In this chapter, you will learn about the scientific evidence for organic evolution and the theories of how life began on earth.

Fossils

The study of fossils provides the strongest evidence of organic evolution. A **fossil** is any trace or remains of an organism that has been preserved by natural processes. By studying fossils, scientists can compare the remains of ancient organisms with organisms living today to see whether or not organic evolution has occurred.

▲ **Figure 28–1**

A Living Fossil. This coelacanth represents a group of lobe-finned fishes that had been thought to be extinct.

▲ **Figure 28–2**
A Fly Fossil Preserved in Amber. This fly became trapped in sticky resin. The resin hardened into amber, preserving the insect.

When an organism dies, it usually decays without leaving any remains. Special circumstances are required for a fossil to form. In the majority of fossils, the soft tissues of the organism have decayed, and only the hard parts, such as bones or shells, have been preserved. In some fossils, however, an entire organism has been preserved with almost no decay.

Fossils in Amber and Ice   The soft tissues of animals usually decay because of the activities of bacteria and fungi. In some circumstances, such as anaerobic conditions or extreme cold, decay does not occur, and entire organisms are preserved.

As you can see in Figure 28–2, entire insect remains can be preserved in amber. *Amber* is a hard, yellow, transparent material formed by the hardening of resin, a sticky substance produced by trees. Insects often become trapped and embedded in the resin, which then hardens into amber.

In the cold Arctic regions, entire animal remains have been preserved in ice for thousands of years. As you can see in Figure 28–3, the remains of extinct woolly mammoths can be so well preserved by the cold that their flesh, skin, and hair are present.

Fossil Bones and Petrifaction   Under some conditions, the hard, mineral parts of animals, such as shells, bones, and teeth, can be preserved for millions of years. Dinosaur bones, some more than 100 million years old, have been found all over the world. Usually only teeth, parts of the skeleton, or skull fragments are preserved. It is uncommon to find a complete fossil skeleton.

Many animal skeletons have been preserved in pools of tar in which the animals were trapped. The La Brea tar pits in Los Angeles, California, contain thousands of fossils. Although most of

Figure 28–3
A Woolly Mammoth. This completely preserved body of a woolly mammoth was uncovered in the Soviet Union by a bulldozer overturning frozen ground. ▶

◀ **Figure 28–4**
Petrified Tree Trunks.

these animals lived less than 25 000 years ago, they include many extinct species, such as the saber-toothed tiger and the woolly mammoth.

In some cases, a dead organism lies in a body of water that contains a high mineral content. Gradually, the original substances of the organism dissolve and are replaced by minerals from the water. In this process, which is called **petrifaction** (peh truh FAK shun), the remains of the organism are turned to stone. Whole trees, estimated to be 150 million years old, have been preserved as stone fossils in the Petrified Forest in Arizona. See Figure 28–4. Many fossil bones actually are petrified replicas of the original bones.

Molds, Casts, and Imprints   By far the greatest number of fossils form on the bottoms of lakes and seas. A dead organism slowly sinks into the sandy or muddy bottom of a lake or ocean. As additional particles of sand or mud accumulate on the bottom, the organism becomes buried. The sand or mud later hardens into rock. Meanwhile, the remains of the organism decay, but its shape is preserved in the rock as a hollow form called a **mold.** Sometimes, a mold becomes filled with minerals, which in turn harden to form rock. The hardened minerals form a **cast,** or a copy of the external form of the original organism.

Impressions made in mud, such as animal footprints, may remain when the mud hardens into rock. The impression is called an **imprint.** Among the largest known imprints are the dinosaur footprints shown in Figure 28–5. Many imprints also have been left by thin structures, such as leaves.

Calculating the Age of Fossils

Most fossils have been found in a kind of rock that is called **sedimentary** (sed uh MEN tuhr ee) **rock.** Most sedimentary rocks form on the bottoms of shallow seas or on ocean bottoms near the shorelines of continents. As a river flows over land, it wears away, or erodes, fine particles of rock. These particles are called *sediments.* When the river waters enter the sea, the sediments slowly settle to the bottom. Gradually, sediments build up on the sea bottom. Various chemical processes, combined with the pressure exerted by the weight of the sediments, slowly harden the material into rock.

Figure 28–5
Dinosaur Footprints. The size of a dinosaur can be estimated relative to the size of its imprints. ▼

Figure 28–6
Deposition of Sediments. Streams flowing into a body of water carry fine rock particles called sediments. These sediments settle to the bottom and may gradually build up to a great thickness. The bodies of dead organisms that settle to the bottom may become fossils embedded in the sediments. The oldest fossils will be in the lowest layers, the youngest in the upper layers. ▶

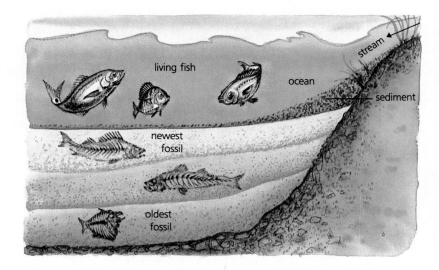

The formation of sedimentary rock may continue for millions of years at any location. Because the size and mineral composition of the sediments will change from time to time, sedimentary rock acquires a layered structure. The oldest layers, those laid down first, are at the bottom. The youngest, or most recent, are at the top. The layers in between are arranged in a time sequence from older to younger. See Figure 28–6.

Critical Thinking in Biology

Identifying Assumptions

People often make assumptions without even knowing it. For example, you may assume that a woman and a boy walking together are mother and son.

Whenever you assume something, you accept it as true without proof or examination. Actions and differences of opinion are often based on assumptions that people have made. For this reason, it is important to learn how to identify assumptions.

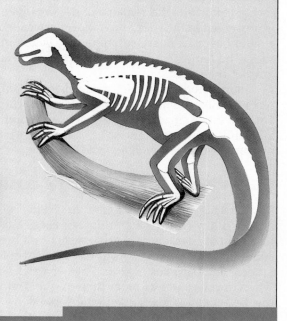

1. A paleontologist used 16 fossil bones to reconstruct an extinct *Iguanodon*, a genus of dinosaurs. The reconstruction is shown on the right. Along with the bones, the paleontologist found an unusual bony spike about four inches long. If you were the paleontologist, where would you place the bony spike?

2. Think about why you placed the bony spike where you did. What assumptions did you make?

3. How valid do you think your assumptions were? Explain.

4. Think About It Describe the thought process you used as you answered question 2.

Relative Dating   Over millions of years, the shifting of the earth's crust has raised some regions that once were under the seas. This shifting has caused some sedimentary layers to be exposed on the side of mountains or plateaus. Sometimes, a river has cut its way through the layers of sedimentary rock. The Grand Canyon, shown in Figure 28–7, is an example of this. If the exposed sedimentary layers have not been greatly disturbed by the motions of the earth's crust, they remain in their original sequence. The oldest are at the bottom, and the youngest are at the top. If these sedimentary layers contain fossils, scientists use the layers to determine when certain organisms existed in relation to others. Any method of determining the order in which events occurred is called **relative dating.** Fossils in the lower layers represent organisms that lived at an earlier time than the fossils in upper layers. This history of life is usually called the **fossil record.**

Absolute Dating   Relative dating does not give the actual age of a rock or a fossil in years. Any method that determines how long ago an event occurred is called **absolute dating.** Many methods of absolute dating have been tried, but most scientists consider radioactive dating the most accurate and reliable method.

The technique of **radioactive dating** is based on the knowledge that certain elements have unstable isotopes. As you read in Chapter 3, the nuclei of these unstable isotopes tend to break down, or decay, changing to different isotopes. During this process, called *radioactive decay,* energy is given off in the form of radiation.

The rate at which a radioactive isotope decays is fixed and unchangeable. The time required for half the atoms of an isotope to decay is called the *half-life* of that isotope. For example, the half-life of carbon-14, which occurs naturally in all organisms, is 5700 years. If a fossil contains carbon-14 and was formed 5700 years ago, only half the original amount of carbon-14 would be left in the fossil today. The other half would have changed to carbon-14's decay product, nitrogen-14. See Figure 28–8. By comparing the ratio of carbon-14 to nitrogen-14, scientists can calculate a fossil's age. This method does not work for fossils older than about 50 000 years because most of the carbon-14 has decayed.

Other isotopes, such as uranium-238 or potassium-40, can be used to date older fossils or rocks in which fossils are embedded. Only *igneous* (IG nee us) rocks—formed when molten material in the crust cooled and hardened—can be dated using radioactive

▲ **Figure 28–7**
Sedimentary Rock. The layers of this rock formation in the Grand Canyon were formed from sediments deposited under water. Shifts in the earth's crust then raised the rock layers without disturbing their order. Each layer is older than those above it.

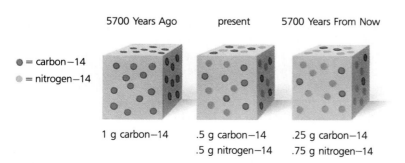

5700 Years Ago   present   5700 Years From Now

● = carbon—14
● = nitrogen—14

1 g carbon—14   .5 g carbon—14   .25 g carbon—14
                  .5 g nitrogen—14   .75 g nitrogen—14

◀ **Figure 28–8**
Half-Life of Carbon-14. The half-life of carbon-14 is 5700 years.

dating methods. These methods do not work with sedimentary rocks because they give the age of the sediments, which existed long before becoming rocks. However, the absolute age of sedimentary rocks can be estimated by the age of igneous rocks that formed above, below, or within them. In this way, scientists can determine the age of fossils found in sedimentary layers.

28-1 Section Review

1. What is a fossil?
2. Name six types of fossils.
3. In what kind of rock are most fossils found?
4. What is the fossil record?

Critical Thinking

5. If you could search for fossils under the ocean floor, where would you look? (*Predicting*)

28-2 Interpreting the Fossil Record

Section Objectives:

- *Define* the terms *correlation* and *index fossil*.
- *Explain* what is meant by the geologic time scale.
- *State* two important conclusions that can be drawn from the fossil record regarding the course of changes in living things over geologic time.
- *Explain* the importance of extinctions.

Fossils supply many clues about the organisms from which they were formed. By examining fossils and the fossil record, scientists have been able to piece together a history of life.

Correlation

Suppose a geologist finds a cliff made of sedimentary rock with five distinct layers of different textures and mineral compositions. Then, a few kilometers away, another exposed cliff face of sedimentary rock is found. Examining the layers of this formation, the geologist notices that the top three layers are exactly the same as the bottom three layers of the first formation. See Figure 28–9. It is reasonable for the geologist to conclude that the layers were originally continuous and were deposited at the same time.

Now suppose that two more layers become visible below the three in the second cliff. The two layers must have been deposited before the other three. Therefore, they must be older. In fact, they

Figure 28–9

Correlation of Separate Rock Formations. Layers C, D, and E on the left have the same composition and thickness as layers C, D, and E on the right. Comparable layers on the left and the right were probably laid down at the same time. Layers F and G on the right are therefore older than layers A and B on the left. ▼

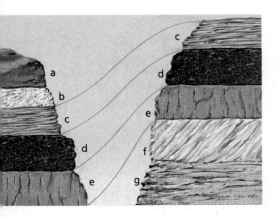

must be older than the bottom three layers of the first cliff. By this process of matching, or **correlation,** scientists can show that certain rock layers in one place are older than certain rock layers in another place. It also follows that the fossils in the older layers are older than fossils in the younger layers.

Correlation of sedimentary rock allows scientists to establish the relative dating of rocks and fossils in different places. Correlation by the comparison of rock layers can produce a fairly extended relative dating of the rocks and fossils in a region. But, the method cannot be used where there are no nearby rocks that are similar. Furthermore, scientists cannot provide a correlation between rocks in a particular region and rocks in another section of the continent or another part of the world.

A study of fossils from many regions has shown that certain types of organisms seem to have appeared, flourished for a time over wide regions of the earth, and then disappeared. Rock layers that have fossils of these organisms must have been formed when

Careers *in Action*

Paleontologist

Dinosaur bones, petrified trees, insects embedded in amber: to a paleontologist, these are clues to the past. A paleontologist is a scientist who studies the traces left behind by once living organisms. These traces, called fossils, are found in many places. In fact, deciding where to look for fossils is a large part of the paleontologist's job.

Once a fossil site has been found, work in the field can be long and hard. Sometimes, the paleontologist must separate fragments of bone from the rock in which they are embedded. In the laboratory, paleontologists analyze the fossils and other clues to piece together the likely structure of an organism that may have become extinct long ago.

Paleontologists work for universities, museums, and oil companies. Most paleontologists specialize in certain types of organisms, such as plants, vertebrates, or invertebrates.

Problem Solving Imagine you are a paleontologist studying the evolution of reptiles. You are about to go on an expedition to look for fossils.

1. Why might it be best to look for fossils in areas that have large formations of sedimentary rock?
2. You find a fossil of what may be the leg of a reptile. You are not sure that it is from the type of reptile that you are studying. You know the age during which your reptile lived. How would you determine if the fossil was from the reptile that you are interested in?
3. Suppose the reptile lived between 150 and 200 million years ago. In what eras and epochs might it have lived?

4. A colleague finds an imprint of a primitive flower in the same stratum of rock that held your fossil. Does this finding then change your answer to question 3? Explain.

■ **Help Wanted:**
Paleontologist. Bachelor's degree in biology required; doctoral degree required for research. Contact: American Geological Institute, 4220 King St., Alexandria, VA 22302.

Figure 28–10

Trilobite Fossils. These two fossilized trilobites were found in Morocco. Trilobites were distant relatives of insects. ▶

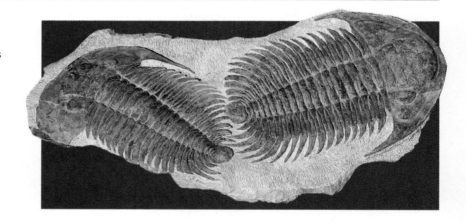

these organisms were in existence. Using the fossils, it becomes possible to match the relative ages of sedimentary rocks in different parts of the world. Fossils that permit the relative dating of rocks within a narrow time span are called **index fossils.** *Trilobites* are good examples of index fossils. See Figure 28–10. Some species of these shelled marine animals only lived between 500 and 600 million years ago. Scientists know, therefore, that sedimentary rock layers containing trilobite fossils must be between 500 and 600 million years old. Index fossils have enabled scientists to find a continuous fossil sequence from the time of the first fossils up to the recent past.

Through a correlation of absolute and relative dating of rocks, geologists have been able to construct a timetable of the earth's history. This timetable is known as the **geologic time scale.** In this time scale, the earth's history is divided into several major divisions called *eras.* Each era is further subdivided into *periods* and *epochs.* Figure 28–11 shows the main subdivisions of the geologic time scale, along with a brief summary of the various types of organisms that appeared, flourished, or disappeared during each time interval.

Patterns of Evolution

When the entire fossil record is studied, some important patterns can readily be seen. One obvious pattern is that the earliest organisms were all relatively simple. As time passed, organisms slowly became more and more complex. As you read in Chapter 5, eukaryotic cells came into existence after the simpler prokaryotic cells. Similarly, multicellular organisms appeared after single-celled organisms, and terrestrial plants and animals arrived later than aquatic species.

Scientists also have observed that the move from simpler species to more complex ones seems to have occurred over thousands or millions of years. Changes in the structure of the horse, for example, have been traced from the first appearance of a horselike mammal through various stages to what is now known as

Figure 28–11 The Geologic Time Scale. ▶

The Geologic Time Scale

Era	Period (or Epoch)		Millions of Years Ago	Plant Life	Animal Life
Cenozoic	Age of Humans	Quaternary — Recent epoch		herbs dominant	modern humans and modern animals
Cenozoic	Age of Humans	Quaternary — Pleistocene epoch	.01	trees decrease; herbs increase	early humans; large mammals become extinct
Cenozoic	Age of Mammals	Tertiary Period — Pliocene epoch	2.5	grasses increase; herbs appear	mammals abundant; earliest humans appear
Cenozoic	Age of Mammals	Tertiary Period — Miocene epoch	12	forests decrease; grasses develop	mammals increase; prehumans appear
Cenozoic	Age of Mammals	Tertiary Period — Oligocene epoch	26	worldwide tropical forests	modern mammals appear
Cenozoic	Age of Mammals	Tertiary Period — Eocene epoch	37	angiosperms increase	early mammals at peak
Cenozoic	Age of Mammals	Tertiary Period — Paleocene epoch	53	modern angiosperms appear	early placental mammals appear; modern birds
Mesozoic	Age of Reptiles	Cretaceous period	65	conifers decrease; flowering plants increase	large reptiles (dinosaurs) at peak, then disappear; small marsupials; toothed birds; modern fishes
Mesozoic	Age of Reptiles	Jurassic period	136	conifers, cycads dominant; flowering plants appear	large reptiles spread; first birds; modern sharks and bony fishes; many bivalves
Mesozoic	Age of Reptiles	Triassic period	190	conifers increase; cycads appear	reptiles increase, first mammals; bony fishes
Paleozoic	Age of Amphibians	Permian period	225	seed ferns disappear	amphibians decline; reptiles increase; modern insects
Paleozoic	Age of Amphibians	Carboniferous period	280	tropical coal forests; seed ferns, conifers	amphibians dominant; reptiles appear; rise of insects
Paleozoic	Age of Fishes	Devonian period	345	first forests; horsetails, ferns	early fishes spread; amphibians appear; many mollusks, crabs
Paleozoic	Age of Invertebrates	Silurian period	395	first land plants	scorpions and spiders (first air-breathers on land)
Paleozoic	Age of Invertebrates	Ordovician period	430	algae dominant	first vertebrates; worms; some mollusks and echinoderms
Paleozoic	Age of Invertebrates	Cambrian period	500	algae, fungi; first plant spores	most invertebrate phyla; trilobites dominant
Precambrian			570	probably bacteria, fungi	a few fossils; sponge spicules; soft-bodied invertebrates
Precambrian			?		

Figure 28–12

Evolution of the Horse. The fossil record provides information on the changes in the structure of the horse over time. During its evolution the horse has increased in size and the number of toes has decreased from four to just one. ▶

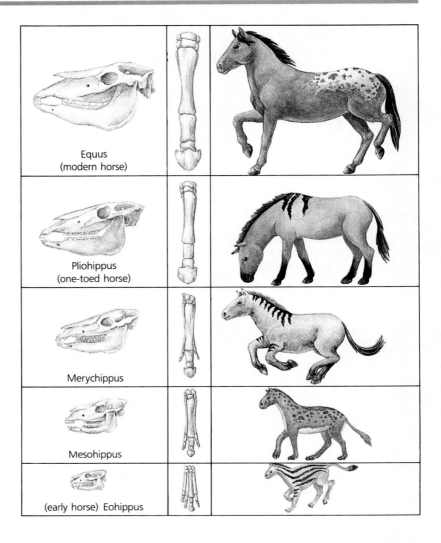

Equus
(modern horse)

Pliohippus
(one-toed horse)

Merychippus

Mesohippus

(early horse) Eohippus

a horse. See Figure 28–12. These changes occur over millions of years in the species, not in any one individual. Other sequences of this kind in the fossil record indicate that later species developed from earlier ones through a series of gradual changes passed on from generation to generation. There are also many interruptions in the fossil record. Species that have not been found in the fossil record are called missing links, or *transitional forms.* Because unusual conditions are needed to form a fossil, it is not surprising that there are gaps in the fossil record. As you have read, most often, the remains of organisms simply decay without leaving behind any traces. Despite its incompleteness, the fossil record is still considered the strongest evidence of organic evolution.

A study of the fossil record also reveals that many fossils come from species no longer living today. In fact, it has been estimated that, of all the species that ever lived, less than 1 percent exist today. When the last individual of a species has died, the species is said to be **extinct.**

The fossil record shows that on several occasions many species became extinct at the same time. The best known example of this is the extinction of the dinosaurs at the end of the Cretaceous period, 65 million years ago. Many other species of plants and animals also became extinct at this time. Some scientists think that the earth may have been struck by a small asteroid or comet, while others believe the extinctions were caused by severe volcanic activity. Whatever the cause, mass extinctions have played a major role in the course of evolution.

28-2 Section Review

1. Where is it not possible to use correlation?
2. What are index fossils?
3. What are the four eras of the geologic time scale?
4. State two patterns of evolution that can be seen in the fossil record.

Critical Thinking

5. Why would the fossil of an organism that lived only in one location be a poor index fossil? (*Reasoning Conditionally*)

28-3 Evidence from Living Organisms

Section Objectives:

- *Distinguish* among homologous, analogous, and vestigial structures.
- *Explain* how similarities in anatomy and embryological development are evidence for evolution.
- *Describe* how similarities in biochemistry show an evolutionary relationship between different species.

The classification system you read about in Chapter 7 is based on similarities and differences in anatomy, embryological development, and biochemistry. The remarkable similarity between some species has led scientists to conclude that similar species have evolved from a common ancestor. In many cases, the fossil remains of these common ancestors have been found. Comparing these remains with living organisms has added to the evidence for organic evolution.

Anatomical Similarities

Comparative anatomy is the study of structural similarities and differences among living things. The presence of certain types of similarities offers evidence for the evolutionary relationships between species.

Science, Technology and Society

Technology: Computer-Simulated Plant Evolution

From fossil evidence, biologists know that early plants looked quite different from plants today. They have developed hypotheses to explain why the plants changed. Now, with the help of a desktop computer, they can "recreate" these hypotheses.

Some scientists hypothesize plants evolved to become more efficient at gathering light. Using the computer, scientists simulate an environment in which three plant species compete against each other. The computer "creates" these plants by varying the pattern of their branches (see photo). Then, it calculates the amount of sunlight that might reach their leaves.

The species with the lowest score is eliminated, and the remaining species disperse spores. Some spores undergo mutations. These new plants enter into the competition. The game is repeated over many generations. The winners are the species that survive to the end.

So far, the computer's simulated pattern of evolution has been consistent with the pattern found in the fossil record.

- **How might learning about the evolutionary history of plants benefit science?**

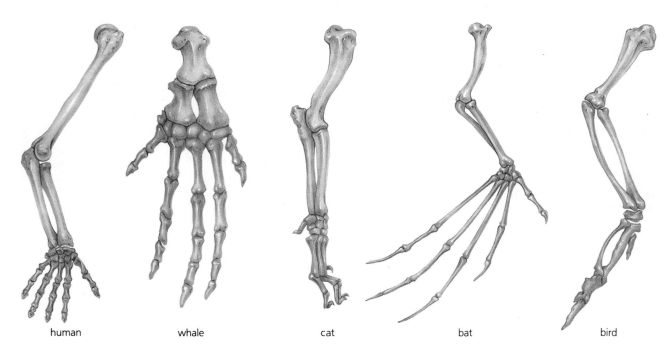

human   whale   cat   bat   bird

▲ **Figure 28–13**

Homologous Structures. Although these limbs function in different ways, they appear to have evolved from the same ancestral structure.

Figure 28–13 shows the structure of the arm and hand of a human, the flipper of a whale, the limb of a cat, the wing of a bat, and the wing of a bird. These structures are quite different in appearance because they are adapted to perform different functions. The human hand is adapted for grasping, and the whale's flipper is adapted for swimming. The wings of bats and birds are both adapted for flying. Even with these different functions, internally the structures of these organs are surprisingly similar. They all have the same number of bones arranged in a similar way. During embryological development of the animal, these structures develop in similar ways. Parts of different organisms that have similar structures and similar embryological development, but have different forms and functions, are called **homologous structures.** Homologous structures are regarded as evidence that some species evolved from a common ancestor.

The human, whale, cat, and bat are all mammals and share the characteristics of mammals. You could expect to find homologous structures among them. On the other hand, there are animals that have similar organs with similar functions but are entirely different kinds of organisms. For example, although birds and insects both have wings, they are different in structure and development. When you examine the internal structure of a bird's wing and an insect's wing, you find no similarity at all. Structures that have similar external forms and functions but different internal structures are called **analogous structures.** If birds and insects have evolved along different pathways, the fact that they have analogous structures, not homologous ones, is understandable. Analogous structures among different animals are regarded as evidence for evolution along different lines.

Another type of evidence for evolutionary relationships is the presence of **vestigial** (ves TIHJ ee ul) **structures** in modern animals. These structures are remnants of structures that were functional in an ancestral form. In modern organisms, vestigial structures are reduced in size and serve little or no function. In the human body, there are more than 100 vestigial structures, including the coccyx, or "tailbone," the appendix, the wisdom teeth, and the muscles that move the nose and ears. The human coccyx is an evolutionary remnant of an ancestral, reptilian tail, and the appendix is the remnant of a large digestive sac. Both whales and pythons have vestigial hind leg bones embedded in the flesh of the body wall. Apparently, whales and snakes evolved from four-legged ancestors.

Embryological Similarities

Comparison of the embryological development of different species can provide additional evidence of evolutionary relationships. Embryos of closely related species show similar patterns of development. Figure 28–14 illustrates various stages in the development of four different vertebrates. In these vertebrates, there are

Figure 28–14

Patterns of Development of Four Vertebrate Embryos. Similarities in the early stages of embryonic development suggest a common ancestor for these vertebrates. ▼

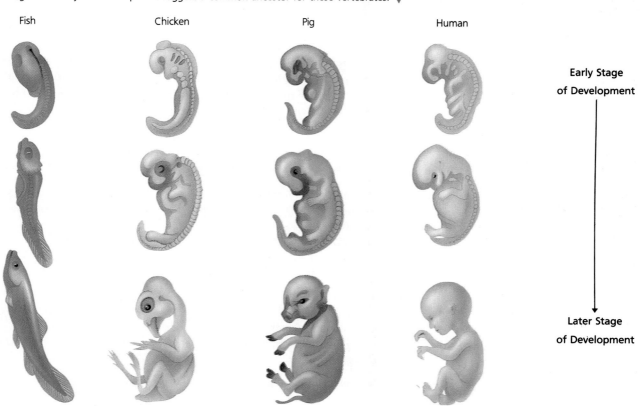

Fish   Chicken   Pig   Human

Early Stage
of Development

Later Stage
of Development

many similarities during the early stages of embryological development. For example, all the embryos have gill slits, two-chambered hearts, and tails. These similarities support the idea that these four organisms have a common evolutionary origin. As development continues, the embryos of each species begin to resemble the adults of their own species. The more closely related the animals, the longer they continue to resemble each other during the stages of development.

Biochemical Similarities

Scientists have discovered that the closer the phylogenetic relationships between organisms based on morphology and anatomy, the more alike the structure of their DNA and protein molecules. For example, the sequences of amino acids in the hemoglobin of closely related species are almost identical.

The degree of evolutionary relationship between different types of organisms also can be estimated with the Nuttall test, which is based on antigen-antibody reactions. (See Chapter 10.) For example, if human blood serum is injected into a rabbit, the rabbit produces antibodies against the proteins in the human blood. If serum from this sensitized rabbit is then mixed with human blood serum, a cloudy precipitate forms, showing an antigen-antibody reaction. The amount of the precipitate can be measured. If serums from a chimpanzee, a baboon, and a pig are then tested individually with serum from the sensitized rabbit, the amount of precipitate formed will be different in each case. The amount of precipitate is an indication of the similarity in protein structure between each of these animals and humans. The greater the amount of precipitate, the greater the similarity in protein structure, and the more closely the animal in question is related to humans.

28-3 Section Review

1. Name three different types of evidence, other than the evidence supplied by the fossil record, that support the theory of organic evolution.
2. Define *homologous structures*.
3. Name two analogous structures.
4. What biochemical similarity is revealed by the Nuttall test?

Critical Thinking

5. Look at Figure 28-14. How do you know that the evolutionary relationship between humans and chickens is closer than the evolutionary relationship between humans and fishes? (*Comparing and Contrasting*)

28-4 The Origins of Life—Early Hypotheses

Section Objectives:

- *Describe* the evidence for spontaneous generation.
- *Explain* how Redi used controlled experiments to disprove the widely accepted hypothesis of the spontaneous generation of maggots.
- *Explain* how Spallanzani's and then Pasteur's experiments finally disproved the hypothesis of the spontaneous generation of microorganisms.

For thousands of years, people believed that living organisms could arise spontaneously, or naturally, in a few days or weeks from nonliving matter. This idea is called **spontaneous generation,** or *abiogenesis* (ay by oh JEN uh sis). Belief in spontaneous generation was based on common observations and intuition. The ancient Egyptians, seeing frogs and snakes coming out of the mud of the Nile River, concluded that these animals were formed from the mud. The Greek philosopher Aristotle reasoned that an "active principle" was responsible for life. This active principle was thought to be present in mud. Some other popular beliefs were that fleas and lice arose from sweat, mice from garbage, and flies from decaying meat.

Early Experiments

In the early 1600s, a Belgian physician named Jan Baptista van Helmont performed an experiment that seemed to support the idea of spontaneous generation. He placed wheat grains in a sweaty shirt. After 21 days, the wheat was gone, and mice were present. Van Helmont reasoned that human sweat was the active principle that changed wheat grains into mice. Even though this was an uncontrolled experiment, his experimental "proof" gained wide acceptance among the scientists of his time.

Then, in the mid-1600s, the Italian physician Francesco Redi struck the first blow against the popular idea of spontaneous generation. It was well known that whenever meat was left exposed to the air, maggots soon appeared on it. Most people were convinced that the maggots developed by spontaneous generation from the decaying meat.

Redi decided to test this idea scientifically. He began by placing many different kinds of meat in open containers. Maggots soon appeared on the meat. See Figure 28–15a. He watched the maggots consume the decaying meat, and he continued to observe them even after the meat was gone. He discovered that the maggots formed pupas, which then developed into flies of various kinds. Redi apparently was the first person to see that maggots developed into flies.

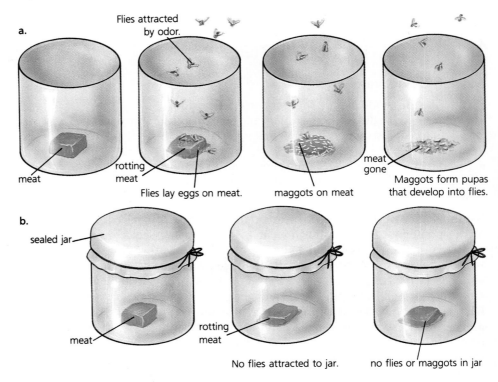

a.

Flies attracted by odor.

meat

rotting meat

Flies lay eggs on meat.

maggots on meat

meat gone

Maggots form pupas that develop into flies.

b.

sealed jar

meat

rotting meat

No flies attracted to jar.

no flies or maggots in jar

▲ **Figure 28–15**

Redi's First Experiment. This experiment showed that meat had to be exposed to the environment to develop maggots.

Redi hypothesized that the maggots developed from eggs laid on the meat by the flies. To test his hypothesis, he placed some pieces of meat in open jars, and he placed other pieces of the same meat in tightly sealed jars. Redi observed that flies entered the open jars and that maggots appeared on the meat. He further observed that no maggots appeared on the meat in the closed jars. See Figure 28–15b.

This experiment proved only that the meat had to be exposed to open air in order to develop maggots. It did not prove that the flies were the source of the maggots. Many scientists of the time claimed that fresh air was necessary for spontaneous generation. By sealing the jars, they argued, Redi had prevented the needed air from reaching the meat.

In order to answer this objection and convince the doubters, Redi performed another set of experiments. In these experiments the containers were covered by fine gauze, as shown in Figure 28–16. The gauze allowed the free circulation of air into the containers but kept the flies out of the containers. Redi observed that the flies attempted to reach the meat by landing on the gauze. The flies also deposited eggs on the gauze, which soon developed into maggots. Still, no maggots appeared inside the jars. Redi had proven conclusively that maggots did not arise spontaneously from decaying meat.

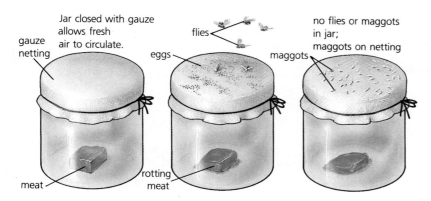

Redi's Second Experiment. Redi's second experiment showed that maggots arise in decaying meat from eggs laid by flies.

Spontaneous Generation of Microorganisms

At about the time that Redi was performing his experiments, Anton van Leeuwenhoek (LAY ven huke) made the startling discovery of microorganisms in a drop of water. Soon, it was found that when hay or soil was placed in sterile water, millions of microorganisms appeared within a few hours. Here, surely, was a clear-cut case of spontaneous generation! The controversy that Redi had almost put to rest flared up again and raged for the next 200 years.

In 1745, John Needham, an English scientist, performed some experiments that reinforced the belief in the spontaneous generation of microorganisms. He boiled flasks of chicken, lamb, and corn broth for a few minutes to kill any microorganisms in them. Then, he sealed the flasks. After several days he opened and examined the flasks and found them full of microorganisms. He repeated the experiment several times and always obtained the same results. Needham and other biologists concluded that the microorganisms developed by spontaneous generation.

About 20 years after Needham did his work, Lorenzo Spallanzani, an Italian scientist, challenged Needham's conclusions. Like Needham, Spallanzani set up flasks of chicken, lamb, and corn broth. However, he boiled the contents of the flasks for a much longer time. No living organisms appeared in the flasks.

Spallanzani claimed that Needham found organisms in his heated flasks because he had not heated them long enough to kill all the organisms originally present. Needham argued that Spallanzani had heated his flasks so long that he had destroyed the "vital principle" in the air that was needed to bring about the generation of new organisms. The debate remained unsettled for almost another 100 years.

Disproving Spontaneous Generation

In 1860, the French chemist Louis Pasteur set out to disprove the theory of spontaneous generation. Pasteur thought that microorganisms and their spores were present in the air and that they became active and reproduced when they entered the nutrient

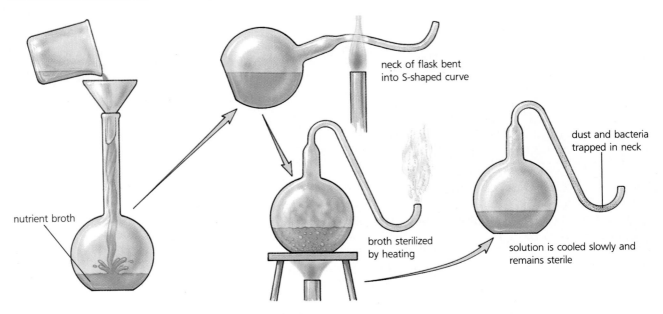

nutrient broth

neck of flask bent
into S-shaped curve

dust and bacteria
trapped in neck

broth sterilized
by heating

solution is cooled slowly and
remains sterile

▲ **Figure 28–17**

Pasteur's Experiment. This experiment showed that microorganisms that developed in a nutrient broth came from spores and microorganisms in the air.

broth. He hypothesized that the presence of air alone could not produce microorganisms in the broth. To test this hypothesis, Pasteur filled flasks with nutrient broth. He then heated the necks of the flasks and drew them out into a long S shape, leaving the ends open. See Figure 28–17. The contents of the flasks were then sterilized by boiling. Fresh air could reach the broth, but microorganisms and their spores were trapped in the long necks of the flasks. As long as the flasks were not disturbed, the contents remained sterile. Microorganisms grew in the flasks only when the flasks were tipped and some of the broth ran into the neck and became contaminated. Pasteur's experiment finally put an end to the idea of spontaneous generation.

28-4 Section Review

1. What is spontaneous generation?
2. What did van Leeuwenhoek discover?
3. Why did Pasteur alter the shape of his flasks in his experiments with spontaneous generation?

Critical Thinking

4. What assumption did Redi make in his first experiment that forced him to conduct his second experiment? (*Identifying Assumptions*)

28-5 The Origins of Life—Modern Hypothesis

Section Objectives:

- *Define* the term *biogenesis.*
- *Describe* the conditions thought to have existed on the primitive earth according to the heterotroph hypothesis.
- *Describe* any experiments that would appear to support the heterotroph hypothesis.

Today, as you read in Chapter 5, most scientists believe in **biogenesis,** the theory that living organisms originate only from other living organisms. But, this theory has one problem: How did the first living things originate on earth? In this section, you will learn about the conditions under which life may have originated on earth.

The Heterotroph Hypothesis

The most widely accepted hypothesis of the origin of life is called the **heterotroph hypothesis.** This hypothesis was formulated by a small group of scientists in the 1920s and 1930s. The scientist most often credited with development of the heterotroph hypothesis was the Russian biochemist A. I. Oparin.

Primitive Conditions on the Earth   Oparin's heterotroph hypothesis assumes that physical and chemical conditions on the earth billions of years ago were very different from those today. See Figure 28–18. For example, the earth's atmosphere now consists almost entirely of nitrogen (N_2) and oxygen (O_2), with a small amount of carbon dioxide (CO_2). Chemists and geologists have evidence showing that the earth's primitive atmosphere consisted of

Figure 28–18

Early Conditions on the Earth Compared with Modern Conditions. On the primitive earth, the composition of the atmosphere was different from the modern atmosphere. In addition, the temperature was higher, and there were more sources of energy for producing chemical change. ▼

Primitive Earth

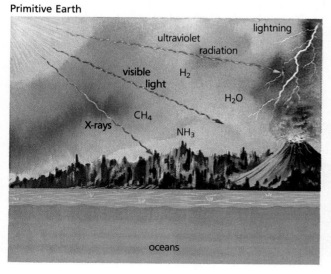

Modern Earth

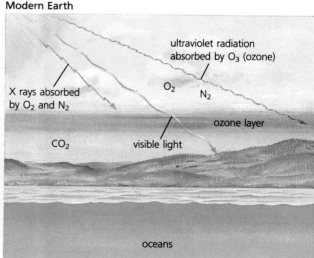

hydrogen (H_2), water vapor (H_2O), ammonia (NH_3), and methane (CH_4). There also is evidence that early temperatures were much higher on the early earth than they are at present. The oceans, when they first formed, were probably not much below the boiling point of water. The oceans of this period have been described as a "hot, thin soup," in which chemical reactions were likely to occur more rapidly than in the cooler waters of the modern earth.

Natural Synthesis of Organic Compounds   Under the primitive conditions just described, simple compounds in the atmosphere and in the oceans could have reacted to form more complex organic compounds. The synthesis of organic compounds from inorganic raw materials requires energy. Many sources of energy are thought to have been present on the primitive earth. There was heat given off by the earth itself; radiation from the decay of radioactive elements in the earth's crust; electrical energy from lightning; and ultraviolet light, visible light, and X rays from the sun. Under these conditions, there would have been enough energy available for the breakdown and formation of chemical bonds. The first nucleotides, amino acids, and sugars could have been formed during this period. There have been experimental results that support this hypothesis.

In 1953, Stanley Miller, working with Harold Urey at the University of Chicago, designed an experiment simulating the conditions of the primitive earth. The specially designed experimental apparatus contained four gases—hydrogen, water vapor, ammonia, and methane. See Figure 28–19. Boiling water in the apparatus forced these gases to circulate past sparking electrodes.

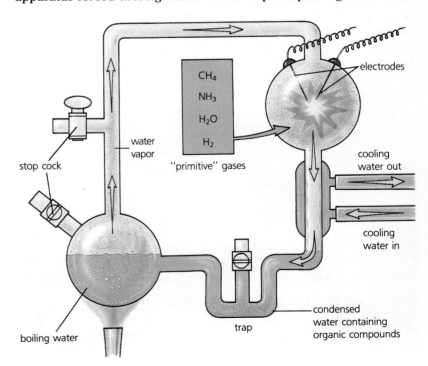

Figure 28–19

Miller's Experiment Simulating Early Conditions on the Earth. A mixture of gases thought to resemble the primitive atmosphere was continuously passed through an electric spark. Water in the apparatus dissolved the new substances produced. After a time, the solution was found to contain many organic compounds. ▶

Miller ran the experiment for a week. At the end of that time, he analyzed the contents of the apparatus and found that it contained urea, various amino acids, hydrogen cyanide, lactic acid, and acetic acid. His experiment clearly demonstrated that organic substances, including amino acids, could have been produced in nature under the conditions assumed for the primitive earth.

The work of an American biochemist, Sidney Fox, showed that, given a supply of amino acids, proteins could also be formed by nonbiological processes. Fox heated a mixture of amino acids at temperatures above 100°C for different lengths of time. Analysis of the resulting compounds revealed the presence of proteins.

Aggregates of Organic Compounds

The heterotroph hypothesis puts forth the idea that protein complexes could develop into nonliving structures that have some of the characteristics of life. Oparin thought that proteinlike substances in the prehistoric oceans may have formed aggregates, or clusters, of large molecules. See Figure 28–20. He called such

Figure 28–20

The Heterotroph Hypotheses. (a) In the early stages, coacervates absorbed organic nutrients from the physical environment. They obtained energy by anaerobic respiration, or fermentation. This released carbon dioxide into the atmosphere. (b) In later stages, autotrophs capable of producing nutrients by photosynthesis appeared. This added oxygen to the atmosphere and led to the development of aerobic respiration. ▼

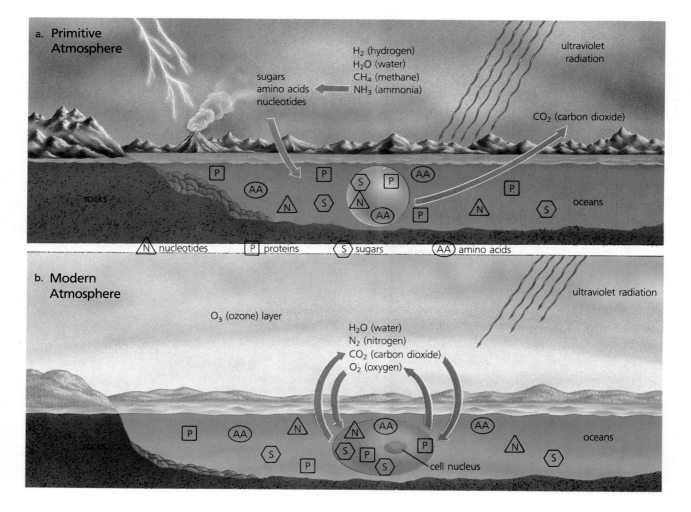

aggregates **coacervates** (koh AS er vayts). According to his hypothesis, these complex structures were surrounded by a "shell" of water molecules, which formed a sort of bounding membrane. The development of a limiting membrane would make it easier for the internal contents of the structure to be chemically different from the external environment. It would also keep various types of molecules in closer contact so that chemical reactions could occur more readily. Coacervates have been formed in laboratories from proteins and other organic molecules.

Oparin believed that within the coacervates, numerous chemical reactions occurred. As coacervates became more complex, they may have developed biochemical systems with the capacity to release energy from various organic nutrients, which were absorbed from the environment. By absorbing material from the environment, coacervates could grow in size. Eventually, they would split in half, and each half would again grow. Such structures would be primitive living things. Oparin called these structures *heterotrophs* because they obtained nutrients from the environment. However, no one yet has been able to produce self-replicating coacervates in the laboratory.

Recently, Graham Cairns-Smith at Glasgow University proposed that the first so-called organisms were inorganic crystals of clay. See Figure 28–21. These materials, which are produced by the weathering of hard rocks, would have been abundant on the early earth. Like all crystals, clay minerals self-assemble in specific patterns that are determined by the materials that compose them. Thus, these early organisms could have carried information.

Occasionally, errors may have occurred during the assembly process, resulting in new, "mutated" forms. Some scientists believe that those forms with favorable properties served as templates for organic molecules. Eventually, these more specialized organic

Figure 28–21
Structure of Clay Crystals. ▶

compounds completely replaced the clay crystals. Currently, scientists are experimenting with clay crystals in order to test this new hypothesis.

Respiration and Photosynthesis

Since the atmosphere of the primitive earth contained no free oxygen, it is thought that the first organisms carried on some form of anaerobic respiration to produce energy. This process would release carbon dioxide into the oceans and atmosphere. As the number of heterotrophs increased, the supply of available nutrients in the environment would decrease. Thus, competitions would arise between existing heterotrophs. Any organism with biochemical machinery that enabled it to use different or more complex nutrients than most other heterotrophs would have a distinct advantage. In this way, organisms containing more and more complex biochemical systems could gradually develop.

Eventually, organisms would develop that could use light energy directly for the synthesis of ATP. These would be the first photosynthetic organisms. In these organisms, the use of light energy for the synthesis of ATP would become coupled with reactions in which carbon dioxide and water were used in the synthesis of carbohydrates. These photosynthetic autotrophs would change the environment further by adding oxygen to the atmosphere.

The presence of oxygen led to the development of organisms with the capacity to carry on aerobic respiration. As you have read in Chapter 6, aerobic respiration is much more efficient than anaerobic respiration, so aerobic organisms became dominant.

The activities of living organisms eventually changed the earth's environment completely. Oxygen produced by organisms blocked the sun's X rays. Ozone (O_3), formed from the oxygen, blocked the intense ultraviolet light. Thus, the high-energy conditions that had originally led to the development of life were destroyed.

28-5 **Section Review**

1. What is biogenesis?
2. What is the name of Oparin's hypothesis of the origin of life?
3. Name the components of the atmosphere thought to have been present on the primitive earth.
4. List the sources of energy thought to have been present on the primitive earth.

Critical Thinking

5. What assumption was made in the Miller and Urey experiment? (*Identifying Assumptions*)

Laboratory Investigation

Examining Fossils

Fossils show us the kinds of organisms that lived in the past. They provide a history of life on earth and, as such, are a primary source of evidence for evolution. In this investigation, you will examine fossils to develop a sense of what they are and what we can learn from them.

Objectives

- *Examine* and study a variety of fossils.
- *Analyze* relationships between fossil organisms and present-day organisms.

Materials

assorted whole fossils
microfossil slides
stereomicroscope
compound microscope

Procedure *

A. Examine the fossils your teacher has put out for you. If possible, use the stereomicroscope to examine large specimens and the compound microscope to examine slides of microfossils, if they are available. Make a small sketch of each fossil. Then, using available reference material, record the following information in a chart like the one below.

1. Name of organism.

2. Era and period during which it lived.

3. Age of fossil. Determine this from the geologic time scale, Figure 28–11 in your text.

4. Type of fossil: (a) preserved organism (original, organic material); (b) mold or cast; (c) imprint; (d) petrified fossil (organic tissue replaced with minerals).

5. Kingdom or phylum to which organism belongs.

6. Present-day organisms, if any, related to fossil.

Analysis and Conclusions

1. Which of the fossils you observed were imprints? Which were actual specimens?

2. Based only on your observations, what information about each organism could you derive from its fossil?

3. Using one of your fossils as an example, describe how it might have been formed.

4. What kinds of living material are likely to be preserved as actual specimens?

5. How old were the oldest fossils you observed in this lab?

6. What similarities and differences exist between any of the specimens you examined and present-day organisms?

Name	Era and Period	Age	Type	Kingdom or Phylum	Living Relatives
trilobite	Paleozoic, Cambrian	500 million years	cast	Arthropoda	arthropods

*For detailed information on safety symbols, see Appendix B.

Chapter Review

Study Outline

28–1 Evidence From the Past

■ The study of fossils provides the strongest evidence for organic evolution.

■ In most fossils, only the hard parts of organisms are preserved. Sometimes, entire organisms may be fossilized in amber or ice. Most fossils are formed when organisms are buried under layers of sediment on the bottoms of lakes and seas.

■ The relative age of fossils found in sedimentary rock can be determined in relation to the order of rock layers.

■ Relative dating determines the order in which events occurred. Absolute dating methods, such as radioactive dating, determine how long ago events occurred.

28–2 Interpreting the Fossil Record

■ The process of correlation involves matching similar layers of rock formations to show the relative ages of rock layers and fossils.

■ Index fossils are fossils of organisms that lived for a well-defined period of time over wide regions of the earth. They permit the relative dating of rocks within a narrow time span.

■ Patterns in the fossil record show that organisms have evolved from simple to more complex forms over thousands or millions of years.

28–3 Evidence from Living Organisms

■ Certain types of anatomical similarities, such as homologous, analogous, or vestigial structures, provide evidence of evolutionary relationships between species.

■ Embryological similarities, and similarities in DNA and protein, provide additional evidence of evolutionary relationships.

28–4 The Origins of Life—Early Hypotheses

■ According to the theory of spontaneous generation, living organisms could arise naturally in a short time from nonliving matter.

■ Redi's and Spallanzani's experiments demonstrated that organisms do not arise spontaneously.

■ Pasteur's experiments with microorganisms conclusively disproved the theory of spontaneous generation.

28–5 The Origins of Life—Modern Hypotheses

■ According to the heterotroph hypothesis, the conditions of the primitive earth produced organic substances that gave rise to heterotrophic organisms.

■ Miller's experiment showed that organic substances could have been produced naturally under the conditions of the primitive earth. Fox's work showed that proteins could form by non-biological means.

■ By releasing oxygen, photosynthetic organisms changed the quality of the atmosphere. The presence of oxygen led to the development of organisms with the ability to carry on aerobic respiration.

Vocabulary Review

geologic evolution (571)

organic evolution (571)

fossil (571)

petrifaction (573)

mold (573)

cast (573)

imprint (573)

sedimentary rock (573)

relative dating (575)

fossil record (575)

absolute dating (575)

radioactive dating (575)

correlation (577)

index fossils (578)

geologic time scale (578)

extinct (580)

homologous structures (582)

analogous structures (582)

vestigial structures (583)

spontaneous generation (585)

biogenesis (589)

heterotroph hypothesis (589)

coacervates (592)

A. Multiple Choice—Choose the letter of the answer that best completes each statement.

1. An impression, such as a footprint, left in mud is called a(n) (a) cast (b) petrifaction (c) mold (d) imprint.

Chapter Review

2. The determination of how long ago an event occurred is called (a) relative dating (b) absolute dating (c) correlation (d) petrifaction.

3. The slow change of the earth itself is called (a) geologic evolution (b) spontaneous generation (c) biogenesis (d) organic evolution.

4. The process whereby the original substances of an organism are replaced by minerals from water is called (a) geologic evolution (b) biogenesis (c) correlation (d) petrifaction.

5. The timetable of the earth's history is known as the (a) fossil record (b) correlation (c) geologic time scale (d) imprint.

6. The human hand and the wing of a bird are examples of (a) vestigial structures (b) analogous structures (c) index fossils (d) homologous structures.

7. Louis Pasteur disproved the theory of (a) spontaneous generation (b) coacervates (c) biogenesis (d) correlation.

B. Sentence Completion—Fill in the vocabulary term that best completes each statement.

8. Any remains or trace of an organism that has been preserved by natural processes is called a(n) _____.

9. A species that is _____ is one that no longer exists today.

10. The remains of organisms, such as trilobites, that permit relative dating within a narrow time span are called _____.

11. The most widely accepted hypothesis of the origin of life is called the _____.

12. _____ is the process of continual change that occurs in species over time.

13. The theory that living organisms originate only from other living organisms is called _____.

14. _____ are nonfunctional remnants of structures in modern animals that were functional in ancestral forms.

15. _____ is formed on the bottoms of shallow seas or on ocean bottoms near shorelines of continents.

Content Review

16. What is the difference between geologic evolution and organic evolution?

17. Explain how fossils help scientists understand organic evolution.

18. Explain the process of petrifaction.

19. What is the relationship between a mold and a cast?

20. Describe how sedimentary rock is formed.

21. How does relative dating differ from absolute dating?

22. How can the absolute age of sedimentary rock be determined through radioactive dating?

23. Explain the process of correlation.

24. Explain why trilobites are good examples of index fossils.

25. What information is illustrated in the geologic time scale?

26. How are gaps in the fossil record explained?

27. How do homologous structures differ from analogous structures?

28. What could you conclude from information that cats and dogs have more homologous structures than do cats and bats?

29. What is the significance of vestigial structures?

30. How do patterns of embryological development provide evidence of evolutionary relationships?

31. How are similarities in biochemical makeup interpreted in the study of evolutionary relationships?

32. Did Redi conclusively disprove the theory of spontaneous generation? Explain your answer, and identify what Redi showed in his experiments.

33. How did van Leeuwenhoek's discovery of microorganisms affect scientists' belief in spontaneous generation?

34. Describe Pasteur's experiment with flasks of nutrient broth. Explain why his experiment finally put an end to the idea of spontaneous generation.

35. According to the heterotroph hypothesis, what were the chemical and physical conditions characterizing primitive earth?

36. How were organic compounds synthesized under the primitive conditions of early earth?

37. How did the activities of living organisms change the earth's environment?

Graphic Organizing

For information on graphic organizers, see Appendix G at the back of this text.

38. **Scale:** Construct a scale of the Paleozoic era. Use the geologic time scale on page 579 as a reference. Plot "millions of years ago" on the scale and label the periods.

39. **Concept Map:** Construct a concept map showing the formation of fossils. Put "Formation of Fossils" in a circle at the top center of your page. Include important concepts from section 28–1. As you link the concepts, do not forget the lines and linking words.

Critical Thinking

40. How do biologists use the fact that fossils exist to support the theory of evolution? (*Reasoning Conditionally*)

41. Why do scientists regard the presence of homologous structures as evidence about evolutionary relationships among organisms? (*Identifying Reasons*)

42. What would serve as an appropriate index fossil for the age of amphibians? For the Cambrian period? For the Cretaceous period? (Refer to the geologic time scale on page 579.) (*Judging Usefulness*)

43. What was the probable effect of the development of photosynthetic autotrophs on anaerobic organisms? What effect did this development have on the evolution of aerobic land animals? (*Predicting*)

Creative Thinking

44. Because evolution occurs so gradually, it is difficult for biologists to predict changes in particular species. Suggest some ways that biologists could overcome this difficulty.

Problem Solving

45. On a fossil-hunting expedition, a biologist finds at a vertical depth of 135 m on an exposed rock face the fossil skeleton of an adult bird that had a wing span of 48 cm. On the same rock face, but at a depth of only 40 m, the biologist finds a fossil skeleton of another adult bird, similar in appearance to the first one except for its wing span, which is only 22 cm. Based on these data, propose a hypothesis as to the relationship of the two birds. Describe the evidence for the hypothesis and the best way to test it.

46. A scientist studying a fossil determines that it contains only one-eighth the amount of carbon-14 that was originally present in it. The half-life of carbon-14 is 5730 years. Approximately how old is the fossil?

47. The 10 fossil snail shells depicted below were removed from successive rock layers. The oldest (1) was fossilized 10 million years ago. The most recent (10) is 3 million years old. Put yourself in the position of a biologist preparing a scientific paper, and attempt to answer the following questions: How are the snail shells similar? How do they differ? How do the shells demonstrate evolution? If specimens 3, 4, 5, and 6 were missing, what might you conclude about the remaining shells?

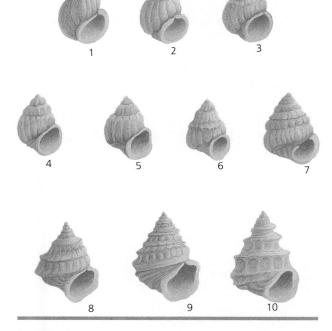

Projects

48. Visit a natural history museum. Observe a fossil display that shows the fossil history of a group of organisms or relates one group of organisms to another. Make notes, sketches, or photographs for a poster, oral report, or written report.

49. Make a cast of leaves, twigs, footprints, or bones, using plaster of Paris or clay. In a short oral report relate this process to fossil formation.

50. Using library resources, research *Archaeopteryx*, a fossil bird, or the *Ginkgo* tree, a "living fossil." Write a report on your findings in the form of a magazine or newspaper article.

The Modern Theory of Evolution

A noisy toucan displays its bill and plumage. Its colors are extravagant, its bill oversized. Each, however, has a function. Its colors attract mates; its large bill discourages predators. Species survive or die out in accordance with how effective their adaptations are for survival. All living things struggle for existence. They compete for food, for territory, for warmth, for the chance to reproduce. Some species blend into their environment, others stand out. Some species change, some stay the same, some disappear. In this chapter you will learn what causes the changes that help species survive.

Guide for Reading

Key words: natural selection, variation, adaptation, population, gene pool, evolution, genetic equilibrium, speciation

Questions to think about:

📖 How did Lamarck's theory of evolution differ from Darwin's theory?

📖 What is the synthetic theory of evolution?

📖 What role does natural selection play in the evolution of species?

29-1 Early Theories of Evolution

Section Objectives:

- *Outline* Lamarck's theory of evolution and describe Weissman's experiment that showed that acquired characteristics are not passed from generation to generation.
- *Explain* the principle of natural selection.
- *List* the six main points of Darwin's theory of evolution and state the chief weakness of Darwin's theory.
- *Distinguish* between gradualism and the theory of punctuated equilibrium.

In the previous chapter, you read some of the scientific evidence for organic evolution. But, the evidence for evolution does not explain how or why it occurs. This chapter deals with theories about how evolutionary change occurs.

Lamarck's Theory of Evolution

One of the first theories of evolution was presented by the French biologist Jean Baptiste de Lamarck in 1809. From his studies of animals, Lamarck became convinced that species were not constant. Instead, he believed they evolved from preexisting species.

▲ **Figure 29–1**

Evolution of Species. The theory of evolution accounts for how this Alaskan walrus and all other species have arisen.

◀ A keel-billed toucan is adapted for life in the rain forests of South America.

▲ **Figure 29–2**
Lamarck's Theory of Evolution. (a) Early giraffes had short necks. (b) When low-growing plants became scarce, giraffes stretched their necks to reach food. (c) The giraffes with stretched necks passed on their long-neck trait to their offspring.

He thought that these evolutionary changes in animals were caused by their need to adapt to changes in the environment.

According to Lamarck's theory, evolution involved two principles. He called his first principle *the law of use and disuse.* According to this principle, the more an animal uses a particular part of its body, the stronger and better developed that part becomes. At the same time, the less a part is used, the weaker and less developed it becomes. An athlete, for example, develops the strength of certain muscles by constant use. Muscles that are not used tend to become smaller and weaker. The second part of Lamarck's theory was *the inheritance of acquired characteristics.* Lamarck assumed that the characteristics an organism developed through use and disuse could be passed on to its offspring.

Using his theory, Lamarck offered the following explanation for the long neck of the giraffe. See Figure 29–2. The ancestors of modern giraffes had short necks and fed on grasses and shrubs close to the ground. As the supply of food near the ground decreased, the giraffes had to stretch their necks to reach leaves farther from the ground. Their necks became longer from stretching, and this trait was passed on to their offspring. In the course of generations, the giraffe's neck became longer and longer, thus giving rise to the modern giraffe.

Modern genetics has shown that traits are passed from one generation to the next by genes in an individual's gametes. As far as we can tell, however, these genes are not affected by an individual's life experiences or activities. Although there have been many experiments looking for evidence of such an effect, all have failed. The most well known of these experiments was performed in the 1870s by the German biologist August Weismann. Weismann cut the tails off mice for 22 generations. In each generation, the mice were born with tails of normal length. The acquired characteristic of shortened tails was not inherited.

Darwin's Observations

The name most closely connected with the theory of evolution is that of Charles Darwin. Darwin was the son of a well-to-do physician. At his father's urging, Darwin began to study medicine, but he did not enjoy the subject and gave it up. He then began to prepare for a career as a minister, but his real interest was in nature study—in observing the natural environment and collecting specimens.

In 1831, the British naval vessel HMS *Beagle* was about to set out on a scientific expedition to chart the coastline of South America and some of the islands of the Pacific Ocean. The expedition also planned to collect specimens of wildlife from the lesser-known regions of this part of the world. Darwin applied for the position of ship's naturalist and was accepted. He was 22 years old when he sailed from England on a voyage that was expected to take 2 years but actually lasted 5 years.

During those years, Darwin collected hundreds of specimens and made detailed observations of the regions through which he traveled. He left the ship several times to make inland journeys, rejoining the *Beagle* later. Darwin had plenty of time for thinking about what he saw. He also read the first volume of *The Principles of Geology* by Charles Lyell, which had been published shortly before Darwin left England. Lyell proposed that the earth was very old, that it had been slowly changing for millions of years, and that it was still changing. His ideas led Darwin to think that perhaps living things also changed slowly over long periods of time.

On his trip, Darwin made several types of observations that supported his idea. He noticed that there was a gradual change in each species as he traveled down the coast of South America. For example, the ostrichlike rheas that live in the latitudes around Buenos Aires are different from those found at the tip of South America. Darwin also observed fossils that were different from the living animals he saw in the same region. At the same time, the fossils had many similarities that suggested they might be related to modern forms.

The most significant of Darwin's observations were those he made on the Galapagos Islands, which lie about 1000 kilometers from the coast of Ecuador in the Pacific Ocean. He found many different species of finches living on these islands. The birds were alike, yet each species was slightly different from those on the next island or in another part of the same island. See Figure 29–3.

Darwin made similar observations about many plants, insects, and other organisms. While species on the Galapagos Islands resembled species on the mainland, they were always different in certain characteristics. See Figure 29–4. Darwin came to believe that these organisms originally had reached the islands from the mainland. Because of their isolation on the islands, the species had opportunities to develop special adaptations to each different region.

Darwin returned to England in 1836 convinced that species evolve. Although he had recorded many observations that supported such a hypothesis, he could offer no explanation of how evolution occurred. Because he could not, he did not publish his ideas at once. Instead, he continued to collect and organize his data and to search for a reasonable theory of how evolution occurs.

Darwin's Theory of Evolution

Shortly after he returned to England, Darwin read *An Essay on the Principle of Population* by Thomas Malthus. This essay greatly influenced Darwin's thinking and became the basis for his theory of evolution. Malthus, a minister, mathematician, and economist, was concerned about the social problems of an increasing human population. Malthus reasoned that the human population tends to increase geometrically (2, 4, 8, 16, . . .). For example, if each pair of parents produced four children, the new generation would have 4

▲ **Figure 29–3**
Darwin's Finches. Two of the finch species native to the Galapagos Islands are shown above.

Figure 29–4
Species Unique to the Galapagos Islands. The blue-footed booby (top) and the land iguana (bottom) are two species unique to the Galapagos Islands. ▼

individuals to replace the two that had produced them. The next generation would have 8, the next 16, and so on. On the other hand, food production could increase only arithmetically (1, 2, 3, 4, . . .) by gradually increasing the amount of land under cultivation. According to this reasoning, the food supply could not keep up with the increase in population. Therefore, to keep a balance between the need for food and the supply of food, millions of individuals had to die by disease, starvation, or war.

Darwin realized that all organisms face the same danger of overpopulation. He was also familiar with the competition and struggle for existence that occurs in nature. In 1838, the idea came to him that organisms with favorable variations would be better able to survive and to reproduce than organisms with unfavorable variations. He called this process **natural selection,** because nature "selects" the survivors. The result of natural selection would be evolution.

Darwin now had an explanation for how evolution occurred. Although many of his friends urged him to publish a book on the subject, Darwin would not be rushed. He insisted on building a strong case for his theory first. Then, in 1858, Darwin received an essay written by Alfred Russel Wallace, an English naturalist working in Indonesia. Wallace had arrived at the same conclusions as Darwin. Not knowing that Darwin had been thinking along the same lines, Wallace had sent the paper to Darwin for his opinion.

Darwin and Wallace agreed that Wallace's essay should be published along with a summary of Darwin's theory. A year later, in 1859, Darwin published his book under the title, *On the Origin of Species by Means of Natural Selection.* Darwin's book was fully supported by examples. His theory of evolution was eventually accepted by most of the leading scientists of his time.

The six main points of Darwin's theory are summarized below.

Overproduction   Most species produce far more offspring than are needed to maintain the population. Species populations remain more or less constant because only a small fraction of offspring live long enough to reproduce.

Competition   Since living space and food are limited, offspring in each generation must compete among themselves and with other species for the necessities of life. Only a small fraction can possibly survive long enough to reproduce.

Variation   The characteristics of the individuals in any species are not exactly alike. They may differ in the exact size or shape of a body, in strength or running speed, in resistance to a particular disease, and so on. These differences are called **variations.** Some variations may not be important. Others may affect the individual's ability to get food, to escape enemies, or to find a mate. These are of vital importance.

Adaptations   Because of variations, some individuals will be better adapted to survive and reproduce than others. In the competition for existence, the individuals that have favorable

Can You Explain This?

Several species of orchid emit a fragrance that is similar to the odor of female insects. Male insects are fooled by the scent and attempt to mate with the orchid's flowers. When the male lands on a flower, it makes contact with the orchid's pollen. Unsuccessful in its mating attempt, the insect flies off in search of a more appropriate mate, carrying the pollen to another orchid flower.

■ *Suggest an explanation for how this ability to deceive insects developed in orchids.*

adaptations to their environment will have a greater chance of living long enough to reproduce. An **adaptation** is any kind of inherited trait that improves an organism's chances of survival and reproduction in a given environment.

Natural Selection   In effect, the environment selects plants and animals with optimal traits to be the parents of the next generation. Individuals with variations that make them better adapted to their environment survive and reproduce in greater numbers than those without such adaptations. Experience has shown that the offspring of better adapted individuals usually inherit these favorable variations.

Speciation   Over many generations, favorable adaptations gradually accumulate in the species and unfavorable ones disappear. Eventually, the accumulated changes become so great that the net result is a new species. The formation of new species is called **speciation** (spee shee AY shun).

Applying Darwin's Theory

Figure 29–5 shows how Darwin's theory would account for the evolution of the modern giraffe. The original giraffe population had short necks and ate grass. However, unlike Lamarck's theory, Darwin's theory assumes some giraffes had longer necks than others. Those with longer necks could eat the lower leaves of trees as well as grass. In times when grass was scarce, the longer-necked animals could obtain more food than the others and, therefore, would be more likely to survive and reproduce. Their offspring would inherit the favorable variation of a longer neck. The longer the neck of a giraffe, the higher it could reach for leaves on the trees and the greater its chances for survival. As a result of natural selection, giraffe necks were slightly longer on the average in each succeeding generation. The modern long-necked animal is the result of this gradual process of evolution.

Overall, Darwin's theory of natural selection gives a satisfactory explanation of evolution. However, there are weaknesses in his theory. For one thing, it does not explain how variations originate and are passed on to the next generation. Also, it does not distinguish between variations caused by hereditary differences and variations caused by the environment, which are not inherited. For example, a plant growing in poor soil may be smaller than a plant of the same species growing in rich, fertile soil. Here, the differences in height are caused by the environment and cannot be inherited.

The Rate of Evolution

At present, scientists do not agree on the rate at which evolution or species formation occurs. According to Darwin's theory, new species arise through the gradual accumulation of small variations. In other words, evolution occurs slowly and continuously over

▲ **Figure 29–5**
Darwin's Theory of Natural Selection. (a) Adult giraffes' necks varied in length. (b) When the environment changed, only the long-necked giraffes could reach food. (c) The short-necked giraffes died, leaving only the long-necked giraffes to reproduce.

Figure 29–6

Gradualism and Punctuated Equilibrium. The gradualism model (left) sees evolution as proceeding more or less steadily through time. The punctuated equilibrium model (right) views evolution as being concentrated in short periods of time (dotted lines) followed by long periods of little or no change. Note that both models result in the same number of species. ▶

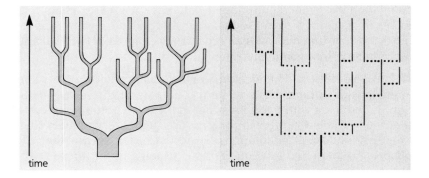

thousands and millions of years. This model is called **gradualism.** The gradualists believe that transitional forms, or links, between species are missing from the fossil record because they were less common. Thus, few of them were preserved.

Steven J. Gould and Niles Eldredge have proposed a different view of evolutionary change, which is known as **punctuated equilibrium.** According to their view, a species remains in equilibrium, or stays the same, for extended periods of time. This view is supported by the fossil record, which shows that each species seems to remain the same for thousands or millions of years. Then, in a relatively short period of time (a few hundred or a few thousand years) according to the fossil record, equilibrium is interrupted by the appearance of a new species. In other words, the long period of equilibrium is interrupted, or punctuated, by a short period of rapid evolution.

The supporters of punctuated equilibrium argue that transitional forms between species are missing because evolution occurs rapidly over a relatively brief period on the geologic time scale. Although the fossil record seems to support this new theory, the mechanisms that could produce new species in such a short interval are unknown. Figure 29–6 compares the Darwinian gradualistic model with the punctuated equilibrium model of evolution.

29-1 Section Review

1. Name the two principles involved in Lamarck's theory of evolution.
2. What is natural selection?
3. List the six main points of Darwin's theory of evolution.
4. Explain how the fossil record supports the theory of punctuated equilibrium.

Critical Thinking

5. Darwin observed similarities and differences between species on the Galapagos Islands and species on the mainland. Why were both similarities and differences necessary for Darwin's conclusion that evolution occurs? (*Reasoning Conditionally*)

29-2 The Synthetic Theory of Evolution

Section Objectives:

- *Define* the term *population genetics* and explain evolution in terms of allele frequencies.
- *Describe* De Vries' contribution to Darwin's theory of evolution.
- *List* the causes of variation in a species according to modern genetic theory.
- *State* the Hardy-Weinberg law and list the conditions under which this law holds true.
- **Laboratory Investigation:** *Test* the Hardy-Weinberg law using traits of students (p. 622).

In Darwin's time, little was known about heredity and genetics. However, modern biologists have combined Darwin's basic theory with the findings of genetics and population biology to form the **synthetic theory** of evolution. According to this theory, evolution happens to populations, not to individuals. Indeed, evolution is now defined as a change in the allele frequency within a population over time. Even with this new definition, however, individuals, not populations, are the units of natural selection.

Population Genetics

The synthetic theory of evolution stresses the importance of populations. See Figure 29–7. A **population** is a group of organisms of the same species living together in a given region and capable of interbreeding. According to the synthetic theory of evolution, *individuals* do not evolve. Their genetic makeup remains the same throughout their lives. However, *populations* do evolve. A population is made up of many individuals, each with its own unique assortment of alleles. As these individuals reproduce and die, the genetic makeup of the population as a whole may change. As its genetic makeup changes from generation to generation, the population evolves. The study of the changes in the genetic makeup of populations is called **population genetics.**

Each individual of a population has a set of alleles that is not exactly the same as the set of any other individual. Still, these individuals do have many of the same alleles. In a population as a whole, there are a certain number of alleles of each kind. Some alleles may be more common than others. For example, every individual in a population may have the alleles for producing a particular enzyme. The *frequencies* of these alleles in the population is 100 percent. On the other hand, only 1 in 100 individuals may have a mutant allele. Its frequency in the population is then 1/100, or 1 percent.

The total of all the alleles present in a population is called the **gene pool.** At any given time, each allele occurs in the population's gene pool with a certain frequency. This frequency may be any-

Figure 29–7

A Population of Penguins. According to the synthetic theory of evolution, it is a population that evolves over time, not individuals. ▼

▲ **Figure 29–8**
Mutations. De Vries formulated his theory of mutation after observing generations of evening primroses.

where from 100 percent to 1 per 10 000 or 1 per 1 000 000. As time goes on, the allele frequencies found in the gene pool may change as the result of natural selection. According to the synthetic theory, **evolution** is the gradual change of the allele frequencies found in a population.

Genetic Sources of Variation

The Dutch botanist Hugo De Vries introduced the concept of mutation at the beginning of this century. As you read in Chapter 26, De Vries based his theory of mutation on research that he conducted over several years with the evening primrose. See Figure 29–8. In the course of his research, De Vries observed that occasionally a plant appeared with a totally new structure or form. This plant would then breed true in later generations. De Vries considered these sudden changes in the hereditary material to be mutations.

De Vries added the idea of mutation to Darwin's theory of evolution. This overcame the question of how new traits could arise, one of the major weaknesses of Darwin's theory. De Vries claimed that the important changes leading to new species occurred as sudden, large changes in heredity that resulted from mutation. According to De Vries, a giraffe with a longer-than-normal neck would have been produced by a mutation. Because the long-necked giraffe and its offspring had an advantage over giraffes with necks of normal length, they survived and multiplied in greater numbers. Eventually, only the long-necked variety was left.

Mutations are not the only source of genetic variation. Recombination, as a result of sexual reproduction and the migration of individuals between populations, also contributes to variation.

Mutations   While gene mutations are a major source of variation in the synthetic theory of evolution, the mutation of any particular gene is a rare event. Out of 10 000 gametes, only one may have a mutated gene. The mutation rate for that gene is said to be 1 per 10 000. On the other hand, each gamete has thousands of genes. Among those thousands of genes, it is very likely that at least one of them has mutated. Thus, a few mutations are likely to be present in every zygote.

Most mutations are recessive. As a result, the mutant trait is usually hidden by the normal, dominant trait. Because of the low frequency of gene mutations, it is rare for mutant alleles to be brought together in the homozygous state. When this does happen, the effect may be harmful or helpful to the individual. However, if environmental conditions change, a harmful mutant allele may suddenly become useful to the species. Natural selection will then tend to gradually increase the frequency of this allele in the population.

Chromosomal mutations, which were described in Chapter 26, are another source of variation in a population. Although these mutations do not produce new genes, they do result in new combinations of genes in an organism. Since most physical traits are controlled by several genes, new gene combinations can give rise to new traits in a later generation.

Genetic Recombination   The formation of new combinations of alleles during sexual reproduction is called **genetic recombination.** Recombination can occur when two gametes undergo fusion to form a zygote. It is crossing-over and independent assortment that bring about recombination during meiosis. Crossing-over involves the exchange of segments between homologous chromosomes. When crossing-over occurs, it results in new allele combinations. Independent assortment provides that alleles on non-homologous chromosomes are randomly grouped, which also results in allele recombination.

Migration   Another source of variation may result from migration into or out of a population. As individuals move into a population, they may bring in genes not already present. When individuals leave a population, they may remove some genes from the population. Migration tends to have its greatest effect on variations in small populations.

Genetic Drift   Another factor that affects small populations is known as genetic drift. **Genetic drift** is a change in the gene pool of a small population that is brought about by chance. In small populations, there is a *chance* that a few individuals have certain alleles that the rest of the population does not have. If these individuals do not mate successfully, the alleles will be lost to the gene pool. For example, imagine that the population of an endangered plant species consists of 80 plants. Suppose that only three plants in this population have a certain allele. If these three plants are killed in a storm before they reproduce, the gene pool is reduced.

Genetic drift usually is harmful to the population because it decreases the variations in the gene pool. In a large population, genetic drift is less likely to occur because there are so many individuals. Therefore, it is unlikely that only a few individuals have an allele that no other individuals possess.

The Hardy-Weinberg Law

In Chapter 25, you read about Mendel's experiments with hybrid pea plants. In these experiments, there were two alleles for each trait that he studied. For example, the allele T produced tall plants, and the allele t produced short plants. The frequencies of T and t in the hybrid plants were equal: 50 percent T and 50 percent t. When the hybrids reproduced, the offspring had a genotype ratio of 1:2:1 and a phenotype ratio of 3:1, but the allele frequencies in the offspring remained the same—50:50. You can check this statement for yourself by counting the T and t alleles in the Punnett square shown in Figure 25–10.

Science, Technology and Society

Issue: Endangered Species

The snow leopard, northern spotted owl, and African elephant are only a few of many species facing a common threat: extinction. Today, the chief causes of extinction are human activities such as poaching (hunting illegally), and destroying natural habitats for industrial or agricultural use. The present extinction rate is over 1000 times higher than in prehistoric times, and it is increasing rapidly.

Environmentalists argue that governments must pass laws to control human activities that result in extinction. All organisms have a right to exist, and many species have potential value as sources of medicine and knowledge.

Other people argue that human needs are more important. Many people depend on the farmland for their food. Others make a living by logging or through other uses of the land. Moreover, competition for survival among species, including humans, is a natural process. Because it is, governments should not interfere.

■ ***Should steps be taken to protect endangered species? Why or why not?***

The condition in which allele frequencies do not change from one generation to the next is called **genetic equilibrium.** To maintain genetic equilibrium, it is not necessary for two alleles in a population to have the same frequency. One may be much more common than the other. Suppose, for example, that there are two alleles for eye color in a certain species. One allele produces white eyes; the other, red eyes. We do not need to worry about dominance or about the eye color of a hybrid. Let us just assume that the allele for white eyes is more common than the allele for red eyes. We will say that 90 percent of the alleles are for white eyes, and 10 percent are for red. As these organisms mate and reproduce, what will happen to the allele frequencies for eye color as generation follows generation? While it might seem logical that the white-eye allele will eventually replace the red-eye allele, in fact, the allele frequency does not change.

In 1908, G. H. Hardy, an English mathematician, and W. H. Weinberg, a German physician, considered this question and came to the same conclusion independently. They showed that segregation and recombination of genes in sexual reproduction could not change allele frequencies by itself. If the frequency of allele p was 90 percent, and the frequency of its allele, q, was 10 percent, random mating would always produce a new generation with the same ratio of 90 percent p and 10 percent q. Their conclusion that sexual reproduction alone does not affect genetic equilibrium is called the **Hardy-Weinberg law.** See Figure 29–9.

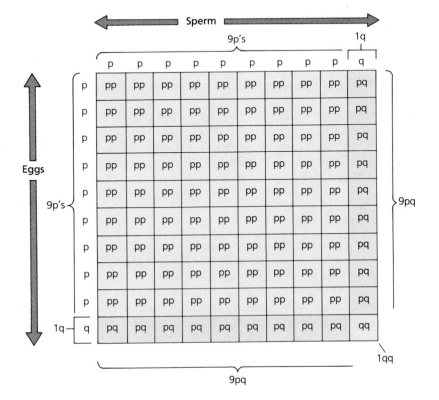

Figure 29–9

Illustrating the Hardy-Weinberg Law. Assume that the frequency of an allele p in a population is 90 percent and that of allele q is 10 percent. Then, out of every 10 sperm cells produced by the population, 9 will carry the p allele and 1 will carry q. The Punnett square shows the results of random fertilizations between these sperm and eggs. Of every 100 gametes formed, 81 are pp, 18 are pq, and 1 is qq. The total number of p's and q's in these 100 gametes is:

$$81\ pp = 162\ p's$$
$$18\ pq = 18\ p's\ and\ 18\ q's$$
$$\underline{1\ qq = 2\ q's}$$
$$Totals:\ 180\ p's\ and\ 20\ q's$$

We see that the ratio of p to q in the offspring generation is 180 to 20, or 9 to 1. This is the same as the ratio in the parent generation. ▶

For the Hardy-Weinberg law to hold true, four conditions must be met.

- The population must be large. In a small population, alleles of low frequency may be lost, or the frequency may change due to genetic drift.

- Individuals must not migrate into or out of the population. Any individuals that do so may change the allele frequencies of the population.

- Mutations must not occur because mutations obviously change the frequencies of the population.

- Reproduction must be completely random. This means that every individual, whatever its genetic makeup, should have an equal chance of producing offspring.

While the first two of these conditions can exist in nature, the last two conditions almost never exist. Populations can be large enough, and migration can be practically zero under certain circumstances. Mutations, however, are always occurring at fixed rates, thus changing allele frequencies. Furthermore, reproduction is not random. Individuals with helpful adaptations are more likely to reproduce because of natural selection. This also results in a change in the allele frequencies.

You may wonder about the usefulness of a law that does not apply to any situation in the real world. The Hardy-Weinberg law is important because it allows us to discover whether or not evolution is occurring in a population. The law tells us that under certain conditions, allele frequencies will remain constant and there will be no evolution. The fact that allele frequencies in a population change tells us that there are external factors causing them to change. In other words, the failure of the Hardy-Weinberg law is a sign that evolution is occurring. The extent of the variation from the Hardy-Weinberg prediction is a measure of how rapid the evolutionary change is.

29-2 Section Review

1. Name the sources of variation within a species according to the synthetic theory of evolution.
2. What is a population?
3. Define the term *gene pool*.
4. List the conditions that must exist for the Hardy-Weinberg law to hold true.

Critical Thinking

5. If mutations occur, then the allele frequencies in a population will change. In a population you are studying, the allele frequencies have changed. Does this prove that mutations have occurred? Why or why not? (*Reasoning Conditionally*)

▲   **Figure 29–10**

Adaptations. The dormouse, like many mammals, hibernates in winter. During hibernation, the animal's metabolic rate and body temperature decrease. This adaptation allows the animal to survive long cold periods when food is scarce.

29-3 Adaptations and Natural Selection

Section Objectives:

■ *Explain* the term *adaptation* and name some different kinds of adaptations.
■ *Define* the terms *camouflage, warning coloration,* and *mimicry.*
■ *Distinguish* among *directional selection, stabilizing selection,* and *disruptive selection.*

Under what conditions will a species evolve? By focusing on allele frequencies, the Hardy-Weinberg law allows scientists to understand the mechanisms of natural selection. In this section, we will look at how natural selection determines which adaptations are favorable for survival.

Types of Adaptations

As you read earlier, an adaptation is any kind of inherited trait that improves the chances of survival and reproduction for an organism. The environment is the selecting force that chooses the best and most useful inherited variations. For example, in a population of plants, there may be a genetic variation in the amount of waxy cutin covering the leaves of the plants. Some plants may be heavily covered with this protective layer, while other plants are only thinly covered. Because cutin protects the plant from drying out, plants with a thick cutin layer will be better able to survive and to produce seeds if the climate becomes very dry. In this case, the cutin is an adaptation that has been "selected" by the environment. After many generations, alleles for this adaptation will accumulate in the gene pool. Eventually, only plants with a heavy cutin layer will remain in the population.

Structural adaptations are adaptations that involve the body of the organism. The wings of birds and insects, for example, are structural adaptations for flight. The fins of fish and the webbed feet of ducks are structural adaptations for swimming. *Physiological adaptations* involve the metabolism of organisms. The protein web made by spiders and the poison venom made by snakes are examples of physiological adaptations. Still other adaptations involve particular behavior patterns. Of course, many adaptations are combinations of various types of adaptations. For example, the mating behavior and migration of birds, the spawning of fishes, and the hibernation of animals involve several types of adaptations. See Figure 29–10.

Many adaptations provide protection. In **camouflage,** the organism blends into the environment, as shown in Figure 29–11. Flounders can become practically invisible against a variety of backgrounds. Fawns are hard to see among the shadows in their usual environment. In **warning coloration,** the colors of the animal

Figure 29–11

Camouflage. The sundial flounder is camouflaged against the sandy ocean floor. ▼

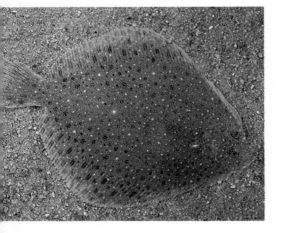

▲ **Figure 29–12**

Mimicry and Warning Coloration. The inedible monarch butterfly (left) warns away potential predators with its bright colors. The edible viceroy butterfly (right) is avoided by predators because it so closely resembles the monarch in color and markings.

actually make it easier to see. This is an advantage for certain insects that birds and other enemies find unpleasant to eat. If a young bird happens to eat one of these insects, it quickly learns to avoid that species in the future. The brightly colored monarch butterfly, which is shown in Figure 29–12, is an example of this kind of warning coloration. In **mimicry,** one organism is protected from its enemies by its resemblance to another species. Birds can eat the viceroy butterfly without suffering unpleasant effects, but they tend to avoid it because it looks like the monarch butterfly. In contrast to mimicry, camouflage helps an organism avoid its enemies by making it difficult to see. As Figure 29–13 shows, shape and structure, as well as coloration, can contribute to an organism's camouflage.

◀ **Figure 29–13**

Camouflage. The lichen katydid is hidden from predators because it resembles the lichen on which it feeds.

Types of Natural Selection

According to the synthetic theory of evolution, natural selection disturbs genetic equilibrium. As a result, the allele frequencies in the population will change. In this way, natural selection determines which adaptations are favorable for a species. There are three main types of natural selection.

Directional Selection   One type of natural selection, in which an extreme phenotype becomes a favorable adaptation, is called **directional selection.** This type of selection usually operates when the environment changes or when species migrate. The new environmental conditions favor the extreme phenotype, causing the population to evolve.

The evolution of long-necked giraffes is an example of directional selection. The red graph in Figure 29–14a shows the continuous variation of neck length in a giraffe population at some time in the past. Most of the giraffes in this population had intermediate neck lengths, although some had short necks and some had long necks. Several alleles for short and long necks probably were present, with low frequencies, in the population. A change in the environment may have given the individuals with the alleles for longer necks a favorable adaptation. In later generations, their offspring made up a larger fraction of the population. Therefore, the

Figure 29–14

Types of Selection. Directional selection favors a relatively rare phenotype. Stabilizing selection favors average phenotypes and acts against extreme phenotypes. Disruptive selection favors two extreme phenotypes and acts against the average phenotype. ▼

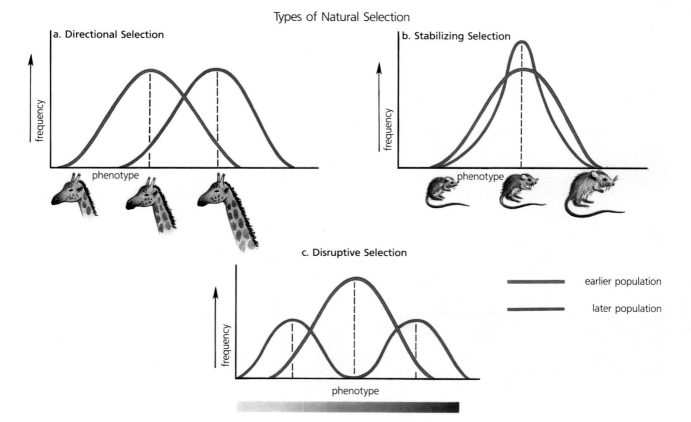

Types of Natural Selection

alleles they carried for long necks would be present at a greater frequency, and the population would have longer necks, on average. This is shown by the blue curve. Notice that the blue curve looks like the red curve, but it is shifted to the right. Thus, directional selection selects in the direction of an extreme phenotype.

Stabilizing Selection   Sometimes, the average phenotype may be a favorable adaptation, and extreme phenotypes are unfavorable. This is called **stabilizing selection.** For example, mice that are too small may not be strong enough to burrow underground in cold weather, while mice that are too large may use too much energy in keeping warm. If the climate becomes colder, fewer large and small mice will survive to reproduce. See Figure 29–14b. Notice that the average size of the population, shown by the dashed line, is the same in the two populations. However, the blue curve representing the population after the climate change is narrower because there is less variation in the population. This is because the frequencies of some alleles carried by the extreme phenotypes have decreased.

Stabilizing selection operates most of the time in most populations. This type of selection limits evolution by keeping allele frequencies relatively constant. In this way, populations of organisms, such as sharks and ferns, have remained stable for millions of years.

Disruptive Selection   A third, rare type of natural selection is called disruptive selection. In **disruptive selection,** two opposite phenotypes are favorable adaptations, while the average phenotypes are unfavorable. As you can see in Figure 29–14c, this creates two subpopulations. For example, a species of crab might show a continuous variation in color from light tan to dark brown. If the environment changed to include both sandy beaches and brown mud, both extremes of coloration could be favorable camouflage against predators. In time, alleles carried by the extreme phenotypes would increase in frequency, resulting in the evolution of two subpopulations. If these populations could not mate with each other, they would be considered two new species.

Can You Explain This?

Several centuries ago, a ship carrying emigrant families was blown off course and wrecked on a deserted tropical island. All of the people were very light skinned. After many years under the tropical sun, however, everyone became dark skinned.

■ *What color skin would babies born to the children of these couples have? What color skin would babies in the fifth generation have? Explain your answers.*

29-3 Section Review

1. Define the term *adaptation.*
2. Name two different types of adaptations and give an example of each.
3. What are the three main types of natural selection?
4. Which type of natural selection tends to prevent evolution?

Critical Thinking

5. How is mimicry of the environment different from camouflage? How is it similar? (*Comparing and Contrasting*)

29-4 Speciation

Section Objectives:

- *Define* the terms *range* and *speciation*.
- *Describe* the processes of isolation and adaptive radiation.
- *Distinguish* between convergent evolution and coevolution.
- 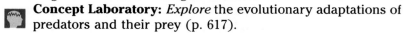 **Concept Laboratory:** *Explore* the evolutionary adaptations of predators and their prey (p. 617).

Under certain circumstances, one species can evolve into two or more species. This formation of new species is called **speciation** (spee shee AY shun).

Speciation and Geographic Separation

Each species is found in a particular region of the earth. This region is called the species' **range.** The characteristics of a species are often different in different parts of its range. Differences in environmental conditions have exerted different selective pressures, leading to different adaptive characteristics. The leopard frog, *Rana pipiens,* for example, has a wide range that extends over most of North America. Across this range, the frogs differ in body size, patterns of coloration, and the temperatures at which their embryos will develop. The species actually consists of separate populations with different gene pools. However, adjacent populations can mate and produce normal offspring. These separate populations are therefore called subspecies, or varieties, of the same species.

The frogs at opposite ends of the range show the greatest differences in characteristics and in gene pools. In fact, frogs from widely separated regions cannot mate successfully. They are still considered to be the same species because there is continuous interbreeding among adjacent subspecies. If, however, a population varies so much from its neighbors that it loses the ability to interbreed, a new species has developed.

Types of Speciation

Isolation   One of the most important factors involved in speciation is isolation. *Isolation* refers to anything that prevents two groups within a species from interbreeding. Isolating a group of organisms separates its gene pool from the gene pool of the rest of the species. Through mutation, genetic recombination, and natural selection, a different gene pool will evolve in each group.

It is generally believed that speciation is a two-step process that first involves geographic isolation, followed by reproductive isolation. **Geographic isolation** occurs when a population is divided by a natural barrier, such as a mountain, desert, river or other body of water, or a landslide caused by an earthquake. See Figure 29–15. As a result, the gene pool of each group becomes isolated, and the two can no longer intermix. Over a period of time, each group will become adapted to its particular environment. When the differences between the isolated groups become great enough, they will no longer be able to interbreed, even if they could get together. The loss of the ability to interbreed by two isolated groups is called **reproductive isolation.**

Reproductive isolation can be produced by several mechanisms. Differences may arise in courtship behavior, times of mating, or the structure of the sex organs. Such changes make it unlikely that mating will occur. Other changes affect events after mating and involve the inability of sperm to fertilize eggs, the death of the embryo early in development, or the development of offspring that are sterile. According to most biologists, if two groups of organisms cannot interbreed successfully, they can be considered different species.

The Kaibab squirrel and Abert squirrel are thought to be cases of speciation by geographic and reproductive isolation. The Kaibab squirrel inhabits the north side of the Grand Canyon, and the Abert squirrel inhabits the south side. It is believed that these two squirrels evolved from a common ancestor. The Grand Canyon, acting as a geographical barrier, divided the ancestral population, which once occupied the entire area. After a long period of geographical isolation, the Kaibab and Abert squirrels evolved. The two squirrels are similar in appearance but are different species because they cannot interbreed.

Polyploidy   Speciation can occur suddenly when abnormal meiosis or mitosis results in polyploidy. As you read in Chapter 25, polyploids are organisms, usually plants, that contain more than the usual number of chromosome sets. When the offspring can interbreed only among themselves, they are considered a new species.

Adaptive Radiation   The process by which a species evolves into a number of different species, each occupying a new environment, is called **adaptive radiation.** This spreading, or radiation, of the organisms into different environments is accompanied by adaptations.

For example, a single ancestral species may have migrated— radiated—into several different environments. If the descendants have few predators and little competition for food, they will be successful. Then, through isolation, genetic variation, and natural selection, they will evolve a variety of adaptations to their new environments. After many generations, they will have evolved into several new species, each having certain adaptive traits. However, their common ancestry is indicated by the traits they share in common.

Darwin's finches are an example of adaptive radiation. In this case, an ancestral type of finch probably arrived in the Galapagos Islands and then radiated into a variety of habitats and ways of life. The initial radiation involved living on the ground and living in trees. Further radiation occurred on the basis of food. Some finches live on the ground and feed on seeds of varying size. Some live in forests and feed on insects in trees. Others feed mainly on cactus or

▲   **Figure 29–15**
Speciation through Geographic Isolation. The spring-dwelling salamander (top) and the cave-dwelling salamander (bottom) can no longer interbreed. These two species are believed to share a common ancestor. The population was divided when one group took up residence in a cave. The differing selective pressures of the two environments resulted in the divergence of the species over time.

▲ **Figure 29–16**
Convergent Evolution. The koala bear is a marsupial that looks very much like a bear.

berries. One species lives in low bushes and feeds on insects. Without competition from other birds, the finches slowly radiated into and adapted to the various types of environment that were present.

Convergent Evolution

As a result of geographic isolation, organisms that are not closely related may develop similar adaptations and come to resemble each other. Natural selection that causes unrelated species to resemble one another is called **convergent evolution.** Convergent evolution produces analogous structures, which you read about in Chapter 28. Bird wings and insect wings are good examples of analogous structures that result from convergent evolution.

Another example of convergent evolution is the similarity between marsupials and their placental counterparts. The marsupial mouse looks much like a placental mouse. There is also a marsupial that resembles a wolf (the Tasmanian wolf) and one that resembles a bear (the koala), which is shown in Figure 29–16. These resemblances are only "skin deep." They evolved because of similar needs in similar environments, leading to the natural selection of analogous structural adaptations.

Coevolution

Two or more species also can evolve in response to each other through cooperative or competitive adaptations. This is called **coevolution.** One example of coevolution is the relationship between flowers and their pollinators. For example, some species of flowers have developed adaptations to attract bees. Bees are active during the day, attracted by bright colors and sweet or minty odors, and usually land on a petal before feeding. Flowers adapted to bees have a sweet or minty odor and are open in the daytime. They have a petal for the bee to land on and are usually bright blue or yellow, because bees cannot see red light. In comparison, bats, which are active at night, feed on the nectar of flowers that are open at night and easily visible in the dark. Coevolution reduces competition between species and benefits both species.

29-4 Section Review

1. What is speciation? Name the two steps involved in speciation.
2. What is adaptive radiation?
3. Give an example of convergent evolution.
4. Define the term *coevolution*.

Critical Thinking

5. What possible outcomes are likely if a species migrates into an area with many predators? Explain your answer. (*Predicting*)

Concept Laboratory

Natural Selection

Goal

To gain an understanding of the process of natural selection.

Scenario

Imagine that you and your classmates are predators, searching for food. Your only source of food is square-shaped organisms. You will eat any square-shaped individual as soon as you see it.

As you perform this experiment, think about how the process of natural selection operates.

Materials

newspaper
solid color
   construction paper
scissors
empty box

Procedure ☞ *

A. Working in groups of four, cut sheets of newspaper into 64 squares of equal size—about 5 cm × 5 cm.

B. Do the same for the colored sheet of paper.

C. Place all the paper squares into a box, shake thoroughly, and dump the squares onto some open, uncut sheets of newspaper. Spread out the squares so they do not overlap.

D. Acting as a predator, each member of your group in turn should quickly look at the spread of squares on the newspaper, grab the square that he or she first sees, then look away. Each person should repeat this procedure five times.

Organizing Your Data

1. Copy the chart below and record how many types of paper square each member of your group captured.

2. Among the members of your group, what differences were there in the number of captured squares of each color? What are some possible explanations?

*For detailed information on safety symbols, see Appendix B.

Student	Newspaper Squares	Colored Squares
1		
2		
3		

3. Calculate the number of squares of each type that were not captured. Graph the data.

4. Compare the results of your group with those of the rest of your class.

Drawing Conclusions

1. What do the two kinds of paper squares represent? What does the open, uncut newspaper represent?

2. Which type of paper square had the lower rate of survival? How would you explain these results?

3. What traits besides coloration might give some individuals of a species an adaptive advantage over others?

4. Based on this investigation, write a brief paragraph explaining how natural selection operates.

Thinking Further

1. Based on your results, what would you expect to happen to the population makeup over the course of many generations?

2. How confident are you that the population would change as you predicted in the previous question? List three factors that might cause the outcome to be different from what you predicted.

3. As you know, predation is only one of many selective forces acting on organisms. Describe a situation in which weather conditions would lead to natural selection in a plant species.

29-5 Observed Natural Selection

Section Objectives:

- *Discuss* industrial melanism and the information gained from the study of the peppered moth in England.
- *Describe* how populations of antibiotic-resistant bacteria and DDT-resistant insects have arisen.

Natural selection may take many thousands of years to produce a change in a population. However, in recent years some excellent examples of natural selection have given scientists an opportunity to study evolution in action. One of these illustrates a kind of adaptation called industrial melanism. **Industrial melanism** is the term used for the development of dark-colored organisms in a population exposed to industrial air pollution.

Industrial Melanism

The peppered moth, *Biston betularia,* is found in wooded areas in England. Before the 1850s, most peppered moths were light in color. Black-colored moths that have a pigment called *melanin* occurred, but they were very rare. During the years from 1850 to 1900, England became heavily industrialized. Where there was a great deal of industry, heavy smoke darkened the tree trunks and killed the light lichens that were growing on them. By the 1890s in these regions, 99 percent of the peppered moths were black in color, while the light-colored variety was rare. In the cleaner, nonindustrial areas of southern England, the light-colored moth continued to predominate. See Figure 29–17.

A hypothesis for the change from light-colored to dark-colored moths can be found in natural selection. The light and dark color of the moth is genetically controlled. The dark color is a mutation that

Figure 29–17

Industrial Melanism. On the light tree (left), the lighter colored peppered moth is better camouflaged. On the dark tree (right), the darker moth is better camouflaged. ▼

occurs at a constant low frequency. During daylight, peppered moths rest on tree trunks. Before England became industrialized, the light-colored moths blended in well with the lichens that covered the tree bark. As a result of this camouflage, birds that feed upon the peppered moth could not easily find the light-colored moths. Dark-colored moths were easily seen and eaten by hungry birds. This situation gave the light-colored moths an obvious reproductive advantage. When the fumes and soot killed the lichens and blackened the trees, however, the light-colored moths were easy to see against the dark background of the tree trunks and became easy prey for birds. Now, the dark-colored moths had a distinct advantage. The blackened trees offered them good camouflage, as shown in Figure 29–17. Through natural selection, more dark moths survived and reproduced than light-colored moths. During the years from 1850 to 1900, a period of 50 generations for peppered moths, the dark-colored moths became the more frequent color in the population.

In the 1950s, H. B. D. Kettlewell and Niko Tinbergen performed controlled experiments that attempted to test the above hypothesis. Light- and dark-colored peppered moths were released in a polluted industrial area and in an unpolluted nonindustrial area. In the polluted area, where the trees were blackened with soot, Kettlewell and Tinbergen recorded more light-colored moths eaten than dark-colored moths. In the unpolluted area, birds ate more dark-colored moths than light-colored moths.

This research on the color change in peppered moths shows that a species can change gradually from one form to another over a period of time. Two details of the evolutionary process are clearly illustrated. One is the presence of variability in the population. Alleles for light color and dark color are in the gene pool. The other is the effect of the changing environment in selecting one color trait over another. The trait that makes the moth best adapted to the environment is preserved.

In the United States, the insects around many major cities are darker in color than the ones in the unpolluted countryside. Interestingly, since the 1950s, air pollution control in England has resulted in an increase in the number of light-colored peppered moths.

Bacterial Resistance to Antibiotics

Antibiotics usually kill bacteria. However, once the use of antibiotics became common, resistant strains of bacteria began to appear. Antibiotics were no longer effective in killing those strains.

Scientists wanted to know how this resistance developed. One possibility was that exposure to an antibiotic caused certain bacterial cells to develop resistance to it. This would be similar to the immunity an individual acquires to a disease organism after recovery from the disease. Another possibility was that in a large population of bacteria, there are always a few individuals with

Biology and You

Q: Can resistance to antibiotics spread from one bacterium to another? If so, will antibiotics soon be useless?

A: Nonresistant bacteria become resistant to antibiotics in many ways. Sometimes rings of DNA, called plasmids, carry antibiotic-resistance genes. Plasmids travel among bacterial cells. When a nonresistant cell acquires a plasmid, it may become resistant to antibiotics. Often resistance is coded for by "jumping genes," which move from chromosome to chromosome. Spontaneous mutation during replication and recombination can also lead to resistance.

During medical treatment, an antibiotic will kill off nonresistant organisms. Resistant strains, however, will survive. Because bacteria reproduce quickly, resistant strains can rapidly spread. This rise in resistant organisms causes antibiotics such as penicillin or streptomycin to be ineffective. One way to reduce this problem is to minimize the use of antibiotics. Another way is to give two unrelated antibiotics at the same time, assuming that resistance to both is unlikely.

■ *Survey five people to find out how many have taken antibiotics in the past three months. Do they know the type taken?*

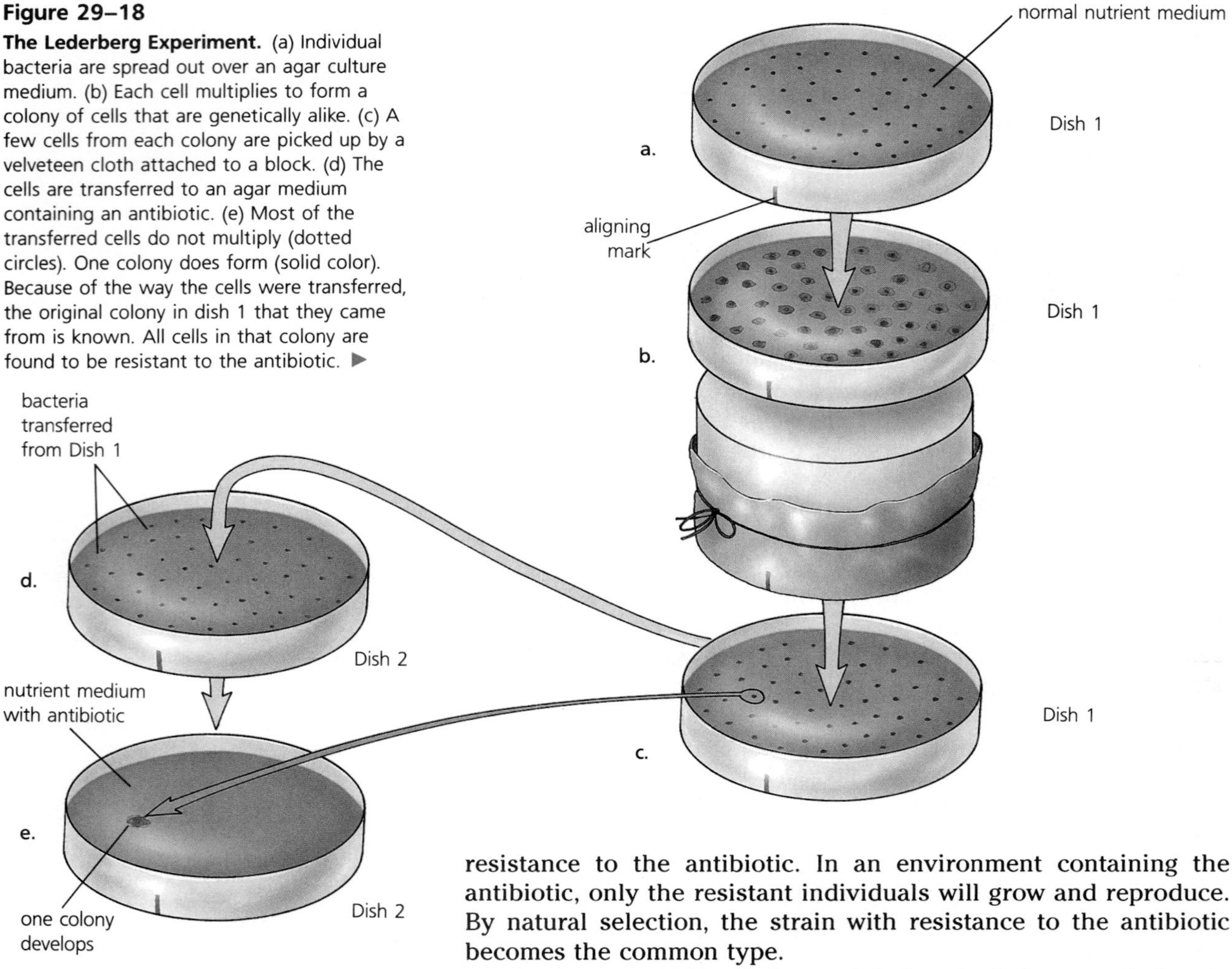

Figure 29-18

The Lederberg Experiment. (a) Individual bacteria are spread out over an agar culture medium. (b) Each cell multiplies to form a colony of cells that are genetically alike. (c) A few cells from each colony are picked up by a velveteen cloth attached to a block. (d) The cells are transferred to an agar medium containing an antibiotic. (e) Most of the transferred cells do not multiply (dotted circles). One colony does form (solid color). Because of the way the cells were transferred, the original colony in dish 1 that they came from is known. All cells in that colony are found to be resistant to the antibiotic. ▶

bacteria transferred from Dish 1

normal nutrient medium

Dish 1

a.

aligning mark

Dish 1

b.

Dish 1

c.

d.

Dish 2

nutrient medium with antibiotic

e.

one colony develops

Dish 2

resistance to the antibiotic. In an environment containing the antibiotic, only the resistant individuals will grow and reproduce. By natural selection, the strain with resistance to the antibiotic becomes the common type.

In the early 1950s, Esther and Joshua Lederberg carried out a series of experiments that showed that the second explanation, natural selection, was the correct one. The Lederbergs worked with the common intestinal bacterium *Escherichia coli,* which is normally killed by the antibiotic streptomycin. The first step of their experiment was to spread a culture of the bacteria very thinly on an agar nutrient medium in a petri dish. See Figure 29–18. This technique, which is called streaking, had the effect of separating the culture into individual bacteria. Each bacterial cell then multiplied on the agar, forming a distinct colony. In each colony, all the cells were genetically alike since they had developed from a single original cell.

The Lederbergs then looked for cells resistant to streptomycin. Since it would have taken too much time to investigate each colony separately, they used a velveteen cloth to pick up bacteria from all the colonies at once. The cloth was then touched to a second agar plate that contained streptomycin, thus transferring bacteria from all the colonies to the agar. Usually, none of the transferred bacteria formed a colony; they could not survive and multiply in the streptomycin environment. Occasionally, however, a colony did grow on the streptomycin plate. When this happened, the Lederbergs could identify the original colony the transferred cells had come from. They could because the velveteen cloth placed the bacteria in the same relative positions on the new agar as the colonies from which they were picked up. It was then a simple matter to test the original colony for streptomycin resistance.

When this test was carried out, it was found that all cells in the original colony were resistant. Remember that these cells had never been exposed to streptomycin. Their resistance was a genetic trait that they already possessed.

The Lederbergs concluded that a few bacteria with resistance to streptomycin had been in the original population. When no antibiotic was present in their environment, these cells had no advantage or disadvantage. However, when the environment was changed to include the antibiotic, the resistant cells had a survival advantage and multiplied, while the normal type died out. The population became 100 percent streptomycin-resistant.

This experiment showed that the change in the environment had not caused the resistance to develop. It had acted only as a selector for organisms that already had the gene for resistance to streptomycin.

Insect Resistance to DDT

When DDT was first introduced, it was an effective killer of insects, including mosquitos. Apparently, however, a small proportion of insects in various insect populations possessed a natural resistance to DDT. When the DDT-sensitive members of a population were killed by spraying, the DDT-resistant insects survived and passed on their natural DDT-resistance to their offspring. Eventually, many insect populations were completely resistant to DDT.

The DDT did *not* create the resistance of the insects. Rather, the DDT acted as the environmental agent for the selection of the resistant strains.

29-5 Section Review

1. What is industrial melanism?
2. Which moths were eaten more easily by birds in industrial areas?
3. Who discovered how antibiotic-resistant bacteria develop?

Critical Thinking

4. Explain why air pollution control in England since the 1950s has led to a greater number of light-colored moths. (*Identifying Causes*)

Laboratory Investigation

The Hardy-Weinberg Law

The Hardy-Weinberg Law, expressed as $p^2 + 2pq + q^2 = 1$, allows the calculation of allele and genotype frequencies in a population. In this investigation, you will determine allele and genotype frequencies for a single human trait.

Objectives

- *Calculate* the frequencies of the dominant and recessive alleles for an inherited trait.
- *Compare* the frequencies of two alleles with the frequencies of their phenotypes.

Materials

calculator

Procedure

A. Tongue rolling is controlled by a single gene. Persons homozygous dominant or heterozygous can roll their tongue. Homozygous recessive individuals cannot. See if you can roll your tongue, as shown below. Following your teacher's instructions, indicate your phenotype on the chalkboard.

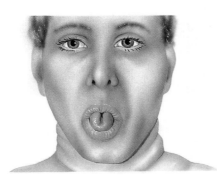

B. Make a table like the one shown above. From the class data, calculate q^2, the fraction of individuals who are homozygous recessive for the trait (those who cannot roll their tongue). Express this as a decimal value. For example, if 3 out of 30 people cannot roll their tongues, then (3/30), or 0.1, of the class are homozygous recessive. Record this value under the q^2 column.

C. Calculate q, the frequency of the recessive allele, by finding the square root of q^2. Record your answer.

D. Determine the frequency of the dominant allele, p, by using the formula $p = 1 - q$. Record your answer in the table.

E. Calculate and record in the table the frequencies of the homozygous dominant (p^2) and heterozygous (2pq) genotypes.

Phenotypes	Numbers of Rollers	
	Numbers of Nonrollers	
fraction with homozygous recessive genotype (q^2)		
frequency of recessive allele (q; $q = \sqrt{q^2}$)		
frequency of dominant allele (p; $p = 1 - q$)		
fraction with homozygous dominant genotype (p^2)		
fraction with heterozygous genotype (2pq)		

Analysis and Conclusions

1. What are the frequencies for the alleles that affect tongue rolling?

2. Is q, the frequency of the recessive allele, larger or smaller than the frequency of people showing the recessive trait? Why?

3. If you tested 10 000 people, do you think the genotype frequencies would be the same as those in your class? Explain.

4. If all Hardy-Weinberg conditions were met, what would be the next generation's allele frequencies?

5. Cystic fibrosis is a recessive, inherited disease whose victims die before they can reproduce. Which Hardy-Weinberg condition would not be met under these circumstances?

6. What do you think would happen to the frequency of the cystic fibrosis allele over several generations?

7. In fact, the allele frequency for cystic fibrosis in the United States seems to be constant over several generations. What could account for this?

8. Evolution is sometimes defined as a change in allele frequencies over time. Comment on this.

Chapter Review

Study Outline

29-1 Early Theories of Evolution

■ Lamarck's theory of evolution was based on the law of use and disuse and the inheritance of acquired characteristics.

■ The modern theory of evolution is based partially on Darwin's concept of natural selection.

■ The six main points of Darwin's theory involve the concepts of overproduction, competition, variation, adaptation, natural selection, and speciation.

29-2 The Synthetic Theory of Evolution

■ The modern, or synthetic, theory of evolution proposes that evolution is the change in allele frequency within a population over time.

■ Genetic variation occurs as a result of gene mutations, genetic recombination, migration, and genetic drift.

■ The Hardy-Weinberg law states that the frequency of alleles within a population remains constant if the population is large, reproduction is random, there are no mutations, and there is no migration.

29-3 Adaptations and Natural Selection

■ Adaptations are inherited traits that improve an organism's chance of survival and reproduction in a given environment.

■ Natural selection can favor an extreme phenotype (directional selection), the average phenotype (stabilizing selection), or two opposite phenotypes (disruptive selection).

29-4 Speciation

■ Speciation, the development of new species, can take place by means of geographic and reproductive isolation, polyploidy, or adaptive radiation.

■ Convergent evolution results in increased resemblance between unrelated species. Coevolution occurs when two or more species evolve in response to each other.

29-5 Observed Natural Selection

■ Studies of industrial melanism in the peppered moth show that natural selection can act in favor of certain adaptations over short periods of time.

■ Resistance of bacteria to antibiotics and of insects to DDT arises from genes that are already present in the population of bacteria or insects.

Vocabulary Review

natural selection (602)

variations (602)

adaptation (603)

speciation (603)

gradualism (604)

punctuated equilibrium (604)

synthetic theory (605)

population (605)

population genetics (605)

gene pool (605)

evolution (606)

genetic recombination (607)

genetic drift (607)

genetic equilibrium (608)

Hardy-Weinberg law (608)

camouflage (610)

warning coloration (610)

mimicry (611)

directional selection (612)

stabilizing selection (613)

disruptive selection (613)

range (614)

geographic isolation (614)

reproductive isolation (615)

adaptive radiation (615)

convergent evolution (616)

coevolution (616)

industrial melanism (618)

A. Matching—Select the vocabulary term that best matches each definition.

1. The study of the changes in the genetic makeup of populations

2. A type of natural selection in which the extreme phenotype is the favorable adaptation

3. A change in the gene pool of a small population that is brought about by chance

4. An adaptation allowing an organism to blend visually into the environment

Chapter Review

5. Process in which there is increased resemblance between unrelated species because of natural selection

6. The formation of new species

7. The total of all the alleles in a population

B. Definition—Replace the italicized definition with the correct vocabulary term.

8. Some *characteristics that differ from the typical characteristics of a species* may help individuals to survive.

9. Darwin proposed that new species are formed by *a process whereby organisms with favorable variations are better able to survive and reproduce than organisms with unfavorable variations.*

10. *The gradual change of allele frequencies* changes the genetic makeup of the whole population over time.

11. The *theory that populations, rather than individuals within a population, evolve* is the modern theory of evolution.

12. The finches that Darwin observed on the Galapagos Islands are an example of the result of *a process by which one species evolves into a number of different species, each occupying a specific environment.*

13. *The proposal that species remain the same for a long period of time and then evolve rapidly during a short time interval* is a new theory regarding the rate of evolutionary change.

14. The *principle that sexual reproduction by itself does not affect genetic equilibrium in a population* enables us to determine whether evolution is occurring in a population.

Content Review

15. Describe Lamarck's theory of evolution.

16. Describe three basic types of observations supporting the idea of evolution that Darwin made during his voyage on the *Beagle.*

17. Describe Darwin's theory of evolution by natural selection.

18. What were the weaknesses in Darwin's theory of evolution?

19. Explain how gradualism and punctuated equilibrium differ.

20. How is evolution defined, according to the synthetic theory?

21. Explain the statement: Individuals do not evolve.

22. How did De Vries explain the appearance of new traits within a species?

23. Why is genetic drift less likely to affect large populations than small ones?

24. In what way is the Hardy-Weinberg law useful?

25. What are adaptations that involve the metabolism of an organism called?

26. What is the difference between camouflage and warning coloration?

27. How does mimicry protect the viceroy butterfly from predators?

28. What type of natural selection might occur as the result of extreme climate change?

29. Describe disruptive selection.

30. When are two groups of organisms considered two different species?

31. What are polyploids?

32. Explain why Darwin's finches may be said to represent an example of adaptive radiation.

33. What is convergent evolution?

34. What general conclusions were drawn from studies of the peppered moth in England?

35. How do antibiotic-resistant bacteria develop?

36. How did insect populations become resistant to DDT?

Graphic Organizing

For information on graphic organizers, see Appendix G at the back of this text.

37. **Line Graph:** Construct a line graph, using the data in the chart at the top of the next page. The data represent the number of trilobite fossils of various lengths found at two different depths of sedimentary rock. In constructing your graph, plot the number of fossils on the vertical axis and the length of the fossil on the horizontal axis. Then, refer to Figure 29–14 to decide which type of natural selection the trilobites underwent as they evolved.

Length of Fossil (cm)	Number of Fossils (Lower Depth)	Number of Fossils (Higher Depth)
3.0	0	0
3.5	0	4
4.0	2	12
4.5	5	18
5.0	12	23
5.5	18	24
6.0	22	23
6.5	24	19
7.0	22	11
7.5	19	6
8.0	13	2
8.5	4	0
9.0	0	0

Critical Thinking

38. Using the evolution of the modern giraffe as an example, compare Darwin's theory with Lamarck's theory of evolution. (*Comparing and Contrasting*)

39. If Lamarck's theory of evolution were correct, what characteristic might you expect in the offspring of two cats whose food was always placed at the top of a tall, smooth, wooden post? (*Predicting*)

40. Classify the following adaptations as examples of camouflage, warning coloration, or mimicry: (a) a chameleon's ability to take on the colors of its background; (b) an edible fish's resemblance to an inedible one; (c) the bright color of a bad-tasting insect. (*Classifying*)

Creative Thinking

41. The dinosaurs, which were successful on earth for millions of years, are now extinct. What factors might have led to their extinction?

42. Write a short story, set thousands of generations in the future, showing some changes that have taken place as a result of evolution.

Problem Solving

43. A biologist is studying industrial melanism. Five hundred red ground beetles and 500 white ground beetles are placed on red clay in a mesh cage containing insect-eating birds. After five days, 475 red beetles and 123 white beetles remain. Calculate the percentage of each type of beetle that survived. What conclusion can be drawn from these results?

44. A botanist identifies two distinct species of violets growing in a field. Also in the field are several other types of violets that, although somewhat similar to the two known species, appear to be different enough in leaf type to be classified as new species. (See the figure below.) Develop a hypothesis as to the origin of these other types of violets. Then design an experiment to determine whether they represent new species.

Viola pedatifida   Viola sagittata   other violets

45. An experiment was performed to determine the resistance of two species of *Anopheles* mosquitoes to the insecticides Malathion and Dieldrin. Ten thousand insects of each species were sprayed with each insecticide. The data from the experiment are presented in the table below. Calculate the percentage of each species that survived. What do the results show about resistance of the insects to each insecticide? Predict what might happen if the offspring of the surviving insects are sprayed with the same concentration of each insecticide.

Insect Species	Insecticide	Number of Surviving Insects
A. culifacies	Malathion	17
	Dieldrin	78
A. stephensi	Malathion	28
	Dieldrin	30

Projects

46. Observe the feeding habits, beak structure, and coloration of the birds in your area. Write a report on the birds' adaptations.

47. Prepare a chart that illustrates the general evolution of marsupial and placental mammals.

48. Prepare a report on one of these career opportunities: cytotechnologist; geographer; environmental engineer.

Biology and Problem Solving

Where did dinosaurs spend the winter?

Bones, teeth, footprints—these are the fossils from which scientists can infer the size and shape of dinosaurs, what they ate, and even how fast they moved. But what about the life style of these huge animals? Did they live individually, in families, or in large groups? Did they remain in one place, roam in a small territory, or travel across great distances? The geographic distribution of fossils suggests some answers to these questions.

Centrosaurus apertus was a dinosaur that lived in the late Cretaceous period, about 75 million years ago. With a horn on its nose and a thick neck frill, it looked something like a rhinoceros. In 1977, a large bed of dinosaur bones, almost entirely those of Centrosaurus, was discovered in Dinosaur Provincial Park in southern Alberta, Canada. The bed contained bones of both male and female animals of all ages—perhaps as many as 250 dinosaurs lay buried there. Scientists believe that all of the animals may have died at the same time.

1. What does this fossil finding suggest about the social organization of Centrosaurus?

2. What might have been the advantage of this kind of social organization?

3. List some possible causes for the death of so many animals at the same time.

Fossils of more than 35 other species of dinosaurs, all dating from about the same period, have also been found in other deposits in Dinosaur Provincial Park.

Unlike the *Centrosaurus* deposit, most of these deposits contained bones from many different species that lived in the area. It is unlikely that great numbers of such large animals could have lived in the same place at the same time. There would not have been enough food in such a limited area.

4. Considering this fact, formulate a hypothesis to explain the presence of so many dinosaur bones in one area.

Fossils of a vegetarian dinosaur, *Pachyrhinosaurus*, have been found both in southern Alberta and at a site 450 miles farther north. A *Pachyrhinosaurus* skull has also been found in the Alaskan Arctic, about 2000 miles away from the southernmost Alberta site.

5. Suggest an explanation for the finding of *Pachyrhinosaurus* fossils at such distant sites.

Seventy-five million years ago, when *Pachyrhinosaurus* lived, the climate in the polar regions was warmer than it is today. With many hours of sunshine each summer day, plants were probably abundant in these regions. These would have provided a great supply of food for vegetarian dinosaurs and other consumers. In turn, these animals would have provided food for the carnivores that preyed upon them.

6. What do you think would have happened to the food supply of the polar regions in winter? Explain.

7. List some possible strategies that dinosaurs might have used to survive polar winters.

8. Based on all of the fossil evidence presented here, which strategy seems to be the most likely one? Explain your answer.

9. What additional evidence would you seek to help determine which strategy was used by dinosaurs to survive polar winters?

Problem Finding

List some questions that you still have about the dinosaurs' pattern of living.

Diversity of Living Things

This African waterhole is host or home to a great diversity of organisms, from the massive water buffalo to the long-legged stork. In this unit, you will see that, despite many differences, organisms can be grouped according to common traits. You will also learn that these similarities often reflect a common evolutionary history.

▲ This magnified image of the silicon-containing skeleton of the radiolarian reveals its pseudopods and spicules.

Monerans, Protists and Viruses

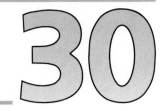

The radiolarian floats like a small space ship in the ocean currents, too small to be seen by the unaided eye. A silicon-containing skeleton surrounds the radiolarian's body, pseudopods and spicules bursting from its center. Long after the radiolarian has died, its skeleton remains, forming part of the ocean floor or of silicon-containing rock. The radiolarian is one of the vast array of organisms you will learn about in this chapter—from the monerans and protists, which are neither plant nor animal, to the viruses, which scientists do not consider to be alive.

Guide for Reading

Key words: Monera, saprobes, parasites, archaebacteria, eubacteria, pathogens, protozoa, algae, virus

Questions to think about:

📖 What are the characteristics of the monerans?

📖 How are the animal-like protists different from the plantlike protists?

📖 Why is it difficult to define viruses as living or nonliving?

30-1 The Monerans

Section Objectives:

- *Describe* the structure, shapes, and functions of bacterial cells.
- *Describe* the metabolic needs of bacterial cells.
- *Distinguish* between the two branches of organisms found in the kingdom Monera.
- *Compare* the harmful and beneficial aspects of bacteria.
- **Laboratory Investigation:** *Observe* the effect of an antibiotic on bacterial growth (p. 660).

The kingdom **Monera** consists of the simplest and most numerous organisms on earth. **Bacteria** is the common name for members of this kingdom, the *monerans*. Although some bacteria cause infections, the vast majority are beneficial. As you will learn, bacteria are important for health, for industry, and even for the air we breathe. Some bacteria can cure, as well as cause, disease.

Characteristics of Bacteria

Bacteria are found almost everywhere—in both fresh and salt water, in soil, in the air, and in plants and animals. In humans, they cover the skin and the linings of the nose and mouth. They are especially numerous in the digestive tract. Bacterial cells are much smaller than eukaryotic cells. In fact, they are the smallest known living cells and can only be seen with the aid of a microscope.

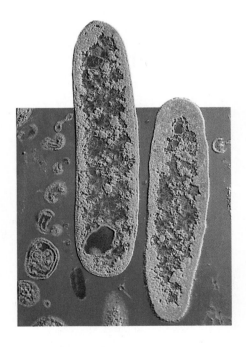

▲ **Figure 30–1**

Disease-causing Bacteria. These rod-shaped bacteria caused the bubonic plague, which killed about one-fourth of the European population in the fourteenth century.

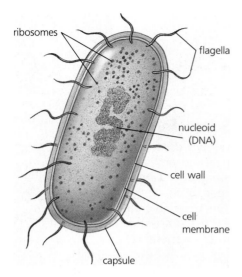

ribosomes

flagella

nucleoid (DNA)

cell wall

cell membrane

capsule

▲ **Figure 30–2**
Structure of a Prokaryote. Many types of bacteria are surrounded by a slimy capsule, in addition to a cell wall.

Cellular Structure The main structural feature that distinguishes moneran cells from the cells of organisms in the other kingdoms is the absence of a distinct, membrane-bound nucleus. Cells that lack a true nucleus are called *prokaryotic cells.* Therefore, all monerans are *prokaryotes.* In contrast, cells that have a membrane-bound nucleus are called *eukaryotic cells.*

As you can see in Figure 30–2, most prokaryotes are surrounded by a protective cell wall. Prokaryotic cell walls, however, have a different structure and chemical composition than plant cell walls. Instead of cellulose, prokaryotic cell walls contain a complex polysaccharide not found in eukaryotic cells. Many prokaryotes also secrete a slimy material that forms a capsule around the outside of the cell wall. The capsule provides additional protection for the cell. For example, many of the bacteria that cause diseases in animals are surrounded by capsules. A capsule prevents the animal's white blood cells and antibodies from destroying the bacterium.

Just inside the cell wall is the cell membrane, which surrounds the cytoplasm. In contrast to eukaryotes, prokaryotes lack organelles, such as mitochondria, endoplasmic reticulum, and chloroplasts, in their cytoplasm. Most of the enzymatic reactions that occur in eukaryotic cells, however, also occur in prokaryotes. Many of these reactions take place on the inner surface of the cell membrane. In bacteria that use oxygen, for example, the reactions of cellular respiration occur on the fingerlike folds of the cell membrane. In bacteria that carry out photosynthesis, membranes containing the photosynthetic pigments fill the cytoplasm.

Small ribosomes are scattered throughout the cytoplasm of prokaryotes. Although prokaryotes lack a true nucleus, they do contain a single, circular chromosome that is found in an area of the cell called the *nucleoid.* This area, which is usually in the center of the cell, contains the hereditary material, DNA. Some bacteria also have *plasmids,* which are smaller circular segments of DNA.

Many bacteria are able to form specialized structures, called **endospores,** within their cytoplasm. Endospores form when conditions for bacterial growth are unfavorable. They consist of a tough protective coat surrounding the nuclear material and a small amount of cytoplasm. In this dormant state, bacteria can survive for years and withstand extreme conditions, such as freezing, boiling, or extremely dry environments. Once conditions become favorable, the endospore again becomes an active, growing bacterial cell. The formation of endospores is an adaptive mechanism for survival. Some disease-causing bacteria can form endospores.

Shape Bacteria can be divided into three major groups according to shape. These shapes are shown in Figure 30–3. A spherical cell is called a **coccus** (KOK us), a rod-shaped cell is called a **bacillus** (buh SIL us), and a spiral or coiled cell is called a **spirillum** (spur IL um). The word *monera* is derived from the Greek word *moneres,* meaning "single." Although many bacteria are single celled, some form colonies in the shape of chains or clumps. Cocci are found as single cells (*monococci*), in pairs (*diplococci*), in chains (*streptococci*), and in grapelike clusters (*staphylococci*). Bacilli also form single cells, pairs (*diplobacilli*), or chains (*streptobacilli*). Spirilla exist only as single cells.

Motility   Many bacteria are capable of movement. Some move with the aid of *flagella*. However, unlike flagella in eukaryotes, in prokaryotes, flagella do not contain microtubules. Instead, the flagella are strands of protein twisted around one another like the strands of a rope. Other bacteria move by gliding on a slimy substance, which they secrete.

Metabolic Needs   Prokaryotes can be described according to the substances they need to live and grow. All prokaryotes require an energy source and a food source. Most bacteria are *aerobic*—they require free oxygen to carry out cellular respiration. Bacteria that cannot live without oxygen are called *obligate aerobes*. Some bacteria, called *facultative anaerobes,* can live in either the presence or absence of oxygen. They obtain energy either by aerobic respiration when oxygen is present or by fermentation when oxygen is absent. Still other bacteria cannot live at all in the presence of oxygen. These are called *obligate anaerobes*. These bacteria obtain energy only by fermentation. Through fermentation, different groups of bacteria produce a wide variety of organic compounds. Besides ethyl alcohol and lactic acid, bacterial fermentation can produce acetic acid, acetone, butyl alcohol, glycol, butyric acid, propionic acid, and methane, the main component of natural gas.

Depending on its source of food, a bacterium is either *heterotrophic* or *autotrophic*. Most bacteria are heterotrophic—they must obtain ready-made food from the environment. Heterotrophic bacteria are either **saprobes** or **parasites.** Saprobes feed on the remains of dead plants and animals and ordinarily do not cause disease. They release digestive enzymes into the organic matter. The enzymes break down the large food molecules into smaller molecules, which are absorbed by the bacterial cells. Parasites live on or in living organisms, absorbing nutrients directly from their host's body. Parasites can cause disease.

Some types of bacteria are autotrophic—that is, they can synthesize the food they require from inorganic substances. Autotrophic bacteria are either *photosynthetic* or *chemosynthetic*. Like plants, many photosynthetic bacteria convert carbon dioxide and water to organic matter and oxygen. Other kinds of bacteria carry out different forms of photosynthesis. For example, some forms can use hydrogen sulfide instead of water.

Only certain groups of bacteria can perform *chemosynthesis*. Like photosynthetic bacteria, these groups synthesize food from carbon dioxide. Unlike photosynthetic forms, they do not use light as a source of energy. Instead, these bacteria use energy from the

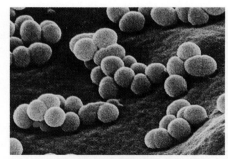
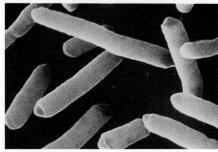
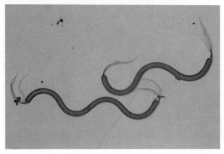

▲ **Figure 30–3**

Shapes of Bacteria. The spherical cocci (top) occur singly, in pairs, in chains, or in clumps. Shown here are clumps of *Micrococcus*. (Magnification 122 000 X). The rod-shaped bacilli (middle) most often appear as single cells but may form chains. This micrograph shows a number of solitary *Enterobacteria*. Spirilla may be S-shaped, like the *Spirillum volutans* shown in this micrograph (bottom), or corkscrew-shaped. (Magnification 700 X)

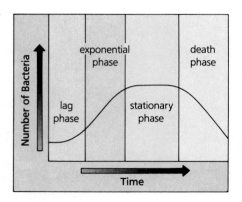

▲ **Figure 30–4**
Growth Curve of a Bacterial Culture.

breakdown of inorganic substances, such as compounds of nitrogen and sulfur, to synthesize food. Both bacterial photosynthesis and chemosynthesis are very important because they cycle the elements and compounds that all living things need.

Reproduction   All bacteria reproduce asexually. Thus, they do not divide by mitosis but by the process of *binary fission.* As you read in Chapter 20, in binary fission the parent cell divides into two identical daughter cells. Some bacteria can reproduce very rapidly. For example, under ideal conditions of food, temperature, and space, some species can divide every 15 to 20 minutes. At this rate, one bacterial cell could produce millions of offspring in just a few days! However, factors such as heat, cold, or lack of food inhibit the rate of reproduction and prevent such rapid growth.

Figure 30–4 shows a typical growth curve for a culture of bacterial cells. (A *culture* is the laboratory growth of a group of bacteria on artificial food material.) It shows the number of bacteria in the culture plotted against time. The growth curve can be divided into four phases. In the *lag phase,* the bacteria are adjusting to their new environment, and growth is slow. In the *exponential phase,* the bacteria are dividing rapidly. In the *stationary phase,* the reproductive rate equals the death rate. In the *death phase,* the bacteria are dying off faster than they are reproducing.

Although bacteria do not reproduce sexually, some species are capable of sharing or exchanging genetic material. This genetic exchange is known as *genetic recombination.* The three methods by which bacteria can exchange DNA are *conjugation, transformation,* and *transduction.* As you learned in Chapter 21, in conjugation one bacterial cell can donate some of its DNA to another cell of the same species. This transfer occurs through a thin, temporary bridge between the two cells. The bacterium receiving the DNA can incorporate it into its own DNA. The DNA exchanged in conjugation may be in the form of plasmids. Some of these plasmids allow bacteria to become resistant to antibiotic drugs. Other plasmids may help bacteria to adapt to stressful environments or to make new nutrients.

Some bacterial cells release DNA when they die and break apart. This DNA can be taken up by other bacterial cells, which can then form new, genetically different cells. This process is called transformation. Transformation occurs in *pneumococcus* bacteria, which you read about in Chapter 27. The type R (rough) bacterial cells were transformed by DNA from dead type S (smooth) bacteria. These bacteria cause pneumonia.

In transduction, DNA is transferred from one bacterium to another by means of viruses. The viruses pick up small amounts of DNA from host bacteria. When they infect other bacterial cells, the new host bacteria can receive and incorporate the DNA into their own.

Recombinant DNA techniques use all three processes to transfer DNA from one organism to another and change the genetic makeup of cells. The technology that uses these techniques, *genetic engineering,* is discussed more fully in Chapter 27.

Types of Bacteria

The kingdom Monera is divided into two main branches: the **archaebacteria** (ar kee bak TIR ee uh) and the **eubacteria.** Both groups include several phyla.

Archaebacteria   Most members of the subdivision archaebacteria live in environments that other organisms could not tolerate. Scientists believe that archaebacteria are among the earliest organisms to inhabit the earth. The name *archaebacteria* is derived from the Greek *archaio,* which means "ancient." The habitats of these primitive bacteria are too harsh for most forms of life today, but they resemble the environments on earth billions of years ago when life was first evolving. Although the archaebacteria are prokaryotes, they are unlike the other monerans in many ways. The chemical makeup of their cell walls is different from that of the eubacteria, the lipids in their cell membranes are also different, and their RNA has a unique structure.

There are three phyla of archaebacteria. The first phylum includes methane-producing bacteria that break down organic matter and produce methane gas as a by-product. They inhabit the intestinal tracts of animals, including humans, and are also found at the bottom of marshes, swamps, and sewage treatment plants. A second phylum includes "salt-loving" bacteria, which inhabit salt lakes such as the Dead Sea and the Great Salt Lake. The third phylum includes bacteria that inhabit hot, acidic environments, such as the hot springs of Yellowstone National Park, where the temperature may reach 80° to 90° Celsius. Figure 30–5 shows some examples of archaebacterial environments.

Eubacteria   The eubacteria include several phyla. Although most eubacteria are heterotrophic, some are autotrophic. These autotrophs are either *phototrophs* (photosynthetic) or *chemotrophs* (chemosynthetic). The main difference between them is that photosynthetic bacteria use light energy to synthesize food, whereas chemosynthetic bacteria use energy from chemical reactions.

The **blue-green bacteria,** a major group of autotrophic eubacteria, are members of the phylum **Cyanobacteria** (sy an oh bak TIR ee ah). Like plants, these bacteria carry out photosynthesis. However, they do not have the chloroplasts of plant cells. Instead, their photosynthetic pigments are located on folded membranes in the cytoplasm. Cyanobacteria contain chlorophyll *a,* as well as a blue pigment, *phycocyanin.* Their blue-green color is due to these two pigments. Some species have additional pigments that color them red, yellow, brown, black, or green. Figure 30–6 shows the structure of blue-green bacterial cells.

Although most blue-green bacteria are found in fresh water, others live in salt water, in soil, and on rocks. The Red Sea owes its name to the occasional appearance of huge numbers of red-colored species of these bacteria. Ponds or lakes that contain a rich supply of organic matter often develop large populations, or *blooms,* of blue-green bacteria. Because they thrive in polluted water, their appearance indicates the presence of organic pollutants.

▲ **Figure 30–5**

Environments Inhabited by Archaebacteria. Although the Great Salt Lake in Utah (top) is too salty to be hospitable to most forms of life, it is well populated by "salt-loving" archaebacteria. Another type of archaebacteria live in the hot, acidic environment of the Hot Springs (bottom) in Yellowstone National Park in Wyoming.

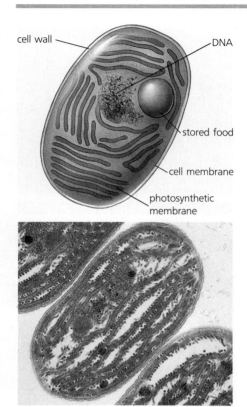

cell wall

DNA

stored food

cell membrane

photosynthetic membrane

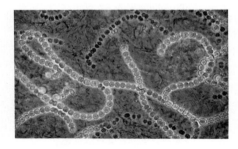

▲ **Figure 30–6**

Blue-Green Bacteria. The structure of a typical blue-green bacterial cell, shown at top, includes a series of folded membranes where photosynthesis occurs. *Spirulina* (middle) is a species of blue-green bacteria sometimes grown commercially as a source of protein. (Magnification 21 600 X) Shown at bottom are filaments of *Nostoc*, a species of nitrogen-fixing blue-green bacteria extremely useful in enriching the soil. (Magnification 256 X)

Although the cells of a few species of blue-green bacteria exist singly, most develop into colonies in the form of plates, clusters, or threadlike filaments. The colonies are usually embedded in a sticky substance. A few species live together with fungi, forming mixed organisms known as *lichens*.

Some blue-green bacteria capture nitrogen from the air and incorporate (or "fix") it into organic compounds. This process, called *nitrogen fixation,* occurs in specialized cells known as *heterocysts.* Nitrogen fixation plays an important role in fertilizing certain crops. The *nitrogen-fixing bacteria* include the blue-green bacteria and a few other kinds. Some kinds of nitrogen-fixing bacteria may live on the roots of plants such as beans and alfalfa, thus providing the plants with a source of nitrogen and reducing the need for artificial fertilizers in nitrogen-poor soils.

Another phylum of autotrophic eubacteria includes the **green-sulfur bacteria** and the **purple bacteria.** These have a form of chlorophyll that differs from the chlorophyll of plants. They are anaerobes and carry out photosynthesis without producing oxygen and without using water as a starting material. These organisms may be colored pink, green, or nearly black. They inhabit the muddy sediments of ponds and seas.

The phylum **Prochlorophyta** (pro klor AH fuh tuh), which was discovered in 1976, also includes photosynthetic bacteria. These one-celled organisms live in close association with certain marine animals. Like plants, these bacteria carry out photosynthesis with both chlorophyll *a* and *b.* Thus, even though the Prochlorophyta are prokaryotes, their pigments and photosynthetic reactions are like those of eukaryotes.

The largest group of heterotrophic eubacteria belong to the phylum **Schizophyta** (skiz AH fuh tuh). Although there are several classes of these bacteria, a universal method of classifying them is the **Gram test,** a staining method developed by the Danish physician Hans Christian Gram. The test divides bacteria into two classes, depending on their reaction to the **Gram stain. Gram-positive** bacteria retain a stain called *crystal violet* and appear purple under a microscope. **Gram-negative** bacteria do not retain this stain, and they appear light pink under a microscope. Figure 30–7 shows gram-positive and gram-negative bacteria as they appear under a microscope.

The chemical nature of the cell walls determines whether bacteria are gram-positive or gram-negative. Because of differences in their cell walls, gram-positive bacteria can be harmed by antibiotics, such as *penicillin.* The cell walls of gram-negative bacteria are not harmed as easily. Thus, infections caused by gram-negative bacteria are more difficult to treat with antibiotics.

The heterotrophic eubacteria live everywhere—in soil, air, food, and water. Some inhabit other living organisms, and others decompose dead organic matter. They include both aerobic and anaerobic forms. Many of them cause human diseases, such as *bacterial pneumonia, diphtheria,* and *tetanus.*

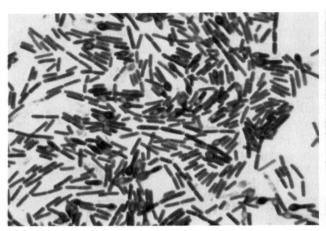
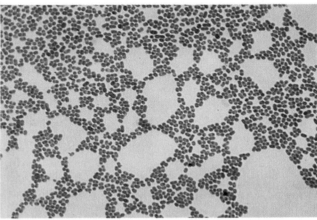

▲ **Figure 30–7**

Gram Stain. The Gram test classifies bacteria into two groups based on differences in their cell walls. When stained, gram-positive bacteria appear purple, as with *Clostridium botulinum* cells (left. (Magnification 1400 X) Gram-negative bacteria appear pink. These *Acinetobacter* cells (right) are gram-negative. (Magnification 2750 X)

Importance of Bacteria

Bacteria and Disease   Harmful bacteria—those capable of causing infections in animals or plants—are called **pathogens.** To cause disease in humans, pathogens must enter the body, overcome the body's immune defenses, and multiply.

Discovery of Bacterial Pathogens   The idea that bacteria can cause disease, known as the **germ theory of disease,** was developed by the French scientist Louis Pasteur in the mid-1880s. Since then, we have learned that bacteria harm the body in three ways. First, the bacteria can become so numerous that they interfere with the functioning of normal cells. Second, they can destroy body cells and tissues. Third, they can produce poisons, or toxins, that kill cells or interfere with their functioning. The third mechanism is the usual cause of harm.

Anthrax, a disease of sheep, cattle, horses, and sometimes humans, was the first disease ever proven to be caused by bacteria. In 1876, the bacterium that causes anthrax was isolated and identified by the German physician Robert Koch. Koch also identified the cause of the disease *tuberculosis* in 1882. Based on the methods he used to identify these bacteria as agents of disease, Koch developed a set of rules to determine whether a specific bacterium is the cause of a specific disease. His rules are:

1. The suspected disease-causing organism should always be found in animals with the disease.

2. The organism must be isolated from the diseased animal and grown in a pure culture. (A pure culture is one containing only one kind of bacterium.)

3. When organisms taken from the pure culture are injected into a healthy animal, they must cause the disease.

4. The organism must be isolated from the experimentally infected animal and grown in pure culture again, and it should be identified as the same organism isolated in step 2.

These rules, which are known as **Koch's postulates,** were developed in the late 1880s and are still used today. They have helped to identify the causes of many bacterial diseases, such as those listed in Figure 30–8.

Pasteur also performed several experiments related to anthrax, and in 1881, he produced a vaccine to prevent this infection. He exposed the anthrax bacteria to temperatures that were high enough to weaken them but not to kill them. He then injected the weakened bacteria into sheep that were healthy. The sheep became slightly ill and then recovered. Later, when the same sheep were injected with unheated bacteria, they did not develop the disease. Pasteur had succeeded in creating a vaccine to prevent this infection. He demonstrated his discovery by performing a series of similar experiments that were viewed by scientists, journalists, and public officials.

Critical Thinking in Biology

Identifying Causes

When you identify causes, you draw a conclusion, based on evidence, about why something has happened. Consider the problem that faced scientists in France in the mid-1800s.

Throughout the country, chickens were dying of chicken cholera. Scientists were able to identify a particular bacterium as the organism responsible for the disease. Each time a culture of young bacteria was injected into healthy chickens, they developed cholera and died.

One day, a researcher injected some bacteria that were several weeks old but still alive into healthy chickens. Previously, only young bacteria had been used. Again, the chickens developed symptoms of cholera. By the next day, however, all the chickens had recovered.

1. Before reading further, list as many possible explanations as you can for why the chickens did not die.

Suppose the scientists conducted the following experiment: A large group of chickens was divided into two similar groups. Group A was injected with young bacteria, while Group B was injected with the same amount of old bacteria. Both groups were kept under identical conditions. Chickens in both groups developed symptoms of cholera. By the next day, all the chickens in Group A had died while all the chickens in Group B had recovered.

2. Based on the results of this experiment, can you rule out any of the explanations you suggested in question 1? Which of your explanations do you consider most likely? Explain.

3. Think About It Explain your thought processes as you answered questions 1 and 2.

Controlling Bacterial Disease   Until this century, bacterial diseases were common causes of death. Around 1900, the work of Pasteur, Koch, and other scientists became known to public health workers in the United States and other developed countries. Knowledge about the germ theory of disease brought about a revolution in hygiene and sanitation. This revolution led to the prevention of many common, serious infections and increased life expectancy dramatically.

A major advance occurred with the development of the first antibiotic, penicillin. An antibiotic is a substance that prevents bacteria from growing. Some antibiotics act by disrupting the formation of bacterial cell walls at the time of cell division. Others interfere with the formation of substances needed by bacterial cells, such as proteins. *Tetracyclines,* which are effective against a wide range of both gram-positive and gram-negative bacteria, act in this manner.

Penicillin was discovered in 1928 by Sir Alexander Fleming, a Scottish bacteriologist. Fleming was growing a culture of bacteria when he noticed that a mold had contaminated the culture and that no bacteria were growing near the mold. The mold was a species of the genus *Penicillium.* The bacteria-inhibiting substance that was secreted by this mold was later to be named penicillin. Although physicians still routinely use penicillin to treat a wide variety of bacterial infections, many other antibiotics are now available.

| Diseases Caused by Bacteria ||
Disease	Bacterium
	Cocci
boils, carbuncles	*Staphylococcus aureus*
gonorrhea	*Neisseria gonorrhoeae*
meningitis	*Neisseria meningitidis*
pneumonia	*Diplococcus pneumoniae*
scarlet fever	*Streptococcus pyogenes*
strep throat	*Streptococcus pyogenes*
	Bacilli
anthrax	*Bacillus anthracis*
botulism	*Clostridium botulinum*
diphtheria	*Corynebacterium diphtheriae*
plague	*Yersinia pestis*
tetanus	*Clostridium tetani*
typhoid fever	*Salmonella typhi*
	Spirilla
cholera	*Vibrio comma*
syphilis	*Treponema pallidum*

◄   **Figure 30–8**
Diseases Caused by Bacteria.

Methods of Preserving Food		
Method	**Food Preserved**	**Why Effective**
freezing	meat, vegetables, desserts	stops growth and reproduction of bacteria
refrigeration	meat, eggs, butter, milk	slows growth and reproduction of bacteria
pasteurization (heating to moderate temperature followed by rapid cooling)	milk, egg products, apple juice, and other beverages	almost all the bacteria are killed by the heat. Cooling slows the growth of remaining bacteria.
drying	meats, grains, flour, starch, fruits, sugar, powdered milk, powdered eggs	removes moisture required by bacteria for growth
canning	vegetables, fruits, meats	high temperature kills all bacteria; sealed container prevents entrance of new bacteria
Preservatives: salt sugar lactic acid (from fermentation) vinegar	meats fruits cucumbers (pickles) and cabbage (sauerkraut) vegetables	high concentrations of sugar and salt osmotically dehydrate bacteria; acid conditions prevent bacterial growth

▲ **Figure 30–9**
Methods of Preserving Food

Bacterial infections can be spread in many ways. Some bacteria are carried through the air on droplets spread by coughing and sneezing. Others are spread by contaminated drinking water. Some are spread by direct contact, such as by shaking hands. *Sexually transmitted diseases* are spread by sexual contact. Still other diseases, such as *Lyme disease* or *typhus,* are spread by insect bites. Bacteria in food are a common cause of illness. The action of bacterial decomposers causes food to spoil and become harmful to eat. For example, *Clostridium botulinum* is an obligate anaerobe, obtaining its energy by fermentation. This organism produces toxins that cause *botulism,* the most dangerous of all types of food poisoning. *Clostridium botulinum* grows in canned foods that have not been heated sufficiently during canning. Even before the germ theory of disease was developed, people were preventing food spoilage by the methods listed in Figure 30–9. All these methods deprive bacteria of some condition they need for growth, such as nutrients, moisture, correct temperature, or oxygen.

Biologic Importance of Bacteria Bacteria play an important role in health as well as in disease. They are normal inhabitants of the human body, and most of them are harmless. Some give off substances that discourage the growth of harmful species. Others help to decompose food wastes in the intestines. Bacteria in the large intestine produce certain vitamins needed by the body. Bacteria cure disease by producing some of the most useful

◀ **Figure 30–10**
Cheese-Making. The production of cheese depends on certain lactic acid producing bacteria that cause milk products to solidify. This woman is coating wheels of cheese with a protective layer of plastic.

antibiotics, such as *streptomycin* and *erythromycin*. Scientists use bacterial cells to study cellular metabolism and molecular biology. The techniques of genetic engineering are based on experiments with the DNA of bacterial cells. Many medically important substances have already been made using genetically engineered bacteria. These include antibiotic drugs and vaccines, hormones such as insulin, and chemicals that fight cancer cells.

The role of bacteria in the environment is equally important. Bacteria are *decomposers*. They break down the tissues of dead animals and plants into organic substances, returning these substances to the environment where other living organisms depend on them for existence. Without bacterial decomposers, these basic substances—oxygen, carbon, nitrogen, phosphorus, and sulfur—would soon be depleted. Thus, bacteria are primarily recyclers, and without them life on earth would die out.

Bacteria are also important in industry and agriculture. They are used in the production of yogurt, buttermilk, cheeses, and vinegar. Chemical companies use bacteria to produce butanol, acetone, and fuels such as methane. Some bacterial species are used as natural pesticides to save crops from insects. Economically important crops, including alfalfa, clover, soybeans, and other beans depend on nitrogen-fixing bacteria for their growth.

30-1 Section Review

1. What are the three shapes of bacterial cells?
2. What are bacteria that cannot live without a supply of oxygen called?
3. Identify the two main branches of monerans.
4. Name the drug discovered by Sir Alexander Fleming.

Critical Thinking

5. Why are bacteria in the human body considered both harmful and beneficial? (*Identifying Reasons*)

30-2 The Protists

Section Objectives:

- *Identify* the three major groups of protists.
- *Compare* the major characteristics of the four types of animal-like protists.
- *Identify* the six main groups of plantlike protists and their distinguishing characteristics.
- *Describe* the general features of the three types of funguslike protists.

The first eukaryotes were probably single-celled organisms much like many modern-day protists. Scientists estimate that the *protists* first evolved about 3 billion years after the monerans were established. Since most protists require oxygen, it is thought that the earliest protists could not have evolved until the blue-green bacteria had been producing oxygen for billions of years.

The kingdom **Protista** contains many species and a great variety of organisms. As eukaryotes, the protists have a membrane-bound nucleus and many different cytoplasmic organelles. They show amazing diversity in cell organization, in methods of reproduction, in metabolic needs, and in habitats. In fact, the members of this kingdom are sometimes defined on the basis of not belonging in any of the other four kingdoms.

Although most protists are unicellular, some are multicellular organisms and may be quite large. Some protists are heterotrophic; others are autotrophic. Reproduction may be sexual or asexual. Their habitats are both aquatic and terrestrial. Because they are so diverse, the members of the kingdom Protista are difficult to classify. They are divided into three main groups: the animal-like, plantlike, and funguslike protists.

Animal-like Protists

The animal-like protists are single-celled or colonial organisms called **protozoa.** They live in fresh and salt water, in the soil, and in the bodies of other organisms. All protozoa are heterotrophic. Some absorb nutrients through their cell membranes, whereas others engulf larger particles of food. The life processes of two common protozoa, ameba and paramecium, are discussed in Unit 2. Most protozoa are motile and are divided into phyla based on their means of locomotion. The major phyla are listed in Figure 30–11.

Sarcodines The *sarcodines* move and capture prey by means of pseudopods, or "false feet." Members of this group belong to the phylum **Sarcodina** (sar kuh DY nuh). The sarcodines are found in both fresh and salt water and in the bodies of animals, where a few cause disease. Reproduction is both asexual and sexual.

The best known of the sarcodines are the amebas. Amebas are unicellular organisms that continually change shape. As explained in Chapter 13, amebas move by means of pseudopods in a type of locomotion known as ameboid movement. They also make use of

Major Divisions of Animal-Like Protists, or Protozoa	
Phylum	**Identifying Characteristics**
Sarcodina (sarcodines) Ameba	move and engulf food by extending pseudopods, which are extensions of the cell surfaces.
Ciliophora (ciliates) Paramecium	use cilia to move and feed.
Zoomastigina (zooflagellates) Trypanosoma	move by means of flagella.
Sporozoa (sporozoans) Plasmodium	nonmotile; mostly parasitic; have complex life cycles.

◀ **Figure 30–11**
Major Divisions of Animal-like Protists, or Protozoa.

pseudopods to surround and engulf food particles. Amebas reproduce asexually by binary fission. They are commonly found in freshwater ponds, lakes, and streams.

Amebas can be parasitic. For example, *amebic dysentery* is a disease caused by a parasitic species of ameba common in tropical areas. It lives in the human large intestine and feeds on the intestinal walls, causing bleeding. Such amebas can sometimes form cysts, which are ameba cells surrounded by a protective capsule. The disease is spread when these cysts pass out of the body with digestive wastes that enter fresh water or soil. A person becomes infected by drinking contaminated water or eating contaminated food.

Some sarcodines are surrounded by protective shells. Among these are the *radiolarians* (rayd ee oh LEHR ee unz) and the *forams,* shown in Figure 30–12. The radiolarians have glassy, silicon-

Figure 30–12

Radiolarians and Forams. A radiolarian (left) is a sarcodine with a delicate shell made of silica. (Magnification 4200 X) A foram (right) has a multichambered shell. The calcium carbonate shells of forams have left an excellent fossil record. ▼

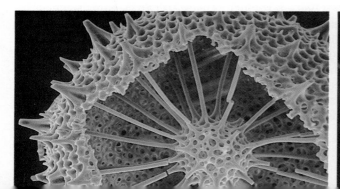

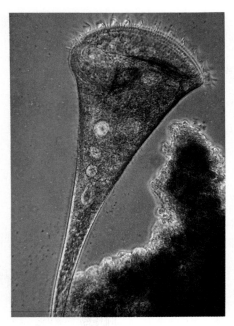

▲ **Figure 30–13**
A Stentor. Ciliates vary in form and in the pattern of cilia. The stentor is trumpet-shaped with cilia arranged in clumps.

Figure 30–14
Zooflagellates. The zooflagellate *Trypanosoma gambiense* causes African sleeping sickness. ▼

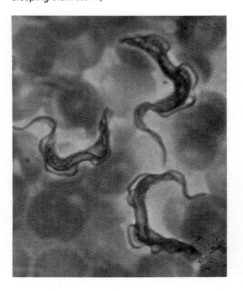

containing shells and long, thin pseudopods. The forams live in many-chambered, snail-like shells containing calcium. Thin projections of cytoplasm protrude through openings in the foram's shell.

Both radiolarians and forams are abundant in the oceans. When these organisms die, their shells drop into the mud of the ocean bottom. In some places, the buildup of tremendous numbers of foram shells has formed huge chalk deposits. The white cliffs of Dover on the English coast were formed in this way. Radiolarian shells make up much of the bottom ooze on some parts of the oceans, and they are also an important part of certain silicon-containing rocks.

Ciliates The *ciliates,* members of the phylum **Ciliophora** (sil ee AHF ur uh), are complex protozoa that live in both fresh and salt water. Ciliates are surrounded by hairlike projections, or cilia. In some species, the cilia are arranged in rows or tufts. In others, such as the stentor, the cilia form clumps. The beating of the cilia moves the organism through the water and propels food and water into its oral groove. The paramecium, a freshwater ciliate, is the most frequently studied member of this phylum.

The intake and digestion of food by the paramecium are described in Chapter 8. Food particles are enclosed and digested in food vacuoles. Some ciliates also have contractile vacuoles, which collect and excrete excess water from the cell.

Unlike amebas, which change shape, ciliates have a rigid outer covering called a **pellicle** that maintains their shape. Beneath the pellicle, some have *trichocysts* (TRIK uh sists), barbed structures that are discharged for defense or to aid in capturing prey.

Asexual reproduction in ciliates is by binary fission. Some also reproduce sexually by the process of conjugation. Both methods of reproduction are explained in Unit 4. Ciliates differ from other protozoa in having two kinds of nuclei, a macronucleus and a micronucleus. The large macronucleus controls cellular metabolism and divides during binary fission. The micronucleus is not necessary for life but is exchanged during conjugation.

Zooflagellates The *zooflagellates* (zoh uh FLAJ uh luhts), of the phylum **Zoomastigina** (zo uh mass tuh JINE uh) move by beating long, whiplike flagella. Some have only one flagellum; others have many. Most are unicellular. Although some zooflagellates are free-living in fresh water, most live in the bodies of animals or the tissues of plants. Zooflagellates reproduce both asexually and sexually.

Among the zooflagellates is *Trypanosoma gambiense* (try pan uh SOHM uh GAM bee enz), which causes African sleeping sickness in humans. This parasite multiplies in the blood where it releases toxins. Symptoms of the disease include weakness, sleepiness, and fever. If left untreated, the victim eventually dies. *Trypanosoma* lives in the blood of domestic and wild animals of Africa. It is spread between animals and humans by the bite of the tsetse fly.

Another zooflagellate is *Trichonympha* (trik uh NIM fuh), which lives in the digestive tract of the termite. Termites do not have the

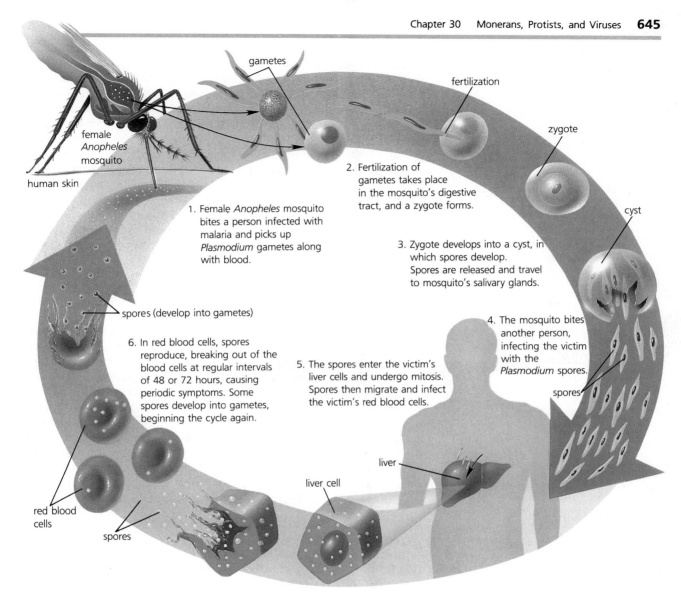

gametes

fertilization

zygote

cyst

female
Anopheles
mosquito

human skin

1. Female *Anopheles* mosquito
bites a person infected with
malaria and picks up
Plasmodium gametes along
with blood.

2. Fertilization of
gametes takes place
in the mosquito's digestive
tract, and a zygote forms.

3. Zygote develops into a cyst, in
which spores develop.
Spores are released and travel
to mosquito's salivary glands.

spores (develop into gametes)

4. The mosquito bites
another person,
infecting the victim
with the
Plasmodium spores.

spores

6. In red blood cells, spores
reproduce, breaking out of the
blood cells at regular intervals
of 48 or 72 hours, causing
periodic symptoms. Some
spores develop into gametes,
beginning the cycle again.

5. The spores enter the victim's
liver cells and undergo mitosis.
Spores then migrate and infect
the victim's red blood cells.

liver

red blood
cells

spores

liver cell

▲ **Figure 30–15**

Major Life Stages of the Malarial Parasite,
Plasmodium.

enzymes necessary to break down wood, but *Trichonympha* does.
The zooflagellate breaks down the wood eaten by the termite, and
both organisms absorb and use the nutrients.

Sporozoans   Members of the phylum **Sporozoa** (spor uh ZOE uh)
are nonmotile and parasitic. They obtain nutrients from the bodies
of their hosts. The members of this phylum, called sporozoans,
cause disease in animals, including humans. Sporozoans get their
name from the fact that they produce spores during the asexual
phase of reproduction. These spores permit the spread of the
parasites. The life cycles of sporozoans are complex and involve
growth and reproduction in more than one kind of host.

The best known sporozoans are members of the genus *Plasmo-
dium* (plaz MOHD ee um), a parasite that causes malaria in humans.
The parasite is transmitted to humans by the bite of the female
Anopheles (uh NAHF uh leez) mosquito. Figure 30–15 traces the life

cycle of *Plasmodium* through the different stages of its reproduction within the human host and the mosquito. Sexual reproduction takes place in the digestive tract of the mosquito, and the asexual phase takes place within the liver and bloodstream of the human host.

At one stage of the life cycle of *Plasmodium,* the spores invade the red blood cells of the human host, multiply there, then break out and invade new cells. The destruction of the red blood cells releases toxic cell wastes into the bloodstream. These waste products cause fever, chills, and other symptoms of malaria.

Malaria is a serious, sometimes fatal, disease. Although it can be treated with drugs, one method of prevention is to eliminate the *Anopheles* mosquito. In spite of the widespread use of pesticides in many countries, millions of people are still infected with malaria, especially in tropical areas.

Plantlike Protists

The plantlike protists, commonly called **algae,** resemble plants in that they are all photosynthetic. Like the protozoa, algae are very diverse. Some are tiny, single-celled organisms with flagella. Others are large, multicellular organisms known as seaweeds.

Like plants, algae have chloroplasts, which contain the photosynthetic pigment chlorophyll. Although all algae have chlorophyll *a,* different groups of algae may have other types of chlorophyll, such as *b* or *c.* Some also have other pigments that give them distinctive colors. Their methods of reproduction vary. They inhabit fresh water, salt water, and moist environments on land. Based on their structure, photosynthetic pigments, and cell wall and food storage substances, algae are divided into six groups.

Euglenoids The *euglenoids* (yoo GLEE noyds) belong to the phylum **Euglenophyta** (yoo gleen AH fuh tuh). These single-celled protists have both plantlike and animal-like characteristics. Like plants, they contain chloroplasts and photosynthetic pigments. However, they do not have cell walls. Like some of the protozoa, euglenoids move by means of flagella. Owing to their pellicles, or flexible protein coverings, they are able to change shape easily as they move. Euglenoids are unusual in that some forms are autotrophic, whereas others are heterotrophic. For these reasons the classification of euglenoids has been controversial, and some biologists still consider them protozoa. Figure 30–16 shows the cellular structure and microscopic appearance of a typical euglenoid, the *euglena,* an organism common in pond water.

The euglena is a single-celled organism having two flagella. The cell has a large, central nucleus and numerous chloroplasts, which contain chlorophylls *a* and *b,* as well as other pigments. The chlorophylls give euglenas their grass-green color. Euglenas are primarily photosynthetic. When light is available, they carry on photosynthesis. However, in the absence of light, they live as heterotrophs, absorbing dissolved nutrients from the environment. Food is stored as a nonstarch carbohydrate in a structure within each chloroplast called a *pyrenoid* (py REE noyd).

Figure 30–16

Euglena. The protist *Euglena* (top) has chloroplasts and carries on photosynthesis. This euglena (bottom) is magnified 560 times. ▼

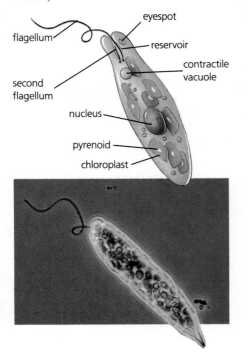

- flagellum
- eyespot
- reservoir
- contractile vacuole
- second flagellum
- nucleus
- pyrenoid
- chloroplast

Major Divisions of Plantlike Protists

	Phylum	Photosynthetic Pigments	Cell Wall Components	Identifying Characteristics
	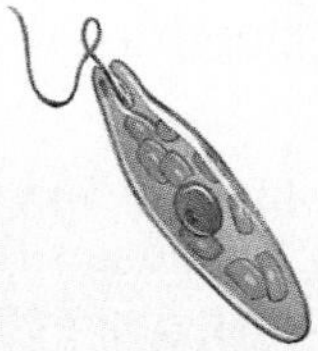 **Euglenophyta** (Euglenoids)	chlorophylls *a* and *c*, other pigments	no cell wall, flexible protein pellicle	when grown in dark, *Euglena's* chloroplasts disappear and it becomes heterotrophic; chloroplasts reappear upon exposure to light.
	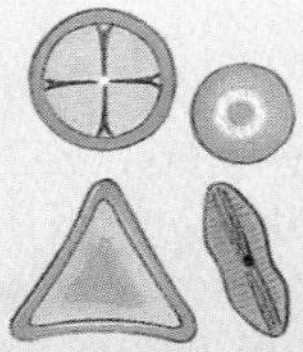 **Chrysophyta** (Golden algae)	chlorophylls *a* and *c*, other pigments	cellulose, some with silica	diatoms have glasslike, ornate shells in two halves
	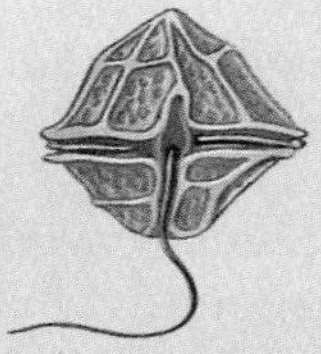 **Dinoflagellata** (Dinoflagellates)	chlorophylls *a* and *c*, other pigments	cellulose and silica	cell walls composed of armorlike plates; some forms are bioluminescent; colorful blooms form red tides.
	 **Chlorophyta** (Green algae)	chlorophylls *a* and *b*, other pigments	cellulose and other substances	unicellular, colonial, and multicellular forms, such as the filamentous *Spirogyra*
	 **Phaeophyta** (Brown algae)	chlorophylls *a* and *c*, other pigments	cellulose	are multicellular, varying in size from microscopic to more than 50 meters in length; include many common seaweeds
	 **Rhodophyta** (Red algae)	chlorophyll *a*, other pigments	cellulose and other substances	most are multicellular; many form a hard crust of calcium carbonate; include many common seaweeds

▲ **Figure 30–17** **Major Divisions of Plantlike Protists.**

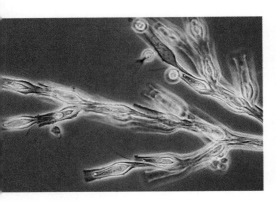

▲ **Figure 30–18**

Golden Algae. The colonial golden alga *Dinoloryon* forms branching colonies in its freshwater habitat.

Euglenas have one large flagellum that is used in movement and one short flagellum that is inactive. The bases of the flagella are within an inpocketing called the *reservoir*. Next to the reservoir is a contractile vacuole, which excretes excess water from the cell into the reservoir. Near the reservoir is the red-orange eyespot, or *stigma*, which is sensitive to light. The stigma enables the euglena to detect light intensity and direction and to position itself for maximum photosynthesis.

Golden Algae   The golden algae belong to the phylum **Chrysophyta** (kris AH fuh tuh). They are mostly unicellular organisms that get their color from large amounts of yellow-brown pigments. Members of this phylum store food as oils or as a starchlike carbohydrate. Some may be flagellated and live in colonies in fresh water.

Of the approximately 10 000 species of golden algae, the most numerous are the **diatoms** (DY uh tahmz). Diatoms are single-celled or colonial organisms that live in both fresh and salt water. Their cell walls are glasslike shells made of silica, with tiny holes for the exchange of gases and other substances. These rigid shells form two halves that fit together like the top and bottom of a pillbox. Each diatom contains a nucleus and one or more chloroplasts. Although diatoms lack flagella, they are capable of a gliding movement. They usually reproduce asexually, but sexual reproduction does occur.

Because of the intricate geometric shapes of their ornate shells, diatoms are quite beautiful. They are tremendously abundant in the oceans, where they serve as food for fish and other aquatic animals. When diatoms die, their shells sink to the ocean floor. In some places, the shells have accumulated in layers hundreds of meters thick, forming rocklike deposits known as *diatomaceous* (dy uh tuh MAY shus) *earth*. Some of these deposits, formed millions of years ago, are now on land. Diatomaceous earth is mined and used in metal polishes, toothpaste, insulation, and filters.

Figure 30–19

Saltwater Diatoms. Diatoms are elaborate, single-celled organisms with glasslike cell walls. ▶

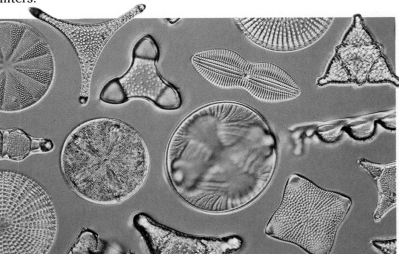

Dinoflagellates   The *dinoflagellates* (dy noh FLAJ uh layts), phylum **Dinoflagellata** (dy noh flag uh LAH tuh), are single-celled algae found mainly in the oceans. Some of them are photosynthetic, storing food as oil or starch. Others lack chloroplasts and are heterotrophic. The photosynthetic dinoflagellates, along with diatoms, serve as the major source of food for many aquatic animals. Reproduction is asexual.

The cell walls of dinoflagellates consist of numerous armorlike plates made up of cellulose and silica. Each organism has two flagella, one running in a beltlike groove around the middle, the other extending from one end. The actions of the flagella cause dinoflagellates to twirl or roll in the water.

In addition to chlorophyll, some of the photosynthetic dinoflagellates contain red and yellow pigments. Sometimes called "fire algae," these organisms vary in color from yellow-green to brown to red. Some of the red ones produce powerful toxins. Blooms (population explosions) of red dinoflagellates in warm, shallow waters cause the water to appear red, a condition called "*red tide*." Shellfish living in areas of red tides should not be eaten because they contain toxins that cause illness in humans. Toxins in the water also kill many fish.

Many dinoflagellates have the property of *bioluminescence* — the ability to produce light. When present in large numbers, and if disturbed, these organisms may cause the ocean to sparkle or glow at night.

Green Algae   The *green algae,* phylum **Chlorophyta** (klor AH fuh tuh), are found in salt and fresh water and in moist places on land. This group has unicellular, colonial, and multicellular forms. Most green algae have a cell wall composed of cellulose, and most contain chlorophylls *a* and *b* in chloroplasts. Some green algae have flagella, which are used in movement.

Chlamydomonas (klam id uh MOH nus), shown in Figure 30–21, is a typical unicellular, freshwater green alga. It has two flagella of equal length, which are used in locomotion. There is one chloroplast and a pyrenoid for starch synthesis. There are two contractile vacuoles near the bases of the flagella. Unlike other green algae, *Chlamydomonas* has a cell wall made up of a compound that contains a carbohydrate and a protein, rather than cellulose.

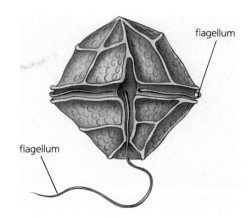

▲ **Figure 30–20**
Structure of a Dinoflagellate. Each dinoflagellate has a pair of flagella for locomotion.

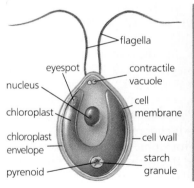
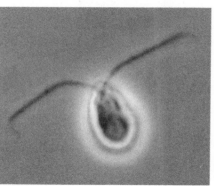

◄ **Figure 30–21**
Structure of *Chlamydomonas*. The unicellular green alga *Chlamydomonas* has two flagella at its anterior end that beat in opposite directions, allowing it to swim rapidly.

▲ **Figure 30–22**
Sargassum. *Sargassum* (top) is a type of brown algae that forms floating masses over wide areas of the Sargasso Sea. Blending into their surroundings, mollusk eggs and a crab are well adapted to life on this strand of sargassum (bottom).

Reproduction is usually asexual. Multicellular green algae include filamentous forms, such as spirogyra and other, more complex, forms.

Brown Algae   The *brown algae,* phylum **Phaeophyta** (fee AH fuh tuh), include many of the common seaweeds. Members of this group are all multicellular and range in size from microscopic to more than 50 meters in length. They are found usually in cold ocean waters. The brown algae contain chlorophyll, which functions in photosynthesis, and other pigments that give them their brown color. Brown algae have cellulose in their cell walls and store food in the form of a polysaccharide or as oil. Their life cycles show an alternation of generations.

Some brown algae that live along the shoreline have rootlike structures that anchor them to rocks. Others found on the ocean surface have gas-filled structures that function as floats. The surface of the Sargasso Sea, an area stretching across the Atlantic Ocean from the West Indies to the coast of Africa, is densely covered with several types of floating brown algae that belong to the genus *Sargassum.*

Red Algae   The *red algae,* phylum **Rhodophyta** (roh DAH fuh tuh), like the brown algae, include many common seaweeds. Red algae are found in warmer waters and at greater depths than brown algae. They usually are attached to rocks or other surfaces. Although most red algae are multicellular, they are not as large as the largest brown algae. The chloroplasts of red algae contain chlorophyll and several other pigments. Many red algae are reddish in color, but others are black, green, yellow, or purple. Red algae have complex life cycles, including an alternation of generations.

The cell walls of the red algae contain cellulose and other substances, including *agar,* which has various industrial uses. It is used as a thickener in foods such as ice cream and as a medium for growing bacteria and fungi in laboratories. *Carrageenan* (kar uh GEE nun), another product of red algae, is used to prevent separation in food mixtures. It is used, for example, in chocolate milk to prevent the milk and the chocolate from separating.

Funguslike Protists

The funguslike protists are similar to the fungi in some ways—namely in appearance and method of nutrition. You will study members of the kingdom Fungi in Chapter 31. *Fungi* are nonphotosynthetic organisms such as yeasts, molds, and mushrooms. The funguslike protists also have some stages in their life cycle that are similar to those of protozoa. For these reasons, their classification has not always been clear-cut. However, most scientists now agree that these organisms are different from fungi in many significant ways, and they classify them in the kingdom Protista.

All funguslike protists are heterotrophic. Most are decomposers (saprobes) that feed on dead and decaying matter in cool, damp environments. A few are parasites. There are three major phyla of funguslike protists, as listed in Figure 30–23.

| Major Divisions of Funguslike Protists ||
Phylum	Identifying Characteristics
Myxomycota (acellular slime molds)	single-celled organisms with many nuclei; feeding stage characterized by ameboid plasmodium.
Acrasiomycota (cellular slime molds)	feeding stage characterized by individual ameboid cells, which group together to form pseudoplasmodium when food is scarce.
Oomycota (water molds and downy mildews)	finely branched, single-celled filaments resemble fungi; water molds are saprobes or parasites; downy mildews are plant parasites.

◀ **Figure 30–23**
Major Divisions of Funguslike Protists.

Acellular Slime Molds   Acellular slime molds, also called *plasmodial slime molds,* are members of the phylum **Myxomycota** (mix uh my KHAT uh). They are single-celled organisms that contain many nuclei.

In the most commonly observed stage of their life cycle, acellular slime molds resemble giant amebas, or slimy masses, as shown in Figure 30–24. This feeding stage is called the *plasmodium.* The plasmodium is ameboid and feeds by engulfing bits of organic matter as it creeps along the forest floor. When conditions for growth become unfavorable, the plasmodium stops moving and develops stalked, spore-producing structures called *fruiting bodies.* Within the fruiting bodies, haploid spores are produced by meiosis. The spores are eventually released and, if they land in a moist, suitable environment, germinate to form flagellated gametes. Two

Figure 30–24
Life Cycle of an Acellular Slime Mold. ▼

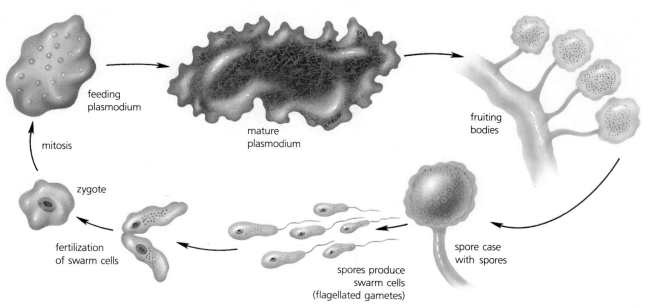

feeding
plasmodium

mature
plasmodium

fruiting
bodies

mitosis

zygote

fertilization
of swarm cells

spores produce
swarm cells
(flagellated gametes)

spore case
with spores

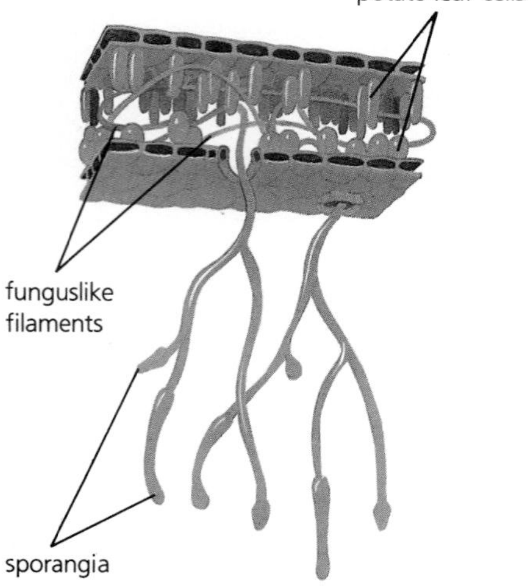

▲ **Figure 30–25**
Growth of a Downy Mildew on a Potato Leaf. The growth of downy mildew on potato plants causes potato blight. As the parasite grows, it destroys the photosynthetic tissue of the leaf.

gametes join to form a diploid zygote, which undergoes mitosis and becomes a feeding plasmodium. Thus sexual reproduction occurs in the acellular slime molds.

Cellular Slime Molds   Slime molds of the phylum **Acrasiomycota** (uh krayz ee oh my KAHT uh) are called cellular slime molds. They live in fresh water, in damp soil, or on decaying vegetation, such as rotting logs. In the feeding stage of their life cycle, the cellular slime molds move about as individual ameboid cells and feed on decaying matter. When food becomes scarce, the individual cells group together to form a mass of cells called a *pseudoplasmodium*. The pseudoplasmodium may migrate for a while. Then it forms fruiting bodies, which produce haploid spores. In favorable conditions, individual, haploid, ameboid cells emerge from the spores, and the feeding stage begins again.

The life cycle of cellular slime molds is less complex than that of the acellular types. Reproduction is asexual, and there is no diploid stage in their life cycle. The other major difference is that, unlike the unicellular feeding plasmodium, the pseudoplasmodium consists of many individual, membrane-bound cells.

Water Molds and Downy Mildews   The water molds and downy mildews are members of the phylum **Oomycota** (oh uh my KAHT uh). These protists, consisting of finely branched, single-celled filaments, look like fungi and also have a funguslike method of nutrition. However, they differ from fungi in the content of their cell walls, their sexual mode of reproduction, and the dominance of the diploid stage in their complex life cycle.

Most water molds are saprobes that live on dead organisms in fresh water. Their filaments form cottony masses. Other water molds are parasites that live on the gills of fish. The downy mildews are parasites that live on plants. One species of downy mildew attacked French vineyards in the 1870s, threatening the wine-making industry. Another species caused late potato blight, which destroyed the potato crops of Ireland from 1845 to 1847. The resulting famine caused over 1 million deaths and the mass migration of nearly half the population of the country.

30-2 Section Review

1. What is the common name for the plantlike protists?
2. Name the substance that gives some protists their glassy cell walls.
3. Are the animal-like protists autotrophic or heterotrophic?
4. What is an acellular slime mold called during its feeding stage?

Critical Thinking

5. What is the main criterion for classifying animal-like protists? Compare this feature in the four types of animal-like protists. (*Comparing and Contrasting*)

30-3 The Viruses

Section Objectives:

- *Describe* the basic structure of a virus.
- *Identify* the criteria that scientists use to classify viruses.
- *Explain* how viruses enter host cells, replicate, and release new viruses.
- *Describe* the body's defenses against viral infection.

Viruses are tiny particles unlike any other living organisms. In fact, scientists do not consider viruses to be living. They are described as somewhere between living cells and nonliving things. Although viruses contain genetic material, they lack all other cell structures necessary for metabolism, reproduction, and growth. A virus consists of genetic material—DNA or RNA—wrapped in a protein coat. Since viruses are not cells, they are not included as members of any of the five kingdoms.

A virus particle cannot reproduce, or *replicate,* unless it is inside a living host cell. Therefore, viruses survive by attacking a living cell, using the cell's machinery to reproduce. This parasitic invasion of plant and animal cells leads to a wide variety of diseases. In humans, viruses cause many infectious diseases that range from the common cold to rabies and acquired immune deficiency syndrome (AIDS).

Size and Structure

The word *virus* is from the Latin word for poison. It originally referred to any poisonous substance. Later, it was used to refer to specific agents of disease.

Viruses are much smaller than cells and were not actually seen until the electron microscope was invented. Scientists knew about viruses about 100 years ago when they began looking for something smaller than bacteria that could cause disease. At that time an epidemic of a disease of tobacco plants, tobacco mosaic disease, occurred in the Soviet Union. In this disease, the tobacco leaves develop a "mosaic" of light- and dark-green patches. Although they could not see the viruses through their microscopes, scientists knew that some "invisible" pathogen was causing the disease. They deduced that the pathogen was not a bacterium, because it would not grow in laboratory cultures, and it could not be killed by alcohol, which is fatal to bacteria. In 1892, the Russian biologist Dmitri Iwanowski discovered the virus that caused tobacco mosaic disease. Martinus Beijerinck, a Dutch scientist, continued experimenting on the disease and was the first to use the term *virus* to describe the pathogen. However, it was not until 1935 that the tobacco mosaic virus was chemically isolated by Wendell Stanley, an American biochemist. Since then, our knowledge about viruses and their ability to cause disease has increased greatly. Through the use of the electron microscope and other scientific advances, many viruses have been isolated and identified.

Figure 30–26

Tobacco Mosaic Virus. These tobacco leaves (top) show the patchwork coloring of a tobacco mosaic virus infection. The rod-shaped virus particles are seen in this scanning electron micrograph (bottom). ▼

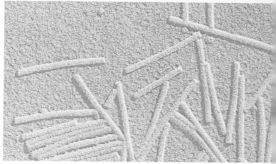

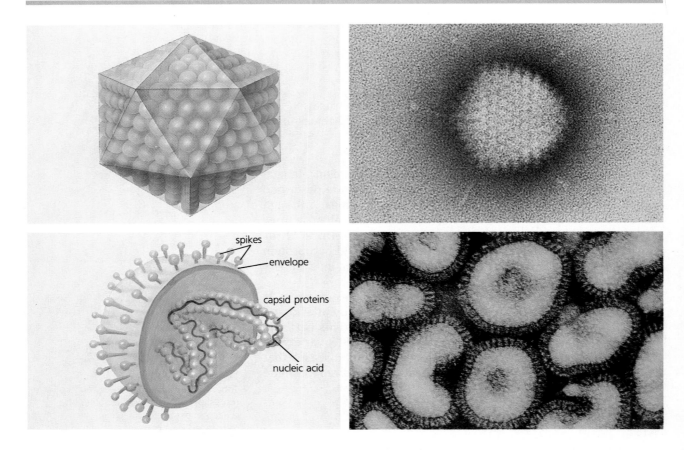

▲ **Figure 30–27**

Viral Shapes and Sizes. At upper left is a diagram of a polyhedron. At upper right is an electron micrograph of the polyhedral-shaped advenovirus. (Magnification 78 500 X) A diagram of an enveloped helical virus is at lower left. An electron micrograph of influenza viruses is at lower right. (Magnification 76 500 X)

Depending on the cells they infect, viruses are frequently classified as plant, animal, or bacterial viruses. Thus, some viruses infect the leaf cells of plants; others infect cells of the human respiratory or digestive tracts. There are many subcategories of each type of virus.

Viruses consist of a nucleic acid core of RNA or DNA surrounded by a protein coat called a **capsid.** In addition to the capsid, some viruses are enclosed in a membrane called an *envelope,* which consists of a portion of the cell membrane or nuclear membrane from the host cell and some viral proteins. The flu virus, for example, has an envelope with protein spikes that enable it to adhere to a host cell.

Viruses vary in their structure and shape. Some are rod-shaped, such as the tobacco mosaic virus. The herpes virus is polyhedral, or many-sided. Still others may be helical, such as the mumps virus. Some examples of different viral shapes and structures are shown in Figure 30–27.

As minute as the typical virus is, there are even smaller and simpler disease-causing viral particles. Scientists have recently isolated *viroids* and *prions.* Viroids, which infect plants, are very short pieces of RNA having no surrounding capsids. Prions, which cause slow, progressive diseases in animals, contain proteins but lack nucleic acid.

Genetic Makeup

Viruses may have either RNA or DNA as their genetic material, but not both. The nucleic acid strands may be single or double, depending on the virus. DNA viruses and RNA viruses have different effects on host cells. Within the host cell, a DNA virus produces RNA, which directs the production of viral proteins. Or the viral DNA may combine with the DNA of a host cell, which then produces new viruses.

RNA viruses work in a different way. Once inside the host cell, an RNA virus may direct the production of proteins by the host cell. Sometimes, the viral RNA may make DNA with the aid of an enzyme called *reverse transcriptase*. That DNA produces new RNA, which, in turn, synthesizes proteins that produce new viruses. Such an RNA virus is called a **retrovirus.** The AIDS infection is caused by a retrovirus.

Reproduction

Bacteriophages, or "*phages*" for short, are viruses that attack bacterial cells. They have a complex structure, consisting of a head and tail portion with long fibers projecting from the tail. Phages are a particularly well-studied group of viruses because their host cells, bacteria, can be easily and quickly grown in laboratory cultures. The following steps show how a phage attacks a bacterial cell, uses it to reproduce, and then destroys it. This is known as the **lytic cycle:**

1. Many phages consist of a tadpole-shaped outer protein coat that transports its nucleic acid by means of a tail. When the phage finds the right host bacterium, it attaches its fibrous tail to the cell and uses an enzyme to eat a hole in the cell wall. The cell wall of a

Figure 30–28
Structure of a Bacteriophage. ▼

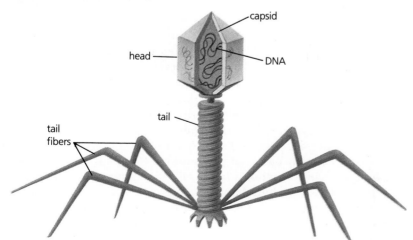

capsid

head

DNA

tail

tail
fibers

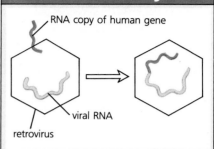

Science, Technology and Society

RNA copy of human gene

viral RNA

retrovirus

Technology: Retroviruses and Gene Therapy

You probably associate viruses with a cold or perhaps with AIDS. But, soon you may think of viruses as tools to treat diseases as diverse as sickle-cell disease and cancer. Scientists are learning to use retroviruses for this purpose.

Retroviruses make a DNA copy of their RNA and insert the copy into a host cell's DNA. Scientists can use this property for gene therapy—a method of treating disorders by replacing a defective gene with a normal one.

To treat sickle-cell disease, for example, the gene coding for normal hemoglobin could be inserted into retroviruses. Bone marrow, where red blood cells are produced, could be removed from a patient and grown in the laboratory. The cultured cells could then be infected with the altered viruses and the marrow reinjected into the patient. Scientists hope enough normal blood cells could be produced to cure the disorder.

■ *Some people oppose gene therapy because the gene for any trait could be transferred in this way. They worry the technology will not be used responsibly. How do you feel about this?*

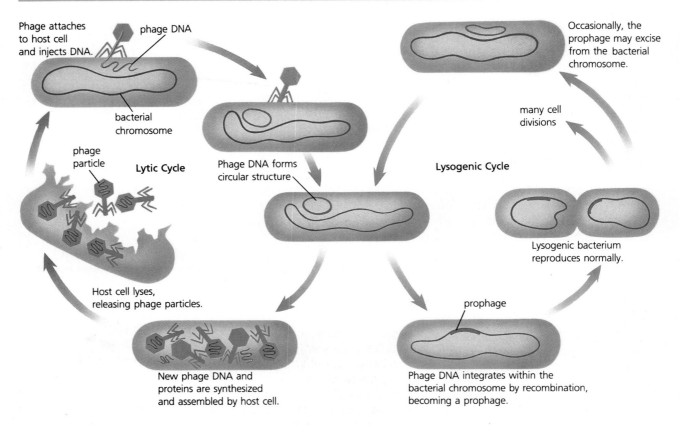

Phage attaches to host cell and injects DNA.

phage DNA

bacterial chromosome

phage particle

Lytic Cycle

Phage DNA forms circular structure

Host cell lyses, releasing phage particles.

New phage DNA and proteins are synthesized and assembled by host cell.

Occasionally, the prophage may excise from the bacterial chromosome.

many cell divisions

Lysogenic Cycle

Lysogenic bacterium reproduces normally.

prophage

Phage DNA integrates within the bacterial chromosome by recombination, becoming a prophage.

▲ **Figure 30–29**

The Lysogenic and Lytic Cycles of a Bacteriophage. Both the lysogenic and lytic cycles begin when the phage injects its DNA into a host cell. In the lytic cycle, the cell is forced to produce new phages. In the lysogenic cycle, the phage's nucleic acid merges with the host's DNA. When the host cell divides, the phage DNA divides with it, thereby producing more cells with viral DNA.

host cell has an area on its surface called a receptor. That area fits together with the capsid shape of a specific virus. For this reason, viruses usually infect only one kind of host cell.

2. The phage then injects its nucleic acid into the host bacterial cell. In some viral infections, the capsid remains outside the host cell. In others, the entire virus enters the cell. Once inside the host cell, the viral RNA or DNA takes over the cell's reproductive machinery. The cell is forced to produce viral nucleic acids and proteins, which are then put together in an assembly-line fashion.

3. Eventually the cell becomes so laden with the new whole viral particles that its wall bursts, or *lyses,* scattering a new generation of viruses that can infect other bacteria. The lytic cycle can be very fast. A whole new generation of phages can be produced in as little as 20 minutes, given optimum conditions.

Sometimes a phage and bacterial cell can coexist for a time without the destruction of the bacterium. This process is known as the **lysogenic cycle.** It occurs when the phage's nucleic acid merges with the bacterial cell's DNA, and the new combination is transmitted through bacterial generations. Once such a virus has attached its DNA to that of a host cell, it is known as a *prophage.* This process can be lethal to the host cell if it is subjected to sudden environmental changes, such as exposure to radiation or chemicals. Changes can activate the phage's nucleic acid, causing production of viral particles and destruction of the host cell. The lysogenic and lytic cycles are compared in Figure 30–29.

As you learned in Chapter 27, viruses sometimes carry DNA from one bacterial cell to another in a process called transduction. This results in the creation of bacterial types that are genetically different.

Viruses and Disease

Viral infections are spread in the same ways as bacterial infections. Viral diseases vary in their severity. If the harm to host cells is slight, the resulting infection is barely noticeable. If the host cells are damaged by the viral attack, a disease with more severe symptoms results. Thus, viruses can damage or destroy cells in their effort to reproduce in specific tissues and organs. Most of us recover completely from colds and the flu, which mainly affect the respiratory tract. The polio virus attacks nerve cells in the brain and spinal cord. As nerve cells cannot be replaced, the infectious spread of the virus through these cells can cause permanent paralysis or death. Other diseases caused by viruses include smallpox, chicken pox, measles, mumps, and AIDS.

Figure 30–30
Viruses that Cause Human Diseases. ▼

Examples of Viruses that Cause Human Disease		
Viral Group	**Nucleic Acid**	**Disease**
papovaviruses	DNA	warts
adenoviruses	DNA	respiratory infections
herpesviruses	DNA	herpes chickenpox shingles infectious mononucleosis Burkitt's lymphoma (a cancer)
poxviruses	DNA	smallpox wartlike skin lesions
picornaviruses	RNA	polio common cold gastrointestinal infections
myxoviruses	RNA	influenza measles mumps
retroviruses	RNA	tumors leukemia AIDS
rhabdoviruses	RNA	rabies
togaviruses	RNA	rubella (German measles)

▲ **Figure 30–31**

The AIDS Quilt. Each of the more than 10 800 panels of the AIDS quilt was made in memory of a person who has died of AIDS. As new panels are contributed, the quilt continues to grow. Shown here in Washington, D.C., the quilt has been displayed in cities across the United States.

The virus that causes AIDS, which is called the human immunodeficiency virus, or HIV, was identified by scientists working independently in France and the United States in 1984. The AIDS virus attacks cells of the immune system called helper T cells. As a result of this damage, the immune response is seriously weakened. The dramatic, devastating effects of AIDS on the immune system are described in Chapter 10.

AIDS was first identified in 1981. As of 1989, over 110 000 people have been diagnosed with AIDS in the United States. Currently, the number of new AIDS cases is doubling every year. Of those diagnosed with AIDS, almost one-half have died. Because their immune systems are so weakened, most AIDS victims die from infections or cancers. It is estimated that as many as 1 to 2 million people may be infected with the AIDS virus, but do not yet show symptoms of the disease. As yet, there is no cure for AIDS, but scientists are actively researching the disease in laboratories around the world. Researchers are working to find out more about the AIDS virus and to develop effective drugs and vaccines.

Viruses have also been implicated in the production of certain cancerous tumors by causing cells to multiply abnormally. Such viruses, known as tumor viruses, may carry genes that cause cancer, or **oncogenes.** They may also stimulate oncogenes already present in the host cell. Tumor viruses alter the genetic makeup of cells by incorporating their nucleic acid with the host cell's DNA. DNA tumor viruses are examples of viruses that direct the constant production of viral DNA and proteins in host cells. This causes rapid cell division and growth. The host cell does not produce new viruses, as it does in other viral infections. Therefore, the host cells are not destroyed but continuously grow and divide.

The human body can protect itself against viruses through its own immune system. Viruses stimulate a response from the immune system of the infected host. This response is the basis for vaccination, in which a weakened or dead virus is used to stimulate production of antiviral substances.

The first vaccine, against smallpox, was developed by Edward Jenner in 1798. In developing this vaccine, Jenner used material from the sores of people infected with cowpox, a disease similar to but milder than smallpox. The cowpox virus was so closely related to the smallpox virus that vaccination with it brought about immunity to smallpox infection. (The word *vaccination* is from the Latin *vacca,* which means "cow.") Smallpox was once one of the most feared diseases in human history, spreading in epidemics to many parts of the world. Now, the use of the smallpox vaccine has completely eliminated the disease.

In 1884, Louis Pasteur developed the first rabies vaccine. He soon used it as a preventive treatment in a nine-year-old who had been bitten by a rabid dog. Rabies is a serious, often fatal, disease that is transmitted to humans by the bite of infected animals, such as dogs and raccoons. In much the same way that he had earlier produced his anthrax vaccine, Pasteur developed weakened forms of the rabies virus. He used the weakened rabies virus in a series of increasingly strong injections. He found that this treatment could prevent rabies in a person exposed to the virus, if administered soon after a bite occurred.

In 1954, Dr. Jonas Salk developed the first polio vaccine, called the Salk vaccine. Salk took active polio viruses and killed them with a poison, formaldehyde. When injected as a vaccine, the killed viruses caused the production of polio antibodies that protected the person from polio. Around this time the United States was in the midst of a polio epidemic. In 1952, there were 57 000 reported cases of polio. Widespread use of the Salk vaccine resulted in a radical drop in the incidence of polio, and by 1957 the United States was free of the epidemic.

In the early 1960s, Dr. Albert Sabin developed a different type of polio vaccine. He treated polio viruses to weaken but not kill them. When fluid containing the weakened viruses is swallowed (in a flavored drink), the person will build antibodies against polio. Today the Sabin vaccine is routinely administered, as it induces a longer-lasting immunity than the Salk vaccine. In parts of the world where widespread vaccination against polio is not enforced, however, the disease still exists. Today, vaccines are available to prevent a wide variety of viral infections, including measles, mumps, and rubella (German measles).

Another natural defense against viruses is *interferon*. Interferon is a protein produced by a virus-infected cell that inhibits virus reproduction. Because human interferon is effective against many kinds of viruses, physicians have been trying to use it to treat viral infections. However, cells produce only minute quantities of interferon, making it impractical to extract. Synthetically produced interferon has only limited use because it can have toxic side effects in humans.

Although vaccines can prevent some viral infections, there is no cure for them once they occur. Antibiotics are not effective against viruses. A few drugs have been developed that can slow viral replication. However, immunity is still the most effective defense against viral diseases.

30-3 Section Review

1. The basic structure of viruses includes which two parts?
2. What is another name for bacterial viruses?
3. Which cycle of viral replication involves the disruption of host cell walls?
4. Dr. Jonas Salk developed the first vaccine for which viral infection?

Critical Thinking

5. Why are antibiotics not useful in treating the common cold? (*Identifying Reasons*)

Biology and You

Q: There are plenty of antibiotics to fight bacterial infections, yet no drugs to treat viral infections such as the common cold. Why not?

A: Viral diseases are difficult to treat because a virus becomes an integral part of the body cell it infects. Once a virus invades a cell, any drug that damages the virus affects the host cell as well. Because bacteria exist as independent cells, antibiotics kill bacteria without damaging body cells.

Some viral diseases are prevented by vaccines. A vaccine for the common cold is impractical, however, since over 200 cold viruses exist. Instead, other methods are being developed.

Doctors are experimenting with a nasal spray that contains interferon, an antiviral agent produced by body cells under viral attack. In experiments, families used the spray whenever a member showed signs of a cold. The spray prevented colds from being passed on, especially colds caused by rhinoviruses, the leading cause of colds. However, more studies need to be done before interferon nasal spray is approved for the public.

Write a one-page story from the perspective of a virus. Describe how you invade and destroy a cell.

Laboratory Investigation

Effect of Antibiotics and Disinfectants on Bacterial Growth

Preventing or slowing the growth of harmful bacteria is an important need. In this investigation, you will test the effect of antibiotics and disinfectants on bacterial growth.

Objective

■ *Compare* the effectiveness of various antibiotics and disinfectants in inhibiting bacterial growth.

■ *Use* standard laboratory materials to culture bacteria.

Materials

2 nutrient agar	disinfectant discs
petri dishes	blank, sterile discs
wax pencil	forceps
antibiotic discs	millimeter ruler

Procedure   *

A. Working in assigned groups, obtain two petri dishes with sterile nutrient agar. Turn the dishes upside down without opening them. With a wax pencil, label the centers of the dishes 1 and 2 and initial them. Then mark off the bottom of each dish into thirds. Label the sections a, b, c.

B. Open dish 1 and gently rub your fingers over the entire surface of the agar. Replace the cover. The same person should repeat this procedure for dish 2.

C. Select two different antibiotic discs. Using forceps, place one disc on section a and another on section b of dish 1. Place a blank, sterile disc on section c as a control. Replace the cover on the petri dish. Record the name of the disc used on each section in a chart like the one above.

D. Using dish 2, repeat step C, this time using two kinds of disinfectant discs and a blank.

E. Incubate the petri dishes at room temperature. Do not remove their covers to examine the bacteria. Examine the petri dishes after days one, two, and five. Without removing the covers, measure the diameter of the clear area surrounding each disc. Record these measurements in your data chart.

Antibiotic/ Disinfectant	Diameter of Clear Area (includes diameter of disc)		
	Day 1	Day 2	Day 3
dish 1 a. b. c.			
dish 2 a.			

F. When you have finished, return the petri dishes to your teacher so that they can be sterilized.

Analysis and Conclusions

1. Why was it important to have the same person inoculate both petri dishes?

2. What was the purpose of using a blank, sterile disc in each dish?

3. How did diffusion play a role in this experiment?

4. Was any antibiotic more effective than the others? What factors might account for the difference?

5. Was any disinfectant more effective than the others?

6. Of all the substances that you tested, which one was most effective in inhibiting bacterial growth? Do your results agree with those of other groups in your class? If not, explain.

7. What additional information about the tested substances could be obtained by monitoring the cultures for a longer period of time?

*For detailed information on safety symbols, see Appendix B.

Chapter Review

30

Study Outline

30–1 The Monerans

■ Monerans, or bacteria, are the simplest, most numerous organisms on earth. Although some bacteria are harmful and cause diseases, most are beneficial. Because they lack a distinct membrane-bound nucleus, monerans are classified as prokaryotes.

■ Most monerans have a cell wall composed of a complex polysaccharide; many are surrounded by a capsule. Monerans lack organelles, except for ribosomes. They contain a circular chromosome and may also have smaller circular segments of DNA, or plasmids.

■ Bacteria may be grouped according to shape: the spherical coccus, the rod-shaped bacillus, and the spiral-shaped spirillum. Some bacteria form colonies.

■ Prokaryotes vary in their ability to live with or without oxygen. They also vary in their food requirements and may be heterotrophic or autotrophic.

■ All bacteria reproduce asexually. Some are capable of exchanging genetic material by conjugation, transformation, or transduction.

■ The branches of the kingdom Monera are archaebacteria and eubacteria. A major group of eubacteria—the cyanobacteria—are photosynthetic autotrophs.

■ Pathogens are heterotrophic bacteria that are disease causing. Much of our understanding of bacterial pathogens can be attributed to the work of Pasteur and Koch. Koch's postulates are a set of rules to determine if a specific bacterium causes a specific disease.

30–2 The Protists

■ The kingdom Protista contains many varied species, and is divided into animal-like, plantlike, and funguslike protists. All protists are composed of eukaryotic cells with a nucleus and cytoplasmic organelles.

■ The animal-like protists, called protozoa, are divided into four phyla, based on their method of locomotion. Sarcodines move by means of pseudopods, ciliates by means of cilia, and zooflagellates by the beating of flagella. Sporozoans are nonmotile.

■ The plantlike protists, or algae, are photosynthetic. They can be divided into six groups. The euglenoids have both plantlike and animal-like characteristics. The unicellular golden algae have yellow-brown pigments. The dinoflagellates have cell walls made up of armorlike cellulose or silica plates. The green algae may be unicellular or multicellular; they contain chlorophyll in chloroplasts. Red algae and brown algae are multicellular; they both include many of the common seaweeds.

■ Many funguslike protists are decomposers; a few are parasites. They include acellular slime molds, cellular slime molds, and water molds and downy mildews.

30–3 The Viruses

■ Viruses are tiny particles that consist of genetic material wrapped in a protein coat. Except for genetic material, viruses lack all cell structures necessary for metabolism, reproduction, and growth.

■ Viruses, which reproduce only inside living cells, attack host cells and gain control of their metabolic machinery. By parasitizing cells, viruses cause many diseases.

■ Viruses vary in shape and may contain either DNA or RNA. DNA and RNA viruses have different effects on the host cell.

■ Viral infections are spread in the same ways as bacterial infections. Once inside a living cell, the virus replicates, destroying the host cell. Viral diseases include colds, flu, polio, measles, mumps, and AIDS.

Vocabulary Review

Monera (631)	eubacteria (635)
bacteria (631)	Cyanobacteria (635)
endospores (632)	blue-green bacteria (635)
coccus (632)	green-sulfur bacteria (636)
bacillus (632)	
spirillum (632)	purple bacteria (636)
saprobes (633)	Prochlorophyta (636)
parasites (633)	Schizophyta (636)
archaebacteria (635)	Gram test (636)

Chapter Review

Gram stain (636)
gram-positive (636)
gram-negative (636)
pathogens (637)
germ theory of disease (637)
Koch's postulates (638)
Protista (642)
protozoa (642)
Sarcodina (642)
Ciliophora (644)
pellicle (644)
Zoomastigina (644)
Sporozoa (645)
algae (646)
Euglenophyta (646)

Chrysophyta (648)
diatoms (648)
Dinoflagellata (649)
Chlorophyta (649)
Phaeophyta (650)
Rhodophyta (650)
Myxomycota (651)
Acrasiomycota (652)
Oomycota (652)
viruses (653)
capsid (654)
retrovirus (655)
bacteriophages (655)
lytic cycle (655)
lysogenic cycle (656)
oncogenes (658)

A. Sentence Completion—Fill in the vocabulary term that best completes each statement.

1. In the phylum _____, folded membranes in the cytoplasm contain pigments for photosynthesis.

2. Cells that lack a distinct nucleus are members of the kingdom _____.

3. Bacteria that retain the stain crystal violet, which makes them purple, are termed _____.

4. An RNA virus that makes DNA from RNA is called a _____.

5. Animal-like protists that move by means of pseudopods are classified in the phylum _____.

6. Harmful bacteria capable of causing infections are called _____.

7. A feeding stage, the plasmodium, is observed among members of the phylum _____.

8. The cell walls of algae in the phylum _____ consist of armorlike plates of cellulose and silica.

9. One of the earliest kinds of bacteria to inhabit the earth are thought to be the _____.

10. Since the malaria parasite is a nonmotile protist, it is classified in the phylum _____.

B. Word Relationships—In each of the following sets of terms, three of the terms have a common characteristic. One term does not belong. Identify the term that does not belong.

11. coccus, bacillus, virus, spirillum

12. pathogen, saprobes, germ theory of disease, Koch's postulates

13. Chlorophyta, Schizophyta, Dinoflagellata, Euglenophyta

14. Oomycota, Sarcodina, Ciliophora, Zoomastigina

Content Review

15. Compare prokaryotic and eukaryotic cells.

16. Summarize both heterotrophic and autotrophic nutrition in bacteria.

17. Draw the growth curve for a culture of bacterial cells, and describe the phases involved.

18. How do bacteria exchange genetic material?

19. What are the habitats of archaebacteria?

20. Under what conditions will a bloom of blue-green bacteria develop?

21. What are Robert Koch's rules to determine if a particular disease is caused by a specific microorganism?

22. Name an important function of bacteria in nature.

23. Briefly describe the life cycle of the malarial parasite, *Plasmodium*.

24. How do euglenoids resemble plant and animal cells?

25. Describe the structure of diatoms, and explain what happens to their shells once they die.

26. List the major characteristics of dinoflagellates.

27. Distinguish between the following three phyla: Chlorophyta, Phaeophyta, and Rhodophyta.

28. Why are the slime molds classified as protists rather than as fungi?

29. Describe the structure and life cycle of acellular slime molds.

30. How do viruses use a host cell?

31. Explain how phages attack bacterial cells.

32. How does the human body combat viral infections?

Graphic Organizing

For information on graphic organizers, see Appendix G at the back of this text.

33. **Scale:** Draw a scale that shows the relative sizes of the following organisms and cells, from smallest to largest: the bacterium *E. coli,* .000001 m; ameba, .0001 m; certain human nerve cells, 1 meter; the marine alga *Acetabularia,* between .01 and .1 m; flu virus, .0000001 m; human red blood cell, .00001 m.

Critical Thinking

34. Compare how viruses and bacteria affect an organism after they infect it. (*Comparing and Contrasting*)

35. Two scientists recently presented opposing views on whether viruses are living. What reasons might they give to support both views? (*Identifying Reasons*)

36. What might happen in a forest ecosystem if all the bacteria suddenly died? (*Predicting*)

Creative Thinking

37. From your knowledge of *Plasmodium's* life cycle, suggest possible plans to eliminate malaria.

38. A veterinarian suspects that a new disease of house cats is caused by a specific type of bacterium. Describe a set of procedures for proving this hypothesis.

Problem Solving

39. In an experiment on bacterial growth, equal amounts of methane-producing bacteria and water were placed in each of four test tubes. One of four sugars (A, B, C, or D) was added to each test tube in equal concentrations. The volume of methane gas (CH_4) liberated from each culture was measured and the data plotted on the graph shown below. What does the graph indicate about the effect of each sugar solution on the growth of the bacteria?

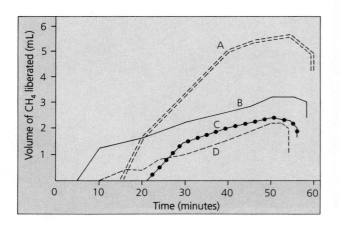

40. Biologists want to determine if euglena can survive without sunlight. The euglena's chloroplasts can be eliminated by treating a euglena culture with antibiotics. Design a controlled experiment to test the ability of euglena to survive by heterotrophic nutrition.

Computer Project

41. When bacteria are placed in a liquid medium, they begin a period of rapid growth, dividing at a regular interval known as the doubling time. The number of bacterial cells doubles repeatedly—one cell divides into two, two into four, and so on. For the bacterium *E. coli,* the doubling time is 20 minutes when conditions are ideal for growth. Another bacterium, *Bacterium X,* has a doubling time of 40 minutes.

In this activity, you will use a computer spreadsheet to answer the following questions about bacterial growth.

A. If you started with one *E. coli,* how many would there be after 4 hours? After 8 hours? How many generations would have passed in each case?

B. If you started with one *Bacterium X,* how many would there be after 4 hours? After 8 hours? How many generations would have passed?

C. Repeat the calculations in A and B above with a starting population of 50 bacteria.

D. So far, it has been assumed that *all* the bacteria survive from one generation to the next. In reality, this is not the case. Suppose that only 80 percent of the bacteria from any one generation survive to reproduce. How would this change the calculations you performed in C? Recalculate and compare the results.

E. Experiment with various survival rates between 0 percent and 100 percent. At what survival rate does no net change in the population result? Refer to the growth curve in Figure 30–4. Which part of the curve illustrates this stage?

F. It is known that viruses that infect *E. coli* reproduce much more rapidly than does *E. coli* itself. In fact, a single virus can give rise to about 100 virus particles in 20 minutes. Suppose there were a bacterial population into which you introduced a single bacterium infected with a single virus. How many virus particles would there be after 4 hours if all survive? How does this compare with the number of *E. coli* present after 4 hours in question A?

G. Problem Finding: What further questions could you answer using this spreadsheet?

▲ Mosses hang from the branches of trees in the Hoh rain forest in Olympia National Park, Washington

Fungi and Plants

31

Forests abound with life. Different plants and fungi find niches in which to live and thrive in complex interrelationships. Mosses cling to rocks and trees and carpet the forest floor. Trees tower and ferns seek the light that is left. Mushrooms and other fungi are scattered in the damp of the forest, extracting nutrients from what is dead and leaving particles to enrich the soil. In this chapter, you will learn about the different groups of plants and fungi and how they maintain their delicate balance.

Guide for Reading

Key words: mycelium, rhizoids, lichen, bryophytes, gymnosperms, angiosperms, monocots, dicots

Questions to think about:

📖 What important role in nature do fungi and bacteria share?

📖 What evolutionary relationship exists between green algae and plants?

31-1 Fungi

Section Objectives:

- *Describe* the general characteristics of fungi.
- *State* the distinguishing characteristics of each phylum, and name two or three of its members.
- *Compare* fungi and lichens.

🔬 **Laboratory Investigation:** *Observe* the structure of bread mold (p. 678).

The fungi, kingdom Fungi, include yeasts, molds, mushrooms, and rusts and smuts. **Fungi** are nongreen organisms that absorb the nutrients they need from the environment. Most fungi are saprobic. They obtain nutrients from the remains of dead plants and animals. To do this, they secrete digestive enzymes onto the food and absorb the digested nutrients. The fungi, along with the bacteria, play a key role in breaking down dead organisms and releasing the substances these organisms contain. Some fungi are parasitic. They obtain nutrients from the organisms on which they live.

General Characteristics of Fungi

Fungi vary in size. Some are microscopic; others weigh several kilograms. The bodies of most fungi consist of threadlike filaments called **hyphae** (HY fee) (singular, hypha). As the hyphae grow, they branch, forming a tangled mass called a **mycelium** (my

▲ **Figure 31–1**

A Fungus. This red mushroom is a member of the kingdom Fungi. There are about 100 000 named species of fungi.

SEE lee um) (plural, mycelia). In some fungi, the cytoplasm in the hyphae is not divided by cell walls. The continuous cytoplasm contains several nuclei. In other fungi, the hyphae are divided by incomplete *septa,* or cross walls. In these fungi, the hyphae are partly divided into compartments, but the cytoplasm is still continuous. Each compartment may contain more than one nucleus. Unlike the cell walls of green plants, the cell walls of most fungi are composed of chitin, not cellulose.

Fungi reproduce both asexually and sexually by means of spores. There are over 80 000 species of fungi. Most of these are grouped into three phyla, based more or less on their pattern of sexual reproduction. These are the conjugation fungi, the sac fungi, and the club fungi. A fourth phylum, called the *imperfect fungi,* includes thousands of fungi that cannot be classified because their pattern of sexual reproduction is unknown.

The Conjugation Fungi

The *conjugation fungi,* phylum **Zygomycota** (zy goh my KOH tuh), produce a special type of thick-walled spore that develops from a zygote during sexual reproduction. These fungi produce another type of spore asexually. Most conjugation fungi are saprobes, but some are parasites on plants, insects, or other fungi. The hyphae of the conjugation fungi lack cross walls, but cross walls do form during the production of gametes and spores. The common bread mold, *Rhizopus,* is a typical member of this group.

Rhizopus grows on the surface of bread and fruit as a cottonlike mass of filaments. See Figure 31–2. The whitish or grayish mycelium consists of several kinds of hyphae. Rootlike hyphae, called **rhizoids,** anchor the fungus, secrete digestive enzymes, and absorb nutrients. Other hyphae, called **stolons,** grow in a network over the surface of the food. The stolons give rise to still another type of

Figure 31–2
Life Cycle of Bread Mold. The bread mold, *Rhizopus,* reproduces sexually by conjugation of different strains of hyphae. *Rhizopus* also reproduces asexually by releasing spores from sporangia. ▼

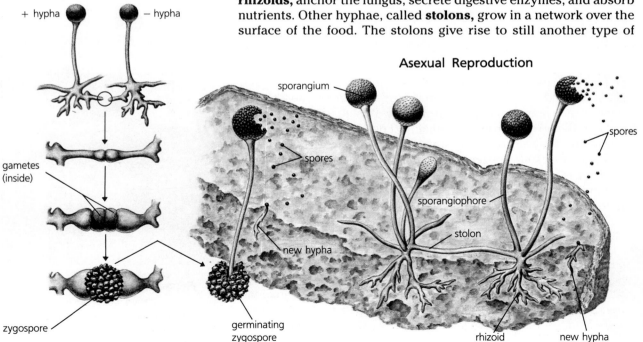

Sexual Reproduction

+ hypha   − hypha

gametes (inside)

zygospore

germinating zygospore

Asexual Reproduction

sporangium

spores

spores

sporangiophore

stolon

new hypha

rhizoid   new hypha

hyphae that grow upward from the surface of the food. These are reproductive hyphae, called *sporangiophores.* At the tip of each sporangiophore, a round spore case, or sporangium, develops. The black dots you see in common bread mold are sporangia. Many spores form within each sporangium. At maturity, when the cases open, the spores are released. Those that land in favorable environments germinate, form new hyphae, and eventually form new mycelia.

Usually, bread mold reproduces asexually by spore formation. Under certain conditions, however, bread mold reproduces sexually by conjugation. The pattern of conjugation is similar to the pattern found in spirogyra, an alga.

Conjugation occurs when hyphae of different strains touch. Once contact has occurred, the tips of both hyphae enlarge, and cross walls form behind the tips. These partitioned-off ends are now gamete-producing structures. The two strains of bread mold are called *plus* and *minus.* One tip contains several plus nuclei, while the other contains several minus nuclei. At the point of contact, the end walls of the two touching hyphae break up, and the nuclei of opposite strains fuse to form a number of diploid nuclei. A hard wall develops around the nuclei and cytoplasm, producing a thick-walled zygospore that can resist harsh environmental conditions. When conditions are favorable, the zygospore germinates. All but one diploid nucleus degenerates. The remaining nucleus undergoes meiosis. Meiosis produces four nuclei, three of which degenerate. The remaining haploid nucleus gives rise to a sporangiophore, which then produces spores asexually.

The Sac Fungi

The *sac fungi,* phylum **Ascomycota** (as koh my KOH tuh), are the largest group of fungi. They include cup fungi, powdery mildews, morels, truffles, blue and green molds, and yeasts.

Sac fungi produce two kinds of spores, each of which can give rise to new organisms under the proper conditions. Spores produced as a result of sexual reproduction are called *ascospores.* Usually eight, but occasionally four, ascospores develop inside a saclike **ascus** (plural, asci), which serves as a sporangium. Spores produced asexually are called *conidia.* Conidia are formed in chains at the tips of specialized reproductive hyphae.

Except for the yeasts, which are unicellular, the sac fungi are multicellular. Cross walls divide hyphae of multicellular sac fungi. Holes in the cross walls permit cytoplasm and nuclei to move from one compartment of the hypha to the next. Each compartment has one to several nuclei.

Most cup fungi are saprobic and grow on dead organic matter. Some cup fungi are shown in Figure 31–3. The visible portion of the fungus is the cup-shaped *fruiting body,* which contains the spore-bearing sacs, or asci. Beneath the surface of the soil is a large mycelium made up of many hyphae.

The unicellular yeasts are not typical of the Ascomycota. Yeasts reproduce by budding and by spore formation (see Chapter

Science, Technology and Society

Technology: Bioremediation

Cleaning up oil spills and toxic chemical dumps is difficult and expensive. Now, scientists are getting help from nature in the form of naturally occurring oil and chemical-eating bacteria and fungi. This process is called bioremediation.

Researchers look for the bacteria and fungi in toxic dumps and oil spills. The organisms either consume the harmful substances or secrete enzymes that break them down. The organisms leave behind carbon dioxide and water or other harmless products.

To determine which organism will break down a particular chemical, scientists collect organisms from polluted sites. In the lab, they feed the organisms different toxic chemicals. Those that live are kept for future use. Lab tests also reveal how to speed up the cleanup, usually by adding nutrients and air to the toxic site.

Bioremediation is cheaper than other methods. Since it does not produce toxic by-products, experts say it is less harmful.

■ *Suppose you were an executive at a chemical company. Prepare a speech offering your views on using company funds for research into bioremediation.*

▲ **Figure 31–3**

Sac Fungi. Cup fungi, a type of sac fungi, grow on decaying matter. Each cup-shaped fruiting body is lined with spore-bearing sacs, or asci.

Figure 31–4

Hunting for Truffles. Truffles, a sac fungus highly prized by cooks, must be rooted out of the ground by pigs or dogs trained specifically for this purpose. ▼

20). In spore formation, the yeast cell itself acts as an ascus. Yeasts are economically important because they are used in the manufacture of bread, alcohol, and alcoholic beverages.

Some Ascomycota cause plant disease, including Dutch elm disease, chestnut blight, and ergot. Ergot is a disease of wheat and rye, which is caused by a parasitic species of Ascomycota. Ergot poisoning results from eating flour made from infected plants. Modern methods of flour production have eliminated this problem.

Truffles and morels are edible ascomycetes that are considered great delicacies. Truffles grow several centimeters below the surface of the soil. They are spherical, brown fruiting bodies that range from about one to seven centimeters in diameter. In France, where truffles have been used in cooking for hundreds of years, pigs and dogs are trained to locate them by their odor. Morels, which are also known as sponge or honeycomb fungi, are common in many areas of the United States. The stalk and distinctive cap are the fruiting body. The asci are located within the folds of the cap.

The Club Fungi

The *club fungi,* phylum **Basidiomycota** (buh sid ee uh my KOH tuh), include most of the large fungi seen in fields and woods. Mushrooms, toadstools, bracket fungi, puffballs, and various parasites, such as rusts and smuts, are club fungi.

In the club fungi, sexual reproduction involves the production of spores called *basidiospores.* These spores are formed in an enlarged, club-shaped reproductive structure, the *basidium,* at the end of a specialized hypha. Some club fungi also produce spores asexually. Incomplete cross walls divide the hyphae of the Basidiomycota. The cells of the hyphae may have one or two nuclei.

The most familiar of the club fungi are the mushrooms. The mushroom is actually a fruiting body, the spore-producing part of the fungus. The main part of the mycelium grows beneath the surface of the ground. This part lives on the remains of plant and animal matter. The mycelium may live for years, slowly growing underground. Only when growing conditions are favorable do mushrooms grow up above the surface.

As shown in Figure 31–5, a mushroom consists of a stalk and a *cap.* Mushrooms begin to develop as small knobs on the underground mycelium. As the cap pushes up through the soil, it is kept closed and protected by a thin membrane that connects the edges of the cap to the stalk. Once the mushroom is above ground, the membrane breaks, and the cap expands. The part of the membrane that remains attached to the stalk is called the *annulus.*

The undersurface of the cap contains many *gills* that radiate out from the center of the stalk like wheel spokes. Each gill is made of many hyphae that are pressed closely together. On the sides of the gills are the basidia, each bearing four basidiospores. One mushroom can produce over 1 billion spores.

Many types of mushrooms are edible, but others are extremely poisonous. It takes an expert to distinguish between edible and

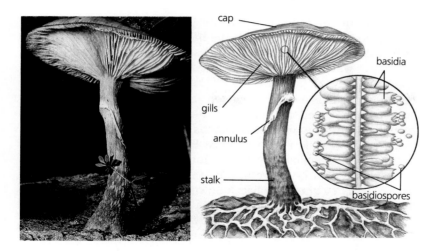

poisonous mushrooms. Never eat mushrooms that you find growing in the wild.

Rusts are club fungi that produce rust-colored spores during one phase of their life cycle. Rusts are parasites on wheat, barley, oats, and other crop plants. Each year they cause millions of dollars worth of damage to these crops. Smuts are similar to rusts. Their name refers to the black and dusty-looking mass of spores that they form within the tissues of the host plant. Smuts attack corn, wheat, oats, barley, and rye.

The Imperfect Fungi

The *imperfect fungi,* phylum **Deuteromycota** (do tur oh my KOH tah), include all fungi that are not known to have a sexual reproductive phase. This may be because the sexual phase does not occur or because it simply has not yet been observed. Because there are no known sexual stages, these fungi are difficult to classify. Therefore, they have been grouped together as "imperfect." If a fungus classified as imperfect is found to have sexual structures, it is reassigned to the appropriate phylum.

Many species of imperfect fungi are beneficial to humans. For instance, the fungus *Penicillium* produces the antibiotic penicillin. Penicillin cures a variety of bacterial diseases, including pneumonia, scarlet fever, and rheumatic fever. Other species of *Penicillium* give flavor to Roquefort and Camembert cheeses. A few imperfect fungi are harmful to humans. Some, for example, cause ringworm and athlete's foot. Others release spores that cause respiratory infections or allergic responses when inhaled.

Lichens

A **lichen** (LY ken) is made up of two organisms—an alga or a blue-green bacterium and a fungus—living together. The algal or bacterial cells are embedded in the mycelium of the fungus. The fungus is usually a sac fungus. Through photosynthesis, the alga or

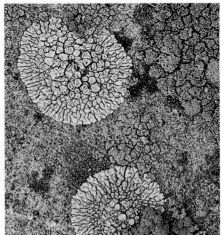

▲ **Figure 31–6**

Lichens. A lichen is made up of both a fungus and an alga. The lichens shown above are crustlike.

the bacterium provides nutrients for the fungus. At the same time, the fungus provides the alga or the bacterium with water, essential elements, and protection from intense light and dryness. Lichens reproduce when fragments break off, blow away, and start growing independently.

Some lichens are crustlike and resemble spots of paint. Some are flat but curled at the edges like leaves. Finally, some are shrublike and have branches. Lichens grow on the bark of trees, on rocks, and on soil. They can exist for months without water. In Arctic regions, lichens serve as food for caribou, musk ox, and other animals. Lichens are usually the first organisms to grow on bare rock. They gradually break down the rock, thus beginning the process of soil formation.

31-1 Section Review

1. What form of nutrition do all fungi have in common?
2. List the four major phyla in the kingdom Fungi.
3. What two organisms make up lichens?

Critical Thinking

4. Suppose that mushrooms appeared repeatedly in only one small part of your yard. What does this indicate about the soil in that area? (*Identifying Causes*)

31-2 Plants

Section Objectives:

- *Discuss* the evolutionary relationship thought to exist between green algae and plants.
- *Name* three kinds of bryophytes.
- *Distinguish* between nonvascular and vascular plants.
- *Describe* the various types of spore-dispersing and seed plants.

The kingdom Plantae is made up of organisms that most people refer to as plants. These include mosses, ferns, conifers, and flowering plants. Plants are multicellular, photosynthetic organisms that are adapted primarily for life on land. Many scientists believe plants have evolved from earlier, algal-like ancestors. Although all plants grow and reproduce on land, they have adapted to many different land environments. Some plants, like the mosses and their relatives, grow and reproduce only in wet or moist environments. Other plants are adapted for living in deserts, the driest of environments. In the following sections, you will read about the structural adaptations that allow plants to live in different environments.

General Characteristics of Plants

The plants, with their complex structural organization, are thought to have evolved from simpler algal-like ancestors. The green algae, which may be unicellular, colonial, or multicellular, are the most similar to plants. For the most part, green algae and plants have the same kinds of chlorophyll, the same food-storage polysaccharide (starch), and the same polysaccharides (cellulose) in their cell walls. These similarities suggest that plants and green algae share a common evolutionary ancestry.

Plants are divided into two groups—the *bryophytes* (BRY uh fyts) and the *tracheophytes* (TRAK ee uh fyts). The bryophytes include mosses, liverworts, and hornworts. These plants do not have conducting, or vascular, tissues. They are short plants and usually grow in areas that have a good supply of water. The tracheophytes include horsetails, ferns, gymnosperms, and flowering plants. Members of this group have well-developed vascular tissues for transport.

Life on land presents a number of problems that do not exist for organisms that live in water. The most immediate problems are obtaining and conserving water. All plants must be able to obtain water, transport it to all their cells, and control its evaporation from their tissues. Plants also need supporting tissues to stand upright against the force of gravity. Many of them need special reproductive mechanisms that enable the sperm to reach the egg without swimming through water.

Nonvascular Plants—The Bryophytes

The division **Bryophyta** (bry AH fuh tuh) includes *mosses, liverworts,* and *hornworts.* These are nonvascular land plants. That is, they have no specialized conducting tissues. The transport of materials through the plant takes place by diffusion, which is slow and inefficient. For this reason, the bryophytes must live where water is plentiful. They are found on forest floors, on damp rocks, in swamps and bogs, and near streams. Without xylem, bryophytes have little in the way of supporting tissues. Most are short, ranging from one to five centimeters in height.

In many bryophytes, some branch filaments of the young plant grow downward and enter the soil, where they function as roots. These rhizoids anchor the plant and absorb minerals and water. Other branches grow upward, forming stemlike shoots and leaves. Since, however, the cells in the rhizoids, shoots, and leaves are all similar, these structures are not true organs.

In the bryophyte life cycle, the haploid gametophyte is the dominant generation. The diploid sporophyte generation is small, short-lived, and dependent on the gametophyte for its nutrition. The life cycle of mosses is shown in Figure 24–4. In mosses and other bryophytes, the sperm must swim to the egg through water. Thus, these plants still display their ancestral origins.

Figure 31–7
Bryophytes. The mosses are the largest class of bryophytes and are found almost everywhere on earth. Like all bryophytes, mosses have no specialized conducting tissues and must live where water is abundant. ▼

General Characteristics of Vascular Plants

The *vascular plants,* division **Tracheophyta** (tray kee AH fuh tuh), are a diverse group that includes most modern-day plants. Whisk ferns, club mosses, and horsetails, as well as the ferns, conifers, and flowering plants, are vascular plants. In the sporophyte generation, all tracheophytes contain the vascular tissues xylem and phloem. In the tracheophytes, the sporophyte generation is dominant and the gametophyte is small and short-lived.

The vascular plants are divided into two groups—the spore-dispersing plants and the seed plants.

Vascular Spore-Dispersing Plants

The spore-dispersing vascular plants include the whisk ferns, club mosses, horsetails, and ferns. Fertilization in these ancient plants requires water.

The Whisk Ferns   The **whisk ferns** are the oldest known vascular plants. Fossil evidence indicates that this group was widespread about 400 million years ago, but there are only a few modern living species. These species, which live only in warm climates, are found from South Carolina to Florida. They are not really ferns. They do not have either true leaves or true roots. The plant body of the sporophyte consists of an underground stem anchored by rhizoids, which absorb water and minerals. Above ground, the stems are green and carry on photosynthesis. As the stems grow, they split into two branches, so that the ends of the stems are Y-shaped. Sporangia form at the tips of some branches. Within the sporangia, haploid spores are produced by meiosis. When the spores are released, some germinate, giving rise to small gametophytes that bear both male and female reproductive organs. These are the antheridia and archegonia, which are explained in Chapter 24, section 24–1. After fertilization occurs, the zygote develops into a new sporophyte.

The Club Mosses   The **club mosses** were one of the dominant forms of plant life during the Carboniferous, or coal-forming, period of the earth about 300 million years ago. There are only a few remaining small genera of club mosses. Some of the prehistoric forms were as large as trees and made up entire forests. Living club mosses are mostly small, reaching about 20 centimeters in height. A few tropical species may reach heights of 90 centimeters and look like bushes.

Club moss sporophytes have true roots, stems, and leaves. Like the whisk ferns, the stems branch so that the ends are Y-shaped. At the tips of some branches, groups of spore-producing structures form. Spores released from them give rise to gametophytes, which bear the reproductive organs. Fertilization results in a zygote that develops into the sporophyte.

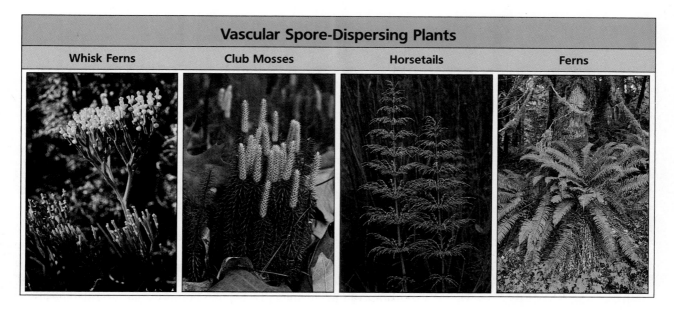

Vascular Spore-Dispersing Plants

Whisk Ferns	Club Mosses	Horsetails	Ferns

Some club mosses are evergreens and are used for holiday decorations. Ground pine and ground cedar are club mosses found in forests in the United States. One species from Mexico is known as the "resurrection plant." When dry, this club moss curls into a gray ball. When water is added, however, the moss opens, forming an attractive green plant.

The Horsetails   The **horsetails** include only about 20 living species of plants. They represent the remains of a group of plants that also flourished during the Carboniferous period. Some ancient horsetails were the size of trees. Over thousands of years, the remains of these and other plants of the Carboniferous period were eventually transformed into coal. Modern horsetails are usually less than one meter in height.

Horsetails are common in shaded woods and around streams, swamps, and ponds. The sporophytes have true roots and leaves. The stems are green and hollow. The leaves grow only at specific points along the stem, forming "collars" of leaves. Cone-shaped, spore-producing structures form at the ends of some stems. Haploid spores, produced by meiosis, are released and give rise to small gametophytes, which bear the reproductive organs. Following fertilization, the zygote gives rise to the sporophyte.

The stems of horsetails contain crystals of silicon. Early settlers often used horsetails for scouring pots and pans, which is why they are also known as *scouring rushes*.

The Ferns   Like the club mosses and horsetails, the **ferns** were most abundant during the Carboniferous period. There are now about 9000 living species of ferns. They are particularly abundant in tropical rain forests, but they are also found in cooler climates. Some tropical tree ferns have a woody, unbranched trunk and may reach heights of more than 15 meters. These ferns may have leaves

▲ **Figure 31–8**
Vascular Spore-Dispersing Plants. There are four groups of vascular spore-dispersing plants alive today. These plants require water for fertilization. The sporophyte generation is always dominant in the life cycle of these plants.

▲ **Figure 31–9**

Ferns. When the fronds of ferns first emerge from the soil, they are curled into the form of a fiddlehead. Some people consider fiddleheads a delicacy in the spring.

4 meters long. The ferns of cooler climates are much smaller. They have horizontal stems, called *rhizomes,* that grow just beneath the surface of the soil. Hairlike roots grow from the rhizomes deeper into the soil. The only visible parts of these ferns are the leaves, or fronds, that grow up from the rhizome. When the leaves first emerge from the soil, they are coiled in a bud called a *fiddlehead.* The fiddlehead gradually uncoils and develops into a mature frond. The fronds usually are divided into tiny leaflets that give them a feathery appearance.

The internal structure of ferns is similar to that of seed plants. Ferns contain xylem and phloem. Their roots have a root cap and show the growth zones found in the roots of higher plants. Because fern stems have no cambium, they show little or no growth in diameter.

Ferns reproduce both sexually and asexually. Sexual reproduction in ferns is discussed in Chapter 24. In asexual reproduction, the rhizome grows through the soil. As it grows, it branches and produces new fronds at the tip of each branch. Over time, older portions of the rhizome die, leaving behind separate rhizomes that continue to grow and to produce fronds.

Vascular Seed Plants

The seed plants have become the dominant and most successful group of plants. There are more than 250 000 species, that range in size from the giant redwood tree to the tiny duckweed, a water plant with leaves only a few millimeters wide. As you can see in Figure

Critical Thinking in Biology

Classifying

The books in a library, the clothes in a dresser, and the albums in a record store are usually organized by some system that makes it easy to find specific items. When you classify items, you arrange them into groups according to some system or organizing principle.

1. Look at the seeds pictured on the right. What characteristic could you use to divide the seeds into two groups?

2. Make a chart showing how the seeds would be grouped. Give each group a title.

3. Choose one of your groups. What characteristic could you use to divide the group into smaller groups? Show how the seeds would be grouped.

4. Aside from the characteristics you used, what other characteristics could be used to classify seeds?

5. Think About It Explain your thought process as you classified the seeds in questions 1 and 2.

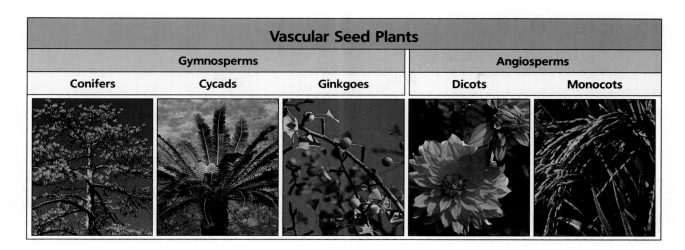

Vascular Seed Plants				
Gymnosperms			Angiosperms	
Conifers	Cycads	Ginkgoes	Dicots	Monocots

31–10, there are two major groups of seed plants—the gymnosperms and the angiosperms. The **gymnosperms** are a diverse group in which the seeds are exposed. That is, they are not contained within a specialized organ. The **angiosperms** are the flowering plants, and their seeds are enclosed within fruits. In both groups, the seed is surrounded by a protective seed coat and contains stored food that nourishes the young seedling until it can function independently. In these plants, water is not needed for fertilization.

The Gymnosperms   The gymnosperms are nonflowering seed plants that usually bear their seeds on the upper surface of scales that form a cone-shaped structure. The gymnosperms have true roots, stems, and leaves. The stems contain cambium, which causes the stem diameter to grow. The gymnosperm plant is the sporophyte generation. The life cycle of gymnosperms is discussed in Chapter 24.

Fossil evidence shows the presence of gymnosperms as early as 350 million years ago. By about 250 million years ago, they were the dominant form of plant life. There are now about 700 living

▲   **Figure 31–10**
Vascular Seed Plants. Seed plants are the most successful of land plants. They include the gymnosperms and the angiosperms. In gymnosperms, the seeds are exposed, but in angiosperms, the seeds are enclosed within fruits.

◀   **Figure 31–11**
Gymnosperms. Gymnosperm cones, like these hemlock, pine, and spruce cones, vary in size and structure.

▲ **Figure 31–12**

Redwood Tree. Some redwood trees are more than 3000 years old.

species of gymnosperms. The conifers are the most important group of gymnosperms. Two other gymnosperm groups are the cycads and the ginkgoes.

The **conifers,** or evergreens, are the best known of the gymnosperms. Members of this group are cone-bearing plants with leaves in the form of needles. In most conifers, the leaves remain green throughout the year. The conifers include pine, spruce, fir, hemlock, redwood, sequoia, cedar, and cypress trees. These trees show wide geographic distribution. In colder regions, they are the dominant trees of the forest. At high altitudes, pine and spruce are most abundant.

Sequoias and redwoods include some of the oldest-living and largest trees in the world. Some are between 3000 and 4000 years old and are more than 90 meters tall. Pine, spruce, and fir trees are widely used as holiday trees, and they and other conifers are used for lumber.

The **cycads** (SY kuds) look like palm trees except that they have cones. Members of this group are slow-growing and may live to be more than 1000 years old. Several species may reach heights of 15 meters. In some, the ovules are the size of large eggs and the cones weigh as much as 45 kilograms. Cycads grow in tropical and semitropical regions. The only cycad found in the United States is *Zamia,* which is found in Florida.

Maidenhair trees, or **ginkgoes,** are the only living representatives of a once numerous group. This species has survived because the trees have been cultivated as ornamental and shade trees. Few survive in the wild. Ginkgo trees, which may reach heights of more than 30 meters, are very hardy. They can survive with limited water and in the presence of air pollution.

The Angiosperms   The angiosperms, the flowering plants, are the most successful of all living plants. This group includes about 250 000 species, many of which are used for food. Flowering plants are found in all types of climates and environments. Some live in the desert where there is almost no water, and others live completely underwater.

In the angiosperms, the flower serves a reproductive function. It contains the structures that produce spores by meiosis. The angiosperm plant is the sporophyte, while the gametophytes are reduced to only a few cells. Fertilization is followed by the development of a seed, which is enclosed in a fruit. The life cycle of angiosperms is discussed in Chapter 24.

The angiosperms are divided into two major groups—the *dicots* and the *monocots.* The seeds of dicots contain two seed leaves, or cotyledons, whereas the seeds of monocots have only one. The monocots include the grasses, palms, lilies, sedges, irises, orchids, and various aquatic plants. The dicots are much more numerous than monocots. Figure 31–13 lists some of the major families of dicots and monocots and representative members of each.

Major Families of Dicots and Monocots		
	Family	**Representative Species**
Dicots	**magnolia**	magnolia and tulip trees
	rose	rose, hawthorn, flowering quince, flowering almond, apple, pear, strawberry, blackberry, raspberry, apricot, cherry, peach, plum
	beech	beech, oak, chestnut trees
	parsley	parsley, carrot, celery, parsnip, dill, caraway, fennel, poison hemlock, anise
	mustard	mustard, cabbage, broccoli, kale, cauliflower, brussels sprout, turnip, horseradish, rutabaga
	heath	heath, heather, rhododendron, mountain laurel, blueberry, huckleberry, cranberry, wintergreen
	pea	pea, soybean, lima bean, peanut, clover, alfalfa, wisteria, sweet pea, black locust, rosewood
	composite	sunflower, dandelion, aster, dahlia, marigold, zinnia, lettuce, artichoke, endive
	nightshade	potato, tomato, tobacco, eggplant, red pepper, petunia
	mallow	hollyhock, okra, cotton
	mint	spearmint, peppermint, lavender, rosemary, thyme, sage
Monocots	**lily**	tiger lily, easter lily, lily of the valley, day lily, onion, leek, chive, garlic, asparagus, tulip, crocus
	grass	rice, wheat, corn, rye, barley, oats, sugar cane, bamboo, buffalo grass, Kentucky bluegrass
	palm	coconut palm, date palm

▲ **Figure 31–13**
Major Families of Dicots and Monocots.

31-2 Section Review

1. From what group of algae did plants probably arise?
2. Name three kinds of bryophytes.
3. Name two tissues that are common to all vascular plants.
4. What are four spore-dispersing tracheophytes?

Critical Thinking

5. What reasons do biologists have for believing that plants and green algae probably have evolved from the same common ancestors? (*Identifying Reasons*)

Laboratory Investigation

Looking at Bread Mold

Molds are familiar examples of fungi. In this investigation, you will examine one type of mold—common bread mold—to become familiar with its features and those of fungi in general.

Objectives

▪ *Identify* the various types of hyphae that make up the body, or *mycelium,* of bread mold.

▪ *Observe* the relationship between the hyphae of a fungus and its food source.

▪ *Observe* the method of asexual reproduction in bread mold.

▪ *Identify* the genus of a bread mold specimen.

Materials

bread mold
stereomicroscope
dropper
dissecting needles
slides
coverslip
tweezers
compound microscope

Procedure *

A. Examine your specimen of bread mold under a stereomicroscope. Note that the body of the mold (the mycelium) is made of cottonlike filaments, or hyphae. Specialized hyphae growing up from the surface of the bread produce dark-colored spores.

B. With dissecting needles, tease apart the filaments of the mycelium. Look for the hyphae, called *stolons,* that extend horizontally over the bread surface. Look, also, for *rhizoids,* highly branched hyphae that grow into the bread. Refer to Figure 31–2 in your text to help you identify these structures.

C. Examine the hyphae that grow upward. These are called *sporangiophores.* At the tip of each sporangiophore is a spore case, or *sporangium.* Within each sporangium, there are many dark-colored spores. These spores give bread mold its color. Refer to Figure 31–2 in your text.

*For detailed information on safety symbols, see Appendix B.

678

D. Place a drop of water on a slide. With tweezers, transfer a small piece of mycelium to the slide and add a coverslip.

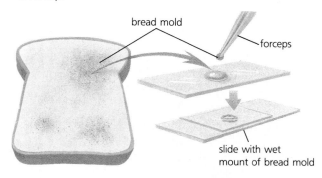

bread mold
forceps
slide with wet mount of bread mold

E. Using a compound microscope, examine the specimen under low power. Draw what you see. Label stolons, rhizoids, sporangiophores, sporangia, and spores.

F. Locate the tip of a stolon and center it in your field of view. Switch to high power. Focus carefully, varying the light intensity and adjusting the condenser, and look for cytoplasmic streaming.

G. Some species of bread mold belong to the genus *Rhizopus,* others to the genus *Mucor.* In species of *Rhizopus,* sporangiophores arise in clusters. In *Mucor* species, the sporangiophores arise singly. Sketch the sporangiophore arrangement in your specimen and name its genus.

Analysis and Conclusions

1. Based on your observations, what would you say is the function of each type of hypha?

2. How is the filamentous nature of the bread mold suited to its method of nutrition?

3. What is the function of the spores?

4. As you observed, bread mold produces its spores at the tips of hyphae that grow into the air. Why do you think forming spores in this location might be an advantage for the mold?

5. Based on your observations, why do you think bread mold is so common?

Chapter Review

Study Outline

31–1 Fungi

■ The kingdom Fungi includes yeasts, molds, mushrooms, rusts, and smuts. All are nongreen and absorb nutrients from the environment. Most are saprobic; some are parasitic.

■ Fungi are made up of a tangled mass of hyphae that form a mycelium. In most fungi, the cell walls are composed of chitin.

■ Fungi are classified according to their method of sexual reproduction. They are divided into four phyla— conjugation fungi, sac fungi, club fungi, and imperfect fungi. All but the imperfect fungi reproduce both sexually and asexually by means of spores.

■ In the conjugation fungi, sexual reproduction occurs by conjugation. In the sac fungi, spores develop within a cup-shaped fruiting body. In the club fungi, spores develop within a club-shaped structure.

■ The imperfect fungi include all fungi in which the pattern of sexual reproduction is unknown.

■ Lichens are made up of an alga or a blue-green bacterium and a fungus living together in symbiotic association. The alga or bacterium provides the fungus with nutrients, and the fungus prevents the alga or bacterium from drying out.

31–2 Plants

■ The kingdom Plantae includes mosses, ferns, conifers, and flowering plants. Plants are multicellular, photosynthetic organisms.

■ The simplest plants, the bryophytes, include the mosses, liverworts, and hornworts. Bryophytes are short plants and must live in a moist area because they do not have vascular tissue.

■ Tracheophytes contain the vascular tissues xylem and phloem. Tracheophytes are divided into two groups— the spore-dispersing plants and the seed plants.

■ The spore-dispersing plants include the whisk ferns, club mosses, horsetails, and ferns. Fertilization in these plants requires water.

■ The seed plants are the most successful group of plants. They include gymnosperms and angiosperms. Seed plants do not require water for fertilization.

■ Gymnosperms are nonflowering plants that bear their seeds exposed on the scales of a cone. The gymnosperms include conifers, cycads, and ginkgoes.

■ Angiosperms are flowering plants that have seeds enclosed within fruits. The angiosperms are divided into two major groups—the dicots and the monocots. Dicots have two cotyledons, whereas monocots have only one.

Vocabulary Review

fungi (665)	Bryophyta (671)
hyphae (665)	Tracheophyta (672)
mycelium (665)	whisk ferns (672)
Zygomycota (666)	club mosses (672)
rhizoids (666)	horsetails (673)
stolons (666)	ferns (673)
Ascomycota (667)	gymnosperms (675)
ascus (667)	angiosperms (675)
Basidiomycota (668)	conifers (676)
Deuteromycota (669)	cycads (676)
lichen (669)	ginkgoes (676)

A. Matching—Select the vocabulary term that best matches each definition.

1. The sac fungi

2. An organism composed of algae or blue-green bacteria and fungi living together in symbiotic association.

3. Seed plants that bear their seeds on scales of cones

4. Threadlike filaments that make up the bodies of most fungi

5. A tangled mass of hyphae

6. Filaments that anchor the plant and absorb minerals and water

Chapter Review

7. Plants that lack specialized conducting tissues

8. Plants with vascular tissue

B. Word Relationships—In each of the following sets of terms, three of the terms have a common characteristic. One term does not belong. Identify the term that does not belong.

9. Deuteromycota, Basidiomycota, club mosses, Ascomycota

10. rhizoids, ascus, Zygomycota, stolons

11. whisk ferns, horsetails, ferns, Zygomycota

12. conifers, ferns, cycads, ginkgoes

13. hyphae, mycelium, angiosperms, Basidiomycota

14. Basidiomycota, Tracheophyta, gymnosperms, angiosperms

Content Review

15. Describe the general structure of fungi.

16. Describe the hyphae that make up the bread mold *Rhizopus.*

17. Describe the conjugation process in the conjugation fungi.

18. In what ways are some Ascomycota harmful or beneficial to humans?

19. How are sac fungi different from club fungi?

20. Describe the structure of a mushroom.

21. How are imperfect fungi different from other types of fungi?

22. How are some species of imperfect fungi beneficial or harmful to humans?

23. What important role do lichens play in nature?

24. Why are green algae considered the evolutionary ancestors of plants?

25. What problems does life on land pose for plants?

26. What are the characteristics of bryophytes?

27. What are the characteristics of tracheophytes?

28. How does the structure of whisk ferns differ from that of horsetails and ferns?

29. Describe asexual reproduction in ferns.

30. In which groups of plants is water not necessary for fertilization?

31. What are the characteristics of conifers?

32. Compare the production of seeds in gymnosperms and angiosperms.

33. Name the two major groups into which the angiosperms are divided. Explain the significance of the names.

34. Classify the following angiosperms as either monocots or dicots: Easter lily, cotton, magnolia, eggplant, onion, corn, tomato, coconut palm, oak, Kentucky bluegrass, rose, peppermint.

Graphic Organizing

For information on graphic organizers, see Appendix G at the back of this text.

35. **Concept Map:** Construct a concept map that will help you describe and understand the tracheophytes. Put "The Tracheophytes" in a circle at the top center of your page. Include important concepts from section 31–2. As you relate the concepts, remember to include all the parts of the map, including the lines and linking words.

Critical Thinking

36. Explain how obtaining nutrients is similar and how it differs in bryophytes and tracheophytes. (*Comparing and Contrasting*)

37. What might happen to bryophytes if they were grown in a dry environment? Explain. (*Predicting*)

38. Suppose that, while walking in the forest, you find three different kinds of fungi. Using a hand lens, you observe the spore-producing bodies of each specimen. The first fungus has a spore case at the end of a single upward-growing hyphae. The second has spores developing inside a saclike structure. The third has spores developing in a club-shaped structure. To which phylum of the kingdom Fungi does each specimen belong? (*Classifying*)

39. Conifers, cycads, grasses, and roses belong to four different plant groups. What characteristics do all of these plants share? How are conifers similar to cycads? How are grasses similar to roses? What characteristics are unique to each of these plants? (*Comparing and Contrasting*)

Creative Thinking

40. The antibiotic penicillin is a natural secretion of a certain kind of fungus, a green mold called *Penicillium*. Penicillin kills bacteria. Why might a mold species have evolved a way of killing bacteria?

41. Many plants contain pigments in addition to chlorophyll. Think of as many ways as you can in which these other pigments might help a plant.

Problem Solving

42. Researchers conducted an investigation to determine the effect of a fertilizer on the growth of dicots (bean plants) and monocots (corn plants). Of the 100 seedlings used in the study, half were bean seedlings, the other half corn. All conditions for growth—such as light, moisture, temperature, and amount of fertilizer—were the same for both groups. The plants were measured each day and the data recorded in the table below. Graph the data. What conclusions can you draw about the effects of the fertilizer on monocots as compared to dicots?

Day	Plant Height (mm)	
	Dicots	**Monocots**
1	2	1
2	3	2
3	4	3
4	6	4
5	8	5
6	10	6

43. How would you improve the investigation involving dicot and monocot seedlings in question 42?

44. Your friend claims that mold grows more rapidly in the daytime as a function of light. Design an experiment to test the validity of that claim. Use bread as a medium for the mold.

45. A mushroom examined under a microscope had five spores in the high-power field. Further examination would reveal that there are about 250 high-power fields on one side of one gill of this mushroom, which has 100 gills in its cap. Approximately how many spores are there on this mushroom?

Projects

46. Design and set up a controlled experiment to determine where mold spores are found in your home. Use pieces of orange skin as a medium for the mold. Expose each piece of orange skin to floor dust from a different part of your home. Place each piece in a separate sandwich bag with several drops of water, and seal the bags completely. Label each bag according to the source of the dust. Put the bags in a dark, warm place for three days. Record your observations daily, and draw a conclusion based on the results. Design a poster for your class illustrating the results in relation to the different parts of your home. Include a floor plan on the poster.

47. Collect examples of local grasses, native flowers, or herbs. Take the entire plant if possible. Be sure not to uproot or disturb any endangered species. Dry the samples as directed by your teacher, and mount them on white paper. Add information about the plants. Display the collection in your classroom.

48. Prepare an oral report for presentation to your class on edible fresh and dried mushrooms sold in local markets. Include information on where and how the mushrooms are grown or collected. Display examples as part of your presentation.

49. Prepare a written report on one of the following career opportunities: (a) mycologist, (b) nursery worker, (c) plant taxonomist, (d) botanical illustrator. If possible, interview a person working in that career. Be sure to prepare your questions in advance. You may wish to inquire about training, future opportunities, and daily routine.

▲ The orange-spotted nudibranch, or sea slug, lives in the shallow waters of the world's oceans.

Invertebrates— Sponges to Mollusks

32

A brightly colored sea slug moves slowly among the riches of an ocean floor, searching out algae and other microorganisms. The sea slug, or nudibranch, is a marine animal without a backbone. It is a relative of the shy octopus, the giant squid, the elusive clam, and the shell-sealed oyster. Less exotic invertebrates include leeches, tapeworms, flukes, earthworms, and jellyfish. In this chapter, you will explore the world, life, structure, and organs of animals with no backbone, from the simple sponge to the complex cephalopod.

Guide for Reading

Key words: invertebrate, sponge, coelenterate, flatworm, roundworm, segmented worm, mollusk

Questions to think about:

📖 How are sponges different from jellyfish and corals?

📖 What are the differences among flatworms, roundworms, and segmented worms?

📖 What are the characteristics of mollusks?

32-1 The Animal Kingdom

Section Objectives:

- *Describe* the basic characteristics of animals.
- *Explain* the difference between vertebrates and invertebrates.
- *Distinguish* between radial symmetry and bilateral symmetry.

Basic Characteristics of Animals

The animal kingdom, kingdom **Animalia,** is the largest of the five kingdoms. Animals are multicellular organisms that must obtain food from their environment. Most have nervous and muscular systems that allow them to move. Most animals reproduce sexually. Some of the simpler forms also reproduce asexually. In some animals, the young have the same basic features as the adult, but in others, the young are different from the adult. In these cases, the young forms are known as **larvae** (LAR vee). The larvae undergo a series of developmental changes that produces the adult form.

The branch of biology that deals with the study of animals is called **zoology** (zoh AHL uh jee). Scientists who study animals are called zoologists (zoh AHL uh jists). Zoologists divide the animal kingdom into about 30 phyla. The nine largest phyla contain the majority of species. It is these phyla that you will study. In this chapter and the next one, you will learn about those animals without backbones—the **invertebrates.** The following two chapters will deal with the **vertebrates,** animals with backbones.

▲ **Figure 32–1**

Invertebrates. These larvae of the sphinx moth are but one example of the many types of invertebrates that exist.

Symmetry

The bodies of most animals show *symmetry* (SIM uh tree). This means the body can be cut into two halves that have matching shapes. A few organisms, including amebas and most sponges, are *asymmetrical* (ay suh MEH trih kul). These organisms cannot be cut into two matching halves.

There are different kinds of symmetry. **Spherical symmetry** is found in a few protists. These organisms are in the shape of a sphere. As you can see in Figure 32–2, any cut passing through the center of the sphere divides the organism into matching halves.

In **radial** (RAY dee ul) **symmetry,** there is a central line, or axis, that runs the length of the animal from top to bottom or from front to rear. Any cross section at right angles to the central axis will show repeating structures arranged around the center like spokes in a wheel. See Figure 32–2. Cross sections at different levels are not alike, but any lengthwise cut down the center divides the animal into matching halves. The hydra shows radial symmetry. One end of the animal has a mouth and tentacles. The other end is closed and rounded. Any lengthwise cut down the center divides the hydra into matching halves, like the halves of a vase. Most animals showing radial symmetry either drift with the water currents or are *sessile,* living attached to a stationary object.

In **bilateral** (by LAT uh rul) **symmetry,** the organism varies from top to bottom and from front to back. See Figure 32–3. The human body shows bilateral symmetry. In this type of symmetry, there is only one way to divide the body into two symmetrical halves. Each half is a mirror image of the other. Bilaterally symmetrical animals have fixed right and left sides. There are special terms that describe positions other than right and left on bilaterally symmetrical animals. **Dorsal** (DOR sul) refers to the upper side or the back of the animal; **ventral** (VEN trul) is the lower, or belly, side of the animal. The front, or head, end of the animal is **anterior,** while the rear, or tail, end is **posterior.**

▲ **Figure 32–2**

Spherical and Radial Symmetry. A radiolarian skeleton (top) displays spherical symmetry. Any cut passing through its center divides it into equal parts. The hydra (bottom) displays radial symmetry. It can be divided into equal halves along many lengthwise planes.

Figure 32–3
Bilateral Symmetry. In bilateral symmetry, there is only one way to cut an organism into two equal parts. ▼

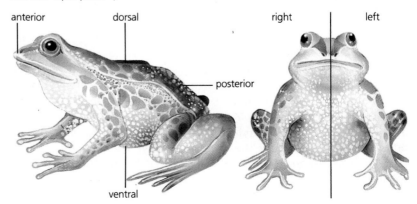

32-1 Section Review

1. How do larvae compare with adult forms?
2. List the three kinds of body symmetry.
3. Give four characteristics that most animals have.

Critical Thinking

4. What type of symmetry is exhibited by the following objects: a basketball, a football, a baseball bat, a fork, and a pair of eyeglasses? (*Classifying*)

32-2 Sponges and Coelenterates

Section Objectives:

- *Describe* the structure of sponges and coelenterates.
- *Explain* how the following life processes are carried out in sponges and coelenterates: nutrition, excretion, respiration, and reproduction.
- *Describe* the two body forms found among coelenterates and name a coelenterate showing each body form.
- *Describe* the life cycle of *Aurelia*.
- **Laboratory Investigation:** *Investigate* the feeding patterns of hydra (p. 702).

General Characteristics of Sponges

The *sponges,* phylum **Porifera** (puh RIF uh ruh), are the simplest multicellular animals. *Porifera* means "pore bearing." Sponges are pierced by many **pores,** or holes, through which water moves continuously. All sponges are aquatic. Most are marine, which means they live in salt water. A few live in fresh water. Although the larvae are free-swimming, adult sponges are *sessile.* They usually live attached to shells or rocks on the ocean floor. There are some sponges that are found in colonies. Some colonies look like plants with individuals branching from a common stem. Other sponges live singly.

Sponges vary widely in size and shape. Most sponges are asymmetrical. Some are the size of a pearl, while others may be the size of a bathtub. Simple sponges are shaped like hollow, upright cylinders or vases. More complex sponges have folds in the body walls. Still other types of sponges have complex systems of canals and chambers within the body walls. Many sponges are gray or black, but others are bright red, yellow, orange, or blue, as you can see in Figure 32-4.

In the past, some types of sponges were used for household cleaning and as bath sponges. Today, however, most commercial sponges are artificially made.

Figure 32–4

Sponges. Sponges are the simplest multicellular animals. There are about 10 000 species of sponges. ▼

Structure and Life Functions of Sponges

Sponges have a simple level of organization. Although their cells show specialization and are present in layers, they do not form true tissues. As Figure 32–5 illustrates, the sponge body is composed of three layers. The outer layer, which consists of thin, flat, epidermal cells, is pierced by numerous pores. These pores allow water, dissolved oxygen, and food particles (microscopic plants and animals) to enter the sponge. The inner layer contains specialized cells called **collar cells.** These cells have a collar of cytoplasm that extends from the cell into the central cavity. Extending through the collar of each cell is a flagellum.

Between the outer and inner cell layers is a middle layer of jellylike material that contains wandering *amebocytes,* amebalike cells. Embedded in the jellylike material of many sponges are small skeletal structures called **spicules** (SPIK yoolz), which are secreted by some of the amebocytes. Spicules provide support and give shape to the sponge. Sponges are classified according to the chemical makeup of their spicules. One group of sponges has spicules composed of calcium compounds; another group has spicules composed of silica. The third group has a network of tough, flexible fibers made of a protein-containing substance called **spongin.**

The pores of the sponge allow water to enter the body of the sponge. Water is drawn into the sponge and circulated in the central cavity by the beating of the flagella of the collar cells. From the central cavity, water passes out of the sponge through the **osculum** (AHS kyoo lum). The osculum is a large opening at the top, or unattached end, of the sponge.

As water passes through the sponge, food particles are captured, ingested, and digested by the collar cells. The amebocytes of the middle layer pick up partly digested food from the collar cells. The amebocytes finish digesting the food and then carry the nutrients to other parts of the sponge.

Wastes diffuse out of the cells into the central cavity of the sponge and leave with the water through the osculum. Gases (oxygen and carbon dioxide) are exchanged by diffusion between the cells and the water. Although sponges have no specialized nerve or muscle cells, some of the cells surrounding the pores respond to harmful substances in the water by closing the pores.

Sponges can reproduce sexually or asexually. In sexual reproduction, some collar cells change into gametes. Both male and female gametes are formed in the same sponge, but self-fertilization does not occur. Mature sperm leave the sponge through the osculum and are drawn into other sponges through their pores. The eggs are found in the jellylike middle layer. After fertilization, the zygote begins cleavage. Eventually, the embryo develops into a free-swimming larva. The larva passes through the inner cell layer and leaves the mother sponge through the osculum. After a time, the larva becomes attached to the ocean floor and develops into an adult sponge.

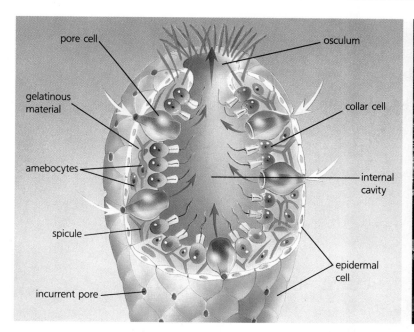

Asexual reproduction usually occurs by budding. Groups of cells on the parent sponge divide to form buds. The buds eventually break off and grow into new individuals. When conditions that are unfavorable for budding or sexual reproduction arise, some fresh-water sponges form asexual reproductive structures that are called **gemmules** (JEM yoolz). The gemmule is a group of cells that are enclosed by a tough outer covering. When environmental conditions become favorable, each gemmule develops into a new sponge. Sponges also have a remarkable capacity for regeneration. They can be cut up into many small pieces, and each piece will grow into a new sponge.

▲ **Figure 32–5**
Structure of a Sponge. Sponges filter food from water pumped through the pores of their body. Water passes out of the osculum, the opening at the unattached end of the sponge.

General Characteristics of Coelenterates

The *coelenterates* (suh LENT uh rayts), phylum **Coelenterata** (suh LENT uh rah tuh), show a more complex level of organization than the sponges. This phylum includes hydras, jellyfish, corals, and sea anemones (uh NEM uh neez). Coelenterates are aquatic. Hydras live in fresh water, but most other coelenterates are marine. There are two general body forms found among the coelenterates. These are shown in Figure 32–6. The **polyp** (PAHL ip) form is usually sessile and has a cylindrical body with a mouth and tentacles at the upper free end. Corals, sea anemones, and hydras are some examples of polyps. The other form, the **medusa** (muh DOO suh), is shaped like an upside-down bowl, with the mouth and tentacles facing downward. The medusa is usually free-swimming. Jellyfish show the medusa body form. Although the two body forms look different, they possess the same basic structure—a hollow sac with a single opening, the mouth, surrounded by tentacles. Most adult coelenterates show radial symmetry.

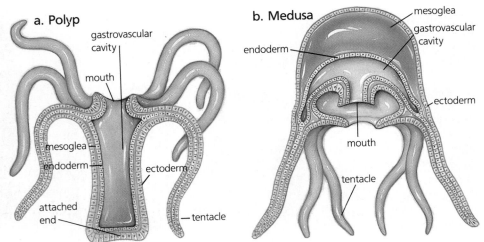

▲ **Figure 32–6**

Body Forms of Coelenterates. The sea nettle (left) is a common jellyfish along the Atlantic coast. Coelenterates have two body forms, the polyp form (middle) which is usually sessile, and the medusa form (right) which is free-swimming.

Structure and Life Functions of Coelenterates

The coelenterates show a tissue level of organization. There are two cell layers, the endoderm (inner layer) and ectoderm (outer layer). These are separated by a jellylike material, composed largely of protein, called the **mesoglea** (mez uh GLEE uh). The ectoderm cells contain contractile fibers. When these fibers contract, the animal moves. However, for the medusas, which are free-swimming, the strength of these contractions is not great enough to overcome the movement of the water. Thus, medusas drift with currents in the water.

Specialized stinging cells called **cnidoblasts** (NYED uh blasts) are characteristic of coelenterates. They are used for defense and for capturing food. Within the cnidoblasts are **nematocysts** (NEM uh tuh sists), which are small, fluid-filled capsules containing a coiled thread. When a cnidoblast on a tentacle is stimulated by pressure, the nematocyst is discharged. The thread uncoils and entangles the prey. Some nematocysts contain poison, which is injected into the prey and paralyzes it. Once the prey is captured, the tentacles of the coelenterate stuff it into the mouth. The structure and function of cnidoblasts in the hydra are discussed in Chapter 8.

The internal body cavity of coelenterates is called the **gastrovascular cavity.** The single opening serves as both a mouth and an anus. Extracellular digestion takes place in the cavity. Enzymes secreted into the cavity by some of the cells of the endoderm begin the process of extracellular digestion. When the food is partially digested, it is engulfed by the endoderm cells, where digestion is completed within food vacuoles. Thus, digestion is both extracellular and intracellular.

No respiratory or excretory system is found in coelenterates. Oxygen is obtained and wastes are excreted by diffusion. The first true nerve cells are found in the coelenterates. The nerve cells form

Corals. Corals are small polyps that grow in colonies. Unlike hydras, they are surrounded by a hard skeleton.

a *nerve net,* which sends impulses in all directions. These animals do not have brains, but the movement of the tentacles shows coordination.

Corals   Many of the structures and life functions of polyps are described in the sections on the hydra in Unit 2. *Corals* (KOR ulz) are small polyps that grow in colonies. See Figure 32–7. Corals are surrounded by a hard, calcium-containing skeleton, which they secrete. In warm, shallow parts of the ocean, islands and large coral reefs are formed by massive colonies of corals. These reefs are among the most productive areas in the world. Coral reefs provide shelter to a great number of different species of invertebrates and fish. In fact, some coral reefs in the Caribbean sea are home to more than 300 species of fish.

Aurelia   *Aurelia* (or EEL yuh) is a common jellyfish. Its life cycle includes both medusa and polyp forms, as shown in Figure 32–8. The jellylike body of the medusa is the form frequently seen on beaches. Protective tentacles hang from the edge of the umbrella-like body. The sexes are separate in *Aurelia,* but the male and female look alike. Sperm from the male medusa are released into the surrounding water. Some sperm cells enter the gastrovascular cavity of a female medusa, where fertilization occurs. Early development occurs while the zygote is attached to the female. The zygote develops into a small, oval-shaped, ciliated larva called a **planula** (PLAN yuh luh). The planula is free-swimming for some time. It then becomes attached by one end to a rock or some other structure on the ocean floor. At this point, the larva develops a mouth and tentacles at the unattached end and becomes a polyp. The polyp grows until it can reproduce asexually by budding to form medusas. During the fall and winter, a series of horizontal divisions make the polyp look like a stack of saucers. One by one, the saucer-shaped structures break off from the top and grow into full-sized medusas.

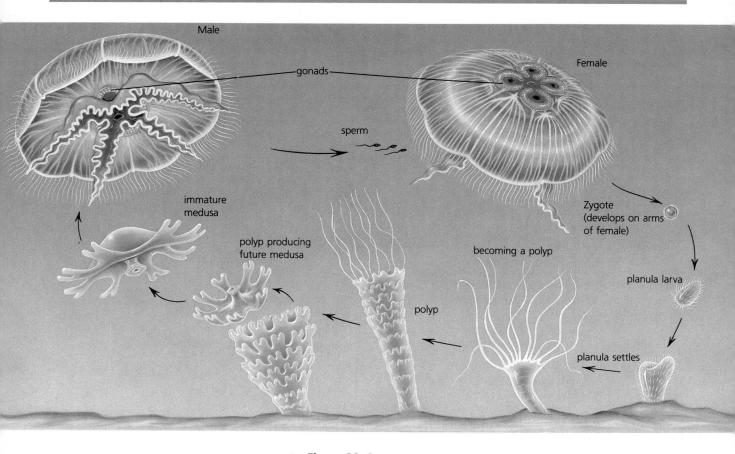

Male

gonads

Female

sperm

immature
medusa

polyp producing
future medusa

Zygote
(develops on arms
of female)

becoming a polyp

planula larva

polyp

planula settles

▲ **Figure 32–8**

Life Cycle of *Aurelia*. The life cycle of *Aurelia* shows an alternation of the polyp and
medusa body forms.

The alternation of the medusa form with the polyp form is
characteristic of some coelenterates. The medusa stage reproduces
sexually by the production of eggs and sperm. It gives rise to the
polyp stage. The polyp stage reproduces asexually by budding and
gives rise to the medusa stage.

32-2 Section Review

1. How does water enter the body of a sponge?
2. What structures provide support to the sponge?
3. Name four coelenterates.
4. Name the two body forms found in the coelenterates.

Critical Thinking

5. Do you think a sponge could survive without amebocytes? Explain.
   (*Relating Parts and Wholes*)

32-3 Flatworms and Roundworms

Section Objectives:

- *Name* the phylum to which flatworms belong and list a representative animal from each of the three classes of flatworms.
- *Describe* the structure and life cycle of planaria, blood flukes, and tapeworms.
- *Describe* the general characteristics and structure of roundworms.
- *Explain* the life cycles of the following roundworms and how they affect humans: trichina, filaria, pinworm, and hookworm.

General Characteristics of Flatworms

The *flatworms,* phylum **Platyhelminthes** (plat ee hel MIN theez), are the simplest animals showing bilateral symmetry. The flatworms are also the simplest invertebrate group showing definite head and tail regions. These animals are called flatworms because their bodies are flattened. There are three major groups of flatworms: free-living flatworms, such as planaria (pluh NEHR ee uh); parasitic flukes; and parasitic tapeworms. Many parasitic flukes and tapeworms have life cycles with more than one host. Usually, the first host is an invertebrate, while the final host is a vertebrate. Free-living flatworms are usually aquatic and are found in both fresh and salt water.

Structure and Life Functions of Flatworms

The body of the flatworm has three distinct tissue layers—ectoderm, mesoderm, and endoderm. These tissues are organized into organs and organ systems. Thus, the flatworms are also the simplest animals with mesodermic layers and organ-system levels of organization.

Planaria   *Planaria,* class **Turbellaria** (ter buh LEHR ee uh), are examples of typical flatworms. Planaria are found in freshwater streams and ponds, where they cling to the bottoms of leaves, rocks, and logs. These animals are gray, brown, or black in color and about 5 to 25 millimeters in length. As you can see in Figure 32–9, the triangular head contains a pair of *eyespots.* Although the eyespots cannot actually detect images, they are sensitive to light, which these animals avoid.

Planaria can move about freely. A piece of liver placed in a stream will be covered with them in a few hours. Moving planaria appear to be gliding over a surface because the underside of the body is covered with microscopic cilia that move the animal. Muscles enable them to change their shape or direction.

Figure 32–9
External Structure of a Planarian. The planarian is a common freshwater flatworm. ▼

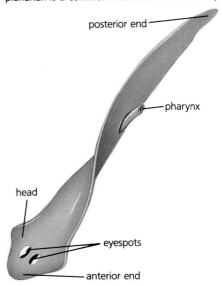

posterior end

pharynx

head

eyespots

anterior end

Figure 32–10

Internal Structure of a Planarian. Flame cells move excess water and liquid wastes into and along the planarian's excretory canal. The nervous system is ladderlike, with two main nerves connected by shorter nerves that run the length of the body. The digestive system consists of a mouth, pharynx, and a highly branched intestine. ▶

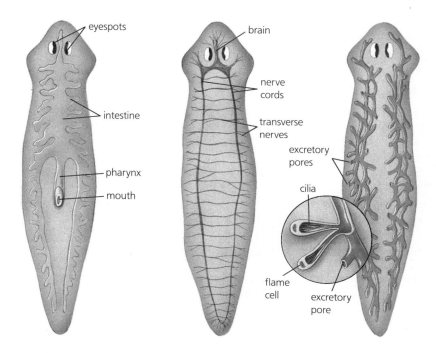

The planarian has a digestive system made up of a mouth, a pharynx, and a highly branched intestine, as shown in Figure 32–10. The muscular **pharynx** is a tube that can be extended through the mouth opening for eating. The mouth is located at the midline on the underside of the body. Planaria feed on living or dead small animals. The pharynx can suck small bits of food into the digestive cavity. Most digestion takes place within food vacuoles in the cells lining the intestine. Digested food diffuses to all cells of the body. Indigestible materials are expelled through the pharynx and mouth.

Planaria have no skeletal, circulatory, or respiratory system. Oxygen and carbon dioxide diffuse into and out of individual cells. A series of tubules that run the length of the body make up the excretory system of planaria. Side branches of the tubules have cells called *flame cells.* These cells remove excess water and liquid wastes from the body and pass them into ducts. The contents of the ducts pass out of the worm through small *excretory pores* on the dorsal surface.

The nervous system includes a small brain beneath the eyespots. From the brain, two nerve cords run the length of the body along either side. Connecting transverse nerves make the nervous system look like a ladder. This ladderlike nervous system enables the planarian to respond to stimuli in a coordinated manner.

Planaria have well-developed reproductive systems. They are *hermaphroditic,* which means that each individual has both ovaries and testes. However, self-fertilization does not occur. Instead, two planaria mate and exchange sperm. Fertilization is internal, and the fertilized eggs are shed in capsules. In a few weeks, the eggs hatch into tiny worms, which grow into adults. The planarian can

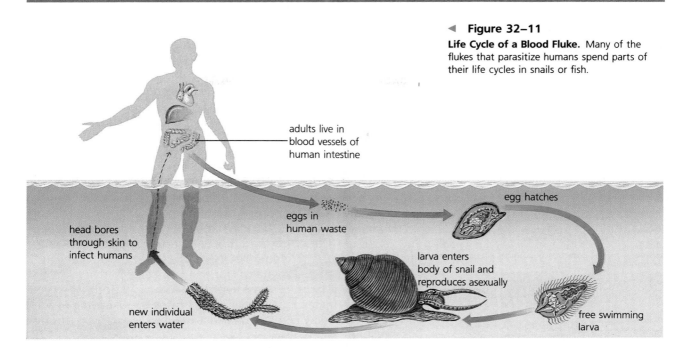

◄ **Figure 32–11**
Life Cycle of a Blood Fluke. Many of the flukes that parasitize humans spend parts of their life cycles in snails or fish.

adults live in blood vessels of human intestine

eggs in human waste

egg hatches

head bores through skin to infect humans

larva enters body of snail and reproduces asexually

free swimming larva

new individual enters water

regenerate an entire animal from a fairly small segment. It can also reproduce asexually by fission, separating its tail end from its head end. Each half regenerates the missing structures.

Flukes   *Flukes* are parasitic flatworms of the class **Trematoda** (trem uh TOHD uh). The body of the fluke is covered with a thick cuticle that protects the parasite from the enzymes of its host once it has entered the host's body. Suckers allow the fluke to attach itself to the tissues of its host. Because the food obtained from the host has already been broken down, flukes do not need well-developed digestive systems.

The blood fluke is a typical fluke. In humans, this parasite causes a disease called *schistosomiasis* (shis tuh soh MY uh sis). The adult fluke is about one centimeter long and lives in the blood vessels of the human intestine. See Figure 32–11. Here, it lays thousands of eggs that pass out of the body with digestive wastes. If the eggs land in water, they hatch into free-swimming larvae. They then enter the bodies of snails, where they reproduce asexually. The new individuals leave the snails and infect streams, rice paddies, and irrigation ditches. Upon contact with humans, the flukes bore through the skin and start their reproductive cycle again. The blood fluke causes loss of blood, diarrhea, and severe pain.

Tapeworms   *Tapeworms* are parasitic flatworms of the class **Cestoda** (ses TOHD uh). The beef tapeworm, which can infect humans, is a long, ribbonlike flatworm. Adults may be from four to nine meters in length. These worms have excretory and nervous systems and highly developed reproductive systems. They lack mouths and digestive systems. Tapeworms live as parasites in the

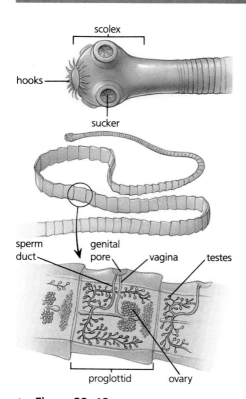

▲ Figure 32–12
Structure of a Tapeworm. The sucker and hooks on the tapeworm's scolex are adaptations for attachment.

intestine and absorb digested food through their skin. The suckers on the knoblike head, or *scolex* (SKOH leks), hold the tapeworm in place. Some tapeworms, such as the pork tapeworm, have hooks as well as suckers. See Figure 32–12.

Below the head and neck are square body segments called **proglottids** (proh GLAHT idz). These segments are produced continuously by budding from the neck region. The proglottids are reproductive structures that produce both sperm and eggs. Periodically, the end segments, filled with over 100 000 fertilized eggs, break off and pass out of the host in the feces. If cattle eat food contaminated with eggs, the eggs develop into larvae in the intestine. The larvae burrow into blood vessels and are carried to the muscle, where they form a dormant capsule.

Humans can become infected when they eat undercooked beef. The capsule surrounding the larva is digested, which releases the small tapeworm. The tapeworm then attaches itself to the wall of the human intestine. Proglottids form and are excreted in the feces. When cattle eat grain or grass infected with proglottids, the cycle begins again. Human tapeworms cause illness by absorbing needed nutrients. They may actually obstruct the passage of food through the intestine.

General Characteristics of Roundworms

The phylum **Nematoda** (nem uh TOHD uh) consists of slender, bilaterally symmetrical *nematodes* (NEM uh tohdz), or *roundworms*. Roundworms have elongated, cylindrical bodies that are tapered at both ends and covered with a tough cuticle. Roundworms range in length from less than one millimeter to more than one meter. Many roundworms are free-living, while others are parasitic. The free-living forms are found in fresh water, salt water, and in soil anywhere from the polar regions to the tropics. They feed on algae, plant sap, and decaying organic matter. Parasitic roundworms live on or in most kinds of plants and animals. The actual number of roundworms present in the environment is tremendous. It has been estimated that 1 million or more roundworms are present in one shovel load of garden soil.

Structure and Life Functions of Roundworms

Roundworms, unlike flatworms, have two openings to their tubular digestive system. Food is taken in through the mouth at the anterior end, and undigested material passes out through the *anus* at the posterior end. Roundworms are the simplest of the animals having complete digestive systems with two openings and a tube-within-a-tube body plan.

Roundworms have no circulatory or respiratory system. They do have a simple excretory system as well as a nervous system. Well-developed muscles located in the body wall allow them to move in a characteristic whiplike fashion.

Roundworms also have well-developed reproductive systems. The sexes are separate, and fertilization occurs within the body of the female. In free-living forms, the fertilized eggs, which are surrounded by a thick shell, are deposited in soil. The newly hatched young resemble the adults.

Most roundworms are parasites, and the diseases caused by them are widespread. Many of those diseases can be controlled by good personal hygiene, proper sanitation, and thorough cooking of food. Some drugs are also useful in controlling these parasites. Trichina, filaria, pinworm, and hookworm are parasitic round-worms that infect humans.

Trichina   The roundworm that causes *trichinosis* (trik uh NOH sis) in humans is called *trichina* (trih KY nuh). Adult trichina worms live in the intestines of hogs. When these worms reproduce, the resulting larvae invade the muscles of the hog. They grow to about one millimeter in length, then curl up, and become enclosed in hard cysts. When pork that has not been cooked well enough to kill the organisms is eaten by a human, digestive enzymes release the larvae from the cysts. The larvae develop into adults in the human intestines and reproduce sexually. The new larvae move through the blood vessels and muscles just as the larvae did in hogs. The movement of the worms through muscle causes intense pain and can cause permanent damage to the muscle. Trichinosis can be prevented easily by cooking pork thoroughly. Hogs become infect-ed when they are fed infected scraps of uncooked meat. Because of better sanitary procedures used today for raising hogs, trichinosis is no longer very common.

Filaria   *Filaria* (fuh LEHR ee uh) cause a disease known as *elephantiasis* (el uh fun TY uh sis). These roundworms are carried by a species of mosquito found in tropical and subtropical regions. Filaria worms are spread to humans by the bite of an infected mosquito. Once in the human body, they invade the lymphatic system. There, they block lymph vessels and cause fluid to accumu-late and tissues to swell. The area of the body in which they lodge often becomes abnormally large, damaging the tissues involved. Within the lymph tissues, the worms reproduce sexually and reproduce larvae that enter the bloodstream. A mosquito becomes infected when it bites an infected person. The larvae mature within the mosquito, and the cycle begins again with the bite of the infected mosquito.

Pinworms   One of the more common parasitic roundworms often found in children is the *pinworm*. Tiny adult pinworms live in the large intestine. The female worms deposit their eggs in the anal region. The presence of the eggs causes itching. When the child scratches, some eggs get on the fingers. Children reinfect them-selves when they put their unclean fingers in their mouths. Pinworms live only a few weeks. Thus, if reinfection can be prevented by cleanliness, the pinworms disappear within a short time.

Hookworm The *hookworm* is a roundworm that infects people in warm climates who walk barefoot on contaminated soil. The hookworm lives in the small intestine, and its eggs leave the body in the feces. When sewage disposal is inadequate, the eggs hatch into larvae on the ground. If people then come into contact with them, the larvae bore through the skin. Once in the body, they are carried to the lungs by the circulatory system. They bore through the lungs, are coughed up, swallowed, and pass again to the small intestine. There, they suck blood from the intestinal wall. Symptoms of hookworm infection include anemia and lack of energy.

32-3 Section Review

1. What type of symmetry do flatworms have?
2. List the three major groups of flatworms.
3. Where are roundworms found?
4. List some parasitic roundworms that infect humans.

Critical Thinking

5. List two similarities and two differences between roundworms and flatworms. (*Comparing and Contrasting*)

32-4 Segmented Worms and Mollusks

Section Objectives:

- *Describe* the general characteristics and structure of segmented worms and mollusks and give representative examples of classes of these phyla.
- *Compare* and *contrast* the structure of the marine worm *Nereis* with the structure of the earthworm.
- *Describe* respiration, nutrition, circulation, excretion, and reproduction in clams.
- *Explain* some of the ways in which gastropods and cephalopods differ from bivalves.

General Characteristics of Segmented Worms

The most familiar worms are those of the phylum **Annelida** (uh NEL uh duh), the *segmented worms*. This phylum includes the earthworm, class **Oligochaeta** (ahl ig oh KEET uh), and the leech, class **Hirudinea** (hir yuh DIN ee uh). The most striking characteristic of the segmented worms, which are also called the *annelids* (AN uh lidz), is the division of the body into separate sections, or segments. Segmented worms are found in both salt and fresh water and on land. Most of these worms are free-living, but a few of them are parasites. Segmented worms can range in length from less than one millimeter to more than two meters.

Structure and Life Functions of Segmented Worms

Segmented worms are bilaterally symmetrical. Their bodies are divided, externally and internally, into segments. These worms are the simplest of the invertebrates that have a closed circulatory system. Like more complex animals, segmented worms have a tube-within-a-tube body plan. The digestive tract, which is lined with endoderm, is the inner tube. It is open at both ends—mouth and anus. The body wall, which is covered with ectoderm, makes up the outer tube. A fluid-filled body cavity is found between the two tubes. This cavity is called a **coelom** (SEE lum) and is lined with mesoderm, as shown in Figure 32–13. The segmented worms are the simplest of the animals that have true coeloms.

Nereis   In most ways, the marine sandworm *Nereis* (NEHR ee is), class **Polychaeta** (pahl ee KEET uh), is similar to the earthworm, which was described in Unit 2. However, there are a few important differences between these two animals.

Nereis lives at tide level, in the intertidal zone. It emerges at night and crawls along the sand or swims in the shallow sea. During the day, it stays in a temporary burrow in mud or sand with its head poking out. Green in color, *Nereis* is composed of about 200 similar segments. The first two segments form a distinct head. The first segment, which is called the *prostomium* (proh STOH mee um), has two short tentacles, two pairs of small eyes, and two other appendages called *palps*. The second segment, which is called the *peristomium* (pehr uh STOH mee um), surrounds the mouth. It has four pairs of tentacles. The structures found on the first two segments serve for finding food and for protection. Except for the first, second, and last segments, a pair of paddlelike extensions called **parapodia** (par uh POHD ee uh) is found on each segment. Parapodia are used for swimming and for creeping over the sand. Parapodia also aid in respiration by providing a surface for gas exchange. Bristlelike *setae* are located on the parapodia. See Figure 32–14.

Nereis eats small animals, which it captures by extending its pharynx out through its mouth. The pharynx has a pair of hard, pointed jaws that grasp the food. As the jaws are pulled back into the mouth, the food is swallowed. The food passes into the esophagus and then to the intestine, where it is digested. Undigested food is eliminated through the anus on the last segment.

Circulation, excretion, and respiration in *Nereis* are basically the same as in the earthworm. The nervous system is also similar.

In *Nereis,* sexes are separate. During the mating season, eggs and sperm develop in the coelom. Eventually, they pass out through the excretory organs (nephridia) or break through the body surface into the sea. Fertilization is external, and the zygote

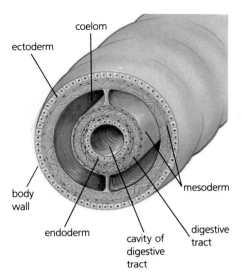

▲   **Figure 32–13**

The Coelom. The coelom is a fluid-filled cavity found between the inner and outer body tubes of annelids and more complex animals.

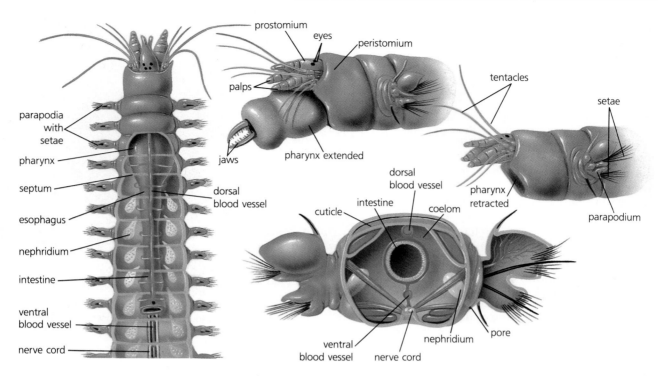

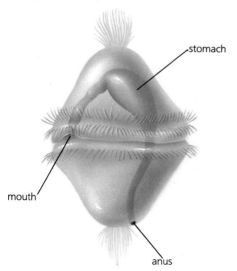

▲ **Figure 32–14**
Structure of the Marine Sandworm *Nereis*.

Figure 32–15

A Trochophore Larva. The life cycle of many mollusks includes the free-swimming, ciliated trocophore larva. ▼

develops into a free-swimming, ciliated **trochophore** (TROH kuh for) **larva,** as shown in Figure 32–15. As the larva develops, the mouth and segments with parapodia appear. Eventually, the young worm settles to the ocean bottom and begins the adult stage of life.

Leeches   *Leeches* are mostly freshwater animals that are parasites of vertebrates. Some are found in moist soil. Most live on the blood of their prey. Segmentation is not easily noticed in leeches. Leeches have suckers at both their anterior and posterior ends. In feeding, the leech attaches itself to its host with its hind sucker. It then attaches the anterior sucker, which contains the mouth and three small jaws. The jaws break through the host's skin. The saliva of the leech contains an enzyme that prevents the host's blood from clotting while it is being sucked up. The leech can ingest many times its own body weight of blood in one feeding. When the leech is full, it drops off the host and remains inactive for long periods. During these periods the blood, which has been stored in the digestive tract, is gradually digested. Leeches are hermaphrodites, but cross fertilization takes place when two leeches exchange sperm. The fertilized eggs develop in water or soil.

General Characteristics of Mollusks

The *mollusks,* phylum **Mollusca** (mahl US kuh) are a highly successful animal group. They are the second largest animal phylum, after the arthropods. Oysters, clams, snails, squids, and octopuses are familiar mollusks. Mollusks are found in salt water, in fresh water, and on land. Members of this group vary greatly in size and shape. See Figure 32–16. Mollusks range from tiny snails 1 millimeter long to giant squids, which can reach 16 meters in length and weigh 2 tons. The giant clam of the South Pacific Ocean can be 1.5 meters long and weigh 250 kilograms.

Many types of mollusks are used by humans for food. Among them are oysters, clams, scallops, mussels, snails, squids, and octopuses. Pearls from oysters are used in jewelry, and mollusk shell is used in buttons and decorative objects. On the other hand, some snails and slugs feed on crops and are highly destructive.

There are three major classes of mollusks: the *bivalves,* class **Bivalvia** (by VALV ee uh), include mollusks with two-part shells, such as clams, oysters, and mussels. The *gastropods* (GAS truh pahdz), class **Gastropoda** (ga STRAHP uh duh), include mollusks with a single shell, such as snails. And the *cephalopods* (SEF uh luh pahdz), class **Cephalopoda** (sef uh LAHP uh duh), include mollusks with little or no shell, such as squids and octopuses. Many marine mollusks have a trochophore larva similar to the trochophore larva of marine segmented worms. This is thought to indicate an evolutionary relationship between the two groups.

▲ **Figure 32–16**
A Mollusk. The nudibranch, or sea slug, has lost its shell during evolution.

Structure and Life Functions of Mollusks

Although adult mollusks vary widely in appearance, they share a number of common characteristics. They are bilaterally symmetrical, and they are composed of three tissue layers. They also have a coelom. All mollusks have a soft body that houses all the organ systems—the digestive system, heart, nervous system, reproductive system, and so on. The foot, mantle, shell, and radula are structures found only in mollusks.

The large, ventral muscular **foot** functions in movement. In clams, the foot is used to burrow or plow through wet sand or mud. The snail uses its foot to creep over rocks or plants. The foot of the squid and octopus is divided into tentacles and covered with suckers. The tentacles are used for seizing and holding prey.

The **mantle** is a fold of skin that surrounds the body organs. In the squid and octopus, the muscular mantle is used for movement. In mollusks with shells, the mantle is a glandular tissue that secretes part of the shell.

The **radula** (RAD joo luh) is a rasping, tonguelike organ found in all mollusks except bivalves. The radula has many rows of teeth and can extend out of the mouth to scrape food from an object and bring it into the digestive system. Some snails use the radula to drill holes in the shells of other mollusks. They then suck out the soft body of the mollusk for food.

The Bivalves   Bivalves, such as clams, scallops, oysters, and mussels, have a shell made up of two parts. The smooth, shiny, innermost layer of the shell, which is just outside the mantle, is called *mother-of-pearl.* See Figure 32–17. In some bivalves, pearls are produced when an irritant, such as sand, gets between the mantle and the shell. The mantle walls off the irritant by secreting mother-of-pearl around it. Eventually, a pearl is formed.

Figure 32–17
Oyster with Pearl. The finest natural pearls are produced by the pearl oysters, but pearls are formed in most bivalves, including freshwater clams. ▼

Figure 32–18
Structure of a Clam. ▶

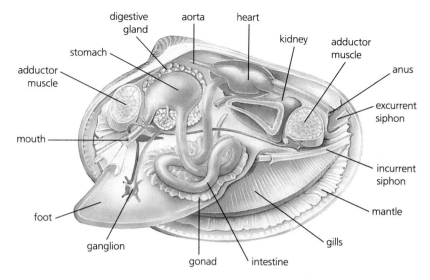

The two halves of the clam's shell can be held firmly closed by two strong *adductor* muscles. See Figure 32–18. When the muscles relax, an elastic hinge keeps the shell open. Usually, the shells are partly open with two tubes extending into the water. One tube, the *incurrent siphon* (SY fun), carries water containing food particles into the mantle cavity. The water is kept in motion by the beating of cilia on the gills. As the water moves over the gills, the exchange of respiratory gases occurs between the blood in the gills and the water. Food particles in the water are trapped by mucus on the gills. The water then flows out of the mantle cavity through the *excurrent siphon.*

Food particles stuck in the mucus on the gills are transported by the cilia into the mouth and then into the rest of the digestive system. Animals that feed by filtering water through their bodies are called *filter feeders.* They feed on organic particles and dead and decaying microscopic organisms in the water.

The clam has an open circulatory system that is made up of a heart and vessels. When the blood reaches the body tissues, it flows out of the vessels and into the body spaces, or *sinuses*, where it bathes the body tissues. From the sinuses, the blood flows into vessels that carry it to the gills. After the exchange of respiratory gases, the blood flows back to the heart.

The clam has a pair of kidneys that remove organic wastes from the blood and empty them into the water through the excurrent siphon. The nervous system consists of three pairs of ganglia connected by nerves to the foot and body organs. Sensory cells enable the clam to respond to chemical changes in the water, to touch, and to light.

In clams, the sexes are separate. Sperm leave the male through the excurrent siphon. They then enter the female through the incurrent siphon. The eggs are held on the gills, where they are fertilized. The young bivalves pass through one or more distinct larval stages before reaching the adult form.

Figure 32–19

A Gastropod. The garden snail, like most gastropods, has a coiled shell. Other gastropods may have cone-shaped shells, and some have no shells at all. ▼

The Gastropods   Snails, whelks, abalones (ab uh LOH neez), conches (KAHN chez), and slugs make up the largest group of mollusks, the gastropods. Most gastropods have a single shell, which is often coiled. A few, such as the slug, lack a shell. Some are aquatic; others are terrestrial.

The common garden snail has a head with tentacles, eyes, and a mouth, as shown in Figure 32–19. The head is connected to the foot. The shell is on top of the foot. For protection in times of danger, all the soft parts of the body can be drawn into the shell. Land snails have simple lungs rather than gills, which are found in aquatic snails. Air is drawn into the mantle cavity, and gas exchange occurs through the mantle.

Land snails usually travel at night when the air is moist. They slide along on a layer of mucus secreted by the foot. To keep from drying out during the day, the snail withdraws into its shell and seals the opening with mucus. The land snail feeds by rubbing its radula against plant material. As the pieces of plant are shredded, they are taken into the mouth.

The Cephalopods   The most advanced mollusks, the cephalopods, include squids, octopuses, and cuttlefish. All cephalopods are marine predators. See Figure 32–20. These animals do not look like other mollusks. The most obvious difference is that most cephalopods have either no shell (the octopus) or a small, internal shell (the squid and cuttlefish). Only a few, such as the nautilus, are enclosed in a shell.

In cephalopods, the mouth is surrounded by tentacles. The tentacles are used to gather food and to manipulate objects. The streamlined bodies of cephalopods permit rapid swimming. Cephalopods swim by expelling a jet of water from their mantle cavity. They have a well-developed nervous system with a large brain. In structure, the eye of the octopus is similar to the eyes of vertebrates, and it works in the same way. In times of danger, some cephalopods, such as squids and octopuses, discharge an inky fluid. This creates a "smoke screen," which distracts the enemy and allows the animal to escape.

▲ **Figure 32–20**

A Cephalopod. Octopuses live on the sea floor where they search for crabs and other food. These animals are believed to be one of the most intelligent of the invertebrates.

32-4 Section Review

1. What is the most striking characteristic that is found in the segmented worms?

2. Where are segmented worms found?

3. Name an organism from each of the three major classes of mollusks.

4. Where are mollusks found?

Critical Thinking

5. Why do taxonomists believe that segmented worms and mollusks are closely related? (*Identifying Reasons*)

Laboratory Investigation

Feeding Behavior of Hydra

The ability to detect and capture food is vital for most animals. Even in simple animals, such as hydra, these activities require sophisticated coordination. In this investigation, you will observe the feeding behavior of hydra.

Objectives

- *Observe* how hydra responds to touch stimuli.
- *Observe* the feeding behavior of hydra.
- *Observe* the digestion of food by hydra.

Materials

brine shrimp culture (colored red)
dropper
glutathione
2 culture dishes
stereomicroscope
toothpick
depression slide
coverslip
compound microscope
beaker
petroleum jelly

Procedure 🧪 🧤 🧽 *

A. Using a dropper, transfer a hydra to a small culture dish half filled with water.

B. While observing the hydra under the stereomicroscope, gently touch the tentacles with a toothpick. Then, gently touch the basal disc. Compare the responses.

C. Add a few brine shrimp to the culture dish. Watch the hydra's behavior, and note the movement of the hydra's tentacles and the way it ingests the shrimp.

D. To observe the ingested brine shrimp, make a hanging-drop preparation. With a toothpick, dab a little petroleum jelly on the corners of a coverslip's upper side. With a dropper, transfer the hydra to the middle of the coverslip. Hold a depression slide over the coverslip, depression side down, and lower it onto the coverslip, as shown below. Turn the slide over so the coverslip faces up.

*For detailed information on safety symbols, see Appendix B.

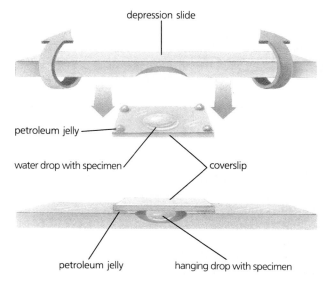

depression slide

petroleum jelly

water drop with specimen

coverslip

petroleum jelly

hanging drop with specimen

E. Using a compound microscope under low power, observe what happens to the ingested shrimp. Record your observations. Then, pour the hydra into the "fed hydra" beaker.

F. Transfer a second hydra to a culture dish filled with water. Using a stereomicroscope, observe this hydra for a few minutes. Add a drop of glutathione solution to the culture dish. Glutathione is a substance found in the body fluids of many animals. Observe, then record, the hydra's response. When you have finished, pour the hydra into the "fed hydra" beaker.

Analysis and Conclusions

1. How did the hydra's response to being touched on its tentacles and on its disc compare? Why might you expect the responses to be different?

2. How did the hydra react to the shrimp? Did it detect the shrimp at a distance or by physical contact?

3. What happens to the ingested shrimp?

4. How did the hydra's response to the glutathione compare to its response to the brine shrimp?

5. On the basis of what you observed, propose a hypothesis to explain how hydra detect prey. Design an experiment that would test this hypothesis.

Chapter Review

32

Study Outline

32–1 The Animal Kingdom

■ Members of the kingdom Animalia are multicellular organisms that must obtain food from their environment.

■ The animal kingdom is divided into two major groups. Vertebrates are animals with backbones; invertebrates are those without backbones.

■ Animals may exhibit spherical, radial, or bilateral body symmetry. A few organisms are asymmetrical.

32–2 Sponges and Coelenterates

■ The sponges, phylum Porifera, are the simplest multicellular animals. The sponge body consists of three layers. The outer layer is pierced by pores through which water containing food and oxygen flows.

■ The spicules, or skeletal structures, of the sponge are embedded in the middle layer. Inner layer cells are specialized for feeding and digestion.

■ Sponges can reproduce either sexually or asexually.

■ The coelenterates, phylum Coelenterata, include hydras, jellyfish, and corals. They are radially symmetrical and have cells that are organized into tissues.

■ The basic structure of coelenterates includes a sac with a central digestive area and a single opening. Coelenterates have two general body forms: the polyp and the medusa.

32–3 Flatworms and Roundworms

■ The flatworms, phylum Platyhelminthes, include planaria, flukes, and tapeworms. They are the simplest animals showing bilateral symmetry and definite head and tail regions.

■ Flatworms' tissues are organized into organs and systems. The digestive tract has a single opening.

■ Roundworms, phylum Nematoda, include trichina, filaria, pinworms, and hookworms. They have bilateral symmetry and are the simplest animals having a complete digestive system with two openings.

■ Although roundworms lack circulatory and respiratory systems, they do have simple excretory and nervous systems and well-developed reproductive systems. Roundworms are generally parasitic.

32–4 Segmented Worms and Mollusks

■ The segmented worms, phylum Annelida, include the earthworm, the *Nereis*, and the leech. They have bilateral symmetry and bodies that are divided into segments.

■ Segmented worms have a tube-within-a-tube body plan. The coelom is a fluid-filled cavity separating the digestive tract from the body wall.

■ Segmented worms are the simplest invertebrates having a closed circulatory system.

■ The mollusks, phylum Mollusca, include clams (bivalves), snails (gastropods), and squids (cephalopods). They are bilaterally symmetrical and have three distinct tissue layers.

■ Mollusks have a muscular foot for movement, a soft body that houses all the organ systems, and a mantle.

Vocabulary Review

Animalia (683)	spicules (686)
larvae (683)	spongin (686)
zoology (683)	osculum (686)
invertebrates (683)	gemmules (687)
vertebrates (683)	Coelenterata (687)
spherical symmetry (684)	polyp (687)
radial symmetry (684)	medusa (687)
bilateral symmetry (684)	mesoglea (688)
dorsal (684)	cnidoblasts (688)
ventral (684)	nematocysts (688)
anterior (684)	gastrovascular cavity (688)
posterior (684)	
Porifera (685)	planula (689)
pores (685)	Platyhelminthes (691)
collar cells (686)	Turbellaria (691)

Chapter Review

pharynx (692)

Trematoda (693)

Cestoda (693)

proglottids (694)

Nematoda (694)

Annelida (696)

Oligochaeta (696)

Hirudinea (696)

coelom (697)

Polychaeta (697)

parapodia (697)

trochophore larva (698)

Mollusca (698)

Bivalvia (699)

Gastropoda (699)

Cephalopoda (699)

foot (699)

mantle (699)

radula (699)

A. Matching—Select the vocabulary term that best matches each definition.

1. Oval, ciliated larva of the jellyfish

2. The largest group of mollusks

3. Animals without a backbone

4. Protein-rich, jellylike material separating two cell layers in coelenterates

5. Fold of skin that surrounds the body organs of mollusks

6. Specialized cells lining the central cavity of the sponge

7. The second-largest animal phylum

8. Square body segments of the tapeworm

9. Fluid-filled body cavity in segmented worms

10. Asexual reproductive structures of the sponge

B. Multiple Choice—Choose the letter of the answer that best completes each statement.

11. Multicellular organisms pierced by many holes through which water flows are part of the phylum (a) Nematoda (b) Porifera (c) Coelenterata (d) Mollusca.

12. The tonguelike organ found in most mollusks is called the (a) parapodia (b) foot (c) radula (d) gemmule.

13. Specialized cells characteristic of the phylum Coelenterata are (a) cnidoblasts (b) trochophore larvae (c) coeloms (d) gemmules

14. The lower or belly side of an animal is referred to as (a) dorsal (b) ventral (c) posterior (d) anterior

15. A mollusk with a shell consisting of two parts is considered a member of the class (a) Cephalopoda (b) Annelida (c) Bivalvia (d) Gastropoda

16. The earthworm belongs to the phylum (a) Nematoda (b) Coelenterata (c) Platyhelminthes (d) Annelida

17. Because it can be divided into equal halves in only one way, the human body is said to show (a) spherical symmetry (b) bilateral symmetry (c) parapodia (d) radial symmetry

18. The sessile body form of the coelenterate is called a (a) polyp (b) medusa (c) planula (d) trochophore larva

Content Review

19. Compare radial and bilateral symmetry and give an example of an organism that shows each.

20. Briefly describe nutrition, gas exchange, and excretion in a sponge.

21. Compare sexual and asexual reproduction in sponges.

22. How do gemmules contribute to the biological success of sponges?

23. Name the three layers of the coelenterate body.

24. What function do cnidoblasts serve?

25. How does respiration occur in coelenterates?

26. Describe the life cycle of *Aurelia*.

27. Explain how feeding and digestion occur in planarians.

28. What are some of the adaptations of tapeworms for parasitic life?

29. What distinguishes the roundworm's body plan from that of the flatworm?

30. What are the causes of tapeworm and trichina infections? What measures can be taken to prevent them?

31. Describe the life cycle of filaria worms.

32. What type of body plan do annelids have?

33. Identify two characteristics that annelids and nematodes have in common.

34. How is *Nereis* well adapted to its environment for finding food, for protecting itself, and for reproducing?

35. Describe the structure and function of the foot, mantle, and radula of mollusks.

36. How do bivalves feed and how do they digest food?

37. What type of circulatory system does the clam have? How does blood flow through the clam?

38. Compare and contrast the characteristics of the land snail with the squid.

Graphic Organizing

For information on graphic organizers, see Appendix G at the back of this text.

39. **Compare/Contrast Matrix:** Construct a compare/contrast matrix to compare flatworms, roundworms, and segmented worms. Use the following characteristics in your comparison: body plan, digestive system (number of openings), circulatory system (presence and type of), separateness of sexes, mode of life (parasitic or free-living).

Critical Thinking

40. What would happen to a land snail if its foot stopped producing mucus? (*Predicting*)

41. Compare and contrast the parasitism of leeches and flukes. (*Comparing and Contrasting*)

42. Predict what would happen if a small piece of hard plastic were placed between the mantle and the shell of an oyster. (*Predicting*)

43. A team of geologists examining hills in Michigan find a thick bed of fossilized coral about 400 million years old. What assumptions can they make about the environment of this area 400 million years ago? (*Identifying Assumptions*)

Creative Thinking

44. From your knowledge of the life cycle of the blood fluke, suggest some methods for preventing infestation by this parasite.

45. Some physicians advocate the use of leeches during certain kinds of surgery and to treat certain injuries. From what you know about leeches, suggest some examples of types of surgery or injuries in which leeches might be used.

Problem Solving

46. A biologist is interested in studying the hydras in a pond containing two types of plants. She notes that the hydras grow on the submerged parts of the plants. How

can she determine if hydras in their natural habitat grow better on one plant species than on the other?

47. A scientist is interested in studying the effects of chemical fertilizers and pesticides on the earthworm population. She chooses two nearly identical fields for her study. One field is treated with chemicals to grow the crop, whereas in the second field the crop is grown without the use of fertilizers or pesticides. She randomly selects five sites in each field. By using a gentle electric shocking device to force the earthworms to the surface, she counts the earthworms in various square-meter areas at each site. Using the data displayed in the table below, calculate the average number of earthworms in one square meter of each field. Interpret the data and then draw a conclusion. What is the control? How could the design of this experiment be improved?

Treated Field		Untreated Field	
Sites	Worms/m²	Sites	Worms/m²
1	7	A	9
2	2	B	6
3	3	C	8
4	4	D	5
5	4	E	7

48. Experiments have shown that glutathione, a peptide of the amino acids glutamic acid, cysteine, and glycine, stimulates feeding in hydras. Hydras eat living daphnia but reject dead daphnia because their glutathione has been destroyed. Hydras that have not been fed for 48 hours, however, will eat dead daphnia that have been dipped into a dilute glutathione solution. Design an experiment to discover whether individual amino acids can stimulate the feeding response. Be sure to develop a hypothesis, design a controlled experiment, and identify the variable being manipulated.

Projects

49. Use reference books to identify the different kinds of coral islands and reefs. Make a model of a reef or create a world map showing where coral formations are located and which ocean currents are near them.

50. Prepare a written report on one of the following topics: (a) Australia's Great Barrier Reef; (b) the Portuguese man-of-war; (c) sponge fishing; (d) mollusks as sources of human food.

▲ Before it became an adult, this butterfly passed through three stages of development.

Invertebrates–Arthropods and Echinoderms

33

An adult butterfly gathers nectar from obliging flowers. In its short life span, this insect will complete a life cycle that has stages as diverse as any in the animal kingdom. Although its wings give it a delicate and lovely appearance, this creature has features common to crabs, crayfish, spiders and all other arthropods—a shell-like external body skeleton and jointed appendages. The arthropods, in all their diversity and complexity, are discussed in of this chapter. In addition, you will learn about another group of invertebrates, the echinoderms.

Guide for Reading

Key words: crustaceans, arachnids, incomplete and complete metamorphosis, entomology, water-vascular system

Questions to think about:

📖 In what ways are insects both helpful and harmful to humans?

📖 Why are echinoderms considered to be closely related to vertebrates?

33-1 Arthropods: Crustaceans

Section Objectives:

- *List* the five classes that make up the phylum Arthropoda, and describe their general characteristics.
- *Describe* the general characteristics of crustaceans, and name four members of this group.
- *Describe* the external structure of the crayfish.
- *Explain* the major life processes in the crayfish: nutrition, excretion, circulation and respiration, nervous regulation, and reproduction.

The phylum **Arthropoda** (ar THRAHP uh duh) ranges from flies, bees, beetles, mosquitoes, butterflies, spiders, and ants to crabs, lobsters, and shrimp. Of all animal groups, *arthropods* are the most biologically successful and the most abundant. There are more arthropod species than all other species of organisms put together. There are about 400 000 known species of plants and about 250 000 known species of animals other than arthropods. But, there are more than 1 million known species of arthropods. Arthropods are found in all regions of the earth and are of the greatest importance to human beings.

The phylum Arthropoda is divided into five classes. These are the crustaceans, centipedes, millipedes, arachnids, and insects. The characteristics of each class will be described separately later in the chapter.

▲ **Figure 33–1.**
A Crustacean. This shrimp is just one of the 1 million known species of arthropods.

Biology and You

Q: I keep hearing about Lyme disease. What causes this illness?

A: Lyme disease is caused by a particular kind of bacteria that is spread by infected deer ticks. When an infected tick bites a person, the bacteria are transmitted into the individual's bloodstream.

While deer ticks feed on many kinds of mammals, they usually pick up the infectious bacteria from the blood of field mice. Woods, marshy areas, high grass, and brush are all high-risk localities for Lyme disease.

The deer tick is about the size of a comma on this page. Because it is so easy to miss and its bite painless, many people never realize they have been bitten. The illness typically begins with an expanding, ring-shaped rash. If Lyme disease is left untreated, severe arthritis, nervous disorders, and heart problems may occur. However, when diagnosed, it can be treated with antibiotic drugs. To prevent Lyme disease, wear light-colored clothing and pants tucked into socks when walking through areas that may be tick infested. If you do recognize symptoms of the disease, call a doctor immediately.

■ *Design an informative poster about Lyme disease to be displayed in your school.*

General Characteristics of Arthropods

In many ways, arthropods are the most advanced invertebrates. They are bilaterally symmetrical and have a small coelom.

Although the phylum consists of a large number of unlike species, all arthropods share certain common features.

Arthropods have jointed legs. The legs are made of several pieces that are connected at hinged joints. These joints are controlled by opposing sets of muscles. Different arrangements of hinged joints allow such different functions as crawling, swimming, hopping, jumping, flying, grabbing, digging, and biting.

Arthropods have exoskeletons composed of protein and **chitin** (KYT un), *a carbohydrate.* The tough, lightweight exoskeleton protects soft body parts. The exoskeleton is also waterproof and prevents water loss, allowing many arthropods to live on land. Because the exoskeleton cannot grow, young arthropods periodically must undergo a process called **molting.** See Figure 33–2. During molting, the exoskeleton is shed and replaced by a new, larger one. Growth takes place before the new exoskeleton hardens. Until the new exoskeleton hardens, the young animal is unable to move or defend itself. Therefore, many arthropods hide until their new exoskeleton has hardened.

Like annelids, arthropods are segmented. However, the body segments are usually modified and fused to form specific body regions. In most arthropods, there is a head, **thorax** (THOR aks), and **abdomen.** The head, which is always composed of six segments, is well developed. It contains a mouth that is specialized for chewing or sucking. The thorax is the middle region of the arthropod, and the abdomen is the posterior region. The number of segments in the thorax and abdomen changes from one group of arthropods to another.

Figure 33–2

A Tarantula Molting. All arthropods molt their exoskeletons periodically. This red-kneed Mexican tarantula has just crawled out of its old exoskeleton. ▼

Arthropods have well-developed nervous systems. There is a distinct brain and a ventral nerve cord located beneath the digestive system. There are a variety of sense organs, including eyes, organs of hearing, sensory cells sensitive to touch, and antennae that are sensitive to touch and chemicals.

Arthropods have an open circulatory system. There is a dorsal tubular heart located above the digestive system. Arteries carry blood away from the heart to the body spaces, where it bathes the tissues directly. The blood eventually reenters the heart through openings in its sides.

General Characteristics of Crustaceans

The class **Crustacea** (krus TAY shuh), the *crustaceans,* includes lobsters, crayfish, crabs, shrimp, waterfleas, sow bugs, barnacles, and many others. Most crustaceans are marine, but some live in fresh water. A few, such as the sow bug, live on land in moist places. Crustaceans vary in size from tiny water fleas to huge crabs with leg spans of 3.5 meters. Microscopic crustaceans are the main source of food for many larger marine animals. All crustaceans have two pairs of antennae on the head.

Figure 33–3

Crustaceans. Representative crustaceans include: (A) Southeast Asian land crab, (B) scarlet reef lobster, (C) blood red shrimp, (D) sow bug, (E) goose barnacles, and (F) waterflea. ▼

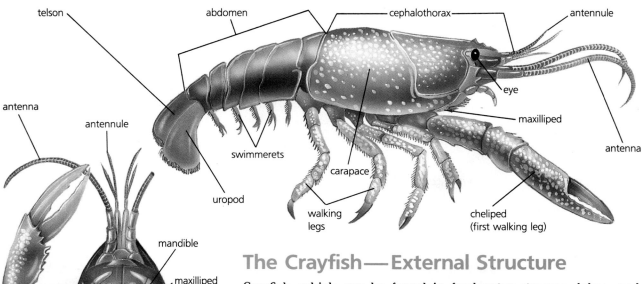

▲ **Figure 33–4**
External Structure of the Crayfish. A ventral view of the crayfish head (bottom) shows the maxilla, mandibles, and maxillipeds. The crayfish has two body sections (top right), the cephalothorax and the abdomen.

The Crayfish—External Structure

Crayfish, which can be found in freshwater streams, lakes, and swamps, are typical crustaceans. They are covered by an exoskeleton hardened with lime. At the joints, where bending occurs, the exoskeleton is softer, thinner, and folded. The crayfish body has two main regions, shown in Figure 33–4. At the anterior end, the segments of the head and thorax are fused to form the **cephalothorax** (sef uh luh THOR aks). The part of the exoskeleton that protects and covers the dorsal and side surfaces of the cephalothorax is called the *carapace* (KAR uh pays). The seven segments behind the cephalothorax form the abdomen. The paddle-shaped last segment of the abdomen is called the *telson.*

The paired appendages of the crayfish have specific functions. Starting at the front, the first pair of appendages are the *antennules* (an TEN yoolz), which function in touch, taste, and balance. Next come the *antennae* (an TEN ee), which are also used for touching and tasting. The **mandibles** (MAN duh bulz), or jaws, crush food by moving from side to side. Two pairs of *maxillae* (mak SIL ee) handle food. Three pairs of *maxillipeds* (mak SIL uh pedz) function in touch and taste and also handle food. The large first legs are called *chelipeds* (KIHL uh pedz). Their grasping claws are used for defense and to catch food. Behind the chelipeds are four pairs of walking legs. On the abdomen are *swimmerets* (swim uh RETS), which are used in swimming. In females, the swimmerets are used to carry the developing eggs. The last pair of appendages are the broad *uropods* (YUR uh pahdz). The uropods, along with the telson, form a fan-shaped tail that is used for backward movement. When the crayfish senses danger, the powerful abdominal muscles whip the tail forward under the abdomen causing the animal to shoot backward.

If a crayfish injures an appendage, it can shed the injured limb at a joint. This process of self-amputation prevents blood loss. Gradually, with each molt, the lost appendage grows back. Regeneration in crayfish is limited to the appendages and eyes.

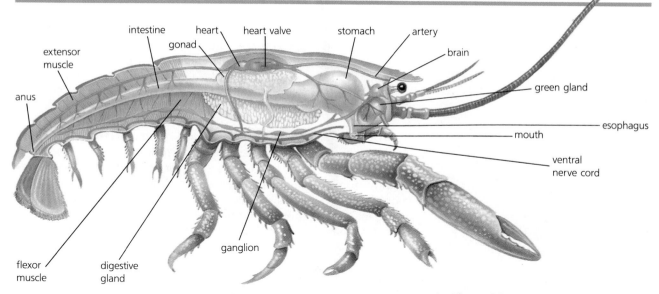

▲ **Figure 33–5**
Internal Structure of the Crayfish.

The Crayfish—Internal Structure and Life Functions

Nutrition The crayfish feeds on dead animals or living animals that it catches with its chelipeds. The food is crushed by the mandibles and passed to the mouth by the maxillae and maxillipeds. The mouth leads into a short esophagus. From there, food passes into the stomach, where it is chewed up by chitinous teeth. The finely ground food particles are digested by enzymes, then passed into the digestive glands and absorbed into the blood. Undigested material passes through the intestine and out the *anus.*

Excretion The excretory organs of the crayfish are called the *green glands.* They are located in the head region. The green glands remove wastes from the blood. These wastes are then excreted from the body through an opening near the base of the antennae.

Circulation and Respiration. The open circulatory system consists of a dorsal heart surrounded by a cavity called the *pericardial* (pehr uh KARD ee ul) *sinus.* Blood in the pericardial sinus enters the heart through three pairs of valves. When the heart contracts, the valves close, and blood is pumped out through arteries to all parts of the body. There are no capillaries or veins. The arteries open into spaces, or sinuses, among the body tissues. There, the blood bathes the cells directly. Oxygen and nutrients from the blood diffuse into the cells, and carbon dioxide and wastes from the cells diffuse into the blood. Eventually, the blood collects in the *sternal* (STERN ul) *sinus,* where it is channeled to the gills. In the gills, it picks up oxygen, gets rid of carbon dioxide, and returns to the pericardial sinus. Dissolved in the plasma of the colorless blood is **hemocyanin** (hee moh SY uh nin), a copper-containing respiratory pigment that aids in the transport of oxygen. Hemocyanin is blue when oxygenated.

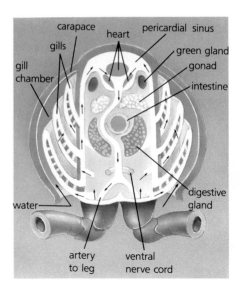

carapace
heart
pericardial sinus
gills
green gland
gonad
gill chamber
intestine
water
digestive gland
artery to leg
ventral nerve cord

▲ **Figure 33–6**
Crayfish Gills. Gas exchange in the crayfish is through the gills, which are located in gill chambers on either side of the thorax. As water flows over the gills, oxygen diffuses into the blood, and carbon dioxide diffuses out.

Figure 33–7
Female Crayfish Carrying Newly Hatched Young. The newly hatched crayfish young remain attached to the mother's swimmerets for several weeks. ▼

The **gills,** where the exchange of respiratory gases occurs, are delicate, plumelike structures. They are located in *gill chambers* on each side of the thorax. The gill chambers are protected by the carapace. Water is kept flowing through the gill chambers by the movement of the second maxillae. Figure 33–6 is a cross section of the thorax, showing the gills in the gill chamber.

Nervous Regulation The nervous system of the crayfish is similar to the annelids' system. The brain, which is in the head, is connected by nerves to the eyes, antennules, and antennae. Extending from the brain, two nerves circle the esophagus and join ventrally to form a double, ventral nerve cord. As the ventral nerve cord runs toward the rear, it enlarges into ganglia in each segment. From the ganglia, nerves branch to the appendages, muscles, and other organs.

The sensory organs of the crayfish are well developed. They include a pair of *compound eyes* located at the ends of movable stalks. Each eye contains about 2000 visual units. Each unit contains a lens system. Unlike the human eye, the eye of a crayfish cannot focus at different distances. While this type of eye is sensitive to movement and offers a wide angle of vision, it produces only a crude image.

The crayfish has two kinds of small sensory hairs that are found on the appendages and other parts of the body. One type of hair is sensitive to touch. The other is sensitive to chemicals and provides information similar to the human senses of taste and smell.

The sense organs of equilibrium, or balance, are found in sacs that are called *statocysts* (STAT uh sists).These sacs are located at the bases of the antennules. Each statocyst contains sensory hairs and grains of sand. When the crayfish moves, the sand grains move, which stimulates some of the sensory hairs. From the stimulated hairs, impulses pass to the brain of the crayfish. The brain interprets the information and initiates impulses that allow the crayfish to adjust its position and to maintain its equilibrium. Each time the animal molts, the sand grains are shed along with the exoskeleton. New sand grains are picked up when the new exoskeleton forms.

Reproduction In crayfish, sexes are separate. Mating takes place in the fall. The male uses his first pair of swimmerets to transfer sperm from his body to the *seminal* (SEM in ul) *receptacle* of the female. The sperm are stored in the receptacle until spring. During the spring, the female lays several hundred eggs that have been fertilized by the stored sperm. The eggs are attached to the female's swimmerets. The waving of the swimmerets back and forth keeps the embryos well supplied with oxygen for development. After five to six weeks, the eggs hatch, but as you can see in Figure 33–7, the young crayfish remain attached to the mother for several more weeks. During this time, the young crayfish begin to molt. Crayfish live for three to five years.

33-1 **Section Review**

1. Name the five major classes of the arthropod phylum.
2. What characteristic do all crustaceans have in common?
3. What are the two main body regions of the crayfish?
4. Hemocyanin functions in what important life process in the crayfish?

Critical Thinking

5. How is the ability to amputate and regenerate injured appendages a survival advantage for the crayfish? (*Judging Usefulness*)

33-2 Arthropods: Centipedes to Arachnids

Section Objectives:

- *Describe* the general physical characteristics of both centipedes and millipedes.
- *Name* four members of the arachnids and describe their general characteristics.

General Characteristics of Centipedes

Centipedes, or "hundred-leggers," belong to the class **Chilopoda** (ky LAHP uh duh). Some centipedes have more than 150 pairs of legs, but 30 to 35 pairs are most common in centipedes. A centipede has a distinct head made up of six segments. The head is followed by a long, wormlike, slightly flattened body made up of many similar segments. Centipedes live on land in dark, damp places, such as under logs or stones.

In the centipedes, all body segments, except the one behind the head and the last two, have one pair of legs. The head has one pair of antennae and various mouthparts. Centipedes feed mainly on insects. The centipede bites its victim with *poison claws,* which are on the first body segment. Small centipedes are harmless to humans. The common house centipede is about 2.5 centimeters long. At night, it searches for food, eating cockroaches, bedbugs, and other insects.

General Characteristics of Millipedes

Millipedes, or "thousand-leggers," belong to the class **Diplopoda** (dih PLAHP uh duh). They do not have 1000 legs, but they may have more than 300 pairs. Like a centipede, a millipede has a distinct head and a long, wormlike body made of many segments. Except for the last two segments, millipedes have two pairs of legs per segment. The head has a pair of antennae and various mouthparts. Unlike centipedes, millipedes do not have poison claws. Millipedes

Figure 33–8

A Centipede and a Millipede. The centipede (top) has one pair of legs per segment, while the millipede (bottom) has two pairs of legs per segment. ▼

also move more slowly than centipedes. They feed mainly on decaying plant material. When they are disturbed, millipedes usually roll themselves into a ball. Many have "stink" glands that give off an unpleasant odor.

General Characteristics of Arachnids

Arachnids, members of the class **Arachnida** (uh RAK nih duh), include spiders, scorpions, ticks, mites, and daddy longlegs. Some arachnids are dangerous to humans and other animals. Mites and ticks live as temporary parasites on the skin of many animals, including humans, dogs, chickens, and cattle. Mites often cause terrible itching. Ticks are carriers of several diseases, including Lyme disease, Rocky Mountain spotted fever, and Texas cattle fever. Scorpions sting with their tail. While the sting is painful, it is usually not fatal to humans.

Spiders are usually harmless. In fact, they are often helpful because they feed on insects. The poisonous spiders of the United States are the black widow and the brown recluse. Spiders rarely bite unless they are disturbed.

Structure and Life Functions of Arachnids

Most arachnids live on land, and many resemble insects. The body of an arachnid is made up of a cephalothorax and an abdomen. Arachnids do not have either antennae or chewing jaws. They have six pairs of jointed appendages, all on the cephalothorax. The first pair of appendages are the fang-like **cheliceras** (kuh LIS uh res), which are used to pierce the prey. The body fluids of the prey are then drawn into the spider's mouth by the action of the *sucking stomach.* Usually, poison glands associated with the cheliceras inject a poison that paralyzes the prey. The second pair of appendages, the **pedipalps** (PED uh palps), is sensitive both to chemicals and to touch. The pedipalps also hold food and are used

Figure 33–9

Arachnids. Arachnids are an extremely successful group containing more than 100 000 species. Some common arachnids include, from top: the scorpion, the spider, the tick, the mite and the daddy longlegs (lower right). ▼

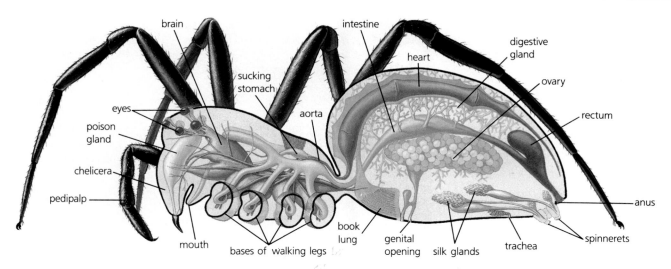

▲ **Figure 33–10**
Structure of an Arachnid. The body of an arachnid is divided into a cephalothorax and an abdomen, and there are four pairs of walking legs. Arachnids are predators, and the sucking stomach is used to draw in the body fluids of prey. The book lungs are specialized structures for gas exchange. Spiders, and some other arachnids, have spinnerets used to spin silk for webs and other purposes.

by the male in reproduction. The next appendages are four pairs of walking legs. You can tell arachnids from insects because insects have three pairs of walking legs.

The respiratory organs of the arachnids are called **book lungs.** Located in chambers on the underside of the abdomen, book lungs are a series of leaflike plates containing blood vessels. Air, drawn into the chambers through slits in the abdomen, circulates between the plates. Gas exchange occurs between the blood in the plates and the air in the chamber. Oxygen and carbon dioxide are transported in the blood between the body cells and the book lungs. Although some insectlike air tubes, or *tracheae,* are present, they play only a minor role in respiration.

In spiders and some other arachnids, the pedipalps of the male are modified for sperm transfer. Following elaborate courtship behavior, the male uses the pedipalps to place the sperm in the seminal receptacle of the female. In spiders, as the female lays the eggs, they are fertilized by the stored sperm, and wrapped in a cocoon. In some species, the female carries the cocoons until the young hatch. An example of this is shown in Figure 33–11. In other species, the eggs in cocoons are deposited on the ground. In still other types of arachnids, sperm are not transferred into the body of the female by the male. Instead, the sperm are enclosed in a case and deposited on the ground. The female then takes up the case into a special body opening.

In spiders and one other small group of arachnids, there are three pairs of **spinnerets** (spin uh RETS) at the end of the abdomen. Spinnerets are used to spin silk produced by silk glands within the abdomen. As the fluid protein is squeezed out of the spinnerets, it hardens into a thread. Spiders use these threads for many purposes. Some use them to construct webs in which they capture prey. Threads are also used to line nests and to make cocoons for the fertilized eggs. Spiders also use threads as a way of raising or lowering themselves from trees and other objects to escape from danger.

Figure 33–11

Female Wolf Spider. The newly hatched young covering this female wolf spider have emerged from the white cocoon attached to her abdomen. ▶

33-2 Section Review

1. How many pairs of legs per body segment do centipedes have? Millipedes?
2. Name three members of the class Arachnida.
3. What do spiders eat?

Critical Thinking

4. How are centipedes and spiders similar? How do they differ? (*Comparing and Contrasting*)

33-3 Arthropods: Insects

Section Objectives:

- *List* several reasons for the success of insects as land animals.
- *Describe* the external structure of the grasshopper.
- *Describe* reproduction and metamorphosis in insects.
- *Name* two members of each of the six major insect orders.
- **Laboratory Investigation:** *Observe* characteristics of arthropods by studying crayfish external anatomy (p. 728).

General Characteristics of Insects

Biologically, *insects*, class **Insecta** (in SEK tuh), are the most successful class of arthropods. There are more than 900 000 known species. Nearly all insects are land animals, although a few live in fresh water and a few in salt water. Insects range in size from tiny beetles 0.25 millimeters long to some large tropical moths with a wingspan of 30 centimeters. Most insects, however, are less than 2.5 centimeters long.

There are several reasons for the remarkable success of insects. The most important of these are listed below.

- Insects are the only invertebrates capable of flying. The ability to fly allows them to travel over great distances in search of food. It also allows them to escape from enemies and to spread into new environments.

- Among the insects, there is tremendous variation in how they are adapted for feeding and reproduction. These adaptations allow insects to exist in all types of environments and to obtain nourishment from many sources.

- Insects have a high rate of reproduction and a short life cycle. A single female can lay hundreds or even thousands of eggs at a time. These eggs develop rapidly and may produce millions of offspring during a year. This increases the ability of insects to adapt.

- Insects are small, which means that they do not need large areas in which to live.

All insects have three separate body regions—the head, thorax, and abdomen. On the head is one pair of antennae, several mouthparts, and in most insects, compound eyes. On the thorax are three pairs of walking legs. In flying insects, the wings are also located on the thorax. The abdomen has as many as 11 segments, with no leglike appendages.

Figure 33–12

Variation Among Insects. Insects are well-adapted for life in a wide variety of environments. This diverse class includes, clockwise from top left: the American cockroach, the bumble bee, the diving beetle, and treehoppers. ▼

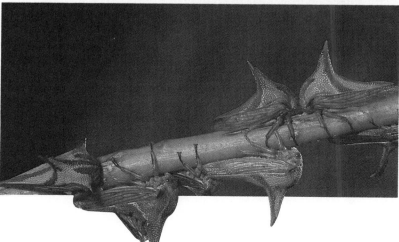

Variations Among Insects

While the grasshopper, which is discussed later in this chapter, shows the general characteristics of the insect class, many insects have other, highly specialized structures. These specialized structures allow them to feed on a particular plant or animal or to live in a particular environment.

Mouthparts The structure of an insect's mouthparts reflects the way in which it obtains food. There are two basic types of mouthparts. Some insects, such as grasshoppers, have chewing mouthparts. Others, such as bees, having sucking mouthparts, which are usually in the form of a tube. Some insects have needlelike projections that enable them to pierce the tissues of animals and plants and suck their juices. In butterflies, the coiled siphoning tube uncoils to suck nectar from flowers. Houseflies have sponging and lapping mouthparts.

Body Form Insects vary greatly in body form. Cockroaches have flattened bodies that are suited for living in cracks and crevices. Beetles have thick, plump bodies. Damsel flies and walking sticks have long slender bodies. Moths are covered with hairs that may serve to protect them from cool evening temperatures. The hairs or bristles on bees help in the collection of pollen.

Legs The legs of insects show many types of modifications. For example, water bugs and some beetles have paddle-shaped legs that are used in swimming. The walking legs of honeybees are modified for the collection of pollen. The forelegs of the praying mantis are modified for grasping prey.

Reproduction and Development in Insects

All insects reproduce sexually. Eggs are produced in the ovaries of the female, and sperm are produced in the testes of the male. In a few insects, eggs hatch directly into miniature adults. The young molt several times, growing larger each time molting occurs. In most species, however, insects undergo a series of distinct changes as they develop from eggs to adults. This series of changes is called **metamorphosis** (met uh MOR fuh sis), and it is controlled by hormones.

Incomplete Metamorphosis The eggs of some insects, such as grasshoppers, crickets, and cockroaches, undergo **incomplete metamorphosis.** In this type of development, the eggs hatch into **nymphs.** The nymph resembles the adult, but lacks certain adult features. As you can see in Figure 33–13, the grasshopper nymph looks like the adult, but it lacks wings and reproductive structures. Nymphs molt several times. With each molt, they become larger and more like the adult. In incomplete metamorphosis, the three stages of development are the egg, nymph, and adult.

Figure 33–13

An Immature Grasshopper. The grasshopper nymph lacks the wings and reproductive structures of the adult. ▼

■ Careers *in Action*

Pest Control Specialist

Insects destroy more trees each year than do forest fires. In addition, some insects spread diseases to other organisms. On the other hand, insects are vital as pollinators of certain crops, as producers of valuable substances, and as destroyers of harmful insects.

It is the job of the pest control specialist to somehow balance getting rid of harmful insects while preserving beneficial ones. Pest control specialists use a variety of methods to destroy insect pests. These methods range from spraying chemical insecticides to releasing natural predators.

A farmer whose crops are being destroyed by beetles or a homeowner whose house is being attacked by termites might call a pest control specialist. The specialist would determine what kind of insect is causing the damage and how many there were. He or she would then devise and carry out a plan for removing the pests while taking into account the effects of the plan on the environment.

Problem Solving Imagine you are a pest control specialist. A fruit-tree grower calls you with a problem—insects are damaging his trees. He wants you to get rid of them. The grower tells you that he relies on bees to pollinate his trees. You inspect the trees and find the leaves have been chewed. You also find the orchard infested with flying insects.

1. Based on your inspection of the trees, what order of insects do you think is causing all the damage?

2. How does the farmer's information about his use of the bees limit your options for eliminating the pests?

3. Propose a plan for destroying the pests.

4. Explain to the farmer why your plan meets his needs.

■ **Help Wanted:**
Pest control specialist. High-school diploma required; two-year associate degree or bachelor's degree recommended. Contact: National Pest Control Association, 8150 Leesburg Pike, Vienna, VA 22180.

Complete Metamorphosis The eggs of most insects undergo **complete metamorphosis.** Moths, butterflies, beetles, bees, and flies are among the insects that undergo complete metamorphosis. The eggs of these insects hatch into segmented larvae. These larvae are known as *caterpillars, maggots* (MAG uts), or *grubs.* During this stage, the larva eats and grows. After several molts, it passes into a resting stage called the **pupa** (PYOO puh). The pupa is surrounded either by a cocoon or by a case made from its outer covering. During the pupal stage, the tissues are reorganized into the adult form. When the changes are complete, the case or cocoon splits open, and the adult emerges. In complete metamorphosis, the four stages of development are the egg, larva, pupa, and adult.

The development of the cecropia (sih KROH pee uh) moth is typical of insects that undergo complete metamorphosis. It includes the egg, larva, pupa, and adult stages. See Figure 33–14. The stages are controlled by the interaction of three hormones—brain hormone, molting hormone, and juvenile hormone.

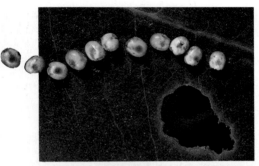

▲ **Figure 33–14**

Development of the Cecropia Moth. The developmental stages of the Cecropia moth include, from top to bottom, eggs, larva, pupa, and adult.

The egg hatches into the larva, in this case a caterpillar. As the caterpillar eats and grows, cells in the brain secrete a hormone that stimulates production of a molting hormone. The molting hormone, produced by an endocrine gland in the thorax, stimulates molting of the exoskeleton. The transformation of the larva into more mature forms is inhibited by a hormone called the juvenile hormone. Juvenile hormone is produced by endocrine glands, which are near the brain. As long as the juvenile hormone is secreted, the larva can molt, but it will not change into the next stage, known as the pupa. At the end of the larval period, the secretion of juvenile hormone decreases. At the time of the next molt, the larva forms a pupa. During the pupal stage, the insect appears inactive, but great changes in body form are occurring. At the end of pupation, the adult moth emerges.

A recent approach to insect control involves the use of substances similar to juvenile hormone. These substances prevent the metamorphosis of larvae into adults, which prevents the insects from reproducing.

Classification of Insects

The branch of biology that deals with the study of insects is called **entomology** (ent uh MAHL uh jee). Scientists who study insects are called *entomologists*. Entomologists divide the class Insecta into 27 orders. Of these 27, only 6 are of major importance. The 6 major orders are the **Hymenoptera** (hy muh NAHP tuh ruh), **Orthoptera** (or THAHP tuh ruh), **Coleoptera** (kohl ee AHP tuh ruh), **Lepidoptera** (lep uh DAHP tuh ruh), **Diptera** (DIP tuh ruh), and **Hemiptera** (heh MIP tuh ruh). Figure 33–16 on page 722 shows the basic characteristics of these and other insect orders.

The Grasshopper—A Representative Insect

Like all insects, the body of the grasshopper is divided into three sections—the *head, thorax,* and *abdomen.* The head is made up of six fused segments. Two large compound eyes, similar to those of the crayfish, are located on the sides of the head. See Figure 33–15. In addition, the grasshopper has three simple eyes located between the compound eyes. The simple eyes do not form images. They are only sensitive to light and dark. On the front of the head is a pair of jointed antennae. The antennae are sensitive to smell and touch.

The mouthparts used in chewing are located outside the mouth. These structures are adapted for eating leafy vegetation. The upper lip and the lower lip hold the food. The mandibles are lined with rough-edged chitinous teeth. In biting off and chewing food, the mandibles move from side to side. Behind the mandibles are the maxillae, which hold the food and pass it to the mandibles. Sensory palps on both the maxillae and the lower lip feel and taste the food. Beneath the lower lip is a tonguelike organ.

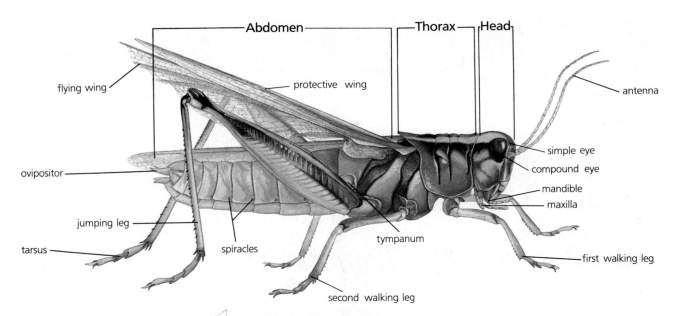

▲ **Figure 33–15**

External Structure of the Grasshopper. The grasshopper body is divided into three sections—the head, the thorax, and the abdomen.

The thorax is composed of three segments. Each segment bears a pair of legs. Each leg has five segments. The last segment is called the *tarsus*, or foot. On the tarsus are pads that enable the grasshopper to cling to smooth surfaces and claws that enable it to climb rough surfaces. The first two pairs of legs are for walking. The third pair of legs is larger than the first two pairs, and it is modified for jumping.

Attached to the last two segments of the grasshopper's thorax are two pairs of wings. The outer pair, known as the fore wings, is hard and serves as a protective covering for the inner pair of wings, called the hind wings. The hind wings, which are flexible, are used in flight. When the hind wings are not in use, they are folded like fans. The thin membranes of the hind wings contain veins that serve to strengthen them.

The abdomen is made up of 10 segments. Along the lower sides of the abdomen and thorax are 10 pairs of **spiracles,** or openings into air tubes. On either side of the first abdominal segments are the *tympana* (TIM puh nuh), the organs of hearing. Each tympanum consists of an oval, flat membrane that vibrates when it is hit by sound waves.

The last segment of the abdomen is modified for reproduction. When grasshoppers mate, the male transfers sperm into the body of the female. The sperm are stored in the seminal receptacle of the female. When the eggs leave the ovary of the female, they enter the oviduct, where fertilization occurs. They then pass out of the body of the female. On the end of the abdomen of the female is a hard, four-pointed organ called the **ovipositor** (OH vee pahz it er). It is used to dig holes in the ground in which the eggs are deposited. The eggs are laid in the fall but do not hatch until spring. For more information about the structure of the grasshopper, see Unit 2.

Classification of Insects

Order	Examples	Mouthparts	Wings	Characteristics/Habitat
Anopleura 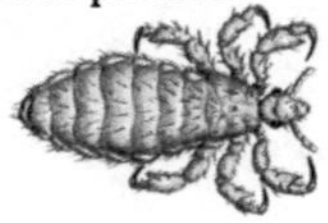	sucking lice	sucking	none	parasites of mammals; suck blood of hosts, including humans. Their bites are irritating, and they spread disease.
Coleoptera 	beetles (diving, bark, carpet, and Japanese beetles, fireflies, ladybirds, june bugs)	chewing	usually 2 pair	largest insect order; found in all habitats. Many feed on plants and are serious pests.
Collembola 	springtails	chewing	none	small insects found in leaf litter, on rotting logs, on beaches, and on surface of pond water. Some species are jumpers.
Diptera 	flies (houseflies, black flies, mosquitos, midges, gnats, horseflies)	sucking	1 pair	found in various habitats; some feed on plants; others are parasites; others feed on insects. Many types are pests. Some damage plants; some transmit animal diseases.
Ephemeroptera 	mayflies	vestigial	2 pairs	found in and around ponds and streams. Adults live only a day or so and do not eat.
Hemiptera 	bugs (water striders, bedbugs, assassin bugs, stinkbugs)	sucking	none or 2 pairs	most terrestrial; some aquatic; few parasitic. Some eat plants, others, insects.
Homoptera 	cicadas, aphids, leafhoppers, spittlebugs, plant hoppers, whiteflies, lac insects	sucking	none or 2 pairs	feed on plants. Many cause serious damage, and some transmit diseases. Lac insects are the source of lac, from which shellac is made.
Hymenoptera 	bees, wasps, ants, sawflies	bee: sucking; wasps, ants, and sawflies: chewing	none	large order whose members live mainly on vegetation, particularly flowers, in various habitats. Some are parasites of other insects. Ants and some wasps and bees are social insects, living in colonies divided into several castes, each serving a function. Honeybees are important pollinators of many types of plants.

▲ **Figure 33–16. Classification of Insects.**

Classification of Insects (continued)				
Order	**Examples**	**Mouthparts**	**Wings**	**Characteristics/Habitat**
Isoptera	termites	chewing	none or 2 pairs	small social insects that feed mainly on wood. Termites damage or destroy buildings and objects made of wood.
Lepidoptera	butterflies and moths	sucking, coiled sucking tube	usually 2 pairs	found on vegetation. The larvae of this group are caterpillars, which feed on plants and often do serious damage. Adults commonly feed on plant nectar and may serve to pollinate the plants they visit. Salivary glands of larvae produce silk used to make cocoon. Silk is produced by silkworm moth larvae.
Mallophaga	chewing lice	chewing	none	parasites of birds and mammals (but not humans).
Odonata	dragonflies, damselflies	chewing	2 pairs	found around water; feed on mosquitos and other insects.
Orthoptera	cockroaches, crickets, grasshoppers, katydids, walking sticks, praying mantis	chewing	usually 2 pairs	large insects found on ground or on low vegetation. Many make noise by rubbing body parts together. Many feed on plants and can do great damage. Cockroaches are pests in buildings.
Siphonaptera	fleas	sucking	none	small parasites on birds and mammals. Fleas are pests, attacking domestic animals and humans. A few types of fleas transmit disease, including bubonic plague.
Thysanura	silverfish, bristletails	chewing	none	small insects. Bristletails found in leaf litter, under logs, etc. Silverfish found in cool, damp places, often pests in buildings.

Economic Importance of Insects

Insects are so widespread and so numerous that they affect almost every part of daily life in some way. Each year, insects cause billions of dollars of damage to crops. Insects spread many plant diseases, such as Dutch elm disease and corn smut. They also transmit animal diseases: mosquitoes carry malaria, yellow fever, and elephantiasis; houseflies carry dysentery and typhoid fever; tsetse (SEET see) flies carry African sleeping sickness; lice carry typhus; and fleas carry plague. Insects also destroy property: termites destroy wood; moths and carpet beetles damage clothing, fabrics, furs, and carpets; silverfish destroy paper; and weevils, cockroaches, and ants ruin food.

Insects also serve some valuable functions. Some insects are necessary for the pollination of important crops. For example, bees pollinate the flowers of apple and pear trees, clover, and berries. They also produce honey. Lac, which is used to make shellac, comes from from lac insects, and silk from silkworm moths.

Some insects destroy other insects that are harmful to humans and property. Ladybird beetles eat scale insects, which injure orange and lemon crops. The praying mantis eats almost any insect it can catch. Wasps, by laying their eggs in caterpillars, eventually kill them. Aquatic bugs eat mosquito larvae. Insects also serve as a source of food for birds, frogs, and fish. Finally, some insects act as scavengers, eating dead plant and animal remains.

Scientists are still searching for ways of controlling harmful insects without harming other insects or animals. Chemical insecticides poison environments, kill harmful and helpful insects, and endanger other animals, including humans. Furthermore, in time, insect populations become resistant to chemicals.

Many scientists believe that biological methods of insect control are safer than chemical insecticides. Biological controls include: sterilizing males and releasing them; developing resistant plants; introducing specific predators and parasites that destroy only harmful insects; and using insect sex attractants, called pheromones, to lure insects into traps.

33-3 Section Review

1. List three reasons for the biological success of insects.
2. What are the stages of complete metamorphosis?
3. List the six major orders of insects.
4. What are two useful products obtained from insects?

Critical Thinking

5. An animal is discovered that has an exoskeleton, sucking mouthparts, head fused with thorax, abdomen, no wings, and four pairs of walking legs. Could you classify the animal as an insect? Why or why not? (*Reasoning Categorically*)

33-4 Echinoderms

Section Objectives:

- *Name* four echinoderms and describe the general characteristics of these organisms.
- *Explain* why echinoderms are considered to be more closely related to the vertebrates than other invertebrate phyla.
- *Describe* the structure and life functions of the starfish.

General Characteristics of Echinoderms

The phylum **Echinodermata** (ih ky nuh der MAH tuh) includes starfish (sea stars), sea urchins, sea cucumbers, and sand dollars. These animals are all marine and live mainly on the ocean floor. Some stay in the same place but most move around. The larvae are bilaterally symmetrical, but the adults are radially symmetrical. *Echinoderms* have a well-developed coelom.

Almost all echinoderms have an internal skeleton that serves as support and protection for the animal. The skeleton consists of hard, calcified plates that are embedded in the body wall. Spiny projections on the plates stick out through the skin. These projections give echinoderms their spiny-skinned appearance.

In all the invertebrates that you have studied so far, the first opening of the digestive system formed in the embryo is the mouth, which is formed from the blastopore (see Chapter 22, section

Figure 33–17
Echinoderms. All echinoderms are marine organisms. Common echinoderms include: (A) the sea urchin, (B) the starfish, (C) the sand dollar, and (D) the sea cucumber. ▼

22–1). The opening for the anus later develops opposite the mouth. In the echinoderms, however, the pattern of development is just the opposite. The blastopore becomes the anus, and the mouth forms at a later stage opposite the anus. This pattern of development is characteristic of vertebrates. It may show, therefore, a possible evolutionary relationship between echinoderms and more complex animals.

The starfish is representative of the phylum, so it can be used to study the structure and life functions of echinoderms. Because it is an echinoderm and not a fish, some people prefer to call it a *sea star.*

Structure and Life Functions of the Starfish

The body of the starfish consists of a central disk from which the arms, or rays, radiate. Most starfish have 5 arms, but some have as many as 20.

Locomotion and food getting in starfish involve a system called the **water-vascular system.** This system is found only in echinoderms. See Figure 33–18. On the dorsal surface of the starfish is an opening called the *sieve plate.* Sea water enters through the sieve plate and enters a system of **canals,** which run into each arm. Connected to the canals in the arms are many small, tubular structures called **tube feet.** Each tube foot has a bulblike structure at one end and a sucker at its tip. The bulbs are within the body of

Figure 33–18

Structure of the Starfish. The water-vascular system is a network of water-filled canals that serves in locomotion and food-getting. Connected to the canals are tube feet, which allow the starfish to crawl along the ocean floor. ▼

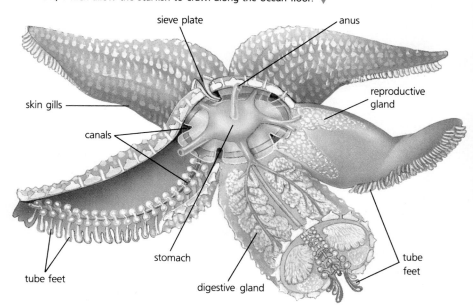

sieve plate · anus · reproductive gland · skin gills · canals · tube feet · stomach · digestive gland · tube feet

the starfish, but the tube feet extend from the ventral, or bottom, surface of the arms. When the bulb contracts, water is forced into the tube foot, causing it to elongate. When the tube foot touches a surface, its sucker holds fast. As the tube foot contracts, or shortens, water is forced back into the bulb, and the starfish is pulled forward. Movement of the animal requires the coordinated action of hundreds of tube feet.

Starfish feed on clams and oysters. They use their water-vascular system to pry open their prey. In feeding, the starfish wraps its arms around both sides of the mollusk, attaches tube feet to each shell, and pulls. Eventually, the mollusk tires, and its shell opens slightly. The stomach of the starfish is then extended out through the mouth, and inserted into the small opening between the mollusk's shell. (The starfish can insert part of its stomach into an opening as small as 0.1 millimeter.) Enzymes secreted by the stomach partly digest the soft body of the mollusk. The food is then taken into the stomach, and the stomach is pulled back into the starfish. Food passes from the stomach into the digestive glands in the arms, where digestion is completed.

Respiration in the starfish occurs by the diffusion of gases across the skin gills and tube feet. *Skin gills* are small, fingerlike structures that extend out from the body surface. They are filled with coelomic fluid. Many materials are distributed by the fluid in the coelom. This fluid bathes the body organs and supplies them with nutrients and oxygen and removes wastes. Excretion takes place by diffusion through the body surface.

Sexes are separate in the starfish. The gametes are shed through openings in the central disk into the water, where fertilization will take place. The fertilized egg develops into a bilaterally symmetrical, free-swimming larva. After several weeks of growth, the larva attaches itself to a solid surface and develops into a small starfish.

Starfish have an amazing ability to regenerate missing parts. An entire new body can grow from as little as a single arm and a tiny part of the central disk.

▲ **Figure 33–19**
Starfish Feeding on a Mussel. Starfish eat many types of shellfish. This starfish has attached its tube feet to the shell of a mussel and is prying it open.

33-4 Section Review

1. Name four echinoderms.
2. Where are echinoderms found?
3. What system of locomotion and food getting is found only in starfish and other echinoderms?

Critical Thinking

4. In a developing sand dollar, which develops first—the mouth or the anus? Explain. (*Reasoning Conditionally*)

Laboratory Investigation

External Anatomy of the Crayfish

The crayfish belongs to the class of arthropods called crustaceans. In this investigation, you will examine the crayfish to become familiar with some of the general features of arthropods.

Objectives

- *Observe* some of the characteristics of arthropods by studying the external anatomy of the crayfish.
- *Relate* the external structure of the crayfish to its way of life.

Materials

preserved crayfish   blunt probes   stereomicroscope

Alternate Approach: A live crayfish or a model or diagram of a crayfish may serve as the basis for an alternative procedure to the one below.

Procedure 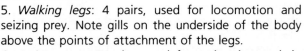 *

A. Using Figure 33–4 in your text as a reference, identify the anterior (head), posterior (rear), dorsal (back), and ventral (underside) regions of the specimen. Sketch a dorsal and a ventral view of your specimen.

B. Locate the head, thorax, and abdomen. The head and thorax are fused into a single structure, the *cephalothorax*. The dorsal part of the skeleton covering the cephalothorax is called the *carapace*. Note that segmentation of the body is evident within each body region, especially the abdomen.

C. Examine the exoskeleton. Note that its dorsal parts are thick and hard, whereas the ventral areas are thinner and less hard.

D. Locate the *rostrum*—the narrow, most anterior and dorsal part of the cephalothorax. On either side of the rostrum is a compound eye at the end of a stalk. Compound eyes, which are characteristic of arthropods, are composed of many lenses. In the crayfish, the eye stalks permit some limited movement of the eye. Observe the compound eye with a stereomicroscope.

E. Examine the abdomen, which consists of six segments. Between the segments, the joints—folded, thin areas of the exoskeleton—permit the abdomen to

*For detailed information on safety symbols, see Appendix B.

bend. At the end of the abdomen is a single flattened structure called the *telson*. The anus is seen as a small opening on the underside of the telson.

F. From anterior to posterior, examine the crayfish's 19 pairs of jointed appendages, noting the number of joints and direction of movement.

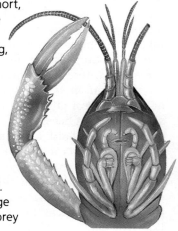

1. *Antennules:* 1 pair, short, double, rodlike, sensitive to chemicals and touch.
2. *Antennae*: 1 pair, long, single, rodlike, sensitive to chemicals and touch.
3. Mouth parts: *mandibles*—1 pair, jaws, crush food; *maxillae*—2 pairs, pass food to mouth; *maxillipeds*—3 pairs, mince and pass food to mouth.
4. *Chelipeds*: 1 pair, large claws, used for seizing prey and for defense.
5. *Walking legs*: 4 pairs, used for locomotion and seizing prey. Note gills on the underside of the body above the points of attachment of the legs.
6. *Swimmerets*: 5 pairs, used for swimming and, in females, for holding eggs. In males, the first two pairs are enlarged and used to transfer sperm.
7. *Uropods*: 1 pair at the end of the abdomen, used with the telson for swimming.

Analysis and Conclusions

1. What sex is your specimen?

2. List the characteristics of arthropods. Which of these were you able to observe?

3. How many joints are on each walking leg? How many sets of muscles would you expect at each joint?

4. List the ways in which the appendages of the crayfish are adapted to their functions.

5. In what way is the thinness between segments of the abdomen and the shape of the telson an adaptation for swimming?

Chapter Review

Study Outline

33–1 Arthropods: Crustaceans

■ Arthropods have jointed legs, a chitinous exoskeleton, and a segmented body. They have a well-developed ventral nervous system with specialized sensory receptors and an open circulatory system.

■ Class Crustacea is characterized by two pairs of antennae on the head.

■ The crayfish, a typical crustacean, has two main body sections, the cephalothorax and the abdomen. A crayfish has several pairs of appendages.

■ Crayfish have a digestive system, excretory organs, an open circulatory system, and a nervous system with a brain, ventral nerve cord, and well-developed sensory organs.

33–2 Arthropods: Centipedes to Arachnids

■ Centipedes, class Chilopoda, have a flattened, segmented body with one pair of legs on most segments.

■ Millipedes, class Diplopoda, have a rounded, segmented body with two pairs of legs on most segments.

■ Class Arachnida includes spiders, scorpions, ticks, mites, and daddy longlegs.

■ The arachnid body is divided into a cephalothorax and an abdomen. Attached to the cephalothorax are a pair of cheliceras, a pair of pedipalps, and four pairs of legs. The respiratory system of the arachnids includes organs called book lungs.

33–3 Arthropods: Insects

■ Class Insecta is the most successful class of arthropods. Insects have three distinct body sections—head, thorax, and abdomen.

■ The head contains a pair of antennae, mouthparts, and compound eyes. Attached to the thorax are three pairs of legs. Some insects have wings attached to the thorax.

■ Most insects undergo distinct changes, or metamorphosis, as they develop from eggs to adults. Metamorphosis may be complete or incomplete.

■ The grasshopper, a typical insect, has mouthparts adapted for chewing and two pairs of wings.

■ Insects are of great economic importance. They are both harmful and helpful.

33–4 Echinoderms

■ All members of phylum Echinodermata are marine. Almost all have a calcified internal skeleton and are spiny skinned.

■ The water-vascular system with its canals and tube feet is unique to echinoderms. It is used for locomotion, feeding, and gas exchange.

■ Echinoderms have no circulatory or excretory systems. Respiration occurs by diffusion of gases across skin gills and tube feet.

Vocabulary Review

Arthropoda (707)

chitin (708)

molting (708)

thorax (708)

abdomen (708)

Crustacea (709)

cephalothorax (710)

mandibles (710)

hemocyanin (711)

gills (712)

Chilopoda (713)

Diplopoda (713)

Arachnida (714)

cheliceras (714)

pedipalps (714)

book lungs (715)

spinnerets (715)

Insecta (716)

metamorphosis (718)

incomplete metamorphosis (718)

nymphs (718)

complete metamorphosis (719)

pupa (719)

entomology (720)

Hymenoptera (720)

Orthoptera (720)

Coleoptera (720)

Lepidoptera (720)

Diptera (720)

Hemiptera (720)

spiracles (721)

ovipositor (721)

Echinodermata (725)

water-vascular system (726)

canals (726)

tube feet (726)

Chapter Review

A. Sentence Completion—Fill in the vocabulary term that best completes each statement.

1. Insects respire through organs in the abdomen and thorax called _____.

2. Young arthropods undergo a process called _____ whereby their exoskeleton is shed periodically, and a new one is formed.

3. Members of class _____ have two pairs of legs attached to every body segment except the last two.

4. Starfish have a system of _____, which run into each arm.

5. The respiratory organs of spiders, _____, consist of leaflike plates containing blood vessels.

6. The study of insects is called _____.

7. The _____, which has a varying number of segments, is the middle region of the arthropod body.

8. When insect development from an egg into an adult includes a pupal stage, the series of changes is called _____.

9. Insect eggs hatch into _____ during incomplete metamorphosis.

B. Word Relationships—In each of the following sets of terms, three of the terms have a common characteristic. One term does not belong. Identify the term that does not belong.

10. Diplopoda, Crustacea, Echinodermata, Arachnida

11. tube feet, spinnerets, book lungs, pedipalps

12. cephalothorax, ovipositor, abdomen, thorax

13. Lepidoptera, Hymenoptera, Chilopoda, Diptera

14. canals, mandibles, water-vascular system, tube feet

15. pupa, molting, cephalothorax, complete metamorphosis

16. Crustacea, hemocyanin, cephalothorax, pedipalps

Content Review

17. Why is molting necessary in the arthropod life cycle?

18. What external feature does the crayfish have that explains why it is classified as a crustacean?

19. Describe the functions of crayfish appendages.

20. Describe the circulatory system of the crayfish.

21. How does the centipede obtain its food?

22. Compare the body structure of the centipede with that of the millipede.

23. Describe the respiratory process of the spider.

24. How do spiders capture their prey?

25. What common arthropod characteristics are shared by spiders, crustaceans, and insects?

26. What characteristics make insects biologically successful organisms?

27. How do the body forms of the insect and the spider differ?

28. Describe the stages of incomplete metamorphosis.

29. Describe the role hormones play in the development of the cecropia moth.

30. Describe the use of each pair of a grasshopper's legs and wings.

31. How are the reproductive processes of the crayfish, spider, and grasshopper similar?

32. In what ways are insects harmful? Helpful?

33. Why are echinoderms thought to be closely related to vertebrates?

34. Describe the structure of the water-vascular system of the starfish.

35. How is the starfish adapted to its environment?

Graphic Organizing

For information on graphic organizers, see Appendix G at the back of this text.

36. **Word Map:** Construct a word map for each of the following terms: (a) crustacean, (b) arachnid, and (c) insect. Be sure to include a definition, examples, and an overarching category in each map.

Critical Thinking

37. What would happen if a spider's silk glands and spinnerets were removed? (*Predicting*)

38. How are complete metamorphosis and incomplete metamorphosis similar? How are they different? (*Comparing and Contrasting*)

39. What could keep an insect larva from developing into a mature form? (*Identifying Causes*)

40. How is the relationship between the crayfish's gills and the crayfish similar to the relationship between the spider's book lungs and the spider? (*Relating Parts and Wholes*)

41. Suppose you found a clamshell on the beach that was only slightly open, and the clam inside was gone. What would you think was the probable cause of the clam's "disappearance"? What evidence would you look for to support your hypothesis? What evidence might you find that would not support your hypothesis? (*Identifying Causes*)

42. The animal in the photo below was identified by a magazine writer as an insect. Was the writer correct or incorrect? Why? (*Evaluating Secondhand Information*)

Creative Thinking

43. A student was given a fossilized extinct animal called a trilobite to study. How might the student determine if the trilobite was an arthropod or an echinoderm? What problems might be encountered in trying to classify the animal?

44. Medical investigators are trying to control an epidemic disease that is believed to be carried by a tick. How might they use knowledge about the structure and life function of arachnids to control the epidemic?

Problem Solving

45. Many insects seem to be attracted to light. Design an experiment to test the hypothesis that fruit flies are attracted to light.

46. Brine shrimp, tiny crustaceans, are found in salty lakes and ponds. The graph below shows the effect of water temperature on the time it takes for brine shrimp eggs to hatch. What can you conclude about brine

shrimp from the graph? How many hours would it take for eggs to hatch at 18°C and at 25°C? Is it valid to predict the amount of time it would take for eggs to hatch at 10°C? Explain.

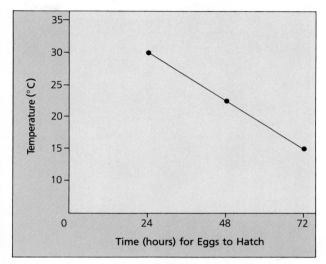

47. A scientist has discovered two fossil animals. Both fossils reveal an exoskeleton and jointed legs. One fossil seems to have a cephalothorax, abdomen, and four pairs of legs. The other seems to have a head, thorax, abdomen, and three pairs of legs. Into what phylum and class should each be classified?

48. Propose a hypothesis to explain why starfish are more numerous on rocks and coral reefs than on sandy areas. After doing this, design an experiment to test the hypothesis.

Projects

49. Collect, mount, and identify local insects and spiders. Display your specimens in class.

50. Sowbugs are crustaceans. Collect some in damp areas under large rocks or rotting logs. Examine them with a stereomicroscope. Design and perform experiments to test their reactions to light, touch, dampness, and temperature. Write a report about your findings.

51. Catch some flies in a glass jar. Using a magnifying glass or stereomicroscope, study the structure of the legs. Based on your observation, propose a hypothesis to explain how flies can walk on walls and ceilings.

52. The novelist Vladimir Nabokov was a well-known lepidopterist. Research and write a short report on the nature and extent of his collection.

▲ A clownfish makes its home among the protective tentacles of a sea anemone.

Vertebrates— Fishes to Reptiles

34

Among the curling tentacles of a sea anemone, a clownfish finds protection from predators. A safe environment is essential for the fish to survive. Yet survival depends on many factors. An internal flotation mechanism keeps the fish level. Powerful muscles maintain the flow of water across the gills, ensuring oxygen intake. A complex digestive system absorbs vital nutrients, which are transported to cells by the circulating blood. In this chapter, you will learn about the unique features and functions of a variety of animals—from saltwater fishes to land-dwelling reptiles.

Guide for Reading

Key words: Chordata, Vertebrata, Urochordata, Cephalochordata, Osteichthyes, Amphibia, Reptilia
Questions to think about:
📖 What characteristics relate the three chordate subphyla?
📖 What are the similarities and differences among fishes, reptiles, and amphibians?
📖 Why is it thought that the amphibians evolved from primitive fishes?

34-1 The Chordates

Section Objectives:

- *List* the three basic characteristics that are present in all chordates.
- *Describe* the structure and life functions of the tunicates and the lancelets.
- *List* the basic characteristics of the vertebrates, and list the eight classes into which they are grouped.

General Characteristics of Chordates

The *chordates*, phylum **Chordata** (kor DAH tah), are divided into three subphyla. The largest of these are *vertebrates*, subphylum **Vertebrata** (vert uh BRAH tah). The other two chordate subphyla are the *urochordates*, subphylum **Urochordata** (yur uh kor DAH tah), and the *cephalochordates*, subphylum **Cephalochordata** (SEF uh loh kor dah tah). The urochordates are also called the *tunicates*. The cephalochordates are commonly called *lancelets*. All the members of these two subphyla live in marine environments, do not have backbones, and are considered to be more primitive than vertebrates.

At some time in their lives, all *chordates* show the following three characteristics. These characteristics distinguish them from all other animals.

▲ **Figure 34–1**

A Vertebrate. This snake belongs to the largest chordate subphylum, Vertebrata.

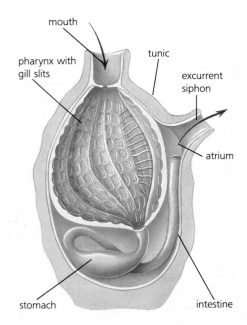

▲ **Figure 34–2**

Internal Structure of a Tunicate. Water passes through the gill slits into the atrium and out of the body through the excurrent siphon. Adult tunicates have gills but lack the hollow nerve cord and the notochord typical of chordates.

Figure 34–3

Blue-and-Gold Adult Tunicates. Tunicates are marine dwellers and grow singly or as colonies on the ocean floor. ▼

■ Chordates have a dorsal, hollow **nerve cord.**

■ Chordates have a flexible, rodlike, internal supporting structure called a **notochord** (NOHT uh kord).

■ Chordates have paired **gill slits** in the throat region, which are used in respiration.

In land-dwelling chordates, the gill slits are seen only in the embryo. In certain other chordates, such as the fishes, the gill slits function in respiration throughout life. The notochord, which is dorsal to the digestive tract, is found in the embryos of all chordates. In the tunicates, the notochord is the only supporting structure. In most vertebrates, it is replaced early in embryonic development by cartilage or bone. The cartilage or bone forms a supporting backbone, or spinal column.

The Tunicates and Lancelets

The tunicates are soft-bodied animals, found only in marine, or saltwater, environments. Adult tunicates are sessile animals that obtain food and oxygen from water that flows through their bodies. See Figure 34–2. Water enters the mouth and then passes into the pharynx. There, it flows through the gill slits in the walls of the pharynx, where gas exchange occurs. The water then passes into a chamber called the atrium and out of the body through the excurrent siphons. The gill slits also trap food particles, which then pass into the digestive system.

Adult tunicates lack a dorsal, hollow nerve cord and notochord. Larval tunicates, unlike adults, have all three chordate characteristics. They are motile and resemble tadpoles. Eventually, the larvae settle to the ocean floor, develop into adults, and lose the three chordate characteristics.

The lancelets are small marine animals. The most common member of this group is *amphioxus* (am fee AHK sus). Lancelets usually live buried in the sand with only their anterior end exposed. Adult lancelets show the three characteristic chordate structures. See Figure 34–4. As in the tunicates, water enters the body through the mouth and passes into the pharynx. From the pharynx, water passes through the gill slits, where gas exchange occurs. Food particles do not pass through the gill slits. Instead, the particles enter the digestive system directly. Water leaves the body through the atrial pore.

The Vertebrates

The vertebrates are the most numerous and complex of the chordates. The basic characteristic distinguishing vertebrates from other chordates is a spinal column made up of vertebrae. The spinal column, or backbone, is the basis for an internal supporting skeleton and allows flexibility and movement. In adult vertebrates, the spinal column surrounds or replaces the notochord.

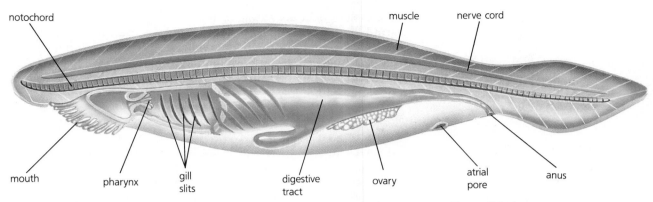

notochord

muscle   nerve cord

mouth   pharynx   gill slits   digestive tract   ovary   atrial pore   anus

▲ **Figure 34–4**

Internal Structure of a Lancelet. Adult lancelets show the characteristics of a chordate. Lancelets have a dorsal, hollow nerve cord, a notochord, and paired gill slits.

In addition to their backbones, vertebrates share a number of other characteristics.

■ The anterior part of the dorsal, hollow nerve cord is enlarged into a brain.

■ The body usually is divided into a head, neck, and trunk. The head contains the brain and various sense organs.

■ In most vertebrates, a tail is present at some stage of development.

■ There is a jointed, internal skeleton.

■ There are two pairs of appendages.

■ There is a heart with two to four chambers. The circulatory system is closed, and the red blood cells contain hemoglobin.

■ In aquatic vertebrates, gas exchange takes place in the gills. In land vertebrates, it occurs in the lungs.

■ There is a large body cavity, or coelom, containing the organs of digestion, excretion, and reproduction, as well as the heart and lungs.

■ The body covering, the skin, is made of at least two layers. The skin often forms other structures, such as glands, scales, feathers, hair, nails, claws, horns, and hoofs.

The vertebrates are divided into eight classes. These are lampreys, hagfishes, cartilaginous fishes, bony fishes, amphibians, reptiles, birds, and mammals. Taxonomists use the plural *fishes* to refer to more than one species.

Fishes, amphibians, and reptiles are *cold-blooded,* or **ectothermic** (ek tuh THER mik), animals. Their body temperature changes with the temperature of their environment. Birds and mammals are *warm-blooded,* or **endothermic** (en duh THER mik) animals. Their body temperature remains fairly constant regardless of the temperature of their environment. The normal body temperature of birds ranges between 40° C and 43° C. In mammals, it ranges from 36.5° C to 39.8° C. Warm-blooded animals are able to survive and to remain active in cold environments. They are able to maintain a constant internal temperature by varying their metabolic rate to produce more or less heat according to their needs. To save energy, some warm-blooded animals hibernate in cold weather. When an animal goes into *hibernation,* the body temperature drops, and the animal becomes inactive. Many cold-blooded animals that live in areas with cold seasons also hibernate during the cold weather when food supplies are low.

34-1 Section Review

1. Name the subphyla of the phylum Chordata.
2. List all the characteristics that distinguish chordates from other animals.
3. What basic characteristic distinguishes vertebrates from the other chordates?
4. What classes make up the vertebrates?

Critical Thinking

5. Will an *amphioxus* lose its gill slits when it becomes an adult? Why or why not? (*Reasoning Categorically*)

34-2 The Fishes

Section Objectives:

- *Name* two classes of jawless fishes and describe the basic characteristics of each class.
- *Name* three members of the cartilaginous fishes and describe the general characteristics of this class.
- *Describe* the general characteristics and life functions of the bony fishes and give three examples of this class.
- *Explain* how a bony fish adjusts the density of its body to maintain its level in the water.

The Jawless Fishes

There are four classes of fishes. The *jawless fishes* include the *lampreys,* class **Cephalaspidomorphi** (sef uh LASS peh duh morf ee), and hagfishes, class **Myxini** (mik SIH nee). These two classes are the most primitive of all living vertebrates. They have long, snakelike bodies and smooth skins with no scales. Lampreys and hagfishes have two single dorsal fins and a tail fin. They do not have the paired fins, true jaws (that is, movable upper and lower jaws), and scales that other fishes have. The skeleton of a jawless fish is made of cartilage. Unlike all other vertebrates, the notochord persists throughout life in hagfishes and lampreys. The sexes are separate, and fertilization occurs externally.

◀ **Figure 34-5**
A River Lamprey. The lamprey attaches itself to other fish with its suckerlike mouth. It then bores a hole in the fish with its teeth and feeds on its blood.

Lampreys   *Lampreys* are found in both fresh and salt water. Most are parasites. They obtain their food by attaching their round, suckerlike mouths to the bodies of other fishes. See Figure 34-5. Once attached, the lamprey uses the teeth on its tongue to gnaw a hole in the body of its victim. The lamprey then sucks the blood and body fluids of the fish.

Lampreys shed their eggs, or *spawn,* in freshwater streams. The eggs are fertilized by the male and hatch into larvae about one centimeter long. The larvae live in the mud of the streams until they mature, which takes three to seven years. The adult lives only one or two years. However, adult lampreys can do serious damage to fish populations.

Hagfishes   *Hagfishes* are found only in salt water. They feed on dead fishes, worms, or other small invertebrates that live on the ocean floor. Hagfishes are also called "slime eels" because their skin glands release large amounts of mucus if they are disturbed. Unlike the lampreys discussed above, hagfishes do not have a larval stage of development.

The Cartilaginous Fishes

The *cartilaginous* (kart ul AJ uh nus) *fishes,* class **Chondrichthyes** (kahn DRIK thee eez), include sharks, rays, and skates. Almost all members of this group are marine. They range in size from small skates less than 1 meter in length to whale sharks 15 meters long. Manta rays may be 6 meters across and weigh as much as 1200 kilograms. In members of this group, the skeleton is made entirely of cartilage, and traces of the notochord are present in the adult. Unlike the jawless fishes, the cartilaginous fishes have movable upper and lower jaws that are equipped with several rows of sharp teeth. These biting jaws enable the cartilaginous fishes to eat a wide variety of food. Like all fishes, members of this class have two-chambered hearts.

Figure 34–6

A Blue-Spotted Stingray. The stingray has a sharp poisonous spine at the base of its tail, which it uses as a weapon. ▶

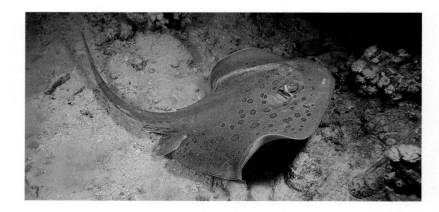

Skates and Rays   *Skates* and *rays* have flattened, winglike bodies with whiplike tails. They live on the ocean floor and feed on worms, mollusks, and crustaceans. As you can see in Figure 34–6, stingrays have poison stingers in their tails, which they use for defense. Electric rays can produce a large electric charge, which they use to stun their prey.

Sharks   *Sharks* are streamlined fishes that swim by moving their trunks and powerful tails from side to side. See Figure 34–7. Swimming forces water through the mouth, over the gills, and out through five to seven pairs of gill slits. The shark obtains the oxygen it needs from this flow of water. If a shark is caught in a net, for example, where it cannot move, it will die from lack of oxygen.

Unlike most fishes, fertilization is internal in the shark. In some species of sharks, the embryos develop within the body of the mother and are born live. In others, the eggs are covered with a leathery coat before they are released from the body of the female.

The shark's sense organs, particularly those for smell and vibration, are well developed. The shark uses its **lateral lines**— lines that extend along each side of the body—to sense vibration. The shark can smell when water enters the nostrils and passes through the *olfactory sacs*. These sacs are sensitive to various chemicals and can detect the presence of food.

Figure 34–7

Sharks. Sharks are well-adapted for predatory life. They are fast swimmers with well-developed sense organs, and powerful jaws. The jaws of the shark are lined with many sharp, triangular teeth that are replaced by other rows of teeth when lost. ▼

Most sharks are meat eaters and active hunters. However, the two largest sharks, the basking shark and the whale shark, are filter feeders. That is, they obtain food by straining microorganisms from the water.

The skin of the shark is covered with embedded, toothlike *placoid* (PLAH koyd) *scales,* which make the skin so tough that it can be used as sandpaper. Unlike the scales of bony fishes, the scales of the shark do not overlap one another.

The Bony Fishes

The *bony fishes,* class **Osteichthyes** (ahs tee IK thee eez), are the largest class of vertebrates. Most of the members of this group have bony skeletons, paired fins, and protective, overlapping scales. They are found in both fresh and salt water all over the earth. Bony fishes vary greatly in size, ranging from the Philippine goby, which is 10 millimeters long, to the swordfish, which may be more than 4 meters long.

Bony fishes differ in structure from the moray eel, which looks much like a snake, to the seahorse. See Figure 34–8. Most, however, are streamlined animals, like the perch. Their fins, which are made of skin webbing, are usually supported by bone or cartilage ribs. Most bony fishes swim by side-to-side movements of the body and tail. The fins aid in maintaining balance and in controlling the direction of movement.

Many adaptations for protection are found among the bony fishes. The pufferfish, for example, is covered with sharp spines. In

Figure 34–8
Bony Fishes. Although most fish resemble the perch in form, the moray eel (left) and the sea horse (right) are fishes too. ▼

▲ **Figure 34–9**
A Spiny Pufferfish. The spines of this pufferfish stand out in response to a threat.

times of danger, it inflates itself with air or water so that its spines stand out. See Figure 34–9. A flying fish has a pair of winglike pectoral fins. To escape its enemies, it leaps out of the water and glides through the air for distances of 100 meters or more. The large South American electric eel can stun or kill its enemies with a strong electric charge.

Bony fishes have a complex nervous system. Ten pairs of cranial nerves extend from the brain, and spinal nerves radiate from the spinal cord. Bony fishes usually have a pair of well-developed eyes, two nostrils, and lateral lines, like those on the shark, on each side of the body. The nostrils are used only for smelling, not for breathing.

There are usually four pairs of gills. These lie on each side of the body under a protective bony flap called the *gill cover,* or **operculum** (oh PER ka lum). Water moves in through the mouth, passes over the gills, and then flows out of the body. The movement of muscles in the mouth and gill covers maintains the flow of water over the gills. A bony fish has a two-chambered heart, consisting of an atrium and a ventricle. Blood travels from the heart to the gills, where oxygen is picked up and carbon dioxide is released. The blood is then distributed through blood vessels to all parts of the body before it returns to the heart.

Most of the body of the fish is muscle. As shown in Figure 34–11, along the ventral side is a small space containing the digestive, excretory, and reproductive organs. The digestive system includes the mouth, pharynx, esophagus, stomach, intestine, liver, gallbladder, pancreas, and anus. The gills are located on the sides of the pharynx. Attached to the short intestine are three tubular structures that are called *pyloric caeca* (SEE kuh). These structures

Figure 34–10
Perch. Most of the fish familiar to us, such as the perch, belong to the class Osteichthyes. ▼

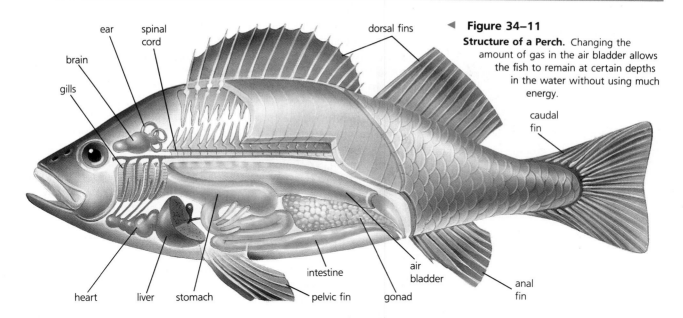

◄ **Figure 34–11**
Structure of a Perch. Changing the amount of gas in the air bladder allows the fish to remain at certain depths in the water without using much energy.

ear   spinal cord   dorsal fins   caudal fin

brain   gills

heart   liver   stomach   intestine   pelvic fin   air bladder   gonad   anal fin

Critical Thinking in Biology

Reasoning by Analogy

Suppose you applied for an after-school job. Before deciding to hire you, the employer might ask to see your high-school record. The employer may reason that the job's responsibilities and the responsibilities you have for your schoolwork are similar in many ways. Thus, she may base her decision on your grades.

This type of reasoning is known as reasoning by analogy. When you reason by analogy, you apply a conclusion about one situation to another situation that you judge to be similar.

1. You are examining a perch, a bony fish, and you discover its swim bladder. Would you expect to find a swim bladder in a whitefish, another bony fish? Explain.
2. Would you expect to find a swim bladder in a shark? Explain.
3. When humans develop bacterial infections, they often run fevers. Scientists do not know if a fever is harmful or beneficial to a human. To investigate this question, some scientists studied lizards with bacterial infections. Healthy lizards like a body temperature of about 38.5°C. Infected lizards seek warmer environments. In one experiment, infected lizards were kept at 34°, 36°, 38°, 40°, and 42°C. Survival increased as the temperature increased. It was greatest at 42°C.

What can you conclude from this experiment? What does this experiment tell you about fevers in humans? Explain your answer.
4. Think About It Describe the thought process you used as you answered question 3.

aid in the absorption of digested materials. Two kidneys filter some nitrogenous wastes from the blood. These wastes, which are in the form of urea, pass through the ureters to the urinary opening and into the water. The gills, however, excrete most of the nitrogenous wastes in the form of ammonia.

All fishes are slightly heavier than water. In order to keep from sinking, they must have some type of flotation device, or they must keep swimming to maintain their level in the water. Most bony fishes have a gas-filled sac called the **swim bladder,** or *air bladder,* in the upper part of their body cavities. The swim bladder acts as a float. That is, it regulates the buoyancy of the fish. By increasing or decreasing the amount of gas in the swim bladder, the fish is able to change the density of its body. The ability to change density allows the fish to remain suspended in the water at any depth. In lungfishes, which are air-breathing fishes, the swim bladder serves as a lung.

As in vertebrates in general, the sexes are separate in bony fishes. The male has testes, and the female has ovaries. Fertilization and development are usually external. The female deposits eggs in the water, and the male then discharges *milt,* a sperm-containing fluid, over the eggs.

34-2 Section Review

1. Describe the feeding habits of lampreys. What are the feeding habits of hagfishes?
2. List the general characteristics of the cartilaginous fishes.
3. What is the function of the olfactory sacs in the shark?
4. List the general characteristics of the bony fishes. What structures aid in maintaining balance and in controlling the direction of movement in bony fishes?

Critical Thinking

5. Which of the following marine animals are bony fishes: perch, skates, moray eels, manta rays, lampreys, sharks, and sea horses? Explain your answer. (*Classifying*)

34-3 The Amphibians

Section Objectives:

- *List* the major characteristics of amphibians and name three members of this class.
- *Describe* the structure and life functions of the frog, including the circulatory and respiratory systems.
- *Describe* metamorphosis in the frog.
- **Laboratory Investigation:** *Observe* the structure and basic features of a vertebrate skeleton (p. 756).

General Characteristics of Amphibians

The *amphibians,* class **Amphibia** (am FIB ee uh), include frogs, toads, salamanders, and newts. Some amphibians live their entire adult lives on land. Others are found only in or around water. In either case, for most amphibians, reproduction and development must take place in water or in a moist place. It is thought that amphibians evolved from primitive, air-breathing fishes. Most scientists believe that amphibians were the very first land-dwelling vertebrates.

In addition to their need for water for reproduction, amphibians share the following characteristics.

- The skin is generally thin and contains mucus-secreting glands.
- There are two pairs of limbs, which are used for walking, jumping, and/or swimming.
- There is a pair of nostrils connected to the mouth cavity.
- The heart has three chambers—two atria and one ventricle.
- The young usually show a distinct larval form and gradually develop adult characteristics.

There are two major groups of amphibians: the tailed amphibians, such as salamanders, and the tailless amphibians, such as frogs. A third, minor group consists of small, tropical, wormlike animals that burrow in moist soil.

Salamanders   The tailed amphibians include salamanders and newts. (Newts are actually a type of salamander.) These animals have long bodies, long tails, and two pairs of short limbs. See Figure 34–12. Most salamanders range from 8 to 20 centimeters in length. However, the Japanese giant salamander, which is the largest living amphibian, may reach a length of 1.5 meters. Salamanders eat fishes, snails, insects, worms, and other salamanders. Some are entirely aquatic, while others live under rocks or logs or in other moist places. They are active only at night. Freshwater salamanders retain their gills, which are used for breathing. The mudpuppy is a well-known salamander that lives in the streams and lakes of the eastern United States. Both the mudpuppy and the *axolotl* (AK suh lot ul), which is a salamander found in the Rocky Mountains and Mexico, are actually larval forms that can reproduce sexually. See Figure 34–13.

Frogs and Toads   The tailless amphibians include frogs and toads. As adults, they have short, squat bodies and lack tails. Their large, powerful hind legs are well suited for jumping.

Toads have dry, rough, warty skins. They can live on land far away from water. Toads burrow or take shelter during the day and come out to feed at night when it is cooler and more humid. Some toads live in the desert, but like most amphibians, they need water for reproduction. During the winter, toads hibernate by burrowing into the ground. During hibernation, life processes slow down, and an animal becomes inactive. See Figure 34–14.

▲ **Figure 34–12**

A Blue-Spotted Salamander. Salamanders live in most northern temperate and tropical regions throughout the world, but most are found in North America.

Figure 34–13

An Axolotl. While most salamanders change form as they develop into adults, the axolotl and the mudpuppy remain in the larval form throughout life. ▼

Figure 34–14

A Texas Toad. Unlike frogs, toads can live far away from water. ▶

Frogs have thin, moist skins that are loosely attached to their bodies. They generally live near ponds, streams, swamps, or other bodies of water. During the winter, they hibernate in the mud at the bottom of pools and streams. Frogs and toads eat insects and worms. The larval forms, called **tadpoles,** eat aquatic plants.

Frogs and toads have many enemies, including snakes, birds, and turtles. Among their protective adaptations are their coloring, which provides good camouflage, and their ability to leap. They often dive underwater to escape their enemies. Glands in their skin produce secretions that are unpleasant tasting or poisonous to their enemies.

Structure and Life Functions of the Frog

The organ systems of the frog are similar to those of most other vertebrates, including humans. For this reason, they are often studied in detail in biology courses.

External Features of the Frog   The frog has a short, broad body with two short forelimbs and two long, muscular hind limbs. See Figure 34–15. The hands have four fingers and are not webbed, while the feet have five webbed toes and are adapted for jumping and swimming. The upper surface of the frog is yellow-green to green-brown in color, and the underside is whitish. This coloring allows the animal to blend in to its surroundings. The skin can change color to some extent to further hide the animal.

Two large, movable eyes protrude from the head. They permit vision in all directions. Each eye is protected by three eyelids—the upper eyelid, the lower eyelid, and the **nictitating** (NIK tuh tayt ing) **membrane.** The nictitating membrane is transparent, which allows the frog to see underwater. Behind each eye is a round eardrum called the **tympanic membrane,** which picks up sound waves from air or water. Two nostrils on the tip of the head enable the frog to breathe air while the rest of the body is floating underwater.

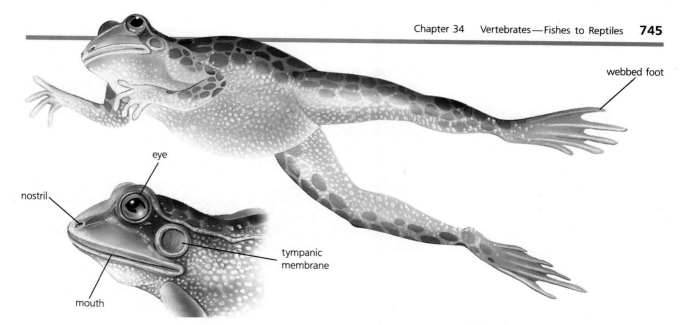

webbed foot

eye

nostril

tympanic
membrane

mouth

▲ **Figure 34–15**
External Structure of the Frog.

The Digestive System   The mouth of the frog is large. The sticky tongue is attached to the front end of the lower jaw. The frog can flip out its tongue rapidly to catch insects in flight. The food sticks to the tongue, which is then pulled back into the mouth. Teeth along the edge of the upper jaw and on the roof of the mouth aid in gripping the food. From the mouth, the food is forced down the opening of the esophagus in the back of the throat.

Food passes down the esophagus into the stomach, where digestion begins. See Figure 34–16. The partially digested food

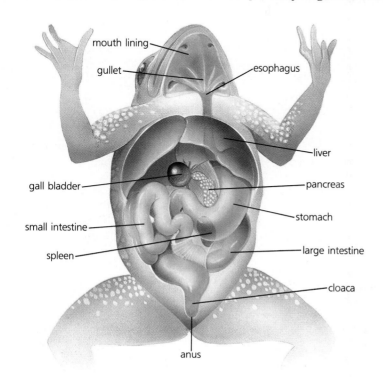

mouth lining

gullet

esophagus

liver

gall bladder

pancreas

small intestine

stomach

spleen

large intestine

cloaca

anus

◄ **Figure 34–16**
Digestive System of the Frog.

then passes through the pyloric valve into the small intestine. The pancreas and liver secrete digestive juices that pass through ducts into the small intestine. Most digestion and absorption of food takes place in the small intestine. Undigested food passes from the small intestine into the large intestine and then into the **cloaca** (kloh AY kuh). The cloaca empties through the anal opening. The cloaca serves as a passageway for urine, eggs, and sperm.

The Circulatory System   The frog has a three-chambered heart made up of two thin-walled atria and one muscular ventricle, as shown in Figure 34–17. Blood leaving the ventricle enters a large blood vessel that branches immediately into two arteries. Each of these arteries divides into many smaller arteries and then into capillaries. Blood from the capillaries is returned to the heart through the veins. Blood from the lungs is carried to the left atrium by the right and left pulmonary veins. This blood is oxygenated only when the frog is breathing air with its lungs. Blood from all the other parts of the body is returned through three large veins into a thin-walled sac. Blood from the sac enters the right atrium. Both the right and left atria empty blood into the ventricle. Thus, blood pumped out by the ventricle is a mixture of oxygenated blood from the left atrium and deoxygenated blood from the right atrium.

The Respiratory System   The respiratory system of the adult frog includes the lungs, the lining of the mouth, and the skin. These structures have thin, moist surfaces and blood vessels.

The frog uses its lungs to meet most of its oxygen requirements. The two lungs are elastic sacs with thin walls. Air is forced

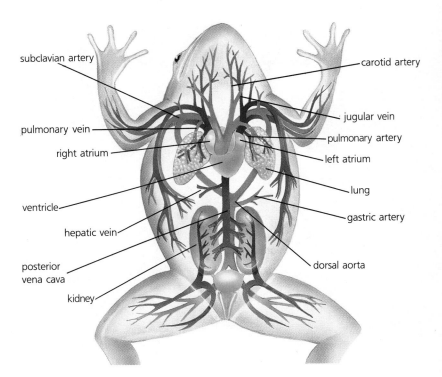

Figure 34–17
Circulatory System of the Frog. ▶

into the lungs by the pumping action of muscles in the floor of the mouth. The floor of the mouth is lowered, and air is drawn into the closed mouth through the nostrils. Then, the nostrils are closed, the floor of the mouth is raised, and air is forced from the mouth, through the glottis, and into the lungs. The exchange of oxygen and carbon dioxide occurs in the capillaries of the lungs.

The thin roof of the mouth also serves as a respiratory surface. The thin, moist skin of the frog can serve as a respiratory surface either in air or in water. This is especially important when the frog remains underwater for long periods of time. When the frog hibernates during winter, its body metabolism is reduced, and skin respiration alone meets the oxygen needs of the animal.

The Nervous System   Like humans, frogs have a central nervous system and a peripheral nervous system. See Figure 34–18. The brain and the spinal cord make up the central nervous system. Ten pairs of cranial nerves connect the brain to parts of the head and abdomen. The spinal cord, encased in vertebrae, is connected to parts of the body by 10 pairs of spinal nerves. The cranial and spinal nerves make up the peripheral nervous system.

The frog's brain is made up of the olfactory lobes, the optic lobes, the cerebrum, the cerebellum, and the medulla. The function of olfactory lobes is smell, and the function of optic lobes is vision. The cerebrum interprets sensory information and controls voluntary muscle action. The cerebellum coordinates movement. The medulla, which connects the brain to the spinal cord, controls many involuntary muscle actions.

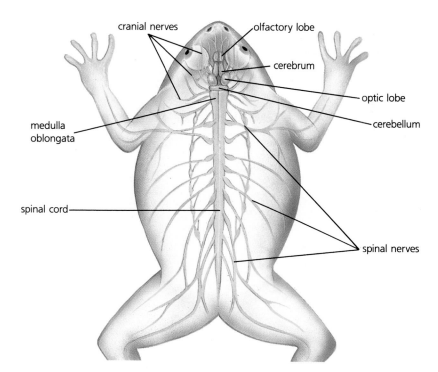

◀ **Figure 34–18**
Nervous System of the Frog.

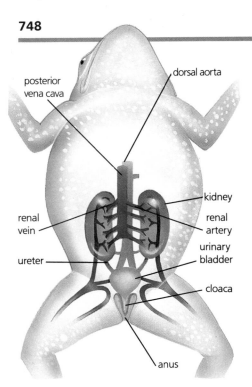

▲ **Figure 34–19**
Excretory System of the Frog.

The sense organs of the frog include eyes, tympanic membranes for hearing, inner ears for balance, taste buds on the tongue, odor-sensitive nerve endings in the nasal passages, and sensory nerve endings in the skin.

The Excretory System   Although most of the carbon dioxide produced by the frog is excreted through the skin, other metabolic wastes are excreted by the kidneys. The kidneys are located in the back of the body cavity on either side of the spine. See Figure 34–19. Wastes filtered from the blood form urine. Urine from each kidney is carried by the ureters to the bladder for temporary storage. From the bladder, the urine passes into the cloaca and out of the body.

The Reproductive System   In the female, the ovaries are located along the back, above the kidneys, as shown in Figure 34–20a. Large numbers of eggs, produced by the ovaries, enter the oviducts, which are coiled tubes. There, the eggs are surrounded by a jellylike substance secreted by the walls of the oviducts. At the base of the oviducts are sacs in which the eggs are stored until they are released from the body through the cloaca.

In the male, the testes are small, yellowish, bean-shaped organs located just above the kidneys. See Figure 34–20b. Sperm

Figure 34–20
Reproductive System of the Frog. In male and female frogs, the reproductive organs are ventral to the kidneys. ▼

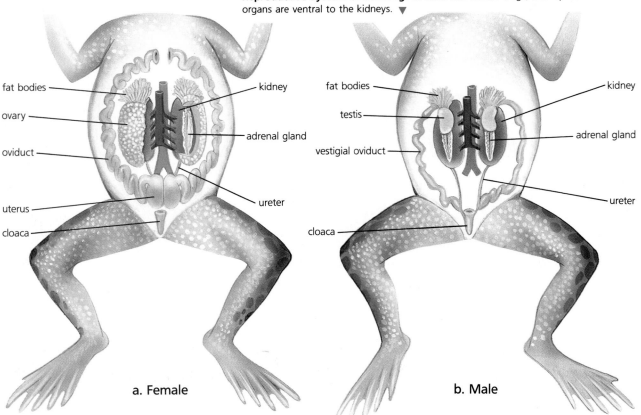

a. Female

b. Male

produced in the testes then pass to the kidneys through microscopic tubules. From the kidneys, the sperm are carried by the ureters to the cloaca. During mating, the sperm are discharged from the male through the cloaca.

Fertilization and Development   Fertilization in frogs is external. During mating, the male clasps the female with his short front legs. This is known as *amplexus* (am PLEK us) (see page 433). As the eggs leave the body of the female, the male releases sperm over them, so that many are fertilized.

After six to nine days, the eggs hatch into tadpoles. See Figure 34–21. They are fishlike, with no legs, a long tail, and gills. The tadpole has a two-chambered heart. Metamorphosis of the tadpole into the adult frog involves the development of legs, the absorption of the tail, the disappearance of gills, and the development of lungs and a three-chambered heart, as well as still other changes. The leopard frog completes metamorphosis in about three months, while bullfrogs take two or three years to complete the process.

Figure 34–21

Development of the Frog. (a) A mass of eggs. (b) Newly hatched tadpoles. (c) Tadpole with limbs. (d) Adult frog. ▼

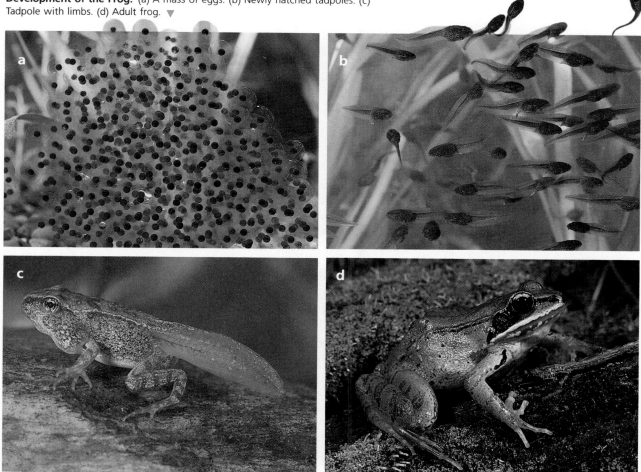

34-3 **Section Review**

1. Name the two main groups of amphibians and give an example of each.
2. What is the function of the nictitating membrane in the frog?
3. What is the tympanic membrane?
4. Name the respiratory surfaces of the frog.

Critical Thinking

5. An ecologist studying a local frog species observed that the population declined in years with droughts in the spring but was not affected by droughts at other times of the year. How would you explain this observation? (*Identifying Causes*)

34-4 Reptiles

Section Objectives:

- *Explain* why reptiles are better adapted to life on land than amphibians.
- *List* the general characteristics of reptiles.
- *Compare* and *contrast* the external structures of crocodiles and turtles.
- *Describe* the major characteristics of lizards and snakes.

General Characteristics of Reptiles

The *reptiles,* class **Reptilia** (rep TIL ee uh), include crocodiles, alligators, turtles, tortoises, lizards, and snakes. Reptiles are well adapted for life on land. Unlike amphibians, they do not require water for reproduction. Fertilization is internal. The fertilized egg is enclosed in a thick, leathery, waterproof shell that protects it from drying out (see page 447). Unlike amphibians, reptiles do not have gills at any stage in their life cycle, and they do not undergo metamorphosis. As you can see in Figure 34–22, when reptiles hatch from eggs, they look like miniature adults.

In addition to their shelled eggs, reptiles share a number of other characteristics.

- The skin of reptiles is dry and covered with scales. This waterproof covering protects them from excessive water loss and from predators.

- Except for snakes, reptiles have two pairs of legs. Most reptiles have five clawed toes on each leg. The legs are adapted for climbing, running, or paddling.

- In most reptiles, there is a three-chambered heart consisting of two atria and one partly divided ventricle. The partial separation of the ventricle lessens the mixing of oxygenated and deoxygenated

◀ **Figure 34–22**
Newly Hatched Wood Turtle. The newly hatched wood turtle, like all reptile young, looks like a miniature adult.

blood in the heart. This in turn increases the amount of oxygen carried to the body cells. Crocodiles and alligators have four-chambered hearts.

▪ Reptiles have well-developed lungs that are protected by a rib cage.

▪ Nitrogenous wastes are excreted mainly as uric acid, so that the urine of many reptiles is a semisolid paste. This is an excellent adaptation for conserving water.

From the fossil record, it appears that reptiles were once a highly successful group. Prehistoric reptiles were a diverse group that included the dinosaurs. There were swimming reptiles, flying reptiles, reptiles that walked on four legs, and reptiles that walked on two legs. Today, however, there are only four orders of living reptiles. One of these orders has only one member, the *tuatara,* which is a lizardlike animal found only in New Zealand. This primitive reptile, shown in Figure 34–23, has an extra eyelike structure on the top of its head.

◀ **Figure 34–23**
A Tuatara. The tuatara is the only living member of the order Rhynchocephalia, a group that has characteristics more primitive than those of the lizards.

Figure 34–24

Alligators. The American alligator is found in the rivers, bayous, and swamps of the southeastern United States. ▶

Crocodiles and Alligators

Crocodiles and alligators are the largest living reptiles. They range in length from 2.5 meters to more than 7 meters. They are found in lakes, swamps, and rivers in tropical regions all over the world. Both alligators and crocodiles have long snouts, powerful jaws with large teeth, and long, muscular tails. See Figure 34–24. The tails are used in swimming. These two types of reptiles look very much alike, but their teeth are arranged differently. Furthermore, the American alligator has a much broader snout than the American crocodile. In both alligators and crocodiles, there are nostrils at the tip of the snout. This allows the animals to lie in water, with only the tip of the snout and the eyes above the surface.

Alligators and crocodiles feed on animals that they capture with their massive, toothed jaws. Crocodiles are more vicious and aggressive than alligators. They will attack large animals, including humans, cattle, and deer. It is the less aggressive alligator that is most commonly found in the southeastern United States. Alligator hides are used for leather goods. Overkilling brought some species close to extinction, but protective laws have allowed the populations to increase again.

Turtles

Turtles are found on land and in both fresh and salt water. Land-dwelling turtles are sometimes called tortoises. As you can see in Figure 34–25, the body of a turtle is enclosed in protective shells. The upper shell is called the *carapace*, and the lower shell is called the *plastron*. For defense, the legs, tail, neck, and head of some turtles can be pulled completely inside the shell. Turtles feed on plants and small animals. They have no teeth, but they grab and tear their food with the hard, sharp edges of their beaks.

Land-dwelling turtles are slow moving. Their short legs have claws that are used in digging. Sea turtles have paddle-shaped legs

▲ **Figure 34–25**

Turtles. The turtle is unique in having a protective shell around its body. If threatened, the turtle will pull its legs, tail, neck and head into its shell.

that are used in swimming. All turtles, including ocean-dwelling turtles, lay their eggs on land in holes that they have dug with their hind legs.

Some species of marine turtles reach lengths of 2 meters and weights of more than 500 kilograms. Some land turtles have reached weights of more than 180 kilograms. Turtles may live to be more than 100 years old.

Lizards and Snakes

Lizards and snakes belong to the same order, but there are many differences between them. Most lizards are four-legged, while snakes have no legs. Lizards have movable eyelids and external ear openings, while snakes have immovable eyelids and no external ear openings. In both lizards and snakes, the skin is covered with scales. Both animals shed their skins periodically. Unlike the scales of the lizard, which are almost uniform in size, the scales of the snake vary in size. While scales on the back and sides of a snake are small, on the belly of the snake, there is a single row of large scales. These scales act as cleats to give the snake traction as it moves across the terrain.

Lizards   Lizards are an extremely diverse group. They are found in deserts, in forests, and in water. Smaller lizards feed on insects, worms, spiders, and snails. Larger ones may also eat eggs, small birds, other lizards, and small mammals. A few lizards feed on plants. Many lizards can shed their tails if seized by an enemy. The tail wiggles, distracting the other animal, and the lizard escapes. A new tail is regenerated in a short time.

The gecko is a small lizard that has sticky toe pads. These pads allow it to walk on vertical surfaces and upside down. It catches insects with a flick of its long, sticky tongue. The American chameleon, the anole, has a remarkable ability to change color and blend in with its surroundings. See Figure 34–26. The Gila monster is a highly colored lizard found in the deserts of the southwestern United States. Its bite is poisonous but rarely fatal to humans. The largest lizard is the Komodo dragon of Indonesia. It weighs over 100

▲ **Figure 34–26**

Lizards. The Komodo dragon (right) is the largest of the lizards. Some species of Anoles (top) have the ability to change color, generally from green to brown. The gecko (bottom) is the most primitive of all lizards.

Figure 34–27

Internal Structure of a Snake. A snake's skeleton is made of a skull and many vertebrae and ribs. Most of the snake's internal organs are long and thin. ▼

kilograms and may be 3 meters long. The Malaysian flying lizard has skin extensions on its sides that enable it to glide from tree to tree.

Snakes Although snakes are feared by many people, only about 200 of the 2500 known species are poisonous. Snakes are actually more helpful than harmful because they kill large numbers of rodents.

Snakes are found on the ground, in trees, and in both fresh and salt water. They are most abundant in tropical areas. The body of a snake consists of the head, trunk, and tail. See Figure 34–27. The trunk contains the body cavity with elongated internal organs. The digestive tract is a straight tube running from the mouth to the anus. The tail is the portion of the body following the anus. The skeleton has a large number of vertebrae and ribs.

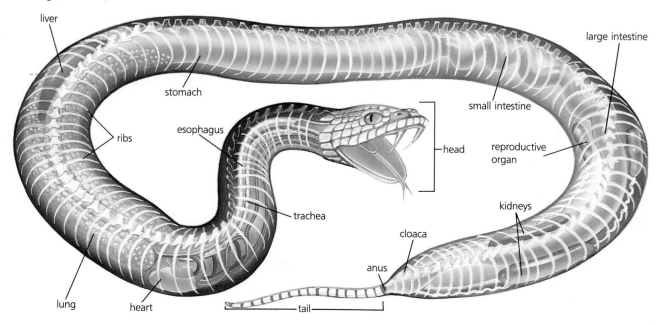

Snakes have special sense organs that are used in hunting for food. The forked tongue of the snake picks up odor-bearing particles. The particles are identified by the *Jacobson's organs* located in the roof of the mouth. Snakes are deaf to airborne sounds, but there are sense organs within the skull that respond to vibrations in the ground. Some snakes are known as *pit vipers*. These snakes have heat-detecting pit organs on their heads, between the nostrils and the eyes. With these organs, pit vipers can accurately track and strike warm-blooded prey, even at night or in deep burrows.

Snakes feed on mice, rats, frogs, toads, insects, fish, and other small animals, depending on where they live. Some snakes swallow their prey alive, while others kill their prey before they swallow it. Large snakes, such as pythons, boas, and king snakes, coil their bodies around their victims and crush or strangle them to death. Some snakes poison their victims.

Snakes can swallow animals that are much larger than they are because the structure of the jaw allows the mouth to open wide. Furthermore, the ribs are unattached at one end, which allows the body cavity to expand. Swallowing is a slow process. The teeth point backward so that the prey cannot pop out of the mouth, and the windpipe is projected forward so that breathing is not blocked. After one large meal, a snake is able to go for weeks or months without eating.

Poisonous snakes have a pair of specialized teeth called *fangs*, shown in Figure 34–28. The fangs are connected to salivary glands, which produce a poison, or venom. Some venoms are *neurotoxins*, or poisons that attack nervous tissues. They can paralyze muscles and affect the action of the heart and lungs. Other venoms, called *hemotoxins*, break down red blood cells and blood vessels. When a snake bites, the fangs carry the venom into the victim. Some snakes have hollow fangs that act like hypodermic needles, injecting the venom. In others, the fangs are grooved, and the venom passes into the victim by capillary action. The poisonous snakes of the United States include rattlesnakes, water moccasins, copperheads, and coral snakes. The best known and most common snake is the garter snake, which is harmless.

▲ **Figure 34–28**
Fangs of a Diamondback Rattlesnake. The rattlesnake, which is most common in the southwestern United States and western Mexico; injects toxin into its victims through hollow fangs. When its jaws are closed, the fangs are folded back against the roof of its mouth.

34-4 Section Review

1. Name four members of the class Reptilia.
2. What are the upper and lower shell of the turtle called?
3. Name the poisonous snakes of the United States.
4. What heat-sensing organs are used by some snakes to detect prey?

Critical Thinking

5. Name two characteristics of reptiles that are important adaptations for life in arid climates. (*Judging Usefulness*)

Laboratory Investigation

Comparing Fish and Frog Skeletons

The skeletons of all vertebrates reflect the basic vertebrate body plan—a skull, a backbone, and paired front and back appendages. In this activity, you will compare the skeletons of two vertebrates—a bony fish and an amphibian. Look for similarities, as well as for evidence of how the basic body plan is adapted to the functions of the two animals, each in its environment.

Objectives

- *Observe* the structure and basic features of a vertebrate skeleton.
- *Compare* the skeletons of an aquatic and a terrestrial vertebrate.

Materials

bony fish skeleton   frog skeleton

Alternate Approach: Diagrams or photographs of fish and frog skeletons may be used as the basis for an alternative to the procedure below.

Procedure

A. Examine each skeleton, noting its major parts—skull, backbone, ribs, and limbs.

B. Compare the skulls of the two animals. Look at the jaws and check their movement.

C. Examine and sketch the backbones. Compare the number, size, and shape of the vertebrae of the two specimens. Observe the bones that stick up from the fish vertebrae—the neural spines. Compare the neural spines of the fish and frog. Draw and label a fish vertebra and a frog vertebra.

D. Compare and record the number, size, and shape of ribs in the fish and frog. Draw a rib of a fish and a frog.

E. Look at the fish's fins. Median fins, which are not paired, include one or two dorsal fins on the dorsal (upper) side of the animal, one caudal (tail) fin, and one anal fin, near the anus on the ventral (bottom) side.

F. Other fins are paired. These include the pectoral fins, located directly behind the skull. They are homologous to the forelimbs of other vertebrates—for example, the arms of a human. The pectoral fins are attached to the pectoral girdle, which is homologous to our

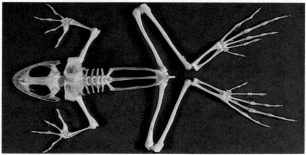

collarbone and shoulder blade. The pelvic fins are located ventral to the pelvic girdle. They are homologous to the hind limbs of other vertebrates. Draw and label the pelvic and pectoral girdles and fins of the fish.

G. Study the pectoral and pelvic girdles and the limbs of the frog. In the hind limbs, you can see the bone pattern typical of land-dwelling vertebrates: one upper leg bone, two lower leg bones (these are often partly fused), a series of "ankle" bones, and a foot with five toes. Observe the same bone pattern in the front limbs of the frog. Draw and label the pectoral and pelvic girdles and the front and hind limbs of the frog.

Analysis and Conclusions

1. Which animal, the fish or the frog, has sturdier bones? Why would you predict this to be the case?

2. Based on the skull sizes you observed, which animal would you infer has the larger brain?

3. How do the vertebrae of the fish and frog compare?

4. Amphibians are thought to have evolved from primitive fish whose skeletons are similar to those of modern-day fish. Describe the changes that would have been necessary for a fish skeleton to have evolved into an amphibian-type skeleton.

Chapter Review

Study Outline

34-1 The Chordates

■ The phylum Chordata includes the subphyla Urochordata, Cephalochordata, and Vertebrata. All chordates have, at some stage in their life, a dorsal, hollow nerve cord, a notochord, and paired gill slits.

■ Tunicates are soft-bodied marine animals that use gill slits for respiration and feeding. Lancelets are small marine animals with completely separate respiratory and digestive systems.

■ Vertebrates have a spinal column consisting of vertebrae. They include fishes, amphibians, reptiles, birds, and mammals.

34-2 The Fishes

■ Jawless fishes include lampreys and hagfishes. They have cartilage skeletons and a notochord that is present throughout life.

■ Cartilaginous fishes include sharks, rays, and skates. They have cartilage skeletons, and movable upper and lower jaws with teeth.

■ Bony fishes have bony skeletons, paired fins, and overlapping scales.

34-3 The Amphibians

■ The amphibians have thin skin with mucus-secreting glands, two pairs of limbs, a pair of nostrils, and a three-chambered heart. They need water for reproduction, and they have a distinct larval form. Development generally involves metamorphosis.

■ Tailed amphibians have long bodies and two pairs of short limbs. Tailless amphibians have short, squat bodies and powerful hind legs.

■ The frog has a well-developed digestive system, a three-chambered heart, and a respiratory system that includes lungs, the lining of the mouth, and the skin.

34-4 Reptiles

■ The reptiles include crocodiles, alligators, turtles, lizards, and snakes.

■ Reptiles do not require water for reproduction. Fertilization is internal. They do not have gills at any stage, and they do not undergo metamorphosis.

■ Reptiles have eggs with shells, dry, scaly skin, a three- or four-chambered heart, and well-developed lungs. Except for snakes, reptiles have two pairs of legs. Nitrogenous waste is excreted mainly as uric acid.

Vocabulary Review

Chordata (733)	lateral lines (738)
Vertebrata (733)	Osteichthyes (739)
Urochordata (733)	operculum (740)
Cephalochordata (733)	swim bladder (742)
nerve cord (734)	Amphibia (743)
notochord (734)	tadpoles (744)
gill slits (734)	nictitating membrane (744)
ectothermic (735)	
endothermic (735)	tympanic membrane (744)
Cephalaspidomorphi (736)	
Myxini (736)	cloaca (746)
Chondrichthyes (737)	Reptilia (750)

A. Definition—Replace the italicized definition with the correct vocabulary term.

1. The *protective bony flap* covers the gills of bony fishes.

2. The frog has a *transparent eyelid* that allows it to see underwater.

3. Members of the class *of cold-blooded vertebrates with dry skin and a three- or four-chambered heart* are well adapted for life on land.

4. A *flexible, rodlike, internal supporting structure* is found in all chordates during some part of their lives.

5. Sharks, rays, and skates belong to the class *of fishes with cartilage skeletons and with scales, fins, and movable jaws.*

6. During some part of their lives, chordates have *paired structures used for respiration.*

Chapter Review

7. The *cavity that serves as an excretory and reproductive passageway* is located between the large intestine and anal opening of the frog.

8. Tunicates belong to the subphylum *of chordates that are soft, sessile, marine animals.*

9. *Rows of sensory organs* along each side of its body allow the shark to sense vibrations.

B. Multiple Choice—Choose the letter of the answer that best completes each statement.

10. Animals whose body temperature remains fairly constant regardless of the temperature of the environment are (a) Reptilia (b) endothermic (c) Amphibia (d) ectothermic.

11. The larval form of a frog or toad is called a (a) notochord (b) cloaca (c) tadpole (d) reptile.

12. Animals with an enlarged brain and a spinal column made up of vertebrae that enclose the nerve cord belong to the subphylum (a) Urochordata (b) Vertebrata (c) Chordata (d) Cephalochordata.

13. The gas-filled sac that regulates buoyancy in most bony fishes is called the (a) lateral line (b) tympanic membrane (c) operculum (d) swim bladder.

14. Chordates that may live on land but that need water for reproduction and development belong to the class (a) Reptilia (b) Osteichthyes (c) Cephalaspidomorphi (d) Amphibia.

15. All chordates at some time in their lives have a dorsal, hollow (a) nerve cord (b) swim bladder (c) notochord (d) cloaca.

16. In frogs, the structure located behind the ear that picks up sound waves is called the (a) nictitating membrane (b) lateral line (c) tympanic membrane (d) operculum.

Content Review

17. Why are tunicates classified as chordates?

18. Describe the structure of lancelets, and explain how gas exchange and nutrition take place.

19. Compare the function of gill slits in tunicates and lancelets.

20. Describe the general characteristics of vertebrates.

21. What is the adaptational advantage of being warm-blooded?

22. Describe the general characteristics of the jawless fishes.

23. What is the advantage to the cartilaginous fishes of having movable upper and lower jaws?

24. What features of electric rays and sharks are adapted for catching prey?

25. Why do sharks need to swim continuously?

26. Describe the respiratory system of the bony fishes.

27. What structural adaptations make bony fishes well suited to swimming?

28. Describe the characteristics of amphibians.

29. Trace the path of food through the frog's digestive system.

30. How does the frog's heart pump both oxygenated and deoxygenated blood to the body?

31. How is respiration in frogs suited to life both in the water and on land?

32. Describe the frog's reproductive system.

33. Describe the characteristics of reptiles.

34. Describe the head and external body features of the alligator and the turtle.

35. What special adaptations of snakes enable them to swallow large prey?

36. When poisonous snakes bite, how do their venoms affect the victims?

Graphic Organizing

For information on graphic organizers, see Appendix G at the back of this text.

37. **Concept Map:** Construct a concept map showing the basic characteristics of the phylum Chordata and its three subphyla. Include the following concepts: urochordates, vertebrates, dorsal hollow nerve cord, soft-bodied, tunicates, spinal column, cephalochordates, lancelets, brain, red blood cells, gas exchange, gill slits, heart, notochord. Include other appropriate concepts of your own choosing. Be sure to use linking words between concepts.

Critical Thinking

38. Why have biologists classified tunicates as chordates rather than as members of the phylum Porifera? (See Chapter 32.) (*Identifying Reasons*)

39. In what ways is an adult whale shark similar to an adult flounder? In what ways are the two different? (*Comparing and Contrasting*)

40. Compare and contrast the ways in which snakes and frogs are adapted for life on land. Consider such features as their water needs, reproduction, respiration, and excretion. (*Comparing and Contrasting*)

41. Reptiles generally inhabit warm climates as opposed to cold regions. Suggest the probable cause for this geographic distribution. (*Identifying Causes*)

Creative Thinking

42. Leatherback sea turtles are the largest of all sea turtles, measuring over six feet long. Although once abundant, they are now endangered due to the commercial marketing of their meat, oil, eggs, leather, and tortoiseshell. Suggest a plan to help preserve them.

43. How would you change the internal and external body design of a land-based reptile if you wanted to enable it to fly?

Problem Solving

44. Pacific sardine fishing is done off the coast of Washington, Oregon, and California. In 1916, there were 4 boats catching sardines, whereas by 1936 there were 137 boats. Because of World War II, only 67 boats were available for fishing from 1939 to 1944. After World War II, the fishing fleet increased to over 100 boats. The graph below illustrates the Pacific sardine catch from 1916 to 1961. Interpret the graph in terms of the change in quantity of sardines caught over the years. Develop a hypothesis as to what caused the great reduction of the total catch after 1944.

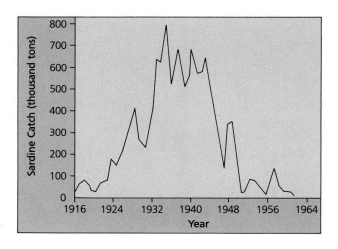

45. A biologist wants to test the effect of an algal herbicide on the bass population of a lake. Bass from various parts of the lake are collected in nets, 152 of them are painlessly tagged, and all are released. Before the herbicide is added, the biologist collects a sample of the bass. Of 213 bass collected, 17 are found to be tagged. A month after the herbicide is added to the lake, the fish population is again sampled. This time 163 bass are collected, of which 13 are found to be tagged. Using the equation

$$\frac{\text{\# of tagged bass in sample}}{\text{total \# of bass in sample}} = \frac{\text{total \# of tagged bass}}{\text{total bass population}}$$

calculate the total bass population before and after the addition of the herbicide. Did the herbicide affect the bass population of the lake?

46. If large tadpoles are placed in a tank with smaller tadpoles, the smaller ones may stop feeding and die, even though food is plentiful. Design an experiment to determine if the agent inhibiting them from eating is a chemical given off by the larger tadpoles.

47. An investigation was conducted to determine how rapidly farm-raised salmon grow. The average weight of 10 salmon was determined at different times after hatching. Construct a graph of the resulting data, which appear in the table below. Draw a conclusion as to the rate at which the salmon gain weight after hatching.

Time after Hatching (weeks)	Average Weight of Salmon (grams)
1	8
2	16
3	24
4	38
6	70
8	120

Projects

48. Some ocean swimmers are fearful of shark attacks. Research this topic and write a short commentary that could be used on television news.

49. Using library resources, write a newspaper article on treating poisonous snake bites.

50. Using field guides, identify some of the reptiles and amphibians native to your area. Present your findings in an oral report to the class.

▲ A gray wolf stalks his prey during a snowstorm on the Alaskan tundra.

Vertebrates—Birds and Mammals

35

A wolf stares intently into the frigid snowscape. The wolf's survival depends on its hunting skills, but the birds and mammals it stalks have skills of their own. Whether predator or prey survives the winter will depend on which better uses its highly developed skills. In this chapter, you will learn more about these two classes of vertebrates—birds and mammals—and their evolutionary link to reptiles.

Guide for Reading

Key words: Aves, air sacs, gizzard, monotremes, marsupials, placental mammals, Neanderthals, Cro-Magnons
Questions to think about:
How are birds like humans?
What are the major groups of placental mammals?

35-1 Birds

Section Objectives:

- *List* the general characteristics of birds.
- *Describe* the structure and functions of feathers.
- *Describe* the internal body systems of birds.

General Characteristics of Birds

Birds, class **Aves** (AY veez), are found in almost all types of environments. Feathers are the one characteristic that distinguishes birds from all other animals. Birds are thought to have evolved from reptiles, and the feathers are thought to be modified scales. Figure 35–1 shows the fossil bird *Archaeopteryx.* Many scientists believe it represents an evolutionary link between reptiles and birds. Contemporary birds do have reptilelike scales on their legs and feet.

In addition to their feathers, birds also share a number of other characteristics.

- The body is usually spindle-shaped and divided into a head, neck, trunk, and tail.

- There are two pairs of limbs. The forelimbs are wings, which, in most birds, are used for flying. The hind limbs are legs that are adapted for perching, walking or swimming, or prey-catching, depending on the type of bird.

- The bones are strong and light, and many bones are filled with air spaces.

- The circulatory system is well developed and includes a four-chambered heart.

▲ **Figure 35–1**

Archaeopteryx. The fossil bird *Archaeopteryx* lived about 150 million years ago. It had some features of both reptiles and birds.

761

▲ **Figure 35–2**

Adaptation in Bird Beaks and Feet. Examples of variation in bird beaks and feet include, from left to right, the Swainson's hawk, the mallard duck, the yellow-bellied sapsucker, and the ostrich.

- The respiratory system is highly efficient and consists of lungs connected to air sacs.

- The mouth is in the form of a horn-covered beak or bill. There are no teeth.

- The excretory system does not include a urinary bladder.

- Fertilization is internal. The large, shell-covered eggs are incubated by the parents, and at hatching, the young are cared for by the parents.

- Birds are warm-blooded. Compared to many vertebrates, their body temperature is quite high.

The beaks and feet of birds show adaptations for different ways of life. The pelican uses its long, sharp beak for catching fish. The cardinal uses its strong beak to crack open seeds. The hooked beak of the hawk allows it to tear its food. The woodpecker uses its beak to bore into trees and extract insects. A duck scoops and strains its food from mud with its beak. The ostrich and other ground-dwelling birds have sturdy feet and toes that enable them to run. Ducks and geese have webbed feet that are useful in swimming. The position of the toes and the presence of sharp claws permit woodpeckers to cling to the sides of trees. Grasping feet with sharp claws or talons are characteristic of falcons, hawks, and perching birds. In these birds, the tendons of the feet are so arranged that the weight of the body forces the toes to grasp the branch when the bird lands on it. This arrangement allows perching birds to sleep without falling off their perches.

Feathers

Feathers are lightweight and flexible, yet incredibly tough. They provide a body covering that protects the skin from wear, supports the bird in flight, and provides insulation from the weather. Feathers grow from follicles in the skin. As the feather grows in the follicle, pigments are deposited in the epidermal cells making up the feather. The color pattern of the feathers is typical of the species. In many species, the male and female have different coloring. Usually, the male is brighter than the female. The difference in coloring between sexes plays a role in the mating behavior of birds.

Figure 35–3 shows a typical feather. The flat area, the *vane,* is supported by a central shaft. The hollow part of the shaft that is attached to the skin follicle is called the *quill.* Each vane consists of many, closely spaced **barbs.** The barbs spread out diagonally from the shaft. Each barb has many **barbules.** The barbules of one barb overlap the barbules of the barb next to it. The barbs are held together by tiny hooks on the barbules themselves. When neighboring barbs become separated, the bird can zip them together with its beak.

Feathers grow only on certain parts of the skin. When fully grown, feathers are not living structures. Usually in the late summer, molting occurs. The feathers are shed and replaced by new feathers. Molting is usually a slow process. No part of the body is ever completely without feathers.

There are several different types of feathers. Elongated *contour feathers* are shown in Figure 35–3. They cover, insulate, and protect the body. Contour feathers that extend beyond the body are called *flight feathers.* Those on the wings support the bird in flight, while those on the tail serve as a rudder for steering. *Down feathers* have short shafts with long barbs. They are soft because the barbules lack hooks. In ducks, geese, and other water birds, down feathers are present beneath the contour feathers. Down feathers provide insulation by trapping air, which helps to save body heat. The winter coats of many birds include down feathers.

An oil gland near the base of the tail in birds helps to make the feathers waterproof. Birds use their beaks to take oil from the gland and spread it over the feathers.

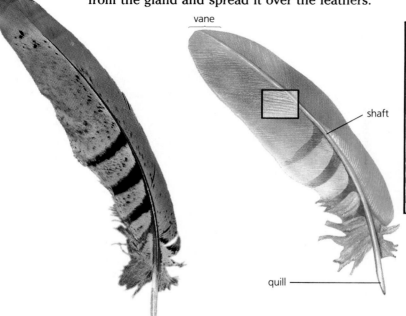

vane

shaft

quill

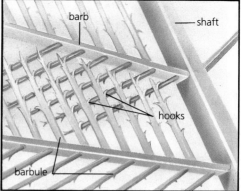

barb

shaft

hooks

barbule

◀ **Figure 35–3**

Structure of a Feather. Elongated contour feathers cover, insulate, and protect the body. Each feather is made up of barbs with overlapping barbules.

Internal Structure of Birds

Respiratory and Circulatory Systems The bird's unusual and highly efficient respiratory system can provide the large amount of oxygen needed for flight. Pouching out from the small lungs are **air sacs.** These are shown in Figure 35–4. The sacs occupy space between the internal organs and even run into the spaces in the larger bones. Air enters the respiratory system through the nostrils and passes down the *trachea,* which divides into two *bronchi.* The bird's *syrinx,* or song box, is located at the point at which the bronchi divide. One bronchus enters each lung. The bronchi pass through the lungs to the posterior air sacs. Thus, oxygen-rich air coming into the respiratory system passes through the lungs to the posterior air sacs without any exchange of respiratory gases. A system of small air tubes leads from the posterior air sacs into the lungs. The air tubes subdivide many times and make close contact with blood capillaries in the lungs. Oxygen-rich air from the posterior sacs is forced through the fine air tubes in the lungs, and gas exchange occurs with the blood. The air, now oxygen-poor, enters the anterior air sacs. From these sacs, it passes back up the trachea and out of the body. The one-way flow of air through the lungs for gas exchange increases the efficiency of the system.

The bird's circulatory system is similar to that of humans. There is a four-chambered heart and a complete separation of oxygenated and deoxygenated blood. Also like humans, birds generate and regulate their body heat internally. In other words, they are endothermic.

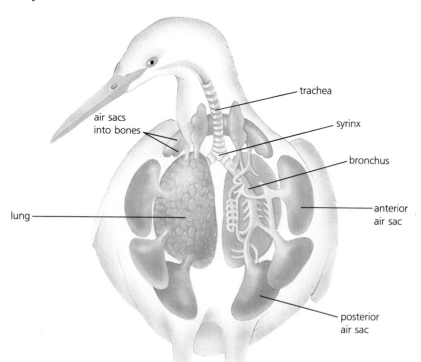

Figure 35–4

The Respiratory System of the Bird. The highly efficient respiratory system of the bird provides it with the large amounts of oxygen needed for flight. Air sacs increase the amount of air inhaled by the bird, and help to make the bird lighter. ▶

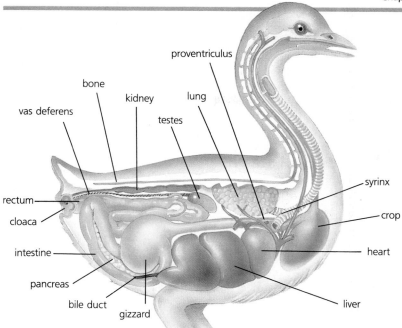

proventriculus
bone
kidney
lung
vas deferens
testes
rectum
cloaca
intestine
pancreas
bile duct
gizzard
syrinx
crop
heart
liver

◀   **Figure 35–5**
The Internal Structure of the Bird.

Digestive System   Because of their high rates of metabolism and the energy needed for flight, birds eat large amounts of food. They feed on seeds, fruits, insects, worms, and in some cases, small reptiles and mammals. Some small birds take in an amount of food equal to 30 percent of their body weight each day.

Food is taken into the mouth, mixed with saliva, and passed down the esophagus to the *crop,* where it is stored and softened. From the crop, it passes into the first portion of the stomach, the *proventriculus* (proh ven TRIK yuh lus), where it is partially digested by gastric juice. It then passes to the **gizzard,** the second part of the stomach. The gizzard is a thick-walled, muscular organ that may contain small stones that have been swallowed by the bird. In the gizzard, the food is ground up and thoroughly mixed with the gastric juices. Next, the food moves into the intestine, where digestion is completed and nutrients are absorbed into the bloodstream. Undigested food enters the short rectum and leaves the body through the *cloaca.* The genital ducts and the ureters from the kidneys also open into the cloaca.

Excretory System   Nitrogenous wastes, in the form of uric acid, are removed from the blood by the kidneys. Since there is no urinary bladder, these wastes pass through the ureters to the cloaca. They combine with the fecal matter in the cloaca to form a whitish, semisolid material.

Nervous System   The brain of the bird is fairly large. The cerebellum, which is involved in muscle coordination, is well developed and allows the bird to perform precise movements in flight. In most birds, the senses of smell and taste are poorly developed, but the senses of sight, hearing, and balance are highly developed.

Science, Technology, and Society

Issue: Animal Experimentation

Millions of animals are used each year in the name of medical and consumer-product research. Some of the most important medical advances of the century were based on experiments done on animals. But, some people would like to limit, or restrict, the use of animals in research.

Should animals, like humans, have a right to life and protection from harm? Animal-rights activists believe they should. They claim that some experiments cause research animals unnecessary pain and that other laboratory techniques could take the place of animal experiments. Besides, they argue, why should animals be made to suffer for research that benefits humans?

Other people believe that animal experimentation is justified. They point to the thousands of human lives that have been saved as a result of animal research. They believe cures for AIDS, cancer, and other diseases can be found only through animal research. They say that animals also benefit from medical breakthroughs. Alternatives to animal use, they say, are available in only a few cases.

■ *Should we continue experimentation on animals? Why or why not?*

Reproductive System   There are no external sex organs in birds. During mating, the sperm are transferred from the male to the female by contact of the cloacas. After fertilization in the female's reproductive tract, a protective shell is deposited around each egg. The female deposits her eggs in a nest, and they are incubated until hatching.

35-1 Section Review

1. Name the single characteristic that distinguishes birds from all other animals.
2. List three different types of feathers.
3. What unique structures are found in the lungs of birds?

Critical Thinking

4. How is the absence of a urinary bladder an adaptation for flight? *(Judging Usefulness)*

35-2 Mammals

Section Objectives:

■ *List* the major characteristics of mammals.
■ *Name* the three different kinds of mammals.
■ *List* the major orders of placental mammals and name some members of each.

Laboratory Investigation: *Compare* amino acids sequences in cytochrome *c* in several kinds of vertebrates. (p. 780).

Concept Laboratory: *Gain* an understanding of the importance of an opposable thumb (p. 772).

General Characteristics of Mammals

The *mammals,* class **Mammalia** (muh MAYL ee uh), include many familiar animals—cats, dogs, bats, monkeys, horses, cows, deer, whales, and humans. Members of this group are found all over the earth in both cold and warm climates. Most are land dwelling, but a few, such as the whale, porpoise, and seal, are found in the oceans. Mammals range in size from the tiny pygmy shrew, which is less than 5 centimeters long and weighs less than 5 grams, to the giant blue whale, which may be 30 meters long and weighs more than 100 000 kilograms.

It is thought that mammals evolved from a group of reptiles that had some mammal-like characteristics. Two characteristics distinguish mammals from all other vertebrates. Mammals nourish their young with milk produced by mammary glands. The body covering of mammals is hair. The amount of hair varies. In whales and porpoises, just a few whiskers are found around the mouth of the animal. In many other mammals, the hair is in the form of a thick coat of fur that covers the body.

Mammals also share several other characteristics. Like birds, they are warm-blooded and have a four-chambered heart. An internal muscular wall, the *diaphragm,* separates the chest cavity from the abdominal cavity. The cerebrum of the brain is more highly developed than in any other group. Mammals are, therefore, the most intelligent of the animals.

Mammals have highly differentiated teeth. The structure and arrangement of the teeth vary from group to group, depending on feeding habits. There are four types of teeth: **incisors** (in SYZ erz), which are for cutting; **canines** (KAY nynz), which are for tearing; and **premolars** (pree MOHL erz) and **molars,** which are for grinding. These are shown in Figure 35–6.

Except for a few egg-laying mammals, all mammals give birth to living young. The number of offspring produced at each birth is fewer than in most other animals. However, because the young are cared for by the parents, they have a better chance of survival.

There are three different kinds of mammals—the **monotremes,** the **marsupials,** and the **placental mammals** (see Chapter 22). The monotremes are egg-laying mammals. They are the most primitive and reptilelike of the mammals. The duckbill platypus and spiny anteater of Australia are the only living monotremes. The marsupials are pouched mammals, such as the kangaroo, opossum, and koala. Marsupials are born at an immature stage and complete their development in their mother's pouch. Marsupials were once more common and widespread than they are today. Except in Australia, wherever placental mammals arose, marsupials eventually died out. Placental mammals are the largest and most successful group of mammals. In the placental mammals, the developing young remain within the uterus of the female until embryonic development is complete. The young are born in a stage of development more advanced than that in the marsupials.

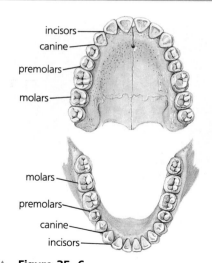

▲ **Figure 35–6**

The Teeth of Mammals. Mammals have differentiated teeth, including incisors, canines, premolars, and molars.

◀ **Figure 35–7**

Marsupials. Opossums are marsupials found in North America.

▲ **Figure 35–8**
Insect-Eating Mammals. These water shrews are feeding on grasshoppers.

Figure 35–9
Rodents and Lagomorphs. Rodents, such as this beaver (top), use their chisel-like incisor teeth to gnaw. The snowshoe hare (bottom), a lagomorph, has powerful hind legs for hopping. ▼

Kinds of Placental Mammals

There are about 15 different orders of placental mammals. Members of these groups range from tiny bats to enormous whales. The major orders of placental mammals are discussed below.

Insect-Eating Mammals Moles, hedgehogs, and shrews are insect-eating mammals, members of the order **Insectivora** (in sek TIV uh ruh). *Insectivores* (in SEK tuh vors) are usually small, mouselike animals. Many live underground. They feed on ants, grubs, beetles, and other insects.

Rodents Mice, rats, beavers, porcupines, squirrels, hamsters, and guinea pigs are all members of the order **Rodentia** (roh DENCH ee uh), the largest order of placental mammals. Their sharp, chisel-like incisor teeth are used for gnawing. These teeth grow throughout the animal's life to replace the ends that wear away. Rodents reproduce rapidly. Some are serious pests, destroying food and carrying disease.

Lagomorphs Rabbits and hares are members of the order **Lagomorpha** (lag uh MOR fuh). Like rodents, they are gnawing animals that feed on plants. Both rabbits and hares have long ears and fluffy tails. Newborn rabbits cannot see or move, and they have no fur. Newborn hares, on the other hand, are covered with fur and become active within a few hours. Both rabbits and hares move by hopping, using their powerful hind legs. They are among the fastest-moving mammals.

Flying Mammals Bats, members of the order **Chiroptera** (ky RAHP tuh ruh), are the only mammals capable of real flight. The wing of a bat consists of four long fingers covered by a membrane of skin. The first finger is used for grasping, as are the small hind legs. At rest, bats hang upside down by their hind legs. Bats are active at night.

While the most common bats feed on insects, some feed on fruit, pollen, or small animals. The vampire bat, which is found in Central and South America, preys mainly on cattle. To obtain blood,

Figure 35–10
Flying Mammals. Bats are the only true flying mammals. ▼

the bat bites off a small piece of skin and then laps up the blood. The total amount of blood lost is small, and the wound itself usually is not serious. However, bat bites can spread diseases.

Bats use a sonarlike system of *echolocation* to find their way around in the dark and to locate their prey. They produce high-frequency sound waves that bounce off any object the waves strike, producing an echo that the bat can hear. The distance to the object is determined by the time between when the sound was made and the echo returns.

Mammals Without Teeth   Anteaters, armadillos, and sloths belong to the order **Edentata** (ee den TAH tuh). These animals have either very small teeth or no teeth at all. Members of this order are found mainly in Central and South America. Anteaters and sloths are covered by long hair, while armadillos are covered by hard plates. Sloths spend much of their time hanging on trees. They feed on leaves and young shoots. Anteaters and armadillos primarily eat ants, termites, and other insects. Both of these animals have long claws and long tongues. They use their claws to break open anthills and termite mounds and their tongues to lick up the insects.

Mammals with Trunks   The order **Proboscidea** (proh buh SID ee uh) includes only African and Asiatic elephants. The muscular trunk of the elephant is formed from a greatly elongated upper lip and nose. The trunk is used to bring food to the mouth. The huge ivory tusks of the elephant are actually greatly enlarged upper incisor teeth. Elephants feed on plants. They are the largest living land animals. To maintain their huge bodies, elephants must feed for as long as 18 hours a day.

Hoofed Mammals   Mammals with feet in the form of hooves are called *ungulates* (UNG yoo lets). The ungulates are divided into two orders according to the number of toes on the hoof. Those with an even number of toes on the hoof belong to the order **Artiodactyla** (art ee uh DAK tuh luh). This order includes pigs, deer, antelopes, sheep, cattle, giraffes, and camels. Ungulates with an odd number of toes belong to the order **Perissodactyla** (puh ris uh DAK tuh luh). This order includes horses, rhinoceroses, and tapirs.

▲   **Figure 35–11**
Mammals Without Teeth. Anteaters use their tongue to gather ants and termites.

Figure 35–12
Mammals with Trunks. Elephants are the only mammals with trunks. ▼

◄   **Figure 35–13**
Hoofed Mammals. The hoofed animals, or ungulates, include goats (left) and zebras (right).

▲ **Figure 35–14**
Meat-Eating Mammals. The grizzly bear is one of many meat-eating mammals, or carnivores.

All ungulates are plant eaters (herbivores) and tend to feed in herds. Their flattened teeth can crush and grind tough plant material. Some ungulates, such as cattle, sheep, camels, and deer, are *ruminants* (ROO muh nents). Their stomachs have four chambers. When grazing, they store large amounts of food in a chamber of the stomach called the **rumen** (ROO men). Later, they bring the food back up into their mouths and chew it thoroughly before swallowing it for a second time.

Meat-Eating Mammals   The order **Carnivora** (kar NIV uh ruh) includes cats, dogs, bears, skunks, and other mammals that eat meat. Some meat eaters (carnivores), such as the bear, eat plant material as well as meat. Most carnivores are strong and fast moving and have sharp claws. Their powerful jaws and large teeth are specialized for seizing, cutting, and tearing meat. They have a well-developed sense of smell. Carnivores are generally intelligent, and much of their hunting behavior is learned.

Aquatic Mammals   Walruses, sea lions, and seals are meat-eating, aquatic mammals that belong to the order **Pinnipedia** (pin uh PEED ee uh). They feed mainly on fish. Their limbs are modified as flippers, and their body shape is adapted for swimming.

Whales, dolphins, and porpoises are members of the order **Cetacea** (see TAYSH ee uh). These animals are well adapted to life in the ocean. Although they are air breathers, they can remain underwater for long periods of time by holding their breath. As you can see in Figure 35–15, the forelimbs of cetaceans also are modified as flippers. There are no hind limbs. Cetaceans swim by moving their powerful tails up and down through the water. Like other mammals, cetaceans give birth to live young, which are fed on milk from the mammary glands.

Porpoises, dolphins, and some whales have teeth and feed on fish. The largest whales, however, feed on plankton, the small organisms that float in the oceans. Plankton is strained from the water by a series of horny plates called *baleen*. Blue whales, which feed on plankton, are the largest animals that have ever lived.

Manatees and dugongs belong to the small order **Sirenia** (sie REEN ee ah). They are plant eaters. Like cetaceans, they lack hind limbs and have flattened tails to help them move through the water.

Figure 35–15
Aquatic Mammals. Aquatic mammals belong to one of three orders. Examples include, from left to right, the sea lion, a member of the order Pinnipedia, the Beluga whale, a Cetacean, and the manatee, which belongs to the order Sirenia. ▼

◀ **Figure 35–16**
A Primate. An orangutan has hands with opposable thumbs.

Primates   Humans, apes, and monkeys are members of the order **Primates** (pry MAYT eez). All *primates* (PRY mayts) have well-developed grasping hands that allow them to handle and manipulate objects. Their fingers and toes have flat nails instead of claws. Humans, apes, and monkeys have **opposable thumbs.** An opposable thumb can be positioned opposite to the other fingers, making it possible to grasp objects in one hand.

Except for humans, gorillas, and baboons, which live on the ground, most primates live in trees. Primates eat both plant material and meat. Primates are the most intelligent of the mammals. Their brains are large and complex, and their sense of sight is well developed.

35-2 Section Review

1. What two characteristics can be used to distinguish mammals from other vertebrates?
2. Name the three different kinds of mammals.
3. What is the largest order of placental mammals?
4. To what order do whales, dolphins, and porpoises belong?

Critical Thinking

5. Why do scientists classify egg-laying *monotremes,* such as the duckbill platypus, as mammals? (*Identifying Reasons*)

Concept Laboratory

The Opposable Thumb

Goal

To gain an understanding of the evolutionary importance of the opposable thumb.

Scenario

Imagine that you are a rehabilitation counselor working with a person who has lost a thumb. To understand the problems faced by your patient, you try a number of activities with your thumb taped to the side of your hand.

As you perform this activity, think about how a single adaptation, such as the opposable thumb, allows an organism to function in ways that otherwise would be difficult or impossible.

Materials

tape	sink	paper clip
book	spoon	scissors
chalk	plastic	deck of
pencil	cup	cards
scrap	comb	newspaper
paper	coin	sneaker

Procedure 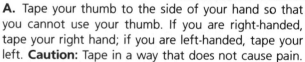 *

A. Tape your thumb to the side of your hand so that you cannot use your thumb. If you are right-handed, tape your right hand; if you are left-handed, tape your left. **Caution:** Tape in a way that does not cause pain.

B. Using only the hand that is taped, perform each of the tasks listed below.

1. Pick up a pencil and write your name.
2. Turn a door knob.
3. Pick up a spoon, and pretend to use it.
4. Pick up a plastic cup, and pretend to use it.
5. Comb your hair.
6. Pick up a coin.

C. Using both your taped and untaped hands, perform the tasks listed below.

7. While holding two pieces of paper in your untaped hand, clip them together with your taped hand.
8. With a pair of scissors in your taped hand, cut a circle out of a piece of paper held in your untaped hand.

*For detailed information on safety symbols, see Appendix B.

9. Shuffle a deck of cards.
10. Hold a deck of cards with your untaped hand, and deal them with your taped hand.
11. Hold a newspaper, and turn a page.
12. Tie a shoe.

Organizing Your Data

1. Make a table like the one shown below. List each task performed, and record whether it was easy, difficult, or nearly impossible to do.

Task	Easy	Difficult	Nearly Impossible
pick up pencil			

2. Compare your results with those of the rest of your class.

Drawing Conclusions

1. Which tasks were the easiest to do? What do these tasks have in common?

2. Which tasks were nearly impossible to do? What do these tasks have in common?

3. Based on your results, explain the evolutionary importance of the opposable thumb.

Thinking Further

1. Do you think that the tasks that were difficult or nearly impossible would become easier with practice? Explain your answer.

2. A *prosthesis* is a device designed to replace the function of a missing body part. What features must a prosthesis have to replace the human thumb?

3. Do your feet have opposable digits? Comment on the significance of this.

4. Monkeys and apes have opposable thumbs on their feet. How is this an adaptation for these animals?

5. Front-facing eyes and bipedal locomotion (walking on hind legs) are two additional physical adaptations that humans have. Explain how you think each of these adaptations has contributed to the survival of humans as a species.

35-3 Human Origins

Section Objectives:

- *List* the characteristics that distinguish humans from other primates.
- *Describe* the characteristics of *Australopithecus, Homo habilis,* and *Homo erectus.*
- *Compare* and *contrast* Neanderthals, Cro-Magnons, and modern humans.

Identifying Human Fossils

The branch of science that attempts to trace the development of the human species is called **anthropology** (an thruh PAHL uh jee). Anthropology deals with human physical, social, and cultural development. It also deals with the study of the origin of humans. Few areas of scientific research have produced more confusion and disagreement than the interpretation of the human fossil record. This record is fragmentary. Often, it consists of a few teeth or scattered pieces of bone. Occasionally, a complete jawbone, skull, pelvis, or thigh bone is found. Putting these bone fragments together to form a complete picture of an organism is like doing a jigsaw puzzle when all the pieces are the same color, many pieces are missing, and you do not know what the finished puzzle is supposed to look like. Dating the fossils is difficult. Their ages must be inferred from the age of the rocks in which they are found. Accurate measurements by absolute dating methods are not always possible.

Humans, apes, and monkeys are all primates that are structurally similar in many ways. However, there are several characteristics that distinguish humans from other primates.

The size and shape of the skull are different from other primates. The human brain is much larger than that of other primates. To hold the brain, the human skull is larger than that of other primates and has a unique shape. The cranium, or brain case, is higher and more rounded than that of other primates. Humans also have relatively flat, vertical foreheads, while other primates have sloping foreheads with bony eyebrow ridges. See Figure 35–17.

Figure 35–17

Comparison of Human and Gorilla Skulls and Jaws. The gorilla's brain capacity is about 450 cm³ while that of the human is about 1450 cm³. ▼

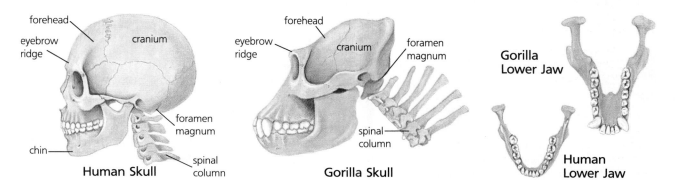

Human Skull   Gorilla Skull   Gorilla Lower Jaw   Human Lower Jaw

The jaws and teeth are different from other primates. Primates other than humans generally have heavy, rectangularly shaped jawbones. The jawbones of humans are lighter and typically U-shaped in front. Also, humans have distinct chins, while the other primates are chinless. Human teeth are generally smaller than those of other primates.

The pelvis and foramen magnum are different. Another unique characteristic of humans is their ability to walk on two legs in an upright position. This is called **bipedal locomotion.** With the exception of birds and kangaroos, most other animals walk on four legs. There are several structural adaptations connected with bipedal locomotion. The pelvis is adapted for upright posture. The back of the pelvis is thick and strong and serves as a site of attachment for the long muscles of the legs. The bowl shape of the pelvis helps to support the internal organs. The S-shaped spinal column places the center of gravity over the pelvis and legs.

The **foramen magnum** (fuh RAY men MAG num) is the opening in the skull where the spinal cord enters. In humans, this opening is under the skull. Its location allows the head to be balanced on top of the spinal column. In other primates, the foramen magnum is toward the back of the skull. This position means that the skull is held forward and tends to face downward. In fossils, the location of the foramen magnum determines whether or not a primate had bipedal locomotion.

Modern humans, and humanlike fossils that exhibited bipedal locomotion, are called *hominids.* The only hominids alive today are humans.

Fossil Evidence of Human Origins

From the fossil record, it is thought that apes began to evolve around 30 million years ago. Based on fossil and biochemical evidence, most scientists believe that humans and apes evolved along different paths from a common ancestor.

Australopithecus   In 1924, workers in a limestone quarry in South Africa found some skulls embedded in the rock. Most of the skulls were those of monkeys, but one skull showed some human characteristics. It looked like the skull of a child about five years old. While the brain size was greater than an ape's, it was smaller than that of a modern human. The position of the foramen magnum indicated that the primate was bipedal. Finally, the teeth were more humanlike than apelike, as can be seen in Figure 35–18. Raymond Dart, a famous anthropologist, named the new species *Australopithecus africanus* (aus truh luh PITH uh kus af ruh KAN us), which means "ape of southern Africa." He believed that **Australopithecus** was more humanlike than apelike and represented an early type of hominid.

In 1936, Dr. Robert Brown discovered the remains of an adult *Australopithecus.* Since then, hundreds of such fossils have been found, and the genus has been divided into several species.

Figure 35–18

Skull of *Australopithecus africanus*. Features of *Australopithecus* indicate a closer link to humans than to apes. ▼

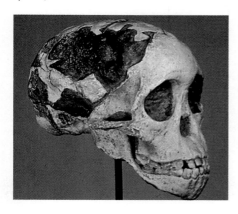

It is believed that *Australopithecus africanus* (also called *A. africanus*) lived between 3 million and 2 million years ago. Members of this group were about 1 meter tall and weighed about 25 kilograms. They could stand and walk upright. In proportion to body size, their brain size was somewhat larger than that of a gorilla.

In 1938, Robert Brown found a fossil of a species that was similar to *A. africanus* but larger and more muscular. He named this massively built but related species *Australopithecus robustus* (roh BUS tus). *A. robustus* was a little over 1.5 meters in height and weighed about 65 kilograms. The jaw of *A. robustus* was heavier than that of *A. africanus,* and the teeth were larger. Extensive pitting in the molars of *A. robustus* suggested that they fed on vegetation containing sand. The teeth of *A. africanus* do not show this pitting, and it is thought, therefore, that they were meat eaters.

A. robustus lived between 2.2 and 1.4 million years ago. Most experts believe that this hominid became extinct and was not an ancestor of modern humans.

In 1974, Donald Johanson and Maurice Taieb discovered the bones of a small female *Australopithecus.* Johanson called her Lucy. Although she had an ape-sized brain and jaw, the pelvis clearly showed she was bipedal. Then, in 1975, they found the fossilized bones of a group of adults and children, apparently all killed at the same time, possibly by a flash flood. All these prehuman fossils are between 3.8 and 2.8 million years old and were named *Australopithecus afarensis.*

At the present time, scientists cannot agree on how many species of *Australopithecus* there were or whether or not they were the ancestors of human beings. Although scientists disagree on these issues, they do agree that bipedal locomotion apparently evolved before an increased brain size.

Homo Habilis   Since 1959, fossils found in East Africa by Louis and Mary Leakey and their son Richard indicate that other hominids lived at the same time as *Australopithecus.* Compared to *Australopithecus,* these hominids had more humanlike teeth and

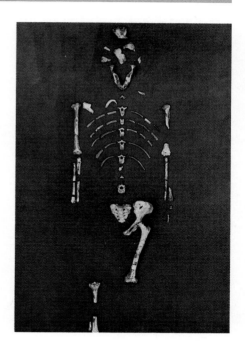

▲ **Figure 35–19**

Australopithecus afarensis. Lucy, a small female *A. afarensis,* lived about 3.6 million years ago.

◄ **Figure 35–20**

The Leakeys. The fossil discoveries made by anthropologists Louis and Mary Leakey (left) and by their son, Richard (right), have been central to our understanding of human evolution.

Figure 35–21
Hand and Skull of *Homo habilis*. The hand shows that *H. habilis* had good manual dexterity. ▶

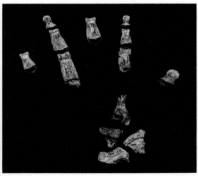

larger brains. Found with these fossils were pebbles chipped to form sharp-edged tools. Because they were toolmakers, these fossils have been classified as *Homo,* the same genus as modern humans. They have been named *Homo habilis* (HOH moh HAB uh lus), which means "handy man." A hand and skull of *H. habilis* is shown in Figure 35–21. *Homo habilis* is thought to have lived between 2.2 and 1.6 million years ago.

Homo Erectus   The first remains showing truly human characteristics appear in the fossil record about 1.5 million years ago. These hominids, thought to be descendants of *Homo habilis,* are classified as *Homo erectus* (uh REK tus). See Figure 35–22. Fossil bones, tools, and living sites of *H. erectus* have been found in Asia, Africa, and Europe. *H. erectus* was the first user of fire. Members of this species lived in groups. Their brain size, although larger than that of *H. habilis,* was significantly smaller than that of modern humans. It is thought that *H. erectus* survived until 350 000 to 250 000 years ago and that they gave rise to the earliest members of the species ***Homo sapiens*** (SAY pee inz), the species of modern humans.

Neanderthals   The **Neanderthals** (nee AN der thals) were an early type of *Homo sapiens.* They are classified as *H. sapiens neanderthalensis.* The Neanderthals first appeared about 130 000 years ago. Their fossils have been found throughout Europe, Africa, and Southeast Asia. These humans were only about 1.5 meters tall but powerfully built. Their faces had heavy, bony eyebrow ridges. According to skull measurements, their brain size was as large as or slightly larger than that of modern humans, but it was shaped differently.

Neanderthals lived in family groups in caves or in simple shelters built of rocks. They used fire and produced a variety of stone tools. It appears that some of these tools were used to scrape hides, which were then used to make clothing. Tools were also used

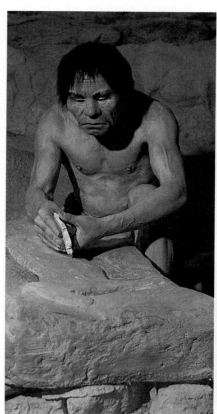

▲ **Figure 35–22**

Early People. These artist's representations show, clockwise from upper right, *Homo erectus,* Neanderthal, and Cro-Magnon figures.

to kill large animals, such as mammoths and woolly rhinoceroses, which were trapped in pits lined with wooden spikes. Fossil evidence indicates that the Neanderthals buried their dead, sometimes with weapons and food.

Cro-Magnons   About 35 000 years ago, the Neanderthals disappeared from the fossil record and were replaced by the **Cro-Magnons** (kroh MAG nuns). Fossil remains of Cro-Magnons have been found in Europe, Asia, Africa, and Australia. Scientists do not know what happened to the Neanderthals. They may have been killed off by the more advanced Cro-Magnons, or possibly the two groups interbred, and the Neanderthals lost their identity.

The skeletons of Cro-Magnons were like those of modern humans, and their brain size was about the same. They lived not only in caves but also in dwellings built of rock, wood, and hides. They made finely chipped stone and bone tools, including axes, knives, awls, chisels, and scrapers. They also made fishhooks, needles, and spear points. They wore clothing sewn from animal skins. On the walls of caves in France are paintings by Cro-Magnons

Figure 35–23
A Cro-Magnon Cave Painting. ▶

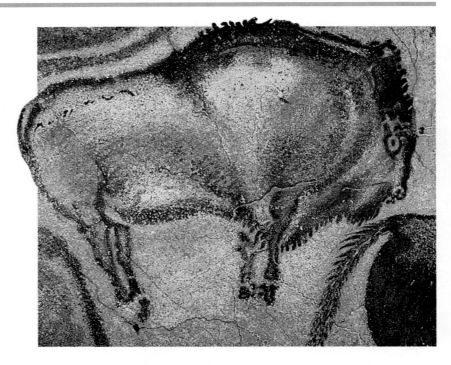

of the animals they hunted. These drawings may have had some spiritual significance. The Cro-Magnons seem to have had a ritual for burying their dead. Often, the body was covered with dye and buried with food and personal articles.

Until the Cro-Magnons, most of the evolutionary changes in the human species apparently were physical changes, such as brain volume. See Figure 35–24. From the Cro-Magnons on, however, changes are thought to have been mainly behavioral and cultural rather than physical. It is believed that the Cro-Magnons migrated to various parts of the world. The different groups gave rise to the three primary human races—Negroid (black), Oriental (yellow), and Caucasoid (white).

Interpreting the Fossil Evidence

Because fossil evidence is limited, the ancestry of humans is still vague. Often, there are large gaps in the fossil record. Therefore, as new fossils are discovered and interpreted, the human evolutionary sequence is revised. Figure 35–25 shows two possible patterns of human evolution that have been proposed by different scientists.

Recent studies by molecular geneticists indicate that as different species evolve from a common ancestor, their DNA sequences change at a fairly constant rate. By comparing the DNA sequences of particular genes in related species whose ages are already known from fossil evidence, the rates of change in the DNA can be assessed. This can be done indirectly by comparing changes in the amino acid sequences of similar proteins. This measurement can serve as a genetic, or *molecular,* clock. It has been shown that there are fewer differences in similar proteins between closely related

Figure 35–24
Brain Volume Increase During Hominid Evolution. The brain size of *Australopithecus* was about three times smaller than that of the earliest *Homo sapiens.* ▼

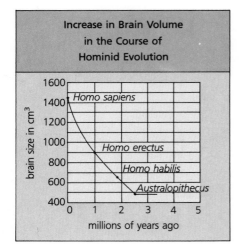

Increase in Brain Volume in the Course of Hominid Evolution

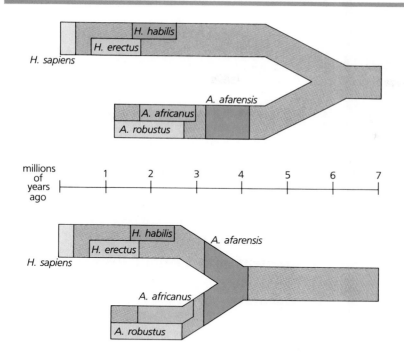

◀ **Figure 35–25**

Two Hypotheses for Human Evolution. The scheme on the top has been proposed by Richard Leakey; the scheme on the bottom has been proposed by Donald Johanson. Gray areas represent gaps for which there is no fossil information.

species than between distantly related species. Thus, the number of base differences in the DNA of related species can be used as a measure of the time at which the species diverged from a common ancestor.

Evidence based on the molecular clock indicates that the split between humans and apes occurred about 5 or 6 million years ago. Fossil evidence suggests that the split occurred between 12 million and 6 million years ago. However, since so few humanlike fossils have been discovered from this period, much about our evolutionary past is not yet known.

35-3 Section Review

1. What is bipedal locomotion?
2. List the names of three species of *Australopithecus*.
3. Which human ancestors or prehumans probably used tools?
4. Which human ancestors are classified as *Homo sapiens?*

Critical Thinking

5. A scientist analyzed the DNA sequence of a single gene in three closely related species. She determined that there were fewer sequence differences between species *X* and *Y* than between either species *Y* and *Z* or species *X* and *Z*. From this, the scientist concluded that species *X* and *Y* were more closely related. What assumption(s) does this conclusion rest on? (*Identifying Assumptions*)

Laboratory Investigation

Comparing Amino Acid Sequences in Vertebrates

The amino acid sequence of a protein may differ between two species. The extent of the difference can be a measure of how long ago in evolution the species diverged from a common ancestor. In this investigation, you will compare the amino acid sequences of the same protein in several vertebrates.

Objectives

- *Compare* the number of amino acid sequence differences in cytochrome *c* among humans, rhesus monkeys, rabbits, and chickens.
- *Construct* a phylogenetic tree based on amino acid sequence data.

Materials

paper and pencil

Procedure

A. Cytochrome *c*, a protein in the electron transport chain in mitochondria, is 104 amino acids long. The figure below shows the amino acids at 13 of the positions in this protein as it exists in humans, rhesus monkeys, rabbits, and chickens. The amino acids at the other 91 positions of cytochrome *c* are identical in all four vertebrates.

B. Count the number of amino acid differences in cytochrome *c* between the following organisms: human and rhesus monkey; human and rabbit; human and chicken.

C. Make a table with three columns, one for each of the vertebrate pairs listed in step B. Record the number of amino acid differences for each pair.

Analysis and Conclusions

1. How many amino acid differences did you find between the cytochrome of the human and rhesus monkey, human and rabbit, and human and chicken?

2. Based on your results, with which organism would you conclude humans have most recently shared a common ancestor? Explain.

3. Did humans share a common ancestor more recently with rabbits or with chickens? Explain.

4. Based on your findings, construct a phylogenetic tree showing the likely sequence in which the four species studied here branched off from each other.

5. Suppose you wanted to check relationships among organisms. What advantages would comparing nucleotide sequences of the DNA that codes for a protein have over comparing amino acid sequences of the protein?

Amino Acid Sequence Number	Human	Rhesus Monkey	Rabbit	Chicken
11	isoleucine	isoleucine	valine	valine
12	methionine	methionine	glutamine	glutamine
15	serine	serine	alanine	serine
44	proline	proline	valine	glutamic acid
46	tyrosine	tyrosine	phenylalanine	phenylalanine
50	alanine	alanine	aspartic acid	aspartic acid
58	isoleucine	threonine	threonine	threonine
83	valine	valine	alanine	alanine
89	glutamic acid	glutamic acid	aspartic acid	serine
92	alanine	alanine	alanine	valine
100	lysine	lysine	lysine	aspartic acid
103	asparagine	asparagine	asparagine	serine
104	glutamic acid	glutamic acid	glutamic acid	lysine

Chapter Review

Study Outline

35–1 Birds

- Similarities between birds and reptiles suggests an evolutionary link between the two groups.

- Birds, class Aves, have wings that are used for flight in most species. The beaks and feet of birds vary widely among species and are adapted for different ways of life.

- Feathers provide a protective body covering for birds, support them in flight, and insulate them from the weather.

- Birds are warm-blooded and have a four-chambered heart. Their respiratory, circulatory, and digestive systems provide the large amounts of oxygen and energy needed for flight.

- Fertilization in birds is internal, and the egg is enclosed in a shell. After birth, the young are cared for by the parents.

35–2 Mammals

- Like birds, it is thought that mammals evolved from a reptilian ancestor.

- Mammals, class Mammalia, nourish their young with milk produced by mammary glands. The body covering of mammals is hair. They are warm-blooded and have a four-chambered heart.

- Monotremes are egg-laying mammals. Marsupials are pouched mammals. All marsupials are born at an immature stage and complete their development in their mother's pouch.

- Placentals are the most successful mammals. Developing young are retained in the mother's uterus until development is complete.

- The several orders of placental mammals include insect-eating mammals, rodents, lagomorphs, flying mammals, mammals without teeth, mammals with trunks, hoofed, meat-eating, and aquatic mammals, and primates.

- Primates have well-developed grasping hands, and many of them have opposable thumbs. They also have large, complex brains and a well-developed sense of sight.

35–3 Human Origins

- Anthropology is the branch of science that attempts to trace the development of the human species.

- Unique characteristics of humans include the size and shape of the skull, jaws and teeth different from those of other primates, and the ability to walk on two legs in an upright position.

- The earliest prehuman primate group for which we have fossil evidence is *Australopithecus*. Other hominids include *Homo habilis, Homo erectus,* Neanderthals, and Cro-Magnons.

- The fossil evidence of human ancestry is limited. Different interpretations of human evolution have been proposed by scientists.

Vocabulary Review

Aves (761)	Proboscidea (769)
barbs (763)	Artiodactyla (769)
barbules (763)	Perissodactyla (769)
air sacs (764)	rumen (770)
gizzard (765)	Carnivora (770)
Mammalia (766)	Pinnipedia (770)
incisors (767)	Cetacea (770)
canines (767)	Sirenia (770)
premolars (767)	Primates (771)
molars (767)	opposable thumbs (771)
monotremes (767)	anthropology (773)
marsupials (767)	bipedal locomotion (774)
placental mammals (767)	foramen magnum (774)
Insectivora (768)	*Australopithecus* (774)
Rodentia (768)	*Homo sapiens* (776)
Lagomorpha (768)	Neanderthals (776)
Chiroptera (768)	Cro-Magnons (777)
Edentata (769)	

Chapter Review

A. Word Relationships—In each of the following sets of terms, three of the terms have a common characteristic. One term does not belong. Identify the term that does not belong.

1. Sirena, Pinnipedia, Cetacea, Edentata
2. molars, barbs, canines, incisors
3. Cetacea, *Homo sapiens*, *Australopithecus*, Cro-Magnons
4. Lagomorpha, Aves, Proboscidea, Artiodactyla
5. bipedal locomotion, opposable thumb, rumen, chin
6. carnivores, marsupials, rodents, bats
7. air sacs, barbules, molars, gizzards

B. Matching—Select the vocabulary term that best matches each definition.

8. Order of mammals having trunks
9. Study of the development of the human species
10. Opening in the skull where the spinal cord enters
11. Species of modern humans
12. Order of placental mammals including humans
13. Second part of the stomach in birds
14. Type of tooth specialized for tearing food
15. Meat-eating order of mammals that includes cats and dogs
16. Egg-laying animals
17. Order of mammals to which bats belong
18. Pouches attached to the lungs in birds

Content Review

19. What are the general characteristics of birds?
20. Draw and label a diagram that shows the fine structure of a contour feather.
21. Explain how birds are adapted for flight.
22. Trace the path of air through the respiratory system of the bird.
23. Why do birds need large amounts of food?
24. Describe the structure and function of the digestive system in birds.

25. What characteristics do birds and mammals share?
26. List four types of mammal teeth and describe their functions.
27. How do monotremes, marsupials, and placentals differ?
28. How does reproduction in birds differ from reproduction in placentals?
29. Compare the feeding structures and habits of the orders Edentata and Carnivora.
30. Describe some adaptations of aquatic mammals for their environment.
31. What makes primates well adapted to life in trees?
32. How do humans differ from other primates?
33. What characteristic is shared by hominids?
34. In what ways was *Australopithecus* more human-like than apelike?
35. Why is *Homo erectus* considered closer to humans than *Homo habilis* is?
36. How were the dwellings of Neanderthals similar to those of Cro-Magnons?
37. What evidence indicates that Neanderthals and Cro-Magnons lived in organized communities?
38. Why do scientists disagree about how humans evolved?

Graphic Organizing

For information on graphic organizers, see Appendix G at the back of this text.

39. **Scale:** The average brain volume of chimpanzees is about 395 cm³; that of gorillas is about 450 cm³. Using this information and the data for *Australopithecus*, *Homo habilis*, *H. erectus*, and *H. sapiens* shown in Figure 35–24, construct a scale from smallest brain volume to largest, placing each of these primates on the scale. Start the scale at 300 cm³ and end it at 1500 cm³. How does the difference in brain volume between gorillas and *Australopithecus* compare with that between *Australopithecus* and *Homo habilis?*

Critical Thinking

40. Since bats are capable of true flight, why are they not classified as birds? (*Classifying*)
41. What assumptions have been made by anthropologists about the intelligence of early hominids? (*Identifying Assumptions*)

42. How is the presence of the rumen in ungulates a beneficial adaptation? (*Identifying Reasons*)

43. Why are the placentals the most successful group of mammals? (*Identifying Reasons*)

44. List the body parts of a bird that contribute to its lightness. (*Relating Parts and Wholes*)

45. Figure 35–22 shows an artist's representations of *Homo erectus,* Neanderthal, and Cro-Magnon people. What would you need to know in order to determine whether these drawings are reliable? (*Evaluating Secondhand Information*)

Creative Thinking

46. Suppose an anthropologist were investigating the remains of our lives 5000 years from now. Look around the room you are in, and list the objects that would provide evidence of intelligent life.

47. Suggest some possible reasons why Cro-Magnons had a burial ritual for their dead.

Problem Solving

48. An investigation was conducted to determine whether latitude affects brood size (number of young produced per mating season) in robins. Using the data in the table below, calculate the average value for each group. Interpret the data and draw a conclusion about robin brood sizes. What further experiments could be done both in the field and in the laboratory to study brood size?

49. Four groups of mice are fed diets that are identical except for the amount of vitamin D. Each of three groups receives a different amount of vitamin D. The fourth group receives none. After four weeks on the diet, the mice are weighed to see whether the amount of vitamin D received has affected their weight. Identify the variable and the control in this experiment.

Projects

50. Prepare an oral report on flightless birds of the world, including those that are endangered and those that are extinct. In your report, discuss how these birds avoid predators.

51. Jane Goodall is a well-known primatologist who has studied chimpanzees in the wild. Write a short biography of Goodall that describes her life and work.

52. Birds and mammals make up the largest number of animals on display in zoos. Interview the director of the zoo closest to your home about the problems of animal maintenance and reproduction that occur with one of their birds or mammals. Choose an animal that is not native to North America. Prepare a poster showing the animal and illustrating the problems of dealing with animals outside their native habitat.

Brood	Number of Young Robins		
	Group A 22° N Latitude	Group B 42° N Latitude	Group C 62° N Latitude
1	3	4	5
2	2	3	7
3	3	5	6
4	4	4	6
5	3	5	7
6	4	5	5
7	3	4	7
8	5	6	6
9	3	5	7
10	3	5	6
11	2	4	6
12	3	4	7

▲ The behavior of this cheetah sends a clear message to an unwanted visitor.

Behavior

A cheetah crouches tensely. Muscles contract in a body coiled for action. Ears lie flat, fangs are bared, and fur bristles along the spine. The long tail swishes nervously. The snarl is sinister, its message unmistakable. And all of this because danger threatens. Organisms respond continually to life situations. These responses, called behavior, allow each organism to adjust to a constantly changing world. In this chapter, you will read about different types of behavior and the purposes they serve.

Guide for Reading

Key words: innate behavior, reflex, instinct, learned behavior, conditioning, imprinting, insight, social behavior

Questions to think about:

📖 How do innate behavior and learned behavior differ?

📖 How do dominance hierarchies, communication, and territoriality benefit a species?

36-1 The Nature of Behavior

Section Objectives:

- *Distinguish* between a behavioral response and a physiological response.
- *Explain* the importance of responding only to particular stimuli and give examples of differing perceptions of stimuli.
- *Describe* the relationship between heredity, structure, the nervous system, and behavior.
- *Explain* the role of biological clocks in certain cycles of behavior.

Stimuli and Behavior

An organism's environment is always changing. These changes may involve one or more external factors, including heat, light, carbon dioxide, oxygen, moisture, and the activities of other organisms. Environmental changes also may involve internal factors, such as thirst or hunger. Any change in the external or internal environment is called a **stimulus** (plural, stimuli).

In a living organism, metabolic processes work best when *homeostasis,* or a stable internal environment, is maintained. Living things maintain homeostasis by physiological or behavioral responses to stimuli. When a hungry dog smells food, it salivates. The smell of food is a stimulus. Secreting saliva is a physiological response. The dog may then actively hunt for food. That response is behavioral. Both responses help the dog to digest food and restore homeostasis. As another example, a person sweats when hot. If this

▲ **Figure 36–1**

Courtship Behavior. The male frigate bird has a bright red, inflatable throat sac used for courtship display.

785

▲ **Figure 36–2**
Perception of Light. Evening primrose as it is seen by a human (top). Evening primrose as it is seen by a honeybee (bottom). Honeybees are able to see ultraviolet light, whereas humans are not. Thus, the same flower is perceived differently by different species.

response does not cool the body, the person may remove outer clothing. Sweating is a physiological response. Removing outer clothing is a form of behavior. **Behavior** is the series of activities performed by an organism in response to stimuli. Although stimuli may bring about physiological responses, such as sweating, these responses usually are not classified as behavior.

Each species has its own pattern of behavior. In flight, a robin usually does not respond to the sight of a rabbit. A hawk, however, may respond by capturing and eating the rabbit.

Behavior aids in the survival of the individual and the species. When a rabbit sees an attacking hawk, it runs first in one direction, and then abruptly changes direction. This behavior may confuse the attacker and allow the rabbit to escape.

Perceiving and Responding to Stimuli

Organisms are exposed to countless environmental stimuli involving sound, light, chemicals, movement, and many other factors. But, one organism may not perceive something that another perceives. The honeybee, for example, can see ultraviolet light, but humans cannot. See Figure 36–2. Because of its form of vision, the bee can distinguish between white light with ultraviolet light and white light without ultraviolet light. Therefore, a bee's "white" and a human's "white" are different.

Many organisms, especially bats, have a keen sense of hearing. Bats feed on moths and other night-flying insects. A bat uses echolocation by bouncing high-pitched sounds off objects. The bat is able to detect the echoes and determine whether an object is an obstacle or moving prey. Having processed this information, the bat responds by directing its flight toward the insect or around the obstacle. Some moths are able to hear the bat sounds. They respond by trying to avoid the bat. See Figure 36–3.

Figure 36–3
Significant Stimulus. A bat responds to a small object that moves, but not to a stationary object. Here, a horseshoe bat pursues a moth in flight. ▶

▲ **Figure 36–4**
Cheetah Chasing Prey. The behavior of an organism is dependent on its body structure.

Another type of environmental factor that can be perceived differently by different organisms involves the sense of smell and chemicals. The female gypsy moth, for example, releases a chemical sex attractant. Male gypsy moths are able to detect minute quantities of this chemical from great distances. In comparison, humans have a poor sense of smell.

Because there are so many stimuli in the environment, an organism must respond only to signs or "messages" that are significant to it. A frog flicks outs its tongue to catch small, dark objects that move. It does not respond to stationary objects. A spider in its web responds only to web vibrations caused by a trapped insect and not to web movements caused by wind. Responding only to particular messages is important when quick action is necessary to capture food or to avoid danger. Each message triggers a specific behavior pattern that is unique to the species involved.

Coordinating Behavior

The behavior of an organism depends on its form and structure. As you can see in Figure 36–4, the cheetah's body structure permits it to run swiftly to catch prey. The robin escapes predators by flying away. Humans and other organisms respond in ways that are suited to their bodies.

Behavior is affected by all the body systems. The nervous system, endocrine system, muscles, and skeleton influence an organism's responses. The more complex an organism's nervous system and structure are, the more complex its behavior is. A frog's behavior, for example, is more complex than a grasshopper's. The behavior of a chimpanzee is more complex than that of a frog. Human behavior is the most complex of all.

▲ **Figure 36–5**

Canada Geese in Flight. Some behavioral activities, such as flying in a "V" formation, occur at the population level.

The Nervous System All multicellular animals, except the sponge, have nervous systems. The nervous system coordinates complex behavior. The simplest nervous systems occur in the coelenterates, which include the hydra, jellyfish, and sea anemone. The members of this group have nervous systems consisting of cells that form a *nerve net* (see Chapter 14). A nerve net produces little or no coordination of activity. In the sea anemone, if a tentacle touches a small piece of food, only that tentacle bends toward the mouth. If the prey is large and struggling, the stimulus spreads over the nerve net, and other tentacles join in the response.

The flatworm exhibits more complex behavior. A flatworm moves by the beating of cilia on its lower body surface and by the rippling and crawling motions of its body. Crawling requires a brain that can act as a center for the coordination of muscles. The flatworm's nervous system includes enlarged ganglia that form a centralized, simple brain. If these ganglia are removed, the crawling activity stops. In higher animals, the brain is more complex. In all animals with brains, the brain coordinates complicated behavioral sequences.

Heredity The development of an organism's body structure, nervous system, and behavior patterns depends on its heredity. DNA carries a "blueprint" for the development of body form and structure. Within this so-called blueprint is a "program" for behavior that uses body form and structure for continued survival. For example, within a spider's DNA are instructions for building the kind of web that its species constructs.

Behavior occurs in groups as well as in individual organisms. An example of group behavior is the flight of geese in a "V" formation, as shown in Figure 36–5. Instructions for this pattern of flight are coded in the DNA of each individual goose. A complex sequence of events links the DNA codes of individuals to their behavior as a group.

As you read earlier, behavior also depends on the ability to perceive and interpret stimuli. Sense organs receive information about environmental conditions. Heredity determines the types of stimuli that can be perceived, how they are interpreted, and the responses that can occur. Heredity also determines the coordinated patterns of muscular movement that form the resulting behavior sequence.

Cycles of Behavior

Although the behavior of an individual organism may vary over time, certain behaviors may be repeated at particular time periods. For example, some activities may occur only at certain times each day. Field mice are active at night and quiet during the day. Morning-glory flowers open during the day and close at night. Hawks hunt by day and owls by night. The activities of these organisms seem to be influenced by a *rhythm,* or cycle, of approximately 24 hours.

Early investigations into biological rhythms were conducted with fiddler crabs. These animals usually become darkly colored during the day and pale at night. To study biological rhythms, some crabs were placed in rooms with complete darkness and constant temperature and humidity. Even in these conditions, a regular cycle of color changes continued. This demonstrated that the cycle involved a physiological process.

Physiological and behavioral cycles occurring over a period of about 24 hours are called **circadian** (sur KAY dee un) **rhythms.** This term comes from the Latin words *circa* (about) and *dies* (day). Circadian rhythms are so regular that sometimes they are said to be controlled by an internal "biological clock." A biological clock, however, usually runs a little faster or slower than a normal clock. It must be adjusted constantly by an external stimulus such as daylight. Suppose, for example, that the inborn cycle of an insect is 25 hours. Each day, the insect's biological clock will adjust by one hour to keep it in phase with the normal day-night cycle. But, if the insect is kept under daylight conditions for a long period, its biological clock will not be adjusted each day. Thus, it gradually becomes out of phase with the actual day-night cycle. If the insect were kept under daylight conditions for 10 days, on day 10, its clock would be 10 hours out of phase with the normal daily cycle.

A similar disruption of a daily cycle is experienced by someone who flies from San Francisco to London. Upon arrival, the person is still oriented to the daily cycle in San Francisco. Most people in London are sleeping when this person is awake, hungry, and ready to work. This phenomenon is commonly called jet lag. In a few days, the individual's biological clock will be reset to London time and the traveler's activities will coincide with the London pattern.

Some animal phenomena follow monthly or even yearly cycles. The menstrual cycle is an example of a monthly cycle in humans.

Biology and You

Q: Are some people born shy? Can you overcome shyness?

A: Have you ever been alone at a party because you were too shy to speak to people? One study has reported that as many as 80 percent of all Americans say they were shy at one time. Shyness can be frustrating and damaging to your social life and your self-esteem.

What causes shyness? Some research shows a link between shyness and genetic makeup. If this is the case, you may be born with a hereditary trait for shyness. This does not mean that you will always be shy. The research shows that people can overcome shyness given the right environment and support. Other research suggests that shyness is not inherited. Instead, it is learned by children in response to their surroundings.

To overcome shyness, one doctor suggests imagining how you might act if you were more outgoing. Another approach is to practice ways of striking up conversations. Whatever the cause, shyness can be overcome.

Write a two paragraph story from the perspective of a shy teenager at a party or some other new social situation.

Another monthly cycle involves the grunion, a small fish that lives along the Pacific coast. Each year from April to June, this fish lays its eggs in the wet sand of the beach during the three or four days of the highest monthly tides. Yearly cycles occur in animals such as the ground squirrel, which hibernates during the colder months of each year. Experiments with ground squirrels kept under constant environmental conditions showed that the animals entered hibernation without environmental stimuli. This indicates that they, too, are influenced by biological clocks.

36-1 Section Review

1. Define the term *behavior*. How does behavior aid an individual or species?
2. List some environmental factors that are important in behavior.
3. What is the role of heredity in behavior?
4. What are circadian rhythms?

Critical Thinking

5. Classify each of the following as a stimulus or a behavior: hunger, a sneeze, running from a predator, thirst, pain, a noise, scratching an itch, a baseball thrown at you, being tickled. (*Classifying*)

36-2 Innate Behavior

Section Objectives:

- *Describe* some innate behavior patterns in plants and protists.
- *Explain* the mechanism and importance of reflexes.
- *Explain* the concept of instinct and give several examples of instincts.
- *Describe* several factors that can influence instinctive behavior.

Innate Behavior in Plants and Protists

Any form of behavior that has not been learned is **innate behavior,** or inborn behavior. Most innate behavior aids survival and reproduction. Innate behavior results from impulse pathways built into the nervous system. Instructions for the impulse pathways are carried in the DNA of the individual. These instructions are all the same for a given species. They are passed from parent to offspring as an inherited trait. An organism cannot "choose" to perform an innate behavior—it occurs automatically.

Innate behavior is found in most organisms. In simple organisms, such as the paramecium, sea anemone, and flatworm, all behavior is almost entirely innate. In more complex animals, only part of the behavior is innate. The remainder is learned behavior, which you will learn about in Section 36–3.

Plants also have inborn responses to stimuli. In plants, as you read in Chapter 19, growth toward or away from stimuli is called a *tropism*. This growth may involve the roots, stems, or leaves. Environmental factors, such as light, gravity, water, heat, and chemicals, act as stimuli for a tropism. Like the sponge, plants do not have a nervous system. Tropisms are controlled by hormones and may take hours or even days to occur.

Some plants exhibit rapid responses. Leaves of the sensitive plant, a type of mimosa, quickly fold and droop after they are touched. The leaves of the Venus flytrap rapidly close when sensitive hairs on its surface are touched. See Figure 17–12. These types of rapid responses in plants are not controlled by hormones. They are brought about by changes in turgor pressure in certain cells.

One characteristic of cells is their sensitivity to stimuli. The ameba and the paramecium do not have a nervous system, but they are able to respond to environmental changes. When a paramecium bumps into an obstacle, it stops. Then, it reverses its ciliary beating and backs away before turning and going forward again. If the paramecium hits the same obstacle, it repeats the back-up and turn. Eventually, it avoids the obstacle.

The ameba avoids intense light in a similar fashion, as you can see in Figure 36–6. It can, however, pursue a food particle until it has engulfed it. If the ameba loses actual contact with the food, it can still sense the location of the food by chemical stimuli.

Any movement by a simple animal or a protist toward or away from a particular stimulus is called a **taxis.** For example, the movement of an ameba away from a strong light is a *negative* taxis. An ameba moving toward food is an example of *positive* taxis.

Figure 36–6

Taxis in Protists. When a paramecium encounters an obstacle (left), it backs away (1——▶2) and then goes forward (3——▶4). If it is still blocked, it repeats the process (4——▶5, 6——▶7) until it avoids the obstacle. An ameba avoids strong light by moving away from the stimulus (middle). It pursues food by moving toward the stimulus (right). ▼

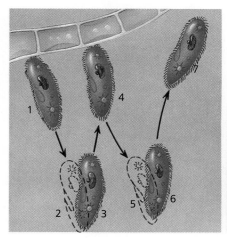
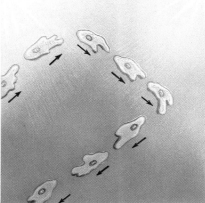
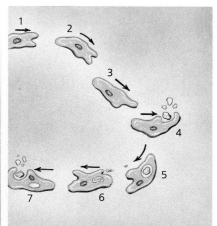

Reflexes

As an animal develops, its neurons develop connections that form a system. For the most part, once the system is formed, neurons do not change their connections with other nerves. As you read in Chapter 15, these fixed pathways in the nervous system are the basis for a type of innate behavior called the **reflex.** A reflex involves a *receptor* (sense organ) that detects stimuli and an *effector* (muscle or gland) that produces a reaction. Nerve cells connecting the receptor and effector through the spine or lower brain complete the nerve pathway, which is called the *reflex arc.* A reflex arc always produces the same response.

Reflex actions are simple, quick, and automatic because they do not require control by the upper brain. You can respond to a stimulus faster if you do not have to process information in your brain and decide what to do. If you touch a hot object, a reflex response causes you to pull back your hand quickly, before you are consciously aware of the pain. Thus, reflexes help an organism to avoid injury by allowing it to react quickly to environmental changes.

Reflexes occur in all animals with a nervous system. A flatworm subjected to electric shock automatically contracts. If a frog's toe is pinched, the leg is always drawn away. If you rub the hind area of a dog's back, it responds by scratching. In complex animals, reflexes account for only a small portion of behavior. In simple animals, reflexes account for a great part of the total behavior.

Sometimes, reflex arcs are linked to produce a complicated series of reflex actions. The behavior of the praying mantis is an example. When a fly or other insect appears in the mantis's field of vision, the mantis turns its head to face the insect. This is the first reflex. Then, the mantis orients its body in the same direction as the head—the second reflex. If the fly is close enough, the mantis instantly strikes, grasping the fly with its front legs. The strike—the third reflex—takes only a fraction of a second. See Figure 36–7.

Figure 36–7

Series of Reflex Actions. When a praying mantis sees a fly, a three-step series of reflex actions occurs. First, the mantis's head turns toward the insect (left). Next, its body orients in line with its head (middle). Finally, the mantis strikes, capturing the fly with its front legs (right). ▼

The stimulus triggering the turning of the mantis's head is the sight of the fly. The subsequent stimulation of receptors in the neck of the mantis results in the turning of the body. The strike follows the stimulus of aligning the head and body. Thus, this series of reflex actions depends on a succession of particular stimuli.

Instincts

Some animal behavior patterns involve a complicated set of un-learned activities that occur in response to a stimulus or series of stimuli. Such a complex, inherited behavior sequence is called an **instinct.** One example of instinctive behavior is nest building in birds. While instincts are quite complicated, they do not have to be learned and are performed automatically. However, an organism performing an instinctive behavior can be consciously aware of what it is doing. In contrast, a taxis and a reflex are simpler behaviors that are performed unconsciously.

Instincts usually are associated with either individual survival or species survival. They often involve activities related to feeding, defense, or reproduction. Instincts provide an animal with ready-made "answers" to its problems of survival because no time is required to learn a response. This is especially important for an animal with a short life span.

Instincts are found primarily in vertebrates and complex invertebrates. Each species displays its own characteristic instinctive behaviors, which are performed by all members of the species. The migration of salmon up the same river in which they were hatched, the communication "dances" of bees, the migration patterns of geese, and the construction of a hanging nest by the Baltimore oriole are instincts unique to those organisms.

Instincts in Invertebrates   Behavior is almost entirely instinctive in most invertebrates. A spider's construction of its intricate web, for example, is an instinct. In fact, the structure of its web can identify a spider as easily as its anatomy. There are four basic types of webs. Within each type, every species exhibits its own unique pattern of construction.

Some spiders perform courtship rituals before mating. A male jumping spider may perform a certain "sidestep" dance in front of his mate. An orb-weaving male spider may "strum" on the strands of a female's web before approaching her. Although these ritualistic acts seem complicated, they are instinctive and appear to communicate the male's intent to mate. Usually, such behavior prevents the male from being eaten by the female.

As you have read, the life cycle of an insect occurs in several stages. Each stage has its own set of instincts. For example, the june beetle larva shuns light. However, adult june beetles are attracted to light, often scraping loudly against the screens of lighted windows.

In insects and spiders, certain complex acts are performed only once, but correctly, without experience. Without being taught, the caterpillar of the gypsy moth climbs upward when hungry. Just

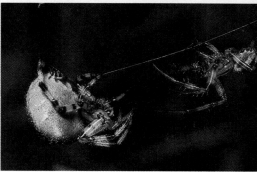

▲ **Figure 36–8**

Instinct. Among spiders, web construction and mating rituals are instinctive. Most orb web spiders (top) weave their web in a single night. Before mating, there is usually a courtship ritual. These two garden spiders (bottom) are about to mate.

1. Egg-laden female appears with head up.

2. Male swims in "zigzag dance."

3. Female swims, with head up, toward male.

4. Male swims toward nest.

5. Female follows male.

6. Male points to nest.

7. Female enters nest.

8. Male "trembles" and nudges female.

9. Female lays eggs in nest.

10. Female leaves nest.

11. Male enters nest and fertilizes eggs.

▲ **Figure 36–9**

Mating Behavior of the Three-Spined Stickleback. The ordered sequence of behavior depends on hormones and is triggered by sight.

before it enters the pupa stage, it spins a particular type of cocoon in a series of steps performed only once in its life. If a female spider builds a silken egg case but misses the opening when laying her eggs, she still completes and guards the empty case.

Instincts in Vertebrates   All vertebrates exhibit instinctive behavior. Fishes, like birds, act instinctively when they build their nests. The male sunfish scoops out a shallow, saucerlike nest in the bottom sediments of a pond or lake. He removes all the pebbles, leaving a layer of sand. After the female lays her eggs in the nest and they are fertilized, the eggs adhere to the sand. Only the male cares for the eggs. He fans the eggs with his tail and drives away predators until the eggs hatch.

In an instinct, one activity may trigger the next. That is, each activity in a behavior sequence may depend on the previous activity. For example, the reproductive behavior of the three-spined stickleback, a fish, is composed of a sequence of ordered stimuli and responses. After establishing a territory (discussed in Section 36–4), the male builds a nest. If an egg-laden female enters his territory, the male performs a "zigzag" courtship dance. The female responds by swimming toward the male. The male then swims toward the nest, and the female follows him. The male points to the entrance to the nest with his head. This act stimulates the female to enter the nest. Her presence in the nest stimulates the male to "tremble" and nudge her at the base of the tail. His trembling and nudging stimulate her to lay eggs. The female leaves the nest, and the male enters and fertilizes the eggs. See Figure 36–9.

Besides nest building, birds exhibit other types of behavior that are primarily instinctive. Many species of birds make an annual round trip, called a **migration,** between their winter feeding grounds in the south and their spring breeding grounds in the north. Every year, these birds follow the same route, sometimes flying thousands of miles in each direction. The golden plover, for example, flies more than 8000 miles each year from its Arctic breeding grounds to southeastern South America and back again. Many other animals, such as the humpback whale and the monarch butterfly, migrate each year.

What prompts these animals to migrate at the same time every year, and how do they know where they are going? They may respond to environmental changes that tell them to start. Research has shown that many migratory birds navigate by using the angles of the sun or stars above the horizon. Certain pigeons and other birds may use the earth's magnetic field to orient themselves.

Hormonal Control of Instincts

Many physiological stimuli, which are related to an organism's metabolism, can trigger instincts. Often, the animal responds to these stimuli with instinctive behavior that tends to restore homeostasis. For example, a lack of food upsets an animal's homeostasis.

This physiological stimulus results in a sensation of hunger. The animal then responds with feeding behavior. Another example is the sensation of thirst, which is triggered by the blood's composition. This physiological stimulus triggers drinking behavior. Once homeostasis is restored, the instinctive behavior ceases.

Reproductive, or sex, hormones are the physiological stimuli that lead to courtship, mating, and care of young. These hormones may act directly on specific organs, such as gonads, causing them to produce eggs or sperm. See Chapter 21. These changes, along with a high level of reproductive hormones, prepare the individual for reproduction. Under these conditions, sight of a mate usually stimulates courtship behavior, which ultimately results in mating.

In some animals, once mating occurs, reproductive behavior stops. In other animals, however, care of the eggs or young is stimulated by the presence of reproductive hormones and the presence of offspring. For example, the hormone *prolactin* influences parental feeding in doves. Doves feed their young "pigeon milk," a thick, white substance produced in the bird's crop. The "milk" is composed of sloughed epithelial cells lining the crop.

Hormones may also influence behavior by acting on the central nervous system. Female canaries, unlike males, usually do not sing. But, if a pellet of male sex hormone (testosterone) is placed under the skin of a female, she will sing a typical canary song until all the hormone is metabolized.

Experiments with thirsty goats have shown that the brain, as well as hormones, is involved in some instincts. When a goat is thirsty, the lack of water in its blood stimulates cells in the hypothalamus. The *hypothalamus* is a portion of the brain to which the *pituitary gland* is attached. A message from the hypothalamus is passed into the pituitary gland, which secretes a hormone. The hormone causes the blood to absorb more water from the filtrate passing through the tubules of the nephrons in the kidneys. (Refer to Chapter 12.) While this physiological means of saving water goes on, the goat looks for water to drink.

The hypothalamus and the pituitary may also control reproductive behavior and parental care of offspring. The hypothalamus or pituitary can be triggered by information from the internal or external environment. For example, in many birds and mammals that reproduce seasonally, increasing daylight stimulates the hypothalamus. In turn, the hypothalamus stimulates the pituitary gland to produce sex hormones. Warmer temperatures also may begin the seasonal production of sex hormones in vertebrates.

Sometimes, a visual stimulus can trigger the hypothalamus. For example, if a female ring dove sees a male, the visual image stimulates cells in her hypothalamus, which activates her pituitary gland. Hormones from the pituitary gland cause the growth of the ovaries and secretion of the female hormone estrogen. The female then lays eggs. Normally, the hormone level of a female ring dove that does not see a male remains low. A female, however, can lay eggs without seeing a male if she is given an injection of hormones.

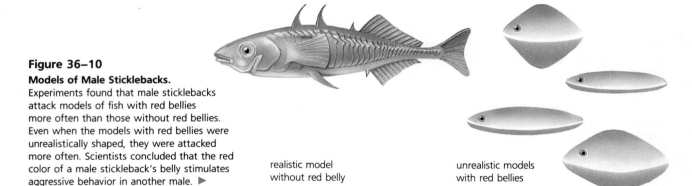

Figure 36–10

Models of Male Sticklebacks.
Experiments found that male sticklebacks attack models of fish with red bellies more often than those without red bellies. Even when the models with red bellies were unrealistically shaped, they were attacked more often. Scientists concluded that the red color of a male stickleback's belly stimulates aggressive behavior in another male. ▶

realistic model
without red belly

unrealistic models
with red bellies

Once internal conditions have been altered by hormones, the animal responds to external stimuli that do not ordinarily affect it. Specific environmental stimuli can then start various forms of reproductive behavior. For example, after hormones have caused the male three-spined stickleback to establish a territory and build a nest, he guards his territory against the intrusion of other males. When other males approach, he becomes aggressive and drives them away. During the mating period, the belly of a male becomes red. It is the red color on the belly of another male that stimulates the aggressive behavior. See Figure 36–10.

36-2 Section Review

1. Define the term *innate behavior.*
2. What is a taxis?
3. What is the chief advantage of instincts?
4. List three instincts that are controlled by hormones.

Critical Thinking

5. How are instincts similar to reflexes? How are they different? (*Comparing and Contrasting*)

36-3 Learned Behavior

Section Objectives:

- *Distinguish* between learned behavior and innate behavior and between classical conditioning and operant conditioning.
- *Explain* habituation and habit formation, and give examples of each.
- *Describe* the process of imprinting and explain its significance.
- *Explain* the concept of insight and relate it to learning.
- **Laboratory Investigation:** *Measure* the progress of learning by trial and error (p. 810).

Experience and Learning

Behavior that changes as a result of experience is called **learned behavior.** Experiences are "stored" in the brain as memory and can be recalled. In learning, a recalled experience is used to help modify behavior in a new situation. Unlike innate behavior, learning permits an animal to choose responses that are appropriate to the stimuli in any new situation. As a result of learning, an animal can adapt to change.

While innate behavior is controlled by genes, learned behavior is controlled only indirectly by genes. Heredity determines the type and complexity of the nervous system, which controls the ability to learn. Animals with more complex nervous systems exhibit a greater capacity to learn. Animals with long life spans and long periods of parental care exhibit mostly learned behavior as adults. In some animals, parents actually teach their offspring certain behaviors. Lions, for example, teach their offspring how to hunt.

Learning has several forms. The simplest type of learning is called **habituation.** In this type of learning, an animal learns to ignore repeated "unimportant" stimuli. Squirrels in a park become accustomed to people and will come quite close to them. Crows learn to ignore a harmless scarecrow in a field.

Humans and other animals can learn to perform many complex activities with little or no thought. A **habit** is a series of actions that is learned first and then becomes automatic through repetition. Dressing, writing, talking, tying shoelaces, typing, and dancing are habits. All habits are performed slowly at first, with concentration and often with difficulty. The habit is learned through repetition until it becomes automatic and requires little or no conscious effort. As the habit continues to develop, the actions become easier and faster to perform and more accurate.

Brushing your teeth is an example of a good habit. The benefits of brushing are best realized if the habit is maintained regularly. A bad habit, such as smoking, is dangerous to your health because its ill effects are repeated again and again. It can be difficult to break a bad habit, just as it can be hard to develop a good habit.

Conditioning

A simple type of learning that changes behavior through forming new associations is called **conditioning.** One type, called *classical conditioning,* was first studied by the Russian physiologist Ivan Pavlov (1849–1936). In classical conditioning, a new stimulus is paired with the stimulus and response of a simple reflex action. The new stimulus, which does not normally cause the reflex response, becomes associated with the response by a process of reinforcement or reward. As a result, the new stimulus can be substituted for the original one.

Dogs normally salivate when they smell or see food. Secretion of saliva is the reflex response. In studies carried out by Pavlov, a bell was rung (new stimulus) every time dogs were given food (old

Stimulus: sight and smell of food

Response: salivation

Stimulus: sight and smell of food and ringing of bell

Response: salivation

Stimulus: after many pairings of food and bell, ringing of bell only

Response: salivation

▲ **Figure 36–11**

Classical Conditioning. Classical conditioning involves the association of a new stimulus with a response. A hungry dog salivates at the sight or smell of food. Pavlov exposed a dog to a ringing bell whenever the dog received food. Eventually, ringing of the bell alone caused the dog to salivate.

stimulus). The association of the new stimulus (bell) with the reflex response (salivation) was reinforced by presenting food immediately after ringing the bell. After a while, the dogs salivated upon hearing the bell without the presence of food. See Figure 36–11.

In Pavlov's experiment, the food was the *unconditioned stimulus.* The ringing bell was the *conditioned stimulus.* Salivation became a *conditioned response.* House pets often become conditioned in a similar manner. Pet owners know that the sound of the refrigerator door or can opener can cause their pets to be quite attentive and to salivate.

Operant conditioning is a type of conditioning in which animals learn to "operate" or do something by associating the action with a reward. The animals may learn to push a lever, open a door, respond to a command, or perform other kinds of actions. In this case, the conditioned response is a behavior that produces the reward. An animal first must be *motivated* to seek the reward by performing the activity. The motivation is usually hunger.

B. F. Skinner, a psychologist, constructed a special cage with a food-release mechanism to study operant conditioning. A rat placed in the cage eventually learns that pressing a bar releases a pellet of food. At first, the rat explores its cage. This leads to a period of trial and error before the rat accidentally performs the correct act—pressing the bar. Food is the reward for pressing the bar. By rewarding the rat for pressing the bar, the rat is more likely to repeat the act. The rat learns to associate the action of pressing the bar with the reward of food. Because of the initial period of exploration, operant conditioning is also called *trial-and-error learning.*

Trainers teach animals to perform tricks by rewarding them for desired behavior and punishing them for undesired behavior. To the animal, the absence of an expected reward is the same as a punishment. Some sea mammals, such as porpoises, sea lions, and whales, will perform tricks at a trainer's command when rewarded with a fish. Hunger is the motivation for doing the trick. During training, the animal is given a fish only after the trick is performed correctly. A poor performance results in no reward.

Imprinting

In 1935 Konrad Lorenz, an Austrian biologist, discovered a simple type of learning that occurs in young animals of certain species. He found that newly hatched ducks, geese, and chickens follow and form a strong attachment to the first moving object they see. If the moving object makes sound, it is more likely to be followed. Forming an attachment to an object or environment soon after hatching or birth is called **imprinting.** This form of learning is rapid and cannot be reversed.

Imprinting in birds occurs several hours after hatching. The young bird will follow the object even over obstacles. Ordinarily, the first object a young bird sees is its mother. Imprinting on the mother bird helps the young bird survive. This behavior creates a bond between the mother bird and her young. It also helps the young birds recognize others of their kind, which is important for mating later in life. See Figure 36–12.

If eggs are hatched under experimental conditions in an incubator, the young birds may imprint on a foreign object, such as a person, a mechanical object, or a dog. The young birds will follow the object when it moves around. Once the period of imprinting has passed, about 36 hours after hatching, the young birds cannot imprint on another object. For example, mature mallard ducks that imprinted on a "decoy" mother duck will court and attempt to mate with the decoy. At maturity, geese and other birds that imprinted on foster parents of another species try to mate with members of the foster species.

Although imprinting was first described in birds, it is now known to occur in other animals. Animals whose young can walk shortly after birth, such as buffalo, deer, sheep, and goats, are capable of imprinting. Puppies that have human contact during their sixth and seventh weeks are easily trained to be good pets. However, if puppies have no human contact until after their fourteenth week, they rarely make friendly pets. Scientists speculate that imprinting plays an important part in the ability of migrating salmon to return to the stream in which they were hatched.

Figure 36–12

Imprinting. Imprinting is a rapid, irreversible form of learning. These young swans are imprinted on their mother (left). Here, young geese have imprinted on Konrad Lorenz, the first living creature they saw (right). ▼

Figure 36–13

Insight. Insight is the ability to plan a solution to an unfamiliar problem. Here, chimpanzees use insight to get bananas suspended from the ceiling. ▶

a.   b.   c.

Insight

Insight is the ability to "create" a solution to an unfamiliar problem without a period of trial and error. The animal surveys a new situation and uses the memory of past learning and experiences to "plan" a reasoned response. Insight is found only in complex vertebrates.

In one of the best-known studies of insight, a hungry chimpanzee is released in a room with boxes scattered on the floor and a bunch of bananas hanging from the ceiling out of reach. The chimp finds that it cannot reach the bananas. Eventually, it piles the boxes up and then climbs them to reach the bananas. Reasoning has resulted in the correct response to a new situation. Chimpanzees and monkeys are often successful at solving problems. Most other animals, including dogs, fail at first.

The ability to reason is most highly developed in humans. At first, young children solve problems by trial and error and by imitation. This helps them master such motor skills as tying shoes. As children mature and learn, they begin to use insight to solve problems. Insight is enhanced by the ability to exchange ideas through speech and written symbols. These skills allow each person to use someone else's experiences to solve problems—an ability that is unique in the animal kingdom.

36-3 Section Review

1. List three types of learned behavior.
2. Which animals exhibit a greater capacity to learn?
3. How does a habit become automatic?
4. What are some benefits of imprinting?

Critical Thinking

5. The day after hatching, a young goose was seen following an adult duck. When the goose matures, will it try to mate with a duck or a goose? Explain. (*Reasoning Conditionally*)

36-4 Social Behavior

Section Objectives:

- *Describe* several forms of helpful and hostile social behavior.
- *Describe* three types of signals used by animals and give an example of each type.
- *Explain* the function of territoriality and the role of communication in maintaining territories.
- *Explain* the organization of a honeybee society and describe how these bees communicate with one another.

Helpful Behavior

Animals often encounter other animals in their environment. Interactions occur between individuals of the same species and between animals of different species. Many environmental factors bring animals together in a group. A street light is a stimulus that attracts a variety of insects. The rich African grassland draws grazing animals such as the zebra, wildebeest, and antelope to feed side by side. A water hole may be a gathering place for many species. These groups of mixed animal species, called *aggregations,* are not social. They are together merely by chance.

Animals belonging to the same species exhibit **social behavior,** which consists of both helpful and hostile interactions. Helpful social behavior includes mating behavior, family interactions, and activities by larger groups. Mating behavior brings a female and a male together, and results in the fertilization of eggs. This form of social behavior may involve courtship as well as actual mating. *Courtship* is a form of communication that signals a readiness to mate and prevents conflict.

Family interactions involve helpful relationships between parents and young. The interactions between members of a family are the basis for the providing of food, shelter, and defense for the young. The interactions usually require certain stimuli. Birds incubate their eggs and later feed their chicks. A gull incubates the eggs only when they are visible and when certain hormones are produced in the parent.

◀ **Figure 36–14**
Courtship Display. This male prairie chicken displays its colorful feathers to attract females for mating.

The needs of the young are often satisfied when their behavior stimulates other forms of behavior among family members. Many birds will not feed their young unless the chicks "communicate" in a form of behavior called *begging*. Begging is usually done by opening the mouth widely. This stimulates the parent to place food in the chick's mouth. Initially, the young are stimulated to beg when the parent's head appears over the edge of the nest or by the jolt of the adult landing on the nest. Sea gull chicks peck at a red spot on the parent's beak when begging. See Figure 36–15. The parent then regurgitates food, picks some up, and presents it to the chick.

Helpful behavior also occurs within larger groups such as herds of animals, schools of fish, and flocks of birds. Individuals can cooperate in groups as long as they can communicate with one another. Information is passed between members of a group by sound signals, visual signals, and chemical signals.

Species that live in groups often depend on the group for survival. A group of animals is more alert than a single individual. When one member senses danger, it communicates it to the whole group. The group then tries to escape the danger. Groups also offer various forms of protection against attack by predators. Male musk oxen form a protective ring around the young and females. Some animals attack in groups as a form of defense. Small birds will attack crows and hawks in groups—a behavior called *mobbing*. Hunting in groups is practiced by wolves, lions, and wild dogs.

Conflict and Dominance Hierarchies

Close association among animals of the same species can result in conflict instead of cooperation. When resources are limited, individuals of a species must compete for food, water, space, and mates. Many animals resolve this competition by aggressive behavior. *Aggression* is threatening or fighting another animal to force it away from something it possesses or is trying to obtain.

There are numerous forms of aggression. Animals may bite, butt, kick, or claw one another. However, aggressive behavior among members of a species seldom causes serious injury or death. See Figure 36–16. More often, "symbolic" threatening behavior results in a "winner." Such displays are instinctive and clearly understood by other species members. In threat displays, animals assume aggressive postures and show the contrasting or brightly colored parts of their bodies. Robins display their red breasts. Fish display colored body parts or puff themselves up.

When actual fights occur, they are mostly symbolic and cause little injury to the rivals. Poisonous male snakes wind themselves together during a bout, and each attempts to butt the other's head with its own. The snake that becomes fatigued first retreats from the fight. Although the snakes have fangs, they never bite each other. Rival male bighorn sheep butt their heads together, resulting in spectacular fights. The sounds of their butting heads can be heard for a long distance.

The animal that loses a fight may simply run away. In some cases, it signals defeat by a *subordination ritual*. A wolf or dog

Figure 36–15
Begging Behavior. A herring gull chick begs for food by pecking at a red spot on its parent's beak. ▼

▲ **Figure 36–16**

Aggressive Behavior. These two male elephants will fight until it is clear which one is dominant. They will not fight to the death

signals submission by presenting its neck to the winner. This display stops further aggression by the victor. Its position of superiority has been established. The animal that wins a fight achieves *dominance,* or better access to contested resources, such as food, water, space, and mates.

In some animals that live in organized groups, or *societies,* fighting establishes a **dominance hierarchy,** or ranking, within the group. Individuals with a high rank have first choice of necessities. In duck and chicken societies, a *pecking order* is established by pecking action. Those birds with the highest standing have uncontested access to food, water, and the roost. Pecking order reduces tension in the group because there is less fighting over who gets what first. Due to their lower position in the group, subordinates must wait their turn to drink or eat. If no food is left, the subordinates go without nourishment. For members of the lowest rank, survival is most difficult.

In baboons, dominance hierarchy is more complicated than a sequence of individuals with decreasing dominance. A group of baboons, called a *troop,* is governed by a clique of dominant males. Any member of the ruling clique that is challenged by an outsider is supported by the other members of the governing group. This helps keep the troop stable by preventing frequent changes in the hierarchy. The clique of dominant males protects the troop against attack by predators.

Communication

Communication plays an important role in both helpful and hostile social behavior. The male and female of many animal species are brought together by signals. The signals sent out by both sexes include visual signals, sound signals, and chemical secretions.

a. b. c.

▲ **Figure 36–17**

Courtship Display. To attract a mate, a male mallard duck displays his brightly colored feathers. Normal swimming (left). "Head up, tail up" display (middle). "Down up" display (right).

Visual signals are common among fish and birds. These signals include movement and posture as well as displays of certain body parts. The zigzag dance of the male stickleback causes the female to approach him and be led to the nest. The male mallard duck courts the female by displaying his brightly colored plumage. See Figure 36–17. The female identifies herself with a signal. She may beg food from the male, which will act as a signal to him not to attack her as he would a male during the reproductive period. Other examples of visual signals during aggression were given in the previous section.

Sound signals are important among insects, frogs, birds, whales, and many other animals. The male *Aedes* mosquito is attracted to the sounds produced by the wings of the female while in flight. Male crickets attract females with sound signals made by rubbing their wings together. Male toads and frogs have characteristic songs that attract the females to the pond or bog. Male birds such as the robin, meadowlark, wood thrush, and vireo produce distinctive high-pitched songs as a signal to the female. Humpback whales can communicate across distances greater than 10 kilometers by using underwater songs. Biologists are still investigating the function of these songs.

Chemical secretions are used by many animals as signals. Such secretions are called **pheromones** (FEHR uh mohnz). These chemicals influence the behavior of other members of the same species. Pheromones can act as sex attractants. The female silkworm moth releases a pheromone so strong that it attracts males from a distance as great as three kilometers. Gypsy moths, cockroaches, and many other insects produce pheromones that act as sex attractants.

Territoriality

A **territory** is an area defended by an individual against intrusion by other members of the same species. Claiming or defending a territory is another aspect of social behavior. Territorial behavior usually takes place during the breeding season and is often limited to males. Maintaining a territory gives an animal the space needed to acquire food, court a mate, and raise a family.

To defend its territory, an individual may spend long periods of time in a conspicuous place. The animal may also "announce" ownership with certain forms of communication. In some birds, such as thrushes, a male will choose an unoccupied area and then sing loudly and vigorously to stake his claim. The loud singing warns away other males but attracts females. A singing duel between competing males often can resolve a boundary dispute. Usually, the loudest-singing male gains the largest territory.

Some mammals use pheromones to mark territory. Deer have pheromone-secreting glands in their hooves. As you can see in Figure 36–18, male antelopes have similar glands close to their eyes. The civet, a cat, has pheromone-secreting glands around its anus. Dogs and wolves mark their "turf" with urine.

Primates vary in their territorial behavior. Although chimpanzee and gorilla troops live on a large range, they show no defense of territory. Different troops may intermingle without causing disturbances. The tree-living howler monkey of Central America displays strong group territoriality. A troop consists of several dozen monkeys. They defend the boundaries of their territory by sessions of howling, which discourages intrusion by other troops. See Figure 36–19. The rhesus, a monkey found in southern Asia, is also territorial and will drive off intruders with active threat displays and, if necessary, ferocious attacks.

Territoriality has several advantages. Due to territories, groups or individuals are evenly distributed throughout the available area. This enables an animal species to use the resources in the environment efficiently. Dividing an area into territories tends to reduce conflict between members of a species. In some cases, especially in birds, position of a territory helps attract a mate.

▲ **Figure 36–18**
Pheromones. A male antelope claims his territory by marking twigs with a secretion produced by a gland near his eye.

◀ **Figure 36–19**
Territoriality. A troop of howler monkeys defends its territory as a group. Their loud howling scares away other troops.

Honeybee Societies

Insect societies are found among termites, ants, and bees in addition to other insects. In an insect society, every effort is directed toward the survival of the entire group. The activities of most insect societies are centered around one female, the *queen*. The queen may live for five years or more. During that time, her only responsibility is to reproduce. All members of the group are offspring of the queen.

A honeybee society involves a complex organization and division of labor. See Figure 36–20. The queen is usually the only reproductive female. She is the focus of hive activity. The colony continues as a group as long as she is healthy and functioning. A queen mates only once with each of several *drones,* or male bees, during a "nuptial" flight. The drones develop from unfertilized eggs and live for only a short time after mating with the queen bee. From the single mating flight, the queen stores enough sperm to fertilize the thousands of eggs she will lay for the colony. Most of these eggs will hatch into sterile females called *worker bees*. These worker bees carry on the essential activities that maintain the hive, including producing honey, feeding larvae, and protecting the hive. See Figure 36–21.

Chemical signals are one form of bee communication. Even the status of the queen's health is passed to members of the hive by means of a chemical signal. This signal originates from the queen as a secretion. It is first passed to workers attending her and then to others in the hive. Chemical signals help maintain the organization of the hive.

Workers at the hive entrance fan hive odor outward, which guides foraging workers home. After a worker locates a food source, it uses visual and other signals to communicate the location of the

Figure 36–20

Honeybee Society. Members of a honeybee hive surround the queen bee. The queen's presence in the hive is established as attending bees lick chemical secretions from her body and pass them to others in the hive. ▶

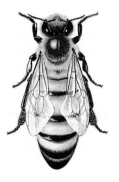
queen

worker

drone

◄ **Figure 36–21**
Types of Bees. The three types of bees that live in a hive are the queen, the worker, and the drone.

■ Careers *in Action*

Zookeeper

Zookeepers take care of the animals in a zoo. They feed and groom the animals and clean and check the safety of the enclosures. In small zoos, keepers work with large groups of animals, such as the mammals or reptiles. In large zoos, they work with smaller groups, such as large cats or marine mammals.

Not only do zookeepers tend to the physical needs of the animals, they also keep a close watch on them to learn about each individual's needs and moods. Changes in the way an animal looks or acts are often signs of illness. Keepers observe the animals to detect any changes in behavior or habits. In this way, keepers assist veterinarians in maintaining the animals' health. Keepers also help to design animal enclosures and assist in demonstrations for zoo visitors.

Problem Solving   You are a zookeeper working with chimpanzees. You notice that the chimpanzees seem to be moping around. To motivate the animals, you decide to make some changes. You know the following about the behavior of chimpanzees in the wild: they spend much of their time swinging in trees; they feed largely on fruits and leaves; they are often seen searching through twigs and leaves on the ground for seeds; and they make tools out of twigs to "fish" for insects in termite mounds.

1. Describe a zoo habitat that could be designed for the chimpanzees that would be more like their natural environment.
2. Based on what you know about their feeding behavior, design a feeding system that would require the chimpanzees to use reasoning.
3. Describe the reasoning process the chimpanzee might use to get the food in the system you designed.

■ **Help Wanted:**
Zookeeper. High-school diploma required; one year of college instruction in zoology recommended. Contact: American Association of Zookeepers, Topeka Zoological Park, 635 Gage Blvd., Topeka, KS 66606.

food to other workers in the hive. The work of Karl von Frisch, a German scientist, helped explain this form of communication between honeybees. In his experiments, he placed sheets of paper smeared with honey near a hive. Eventually the bees discovered the honey. Von Frisch noticed that when one bee discovered the honey, others would come in a short time. Somehow the other bees were informed by the first bee.

Von Frisch set up a hive with glass sides so he could observe the scout bees returning from a new food supply. A bee that landed at the new food supply was marked with a little paint to help identify her. Upon returning to the hive, the bee fed several other workers. Then she performed a "dance" on the inner wall of the honeycomb. Other workers near her became excited and followed with antennae held close to her. One by one, the other workers left the dancer and in a short time appeared at the location of the food. The dance pattern, called the *round dance* by von Frisch, consists of circling first in one direction and then in the other direction. See Figure 36–22. The round dance is repeated many times. This dance seems to inform other workers that food is nearby, while the bee passes on the scent of the flower to the workers.

To determine if the round dance told other workers the direction of the food, von Frisch placed four dishes of sugar water

Figure 36–22

Bee Communication. The round dance informs the bees of a hive that food is nearby (left). The waggle dance indicates the direction and distance of the food from the hive (right). ▼

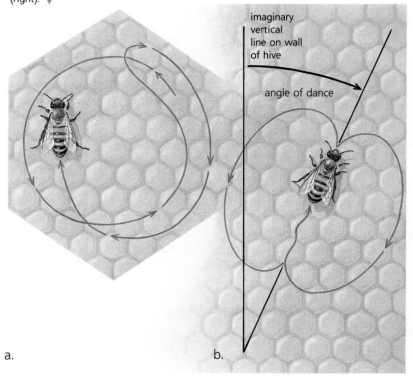

imaginary vertical line on wall of hive

angle of dance

a.

b.

scented with honey 10 meters from the hive—one dish north, one south, one east, and one west. Although von Frisch observed that a worker might find only one sample, he also noticed that equal numbers of bees appeared at each dish. He concluded that the round dance did not communicate direction. Other experiments were performed, which showed that the dance did not communicate distance. The round dance, then, seems only to tell the bees that food is nearby.

Von Frisch set up another experiment. He used two dishes of sugar water scented with lavender oil. He placed one dish at 10 meters from the hive and the other at 300 meters from the hive. The foraging bees that discovered the dish at 10 meters returned to the hive and danced the round dance. Workers returning from the dish 300 meters away performed a different dance. Von Frisch called this dance the *waggle dance*. In this dance, the bee runs along the wall of the hive in a straight line for a short distance while wagging her abdomen from side to side. She then circles back, runs forward again, circles in the opposite direction, and runs forward again. This "figure eight" motion is repeated several times. See Figure 36–22b. The distance of the food is communicated by the number of straight runs and the number of waggles given in a 15-second time period. As distance increases, the number of straight runs decreases while the number of waggles increases. The waggle dance appears to be used to communicate food distances greater than 50 meters.

The direction of the food source is indicated by the direction of the waggle dance on the honeycomb wall. If the food is located in the direction of the sun, the dancer travels vertically up the wall of the honeycomb on the straight run. If the food is in a direction away from the sun, the dance is directed down the honeycomb wall. If the food is located at some angle between the line from the hive to the sun, the straight run is oriented at the same angle relative to a vertical line. Bees are thus able to translate information concerning the distance and direction of food into the speed and angle of a dance.

36-4 **Section Review**

1. Define the term *social behavior*.
2. What is a dominance hierarchy?
3. What signals bring together the male and female of a species during courtship?
4. What are two insects that form complex societies?

Critical Thinking

5. What are some advantages of "symbolic" threatening behavior? (*Judging Usefulness*)

Laboratory Investigation

Learning by Trial and Error

Learned behavior is behavior that develops as a result of experience. Of all the ways humans learn, trial and error is one of the most time-consuming. In this investigation, you will participate in trial-and-error learning and observe the effect of practice on the time it takes to reach a solution.

Objectives

- *Solve* a problem by trial and error.
- *Observe* the relationship between practice and the time it takes to solve a problem by trial and error.

Materials

puzzles 1 and 2
watch with second hand
graph paper

Procedure

A. Make a data chart like the one shown below. Working in pairs, one student should take the envelope marked puzzle 1. When your partner tells you to begin, remove the pieces and arrange them so they form a perfect square. Record the time it took to complete the puzzle. Mix the pieces and return them to the envelope.

Trial	Time to Complete Task (Minutes)	
	Puzzle 1	Puzzle 2
1.		
2.		
3.		

B. Repeat step A two more times and record your results in the trial 2 and trial 3 entries of your data chart. Wait two minutes between trials.

C. Now take an envelope marked puzzle 2. On your partner's signal, remove the pieces and arrange them so that they form the letter *T*. Record the time it took to complete the puzzle in your data chart. Repeat this two more times, following the procedure outlined in steps A and B.

D. Partners should switch roles and repeat steps A through C.

E. Using the format shown below, make bar graphs that show the time it took to complete each puzzle in each trial.

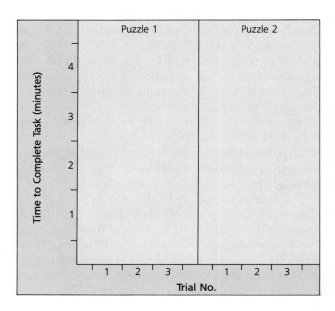

Analysis and Conclusions

1. How did practice affect the amount of time it took you to complete each puzzle?

2. What was the purpose of waiting two minutes between trials?

3. How did the time to complete the task vary from puzzle 1 to puzzle 2? Can you explain why?

4. How did the time to complete the puzzles differ between the first-round and second-round partners?

5. What factors favor the possibility that the second-round partner will solve the puzzles faster than the first-round partner? Experimentally, how could you eliminate this factor?

6. How does each instance of solving a puzzle affect how you go about solving the puzzle in a subsequent trial?

Chapter Review

36

Study Outline

36–1 The Nature of Behavior

■ Behavior is a series of activities performed by organisms in response to stimuli.

■ Living things respond only to pertinent stimuli with a behavior pattern unique to the species involved. The unique behavior depends on the organism's form and structure and on information coded in the DNA.

■ Some inherited behavior cycles are controlled by an internal "biological clock" that varies over the day, month, or year.

36–2 Innate Behavior

■ Innate behaviors, such as reflexes and instincts, aid in survival and reproduction. In some simple organisms, behavior is entirely innate.

■ Reflexes are simple, quick, and automatic. They result from a fixed pathway in the nervous system.

■ Instincts are complex innate behavior sequences associated with individual or species survival.

■ Physiological factors such as hunger, thirst, and hormone levels influence instincts. External factors, including visual images or the length of day, may also trigger instincts.

36–3 Learned Behavior

■ Learned behavior changes as a result of experience. The experiences are stored in the memory and recalled at a later time, helping the animal to adapt to change.

■ Habituation, conditioning, and imprinting are simple forms of learned behavior.

■ Insight, a more complex learned behavior, is the ability to solve an unfamiliar problem without a period of trial and error.

36–4 Social Behavior

■ Social behavior consists of interactions among animals of the same species. The interactions may include courtship, mating behavior, family interactions, and large-group interactions.

■ Conflict arising among individuals in a large group can result in a ranking of the group's members. This ranking is called a dominance hierarchy.

■ Communications in the form of visual signals, sound signals, and chemical secretions between males and females of many animal species affect social behavior.

■ A member of a species may claim a territory to provide it with food and a place to raise a family. The individual will defend the territory against intrusion by others of the same species.

■ Insect societies are found among ants, termites, and bees. Honeybee societies are highly organized. Worker bees communicate the location of food by a round dance or a waggle dance.

Vocabulary Review

stimulus (785)	habit (797)
behavior (786)	conditioning (797)
circadian rhythms (789)	imprinting (799)
innate behavior (790)	insight (800)
taxis (791)	social behavior (801)
reflex (792)	dominance hierarchy (803)
instinct (793)	
migration (794)	pheromones (804)
learned behavior (797)	territory (804)
habituation (797)	

A. Definition—Replace the italicized definition with the correct vocabulary term.

1. Birds exhibit an *innate complex behavior sequence* during nest-building activities.

2. Nerve cells between the receptor and effector complete a *fixed pathway in the nervous system*.

3. A behavioral response is a reaction to a *change in the environment*.

4. A *home area defended against members of the same species* gives the resident space to raise a family.

Chapter Review

5. *A form of attachment that develops soon after hatching* causes newly hatched birds to follow the first moving object they see.

6. Through fighting, animals living in organized groups establish a *ranking.*

7. Chimpanzees are able to solve unfamiliar problems by *creating a solution without trial and error.*

8. *Behavioral cycles that occur over a 24 hour period* are controlled by "biological clocks."

B. Multiple Choice—Choose the letter of the answer that best completes each statement.

9. A form of behavior in which animals learn to ignore repeated unimportant stimuli is called (a) habit (b) conditioning (c) habituation (d) imprinting.

10. A type of instinct in which a bird makes an annual round trip is called (a) migration (b) reflex (c) taxis (d) conditioning.

11. A female gypsy moth may attract a mate with (a) habit (b) conditioning (c) a pheromone (d) a circadian rhythm.

12. A form of behavior that changes as a result of experience is a(n) (a) response (b) instinct (c) social behavior (d) learned behavior.

13. A form of behavior inherited by offspring is called (a) innate behavior (b) imprinting (c) conditioning (d) insight.

14. By moving toward or away from a stimulus, a protist exhibits a form of behavior known as (a) taxis (b) migration (c) reflex (d) conditioning.

Content Review

15. Why do organisms respond only to specific stimuli?

16. How does a bat determine whether an object is an obstacle or moving prey?

17. Describe evidence supporting the existence of internal biological clocks.

18. What is the approximate length of the cycle of a circadian rhythm?

19. Describe the functions of the receptor and the effector in producing a reflex.

20. Why is instinct an important adaptation for survival in animals that live only a short time?

21. Describe the relationships among a single reflex, a series of reflexes such as those exhibited by a praying mantis capturing food, and an instinct.

22. What physiological stimuli lead an animal to courtship, mating, and care of young?

23. Explain the difference between innate behavior and learned behavior.

24. Briefly describe Pavlov's conditioning of dogs.

25. What is a conditioned response?

26. How does insight differ from instinct?

27. Compare learning in humans and in chimpanzees.

28. Describe the types of helpful social behavior.

29. What is the advantage of a dominance hierarchy as a type of social organization?

30. How does the subordination ritual limit further aggression?

31. List two purposes of sound signals used by animal groups.

32. In what ways is the establishment of territories a useful form of social behavior?

33. Describe the social organization of a beehive.

Graphic Organizing

For information on graphic organizers, see Appendix G at the back of this text.

34. **Concept Map:** Construct a concept map of learned behavior using the following concepts: learned behavior, conditioning, insight, imprinting, habit, habituation, classical conditioning, and operant conditioning. Select 10 additional concepts to add to the map that help explain the different types of learned behavior. As you link the concepts, do not forget the lines and linking words.

Critical Thinking

35. The migration of birds and the nest building of fishes are inherited behavior patterns known as instincts. From these examples, make a general statement about why some behaviors are passed on from generation to generation. (*Generalizing*)

36. How are genetically determined behaviors different from more complex learned behaviors? (*Comparing and Contrasting*)

37. Some biologists think that lower vertebrates such as fishes are capable of learning. What assumptions must these biologists have made? (*Identifying Assumptions*)

38. Order the following list of administrative and teaching staff at your school to show the dominance hierarchy within this group: student teacher, principal, teacher, department head, school board member, and assistant principal. (*Ordering*)

39. List one similarity and one difference between Pavlov's concept of classical conditioning and Lorenz's concept of imprinting. (*Comparing and Contrasting*)

Creative Thinking

40. If a plant suddenly developed the ability to learn, what types of behavior would you expect of it?

41. Some people think that because a person cannot read or write, that person cannot learn. Think of ways in which this assumption might be incorrect in certain cases, and give examples.

Problem Solving

42. Biologists have evidence that the chicks of herring gulls beg food from their parents by pecking at a red spot on the beaks of the adult birds. Design a controlled experiment to determine whether chicks peck only a red spot.

43. The ability of an animal to learn can be tested in a maze. A maze consists of a route with several choices. The choices are like forks in a road. By choosing the correct turn at each fork, the animal will complete the maze. In a learning experiment, two groups of rats—a control group and a test group—were tested daily for fourteen days. The test group received a reward of food upon completing the maze. The control group received no food. The errors, or wrong turns, for each group were counted, and the averages were graphed, as shown below. From this data, what can you conclude about learning and the use of rewards?

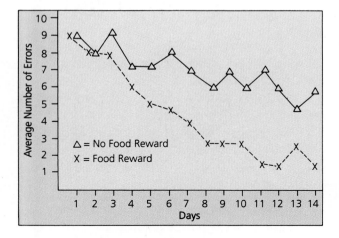

Projects

44. Find a spider in its web. Determine the responses of the spider to a variety of stimuli, such as sound, light, movement of the web, insect in the web, and so forth. List each stimulus and response on a chart. For further reference, read Jean-Henri Fabre, *The Life of the Spider* (Horizon Press, 1971) or B. J. Kaston, *How to Know the Spiders,* 3d ed. (William C. Brown & Co., 1978).

45. Write a report on the life and contributions of one of the following scientists: Konrad Lorenz; Karl von Frisch; Niko Tinbergen; Jane Goodall.

46. Visual signals are important indicators of animal behavior. In many pets, the various positions of the head, ears, and tail have definite meanings. Observe a dog, cat, or other pet, and gather additional information at the library. Prepare a poster for class display that illustrates the positions of the head, ears, and tail as visual signals of a pet's behavior.

Biology and Problem Solving

How do elephants communicate with each other?

A female elephant is pregnant for almost two years. She spends an additional two years nursing her calf. During this four-year period, the female lives independently and often far away from male elephants.

After this time, the female elephant is ready to breed again. Her breeding period is short—only a few days out of every four or five years. Yet the distant males are able to find her quickly once she has entered her reproductive period. A human observer would have noticed no signal sent by the female. Yet, somehow, she has let the males know where to find her.

How do male and female elephants locate one another for reproduction? This question has puzzled scientists for many years.

1. List some ways in which elephants might find each other during the female's breeding period.

Scientists studying this question have made some observations suggesting that elephants may respond to signals that humans cannot perceive.

Observation A: A herd of 100 elephants is grazing peacefully. Suddenly all become motionless, standing with ears raised for almost a minute.

Observation B: For several days no elephants are seen at a water hole. Then, several groups arrive simultaneously, appearing from different directions.

In neither situation did human observers perceive any signals. Furthermore, such situations have been observed even when the wind direction prevented the elephants from using their sense of smell.

2. What are some possible explanations for the synchronized behavior of the elephants?

To what signals were the elephants responding in these situations? One scientist who has worked to answer this question is Katharine Payne. One day when Dr. Payne was observing a group of elephants in a zoo, she felt a deep throbbing in the air. She associated the throbbing with distant thunder or with the playing of the largest pipe of an organ. At the same time, she noticed a fluttering of the skin on the elephants' foreheads. Perhaps, Dr. Payne hypothesized, the elephants were making very low-pitched sounds— sounds in a range of frequency that humans cannot hear. Perhaps, she reasoned, elephants were using such sounds to communicate with each other.

To test this hypothesis, Dr. Payne and her associates set up microphones near elephants in a zoo and recorded sounds for a month. The electronic printouts from this investigation showed 400 calls, three times as many as the scientists had heard. Many of the sounds were below the range of human hearing, in what is known as *infrasound*.

3. Did Dr. Payne's findings with zoo elephants support her hypothesis that elephants use low-pitched sounds to communicate? Explain.

4. Design an experiment by which Dr. Payne could test her hypothesis with elephants in the wild, rather than in a zoo.

Dr. Payne and her associates conducted many experiments with wild elephants during expeditions in southwest Africa. In one experiment, two scientists were positioned one mile away from two male elephants that were being observed by Dr. Payne. The associates played a recording of an infrasound elephant call, noting the starting time. They did not tell Dr. Payne what the call was or when they began playing it.

Dr. Payne observed both elephants drinking and bathing at the water hole. Then, although she heard nothing, Dr. Payne saw both elephants lift their heads and spread their ears. They then set off in the direction of the call, not even stopping to touch the water.

The call that Dr. Payne's associates had played was the infrasound mating call of a female elephant. The time at which Dr. Payne noted that the elephants lifted their heads and spread their ears was the same time at which her associates had played the recording.

5. What can you conclude from this experiment?

6. Why was it important that Dr. Payne not know the type of call and the time it was played?

Repeated experiments showed that the elephants responded with changed behavior and with calls of their own when recorded calls were played by the scientists.

Another scientist studying elephant behavior has used radio collars to track the travels of female elephants for the past 10 years. He has found that family groups often coordinate the direction in which they wander, even when the groups are far apart.

7. How might Dr. Payne work with this scientist to learn more about elephant communication?

Problem Finding

List some questions that you still have about elephant communication.

Ecology

Tidal waters stretch as far as the eye can see. The scene is peaceful, beautiful, and barren. Or is it barren? Actually, here as on every shore, life abounds—in the waters, among the rocks, on and under the sand. There are fish, crustaceans, worms, plants, and one-celled organisms, all living active lives and all dependent on one another for survival. This is what ecology is all about—species of organisms so interrelated that a natural and continuing balance results. The seashore is but one ecological setting. There are many more. In this unit, you will learn that the earth is a delicately balanced system with limited space and resources. You will examine the interactions between organisms and their environment and see how these interactions result in a flow of energy and a cycling of materials.

▲ This crab spider is making a meal of a small, iridescent fly.

Organization in the Biosphere

37

It is morning in the tropical valley. Birds call, insects buzz, and raindrops rattle onto huge, shiny leaves and drip to the ground. The crab spider sits on a purple orchid, waiting. Soon, the orchid's sticky stamens attract a fly. The fly begins to feed on the nectar, but its meal is brief. In a flash, the spider pounces. The fly becomes food for the spider. How are the fly, the spider, the orchid, and the rain connected? In this chapter, you will learn how living and nonliving things interact to permit life on our planet.

Guide for Reading

Key words: ecology; population; ecosystem; food chain; food web; nitrogen, carbon, and oxygen cycles; succession; climax community

Questions to think about:

📖 What nonliving factors affect living things?

📖 How are materials cycled between living things and the environment?

37-1 Abiotic Factors in the Environment

Section Objectives:

- *List* the abiotic factors in the environment.
- *Describe* how light, temperature, and precipitation vary with position on the earth's surface.
- *Describe* the process of soil formation.

All types of living organisms have adaptations that allow them to survive in a particular environment. They may show adaptations for food getting and reproduction, as well as for avoiding predators. Living organisms are affected by physical factors in their environment, such as the availability of water, changes in temperature, the amount of light present in the environment, and the composition of the soil. The physical environment is also affected by the organisms that live in it. For example, some organisms help to break rock into soil. The growth of plants contributes to the filling in of ponds. Finally, organisms are affected by other organisms living in the same area. The branch of biology that deals with the interactions among organisms and between organisms and their environment is called **ecology** (ih KAHL uh jee).

In studying the interaction between organisms and their environment, both the living and nonliving factors must be considered. The **biotic** (by AHT ik), or living, **factors** include all the living organisms in the environment and their effects, both direct and

▲ **Figure 37–1**

Environmental Adaptatations. In order to survive in a desert environment, these plants have had to adapt to conditions of little water and large changes in temperature.

indirect, on other living things. The **abiotic** (ay by AHT ik), or nonliving, **factors** include water, oxygen, light, temperature, soil, and inorganic and organic nutrients.

Abiotic factors determine what types of organisms can survive in a particular environment. For example, in deserts there is little water available, and the temperature can change daily from very hot to cold. Only plants that are adapted to these conditions, such as sagebrush and cactus, can survive. Other types of plants, such as corn, oak trees, and orchids, grow in other environments with different abiotic conditions to which they are adapted.

Light

The energy for almost all living things on earth comes directly or indirectly from sunlight. The amount of sunlight striking a given area of the earth's surface changes with the latitude of the area. *Latitude* is the distance north or south of the equator. Both the *intensity,* or strength, of sunlight and the *duration,* or length, of daylight also vary with latitude. As shown in Figure 37–2, areas around the equator receive sunlight of the strongest intensity, while areas around the North and South Poles receive light of the weakest intensity. Areas at the equator receive about 12 hours of daylight throughout the year. At the North and South Poles, the sun does not rise above the horizon for the six winter months of each year. During the summers, the sun never sets. In regions between the equator and the poles, day and night lengths vary with the season. These variations in the amount of sunlight are caused by the daily rotation of the earth, the movement of the earth around the sun, and the tilt of the earth's axis.

The intensity and duration of sunlight affect the growth and flowering of plants. Some plants require high light intensity; others require low light intensity. Some require long days for flowering; others require short days. In many animals, migration, hibernation, and reproduction are influenced by day and night lengths.

Figure 37–2
Intensity of Sunlight at Different Latitudes. In this diagram, each ray represents the same amount of light. Note that near the equator a larger amount of surface is struck by more rays than is struck near the poles. ▶

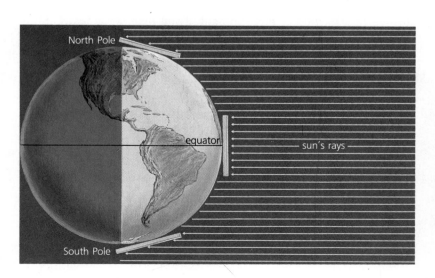

Light conditions also vary in aquatic environments. Light is absorbed as it passes through water. Thus, the amount of light present decreases with increasing depth. The layer of water through which light penetrates is called the *photic* (FOH tik) *zone.* About 80 percent of the earth's photosynthesis takes place in this zone. Below the photic zone is the *aphotic* (ay FOH tik) *zone,* where there is no light. Except for a few chemotrophs, organisms in the aphotic zone are heterotrophs that derive their energy from organisms that drift or migrate down from the photic zone.

Temperature

Temperature patterns on the earth's surface vary with latitude and with altitude. *Altitude* is the vertical distance above sea level. The temperature pattern of a region may be affected by the presence of nearby major geographic features, such as a mountain or an ocean. As the altitude rises, the temperature falls. Thus, the tops of mountains may be snow-covered even in warm regions.

The warmest average temperatures on the earth's surface are around the equator. Traveling north or south of the equator, the average temperatures drop. The North and South Poles are the coldest regions on earth.

Water

The release of water from the atmosphere as rain, snow, dew, and fog is called *precipitation.* The annual amount of precipitation varies from one region to another on the earth's surface. Annual precipitation patterns are related to latitude and to altitude. They are also influenced by local features, such as mountains and large bodies of water. Areas around the equator are hot and humid with heavy rainfall throughout the year. Because of the pattern of airflow over the earth's surface, most deserts are found around latitudes 30° north and 30° south of the equator. In these regions, there is a brief rainy season and almost no rain at all during the rest of the year. Still further north and south of the equator are *temperate* regions with hot summers and cold winters. Rainfall is fairly abundant in these regions. The polar regions are cold, and precipitation is in the form of snow.

Soil and Minerals

Soil consists of inorganic and organic materials. The inorganic material is mostly rock particles broken off from larger rocks by the action of water and wind—a process known as *weathering.* Alternate freezing and thawing of water helps to crack the rock and break off pieces. Soluble minerals in the rock dissolve in water, breaking down the rock still further. Lichens and other organisms also help to break down rock. When these organisms die, their remains are mixed with the rock particles, thus adding organic matter to the developing soil. Plants may take root in the soil. When they die, their remains add more organic matter to the soil.

The minerals present in soil depend partly on the type of rock from which the soil was formed and partly on the types of organisms living in the soil. The amount of precipitation determines the extent to which minerals will be retained in, or washed out of, the soil.

As soil development slowly proceeds, three distinct layers form. See Figure 37–3. The uppermost layer, which is called *topsoil,* includes organic matter and various living organisms. Plant "litter," such as fallen leaves and twigs, overlies the topsoil and gradually blends into it. The dark, rich organic matter in the topsoil is called **humus** (HYOO mus). Humus is formed from the decay of dead plants and animals. The living organisms found in the topsoil include plant roots, earthworms, insects, and many other animals and protists. The organisms of decay—bacteria and fungi—are also found in this layer.

Beneath the topsoil is a layer of *subsoil.* The subsoil is made of rock particles mixed with inorganic compounds, including mineral nutrients. Water-soluble materials from the topsoil are carried downward into the subsoil by the downward movement of water. The bottommost layer of the soil is made of bits of rock broken off from the parent bedrock below.

There are many types of soils. They are classified according to their organic content, mineral composition, pH, and size of the rock particles. Sandy soil has the largest particles, silt has particles of intermediate size, and clay is made up of small particles. Water drains too quickly through sand and too slowly through clay for good plant growth. In general, the best soils for plants to grow in consist of a mixture of clay and larger particles. However, different types of plants grow well in different types of soil. Some plants require well-drained, sandy soils, while other plants do best in soils that are clayey. Also, some plants thrive in acid soils, but others need alkaline soils.

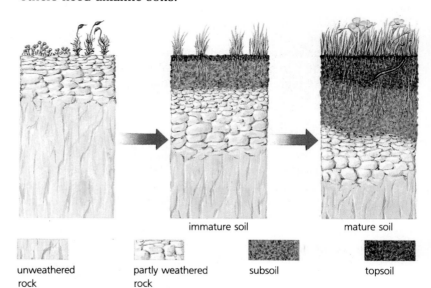

immature soil   mature soil

unweathered rock   partly weathered rock   subsoil   topsoil

Figure 37–3

Three Stages in the Development of Soil. The forces of weathering and the action of certain small organisms gradually break down bedrock into fine particles. The products and remains of organisms add organic matter to the rock particles, forming soil. ▶

37-1 **Section Review**

1. What is ecology?
2. List some common abiotic environmental factors.
3. What are some geographic features that influence temperature patterns?
4. Which environmental factors cause the weathering of rock during soil formation?

Critical Thinking

5. First compare and then contrast topsoil with subsoil. (*Comparing and Contrasting*)

37-2 Biotic Relationships in Ecosystems

Section Objectives:

- *Explain* the terms *population, community, ecosystem, biosphere,* and *autotrophic* and *heterotrophic nutrition.*
- *List* the different types of symbiotic relationships and *describe* each of them.
- *Describe* the feeding relationships in an ecosystem in terms of competition, food chains, and food webs.
- *Explain* the flow of energy in an ecosystem using the concepts of pyramids of energy and biomass.
  Concept Laboratory: *Gain* an understanding of the interactions that occur within an ecosystem (p. 827).

Populations, Communities, and Ecosystems

In studying organisms in nature, ecologists often look at a particular group of organisms in a particular type of natural setting. The simplest grouping of organisms in nature is a **population.** A population includes all individuals of a particular species within a certain area. All the black oak trees in a forest make up a population. All the bullfrogs in a pond make up a population. Populations can also be considered as parts of larger groups. All the populations of different organisms within a given area make up a **community.** For example, all the frogs, fish, algae, plants, and other living things in and around a pond make up a pond community.

An **ecosystem** includes a community and its physical environment. In an ecosystem, both the biotic and abiotic factors are included. There is an ongoing exchange of materials between the nonliving and living parts of an ecosystem. All the ecosystems of the earth are linked to one another. Organisms move from one ecosystem to another. Water and other inorganic substances pass from one ecosystem to another. Also, organic compounds, with their stored energy, are transferred between ecosystems.

The Biosphere

The portion of the earth in which living things exist is known as the **biosphere.** Compared to the diameter of the earth, the biosphere is a thin zone. It is about 20 kilometers in thickness, extending from the ocean floor to the highest point in the atmosphere where life is found. The biosphere includes portions of the *lithosphere* (the solid part of the earth's surface), the *hydrosphere* (the water on and under the earth's surface as well as the water vapor of the air), and the *atmosphere* (the mass of air surrounding the earth).

Autotrophic and Heterotrophic Nutrition

An ecosystem includes all kinds of organisms—microorganisms, plants, and animals. These organisms interact in many ways, but their nutritional and energy relationships are among the most important.

Autotrophs are organisms that can make their own food using carbon dioxide. Most autotrophs carry on photosynthesis. A few, however, carry on chemosynthesis. Directly or indirectly, autotrophs provide the food for heterotrophs—those organisms that cannot synthesize their own food.

Heterotrophs are divided into several groups according to what they eat and how they obtain their food. Heterotrophs include herbivores, carnivores, omnivores, and saprobes. **Herbivores** (HER buh vorz) are animals that feed only on plants. Rabbits, cattle, horses, sheep, and deer are herbivores. **Carnivores** (KAR nuh vorz) are animals that feed on other animals. Some carnivores are predators, and some are scavengers. **Predators** (PRED uh terz), such as lions, hawks, and wolves, attack and kill their prey and feed on their bodies. **Scavengers** (SKAV en jerz) feed on dead animals they find. Vultures and hyenas are scavengers. **Omnivores** (AHM nih vorz) are animals that feed on both plants and animals. Humans and bears are omnivores. **Saprobes** are organisms that obtain nutrients by breaking down the remains of dead plants and animals. Many bacteria and fungi function as saprobes.

Symbiotic Relationships

Symbiotic relationships are relationships in which two different organisms live in close association with each other to the benefit of at least one of them. There are three types of symbiotic relationships: mutualism, commensalism, and parasitism.

In **mutualism** (MYOOCH uh wuh liz um), both organisms benefit from their association. For example, termites have cellulose-digesting microorganisms living in their digestive tracts. Without

▲ **Figure 37–4**
Herbivores. Giraffes are just one of many kinds of herbivores that inhabit the African tropical grasslands.

these microorganisms, termites could not get nutrients from the wood they eat. In turn, the termites provide the microorganisms with food and a place to live.

Lichens consist of algal or blue-green bacterial and fungal cells. Both types of cells benefit from this association. It allows them to live in environments in which neither could survive alone. Through photosynthesis, the algae or blue-green bacteria produce food for themselves and for the fungi. The fungi provide moisture and the structural framework and attachment sites in which the algae or bacteria grow.

Peas, clover, and alfalfa are *legumes*. Legumes have nodules on their roots in which certain bacteria grow. The bacteria convert nitrogen gas from the air in the soil into forms usable by the plants. In this relationship, the plants are supplied with the nitrogen compounds they need, while the bacteria are given an environment in which they can grow and reproduce. Legumes are part of the nitrogen cycle described in section 37–3.

In **commensalism** (kuh MEN suh liz um), one organism benefits from a symbiotic relationship, and the other is not affected. For example, pilotfish are small fish that live with sharks. They eat the scraps left over from the shark's feeding. Thus, the shark provides the pilotfish with food. As far as is known, the pilotfish neither helps nor hurts the shark. Barnacles may attach themselves to the large body surface of a whale. Barnacles are sessile and rely on water currents to bring them food. The movements of the whale provide them with a constantly changing environment and food supply. The whale is not affected by the presence of the barnacles.

In **parasitism** (PAR uh suh tiz um), one organism benefits from a symbiotic relationship, and the other is harmed. The organism that benefits is called the *parasite,* while the organism that is harmed is the *host.* Some parasites cause only slight damage to their hosts, while others kill the host. Tapeworms, for example, are parasites that live in the digestive tracts of various animals. There,

▲ **Figure 37–5**

Mutualism. These termites have microorganisms within their intestines that allow the termites to obtain nutrients. The microorganisms benefit from the presence of food and a place to live.

◀ **Figure 37–6**

Commensalism. The pilotfish around this shark eat leftover scraps from the shark's feeding. The shark is neither helped nor harmed by the presence of the pilotfish.

they are provided with nutrients and an environment in which to grow and reproduce. However, the host is harmed by the presence of the tapeworms. The loss of nutrients and tissue damage caused by the worm can cause serious illness. There are also parasitic plants that grow on other plants. Two examples of plant parasites are mistletoe and Indian pipe.

Symbiotic relationships, particularly those involving mutualism or commensalism, are not always permanent. Also, it is not always possible to say whether an organism is helped or harmed by such a relationship. For example, in many environments the algal cells of a lichen can survive well without the fungal cells. The fungal cells, on the other hand, may not be able to survive alone.

Competition in Ecosystems

Each type of organism within an ecosystem has a particular part of the environment in which it lives. This is its **habitat.** For example, the habitat of a slime mold is the damp floor of a forest. Because of the complex interactions that occur within an ecosystem, each species also plays a particular role in an ecosystem. The role of a species in an ecosystem is its **niche.** An organism's habitat is part of its niche, but only part. Also included are how, when, and where it obtains nutrients, its reproductive behavior, and its direct and indirect effects on the environment and on other species within the ecosystem.

In a balanced ecosystem, each species occupies its own niche. It occupies a particular territory (its habitat) and obtains nutrients in a particular way. Competition arises when the niches of two species overlap. The greater the overlap—the more requirements the two species have in common—the more intense the competition. Competition between two different species is called **interspecific competition.** As the resources being competed for become

Figure 37–7
Habitat of a Slime Mold. This white slime mold has grown around a small twig on the forest floor. Slime molds are found in cool, shady, moist places in the woods. ▶

Concept Laboratory

Nature's Interactions — A Model

Goal

To gain an understanding of the relationships and interactions within an ecosystem.

Scenario

Imagine you are an ecologist studying an ecosystem. To analyze the ecosystem, you decide to simulate the natural ecosystem by setting up a mini-ecosystem in a glass jar.

As you work, think about how organisms interact with each other and with their environment.

Materials

large glass jar
boiled water
soil, sand, gravel, or mud
plants, plant seeds, and spores
invertebrate animals (earthworms,
    sow bugs, insects, slugs, snails, etc.)
cheesecloth
rubber band

Procedure 🔲 🧪 ♻️ 🧤 🤲 🪨 *

A. Choose an ecosystem, such as a pond, forest, or field, that you would like to simulate. Design a mini-ecosystem that will support at least three types of plants and three types of animals from that ecosystem. Write up your plans and submit them to your teacher for approval.

B. After obtaining approval, add soil characteristic of the ecosystem to the jar.

C. Add plants and/or sow seeds and spores. Allow time for seeds and spores to germinate.

mosses
fungus
soil
cheesecloth
rubber band
flowering plant
leaf litter
earthworms

*For detailed information on safety symbols, see Appendix B.

D. Add several types of animals to the jar. Then, cover the jar with several layers of cheesecloth held in place with a rubber band, and place the jar in a well-lit area.

E. Observe the jar daily, adding water as it is required. Record the number of species of plants and animals, and note their behavior and growth. Look for interactions between organisms and for any changes that may occur.

Organizing Your Data

1. What is the energy source in your ecosystem?

2. How many different species did you observe in your ecosystem?

3. Which organisms increased in number? Which decreased in number? What factors may have caused these changes?

4. Describe the niche of any one of the organisms in the ecosystem you designed.

Drawing Conclusions

1. What are the general conditions necessary to maintain an ecosystem?

2. Write out the food chain for your ecosystem. Identify producers, consumers, and decomposers.

3. Using a diagram, show how carbon is recycled within your ecosystem.

4. What evidence is there of competition among the organisms in your ecosystem?

Thinking Further

1. What factors in your ecosystem would limit population growth?

2. How do you think adding more plants would affect your ecosystem?

3. What effect would sealing the jar with a lid have on your ecosystem? Explain your answer.

4. How is the ecosystem you have designed different from an actual ecosystem found in nature? In what way is it similar?

5. How can studying ecosystems be of benefit as humans begin to populate space?

more scarce, the competition becomes more intense. Eventually, one of the species is eliminated from the ecosystem, leaving the more successful species to occupy the niche.

Competition also occurs between members of the same species. This is called **intraspecific competition.** The intensity of the competition between members of the same species is affected by population density and the availability of resources. If conditions become harsh, those individuals with the most helpful adaptations will survive. The less-well-adapted individuals will not.

Producers, Consumers, and Decomposers

In all but a few small ecosystems, the autotrophs are plants and other photosynthetic organisms. They trap energy from sunlight and use it for the synthesis of sugars and starch. These substances can be changed to other organic compounds that are needed by the plant, or they can be broken down for energy. Heterotrophs can only use the chemical energy stored in organic compounds for their life processes. These organic nutrients must be obtained from the bodies of other organisms—either plants or animals. Because autotrophs are the only organisms in an ecosystem that can produce organic compounds (food) from inorganic compounds, they are called **producers.** Since heterotrophs must obtain nutrients from other organisms, they are called **consumers.**

Saprobes play an important role in an ecosystem. They function as *organisms of decay,* or **decomposers.** They break down the remains of dead plants and animals, releasing substances that can be reused by other members of the ecosystem. In this way, many important substances are recycled in an ecosystem.

Food Chains and Food Webs

Within an ecosystem, there is a pathway of energy flow that always begins with the producers. Energy stored in organic nutrients synthesized by the producers is transferred to consumers when the plants are eaten. Herbivores are the primary consumers, or *first-order consumers*. The carnivores that feed on the plant-eating animals are secondary consumers, or *second-order consumers*. For example, mice feed on plants and are first-level consumers. The snake that eats the mice is a second-level consumer, while the hawk that eats the snake is a third-level consumer. Since many consumers have a varied diet, they may be second-, third-, or higher-level consumers, depending on their prey. Each of these feeding relationships forms a **food chain,** a series of organisms through which food energy is passed. A simple food chain is shown in Figure 37–8.

Feeding relationships in an ecosystem are never just simple food chains. There are many types of organisms at each feeding level, and there are always many food chains in an ecosystem. These food chains are connected at different points, forming a **food web.** One of these is shown in Figure 37–9.

▲ **Figure 37–8**
A Simple Food Chain. The grass is a producer, the field mouse is a first-order consumer, and the owl is a second-order consumer. The arrows show the flow of energy in the food chain.

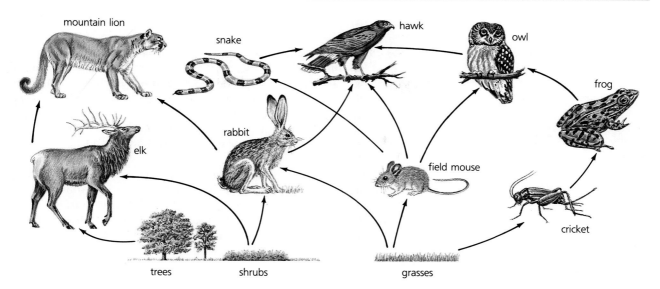

At every level in an ecosystem, there are organisms that act as decomposers. The decomposers make use of the wastes and remains of all organisms in the system. They use the energy they find in these materials for their own metabolism. At the same time, they break down organic compounds into inorganic compounds and make substances available for reuse. The decomposers are the final consumers in every food chain and food web.

▲ **Figure 37–9**

A Simple Food Web. Usually, each organism is part of several different food chains.

Critical Thinking in Biology

Predicting

When you say, "The Birds will win the pennant this year," you are making a prediction. Predicting is determining the probable outcome of an event before it occurs.

Consider the ecosystem inhabited by the organisms in the food web shown in Figure 37–9. Suppose a disease caused most of the rabbits in the ecosystem to die.

1. Predict what effects this would have on (a) the hawk population; (b) the amount of grass available, and (c) the field mouse population. Explain the basis for your prediction.
2. Which, if any, of these effects is certain to occur? Explain.
3. Why do predictions about changes in ecosystems contain a high degree of uncertainty?
4. Given the uncertainty of ecological predictions, what advice would you give to scientists working on acid rain, global warming, and other environmental problems?
5. Think About It Describe your thought process as you answered question 1.

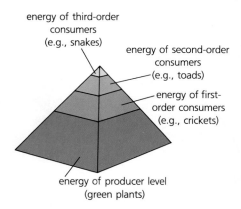

energy of third-order
consumers
(e.g., snakes)

energy of second-order
consumers
(e.g., toads)

energy of first-
order consumers
(e.g., crickets)

energy of producer level
(green plants)

▲ **Figure 37–10**
**The Pyramid of Energy in an
Ecosystem.** At each level in an ecosystem,
the energy available is only about 10 percent
of the energy at the level below it.

Pyramids of Energy and Biomass

The amount of energy available in a food web decreases with each higher feeding level. This happens because only a small fraction of the energy taken in as food becomes stored as new tissue. Much of the food eaten is not digested and absorbed. Furthermore, a large part of the energy in the food is used for respiration and maintenance. This energy is lost as heat. As a result, only about 10 percent of the energy taken in at any feeding level is passed to the next feeding level.

The amount of energy available in an ecosystem is commonly shown in the form of a **pyramid of energy.** The greatest amount of energy is present in the producers—the base of the pyramid—and the least energy is present at the top of the pyramid—the highest-level consumers. Because the amount of available energy decreases so steeply, there are usually no more than four or five feeding levels in an ecosystem.

Figure 37–10 is an example of a pyramid of energy. In a field, green plants are the producers, trapping the sun's energy for photosynthesis. For every 1000 calories of energy absorbed by the plants, only 100 are stored. Thus, 100 calories are available to a first-order consumer, such as a cricket, that feeds on the plants as they grow. As a second-order consumer, a toad that eats the cricket receives only 10 of the original 100 calories consumed by the cricket. At the top of the energy pyramid, a snake that eats the toad receives only 1 calorie of the 10 calories received by the toad. This amount is only 0.001 percent of the energy originally absorbed by the plants at the base of the pyramid.

Since the total amount of energy available decreases with each higher feeding level, the total mass of living organisms that can be supported at each level decreases, too. This relationship can also be represented by a pyramid. The relationship, known as the **pyramid of biomass,** shows the relative mass of the organisms— the *biomass*—at each feeding level. The greatest amount of biomass is in the lowest level, the producers. The least is found in the highest level of consumers.

37-2 Section Review

1. What is a population?
2. List three types of symbiotic relationships.
3. Which kind of organisms are always found at the base of a food chain?
4. Which level of an energy pyramid contains the most energy?

Critical Thinking

5. Can two species occupy the same habitat? Can they occupy the same niche? Explain. (*Reasoning Conditionally*)

37-3 Cycles of Materials

Section Objectives:

■ *Describe* each of the following biogeochemical cycles: the nitrogen cycle, the carbon and oxygen cycles, and the water cycle.

Laboratory Investigation: *Observe* nitrogen-fixing nodules in legumes (p. 840).

In all ecosystems, materials revolve between living things and the environment. Organisms incorporate certain substances from the environment into their bodies. When these organisms die, their bodies are broken down by decomposers, and the substances returned to the environment. If these substances were not returned to the environment, their supply would eventually become exhausted. The cycles of materials between living things and the environment are called *biogeochemical cycles*. Nitrogen, carbon, oxygen, and water are among the substances involved in such cycles.

The Nitrogen Cycle

Nitrogen is an important element in living things. It is a basic component of amino acids, which form proteins, and of nucleotides, which form nucleic acids. Nitrogen gas (N_2) makes up almost 80 percent of the earth's atmosphere. However, most organisms cannot use N_2 directly. They must use nitrogen compounds. Most plants can use nitrogen only in two inorganic forms, ammonia (NH_3) and nitrate (NO_3^-). Usually, nitrate is the major source of nitrogen for plants. From nitrate and ammonia, plants can make proteins and nucleic acids. Animals lack this ability. They can use nitrogen only in an organic form. Thus, animals must ingest plants or other animals to meet their nitrogen needs.

Nitrogen in the wastes and in the remains of organisms must be made available to living plants for reuse. This is accomplished through the activity of decomposers that break down the complex organic compounds in plant and animal remains. During decomposition, most of the nitrogen in organic compounds is released as ammonia. Some of this may be taken up directly by plants, but most is converted by **nitrifying bacteria** to nitrite (NO_2^-) and finally to nitrate. The nitrate then is available for uptake again by plants.

Not all nitrate in the soil and in water remains as nitrate until taken up by plants. **Denitrifying bacteria** get energy for their life processes by converting nitrite and nitrate to nitrogen gas (N_2). This form of nitrogen, which is released into the atmosphere, cannot be used by plants and animals. However, nitrogen gas can be changed to a form available to plants. A few kinds of bacteria, including blue-green bacteria, convert nitrogen gas directly to ammonia through a process called **nitrogen fixation.** Some of these **nitrogen fixers** are free-living. The ammonia they produce is used to synthesize their own nitrogen-containing compounds. Other nitrogen-fixers are symbiotic. They fix nitrogen only when living in close association with a host plant. Figure 37–11 shows the

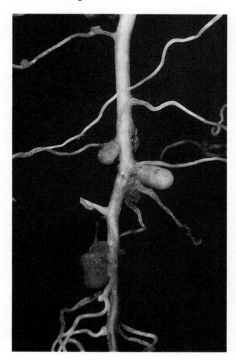

Figure 37–11

Root Nodules of a Legume. The nodules on this legume root contain nitrogen-fixing bacteria that provide the plant with usable forms of nitrogen. ▼

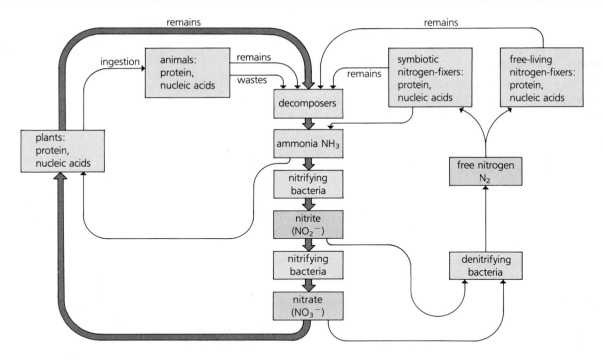

▲ **Figure 37–12**

The Nitrogen Cycle. The nitrogen cycle is a complex pathway by which nitrogen moves through an ecosystem. Most nitrogen enters living organisms through nitrogen-fixing bacteria that oxidize it to nitrites and nitrates. Nitrogen is returned to the soil through animal waste products and the decay of organisms.

bacteria-containing nodules on the roots of one of the legume plants. In these symbiotic associations, the nitrogen fixers utilize the ammonia themselves. They also supply some ammonia directly to the host plant. When the nitrogen fixers die, their nitrogen is recycled through decomposition.

Figure 37–12 shows the various pathways of the nitrogen cycle. The nitrogen cycle keeps the level of usable nitrogen in the soil fairly constant. The nitrogen cycle also occurs in lakes, streams, and oceans. Most of the nitrogen being cycled remains in compound form. Only a small fraction is cycled through the atmosphere.

The Carbon and Oxygen Cycles

Carbon, in the form of carbon dioxide, makes up about 0.03 percent of the atmosphere. Carbon dioxide is also found dissolved in the waters of the earth. In the course of photosynthesis, carbon dioxide from the atmosphere is incorporated into organic compounds, a process known as *carbon fixation*. Some of these organic compounds are broken down during cellular respiration by photosynthetic organisms. This releases carbon dioxide into the atmosphere. If the plants or other photosynthetic organisms are eaten by animals, the carbon compounds pass through a food web. At each level, some are broken down by cellular respiration, releasing carbon dioxide into the atmosphere. Finally, the remains of dead plants and animals and animal wastes are broken down by decomposers, which also releases carbon dioxide.

In the **carbon cycle,** carbon dioxide is removed from the atmosphere by photosynthesis, and it is returned to the atmosphere by cellular respiration. The carbon cycle is shown in Figure 37–13. These two processes are normally in balance, maintaining a relatively constant level of carbon dioxide in the atmosphere. However, the burning of fossil fuels (oil, coal, and natural gas) also releases carbon dioxide. Because of the increasing use of these fuels, there has been a gradual increase in the carbon dioxide content of the atmosphere since the mid-1800s. The long-term effects of this change are not known. However, some scientists think that it will result in an increase in temperature on the earth's surface. This would occur because the atmospheric carbon dioxide absorbs heat from the earth that would otherwise be radiated away into space. This phenomenon is known as the "greenhouse effect."

Oxygen makes up about 20 percent of the earth's atmosphere. During photosynthesis, water molecules are split into hydrogen and oxygen. The hydrogen is used in the formation of carbohydrates, and the oxygen is released into the atmosphere. Animals, plants, and many protists use oxygen in cellular respiration and release carbon dioxide. Thus, in the **oxygen cycle,** oxygen is released into the atmosphere by the process of photosynthesis and removed from the atmosphere by cellular respiration.

Figure 37–13
The Oxygen and Carbon Cycles. The cycling of carbon and oxygen in an ecosystem results from photosynthesis and respiration. ▼

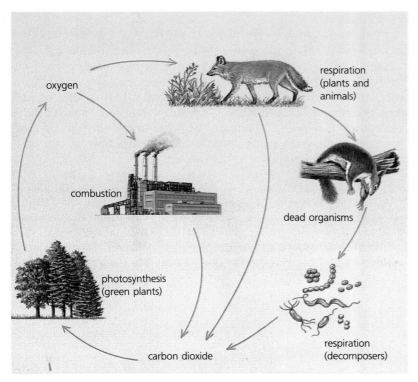

Science, Technology and Society

Issue: Greenhouse Effect

For years, scientists have been worried about the greenhouse effect. Many human activities, especially the burning of fossil fuels, release carbon dioxide into the air. Carbon dioxide traps the sun's heat, which is reflected back to the earth. This is causing the average temperature of the earth to rise gradually.

Some scientists fear that global warming may lead to drought and famine. As the earth warms, the melting of icecaps may cause floods in major coastal cities. If we do not prevent the greenhouse effect soon, they claim, it will be too late. Thus, they want to look for solutions and explore alternatives to fossil fuels now.

Other scientists claim that the greenhouse effect may have been blown out of proportion. Scientific data is somewhat contradictory and does not explain all the factors affecting climate. They argue that global warming is a gradual process and no quick decisions should be made. Time and money should be spent on research now and not on taking action.

■ *Should we take immediate action against the greenhouse effect? Why or why not?*

Figure 37–14

The Water Cycle. Water from the earth's surface enters the atmosphere in the form of water vapor through the processes of evaporation and transpiration. It returns to the surface through condensation and precipitation. ▶

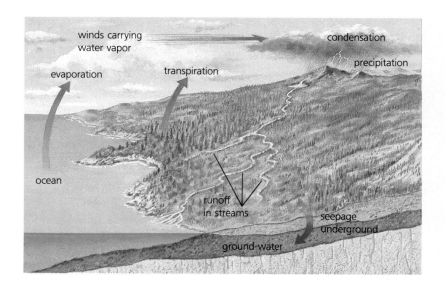

The Water Cycle

The cycling of water on the earth is almost entirely a physical process. Wherever water is exposed to the air, it evaporates—escapes into the air in the form of water vapor. Plants contribute to this loss of water to the air by the process of transpiration. However, there is a limit to the amount of water vapor the air can hold. Through various physical processes, excess water vapor condenses to form clouds and falls back to the earth's surface as precipitation.

This cycling of water between the surface of the earth and the atmosphere is called the **water cycle.** It is shown in Figure 37–14. Unlike the other cycles we have examined, no chemical changes are involved. Nor do any truly biological processes enter into it. It is true that some water is broken down chemically to hydrogen and oxygen during photosynthesis. This water is restored by cellular respiration. However, the amount of water involved in the photosynthesis-respiration cycle is only a small fraction of the total amount that passes through the water cycle.

37-3 Section Review

1. Name three biogeochemical cycles.
2. What kind of organisms return soil nitrogen to the atmosphere?
3. What carbon compound is released as the result of burning fossil fuels?

Critical Thinking

4. List two similarities between the nitrogen cycle and the carbon cycle. (*Comparing and Contrasting*)

37-4 Maintenance and Change in Ecosystems

Section Objectives:

- *Describe* the conditions necessary for a stable, self-sustaining ecosystem.
- *Explain* the terms ecological succession, dominant species, climax community, primary succession, and secondary succession.
- *Compare* succession on land, which leads to development of a forest community, with succession in lakes and ponds.

Maintenance in an Ecosystem

For an ecosystem to be stable and self-sustaining, certain conditions must exist. First, there must be a constant source of energy. For almost all ecosystems on earth, the source of energy is light from the sun. Only a few ecosystems are based on chemosynthesis. In these ecosystems, the producers derive energy for the synthesis of organic compounds from chemical reactions involving various inorganic compounds. Second, there must be organisms within the ecosystem that can use incoming energy (light) for the synthesis of organic compounds. This role is filled by green plants and algae, which are the producers of the ecosystem. Third, there must be a cycle of materials between living organisms in the ecosystem and the environment. The producers incorporate inorganic compounds from the environment into organic compounds, which may then pass through a food chain or food web. Eventually, however, the decomposers break down the remains of dead organisms, releasing the inorganic substances back into the environment for reuse.

Ecological Succession

Although ecosystems appear stable, they do undergo change. Change occurs because the living organisms present in the ecosystem alter the environment. Some of the changes tend to make the environment more suitable for new types of organisms and less suitable for the existing organisms. Thus, the original organisms in an ecosystem are slowly replaced by other types. A new community replaces the original community in the ecosystem. Over time, this community is gradually replaced by still another community. The process by which an existing community is slowly replaced by another community is called **ecological succession.** In land environments, ecological succession usually depends upon the types of plants that are present at any given time. Plants determine the type of community that develops because plants are the producers. The types of animals that can survive in the community depend, directly or indirectly, on the types of plants.

During each stage of ecological succession, only a few species have a great effect on the environment and on other members of the

Early Plant Forms

Intermediate Forms

Climax Forest

maples

beeches

oaks

pines

shrubs and
grasses

ferns

lichens and
mosses

soil and humus

rock

▲ **Figure 37–15**

Stages of Primary Succession. The sequence of stages is shown from left to right. In an actual succession, only one stage is present at any given time.

community. These species are called the **dominant species.** The conditions imposed on the environment by the dominant species determine the types of other species that can survive in each successive community.

Succession of one community by another goes on until a mature, stable community develops. Such a community is called a **climax community.** In an ecosystem with a climax community, the conditions continue to be suitable for all the members of the community. The climax community remains until it is upset by a catastrophic event, such as a fire, flood, or volcanic eruption. After the destruction of a climax community, succession begins again and continues until a new climax community develops. Such catastrophic events are not necessarily bad. For example, some conifer species require a fire in order to release their seeds.

When succession occurs in an area that has no existing life, for example, on a bare rock, it is called **primary succession.** Succession that occurs in an area in which an existing community has been partially destroyed and its balance upset is called **secondary succession.**

Succession on Land   Primary succession occurs on land in areas that are nearly lifeless. The process is shown in Figure 37–15. Such conditions exist on rocky cliffs, sand dunes, newly formed volcanic islands, and newly exposed land areas. Primary succession is a slow process because it must begin with the formation of soil.

Soil forms slowly over thousands of years. By the process of weathering, large rocks are gradually broken into smaller and smaller pieces. Eventually, some of the rock is broken down into small particles. The first organisms to inhabit an area are called *pioneer organisms*. These organisms include bacteria, fungi, and lichens. They break down the rock still further and add organic matter to the developing soil. Lichens are adapted to exposed conditions. They become attached to irregularities in the surface of the rock by rootlike rhizoids. They secrete acids that dissolve the rock. Some lichens die, and their remains are added to the soil. Mosses appear in areas where a little soil has accumulated. The mosses may shade the lichens, causing them to die and thus adding more organic material to the still primitive soil.

Eventually, grasses and annual plants grow in the areas where organic material has accumulated. When these plants die, the soil becomes richer. Small shrubs begin to grow, and their roots break rocks apart. The shrubs may shade the grasses, killing them. Tree seedlings may take root. The trees eventually shade out the shrubs. The seedlings that grew among the shrubs may have required a fair

■ Careers *in Action*

Forestry Technician

Under the direction of a forester, a forestry technician helps manage forest lands and resources. Because forests are used in many ways, forestry technicians perform a variety of activities.

Forestry technicians estimate timber yields and supervise some logging and tree replanting operations. They also help maintain the lands for campers, hikers, and others. By checking trees for disease and insect damage and by pruning and clearing unwanted growth, technicians maintain the health of forests.

An important part of the forestry technician's job is to prevent fires. When fires do occur, technicians help to supervise fire-fighting crews. Forestry technicians may also assist in research, work to prevent damage from floods and erosion,

maintain and repair tools, and prepare educational programs for the public.

Problem Solving In 1988, a huge fire raged in Yellowstone National Park. Suppose you were a forestry technician at the park. The head forester has decided to let the fire burn in certain climax areas. The people in a nearby town are worried that the areas will be charred forever. The forester asks you to discuss these concerns with the townspeople.

1. Are the townspeople right? Why or why not?
2. Prepare an explanation of what is likely to happen to the burned areas in the future.
3. The people are concerned about erosion in burned areas and possible flooding in the town from heavy rains. What reasons would they have for these concerns?

4. The townspeople are worried that the animals living in the forests will escape the fire and become pests in the town. Do you think this is likely? Explain your answer in terms of habitats.

■ **Help Wanted:**
Forestry technician. High-school diploma required; one- or two-year college forestry course recommended. Contact: The American Forestry Association, 1319 18th St., N.W., Washington, D.C. 20036.

amount of sunlight. Thus, when they become mature trees, there may not be enough sun on the forest floor for seedlings of the same type to survive. However, seedlings of other trees may grow well in the shade. In this way, one community of trees will be succeeded by another community with different trees. After thousands of years, a climax community will develop. Climax communities are usually described in terms of their dominant plant forms.

The dominant plants of a climax community are determined by the physical factors of the environment. Where there is adequate rainfall and suitable soil, the climax community is likely to be a forest. If there is not enough water for a forest, the climax community can consist of grasses or some other type of plant.

Animal life changes with the plant communities. For example, as a succession proceeds toward a forest community, animals that live among grasses and shrubs will be replaced by animals that live on the forest floor and at varying levels in the trees.

Secondary succession occurs in areas in which the climax community has been destroyed. For example, a forest may be cut down in order to clear the land for farming. If, after being farmed for awhile, the land is left untended, a new succession will begin. Eventually, it, too, will end in another forest climax community. In secondary succession, the area already has existing soil. Since the sequence does not begin with soil formation, the process is faster than primary succession. A climax community may become reestablished after a few hundreds of years, rather than the thousands originally needed for the primary succession. After the forest fires in 1988, the process of succession has begun to occur in Yellowstone National Park.

Succession in Lakes and Ponds Lakes and ponds may also pass through stages of ecological succession, eventually developing into a forest climax community as shown in Figure 37–17. The

Figure 37–16

Secondary Succession. In 1988, forest fires swept through Yellowstone National Park in Wyoming, destroying much of the vegetation. As seen in this recent photo, the process of secondary succession has already begun. ▶

a

b

c

d

process begins when sediment, fallen leaves, and other debris gradually accumulate on the lake bottom, decreasing its depth. Around the edges of the lake, sphagnum moss and many of the rooted plants, such as cattails, reeds, and rushes, grow out into the shallower water. They gradually extend the banks inward, decreasing the size of the lake. As the lake fills in, it becomes rich in nutrients that can support a large population of organisms. The increased number of plants and animals contribute organic material to the sediment, which hastens the filling-in process. As succession continues, the lake becomes a marsh. Still later, the marsh fills in, forming dry land. Land communities replace aquatic forms. Over time, the filled area becomes part of the surrounding community.

▲ **Figure 37–17**
Ecological Succession in a Pond. A pond may gradually change to dry land that supports a forest community.

37-4 **Section Review**

1. By what process do ecosystems change?
2. What is any group of organisms that is the first to inhabit an area?
3. How does secondary succession differ from primary succession?

Critical Thinking

4. A fire has wiped out an old maple forest near your home. Predict the dominant plants that will inhabit the area 400 years after the fire. (*Predicting*)

Laboratory Investigation

Investigation of Nitrogen-Fixing Nodules

Only some types of bacteria can fix nitrogen. These organisms convert nitrogen gas (N_2) into a form that can be incorporated into organic molecules. In this investigation, you will study *Rhizobium,* a genus of bacteria that fixes nitrogen while living symbiotically in the roots of legumes.

Objectives

- *Observe Rhizobium* in prepared slides of legume root nodules.
- *Observe* legume root nodules of living plants.

Materials

microscope
prepared slides of
  legume nodules
living legume plants
  with root nodules

scalpel
paper towels
toluidine blue
dropper
dissecting microscope

Procedure 🏺 🧪 🧤 *

A. Examine a prepared slide of a legume nodule under low power. Notice that there are many closely packed, thin-walled, blue-stained cells in the center of the nodule. Inside these parenchyma cells live the bacteria that fix nitrogen.

B. Look in some of the parenchyma cells for a nucleus, which appears as a round blue body, and a vacuole, which appears as a large empty space.

C. Switch to high power. Notice the small, red-stained structures inside the parenchyma cells. These are the bacteria.

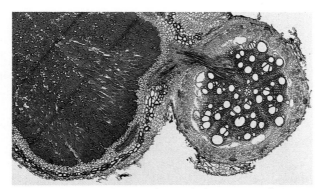

*For detailed information on safety symbols, see Appendix B.

D. Return to low power. Look at the rest of the nodule. The parenchyma cells are surrounded by thick-walled supporting cells. These form the outside of the nodule. Vascular bundles appear as circular clusters of cells in this region of the nodule.

E. Draw a diagram of the root nodule as you see it through the microscope. Label parenchyma cells, cell walls, nuclei, vacuoles, bacteria, supporting cells, xylem, and phloem.

☠ 🤚 **F.** Look at the roots of a live legume and examine the nodules on the roots. Remove one nodule from the plant, noting how the nodule is attached to the root. Cut the nodule open with a scalpel, and examine it with a dissecting microscope. Add a drop of toluidine blue. Try to match the structures you see in the live nodule with the ones you saw on the prepared slide.

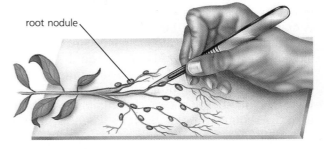

root nodule

G. Draw a diagram of a legume root showing the structural relationship between a nodule and the root.

Analysis and Conclusions

1. Why might you predict that legumes will grow well in soils poor in nitrogen?

2. Trace the path of nitrogen gas from the atmosphere to the inside of the bacteria.

3. Explain how the nitrogen fixed in a root nodule can end up in a protein in a cell in your body.

4. What is the significance of the root nodules having vascular tissue?

5. How might a farmer or gardener take advantage of the nitrogen-fixing properties of legumes?

Chapter Review

37

Study Outline

37-1 Abiotic Factors in the Environment

■ Ecology deals with the interactions among organisms and between organisms and their environment. The environment consists of living, or biotic, factors and nonliving, or abiotic, factors.

■ Abiotic factors include light, temperature, water, soil, and minerals. They determine the kinds of organisms that can survive in an environment.

37-2 Biotic Relationships in Ecosystems

■ Living things in the environment are divided into populations that include individuals of a certain species within a given area. All the populations in a given area make up a community.

■ An ecosystem consists of a community and its physical environment. The biosphere is the part of the earth in which living things exist.

■ Autotrophs are organisms that synthesize their own nutrients. Heterotrophs, which feed on other organisms, include herbivores, omnivores, and carnivores. Predators and scavengers are carnivores.

■ In symbiotic relationships, such as mutualism, commensalism, and parasitism, two species of organisms live in close association.

■ The role of a species in an ecosystem is its niche. Competition between two species is called interspecific competition. Intraspecific competition occurs among members of the same species.

■ In an ecosystem, autotrophs are the producers, heterotrophs the consumers, and saprobes the decomposers. The path of energy, called a food chain, flows from producers to consumers and finally to decomposers. Interconnections between food chains form a food web.

■ The amount of energy in an ecosystem is greatest at the producer level and decreases with each higher feeding level of the food web, forming an energy pyramid. The relative mass of organisms at each level also decreases, forming a pyramid of biomass.

37-3 Cycles of Materials

■ In all ecosystems there exist biogeochemical cycles, which are exchanges of materials between organisms and their environment.

■ Biogeochemical cycles include the nitrogen cycle, the carbon cycle, the oxygen cycle, and the water cycle.

37-4 Maintenance and Change in Ecosystems

■ Ecosystems remain stable and self-sustaining if there are a constant source of energy, the presence of autotrophs, and the cycling of materials between living organisms and the environment.

■ Ecosystems undergo change called ecological succession in which an existing community is replaced by another community. The final stage of succession is called the climax community.

Vocabulary Review

ecology (819)
biotic factors (819)
abiotic factors (820)
humus (822)
population (823)
community (823)
ecosystem (823)
biosphere (824)
herbivores (824)
carnivores (824)
predators (824)
scavengers (824)
omnivores (824)
saprobes (824)
symbiotic relationships (824)
mutualism (824)
commensalism (825)

parasitism (825)
habitat (826)
niche (826)
interspecific competition (826)
intraspecific competition (828)
producers (828)
consumers (828)
decomposers (828)
food chain (828)
food web (828)
pyramid of energy (830)
pyramid of biomass (830)
nitrifying bacteria (831)
denitrifying bacteria (831)
nitrogen fixation (831)
nitrogen fixers (831)

Chapter Review

carbon cycle (833)

oxygen cycle (833)

water cycle (834)

ecological succession (835)

dominant species (836)

climax community (836)

primary succession (836)

secondary succession (836)

A. Sentence Completion—Fill in the vocabulary term that best completes each statement.

1. Relationships in which two different kinds of organisms live in close association so that at least one benefits are called _____.

2. An ecosystem includes a _____ and its physical environment.

3. A food web results from interconnected feeding relationships in _____.

4. The relative mass of organisms at each feeding level can be shown as a _____.

5. The mature, stable community that finally develops as a result of ecological succession is called the _____.

6. In the carbon cycle, carbon dioxide is removed from the atmosphere by photosynthetic organisms called _____.

7. Competition between two different species is referred to as _____.

8. In the nitrogen cycle, certain bacteria called _____ convert ammonia to nitrates.

9. The amount of energy available in an ecosystem is commonly illustrated by a _____.

B. Analogies—Select the vocabulary term that relates to the single term in the same way that the paired terms are related.

10. parasitism **is to** tapeworm as _____ **is to** lichen

11. population: community as ecosystem: _____

12. consumers: heterotrophs as decomposers: _____

13. abiotic factors: light as _____: plants

14. herbivores: plants as _____: animals

Content Review

15. Discuss the effect of latitude on the intensity and duration of sunlight.

16. Describe the process of soil formation and the three layers that eventually form.

17. What is an ecosystem?

18. Name and describe the three main parts of the biosphere.

19. Compare a predator with a scavenger, and give examples of each.

20. What are the kinds of symbiotic relationships? Give one example of each.

21. How does a niche differ from a habitat?

22. What happens when two different species occupy the same niche?

23. Summarize the parts of a simple food chain.

24. Outline the nitrogen cycle.

25. Why do some farmers alternate legumes, such as clover plants, with their regular cash crops?

26. How is the carbon cycle related to the oxygen cycle?

27. Under what conditions does an ecosystem tend to be stable and self-sustaining?

28. Describe the stages of primary succession both on land and in a pond.

29. What is secondary succession?

Graphic Organizing

For information on graphic organizers, see Appendix G at the back of this text.

30. **Flow Chart:** Referring to the pyramid of energy shown on page 830, construct a flow chart to show the order of the feeding levels. Begin with the level that contains the greatest amount of energy, and end with the one that contains the least. As you move up the pyramid of energy, only 10 percent of the energy taken in at any feeding level is passed on to the next. Assume the producers supply 2697 calories. Use your calculator to compute the amounts of energy in the three feeding levels.

Critical Thinking

31. What would happen to an ecosystem if all the decomposers were destroyed? (*Predicting*)

32. Suppose the intensity of sunlight were drastically reduced for several months by smoke and ash from an erupting volcano such as Mt. St. Helens. Predict how each of the following members of nearby ecosystems would be affected: sweet grass, rabbits, and hawks. (*Predicting*)

33. All consumers, either directly or indirectly, depend on producers for their food. Humans are consumers. Therefore, humans depend on producers for their food. Is this a valid argument? Explain your answer. (*Reasoning Categorically*)

Creative Thinking

34. Suggest some characteristics that plants would need in order to grow successfully in far northern latitudes.

35. What special adaptations might an animal need to survive in areas that (a) were always very hot, (b) were always very cold, and (c) varied in temperature each day from very hot to very cold?

Problem Solving

36. An ecologist samples a shrubby meadow to determine the populations of certain key organisms. Data from the investigation are recorded in the table shown below. On a piece of graph paper, construct a pyramid showing the numbers of organisms at each feeding level for each of the three months. What is the relationship between population size and time of the year? How could you explain this relationship?

Type of Organism	Number of Organisms		
	May	**July**	**September**
grasshoppers	100	500	150
birds	25	100	10
shrubs and grass	700	2000	600
spiders	75	200	50

37. The graph in the next column shows changes in two populations, the snowshoe hare, a large relative of the rabbit, and the lynx, a small wildcat. Analyze the graph and develop a hypothesis to explain the changes in the two populations.

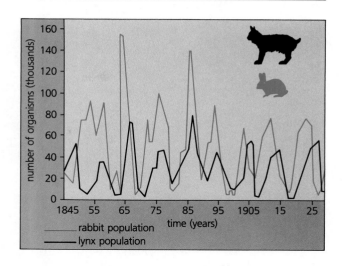

rabbit population
lynx population

Computer Project

38. Your goal as a wildlife resource manager is to introduce wolves to maintain a deer population at a constant level. You will use a computer spreadsheet to estimate the number of wolves to be introduced. (Round all numbers off to the nearest whole number.) Currently, there are 2000 deer in the population. On average, each wolf eats 30 deer per year, and the surviving deer population increases by 40 percent annually.

A. If no wolves are introduced, how many deers will there be in 1 year? 5 years? 10 years?

B. If you introduce 10 wolves, how would this change your answers in question A? Assume that the number of wolves does not change.

C. How many wolves should be introduced to keep the deer population nearly constant for five years?

D. Suppose the wolf population increases at a 10 percent rate each year. Calculate how many wolves should be introduced in the first year to keep the deer population as constant as possible for five years. What happens after five years?

E. Suppose that a second predator were introduced along with the wolves. This predator increases its population at a 20 percent rate each year but only eats 10 deer per year. If you introduce 10 wolves (increasing at a rate of 10 percent each year), calculate how many of the new predators you would need to add to keep the deer population fairly constant over five years.

F. Problem Finding: What further questions could you answer using this spreadsheet?

▲ Sand dunes dominate the landscape at Death Valley National Monument in California.

Biomes of the Earth

Formed into ripples by the wind, oceans of sand spread out to the horizon. With almost no rainfall, and with generally poor soil, the deserts of the world would seem incapable of supporting life; yet certain plants and animals have adapted to these harsh, arid conditions. The desert is just one of the earth's major biomes—large geographic regions with specific climates and particular types of plant and animal communities. In this chapter you will learn about all of the major biomes and the plants and animals that live in them.

Guide for Reading

Key words: biome, tundra, taiga, tropical rain forest, benthos, plankton, littoral zone

Questions to think about:

📖 How does increasing altitude in mountainous regions produce effects similar to increasing latitude?

📖 What different types of problems do organisms encounter in adapting to salt water and freshwater biomes?

38-1 Terrestrial Biomes

Section Objectives:

- *List* the major terrestrial biomes of the earth and specify the location of each.
- *Describe* the climatic factors that characterize each of the terrestrial biomes.
- *Name* the dominant plant and animal types of each biome.
  - **Laboratory Investigation:** *Compare* the water-holding capacity of three different soil types (p. 858).

The climate and other physical conditions of an area determine the type of climax community that can develop in that area. Areas that are similar in climate and other physical conditions develop similar types of climax communities. The term **biome** (BY ohm) refers to a large geographical region that has a particular type of climax community. In the case of a land, or terrestrial, biome, the climax community is defined by the dominant type of plant life found there. For example, one biome may consist of climax communities of grasses. Another biome may contain evergreen trees. The species may vary from one part of a biome to another, but the general type of plant life, or vegetation, is the same throughout the biome. The major terrestrial biomes are the tundra, taiga, deciduous forest, grassland, desert, and tropical rain forest. In the sections that follow, you will read more about these major biomes and learn about their vegetation and inhabitants.

▲ **Figure 38–1**

A Mountain Environment. The physical conditions of this environment determine the types of organisms that thrive there.

The term *biome* is also applied to communities that develop in aquatic environments. Ecologists refer to freshwater biomes, or communities of organisms inhabiting lakes and streams, and to saltwater, or marine, biomes.

The Tundra

The **tundra** (TUN druh) is a region that lies south of the ice caps of the Arctic and extends across North America, Europe, and Siberia. In the Southern Hemisphere, the areas that would be tundra are oceans. The tundra is characterized by a low average temperature and a short growing season of about 60 days. During the long, cold winters, the ground is completely frozen. During the short summer, only the topmost layer of soil thaws. The layers beneath this layer remain frozen. These layers are called **permafrost** (PER muh frost). The average precipitation in the tundra is only about 10 to 12 centimeters a year. However, because of the low rate of evaporation, the region is wet with bogs (see page 857) and ponds during the warm season.

Vegetation in the tundra is limited to lichens, mosses, grasses, sedges, and shrubs. There are almost no trees because of the short growing season and the permafrost. The vegetation that does grow

Figure 38–2

The Tundra. Caribou live in large herds that migrate across the Alaskan tundra. Many other large grazing animals and carnivores live in the tundra. ▼

◀ **Figure 38–3**
The Taiga. Taiga vegetation, such as this Alaskan forest, is dominated by pine, fir, and spruce trees.

in the tundra supports a limited number of animal species. Animals found in the tundra include reindeer, musk oxen, caribou, wolves, Arctic hares, Arctic foxes, lemmings, snowy owls, and ptarmigans, which are a type of bird.

During the warm season, a great number of flies and mosquitos appear in the region. Various types of birds, including sandpipers, ducks, and geese, migrate to the tundra during this season. Here, these migratory birds can nest and breed in safety, because of the relative absence of their usual predators. During their breeding season, the birds feed on the growing vegetation and the abundant supply of insects.

The Taiga

Moving south across the tundra, the vegetation slowly changes. Groups of stunted trees begin to appear in sheltered places. Farther south, the trees become larger and closer together, at last giving way to evergreen forests. This belt of evergreen forest, which extends across North America, Europe, and Asia, is the **taiga** (TY guh). The taiga has cold winters during which the ground is covered by deep snow. However, the growing season is longer than that of the tundra—about 120 days. The summer days are warmer than in the tundra, and the ground thaws completely. Precipitation is greater than in the taiga, averaging between 50 and 100 centimeters a year. As in the tundra, there are many ponds and bogs. Pines, firs, and spruce are the dominant vegetation, although some deciduous trees, which shed their leaves, are also present. These include willows and birches. There are also shrubs and herbaceous plants.

Figure 38–4

A Temperate Deciduous Forest. Temperate deciduous forest covers much of the eastern United States. In the summer, the broad-leaved deciduous trees shade the forest floor. In the spring, many species of herbs and other low-growing plants cover the forest floor before the trees develop leaves. ▶

Animals of the taiga include moose, wolves, bears, lynx, deer, elks, wolverines, martens, snowshoe hares, porcupines, and various rodents, birds, and insects.

The Temperate Deciduous Forest

The areas south of the taiga have varying amounts of rainfall, so there is no single type of biome that can be said to stretch across these latitudes. In eastern North America and Europe, these areas are regions of **temperate deciduous forest.** In temperate deciduous forest biomes the summers are usually hot and humid, and the winters are cold. Rainfall averages between 75 and 150 centimeters a year.

The species present in deciduous forests vary with the local amount of rainfall. Common trees of deciduous forests include oak, maple, hickory, beech, chestnut, and birch. Smaller trees and shrubs are also present, as well as herbaceous plants, ferns, and mosses.

Animals of the deciduous forest include wolves, gray foxes, bobcats, deer, raccoons, squirrels, and chipmunks, as well as a wide variety of birds and insects.

Grasslands

Grasslands, or prairies, are found in the interiors of North America, Asia, South America, and Africa. They occur in both temperate and tropical climates where rainfall ranges from 25 to 75 centimeters a year. This quantity of rainfall cannot support a deciduous forest, and grasses become the dominant form of vegetation. See Figure 38–5. The soil of the grasslands is often deep and rich, and these areas have become the most productive farmlands of the earth.

▲ **Figure 38–5**
Grasslands. The natural vegetation of the grasslands includes many species of grass and wildflowers. Grasshoppers are a common sight.

The natural vegetation of the grasslands includes many species of grasses and wild flowers. In wetter areas, near rivers, the vegetation may be dense and include various shrubs.

Animals of the North American grasslands include coyotes, badgers (carnivorous, burrowing mammals), rattlesnakes, prairie dogs, jackrabbits, and ground squirrels. In the past, great herds of bison and pronghorn antelope were common. Now, most have been replaced by domesticated cattle and sheep. In Africa, the grasslands are populated by zebras, giraffes, gazelles, and other large grazing animals. Predators, such as lions, that feed on the grazers are also present. There are fewer types of birds in the grasslands than in the deciduous forest. In the North American grasslands, there are meadowlarks, ringnecked pheasants, prairie chickens, hawks, and owls. There are many insects, but the grasshopper populations, in particular, may be huge.

Deserts

Deserts occur in regions that are too dry to support grasses. The soil is sandy and poor. Rainfall is usually less than 25 centimeters in one year. In North America, there is a desert extending from Mexico, north to the eastern part of Washington. Huge areas of desert are also found in South America, Africa, Asia, and Australia. Temperatures in the desert vary widely in the course of a day. While the day is hot, the temperature may drop as much as 30° C at night. Some deserts have almost no vegetation at all, while others have a variety of plants.

▲ **Figure 38–6**

Fennec Fox in African Desert. The fennec is a small desert fox found in the North African desert. Most small animals avoid the harsh conditions of the desert by living in burrows and feeding at night.

Desert plants have special adaptations for the conservation of water and for the completion of their reproductive cycles. Most have widespread, shallow roots that allow them to absorb the maximum amount of water when it is available. Many desert plants, such as cacti, store water in their tissues. Many live only a short time. They sprout, flower, and produce seeds during brief rainy periods, which may last only a few days. Cactus, yucca, mesquite, sagebrush, and creosote bush are characteristic plants of the North American deserts.

Like desert plants, desert animals have a wide variety of adaptations for survival in the harsh environment. Most are active at night and spend the hot days in burrows in the ground or hidden in any available shade. Many desert rodents can survive with little drinking water. They get by on the water produced by cellular metabolism and the water present in the plants they eat. The fennec, a small desert fox, spends its days in a burrow, coming out only at night to feed on birds and other small animals. Its long ears provide a surface area for getting rid of excess body heat. Also found in the desert are snakes, lizards, spiders, and insects.

Tropical Rain Forests

Tropical rain forests are found in areas around the equator. In these regions, the climate is uniform throughout the year. There is a constant supply of rainfall, which may total between 200 and 400 centimeters a year. Rain occurs nearly every day, and the humidity is high. Temperatures remain constant at about 25° C throughout the year. Tropical rain forests contain an enormous variety of plants and animals. At present, the habitats of these animals are disappearing at an alarming rate as the rain forests are destroyed. (See page 379.)

Figure 38–7

Tropical Rain Forests. Tropical rain forests contain more plant and animal species than any other biome. The green forest of huge trees supports a wide variety of plants and animals in its branches, high above the forest floor. ▶

Within a tropical rain forest, the tree cover is so dense that little light reaches the ground, as you can see in Figure 38-7. The treetops form a canopy about 50 meters high. Below the canopy are shorter trees that can grow in the shade. The trees of the rain forest have shallow root systems that allow them to absorb nutrients from the thin layer of wet soil. Many have braces, or buttresses, that extend from the trunks to the ground. Like prop roots, these buttresses help to keep the tree standing upright.

Organic materials decay quickly in this warm, humid environment. Minerals released by decomposition are rapidly taken up again by the plants. Materials not absorbed by the plants are washed away by the frequent rains. Little organic matter is stored in the soil. Most of the nutrients in a tropical rain forest are found within organisms. Because of the poor soil, land cleared in this area cannot support crops for more than one or two years.

Among the 100 or more different species of trees found in a rain forest, there are many with large, broad leaves. In addition to trees, there are thick vines, called *lianas,* that are attached to the tree trunks and grow up through the treetops. The roots of these vines are in the ground. There are also many *epiphytes* (EP uh fyts).

▲ **Figure 38–8**

Epiphytes. Epiphytes grow on the branches of other plants. Bromeliads are epiphytes with leaves at their bases that absorb water.

Critical Thinking in Biology

Ranking

When you rank items or information, you arrange them in order, according to how well they serve a particular function.

Suppose you were a field biologist about to begin a six-month study of bird courtship behavior in the tropical rain forest of South America. When your study is complete, you hope to share your research findings with other scientists through talks and written articles.

Your work in the rain forest will involve hiking through rugged terrain. All your equipment must be lightweight, durable, and easy to use. You can take along only one of the following instruments to observe and/or record the birds' behavior: a pair of binoculars, a camera with a telephoto lens, or a video camera (see photo).

1. Using your knowledge of the three instruments, list the pros and cons of each for the field biologist.

2. Based on your list of pros and cons, rank the three instruments from the most useful to the least. Explain your reasoning.

3. Rank the three instruments according to how useful they would be in your personal life. Explain your reasoning.

4. Thinking About It Explain your thought process as you ranked the instruments in questions 1 and 2.

Epiphytes are plants that grow on other plants but are not parasites. Various orchids, cacti, and ferns are epiphytes. The roots of some epiphytes absorb moisture from the air. Others have leaves that form cups at their bases. Water-absorbing structures pick up the moisture trapped in the leaves. Plants that are tolerant of almost complete shade cover the floors of tropical rain forests.

A wide variety of animal species inhabit tropical rain forests. Many of them have adaptations that allow them to live at a particular level in the trees. Monkeys, bats, squirrels, and parrots and other birds feed on fruits and nuts in the treetops. Flying squirrels leap from one tree to another. Snakes and lizards live in the branches of the trees. Rodents, tapirs, antelope, deer, and other large animals live on the forest floor. Spiders and insects are present at all levels. There are also ants, termites, bees, butterflies, and moths.

▲ Figure 38–9
Animals of the Rain Forest. Rainbow lorikeets are among the many animals that live in the trees of the rain forest.

Effects of Altitude on Climax Vegetation

With the exceptions of the tundra and the taiga, the terrestrial biomes of the earth are distributed in irregular belts around the earth, more or less in sequence according to latitude. Their distribution is shown in Figure 38–10. Because increasing altitude usually produces climatic effects similar to increasing latitude,

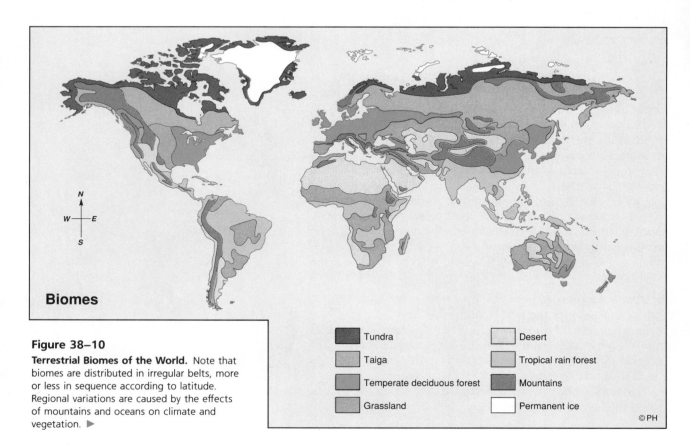

Biomes

Figure 38–10
Terrestrial Biomes of the World. Note that biomes are distributed in irregular belts, more or less in sequence according to latitude. Regional variations are caused by the effects of mountains and oceans on climate and vegetation. ▶

Tundra

Taiga

Temperate deciduous forest

Grassland

Desert

Tropical rain forest

Mountains

Permanent ice

©PH

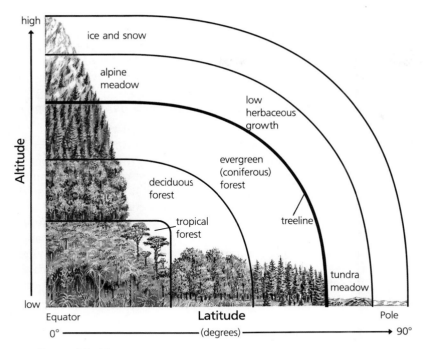

▲ **Figure 38–11**
Effects of Altitude on Climax Vegetation. The climactic effects of altitude and
latitude roughly correspond. For example, note that plant life near the top of a
mountain resembles that of a tundra.

mountainous regions are often left out of the biome classifications.
The sides of mountains may show a succession of plant communities that change with increasing altitude. These communities will
have many characteristics in common with those of specific
biomes. For example, the climatic conditions and types of plant life
near the top of a mountain may resemble those of the tundra. Lower
down, evergreen forests that are characteristic of the taiga will
appear. This relationship between higher altitude and higher
latitude is shown in Figure 38–11.

38-1 Section Review

1. Define the term *biome*.
2. What vegetation characterizes the taiga?
3. In what kinds of climate do grasslands usually develop?
4. List three types of plants common to tropical rain forests.

Critical Thinking

5. Order the terrestrial biomes from least rainfall received to most rainfall
   received. (*Ordering*)

38-2 Aquatic Biomes

Section Objectives:

- *Compare* the physical factors in aquatic biomes with those in terrestrial biomes.
- *Describe* marine and freshwater biomes, and *name* some representative organisms of each.

The problems of life in aquatic biomes are different from the problems in terrestrial biomes. For one thing, in aquatic biomes, water is always present. However, in fresh water, organisms must excrete excess water, and in salt water, excess salt may be excreted by organisms. Temperature changes in the course of a year are much less in aquatic environments than they are on land. Temperatures in the oceans show the least change, while those in lakes and ponds show more change. Other physical factors that affect living things in aquatic biomes are the amounts of oxygen and carbon dioxide dissolved in the water, the availability of organic and inorganic nutrients, and light intensity.

The Marine Biome

Since all the oceans of the earth are interconnected, they are said to form a single marine, or saltwater, biome. Conditions and life forms vary from one region of the marine biome to another but without the clearcut differences of the terrestrial biomes.

Characteristics of the Marine Biome   The marine biome covers more than 70 percent of the earth's surface. Because of the heat capacity of water, the oceans absorb solar heat energy during warm seasons and hold it during cold seasons. As a result, ocean temperatures remain fairly stable. The oceans also have a stabilizing effect on average temperatures of land areas. Temperatures on the earth would vary much more than they do if the oceans did not exist. The aquatic environment of the oceans is stable in other respects, too. In any given region, the supply of nutrients and the concentration of dissolved salts remain relatively constant although they may vary from region to region.

Although environmental conditions tend to remain constant in any particular region of the marine biome, they can vary from region to region within the biome. The salt content in particular varies from one place to another. It is lower where large rivers bring fresh water into the ocean, and it is higher where atmospheric temperatures cause rapid evaporation. In general, salt concentrations in the ocean are similar to those in living cells. Marine organisms, therefore, do not usually have the problem of water balance that freshwater organisms have.

Although temperatures remain fairly constant in any given part of the marine biome, there is a variation with latitude. Ocean temperatures vary from near 0° C in the polar regions to 32° C near the equator.

Figure 38–12
Tropical Ocean Reef. This Australian coral reef, like all tropical ocean reefs, supports a wide variety of marine life forms. At deeper depths, conditions are not as favorable for life, and far fewer marine species are found. ▼

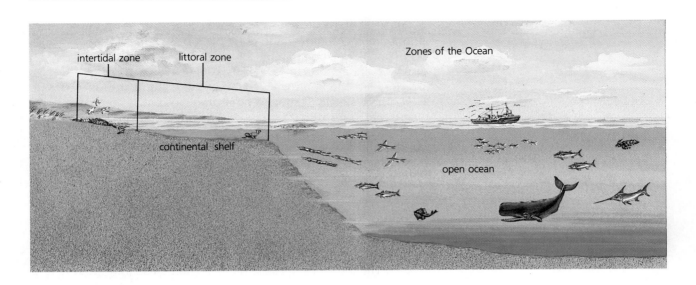

intertidal zone   littoral zone

Zones of the Ocean

continental shelf

open ocean

Organisms of the Oceans

The marine biome supports a great variety of life forms. Some marine organisms are sessile and live attached to the ocean floor or to other fixed surfaces. Sessile organisms include sponges, sea anemones, corals, and barnacles. Organisms that live on the ocean floor are called **benthos.** Starfish, which are sometimes called seastars, clams, worms, snails, and crabs are benthic organisms. Small organisms that float near the surface and are carried by ocean currents are called **plankton.** Planktonic organisms include protozoa and algae, tiny crustaceans called copepods, the larvae of various animals, small jellyfish, and worms. Photosynthetic planktonic organisms, called **phytoplankton,** are the major producers of the oceans. Nonphotosynthetic planktonic protists and planktonic animals are called **zooplankton.** Both phytoplankton and zooplankton make up the lowest level of the complex marine food web. Many types of animals, from small worms to whales, feed on plankton. Free-swimming organisms that live in the oceans are called **nekton.** Nektonic organisms include squid, fishes, turtles, seals, and whales.

Zones of the Oceans

As you can see in Figure 38–13, the oceans are divided on the basis of depth into several zones. The **intertidal zone** is the area along the shoreline that is covered by water at high tide and uncovered at low tide. Various types of seaweeds—red and brown algae—are abundant in this zone. On sandy beaches, clams, crabs, sand fleas, and worms live in the sand. Many types of birds live along the shore, including gulls, terns, and sandpipers. On rocky coasts, algae, barnacles, mussels, and starfish cling to the rocks.

Beyond the intertidal zone is the **littoral zone,** which includes the shallow waters above the continental shelf. The gently sloping continental shelf extends out from the edge of the continent for about 300 kilometers. This zone contains nutrients carried into the oceans by rivers and streams. Because the water is shallow, light

▲ **Figure 38–13**
Zones of the Ocean. Oceans are divided into several zones, according to depth of the water.

reaches all the way to the ocean floor in the littoral zone. The littoral zone contains many different forms of life. In many places there are large populations of algae, as well as fishes, oysters, mussels, crabs, barnacles, worms, and sea cucumbers.

Beyond the continental shelf is the zone of the open ocean. Here, the water is deep, and light does not reach the ocean floor. The upper layer of the open ocean is filled with plankton. In this zone, there are also large animals, including sharks, porpoises, squids, and whales. Below the zone of photosynthesis, the organisms are heterotrophs. They feed on organic matter that drifts downward from the photosynthetic zone above.

Freshwater Biomes

Freshwater biomes can be divided into two types—running water (streams) and standing water (lakes, ponds, swamps, and bogs). The volume of water in these biomes is much smaller than that of the marine biome. As a result, the temperature variations are larger. Organisms living in fresh water must be able to adapt to greater seasonal variations than those living in the ocean. They also have the problem of maintaining water balance. In a freshwater environment, water enters living cells by osmosis, as explained in Chapter 5, Section 5–3. Freshwater organisms usually need a mechanism for removing excess water by active transport. The contractile vacuoles of the ameba and paramecium are examples of such mechanisms.

Acid rain is an increasingly serious problem in freshwater biomes. (See Chapter 3.)

Streams   In fast-moving streams, the bottom is made of rocks and gravel. Most organisms are found in calmer, shallow areas near the banks of the stream. Here, algae grow on rocks, and there are many insects and insect larvae. Fish and microscopic floating algae are found both in running water and in the calmer pools. Where streams are slow moving, muddy sediment accumulates on the bottom. Many animals, including aquatic insects and their larvae, worms, snails, and crayfish, live in the bottom mud. Raccoons, birds, and other animals that live along the banks catch fish and other animals from the stream.

Lakes and Ponds   Lakes and ponds are bodies of standing water. Usually, lakes are larger than ponds and so deep that light does not reach the bottom in all parts. Ponds are usually shallow enough for light to reach the bottom throughout.

Around the shores of a lake is a zone of shallow water in which light reaches the bottom. In this zone, cattails, bulrushes, and other plants grow above the surface of the lake. These plants have roots in the lake bottom. In deeper water, out from the shore, there are floating plants, such as water lilies, which are also rooted in the bottom. Many types of animals are found in the bottom in the shallow water zone. There are insect larvae, crayfish, worms, hydra, clams, and snails. Free-swimming animals include diving beetles, mosquito larvae, giant water bugs, fish, frogs, salamanders, turtles,

Figure 38–14

Streams. These racoon kittens live along the stream and fish for food from its banks. ▼

◀   **Figure 38–15**
Lakes and Ponds. Floating plants, such as water lilies, often are seen on the surface of a pond.

and snakes. On the surface, there may be water striders, water boatmen, and whirligig beetles. Life in ponds is much the same as life in the shallow waters of lakes.

In the deep, open waters of the lake where light does not reach the bottom, the main producers are microscopic algae (phytoplankton) that float near the surface. Zooplankton is also present. These microscopic, heterotrophic organisms are the primary consumers in a complex food web. The planktonic organisms are eaten by small fish, which, in turn, are eaten by larger fish, and so on.

Swamps and Bogs   Swamps are low, wetland areas in which the vegetation includes shrubs and trees. Fresh and saltwater swamps and marshes are often called *wetlands*. Some of these regions, especially those in coastal areas, are threatened by development. Many types of plants and animals are found in wetlands. Wetlands are also important nesting sites for water birds.

Bogs are shallow bodies of water that contain sphagnum moss. The moss and other factors create an acid environment in which the rate of decay is slowed. This reduces the cycling of nitrogen through the ecosystem. Several plants common in bogs are insectivorous. These include pitcher plants and sundews.

Figure 38–16

Swamps and Bogs. Insectivorous plants, such as these pitcher plants, thrive in the shallow water of bogs. ▼

38-2 Section Review

1. Name the two basic kinds of aquatic biomes.
2. How do plankton move from one place to another in the ocean?
3. What are the two basic types of freshwater biomes?

Critical Thinking

4. Predict what would happen to a marine animal placed in freshwater. (*Predicting*)

Laboratory Investigation

Physical Properties of Soil

Soil possesses many physical properties that affect the plants growing in it. In this investigation, you will measure two physical properties of soil—water-holding capacity and porosity (pore space).

Objectives

- *Measure* the water-holding capacity and porosity of soil.
- *Compare* the water-holding capacity and porosity of various types of soil.

Materials

topsoil	250 mL beakers
sand	100 mL graduated
clay soil	cylinders
local soil sample	25 mL graduated
stereomicroscope	cylinders
funnels	water
ring stand	wax pencil
iron ring	stirring rods
filter paper	

Procedure 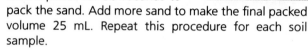 *

A. Examine each soil sample under a stereomicroscope. Observe the size and shape of the particles. Draw a diagram of the particles you observe.

B. For each soil sample, set up a funnel with filter paper above a 250 mL beaker. Into each funnel, place 25 mL of a soil sample. Label each beaker according to the soil sample above it.

C. Note the time. Then, pour exactly 100 mL of water into each funnel. After 10 minutes, measure the amount of water in each beaker and record the values in a chart like the one below.

D. Calculate the water-holding capacity (the amount of water retained by each sample) by subtracting the volume of water in each beaker from 100 mL. Record these values in your chart.

E. Add sand to a 25 mL graduated cylinder. Tap the cylinder with a pencil for 30 seconds to pack the sand. Add more sand to make the final packed volume 25 mL. Repeat this procedure for each soil sample.

F. Pour 50 mL of water into a 100 mL graduated cylinder, and add the packed sand. Stir and wait five minutes. Measure, then record, the final volume of the soil-water mixture. Repeat this for each sample.

G. Calculate the porosity of each sample by subtracting the volume of the soil-water mixture from 75 mL (the combined volume of the soil and water before mixing). Record these values on your chart.

Analysis and Conclusions

1. Which soil has the largest water-holding capacity? Which has the smallest?

2. How does your local soil sample compare with sand and with clay with respect to the two properties?

3. Based on your observations, how do soil particle size and shape affect water-holding capacity and porosity?

4. How do you think water-holding capacity and pore space are related?

5. What would you predict to be the water-holding capacity and porosity of a soil made up of primarily small, jagged-edged particles?

Soil Sample	A Volume of Water in Beaker After 10 Minutes	B Water-Holding Capacity (100 mL– mL from Column A)	C Volume of Soil-Water Mixture	D Porosity (75 mL–mL from Column C)
sand				

* For detailed information on safety symbols, see Appendix B.

858

Chapter Review

Study Outline

38–1 Terrestrial Biomes

■ A biome is a large geographical region that has a unique climax community. The climax community is determined by climate and other physical conditions. Biomes are divided into two major types—terrestrial and aquatic.

■ Terrestrial biomes have a climax community defined by the dominant type of plant life. The major terrestrial biomes located between the poles and the equator are tundra, taiga, temperate deciduous forest, grassland, desert, and tropical rain forest.

■ The tundra biome occurs south of the arctic ice cap. The tundra is characterized by a low average temperature with a short growing season, permafrost, low rainfall, and vegetation limited to lichens, mosses, grasses, sedges, and shrubs.

■ The taiga is characterized by cold winters, warm summers, and a higher rainfall and longer growing season than the tundra. The dominant vegetation consists of evergreen spruce, pine, and fir trees.

■ The temperate deciduous forest has cold winters, hot, humid summers, and abundant rainfall. The dominant vegetation includes a variety of deciduous trees such as oak and beech.

■ Grasslands occur where rainfall will not support a temperate deciduous forest. The dominant vegetation includes a variety of grasses and wildflowers growing in a deep, rich soil.

■ In deserts, rainfall is slight, and plants with water-conserving adaptations are common. Some plants have short reproductive cycles. The days can be very hot and the nights extremely cold.

■ Tropical rain forests are located near the equator. The climate is uniform, with abundant daily rainfall. Trees form a dense canopy over the moist, shallow soil.

■ Increasing altitude produces climatic effects similar to those produced by increasing latitude. Mountainsides, therefore, may show a succession of plant communities similar to those of specific terrestrial biomes.

38–2 Aquatic Biomes

■ The marine biome includes all the oceans of the world. The marine environment remains largely constant, although the temperature varies from the polar regions to the equator.

■ Marine organisms are classified as benthos (which live on the ocean floor), plankton (which float near the surface), and nekton (which are free-swimming).

■ The oceans can be divided into three zones—intertidal zone (shoreline), littoral zone (above the continental shelf), and open ocean.

■ Freshwater biomes are divided into two types—running water, and standing water. Running water includes streams and rivers. Lakes and ponds are bodies of standing water, as are swamps and bogs.

Vocabulary Review

biome (845)	benthos (855)
tundra (846)	plankton (855)
permafrost (846)	phytoplankton (855)
taiga (847)	zooplankton (855)
temperate deciduous forest (848)	nekton (855)
grasslands (848)	intertidal zone (855)
deserts (849)	littoral zone (855)
tropical rain forests (850)	

A. Matching—Select the vocabulary term that best matches each definition.

1. Small, floating organisms that are the major producers of the oceans.

2. A large geographical region with a particular climax community

3. A layer of soil that remains frozen throughout the year

4. Area of relatively shallow water above the continental shelf

5. Free-swimming organisms that live in the ocean

Chapter Review

B. Multiple Choice—Choose the letter of the answer that best completes each statement.

6. A climax forest in which leaves are shed in the fall is found in the (a) taiga (b) desert (c) temperate deciduous forest (d) grassland.

7. Sparse rainfall and extreme daily temperatures occur in the (a) taiga (b) desert (c) tropical rain forest (d) temperate deciduous forest.

8. A warm, uniform climate and daily rainfall occur in the (a) grassland (b) desert (c) tropical rain forest (d) taiga.

9. A climax forest of pine, fir, and spruce trees, inhabited by snowshoe hare, lynx, bear, wolves, and moose, is a (a) tundra (b) temperate deciduous forest (c) desert (d) taiga.

10. The biome with rich, deep soil but no trees is a (a) grassland (b) tropical rain forest (c) temperate deciduous forest (d) taiga.

Content Review

11. What physical factors are unique to the tundra?

12. Why are there few trees in the tundra?

13. Describe the physical characteristics of the taiga.

14. Much of the eastern United States is in the temperate deciduous forest biome. Why does this biome not extend farther west?

15. Identify the vegetation and animal life of temperate deciduous forests.

16. What characteristics of grasslands make these regions good farmland?

17. How have the natural plants and animals of the grasslands changed in the United States?

18. Describe the physical characteristics of the desert.

19. How are desert plants and animals adapted to desert conditions?

20. Describe the physical characteristics of tropical rain forests.

21. Why is the soil of the tropical rain forest poor?

22. How does altitude affect climax vegetation?

23. List examples of benthos, plankton, and nekton.

24. What role is played by photosynthetic plankton in the marine biome?

25. Describe the zones of the oceans.

26. Why are temperature variations larger in fresh water than in the marine biome?

27. How do the vegetation and animal life of streams compare with those of lakes and ponds?

28. What physical factors that dominate the stream environment are absent in a lake?

29. Summarize the general characteristics of swamps and bogs.

Graphic Organizing

For information on graphic organizers, see Appendix G at the back of this text.

30. **Bar Graph:** Using the rainfall information in your textbook, construct a bar graph of annual rainfall for the terrestrial biomes. Since rainfall is given as ranges, indicate both the low and high figures on the bar for each biome.

Critical Thinking

31. How do you think conditions for marine algae that live in the intertidal zone differ from those for marine algae in the littoral zone? How are they alike? (*Comparing and Contrasting*)

32. Successive biology classes studied a water bird population for four years. Their data showed a sharp decrease in the water bird population. List as many possible causes as you can for this decrease.

Upon investigation, it was discovered that a local wetland area had been filled in to build a shopping mall five years before. Might the shopping mall be the cause of the decrease in the water bird population? Explain. (*Identifying Causes*)

33. Why would bats be good pollinators in the desert biome? (*Identifying Reasons*)

34. Indicate the suitability of each terrestrial biome for (a) growing crops, (b) building new cities, and (c) maintaining wildlife preserves. (*Classifying*)

Creative Thinking

35. Tropical rain forests are ecologically valuable, yet they are being destroyed at an alarming rate. What can you do to slow down or stop their destruction? List as many ideas as you can.

36. In the benthic zones of the oceans far below the depth to which sunlight penetrates, many types of species thrive. Suggest some special characteristics that would enable species to inhabit the benthos.

Problem Solving

37. Deciduous trees, such as maples and oaks, grow on the southern slopes of hills in the taiga biome. A student set up a laboratory investigation using local soil, maple seedlings, and several sunlamps to simulate the physical factors of the southern slopes. The purpose of the investigation was to determine what conditions enabled the deciduous trees to grow within the taiga biome. Critique the student's experimental design.

38. An ecologist studied the climate on the mountain range in Central America shown below. She assigned a student to collect temperature and precipitation data in each zone identified on the basis of its vegetation. The student's data are recorded in the table. Do you think the data are correct or incorrect? Why?

Zone 4

Zone 3

Zone 2

Zone 1

			Month			
			May	**June**	**July**	**August**
Z	1	T	22	24	25	23
		P	1.5	2	3	6
O	2	T	28	30	31	30
		P	10	8	10	8
N	3	T	19	19	20	18
		P	0.5	1	1	2
E	4	T	31	32	33	33
		P	28	29	32	31

T = Temperature, °C   P = Precipitation, cm

Projects

39. In order for aquatic insects to live in a fast-moving stream, they must have adaptations to keep from being washed away. The insects have structures or body modifications that enable them to cling to stones at the bottom of the stream. Prepare a class presentation about the ability of some aquatic insects to keep their position in a swift current. If possible, go to a stream and collect some aquatic insects for class display. Use illustrations to support your presentation.

40. Desert plants are adapted to survive under extreme conditions of temperature and moisture. Prepare a report about some of these plants and their adaptations for survival. Set up a classroom display of cacti and other desert plants.

41. A city is an environment created by the activities of its people. A city is also influenced by the climatic conditions of the biome in which it is located. Many kinds of plants and animals inhabit a city. Survey an urban area near your home to find common plants and animals. You could survey a park or an empty lot. Make a list of the plants and animals you find, using a field guide as a reference. Set up a poster display of the common organisms you discovered.

42. Write a report about one of these three career opportunities: range manager; game warden; demographer.

▲ This endangered peregrine falcon makes its home on a Denver skyscraper.

Human Ecology

39

A peregrine falcon surveys its new home. It has become a city dweller. Endangered and displaced from their natural habitat, some of these wild birds now live among another species whose quality of life is threatened by environmental damage: human beings. People have dramatically altered the delicate biosphere on which all living things depend. In this chapter, you will learn how cities, farms, and factories threaten various ecosystems. You will also see how concerned people are working to rescue endangered species and to restore the environment.

Guide for Reading

Key words: carrying capacity, urbanization, biodegradable, eutrophication, biological magnification, renewable resources, nonrenewable resources

Questions to think about:

📖 How have ecosystems been disrupted by human activities?

📖 How can the environment be restored?

39-1 Causes of Environmental Damage

Section Objectives:

- *Discuss* the importance of human population control in improving the quality of human life.
- *Describe* the problems that can arise from the movement of populations to the cities and from the importation of organisms into new environments.
- *Describe* how poor farming practices and the use of pesticides have damaged the environment.
- *List* the major types of pollution and describe the ways in which they damage the environment and/or human health.

📝 **Laboratory Investigation:** *Rate* pollution levels in your community (p. 880).

In the past, most people did not worry about the effects of human activities on the environment. Forests were cut down, rivers were dammed, and soil erosion was not prevented. Wastes from mines and other industries were dumped on the land, into waterways, and into the air. Within the last 30 years or so, however, more and more people have come to realize that the environment cannot be used thoughtlessly any longer. Human activities have damaged the environment, and the damage may be dangerous and permanent. In response to this awareness, many agencies are now devoted to restoring the environment. Rivers that had been so polluted that

▲ **Figure 39–1**

Environmental Damage. Although forests may be replanted, poor tree-cutting practices can damage the forest ecosystem.

▲ **Figure 39–2**

Lake Erie. Several years ago, pollution seriously threatened fish and other life in Lake Erie. Recent cleanup efforts have been effective.

they contained no fish have now been cleaned up. Lake Erie, for example, is one lake that has been saved. In some cities where the air was dangerously polluted, it is now somewhat cleaner.

Human ecology deals with the relationship between humans and their environment. In this chapter, we will discuss some of the important aspects of this relationship.

Human Population Growth

Many of today's most serious environmental problems are related to the growth in the human population in recent decades. In 1850, the population of the earth was estimated to be about 1 billion. Within 80 years, by about 1930, the population had doubled, reaching 2 billion. It doubled again, reaching 4 billion, by the mid-1970s. In 1990, it is estimated that there will be 5.1 billion people in the world. As you may have noticed, the time needed for the doubling of the world's population has become shorter. By the year 2000, the human population may have reached 6.2 billion.

As in other natural populations, if the human population continues to grow indefinitely, it will eventually reach a point at which the environment cannot support it any longer. A lack of food, water, space, or some other basic necessity acts as a limiting factor for every population. A **limiting factor** is some factor, such as a lack of food, that halts any further growth in a population. The size of a population that can be supported by the environment is called the **carrying capacity** of the environment. At some point in the future, human population growth must stop because the earth will reach its carrying capacity and will not be able to support any more humans.

A population remains the same size if the birth rate and death rate are equal and no changes result from migration. In this century, the death rate in the industrialized countries has declined sharply because of improvements in medical care, food production, and sanitation. In many of these countries, however, there has also been a decline in the birth rate. This has led to a stable, but older, population. In a few countries, the birth rate has dropped below the death rate, resulting in a shrinking population. In the underdeveloped countries, the birth rate remains high. At the same time, the death rate in many of these countries has dropped because of improved living conditions. Thus, the continued high birth rate is causing a rapid growth in the populations. Unfortunately, food production is not increasing as fast as the population. Any failure in crop production in these countries can result in malnutrition and starvation.

Directly or indirectly, the problem of human population growth affects everyone. If food becomes the limiting factor in human population growth, then starvation will become the major means of population control in many parts of the world. One way to avoid this situation is to reduce the birth rate to a level that maintains the population at its present size. This level is called the level of *reproductive replacement*. At this level, the number of births equals the number of deaths over a period of time.

Population control, alone, however, will not guarantee adequate supplies of needed materials. If the industrialized nations of the world reduced their levels of wasteful consumption, more resources would be available for use by the underdeveloped countries. The careful use of conservation measures in farming, lumbering, mining, and water use could ensure an increase in the production of food. Conservation measures could also ensure constant levels of other needed materials. The use of fertilizers and irrigation, as well as the development of crops with higher yields, could increase the food supply. New sources of food could be developed—from the oceans, for example. Most ecologists feel, however, that no matter what steps are taken, the food supply will eventually become inadequate if the current rate of population growth continues.

Disruption of Existing Ecosystems

Population increases and technological advances have resulted in the careless destruction or disruption of many ecosystems. As the population has increased, patterns of land use have changed. There has been a shift from rural (farming) areas to cities. The movement of the population to cities, or **urbanization,** has resulted in the destruction of productive farmland. As the cities have grown, farms have been turned into housing developments and shopping centers. The growth of cities has also destroyed or endangered other ecosystems, such as wetlands, which previously had been untouched. These changes have destroyed the natural habitats of many species of plants and animals.

Biology and You

Q: Why do children in photos of famine-stricken areas have large bellies? They don't look like they are starving.

A: Although these children have swollen bellies, they are starving. Their bellies swell from a buildup of gas. Malnutrition disturbs the bacteria that live in the intestines. The bacteria grow out of control and produce large amounts of gas, causing the belly to bulge.

The body needs nutrients and fuel to function. When a person does not get enough of these, he or she is malnourished. In famine-stricken areas, diets often are lacking in calories, in protein, or in both. Children usually are the first to be affected by malnutrition because their bodies need extra calories and protein during the growing phases of childhood.

Malnutrition occurs all over the world and is a serious public-health problem. Some long-term effects of severe malnutrition are irreversible. Brain damage and stunted growth are common. With enough food and care, a starving child will regain weight, but some physical problems may remain.

Write a letter to your local newspaper outlining what you think should be done to end world hunger.

Figure 39–3
Destruction of Wetlands. Wetlands, like many other ecosystems, are easily destroyed by careless development. ▶

In a balanced ecosystem, the number of organisms at each level in a food chain is controlled by the number of organisms at the next higher feeding level. For example, the number of insects in a region is controlled by the number of organisms that feed on them. When an organism is removed from its ecosystem and introduced into a new ecosystem, there may be no predators to control its numbers. This has happened in a number of cases in which insects and other organisms have accidentally or intentionally been introduced into new environments. The Japanese beetle, the fungus that causes Dutch elm disease, and the gypsy moth were all imported into North America. With few natural enemies, they have spread, doing great damage to plants.

Poor Farming Practices

Overfarming and Overgrazing

In a natural ecosystem, dead plants cover the ground. They decompose and form rich humus that is added to the soil. In farmland, the crops are harvested each year, and most of the plant parts are removed from the fields. When these plant parts are removed, the nutrients these plants have absorbed from the soil are also removed. If these nutrients are not returned, the soil becomes less fertile, and crop yield drops. In the past when this happened, the fields were abandoned. When fields are left without a cover of vegetation, heavy rains or winds can carry away the topsoil. The removal of soil by wind and water is called **erosion.** Wind action blows the soil away, occasionally causing dust storms. The dust bowls of the 1930s in Oklahoma and other states were caused by overfarming and wind action. After rainstorms, water running over the surface of the land carries soil into nearby streams. Eventually the land becomes useless for cultivation.

Figure 39–4
Soil Erosion. Without a protective covering of vegetation, topsoil is easily washed or blown away. Strong winds result in dust storms such as the one in this Texas desert. ▼

In many areas, overgrazing by herds of cattle and sheep has left former grasslands without a cover of vegetation. In these areas, too, the end result of thoughtless farming practices has been the loss of topsoil.

Misuse of Pesticides   Pesticides, used indiscriminately, have contaminated the air and water in many places. They have also disrupted food chains by killing organisms that are not pests. After years of use, some widely used pesticides have been found to be dangerous. DDT, for example, was used for many years before it was found to be highly poisonous to many animals, including humans. DDT sprayed on plants was washed off by rainwater and carried into streams and rivers. Eventually it entered the oceans, where it was taken up by the plankton. DDT is not readily broken down. Eventually it became concentrated in the bodies of the higher-level consumers. Following the path of DDT through the ecosystem well illustrates the interconnected nature of the world's ecosystems. DDT has been found in the bodies of polar bears, in the ice of the Antarctic, and even in the fatty tissues of humans.

Frequently, the organisms the pesticides were intended to kill become resistant to them. As you read in Chapter 29, this type of resistance is an inherited trait, which makes the pests even more difficult to control.

Pollution

Adding anything to the environment or affecting the environment in a way that makes it less fit for living things is called **pollution.** Pollution of the environment takes many forms and has increased with population growth and industrial development. One example of pollution is **noise pollution,** loud sounds that can cause hearing loss. Other examples are sewage and industrial wastes dumped into streams and rivers. Gaseous wastes from cars and trucks, the burning of fuels, and industrial gases have polluted the air. The land has been polluted by tremendous quantities of solid wastes generated by industry and by the population in general. Some of those industrial wastes are highly toxic.

Water Pollution   Industrialized countries use enormous quantities of water each day. The most recent data indicate that 693 liters (183 gallons) of water per person are consumed every day in the United States. Much of the available water, however, is polluted. There are six major sources of water pollution: organic wastes, inorganic chemicals, disease-causing microorganisms, changes in water temperature, oil spills, and radioactive wastes.

Many *organic wastes* are materials from plants and animals. Usually, these materials are **biodegradable** (by oh duh GRAYD uh bul). That is, they can be broken down by bacteria and other decay organisms into simpler substances. Sewage and wastes from canning, brewing, meat packing, and paper mills are major sources of organic materials in waterways. If organic wastes are added in small quantities, bacteria and other decay organisms break them down,

Figure 39–5
Noise Pollution. Construction sites are just one of many sources of noise pollution in urban areas. ▼

Science, Technology and Society

Technology: Biodegradable Plastics

Plastic waste makes up almost one-third of this country's garbage. It threatens to clog landfills and litter beaches. Plastics, unlike many wastes, do not degrade quickly in a landfill. But, chemists may have the answer to the problem: plastics that decay in sunlight and soil.

Plastics normally take 200 years to break down. They do not degrade quickly because they are composed of polymers, long chains of repeating molecules linked together. These chains are so tightly bound that they cannot be penetrated by decay organisms.

For chemists, the challenge was to find a way to weaken the chains without robbing the plastic of its strength. They achieved this by inserting cornstarch, a polymer that microorganisms can penetrate, into the plastic polymer. The organisms eat the cornstarch, breaking apart the molecule chains. Although these biodegradable plastics are expensive to produce, several states have laws requiring that the plastic yokes that hold six-packs together be made of them.

■ *Do you think states should require the use of biodegradable plastics? Why or why not?*

keeping the water clean. However, the breakdown of these materials uses oxygen from the water. If sewage and other organic materials are discharged into the water in large quantities, the oxygen content of the water is seriously reduced. The lack of oxygen kills off fish and other aquatic organisms.

Some organic wastes are plant nutrients. When large amounts of these substances are present, they stimulate the growth of algae and aquatic plants. In lakes, the presence of nutrients can speed up the process of succession. As the organisms die, material is added to the lake bottom, which reduces its depth. Growth around the shores further reduces the size of the lake. This accelerated aging process is called **eutrophication** (yoo truh fuh KAY shun).

When nutrients cause an explosive growth in the algae populations, only the topmost layer of algae receives enough light and oxygen for survival. The lower layers die. When they decay, the oxygen content of the lake is reduced, which kills various other forms of aquatic life.

Various pesticides, fertilizers, and detergents, as well as other synthetic organic chemicals, are poisonous to aquatic life. At the same time, these fertilizers and detergents contain plant nutrients. The net effect is to upset the natural balance of an ecosystem and possibly to destroy it.

Inorganic chemicals are dumped into waterways by mining and other industrial processes. Some of these wastes contain metals, particularly mercury and lead, that are poisonous to humans and other animals.

When dumped into waterways, mercury, lead, and some pesticides are picked up first by small aquatic plants and algae. These in turn are eaten by first-level consumers. Because these consumers eat large amounts of plants and algae, the toxic substances build up in their bodies. In the food web, larger second-level consumers eat many first-level consumers. The toxic substances then build up in higher concentrations in the bodies of the second-level consumers. As the food chain proceeds, each higher level of consumers accumulates larger quantities of poisonous substances. This process is called **biological magnification.** The animals at the end of the food chain, where the concentrations are highest, are most harmed by the pesticide or chemicals. In some cases, this has been humans. People in Japan who rely heavily on fish in their diet have suffered mercury poisoning from mercury that had been dumped into the ocean. Through biological magnification, the mercury accumulated in high concentrations in the bodies of large food fish.

Disease-causing microorganisms may enter the water from untreated sewage and wastes. Sewage contamination can be detected by testing for the bacterium *Escherichia coli.* These organisms, as well as other infectious bacteria and viruses, live in the intestines of warm-blooded animals and are found in their wastes.

Changes in the water temperature in streams and rivers can kill the fish and other organisms living there. This type of pollution, called **thermal pollution,** occurs when water is taken from a stream and used for cooling industrial equipment. The cold stream water is

▲ **Figure 39–6**
Oil Spill. In 1989, the Exxon Valdez oil tanker spilled about 240 000 barrels of oil off the coast of Alaska. This photograph shows seals coated by oil from the spill. The long-term effects of the spill on the ecosystem will only become apparent over time.

run through pipes next to pipes containing hot water. Heat is transferred from the hot water to the cold water, and the stream water, now heated, is returned to the waterway. In addition to the direct effects it has on living organisms, the warmed stream water holds less oxygen than the cold water. Nuclear power plants, in particular, require great amounts of water for cooling.

Other forms of water pollution involve oil spills. Oil is toxic to all forms of aquatic life and even kills many types of bacteria. Water birds die when they ingest the oil in trying to clean it off their feathers. The Valdez oil spill in Alaska was the worst of these oil spills to date.

Radioactive wastes are produced by nuclear reactors, mining, and the processing of radioactive materials. Exposure to relatively small amounts of radioactivity can be harmful.

Air Pollution   Industrialized countries with large urban populations and many cars face serious air pollution problems. In the United States alone, more than 200 million metric tons of pollutants are released into the atmosphere each year.

Some pollutants are **aerosols.** Aerosols have tiny solid particles or liquid droplets that remain suspended in air. Dust and smoke are aerosols. There are also artificial aerosols such as hair spray. Aerosol particles scatter sunlight, reducing the amount of light reaching the earth's surface, and thus lowering the surface temperature.

Some pollutants are gases that mix with the air. Sulfur dioxide (SO_2) is a pollutant produced by the burning of coal and oil that contain sulfur. In the atmosphere, sulfur dioxide may react chemically to form sulfuric acid, a harsh irritant to the respiratory system.

Sulfuric acid dissolved in rainwater is a major component of acid rain, which can gradually destroy stone buildings and other structures. It can also lower the pH of lakes and ponds, killing many of the organisms they contain or affecting their ability to reproduce. (See page 48.)

Hydrogen sulfide (H_2S) is a pollutant produced by several industrial processes, including the refining of oil and the manufacture of paper pulp. This gas, which has an odor like rotten eggs, is a nuisance in low concentrations but can be poisonous in high concentrations.

Carbon monoxide (CO) is produced by the burning of gasoline, coal, and oil. It combines easily with the hemoglobin of the red blood cells and reduces the hemoglobin's capacity to carry oxygen. In low concentrations, carbon monoxide can cause drowsiness and slow reaction times. In high concentrations, it causes death.

Nitrogen oxide (NO) and nitrogen dioxide (NO_2) are produced by the burning of gasoline, oil, and natural gas. When nitrogen dioxide is exposed to sunlight, it turns a dirty brownish color, darkening the air. Reactions in the atmosphere between nitrogen oxide, oxygen, and ultraviolet light produce ozone (O_3), which is a pollutant at lower levels. However, the ozone layer is a necessary part of the upper atmosphere. This layer protects us from damaging ultraviolet radiation.

Hydrocarbons, which are compounds of hydrogen and carbon, are produced by the burning of gasoline, coal, oil, natural gas, and wood. Several related compounds, such as formaldehyde and acetaldehyde, irritate the eyes, nose, and throat, but most are not dangerous at existing levels. However, when hydrocarbons react with nitrogen oxides in the presence of sunlight, they form what is known as *photochemical smog*, the type of smog found in Los Angeles. This type of smog, which occurs in dry, warm climates, is highly irritating to the lungs and eyes. It also damages plants. Los Angeles has recently enacted legislation designed to reduce this problem in the 1990s.

Usually, the air layer closest to the earth's surface is the warmest layer, and the air temperature drops with increasing altitude. Under these conditions, the less dense, warmer air rises,

Figure 39–7

Air Pollution in Mexico City. In cities, the major cause of air pollution is the burning of fossil fuels by industries and automobiles. ▼

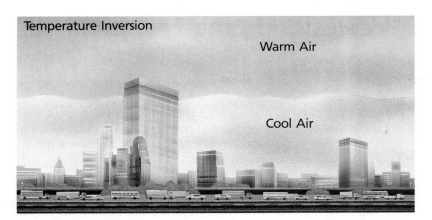

Temperature Inversion

Warm Air

Cool Air

◀ **Figure 39–8**
Temperature Inversion. In a temperature inversion, a layer of cool air becomes trapped beneath a layer of warm air.

carrying pollutants away from the earth's surface. In a **temperature inversion,** however, a layer of cooler, denser air becomes trapped below a layer of warmer air, as shown in Figure 39–8. The warm air acts as a lid, preventing the upward movement of air from the earth's surface. Pollutants accumulate in the cool layer. The condition lasts until the air masses move away.

Land Pollution   As cities have grown, land pollution has become a growing problem. Even small cities produce many tons of solid waste, or refuse, every day. Two ways of disposing of refuse are sanitary landfills and incineration (burning).

A *sanitary landfill* is a large area where the refuse is dumped into a trench. A liner is used to prevent seepage into the water supply. The refuse is compacted and covered over with dirt. However, as cities have grown, it has become increasingly difficult to find land for this purpose.

Burning refuse in large furnaces, or incinerators, is another method of disposal. However, incinerators must be equipped with pollution control devices, or they release large quantities of pollutants into the air. After burning, the ashes must still be disposed of in sanitary landfills. Some cities are using the steam produced by the burning of refuse to generate electricity.

39-1 Section Review

1. What is human ecology?
2. Define the term *carrying capacity*.
3. What problem may arise when an organism is imported into a new environment?
4. List some of the major sources of water pollution.

Critical Thinking

5. Rank the major types of pollution from the most difficult to clean up or control to the least difficult.   *(Ranking)*

Science, Technology and Society

Issue: Nuclear Wastes

Tons of radioactive waste from nuclear power plants sit in temporary storage facilities. Finding a suitable permanent site is difficult. Much of the waste will remain dangerous to humans and the environment for thousands of years.

Many people fear the storage of nuclear waste near their homes. Studies have shown that exposure to radiation can cause cancer and birth defects. What if the waste leaks from the storage site? How can wastes be transported to the site safely? Would catastrophes, such as earthquakes, cause radioactive substances to leak into groundwater? These people argue that the safety of waste sites cannot be guaranteed.

Others claim that building a permanent facility would create jobs and bring money into a community. The location of the site would be determined by years of scientific research. Such a facility would be constructed under strict guidelines to ensure safety. Besides, they argue, temporary storage sites are more dangerous than permanent sites would be.

■ **What should be done about the storage of nuclear wastes?**

39-2 Restoring the Environment

Section Objectives:

■ *Describe* some of the efforts that are being made to control pollution.
■ *Explain* how each of the following techniques is used in soil conservation: cover crops, strip cropping, terracing, contour farming, windbreaks, dams, crop rotations, and fertilizers.
■ *Discuss* methods of forest and wildlife conservation.
■ *Describe* several biological methods of pest control.

Many countries have introduced programs to improve the lives of their citizens and to halt the deterioration of the environment. Some programs are aimed at family planning to slow the rate of population growth. Other programs deal with disease control and sanitation. In terms of the environment itself, there are national and international programs aimed at pollution control, conservation of natural resources, and preservation of existing species. Various methods of maintaining and restoring the environment are discussed in the following sections.

Controlling Pollution

In some countries where much of the air pollution is caused by exhaust from automobiles, the introduction of emission controls and the use of unleaded gasoline have reduced pollution. Waste gases from industrial processes are treated in many different ways to remove the most serious pollutants before the gases are released into the atmosphere. Some areas are now banning the use of aerosol sprays to help protect the atmosphere.

Sewage treatment plants that use bacteria to break down sewage before it is released into waterways have greatly reduced water pollution. Where these plants have been built, the waterways are much cleaner. However, many additional plants are needed. Although some of the most toxic industrial wastes are no longer dumped into waterways, others are still pouring into lakes, streams, and rivers. The dumping of wastes into the oceans by coastal nations, combined with wastes carried into the oceans by rivers, threatens the future productivity of the oceans.

Some efforts are being made to reclaim land that has been strip-mined. Other efforts are being made to control the dumping of wastes on land. In many areas, there are now special sites reserved for the disposal of toxic wastes. Because these wastes remain toxic for long periods of time, there is always a problem with leakage and the contamination of surrounding areas. The disposal of radioactive wastes from nuclear reactors is proving to be an especially difficult problem. These wastes must be stored so that the radioactivity is safely absorbed by surrounding materials and cannot leak into the environment for the thousands of years during which the wastes remain radioactive.

Conserving Natural Resources

Natural resources are materials in the environment that are used by humans either for their life processes or for cultural activities. There are two types of natural resources—renewable and nonrenewable. **Renewable natural resources** include air, water, soil, sunlight, and living things. Natural events replace these resources as they are used. However, careless human activities can disrupt the natural events that replace renewable resources. **Nonrenewable natural resources** are resources that can be taken from the earth only once. Coal, oil, natural gas, metals, and minerals are nonrenewable natural resources. The conservation of both types of resources is essential.

Some products made from nonrenewable resources can be reprocessed and used again for their original purpose. This is called **recycling.** Some states have passed laws requiring the recycling of refuse. Figure 39–9 shows the current status of such legislation.

Conserving Soil   Although soil is a renewable resource, the process of soil formation is very slow. It may take thousands of years to yield a few centimeters of topsoil. Therefore, it is important to prevent the loss of soil through erosion and the loss of nutrients. There are a number of techniques used in soil conservation.

To prevent soil erosion, some farmers use cover crops. **Cover crops** are crops planted to cover a whole field. These crops have fibrous roots that form a dense mat in the soil. This mat prevents

Figure 39–9

Recycling. Glass, metal, plastic, and paper may be recycled. Many states have laws requiring recycling containers, and some communities have recycling centers. ▼

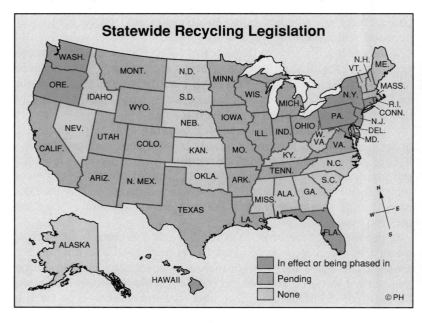

Statewide Recycling Legislation

In effect or being phased in
Pending
None

©PH

Figure 39–10
Some Soil Conservation Practices. ▶

strip cropping

terracing

contour plowing

windbreak

soil erosion. Commonly used cover crops include clover, alfalfa, oats, and wheat. Crops that are planted in rows, such as corn, beans, and cabbage, do not prevent erosion of the exposed soil between the rows of plants.

When row crops are planted, strip cropping may be used to protect the soil. **Strip cropping** is a conservation practice in which cover crops are planted between strips of row crops. Thus, no exposed soil is left open to erosion.

Terracing is used on the sides of hills. Flat areas, or terraces, are dug in the hillside, providing areas for planting. Each terrace has a boundary made up of an earth bank held in place by plants and rocks. Terracing prevents surface water from running directly down the hill, carrying the soil with it.

On uneven landscapes, **contour farming** may help to prevent erosion. In contour farming, rows are plowed across slopes, following the contour of the land. The mounds formed by the plow and the plants prevent water from running straight down the slopes.

Windbreaks are used to prevent wind erosion. Windbreaks usually consist of rows of trees. Poplar trees are commonly used for this purpose.

In areas that are already eroded, **dams** are often built to slow down the running of water and to reduce or to prevent further erosion. Dams are also a major means of water conservation. Large amounts of water collect behind the dam. This collected water can be used for drinking, irrigation, and recreation, as well as for the generation of electricity.

Since each plant has its own mineral requirements, planting the same crops every year can remove valuable nutrients from the soil. **Crop rotation** involves growing different crops in succeeding years. Planting different crops prevents the reduction of soil nutrients, which is known as *soil depletion*. Legumes, such as clover, are rotated along with other crops to restore nitrates to the depleted soil.

Fertilizers also are used to replace essential soil materials removed by crops. Both natural fertilizers, such as manure, and commercial chemical fertilizers are widely used.

Conserving Forests   Forests, which supply a wide variety of materials for human use, are another renewable natural resource. In addition to furnishing wood, trees are used for the production of paper, charcoal, turpentine, and rayon. Forest soils hold large quantities of water, and the trees and undergrowth of the forest prevent soil erosion. However, like soil, the replacement of forests is a slow process, and poor tree-cutting practices can cause permanent damage to the forest ecosystem.

As populations have grown, the need both for cleared land and for forest products has increased. As a result, in many parts of the world, especially the tropics, the amount of forest land is shrinking. In an attempt to raise productivity and ensure future supplies, a variety of conservation practices are being applied to remaining forests.

Sustained-yield tree farming involves cutting down trees in only certain areas of a forest. This leaves surrounding areas untouched. In *block cutting,* square areas of forest are cut. Reseeding of the cut section takes place naturally by seeds coming from the surrounding forest. In *strip cutting,* strips of trees are cut between strips of untouched forest. In *selective harvesting,* certain trees are marked and cut, leaving the others undisturbed to grow and mature.

To replace trees lost in cutting, **reforestation** programs plant seeds or seedlings of a particular type in cut areas. The seeds or seedlings planted are usually fast-growing, disease-resistant types

◄   **Figure 39–11**

Reforestation. In the tropics, forests are being cleared at an increasing pace for agriculture and cattle farming. In Brazil, as in many other areas of the world, efforts are being made to replant these deforested areas.

Endangered Species		
	Number of Species	
	Extinct Since 1600	**Endangered Today**
amphibians	1	46
reptiles	20	191
birds	109	634
mammals	73	458

▲ **Figure 39–12**
Endangered Species. By the end of this century, thousands of species may have become extinct. Examples of endangered species include the cheetah, the leatherback sea turtle, the golden coqui frog, and the California condor.

that will produce a good quality of lumber. In all forestry programs, undesirable, diseased, or dead trees are removed to allow space for growth of good timber.

Conserving Wildlife The growth of cities and suburbs has led to the destruction of the natural habitats of many types of plants and animals. Furthermore, hunting has brought about the extinction of a few species and the near-extinction of many others. The passenger pigeon was a bird that was found in large numbers in North America until the mid-1800s. A prime target of bird hunters, the huge flocks were killed off, and the last known passenger pigeon died in the early 1900s. Many other species of birds, as well as whales and other animals, are now in danger of becoming extinct. Among them are the golden coqui frog of Puerto Rico, the leatherback sea turtle, the California condor, and the cheetah. Figure 39–12 shows the numbers of species of amphibians, reptiles, birds, and mammals that are in danger of extinction today. Compare these figures with the numbers that have become extinct since 1600. Sadly, the rate of extinctions is rising. Several plant species have also become extinct, and others are endangered.

Concern over the possible extinction of various species of animals and plants has resulted in wildlife conservation practices that are being carried out in many areas. Hunting and fishing laws now restrict the sex, size, and number of prey that can be taken and limit the hunting season. In many areas, game and bird preserves, where no hunting is allowed, have been established.

In some areas, game fish are being bred in fish hatcheries. These fish are used to restock heavily fished lakes and streams. This keeps the populations at a reasonable level but also allows recreational fishing.

Finally, restricting the use of pesticides and herbicides has limited the number of accidental deaths caused by these chemicals.

Legal protection for endangered species has made it possible for some species that were near extinction to begin to show a population increase. This is true of the bison, egret, and whooping crane. Other endangered species are still decreasing in numbers.

Controlling Pests Biologically

Although chemical pesticides have helped to control insect damage to agricultural crops, they have also created some serious ecological problems. As you have read, many chemical pesticides are not readily broken down in nature, so they build up in the environment, harming plants and animals. In response to this problem, scientists have developed some chemical pesticides that break down within a few days to harmless substances. These pesticides will not build up. On the other hand, they have to be applied more frequently, which makes them more expensive and difficult to use.

As an alternative to chemical pesticides, various biological methods of pest control have been discovered. These methods are more specific than the chemical pesticides and have much less of an effect on the environment in general.

In some areas, natural enemies have been imported to control certain pests. For example, ladybugs have been used to control aphids, and a wasp that preys on the alfalfa weevil has been used to control that insect pest. In introducing one organism to control the population of another organism, it is important to know whether or not there is any way to control the population of the introduced organism.

Sometimes, various insect larvae, including gypsy moth caterpillars and mosquito larvae, can be controlled when they are infected with a particular type of bacterium. Viruses have been used against worms that attack vegetable crops.

Crop rotation, which helps to prevent soil depletion, also helps to control pests. Planting different crops in succeeding years removes the favored food source of a pest organism and thereby decreases the pest population.

Figure 39–13

Biological Control of Insects. Several methods of insect biological control have been developed. Scientists continue to experiment with new ways to control insect damage to agricultural crops. ▶

Scientists have also developed ways to use insecticides without contaminating the environment. **Pheromones** (FEHR uh mohnz) are a type of animal secretion that serves as a sex attractant between members of a species. Scientists have developed pheromones that are used to lure insects into traps where they are exposed to contact poisons. Although the insects are killed, the environment is not polluted.

Another method of insect control involves the release of sterile males into the population. The males, which have been sterilized by exposure to radiation, mate with the females but no offspring result.

Key to the Future

Pollution is a pressing societal issue, one that directly affects the future survival of all the earth's living organisms. Yet, although there are various technologies to stop pollution, it is difficult to devise an overall plan to deal with environmental problems. Any solution usually requires social and cultural changes that are not easily brought about. The views and attitudes of any society are based on the values important to the society. For this reason, environmental problems cannot be left only to the scientist. Every member of every community should be involved in these decisions.

Some people are already trying to preserve and improve the environment voluntarily, but many more people need to become involved in these efforts. Unfortunately, some people show little concern about their surroundings, and continue their destructive habits, while other people do not have enough information to make the correct choices.

The actions of individuals are only a small part of the problem. It is people in groups—in towns, cities, and industries—who produce the greatest effects on the environment. Often, industries will not voluntarily reduce their polluting activities because the costs are high, a factor which reduces profits. Some cities and states that have industries are hesitant to set strict pollution standards for fear of placing heavy financial burdens on the companies. If costs become too high, the industry may move out of the community, resulting in unemployment and lower tax revenues. Also, the major causes of pollution are not limited to industries. Cities themselves often degrade the environment with inefficient sewage plants and poorly managed landfills. The reasons are the same—cost factors. Efficient operation increases cost, which in turn affects taxes, and this touches all members of the community. Yet, well-informed individuals know that failure to stop pollution will also endanger the community in the long run.

One method of controlling pollution is through legislation. At present there are many laws regulating pollution in the United States and Canada. However, laws that regulate water pollution, air pollution, landfills, heat, and noise often do not prevent pollution. Some states are now proposing new economic incentives in the form of lower taxes to industries that reduce pollution.

Perhaps the most promising way to influence people is by educating them. In successful environmental education, individuals obtain the essential facts, identify the relevant values, analyze the effects of different values in decision-making processes, and predict the consequences of various choices. Once people realize that certain consequences are unacceptable, they may recognize that individual and collective behavior must be changed. People must learn to "think globally." Countries need to work together to solve many interrelated environmental problems.

Well-informed citizens know that the problem of pollution *must* be solved if the earth is to remain a safe, stable place for future inhabitants. They understand that informed voters in a democratic society can bring out an enlightened government. A government supported by its people could enforce strong laws that are based on an understanding of earth's delicately balanced biosphere. With such insight, the life-sustaining conditions of our earth could be preserved for all the living organisms that will follow us.

▲ **Figure 39–14**

An International Conference. In March of 1989, a two-day conference on the protection of the global environment was held in Belgium. Twenty-four nations participated in the conference.

39-2 Section Review

1. What are the two major types of natural resources?
2. List three methods of soil conservation.
3. Name two conservation practices being used to protect forests.
4. How are pheromones used to control insects biologically?

Critical Thinking

5. Compare renewable and nonrenewable resources.   (*Comparing and Contrasting*)

Laboratory Investigation

Community Pollution Issues

Pollution is not as far removed from our lives as we might think. In this investigation, you will evaluate pollution problems in your community and the efforts, if any, being made to deal with them.

Objectives

- *Identify* factors within the community that contribute to water, air, and land pollution.
- *Assess* the degree of pollution occurring within your community and evaluate the community effort to reduce or prevent it.

Materials

notebook
pencil
resource materials

Procedure

A. As a group, choose one type of pollution (land, water, or air) to evaluate. In a chart, list specific examples that are occurring in your community of that type of pollution.

B. Each group member should investigate at least one of the pollution problems listed in the chart. Research background information about each problem and the measures, if any, that have been, or will be, taken to solve it.

C. Using the Problem Points guidelines below, rate each pollution problem and record the ratings in the chart.

Problem Points	Guidelines
0	no problem locally.
1	some problem, coming from other community.
2	mild problem that could become significant.
3	moderate problem with potential to become hazardous.
4	severe problem, becoming hazardous.
5	critical problem, hazards existing.

D. Using the Solution Points guidelines below, rate your community's past and present efforts to deal with each of the pollution problems. Record the ratings in your chart.

Solution Points	Guidelines
0	no pollution awareness or antipollution actions
1	some community awareness; no community activities or legislation
2	definite awareness; citizens groups; some movement toward legislation
3	strong awareness; much discussion and news coverage; legislation in process
4	strong public and private community awareness; legislation in place
5	community organized to deal with and prevent pollution; legislation enforced

E. Add all the problem points and solution points. Compare your findings with the findings of the other groups.

Analysis and Conclusions

1. Using the total problem points and solution points from your group, describe the strengths and weaknesses in your community's pollution status.

2. What issues does your community deal with effectively? What issues, if any, is it neglecting?

3. What might a rating of 15 problem points and 12 solution points indicate about a community?

4. Solution points depend upon community awareness, political action, and individual responsibility. Explain each of these and give an example.

5. What would you suggest to reduce the pollution problems you have studied? Why might some people not favor all of your suggestions?

6. Identify five actions an individual can take that will help to prevent further pollution.

Chapter Review

Study Outline

39–1 Causes of Environmental Damage

■ There is a growing awareness that humans have damaged the environment. It can no longer be used thoughtlessly.

■ The rapidly growing human population places a strain on the earth's resources.

■ Movement of human population to growing cities has resulted in loss of farmland and ecosystems important to many plants and animals.

■ Overfarmed and overgrazed land has been lost by erosion.

■ Pollution has increased with human population growth and industrial development. The release of industrial wastes and sewage into streams and rivers causes water pollution.

■ Gaseous wastes from fossil fuels, cars, and industries pollute the air. Solid wastes from many sources result in serious disposal problems.

39–2 Restoring the Environment

■ Efforts are being made to control pollution by sewage treatment, emission controls, and sanitary disposal of solid wastes. Some natural resources can be conserved by recycling.

■ Many techniques are used to conserve soil. Various farming practices can be used to prevent soil erosion by water and wind. Crop rotation and fertilizers can be used to prevent soil depletion.

■ Although renewable, forests are not always replanted as soon as they are cut down. Sustained-yield tree farming and reforestation programs should be used.

■ Because the natural habitats of many plants and animals are being destroyed, many species are endangered. Some have already become extinct. Awareness of the danger is growing, and the use of conservation practices is helping save some species.

■ Wherever possible, long-lasting chemical pesticides are being replaced with biological controls.

Vocabulary Review

human ecology (864)
limiting factor (864)
carrying capacity (864)
urbanization (865)
erosion (866)
pollution (867)
noise pollution (867)
biodegradable (867)
eutrophication (868)
biological magnification (868)
thermal pollution (868)
aerosols (869)
temperature inversion (871)
renewable natural resources (873)

nonrenewable natural resources (873)
recycling (873)
cover crops (873)
strip cropping (874)
terracing (874)
contour farming (874)
windbreaks (874)
dams (874)
crop rotation (875)
fertilizers (875)
sustained-yield tree farming (875)
reforestation (875)
pheromones (878)

A. Sentence Completion—Fill in the vocabulary term that best completes each statement.

1. The situation in which cooler, dense air is trapped below warmer air is called a(n) _____.

2. Toxic materials accumulate at higher levels of a food chain by a process called _____.

3. Anything that makes the environment less fit for living things is called _____.

4. Some nonrenewable resources can be reprocessed for use by _____.

5. Materials that are _____ can be broken down by bacteria.

6. A shortage of food or water acts as a(n) _____.

7. Movement of human populations into the cities is called _____.

8. Water used for cooling industrial equipment such as that in nuclear power plants can result in _____.

Chapter Review

9. _____ consist of tiny particles that remain suspended in air.

10. Planting different crops in succeeding years is called _____.

B. Word Relationships—In each of the following sets of terms, three of the terms have a common characteristic. One term does not belong. Identify the term that does not belong.

11. strip cropping, contour farming, terracing, eutrophication

12. recycling, cover crops, pheromones, dams

13. carrying capacity, noise pollution, thermal pollution, aerosols

14. reforestation, terracing, sustained-yield tree farming, renewable natural resources

Content Review

15. What could happen if human population growth continues at the present rate?

16. How have changes in birth and death rates affected population growth in industrialized and nonindustrialized countries?

17. How has the shift of populations from rural to urban areas affected the environment?

18. How have poor farming practices ruined former farmlands?

19. Describe some of the problems that can arise from the use of chemical pesticides.

20. Describe the harmful effects of organic wastes as water pollutants.

21. Name the major air pollutants, and describe how each is harmful.

22. Describe the methods used for refuse disposal.

23. Explain how refuse can be used to produce energy.

24. Explain the difference between renewable and nonrenewable natural resources, and give examples.

25. How can soil depletion be reduced?

26. Describe five farming techniques used to prevent soil erosion.

27. Although soil and forests are renewable natural resources, it is important that they be conserved. Why?

28. List five wildlife conservation practices.

29. What are some of the advantages of biological pest control over chemical pesticides?

30. List four examples of biological pest control.

31. What is the key to getting people to "think globally" and to become interested in conservation?

Graphic Organizing

For information on graphic organizers, see Appendix G at the back of this text.

32. **Concept Map:** Construct a concept map that shows the major types of water pollution. Include at least 15 concepts. Be sure to include linking words between related concepts.

Critical Thinking

33. Suppose a pond suddenly becomes overgrown with algae and plants. Suggest some possible causes for the growth. What questions would you ask to determine the most probable cause for the sudden algae growth? (*Identifying Causes*)

34. Discuss farming methods that destroy resources and other farming methods that could conserve resources. (*Comparing and Contrasting*)

35. PCBs are chemicals that pollute water. The Food and Drug Administration has declared that food with more than 2 parts per million (ppm) of PCBs should not be eaten because it presents an unacceptable risk of causing cancer. Suppose some lobsters have been tested as having 0.2 ppm of PCBs in the meat and 40 ppm of PCBs in the green gland, often called tomalley. Some people eat this green gland. Should these lobsters be sold for human consumption? Why or why not? (*Making Ethical Judgments*)

36. Suppose you live in a small town whose economy depends largely on one factory. The factory generates pollutants that may cause health risks to children, the elderly, and sick people. If the factory is forced to clean up its emissions, it will no longer be profitable and will be closed down by the large corporation that owns it. If you were the mayor of the town, would you allow the factory to continue to operate, or would you force the owners to clean up the emissions? Explain your answer. (*Making Ethical Judgments*)

Creative Thinking

37. Some environmentalists have raised the possibility of a "greenhouse effect" as a consequence of air pollution. They believe that the build-up of carbon dioxide in the atmosphere may increase heat at the earth's surface. Propose some possible consequences of a worldwide increase in temperature.

38. Think of as many ways as you can for your school to decrease its negative impact on the environment.

Problem Solving

39. Your father notices cracks in the tires of his car. A chemistry teacher tells you that ozone, an air pollutant, can cause tires to deteriorate. Design a controlled experiment to test the validity of that statement.

40. The human population 2000 years ago was approximately 250 million people. By 1850, the population had reached one billion. It had doubled by 1930. Presently, the population has exceeded 5 billion people and is increasing at a rate of about 1.5 million people per week. Calculate the projected population for each of the next five years. Assume that the population in 1991 is 5.1 billion.

41. Carrying capacity affects the size of all natural populations living in an area. The graph below shows the relationship between the carrying capacity of an island and the number of deer living on it over a 45 year period. How would you explain the decline in carrying capacity around 1925?

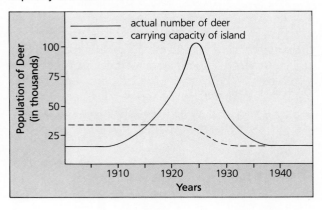

42. A population of wolves lived on the same island as the deer in problem 41. The table below contains data about the wolf population. Using the graph in 41 and information from the table, develop a hypothesis to explain the change in the deer population and carrying capacity.

Year	Wolf Population
1905	25
1910	20
1915	15
1920	10
1925	0

Projects

43. Sample the airborne particulate matter in the area of your home. Coat four microscope slides with a very thin layer of petroleum jelly. Then place the slides in the open air on windowsills around your house. After 24 hours, cover each exposed slide with a second microscope slide. Carefully return the slides to school, and view them with a microscope. Describe your observations. (Use a field guide to help identify types of pollen you may have collected.)

44. Gather information about local industries that process recycled materials for use in their products. Interview industry representatives as part of your information gathering. Include the information in a written report about the status of recycling in your community.

45. A farming community has the opportunity to attract a small paint factory that will bring many needed jobs to the area. Large quantities of water are used to manufacture paint, and there is an occasional need to dispose of chemical wastes. As the town mayor, you are asked to outline a solution to the environmental problems that would be created by the new industry. Research the paint industry in the library, and present your plan to the class.

Unit 8

Biology and Problem Solving

What can be done to control the gypsy moth threat?

Imagine that you have been elected Parks and Woodlands chairperson for the council that governs your county. You have just received a letter from the governor's office saying that gypsy moths are expected to migrate into your county this spring.

From the library and from interviews with ecologists, you learn that, in North America, there are no predators that control the gypsy moth population. You also find out that, without this control, many trees, particularly oak and hickory trees, are destroyed as the gypsy moth larvae feed on the leaves. You think about the many parks and wooded areas in your county. "What can we do to save our trees?" you ask yourself.

1. Describe the problem you face as Parks and Woodlands chairperson.

You learn that other counties have used several measures to prevent destruction of the trees. Some steps that can be taken are:

■ Citizens can act individually to protect trees. When citizens place a band of material around each tree trunk, gypsy moth larvae become trapped and cannot crawl up the tree to the leaves. One banding method requires that the moths be picked off the band by hand each day and destroyed.

■ A chemical insecticide can be sprayed from airplanes flying over the county. This insecticide is effective and disappears fairly rapidly, although some may run off with rainwater into streams and lakes.

■ A spray containing a type of bacteria that naturally feed on gypsy moth larvae can be sprayed onto trees. These naturally occurring bacteria are believed to be otherwise harmless to the environment. The spray, however, is very expensive.

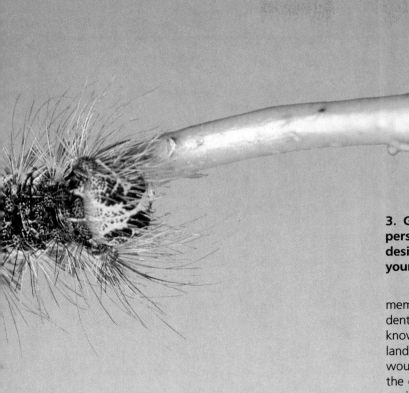

■ Do nothing. Some ecologists recommend "letting nature run its course" and allowing major changes to occur in the ecosystem.

2. For each of the four alternatives, predict the short-term and long-term consequences for the environment.

You know that your primary responsibility as Parks and Woodlands chairperson is to protect your county's woodland areas. These areas have been set aside for the recreational use and enjoyment of all of the county's citizens.

3. Given your role as Parks and Woodlands chairperson, rank the four alternatives from the most desirable to the least. Explain how you arrived at your ranking.

Now, you think about how the other council members might react to this problem. You are confident that they want to serve their citizens well. They know that the people in the county value their woodland areas. But, they also are well aware that citizens would strongly oppose a budget increase. You fear that the other council members might not agree with your position on this issue.

4. Prepare a short speech designed to persuade council members that they should support your top-ranked alternative. Make sure your speech addresses the concerns they might have.

Problem Finding

Suppose you were one of the other council members and you had just heard the speech given by the Parks and Woodlands chairperson. What questions would you still need answered before you take a position on this issue?

Appendix A: Care and Use of the Microscope

One of the biologist's most important tools is the microscope. Microscopes allow biologists to study structures so small they cannot be seen with the unaided eye. The type of microscope used in most biology classrooms is the compound microscope.

ocular (eyepiece): contains lens

coarse adjustment: moves body tube

fine adjustment: moves body tube

arm: supports body tube

clip: holds slide

base: supports microscope

body tube: separates ocular from the nosepiece

nosepiece: hold objectives

high-power objective: contains lens

low-power objective: contains lens

stage: supports slide

diaphragm: controls amount of light

mirror

Learning the Parts of a Microscope

Figure 1 shows the parts of the compound microscope. You should become familiar with all the parts and the function of each part. At the top of the microscope is the *ocular,* or *eyepiece, lens.* This is the part into which you look. Usually, the ocular lens has a magnifying power of 10X or less. A magnifying power of 10X means that the lens makes an object appear 10 times larger than it really is.

Beneath the ocular is the *body tube.* At the bottom of the body tube is a revolving *nosepiece* containing two more lenses. These are called the objective lenses. The shorter objective lens is the *low-power objective,* which usually has a magnification of 10X. The longer objective is the *high-power*

objective, and it usually has a magnification of about 40X. The magnification of each objective lens will be marked on it. Powers of magnification may vary.

When you are observing specimens with the microscope, you should know how much they are being magnified. To compute the total magnification, multiply the magnifying power of the ocular lens (10X) by the magnifying power of the objective lens (either 10X or about 40X). For example, if you are looking at an object with the low-power objective, the magnification would be 10 times 10, which equals 100X. Thus the object would appear to be 100 times larger than its natural size.

Preparing the Microscope for Use

To prepare a microscope for use, follow these steps:

1. Remove the microscope from its storage case by grasping the arm of the microscope with one hand and placing your other hand under the base. Carry the microscope upright to your lab table.
2. Gently set the microscope down on its base on the lab table. Position it so that the ocular and the arm are toward you. Make sure that the base of the microscope rests evenly on the table and that the front edge of the base is at least 10 centimeters from the edge of the table.
3. Clean the ocular and the objective lenses with a piece of lens paper. Touch the lenses only with lens paper; anything else may damage them.
4. Use the coarse adjustment knob to raise the body tube to its highest point above the stage. Then, turn the revolving nosepiece until the low-power

objective clicks into place. When in place, the objective lens should be directly above the opening in the stage.

5. Look through the ocular with both eyes open. Switch on the illuminator lamp or turn the mirror until you see an evenly lit field of white light. If the field of light is too dark or too light, open or close the diaphragm. CAUTION: If your microscope has a mirror, never use direct sunlight as a light source. It can injure your eyes.

Observing a Prepared Slide

To observe a prepared slide, follow these steps:

1. Place a prepared slide on the stage so that the stage clips hold down the left and right sides of the slide and the specimen is on top. Move the slide so that the specimen is centered beneath the low-power objective.

2. Looking at the side of the microscope, slowly lower the low-power objective with the coarse adjustment knob until it is just above, but not touching, the slide. CAUTION: Do not let the low-power objective touch the slide, as it may damage the objective or break the slide.

3. With both eyes open, look through the ocular. Move the slide back and forth with your fingers until the specimen is centered. If the specimen looks out of focus, turn the coarse adjustment knob slowly to raise the body tube. When the specimen comes into view, stop turning. For fine focusing, use the fine adjustment knob. Always look through the eyepiece as you adjust the focus.

4. If necessary, adjust the diaphragm to increase or decrease the amount of light.

5. To view the specimen with the high-power objective (40X), lift your head away from the ocular and revolve the nosepiece until the high-power objective clicks into place. (Do *not* look through the ocular while switching from low to high power.) Then look through the ocular and use the fine adjustment knob to bring the specimen into focus. CAUTION: Do not let the high-power objective touch the slide.

6. After using the microscope, clean the lenses with lens paper. If necessary, also clean the stage.

Preparing a Wet Mount

To make a wet mount, follow these steps:

1. Obtain from your teacher a clean microscope slide and a clean coverslip.

2. Place the specimen in the middle of the microscope slide. Make sure the specimen is thin enough for light to pass through it.

3. With an eyedropper, place a drop of water on the specimen.

4. Hold the coverslip by its thin edges between your index finger and your thumb, and lower it at a 45° angle into the drop of water. After the water spreads along one edge of the coverslip, slowly lower it until it lies flat on top of the slide. This is called a wet mount. If air bubbles are trapped beneath the coverslip, lift it up and lower it again, or tap it gently with the eraser end of a pencil.

5. Remove any excess water with a piece of paper towel. If the specimen starts to dry out, add another drop of water at one edge of the coverslip.

6. To view the wet mount, follow the same steps as for viewing a prepared slide.

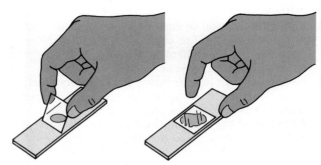

Staining a Specimen

To stain a microscope specimen, follow these steps:

1. Obtain from your teacher a clean microscope slide and a clean coverslip.

2. Place the specimen in the middle of the slide.

3. With an eyedropper, place one drop of water on the specimen.

4. Holding the coverslip by its thin edges, lower it at a 45° angle into the water on the slide. Continue lowering it gently until it lies flat on the slide.

5. Using the eyedropper again, add a drop of stain along one edge of the coverslip. Touch a small piece of paper towel to the opposite edge so as to draw the stain under the coverslip. Stains are used to make the structures of a specimen easier to see.

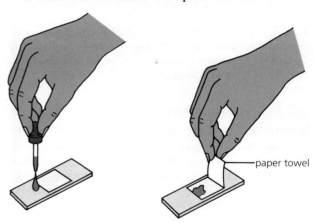

paper towel

Appendix B: Safety in the Laboratory

Laboratory work is an important part of your biology class. The work, however, may require you to handle strong chemicals, sharp instruments, hot materials, or expensive equipment. To prevent accidents, you need to act with care and observe all safety rules at all times.

General Safety Rules

Review these safety rules before starting a lab activity:

1. Know the location of emergency equipment, such as the first-aid kit, fire alarm, or fire extinguisher. Only use the equipment, however, if you are instructed to do so by your teacher.
2. Be familiar with how to leave the laboratory safely in an emergency. Be sure you know a safe exit route in case of a fire or an explosion.
3. Be prepared for your lab exercise when you arrive in the laboratory. Read the directions for the lab activity beforehand, so that you will know what to do.
4. Do not bring any food or drink into the laboratory. Do not eat or drink out of any lab equipment.

Personal Safety Rules

Each person should follow these additional safety rules:

1. Inform your teacher of any medical problems that may affect your safety in doing lab work. You should report allergies, asthma, sensitivity to certain chemicals, epilepsy, or a heart condition.
2. Wear clothing that will not present a safety hazard in the laboratory. Make sure articles of clothing will not interfere with laboratory work or be a fire hazard. Secure or take off a loose tie or jacket. Roll up long sleeves. Remove dangling jewelry.
3. Tie back or cover long hair.
4. Wear regular glasses instead of contact lenses. If you must wear contact lenses, notify your teacher and wear safety goggles to prevent accidents or loss of a contact lens.

Lab Safety Procedures

When working in the lab, follow these safety rules:

1. Understand the correct laboratory procedure to be used and be aware of possible hazards. Perform only those lab activities assigned and explained by your teacher. Listen carefully to your teacher's instructions and follow them exactly.

2. Keep work areas clean and neat at all times. Only textbooks, lab manuals, or notebooks should be in the work area. Allow enough time to clean and dry all work areas before the end of the lab session. Wash your hands thoroughly after the lab is completed and the area cleaned.
3. Report all accidents or unusual occurrences to your teacher immediately.

Safety Symbols

You should recognize the following safety symbols and know what they mean. **In the event of an accident or unusual occurrence, remember to report it to your teacher immediately.**

Safety Clothing Wear a lab apron or lab coat whenever working with materials that could damage your skin or clothing. For some activities, you may also need protective gloves. Wear safety clothing at all times during a specified laboratory activity.

Safety Goggles Use safety goggles in any lab activity involving chemicals, flames, or the possibility of broken glassware. Wear goggles at all times during these activities. Place them over eyeglasses to protect the sides of the eyes.

Fire Safety Tie hair back, roll up sleeves, and remove or secure loose clothing, such as ties, scarves, or jackets, that may get too close to a flame. Keep all flammable items away from the area of a flame or from an electrical source or appliance. Follow your teacher's instructions on how to light a burner safely. If you use matches, soak them in water before throwing them away. Never use organic solvents, such as ethanol or acetone, in the same room as a burner. Turn burners completely off when you have finished with them. When heating a test tube, point the open end away from yourself and others. Do not heat material in a stoppered test tube. If you do get burned, cool the area with cold (not ice) water and cover with a sterile dressing. If the burn results from chemicals, rinse the burned area under cold running water for 20 minutes. If the burn breaks the skin, if it is on the hands or face, or if it covers a large area, seek medical care. Do not break any blisters that may develop.

Corrosive Substance Avoid letting any corrosive substance, such as a strong acid or base, touch your skin, eyes, or clothing. Wear protective clothing and goggles. Do not inhale vapors. Use a fume hood if vapors are dangerous. Never smell or taste chemicals in the laboratory. When mixing a strong acid with water, pour the acid into the water to keep it from splattering. Never pour the water into the acid. When labeling the contents of a container, label the empty container before adding the material. Dispose of all hazardous chemicals in a toxic-waste-disposal container. If a corrosive substance spills on you, immediately rinse the area under cold running water for 20 minutes. Get medical help if needed.

Breakage Handle breakable equipment, such as glassware, carefully. Do not use cracked, chipped, scored, or badly scratched glassware. Always lubricate glassware, such as tubing, with water or glycerin before trying to insert it into a stopper. Wear protective gloves and use a twisting motion when inserting lubricated glassware into a stopper. Do not plunge hot glassware into cold water or let empty glassware get too hot. Never leave any glassware unattended while it is being heated. Place hot glassware on an insulated surface. Remember that glass may remain hot for quite a while after it has been removed from a heat source, even though it may not look hot; handle it with caution. If glassware breaks, clean up the pieces immediately but not with your bare hands. Throw away the pieces in a special container marked "broken glass." If you are cut, rinse the area with soap and water and apply a sterile dressing. Use direct pressure and elevate the injured part to stop the bleeding. If the bleeding cannot be controlled, get medical help.

Hand Safety Dissect specimens in dissecting pans and never while holding the specimen in your hands. Make sure the specimen is pinned down firmly in a dissecting tray before beginning work. Direct sharp-edged instruments away from yourself and others. Use scissors instead of a scalpel whenever possible. Do not use too much force when working with a sharp instrument, such as a scalpel. Cut in a direction away from yourself and others. Do not use chipped or cracked glassware. Use protective gloves when working with hot materials or corrosive substances. If you cut yourself, wash the wound thoroughly with soap and water, control the bleeding, and place a sterile dressing over the wound. If you are burned, cool the burned area with cold (not ice) water and cover with a sterile dressing. If you cannot control the bleeding, or if the burn is large, deep, or on the face or hands, seek medical attention.

Dangerous Vapors Use a fume hood in the presence of dangerous vapors and avoid inhaling them. When testing an odor, use a wafting motion of your hand to direct the vapor toward your nose. Never directly inhale laboratory vapors. If you begin to get a headache or feel dizzy or weak, leave the room or open a window to get fresh air.

Explosion Danger Avoid the danger of explosion by studying laboratory instructions and following them exactly. Never look into the top of a container that is being heated. Do not bring any substance into contact with a flame unless instructed to do so. When heating a substance, point the mouth of the test tube away from yourself and others.

Poison Do not let a poisonous substance come into contact with your skin, do not swallow any of the substance, and do not breathe any of its vapors. If you accidentally come into contact with a poison, have someone call the local poison control center immediately, and follow their instructions exactly.

Electrical Shock Avoid electrical shock by reading laboratory instructions carefully and following them exactly. Disconnect all electrical equipment when not in use. Do not use any electrical equipment around water or when the equipment is wet.

Biohazard Assume that any microorganisms, such as bacteria, might be hazardous. Treat microorganisms as directed by your teacher. Do not use preserved specimens showing any signs of decay for any type of lab observation or dissection. Avoid direct contact with a culture of microorganisms, and wash your hands thoroughly with soap and water after working with microorganisms.

Plant Safety Use a reliable field guide when collecting plants and fungi outdoors. Never eat any part of an unknown plant or fungus. Know how to recognize the poisonous plants and fungi in your area and avoid contact with them.

Animal Safety Before handling an animal, get professional advice about how to do so correctly. If you are bitten or scratched, wash the area thoroughly with soap and water, control the bleeding, and place a sterile dressing on the wound. Check with a physician to see whether you need a tetanus booster, especially if you have received a puncture wound. If you think the animal that injured you was poisonous, or if you show signs of an allergic reaction or infection, get medical attention immediately.

Animal Welfare Treat all animals in as humane a way as possible. Handle animals gently to avoid producing undue excitement or trauma. Avoid subjecting animals to stressful conditions, such as exhausting exercise or painful stimuli. Understand the purpose and procedure of an activity involving live animals before you begin work.

Hazardous Waste Disposal Dispose of all potentially hazardous materials as directed by your teacher. Do not place hazardous materials in the wastebasket or wash them down the sink.

Appendix C: Humane Treatment of Animals

Some of the most exciting parts of a biology course are those that deal with living animals. Some classrooms have a permanent collection of living animals. Others use living animals only occasionally as part of laboratory activities. It is the moral and ethical responsibility of the teacher and students to give adequate care to all living animals and to treat them humanely. In all your dealings with living animals, you should show a respect for life.

To insure humane treatment of animals in the laboratory or classroom, follow these steps:

1. Be sure that a proper environment can be created and maintained for an animal before it is brought into the classroom. Research the animal's needs and make sure you can meet them.
2. In general, avoid bringing wild animals into the classroom. Most wild animals cannot adapt to a classroom or laboratory setting. Exceptions are small animals, such as earthworms, spiders, insects, fish, and some amphibians. Do not collect any animals that are members of an endangered species. Do not collect animals that are sick or dying. Handle wild animals as infrequently as possible. After a period of observation, return the animals to their natural habitats.
3. Obtain healthy animals from pet stores or biological supply houses.

4. Make sure you know how to treat each kind of animal. Get directions from your teacher before handling a live animal. You need to know how to handle an animal properly in order to prevent injury to it and to yourself. Do not make loud noises, tap on the animals' cages or containers, or otherwise disturb or frighten the animals.
5. Keep records of who is feeding the animals and how much and how often they are fed. Make sure that the animals' nutritional needs are met but not exceeded.
6. Make sure that the areas in which the animals are housed are cleaned regularly.
7. Arrange for the animals to be cared for when school is not in session, such as during weekends and vacations.
8. Do not cause any animal pain or injury. Experiments with live animals should have humane objectives and be closely supervised by the teacher. Follow your teacher's instructions carefully.
9. **Discuss your attitudes and feelings about dissection of preserved animals with your teacher. If you are strongly opposed to dissecting animals, alternatives are available to you. Your textbook and lab manual present these alternatives. Some of them are: (1) computer simulations, (2) anatomical models and charts, (3) laser discs, (4) films, (5) filmstrips, (6) transparencies, (7) videotapes, and (8) books.**

Appendix D: Careers in Biology

Biology offers a wide range of different careers in which a variety of interests may be pursued. Some jobs in biology-related fields require only on-the-job training, while others require up to eight years of college and post-college education. Below is a list of some biology-related careers along with a brief description of each and the address of organizations providing more information. Further information can be found in the "Occupational Outlook Handbook" published by the U.S. Department of Labor. Your school guidance counselor may also be able to provide you with information.

Careers Requiring a High School Diploma or a Two-Year Degree

Animal lab assistants care for animals treated by veterinarians or used in biomedical research. They feed the animals, maintain records on them, collect specimens, perform laboratory procedures, and sometimes assist in breeding and medical procedures. **WHERE TO WRITE:** American Veterinary Medical Association, 930 N. Meacham Rd., Schaumburg, IL 60196.

Biomedical engineer technicians install, operate, repair, and maintain electronic biomedical equipment used in the diagnosis and treatment of diseases. They also teach health workers how to operate the instruments. **WHERE TO WRITE:** Biomedical Engineering Society, P.O. Box 2399, Culver City, CA 90230.

Biomedical photographers provide the photographs that illustrate magazines, books, articles, and teaching materials in the biomedical sciences. Their work may range from photographing bacteria under a microscope to photographing whales in the ocean. **WHERE TO WRITE:** Professional Photographers of America, 1090 Express Way, Des Plaines, IL 60018.

Dental technicians construct a variety of dental appliances, such as dentures, bridges, crowns, and inlays, based on a dentist's instructions. **WHERE TO WRITE:** American Dental Association, 211 E. Chicago Ave., Chicago, IL 60611.

Emergency medical technicians are trained to deal with medical emergencies. Arriving at the scene of an emergency in an ambulance, they treat patients in shock, perform cardiopulmonary resuscitation, control bleeding, and administer other emergency care until the patient can be given hospital care. **WHERE TO WRITE:** National Registry of Emergency Medical Technicians, 1395 E. Dublin-Granville Rd., Columbus, OH 43229.

Fish culturists supervise the day-to-day operations of fish hatcheries, under the direction of fish biologists. They monitor the breeding, development, feeding, and health of the fish, and oversee the crews that maintain hatchery equipment. **WHERE TO WRITE:** American Fisheries Society, 5410 Grosvenor Lane, Bethesda, MD 20014.

Floral designers design, create, and often sell ornamental flower arrangements. They help customers choose the appropriate flowers for a particular occasion, and some grow the plants and flowers they use. **WHERE TO WRITE:** American Society for Horticultural Science, 701 North Saint Asaph St., Alexandria, VA 22314.

Histological technicians work in hospital pathology laboratories. They prepare tissue specimens for examination by a pathologist using methods such as staining, mounting on slides, and dehydration. **WHERE TO WRITE:** National Society for Histotechnicians, P.O. Box 36, Lanham, MD 20706.

Medical assistants help physicians manage their offices efficiently. They function as receptionists, administrators, and secretaries and assist physicians with patient examinations. **WHERE TO WRITE:** American Medical Assistants, Inc., 1 E. Wacker Dr., Chicago, IL 60601.

Medical laboratory assistants work under the supervision of a medical technologist in a clinical laboratory. They perform routine laboratory procedures on specimens of body tissue and fluids to aid in the diagnosis and treatment of disease. **WHERE TO WRITE:** American Society for Medical Technology, 330 Meadowfern Dr., Houston, TX 77067.

Medical laboratory technicians conduct a wide range of routine and specialized laboratory tests to help physicians diagnose and treat disease. They prepare and analyze specimens of body fluids and tissues under the supervision of a laboratory supervisor. **WHERE TO WRITE:** American Medical Technologists, 710 Higgins Rd., Park Ridge, IL 60068.

Museum technicians help museum curators and others prepare, construct, and maintain exhibits. They may also assist in preserving specimens and interpreting historical research. **WHERE TO WRITE:** American Association of Museums, 1055 Thomas Jefferson St. NW, Washington, D.C. 20007.

X-ray technologists prepare patients for x rays, operate x-ray equipment, and prepare radioactive materials. They assist physicians in performing other radiologic procedures used for diagnosing medical conditions or treating specific diseases. **WHERE TO WRITE:** American Society of Radiologic Technologists, 1500 Central Ave. SE, Albuquerque, NM 87123.

Careers Requiring a Bachelor's Degree

Dental hygienists provide preventive dental services by cleaning teeth, applying compounds that prevent decay, and instructing patients in proper oral hygiene. In addition, they may assist dentists by taking patient histories and preparing diagnostic aids. **WHERE TO WRITE:** American Dental Hygiene Association, 444 N. Michigan Ave., Chicago, IL 60611.

Foresters develop and manage forests for their many uses, from timber production to recreation. They plan and supervise the growth and harvesting of trees, and they protect forests from fires, floods, insects, and disease. **WHERE TO WRITE:** Society of American Foresters, 5400 Grosvenor Lane, Bethesda, MD 20814.

Horticulturists cultivate orchard and garden plants. They breed, grow, and distribute flowers, fruits, vegetables, trees, and bushes. Using a knowledge of genetics, they develop new plants with such traits as high crop yield or physical beauty. **WHERE TO WRITE:** American Society for Horticultural Science, 701 North Saint Asaph St., Alexandria, VA 22314.

Medical illustrators draw and diagram medical instruments, procedures, and anatomical structures for textbooks and medical supply companies. **WHERE TO WRITE:** Association of Medical Illustrators, 2692 Huguenot Springs Rd., Midlothian, VA 23113.

Pharmacists prepare and dispense medications prescribed by physicians and dentists. They must know the compositions and interactions of drugs to predict the effects of a particular medication. They also advise customers on medications and health aids. **WHERE TO WRITE:** American Pharmaceutical Association, 2215 Constitution Ave., Washington, D.C. 20037.

Physician's assistants examine and treat patients under the guidance of a physician. They collect patient histories, perform examinations, give injections, treat wounds, perform laboratory tests, and assist physicians as needed. **WHERE TO WRITE:** American Academy of Physician's Assistants, 2341 Jefferson Davis Highway, Suite 700, Arlington, VA 22202.

Registered nurses plan, administer, and supervise patient nursing care. They monitor and record patients' symptoms and progress and give medications ordered by physicians. They may also supervise nursing assistants and teach. **WHERE TO WRITE:** National League for Nursing, 10 Columbus Circle, New York, NY 10019.

Careers Requiring a Master's or Doctorate Degree (Ph.D.)

Biochemists study the chemistry of organisms to determine what compounds make up living things and how these compounds interact in processes such as metabolism, growth, and reproduction. Some study the effects of foods, drugs, and toxins, while others develop methods for diagnosing and treating disease. **WHERE TO WRITE:** American Society of Biological Chemists, 9650 Rockville Pike, Bethesda, MD 20014.

Dentists examine, diagnose, and treat patients with diseases, injuries, and other problems related to their teeth and mouths. They restore and repair defective teeth and replace missing ones. **WHERE TO WRITE:** American Dental Association, 211 E. Chicago Ave., Chicago, IL 60611.

Geneticists study the process of inheritance and the structures and chemicals involved. Some geneticists do applied research and study genetic diseases and plant and animal breeding, while others do basic research and study biological processes at the molecular level. **WHERE TO WRITE:** Genetics Society of America, 9650 Rockville Pike, Bethesda, MD 20814.

Microbiologists study the growth, structure, and functions of bacteria, viruses, and other microorganisms using sophisticated laboratory tools. Some specialize in identifying and classifying the microorganisms in patients' fluid and tissue specimens. **WHERE TO WRITE:** American Society of Microbiologists, 1913 I St. NW, Washington, D.C. 20006.

Pathologists are specialized physicians who interpret the nature of diseases and the changes they produce in the body. They examine tissues and body fluids to determine the levels of various biochemicals, the presence of certain types of cells and infectious organisms, and the status of the immune system. **WHERE TO WRITE:** American Medical Association, 535 N. Dearborn St., Chicago, IL 60610.

Physicians are responsible for the overall health care of patients. They examine, diagnose, and treat patients, perform routine and emergency medical procedures, and advise patients on ways to stay healthy. **WHERE TO WRITE:** American Medical Association, 535 N. Dearborn St., Chicago, IL 60610.

Veterinarians examine, diagnose, and treat animals with diseases and other medical conditions. They perform surgery and laboratory tests, and administer medications and immunizations. Some specialize in the care of small domestic animals, while others specialize in the care of large farm animals. **WHERE TO WRITE:** American Veterinary Medical Association, 930 N. Meacham Rd., Schaumburg, IL 60196.

Appendix E: The SI System of Measurement

The abbreviation SI refers to the International Metric System (formally, Système Internationale d'Unités). This widely used system of measurement is conveniently based on units of 10. Tables of SI base units, their multiples and submultiples, and selected Metric/English conversions are presented below. Also included are notes on temperature and temperature conversions, and heat energy.

SI Base Units			
To Measure	**SI Base Unit Used**	**Common Multiples and Submultiples**	**Approximate Size**
Length	meter (m)	meter = 1 m kilometer 1 km = 1000 m centimeter 1 cm = 0.01 m millimeter 1 mm = 0.001 m micron 1 μm = 10^{-6} m (micrometer) nanometer 1 nm = 10^{-9} m	1 m—height of a doorknob 1 km—length of 5 city blocks 1 cm—width of a paper clip 1 mm—thickness of a dime 1 μm—diameter of a bacterium 1 nm—length of an amino acid
Mass	gram (g)	gram = 1 g kilogram 1 kg = 1000 g milligram 1 mg = 0.001 g	1 g—mass of a paper clip 1 kg—mass of a pair of adult shoes 1 mg—mass of a grain of rice
Volume	liter (L)	liter = 1 L milliliter 1 mL = 0.001 L cubic centimeter 1 cm³ (also cc) $\approx$ 1 mL	1 L—capacity of a water pitcher 1 mL (1 cm³)—capacity of an eye dropper

Metric Prefixes			
Prefix	**Symbol**	**Meaning**	**Multiples and Submultiples**
atto-	a-	quintillionth	10^{-18}
femto-	f-	quadrillionth	10^{-15}
pico-	p-	trillionth	10^{-12}
nano-	n-	billionth	10^{-9}
micro-	μ-	millionth	10^{-6}
milli-	m-	thousandth	10^{-3}
centi-	c-	hundredth	10^{-2}
deci-	d-	tenth	10^{-1}
deka-	dk-, da-	ten	10
hecto-	h-	hundred	10^{2}
kilo-	k-	thousand	10^{3}
mega-	M-	million	10^{6}
giga-	G-	billion	10^{9}
tera-	T-	trillion	10^{12}

Some Metric/English Conversions		
To Obtain	**Multiply**	**By**
Feet	Meters	3.2808
Miles	Kilometers	0.6214
Ounces	Grams	0.0353
Pounds	Kilograms	2.2046
Liquid quarts	Liters	1.0567
Fluid ounces	Milliliters	0.0338
Meters	Feet	0.3048
Kilometers	Miles	1.6093
Grams	Ounces	28.3495
Kilograms	Pounds	0.4536
Liters	Liquid quarts	0.9463
Milliliters	Fluid ounces	29.5735

Temperature

The SI base unit for temperature is the kelvin (K). However, as a matter of convenience, the degree Celsius (°C) is more commonly used.

Some Sample Temperatures			
Scale	**Water Freezes**	**Water Boils**	**Range**
K	273	373	100
°C	0	100	100
°F	32	212	180

Heat Energy

1 calorie (cal) = The amount of heat needed to raise the temperature of 1 g of water 1 °C (also called the small calorie or gram calorie)
1 cal = 4.2 joules

1 Calorie (Cal) = 1 kilocalorie, or 1000 cal (also called the large calorie or kilogram calorie)
1 Cal = 4184 joules

Temperature Conversion Formulas
°C = K − 273   K = °C + 273
°C = 5/9(°F − 32)   °F = (9/5)°C + 32
°F = 9/5(K − 273) + 32   K= 5/9 (°F − 32) + 273

Appendix F: Classification of Organisms

Kingdom Monera

Monerans are the simplest organisms. Most are single-celled, although some form colonies. Their cells are prokaryotic, so they lack nuclear membranes, mitochondria, chloroplasts, and other membranous organelles. Reproduction is mainly asexual by fission.

Subkingdom Archaebacteria
Archaebacteria live in harsh environments that other organisms cannot tolerate. They are distinguished from other monerans by the chemical makeup of their cell walls and cell membranes and the unique structure of their RNA.

Phylum Methanogenic Bacteria These archaebacteria break down organic matter and produce methane gas. They are found in the intestinal tracts of animals and in marshes, swamps, and sewage treatment plants.

Phylum Halophilic Bacteria These are salt-loving archaebacteria. They are found mainly in salt lakes.

Phylum Thermoacidophilic Bacteria These archaebacteria live in very hot, acidic environments.

Subkingdom Eubacteria
Eubacteria include all bacteria except the archaebacteria. Most eubacteria are heterotrophic; the rest—the autotrophic eubacteria—are either phototrophs or chemotrophs.

Phylum Anaerobic Phototrophic Bacteria These bacteria carry out photosynthesis without using water as a starting material and without producing oxygen. This phylum includes green-sulfur bacteria and purple bacteria.

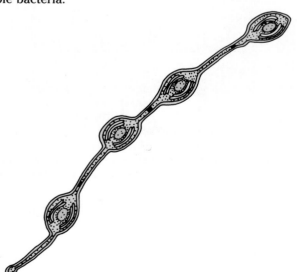

Phylum Cyanobacteria Members of this phylum, the *blue-green bacteria,* carry out photosynthesis using chlorophyll *a* and the blue pigment phycocyanin. Cyanobacteria include both unicellular and colonial species; some perform nitrogen fixation. **EXAMPLE:** *Nostoc.*

Phylum Prochlorophyta These bacteria carry out photosynthesis with both chlorophyll *a* and *b.* They live in association with certain marine animals. **EXAMPLE:** *Prochloron.*

Phylum Schizophyta This is a diverse group of heterotrophic eubacteria that includes both saprobes and parasites, and both aerobic and anaerobic forms. Many cause diseases. **EXAMPLES:** *E. coli, Rickettsia, Streptococcus, Chlamydia, Mycoplasma, Treponema, Actinomyces.*

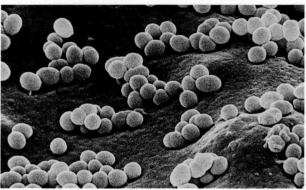

Kingdom Protista

Protists are unicellular, colonial, and multicellular organisms. The cells of protists are eukaryotic with a membrane-bound nucleus and other organelles. Protists may be animal-like, plantlike, or funguslike.

Phylum Sarcodina The unicellular, animal-like *sarcodines* move and capture prey by means of pseudopods. They are found in both fresh water and salt water and in the bodies of animals. A few cause disease. Reproduction is both asexual and sexual. Of the sarcodines, amebas are surrounded only by a cell membrane, while radiolarians and forams have protective shells. **EXAMPLES:** *Ameba, Entamoeba, Globigerina, Pelomyxa.*

Phylum Ciliophora *Ciliates* are complex, animal-like protists that move by means of cilia. Their cells contain a macronucleus and a micronucleus. They are found in both fresh water and salt water. Reproduction is both asexual and sexual. **EXAMPLES:** *Paramecium, Stentor, Tetrahymena, Vorticella.*

Phylum Zoomastigina The *zooflagellates* move by means of one or more flagella. Most are unicellular and live inside animals or plants. A few are free-living in fresh water. Reproduction is both asexual and sexual. **EXAMPLES:** *Trypanosoma, Trichonympha.*

Phylum Sporozoa The *sporozoans* are nonmotile, spore-producing protists. They are parasites with a complex life cycle that includes more than one kind of host. **EXAMPLES:** *Plasmodium, Toxoplasma.*

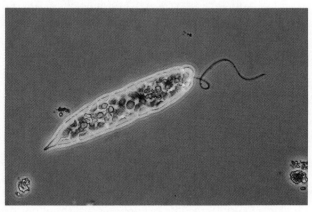

Phylum Euglenophyta The *euglenoids* are unicellular, photosynthetic protists. They do not have cell walls but have pellicles, or flexible protein coverings. They move by means of flagella and are found mainly in fresh water. Reproduction is asexual. **EXAMPLE:** *Euglena.*

Phylum Chrysophyta The *golden algae* are a diverse group of photosynthetic protists that contain yellow-brown pigments and store food as oils or starchlike carbohydrates. They are mostly unicellular; some are colonial. They are found in both fresh water and salt water. The most numerous kind, the diatoms, have shells made of silica. Reproduction is both asexual and sexual. **EXAMPLES:** *Botrydium, Chrysameba, Pinnularia.*

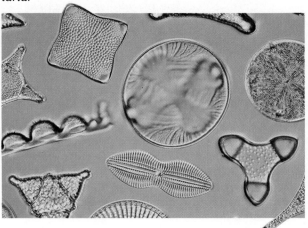

Phylum Dinoflagellata The *dinoflagellates* are unicellular algae with two flagella and armorlike cell walls containing cellulose and silica. Some are photosynthetic; others are heterotrophic. Most live in salt water. Reproduction is asexual. **EXAMPLES:** *Gonyaulax, Gymnodinium.*

Phylum Chlorophyta The *green algae* are unicellular, colonial, or multicellular protists. Some have flagella. Most contain chlorophylls *a* and *b* and cell walls made of cellulose. They are found in salt water and fresh water and in moist places on land. **EXAMPLES:** *Ulothrix, Chlamydomonas, Chlorella, Volvox, Spirogyra, Ulva.*

Phylum Phaeophyta The multicellular, photosynthetic *brown algae* include many common seaweeds found in ocean waters. The brown algae contain a brown photosynthetic pigment in addition to chlorophyll. They have cellulose in their cell walls and store food in the form of polysaccharides or oils. Their life cycles show alternation of generations. **EXAMPLES:** *Sargassum, Macrocystis, Fucus, Laminaria.*

Phylum Rhodophyta The *red algae* are mostly multicellular, marine protists that include many common seaweeds. They contain chlorophyll as well as other pigments. They have complex life cycles, including alternation of generations. **EXAMPLES:** *Porphyra, Polysiphonia, Chondrus.*

Phylum Myxomycota *Acellular slime molds* are unicellular organisms with many nuclei. They are found on damp soil, on rotting logs, and in leaf litter. Reproduction is sexual. The life cycle includes an ameboid plasmodium and spore-producing fruiting bodies. **EXAMPLE:** *Physarum.*

Phylum Acrasiomycota The *cellular slime molds* are found in fresh water, on damp soil, or on decaying vegetation. Reproduction is asexual. The life cycle includes a pseudoplasmodium, or mass of many individual, membrane-bound cells. The pseudoplasmodium can form fruiting bodies that produce haploid spores. Their feeding stage consists of independent ameboid cells. **EXAMPLE:** *Dictyostelium.*

Phylum Oomycota *Water molds* and *downy mildews* consist of finely branched, single-celled filaments. Most water molds are saprobes that live in fresh water. Some are parasites of fish. Their cell walls contain cellulose, and they have a complex life cycle dominated by a diploid stage. The downy mildews are plant parasites. **EXAMPLE:** *Saprolegnia.*

Kingdom Fungi

Fungi are eukaryotic, heterotrophic organisms that absorb nutrients from their environment. A few are unicellular, but most are multicellular and are made up of masses of threadlike filaments, hyphae. In most, the cell walls are made up of chitin. Reproduction is either asexual or sexual by means of spores.

Phylum Zygomycota *Conjugation fungi* reproduce sexually by conjugation. The also reproduce asexually. Their hyphae lack cross walls. Most of these fungi are saprobes; some are parasites. **EXAMPLES:** *Rhizopus, Phycomyces.*

Phylum Ascomycota The *sac fungi* are the largest group of fungi. They reproduce sexually by ascospores, which develop inside the ascus, a saclike structure. They also reproduce asexually by conidia. Except for unicellular yeasts, sac fungi are multicellular with incomplete cross walls dividing the hyphae. This group includes cup fungi, yeasts, powdery mildews, truffles, morels, and blue and green molds. **EXAMPLES:** *Saccharomyces, Neurospora.*

Phylum Basidiomycota *Club fungi* have a clublike reproductive structure called a basidium. Club fungi may reproduce either asexually, or sexually by basidiospores. Incomplete cross walls divide the hyphae of club fungi. This group includes mushrooms, bracket fungi, puffballs, rusts, and smuts. **EXAMPLES:** *Amanita, Lycoperdon, Phragmidium.*

Phylum Deuteromycota In the *Fungi Imperfecti,* the pattern of sexual reproduction is unknown. They can reproduce asexually by conidia, however. Many species are beneficial to humans. A few cause diseases, such as athlete's foot and candidiasis. **EXAMPLES:** *Trichophyton, Candida, Aspergillus, Penicillium.*

Kingdom Plantae

Plants are generally nonmotile, multicellular, photosynthetic organisms. Their cells contain plastids and are surrounded by cell walls. Chlorophylls and carotenoids are present in chloroplasts. Most plants have specialized tissues and organs. Reproduction may be asexual or sexual. Sexual reproduction in plants involves an alternation of generations with multicellular diploid (sporophyte) and multicellular haploid (gametophyte) stages.

Division Bryophyta The bryophytes are small multicellular plants that lack xylem and phloem, and do not have true leaves, stems, or roots. Bryophytes are found usually in moist areas, and water is needed for fertilization. The gametophyte generation is dominant; the sporophyte generation is reduced in size and is dependent on the gametophyte. Dispersal is by means of spores.

Class Musci In mosses, the gametophyte generation consists of small, erect plants that have tiny leaf-like structures arranged spirally around a stalk. **EXAMPLES:** *Sphagnum, Polytrichum.*

Class Hepaticae In liverworts, the gametophytes consist of flattened or leaflike structures. **EXAMPLES:** *Marchantia, Riccia.*

Class *Antherocerotae* In hornworts, the gametophyte is thalluslike; the sporophyte is cylindrical.
EXAMPLE: *Anthoceros.*

Division Tracheophyta The vascular plants contain
xylem and phloem and have true leaves, stems, and
roots. The sporophyte generation is dominant, and
the gametophyte generation is greatly reduced. Chlorophylls *a* and *b* are present, and food is stored as
starch in plastids. Tracheophytes include most modern-day plants.

Subdivision Psilophyta In whisk ferns, the highly
branched vascular stems lack true leaves and roots.
Whisk ferns are rare plants that are found mainly in
warm regions. **EXAMPLE:** *Psilotum.*

Subdivision Lycophyta Club mosses have small
leaves arranged in a spiral. Spores are found at the
tips of some branches in conelike strobili. **EXAMPLES:**
Lycopodium, Selaginella.

Subdivision Sphenophyta Horsetails have small
leaves arranged in whorls at specific points along the
stems. Conelike strobili form at the ends of some
stems. **EXAMPLE:** *Equisetum.*

Subdivision Pterophyta This group includes the
ferns, gymnosperms, and angiosperms. These plants
have relatively large expanded leaves or are derived
from plants with leaves of that type.

Class *Filicineae* Ferns have fronds that grow up from
horizontal underground stems. Sori are found on the
backs of the leaflets of some fronds. The independent
gametophyte generation consists of a small prothallus. The sperm are motile, and water is required for
fertilization. **EXAMPLES:** *Polypodium, Dryopteris, Os-
munda.*

Class *Gymnospermae* Gymnosperms are the plants
that bear seeds not enclosed in a fruit. The tiny
gametophytes grow on the dominant sporophyte.
Sperm are enclosed in a pollen tube, so water is not
needed for fertilization.

   Subclass Coniferophyta Conifers, or evergreens, are
   cone-bearing trees with needlelike or scalelike leaves.
   EXAMPLES: pines *(Pinus)*, spruce *(Picea)*, hemlocks
   (Tsuga), firs *(Abies)*, redwoods *(Sequoia).*

   Subclass Cycadophyta Cycads are tropical, palmlike
   gymnosperms. **EXAMPLES:** *Cycas, Zamia.*

   Subclass Ginkgophyta Ginkgo, or maidenhair, trees,
   have fan-shaped leaves. The only surviving species is
   Ginkgo biloba.

Class *Angiospermae* Angiosperms are flowering
plants whose seeds are enclosed in an ovary that
ripens into a fruit. The tiny gametophytes grow on the
dominant sporophyte. Sperm are enclosed in a pollen
tube so that water is not needed for fertilization.

Subclass Monocotyledonae Monocots have an embryo
with a single cotyledon, leaves with parallel veins,
flower parts in threes or sixes, and vascular bundles
scattered throughout the stem tissue. They are primarily herbaceous plants. **EXAMPLES:** grasses, including rye *(Secale)*, corn *(Zea)*, and wheat *(Triticum)*;
lilies *(Lilium)*, tulips *(Tulipa)*; orchids *(Orchis).*

Subclass Dicotyledonae Dicots have an embryo with
two cotyledons, veins of leaves in the form of a
network, flower parts in fours or fives, and vascular
tissue organized in a concentric ring. They include
both herbaceous and woody plants. **EXAMPLES:** oaks
(Quercus, Lithocarpus), maples *(Acer)*, magnolias
(Magnolia), cucumbers *(Cucumis)*, carrots *(Daucus)*,
roses *(Rosa).*

Kingdom Animalia

Animals are multicellular, heterotrophic organisms
with specialized tissues. Most are motile. Their cells
are eukaryotic and lack cell walls. Reproduction is
mainly sexual.

Phylum Porifera *Sponges* are sessile, aquatic animals; most are marine. Their asymmetrical bodies
have two cell layers and are pierced by pores; they are
stiffened by skeletal elements called spicules. Reproduction is both asexual and sexual. **EXAMPLES:** *Gran-
tia, Scypha, Euplectella.*

Phylum Coelenterata In *coelenterates*, the radially
symmetrical body is saclike and is made up of two cell
layers. There are two body forms—the polyp and the
medusa. The digestive cavity has a single opening
surrounded by tentacles containing stinging cells.
Coelenterates are all aquatic; most are marine. Some
coelenterates, such as corals, are colonial. Reproduction is sexual in the medusa stage and asexual in the
polyp stage. **EXAMPLES:** hydra *(Hydra)*, jellyfish *(Aurelia, Obelia, Physalia)*, corals *(Gorgonia)*, sea anemones *(Actinia).*

Phylum Platyhelminthes The *flatworms* have flattened, bilaterally symmetrical bodies made up of three tissue layers. The digestive system has only one opening.

Class Turbellaria These are free-living flatworms with eyespots. EXAMPLES: *Planaria, Dugesia.*

Class Trematoda Flukes are parasitic flatworms, usually with suckers. EXAMPLES: *Schistosoma, Fasciola.*

Class Cestoda Tapeworms are parasitic flatworms without digestive systems. EXAMPLE: *Taenia.*

Phylum Nematoda Roundworms have long, cylindrical bodies; there is a digestive system with both a mouth and an anus. Most are parasitic. EXAMPLES: *Ascaris, Necatur, Trichinella.*

Phylum Annelida Segmented worms have a body made up of many similar segments. They have a well-developed coelom, a complete digestive tract, a closed circulatory system, and a ventral nervous system.

Class Polychaeta These include mostly marine worms. They have a well-developed head. EXAMPLE: *Nereis.*

Class Oligochaeta This class includes terrestrial and aquatic worms. Members of this group, including earthworms, have poorly developed heads. EXAMPLES: *Lumbricus, Tubifex.*

Class Hirudinea Leeches are parasitic annelids with suckers at one or both ends of the body. EXAMPLE: *Hirudo.*

Phylum Mollusca Mollusks have a soft, unsegmented body, often with a muscular foot, mantle, and radula. The digestive, circulatory, and nervous systems are well developed.

Class Bivalvia Bivalves include clams, oysters, mussels, and scallops. These mollusks have a two-part, hinged shell and no head or radula. EXAMPLES: *Mytilus, Pecten, Teredo.*

Class Gastropoda Gastropods include snails, slugs, and whelks. These mollusks have a head with tentacles; most have a spiral shell. EXAMPLES: *Limax, Helix, Busycon.*

Class Cephalopoda Cephalopods have a large head surrounded by arms, or tentacles. The octopus has no shell, the squid has an internal shell, and the nautilus has an external shell. The nervous system is particularly well-developed. EXAMPLES: octopus *(Octopus),* squid *(Loligo),* nautilus *(Nautilus).*

Phylum Arthropoda Arthropods have a segmented body with paired, jointed appendages and an exoskeleton composed of chitin.

Class Crustacea Crustaceans have two pairs of antennae. Most are aquatic and respiration is by gills. EXAMPLES: lobsters *(Homarus),* crabs *(Cancer),* crayfish *(Cambarus),* water fleas *(Cyclops, Daphnia).*

Class Chilopoda Centipedes have one pair of antennae, many body segments, and one pair of legs on most body segments. EXAMPLE: *Scolopendra.*

Class Diplopoda Millipedes have one pair of antennae, many body segments, and two pairs of legs on most body segments. EXAMPLE: *Glomeris.*

Class Arachnida Arachnids have no antennae, two body regions, four pairs of legs, and book lungs. EXAMPLES: spiders *(Argiope),* scorpions *(Chelifer).*

Class Insecta Insects have one pair of antennae, three body regions, three pairs of legs, and tracheal respiration; many have two pairs of wings. Group includes flies, ants, beetles, fleas, lice, bees, and roaches. (See pages 722–723 for a table of insect classification.)

Phylum Echinodermata Echinoderms have a water-vascular system, an internal skeleton, and a spiny skin. Adults are radially symmetrical. All are marine. **EXAMPLES:** starfish *(Asterias),* sea urchins *(Arbacia),* sea cucumbers *(Cucumaria),* sand dollars *(Echinarachnius).*

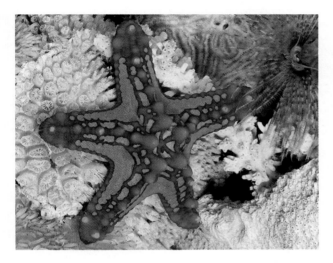

Phylum Chordata At some stage of development chordates have a notochord, paired gill slits, and a dorsal, hollow nerve cord.

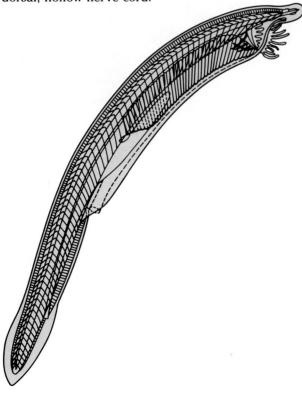

Subphylum Urochordata Adult tunicates are soft, saclike, sessile, marine animals. Larvae are free-swimming. **EXAMPLES:** *Ciona, Appendicularia.*

Subphylum Cephalochordata Lancelets are small, fishlike, marine animals. A notochord is present in adults. They have prominent gill slits. **EXAMPLE:** *Branchiostoma.*

Subphylum Vertebrata Vertebrates have an enlarged brain and a spinal column made up of vertebrae that enclose the dorsal nerve cord.

Class **Cephalaspidomorphi** The lampreys have skeletons made of cartilage, snakelike bodies, and no scales. They lack true jaws and retain their notochord throughout their life. **EXAMPLE:** *Petromyzon.*

Class Myxini The hagfishes, another type of jawless fish, are found only in salt water. They also have cartilaginous skeletons, snakelike bodies, and smooth skins without scales. **EXAMPLE:** *Myxine.*

Class **Chondrichthyes** The cartilaginous fishes have a skeleton composed of cartilage, movable jaws, scales, and fins. **EXAMPLES:** sharks *(Squalus),* skates *(Raja).*

Class Osteichthyes The bony fishes have a skeleton made of bone, movable jaws, overlapping scales, paired fins, and an air bladder. **EXAMPLES:** salmons and trouts *(Salmo),* carps *(Cyprinus),* perches *(Perca),* codfish *(Gadus).*

Class Amphibia Most amphibians have moist, smooth, scaleless skin and four limbs. Water is needed for reproduction. Aquatic larvae have gills and undergo metamorphosis. Adults are usually terrestrial and have lungs and a three-chambered heart. **EXAMPLES:** frogs *(Rana),* toads *(Bufo),* salamanders *(Necturus, Triturus).*

Class Reptilia The reptiles have dry skin and a scale-covered body with four limbs (absent in snakes). Fertilization is internal. Their eggs have a leathery shell and protective membranes. Most reptiles live and reproduce on land. They have lungs and a three-chambered heart with a partially divided ventricle. **EXAMPLES:** turtles *(Chelydra, Terrapene)*, crocodiles *(Crocodylus)*, alligators *(Alligator)*, snakes *(Crotalus)*.

Class Aves Birds have feathers and their front limbs are wings. Their eggs have a hard shell. They have a four-chambered heart and are warm-blooded. **EXAMPLES:** robins and thrushes *(Turdus)*, chickens *(Gallus)*, ducks *(Anas)*, sparrows *(Passer, Melospiza)*, starlings *(Sturnus)*.

Class Mammalia Mammals nourish their young with milk produced by mammary glands. Their body covering is hair or fur. They have a four-chambered heart and are warm-blooded.

Subclass Prototheria Monotremes are egg-laying mammals. **EXAMPLES:** duckbill platypus *(Ornithorhynchus)*, spiny anteater *(Tachyglossus)*.

Subclass Metatheria Marsupials are pouched mammals, found mainly in Australia. **EXAMPLES:** kangaroos *(Macropus)*, opossums *(Didelphis)*, koalas *(Phascolarctos)*.

Subclass Eutheria The placental mammals include most living mammals. Developing embryos receive nourishment from the mother's circulatory system by means of a structure called the placenta.

Order Insectivora moles *(Scalopus)*, shrews *(Sorex)*

Order Rodentia rats *(Rattus)*, mice *(Mus)*, squirrels *(Sciurus)*

Order Lagomorpha rabbits *(Sylvilagus)*, hares *(Lepus)*

Order Chiroptera bats *(Myotis)*

Order Pinnipedia walruses *(Odobenus)*, sea lions *(Zalophus, Otaria)*, seals *(Arctocephalus, Callorhinus, Phoca)*

Order Cetacea whales *(Balaena)*, dolphins *(Delphinus)*, porpoises *(Phocaena)*

Order Sirenia manatees *(Trichechus)*, dugongs *(Dugong)*

Order Edentata anteaters *(Myrmecophaga)*, armadillos *(Dasypus)*

Order Proboscidea elephants *(Elephas, Loxodonta)*

Order Artiodactyla camels *(Camelus)*, sheep *(Ovis)*, pigs *(Sus)*, cattle *(Bos)*

Order Perissodactyla horses *(Equus)*, rhinoceroses *(Rhinoceros)*

Order Carnivora cats *(Felis)*, dogs *(Canis)*, bears *(Ursus)*, raccoons *(Procyon)*

Order Primates humans *(Homo)*, chimpanzees *(Pan)*, orangutans *(Pongo)*, monkeys *(Macacus)*

Appendix G: Graphic Organization

In your study of biology, you will learn many new facts to increase your understanding of the world around you. But, more than simply a collection of facts, the science of biology is a dynamic field in which, each day, scientists combine new information with prior knowledge to gain new insights into the living world.

How do scientists do this? They train themselves to step back from the individual facts and look at the "big picture." Have you ever seen an aerial view of a city at rush hour? If so, then you understand how useful the "big picture" can be in making sense of something complex and ever-changing.

One way to develop a "big picture" of each topic that you study is to organize information in a visual fashion, using what is known as a **graphic organizer.** Think of a graphic organizer as a visual representation of a topic that emphasizes relationships between concepts or data. A good graphic organizer can show at a glance key information and relationships that are not apparent in a verbal description. By constructing and using graphic organizers, you will improve your ability to understand, summarize, and interrelate biological information. Some graphic organizers that you may find useful in this course are explained below.

Flow Chart

A biological process or sequence of events can be depicted in a flow chart like the one below.

Problem: To start a car

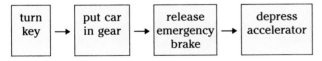

Flow charts can be used to outline the procedures of an experiment or to break down a complex process into its component steps.

Compare/Contrast Matrix

To compare and contrast two or more items, the graphic organizer shown in the next column can be used. Use this organizer to clarify differences between similar items and to analyze opposing viewpoints on a controversial issue.

Problem: How do compact disks compare with vinyl records?

Characteristic	Compact Disk	Record
cost	$15–$20	$10
sound quality	excellent	good
durability	excellent	fair
availability	good and improving	good, but decreasing

Word Map

When a new term is introduced, it may help to define it visually, in relation to other terms. This can be done in the following manner.

Problem: To define the term *bicycle*

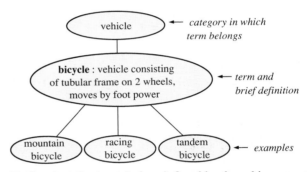

Notice that the term being defined is placed in an oval in the center of the map and a brief definition is given. The general category in which the term belongs is written above it, and specific examples are written below.

Depending on the term, a slightly different type of word map may be useful. In this type of word map, instead of including examples, components are listed in the third level of ovals, as shown below.

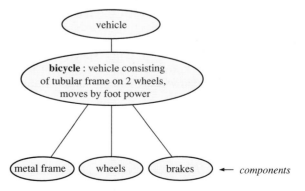

Word maps are useful for gaining a complete understanding of an unfamiliar term and for exploring how a term relates to both larger and smaller concepts.

Concept Map

A concept map is a useful graphic tool for organizing information on an entire topic. Concept maps emphasize relationships between the important concepts of a topic in a way that can be easily modified as you gather additional information.

A concept map is constructed by placing *concept words* (usually nouns) in ovals and connecting them with *linking words,* which are written along lines extending between the ovals. The most general concept word is placed in an oval at the top of the map, and the words become more specific as you move downward. This is shown in the concept map below.

Problem: To organize information on the topic of *mammals.*

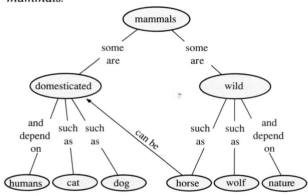

Note how the concept words flow from general to specific in the example above. Note also how each set of linking words describes the relationship between the two concept words it connects. If you follow any string of concept and linking words down the map, it reads like a sentence. There is one set of linking words, however, that does not follow the downward flow—the words *can be* between the concepts *horse* and *domesticated.* These linking words connect a concept from one branch of the map to a concept on another branch of the map. This special type of linking word is often known as a *cross-linkage.* Cross-linkages are used to show more complex interrelationships among concepts. Some cross-linkages are designated with arrows to show the direction of the connection.

To construct your own concept map, follow these guidelines.

1. List all the concepts to be mapped.
2. Pick out the main concept. Rank the remaining concepts from most general to most specific. Group together related concepts.
3. Arrange the concepts in a downward, branching structure. (It may be helpful to write each term on a separate index card and then arrange the cards in a branching structure.)
4. Link related concepts and choose appropriate linking words.
5. Look for places where cross-linkages can be drawn.

Concept maps are useful for summarizing or for clarifying information on a complex topic (such as some sections or chapters in your textbook). They are also helpful study aids that can be used to review a topic for an examination. In addition, concept maps may help you to organize your notes before you write a report.

Concept maps are dynamic tools. Once you have constructed one, you can alter it or add to it as you learn more about a particular topic. Furthermore, by linking individual concept maps together, you can begin to develop an overall picture of how broad topics are interrelated.

Scale

A scale can be used to put several items in a sequence with respect to a single characteristic, such as size or age. The degree to which an item exhibits the characteristic determines how far from the scale's end points the item is placed. In constructing a scale, it may be helpful to include appropriate units of measurement along the scale. For example, decibel units could be written along the scale shown below. A scale is useful for ordering items based on a specific characteristic or for showing a chronology of events, as in a timeline.

Problem: Order these everyday sounds from softest to loudest.

Line Graph

One way to depict a numerical relationship between two variables is with a line graph, as in the following example.

Problem: What is the relationship between plant size and light exposure?

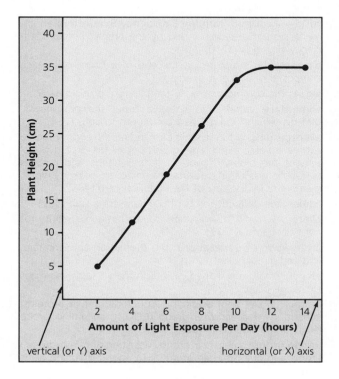

vertical (or Y) axis   horizontal (or X) axis

Line graphs can be used to depict data collected in experiments where one quantity changes in response either to time or to changes in another quantity. The data are plotted as individual points, which are then connected to form a continuous line.

Bar Graph

Bar graphs are useful for depicting data collected in a number of discrete but related categories. You can see at a glance how the categories compare with respect to the characteristic for which the data were collected. For each category, the data are plotted as a bar of a particular height, which is determined by a numerical scale along the vertical axis.

By presenting related sets of data side by side, a bar graph makes comparing data easy. This is shown in the following example.

Problem: How much precipitation fell in each summer month?

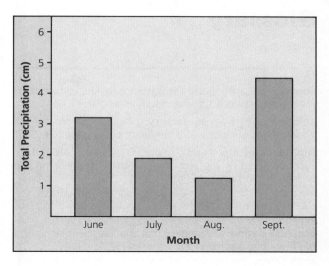

Circle Graph

In addition to bar graphs, circle graphs can also depict related sets of data in discrete categories. Unlike bar graphs, however, circle graphs can be used only when *all* of the categories that comprise a particular subject are represented. Circle graphs are sometimes called pie graphs because they resemble a pie cut into slices, as you can see in the example below.

Problem: How do I spend my time on a typical weekday?

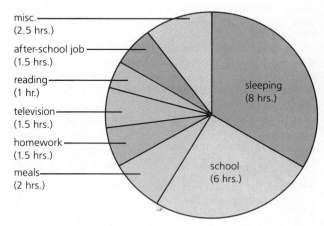

Circle graphs are useful for showing how something is divided into components, when each component has a numerical value that describes it. Each triangular section, or slice of the pie, represents one component. The size of each section is determined by its numerical value. Together, all of the sections should fill the circle completely.

Glossary

abdomen: in *arthropods,* the posterior region; in *mammals,* the region between the diaphragm and pelvis (709)

abiotic (ay by AHT ik) **factors:** physical factors of the environment, such as water, air, light, or temperature (820)

ABO blood group: a group of blood types (A, B, AB, and O), each identified by the presence of certain antigens on the surface of red blood cells (206)

abscisic acid: an organic compound that influences the shedding of leaves and the seasonal slowing down of plant activities (382)

absolute dating: any method that enables scientists to find out how long ago an event occurred (575)

absorption: the passage of materials across a cell membrane into the cell; the process by which usable materials are taken into an organism (155)

absorption spectrum: the different colors, or wavelengths, of light absorbed by each pigment in photosynthesis (341)

acetyl CoA: a compound produced during pyruvic acid breakdown by the combination of an acetyl group with coenzyme A (115)

acid: any compound that produces hydrogen ions in water solution (48)

Acquired Immune Deficiency Syndrome: *see* **AIDS** (210)

Acrasiomycota: a protist phylum; cellular slime molds; life cycle includes a stage as separate ameboid cells and a stage as a pseudoplasmodium—an aggregate of cells (652)

active immunity: a type of acquired immunity in which the body produces its own antibodies to attack a particular antigen (205)

active site: the region on the surface of an enzyme where substrate molecules attach, thus enabling a reaction (69)

active transport: a process in which the movement of materials across a cell membrane requires the expenditure of cellular energy (96)

adaptation: any kind of inherited trait that improves the chances of survival and reproduction for an organism (603)

adaptive radiation: the process by which a species evolves into a number of different species, each occupying a new environment (615)

addition: the breaking-off of a segment of a chromosome and its attachment to the homologous chromosome (547)

adenine: a nitrogenous base found in DNA and RNA (526)

adenosine diphosphate: *see* **ADP** (108)

adenosine triphosphate (uh DEN uh seen try FAHS fayt)**:** *see* **ATP** (109)

adhesion (ad HEE zhun)**:** the attraction between molecules of two different substances (59)

ADP (adenosine diphosphate): the lower-energy compound remaining after one phosphate group is removed from ATP (108)

adrenal glands: the endocrine glands that secrete hormones that help the body deal with stress (324)

aerobic respiration: respiration requiring the presence of free oxygen, in which glucose is completely oxidized to carbon dioxide and water (112)

aerosols: very small solid particles or liquid droplets suspended in air (869)

AIDS—Acquired Immune Deficiency Syndrome: an immune system disease; caused by the human immunodeficiency virus (HIV) (210)

air sacs: the structures at the end of a bronchiole; each composed of a cluster of alveoli (223, 764)

algae: the common term for the plantlike protists (646)

alimentary canal: the digestive tube; the passageway through which food moves from mouth to anus (158)

allantois (uh LAN tuh wis)**:** in *bird* and *reptile eggs,* a saclike extraembryonic membrane; grows out of the digestive system of the embryo; controls gas exchange and collects metabolic wastes; in *placental mammals,* an extraembryonic membrane that is part of the umbilical cord (448)

alleles: the different forms of the gene for a trait (502)

allergy: a rapid overreaction to an antigen that is not normally harmful (212)

alternation of generations: the alternation between haploid and diploid plant forms (474)

alveoli: small, cup-shaped cavities in the air sacs where gas exchange occurs (226)

amino (uh MEE noh) **acids:** the structural unit of proteins; consisting of a carboxyl group (COOH), an amino group (NH_2), and a side chain (66)

amino group: a chemical group consisting of two hydrogen atoms and one nitrogen atom (NH_2), found in amino acids (66)

amniocentesis: a technique used to detect genetic disorders in a fetus; a long needle is used to draw out fluid from the amniotic sac in which the fetus lives (553)

amnion: in both shelled eggs and mammals, a fluid-filled extraembryonic sac that surrounds the embryo, provides a watery environment, and protects the embryo (448)

amniotic fluid: the fluid that fills the amnion (465)

Amphibia (am FIB ee uh)**:** a class of water- and land-dwelling vertebrates having moist, smooth skin and four limbs; need water for reproduction; amphibians (743)

anaerobic respiration: respiration in the absence of free oxygen, in which glucose is partially oxidized (112)

anal pore: the opening through which indigestible wastes are ejected from a paramecium (156)

analogous structures: structures found in different types of organisms that are similar in function or outward appearance but dissimilar in basic structure or embryological development (582)

anaphase: the stage of mitosis during which the daughter chromosomes move to opposite poles (400)

anemia (uh NEE mee uh)**:** a condition in which a person has too few red blood cells or an insufficient amount of hemoglobin (197)

angiosperms (AN jee uh spermz): the flowering plants (473, 675)

Animalia (an uh MAL ee ah): one of the five kingdoms, composed of multicellular, usually motile, heterotrophic organisms; animals (135, 683)

Annelida: a phylum of animals with bodies made up of many similar segments; segmented worms (138, 696)

anterior: pertaining to the front, or head, end of a bilaterally symmetrical animal (684)

anther: the saclike structure of a stamen in which pollen grains are produced (481)

antheridium: the male gametophyte in mosses (476)

anthropology (an thruh PAHL uh jee): the branch of science that traces the development of the human species (773)

antibodies: proteins in the blood that bind to help destroy foreign substances in the body (196)

anticodon: a sequence of three bases of a tRNA molecule that pairs with the complementary three-nucleotide codon of an mRNA molecule during protein synthesis (531)

antigen (ANT uh jin): any substance that can cause an immune response (202)

anus: the opening of the digestive tube through which undigested materials are eliminated from the body (158)

aorta: the largest artery of the body (180)

appendicitis: inflammation of the appendix (165)

appendicular skeleton: the division of the human skeleton that includes the arms, legs, pectoral girdle, and pelvic girdle (260)

appendix: a small, fingerlike pouch found where the small intestine joins the large intestine (165)

Arachnida (uh RAK nih duh): a class of arthropods having no antennae, two body regions, four pairs of legs, and book lungs; includes spiders and scorpions; arachnids (714)

archaebacteria: bacteria that live in environments that other organisms could not tolerate (635)

archegonium: the female gametophyte in mosses (476)

arteries (AR tuh reez): blood vessels that carry blood away from the heart to the tissues and organs of the body (177)

Arthropoda: the phylum of animals having a segmented body with paired, jointed appendages and an exoskeleton; arthropods (138, 707)

Artiodactyla (art ee uh DAK tuh luh): an order of hoofed mammals having an even number of toes; includes camels and deer (769)

Ascomycota (as koh my KOH tuh): the largest phylum of fungi, both aquatic and terrestrial, which produce spores in asci; sac fungi (667)

ascus: serves as a sporangium; saclike (667)

asexual (ay SEK shuh wul) **reproduction:** reproduction with only one parent; offspring are identical to parent (8, 398)

assimilation (uh sim uh LAY shun): the incorporation of materials into the body of the organism (7)

asters: star-shaped structures formed during mitosis or meiosis in animal cells by fibers from the centrioles (400)

atomic number: the number of protons in the nucleus of an atom; the number that identifies an element (35)

atoms: the smallest particles of an element that have the properties of that element; consist of electrons, protons, and neutrons (34)

ATP (adenosine triphosphate): the compound in which energy released by cellular respiration is stored (109)

atria (AY tree uh): the two upper, thin-walled chambers of the heart (179)

atrioventricular node: *see* **A-V node** (181)

auditory canal: the passage leading from the outer ear to the middle ear (302)

auditory nerve: the nerve that carries impulses from the inner ear to the brain during the process of hearing (304)

auricles (OR ih kulz): *see* **atria** (179)

Australopithecus: a genus of fossil bipedal mammals found in southern Africa, showing more human than ape characteristics (774)

autoimmune disease: any disease in which the immune system of an individual fails to recognize certain of the body's cells as "self" and produces antibodies to combat them (212)

autonomic nervous system: a division of the peripheral nervous system consisting of motor fibers from the brain and spinal cord that serve the internal organs of the body; not under voluntary control (296)

autosomes: the chromosomes other than the sex chromosomes (518)

autotrophs: plants and other organisms that make their own food from inorganic substances (137, 150)

auxins: the hormones that affect the growth of all types of plant tissues (381)

Aves (AY veez): a class of warm-blooded vertebrates that have feathers, a four-chambered heart, front limbs as wings, and eggs with hard shells; birds (761)

A-V node (atrioventricular node): a small bundle of muscle cells at the base of the right atrium; triggers an impulse that causes contraction of the ventricles (181)

axial skeleton: the division of the human skeleton that includes the skull, vertebrae, ribs, and breastbone (260)

axon: a long, thin fiber that carries nerve impulses away from the cell body of a neuron (273)

B

bacillus: a rod-shaped bacterium (632)

bacteria (sing. bacterium): unicellular organisms with prokaryotic cells; belonging to the kingdom Monera (631)

bacteriophages: viruses that attack bacterial cells (655)

ball-and-socket joints: a type of joint that permits movement in all directions (262)

barbs: thin, hairlike branches growing diagonally from the shaft of a feather (763)

barbules: threadlike branches of a feather made up of many tiny hooks, which provide support to adjacent barbs (763)

base: any compound that produces excess hydroxide ions when dissolved in water (48)

Basidiomycota (buh sid ee uh my KOH tuh): a phylum of terrestrial fungi that produce spores in a clublike structure called a basidium; club fungi (668)

B cells: a type of lymphocyte that is produced and matured in the bone marrow, then released into the circulatory and lymphatic systems (203)

behavior: the series of activities performed by an organism in response to stimuli (786)

benthos: organisms that live on the ocean floor (855)

bilateral symmetry: a type of symmetry in which there is only one longitudinal section that will divide the organism into two parts that are mirror images of each other (684)

bile: a fluid secretion of liver cells that aids in the breakdown of fats (164)

binary fission: the simplest form of asexual reproduction, in which a unicellular parent organism divides into two approximately equal cells (405)

binomial (by NOH mee ul) **nomenclature:** two-word system of identifying an organism (129)

biodegradable: organic wastes that can be broken down by bacteria and other organisms into simpler substances (867)

biogenesis: the theory that living organisms only originate from other living organisms (589)

biological magnification: the accumulation of substances in larger and larger quantities in the bodies of organisms at each higher level of a food chain (868)

biology: the study of living things (3)

biome: a large geographical region that has a particular type of climax community (845)

biosphere: the portion of the earth in which living things exist (824)

biotic (by AHT ik) **factors:** all the living organisms in an environment and their effect on other living things (819)

bipedal locomotion: walking upright on two legs (774)

Bivalvia: a major class of the phylum Mollusca; includes mollusks with two-part shells such as clams, oysters, and mussels (699)

blastocoel (BLAS tuh seel): the fluid-filled cavity in a blastula (442)

blastopore: the opening in a gastrula created by the gastrulation process; becomes an opening to the digestive system in the adult organism (443)

blastula (BLAS chuh luh): a stage of development in which the embryo consists of a single layer of cells surrounding a fluid-filled cavity (442)

blue-green bacteria: members of the phylum Cyanobacteria of autotrophic eubacteria (635)

bone: a type of connective tissue made up of living cells, connective tissue fibers, and inorganic compounds (259)

book lungs: respiratory organs of the arachnids, consisting of leaflike plates in which gas exchange occurs (715)

Bowman's capsule: a double-walled, cup-shaped structure surrounding the glomerulus (244)

brain: a group of specialized nerve cells that control and coordinate the activities of a nervous system (272)

brainstem: the pons, the medulla oblongata, and the midbrain (291)

bronchi: two cartilage-ringed tubes that branch off the trachea and enter the lungs (226)

bronchial tubes: the branches of the bronchi (226)

bronchioles: the finest branches of the bronchial tubes; each ends in an air sac (226)

Bryophyta (bry AH fuh tuh): a phylum of land plants lacking specialized conducting tissues; these mosses, liverworts, and hornworts grow in moist areas; bryophytes (671)

budding: a type of asexual reproduction in which the parent organism divides into two unequal parts (406)

bulb: a short underground stem with thickened storage leaves; plants develop by vegetative reproduction (409)

C

calorie: the amount of heat that will raise the temperature of 1 g of water 1°C (the large Calorie, used to measure the energy content of foods, is equal to 1000 calories) (151)

calyx: the complete circle of sepals in a flower (481)

cambium: in woody plants, a meristematic tissue (356)

camouflage: a protective adaptation that enables an organism to visually blend into the environment (610)

canals: in echinoderms, a system of channels connecting the sieve plate to the tube feet; part of the water vascular system (726)

canines: a type of tooth specialized for tearing food (767)

capillaries (KAP uh ler eez): microscopic blood vessels (176)

capillarity: see **capillary action** (59)

capillary action: the upward movement of a liquid in a tube of narrow diameter (59)

capsid: in viruses, the protein coat that surrounds the nucleic acid core of RNA or DNA (654)

carbohydrates (kar boh HY drayts): organic compounds of carbon, hydrogen, and oxygen in which the ratio of hydrogen to oxygen to carbon is 2:1:1 (60)

carbon cycle: the pathways by which carbon is circulated through the biosphere (833)

carbon fixation: the process in which carbon dioxide is incorporated into organic compounds by the processes of photosynthesis and chemosynthesis (346)

carboxyl (kar BAHK sul) **group:** the characteristic chemical group (COOH) of organic acids (62)

Carnivora (kar NIV uh ruh): an order of flesh-eating mammals; includes cats, dogs, bears, and seals; carnivores (770)

carnivores (KAR nuh vorz): animals that feed on other animals (824)

carrying capacity: the maximum size of a population that can be supported by an environment (864)

cartilage: a type of flexible connective tissue (260)

cast: a type of fossil formed when a mold becomes filled with minerals and then hardens, producing a copy of the external features of an organism (573)

catalyst (KAT uh list): a substance that increases the rate of a chemical reaction without being changed itself (68)

cell body: the part of a nerve cell that contains the nucleus; cyton (273)

cell membrane: the structure that encloses the interior of a cell and controls the passage of materials into and out of the cell; plasma membrane (82)

cell plate: a structure formed during cytokinesis in a plant cell that divides the cell in half, forming part of the new cell walls of the daughter cells (402)

cells: the basic units of structure and function in living things; the smallest units in living things; the smallest units in living things that show the characteristics of life (78)

cellular respiration: the process by which the energy stored in food inside the cells is released (108)

cell wall: the rigid structure that encloses the cells of plants and various microorganisms; often composed of polysaccharides (82)

central nervous system: the division of the nervous system that includes the brain and spinal cord (282)

centrifugation (sen truh fyoo GAY shun): a process in which materials of different densities are separated from each other by suspending them in a liquid (25)

centrioles: cylindrical organelles found near the nucleus in animal cells that are involved in mitosis (88, 400)

centromere: the region of attachment of two sister chromatids (400)

Cephalaspidomorphi (sef uh LASS peh duh morf ee): the class of jawless fishes that includes the lampreys and hagfishes; the most primitive of all living vertebrates (736)

Cephalochordata (sef uh loh kor DAH tah): a subphylum of chordates; small, fishlike, marine animals with a notochord present in adults; lancelets (733)

Cephalopoda: a major class of the phylum Mollusca; includes mollusks with little or no shell such as squids and octopuses (699)

cephalothorax (sef uh luh THOR aks): the anterior portion of some arthropods, made up of the fused segments of the head and thorax (710)

cerebellum: a part of the human brain below the rear part of the cerebrum; coordinates voluntary movements (293)

cerebral cortex: the outer layer of the cerebrum; the gray matter of the brain (292)

cerebral hemispheres: the two halves of the cerebrum, which are partially separated by a deep groove (291)

cerebrum: the largest part of the human brain (291)

cervix: the narrow neck of the uterus (460)

Cestoda: the class of parasitic flatworms without digestive systems (693)

Cetacea (see TAYSH ee uh): an order of nearly hairless marine mammals having paddlelike forelimbs; includes whales, dolphins, and porpoises; cetaceans (770)

cheliceras: in arachnids, a pair of fanglike appendages used to pierce prey; the first pair of appendages (714)

chemical bond: a force of attraction between atoms that holds them together in compounds (40)

chemical formula: a written representation of a compound in which a chemical symbol replaces each element (43)

chemical reaction: the process in which chemical bonds are broken and the atoms form new bonds, producing new substances (44)

chemoautotrophs: the autotrophs that carry on chemosynthesis (340)

chemosynthesis: a form of autotrophic nutrition in which energy for synthesizing organic compounds is obtained from inorganic compounds rather than from light (349)

Chilopoda (ky LAHP uh duh): a class of arthropods with one pair of antennae, many body segments, and one pair of legs on most body segments; centipedes (713)

Chiroptera (ky RAHP tuh ruh): an order of mammals capable of real flight; bats (768)

chitin: the polysaccharide that makes up the exoskeleton of arthropods (254, 708)

chlorophylls: the major photosynthetic pigments of plants and algae (342)

Chlorophyta (klor AH fuh tuh): a phylum of mainly aquatic plantlike protists; green algae (649)

chloroplasts: the plastids that contain chlorophyll; the sites of photosynthesis in eukaryotic cells (89, 343)

cholesterol (kuh LES tuh rohl): an essential compound found in most animal tissues; plays a role in the buildup of fatty deposits in arteries (63)

Chondrichthyes (kahn DRIK thee eez): the cartilaginous fishes including sharks, rays, and skates (737)

Chordata (kor DAH tah): a phylum of animals having at some stage of development a notochord, paired gill slits, and a dorsal, hollow nerve chord; chordates (138, 733)

chorion: the membrane that surrounds the embryo and the other extraembryonic membranes in mammals, birds, and reptiles (448, 464)

chorionic villus sampling: a technique used by physicians to detect genetic disorders; a sample of the chorion, a part of the placenta, is removed for genetic analysis (553)

choroid coat: the darkly pigmented middle layer of the eye (300)

chromatid: one of the two strands of a doubled chromosome (400)

chromatin: the material of which chromosomes are composed (398)

chromatography (kroh muh TAHG ruh fee): any technique that separates substances in a mixture on the basis of their chemical properties (26)

chromoplasts: the plastids that contain pigments other than chlorophyll (89)

chromosomal mutation: a change in chromosome structure, resulting in new gene combinations (546)

chromosomes: rodlike structures in cells that undergo division and that contain hereditary information of the organism (398)

Chrysophyta: a phylum of algaelike protists; chrysophytes are mostly unicellular and contain large amounts of yellow-brown pigment (648)

chyme: the thin, soupy liquid produced from food by the stomach (163)

cilia: short, hairlike organelles at the surface of a cell, with the capacity for movement (88, 255)

Ciliophora: a phylum containing the most complex protozoa, which have an abundance of hairlike cilia; ciliates (644)

circadian rhythms: the physiological or behavioral cycles that repeat approximately every 24 hours (798)

circulatory system: a system that links the cells of a complex organism with its environment (173)

class: a group of related orders (128)

cleavage: in a fertilized egg, the first series of cell divisions that occur without growth and continue until the cells of the embryo are reduced to the size of the cells of an adult organism (442)

climax community: a mature, stable community that is the final stage of ecological succession (836)

cloaca (kloh AY kuh): in *reptiles, birds, amphibians,* and in many *fishes,* the cavity into which both the intestinal and genitourinary tracts empty; in some *vertebrates,* a cavity that serves as a duct for excretion, respiration, and reproduction (745)

clone: a group of individual organisms that have identical genetic makeup (556)

closed circulatory system: a system in which the blood is always contained within tubes or vessels in the body (175)

clotting: the solidification of blood (199)

club mosses: small, spore-dispersing tracheophytes with true roots, stems, and leaves (672)

cnidoblasts: the stinging cells found in coelenterates; used for defense and capturing food (688)

coacervates (koh AS er vayts): according to the heterotroph hypothesis, an aggregate of large proteinlike molecules; thought to have developed into the first forms of life on the primitive earth (592)

coccus: a spherical bacterium (632)

cochlea: the organ of hearing, found in the inner ear, consisting of coiled, liquid-filled tubes (304)

codominance: a type of inheritance in which two dominant alleles are expressed at the same time without blending of traits (510)

codon: a group of three bases in an mRNA molecule that specifies a particular amino acid (530)

Coelenterata: a phylum of simple aquatic invertebrates that have a mouth, tentacles, and cnidoblasts; includes hydras, jellyfishes, and corals (138, 687)

coelom: The fluid-filled body cavity between the body wall and the digestive tube in most multicellular animals (697)

coenzymes (koh EN zymz): the nonprotein, organic substances necessary to the functioning of particular enzymes (71)

coevolution: when two or more species can evolve through cooperative or competitive adaptations (616)

cohesion (koh HEE zhun): the force of attraction between molecules of the same substance (58)

Coleoptera (kohl ee AHP tuh ruh): the largest insect order, having front wings modified to form a horny covering for hind wings; beetles (720)

collagen: a fibrous structural protein that is a constituent of connective tissue (259)

collar cells: the flagellated cells found in the inner layer of a sponge (686)

colloidal dispersion (kuh LOYD ul dis PER zhun): a mixture in which the solute particles are larger than molecules or ions, but are too small to settle out (47)

color blindness: a sex-linked trait in which an individual cannot perceive certain colors (520)

commensalism: a type of symbiotic relationship in which one organism benefits from the association and the other is not affected (825)

community: all the populations of different organisms within a given area (823)

complement system: a series of enzymes in blood that catalyze reactions that help destroy invading cells (204)

complete metamorphosis: the type of development in most insects; involves the stages of larva, pupa, and adult (719)

compound leaf: a leaf in which the blade is divided into several parts, or leaflets, that are attached to a petiole (367)

compound microscope: a microscope with two lenses (20)

compounds: substances made of two or more kinds of atoms combined in different proportions (34)

concentration gradient: the difference in concentration between a region of greater concentration and one of lesser concentration (92)

conditioning: a simple form of learning in which behavior is changed through association (797)

cones: in the *retina of the eye,* the structure responsible for color vision; in *gymnosperms,* the seed-bearing or pollen-bearing structures (301)

conifers: subclass Coniferophyta of the gymnosperms; cone-bearing seed plants with needlelike or scalelike leaves (676)

conjugation: a form of sexual reproduction found in protists; individual organisms appear to be identical but are of different mating types (425)

connective tissue: a type of tissue that supports other body tissues and binds tissues and organs together (99)

consumers: the heterotrophs; the organisms that obtain nutrients from other organisms (828)

contour farming: a method of farming in which rows are plowed horizontally across slopes, following the contour of the land, and acting to reduce the flow of water down the slopes (874)

control: the setup in an experiment in which no factor was changed (15)

controlled experiment: an experiment set up in duplicate in which a single factor is changed in one setup but not the other (15)

convergent evolution: the evolution of outward similarities in organisms that are not closely related, because they have to meet similar problems in their habitats (616)

cork: a protective plant tissue that covers the surface of woody stems and roots (357)

cork cambium: the meristematic tissue that produces cork (356)

corm: a short, stout underground stem that contains stored food (409)

cornea: the transparent part of the sclera in the front of the eye through which light enters (300)

corolla: the complete circle of petals in a flower (481)

coronary (KOR uh ner ee) **circulation:** the branch of the systemic circulation that supplies the heart muscle with blood (187)

corpus callosum: the bridgelike connection between the hemispheres of the brain (292)

corpus luteum: a progesterone-secreting yellow body in the ovary, formed when luteinizing hormone causes a ruptured follicle to fill with cells (462)

correlation: a process by which geologists determine the relative ages of rock layers and fossils in a local region (577)

cortex: the outer region or layer of a plant or animal organ or structure (362)

corticosteroids: a type of hormone produced by the adrenal cortex, synthesized from cholesterol (325)

cotyledon: a modified leaf of a seed plant embryo; often provides nourishment for the developing seedling (484)

covalent (koh VAY lent) **bond:** a chemical bond that is formed by the sharing of electrons (40)

cover crops: crops planted over a whole field instead of in rows; used to prevent soil erosion (873)

cranial nerves: the nerves connected to the brain (295)

cranium: the upper part of the skull, which houses and protects the brain (260)

Cro-Magnons (kroh MAG nuns): a type of prehistoric human, considered to be the same as modern humans; replaced the Neanderthals about 35 000 years ago (777)

crop: in birds and many invertebrates, a thin-walled organ that temporarily stores food from the esophagus (158)

crop rotation: a method of farming in which different crops are grown on a field in successive years to prevent the reduction of soil nutrients (875)

crossing-over: the process in which pieces of homologous chromosomes are exchanged during synapsis in the first meiotic division (522)

Crustacea (krus TAY shuh): a class of mostly aquatic arthropods having two pairs of antennae; includes lobsters, crabs, shrimp, and barnacles; crustaceans (710)

cuticle: the layer of cutin that covers plant epidermis (356)

cutting: any vegetative part of a plant used to produce a new plant by artificial vegetative reproduction (410)

Cyanobacteria: the phylum of prokaryotes that carries out photosynthesis; includes blue-green bacteria (635)

cycads (SY kuds): subclass Cycadophyta of the gymnosperms; tropical plants resembling ferns or palm trees (676)

cytokinesis (sy toh kih NEE sis): the division of the cytoplasm of the cell after mitosis or meiosis; the cell divides into two parts, each containing one of the newly formed nuclei and half of the other contents of the parent cell (398)

cytokinins: a type of plant hormone that stimulates cell division and growth (382)

cytoplasm: the watery material between the nucleus and the cell membrane of a cell (85)

cytosine: a nitrogenous base found in DNA and RNA (526)

D

dams: barriers built to hold back flowing water (874)

dark reactions: the series of reactions in photosynthesis in which carbon fixation occurs and for which light is not required (344)

decomposers: organisms of decay (828)

dehydration synthesis (dee hy DRAY shun SIN thus sis): a type of reaction in which two molecules are bonded together by the removal of a water molecule (61)

deletion: a type of chromosomal alteration in which a portion of a chromosome and the genes it contains is lost (547)

dendrites: the short, branched parts of a neuron specialized for receiving nerve impulses and transmitting them to the cell bodies (273)

denitrifying bacteria: anaerobic bacteria that convert nitrates and nitrites to nitrogen gas, which is released into the atmosphere (831)

deoxyribose: a five-carbon sugar found in DNA (526)

dermis: the layer of skin beneath the epidermis, consisting of elastic connective tissue (247)

deserts: a type of biome in which there is too little rainfall to support trees or grasses; may show great variation in temperature between day and night (849)

Deuteromycota (do tur oh my KOH tuh): the phylum that includes all fungi that are not known to have a sexual reproductive phase; *Penicillium* (669)

development: the orderly series of changes that occur as an organism grows to maturity (441)

diabetes mellitus: a condition caused by an insufficient concentration of insulin in the blood (326)

diaphragm: the muscle that forms the floor of the chest cavity (225)

diastole (dy AS tuh lee): the period of relaxation of the heart (180)

diatomic (dy uh TAHM ik) **molecule:** a molecule formed when two atoms of the same element combine in a covalent bond, such as O_2 (40)

diatoms: members of the phylum Chrysophyta; unicellular, silica-shelled organisms found largely in salt water (648)

dicots: the plants whose seeds have two cotyledons (485)

differentiation: the series of changes that transforms unspecialized embryonic cells into the specialized cells, tissues, and organs that make up an adult organism (361, 444)

diffusion: the movement of molecules or particles from an area of greater concentration to an area of lesser concentration (92)

digestion: the breakdown of complex food materials into simpler forms that an organism can use (6, 155)

dihybrid cross: a genetic cross in which two pairs of contrasting traits are studied (508)

Dinoflagellata: a phylum of single-celled algae found mainly in the oceans (649)

dipeptide: the molecule formed when two amino acids are joined by a peptide bond (67)

diploid: having two sets of chromosomes, or all the homologous chromosomes that are characteristic of the species (419)

Diplopoda (dih PLAHP uh duh): a class of arthropods with one pair of antennae, many body segments, and two pairs of legs on most body segments; millipedes (713)

Diptera (DIP tuh ruh): a widespread insect order having one pair of functional, membranous wings; includes flies (720)

directional selection: a type of natural selection in which an extreme phenotype becomes a favorable adaptation (612)

disaccharide (dy SAK uh ryd): the molecule formed by the joining of two simple sugars (61)

disjunction: the separation of homologous chromosomes during anaphase I of meiosis (420)

disruptive selection: a rare type of natural selection in which two opposite phenotypes are favorable adaptations and the average phenotypes are unfavorable; creates two subpopulations (613)

DNA (deoxyribonucleic [dee AHK see ry boh noo klay ik] **acid):** the nucleic acid found in all the cells of an organism; the hereditary material passed on during reproduction (64)

dominance hierarchy: a ranking within a group of animals that is established through fighting or displays of aggression (803)

dominant: the inherited characteristic that appears in an organism (500)

dominant species: the species that exert the greatest effects on the environment and on other members of the community (836)

dormancy: a period during which growth and other metabolic activities stop or are severely reduced (487)

dorsal: pertaining to the upper side or the back of a bilaterally symmetrical animal (684)

double fertilization: in flowering plants, the fertilization of the egg and of the two polar nuclei to form the diploid zygote and the triploid endosperm nucleus, respectively (483)

Down syndrome: a disorder that results from an extra number 21 chromosome (552)

E

Echinodermata (ih ky nuh der MAH tuh): a phylum of marine animals having a water-vascular system, an internal skeleton, and a spiny skin; echinoderms (725)

ecological succession: the process by which an existing community is slowly replaced by another community (835)

ecology: the branch of biology that deals with the interactions among organisms and between organisms and their environment (819)

ecosystem: a community and the physical environment that it occupies (823)

ectoderm: the outer layer of cells in a simple animal or embryo; one of the germ layers of an animal embryo (443)

ectothermic (ek tuh THER mik): having a body temperature that varies with the temperature of the environment; cold-blooded (735)

Edentata (ee den TAH tuh): an order of mammals having only molars or no teeth at all; includes anteaters, sloths, and armadillos (769)

effector: a muscle or gland; responds to a stimulus (272)

ejaculation: a process by which involuntary muscular contractions force the semen through the urethra and out of the body (459)

electrons (ih LEK trahnz): negatively charged particles found in the spaces outside the nuclei of atoms; have much less mass than protons or neutrons (35)

electron transport chain: a series of oxidation-reduction reactions in which most of the energy produced from the breakdown of glucose is transferred to ATP (116)

electrophoresis (ih lek truh fuh REE sis): a technique for separating substances made up of particles that have an electrical charge (26)

elements: substances made entirely of one kind of atom (34)

elongation zone: in plants, a region behind the meristematic zone of the root, in which the cells produced in the meristematic zone grow longer (361)

embryo: a multicellular organism in the early stages of development (441)

embryonic induction: the process by which one group of cells (the organizer) induces another group of cells to differentiate (446)

embryo sac: the mature female gametophyte of a flowering plant (482)

endocrine glands: the ductless glands (314)

endocrine system: the system of the body that regulates overall metabolism, homeostasis, growth, and reproduction (314)

endocytosis: the process of transporting material into a cell by means of a vesicle (97)

endoderm: the inner layer of cells in a simple animal or embryo; one of the germ layers of an animal embryo (443)

endodermis: in plants, the innermost layer of the cortex of the root (362)

endoplasmic reticulum: a system of fluid-filled canals enclosed by membranes (85)

endoskeleton: a skeleton composed of bone and/or cartilage located within the body walls (254)

endosperm: the tissue that develops from the endosperm nucleus, often serving as a food supply for the plant embryo (483)

endospores: structures containing genetic material formed by bacteria and found in their cytoplasm when conditions for growth are unfavorable (632)

endosymbiosis: the condition in which one organism lives inside the cell of another to the benefit of both (89)

endothermic (en duh THER mik): having a body temperature that remains relatively constant regardless of the temperature of the environment; warm-blooded (735)

enhancer: a section of DNA that controls the access of an enzyme to a promoter (538)

entomology: the branch of biology that deals with the study of insects (720)

enzymes (EN zymz)**:** protein catalysts that are necessary for most of the chemical reactions that occur in living cells (68)

epicotyl: the part of a plant embryo above the point of attachment of the cotyledons; gives rise to the terminal bud, leaves, and stem (485)

epidermis: in *plants,* a protective tissue that forms the outer layer of leaves, green stems, and roots; in *animals,* the outer layer of skin consisting of layers of tightly packed epithelial cells (246, 356)

epididymis: a storage area for sperm on the upper, rear part of the testis (458)

epiglottis: a flap of tissue that covers the trachea during swallowing, so that food passes only into the esophagus (162)

epinephrine: a hormone produced by the adrenal medulla; a neurotransmitter produced by some nerve cells; adrenaline (324)

epithelial tissues: the tissues that cover body surfaces and line body cavities and organs (99)

erosion: the removal of soil by the action of wind and/or water (866)

esophagus (pl. **esophagi**)**:** the tube that is the passageway for food from the mouth to the stomach (158)

estrogen: a hormone secreted by the ovaries that promotes development of female secondary sex characteristics and regulates the reproductive cycle (327)

ethylene: an organic compound that stimulates flowering in some plants and hastens the ripening of fruit (382)

eubacteria: the more recent of the two lineages of prokaryotes, which includes cyanobacteria (635)

Euglenophyta: a phylum of protists; euglenoids show both plantlike and animal-like characteristics (646)

eukaryotic cells: cells containing membrane-bound nuclei (79)

Eustachian tube: the tube extending between the middle ear and the throat that equalizes the pressure between the middle ear and the environment (303)

eutrophication (yoo truh fuh KAY shun)**:** an accelerated aging process in a lake or pond, in which the body of water fills in with plant remains and is reduced in size (868)

evolution: the gradual change of allele frequencies found in a population (606)

excretion: the removal of waste substances from an organism (7, 237)

exhalation: the phase of breathing in which air is pushed out of the lungs (227)

exocrine glands: the glands that discharge their secretions into ducts (314)

exocytosis: movement of materials out of the cell by the reverse of endocytosis (97)

exon: a segment of DNA that codes for amino acids that will become part of a protein (537)

exoskeleton: a skeleton found on the outside of the body, enclosing the soft parts (254)

extensor: a muscle that extends a joint (264)

external fertilization: the process in which eggs are fertilized outside the body of the female (432)

extinct: the end of a species when the last individual of that species has died (580)

extraembryonic membrane: in shelled eggs of reptiles and birds, any one of four membranes outside the embryo but inside the shell (448)

F

F_2 (second filial) generation: the second generation produced in a breeding experiment (499)

facilitated diffusion: a process by which certain molecules diffuse quickly across a cell membrane (92)

family: a group of related genera (128)

farsighted: a vision condition that occurs because the eyeball is too short (301)

fatty acid: an organic molecule with two parts: a carbon chain, and one or more carboxyl groups (COOH); one of the end products of the digestion of fats (62)

feces: undigested and indigestible food material that is solidified in the large intestine and then eliminated (167)

fermentation (fer men TAY shun)**:** following glycolysis, the conversion of pyruvic acid to an end product with no further release of energy (113)

ferns: a type of spore-dispersing tracheophyte; Filicineae (476, 673)

fertilization: the fusion of the nuclei of the male and female gametes (418)

fertilizers: materials used to provide or replace soil nutrients (875)

fetoscopy: a technique that allows direct observation of the fetus and surrounding tissues (554)

fetus: the developing baby after about the second month of pregnancy (464)

fiber: indigestible materials, such as cellulose, found in the cell walls of fruits (153)

filament: the starlike part of a stamen that supports the anther (481)

first filial (F_1) generation: the first generation produced in a breeding experiment (499)

flagella: long, hairlike organelles at the surface of a cell, with the capacity for movement (256)

flexor: a muscle that bends a joint (264)

follicles: the structures in the ovaries in which the mature eggs develop (459)

food chain: a series of organisms through which food energy is passed (828)

food web: food chains that are interconnected at various points (828)

foot: a large, ventral, muscular structure that is used for locomotion in mollusks (699)

foramen magnum (fuh RAY men MAG num)**:** the opening in the skull where the spinal cord enters (774)

fossil: any trace or remains of an organism that has been preserved by natural processes (132, 571)

fossil record: the history of life as determined by the relative age of fossils (575)

fraternal twins: two individuals formed when two eggs are fertilized at the same time; twins that are genetically different (467)

fruit: a structure that develops from the ovary and associated flower parts after fertilization; contains seeds (481)

fundamental tissues: the tissues involved in the production and storage of food and in the support of plants; parenchyma, collenchyma, and sclerenchyma (358)

Fungi (FUN jy): one of the five kingdoms; its members are saprobic or parasitic, mostly multicellular, and usually consist of filaments (135, 665)

G

gallbladder: the organ that stores bile produced by the liver (165)

gametes: the haploid cells that fuse with other haploid cells to form zygotes; the sperm cells or egg cells (417)

gametogenesis: the process by which gametes develop in the gonads (429)

gametophyte (guh MEET uh fyt): a gamete-producing plant (474)

ganglion: a group of cell bodies and interneurons that switch, relay, and coordinate nerve impulses (282)

gas exchange: the process by which protist and animal cells get rid of excess carbon dioxide (219)

gastric juice: the digestive secretion of glands in the stomach, containing hydrochloric acid and pepsin (162)

Gastropoda: a major class of the phylum Mollusca; includes mollusks with a single shell, such as snails (699)

gastrovascular cavity: the internal body cavity of a coelenterate (157, 688)

gastrula: in animals, an early stage of embryonic development during which the second germ layer is formed (443)

gastrulation: the process in which the cells on one side of a blastula move inward to form the two-layered gastrula (443)

gemmules: asexual reproduction structures that form on some freshwater sponges (687)

gene: a distinct unit of hereditary material found in chromosomes; a sequence of nucleotides in DNA that codes for a particular tRNA, rRNA, or polypeptide (502)

gene mutation: a change in the sequence of the bases in a gene, which changes the structure of the polypeptide that the gene codes for (546)

gene pool: the total of all the alleles in a population (605)

gene therapy: the process of correcting genetic defects by transferring normal genes to cells that lack them (558)

genetic drift: a change in the gene pool of a small population that is brought about by chance (607)

genetic engineering: the process of producing altered DNA, usually by breaking a DNA molecule and inserting new genes (558)

genetic equilibrium: the condition in which allele frequencies do not change from one generation to the next (608)

genetic recombination: the formation of new combinations of alleles during sexual reproduction (607)

genetics: the branch of biology that studies the ways in which hereditary information is passed on from parents to offspring (498)

genotype: the genetic makeup of an individual (503)

genus (JEE nus): a group of closely related species (127)

geographic isolation: the first stage of speciation, in which a population of organisms is prevented from interbreeding with other populations of that species by a natural barrier (614)

geologic evolution: the process of continual change that the earth undergoes (571)

geologic time scale: a timetable of the earth's history constructed by geologists (578)

geotropism: the growth of a plant in response to the stimulus of gravity; roots of a plant generally grow down in the direction of the gravity force (383)

germ layers: the three embryonic cell layers—the ectoderm, mesoderm, and endoderm—that give rise to all tissues and organs of animals (444)

germ theory of disease: the idea that bacteria and other microorganisms can cause disease (637)

gestation: the length of pregnancy (465)

gibberellins: the hormones that affect plant growth as well as the development of fruits and seeds (382)

gills: in aquatic animals, thin layers of tissue, richly supplied with blood vessels, that are the respiratory organs (223, 712)

gill slits: a structure found in pairs in the throat region of all chordates during some part of their lives (734)

ginkgoes: subclass Ginkgophyta of the gymnosperms; the maidenhair trees with fan-shaped leaves—the only surviving species (676)

gizzard: in birds and many invertebrates, a thick-walled grinding organ that crushes food released from the crop (158, 765)

glands: organs made up of epithelial cells that specialize in the secretion of substances needed by the organism (314)

gliding joint: a joint that permits limited flexibility in all directions (262)

glomerulus (pl. **glomeruli**): a group of capillaries in the nephron of a kidney (244)

glucagon: a hormone secreted by the pancreas that increases the blood-glucose level (325)

glycolysis (gly KAHL uh sis): the process of breaking down the glucose molecule into two three-carbon pyruvic acid molecules (112)

Golgi bodies: the organelles consisting of stacks of membranes forming flattened sacs in the cytoplasm, which serve as storage centers for proteins synthesized by cells (86)

gonads: in animals, the specialized organs in which gametes develop (326, 428)

gradualism: Darwin's theory of evolution, in which new species arise through gradual changes in their characteristics, and thus evolution occurs very slowly over millions of years (604)

grafting: a type of artificial vegetative propagation accomplished by permanently joining a part of one plant to another plant (411)

gram-negative: bacteria that do not retain crystal violet stain and that appear light pink under a microscope (636)

gram-positive: bacteria that retain crystal violet stain and that appear purple under a microscope (636)

Gram stain: a process of staining used to divide bacteria into two classes (636)

Gram test: a universal method of classifying bacteria (636)

grana: stacks of thylakoids in plant chloroplasts (343)

grasslands: a type of biome in which there is not enough rainfall to support trees and the dominant form of vegetation is grasses; prairies (848)

green-sulfur bacteria: autotrophic eubacteria that carry out photosynthesis without producing oxygen (636)

growth: the process by which living organisms increase in size (7)

guanine: a nitrogenous base found in DNA and RNA (526)

guard cells: in leaves, pairs of specialized epidermal cells that regulate the opening and closing of the stomates (368)

gullet: in the paramecium, the part where food particles enter the cell (156)

guttation: the formation of water droplets at the edges, or tips, of leaves as a result of root pressure (378)

gymnosperms (JIM nuh spermz)**:** a seed plant whose seeds are not enclosed within a fruit (473, 675)

H

habit: learned behavior that becomes automatic (797)

habitat: a particular part of the environment where an organism lives (826)

habituation: the simplest type of learning; the animal learns not to respond to repeated "unimportant" stimuli (797)

haploid: having only one chromosome from each pair of homologous chromosomes (419)

Hardy-Weinberg law: the principle that sexual reproduction alone does not affect allele frequencies in a population (608)

Haversian canal: a cavity in bone that contains the blood vessels and nerves that serve the osteocytes (259)

heart: the organ that pumps blood through the circulatory system (174)

helix: a shape like a coiled spring, used to describe the structure of DNA molecules (527)

Hemiptera (heh MIP tuh ruh)**:** a large order of insects— mostly terrestrial, some aquatic; bugs (720)

hemocyanin (hee moh SY uh nin)**:** a copper-containing respiratory pigment that aids in the transport of oxygen in some vertebrates (711)

hemoglobin: a substance that increases the amount of oxygen the blood can carry (175, 197)

hemophilia (hee muh FIL ee uh)**:** a hereditary disease in which one or more of the clotting factors are missing from the blood (200)

hepatic-portal (heh PAT ik PORT ul) **circulation:** the branch of the systemic circulation that carries blood to the liver from the digestive tract (188)

herbaceous stems: plant stems that are soft, green, and juicy (363)

herbivores (HER buh vorz)**:** animals that feed only on plants (824)

hermaphrodites: the individual organisms that possess both testes and ovaries (428)

heterotroph hypothesis: the hypothesis that the first organic compounds were formed by natural chemical processes on the primitive earth and that the first lifelike structures developed from coacervates and were heterotrophs (589)

heterotrophs: the organisms that cannot synthesize their own food and must obtain it ready-made (137, 150)

heterozygous: having two different alleles for a trait (503)

hinge joints: a type of joint that permits back-and-forth motion, e.g., at the elbow and knee (262)

Hirudinea: a major class of the phylum Annelida; includes the leeches (696)

histones: small groups of proteins around which DNA is wrapped to form chromatin (398)

HIV—human immunodeficiency virus: the virus that causes AIDS (210)

homeostasis (hoh mee oh STAY sis)**:** the condition of a stable internal environment in an organism (6)

homeotic genes: in fruit flies, the genes that control key events in the flies' development (539)

homologous chromosomes: a pair of chromosomes having the same size and shape and carrying alleles for the same traits (418)

homologous structures: structures found in different kinds of organisms that have the same basic arrangement of parts and a similar pattern of embryological development (582)

Homo sapiens: the species of modern humans (776)

homozygous: having two identical alleles for a trait (503)

hormones: the secretions of the endocrine glands (315)

horsetails: the spore-dispersing members of the tracheophytes, each having a hollow stem and a collar of small leaves (673)

human ecology: the study of the relationship between humans and their environment (864)

human immunodeficiency virus: *see* **HIV** (210)

humus: the dark, rich organic matter in topsoil formed from the decay of dead plants and animals (822)

hybrids: individuals that are heterozygous for particular traits; individuals produced by a cross between members of two closely related species (499)

hybrid vigor: superior characteristics that are often found in hybrids produced by a cross of two closely related species; heterosis (556)

hydrolysis (hy DRAHL uh sis)**:** the process by which molecules are broken apart by the addition of water molecules (62)

hydrotropism: growth of plant roots toward water (383)

Hymenoptera (hy muh NAHP tuh ruh)**:** a large, varied order of colonial insects; includes bees, wasps, and ants (720)

hypersecretion: an excess of a hormone (319)

hypertonic solution: a solution whose concentration of solutes is higher than that of a cell placed in it (96)

hyphae (HY fee)**:** threadlike filaments, many of which make up the bodies of most fungi (665)

hypocotyl: the part of a plant embryo between the radicle and the point of attachment of the cotyledons (485)

hyposecretion: a deficiency of a hormone (319)

hypothalamus: the part of the brain located below the thalamus; controls body temperature, blood pressure, and emotions (291, 321)

hypothesis: a possible explanation for an observed set of facts (14)

hypotonic solution: a solution that contains a lower concentration of dissolved substances than that of a cell placed in it (95)

identical twins: two individuals formed when one fertilized egg divides at an early stage of development, producing two organisms with the same genetic makeup (467)

immovable joints: a type of joint in which the bones are fitted tightly together and cannot move (262)

immune response: the reaction of the immune system to the presence of foreign cells or molecules (202)

immunity: the ability of the body to resist a disease (202)

implantation: the fastening of the embryo to the wall of the uterus (463)

imprint: a type of fossil formed when an impression made in mud by a living thing is preserved when the mud is transformed into rock (573)

imprinting: in some animals, the forming of an attachment to an organism, object, or other environmental factor soon after hatching or birth (799)

impulses: messages carried by the nerve cells of a nervous system; regions of electrical and chemical changes that pass along the cell membrane of a neuron (272)

inbreeding: a breeding method in which closely related individuals are mated to retain or strengthen certain desirable traits (555)

incisors (in SYZ erz)**:** a type of tooth specialized for cutting food (767)

incomplete dominance: a type of inheritance in which two contrasting alleles contribute to the individual a trait not exactly like either parent; blending inheritance (509)

incomplete metamorphosis: a type of development in some insects in which there is no larval stage (718)

index fossils: fossils that permit the relative dating of rocks within a narrow time span (578)

indicator: a substance that changes color when the pH goes above or below a certain value (50)

industrial melanism: the development of dark-colored organisms in a population exposed to industrial air pollution (618)

inferior vena cava (VEE nuh KAY vah)**:** veins, some of the largest in the body, that return blood from the lower regions of the body into the right atrium of the heart (187)

inflammatory response: the result of a pathogen activating the second line of defense (201)

ingestion: the taking in of food from the environment (6)

inhalation: the phase of breathing in which air is drawn into the lungs (227)

innate behavior: behavior determined by heredity (790)

inorganic compounds: any type of compound that is not an organic compound; usually do not contain carbon (58)

Insecta (in SEK tuh)**:** a class of arthropods having no antennae, three body regions, three pairs of legs, and tracheal respiration; insects (716)

Insectivora (in sek TIV uh ruh)**:** an order of generally small, primitive mammals that feed mainly on insects; includes moles, shrews, and hedgehogs (768)

insight: the ability to create a solution to an unfamiliar problem without a period of trial and error (800)

instinct: a complex, inborn behavior pattern (793)

insulin: a hormone secreted by the pancreas that lowers blood glucose levels (315)

intercellular (in ter SEL yoo ler) **fluid:** a fluid that helps move materials between the capillaries and the body cells (188)

interferon: a protein produced by body cells in response to attack by viruses (202)

internal fertilization: the process in which eggs are fertilized inside the body of the female (433)

interneurons: the neurons that relay impulses from one neuron to another in the brain and spinal cord (274)

interphase: the stage of the cell reproductive cycle lasting from the end of one mitotic cycle to the beginning of the next (399)

interspecific competition: competition between members of two different species in an ecosystem (826)

intertidal zone: the biome along the ocean shoreline that is covered by water at high tide and uncovered at low tide (855)

intestinal juice: a secretion of the cells of the walls of the small intestine, containing digestive enzymes (164)

intestine: the organ in which most digestion and the absorption of food occurs (158)

intraspecific competition: competition between members of the same species in an ecosystem (828)

intron: a segment of DNA that does not code for amino acids of a protein (537)

inversion: a type of chromosomal mutation in which a piece of chromosome is rotated, resulting in reversal of the order of the genes in that segment (547)

invertebrate: an animal without a backbone (683)

in vitro fertilization: the process of fertilization in a glass laboratory dish (463)

ion (I ahn)**:** an atom or group of atoms with an excess electrical charge (41)

ionic (i AHN ik) **bond:** the force of attraction between two ions in a chemical compound (42)

iris: the round, colored part of the eye formed from the choroid layer; controls the size of the pupil (300)

irritability: the capacity of a cell or organism to respond to stimuli (272)

islets of Langerhans: the endocrine portion of the pancreas, consisting of clusters of hormone-secreting cells (325)

isotonic solution: a solution that contains the same concentration of dissolved substances as does a living cell placed in it (95)

isotopes (I suh tohps): atoms that differ from other atoms of the same element by the number of neutrons in their nucleus (37)

J ─────────────────────

joint: a point in the skeleton where bones meet (262)

K ─────────────────────

Kaposi's sarcoma (KAH poh seez sar KOH muh): cancer of the blood vessels; common cause of death in AIDS patients (211)

karyotyping: a technique for examining the chromosome makeup of an individual (552)

kidneys: a pair of organs in vertebrates that excrete nitrogenous wastes and regulate the blood's chemical balance; produce urine (242)

kilocalorie: the amount of heat needed to raise the temperature of 1 kg of water 1°C; 1000 calories or 1 Calorie (151)

kingdom: a group of related phyla; the largest category in classification systems (128)

Koch's postulates: a set of rules to determine the specific bacterium causing a particular disease (638)

Krebs cycle: the series of chemical reactions that begins with the acetyl coenzyme A (CoA) formed from pyruvic acid (115)

L ─────────────────────

labor: the slow, rhythmic contractions of the uterine muscles during childbirth (466)

lacteals: the small lymph vessels found in the center of a villus (164)

Lagomorpha (lag uh MOR fuh): an order of plant-eating mammals having short tails and two pairs of upper incisors, one behind the other; includes rabbits and hares (768)

large intestine: the final section of the digestive tract; reabsorbs water, absorbs vitamins, and eliminates undigested and indigestible material (165)

larva (pl. **larvae**): an early developmental stage of some animals after hatching; must undergo metamorphosis to reach the adult form (683)

larynx: the voice box; connects the pharynx with the trachea (225)

lateral bud: a bud in the upper angle where a leaf joins a stem; axillary bud (367)

lateral lines: lines that extend along each side of a shark to help sense vibration (738)

law of conservation of mass: the principle stating that mass can neither be created nor destroyed (44)

law of dominance: the principle of genetics stating that when organisms pure for contrasting traits are crossed, all their offspring will show the dominant trait (500)

law of independent assortment: the principle of genetics stating that different traits are inherited independently of one another (508)

law of probability: the principle stating that if there are several possible events that might happen, and no one of them is more likely to happen than any other, then they will happen in equal numbers over a large number of trials (504)

law of segregation: the genetic principle stating that the alleles of a gene occur in pairs and are separated from each other during meiosis and are recombined at fertilization (501)

layering: a type of artificial vegetative propagation, accomplished by covering part of a growing plant with soil (411)

learned behavior: behavior that develops as a result of experience (797)

leaves: the usually thin, flat outgrowths of stems; carry out photosynthesis (355)

lens: in the eye, the structure behind the iris that focuses light on the retina (300)

lenticels: the openings in cork tissues that allow the exchange of respiratory gases between the atmosphere and the plant tissues (367)

Lepidoptera (lep uh DAHP tuh ruh): an insect order having two pairs of broad, membranous wings; butterflies and moths; (720)

leucoplasts: the colorless plastids in which starch or other plant nutrients are stored (89)

leukocytes: the nucleated blood cells that protect the body from disease-causing organisms, such as bacteria and viruses; white blood cells (197)

lichen (LY ken): an organism consisting of an alga or a blue-green bacterium and a fungus living together in symbiosis; they grow on soil, rocks, and tree trunks (669)

ligaments: the tough, fibrous bands of connective tissues that hold the bones together at movable joints (262)

light microscope: any device that enables us to see small details in an object by enlarging the object (19)

light reactions: in photosynthesis, a series of reactions requiring light in which water or some other compound is oxidized and ATP and $NADPH_2$ are produced (344)

light system: the mirror and diaphragm in a microscope (22)

limiting factor: a condition of the environment that limits the growth of a population, such as availability of food, water, space, or some other necessity (864)

linkage group: all of the genes that are on the same chromosome (521)

lipids (LIP idz): the organic compounds other than carbohydrates, consisting of carbon, hydrogen, and oxygen; fats, oils, or waxes (62)

littoral zone: the biome between the intertidal zone and the continental shelf; the water is relatively shallow, and light reaches the ocean floor (855)

liver: an organ that secretes bile and removes toxic substances from the blood (161)

lungs: in vertebrates, the organs specialized for the exchange of gases between the blood and the atmosphere (224)

lymph: intercellular fluid and proteins are called *lymph* once they are inside the lymphatic system (189)

lymphatic (lim FAT ik) **system:** a system of vessels that return excess fluid and proteins from the intercellular spaces to the blood (188)

lymph nodes: lymphatic glands that play an important role in the body's defense against disease (189)

lymphocytes (LIM fuh syts): a type of white blood cell that recognizes and destroys antigens present in the body tissue (197)

lysogenic cycle: the process that occurs when the phage's nucleic acid merges with the bacterial cell's DNA, and the new combination is transmitted through bacterial generations (656)

lysosomes: the small, saclike structures surrounded by single membranes and containing strong digestive or hydrolytic enzymes (86)

lytic cycle: the process in which a phage attacks a bacterial cell, uses it to reproduce, and then destroys it (655)

M

macrophages: giant white blood cells that can ingest large numbers of bacteria (201)

magnification: the ratio of the image size to the object size (19)

Malpighian tubules: the excretory organs of grasshoppers and other insects (240)

Mammalia (muh MAYL ee uh): a class of warm-blooded vertebrates that have a four-chambered heart, are covered with hair or fur, and nourish their young with milk; mammals (766)

mandibles (MAN duh bulz): the jaws of crustaceans that crush food by moving from side to side (710)

mantle: a fold of skin that surrounds the body organs in mollusks and contains glands that secrete the shell of shelled mollusks (699)

marrow: the soft tissue that fills the hollow spaces in bone (260)

marsupials: nonplacental mammals in which the fetus is born at a very immature stage and completes its development in a pouch on the mother's body (767)

mass number: the sum of the protons and neutrons in the nucleus of an atom (36)

maturation zone: the region behind the elongation zone of the root in which cells differentiate (361)

mechanical system: the structural parts of a microscope that hold the specimen and lenses and permit focusing of the image (20)

medulla oblongata: the part of the brain beneath the cerebellum and continuous with the spinal cord; controls involuntary activities (293)

medusa: the body form of free-swimming coelenterates (687)

meiosis: cell division in diploid cells that results in haploid cells; reduction division (418)

menstrual cycle: the hormone-controlled cycle in the human female, lasting about a month, in which an egg matures and is released from the ovary and the uterus prepares to receive it (460)

menstruation: the last stage of the menstrual cycle, marked by the shedding of some of the uterine lining, the unfertilized egg, and a small amount of blood through the vagina, which occurs about once a month in the human female (462)

meristematic tissues: plant tissues whose cells undergo, or are capable of, repeated cell division; meristems (356)

meristematic zone: a region of actively dividing cells just behind the root cap (361)

meristems: in plants, a region or tissue composed of cells that undergo or are capable of repeated cell division (356)

mesoderm: the germ layer between the endoderm and ectoderm (443)

mesoglea: the jellylike material, composed largely of protein, found between the ectoderm and endoderm layers of coelenterates (688)

mesophyll (MEZ uh fil): a layer of photosynthetic tissue found between the epidermal layers of a leaf (368)

messenger RNA (mRNA): the type of RNA that carries the code for a polypeptide from DNA to the ribosomes where it is translated (530)

metabolism (muh TAB uh liz um): all the chemical reactions occurring within the cells of an organism (8)

metamorphosis: the series of changes that certain types of organisms undergo as they develop from a larva or nymph to an adult (718)

metaphase: the stage of mitosis or meiosis during which the centromeres of the chromosomes are lined up at the equatorial plane (400)

microdissection (my kroh dis EK shun): dissections using tiny instruments to perform operations on living cells (26)

microfilaments: long, threadlike strands found in the cytoplasm of some cells; involved in movement (87)

micropyle: a small opening in the ovule through which the pollen tube grows (479)

microtubules: the long, cylindrical organelles found in cilia and flagella (87)

migration: an annual round trip made by many species between their winter feeding grounds in the south and their spring breeding grounds in the north (794)

mimicry: a protective adaptation in which one species is protected from its enemies by its resemblance to another species (611)

minerals: chemical elements that organisms need for normal functioning (149)

mitochondrion (pl. **mitochondria**): an oval membrane-enclosed organelle in which most of the reactions of cellular respiration occur (86)

mitosis: the process by which the nucleus of a cell divides, while maintaining the chromosome number (398)

mixture: a combination of substances that are physically mixed without forming new chemical bonds (45)

molars: a type of tooth specialized for grinding food (767)

mold: a type of fossil formed when sediment, in which an organism is embedded, hardens, preserving the shape of the organism after its remains decompose (573)

molecule: an uncharged group of atoms held together by covalent bonds; the smallest particle that retains the properties of a covalent compound (39)

Mollusca: a phylum of invertebrates having soft, unsegmented bodies, often enclosed in a mantle; mollusks (698)

molting: in arthropods, shedding of the exoskeleton (709)

Monera (muh NER uh): the kingdom that includes the simplest one-celled prokaryotic organisms—the bacteria; monerans (134)

monocots: the flowering plants whose seeds have one cotyledon (485)

monohybrid cross: a genetic cross in which only one pair of contrasting traits is studied (507)

monosaccharides (mahn uh SAK uh rydz): the simplest type of carbohydrates; the simple sugars (60)

monotremes: egg-laying mammals; includes the duckbill platypus and spiny anteater (767)

morula (pl. **morulae**): an early stage of animal development in which the embryo consists of a solid ball of cells formed by cleavage of the fertilized egg (442)

mosses: any of a class of bryophytes; small, simple, green plants that grow in moist environments; Musci (475)

multiple alleles: three or more different forms of a gene, each producing a different phenotype (511)

multiple-gene inheritance: the type of inheritance in which two or more pairs of genes affect the same characteristic; polygenic inheritance (523)

muscle tissues: the tissues consisting of cells that have the capacity to contract and exert a pull (100)

muscle tone: the state of partial contraction in which all muscles are kept (265)

mutagens: factors in the environment that cause mutations (546)

mutation: the appearance of a new allele on a chromosome (545)

mutualism: a symbiotic relationship in which both organisms benefit from their association (824)

mycelium (my SEE lee um): a tangled mass of grown-out hyphae (665)

myelin: layers of white, fatty substance produced by Schwann cells on some axons (273)

myofibrils: the small fibers that are arranged in a bundle making up the larger muscle fiber (263)

Myxini (mik SIH nee): a class of vertebrates that lack paired fins, true jaws, and scales of other fishes, and in which the notochord persists throughout life (736)

Myxomycota: a phylum of protists with characteristics of both protozoa and fungi during their life cycles; slime molds (651)

N

nasal passages: the hollow spaces in the nose through which air flows from the nostril to the pharynx (225)

nastic movement: a plant movement that is in response to a stimulus but independent of the direction of the stimulus (383)

natural selection: the process whereby organisms with favorable variations survive and produce more offspring than less well-adapted organisms (602)

Neanderthals (nee AN der thals): an early type of *Homo sapiens* that first appeared about 100 000 years ago (776)

nearsighted: a vision condition that occurs because the eyeball is too long (301)

nectar: a sugary liquid produced by flowers (482)

negative feedback: chemical response to chemical regulation of glandular secretions that opposes the original change (316)

nekton: free-swimming marine organisms (855)

nematocysts: the capsules within cnidoblasts containing coiled, hollow threads that are discharged when the cnidoblasts are stimulated (688)

Nematoda: the phylum consisting of slender, bilaterally symmetrical roundworms (694)

nephridia (nih FRID ee uh): the organs of excretion in the earthworm and other annelids (238)

nephrons: the functional units of the kidney (243)

nerve cord: in chordates, a dorsal hollow structure (734)

nerve net: a type of nervous system found in the hydra; the nerve cells are formed into an irregular network through which coordinated movement can occur (282)

nerves: bundles of axons or dendrites that are bound together by connective tissues (274)

neuromuscular junctions: the junctions between motor neurons and muscle fibers (280)

neuron: a cell specialized for the transmission of impulses: a nerve cell (273)

neurotransmitters: a substance released from the synaptic knob into the synaptic cleft that initiates impulses in adjacent neurons (279)

neutralization (noo truh luh ZAY shun): the reaction of an acid with a base to produce a neutral solution (49)

neutrons (NOO trahnz): particles in the nuclei of atoms that have no electrical charge; have roughly the same mass as protons (34)

niche: the role of a species in an ecosystem (826)

nictitating (NIK tuh tayt ing) **membrane:** a transparent kind of eyelid present in many vertebrates, allowing them to see underwater (744)

nitrifying bacteria: bacteria that can convert ammonia to nitrite and nitrate (831)

nitrogen fixation: a process by which nitrogen-fixing organisms produce nitrogen compounds from the gaseous nitrogen of the atmosphere (831)

nitrogen fixers: bacteria and blue-green algae that can produce nitrogen compounds from the gaseous nitrogen of the atmosphere (831)

noise pollution: loud sounds that can cause hearing loss (867)

nomenclature (NOH men klay chur)**:** a system for naming things (129)

nondisjunction: the failure of homologous chromosomes to separate normally during meiosis, producing gametes or spores with one more or one less chromosome than normal (547)

nonrenewable natural resources: resources that can be used only once, such as coal, oil, and minerals, and cannot be replaced (873)

norepinephrine: an excitatory neurotransmitter; noradrenaline (324)

notochord (NOHT uh kord)**:** a flexible, rodlike, internal supporting structure found in all chordates during some part of their lives (734)

nuclear envelope: the membrane that surrounds the nucleus (84)

nucleic acids: compounds that contain phosphorous and nitrogen in addition to carbon, hydrogen, and oxygen; DNA or RNA (64)

nucleoli: dense, granular bodies that are found in the nucleus of cells and are a site of RNA production (85)

nucleotides: the base units of nucleic acids, each containing a sugar, a phosphate group, and one of four nitrogenous bases (64, 526)

nucleus: in a *eukaryotic cell,* a large, membrane-enclosed organelle that contains the cell's DNA; in an *atom,* containing protons and neutrons (77)

nutrients (NOO tree unts)**:** the substances that an organism needs for energy, growth, repair, or maintenance (6, 149)

nutrition: the process by which organisms take in food and break it down so it can be used for metabolism (149)

nymphs: the young of an insect, such as a grasshopper, hatched from an egg in incomplete metamorphosis; resembles the adult insect but lacks some of the adult features (718)

O

olfactory cells: receptors for smell located in the mucous membrane lining the upper nasal cavities (306)

Oligochaeta: a major class of the phylum Annelida; includes the earthworm (696)

omnivores (AHM nih vorz)**:** a heterotroph that feeds on both plants and animals (824)

oncogenes: any of various genes that, when activated, may cause normal cells to become cancerous (539, 658)

one gene-one polypeptide hypothesis: the hypothesis that every gene directs the synthesis of a particular polypeptide chain; originally called the one gene-one enzyme hypothesis (529)

oogenesis: the formation of eggs in the ovaries (429)

Oomycota (oh uh my KOH tuh)**:** a phylum of protists that look like fungi but differ from fungi in the content of their cell walls and in their mode of sexual reproduction; water molds (652)

ootid: the large, haploid daughter cell produced by the meiotic division of the secondary oocyte; matures into an egg (429)

open circulatory system: the blood flows directly into body tissues (176)

operculum (oh PER kah lum)**:** a protective flap that covers the gills of bony fishes; gill cover (740)

operon: in prokaryotes, such as bacteria, the promoter, the operator, and their associated structural genes (537)

opposable thumb: a thumb positioned opposite to the other fingers, making it possible to grasp objects; found in primates (771)

optical system: the lenses of a compound microscope (20)

optic nerve: the nerve that carries impulses from the receptors in the retina of the eye to the brain (300)

oral groove: the opening in the paramecium through which food is ingested (156)

order: a group of related families (128)

organ: a group of tissues that work together to perform a specific function (100)

organelles: the specialized structures in the cytoplasm of cells that carry out specific functions (82)

organic compounds: compounds that contain carbon and usually hydrogen; most occur naturally only in the bodies and products of organisms (57)

organic evolution: the process of continual change that occurs in species over time (571)

organism (OR guh nih sum)**:** a living thing (4)

organ system: a group of organs that work together to perform a specific function (101)

Orthoptera (or THAHP tuh ruh)**:** a large order of insects that exhibit incomplete metamorphosis, have biting mouthparts, and inhabit the ground or low vegetation; includes cockroaches and grasshoppers (720)

osculum: an opening at the unattached end of a sponge that serves as the excurrent opening (686)

osmosis: the diffusion of water across a semipermeable membrane from a region of high concentration of water to a region of low concentration of water (93)

osmotic pressure: the increase in pressure resulting from the flow of water in osmosis (94)

ossification: the process by which cartilage is replaced by bone in the skeletons of most vertebrates (260)

Osteichthyes (ahs tee IK thee eez)**:** a class of vertebrates having a bony skeleton, movable jaws, overlapping scales, paired fins, and an air bladder; bony fishes (739)

osteocytes: the bone-forming cells entrapped in small cavities within the bone substance (259)

ova (pl. **ovum**)**:** egg cells (428)

oval window: the membrane between the middle and inner ear, connected to the eardrum by three small bones (303)

ovary (pl. **ovaries**)**:** in *animals,* the female gonad, which produces egg cells; in *flowering plants,* the basal part of the pistil, which contains ovules and, later, seeds, and which develops into a fruit (428, 459, 481)

oviduct: a tube that carries the egg away from the ovary (459)

ovipositor (OH vuh pahz it er): a hard, four-pointed organ that is used to dig holes in the ground in which eggs are deposited (721)

ovulation: the release of an egg from an ovary (459)

ovule: in seed plants, a structure within the ovary that contains a female gametophyte and that develops into a seed after fertilization (478)

oxidation (ox sih DAY shun): any chemical change in which an atom or a molecule loses electrons (110)

oxidation-reduction reaction: a reaction in which one substance is oxidized and another substance is reduced (110)

oxygen cycle: the pathways of oxygen in the biosphere (833)

oxygen debt: the amount of oxygen needed to get rid of lactic acid (119)

P

palisade mesophyll: the upper portion of the mesophyll (368)

pancreas: an organ that is both an exocrine gland and an endocrine gland and that secretes digestive juice and the hormones insulin and glucagon (161)

pancreatic juice: the digestive secretion of the pancreas containing sodium bicarbonate, amylase, proteases, and lipases (164)

parapodia: paired, paddlelike extensions on each segment of some annelids; used for swimming and creeping (697)

parasites: a type of heterotroph; obtain nutrients from living organisms in or on which they live (633)

parasitism: a symbiotic relationship in which one organism benefits from the association and the other is harmed (825)

parasympathetic nervous system: the division of the autonomic nervous system that slows down the functioning of various body systems (296)

parathyroid glands: the four small glands embedded in back of the thyroid that secrete parathormone (323)

parent (P) generation: the starting generation in a breeding experiment (499)

parthenogenesis: the development of an unfertilized egg into an adult animal without fusion with sperm (434)

passive immunity: a type of immunity that is acquired when a person is given antibiotics, obtained from the blood of either another person or an animal, to attack a particular antigen (205)

passive transport: a process by which materials move across cell membranes without the expenditure of cellular energy (96)

pathogens: viruses, bacteria, and other microorganisms that cause disease (201, 637)

pedicel: the stalk that bears a single flower (480)

pedigree chart: a diagram that shows the presence or absence of a particular trait in each member of each generation (549)

pedipalps: in arachnids, the second pair of appendages; sensitive both to chemicals and to touch (714)

pellicle: a grooved, flexible, proteinaceous structure found inside the cell membrane of some protists (644)

pepsin: a protein-digesting enzyme in gastric juice (162)

peptide (PEP tyd) **bond:** the bond formed between two amino acids by dehydration synthesis (67)

pericardium (per uh KARD ee um): a tough membrane that covers and protects the heart (179)

periosteum: a tough membrane covering the outside of bones, except at joints (259)

peripheral nervous system: the division of the nervous system that includes all the neurons and nerve fibers outside the brain and spinal cord (282)

Perissodactyla (puh ris uh DAK tuh luh): an order of hoofed mammals having an uneven number of toes on each foot; includes horses and rhinoceroses (769)

peristalsis: the alternate waves of contraction and relaxation in the walls of the alimentary canal (162)

permafrost: the lower layers in the tundra that remain frozen throughout the year (846)

petals: the usually showy flower structures located between the sepals and the stamens (481)

petiole: the structure that attaches the leaf to the stem of a plant (367)

petrifaction (peh truh FAK shun): the process by which the body of a dead organism is slowly replaced by dissolved minerals (573)

pH: a unit of measurement that indicates the concentration of hydrogen ions in a solution (50)

Phaeophyta (fee AH fuh tuh): a plant phylum that includes multicellular seaweeds and kelps; brown algae (650)

phagocytosis: the process in which large particles or small organisms are ingested into a cell (97)

pharynx (FA rinks): the throat (158, 225, 692)

phase-contrast microscope: a form of compound microscope that allows the details within living specimens to be seen without straining (24)

phenotype: the physical traits that appear in an individual as a result of its genetic makeup (503)

pheromones (FEHR uh mohnz): a type of animal secretion that serves as a means of communication between members of the same species (804, 878)

phloem: the tissue that conducts food and other dissolved materials throughout the body of a vascular plant (357)

photoautotrophs: organisms that use light energy to drive the reactions needed to make food; photosynthetic organisms (340)

photon: a particle of light (341)

photoperiodism: the response of a plant to the changing duration of light and darkness during the year (384)

photosynthesis: the process by which organic compounds are synthesized from inorganic carbon, i.e., CO_2, in the presence of light in most autotrophic organisms (340)

phototropism: the growth of a plant in response to the stimulus of light; the stem of a plant grows toward light or away from it (383)

phylogeny: the evolutionary history of a species or a group of organisms (130)

phylum (FY lum): one of the largest or most inclusive groups within a kingdom (128)

phytoplankton: photosynthetic organisms that float near the surface of water (855)

pigment: a substance that absorbs only certain wavelengths of light; a substance that has color (341)

pineal gland: a pea-sized gland attached to the base of the brain that produces melatonin (328)

Pinnipedia (pin uh PEED ee uh): an order of aquatic mammals having flippers; seals and walruses (770)

pinocytosis: the process in which liquids or very small particles from the surrounding medium are taken into a cell by the formation of a vesicle (97)

pistil: the part of a flower that contains the ovules and through which pollen tubes grow (481)

pith: the center of a herbaceous dicot system, made up of parenchyma cells (365)

pituitary gland: the endocrine gland attached to the hypothalamus that controls the activities of many other endocrine glands in the body (320)

pivot joints: a type of joint that permits rotation from side to side as well as up-and-down movement (262)

placenta: in mammals, a temporary organ through which the fetus receives food and oxygen from the mother's body and gets rid of wastes (450, 465)

placental mammals: mammals in which a placenta forms during development of the embryo (767)

plankton: organisms that float in a body of water (855)

Plantae (PLAN tee): one of the five kingdoms; members are mostly multicellular and photosynthetic; plants (135)

planula: the small, ciliated larva of many coelenterates (689)

plasma (PLAZ muh): the liquid part of blood (196)

plasmids: the small, circular segments of DNA that are found in bacteria and that stay separate from the bacterial chromosomes; used in genetic engineering (559)

plasmolysis: the shrinking of cytoplasm resulting from loss of water by osmosis in a cell placed in a hypertonic solution (96)

plastids: membrane-enclosed organelles found in the cells of some protists and almost all plants; includes chloroplasts, chromoplasts, and leucoplasts (89)

platelets: small, round or oval blood fragments that trigger the blood-clotting process (198)

Platyhelminthes: the simplest animals showing bilateral symmetry (691)

pleura: a two-layered membrane that encloses the human lung (225)

point mutation: a type of gene mutation in which only a single nucleotide in the gene has been changed (547)

polar molecule: a molecule with regions of partial negative and partial positive charges (58)

polar nuclei: the two nuclei found within the embryo sac (482)

pollen grain: the male gametophyte of seed plants (478)

pollen tube: the tubelike outgrowth of the pollen grain through which the sperm nuclei pass to the ovule (479)

pollination: the transfer of pollen from an anther to a stigma of a flower (478)

pollution: the addition of anything to the environment that makes it less fit for living things (867)

Polychaeta: a class that includes the marine sandworm *Nereis;* similar to the earthworm (697)

polymers (PAHL uh merz): large molecules consisting of chains of repeating units (61)

polyp: the sessile body form of a coelenterate (687)

polypeptide: a chain of amino acids joined by peptide bonds (67)

polyploidy: a condition in which the cells have some multiple of the normal chromosome number (547)

polysaccharides (pahl ee SAK uh ryds): a long chain of repeating sugar units formed by joining simple sugars by dehydration synthesis (61)

pons: a part of the brain that serves as a relay system linking the spinal cord, medulla oblongata, cerebellum, and cerebrum (291)

population: the simplest grouping of organisms in nature (605, 823)

population genetics: the study of the changes in the genetic makeup of populations (605)

pores: the tiny openings found in plant leaves and animal skins through which fluids are absorbed or discharged (685)

Porifera: a phylum of the simplest multicellular animals; aquatic, immobile animals with an outer layer pierced by many pores; sponges (685)

positive feedback: feedback from chemical regulation of glandular secretions that reinforces the original change (316)

posterior: pertaining to the rear or tail end of a bilaterally symmetrical animal (684)

predators (PRED uh terz): the carnivores that capture and feed on prey (824)

pregnancy: in mammals, the period during which the developing embryo is carried in the uterus (463)

premolars: a type of tooth specialized for grinding food (767)

pressure-flow theory: an explanation for the movement of liquid in plant phloem from an area of high concentration to one of low concentration (379)

primary immune response: the body's initial response to an antigen; does not produce measurable amounts of antibodies (203)

primary root: the first structure to emerge from a sprouting seed (359)

primary succession: succession that occurs in an area that had no previously existing life (836)

Primates (pry MAYT eez): an order of mammals having grasping hands and flexible feet, each with five digits; includes humans, apes, monkeys, and lemurs (771)

primitive gut: the cavity within the gastrula of an embryo that eventually forms the digestive tract (443)

Proboscidea (proh buh SID ee uh): an order of mammals having tusks and long, flexible, tubelike snouts; includes elephants (769)

Prochlorophyta: one-celled organisms that live in close association with certain marine animals (636)

producers: organisms, such as green plants, that produce organic compounds from inorganic compounds; the autotrophs (828)

products: the new substances produced by chemical reactions (44)

progesterone: a hormone secreted by the ovaries that helps to regulate the menstrual cycle and maintains the uterus during pregnancy (327)

proglottids: the segmentlike divisions of a tapeworm's body (694)

prokaryotic cells: cells lacking distinct membrane-bound structures; monerans; bacteria or blue-green algae (79)

prophase: the stage of mitosis or meiosis in which the chromosomes and spindle appear and the nuclear membrane disappears (400)

prostaglandins: local hormones that produce their effects on the cells in which they are synthesized, without entering the bloodstream (315)

proteins (PROH teenz)**:** organic compounds consisting of one or more chains of amino acids; contain nitrogen as well as carbon, hydrogen, and oxygen (66)

Protista (proh TIST ah)**:** one of the five kingdoms; includes simple, mostly unicellular, eukaryotic organisms; protists (134, 642)

protons (PROH tahnz)**:** positively charged particles found in the nuclei of all atoms; have roughly the same mass as neutrons (34)

protozoa (sing. **protozoan**)**:** protists that are usually motile, e.g., amebas and paramecia (642)

pseudopods: in certain cells, the temporary projections of cell surfaces that enable cells to move and engulf particles (255)

pulmonary artery: the two-branched vessel that brings blood from the right ventricle of the heart to the lungs (180)

pulmonary circulation: this pathway in the body carries blood between the heart and the lungs (184)

pulse: the expansion and relaxation that can be felt in an artery each time the left ventricle of the heart contracts and relaxes (182)

punctuated equilibrium: a theory of evolution stating that a species remains the same for a long time and then evolves rapidly during a short time interval (604)

Punnett square: a diagram, used in genetics, to show the results of a cross (504)

pupa (PYOO puh)**:** the resting stage that larvae pass into after several molts (719)

pupil: the opening in the center of the iris of the eye, which allows light to enter the eye (300)

purple bacteria: autotrophic eubacteria that carry out photosynthesis without producing oxygen (636)

pyramid of biomass: the relative mass of organisms at each feeding level in an ecosystem (830)

pyramid of energy: the amount of available energy in an ecosystem (830)

R

radial symmetry: a type of symmetry in which any section through and parallel to the central axis of the organism divides it into similar halves (684)

radicle: the root portion of a seed embryo (485)

radioactive dating: a dating method based on the rate of disintegration of radioactive isotopes; used to determine the age of rocks and fossils (575)

radioactivity (ray dee oh ak TIV uh tee)**:** the process in which the nucleus of an atom gives off radiation or charged particles; changes the atom to another isotope or a different element (38)

radioisotopes (ray dee oh I suh tohps)**:** radioactive isotopes (38)

radula: a rasping, tonguelike organ in mollusks (699)

range: the particular region of the earth where a species is found (614)

reactants (ree AK tuntz)**:** the substances that take part in chemical reactions (44)

receptacle: the expanded end of the pedicel, to which the flower parts are attached (480)

receptors: in a nervous system, the specialized structures sensitive to certain types of stimuli; sense organs (272)

recessive: the inherited characteristic often masked by the dominant characteristic and not seen in an organism (500)

recombinant DNA: DNA that has been altered by genetic engineering (558)

rectum: a structure in which undigested food (feces) is stored prior to elimination from the body (159)

recycling: the process of reusing materials rather than discarding them as waste (873)

red blood cells: red cells that carry oxygen and carbon dioxide; erythrocytes (196)

reduction: chemical reaction in which a substance gains an electron (110)

reflex: an involuntary, automatic response to a given stimulus, not involving the brain (297, 792)

reflex arc: the pathway over which the nerve impulses travel in a reflex (298)

reforestation: a program that plants seeds or seedlings to replace trees lost in cutting (875)

refractory period: the brief recovery period during which the cell membrane of a neuron cannot carry impulses (277)

regeneration: the ability of an organism to regrow lost body parts (408)

regulation: all the activities that help to maintain an organism's homeostasis (7)

relative dating: any method of determining the order in which events occurred (575)

releasing factors: hormones that are produced by the hypothalamus and that control the release of a hormone from the anterior pituitary (321)

renal (REEN ul) **arteries:** the vessels that bring blood to the kidneys (244)

renal circulation: the branch of the systemic circulation that carries blood to and from the kidneys (188)

renal veins: the vessels through which blood flows from the kidneys (244)

renewable natural resources: a natural resource, such as air, water, soil, sunlight, and living organisms, that can be replaced by natural processes (873)

reproductive isolation: the loss of the ability to interbreed by two isolated groups (615)

Reptilia (rep TIL ee uh)**:** a class of cold-blooded vertebrates with dry skin, scales, four limbs (except for snakes), lungs, and a three-chambered heart; fertilization of eggs is internal; reptiles (750)

resolution: the ability of a microscope to show two points that are close together as separate images (23)

respiration (res puh RAY shun)**:** the process of releasing energy in a complex series of chemical reactions (6)

respiratory pigments: colored substances in the blood of most multicellular animals which carry oxygen and carbon dioxide between the respiratory surface and the body cells (221)

respiratory surface: a moist surface through which the exchange of respiratory gases takes place (220)

restriction enzymes: the proteins used to cut DNA molecules at specific places so that scientists can isolate pieces with the desired genes (558)

retina: the innermost layer of the eye, on which an image is projected by the lens (300)

retrovirus: an RNA virus; a virus whose genetic material is RNA (655)

Rh factor: a group of antigens found on the surface of red blood cells (208)

rhizoids: rootlike hyphae that anchor a fungus, secrete digestive enzymes, and absorb nutrients (475, 666)

rhizome: a thick, horizontal stem containing stored food, which forms new plants by vegetative reproduction (410)

Rhodophyta (roh DAH fuh tuh)**:** a phylum of plants that includes unicellular organisms and multicellular seaweeds; red algae (650)

ribosomal RNA (rRNA): a type of RNA transcribed from DNA in the nucleolus and found in the ribosomes (532)

ribosomes: the organelles that are the sites of protein synthesis in cells (85)

RNA (ribonucleic [ry boh noo KLAY ik] **acid):** the nucleic acid that is transcribed from DNA (64)

Rodentia (roh DENCH ee uh)**:** the largest order of placental mammals, having sharp incisors for gnawing; includes rats, mice, and squirrels; rodents (768)

rods: structures in the retina of the eye responsible for black-and-white vision (301)

root cap: a thimble-shaped group of cells that form a protective covering for the root tip (361)

root hairs: hairlike extensions of root epidermal cells that increase the surface area for absorption (362)

root pressure: the osmotic pressure in the xylem of a root (377)

roots: the structures containing vascular tissues adapted for anchoring plants and absorbing water and dissolved substances (355)

rumen (ROO men)**:** in ruminants, the chamber of the stomach in which food is stored (770)

runner: a horizontal stem with long internodes that forms independent plants by vegetative reproduction; a stolon (410)

S

saliva: the secretion of the salivary glands (159)

salivary amylase: the enzyme in saliva that hydrolyzes starch into maltose; ptyalin (161)

salivary glands: the glands that secrete saliva into the mouth (159)

salt: an ionic compound produced by the neutralization reaction between an acid and a base (49)

S-A node (sinoatrial node): a small group of specialized muscle cells in the wall of the right atrium; stimulates contraction of the heart; pacemaker (181)

saprobes: the organisms that obtain nutrients by breaking down the remains of dead plants and animals (633, 824)

Sarcodina: a phylum of protists composed of protozoa that move and capture prey by using pseudopods; sarcodines (642)

saturated (SATCH uh rayt ed) **fats:** fats formed from fatty acids in which all carbon-to-carbon bonds are single bonds (63)

scanning electron microscope: an electron microscope that uses an electron beam focused to a fine point and passed back and forth over the surface of the specimen (25)

scavengers: carnivores that feed on dead animals they find (824)

Schizophyta: the largest group of heterotrophic eubacteria (636)

Schwann cells: cells that surround some axons and form myelin (273)

scientific law: a statement that describes some aspect of a phenomenon that is always true (17)

scientific method: a universal approach to scientific problems, consisting of defining the problem, formulating a hypothesis, testing the hypothesis, and recording and reporting observations (13)

sclera: the tough, fibrous, white outer layer of the eye (299)

scrotum: a sac of skin outside the body wall in which the testes are located (458)

sebaceous glands: the glands in the skin that produce oily secretions (247)

secondary response: a rapid response by the immune system when it encounters the same antigen a second time in the body; produces high levels of antibodies within two days (203)

secondary roots: the roots that branch off from primary roots (359)

secondary sex characteristics: characteristics—such as body hair, muscle development, broadened pelvis, or voice depth—controlled by the male and female sex hormones, but not essential to the reproductive process (457)

secondary succession: succession that occurs in an area in which an existing community has been partially destroyed and its balance upset (836)

secretion: in cells, the process by which materials made by the cell are released to the outside of the cell (86)

sedimentary (sed uh MEN tuhr ee) **rock:** a type of rock formed from layers of particles that settled to the bottom of a body of water, often containing fossils (573)

seed: in seed plants, the structure formed from the ovule following fertilization; contains the plant embryo, stored nutrients, and a seed coat (478)

seed coat: a tough, protective covering around a seed that develops from the wall of the ovule (478)

selection: a technique in which only those animals and plants with the most desirable traits are chosen for breeding (555)

selectively permeable: a characteristic of a cell membrane that allows some substances to pass freely through the membrane, while others can pass through to a slight extent or not at all (84)

semen: the mixture of sperm and fluids released during ejaculation (459)

semicircular canals: a system of loop-shaped tubes in the inner ear that enable the body to maintain balance (304)

sensory neurons: nerve cells that carry impulses from receptors toward the spinal cord and brain (274)

sepals: the leaflike structures at the base of flowers (480)

setae: tiny bristles on the body segments of annelids, used in locomotion (257)

sex chromosomes: the two unmatched chromosomes that determine the sex of an individual; represented as X and Y (518)

sex-linked trait: a trait that is controlled by a gene found on one of the sex chromosomes (520)

sexual reproduction: a form of reproduction in which a new individual is produced by the union of the nuclei of two specialized sex cells, i.e., gametes, usually from two separate parent organisms (8, 398)

SI: the International System of Units; the metric system (17)

sickle-cell disease: a recessive inherited disorder of the blood in which the red blood cells have an abnormal sickle shape (550)

simple leaf: a leaf with only one blade and one petiole (367)

simple microscope: a magnifying glass (20)

sinoatrial node: *see* **S-A node** (181)

Sirenia (sy REEN ee ah): an order of large, vegetarian sea mammals; manatees and dugongs (770)

skeletal muscle: muscle that is attached to bone and is involved in locomotion and voluntary movement; striated muscle (263)

small intestine: the part of the digestive tract where most chemical digestion and almost all absorption occurs (164)

smooth muscle: muscle tissue made up of individual cells, not marked by striations, and not under voluntary control (265)

social behavior: animal behavior consisting of both helpful and hostile interactions (801)

sodium-potassium pump: an active transport mechanism that pumps sodium ions out of, and potassium ions into, a nerve cell; sodium pump (276)

solutes (SAHL yoots): substances that are dissolved in solvents (46)

solution: any homogeneous mixture (46)

solvent: the substance, usually a liquid, that makes up the bulk of a solution (46)

somatic cells: the body cells, as distinguished from the sex cells (418)

somatic nervous system: the division of the peripheral nervous system that contains sensory and motor neurons that connect the central nervous system to skeletal muscles, skin, and sense organs (296)

speciation: the formation of new species (603)

species (SPEE sheez): a group of organisms that are structurally similar and that pass these similarities on to their offspring (128)

spectrophotometry: a method of identifying and quantifying a substance by measuring the amount of light at different wavelengths that it absorbs (26)

spermatogenesis: the formation of sperm in the testes (429)

spermatogonia: diploid cells in the testes, from which sperm are formed through meiosis and differentiation (430)

sperm cells: the male gametes (428)

spherical symmetry: a type of symmetry in which any cut passing through the center of the organism divides it into matching halves (684)

sphincter: a ring of muscle that acts as a valve (162)

spicules: the small skeletal structures embedded in the middle layer of sponges that provide support and give shape to the organism (686)

spinal column: in vertebrates, the series of vertebrae connected by cartilage disks that surrounds and protects the spinal cord; the backbone (260)

spinal cord: the cord of nervous tissue in vertebrates that extends down from the brain, running through the vertebrae of the spinal column (294)

spinal nerves: the nerves connected to the spinal cord (295)

spindle: a structure formed by fibers during mitosis (400)

spinnerets (spin uh RETS): the organs in spiders and other small arachnids that are used to spin silk (715)

spiracles: several paired openings through which air enters and leaves the body of terrestrial arthropods (223, 721)

spirillum: a spiral or coiled bacterium (632)

spongin: protein-containing substance that makes up the fibers in the skeleton of some sponges (686)

spongy mesophyll: the lower portion of mesophyll in a leaf, consisting of irregularly shaped, chloroplast-filled cells separated by large air spaces (369)

spontaneous generation: the idea that living things regularly arise from nonliving matter; abiogenesis (585)

spores: the specialized reproductive cells that can give rise to new organisms (407)

sporophyte (SPOR uh fyt): a spore-producing plant (474)

Sporozoa: a phylum of protists composed of nonmotile, parasitic protozoans; sporozoans (645)

stabilizing selection: a type of natural selection in which the average phenotype may be a favorable adaptation, and the extreme phenotypes are unfavorable (613)

stamen: the organ of a flower that bears pollen grains (481)

starch: a polysaccharide that is the main food storage compound in plants (61)

stems: the leaf-bearing structures of vascular plants (355)

stereomicroscope: a microscope used in studying the external, or surface, structure of specimens; has low magnifying power, ranging from 6X to 50X, thus providing a three-dimensional image of the specimen (24)

stigma: in a *pistil,* the enlarged, sticky knob on top of a style that receives the pollen; in *protists,* an eyespot (481)

stimulus (pl. **stimuli**): any factor that causes a receptor to trigger impulses in a nerve pathway, resulting in a change of activity (272, 785)

stolons: in fungi, hyphae that grow in a network over the surface of food (666)

stomach: the organ of the digestive tract in which food is temporarily stored and partially digested (159)

stomates (STOH mayts): the openings in the epidermis of leaves that allow the exchange of respiratory gases between the internal tissues of the leaf and the atmosphere (364)

strip cropping: a conservation practice in which cover crops are planted between strips of row crops, leaving no soil open to erosion (874)

stroma: in plant chloroplasts, the regions between the grana (343)

structural formula: a kind of chemical formula that shows how atoms in a molecule are bonded to one another (43)

style: in flowers, the part of the pistil between the stigma and the ovary (481)

substrate (SUB strayt): the substance that an enzyme acts upon (68)

superior vena cava (VEE nuh KAY vuh): veins, some of the largest in the body, that return blood from the head, arms, and chest into the right atrium of the heart (187)

suspension (suh SPEN shun): a mixture that separates when left still (46)

sustained-yield tree farming: a method of forest conservation in which trees are cut down in certain areas of a forest, leaving surrounding areas untouched (875)

sweat glands: a gland composed of a tiny coiled tube that opens to the surface of the skin and secretes perspiration (247)

swim bladder: in bony fishes, a gas-filled sac that regulates the buoyancy of the fish; air bladder (742)

symbiotic relationships: relationships in which two different organisms live in close association with each other to the benefit of at least one of them (824)

sympathetic nervous system: the division of the autonomic nervous system that generally accelerates body activities (296)

synapse: the region where nerve impulses pass from one neuron to another (274)

synapsis: the pairing of replicated homologous chromosomes during prophase I of meiosis (420)

synthesis (SIN thuh sis): the chemical combining of simple substances to form more complex substances (7)

synthetic theory: a modern theory of evolution stating that populations evolve, rather than the individuals within populations (605)

systemic circulation: a pathway that carries blood between the heart and the rest of the body, excluding the lungs (184)

systole (SIS tuh lee): the period of contraction of the heart (180)

T

T cells: lymphocytes that are produced in the bone marrow and matured in the thymus gland and are then released into the circulatory and lymphatic systems (203)

tadpoles: larval form of frogs and toads (744)

taiga (TY guh): a biome with cold winters and warm, moist summers; climax flora are evergreen vegetation (847)

taste buds: the taste receptors on the tongue (306)

taxis: any movement by a simple animal or a protist toward or away from a particular stimulus (791)

taxonomic key: a procedure that is a series of instructions biologists use to identify an organism (136)

taxonomy (tak SAHN uh mee): the branch of biology that deals with the classification and naming of living things (125)

telophase: the stage of mitosis during which the chromosomes uncoil, the spindle and asters disappear, and the nuclear membrane reforms (400)

temperate deciduous forest: a biome in which the climax vegetation is deciduous trees; characterized by hot, humid summers and cold winters (848)

temperature inversion: a situation in which a layer of cooler, denser air becomes trapped below a layer of warmer air (871)

tendons: strong bands of connective tissues that attach skeletal muscle to bone (264)

terminal bud: the bud at the tip of a plant stem (366)

terracing: a method of altering land for cultivation, in which flat areas are cut into the sides of a hill to prevent soil erosion from water running over the surface (874)

territory: an area defended by an individual against intrusion by other members of the same species (804)

test cross: a genetic cross in which a test organism showing the dominant trait is crossed with one showing the recessive trait; used to determine whether the test organism is homozygous dominant or heterozygous (506)

testes: the male gonads, which produce sperm and secrete male sex hormones (428, 457)

testosterone: a male sex hormone secreted by the testes; stimulates development of the male reproductive system and promotes male secondary sex characteristics (327)

tetrad: the group of four chromatids formed in prophase I of meiosis (420)

thalamus: a relay center between various parts of the brain and the spinal cord (291)

theories: explanations based on facts that apply to a broad range of phenomena (17)

theory of evolution: theory of gradual change in species over time (130)

thermal pollution: a type of pollution in which warmed water, which has been used to cool industrial equipment, is returned to a stream or river; the change in water temperature kills fish and other organisms (868)

thigmotropism: growth of a plant in response to the stimulus of touch, as when tendrils of a grapevine wind themselves around the stem of another plant (383)

thorax (THOR aks)**:** the head in most arthropods (709)

threshold: the minimum sensitivity level of a nerve cell; impulses below this level do not initiate responses (277)

thylakoids: the photosynthetic membranes in the chloroplast which are arranged in the shape of flattened sacs (343)

thymine: a nitrogenous base found in DNA (526)

thymus: a gland located in the upper part of the chest cavity that is involved in immunity (327)

thyroid gland: the endocrine gland located in front of the trachea; secretes thyroxine and calcitonin (322)

thyroxine: an iodine-containing hormone that is secreted by the thyroid and that regulates the rate of metabolism in the body (317)

tissue: in multicellular organisms, a group of cells that are similar and are organized into a functional unit; usually integrated with other tissues to form an organ (99)

tissue culture: a technique of maintaining living cells or tissues in a culture medium outside the body (26)

trachea: the tube through which air passes from the pharynx to the lungs (225)

tracheal tubes: a system of branching air tubes that carries air directly to all the cells of the body (222)

Tracheophyta (tray kee AH fuh tuh)**:** the phylum of vascular plants; contain the water-conducting and food-conducting tissues—xylem and phloem; tracheophytes (672)

transcription: the copying of a genetic message from a strand of DNA into a molecule of RNA (530)

transduction: the process in which pieces of DNA are transferred by viruses from one bacterial cell to another (561)

transfer RNA (tRNA): the type of RNA that carries a particular amino acid to mRNA at the ribosome in protein synthesis (531)

transformation: the transfer of DNA from dead, ruptured bacteria to living bacterial cells (560)

translation: the process by which the information coded in RNA is used for the assembly of a particular amino acid sequence (534)

translocation: in *plants,* the movement of dissolved materials; in *genetics,* the transfer of a chromosome segment to a nonhomologous chromosome (379, 547)

transmission electron microscope: a microscope that uses electron beams rather than light and electromagnetic lenses rather than glass lenses (24)

transpiration: the evaporation of water vapor from plant surfaces (375)

transpirational pull: the chief process by which water moves through the xylem of a plant (378)

transport: the process by which substances move into or out of cells or are distributed within cells (6, 173)

Trematoda: a class of the phylum Platyhelminthes; the parasitic flatworms; flukes (693)

trochophore larva: a type of free-swimming, ciliated larva (698)

tropical rain forests: a biome found around the equator in which there is a constant supply of rainfall and the temperature remains at about 25 °C throughout the year (850)

tropism: a growth response in a plant caused by a stimulus that comes primarily from one direction (382)

tube feet: in echinoderms, water-filled tubes, each ending in a suction disk; used in locomotion, feeding, and respiration (726)

tuber: an enlarged portion of an underground stem that can grow into a new plant by vegetative reproduction (409)

tundra: a region that lies south of the ice caps of the Arctic and extends across North America, Europe, and Siberia (846)

Turbellaria: the class of free-living flatworms with eyespots; *Planaria* (691)

tympanic membrane: a delicate membrane stretched across the inner end of the auditory canal of the ear; eardrum (303, 744)

typhlosole: a longitudinal fold in the intestinal wall of some animals, such as the earthworm, that increases the surface area of the intestine (158)

U

ultrasound: a procedure used to determine the size and position of a developing fetus; high-frequency sound waves are reflected off the fetus to produce an image that can be studied (554)

umbilical cord: in placental mammals, the structure that connects the fetus and the placenta (450, 465)

universal donors: individuals with type O blood (209)

universal recipients: individuals with type AB blood (209)

unsaturated fats: fatty acid molecules that have one or more double or triple carbon-to-carbon bonds (63)

urbanization: the transformation of a rural area to a city environment (865)

urea (yuh REE uh)**:** a nitrogenous waste formed from ammonia and carbon dioxide (239)

ureter (YUR et ur)**:** a tube that carries urine from a kidney to the bladder (242)

urethra: the tube that carries urine from the bladder to the outside of the body (243, 458)

uric (YUR ik) **acid:** a dry, nitrogenous waste product excreted by birds, reptiles, and insects (240)

urinary bladder: a saclike organ where urine is stored before being excreted (242)

urinary (YUR uh ner ee) **system:** the system involved in the production and excretion of urine, including the kidneys, bladder, and associated tubes (242)

urine (YUR en): an excretory liquid composed of water, urea, and salts (239)

Urochordata (yur uh kor DAH tah): a subphylum of chordates that are soft, saclike, sessile, marine animals; tunicates (733)

uterus: the thick, muscular, pear-shaped organ in the female mammal in which the embryo develops (449, 460)

V

vacuoles: a membrane-enclosed, fluid-filled cavity in a cell (88)

vagina: the structure leading from the uterus to the outside of the body in mammalian females; the birth canal (460)

valves: flaplike structures that allow blood to flow in only one direction—toward the heart (178)

variable: a single factor that is changed in a controlled experiment (15)

variations: the characteristics in individuals that differ from the typical characteristics of other individuals of the same species (602)

vascular bundle: a structure within a stem containing parallel strands of xylem and phloem; may also contain cambium (364)

vascular cambium: the meristematic layer of cells that causes growth in width of a stem or root (356)

vascular cylinder: the central core of the root, which contains xylem and phloem; central cylinder (362)

vascular tissues: the xylem and phloem of a plant; conducting tissues (357)

vas deferens: the tubes that carry sperm from each testis to the urethra (458)

vegetative reproduction: the process in which undifferentiated plant cells first divide mitotically and then differentiate to produce an independent plant; vegetative propagation (408)

veins (vaynz): vessels that return blood from the body tissues to the heart (178)

venation: the arrangement of veins in a leaf (369)

ventral: pertaining to the lower or belly side of a bilaterally symmetrical animal (684)

ventricles (VEN trih kulz): the two lower, thick-walled chambers of the heart (179)

vertebrae: the bones of the spinal column that surround and protect the spinal cord (260)

Vertebrata (vert uh BRAH tah): the largest subphylum of chordates having an enlarged brain and a spinal column made up of vertebrae that enclose the dorsal nerve cord; vertebrates (733)

vertebrates: the animals with backbones, including mammals, fishes, birds, reptiles, and amphibians; see **Vertebrata** (683)

vestigial (ves TI jee ul) **structures:** nonfunctional structures in an organism that are a remnant of structures that were functional in some ancestral form of the organism (583)

villi (sing. **villus**): small, fingerlike projections of the lining of the small intestine (164)

viruses: extremely small particles of DNA and RNA surrounded by protein coats; capable of reproducing themselves inside living cells (653)

vitamins: organic molecules required in the diet in very small amounts; exist as coenzymes or are converted into coenzymes in the cell (150)

vocal cords: two pairs of membranes that are stretched across the interior of the larynx, and which, when air passes over them, can be controlled to make sounds (225)

W

warning coloration: a protective adaptation in which the bright colors of an organism make it easily recognized (610)

water cycle: the cycling of water between the surface of the earth and the atmosphere (834)

water-vascular system: a system found only in echinoderms, for locomotion and food getting (726)

whisk ferns: the oldest known vascular plants (672)

white blood cells: see **leukocytes** (197)

windbreaks: rows of trees used to prevent wind erosion of the soil (874)

woody stems: a type of stem containing wood (364)

X

X chromosomes: one of the two types of sex chromosomes (518)

xylem: the tissue that conducts water and minerals from the roots upward through the plant and helps to support the plant (357)

Y

Y chromosomes: one of the two types of sex chromosomes (518)

yolk sac: in *shelled eggs*, the extraembryonic membrane that surrounds the yolk, containing blood vessels that transport food to the embryo; in *mammals*, an extraembryonic membrane that forms part of the umbilical cord (448)

Z

zoology: the branch of biology that deals with the study of animals (683)

Zoomastigina: the most primitive of the animal-like protists; most are unicellular (644)

zooplankton: small animals and nonphotosynthetic protists that float near the surface of water (855)

Zygomycota (zy goh my KOH tuh): a phylum of terrestrial fungi that produces two types of spores, one for sexual reproduction and the other for asexual reproduction; conjugation fungi (666)

zygospore: a zygote covered by a thick, protective wall (426)

zygote: the diploid cell resulting from fusion of two gametes (418)

Index

Acknowledgments

PHOTO CREDITS

KEY TO PHOTO SOURCE ABBREVIATIONS
Animals Animals: AA; Peter Arnold Inc.: PA; Biological Photo Service: BPS; Bruce Coleman Inc.: BC; Color Pic Inc.: CP; Grant Heilman Photography: GH; Offshoot Stock: OS; Photo Researchers Inc.: PR; Tom Stack & Associates: TSA; Tony Stone Worldwide: TSW; Visuals Unlimited: VU.

KEY TO PHOTO POSITION ON TEXT PAGE
T=Top; M=Middle; B=Bottom; L= Left; R=Right.

Front Cover: Ron Kimball Photography
Back Cover: T, Stephen Thompson/Uniphoto; **L, R,** Superstock; **BL,** Comstock

Frontis: **i, ii-iii, v T,** Ron Kimball Photography; **v B,** Mark Heifner/The Stock Broker; **vi,** Woodbridge Williams/After Image; **vii T,** B. Dittrich/The Image Bank; **vii B,** Art Wolfe/Allstock; **viii T,** Breck P. Kent/AA; **viii B,** Michael Fogden/AA; **ix T,** Carl Roessler/AA; **ix B,** Michael Fogden/AA; **x T,** Gerry Ellis/Ellis Wildlife Collection; **x B,** Michael C. T. Smith/Allstock; **xi T,** Gerry Ellis/Ellis Wildlife Collection; **xi BL,** Chris Harvey/TSW; **xi BR,** Tom Walker/Photri; **xii T,** Mitchell Warner/PA; **xii B,** Jeffrey Rotman/PA; **xiii T,** Sean Morris/AA (Earth Scenes); **xiii M,** Wayne Lynch/Masterfile; **xiii B,** Chris Harver/TSW; **xiv T,** Y. Arthus-Bertrand/PA; **xiv M,** Walter Urie/Westlight; **xiv B,** Zig Leszczynski/AA; **xv,** Kevin Schafer/Allstock.

UNIT 1 **xvi–1,** Fred Bavendam/Allstock; **2, 10,** Jeffrey Rotman; **3,** Stephen Krasemann/PR; **4 L,** Barbara Miller/BPS; **4 R,** E. R. Degginger/BC; **5,** Alfred Pasieka/BC; **6,** Fred Bodin/OS; **7 T,** Runk/Schoenberger/GH; **7 B,** Comstock; **8 T,** Runk/Schoenberger/GH; **8 B,** Nick Bergkessel/PR; **12, 29,** Carl R. Sams II; **12,** Bonnie Kamin/M. L. Dembinsky Jr; **14 L,** Kevin McCarthy/OS; **14 M,** Runk Schoenberger/GH; **14 R,** PA; **15,** Nancy Sheehan; **16,** John Kennard; **18 L,** Norman Tomalin/BC; **18 R,** Kevin Galvin/BC; **20 T,** E. R. Degginger/CP; **20 B,** Beth Ullman/Taurus; **21,** The Bettmann Archive; **23 T,** James Webb/BC; **23 BL, BM,** Bruce Iverson; **24 T, M,** Biophoto Assoc./PR; **24 BL,** Leonard Lessin/PA; **24 BR,** PA; **25,** Jeremy Burgess/PR; **26,** E. R. Degginger/CP; **27,** SPL/PR; **32, 53,** Fritz Goro/Life Picture Svc; **33,** Skip Moody/M. L. Dembinsky Jr.; **34 T, M,** E. R. Degginger/CP; **38,** Julie Houck; **39,** Phototake; **43,** E. R. Degginger/CP; **44,** Norman Tomalin/BC; **45,** Bruce Iverson; **47 L, R,** E. R. Degginger/CP; **49,** M. L. Miller/Picture Group; **50,** E. R. Degginger/CP; **51,** Bob & Clara Calhoun/BC; **56, 73,** John M. Burnley/PR; **57,** Mike Monamee/PR; **59,** Bruce Iverson; **64,** Blair Seitz/PR; **67,** Lawrence Livermore Nat. Lab; **76, 103,** Jeremy Burgess/PR; **77,** Peter Parks/Oxford Scientific; **78 T,** W. Amos/BC; **78 B,** Paul W. Johnson & J. McN. Sieburth/BPS; **79 L,** Lee Simon/PR; **79 B, 83 T,** Don Fawcett; **83 B,** Biophoto Assoc./PR; **84,** Don Fawcett/PR; **85,** Warren Rosenberg, Iona College/BPS; **86,** M. Powell/VU; **87 T,** K. G. Murti/VU; **87 BL,** M. Schliwa/VU; **87 BR,** Gopal Murti/PR; **88 T,** David Phillips/VU; **88 BL,** Don Fawcett/PR; **88 BR,** David Phillips/VU; **89,** GH; **97,** Don Fawcett; **99,** Kim Taylor/BC; **101 TL, TR,** Fred Hossler/VU; **101 M,** G. W. Willis, Ochsner Med. Instit./BPS; **106, 121,** John Shaw/TSA; **107,** Paulo Bonino/PR; **108,** John Kennard; **115,** M. Schliwa/VU; **118,** Deni McIntyre/PR; **124, 141,** Lawrence Schiller/PR; **125,** Alvis Upitus/The Image Bank; **126,** Larry Lefever/GH; **128 T,** Tom McHugh/PR; **128 M,** Stephen J. Krasemann/PA; **128 B, 132 L,** Tom McHugh/PR; **132 R,** E. R. Degginger/BC; **134 T,** T. E. Adams/VU; **134 M,** Henry C. Aldrich/VU; **134 B,** Tom Branch/PR; **135 T,** GH; **135 B,** Kevin McCarthy/OS; **136,** E. R. Degginger/CP; **137,** M. I. Walker/PR; **138 T,** M. Abbey/VU; **138 M,** Richard Campbell, Univ. of Calif./BPS; **138 B,** Hans Pfletschinger/PA; **139,** Jay Baumgarner/Photo Nats; **144 TL, 145 TR,** Kim Taylor/BC; **144 TM,** Robert Randolph/Uniphoto; **144 TR,** Roger Wilmshurst/BC; **144 B,** Les Moore/Uniphoto; **145 TL, MR,** Kenneth Fink/BC.

UNIT 2 **146–147,** Erwin & Peggy Bauer/BC; **148, 169,** Merlin Tuttle/PR; **149,** Maslowski Photo/VU; **153 all,** David Dempster/OS; **154,** Nancy Sheehan; **155,** Terry Domico/Earth Images; **156 TL, TM, TR,** Michael Abbey/PR; **156 B,** Eric Grave/PR; **157 all,** Kim Taylor/BC; **159,** William Weber/VU; **162,** Jim Caccavo/Stock Boston; **164,** David Phillips/VU; **165,** Jeffrey Myers/Stock Boston; **167,** Charles Gupton/TSW; **172,** Michael Fogden/BC; **173,** Steve Allen/PA; **173 TR,** Jacques Jangoux/PR; **178,** from *Tissues and Organs;* **179,** John Cunningham/VU; **180,** Custom Medical Stock Photo; **181,** VU; **182,** Manfred Kage/PA; **187,** Hans Halberstadt/PR; **194, 195, 215,** David Phillips/VU; **196,** Martin Rotker/Taurus Photos; **197,** S. M. Phillips/VU; **198,** Omikron/PR; **200,** Bob Daemmrich; **203,** David Phillips/VU; **205,** Jeremy Burgess/PR; **206,** North Wind Picture Archives; **207,** Doug Handel/The Stock Market; **208,** Bob Daemmrich; **211,** Bonnier Fakta; **212,** Sylvia Harle/Custom Medical Stock; **213,** A. Leipens/PR; **218, 233,** J. W. Mowbray/PR; **219,** Tom Ulrich/TSW; **222,** E. R. Degginger/CP; **223,** Vail Tyler/OS; **225,** Michael Gabridge/VU; **226 TL,** Manfred Kage/PA; **226 TR,** O.

Auerbach/VU; **226 BL,** Chris Collins/The Stock Market; **229,** American Cancer Society; **236 BKGD,** Bob Daemmrich/Stock Boston; **236 runner, 237,** David Madison/duomo; **238,** Biophoto Assoc./PR; **245 T,** Biophoto Assoc./PR; **245 B,** PA; **246,** Stan Elams/VU; **252, 267,** John Cancalosi/TSA; **253,** Fred McConnaughey/PR; **254 T,** Owen Franken/Stock Boston; **254 B,** Kevin McCarthy/OS; **256 T,** Kessel & Shih/PA; **256 B,** Manfred Kage/PA; **258,** Stephen Dalton/PR; **259,** Lawrence Migdale/Stock Boston; **260,** Manfred Kage/PA; **264,** David Madison/BC; **265,** Manfred Kage/PA; **270, 285,** Steven Kaufman/BC; **271 TR,** Len Lee Rue III/PR; **271 B,** F. McConnaughey/BC; **274 B,** from *Tissues and Organs;* **288, 309,** Thomas Roddies/Sygma; **289,** George Michael/BC; **292,** Dan McCoy/Rainbow; **301,** Runk/Schoenberger/GH; **302,** Ralph Eagle/PR; **304,** John Coletti/Stock Boston; **307,** from *Tissues & Organs;* **312, 331,** Sharon Cummings/M. L. Dembinsky Jr.; **313 TR,** Len Lee Rue III/PR; **313 BR,** Greg Dimijian/PR; **315,** Bill Dobbins/Allsport; **322,** PR; **325,** Larry Lawfer/OS; **326,** OS; **327 T,** Sylvain Legrand/PR; **327 B,** Jean Marc Barey/PR; **328,** Kevin McCarthy/OS; **329,** Jeff Zaruba/The Stock Market; **334 T,** Lou Jones/Uniphoto; **334 B,** Robert Kligge/The Stock Market; **334–335 M, 335 T,** Gary Cralle/The Image Bank; **335 M,** Bruce Berman/The Stock Market; **335 B,** SPL/PR.

UNIT 3 **336–337,** David Muench/Allstock; **338, 351,** David Muench; **339,** C. E. Jeffree/Oxford Scientific; **340,** Bettmann Archive; **341,** Runk/Schoenberger/GH; **342,** John D. Cunningham/VU; **348 T,** W. H. Hodge/PR; **348 B,** Norma J. Lang, Univ. of Calif./BPS; **349 L,** E. R. Degginger/BC; **349 R,** Bob Gossing/BC; **354, 371,** David Muench; **355,** Russell Schleipman/OS; **357 T, B,** John D. Cunningham/VU; **358 T,** Randy Moore/VU; **358 TL, ML,** Ed Reschke; **358 BL,** John D. Cunningham/VU; **359,** William E. Ferguson; **360 TL,** John Kaprielian/PR; **360 TR,** John D. Cunningham/VU; **360 BL,** David Newman/VU; **360 BR,** E. R. Degginger/CP; **361,** Ed Reschke; **362,** Larry Lefever/GH; **363 T,** Stanley L. Flegler/VU; **363 B,** N. C. Schenck/VU; **364 L, R,** Ed Reschke; **365 T,** Nathan Benn/Stock Boston, Inc.; **365 B,** John D. Cunningham/VU; **366,** W. Gilbert Grant/PR; **369 L,** Dale E. Boyer/PR; **370 R,** Charles Krebs/Allstock; **374, 389,** Mike Chuang/FPG International; **375,** John Gerlach/VU; **376,** Jeremy Burgess; **377,** Gilbert Grant/PR; **378,** Robert P. Carr/BC; **379,** E. R. Degginger/CP; **380 L,** GH; **380 R,** David Newman/VU; **381 L,** Robert P. Comport/AA; **381 R,** Runk/Schoenberger/GH; **383,** GH; **384 L, R,** John D. Cunningham/VU; **385 L,** Richard D. Poe/VU; **385 R,** Brock May/PR; **386,** Linda K. Moore/Rainbow; **387,** Peter Pears/TSW; **392 T,** Superstock; **392 BL,** Hans Pfletschinger/PA; **392 BR,** D. Cavagnaro/PA; **393 T,** Ray Coleman/PR; **393 M,** Norman O. Tomalin/BC; **393 B,** Hans Pfletschinger/PA.

UNIT 4 **394–395,** Frans Lanting/Minden Pictures; **396, 413,** CNRI/PR; **397,** David Phillips/VU; **398 T,** Ed Reschke; **398 B,** Runk/Schoenberger/GH; **401 T,** Ed Reschke/PA; **401 ML, MR, B, 402,** Ed Reschke; **403,** Blair Seitz/PR; **406,** Biophoto Assoc./PR; **407,** Kim Taylor/BC; **408,** T. E. Adams/VU; **409 T,** Ed Reschke; **409 BL, BR, 410 L,** Runk/Schoenberger/GH; **410 R,** GH; **411,** N. Pecik/VU; **416, 437,** Hans Reinhard/PR; **417,** SPL/PR; **418 L, R,** CNRI/PR; **425,** SPL/PR; **426,** M. I. Walker/PR; **427,** Ed Reschke; **428,** Hans Pfletschinger/PR; **431,** Manfred Kage/PR; **432,** Francis Leroy/PR; **433,** Zig Leszczynski/AA; **434,** C. A. Hasenkanpf, Univ. of Calif./BPS; **435,** Donald Specker/AA; **436,** E. R. Degginger/CP; **440, 441, 453,** Michael Fogden/AA; **445,** Cincinnati Zoo; **447 T,** Dwight R. Ruhn/BC; **447 BL,** Manfred Danegger/PA; **447 BR,** Kerry T. Givens/TSA; **449 all,** Kim Taylor/BC; **450,** Tom McHugh/PR; **451,** J. Cancalosi/PR; **456, 469,** Superstock; **457,** Dick Dempster/OS; **463,** Francis Leroy/PR; **464,** Lief Skoogfors/Woodfin Camp; **465,** Erika Stone/PR; **466 T,** Tom McCarthy/Picture Cube; **466 B,** Herb Snitzer/Stock Boston; **467,** Heather & Hollie Leffel/Woodfin Camp; **472, 489,** Norman Tomalin/BC; **473,** Robert Bornemann/PR; **475,** Eric & David Hosking/PR; **476 T,** Richard L. Carlton/PR; **476 B,** John D. Cunningham/VU; **477 T,** Biophoto Assoc./PR; **477 B,** Stan Elems/VU; **478,** George Kleiman/PR; **479,** Brooking Tatum/VU; **480,** Pat Lynch/PR; **481 L,** R. Litchfield/PR; **481 R,** G. F. Leedale/PR; **482,** Stephen P. Parker/PR; **484 TL, TML, TMR;** Runk/Schoenberger/GH; **484 TR,** Larry Lefever/GH; **484 T,** N. Pecnik/VU; **484 M,** A. W. Ambler/PR; **484 B,** John L. Pontier/AA; **487,** E. R. Degginger/CP; **492,** Comstock; **493 TL, B,** Jeffrey Reed/Stock Shop/Medichrome; **493 TR,** David York/Stock Shop/Medichrome.

UNIT 5 **494–495,** Charles Krebs/Allstock; **496, 513,** D. & R. Sullivan/BC; **497 T,** Ralph A. Reinhold/AA; **497 B,** Leonard Lee Rue, III/PR; **498,** Jane Grushow/GH; **505,** Runk/Schoenberger/GH; **510 L, R,** Larry Lefever/GH; **512,** Carolina Biological Supply; **516, 541,** Dan McCoy & R. Langridge/Rainbow; **517,** Darwin Dale/PR; **520,** John Running/Stock Boston, Inc.; **544, 563,** Tom McHugh/PR; **545,** Superstock; **547,** William E. Ferguson; **548,** Jim Brandenburg/Brandenburg Stock; **550 L, R,** Bill Longcore/PR; **552 L,** Leonard Lessin/PR; **552 R,** Cathlyn Melloan/TSW; **553,** Julie Houck; **554,** Charles Gupton/TSW; **555,** Richard Kolar/AA; **557 T,** Dave Schaefer/Schaefer Photography; **557 B,** Runk/Schoenberger/GH; **558,** Bruce Hoertel; **560,** ARS-USDA; **566–567,** Gunter Ziesler/PA; **566 T, 567 B,** Karl Ammann/BC.

UNIT 6 **568–569,** Frans Lanting/Minden Pictures; **569 inset,** Wolfgang

Kaehler; **570, 595,** E. R. Degginger/CP; **571,** Peter Scoones/Woodfin Camp; **572 T,** E. R. Degginger/CP; **572 B,** Tass/Sovfoto; **573 T,** Walter H. Hodge/PA; **573 B,** R. T. Bird/American Museum of Natural History; **575,** Jeff Foott/TSA; **577,** William E. Ferguson; **578,** John Cancalosi/TSA; **581,** James Kilkelly; **592,** Walter D. Keller/University of Missouri; **594 T,** John Cancalosi/TSA; **594 B,** Kevin Schafer/TSA; **598, 623,** Kevin Schafer/PA; **599,** William W. Bacon, III/PR; **601 T,** BC; **601 TM,** Miguel Castro/PR; **601 B,** Fred Bavendam/PA; **601 BM,** Keith Gunnar/BC; **602,** J.A.L. Cooke/AA; **605,** Art Wolfe/Allstock; **606,** John Bova/PR; **607,** R. S. Virdee/GH; **610 T,** Kim Taylor/BC; **610 B,** Runk/Schoenberger/GH; **611 TL,** George K. Bryce/AA; **611 TR,** Ralph Reinhold/AA; **611 B,** Michael P. L. Fogden/BC; **613 T,** Jack Fields/PR; **615 T,** Joe McDonald/BC; **615 B,** E. J. Maruska/VU; **616,** J. Cancalosi/TSA; **617,** Superstock; **618 L,** M.W.F. Tweedie/BC; **618 R,** M.W.F. Tweedie/PR; **619,** Patrick L. Pfister/Stock Boston, Inc.; **626, 627,** Vladimir Krb/American Museum of Natural History.

UNIT 7 **628–629,** John Flannery/BC; **630, 661,** Manfred Kage/PA; **631,** CRI/PR; **633 T, M,** David M. Phillips/VU; **633 B, 635 T,** John D. Cunningham/VU; **635 B,** F.C.F. Earney/VU; **636 T,** Paul W. Johnson & J. McN. Sieburth/BPS; **636 B,** Sinclair Stammers/PR; **637 L,** M. Abbey/PR; **637 R,** Leon J. Le Beau/BPS; **638,** Bettmann Archive; **641,** John Colwell/GH; **643 L, R,** Manfred Kage/PA; **644 T, B,** Eric V. Grave/PR; **646,** Paul W. Johnson/BPS; **648 T,** J. R. Waaland, Univ. of Washington/BPS; **648 B,** Manfred Kage/PA; **649,** J. R. Waaland, Univ. of Washington/BPS; **650 T,** Heather Angel/Biofotos; **650 B,** Robert C. Hermes/PR; **653 T,** R. Knauft/Biology Media/PR; **654 T,** M. Wurtz/PA; **654 B,** K. G. Murti/VU; **658,** Alon Reininger/Woodfin Camp; **659,** Michael P. Gadomski/PR; **664, 679,** David Muench; **665,** Rod Planck/TSA; **667,** Joe Baraban; **668 T,** Michael P. L. Fogden/BC; **668 B,** Adam Woolfit/Woodfin Camp; **669,** Terry G. Murphy/AA; **670 T,** GH; **670 B,** Doug Sokell/TSA; **671, 673 L,** Heather Angel/Biofotos; **673 ML,** Rod Planck/TSA; **673 MR,** Heather Angel/Biofotos; **673 R,** Brian Parker/TSA; **674 T,** John Shaw/TSA; **674 B,** Julie Houck; **675 TL,** Luis Castaneda/Image Bank; **675 TML,** Ed Reschke/PA; **675 TM,** E. R. Degginger/CP; **675 TMR,** E. R. Degginger/AA; **675 TR,** Heather Angel/Biofotos; **675 BL,** E. R. Degginger/CP; **675 BM, BR,** Michael P. Gadomski/PR; **676,** Thomas Kitchin/TSA; **682, 703,** F. Stuart Westmoreland/Allstock; **683,** Runk/Schoenberger/GH; **684 T,** Manfred Kage/PA; **684 B,** Heather Angel/Biofotos; **685,** Runk/Schoenberger/GH; **687,** Carl Roessler/TSA; **688,** Runk/Schoenberger/GH; **689 TL,** Bob Evans/PA; **689 TR,** Dave Fleetham/TSA; **699 T,** Tom McHugh/PR; **699 B,** Breck P. Kent/PA; **700,** Milton Rand/TSA; **701,** Jane Burton/BC; **706, 729,** J. Carmichael, Jr./Image Bank; **707,** Zig Leszczynski/AA; **708 T,** E. R. Degginger/AA; **708 B,** Raymond A. Mendez/AA; **709 T,** E. R. Degginger/CP; **709 TR,** E. R. Degginger/AA; **709 ML,** E. R. Degginger/CP; **709 MR,** Gary Retherford/PR; **709 BL,** Breck P. Kent/AA; **709 BM,** Fred Bavendam/Allstock; **709 BR,** E. R. Degginger/CP; **712,** Breck P. Kent/AA; **713 T,** E. R. Degginger/CP; **713 B,** Michael P. L. Fogden/BC; **714 T,** E. R. Degginger/CP; **714 ML,** John Cancalosi/TSA; **714 BL,** Jeff Foott/TSA; **714 BM,** Sean Morris/AA; **714 BR,** Russ Lappa/Picture Cube; **716,** G. C. Kelley/PR; **717 TL, TR,** Runk/Schoenberger/GH; **717 BL,** Richard Kolar/AA; **717 BR,** E. R. Degginger/CP; **718,** Jack Wilburn/AA; **719,** William E. Ferguson; **720 T, TM,** E. R. Degginger/CP; **720 B,** E. R. Degginger/AA; **720 BM,** Paul Metzger/PR; **725 L,** E. R. Degginger/CP; **725 TR,** Brian Parker/TSA; **725 MR,** E. R. Degginger/AA; **725 BR,** E. R. Degginger/CP; **727,** Thomas Kitchin/TSA; **731,** Raymond A. Mendez/AA; **732, 757,** Mark Newman/TSA; **733,** James H. Carmichael/TSA; **734,** F. Stuart Westmoreland/TSA; **737,** Heather Angel/Biofotos; **738 T,** Dave Fleetham/TSA; **738 BL,** Ron & Valerie Taylor/BC; **738 BR,** Norbert Wu/PA; **739 L,** Ian Took/Biofotos; **739 R,** Tom McHugh/PR; **740 T,** W. Gregory Brown/AA; **740 B,** Colin Milkins/AA; **741,** E. R. Degginger/CP; **743 T,** Zig Leszczynski/AA; **743 B,** Michael P. L. Fogden/BC; **744,** Zig Leszczynski/AA; **749 TL, TR, BL,** Hans Pfletschinger/PA; **749 BR,** Zig Leszczynski/AA; **751 T,** Zig Leszczynski/AA; **751 B,** Heather Angel/Biofotos; **752,** Zig Leszczynski/AA; **753 L,** Breck P. Kent/PA; **754 TL,** E. R. Degginger/CP; **754 R,** Tom McHugh/PR; **754 B,** Zig Leszczynski/AA; **755 T,** E. R. Degginger/CP; **755 B,** Tom McHugh/PR; **756 T,** Runk/Schoenberger/GH; **756 B,** D. J. Lyons/BC; **760, 781,** Tom & Pat Leeson; **761,** Breck P. Kent/AA; **762 L,** GH; **762 ML,** Mary Clay/TSA; **762 MR,** Thomas Kitchin/TSA; **762 R,** Laura Riley/BC; **763,** Heather Angel/Biofotos, **766,** Richard Dunoff/Stock Market; **767,** E. R. Degginger/BC; **768 T,** Stouffer Prod. Ltd./AA; **768 M,** Jim Brandenburg/BC; **768 BL,** Rod Planck/TSA; **768 BR,** Stephen Dalton/PR; **769 T,** Breck P. Kent/AA; **769 M,** Gordon Langsbury/BC; **769 BL,** Mary Clay/TSA; **769 BR,** Martin W. Grosnick/BC; **770 T,** Phil Degginger/CP; **770 BL,** Alan G. Nelson/AA; **770 BM,** E. R. Degginger/CP; **770 BR,** Harvey Barnett/PA; **771,** Adrienne T. Gibson/AA; **772,** E. R. Degginger/CP; **774,** American Museum of Natural History; **775 T,** Cleveland Museum of Natural History; **775 BL,** Des & Jen Bartlett/BC; **775 BR,** Delta Willis/BC; **776 L, R,** M. Crabtree/AAAS; **777 TL,** Tom McHugh/PR; **778,** Douglas Mazonowicz/Gallery of Prehistoric Art; **784, 811,** Peter Lamberti/TSW; **785,** John K. Nakata/Terraphoto Graphics/BPS; **786 T, M,** Heather Angel/Biofotos; **786 B,** Stephen Dalton; **787,** Y. Arthus-Bertrand/PA; **788,** S. Nielsen/BC; **789,** Billy E. Barnes/Stock Boston; **793 T,** E. R. Degginger/CP; **793 B,** Hans Pfletschinger/PA; **799 L,** G. Ziesler/PA; **799 R,** Nina Leen/Life Picture Service; **801,** C. W. Schwarz/AA;

802, J. H. Robinson/AA; **803,** Clem Haagner/BC; **805,** Raymond A. Mendez/AA; **806,** C. Allen Morgan/PA; **807,** Dan Guravich/PR; **814, 815,** Tom P. Kahl/BC.

UNIT 8 **816–817,** David Mauzy/Allstock; **818, 841,** James Carmichael/The Image Bank; **819,** Phil Degginger/The Image Bank; **824,** E. R. Degginger/CP; **825 T,** Norm Thomas/PR; **825 B,** Carl Roessler/AA; **826,** Ken Brate/PR; **829,** Len Lee Rue III/BC; **831,** E. R. Degginger/CP; **833,** Gerhard Gscheidle/PA; **837,** Gabe Palmer/The Stock Market; **838,** Joe McDonald/BC; **840,** E. R. Degginger/CP; **844, 845, 859,** David Muench; **847,** Johnny Johnson/AA; **847,** Charlie Ott/PR; **848,** Michael Gadomski/PR; **849 T,** Jack Ryan/PR; **849 B,** Breck P. Kent/AA; **850 T,** Jeff Rotman/PA; **850 B,** Gregory Dimijian/PR; **851 T,** Patti Murray/AA; **851 B,** Nancy Sheehan; **852,** John Cancalosi/TSA; **854,** Carl Roessler/AA; **856,** E. R. Degginger/CP; **857 T,** Heather Angel/Biophotos; **857 B,** Rod Planck/TSA; **862, 881,** Wendy Shattil/TSA; **863,** M. J. Balick/PA; **864,** Runk/Schoenberger/GH; **865,** Anthony Suau/Black Star; **866 T,** Robin Jane Solvano/BC; **866 B,** Jim Brandenberg/BC; **867,** Wendell Metzen/BC; **868,** Franz Kraus/Picture Cube; **869,** B. Nation/Sygma; **870,** Stephanie Maze/Woodfin Camp; **872,** Will McIntyre/PR; **873,** Steve Elmore/TSA; **875,** Luiz Claudio Maurio/PA; **876 L,** Kenneth Fink/BC; **876 TR,** Y. Arthus-Bertrand/PA; **876 BL,** Raymond Mendez/AA; **876 BR,** Emily Harsie/BC; **878 L,** Mark Sherman/PR; **878 R,** William E. Ferguson; **879,** Ruod Lubbus/Sipa Press; **884–885, 884 B,** GH; **885 M,** Runk/Schoenberger/GH; **885 B,** Larry Mulvehill/PR.

APPENDIX F **894,** David M. Phillips/VU; **895 T,** J. R. Waaland/Univ. of Wash./BPS; **895 B,** Manfred Kage/PA; **895 M,** Robert C. Hermes/PR; **896 L,** Michael Fogden/BC; **896 R,** Rod Planck/TSA; **897,** E. R. Degginger/AA; **898,** Tom McHugh/PR; **899 T,** E.R. Degginger/CP; **899 B,** Dave Fleetham/TSA; **900,** Mary Clay/TSA.

ILLUSTRATION CREDITS

Edmond S. Alexander/Alexander and Turner: 158, 190, 232, 257B, 308RT, 308RB, 329, 330L, 346T, 350, 518T, 698T, 698B, 658TR.

Sally J. Bensusen/Visual Science Studio: 159, 175, 176, 222, 239, 240, 282B, 283, 715, 721, 807, 808.

Kent Boughton: 228.

Carmella M. Clifford: 161, 227, 246, 262T, 278, 284BL, 284TR, 300, 302, 303, 305, 306T, 306B, 307, 370, 767, 773.

Barbara Cousins: 255, 316, 318T, 391R, 491.

Paul Foti/Boston Graphics, Inc.: 886, 887, 894, 898, 899, 903.

Howard Friedman: 22, 72BL, 91, 93B, 94, 576, 589, 693, 820, 822, 827, 828, 829, 833, 834, 836, 839, 853, 855.l

Andrew Grivas: 157, 174B, 221B, 282T, 318B, 356L, 356R, 357, 358, 361, 379, 491, 604, 688, 690, 697, 791R.

Steven J. Harrison/Fine Line Studio: 150, 344, 586, 587, 588, 590, 620.

Jackie Heda: 452B, 574BR, 574BL, 684L, 684BL, 684BR, 687, 700, 710, 711, 712, 726, 728, 735, 741, 745T, 745B, 746, 747, 748T, 748B, 754, 763, 764, 765, 794, 796.

Floyd E. Hosmer: 81, 92T, 92B, 123, 179T, 189, 263T, 263B, 273, 274, 275, 276, 277, 287, 295, 298, 391L, 597, 625, 654, 655B, 678, 702.

Keith Kasnot: 83T, 83B, 84B, 85, 86, 87, 89, 160, 163, 185, 186, 188, 224, 241, 243, 244, 314, 368, 399, 401, 402, 405, 406T, 406B, 464, 575, 583, 652.

Elizabeth McClelland: 580, 666, 669, 722, 723, 792, 798, 800, 804, 805.

Robert Margulies/Margulies Medical Art: 84T, 93T, 95, 96, 156, 164, 174T, 221T, 256, 257T, 279, 281, 376, 426, 442, 443TL, 443TR, 443B, 444, 448, 632, 636, 643, 646, 647, 649BL, 649TR, 655T, 691, 692, 694, 734, 791L.

Martucci Studio/Jerry Malone: 15TR, 35, 36, 37T, 37B, 39, 40, 41, 42, 43T, 43B, 44, 46, 50, 52BR, 54, 55, 58, 59, 60, 61T, 61B, 62T, 62B, 63T, 63B, 65, 66T, 66B, 69, 70T, 70B, 71, 80, 104, 105L, 109T, 109B, 113, 114, 116, 117, 118, 119, 127, 128, 131, 170, 171, 193L, 207, 231, 242, 251, 269, 311R, 323T, 340, 341, 343, 345, 346B, 353L, 461, 498, 499, 500, 501, 503, 504T, 504B, 506T, 506B, 507, 508TL, 508B, 509TR, 509BR, 510, 518B, 519T, 519M, 519B, 521, 522, 523, 527, 528, 530, 531T, 531B, 533L, 534T, 534B, 536, 537, 540L, 543R, 547, 549, 551, 559, 562, 565R, 608, 634, 663L, 731, 759L, 778, 779, 810R, 813, 830, 832, 843, 883.

Fran Milner: 178T, 178B, 179B, 181, 198, 199, 204, 260, 261, 262B, 265, 290B, 290L, 291L, 291R, 292, 293, 294, 296, 321, 322, 323B, 324, 419, 421, 425, 429, 430, 431, 458, 459, 460, 468L, 468R, 526, 582.

Sanderson Associates: 852, 873.

Lois Sloan: 475, 538, 574T, 600, 603, 861L, 874.

Gary Torrisi: 474, 525, 591, 871.

Cynthia Turner/Alexander and Turner: 28L, 102L, 102R, 120, 168L, 266, 366, 367, 378, 388, 436, 476, 477, 479, 480, 481T, 481B, 482, 483, 485, 486T, 486B, 533MR, 533TR, 612, 622L, 645, 651, 656, 660L, 840R.

Any photo or illustration acknowledgement inadvertently omitted will be amended upon notification.

W9-ARW-963

Happiest Birthday
to my treasured
friend, Roberta!

Esther

OTHER BOOKS BY PERLA MEYERS

The Seasonal Kitchen
The Peasant Kitchen
From Market to Kitchen

THE AMERICAN HARVEST SERIES:
The Spring Garden
The Early Summer Garden

PERLA MEYERS'

ART OF

SEASONAL

COOKING

EDITED BY JUDY KNIPE

PERLA MEYERS

SIMON & SCHUSTER NEW YORK LONDON TORONTO SYDNEY TOKYO SINGAPORE

Simon & Schuster
Simon & Schuster Building
Rockefeller Center
1230 Avenue of the Americas
New York, New York 10020

Copyright © 1991 by Perla Meyers

All rights reserved
including the right of reproduction
in whole or in part in any form.

SIMON & SCHUSTER and colophon are registered trademarks
of Simon & Schuster Inc.

Designed by Bonni Leon
Illustrated by John Pitman
Manufactured in the United States of America

10   9   8   7   6   5   4   3   2   1

Library of Congress Cataloging in Publication Data
Meyers, Perla.
    [Art of seasonal cooking]
    Perla Meyers' art of seasonal cooking / Perla Meyers.
      p.   cm.
    Includes index.
    1. Cookery, American.   I. Title.   II. Title: Art of seasonal
cooking.
TX715.M6254   1991
641.5973—dc20   91-14009
                                        CIP

ISBN 0-671-64984-1

To my mother,

who always supported me in my love of food
and who nourished
my passion for cooking in every season.

First and last, immeasurable thanks to my friend and assistant, Michael di Beneditto, for countless hours of planning, testing, tasting, and then retesting. Without his generosity, enthusiasm, and integrity this book could never have been completed.

Thanks as well to Judy Knipe, who helped to shape and strengthen the book and whose candor and patience encouraged me to find my own best voice.

I'm grateful to my editor, Kerri Conan, for her invaluable finishing touches and her care in guiding the manuscript through its final stages.

I'm especially grateful to the many chefs, both here and in Europe, who have so generously opened their kitchens and have inspired me with their talent and creativity.

Thanks to my former editor, Carol Lalli, for believing in my work and for enthusiastically supporting this book in its early development.

CONTENTS

INTRODUCTION

When my first book, *The Seasonal Kitchen,* was published in 1973 it seemed to many people almost revolutionary. During my book tour of more than fifteen cities, I was invariably asked the same questions: What is wrong with frozen beans, corn, peas? Why not use garlic powder or a touch of onion salt? Why bother with a homemade dressing when the bottled stuff is so convenient? These questions were valid then. For decades Americans had been exposed to convenience foods, and with modern transportation being far more advanced here than in Europe, it was possible to prolong the shelf life of perishable foods to the point that seasonal eating had little meaning—a pity in a country that offers the consumer fresh produce twelve months out of the year.

I, on the other hand, had never known these foods. I grew up in Spain at a time when it was still one of the least advanced countries in Europe. With rudimentary refrigeration and little access to nonregional or even local produce, the buying and preparation of food was an entirely different matter.

My mother shopped daily, and the market was not just a place to buy food but also a diversion and a way of life. Much of the social contact took place there, and each of the vendors knew us. When my mother sent me back to the market to get four soles that she had forgotten to buy, Mercedes, the fish lady, knew exactly how to prepare them. She would garnish the package with sprigs of fresh parsley, which, to this day, is given free with a purchase of seafood or meat. Mercedes would then send me on my way with best wishes to the family. I did not have to pay; there were no charge accounts, since Mercedes knew my mother would be back. Her first question was always, "So how were the soles?" Shopping for food there was a truly pleasurable experience, the likes of which many Americans, alas, have never known. So, *The Seasonal Kitchen* was not revolutionary to me—it was simply a natural extension of my early life, the only way I could conceive of preparing food. It was not a question of whether seasonal cooking was the upcoming trend, it was just common sense, since the best results in cooking are obtained by using fresh food at its prime.

Ironically, I was ahead of my time. In the eighties *freshness* became the buzz word in cooking literature. During that time supermarkets enlarged their produce areas and stopped prepackaging vegetables. Small produce stores sprouted up in large cities, as did country farmstands and produce markets. "Gourmet" vegetables such as ginger and snow peas became as common as onions and carrots, and the seasonality of asparagus and strawberries could actually be felt as you entered the supermarket. Restaurants started to feature fresh foods from afar. Day-old wild chanterelles from the Northwest marked the beginning of fall; vine-ripened tomatoes from Florida appeared in the winter, not to mention fresh New Zealand lamb and imported Italian chicory lettuce and tiny green beans. Suddenly, fresh herbs like basil were as easy to find as parsley and seemed to stretch the season from early spring to late fall. Indeed, stretching the season was the theme of the eighties, and it has continued into the nineties.

Now everyone, from the restaurant chef to the home cook, is in pursuit of capturing global freshness. Asparagus are available almost year-round, as are Belgian endives, snow peas, and a variety of wild mushrooms, to say nothing of Israeli tomatoes and melons, and strawberries from New Zealand. But try to get a crisp young turnip with its greens attached or the wonderful curly Savoy cabbage in its true season and you almost have to travel to the very source. We cannot turn back the clock—nor would I want to. The abundance of fresh fruits and vegetables that spans several seasons makes our lives as cooks and chefs more exciting and creative. I can't imagine a winter now without fennel, broccoli rape, or celery root, and when it comes to salads, even radicchio and frisee are almost commonplace. And how about fruits such as fresh figs, golden raspberries, fresh currants, and even sour cherries? All these have expanded our cooking repertoire and have helped to make this book a reality.

Much of *The Seasonal Kitchen* was based on the regional cooking of Spain, Italy, France, and central Europe, where native cuisines have been shaped by nature's calendar. These dishes were, in a way, almost classic, not only because they were seasonal but because they made sense and simply tasted good. To me there is a joy in cooking and sampling certain dishes again and again with each passing season, and I love to rediscover familiar flavors and textures.

What has happened to the simple concept of these seasonal dishes? What has happened to real cooking? So much of what we taste today is merely clever and involves an assembly of ingredients rather than the true marriage and harmonious combination of flavors. We have seen lots of wild experimentation in the name of freshness and at the expense of old-fashioned concepts that once produced dishes that we never tired of.

In the years since *The Seasonal Kitchen* was published I feel that I have grown as a cook, focusing on American ingredients and regional produce. I have tried to lighten my food and its seasoning, and it is this food that I present here. I don't believe that all that was good in the seventies and the eighties should be discarded in favor of new and fashionable food. I am most concerned with creating dishes that are alive and full of taste, sparkling with colors, shape, and texture. I look for food with character that delivers. Some of it is quite modern and may sometimes call for a more updated and contemporary presentation, but at heart the dishes are simple and easy to make. In developing them, I kept in mind the basic limitations facing a home cook who is trying to juggle a career and a home life that includes cooking and entertaining.

Mostly I try to be true to the seasons as I know them. Anyone who has traveled in this country realizes that the United States is composed of five major regions, each with its own climate and life-style. What is truly seasonal in one region may not be so in another. A red snapper braised with vine-ripened fresh tomatoes and basil is an early spring dish in Florida, while freshly caught and grilled bluefish is at its best sampled off the coast of New England in the summer. Picking fresh cherries off the fruit-laden trees is as common in Oregon during the summer as it is to pick ripe apricots in Fresno, California, or to buy raspberries by the flat in Seattle. But although I teach cooking and work with local ingredients in many parts of the country, the recipes reflect the seasons of the Northeast, where I live. These can easily be adapted to your region if you are familiar with your seasons and learn to shop spontaneously. In fact, to be truly seasonal, you should be tuned in to the geography of your region and be able to adapt the recipes to your life-style. Even in an age of shrinking distances and increased global availability it is important to learn to distinguish a food that is fresh and seasonal from one that is merely fresh. An early fall apple is far superior to one that has been stored throughout the winter; the sweet, mellow taste of a real spring onion cannot be duplicated at any other time of year.

Even in summer, the season we all experience most directly, I often hear the question "Is this local corn?" when the corn stalks are still less than two feet high. It's like asking for a ripe local melon when the spring daffodils are in full bloom and the trees are just starting to bud.

Learning when foods are truly seasonal is really worthwhile. It is something that is ingrained in every Mediterranean cook because the seasonal changes in that region are still experienced joyously. Any Italian will tell you when fresh porcini or shell beans are in season; most French know that scallops are at their best in the fall, and the Spanish cherish their grilled sardines in the summer and their wonderful fava beans in the spring. It is equally important for the American cook to take seasonality into consideration when planning even the simplest of meals. Personally, I feel that if I am going to spend time in the kitchen I may as well work with the best tools and the best possible ingredients.

The recipes in this book are a result of many years of experimenting and tasting. Many were inspired by dishes, techniques, and garnishes I saw in restaurant kitchens in Europe and America, and others come from my mother's and grandmother's old notebooks. Some are foods I grew up with and still love, such as the Catalan bean ragout or fava beans à la Basquaise or a roast chicken in vinegar sauce. Many are inspired by people and chefs I have had the pleasure of working and cooking with, such as Pierre Gagnaire, Jacques Chibois, Alain Passard, Jacques Lameloise, Gina from the Due Sorrelle in Venice, and Pepe, a wonderful man and owner of the Caballito Blanco, a tiny bistro in Barcelona. But most of all, the recipes were inspired by the seasons, those exact moments in time when I was looking at some fresh vegetable, fish, fruit, or berry trying to decide how to prepare it and calling on my creativity as a cook to come up with a suitable recipe.

This led me to organize this book by season, with five specific menu suggestions for each one and then a selection of other seasonal recipes that offer the cook the means to create many other menus tailored to individual lifestyles.

I have tried to make the recipes as flexible as possible so any cook can make a given dish with some variation in more than one season. While some dishes take minutes to prepare, others require slow, leisurely cooking to develop and deepen their flavor.

Some dishes are made with little amounts of butter and cream, while others call for more traditional amounts. I believe that for a dish to retain its integrity, it should be enjoyed without guilt as long as you do not indulge in fats and excess calories at every meal. All the butter enrichments called for in the recipes are optional, but it is useful to remember that if a dish serves four to five people and is enriched with three tablespoons of butter, the amount of butter consumed by each person is less than one tablespoon. I believe, above all, that a menu should be well balanced, and that while one dish may be somewhat rich, the others should be light and flavorful.

I am not concerned with the ins and outs of food. A classic beurre blanc is a sauce that will never go out of style, and a spoonful does magic to a piece of grilled or sautéed fish. A savory tart that combines leeks and bacon and is bound by a flavorful custard is old-fashioned French ambrosia and so is a buttery potato puree. To me there is only good and bad food. Not Nouvelle or Bistro or Cal-Italian or Southwestern. I do not consider myself a French, Austrian, or Spanish cook, although my background is a mixture of all three. Indeed, I love many kinds of food, both old-fashioned and new, and from every type of cuisine—as long as it tastes good. When I was a child, my father always used to tell me, "I love everything that begins with an *a*—a good piece of fish, a good piece of bread, a good stew, a good soup. . . ." At the time it was just his little joke, but today it is my reality. For me, *good* always means fresh, good quality and full of flavor, and that ideal has been my touchstone as a cook, as a writer, and as a teacher of cooking.

I admit that I envy those cooks who can get seafood caught only hours before it arrives at the market, or chickens that are truly free-range, or fresh farm eggs. Above all, I greatly admire those who have time and are willing to spend hours tending to their vegetable gardens in pursuit of the simple pleasure of tasting a home-grown vegetable.

Considering the limitations of flavor in today's ingredients, I feel that I'm particularly lucky. I have access to so much because I live in New York and have a small garden in New England, but mostly I am lucky to have been exposed by my parents to so many kinds of foods and cuisines. It is this exposure that developed my curiosity and my need to seek out the best ingredients at each time of the year and prompted my continued fascination with cooking from season to season.

THE

SPRING

KITCHEN

When I was growing up in Spain, spring began for us in early March with a burst of brilliantly colored flowers—wildly tinted primrose, fragrant mimosa, and spectacular lilac. The first mild, sunny days were a sure sign that it was time to start planting the spring crop, and we made a mad rush for the garden.

As a child it delighted me, just as it does now, to see the earliest vegetables tenderly poking out of the ground: delicate little radish leaves; the first row of pale green lettuce just beginning to show its various shapes and colors; vibrant green sorrel, its leaves unruffling in the sun.

Spring was exuberantly welcomed in Barcelona's great outdoor flower market and in the many neighborhood food markets, their stands heaped with peas, tiny artichokes, glistening fava beans, and creamy white asparagus. I reveled in pearly new potatoes, jewellike turnips, and succulent baby carrots that were so tender I could easily eat a bunch while helping my mother with her daily shopping.

Later, as a student in Geneva, I'd take my weekly stroll through the Saturday market in the old town, which by April had come alive in celebration of the season. The mountains were still capped by snow and the air still brisk, but a heavenly whiff of spring was wafted in by breezes from the lake. What the market lacked in Mediterranean flair it made up for in exquisite quality and presentation. Local farmers brought in their first fresh goat cheeses; mounds of spongy, earthy morels; and crates of corn lettuce, with its beautifully shaped leaves that so perfectly deserve the name lamb's tongue. The beets were so pristine that it seemed unlikely they had ever touched the soil, the lacy fronds of dill so decorative that they'd have served equally well in a vase or in a chicken soup.

My appetite for spring foods was reawakened, and I wanted to buy everything. As I walked among the market stands I would revise the week's menus to accommodate each glorious new offering I encountered.

Nowadays, of course, we can buy spring produce by the time winter is half over. Good asparagus arrives as early as mid-February, and peas can be found in many markets in late winter. Fresh hothouse herbs—basil, thyme, rosemary, among others—may not have the bright flavor of the home-grown variety, but they taste infinitely more vivid than dried herbs, and they're available almost year-round in supermarkets.

But even though supermarkets are undeniably convenient, and early asparagus tastes perfectly delicious, for those of us who are gardeners it's spring only after we've spent a few warm days turning the soil and then watching the first seedlings pop up.

Spring has always seemed to me at its most poignant and most beautiful in the Pacific Northwest, where lush, spectacularly colored rhododendrons come into bloom just as the finest green asparagus, the freshest of salmon, and snowy white halibut arrive at the market along with such exotic greens as the vines of fresh peas, wild leeks, and fiddlehead.

Spring passes quickly, and your seasonal menus should be uncomplicated, with a focus on the clear, direct flavors of young vegetables and fruits as they come to market. This is the time for one of the great vegetables, the brilliantly green asparagus, which always astonishes by its wonderful flavor and endless versatility.

Fresh peas are never so good as in early spring. A plateful of young peas tossed in the pan juices of a roast chicken or squab can be a memorable feast. I often begin a spring dinner with a plate of freshly cooked peas, either combined with diced sautéed prosciutto or lightly sugared and tossed with mint leaves.

But the quintessential spring meal has always been a simply roasted leg of lamb accompanied by braised buttery fava beans.

Spring is also the time to begin making lighter desserts. Enjoy the last of the good citrus fruit, and welcome the wonderful strawberry in a variety of ways. If you keep your meals simple, zesty, and seasonal, you'll have more time to spend in the garden, preparing for summer's abundance.

SPRING MENUS

1

Spinach Fritters with Herbed Sour Cream and Smoked-Salmon
Sauce
Braised Veal Shanks with Peas and Mushrooms Primavera
Crème au Grand Marnier with Strawberries

2

Steamed Asparagus in Bright-Green Provençal Mayonnaise
Roast Cornish Hens in Garlic Cream with Pan-Sautéed Escarole
Terrine of Citrus Fruit with Strawberry Sauce

3

Risotto with Shrimp and Radicchio Treviso
Sauté of Salmon with Asparagus Vinaigrette
Sweet Spring Mango Soufflés

4

Globe Artichokes in Curry-Lemon Vinaigrette
Chicken Breasts in Sorrel-Cream Primavera
Fallen Ricotta and Orange Soufflés

5

Asparagus, Leek, and Potato Soup du Jour
Fillets of Beef in Emerald Sauce with Crisp Fried Onion Rings
Very Lemony Meringue Custards

SPRING APPETIZERS

Pickled Halibut with Spicy Peppers and Citrus Zest

Fried Sardines with Wilted Spinach and Lemon Vinaigrette

Double Asparagus and Smoked-Salmon Tart

Stir-Fry of Calamari with Red Peppers and Lemon

Warm Shrimp Salad with Crisp Vegetables Niçoise

Pan-Grilled Shrimp Salad in Sesame-Scallion Vinaigrette

Twice-Baked Cheese Soufflés with Peas and Asparagus

Goat Cheese Pots de Crème with Peas and Spring Herbs

Globe Artichokes with Creamy Goat Cheese and Shallot Dressing

Asparagus with Beet Vinaigrette

Fava Beans à la Basquaise

Ziti with Asparagus, Peas, and Lemon Cream

SPRING SOUPS

Creamy Asparagus Soup Printanier

Cream of Broccoli and Tangy Lemon Soup

Viennese Kohlrabi Soup with Garden Vegetables

Sorrel, Tomato, and New Potato Soup

SPRING MAIN COURSES

Fillets of Salmon on a Bed of Cabbage with Crème Fraîche—Dill Sauce

Fricassee of Salmon with Baby Vegetables Lameloise

Shad Roe in Parsley Butter with Braised Spring Onions

Oven-Braised Shad with New Potatoes

Fillets of Sea Bass with a Vegetable Pilaf and Chive-Butter Sauce

Braised Red Snapper Fillets in a Fresh Tomato and Dill Sauce with Sautéed Cucumbers

Grilled Brochettes of Monkfish in Tangerine and Spice Marinade

Medallions of Maryland Crab and Shrimp Mousse in Curry-Lemon Sauce

Sautéed Soft-Shell Crabs in Shallot and Mustard Sauce on a Bed of Roasted Vegetables

Spicy Steamed Mussels with Sautéed Sausage and Orange-Saffron Sauce

Fettuccine in Saffron Cream with a Sauté of Shrimp and Scallops

Roast Chicken in Vinegar and Garlic Sauce

Roast Squabs with a Fricassee of Pearl Onions and Fresh Garden Peas

Casserole of Lamb with Carrots, Pearl Onions, and Rosemary

Roast Leg of Lamb with Garlic Confit in Tomato Fondue

Roast Veal with Glazed Fennel and Tomato Jam

SPRING VEGETABLES

Braised Beets with Balsamic Vinegar and Crème Fraîche

Sauté of Cucumbers with Peas and Mint

Tuscan Braised Green Beans in Parmesan Cream

Gratin of Kohlrabi and Leeks

Gratin of New Potatoes Niçoise

Potato Puree with Glazed Scallions

Butter-Braised Spinach with Glazed Garlic Cloves

Pan-Seared Snow Peas in Parsley and Shallot Butter

SPRING SALADS

Catalan Broccoli Salad

Salad of Carrots in Ginger, Mustard, and Fresh Cilantro
Vinaigrette

Endive, Radish, and Watercress Salad in Double-Mustard Dressing

Marinated White Mushroom Salad in Crème Fraîche and Chive
Vinaigrette

Peasant Salad with Provençal Vinaigrette

S P R I N G   D E S S E R T S

Fallen Chocolate Soufflé Cake with White Chocolate Sauce and
Raspberries

Sugar-Laced Crepes with a Strawberry and Mascarpone Mousse

Spiced Oranges in Red Wine and Cinnamon Syrup

Pineapple Compote in Wine and Orange Syrup

Spring-Berry Bread Pudding

Raspberry Pots de Crème

Stuffed Strawberries in Pineapple-Mint Coulis

Caramelized Meringue Puffs in Strawberry Sauce

Spinach Fritters with Herbed Sour Cream and Smoked-Salmon
Sauce
■
Braised Veal Shanks with Peas and Mushrooms Primavera
■
Crème au Grand Marnier with Strawberries

Just like the summer kitchen with its superstar vegetables, the spring kitchen offers us its own very special, if somewhat more subdued, delicacies. Peas, asparagus, tender young spinach, lettuce, sorrel, and, of course, strawberries should be the main components of your spring meals.

I particularly like this rather traditional, homey menu because it is so in keeping with the kind of seasonal food I grew up with and still enjoy serving to family and friends. For starters, even the vivid green color of the spinach fritters makes this appetizer feel like spring. Serve it on individual plates and garnish according to your budget and taste. The Provençal Mayonnaise (page 34) is probably the perfect choice, especially if you've made a batch of it for another use and have it on hand. Another interesting presentation is to top each fritter with a pan-grilled sea scallop and a drizzle of lemon butter. For something even more elaborate, garnish each plate with a few pencil-thin steamed asparagus, some finely sliced prosciutto, and a shower of freshly grated Parmesan.

Any of these presentations would set the mood for one of the season's best main courses: braised veal shanks. The shanks are known primarily as the traditional cut for the hearty osso buco, which is usually served with a Milanese saffron-flavored risotto, but they are equally delicious when braised gently in a flavorful stock with fresh rosemary, peas, and sautéed mushrooms. The natural sweetness of young peas complements the rich flavor of the veal shanks, making this a seasonal winner. Buttery-crisp rosti potatoes (page 382) are the perfect side dish, but since the appetizer in this menu is rather filling, you can also keep it quite simple by serving good crusty bread for mopping up the delicious sauce. I would follow the veal with a crisp, well-seasoned Belgian endive and arugula salad and then move on in a leisurely way to dessert.

Since strawberries are the spring star fruit in most parts of the country, they are a natural choice for this season's dessert repertoire.

A true vine-ripened berry needs little more than a touch of sugar and a pitcher of cream. You may also serve berries simply tossed in fresh orange juice, but for more elegant entertaining, the Grand Marnier–flavored Bavarian cream heaped with berries and a shiny red currant glaze is a perfect ending to a lovely spring meal.

SPINACH FRITTERS WITH HERBED SOUR CREAM AND SMOKED-SALMON SAUCE

serves 6–8

SAUCE
1½ cups sour cream
2 tablespoons finely minced fresh chives
3 tablespoons finely minced fresh dill
Finely grated rind of 1 large lime
3 ounces smoked salmon, finely minced
Salt and freshly ground white pepper

FRITTERS
2 cups fresh spinach, stemmed and washed
2 large eggs

1 tablespoon grated onion
1 cup buttermilk
1 cup all-purpose flour
1 teaspoon baking powder
4 tablespoons unsalted butter, melted
Salt and freshly ground white pepper
5–6 tablespoons Clarified Butter (page 478)
Freshly ground black pepper

GARNISH
Sprigs of fresh dill and chives

1. Start by making the smoked-salmon sauce: Combine the sour cream, chives, dill, grated rind, and minced salmon in a small bowl. Stir until well blended and season with salt and pepper to taste. Cover the sauce and refrigerate at least 1 hour before using.

2. Preheat the oven to 200°F.

3. To prepare the fritters: Place the spinach in a small saucepan with the water that clings to its leaves, cover, and cook over medium heat for 2 or 3 minutes or until the spinach is just wilted. Drain and, when cool enough to handle, squeeze out the excess liquid.

4. Transfer the cooked spinach to the workbowl of a food processor or blender, add the eggs, and process until the spinach is finely chopped. Add the onion, buttermilk, flour, and baking powder, and pulse on and off to combine. Transfer the batter to a mixing bowl, add the melted butter, and whisk until well blended. Season with salt and pepper.

5. In a medium nonstick skillet melt 2 teaspoons of the clarified butter over medium heat. Add the batter to the hot skillet using 1 heaping tablespoonful for each fritter; do not crowd the pan. You will have to cook the fritters in batches, 4 or 5 per batch. Cook until the fritters are nicely browned on both sides.

6. Remove the fritters from the skillet and transfer to a baking sheet. Place in the preheated oven, cover, and keep warm while making the remaining fritters.

7. Continue cooking the fritters, adding 2 teaspoons of clarified butter to the skillet for each batch. After a couple of batches, wipe the skillet clean with paper towels and begin again.

8. To serve, sprinkle the warm fritters with a grinding of black pepper. Arrange 3 or 4 fritters on each serving plate and garnish with a dollop of the smoked-salmon sauce and sprigs of dill and chives. Serve at once.

Note: If pressed for time, use 4½ ounces of frozen chopped spinach, thawed, drained, and squeezed. You need not cook the frozen spinach. Omit step 3 and proceed with the recipe.

The fritter batter can be made the night before, omitting the melted butter. Bring the batter back to room temperature the next day and whisk in the melted butter.

BRAISED VEAL SHANKS WITH PEAS AND MUSHROOMS PRIMAVERA

serves 6

6 tablespoons unsalted butter
1 tablespoon olive oil
6 pounds veal shanks, cut crosswise into 2-inch pieces
Salt and freshly ground black pepper
6 large cloves garlic, peeled and sliced
6 small sprigs of fresh thyme
2 small sprigs of fresh rosemary
⅔ cup dry white wine
1½ cups Brown Chicken Stock (page 467)

8 small onions, peeled, or 10 white onions, peeled
1 pound Cremini mushrooms, stemmed and cubed
1½ cups shelled fresh peas
1 Beurre Manie (page 478)

GARNISH
3 tablespoons finely minced fresh parsley

1. In each of 2 (12-inch) cast-iron skillets, melt 2 tablespoons of butter with olive oil. Add the veal shanks in 1 layer and brown over medium-high heat on both sides. When the veal has browned, remove to a side dish, season with salt and pepper, and reserve.

2. Add the garlic, thyme, rosemary, and wine. Bring to a boil, reduce the wine to a glaze, and return the veal shanks to each skillet. Add enough stock to come halfway up the sides of the shanks, and divide the onions between the 2 skillets. Cover tightly with foil and then the lid. Braise the veal on the lowest possible heat for 1 hour, or until very tender when pierced with a fork. Reserve.

3. While the veal is braising, sauté the mushrooms. In a 10-inch skillet, melt the remaining butter and oil over medium-high heat. Add the mushrooms, season with salt and pepper, and cook until nicely browned. Add the mushrooms to the veal during the last 10 minutes of braising.

4. In a medium saucepan bring lightly salted water to a boil, add the peas, and cook for 5 minutes, or until just tender. Drain the peas and add them to the veal.

5. When the veal is done, transfer it to a deep serving platter with the mushrooms and peas. Combine the pan juices in one skillet, bring to a boil, and whisk in bits of beurre manie. Cook until the sauce coats a spoon. Taste the sauce and correct the seasoning, spoon the sauce over the veal, and garnish with minced parsley. Serve hot, accompanied by crusty bread and buttered noodles.

Note: Cremini mushrooms are far tastier and have a better texture than all-purpose mushrooms; they are now available in many supermarkets, where they are often called brown mushrooms. If you cannot find them, substitute all-purpose mushrooms.

For an interesting and colorful variation, substitute 1 large diced zucchini (using the skin part only and discarding the pulp) and 1 red pepper—cored, seeded, and diced—for the mushrooms. Braise the zucchini and pepper in a little olive oil and ⅓ cup water or broth until tender and add to the veal shanks just before serving.

CRÈME AU GRAND MARNIER WITH STRAWBERRIES

serves 6–8

½ cup Grand Marnier
1 tablespoon unflavored gelatin
5 large egg yolks
¾ cup minus 2 tablespoons granulated
  sugar
1½ cups warm milk
1 teaspoon vanilla extract
1 cup heavy cream, whipped

2 pints fresh small strawberries, washed
  and hulled
2 (10-ounce) packages frozen
  strawberries, thawed and drained;
  reserve juice from 1 package
¾ cup red currant jelly
A few drops of lemon juice

1. Place ¼ cup of the Grand Marnier in a small, heavy saucepan and sprinkle the gelatin over it. Let stand for 5 minutes, or until the gelatin has softened. Place the pan over low heat until the gelatin has completely dissolved; keep warm.

2. In the top of a double boiler combine the yolks, sugar, and warm milk. Set over simmering water and whisk the mixture until it thickens.

3. Pour the gelatin into the hot custard, add the vanilla, and whisk until the gelatin has completely dissolved. Transfer the mixture to a stainless-steel bowl and chill until it is almost set. Watch carefully so that the custard does not gel. Fold in the whipped cream and transfer the custard to a glass serving bowl. Smooth the top evenly with a spatula and chill for about 3 to 4 hours, or until completely set.

4. About 2 hours before serving, arrange the fresh strawberries over the custard, stem side down. (If the berries are large, slice and arrange in a decorative pattern.)

5. Place the frozen strawberries in the workbowl of a food processor together with the reserved juice and process until smooth. Pass through a fine sieve into a medium bowl and divide the puree in half.

6. Combine the currant jelly and 3 tablespoons of Grand Marnier in a small saucepan. Place over low heat and stir until the jelly has completely melted.

7. Add the currant-jelly mixture to half of the pureed strawberries and pour the mixture over the berries on the custard. Chill.

8. Add a few drops of lemon juice and the remaining Grand Marnier to the remaining pureed strawberries, pour the mixture into a bowl, and serve the custard lightly chilled with the strawberry sauce on the side.

Steamed Asparagus in Bright-Green Provençal Mayonnaise

Roast Cornish Hens in Garlic Cream with Pan-Sautéed Escarole

Terrine of Citrus Fruit with Strawberry Sauce

Here is a more ambitious spring menu, but although it requires more advance preparation, there is little or no last-minute work in the kitchen.

The Provençal mayonnaise is one of my favorite all-purpose sauces. It can easily double as a dip for raw vegetables and is delicious when served with poached or sautéed salmon or tossed with small steamed scallops and shrimp. It is equally delicious when served with steamed broccoli, sliced ripe avocados, or as a topping to a whole poached head of cauliflower. The sauce can be made days in advance and will continue to develop flavor as it sits. Because the mayonnaise is good with either warm or chilled asparagus this combination has become my spur-of-the-moment seasonal appetizer.

In warm weather, it is a nice change of pace to grill the Cornish hens, smoking them lightly by adding some green lilac or applewood to the coals. The garlic cream can be made separately several hours or even a day ahead of time and the hens grilled at least one or two hours before serving, covered with foil, and kept warm in a low oven.

Escarole, with its robust taste and texture, is an especially well-suited accompaniment to the crisp, juicy hens. This earthy green, available year-round, is often taken for granted and rarely used creatively, yet its slightly bitter flavor works beautifully with the sweet but assertive garlic cream, adding a nice balance and an interesting touch to this main course.

Although simply sugared ripe berries are always a welcome ending to this type of seasonal menu, I also like to take advantage of the last truly flavorful citrus fruit. Here ruby-red grapefruits and juicy navel oranges are bound by a light gelatin citrus base and molded into a loaf pan. The result has a lovely pâtélike effect. Served on individual plates in a pool of either strawberry or raspberry sauce with a garnish of tiny mint leaves, it is one of the most appealing endings to any meal. Since the terrine must be frozen, it also has the advantage that you can make it days ahead of time, something that I find to be most important at a time of year when I want to spend as much time as I can outdoors.

STEAMED ASPARAGUS IN BRIGHT-GREEN PROVENÇAL MAYONNAISE

serves 6

MAYONNAISE

2 cups tightly packed fresh spinach leaves, cooked, squeezed, and chopped

3 tablespoons finely minced scallions

3–4 tablespoons finely minced fresh parsley

2 large cloves garlic, peeled and crushed

3–4 flat anchovy fillets, finely minced

½ cup Crème Fraîche (page 491) or sour cream

1 cup Mayonnaise (page 477)

Juice of 1 large lemon

Salt and freshly ground black pepper

1½ pounds fresh asparagus, peeled, steamed, and cooled

GARNISH

12–18 medium shrimp, cooked and marinated in a little Provençal Vinaigrette (page 480)

2 tablespoons finely minced fresh chives

2 hard-boiled egg yolks, sieved

1. To make the Provençal mayonnaise: Combine the spinach, scallions, parsley, garlic, and anchovies in the workbowl of a food processor or blender, add a little of the crème fraîche or sour cream, and process until smooth. Add the mayonnaise, remaining crème fraîche, and lemon juice and process again until smooth. Season with salt and a large grinding of black pepper and chill for 4 to 6 hours, or overnight.

2. To serve, arrange the asparagus on a rectangular platter in a single, even layer. Spoon the sauce over the stems, leaving the tips exposed, and garnish the sauce with the shrimp, chives, and sieved egg yolks. Serve lightly chilled.

ROAST CORNISH HENS IN GARLIC CREAM WITH PAN-SAUTÉED ESCAROLE

serves 6–8

4 Cornish hens, 1–1¼ pounds each
1½ teaspoons dried thyme
1½ teaspoons dried marjoram
1 tablespoon Dijon mustard
1½ teaspoons imported mild paprika
Coarse salt and freshly ground black pepper
2 tablespoons unsalted butter
5 tablespoons olive oil
2 small onions, peeled and cut in half
1¼ cups Chicken Stock (page 470)

8 large cloves garlic, peeled
1½ cups heavy cream
1 large head escarole, washed, dried, and torn into 1-inch pieces
8 fresh shiitake mushrooms, sliced and sautéed in a little olive oil, optional
1 Beurre Manie (page 478)

GARNISH
Sprigs of watercress
8 medium cloves garlic, roasted (page 482) and peeled, optional

1. Preheat the oven to 400° F.

2. Dry the Cornish hens thoroughly with paper towels and truss them. In a small bowl combine the thyme, marjoram, mustard, and paprika. Mix well and rub all over the hens. Season with coarse salt and black pepper.

3. Melt 2 tablespoons of the butter with 1 tablespoon of the oil in a large, heavy flameproof baking dish over medium-high heat. Add the hens and brown on all sides, regulating the heat so as not to burn them.

4. Add the onions to the baking dish, cut side down, together with ¼ cup of the stock. Place the hens on their sides and set the baking dish in the center of the preheated oven. Roast for 35 to 40 minutes, adding a little stock to the pan every 10 minutes and basting them with the juices. Turn the hens once during roasting. They are done when the juices run pale yellow or when the inner temperature reaches 180° F on a meat thermometer.

5. While the hens are roasting prepare the garlic cream. Bring water to a boil in a small saucepan, add the garlic cloves, and blanch for 1 minute. Drain the cloves and discard the water. Add more water to the saucepan and bring to a boil. Add the blanched garlic cloves and cook for 10 to 15 minutes, or until tender. Drain the garlic cloves again and add them to the workbowl of a food processor with the cream. Process until smooth and reserve.

6. Heat 2 tablespoons of oil in a large casserole over medium heat. Add half of the escarole and cook, tossing in the oil, until the leaves have wilted. Remove and set aside. Repeat with the rest of the escarole. Return first batch to the

casserole, add the optional shiitake mushrooms, correct the seasoning, and keep warm.

7. When the hens are done, transfer them to a cutting board and set aside while completing the sauce.

8. Remove the onions from the pan juices and discard. Thoroughly degrease the pan juices and return them to the baking dish. Add the remaining stock, bring to a boil, and reduce by half. Add the garlic cream, bring to a boil, and reduce slightly. Whisk in bits of beurre manie until the sauce coats a spoon. Taste and correct the seasoning.

9. Remove the strings from the hens, cut in half, and transfer to individual serving plates. Spoon some of the sauce around the hens and garnish with sprigs of watercress, the escarole mixture, and optional roasted garlic cloves. Serve at once.

TERRINE OF CITRUS FRUIT WITH STRAWBERRY SAUCE

serves 6–8

Terrine of citrus fruit is a relatively new dessert, which I first sampled at Gerard Boyer's restaurant in Reims. It is at its most appealing when molded into a loaf shape and then sliced like a pâté, but it can also be frozen in attractive individual molds, a practical alternative if you don't plan to serve the entire recipe at one meal. Because the quality of the fruit is so important, make the terrine in early spring, when both California navel oranges and pink grapefruits are still in good supply. You can make several loaves at a time and freeze them for warm-weather entertaining.

Flavorless oil (preferably almond oil) for mold
1½ tablespoons unflavored gelatin
2 cups fresh orange juice, strained
2 pink grapefruits, about 1½ pounds total

4 navel oranges, about 2 pounds total
1 recipe Strawberry Sauce (page 220)

GARNISH
Sprigs of fresh mint

1. Lightly oil the inside of a 1½ quart terrine, preferably porcelain.

2. Combine the gelatin and orange juice in a small saucepan and let sit for 5 minutes, or until the gelatin has softened.

3. Peel the grapefruits and oranges with a sharp knife, removing all traces of the rind and white pith. Cut each segment away from the membrane on either side, then cut each segment in half crosswise. Set aside.

4. Heat the orange juice and gelatin mixture over medium-low heat, stirring constantly until the gelatin has completely dissolved. Remove from the heat and transfer the mixture to a large stainless-steel or glass bowl. Chill until the mixture begins to thicken but is not set. (The thickened juice will keep the fruit suspended throughout the gelatin mixture instead of letting it sink to the bottom of the terrine.)

5. Fold the fruit into the gelatin mixture and pour into the oiled terrine. Cover and chill overnight.

6. The next day remove the terrine from the refrigerator and run a hot knife around the inside of the terrine to loosen the molded fruit. Invert onto a sheet of plastic wrap and shake to loosen the mold from the terrine. Wrap completely in the plastic wrap and place in the freezer for 3 to 4 hours or until just frozen.

7. To serve, cut the terrine into ½-inch slices and place 1 slice on each individual dessert plate. Let sit for a few minutes, then wipe excess liquid from the plate with a paper towel. Surround with strawberry sauce and garnish with sprigs of mint.

Risotto with Shrimp and Radicchio Treviso

■

Sauté of Salmon with Asparagus Vinaigrette

■

Sweet Spring Mango Soufflés

Few dishes capture the intensity of flavors as well as a risotto. When well prepared, this superb rice dish is one of the most satisfying starters to any meal. Here is a risotto that is the signature dish of the tiny Café Orientale in Venice. I always order it, and even when it is not on the menu, the cook, a wonderful woman whose two sons own and run the restaurant, will go out of her way to make it for me.

Spring is the time for such favorites as peas, asparagus, young spinach, and the first buttery greens, all of which lend their delicate flavor to this risotto. You may substitute escarole for the radicchio and, for the shrimp, either small scallops or cooked mussels, using some of the mussel poaching liquid instead of chicken stock.

The risotto is a good opener for the flash-sautéed salmon, served with a mellow, pale green asparagus vinaigrette that is drizzled in a free-form manner. Small bundles of additional asparagus spears and parsleyed new potatoes will complement this light but full-flavored spring main course.

Fresh mangoes are without a doubt among the most delicious of all late-spring fruits. Unfortunately, *fresh* does not mean ripe, and a great part of the country does not get this tree-ripened fruit. A mango, simply sliced and served with good-quality store-bought or homemade vanilla ice cream, is a most satisfying dessert, but for an impressive finale to a meal, this richly flavored soufflé is well worth the time.

■ ■

RISOTTO WITH SHRIMP AND RADICCHIO TREVISO

serves 6

½ pound medium shrimp, peeled
4 tablespoons unsalted butter
2 teaspoons olive oil
Salt and freshly ground black pepper
1½ tablespoons finely minced shallots
1½ cups Italian rice, preferably Arborio
5 cups hot Chicken Stock (page 470)

2 cups trimmed and shredded radicchio
2 cups Boston lettuce, washed, dried,
   and torn into 1½-inch pieces
1 cup cooked fresh peas
½ cup heavy cream
⅓ cup freshly grated Parmesan cheese

1. Dry the shrimp thoroughly on paper towels. In a heavy 3-quart casserole, melt 2 tablespoons of the butter together with 1 teaspoon of the olive oil over medium-high heat. Add the shrimp and cook for 2 to 3 minutes or until they turn just pink. Season with salt and pepper and transfer to a side dish with a slotted spoon. Reserve.

2. Reduce the heat, add the shallots to the casserole, and cook for 1 minute without browning. Add the rice and toss with the butter and shallots until the grains have become opaque. Season with salt and pepper. Add 3 cups of the stock and bring to a boil. Reduce the heat to low and simmer the rice, covered, for 15 minutes.

3. While the rice is cooking prepare the radicchio: In a medium skillet, melt the remaining butter and oil over medium heat. Add the radicchio and cook, stirring often, until just wilted. Season with salt and pepper and set aside.

4. Uncover the rice and add the remaining stock, ⅓ cup at a time, stirring constantly until it is absorbed by the rice, about 10 minutes longer. Each grain of rice should be soft on the outside but still chewy in the center, and the risotto should be quite moist and creamy. Add the lettuce and stir until it just wilts in the risotto. Add the reserved radicchio, shrimp, and peas and just heat through.

5. Add the cream and Parmesan and correct the seasoning, adding a large grinding of pepper. Serve at once accompanied by a crusty loaf of French bread.

SAUTÉ OF SALMON WITH ASPARAGUS VINAIGRETTE

serves 6

1½ pounds young asparagus
2 tablespoons finely minced fresh
  parsley
2 tablespoons finely minced scallions
⅓–½ cup Crème Fraîche (page 491)
3 tablespoons sherry vinegar
½ cup extra-virgin olive oil
Salt and freshly ground white pepper
2–3 tablespoons heavy cream, optional
2 tablespoons julienne of fresh sorrel
  leaves, optional

2½–3 pounds fresh salmon fillets, with
  skin left on
All-purpose flour for dredging
2 tablespoons unsalted butter

GARNISH
1–2 tablespoons red salmon caviar
Fresh lemon juice
Sprigs of fresh dill or watercress

1. Peel the asparagus stalks. Cut the tips off with about 2 inches of stalk and reserve. Dice the remaining stalks; you should have about 2 cups. In a saucepan bring 3 cups of salted water to a boil. Add the diced asparagus stalks and cook until very tender. Drain and set aside.

2. In the workbowl of a food processor combine the parsley, scallions, cooked asparagus stalks, and a little of the crème fraîche. Process until smooth. Add 2 tablespoons of the vinegar and 6 tablespoons of the olive oil and process again until smooth. Season the vinaigrette with salt and pepper and, if too thick, thin with a little heavy cream, if desired. Transfer the vinaigrette to a bowl, add the optional julienne of sorrel, and chill for at least 1 to 2 hours.

3. In a saucepan bring 3 cups of salted water to a boil. Add the asparagus tips and poach until barely tender; do not overcook. Drain and transfer to a side dish. Sprinkle with the remaining vinegar and olive oil and chill until serving.

4. Cut the salmon on the diagonal into 6 scallops ½ inch thick. Dry the scallops thoroughly on paper towels. Season with salt and pepper and dredge very lightly in flour. In a large nonstick skillet melt the butter over medium-high heat. Add the salmon without crowding the pan and sauté until nicely browned on both sides; do not overcook. Remove and set aside.

5. Divide the vinaigrette among 6 individual serving plates. Place a salmon scallop in the center of each plate, and garnish with 3 or 4 well-drained asparagus tips. Sprinkle each portion with a little salmon caviar and drops of lemon juice. Garnish with sprigs of dill or watercress and serve at once.

SWEET SPRING MANGO SOUFFLÉS

serves 6

Butter and granulated sugar for soufflé
dishes
2 ripe mangoes, peeled, pitted, and
diced
2 large egg yolks
½ cup granulated sugar

½ teaspoon vanilla extract
4 large egg whites

GARNISH
Confectioners' sugar
Honey Custard Ice Cream (page 233),
optional

1. Preheat the oven to 400° F. Generously butter the insides of 6 (1-cup) soufflé dishes. Sprinkle with sugar and shake out the excess.

2. Place the diced mango in the workbowl of a food processor and puree until smooth. You should have about 1 cup of puree. Reserve.

3. In a medium mixing bowl combine the egg yolks, ¼ cup of the sugar, and the vanilla extract. Beat the mixture until fluffy and pale yellow, fold in the mango puree, and set aside.

4. In another mixing bowl, combine the egg whites with the remaining sugar and beat until stiff and shiny.

5. Fold one quarter of the beaten egg whites into the mango mixture, gently but thoroughly. Then fold the lightened mango mixture into the remaining whites, again gently but thoroughly.

6. Divide the soufflé mixture evenly between the prepared soufflé dishes. Place the soufflés in the center of the preheated oven and bake for 7 to 9 minutes, or until the tops are nicely browned. The soufflés should still be a bit runny in the center.

7. Remove the soufflés from the oven and sprinkle heavily with confectioners' sugar. Serve immediately with Honey Custard Ice Cream.

4

Globe Artichokes in Curry-Lemon Vinaigrette

Chicken Breasts in Sorrel-Cream Primavera

Fallen Ricotta and Orange Soufflés

Artichokes and asparagus are, to my mind, the best starters to a spring meal. Simply poached and then served hot with hazelnut butter or at room temperature with a zesty anchovy-garlic cream, artichokes have a presence that few other vegetables can command. Here they are served with a cool curry mustard sauce that complements the naturally delicate flavor of this unique vegetable.

Follow with a simple sauté of chicken breasts in a sauce enhanced by the tangy taste of young sorrel. A sprinkling of braised julienned beets enlivened with a touch of chives or mint makes a wonderful accompaniment to this dish, adding color and texture.

After a light main course, I often serve a slightly more filling dessert, such as the ricotta puddings. Prepare the pudding base well ahead of time, but fold in the beaten egg whites just before sitting down to dinner. The soufflés can also be baked ahead of time and served as fallen soufflés. They will unmold easily and look lovely placed in a pool of chilled fruit sauce with a garnish of seasonal berries and a sprig of mint.

GLOBE ARTICHOKES IN CURRY-LEMON VINAIGRETTE

serves 4

4 large globe artichokes
1 lemon, halved
Salt

VINAIGRETTE
1 teaspoon egg yolk
1 teaspoon curry powder (preferably Madras)
1 teaspoon ginger, crushed through a garlic press

1½ tablespoons sherry vinegar
2 teaspoons Dijon mustard
1 large clove garlic, peeled and mashed
¾ cup olive oil
Juice of ½ lemon
3 anchovy fillets, drained and finely minced, optional
2 tablespoons finely minced scallions or chives, optional

1. Cut the stems off the artichokes. With a sharp knife cut the tips off the leaves. Immediately rub the cut parts with 1 lemon half.

2. In a large casserole, bring salted water to a boil. Add the remaining lemon half and artichokes and cook for 25 minutes, or until tender. Test by piercing the artichoke bottoms with the tip of a sharp knife.

3. While the artichokes are cooking, make the vinaigrette: In a small, heavy bowl combine the egg yolk, curry powder, ginger juice, vinegar, mustard and garlic and whisk the mixture until well blended. Start adding the oil, drop by drop, whisking constantly until the dressing is thick and smooth. Add the lemon juice and the optional anchovies, and scallions or chives. Taste the sauce; if you are using anchovies it probably will not need salt, just a large grinding of black pepper. Set the sauce aside.

4. When the artichokes are tender, drain and carefully spread the leaves out from the center. Scoop out the fuzzy choke with a small spoon and fill each artichoke with a little vinaigrette. Place on a serving platter and serve at room temperature, with any remaining sauce on the side.

Note: To extract the ginger juice, peel a very fresh piece of ginger about the size of a walnut and force it through a garlic press. You can now find ginger presses which look just like a garlic press, but with larger holes. This vinaigrette is equally delicious served as a sauce with grilled scallops, shrimp, or fish steaks. Since the sauce keeps for 10 days in the refrigerator, you can easily make a double batch and use it on another vegetable or on seafood at another time.

CHICKEN BREASTS IN SORREL-CREAM PRIMAVERA

serves 6

1 cup Crème Fraîche (page 491)
2 teaspoons fresh lemon juice
2 cups sorrel leaves, stemmed, washed, cooked, and drained (page 68)
5 tablespoons unsalted butter
¾ pound small button mushrooms, wiped
4 small whole chicken breasts, cut in
half, skin removed, and flattened slightly
Salt and freshly ground white pepper
All-purpose flour for dredging
1 tablespoon peanut oil
⅓ cup Chicken Stock (page 470)
1 Beurre Manié (page 478)
2 tablespoons finely minced fresh chives

1. A few hours ahead, combine the crème fraîche and lemon juice in a small bowl and let the mixture stand at room temperature for 2 hours.

2. Transfer the lemon-cream mixture to the workbowl of a food processor together with the cooked sorrel and process until smooth. Set aside.

3. In a heavy, medium skillet, melt 2 tablespoons of the butter. Add the mushrooms and cook over medium-high heat for 2 to 3 minutes, shaking the pan back and forth, until the mushrooms are nicely browned on all sides. Reserve.

4. Season the chicken breasts with salt and pepper. Dredge lightly in flour, shaking off the excess. In a large cast-iron skillet, heat the remaining butter and the oil over medium-high heat. Add the chicken breasts and cook for about 3 to 4 minutes or until nicely browned on both sides. Remove the chicken breasts to a side dish and reserve.

5. Remove all but 1 tablespoon of fat from the pan and add the stock. Bring to a boil over high heat and reduce the stock to 2 tablespoons, stirring often. Whisk in the sorrel-cream, season with salt and pepper, and cook the sauce until it is slightly reduced. Whisk in bits of beurre manie, whisking constantly until the sauce lightly coats a spoon.

6. Return the chicken breasts to the pan and simmer them in the sauce for another 4 to 5 minutes or until just done. Do not overcook. Taste and correct the seasoning.

7. Transfer the chicken breasts to a serving platter. Spoon the sauce over them and garnish with the sautéed mushrooms. Sprinkle with chives and serve hot.

FALLEN RICOTTA AND ORANGE SOUFFLÉS

serves 6

Butter and granulated sugar for soufflé dishes
4 large eggs, separated
¼ cup granulated sugar
1 teaspoon vanilla extract
2 tablespoons lightly toasted almonds, finely ground

Finely grated rind of 1 orange
1 cup whole-milk ricotta
⅓ cup confectioners' sugar

GARNISH
Confectioners' sugar

1. Generously butter the insides of 6 (4-ounce) ramekins. Sprinkle with granulated sugar, shake out the excess, and set aside.

2. Preheat the oven to 475° F.

3. In a large mixing bowl combine the egg yolks, granulated sugar, and vanilla. Beat with an electric hand mixer until fluffy and pale yellow. Add the almonds, orange rind, and ricotta and fold in gently but thoroughly. Reserve.

4. In another large mixing bowl, combine the egg whites and confectioners' sugar and beat until the whites are shiny and hold firm peaks. Fold one quarter of the egg whites into the ricotta mixture to lighten it. Fold the lightened ricotta mixture into the remaining egg whites, gently but thoroughly.

5. Spoon the soufflé mixture into the prepared ramekins. Place the ramekins in a heavy rectangular baking dish and add enough boiling water to the baking dish to come halfway up the sides of the ramekins. Place the baking dish in the center of the preheated oven and bake for 10 to 12 minutes, or until the tops of the soufflés are golden brown. Remove the soufflés from the oven and let cool slightly on wire racks.

6. When the ramekins are cool enough to handle but the soufflés are still warm, unmold each soufflé onto an individual dessert dish, dust heavily with confectioners' sugar, and serve at once.

Note: The soufflés may also be served at room temperature or chilled. For a variation, surround each unmolded soufflé with Strawberry Sauce (page 220) and garnish with fresh blueberries and mint.

Asparagus, Leek, and Potato Soup du Jour

Fillets of Beef in Emerald Sauce with Crisp-Fried Onion Rings

Very Lemony Meringue Custards

In the Northeast, spring is our most unpredictable season. Although the trees start to bud, and everything seems to grow and bloom almost overnight, there are a number of chilly days that make you wonder whether spring is really here. This hearty, cozy menu is meant for just such a day. A fresh asparagus soup is a must in the spring kitchen. Whether made simply in a bistro style, such as this, or prepared in a more elegant, richer classic version with a creamy velouté base, it is truly a seasonal treasure that is a perfect starter for casual or elegant entertaining.

Follow the soup with fillets of beef in an updated classic preparation that, when properly cooked, can hold its own against the best of nouvelle cuisine main courses. Its success depends on just three ingredients: good beef, excellent herb mustard, and perfectly fried onions. Weather permitting, you can, of course, grill the fillets and then finish cooking them in a skillet. If you don't feel like adding the fried onion rings, which would be a shame, you can oven-grill the onions by enclosing them in heavy-duty foil with a touch of virgin olive oil and some minced fresh herbs, and cook the onions at the same time as the fillets. I would opt for homemade shoestring potatoes as an accompaniment to this main course or the buttery puree of potatoes with glazed scallions (page 103).

It is almost de rigueur to follow a main course of beef with something light and fruity, and the lemon meringue custards always seem to hit the spot. These sabayonlike custards are extremely versatile and easy to make. You may tuck a few raspberries or a sliced strawberry in the bottom of each ramekin or add a few tiny lemon thyme leaves—yes, thyme—to the custard. Either way this simple dessert makes a lively and satisfying ending to a cold-weather spring meal.

ASPARAGUS, LEEK, AND POTATO SOUP DU JOUR

serves 6

4 tablespoons unsalted butter
2 medium leeks, trimmed of all but 2 inches of greens, thinly sliced and rinsed
2 medium all-purpose potatoes, peeled and diced
1 pound fresh asparagus, trimmed, stalks peeled and diced, and tips reserved

6 cups Chicken Stock (page 470)
½ cup heavy cream
Salt and freshly ground white pepper
2 tablespoons unsalted butter, for enrichment, optional

GARNISH
4 tablespoons finely minced fresh chives

1. In a large, heavy casserole, melt the butter over low heat. Add the leeks, potatoes, and asparagus stalks and cook for 2 to 3 minutes. Add the chicken stock, bring to a boil, reduce heat, and simmer for 25 minutes, covered, or until the vegetables are tender. Cool the soup slightly.

2. With a slotted spoon, remove 1 cup of the vegetables and reserve. Transfer the remaining soup to the workbowl of a food processor or blender and process until smooth.

3. Return the pureed soup to the casserole and place over medium heat. Add the asparagus tips and cook for 4 to 5 minutes or until just tender. Add the reserved cooked vegetables and cream and simmer 5 minutes longer. Taste and correct the seasoning, adding the optional butter and the chives. Serve the soup hot or lightly chilled.

FILLETS OF BEEF IN EMERALD SAUCE WITH CRISP-FRIED ONION RINGS

serves 6

SAUCE
3 tablespoons finely minced shallots
1 cup Chicken Stock (page 470)
3 tablespoons Maille mustard with green herbs
16 tablespoons (½ pound) unsalted butter, chilled and cut into 16 pieces
Salt and freshly ground white pepper

FILLETS
6 fillets of beef, cut ¾ inch thick, about 2–2¼ pounds total
3 tablespoons unsalted butter
2 teaspoons peanut oil
Coarse salt and coarsely ground black pepper

GARNISH
Fresh watercress leaves
Crisp Fried Onion Rings (recipe follows)

1. Start by making the sauce: In a heavy 2-quart saucepan, preferably copper, combine the shallots and chicken stock. Bring to a boil and reduce to 2 tablespoons, whisking from time to time.

2. Remove the saucepan from the heat and whisk in the mustard. Return to very low heat and add the butter, 1 tablespoon at a time, whisking constantly, until it has been completely incorporated and the sauce is quite smooth and creamy. Season with salt and a large grinding of white pepper. Strain the sauce through a fine sieve and transfer to the top of a double boiler. Cover and set over lukewarm water until ready to serve.

3. Dry the fillets thoroughly on paper towels. In a large, heavy skillet, preferably cast iron, melt the 3 tablespoons of butter together with the oil over high heat. Add the fillets, without crowding the pan, and brown quickly on each side. Reduce the heat slightly and season with coarse salt and pepper. Cover and continue to cook for 2 minutes on each side for medium-rare.

4. When the fillets are done transfer them to paper towels and let sit for about 15 seconds. Dip each fillet into the green mustard sauce to coat completely, and place each in the center of an individual serving plate. Garnish with leaves of watercress and surround with Crisp Fried Onion Rings.

CRISP-FRIED ONION RINGS

serves 6

1½ cups all-purpose flour
1¼ cups beer
3 large egg whites, lightly beaten
½ teaspoon coarse salt

1 quart peanut or corn oil
2–3 medium onions, peeled and cut into
  ¼-inch slices

1. In a large mixing bowl combine the flour, beer, egg whites, and salt. Whisk until smooth and set aside.

2. Place the oil in a deep pan or deep-fat fryer and heat to 375° F.

3. Separate the onion slices into rings and dip the rings into the batter, letting the excess drip off. Drop into the hot oil and fry about 10 rings at a time until they are golden brown, about 2 to 3 minutes, turning often. Remove with a slotted spoon to paper towels. Sprinkle with coarse salt and continue frying the remaining onion rings, making sure to let the oil return to 375° F before adding the next batch.

VERY LEMONY MERINGUE CUSTARDS

serves 8

CURD
3 large eggs
3 large egg yolks
¾ cup granulated sugar
¾ cup fresh lemon juice
Zest of 1 lemon, finely grated, blanched,
    and drained

10 tablespoons unsalted butter, cut into
    10 pieces, chilled

MERINGUE
6 large egg whites
Pinch of salt
⅔ cup granulated sugar
Confectioners' sugar

1. Start by making the curd: In a large stainless-steel mixing bowl, combine the eggs and egg yolks. Whisk in the sugar, lemon juice, and lemon zest until well blended. Transfer the mixture to a heavy 2-quart saucepan, preferably copper. Place over medium heat, add the butter, and whisk constantly until the butter is completely incorporated and the mixture is quite thick. Make sure the mixture does not come to a boil or the eggs will curdle.

2. Immediately remove the lemon curd from the heat and divide between 8 (4-ounce) porcelain ramekins. Place a piece of plastic wrap directly on the surface of each curd and chill for 6 hours or overnight.

3. Preheat the broiler.

4. To prepare the meringue: In a large mixing bowl combine the egg whites and pinch of salt. Beat with an electric mixer until the whites hold soft peaks. Add the granulated sugar, 2 tablespoons at a time, beating constantly, until the mixture is stiff and shiny.

5. Using a pastry bag fitted with a ½-inch star-tip nozzle, pipe the meringue in a large rosette over each curd, covering the surface completely. Sprinkle heavily with confectioners' sugar and place under the broiler, 6 inches from the source of heat, until golden brown. Be careful not to burn the meringue. Serve at once.

PICKLED HALIBUT WITH SPICY PEPPERS AND CITRUS ZEST

serves 4–6

Halibut is a mild fish with a firm texture that lends itself well to this preparation, which is much like a seviche. The success of this dish largely depends on the freshness of the fish. I often serve this cool, zesty salad with warm parsleyed new potatoes as a light main course.

2 pounds very fresh halibut steaks, cut 1 inch thick
Juice of 1 large lime
Juice of 1 large lemon
½ red bell pepper, cored, seeded, and finely diced
½ yellow bell pepper, cored, seeded, and finely diced
3 tablespoons finely diced red onion
¼ cup tiny fresh cilantro leaves
1 fresh green chili pepper, thinly sliced
6 tablespoons extra-virgin olive oil

1 tablespoon fine julienne of lemon zest
1 tablespoon fine julienne of orange zest
Coarse salt and freshly ground black pepper

GREENS
Bibb or Boston lettuce leaves, rinsed and dried
Radicchio leaves
2 Belgian endives, cored and separated into leaves
Small fresh cilantro leaves

1. Cut the halibut into ¾-inch cubes, removing all skin and bones. Place the fish in a shallow glass or stainless-steel bowl and add the remaining ingredients. Toss gently, cover, and chill for 4 to 6 hours.

2. Before serving, taste the marinade and season with coarse salt and pepper to taste. The marinade should be rather spicy.

3. To serve, make a bed of Bibb lettuce, radicchio, and endive on 4 to 6 individual serving plates. Divide the pickled halibut between the plates and garnish with fresh cilantro leaves. Spoon some of the marinade over the greens. Serve chilled but not cold.

FRIED SARDINES WITH WILTED SPINACH
AND LEMON VINAIGRETTE

serves 4

A fresh sardine, either grilled or pan-fried, is one of those delicious but underappreciated morsels of spring. Buy only the freshest sardines—they must glisten and have their heads attached. Unfortunately, sardines are not always available outside large East Coast cities such as Boston and New York, and if this is true in your area, substitute smelts for the sardines. Although they are less flavorful, they will benefit greatly from the zesty lemon vinaigrette.

VINAIGRETTE
Juice of 1 lemon
2 tablespoons finely minced shallots
Salt and freshly ground black pepper
8 tablespoons olive oil

SARDINES
12 small fresh sardines, cleaned
Coarse salt and freshly ground black pepper

3–4 tablespoons extra-virgin olive oil

SPINACH
3 tablespoons extra-virgin olive oil
4 cups fresh spinach, preferably New Zealand, stemmed, washed, and dried well on paper towels
1 large clove garlic, finely minced
1 teaspoon light soy sauce

1. Start by making the vinaigrette: In a small bowl combine the lemon juice and shallots. Season with salt and pepper and add the oil in a slow, steady stream, whisking constantly until smooth and creamy. Reserve.

2. Dry the sardines thoroughly on paper towels and season with coarse salt and black pepper. In a large nonstick skillet, heat the olive oil over medium heat. Add the sardines without crowding the pan and cook for 2 to 3 minutes on each side, until nicely browned and crisp. Remove from the heat and reserve.

3. To prepare the spinach: Heat the oil in a large skillet over medium heat. Add the spinach and toss quickly in the oil until just wilted. Add the garlic and soy sauce, toss well, and immediately divide between 4 warm serving plates. Top each portion of spinach with 3 fried sardines and drizzle with the vinaigrette. Serve at once.

DOUBLE ASPARAGUS AND
SMOKED-SALMON TART

serves 6

With pencil-thin asparagus creating a lovely design on the top, this sophisticated and beautiful tart makes a wonderful first course for a somewhat formal meal. You can also serve it as the main course of a light seasonal lunch.

Or you can turn the tart into a simpler spur-of-the-moment savory flan by omitting the pastry shell. Heavily butter a ten-inch porcelain quiche pan, and sprinkle it generously with bread crumbs. Add two additional eggs to the custard mixture and, when the custard is almost set, top with the asparagus, arranged like spokes of a wheel.

Salt
1 pound asparagus, trimmed
3 large eggs
1½ cups heavy cream
3 ounces smoked salmon, cut into
   ¼-inch dice

2 tablespoons finely minced fresh chives
2 tablespoons finely minced fresh dill
Freshly ground white pepper
1 partially baked 9-inch Basic Tart Shell
   (page 492), in a pan with a
   removable bottom

1. Bring salted water to a boil in a medium saucepan. Cut off the asparagus tips, add them to the boiling water, and cook for 2 minutes or until tender. Remove with a slotted spoon and reserve.

2. Peel the asparagus stalks with a vegetable peeler and cut into 1-inch pieces. Bring the water back to a boil, add the stalks, and cook for 6 to 7 minutes or until very tender. Drain thoroughly and mince finely. Set aside.

3. Preheat the oven to 350° F.

4. In a large mixing bowl combine the eggs and cream. Whisk until well blended. Fold in the salmon, asparagus tips, minced asparagus stalks, chives, and dill. Season the mixture carefully with salt and pepper and pour into the partially baked tart shell.

5. Place the tart in the center of the preheated oven and bake for 30 to 40 minutes, or until the custard is set and the top is nicely browned; a knife inserted in the center should come out clean.

6. Remove from the oven and serve warm or at room temperature as an appetizer or with a fresh green salad as a main course for lunch.

Variation: Instead of smoked salmon, use flaked poached fresh salmon and 1 cup of diced mushrooms that have been sautéed in a little butter. Add herbs such as dill, chives, or parsley to the tart or flan for even more depth of flavor.

STIR-FRY OF CALAMARI WITH RED PEPPERS AND LEMON

serves 4

This quick and flavorful sauté of calamari is one of my favorite tapas. You can find it in many of the tapas bars throughout Spain where the dish is made to order in a matter of minutes and served piping hot in small earthenware gratin dishes with crusty bread. A glass of dry sherry is the perfect accompaniment to this spicy starter.

4–6 tablespoons extra-virgin olive oil
2 small dry red chili peppers, crushed
1 large red bell pepper, cored, seeded, and cut into ¾-inch cubes
1 pound small calamari, cleaned and cut crosswise into ¼-inch rings
1 small lemon, unpeeled, cut crosswise
into ⅛-inch slices, and each slice cut into quarters
Coarse salt and freshly ground black pepper
2 tablespoons tiny fresh cilantro leaves, optional

1. In a large, heavy-bottomed saucepan, heat 2 to 3 tablespoons of the olive oil over high heat. Add 1 chili pepper and half of the red bell pepper and cook for 1 to 2 minutes, or until barely tender.

2. Add half of the calamari and stir-fry for 1 minute, or until just opaque. Immediately add half the lemon and toss for 30 seconds, or until the lemon is just heated through. Transfer to a serving dish and cook the remaining pepper cubes and calamari in the same manner, using the remaining olive oil, chili pepper, and lemon. Add to the serving dish, sprinkle with coarse salt, black pepper, and optional cilantro leaves, and serve at once.

Note: This dish must be done in batches so that the calamari does not overcook.

WARM SHRIMP SALAD WITH CRISP
VEGETABLES NIÇOISE

serves 6

Here is a deceptively simple Provençal appetizer that can be made only with the freshest and youngest spring vegetables. By deceptively, I mean that the key to the success of this refreshing dish lies in the quality of all the ingredients: the olive oil, which must be superb; the asparagus, which must be perfectly crisp and young; the fennel and the tiny, fresh lima beans. If these last are not available, make the dish without them instead of using frozen or older beans. I recommend a French extra-virgin oil, such as the extra-virgin James Plagniol or Hilaire Fabré or à l'Olivier—a cold-pressed oil that is both rich and fruity. These oils are full of flavor but will not overwhelm the delicate taste of either the shrimp or the vegetables.

½ pound fresh asparagus, peeled and trimmed
1 cup (about 5 ounces) finely diced fennel bulb
2 small ripe tomatoes, unpeeled, seeded, and diced

½ pound shelled baby lima beans
Salt and freshly ground black pepper
½ cup extra-virgin olive oil
Juice of 1 large lemon
1½ pounds medium shrimp, peeled with tails left on

1. Cut off the tips of the asparagus and place in a medium mixing bowl. Slice the stalks very thinly on the bias and add to the bowl with the fennel, diced tomatoes, and lima beans. Season with salt and pepper to taste and toss the vegetables with ¼ cup of the olive oil and the lemon juice. Set aside to marinate for 1 hour.

2. Dry the shrimp thoroughly on paper towels. Heat the remaining oil in a large heavy skillet over high heat. When the oil is very hot, add the shrimp and sauté quickly until bright pink and lightly browned. Season with salt and pepper and keep warm.

3. To serve, divide the marinated vegetables between 6 salad plates and arrange the sautéed shrimp over them. Serve immediately.

PAN-GRILLED SHRIMP SALAD IN SESAME-SCALLION VINAIGRETTE

serves 4–5

I must confess that I've resisted the culinary trend toward "East meets West" which, just for the sake of novelty, often combines ingredients that have no natural affinity for one another.

But I do love the flavors of some Oriental ingredients, in particular, fresh ginger and toasted sesame oil, and I have used them, singly and together, in a number of dishes. In combination, they contribute a lot to this full-flavored hot-skillet salad. The sesame vinaigrette is also wonderful with a salad of oven-grilled bell peppers, zucchini, and shiitake mushrooms. In summer, try it with cubed grilled swordfish that has been briefly marinated in a little lemon juice and toasted sesame oil, or use it as a sauce for grilled tuna.

SALAD

6 tablespoons Sesame-Scallion Vinaigrette (recipe follows)

2 Belgian endives, cored and leaves separated

1 small head radicchio, leaves separated

2 small heads Bibb lettuce, leaves separated, rinsed and dried

1 pound medium shrimp, peeled, with tails left attached

Coarse salt and freshly ground black pepper

2–3 tablespoons extra-virgin olive oil

2 teaspoons finely minced garlic

2 teaspoons finely minced fresh ginger

2 tablespoons finely minced scallions

GARNISH

1 tablespoon lightly toasted sesame seeds

1. Start by making the salad: Place the dressing on the bottom of a large salad bowl. Top with the endive, radicchio, and Bibb leaves but do not toss. Cover and refrigerate until serving.

2. Dry the shrimp thoroughly on paper towels. Sprinkle with coarse salt and black pepper. In a large, heavy skillet, heat the olive oil over high heat. When hot, add the shrimp and toss in the hot oil until just pink and lightly browned. Remove the skillet from the heat and add the garlic, ginger, and scallions. Toss until the mixture is fragrant and the scallions are wilted. With a slotted spoon, transfer the shrimp to a side dish and keep warm.

3. Toss the salad greens with the dressing and place them in a circle on 4 to 5 individual salad plates. Divide the warm shrimp evenly and place in the center of each plate. Sprinkle the shrimp with the toasted sesame seeds and serve at once with a crusty loaf of French bread.

SESAME-SCALLION VINAIGRETTE

makes about 1 ½ cups

1 medium clove garlic, peeled and crushed
1 teaspoon Dijon mustard
2 tablespoons sherry vinegar
2 tablespoons light soy sauce
1 teaspoon egg yolk

¾–1 cup peanut oil, preferably cold-pressed
¼ cup Oriental sesame oil
¼ cup finely minced scallions
Freshly ground black pepper

1. In a medium stainless-steel mixing bowl, combine the garlic, mustard, vinegar, soy sauce, and egg yolk. Whisk until well blended.

2. Add the peanut oil, drop by drop, into the egg-soy mixture, whisking constantly, as if making a mayonnaise, until all the peanut oil has been added. Be sure to add the oil slowly or the mixture will not emulsify. Add the sesame oil in a slow, steady stream, again whisking constantly, until the mixture resembles a loose mayonnaise.

3. Add the scallions and a large grinding of black pepper. Keep the vinaigrette covered at room temperature if serving it the same day. If you've prepared the vinaigrette a few days in advance, keep in a tightly covered jar in the refrigerator. Bring back to room temperature before serving, whisking the vinaigrette until smooth.

TWICE-BAKED CHEESE SOUFFLÉS WITH PEAS AND ASPARAGUS

makes 12

One of the first dishes I learned in cooking school more than twenty-five years ago was cheese soufflé. Although I have a passion for all dessert soufflés, savory soufflés, especially those flavored with cheese, have never been among my favorites. I find them heavy and not in keeping with the way we eat today. So when I came across these delicious soufflés in a restaurant in Basel, I was truly amazed by their superb texture and flavor. The technique may sound complex, but while it is rather unusual, it is really quite simple, and, of course, the soufflé base can be made early in the day.

I usually follow the soufflés with a simple main course, such as the Roast Veal with Glazed Fennel and Tomato Jam (page 96), which can be made in advance, leaving you with little else to do other than the last-minute preparation of the soufflés. The pineapple compote in wine and apple syrup would make an excellent do-ahead dessert.

6 tablespoons unsalted butter
6 tablespoons all-purpose flour
2 cups whole milk
3 tablespoons finely minced fresh chives
6 large egg yolks
Salt and freshly ground white pepper
2½ cups hot heavy cream
8 large egg whites

¾ cup finely grated Gruyère cheese
12 asparagus spears, peeled, cooked, and thinly sliced on the bias
½ cup cooked fresh peas

GARNISH
3–4 tablespoons finely minced fresh chives

1. Preheat the oven to 400° F. Place 12 flan rings, each 3 inches in diameter, in the freezer until needed.

2. Melt the butter in a medium saucepan over medium heat. Add the flour, whisk until well blended, and cook for 2 minutes without browning. Add the milk all at once and whisk constantly until the mixture comes to a boil.

3. Remove the saucepan from the heat. Add the chives and egg yolks and whisk until well blended. Season the soufflé base with salt and pepper and set aside.

4. Remove the flan rings from the freezer and generously coat the insides of the rings with the remaining butter. Place them on a buttered baking sheet and reserve.

5. Pour about 3 tablespoons of hot cream into each of 12 individual round au gratin dishes, about 6 inches in diameter. Set aside.

6. Using an electric mixer, beat the egg whites with a pinch of salt until the peaks are firm. Fold the whites gently but thoroughly into the soufflé base and spoon the mixture into the flan rings, mounding it slightly.

7. Place the baking sheet in the center of the preheated oven and bake for 4 to 5 minutes, or until the tops of the soufflés are lightly browned. Remove from the oven and let sit for 8 to 10 minutes, or until the soufflés fall back and pull away from the sides of the flan rings. With the aid of a spatula and using a towel to protect your hands, lift each soufflé from the baking sheet with its flan ring and place in the center of each gratin dish. Carefully remove the ring and sprinkle each soufflé with 1 heaping tablespoon of Gruyère cheese.

8. Return the soufflés to the center of the preheated oven and bake for an additional 4 to 5 minutes, or until puffed and golden brown. Remove from the oven and sprinkle each gratin dish with some sliced asparagus, cooked peas, and fresh chives. Serve at once.

GOAT CHEESE POTS DE CRÈME WITH PEAS AND SPRING HERBS

serves 6

In recent years food terminology has changed so much that it is sometimes hard to tell what a dish is supposed to be. Not so long ago a pot de crème was strictly a dessert, and the words immediately brought to mind lovely individual porcelain ramekins filled with custard that was flavored with vanilla, chocolate, coffee, or a liqueur. Now, you will find blue cheese, garlic, lettuce, and other fanciful pots de crème served as appetizers and as savory garnishes to a main course. But all pots de crème still have one thing in common, and that is the use of eggs and a little cream as a binder. These spring custards are an easy way to begin a spring lunch, dinner, or informal supper. Be sure to use young fresh peas, and, if you cannot get them, make the pots with wilted Bibb lettuce and a touch of herbs. Do not use frozen peas—they are too watery, tasteless, and not worth bothering with.

½ cup shelled fresh peas
1 tablespoon granulated sugar
Salt and freshly ground white pepper
6 tablespoons creamy goat cheese, such as Montrachet, Bucheron, or a domestic chèvre
4 extra-large eggs
1⅓ cups heavy cream
2 tablespoons freshly grated Parmesan cheese

2 tablespoons finely minced scallions
2 tablespoons finely minced fresh dill or tarragon
6 small fresh mint leaves, whole or minced

GARNISH
6 tiny sprigs of fresh dill

1. Preheat the oven to 350° F.

2. Bring salted water to a boil in a small saucepan. Add the peas and sugar and season with salt and pepper. Simmer the peas until they are just tender, about 3 to 5 minutes. Drain and reserve.

3. Combine the goat cheese, eggs, cream, and Parmesan in the workbowl of a food processor and process until smooth. Transfer the mixture to a mixing bowl. Fold in the scallions and dill or tarragon, taste and correct the seasoning, and reserve.

4. Butter 6 (5-ounce) porcelain ramekins and divide the peas evenly among them. Spoon the cheese mixture into the ramekins and top each with a mint leaf.

5. Place the ramekins in a large baking dish and fill the dish with enough boiling water to come halfway up the sides of the ramekins. Place in the center of the preheated oven and bake for 25 to 30 minutes or until the custard is set and lightly browned. Remove the ramekins from the water bath and cool for 10 minutes.

6. Garnish the pots de crème with tiny sprigs of dill. Serve accompanied by French bread.

GLOBE ARTICHOKES WITH CREAMY GOAT CHEESE AND SHALLOT DRESSING

serves 6

In most parts of the country, the artichoke season really seems too short. And particularly in the Northeast and the Midwest, artichokes are always too expensive to allow for some of the more extravagant preparations that are "peasant fare" in Mediterranean Europe, where this vegetable is so plentiful that only the hearts are used and the leaves discarded. For those of us to whom the artichoke is still something of a delicacy, a simple brown butter or a well-flavored mayonnaise is all that this glorious vegetable needs, but for a nice change of pace I like to serve it with this zesty goat cheese and shallot vinaigrette.

1 tablespoon sherry vinegar
1 teaspoon balsamic vinegar
½ cup mild goat cheese, such as
    Montrachet, crumbled
1 tablespoon finely minced shallots
¼ cup heavy cream
½ cup extra-virgin olive oil

1 teaspoon tiny fresh thyme leaves
Salt and freshly ground black pepper
6 medium globe artichokes
1 large lemon, cut in half

GARNISH
Sprigs of fresh parsley

1. In the workbowl of a food processor combine the vinegars, goat cheese, shallots, heavy cream, and olive oil and process until smooth and creamy. Add the thyme and season with salt and a large grinding of black pepper. Set aside.

2. Cut the stems off the artichokes. With kitchen scissors, cut the pointed tips off the leaves and immediately rub the cut parts with the halved lemon.

3. Bring salted water to a boil in a vegetable steamer. Add the artichokes, cover, and steam for about 20 minutes or until tender. Test by piercing the bottoms with the tip of a sharp knife.

4. Drain the artichokes and, when cool enough to handle, carefully spread the leaves out from the center. Scoop out the fuzzy choke with a small spoon and fill each artichoke cavity with a little dressing. Place on a serving platter, garnish with sprigs of parsley, and serve the remaining dressing on the side.

Note: The goat cheese and shallot dressing can be made in quantity. It will keep, refrigerated, for several days. Serve the dressing with steamed broccoli, roasted peppers, or tossed with cooked new potatoes.

ASPARAGUS WITH BEET VINAIGRETTE

serves 4–6

I am always surprised at how many people dislike the flavor of beets or object to them because of their intense color and ability to discolor other foods. While this delicious vegetable certainly has some overpowering characteristics, especially its color, its sweet and mellow taste certainly makes up for it. I often use the coloring properties of beets to advantage, especially when I want to transform a simple mayonnaise or vinaigrette into a brilliant ruby-red dressing.

In this elegant presentation, the beet vinaigrette covers the entire plate and is then topped with bright-green asparagus. A nice finishing touch is to arrange three or four pan-seared sea scallops around the plate. For a more robust version, I often wrap each asparagus spear in a paper-thin slice of mild prosciutto, smoked Westphalian ham, or smoked salmon and serve the dish as the main course at a weekend lunch.

2 pounds asparagus, trimmed and
  peeled
1 cup small broccoli florets
½ teaspoon egg yolk
Juice of ½ lemon

1½ tablespoons sherry vinegar
1 teaspoon grated raw onion
¾ cup walnut oil
1 cup cooked beets, diced
Salt and freshly ground black pepper

1. Place the asparagus in a vegetable steamer, set over simmering water, and steam, covered, until just tender, about 5 to 7 minutes. Remove the asparagus and run under cold water to stop further cooking. Place on a double layer of paper towels to drain. Reserve.

2. Place the broccoli florets in the steamer, and steam, covered, for 4 to 5 minutes, or until just tender. Remove and run the florets under cold water to stop further cooking. Place on paper towels to drain and set aside.

3. To make the vinaigrette: In a medium mixing bowl combine the egg yolk, lemon juice, vinegar, and raw onion. Whisk until well blended. Add the walnut oil, drop by drop, whisking constantly, until the mixture starts to emulsify. Continue adding the oil in a slow, steady stream, whisking constantly, until all the oil has been added and the vinaigrette is creamy and has the consistency of a loose mayonnaise. Reserve.

4. Place the sliced cooked beets in the workbowl of a food processor, add 3 tablespoons of the vinaigrette, and process until very smooth. Fold the beet

puree into the remaining vinaigrette, season with salt and pepper, and set aside.

5. To serve, spoon some of the beet vinaigrette into the center of 4 to 6 individual serving plates, covering the surface of each plate completely. Place 4 to 6 asparagus spears in the center of the vinaigrette and garnish the edge of the vinaigrette with 3 to 4 broccoli florets. Serve with a crusty loaf of French bread.

Variation: Substitute poached leeks for the asparagus. If you cannot get very thin leeks, cut them in half lengthwise before serving.

FAVA BEANS À LA BASQUAISE

serves 4–5

In Spain the one sure sign of spring is the appearance of fresh fava beans in the local markets. The season is relatively short, and because of this the enormous pale-green fuzzy-looking pods seem to call for our immediate attention. The pods contain delicious, tender beans that are an important ingredient in some of the great classic vegetable dishes of Basque and Catalan cooking. But while the Catalan version of this recipe calls for salt pork and blood sausage, the Basque version is often made with just tomatoes, onions, and red peppers. Over the years, I have developed my own version of this great dish which leaves plenty of room for creative interpretations. Remember that the key to cooking fava beans properly lies in long, slow braising. This is not a vegetable that benefits from being cooked al dente; instead, it needs to develop flavor through gentle simmering in the company of onions and tomatoes.

¼ pound meaty slab bacon
1 tablespoon unsalted butter
2 tablespoons extra-virgin olive oil
1 large Spanish onion, peeled and diced
2 large garlic cloves, peeled and finely minced
3 large ripe tomatoes, peeled, seeded, and chopped or 1 (2-pound) can Italian plum tomatoes, thoroughly drained and coarsely chopped (see Note)

½ pound boneless smoked pork shoulder butt, cut into 1-inch cubes
3 pounds fresh fava beans, shelled
Salt and freshly ground black pepper
2–3 cups beef bouillon
1 large sprig of fresh thyme plus 1 bay leaf
6 ounces chorizo sausage, cut crosswise into ¼-inch slices, optional

1. In a medium saucepan bring the water to a simmer, add the bacon, and cook for 3 minutes. Drain and dry thoroughly with paper towels.

2. In a heavy casserole, melt the butter over medium heat, add the bacon, and sauté until almost crisp. Remove the bacon with a slotted spoon to a side dish and reserve.

3. Remove all but 2 tablespoons of fat from the casserole. Add the olive oil and, when hot, add the onion and garlic, lower the heat, and cook the mixture for about 10 to 15 minutes or until soft and nicely browned. Be careful not to burn the onions, which must caramelize without burning.

4. Add the tomatoes, pork butt, and bacon, bring to a simmer, and cook the mixture for 10 minutes longer.

5. Add the fava beans, season with salt and pepper, and add enough bouillon to barely cover the beans or about 2 cups. Add the thyme and bay leaf, cover the casserole tightly, and simmer over low heat for 1 hour to 1 hour and 15 minutes, adding some broth every 15 minutes, as necessary. Add the optional chorizo for the last 15 minutes of braising, or when the fava beans are tender.

6. When the fava beans are done, taste and correct the seasoning. Serve the beans hot as an appetizer or as a vegetable with a roast leg of lamb, sautéed lamb chops, or roast chicken.

Note: The pork butt and bacon give this dish its gutsy character and flavor, but you can discard them once the fava beans are cooked. When using canned tomatoes, be sure to choose a variety in which the tomatoes are whole, preferably those that are from San Marzano (see page 427). Once drained, use 6 of the tomatoes and keep any leftovers for a soup or stock.

ZITI WITH ASPARAGUS, PEAS, AND LEMON CREAM

serves 4

It was a lovely sunny day, and my mother and I were having lunch in a tiny trattoria in a deserted square in Gubbio, one of the most picturesque little hill towns in Umbria. A large Italian family at the next table was having their pasta course, and it looked extremely appealing. The dish did not seem to be on the menu, and when I asked the owner whether we could order it, he hesitated for a moment and then asked if we would be willing to wait a bit before we ate. Oh yes, we said, seeing the intense satisfaction our neighbors radiated as they ate their pasta. And so he dispatched his son, who returned forty minutes later with a basket of their early home-grown asparagus. Very soon a large bowl of homemade fettuccine arrived at the table. It was heaped with the asparagus, moistened with cream, and had a bonus of tiny freshly cooked fava beans. The unexpected perfume of grated lemon zest made the dish even more memorable.

I serve this dish at home now, made with various kinds of fresh and dried pasta, either as an appetizer or a light main course, and I often use fresh peas and some arugula, which are equally successful.

½ cup shelled fresh peas, about ½ pound unshelled

½ pound asparagus, trimmed, peeled, and cut on the diagonal into 1½-inch pieces

8 tablespoons unsalted butter

1¼ cups heavy cream

Juice of 1 lemon

1 teaspoon finely grated lemon rind

Salt and freshly ground white pepper

½ pound ziti

3 hearts of Bibb lettuce, separated into leaves

½ cup freshly grated Parmesan cheese

1. Place the peas in a vegetable steamer, set over simmering water, and steam, covered, for 3 to 5 minutes or until just tender. Remove and run under cold water to stop further cooking. Drain and set aside.

2. Add the asparagus to the vegetable steamer. Cover and steam for 3 to 5 minutes or until just tender. Run under cold water, drain, and reserve.

3. In a large, heavy skillet melt the butter over medium heat and whisk in the cream. Bring to a boil, reduce the heat, and simmer until reduced by one third. Add the lemon juice and rind. Season with salt and white pepper and keep warm.

4. Bring plenty of salted water to a boil in a large casserole. Add the ziti and cook until just tender, al dente. Add 2 cups of cold water to stop further cooking. Drain thoroughly, and add to the lemon cream together with the peas, asparagus, and Bibb lettuce and simmer until the sauce lightly coats the pasta and the lettuce has just wilted.

5. Add the Parmesan and toss gently. Taste and correct the seasoning and serve at once.

CREAMY ASPARAGUS SOUP PRINTANIER

serves 5–6

1 pound fresh asparagus
2 small leeks, trimmed of all but 2 inches
  of greens
5 tablespoons unsalted butter
6 cups Chicken Stock (page 470)
1 pound fresh sorrel leaves, stemmed
  and washed (page 68)
2½ tablespoons all-purpose flour

Salt and freshly ground white pepper
1 cup heavy cream
1 cup cooked fresh peas
1 heart of Boston lettuce, leaves
  separated and washed

GARNISH
3 tablespoons finely minced fresh chives

1. Peel the asparagus stalks and cut into 1-inch pieces, reserving the tips separately.

2. Cut the leeks crosswise into thin slices. Rinse under warm water to remove all sand, drain well, and reserve.

3. In a heavy 3½-quart casserole melt 2 tablespoons of the butter over low heat. Add the leeks and a little stock, cover, and simmer until just tender. Add the remaining stock and the asparagus stalks. Bring to a boil and cook until the asparagus is tender. Add two thirds of the sorrel leaves and cook for 5 minutes or until tender. Cool the soup slightly, transfer to the workbowl of a food processor, and process until smooth. Set aside.

4. Heat the remaining butter in the casserole over low heat, add the flour, and whisk until well blended. Add the pureed soup and bring to a boil. Reduce the heat, add the asparagus tips, and cook until just tender. Add the cream and peas, and season to taste with salt and pepper.

5. Just before serving, add the remaining sorrel and lettuce leaves and heat the soup without letting it come to a boil. Serve the soup hot, garnished with a sprinkling of fresh chives.

Note: This soup is best when prepared a day in advance and reheated.

CREAM OF BROCCOLI AND TANGY
LEMON SOUP

serves 6

Broccoli is one of my favorite vegetables. Here lemon juice provides a zesty flavor that makes the soup an excellent starter to a light spring meal. The soup freezes beautifully and can be flavored with a sprinkling of your first summer herbs, such as tarragon, dill, or chives.

1 large bunch broccoli, about 1½ pounds
6 cups Chicken Stock (page 470) (see Note) or bouillon
4 tablespoons unsalted butter
1 cup finely minced scallions

2 tablespoons all-purpose flour
½ cup heavy cream
Juice of ½ lemon, or more to taste
2 large egg yolks

GARNISH
Finely minced fresh parsley

1. Trim the broccoli stalks, removing all the leaves. Peel and thinly slice the stalks crosswise and separate the tops into small, even florets.

2. Combine the chicken stock and the broccoli stalks and florets in a large saucepan. Bring to a boil, reduce the heat, and simmer until very soft.

3. Cool and strain the soup. Transfer the broccoli to the workbowl of a food processor or a blender and process until smooth. Whisk the broccoli puree into the strained broth. Reserve.

4. In a large, heavy saucepan melt the butter over medium heat. Add the scallions and cook until soft but not browned. Add the flour and blend well. Immediately add the broccoli mixture, bring to a boil, reduce the heat, and simmer for 10 minutes. Season with salt and pepper.

5. Combine the cream, lemon juice, and egg yolks in a small bowl and whisk until smooth. Add the cream mixture to the hot soup and just heat through. Do not let the soup return to a boil or the yolks may curdle. Taste and correct the seasoning. Cool the soup to room temperature, then chill 2 to 4 hours before serving. Garnish with parsley.

Note: If the soup thickens too much after chilling, add an additional 1 cup of chicken stock. When I serve the soup hot, I add tiny broccoli florets steamed separately, for additional texture.

VIENNESE KOHLRABI SOUP WITH
GARDEN VEGETABLES

serves 4–5

A relative newcomer to the supermarket produce section, kohlrabi is a delicious vegetable that can be used in many interesting ways. Be sure to buy small bulbs with fresh, crisp-looking tops and use it within two to three days of purchase.

2 pounds kohlrabi, 5–6 medium bulbs
2 large leeks, trimmed of all but 2 inches of greens
4 tablespoons unsalted butter
2 medium carrots, peeled and thinly sliced

6 cups Chicken Stock (page 470)
1 cup fresh shelled peas
2 cups fresh spinach leaves, stemmed and well washed
Salt and freshly ground white pepper

1. Peel the kohlrabi with a sharp knife and cut into ½-inch cubes. Reserve.

2. Cut the leeks lengthwise and then crosswise into thin slices. Rinse thoroughly under warm water, drain, and set aside.

3. In a large casserole melt the butter over low heat, add the leeks and carrots, and sauté for 2 to 3 minutes without browning. Add the kohlrabi and stock, bring to a boil, reduce the heat, and simmer, partially covered, for 30 minutes. Add the peas and continue to cook until they are just tender.

4. Add the spinach and cook until just wilted. Taste and correct the seasoning. Serve the soup hot or slightly chilled.

Kohlrabi

I keep hoping that kohlrabi will be the next "hot" vegetable. Unfortunately, it does not look as if this will happen in the near future, although I think kohlrabi is the perfect solution to tired vegetables. For those home cooks and gardeners who love and want variety, kohlrabi is a great choice in spring. This somewhat unique vegetable is a cross between a turnip and a cabbage with the texture of a sweet radish. When you do see them truly fresh with crisp green tops and tender skins, and not too large, don't hesitate and buy them. Kohlrabi can be substituted in any preparation calling for turnips. (Like turnips, they must be peeled.) They love the company of carrots, peas, and potatoes. Oven-glaze some with a touch of sugar and butter or simply boil a potful and toss in parsley or dill with a touch of sour cream.

■ ■ ■

Sorrel

Sorrel became the star green in France in the seventies after long years when it had been relegated to the back of the vegetable patch, seemingly forgotten. But the restaurant chefs' love affair with sorrel has since tapered off and in Europe this delicious green is once again facing obscurity—but not in the United States.

Sorrel has only recently become available here commercially, and its popularity is beginning to grow. In its season, which starts in late spring, it is well worth seeking out. Most sorrel is grown in hothouses and sold in small bunches. It is far less expensive to buy it by the pound, and if you tell the produce manager that you will take a whole box, he will happily oblige you. Stem and wash the sorrel leaves thoroughly in warm water, place them in a heavy saucepan, and melt down with only the water that clings to the leaves. Drain the sorrel well and freeze in 1-pint containers or zip-lock bags. You can now enjoy sorrel soups, sauces, and purees on the spur of the moment all through the spring and well into the summer.

If you are a gardener, plant at least 2 sorrel plants. You will not regret it, because this good-natured, easy-to-grow perennial will come up every year, and you'll find that garden-grown sorrel is far more flavorful and pungent than the store-bought variety.

■ ■ ■

SORREL, TOMATO, AND NEW POTATO SOUP

serves 4–6

Traditionally, sorrel has been used in two well-known soups: the Russian schav, often called a white borscht, and the French potage à l'Oseille, which is a creamy, rich velouté. But there is no reason to limit the use of sorrel to these two. Here is a homey peasant soup that also includes the first flavorful ripe tomatoes. However, if these still look and taste like waxballs, substitute another vegetable such as asparagus or Swiss chard. Canned tomatoes are not a good choice here since their acidity, coupled with the natural tartness of sorrel, will result in an overly tangy, rather sour soup.

2 tablespoons unsalted butter
1 tablespoon olive oil
2 medium onions, peeled and finely minced
4 large ripe tomatoes, peeled, seeded, and chopped
4 cups Chicken Stock (page 470) or bouillon
Salt and freshly ground black pepper
1 medium leek, trimmed of all but 2 inches of greens, finely diced and well rinsed
2 medium red potatoes, peeled and diced
3 cups sorrel leaves, stemmed, rinsed and cut into fine julienne

GARNISH
Bowl of Crème Fraîche (page 491) or sour cream

1. In a heavy 3-quart casserole, melt the butter together with the oil over medium heat. Add the onions and cook until soft but not browned. Add the tomatoes and stock, and season with salt and pepper. Bring to a boil, reduce the heat, and simmer, partially covered, for 25 minutes.

2. Cool the soup slightly, transfer to the workbowl of a food processor or blender, and process until smooth.

3. Return the soup to the casserole, add the leek and potatoes, and simmer for 20 minutes longer, or until the potatoes are tender but not falling apart.

4. Add the sorrel leaves and simmer for 2 minutes, or just until wilted. Taste the soup and correct the seasoning. Serve hot or at room temperature accompanied by crusty bread and a bowl of crème fraîche or sour cream.

FILLETS OF SALMON ON A BED OF CABBAGE WITH CRÈME FRAÎCHE–DILL SAUCE

serves 4

I can think of no other fish that is cooked more creatively these days than salmon. Probably because of its wide appeal and easy availability, salmon tops the list of most restaurant chefs who seem to prepare this versatile fish in exciting and experimental ways.

The combination of salmon and cabbage is just one example of this trend. Although in Alsace and in Scandinavia, various types of fish are often teamed with cabbage and dill, it is still a rather new marriage of ingredients, but one that I feel has real staying power. Serve the fillets with simply cooked buttered and parsleyed new potatoes or possibly the Vegetable Pilaf on page 76.

4–5 cups shredded Savoy or green cabbage
1¾ cups Fish Stock (page 472)
2 tablespoons finely minced shallots
¾ cup Crème Fraîche (page 491)
9 tablespoons unsalted butter
3 teaspoons fresh lemon juice

3–4 tablespoons finely minced fresh dill
Salt and freshly ground white pepper
4 salmon fillets, skinless, about 6–7 ounces each

GARNISH
Sprigs of fresh dill

1. Arrange the cabbage in a vegetable steamer. Place over simmering water, cover, and steam for 3 to 4 minutes, or until just tender. Do not overcook. Remove, press lightly to remove excess moisture, and set aside on paper towels.

2. In a medium saucepan combine ¾ cup of the fish stock and shallots. Bring to a boil, reduce the heat, and simmer until reduced to ¼ cup.

3. Add the crème fraîche, return the sauce to a boil, and reduce by one third. Remove from the heat and whisk in 6 tablespoons of the butter, 1 tablespoon at a time, until all of it has been absorbed and the sauce is smooth. Add the lemon juice, dill, and salt and pepper to taste. Return the sauce to the lowest heat possible, cover, and keep warm.

4. Add the remaining fish stock to a large flameproof baking dish and place over high heat. Bring to a boil and add the salmon fillets. Reduce the heat, cover the dish with buttered foil, and cook the fillets 3 to 4 minutes on each side, turning them carefully with a spatula.

5. While the salmon is poaching, finish preparing the cabbage: In a medium skillet melt the remaining butter over medium-low heat. Add the cabbage, season with salt and pepper, and just heat through. Transfer to a large serving platter, cover, and keep warm.

6. Remove the salmon from the poaching liquid with a slotted spatula and place on top of the bed of cabbage. Spoon the dill sauce over the salmon and garnish with sprigs of fresh dill. Serve at once.

FRICASSEE OF SALMON WITH BABY VEGETABLES LAMELOISE

serves 6

Visually, I find this to be one of the most pleasing dishes of the spring season. At Lameloise in Burgundy, the cubed salmon is sautéed to crisp perfection, then served in shallow soup bowls as you would a fricassee, with a colorful array of baby vegetables—from baby turnips and radishes to miniature spinach leaves and the tiniest of carrots, all bound by the superb Lameloise sauce. The entire effect is fresh and utterly seasonal.

In this country, baby vegetables can often be disappointing. Because of lack of turnover they lose their fresh taste, and until recently they have been quite expensive. Now, however, a good supply of flavorful baby zucchini and other squashes has begun coming in from Central America, and they are fresh and well priced. If you do not find the baby vegetables called for in this dish, simply use standard-size vegetables and cut them into ¾-inch cubes.

12 baby zucchini, cut in half lengthwise
12 baby yellow squash, cut in half lengthwise
24 baby carrots, trimmed and peeled
1½ pounds fresh spinach, stemmed and well washed
1½ pounds fresh salmon fillets, preferably center cut and skin removed

Coarse salt and freshly ground black pepper
2 cups Lameloise Sauce (page 479)
6 tablespoons unsalted butter
2 teaspoons olive oil

GARNISH
2 small tomatoes, seeded, diced, and tossed with a touch of olive oil and finely minced fresh chives

1. Bring water to a boil in a vegetable steamer. Add the zucchini and squash, cover, and steam until just tender. Transfer to a side dish and reserve. Steam the carrots in the same manner and reserve. Add the spinach to the steamer, cover, and steam until just wilted.

2. Cut the salmon into ¾-inch cubes and remove any bones. Season with coarse salt and pepper and reserve.

3. Place the Lameloise sauce in a small saucepan and just heat through. Taste and correct the seasoning, cover, and keep warm.

4. In each of 2 nonstick skillets, melt 2 tablespoons of butter and 1 teaspoon oil over high heat. When the butter is very hot, add the salmon cubes. Do not crowd the pans. Sauté the salmon until very crisp and nicely browned but still pink in the center; do not overcook. Set the salmon aside on a double layer of paper towels.

5. Wipe out the skillets with paper towels and melt the remaining butter over low heat. Add the steamed zucchini, squash, and carrots and sauté until just heated through. Season with salt and pepper and reserve.

6. To serve, place a dollop of steamed spinach in each of 4 to 6 individual shallow soup plates. Top with the sautéed salmon and mixed vegetables. Spoon some of the warm sauce over the fish and vegetables, and garnish with diced marinated tomatoes.

SHAD ROE IN PARSLEY BUTTER WITH BRAISED SPRING ONIONS

serves 4

Sautéed shad roe is one of my favorite spring treats. Like soft-shell crabs, the roe is in season for only a brief time, so I try to take advantage of this delicacy by preparing it in many interesting but quite simple ways that enhance the exquisite flavor of the roe. Shad roe should be treated with care. It must never be overcooked and is best served with a spring garnish, such as braised scallions, steamed asparagus, or braised Bibb lettuce.

8 medium scallions, trimmed of all but 3 inches of greens
8 tablespoons unsalted butter
Salt
1 teaspoon granulated sugar
4 pair shad roe

All-purpose flour for dredging
Freshly ground black pepper
3 tablespoons finely minced fresh Italian parsley
Juice of 1 large lemon

1. Place the scallions in a deep 10-inch skillet and add just enough water to cover. Add 2 tablespoons of the butter and season with a pinch of salt. Bring to a boil, reduce the heat, and simmer the scallions for 3 to 4 minutes, or until just tender.

2. Discard all but 2 tablespoons of the water. Add the sugar and continue to simmer until the scallions are nicely glazed and all the water has evaporated. Set aside.

3. Dredge the shad roe lightly in flour, shaking off the excess.

4. In a large, heavy nonstick skillet, melt 3 tablespoons of the butter over medium heat. Add the roe, reduce the heat, and sauté, covered, for 2 to 3 minutes. Turn the roe, season with salt and pepper, and continue to sauté, partially covered, over low heat for another 2 to 3 minutes, or until springy to the touch. Transfer the shad roe carefully to warm individual plates.

5. Add the remaining butter, parsley, and lemon juice to the pan juices and whisk until well blended. Spoon the herb butter over the shad roe and garnish with the braised scallions. Serve immediately.

Note: If the butter has burned after sautéeing the roe, discard it and melt an additional 6 tablespoons of butter in a small saucepan over moderate heat. When the butter turns a hazelnut brown, add the lemon juice and parsley, whisking constantly. Spoon the parsley butter over the shad roe and serve at once.

OVEN-BRAISED SHAD WITH NEW POTATOES

serves 4–5

Shad fillets are becoming a delicacy since the complex art of boning this spring fish is fast disappearing. The short-seasoned shad is widely available in the Northeast and is now shipped daily to top restaurants across the country as well as to quality seafood markets.

Since shad is riddled with tiny bones, make sure you are buying fillets that have been properly boned. The fish should look glossy, moist, and plump. Do not postpone planning some of your meals around shad, since promptly on Mother's Day it disappears from the market until the following spring.

For the simplest of preparations, braise the fillets and serve them with Butter-Braised Spinach with Glazed Garlic Cloves (page 104) or Sauté of Cucumbers with Peas and Mint (page 99). Shad is also delicious pan-sautéed in a little clarified butter and served in a pool of Asparagus Vinaigrette (page 40).

2 pounds boneless shad, skin removed and cut into 6-ounce fillets
1 lemon, thinly sliced
4 tablespoons extra-virgin olive oil

POTATOES
¾ pound new potatoes, unpeeled and cut into ¼-inch slices
1 large red onion, peeled, quartered, and thinly sliced
2 tablespoons finely minced fresh thyme
1 teaspoon fresh oregano

3 tablespoons extra-virgin olive oil
2 large cloves garlic, peeled and thinly sliced
Coarse salt and freshly ground black pepper
½ cup Fish Stock (page 472) or Chicken Stock (page 470)
6 tablespoons unsalted butter
2 tablespoons finely minced shallots
½ cup dry white wine
Salt and freshly ground black pepper
4 tablespoons finely julienned fresh basil

1. Start by marinating the shad: Place the shad in a rectangular glass or porcelain dish, top with the sliced lemon, and drizzle with the olive oil. Cover and refrigerate for 2 to 4 hours.

2. Preheat the oven to 375° F.

3. While the shad is marinating, prepare the potatoes: In a mixing bowl, toss the sliced potatoes with the onion, thyme, and oregano. In a flameproof rectangular baking dish large enough to hold the potatoes in one layer, heat 2 tablespoons of the olive oil over low heat, add the sliced garlic, and sauté for 1 or 2 minutes, or until the garlic is lightly colored. Immediately remove from the heat and place the potato mixture evenly in the pan. Drizzle with the remaining olive oil and season with coarse salt and a large grinding of black pepper. Add the stock and cover loosely with buttered foil.

4. Bake the potato mixture for 35 to 40 minutes, or until the potatoes are tender and lightly browned. Uncover during the last 10 minutes of baking and add a little more stock if the mixture seems dry. Remove from the oven and keep warm.

5. Reduce the oven temperature to 350° F.

6. Remove the shad from the marinade and reserve the fish and marinade separately. In another flameproof rectangular baking dish, melt 3 tablespoons of the butter over low heat, add the shallots and white wine, and place the shad fillets over the shallot mixture. Season with salt and pepper, sprinkle some of the reserved marinade over the shad, and cover with buttered parchment paper.

7. Bake the shad for 4 to 5 minutes, or until the fillets just turn opaque; do not overcook. Remove the dish from the oven and, with a spatula, carefully remove the shad to a side dish. Reserve.

8. Place the baking pan directly over medium-high heat and reduce the juices until they become syrupy. Remove the pan from the heat and whisk in the remaining 3 tablespoons of butter, 1 tablespoon at a time, whisking constantly until the mixture is emulsified. Whisk in the basil, taste the sauce, and correct the seasoning.

9. With a spatula place an individual serving of potatoes on each dinner plate. Place 2 shad fillets on each serving of potatoes and spoon some of the sauce over each portion. Serve immediately.

Note: The potatoes can be made a day ahead and reheated either in a low oven or in the microwave. Truly fresh shad does not require long marinating, and if you are pressed for time, omit the first step altogether. You can add a dash of lemon juice to the sauce to achieve the flavor of the marinade.

Rather than cutting the basil into julienne, I often puree it with a touch of cream and whisk it into the sauce. You can garnish the dish with Italian parsley or tiny cilantro leaves instead.

FILLETS OF SEA BASS WITH A VEGETABLE PILAF AND CHIVE-BUTTER SAUCE

serves 4

I still remember a time when fish dishes, especially those served with a cream sauce, were accompanied only by parsleyed new potatoes. Now all this has changed, and there is a new freedom when it comes to the preparation and accompaniments of fish. I truly welcome this change because seafood should be approached with the same creativity many cooks reserve for beef, poultry, and veal. Here the full-flavored fillets of sea bass, one of the best fish we have in our waters, are served with a vegetable pilaf that can be made as much as a day ahead of time and reheated.

You may add two to three tablespoons of Tomato Fondue (page 474) to the butter sauce and substitute tarragon for the chives. Either way, it is a lovely spring main course.

PILAF
1 medium zucchini, trimmed
3 tablespoons unsalted butter
3 medium scallions, trimmed of all but 2 inches of greens and finely sliced
1 medium carrot, peeled and cut into ¼-inch dice
3 tablespoons finely diced red bell pepper
3 tablespoons water
1 cup Arborio rice
Salt and freshly ground white pepper
2 cups Fish Stock (page 472) or Chicken Stock (page 470)
3 tablespoons finely minced fresh parsley

FILLETS
9 tablespoons unsalted butter
Salt and freshly ground white pepper
2 tablespoons finely minced shallots
4 fillets of sea bass, about 6 ounces each
1½ cups Fish Stock (page 472)
3 tablespoons Crème Fraîche (page 491) or heavy cream
3–4 tablespoons finely minced fresh chives
2 teaspoons fresh lemon juice

GARNISH
Sprigs of fresh chives
Steamed sugarsnap peas or snow peas

1. Start by making the pilaf: Using a sharp knife, remove the skin of the zucchini with ¼ inch of the pulp. Cut the strips into ¼-inch dice and reserve.

2. In a heavy saucepan, melt the butter over low heat. Add the scallions, zucchini, carrot, and red pepper. Add the water, cover, and simmer the vegetables for 3 to 4 minutes, or until barely tender.

3. Add the rice, toss gently with the vegetable mixture, and season with salt

and pepper. Add the stock and bring to a boil. Reduce the heat and simmer, covered, for 25 to 30 minutes, or until the rice is just tender.

4. Fold in the parsley and pack the mixture into 4 lightly oiled custard cups or timbale molds, about 3 ounces each. Set the cups in a pan of hot water, cover, and set aside.

5. Preheat the oven to 425° F.

6. To prepare the fillets: Butter a heavy flameproof baking dish large enough to hold the bass fillets in one layer, using 3 tablespoons of the butter. Sprinkle the bottom of the dish with salt, pepper, and the shallots.

7. Season the fillets with salt and pepper and place on top of the shallots in the baking dish. Spoon 2 or 3 tablespoons of the fish stock around the fillets, cover with buttered foil or parchment paper, and place in the center of the preheated oven. Braise the fillets for 3 minutes. Carefully turn them with a spatula and braise for another 3 minutes or until opaque throughout; do not overcook.

8. Remove the baking dish from the oven and carefully transfer the fillets to a plate. Remove and discard the skin and cover the fillets while you finish the sauce.

9. Add the remaining stock to the baking dish, bring to a boil on top of the stove, and reduce by half.

10. Strain the sauce into a medium saucepan, place over very low heat, and whisk in the remaining butter, 2 tablespoons at a time. The sauce should be smooth and velvety. Add the crème fraîche, chives, and lemon juice and correct the seasoning.

11. To serve, unmold the rice timbales onto each of 4 warm individual serving plates. Add any accumulated juices from the bass fillets to the sauce and place 1 fillet on each plate. Spoon a little sauce over each fillet and garnish the plate with sprigs of chives and steamed sugarsnap peas or snow peas. Serve at once.

Note: The rice can be cooked hours ahead of time and the molds kept warm in a warm, but not hot, water bath. Steam the peas ahead of time and then reheat in a little butter while the pan juices are reducing.

Other fish, in particular mahi mahi, red snapper, and monkfish, work well in this recipe.

You will have quite a bit of leftover pilaf. It can easily be reheated the next day in a low oven or the microwave, or you can turn it into a quick rice salad by tossing it with Provençal Vinaigrette (page 480).

BRAISED RED SNAPPER FILLETS IN A FRESH TOMATO AND DILL SAUCE WITH SAUTÉED CUCUMBERS

serves 6

One of the great pleasures of being in Florida in the spring is that truly vine-ripened tomatoes and delicious red snapper are in season together at a time when the rest of the country has almost forgotten how ripe tomatoes taste. I find the marriage of seafood with tomatoes unbeatable, and the glazed and still slightly crisp cucumbers add a nice finishing touch. Several herbs—especially dill, chives, and chervil, or a mixture of all three—can be used as well, but I prefer the unadulterated flavors of the sweet fresh fish, ripe tomatoes, and buttery cucumbers.

Other fish, such as sea bass, white snapper, or mahi mahi, can be prepared successfully in the same way.

SAUCE
3 tablespoons unsalted butter
2 tablespoons finely minced shallots
3 large ripe tomatoes, peeled, seeded, and chopped
Salt and freshly ground black pepper

SNAPPER
2 whole red snapper, about 2 pounds each, filleted and cut into 6-ounce serving pieces (reserve bones for making stock)
4 tablespoons unsalted butter
3 tablespoons finely minced shallots

½ cup Fish Stock (page 472)

CUCUMBERS
3 tablespoons unsalted butter
2 seedless cucumbers, unpeeled, cut in half lengthwise, and cut crosswise into ¼-inch slices
Pinch of granulated sugar
⅓ cup water
2–3 tablespoons finely minced fresh dill
Salt and freshly ground black pepper

1 Beurre Manie (page 478)
4 tablespoons unsalted butter, for enrichment, optional

1. Start by making the sauce: Melt the butter in a medium skillet, add the shallots, and sauté over medium heat until soft but not brown, about 1 minute. Add the tomatoes, season with salt and pepper, and cook until all the tomato liquid has evaporated and the mixture is thick.

2. Cool the tomato mixture slightly and transfer to the workbowl of a food processor or blender. Puree until smooth and reserve.

3. Preheat the oven to 400° F.

4. To prepare the snapper: Season both sides of the snapper fillets lightly with salt and heavily with pepper. In a large, heavy baking dish, melt the butter over low heat and add the shallots. Add the snapper fillets and turn to coat well with the butter mixture. Add a little stock and cover with buttered foil. Place in the center of the preheated oven and bake for 10 minutes, or until the fish is opaque and just barely flakes; do not overcook.

5. While the fish is baking prepare the cucumbers: Melt the butter in a large iron skillet over medium heat. Add the cucumber slices, season with salt and pepper, and sprinkle with the sugar. Add the water, cover, and simmer until the cucumbers are just tender, 5 to 7 minutes. Uncover the pan, raise the heat, and cook until all the water has evaporated and the cucumbers are glazed. Add the minced dill and reserve.

6. When the snapper is done, remove from the oven. Transfer the fillets to a warm serving platter and set aside.

7. Strain the pan juices and transfer to a skillet, add the pureed tomato mixture, and place over medium-low heat. Bring to a simmer and, if the sauce seems too thin, whisk in bits of beurre manie and cook until the sauce lightly coats a spoon. Whisk in the optional butter for enrichment, add the cucumbers, and just heat through.

8. Spoon the tomato-dill sauce over the snapper fillets, surround with the cucumbers, and garnish with sprigs of fresh dill.

GRILLED BROCHETTES OF MONKFISH IN TANGERINE AND SPICE MARINADE

serves 4

MARINADE
2 teaspoons finely minced fresh ginger
1 teaspoon finely minced fresh garlic
Grated rind of 1 tangerine
Juice of 1 tangerine
Juice of 1 lime
2 tablespoons light soy sauce
2 tablespoons olive oil
1 tablespoon Oriental sesame oil
2 tablespoons finely minced fresh
  cilantro
1 jalapeño pepper, with seeds, finely
  minced

1¼ pounds trimmed monkfish, cut into
  1½-inch cubes, to make 16 cubes

BROCHETTES
1 large onion, peeled and cut into
  1-inch squares
½ medium red bell pepper, cored,
  seeded, and cut into 1-inch squares
½ medium yellow bell pepper, cored,
  seeded, and cut into 1-inch squares
½ medium green bell pepper, cored,
  seeded, and cut into 1-inch squares
Coarse salt and coarsely ground black
  pepper
Vegetable oil for brushing on grill

GARNISH
1 ripe avocado, peeled, halved, seeded,
  and cut into ¼-inch slices
Juice of 1 lime
2 tablespoons olive oil
Salt and freshly ground black pepper
Radish sprouts
Sprigs of fresh cilantro, optional

1. Combine all the ingredients for the marinade in a shallow bowl. Add the monkfish and marinate for 2 to 3 hours in the refrigerator, turning several times in the marinade.

2. Prepare the charcoal grill.

3. To assemble the brochettes: Remove the monkfish from the marinade and set the marinade aside. Using 4 (10-inch) wooden skewers, thread the monkfish alternately with the onions and peppers: 4 pieces of fish, 3 pieces of onion, and 1 piece each of red, yellow, and green pepper per skewer. Sprinkle with coarse salt and pepper.

4. Brush the grill with vegetable oil. Place the brochettes over the hot coals and grill until lightly browned on each side, about 4 minutes total, or until just done, brushing with some of the reserved marinade. Be careful not to overcook.

5. Remove the brochettes from the grill and slide the fish and vegetables carefully off the skewers onto one side of each individual serving plate. Arrange the avocado slices on the other side of the plates. Sprinkle the avocado with a little lime juice, olive oil, and salt and pepper. Garnish with a few radish sprouts and an optional sprig of fresh cilantro. Serve at once.

Note: The brochettes may be broiled instead of grilled. Place over a pan on the broiler rack 6 inches from the source of heat. Broil 2 minutes on each side, for a total of 4 minutes, or until just done. Do not overcook.

MEDALLIONS OF MARYLAND CRAB AND SHRIMP MOUSSE IN CURRY-LEMON SAUCE

serves 6

Not so long ago, crab cakes were a regional culinary highlight, sought out by every visitor to Maryland. The eternal question for all those entering Baltimore was (and still is): "Where can I get a good crab cake?"

Today crab cakes are an all-American favorite, and their popularity has inspired many restaurants to come up with variations on the basic theme. I've had some good ones and others that were barely edible, but when they are really good, crab cakes are among the best dishes in American cookery. These medallions are a takeoff on the classic crab cakes, an amusing version in which lump crabmeat is bound by a light and flavorful shrimp mousse and then baked in flan rings. Since the recipe makes a large amount, you can prepare the flans in advance and reheat them the next day in a low oven without loss of taste and texture.

Vary your sauce according to your mood and the season. A simple brown hazelnut butter is an ideal accompaniment, but so is a bowl of crème fraîche or sour cream flavored with minced jalapeño pepper, a touch of finely grated lemon zest, and a few tiny cilantro leaves. Serve with greens—spinach is a good choice—that have been sautéed in olive oil.

MEDALLIONS
2 tablespoons unsalted butter
1 bunch scallions, trimmed of all but 2 inches of greens and finely chopped
Salt and freshly ground white pepper
½ pound medium shrimp, peeled
2 large eggs
1⅓ cups cold heavy cream

1 pound fresh crabmeat, flaked and all shells removed
1 tablespoon Dijon mustard
½ teaspoon Tabasco sauce
2 tablespoons finely minced fresh parsley

(cont.)

2 teaspoons unsalted butter
⅔ teaspoon good-quality imported
   curry powder
½ cup water
¼ cup heavy cream
4 large egg yolks

8 ounces unsalted butter, melted and
   kept warm
Salt and freshly ground white pepper
Pinch of cayenne
Lemon juice to taste

1.   Preheat the oven to 375° F.

2.   Start by making the crab medallions: In a medium skillet melt the butter over low heat. Add the scallions and cook for 1 minute or until just wilted. Season with salt and pepper and set aside to cool.

3.   In the workbowl of a food processor, combine the shrimp and eggs and process until smooth. With the machine running, pour the cream slowly through the feed tube. When the cream has been absorbed completely, transfer the shrimp mixture to a large mixing bowl, add the wilted scallions, crabmeat, mustard, Tabasco, and parsley. Season with salt and pepper and fold gently, until thoroughly combined.

4.   Lightly oil 10 to 12 (3-inch) flan rings. Place the rings on a nonstick baking sheet and fill each ring to the top with some of the crab mixture. Smooth the mixture evenly with a spatula and place the baking sheet in the center of the preheated oven. Bake for 5 to 7 minutes, or until the flans begin to set. Turn each flan and its flan ring over with the aid of a spatula and bake for an additional 2 minutes or until the flans are completely set. Do not overcook. When the flans are done, remove them from the oven and reserve.

5.   To prepare the butter sauce: In a heavy 2½-quart saucepan, preferably copper, melt the 2 teaspoons of butter over low heat. Add the curry powder and cook until fragrant, about 1 minute, whisking constantly. Immediately remove from the heat, add the water and cream, then whisk in the egg yolks. Return the saucepan to the heat and whisk just until the egg yolks have thickened. Do not cook too long, otherwise the egg yolks will curdle. Remove the pan from the heat and whisk in the melted butter. Season with salt, pepper, cayenne, and lemon juice to taste.

6.   To serve, remove the flan rings from the medallions and place 1 or 2 medallions on each individual serving plate. Spoon some of the sauce over each portion and surround with warm sautéed greens.

SAUTÉED SOFT-SHELL CRABS IN SHALLOT AND MUSTARD SAUCE ON A BED OF ROASTED VEGETABLES

serves 4

There are few foods that can better celebrate spring than soft-shell crabs. Their short season deserves your full attention, and while they are always delicious when simply sautéed and served in a lemon-spiked brown butter, there is also room for variations that bring out the wonderfully distinctive flavor of the crabs.

8 soft-shell crabs, cleaned
Half-and-half for marinating

VEGETABLES
4–6 tablespoons extra-virgin olive oil
2 large cloves garlic, peeled and sliced
2 medium red bell peppers, cored, seeded, and cut into eighths
2 medium yellow bell peppers, cored, seeded, and cut into eighths
3 medium zucchini, trimmed and cut into ½-inch slices
2½ tablespoons finely minced fresh thyme

SAUCE
1 medium shallot, peeled and finely minced

½ cup dry white wine
½ cup Fish Stock (page 472) or bouillon
2 tablespoons Dijon mustard
1¼ cups heavy cream
Freshly ground white pepper
2 tablespoons finely minced fresh chives

Coarse salt and freshly ground black pepper
Instant flour for dredging
3 tablespoons Clarified Butter (page 478)
2 teaspoons olive oil

GARNISH
Tiny fresh parsley leaves

1. Start by marinating the crabs: Place the cleaned crabs in a shallow baking dish and cover with half-and-half. Let the crabs marinate for at least 1 hour.

2. Preheat the oven to 375° F.

3. While the crabs are marinating, roast the vegetables: In a heavy rectangular roasting pan or 12-inch cast-iron skillet, heat the olive oil over low heat. Add the garlic, bell peppers, zucchini, and thyme and toss lightly until well coated with the oil. Place the pan or skillet in the center of the preheated oven and roast the vegetables for 45 minutes to 1 hour, tossing them gently with a wooden spoon 2 to 3 times during roasting.

4. While the vegetables are roasting, prepare the sauce: In a heavy 1-quart

saucepan, combine the shallot, wine, and fish stock. Bring to a boil, reduce the heat, and simmer the mixture until it is reduced by half. Add the mustard and cream and whisk until well blended. Season with a dash of pepper and simmer until reduced to ¾ cup. Add the chives and keep the sauce warm.

5.   Just before serving, remove the crabs from the half-and-half. Dry thoroughly on paper towels and season lightly with coarse salt and pepper. Dredge lightly in the instant flour, shaking off the excess, and set aside.

6.   Heat the clarified butter together with the oil in 2 (10-inch) nonstick skillets over medium-high heat. Place 4 soft-shell crabs in each skillet and cover the skillets lightly with aluminum foil. Sauté the crabs for 2 minutes on each side, remove to a side dish, and keep warm.

7.   To serve, place 2 soft-shell crabs in the center of each individual serving plate. Surround with some roasted vegetables, top each crab with a dollop of sauce, and garnish with whole parsley leaves. Serve at once.

SPICY STEAMED MUSSELS WITH SAUTÉED SAUSAGE AND ORANGE-SAFFRON SAUCE

serves 4–6

Clams braised in a rich sofrito of tomatoes, onions, and chorizo sausage is a popular dish all over the Iberian peninsula. Because I find that American littleneck clams are overwhelmed by the sauce, I have turned to mussels, which are somewhat more assertive and stand up nicely to the flavor-packed sauce. Saffron and citrus add zest and a piquant taste to this easy main course. Serve it in deep bowls with crusty French bread and for those who like the taste of this herb, a sprinkling of cilantro leaves.

4 pounds small mussels
1 pound sweet Italian fennel sausage
3 tablespoons extra-virgin olive oil
2 tablespoons finely minced shallots
1 small dry red chili pepper, crushed

3 large cloves garlic, peeled and finely minced
2 tablespoons fresh thyme leaves, minced
2 teaspoons fresh oregano leaves, minced

1 teaspoon powdered saffron or saffron threads
3 large ripe tomatoes, peeled, seeded, and chopped
Juice of 2 large oranges
1 tablespoon grated orange zest
½ cup dry white wine

Salt and freshly ground black pepper

GARNISH
2 tablespoons finely minced fresh Italian parsley
Small fresh cilantro or basil leaves

1. Wash the mussels well, scrubbing with a stiff brush. Discard any open mussels. Set aside.

2. Remove the sausage from its casing and break into small pieces. Reserve.

3. In a large, heavy casserole, heat the oil over medium heat. Add the shallots, chili pepper, and garlic and sauté the mixture for 1 minute without browning. Add the sausage meat and cook until lightly browned, breaking up the pieces with a fork.

4. Add the thyme, oregano, saffron, and tomatoes, bring to a simmer, and cook until most of the tomato juices have evaporated. Add the orange juice and zest and continue simmering the sauce until it is reduced by half.

5. Transfer the sauce to the workbowl of a food processor or blender and process until smooth. Return to the casserole and set aside.

6. In another casserole, combine the wine and mussels. Bring to a boil, cover, and cook for 3 to 5 minutes, or until the mussels have opened. Discard any mussels that have not opened. Transfer the mussels to a deep serving platter and reserve.

7. Strain the mussel juices through a double layer of cheesecloth into a bowl and add 1 cup of it to the reserved sauce. Reheat the sauce over low heat and correct the seasoning, adding a large grinding of black pepper. Pour the sauce over the mussels, sprinkle with minced parsley and cilantro or basil leaves, and toss gently with two forks. Serve the mussels immediately, accompanied by crusty French bread.

Note: The sauce can be made a day ahead, and the mussels can be steamed open several hours ahead of time and reheated in the sauce.

FETTUCCINE IN SAFFRON CREAM WITH A SAUTÉ OF SHRIMP AND SCALLOPS

serves 4–5

1 large carrot, trimmed and peeled
1 medium zucchini, trimmed
1 large leek, trimmed of all but 1 inch of greens
4 ounces snow peas, strings removed and trimmed
1 teaspoon saffron threads
½ cup Chicken Stock (page 470) or Fish Stock (page 472)
¾ cup heavy cream or Crème Fraîche (page 491)

Coarse salt and freshly ground white pepper
3 tablespoons olive oil
½ pound small to medium shrimp, peeled
½ pound bay scallops or small sea scallops
6 tablespoons unsalted butter
½ pound fresh fettuccine
2 tablespoons finely minced fresh chives
2 tablespoons finely minced fresh parsley

1. Start by preparing the vegetables: Cut the carrot into ¼-inch × 1-inch matchsticks and set aside.

2. With a sharp knife, remove the zucchini skins, including about ¼ inch of the pulp. Cut skins into ¼-inch × 1-inch matchsticks; discard the seedy pulp. Set aside.

3. Cut the leek into the same size matchsticks. Place the leek in a strainer, rinse thoroughly under warm water, drain well, and reserve.

4. Place all the vegetables in a vegetable steamer and set over simmering water. Cover and steam until just tender. Do not overcook. Remove to a side dish and reserve.

5. In a small saucepan, combine the saffron and stock. Bring to a simmer and reduce to 3 tablespoons. Add the heavy cream or crème fraîche and just heat through. Season the saffron cream with salt and pepper and keep warm.

6. In a heavy 10-inch nonstick skillet, heat half the olive oil until it is almost smoking. Add the shrimp and sauté on both sides until nicely browned. Remove with a slotted spoon to a side dish, season with coarse salt and pepper, and reserve.

7. Add the remaining olive oil to the skillet and when hot, add the scallops and sauté until nicely browned on both sides. Do not overcook. Season the scallops with salt and pepper.

8.   Return the shrimp to the skillet together with the steamed vegetables and 2 tablespoons of the butter. Simmer the mixture for 1 minute. Add the saffron cream and just heat through. Taste and correct the seasoning and set aside.

9.   In a large casserole, bring plenty of salted water to a boil. Add the fettuccine and cook 3 to 4 minutes, or until just done, al dente.

10.   Add 2 cups of cold water to the casserole to stop further cooking. Drain the fettuccine well, return to the casserole, and fold in the remaining butter together with the chives and parsley. Season with salt and pepper. Add the scallop and shrimp mixture to the fettuccine, toss lightly with 2 forks, and transfer to a large serving dish. Serve immediately.

Saffron

There is no substitute for the unique taste of saffron. Known as the world's most expensive spice, saffron is harvested from the stigmas of autumn crocus. The reason for the high price is that these stigmas have to be individually plucked by hand. The spice must then go through a long process of drying before being bottled or powdered.

At one time saffron was the key ingredient to three classic Mediterranean dishes, the Spanish paella, the French bouillabaise, and the Italian risotto Milanese, but in recent years saffron has been used more creatively and it is now enjoying an unprecedented popularity in spite of its cost. In pasta, sauces, and rice dishes, and even as a colorant for potatoes, saffron seems to be popping up everywhere, turning these foods a brilliant gold.

Buy saffron threads only and be wary of inexpensive saffron. The best comes from Spain. Egyptian and Indian saffron, especially the powdered variety, is almost always adulterated.

For some preparations—such as pasta dough—you will do better using powdered saffron. Here the quality is not that important since very little of the saffron flavor comes through in the dough. Still, I usually choose either the Spanish or Italian packages, which have an intense saffron aroma. You can use saffron threads instead and, if you do, it is best to pound them lightly in a wooden mortar and pestle and then infuse ½ to 1 teaspoon in ½ cup of hot water or broth for about 10 minutes.

Do not use more than a recipe calls for. Good saffron is powerful and when used in excess becomes bitter. However, be sure the color of the infusion is a brilliant, golden red. If pale or too dark, the saffron is stale.

Saffron is particularly good with rice, pasta, all seafood, tomatoes, yogurt, crème fraîche, and a variety of cakes and breads. In many North African and Indian dishes there is simply no substitute.

▪▪▪

ROAST CHICKEN IN VINEGAR AND GARLIC SAUCE

serves 4–6

It is hard to believe that only a short time ago sherry vinegar was almost completely unavailable in this country, making it impossible to cook such simple, gutsy dishes as this roast chicken in a light vinegar sauce. Cooking with sherry vinegar is extremely popular in Spain, where it is used in many fish preparations. A crisp roasted chicken, with its pan juices slightly deglazed and flavored with whole garlic cloves, summer herbs, and possibly an uncooked tomato concassé, is one of spring's easiest main courses. It is in fact a wonderful dish to serve at any time of the year, but I like to make it truly seasonal by accompaniments such as sautéed snow peas or simply oven-roasted new potatoes tossed in a basil-infused olive oil.

2 small whole chickens, about 2¾
  pounds each
Salt and freshly ground black pepper
1 teaspoon imported paprika
1 teaspoon dried thyme
4 teaspoons Dijon mustard
2 large cloves garlic, peeled and
  mashed
2 tablespoons unsalted butter
1 teaspoon olive oil
16 large garlic cloves, peeled

⅓ cup sherry vinegar
1 cup Chicken Stock (page 470)
1 cup Crème Fraîche (page 491)
  combined with 2 teaspoons tomato
  paste
1 tablespoon cognac
1 Beurre Manie (page 478), optional

GARNISH
2 ripe medium tomatoes, diced
Finely minced fresh parsley and chives

1. Preheat the oven to 375° F.

2. Dry the chickens thoroughly on paper towels, then season with salt, pepper, paprika, thyme, 2 teaspoons Dijon mustard, and the mashed garlic. Truss and set aside.

3. In a large ovenproof dish, heat the butter and olive oil over medium-high heat. Add the chickens and brown lightly on both sides. Add the 16 garlic cloves to the dish together with the sherry vinegar and reduce to a glaze. Add a little chicken stock to the baking dish and place in the center of the preheated oven. Baste the chickens every 10 minutes, adding a little stock to the baking dish. Continue to roast until the juices run pale yellow, about 1 hour.

4. When done, transfer the chickens to a roasting pan and set aside.

5. Pass the pan juices through a fine sieve into a small saucepan, pressing down on the garlic cloves to extract all of their juices. Thoroughly degrease the pan juices, return to the baking dish, and set over medium heat. Bring to a boil and whisk in the crème fraîche–tomato mixture together with the remaining Dijon mustard and cognac and simmer until it lightly coats a spoon. (When using homemade crème fraîche you may need to add a little beurre manie.) Add the diced tomatoes to the sauce and keep warm.

6. Remove the trussing strings from the chickens and if necessary run under the broiler to brown evenly. Carve the chickens.

7. Spoon a little of the sauce onto each serving plate. Top with the chickens and sprinkle with parsley and chives. Serve at once accompanied by a crusty loaf of French bread.

Note: Do not attempt to make this dish with anything but sherry vinegar since the result will be an acidic and harsh sauce. Sherry vinegar becomes rather sweet in cooking and does not leave a vinegary aftertaste.

Tomato Concassé

Tomato concassé is a pretty garnish of diced raw tomatoes and fresh herbs that adds zest and color to many dishes. Use either peeled or unpeeled tomatoes, depending on their ripeness. A peeled tomato that is not truly ripe will look pink, not red, so it is better not to peel it. For an unpeeled tomato, use a serrated knife and cut the flesh off on all four sides, discarding the seeds and the center of the tomato. Cut the flesh into fine dice.

Ripe tomatoes should be blanched in boiling water for about 10 seconds, then peeled. Cut the peeled tomato in half crosswise: It will reveal perfect little pockets of seeds. Gently squeeze out the seeds, then cut the tomato into ¼- to ½-inch pieces, depending on their use.

Combine the diced tomato with extra-virgin olive oil, a splash of sherry vinegar, and some fresh herbs such as basil, cilantro, chives, dill, or parsley, or a combination of 2 or 3 herbs. Do not salt the concassé until just before serving or the tomatoes will soften and become very watery. Make the concassé as you need it; it will not keep for more than 3 to 4 hours.

A spoonful of tomato concassé is good with a frittata, roast chicken, sautéed salmon, or various grilled fish; or as a topping to many pasta dishes or cool summer soups.

■ ■ ■

ROAST SQUABS WITH A FRICASSEE OF PEARL ONIONS AND FRESH GARDEN PEAS

serves 6

Fresh squabs are now available year-round, and I like cooking them in a variety of ways, but come spring I choose a traditional preparation in which the squabs are combined with fresh peas and pearl onions. As with so many classic preparations, this dish is reappearing on menus of many trendy restaurants with slight variations, mainly that the squabs are grilled rather than roasted. But it is the old-fashioned recipe that I love and find perfect for the home cook.

Here, the natural sweet, slightly gamey flavor of the squab is enhanced by the sweet taste of peas and glazed pearl onions. Serve the squabs with braised kohlrabi, a lemon- and chive-flavored risotto, or for a comforting, homey meal, the Potato Puree with Glazed Scallions (page 103) is a lovely choice as well.

1 pint pearl onions
5 tablespoons unsalted butter
1 tablespoon peanut oil
1 teaspoon granulated sugar
Salt and freshly ground black pepper
1 tablespoon tomato paste
2 tablespoons dark-brown sugar
1½ tablespoons sherry vinegar
1 small sprig plus 2 tablespoons finely minced fresh thyme

1 large clove garlic, peeled and thinly sliced
1½–2 cups Brown Chicken Stock (page 467)
4 squabs, about 1–1¼ pounds each
1 pound fresh peas, shelled, about 1 cup
1 Beurre Manie (page 478)

1. Drop the onions into boiling water for 30 seconds. Drain and, when cool enough to handle, peel the onions with a sharp knife. Make a tiny x through the root end and set aside.

2. In a 10-inch cast-iron skillet, melt 2 tablespoons of butter together with 1 teaspoon of oil over medium heat. Add the onions and sauté until lightly browned, about 5 minutes.

3. Sprinkle with the granulated sugar and season with salt and pepper. Continue to sauté, shaking the pan back and forth until the onions are nicely browned and somewhat caramelized, being careful not to burn them.

4. Add the tomato paste, brown sugar, vinegar, sprig of thyme, and garlic and stir to coat the onions well. Add ½ cup of the stock, bring to a boil, reduce the heat, and cover the skillet tightly. Simmer for 30 minutes, or until the onions are tender, adding a little more stock if the pan juices run dry. Set the onions aside.

5. Preheat the oven to 375° F.

6. Truss the squabs and season with salt, pepper, and minced thyme.

7. In a 12-inch cast-iron skillet or rectangular baking dish, melt the remaining butter and oil over high heat. Add the squabs and brown nicely on all sides. Add 3 tablespoons of stock to the skillet and place in the center of the preheated oven. Roast the squabs for 30 minutes, or until medium-rare, adding a little stock to the skillet and basting with the pan juices every 10 minutes. Do not overcook.

8. While the squabs are roasting, cook the peas. Bring salted water to a boil in a small saucepan. Add the peas and a pinch of sugar. Cook for 5 to 7 minutes or until just tender. When done, drain the peas and add to the onion mixture. Reserve.

9. When the squabs are done, transfer to a serving platter. Discard the trussing strings and set aside.

10. Degrease the pan juices thoroughly and return them to the skillet. Place over low heat. Add the onion and pea mixture and any remaining stock. Bring to a simmer and whisk in bits of beurre manie, just enough to lightly coat a spoon. Taste and correct the seasoning.

11. Spoon the sauce around the squabs and serve immediately, accompanied by a carrot and sweet potato puree or a gratin of turnips.

CASSEROLE OF LAMB WITH CARROTS, PEARL ONIONS, AND ROSEMARY

serves 4–5

This flavorful ragout is similar to the classic French navarin, a traditional full-bodied ragout in which young spring lamb is braised with seasonal root vegetables. Since it is essentially peasant fare, there are many variations of this dish, depending on the region and local availability of produce. Although quite French in nature, this flavorful dish is also an extremely popular spring main course in northern Spain. My mother used to make a wonderful navarin as soon as both baby lamb and the first young spring peas appeared in the market.

3–3½ pounds boneless shoulder of lamb, cut into 1½- to 2-inch cubes
4½ tablespoons olive oil
Salt and freshly ground black pepper
½ pound slab bacon, cut into ¼-inch dice
¼ cup finely minced shallots
¾ cup dry white wine
1 tablespoon all-purpose flour
½ teaspoon granulated sugar
2¼ cups Brown Lamb Stock (page 469)
1 tablespoon fresh thyme leaves or 1 teaspoon dried

1 tablespoon fresh rosemary leaves or 1 teaspoon dried
16 small white onions
2 large carrots, peeled and cut into 1 × ½-inch pieces
3 large egg yolks
2 tablespoons fresh lemon juice

GARNISH
¼ cup finely minced fresh Italian parsley or cilantro
Pilaf of rice

1. Preheat the oven to 350° F.

2. Dry the lamb cubes thoroughly with paper towels.

3. In a large cast-iron skillet heat 3 tablespoons of the olive oil over medium heat. Add the lamb cubes a few at a time and sauté until nicely browned on all sides. Remove the sautéed lamb to a dish, season with salt and pepper, and set aside.

4. Bring water to a boil in a small saucepan, add the bacon, and blanch for 3 minutes. Remove with a slotted spoon to a double layer of paper towels and set aside.

5.   Discard all but 2 tablespoons of fat from the skillet in which the lamb was browned. Add the blanched bacon and cook for 2 minutes over medium heat until lightly browned but not crisp. Add the shallots and cook for 1 minute without burning. Add the white wine and cook for 5 minutes, or until it is reduced to a glaze.

6.   Return the lamb to the skillet and sprinkle with the flour and ¼ teaspoon of the sugar. Toss until it is well coated with the flour and lightly glazed. Transfer the lamb and shallot mixture to a heavy flameproof 4-quart casserole.

7.   Add 2 cups of the stock, the thyme, and the rosemary to the casserole and bring to a boil over high heat. Immediately remove the casserole from the heat and cover it tightly. Braise the lamb in the center of the preheated oven for 1 hour and 20 minutes to 1 hour and 40 minutes, or until tender.

8.   While the lamb is braising, bring water to a boil in a medium saucepan. Add the onions, blanch for 1 minute, and drain. When cool enough to handle, peel and set aside.

9.   Heat 1½ tablespoons of olive oil in a 10-inch skillet over medium heat. Add the onions, carrots, and ¼ teaspoon of sugar. Shake the pan back and forth and cook for 10 minutes or until the onions and carrots are nicely browned. Set aside.

10.   Fifteen minutes before the meat is done, add the reserved onion-carrot mixture. When the meat is cooked, transfer it and the onions and carrots to a dish with a slotted spoon and set aside.

11.   Carefully degrease the pan juices. Return the juices to the casserole and place over medium-low heat. Add the remaining stock. In a small bowl combine the egg yolks and lemon juice and whisk into the pan juices. Just heat through. Do not let the mixture come to a boil or the egg yolks will curdle.

12.   Return the lamb and vegetables to the casserole. Season with salt and a large grinding of pepper and toss to coat well with the sauce. Transfer the ragout to a serving dish, garnish with the minced parsley, and serve hot over braised rice.

ROAST LEG OF LAMB WITH GARLIC CONFIT IN TOMATO FONDUE

serves 6

1 leg of lamb, about 6–6½ pounds
Salt and freshly ground black pepper
1 teaspoon dried marjoram
1 teaspoon dried thyme
1 teaspoon dried rosemary
2 teaspoons imported paprika
1 tablespoon Dijon mustard
2 large cloves garlic, peeled and mashed
4 flat anchovy fillets, drained and finely minced

CONFIT
¾ cup olive oil
2 large heads garlic, separated into cloves and peeled

3 tablespoons unsalted butter
1 tablespoon extra-virgin olive oil
1 medium onion, peeled and cut in half crosswise

2 cups hot Brown Chicken Stock (page 467) or bouillon

FONDUE
2 tablespoons extra-virgin olive oil
2 large shallots, peeled and finely minced
4 large ripe tomatoes, peeled, seeded, and chopped
Salt and freshly ground black pepper
1 tablespoon fresh thyme

OPTIONAL
2 teaspoons arrowroot mixed with a little stock
1 large clove garlic, peeled and mashed

GARNISH
Sprigs of fresh watercress

1. Season the lamb with salt, pepper, marjoram, thyme, and rosemary and rub with the paprika and mustard. Make tiny slits along the bottom of the roast and insert bits of the garlic and anchovy fillets. Set the lamb aside at room temperature for 2 hours.

2. While the lamb is marinating, prepare the confit: In a small, heavy saucepan, combine the olive oil and the garlic cloves. Place over very low heat and stew for 1½ hours, or until the garlic is pale beige and very soft. Remove from the heat and let cool in the oil.

3. Remove the garlic from the oil and transfer to the workbowl of a food processor and process until smooth. Reserve.

---

4. Preheat the oven to 375° F.

5. In a large flameproof baking dish, melt the butter and oil over medium heat. Place the lamb together with the onion halves, cut side down, in the baking dish and spoon some of the melted butter over the lamb. Set the baking dish in the center of the preheated oven and roast for 1 hour and 20 minutes, or 12 minutes per pound, basting it every 10 minutes with a little hot stock.

6. While the lamb is roasting, prepare the fondue: In a heavy 10-inch skillet, heat the oil over medium heat, add the shallots, and cook for 1 minute without browning. Add the tomatoes, salt, pepper, and thyme. Cook until reduced to a thick puree. Add the reserved garlic puree and set aside.

7. When the lamb is done, remove from the oven, transfer it to a serving platter, cover, and keep warm.

8. Carefully degrease the pan juices and return them to the baking dish. Place over high heat and add any remaining stock to the pan. Bring to a boil and reduce to ½ cup. Add the tomato-garlic fondue and heat through.

9. Whisk in a little of the optional arrowroot mixture if necessary, just enough to lightly thicken the sauce. Taste and correct the seasoning, adding the optional mashed garlic.

10. Garnish the roast with sprigs of watercress and serve garlic sauce on the side.

Garlic Confit

A garlic confit is an updated name for an old Mediterranean classic. Stewed garlic cloves have been part of the peasant cooking of Spain, France, and probably some parts of Italy for many years. The peeled garlic cloves are covered with a fruity olive oil and stewed gently over very low heat until golden brown and tender. Once stewed, the garlic can be transferred to a jar, where it will keep for several months.

Add these tender, sweet cloves to the pan juices of a roast, to a variety of pasta dishes, or to a platter of roasted vegetables, or use as a topping to pizza. The stewed garlic is so mild, you may want to use some crushed cloves with various salad dressings, adding texture as well as taste.

■■■

ROAST VEAL WITH GLAZED FENNEL AND TOMATO JAM

serves 4

A veal roast always seems to conjure up the image of an elegant dish for entertaining, but made with the shoulder butt, it is inexpensive and quite accessible, and it allows for endless seasonal variations. Here the roast becomes the centerpiece for a glazed fennel and tomato jam, a combination that is rather unusual since fennel and good vine-ripened tomatoes share the season for a very short time in most parts of the country. If you cannot get good tomatoes, it is preferable to use the Tomato Fondue on page 474. For a simple but satisfying variation, cut the fennel bulbs into eighths and place them around the veal so that they roast together. As soon as the fennel is tender, remove it to a side dish and then reheat it in the pan juices just before serving.

1 boned shoulder of veal, about 2½–3 pounds, rolled and tied
1 large clove garlic, peeled and mashed
2 teaspoons dried thyme
½ teaspoon imported paprika
2 tablespoons unsalted butter
1 teaspoon peanut oil
Coarse salt and freshly ground black pepper
2 small onions, peeled and cut in half crosswise
2½ cups Brown Chicken Stock (page 467)

JAM
1 tablespoon plus 2 teaspoons olive oil
1 tablespoon finely minced shallots

3 ripe medium tomatoes, peeled, seeded, and chopped
Salt and freshly ground black pepper
2 tablespoons unsalted butter
3 medium fennel bulbs, trimmed, cored, and thinly sliced
1 tablespoon granulated sugar
3 tablespoons water

1 Beurre Manie (page 478)

GARNISH
Sprigs of fresh watercress

1. Preheat the oven to 375° F.

2. Dry the veal thoroughly on paper towels. Make tiny incisions on the bottom of the roast and insert pieces of the mashed garlic. Season with the thyme and paprika and set aside.

3. In a heavy flameproof roasting pan, melt the butter and oil over medium-high heat. Add the veal and brown nicely on all sides. Season with salt and pepper and add the onions to the pan, cut side down, together with ¼ cup stock. Place the pan in the center of the preheated oven and roast for 1 hour and 45 minutes, or until a meat thermometer registers 160° F. Baste the roast every 10 to 15 minutes by adding some stock to the pan, then spooning some of the pan juices over the veal.

4. While the veal is roasting, prepare the jam: In a medium saucepan heat 1 tablespoon of the olive oil over low heat. Add the shallots and cook for 1 minute until soft but not browned. Add the tomatoes, season with salt and pepper, and cook, stirring often, until all the tomato liquid has evaporated. Set aside.

5. In a heavy 10-inch skillet, melt the butter together with the remaining oil over medium-high heat. Add the fennel, season with salt, pepper, and sugar and cook for 3 to 4 minutes, stirring often, until the fennel begins to brown. Reduce the heat, add the water, and simmer, covered, for 6 to 8 minutes, or until the water has evaporated and the fennel is golden brown.

6. Add the reserved tomato mixture to the fennel, fold gently, and just heat through. Cover and keep warm.

7. When the veal is done, remove from the oven and transfer to a carving board. Remove the trussing string and let the meat rest for 5 minutes.

8. Degrease the pan juices thoroughly and discard the onions. Add 2 tablespoons of the juices to the fennel and tomato mixture and return the remaining juices to the baking dish. Add the remaining stock, bring to a boil, and whisk in bits of beurre manie until the sauce lightly coats a spoon. Taste and correct the seasoning and keep warm.

9. To serve, cut the veal roast crosswise into thin slices. Place 3 to 4 slices on each of 4 individual serving plates, surround with the fennel and tomato jam, and spoon a little sauce over each portion of veal. Garnish with sprigs of watercress and serve at once.

Note: The fennel and tomato jam is an excellent vegetable dish that I often serve as an appetizer or as an accompaniment to sautéed chicken breasts or pan-seared salmon fillets.

BRAISED BEETS WITH BALSAMIC VINEGAR
AND CRÈME FRAÎCHE

serves 5–6

6–8 medium beets, trimmed of all but 2
  inches of tops
1 tablespoon red wine vinegar
1 tablespoon balsamic vinegar
1 teaspoon granulated sugar

4 tablespoons unsalted butter
4 tablespoons finely minced scallions
2 tablespoons water
¾ cup Crème Fraîche (page 491)
Salt and freshly ground black pepper

1. Place the beets in a large saucepan and add enough water to cover them by 2 inches. Bring to a boil, reduce the heat, and simmer for 35 to 45 minutes, or until tender when pierced with the tip of a sharp knife.

2. Remove the beets and, when cool enough to handle, peel them with your hands; the skins should slip off easily. Cut into ½-inch julienne strips and reserve.

3. In a small bowl combine the vinegars and sugar, whisk until the sugar is completely dissolved, and set aside.

4. Melt 2 tablespoons of the butter in a 10-inch cast-iron skillet over low

---

Beets

One of the most common objections I hear about beets is how long they take to cook. Much of the problem comes from the fact that store-bought beets are often fully mature and not quite as tender as when bought in late spring or summer at a local farm or produce stand. Remember that beets can also be baked just like potatoes, requiring no attention whatsoever.

To prepare beets for cooking, remove all but ½ to ¾ inch of the beet tops. (Save these; they make a delicious stir-fried green.) Rinse the beets well, place in a baking dish, and bake at 350° F for an hour and 30 minutes or until they are easily pierced with the tip of a sharp knife. Do not prick beets too often as they will start to "bleed" and lose their glorious red color. Although beets can also be cooked in the microwave, I prefer the old-fashioned boiling or baking method. Cooked beets should be refrigerated unpeeled. They will keep for 4 to 6 days.

■ ■ ■

---

heat. Add the scallions and 2 tablespoons of water, cover the skillet, and simmer for 3 to 5 minutes, or until very tender. Transfer the scallions to a side bowl and reserve.

5. Add the remaining butter to the skillet and, when melted, add the beets and cook over low heat for 3 to 4 minutes to coat well with the butter. Add the vinegar mixture and simmer until completely absorbed by the beets. Add the crème fraîche and reserved scallions and simmer until the crème fraîche is reduced and coats the beets. Season with salt and pepper and serve hot as an accompaniment to veal chops, pan-seared salmon, or roast or sautéed chicken.

SAUTÉ OF CUCUMBERS WITH PEAS AND MINT

serves 6

Salt
2½ teaspoons granulated sugar
2 pounds fresh, unshelled peas (about 2 cups shelled)
2 large gourmet cucumbers, unpeeled

4 tablespoons unsalted butter
¼ cup water
Freshly ground white pepper
3–4 tablespoons finely minced fresh mint

1. In a large saucepan bring salted water to a boil. Add 1 teaspoon of sugar and the shelled peas and simmer for 5 to 7 minutes or until just tender. Drain the peas and rinse under cold water to stop further cooking and to retain the peas' bright-green color. Set aside.

2. Cut the cucumbers in half lengthwise. Scoop out the seeds with a small spoon and discard. Cut the cucumbers crosswise into ¼-inch slices (half moons) and reserve.

3. Melt the butter in a 12-inch skillet over medium heat. Add the cucumbers and the remaining sugar and sauté for 2 minutes, tossing the cucumbers constantly in the butter. Add the water and season with salt and pepper. Bring to a boil, reduce the heat, and simmer, covered, for 5 minutes, or until tender. Add the peas to the skillet during the last minute or two of cooking. Remove the cover and cook until all the water has evaporated and the cucumbers and peas are glazed.

4. Add the mint to the skillet. Taste and correct the seasoning and serve at once.

TUSCAN BRAISED GREEN BEANS IN PARMESAN CREAM

serves 4–6

1¼ pounds young green beans, trimmed
Salt
3 tablespoons unsalted butter
½ cup heavy cream

Juice of ½ lemon
Freshly ground black pepper
3 tablespoons coarsely grated fresh Parmesan cheese

1. Drop the beans into plenty of salted boiling water and cook for 3 minutes, or until just tender. Drain the beans, run under cold water to stop the cooking, and cut into 1-inch pieces. Reserve.

2. In a large skillet melt the butter over low heat. Add the beans and simmer, uncovered, for 2 to 3 minutes. Add the cream and lemon juice, season with salt and pepper, and simmer until the cream is reduced and has been absorbed by the beans. Add the Parmesan and toss the beans gently until they are well coated with the Parmesan cream. Transfer to a serving bowl and serve immediately.

Note: The beans can be cooked or steamed well in advance, or even the day before, and reheated gently in the cream sauce. For an interesting variation in texture and color, combine wax and green beans and 2 small carrots cut into 1-inch matchsticks and steamed separately.

GRATIN OF KOHLRABI AND LEEKS

serves 6

Kohlrabi has yet to win the popularity in America that it enjoys in central Europe. I find its neglect quite amazing since this wonderful spring vegetable is both versatile and delicious. Even more puzzling is that kohlrabi has become so readily available that I often wonder who uses it.

Every recipe that is suitable for turnips works with kohlrabi. Braised, sautéed, caramelized, or combined with other vegetables such as carrots, potatoes, and peas, it has an adaptable personality that makes it one of my seasonal favorites.

2½ pounds trimmed kohlrabi, peeled and cut into ⅛-inch slices
Salt and freshly ground white pepper
6 medium leeks, trimmed of all but 1 inch of greens

5 tablespoons unsalted butter
¼ cup water
⅓–½ cup Chicken Stock (page 470) or bouillon

1. Preheat the oven to 375° F.

2. Place the slices of kohlrabi in a vegetable steamer, set over simmering water, and steam, covered, for 3 minutes. Remove the kohlrabi from the steamer, season with salt and pepper, and set aside.

3. Cut the leeks crosswise into ¼-inch slices, place in a colander, and run under warm water to remove all the sand. Drain and reserve.

4. Melt 3 tablespoons of the butter in a heavy 10-inch skillet over medium heat. Add the leeks and toss in the butter for 1 minute. Season with salt and pepper. Add the water, reduce the heat, and braise the leeks, covered, for 5 to 7 minutes, or until just tender. Uncover the skillet and if any water remains, cook until it has evaporated. Transfer the leeks to a side dish and reserve.

5. Place half the kohlrabi in a buttered rectangular baking dish in an even layer. Top with the cooked leeks and cover completely with the remaining kohlrabi. Add ⅓ cup of stock and dot the gratin with the remaining butter. Cover with buttered foil, place in the center of the preheated oven, and bake until the kohlrabi is tender and the stock has been absorbed, about 20 to 25 minutes. If the liquid is absorbed too quickly, add the remaining stock.

6. Remove the gratin from the oven and serve hot as an accompaniment to roast Cornish hens or other roasted poultry.

GRATIN OF NEW POTATOES NIÇOISE

serves 4–5

1 pound small red new potatoes, unpeeled
5 tablespoons extra-virgin olive oil
2 large red onions, peeled, quartered, and thinly sliced
3 tablespoons finely minced fresh thyme
Salt and freshly ground black pepper

2 large cloves garlic, peeled and thinly sliced
½–¾ cup beef bouillon
3 medium ripe tomatoes, sliced crosswise
½ cup black oil-cured olives, pitted and cut in half

1. Preheat the oven to 350° F.

2. Cut the potatoes into ¼-inch slices. Set aside.

3. In a large, heavy skillet heat 2 tablespoons of the oil, add the onions and 2 tablespoons of the thyme, and cook the mixture over medium heat for 3 to 4 minutes, or until the onions start to soften.

4. Add the potatoes and fold them into the onion mixture. Season with salt and pepper and reserve.

5. In a flameproof rectangular or oval baking dish heat 2 tablespoons of the oil, add the garlic, and spoon the potato mixture into the dish. Add ½ cup of bouillon and top with overlapping slices of tomatoes. Add the remaining thyme and drizzle with the remaining oil.

6. Place the dish in the oven and bake for 45 minutes, or until the potatoes are tender. You may need to add the remaining bouillon depending on the juiciness of the tomatoes.

7. When the potatoes are almost done, sprinkle the gratin with the olives and continue to bake until done. Serve hot or at room temperature as an accompaniment to a roast leg of lamb, grilled beef, a grilled veal roast, or roast chicken.

POTATO PUREE WITH GLAZED SCALLIONS

serves 5–6

2 pounds all-purpose potatoes,
   preferably russet
Salt
8 scallions, trimmed and minced
5 tablespoons unsalted butter

Freshly ground black pepper
1 teaspoon granulated sugar
¼–⅓ cup chicken bouillon
½ cup sour cream
2 tablespoons finely minced fresh chives

1.   Peel the potatoes and cut into ¾-inch cubes. Place the potatoes in a large saucepan, add water to cover, and salt the water lightly. Bring to a boil and cook the potatoes until tender, about 15 minutes.

2.   While the potatoes are cooking prepare the scallions: In a medium skillet, melt 3 tablespoons of the butter over low heat. Add the scallions, season with salt, pepper, and sugar, and add ¼ cup of bouillon. Cover the skillet tightly and braise the scallions for 10 minutes or until very tender. If the juices run dry, add the remaining broth. Set aside.

3.   When the potatoes are done, drain thoroughly and pass through a food mill or potato ricer into a bowl. Add the remaining butter, sour cream, chives, and glazed scallions. Blend the mixture thoroughly and season with salt and pepper. Serve hot as an accompaniment to a ragout of veal, a sauté of chicken, or roast pork tenderloins.

Note: Finely minced leeks can be substituted for the scallions and chives. When serving the potato puree as an accompaniment to seafood, substitute fruity olive oil for the butter in the puree.

Depending on the quality of potatoes, you may need less or more sour cream. If you cannot get russet potatoes, you may use baking potatoes, but they are somewhat mealier and the result will be a drier puree that will need more butter or sour cream, or both.

BUTTER-BRAISED SPINACH WITH GLAZED GARLIC CLOVES

serves 4

I once read in an English cook book that spinach can absorb an unlimited amount of butter and that, indeed, the success of a spinach puree depends largely on how much butter is used in its preparation. These days, no one would dare to use unlimited amounts of butter in cooking *any* dish, no matter how delicious the result. Fortunately, I have found that spinach can do nicely with just a little of the good stuff and, when accented with glazed garlic cloves, the success of this simple vegetable dish can be achieved without any feelings of guilt.

12 garlic cloves, peeled
1½ pounds fresh spinach, stemmed and washed

5 tablespoons unsalted butter
Salt and freshly ground black pepper

Spinach

Although spinach seems to be available year-round, it is very much a seasonal cool-weather green. For those living in colder regions spinach means the curly, crinkly, fleshy variety that is so flavorful and lovely to look at. In warmer climates such as in California, Texas, and Arizona, you will usually find the flat-leafed spinach that can take the hot weather; it is a variety called New Zealand spinach that looks more like sorrel and is quite mild and not as interesting as true spinach. Cooked, this spinach is rather bland; raw, it is excellent used in salads by itself or in combination with other greens. I find it pairs especially well with Belgian endives. It stands up to hearty vinaigrettes such as the Creamy Goat Cheese and Shallot Dressing on page 60 and the Curry-Lemon Vinaigrette on page 44.

Spinach is also excellent simply pan-sautéed in virgin olive oil with some sliced garlic but can also be steamed in just the moisture that clings after washing.

Try to find unbagged spinach. Although it is more gritty and will take longer to clean, a truly fresh bunch of spinach is so much better than the packaged sort where many of the leaves are broken and a lot of the stems are included. Raw spinach should be used within a day or two of purchase. Steamed or braised it will keep for several days.

■■■

---

1. In a small saucepan, bring water to a boil, add the garlic, and boil for 1 minute. Drain well. Add fresh water to the saucepan and bring to a boil. Add the blanched garlic cloves and cook until tender, about 10 to 15 minutes. Drain and reserve.

---

2. Place the spinach in a vegetable steamer, set over simmering water, and steam, covered, for 2 to 3 minutes or until just wilted. Do not overcook. Remove from the steamer and set the spinach aside.

3. In a large nonstick skillet, melt the butter over medium-low heat, add the garlic cloves, and sauté for 1 to 2 minutes or until lightly browned. Add the steamed spinach, season with salt and pepper, and just heat through. Serve at once as an accompaniment to veal scaloppine or sautéed fish.

PAN-SEARED SNOW PEAS IN PARSLEY AND SHALLOT BUTTER

serves 4–5

3 tablespoons unsalted butter, softened
1½ tablespoons finely minced shallots
2 tablespoons finely minced fresh parsley
2 tablespoons extra-virgin olive oil
1 pound snow peas, ends trimmed

2 large cloves garlic, peeled and thinly sliced
2 tablespoons dry sherry
¼ cup chicken stock or bouillon
Coarse salt and freshly ground black pepper

1. In a small bowl combine the butter, shallots, and parsley, and with a fork blend the mixture into a smooth paste. Chill.

2. In a large nonstick skillet heat the oil, add the snow peas, and sauté over fairly high heat until they start to brown. Add the garlic cloves and sherry and continue sautéeing until they are nicely browned, shaking the pan back and forth to make sure that the snow peas do not burn.

3. Add the chicken stock and season with salt and pepper. Cover the skillet and simmer the snow peas for another 3 minutes. Do not overcook.

4. Remove the pan from the heat, add the shallot mixture, and just let it melt into the peas.

5. Transfer to a deep serving bowl and serve immediately.

Note: Leftover snow peas can be heated successfully in the microwave as long as they have not been overcooked.

CATALAN BROCCOLI SALAD

serves 6

3 flat anchovy fillets, drained and finely
   minced
3 tablespoons extra-virgin olive oil
1 tablespoon red wine vinegar
1 large clove garlic, peeled and mashed
⅔ cup Mayonnaise (page 477)
3 tablespoons Danish blue cheese or
   mild blue cheese, crumbled, about 2¼
   ounces
Salt and freshly ground black pepper
1 tablespoon tiny capers, drained

1½ pounds fresh broccoli
1 small red onion, peeled, cut into
   quarters, and finely sliced

GARNISH
1 small red bell pepper, cored, seeded,
   and cut into fine julienne
10–12 small black oil-cured olives,
   preferably Niçoise
1 hard-boiled egg, peeled and minced

1. In the workbowl of a food processor combine the anchovies, olive oil, vinegar, and garlic and process for 15 seconds, or until smooth.

2. Add the mayonnaise and blue cheese to the workbowl and process for 15 seconds, or until thoroughly combined with the anchovy vinaigrette. Season with salt and pepper. Transfer the dressing to a small bowl, fold in the capers, cover, and refrigerate until needed.

3. Trim all but 2½ inches of stalks from the broccoli. Separate the broccoli into spears and, with a vegetable peeler, peel the stalks. Set the trimmed spears aside.

4. Bring salted water to a boil in a large saucepan. Add the broccoli spears and cook for 3 to 4 minutes, or until the stalks are tender. Transfer to paper towels and set aside to cool.

5. Place the cooled broccoli in a shallow serving dish in a decorative pattern, sprinkle with the red onion, and pour the chilled dressing over the vegetables. Garnish the salad with the red pepper, black olives, and minced hard-boiled egg. Serve chilled, but not cold, as an appetizer.

SALAD OF CARROTS IN GINGER, MUSTARD, AND FRESH CILANTRO VINAIGRETTE

serves 6

12–14 medium carrots, about 2 pounds, peeled, trimmed, and cut into ¼-inch matchsticks
1 tablespoon cider vinegar
2 tablespoons fresh lemon juice
½ teaspoon fresh ginger, mashed through a garlic press
2 teaspoons granulated sugar

1 tablespoon Dijon mustard
⅓ cup olive oil
2–3 tablespoons finely minced fresh cilantro
Salt and freshly ground black pepper

GARNISH
Sprigs of fresh cilantro

1. Bring water to a boil in a vegetable steamer, add the carrots, cover, and steam for 6 to 7 minutes, or until tender. Remove from the steamer and immediately run under cold water to stop further cooking. Drain and dry thoroughly on paper towels. Set aside.

2. In a medium salad bowl combine the vinegar, lemon juice, ginger, sugar, and mustard and whisk until the sugar has dissolved. Add the olive oil in a slow steady stream, whisking constantly until the vinaigrette is well blended and creamy. Add the cilantro and season with salt and pepper.

3. Add the carrots and toss to coat evenly. Cover and refrigerate 4 to 6 hours.

4. Thirty minutes before serving, bring the salad back to room temperature and correct the seasoning, adding a large grinding of black pepper. Garnish with sprigs of fresh cilantro and serve as part of a buffet or antipasto table.

Note: You may substitute 2 to 3 tablespoons finely minced fresh mint for the fresh cilantro.

ENDIVE, RADISH, AND WATERCRESS SALAD IN DOUBLE-MUSTARD DRESSING

serves 6

Juice of 1 large lemon
¼ teaspoon dry mustard
1 tablespoon Dijon mustard
2 teaspoons granulated sugar
4 tablespoons finely minced scallions
6–8 tablespoons extra-virgin olive oil
Salt and freshly ground white pepper
2 cups thinly sliced radishes

¾ cup Gruyère or imported Swiss cheese, cut into fine julienne, about 1½ inches long
6 Belgian endives, cored and cut into thick julienne
1 large bunch watercress, stemmed, washed, and dried on paper towels

1. In a large salad bowl combine the lemon juice, mustards, sugar, and scallions and whisk until smooth. Add the olive oil in a slow steady stream, whisking constantly until the dressing is creamy. Season with salt and pepper to taste.

2. Top the dressing first with the radishes, then the cheese, endives, and watercress, but do not toss. Cover and refrigerate until serving time.

3. Remove the salad from the refrigerator 15 minutes before serving, toss gently, and correct the seasoning, adding a large grinding of pepper. Serve with French bread and sweet butter.

MARINATED WHITE MUSHROOM SALAD IN CRÈME FRAÎCHE AND CHIVE VINAIGRETTE

serves 4

In truth, I find the flavor of cooked, all-purpose, cultivated mushrooms rather dull—almost tasteless. Nevertheless, I do like their texture and often add them to cooked dishes that call for a variety of other types of mushrooms. They are especially good combined with dried porcini, which have a unique taste that does not allow for an additional assertive flavor. When used raw, however, and combined with a tangy vinaigrette, the all-purpose mushroom comes to life, and the result is a very refreshing spring salad that I like to serve as an easy and quick starter to a warm-weather menu or as an accompaniment to poached salmon steaks, carpaccio of veal, pan-seared tuna, or grilled chicken.

Juice of ½ lemon
1½ tablespoons sherry vinegar
2 tablespoons finely minced shallots
6 tablespoons olive oil
12 ounces all-purpose mushrooms, trimmed, wiped, and cut into ½-inch dice

Salt and freshly ground black pepper
¾ cup Crème Fraîche (page 491)
2–3 tablespoons finely minced fresh chives

1.   In a medium mixing bowl combine the lemon juice, vinegar, and shallots and whisk until smooth. Add the olive oil in a slow steady stream, whisking constantly until the vinaigrette is creamy. Add the mushrooms and toss to coat with the vinaigrette. Season with salt and pepper, cover, and marinate the mushrooms at room temperature for 30 minutes.

2.   Fold in the crème fraîche and chives, then taste and correct the seasoning. Cover and refrigerate until lightly chilled.

PEASANT SALAD WITH PROVENÇAL VINAIGRETTE

serves 4

Here is a salad that seems to have been part of my cooking repertoire forever. It is somewhat similar to a salade Niçoise in that it calls for tuna, hard-boiled eggs, and black olives, but otherwise, it is simply a delicious mixture of greens, fresh mushrooms, and a red onion, dressed in a pungent mustard and garlic vinaigrette. I make the dressing in large quantities and refrigerate it in a covered jar, to which I add lightly crushed garlic cloves. Once the dressing is brought to room temperature, I shake the jar, and the dressing emulsifies beautifully.

Nothing is a must in this salad. You can add or omit different greens and turn it into a main-course salad with the addition of the tuna, hard-boiled eggs, some roasted new potatoes, or serve it just as a mixed salad. I always like to remind my students that a salad bowl is not a catchall, and that the balance of greens and their affinity to this type of hearty dressing has to be taken into consideration when making it.

1 recipe Provençal Vinaigrette (page 480)
1 small red onion, peeled and thinly sliced
6–8 medium all-purpose mushrooms, trimmed, wiped, and thinly sliced
3 cups fresh spinach leaves, stemmed, washed, and dried
1 heart of romaine lettuce, washed and dried on paper towels
2 Belgian endives, cored and cut into julienne strips

1 small head radicchio, torn into 2-inch pieces, optional
1 bunch arugula leaves, washed and dried on paper towels
6 ripe cherry tomatoes, cut in half
2 or 3 small hard-boiled eggs, peeled and cut in half
4 ounces light tuna packed in oil, drained and flaked
8 small black oil-cured olives
1 large red bell pepper, roasted (page 483), peeled, and thinly sliced
Salt and freshly ground black pepper

1. In a large salad bowl combine the vinaigrette, red onion, and mushrooms. Toss gently and marinate at room temperature for 30 minutes to 1 hour.

2. Place the spinach, romaine, Belgian endives, optional radicchio, and arugula on top of the marinated vegetables, in that order. Do not toss. Top the greens with the remaining ingredients except the salt and pepper, cover, and refrigerate until serving.

3. Remove the salad from the refrigerator 20 minutes before serving. Season with salt and pepper, toss gently, and serve at once, accompanied by crusty French bread and a selection of regional cheeses.

FALLEN CHOCOLATE SOUFFLÉ CAKE
WITH WHITE CHOCOLATE SAUCE
AND RASPBERRIES

serves 8

Nothing could be simpler or more delicious than this rich, dense chocolate mousse cake. It can be assembled in a matter of minutes and gets better every day, although in my house it does not seem to last very long. What I like best about the cake is its adaptability to all types of seasonal accompaniments, in particular roasted or poached pears, a pear coulis, white or mocha-flavored chocolate sauce, all spring berries, homemade ice creams, or as my mother always says, "Just cover it with Schlaag," the Viennese word for whipped cream.

7 ounces extra-bittersweet chocolate,
   broken into pieces
7 ounces unsalted butter
2 tablespoons strong prepared coffee
5 large eggs, separated
½ cup granulated sugar
2 tablespoons unsweetened cocoa,
   preferably Droste's

Pinch of salt

GARNISH
Heavy dusting of cocoa
White Chocolate Sauce (page 488)
Fresh raspberries

1. Preheat the oven to 350° F.

2. Butter the bottom and sides of a 9½-inch springform pan. Line the bottom with a circle of parchment paper and butter the paper. Reserve.

3. In the top of a double boiler combine the chocolate, butter, and coffee. Set over simmering water and stir until melted and smooth. Transfer the chocolate mixture to a large mixing bowl and let cool slightly.

4. Add the egg yolks and whisk until smooth. Sift the sugar and cocoa into the chocolate mixture and whisk until well blended. Set aside.

5. In another large mixing bowl combine the egg whites and pinch of salt and beat until soft peaks form. Gently fold a third of the whites into the choc-

olate mixture. Fold the lightened chocolate mixture into the remaining whites, gently but thoroughly. Pour the batter into the prepared pan and set on the lower rack in the center of the preheated oven. Bake for 30 minutes.

6. Remove the cake from the oven and place on a wire rack. Immediately loosen the collar of the springform pan but do not remove. Place the bottom of another 9½-inch springform pan or the bottom of a 9-inch removable-bottom tart pan on top of the cake and lightly press to deflate. Let the cake cool in the pan.

7. When the cake is cool, remove the collar and carefully invert the cake onto a serving plate. Remove the bottom of the springform pan and carefully peel off the parchment paper.

8. Serve the cake cut into wedges with a heavy dusting of cocoa and a pool of white chocolate sauce and a sprinkling of fresh raspberries.

SUGAR-LACED CREPES WITH A STRAWBERRY AND MASCARPONE MOUSSE

serves 4

Mascarpone is a buttery, mild cream cheese imported from Italy, now available in many supermarkets and fancy grocery stores throughout the country. It is one of the key ingredients in tiramisu, a dessert that has become very popular of late. Before the advent of tiramisu, mascarpone was usually served with sugared berries, to which this cheese has a wonderful affinity.

Sweet crepes are among the most versatile of dessert classics and should be part of every cook's repertoire. They are not difficult or time-consuming to make, and they freeze well. Here, these delicate pancakes are filled with a light combination of strawberries, mascarpone, and whipped cream and then topped with a warm orange-flavored sauce. This is a truly memorable dessert.

If you are pressed for time, use the filling as a dessert by itself. Simply spoon it into individual dessert dishes or coupes, sprinkle with additional assorted berries, and add a sprig of mint—the perfect ending to a homey supper.

FILLING
¼ cup granulated sugar
2 cups thinly sliced fresh strawberries
½ pound mascarpone cheese
1 cup heavy cream, whipped

SAUCE
8 tablespoons unsalted butter
¼ cup granulated sugar
¼ cup Grand Marnier
2 tablespoons cognac
2 teaspoons finely grated orange rind

12 Dessert Crepes (page 493)

1. Start by making the filling: Combine the sugar and strawberries in a bowl, toss gently, and let the strawberries macerate for 30 minutes at room temperature.

2. Place the mascarpone in a large mixing bowl and beat with an electric beater until fluffy. Fold in the whipped cream and macerated strawberries gently but thoroughly. Cover and chill until needed.

3. Preheat the oven to 325° F.

4. To make the sauce: Melt the butter in a medium saucepan over low heat. Add the sugar, Grand Marnier, cognac, and rind and whisk until the sugar is dissolved. Keep warm.

5. Spread each crepe with some of the mascarpone-strawberry filling. Fold the crepes into quarters and place in a buttered baking dish, slightly overlapping, in a single layer. Spoon the sauce over the crepes. Cover the dish with foil and place in the center of the preheated oven. Bake for 10 to 15 minutes. The filling should still be cool, but the sauce and crepes should be heated through. Serve immediately.

SPICED ORANGES IN RED WINE AND CINNAMON SYRUP

serves 6

6 medium California navel oranges
3 cups dry red wine
1½ cups granulated sugar
1 (3-inch) piece cinnamon stick
1 (3-inch) piece of vanilla bean, split
   lengthwise
2 whole cloves
1 (3-inch) strip of orange zest

3 tablespoons Cointreau or other
   orange-flavored liqueur, optional

GARNISH
Tiny whole fresh strawberries or
   raspberries
Tiny fresh mint leaves

1. Using a sharp paring knife, peel the oranges, removing all the rind and white pith. Slice the oranges crosswise into ¼- to ⅓-inch slices. Remove all the pits with the tip of the knife and arrange the orange slices in a decorative pattern in a shallow serving dish. Set aside.

2. In a medium stainless-steel saucepan combine the wine, sugar, cinnamon stick, vanilla bean, cloves, and orange zest. Bring to a boil and stir to dissolve the sugar. Reduce the heat and simmer the mixture until reduced to 1 cup. Remove from the heat, strain into a bowl, and set aside until completely cooled.

3. Add the optional Cointreau to the syrup and pour over the orange slices. Cover and refrigerate for 2 to 4 hours, or until the oranges have taken on a red color.

4. To serve, place the marinated orange slices on individual dessert plates, spoon some of the red wine syrup around the slices, and garnish with tiny whole strawberries or raspberries and leaves of mint.

Note: The entire dish can be made the night before and covered and refrigerated. You may also serve a small scoop of good-quality vanilla ice cream with the oranges.

PINEAPPLE COMPOTE IN WINE AND ORANGE SYRUP

serves 4–6

Zest of 1 large navel orange, cut into ¼-inch julienne strips
Zest of 1 large lemon, cut into ¼-inch julienne strips
1 bottle blush or rosé wine
1⅓ cups granulated sugar
8 whole black peppercorns
1 large ripe pineapple, peeled and cut into 1-inch cubes

2 tablespoons dark rum, optional
1 ripe banana, peeled and sliced, optional
1 ripe kiwi, peeled and thinly sliced, optional

GARNISH
Tiny fresh mint leaves

1. Bring water to a boil in a small saucepan, add the orange and lemon zest, and blanch for 3 to 4 minutes, or until tender. Drain and set aside.

2. In a 3-quart casserole, combine the blush or rosé wine with the sugar and peppercorns, bring to a boil, and cook for 1 to 2 minutes, or until the sugar is completely dissolved. Add the pineapple, reduce the heat, and simmer, uncovered, for 20 minutes.

3. Remove the pan from the heat and transfer the pineapple and its poaching liquid to a serving bowl and let cool completely.

4. Add the optional rum, banana, and kiwi and serve lightly chilled, garnished with mint leaves.

SPRING-BERRY BREAD PUDDING

serves 10

4 tablespoons unsalted butter, softened
   at room temperature
12 slices good-quality white bread,
   crusts removed
4 large eggs
2 large egg yolks
1 cup granulated sugar
1½ teaspoons vanilla extract
Finely grated rind of 1 large lemon

3 cups light cream
1 pint fresh blueberries, rinsed
½ pint fresh raspberries

GARNISH
Heavy dusting of confectioners' sugar
Fresh raspberries and blueberries
Tiny fresh mint leaves
Crème Anglaise (page 490), optional

1. Butter the bottom and sides of a 2½-quart rectangular baking dish or pudding dish, using 1 tablespoon of the butter. Set aside.

2. Coat each side of the bread with the remaining butter. Reserve.

3. In a large mixing bowl combine the eggs, egg yolks, sugar, vanilla, and lemon rind and whisk until fluffy and pale yellow. Add the cream and whisk until smooth.

4. Arrange 6 slices of the bread in 1 layer on the bottom of the baking dish, trimming the bread to fit evenly, if necessary. Top with the blueberries and raspberries and place the remaining bread in a single layer on top of the berries. Pour the cream mixture over and let stand at room temperature for 2 hours, spooning the cream mixture over the bread from time to time.

5. Preheat the oven to 350° F.

6. Place the baking dish inside a larger baking dish. Fill the larger dish with enough boiling water to reach halfway up the sides of the smaller dish. Place in the center of the preheated oven and bake for 45 to 50 minutes, or until the top has browned lightly and a knife inserted in the center comes out clean.

7. Remove the pudding from the oven and the water bath. Let cool completely.

8. Serve at room temperature, heavily sprinkled with confectioners' sugar. Garnish with berries and mint and serve cut into squares with optional crème anglaise.

RASPBERRY POTS DE CRÈME

serves 6

¾ cup whole milk
½ cup heavy cream
2 large eggs
2 large egg yolks
⅓ cup granulated sugar
1 teaspoon vanilla extract

¾ pint fresh raspberries
¾ cup red currant jelly
2 tablespoons Grand Marnier

GARNISH
Tiny fresh mint leaves

1. Preheat the oven to 350° F.

2. In a small, heavy saucepan combine the milk and cream and heat the mixture over medium-low heat without letting it come to a boil. Keep warm.

3. Combine the eggs, yolks, sugar, and vanilla in a medium mixing bowl and whisk until fluffy and pale yellow. Add the warm milk-cream mixture and whisk until well blended. Pour the mixture into 6 (4-ounce) ramekins.

4. Place the ramekins in a heavy rectangular baking dish and fill the baking dish with enough boiling water to come halfway up the sides of the ramekins.

5. Place the baking dish in the center of the preheated oven and bake the custards for 15 minutes, or until the tip of a knife inserted in the center comes out clean. Do not overcook. Remove the ramekins from the oven and cool on wire racks.

6. When completely cooled, garnish each ramekin with a single layer of raspberries to cover the top completely. Set aside.

7. In a small saucepan, melt the currant jelly over low heat, add the Grand Marnier, and heat through. Remove the glaze from the heat and let cool slightly. Spoon enough glaze over the raspberries to cover them completely, and chill for 2 to 4 hours before serving.

8. Just before serving, garnish each ramekin with tiny leaves of mint and serve lightly chilled.

STUFFED STRAWBERRIES IN
PINEAPPLE-MINT COULIS

serves 6

Here is a lighthearted, somewhat whimsical idea. I still remember the flavor of that first strawberry I popped into my mouth at Jacques Chibois' restaurant in Cannes. When I suddenly realized that it was a strawberry stuffed with a velvety Grand Marnier mousse, I felt like eating the whole platter.

When making this dessert, be sure to choose even-sized berries and plan to serve five per person. Although the crème Chiboust is a lovely filling that also doubles as a topping to other berries and can be frozen for later use, you can fill the berries simply with sugared mascarpone, to which you might add finely minced fresh berries and a touch of mint. Do not fill the berries more than three to four hours ahead of time or they will become mushy. Either a strawberry sauce or a coulis made with frozen raspberries is a lovely accompaniment to the strawberries.

COULIS
¾ cup granulated sugar
¾ cup water
⅓ cup Sauternes
1½ teaspoons unflavored gelatin
1 cup fresh mint sprigs
1 small ripe pineapple, peeled, cored, and cut into cubes

CRÈME CHIBOUST
3 large egg yolks
2 tablespoons cornstarch

⅓ cup granulated sugar
1 cup less 2 tablespoons hot milk
1 teaspoon vanilla extract
2 tablespoons Grand Marnier
1 cup heavy cream, whipped

30 fresh, whole, medium strawberries, washed but not hulled

GARNISH
Tiny fresh mint leaves

1. Start by making the coulis: In a medium saucepan, combine the sugar, water, and Sauternes. Bring to a boil over high heat and stir to dissolve the sugar. Add the gelatin and stir until dissolved. Add the mint and remove from the heat. Let cool.

2. Place the pineapple in the workbowl of a food processor, add the mint syrup, and process until smooth. Pass the mixture through a fine sieve into a bowl and chill until serving time, stirring once or twice.

3.   To make the crème Chiboust: In a medium mixing bowl combine the egg yolks and cornstarch and whisk until well blended. Add the sugar and whisk until fluffy and pale yellow. Add the hot milk all at once and transfer to a heavy-bottomed 2-quart saucepan.

4.   Place the saucepan over medium-low heat and whisk constantly until the mixture is smooth and quite thick. Immediately transfer to a stainless-steel bowl and whisk in the vanilla and Grand Marnier. Place a piece of plastic wrap directly on the surface of the custard to cover completely. Chill until completely cold. Gently fold in the whipped cream and chill until needed.

5.   Cut a ¼-inch slice off the pointed end of each strawberry and discard. Cut ¼ inch off the stem ends and reserve these caps.

6.   With a very small melon-ball cutter, scoop out the exact center of each strawberry, creating a cavity no larger than the diameter of the melon-ball cutter. Stand the strawberries cavity side up and reserve.

7.   Fill a pastry bag, fitted with a ½-inch plain nozzle, with crème Chiboust. Pipe some of the cream into each strawberry cavity to mound slightly and top with the reserved cap. The strawberries can be filled several hours before serving and kept in the refrigerator.

8.   Place 5 strawberries on each of 6 individual dessert dishes and spoon a little pineapple-mint coulis into each dish. Garnish with mint leaves and serve at once.

Note: The remaining crème Chiboust can be served as a dessert by itself in individual ramekins, topped with fresh raspberries or grated chocolate.

CARAMELIZED MERINGUE PUFFS IN STRAWBERRY SAUCE

serves 6

Here is an updated version of one of the great French classic desserts, oeufs à la neige, or snow eggs. It was one of the last recipes I learned in cooking school, and it remained in my entertaining repertoire for years, when it was abandoned in favor of easier, more quickly assembled desserts. Recently I returned to this delicious and versatile dessert, once I realized that it could be made in two stages.

CUSTARD
5 large egg yolks
⅓ cup granulated sugar
1½ cups hot whole milk
1 teaspoon vanilla extract

STRAWBERRIES
2 pints fresh strawberries, washed and hulled
3 tablespoons granulated sugar
2 tablespoons orange-flavored liqueur

PUFFS
5 large egg whites
½ cup granulated sugar

CARAMEL
½ cup granulated sugar
2 tablespoons water

GARNISH
Tiny fresh mint leaves

1. Start by making the custard, which can be prepared up to 1 week ahead of time: In a heavy stainless-steel bowl, combine the egg yolks and sugar and whisk until fluffy and pale yellow. Add the hot milk and whisk until well blended. Transfer the mixture to the top part of a double boiler, place over simmering water, and whisk constantly until the custard heavily coats a spoon. Immediately transfer the custard to a bowl and add the vanilla extract. Place a piece of plastic wrap directly on the surface to cover completely and chill.

2. Dice the strawberries, reserving about 6 to 8 perfect ones for garnish, and reserve.

3. Combine the diced strawberries, sugar, and liqueur in the workbowl of a food processor and process until smooth. Strain through a fine sieve and fold the puree into the chilled custard. Return to the refrigerator, cover, and chill until serving.

4. Three to four hours before you serve the dessert, prepare the puffs: Place the egg whites in a large copper bowl, add 1 tablespoon of the sugar, and beat the whites with a small electric hand beater on high speed. When the egg

whites start to froth, add the remaining sugar, 1 tablespoon at a time. Continue beating the whites until they are firm and glossy, but not dry. Set aside.

5. In a deep 12-inch skillet or sauté pan, add water to the depth of 1 inch and place over high heat. Bring to a boil and reduce to a simmer. With a medium-sized soup ladle, scoop up the meringue and smooth the scoops with a cake spatula. Carefully transfer the scoops to the simmering water and cook for 50 seconds. With 2 spoons, carefully turn over and cook for another 50 seconds. Do not overcook or the puffs will get water-logged.

6. Transfer the cooked puffs to a linen kitchen towel to drain and continue making meringue puffs until all the meringue mixture is used.

7. Pour the strawberry custard into a deep oval or round serving dish. Carefully place the puffs on top of the custard using a spatula and set aside.

8. To make the caramel: In a small saucepan combine the sugar and water. Bring to a boil and cook the mixture without stirring until the sugar caramelizes and turns a light hazelnut brown. Watch carefully so that the sugar does not become too dark. Remove the saucepan from the heat and let the caramel cool slightly. Do not let cool completely or the caramel will become quite hard. When it has become thick and syrupy, drizzle the caramel over the meringue puffs in a decorative pattern and garnish the platter with the reserved berries and tiny leaves of fresh mint. Serve cool, but not chilled.

Note: You can substitute raspberries for the strawberries and garnish with a mixture of fresh berries, such as blueberries, blackberries, red currants, or loganberries.

You may beat the egg whites in a heavy-duty electric mixer, but be careful not to overbeat them. It is best to finish beating the egg whites by hand using a large whisk.

THE

SUMMER

KITCHEN

Several years ago I spent the summer months in a small Mediterranean fishing village. The tiny narrow streets were filled with the snarling traffic of summer vacationers working their way down the steep hill to the beach. Every day a farmer on a large tricycle would make his way to a tiny space on one side of the street. It was heaped with brilliant summer produce, and both the locals and summer vacationers would patiently line up to buy their vegetables and herbs. The little tricycle was a symbol of summer for me; rich and colorful, it spelled the essence of the season, a season filled with intensity and great flavors rather than variety. Because we were staying at a hotel, it was a challenge trying to figure out how I could make use of this wonderful bounty within the limitations of our room.

Whether at the beach or in the country, the sheer richness of this relatively short season is hard to resist. Each year I find it impossible not to get carried away by the colorful farmstands and produce markets, where everything seems to vie for my attention: crisp beans, shiny, firm zucchini, brilliant green peppers, moist lettuces, feathery carrots, bushels of vine-ripened tomatoes, and sacks of freshly picked corn. Summer is a concentrated time of total excitement, a never-ending discovery of simple new dishes and new preparations. And, at the same time, it is a time to return to some of the familiar undemanding and refreshing summer classics.

The carefree spirit and ease of summer living extends into the kitchen in a unique way. Somehow the season's menus seem to fall into place naturally, with more spontaneity and little advance preparation. Because both fruits and vegetables are at their peak they require little in the way of adornment. All you need for a perfect summer lunch is a well-seasoned salad, possibly topped with a julienne of pan-seared chicken breast or a crumbling of goat cheese. A few sliced tomatoes dressed with some basil vinaigrette, a bowl of oil-cured olives, and some crusty bread will round off this kind of casual lighthearted meal. Then, for dinner, I may add a zucchini and pepper frittata or some fresh pasta tossed with a julienne of braised summer vegetables, finishing with a piece of fresh fruit. Nothing is a must; this is one time of year when meals do not need to follow the traditional theme of an appetizer, main course, and dessert.

It is important to keep in mind that while summer foods can be kept simple, simplicity does require perfection. The best vinaigrette won't change the taste of an unripe tomato, nor can the best homemade ice cream mask the lack of flavor of an underripe peach. Even the best summer produce requires careful seasoning and will benefit from an imaginative preparation. It is also important to remember that undercooked vegetables are as uninteresting as overcooked ones, and that cool foods are best served slightly chilled and not icy cold. Yet the possibilities of creating more substantial and memorable meals are here as well. A chicken and green bean salad tonnato or Galician summer vegetable soup followed by either a sauté of shrimp with tomato fondue and Indian spices or charcoal-grilled beef tartare are just the kind of appealing examples that add interest and diversity to summer menus. Since your main course may be rather simple and straightforward, this is a good time for desserts to reflect the season in a creative way. A blueberry custard tart, a summer pudding, or a berry and fruit compote served with ice cream can highlight your menus and make them truly memorable.

In the past few years the grill has had a tremendous impact on summer cooking, and it has become all too easy to resort to grills and marinades at every meal. But, this method of cooking can get pretty monotonous after a while. I use the grill selectively and with some subtlety, not allowing every food to take on the same flavor. If I plan on a main course to be prepared on the grill, I usually serve it surrounded by a variety of my favorite vegetables, such as a sauté of summer squash in thyme and sour cream, an interesting green bean salad tossed in a tuna vinaigrette, and a roast pepper chutney. Many vegetables are, of course, wonderful prepared on the grill, and something more unusual, such as smoked eggplant fritters, makes for a wonderful starter, allowing you to take full advantage of the hot coals.

In addition to the grill and the ever-popular summer pasta dishes, this is a good time to experiment with a variety of homemade pizzas and summer risottos. Since Italian cooking has had such an influence on the American kitchen, we can continue to look toward this cuisine for its many imaginative preparations of vegetables, soups, and light main courses. To me, everything seems to fall into place naturally in the summer kitchen, and I like to draw on its full potential with creativity. Above all, I try to make the most of this brief season and enjoy the pure and refreshing flavors that say summer.

1

Creamy Beet Soup with a Concassé of Mushrooms,
Crème Fraîche, and Dill
Grilled Lamb on a Bed of Sweet-and-Sour Red-Onion Jam
Charcoal-Roasted Eggplant Fritters with Yogurt and Mint Sauce
Purple Plums in White-Wine Sabayon

2

Spicy Summer Black-Bean Salad
Sauté of Swordfish with Pepper and Anchovy Marmalade
Baked Peaches in Praline Caramel

3

Yellow Squash Soup with a Garnish of Summer Vegetables
Sauté of Chicken Bressane
Gratin of Raspberries and Blueberries in Ginger Cream

4

Risotto of Leek and Tomatoes Cipriani
Lilac-Smoked Cornish Hens with Grilled Eggplant Aioli
Tricolor of Melons and Kiwifruit in Lime Caramel

5

Sicilian Braised Vegetables in Sweet Balsamic Dressing
Spice-Charred Grilled Rib Eye with Black Radish
and Mustard Sauce
Red-Potato Salad in Sour Cream Vinaigrette
Drunken Blueberry Fool with Cassis

SUMMER APPETIZERS

Chicken Salad in Sweet Mustard Vinaigrette with
Beets and Cucumbers

Herbed Zucchini and Potato Frittata

Mozzarella Toasts in Tomato Vinaigrette

Eggplant Sandwiches à l'Antiboise

Niçoise Fusilli Salad

Multicolored Pepper, Onion, and Two-Cheese Pizza

Stir-Fried Swiss Chard, Bacon, and Pine-Nut Pizza

Pesto, Tomato, and Summer-Squash Pizza

SUMMER SOUPS

Spicy Corn and Zucchini Soup

Galician Summer Vegetable Soup

Pistou of Red Beans, Tomato, and Summer Squash

Cream of Tomato and Tangy Lemon Soup

Creamy Gazpacho with Lime-Marinated Bay Scallops

Avocado Soup à la Mexicaine with Traditional Salsa Topping

SUMMER MAIN COURSES

Brochette of Swordfish with Leek and Tomato Compote and
Balsamic Vinegar Sauce

Penne with Pan-Seared Peppered Tuna, Tomatoes, and Gremolata

Cedar-Roasted Salmon with Chive-Butter Sauce

Salmon and Lobster Fritters in Sorrel Cream

Pernod-Scented Mussels in Tomato, Mint, and Saffron Sauce

Fusilli with Sautéed Shrimp, Tomato, and Ginger Fondue

Sauté of Shrimp with Tomato Fondue and Fragrant Indian Spices

A Southern Stew of Chicken, Okra, and Tomatoes

Sautéed Chicken with Roasted Garlic, Rosemary, and Peppers

Sautéed Chicken Breasts with Tomatoes, Basil, and a
Confit of Garlic

Grilled Quail with Red Onion—Candied Ginger—Thyme Jam

Grilled Turkey Breast with Spicy Yogurt Paste and
Two-Bean Fricassee

Grilled Flank Steaks in Spicy Chili Marinade

Charcoal-Grilled Tartare of Beef with Capers and
Chive Essence

Grilled Veal Roast à la Florentine

Corn, Polenta, and Jalapeño Pepper Fritters

Braised Okra in Spicy Middle Eastern Tomato Sauce

Creamy Sauté of Corn and Yellow Squash

Sautéed Young Green Beans with Ginger,
Garlic, and Plum Tomatoes

Grilled Stuffed Peppers à la Basquaise

Moroccan Stuffed Baked Tomatoes

Braised New Potatoes with Cucumbers, Tomatoes,
and Summer Herbs

Potato and Farmstand Vegetable "Paella"

Thyme-Braised Zucchini in Sour Cream

Lemon- and Parsley-Scented Braised Bulgur

SUMMER SALADS

Sweet Steamed Scallop Salad in Tomato and Basil Vinaigrette

Corsican Eggplant Salad with Onions, Olives, and Capers

Red and White Radish Salad with Arugula and Yogurt Vinaigrette

Vine-Ripened Tomato Salad in Basil-Yogurt Dressing

Farmstand Zucchini, Cherry Tomato, and Tuna Salad
in a Basil Vinaigrette

Two-Tone Zucchini and Pepper Salad Parma

Zucchini Salad with Pink-Lemon Mayonnaise

Cool Rice Salad with Fragrant Indian Spices

SUMMER DESSERTS

Banana-Caramel Sorbet with Chocolate Sauce

Coeur à la Crème with Summer Berries

Blueberry-Ginger Custard Tart

Bing Cherry Tart with Hazelnut-Butter Custard

Kir-Poached Cherries with Praline Meringue

Poached Figs in Port and Red Wine Syrup

Mango and Lemon Bavarians

Compote of Peaches and Blueberries

Pecan-Roasted Peaches in Sauternes

Kirsch-Soaked Pineapple in Pineapple Zabaglione

Field Strawberries in Crème Rouge

Chocolate Mousse Ice Cream

Honey Custard Ice Cream

Creamy Beet Soup with a Concassé of Mushrooms,
Crème Fraîche, and Dill
∎
Grilled Lamb on a Bed of Sweet-and-Sour Red-Onion Jam
∎
Charcoal-Roasted Eggplant Fritters with Yogurt and Mint Sauce
∎
Purple Plums in White-Wine Sabayon

For many of us, summer entertaining means cooking on the grill. Personally, I never tire of it, and over the years I have learned to master the grill with increasing creativity, using a variety of woods and marinades and developing a better understanding of heat control, which is the key element to successful grilling.

When choosing a menu in which the main course will be grilled, use that source of heat for as many other elements of the meal as possible without making everything taste the same. In this menu, the eggplants are grilled directly over the burning coals and the tender pulp turned into light fritters that are enhanced by a slightly smoky taste. I usually grill one or two additional eggplants that can be made into fritters two or three days later. Spread with goat cheese and thyme, and topped with a light tomato sauce, the fritters are a delicious light main course.

For many grill buffs, butterflied leg of lamb is a summer favorite. I like to marinate the lamb in red wine to tenderize it and add a more interesting taste to a cut of meat which can be rather gamy in the summer. The lamb is complemented beautifully by a red-onion marmalade, in which the onions are melted down by slow, even cooking. I usually double or triple the recipe, and keep it covered with olive oil for up to two months in the refrigerator. The jam is an excellent accompaniment to grilled meats, seafood, and poultry, and it makes a terrific topping for pizzas.

The starter and dessert for this menu can be prepared in advance, which will leave you plenty of time to concentrate on the main course. The cool, creamy beet soup, enhanced by the tangy flavor of lemon, is a light, refreshing beginning. After the substantial and richly flavored lamb, the compote of plums adds a nice, and not-too-sweet, finishing touch.

∎∎∎

CREAMY BEET SOUP WITH A CONCASSÉ OF MUSHROOMS, CRÈME FRAÎCHE, AND DILL

serves 6

The first time I tasted this brilliant magenta soup was in a small Russian restaurant in Paris. It was served topped with a light mushroom dumpling, which provided an unusual combination of flavors. I've since transformed the garnish into this simple, delicate, but intensely flavored mushroom relish. The concassé can also be used with other cream soups and as a filling for tomatoes. It is an unusual and very interesting accompaniment to a platter of thinly sliced prosciutto.

CONCASSÉ
1½ tablespoons fresh lemon juice
5 tablespoons olive oil
¾ cup Crème Fraîche (page 491)
Coarse salt and freshly ground black
   pepper
12 ounces mushrooms, wiped, stemmed,
   and cut into ¼-inch dice
3 tablespoons finely minced fresh chives
2 tablespoons finely minced fresh dill

SOUP
Salt
6 medium fresh beets, trimmed of all but
   2 inches of greens, roots attached

7 cups Chicken Stock (page 470)
4 tablespoons unsalted butter
3 tablespoons all-purpose flour
Freshly ground white pepper
¾ cup heavy cream
Juice of 1–2 lemons
3 tablespoons finely minced fresh dill

GARNISH
Sprigs of fresh dill

1. Start by making the concassé: In a medium mixing bowl combine the lemon juice and olive oil and whisk until creamy. Add the crème fraîche and whisk until smooth. Season the vinaigrette with salt and pepper.

2. Fold the mushrooms, chives, and dill into the vinaigrette, gently but thoroughly, and let marinate at room temperature for 1 hour to develop flavor.

3. While the mushrooms are marinating, prepare the beet soup: In a medium casserole, bring lightly salted water to a boil. Add the beets, reduce the heat, and cook, partially covered, for 45 minutes, or until tender when pierced

with the tip of a sharp knife. Drain the beets and, when cool enough to handle, peel, trim, and cube. The skins should slip right off.

4. Place the beets in the workbowl of a food processor, add 1 cup of the stock, and puree until smooth. Set aside.

5. In a heavy 4-quart casserole, melt the butter over medium heat. Add the flour and cook, whisking constantly, for 1 to 2 minutes without browning. Add the remaining stock all at once and whisk until smooth. Bring to a boil, add the pureed beets, and season with salt and white pepper. Reduce the heat, partially cover, and simmer for 20 minutes.

6. Add the heavy cream and lemon juice and correct the seasoning. Simmer for 10 minutes longer without letting the soup come to a boil and whisk in the dill. Serve the soup at room temperature, ladled into individual soup bowls, and garnish with a dollop of the mushroom concassé and a sprig of dill.

GRILLED LAMB ON A BED OF SWEET-AND-SOUR RED-ONION JAM

serves 6

1 leg of lamb, 6½–7 pounds, boned and butterflied

MARINADE
Juice of 4 limes
8 tablespoons extra-virgin olive oil
4 tablespoons light soy sauce
2 large cloves garlic, peeled and crushed
2 large onions, peeled and sliced
2 tablespoons fresh oregano leaves
2 tablespoons fresh thyme leaves
2 tablespoons fresh rosemary leaves
1 teaspoon coriander seeds, toasted and crushed

JAM
⅓ cup extra-virgin olive oil
6–8 medium red onions, peeled and thinly sliced (about 6 cups)
2 tablespoons finely minced fresh thyme
Salt and freshly ground black pepper
2 tablespoons sherry vinegar
¼ cup dark-brown sugar

GARNISH
Sprigs of watercress

1.   Start by marinating the lamb: Combine all the marinade ingredients in a mixing bowl, mix well, and pour into a heavy zip-lock plastic bag. Add the lamb, close the bag, and place in a baking dish. Refrigerate overnight, turning the bag several times.

2.   To prepare the jam: In a 12-inch cast-iron skillet, heat the oil over high heat. Add the onions and cook for 2 to 3 minutes, or until they start to brown. Reduce the heat to low, add the thyme, and season with salt and pepper. Cover the skillet and cook the onions for 45 to 50 minutes, or until they are very soft.

3.   While the onions are braising, remove the lamb from the refrigerator and bring back to room temperature. Reserve.

4.   When the onions are very soft, uncover the skillet, raise the heat, and add the vinegar and sugar. Continue to cook the onions for 10 minutes longer, or until they are nicely glazed. Taste and correct the seasoning and keep warm.

5.   Prepare the charcoal grill.

6.   When the coals are white-hot, place the lamb on the hot grill directly above the coals. Grill the lamb for 20 to 25 minutes for medium-rare. The inner temperature should register 135° F on an instant meat thermometer inserted into the thickest part of the lamb. Turn the lamb once during grilling and baste often with the marinade.

7.   When the lamb is done, transfer to a carving board, let rest for 5 minutes, and cut crosswise into thin slices. Place a dollop of red-onion jam in the center of each individual serving plate and top with 3 or 4 slices of the grilled lamb. Garnish with sprigs of watercress and serve at once.

CHARCOAL-ROASTED EGGPLANT FRITTERS WITH YOGURT AND MINT SAUCE

serves 6

SAUCE
1 cup plain yogurt
1 (½-inch) piece fresh gingerroot, peeled and mashed through a garlic press
1 large clove garlic, peeled and mashed through a garlic press
½ teaspoon ground cumin
2 tablespoons finely minced fresh mint
Salt and freshly ground white pepper

FRITTERS
1 small eggplant, about 1–1¼ pounds
1 tablespoon grated raw onion
½ cup all-purpose flour
1 teaspoon baking powder
1 large egg, lightly beaten
Salt and freshly ground white pepper
Corn oil for frying

1. Start by making the sauce: Combine all the ingredients in a small bowl and mix well. Cover and chill until serving time.

2. Prepare the charcoal grill: When the coals are hot, place the whole eggplant directly on the coals and grill until evenly charred on all sides and soft in the center, about 10 to 15 minutes. Remove the eggplant from the grill and set aside until completely cool.

3. When cool enough to handle, peel the eggplant, removing all the charred skin. Finely chop the pulp by hand and transfer to a mixing bowl. Add the onion, flour, baking powder, and egg and season with salt and pepper. Mash the ingredients together with a fork until thoroughly combined. Set aside.

4. In a 10-inch skillet, heat ¼ inch of corn oil over medium heat. When hot, drop the fritter batter by tablespoonfuls into the hot oil without crowding; fry the fritters in 2 batches. Cook for 1 to 2 minutes on each side, or until nicely browned, adjusting the heat so that the fritters do not brown too quickly. Remove with a slotted spoon to paper towels to drain. Serve hot, accompanied by the yogurt and mint sauce.

Note: You can char the eggplant on a gas burner or electric stovetop instead of a grill. Place the eggplant directly on a gas burner over a medium flame or directly on an electric burner over medium heat and turn to char evenly, cooking for about 15 to 20 minutes.

PURPLE PLUMS IN WHITE-WINE SABAYON

serves 6

PLUMS
1½ cups granulated sugar
2 cups Sauternes
1 (3-inch) piece of cinnamon stick
3 whole cloves
30 Italian prune plums or small purple plums, washed but not peeled or stoned
2 slices lemon

SABAYON
⅓ cup granulated sugar
5 large egg yolks

1 cup reserved plum poaching liquid
1 cup heavy cream, whipped

GARNISH
Tiny fresh mint leaves
Crisp Brown Sugar and Almond Cookies (page 450)

1. Start by poaching the plums: In a large saucepan combine the sugar, wine, cinnamon stick, and cloves. Bring to a boil and cook over high heat until the sugar is dissolved.

2. Add the plums and lemon slices. Reduce the heat and simmer, partially covered, about 5 to 8 minutes or until the plums are just tender; do not over-cook.

3. Remove the lemon slices and let the plums cool completely in their poaching liquid. Set 1 cup of the liquid aside.

4. To make the sabayon: In a medium mixing bowl combine the sugar and egg yolks and whisk until fluffy and pale yellow. Add the reserved plum poaching liquid and whisk until well blended.

5. Transfer the mixture to the top of a double boiler. Place over simmering water and whisk constantly until thick and creamy. The sabayon should heavily coat a spoon.

6. Immediately remove the pan from the heat and transfer the sabayon to a large mixing bowl. Place in the refrigerator until completely cool, whisking often. Fold in the whipped cream gently but thoroughly, and chill until serving.

7. To serve, drain the poached plums and place 5 of them in each of 6 individual dessert bowls. Spoon some of the sabayon over each portion and garnish with leaves of mint and brown sugar cookies.

Spicy Summer Black-Bean Salad
■
Sauté of Swordfish with Pepper and Anchovy Marmalade
■
Baked Peaches in Praline Caramel

Summer is the only season that evokes the same responses to food, regardless of region. We love cooking and eating outdoors, and we don't hesitate to combine freely a variety of flavors into one menu, using a number of different cuisines. Here's a truly American menu, inspired by the best of several international favorites. The beans are somewhat Mexican in derivation, the swordfish with a pepper and anchovy marmalade could be found in any part of Mediterranean Europe, and the baked peaches are a quintessentially American fruit dessert.

It's best to serve this informal meal buffet-style, with two or three additional salads which you can improvise spontaneously by choosing what looks best and freshest at your local farmers market or vegetable stand. For a color, flavor, and texture contrast with the black-bean salad, I might make a cucumber salad in a yogurt and garlic dressing and a zucchini and tomato salad in a basil vinaigrette. As a nice change from the swordfish, why not experiment with other fish, such as tuna, halibut, mako shark, or lake trout. Of course, a platter of juicy, freshly picked corn would round off this meal beautifully.

As for dessert, an all-American peach or nectarine cobbler made with real tree-ripened fruit would fit this menu perfectly. Add to this a homemade ginger or vanilla ice cream and you have the makings of a great summer meal.

SPICY SUMMER BLACK-BEAN SALAD

serves 6

In the world of legumes, black beans are by far the most striking in their color and, to me, the most flavorful. I used to know them only as a main ingredient of Cuban black-bean soup, but now their distinctive taste has earned them a well-deserved popularity, and has inspired restaurant chefs and home cooks to use them in a variety of hot and cold salads.

2 teaspoons Dijon mustard
4 tablespoons fresh lemon juice
3 tablespoons light soy sauce
1 large clove garlic, peeled and mashed
2/3 cup extra-virgin olive oil
1 tablespoon finely minced jalapeño pepper
1/2 cup finely diced red onion

5 cups cooked black turtle beans, see Cooked White Beans (page 484)
Salt and freshly ground black pepper

GARNISH
Tiny fresh cilantro leaves
Ripe cherry tomatoes, cut in half

1.   In a large serving bowl combine the mustard, lemon juice, soy sauce, and garlic and whisk until well blended. Add the olive oil in a slow steady stream, whisking constantly until the vinaigrette is creamy. Add the jalapeño pepper, red onion, and cooked black beans. Fold gently and season with salt and pepper. Cover and chill for 2 hours.

2.   Thirty minutes before serving, bring the salad back to room temperature. Taste and correct the seasoning. Garnish with leaves of cilantro and cherry tomatoes.

SAUTÉ OF SWORDFISH WITH PEPPER AND ANCHOVY MARMALADE

serves 6

1 can flat anchovies, drained
1/2 cup whole milk
3 large red bell peppers, roasted, cored, seeded, and peeled (page 483)
3 large yellow bell peppers, roasted, cored, seeded, and peeled (page 483)
6 tablespoons extra-virgin olive oil
2 cloves garlic, peeled and mashed
2 tablespoons tiny capers, drained
1/2 teaspoon red wine vinegar

Coarse salt and freshly ground black pepper
2 1/2 pounds fresh swordfish, cut into 6 steaks, each about 1/2 inch thick and each weighing about 6 ounces
All-purpose flour for dredging

GARNISH
Sprigs of fresh Italian parsley
Tiny black oil-cured olives

1.   Place the anchovies in a small bowl, cover with the milk, and let soak for 1 hour. Drain, finely mince the anchovies, and set aside.

2. Cut the red and yellow peppers into thin, ½-inch strips and reserve.

3. In a heavy 2-quart saucepan, heat 3 tablespoons of the olive oil over low heat. Add the garlic and minced anchovies and simmer the mixture until the anchovies have melted. Add the peppers and simmer for 4 to 5 minutes or until just tender.

4. Add the capers and vinegar and season with salt and pepper. Cover and keep warm. This can be done several hours before serving, or even the day before.

5. Season the swordfish steaks with coarse salt and pepper and dredge lightly in flour, shaking off the excess.

6. In a large nonstick skillet, heat 1½ tablespoons of oil over very high heat. The oil should just "film" the bottom of the skillet. When almost smoking, add 3 swordfish steaks, without crowding the skillet, and sauté for 2 minutes on each side, or until just done; do not overcook. Transfer the steaks to a serving platter and keep warm.

7. Add the remaining oil to the skillet and, when almost smoking, add the remaining swordfish steaks and sauté for 2 minutes on each side. Transfer to the platter with the first batch of swordfish.

8. Top each steak with a dollop of the marmalade and garnish each with a sprig of parsley and an olive. Serve immediately.

Note: Any leftover marmalade makes a delicious filling for an omelet. It is also a wonderful topping for small pizzas or tart shells.

Other fish steaks, in particular tuna, mako shark, and salmon, can be substituted for swordfish in this preparation.

BAKED PEACHES IN PRALINE CARAMEL

serves 6

½ cup granulated sugar
3 tablespoons water
½ cup all-purpose flour
½ cup Praline Powder (page 489)
5 tablespoons unsalted butter
6 large ripe freestone peaches cut in
    wedges
2 tablespoons cornstarch

GARNISH
Good-quality vanilla ice cream, Honey
    Custard Ice Cream (page 233), or
    sugared Crème Fraîche (page 491)

1.   Preheat the oven to 375° F. Choose a heavy baking dish that will hold the peaches in a tight-fitting layer and reserve.

2.   In a small saucepan combine the sugar and water. Bring to a boil over high heat and stir once to dissolve the sugar. Continue to boil until the mixture turns a deep hazelnut-brown caramel. Immediately pour the caramel into the baking dish, turning the dish so that the caramel completely coats the entire bottom. Set aside to cool.

3.   In the workbowl of a food processor combine the flour, praline powder, and butter and pulse for 15 to 20 seconds or process until the mixture becomes quite crumbly. Reserve.

4.   Bring plenty of water to a boil in a medium saucepan. Add the peaches and blanch for 30 seconds. Immediately remove from the water with a slotted spoon and cool slightly.

5.   Peel the peaches (the skins should slip off quite easily), cut in half, and discard the pits. Cut each peach half into 4 wedges and toss with the cornstarch.

6.   Place the peaches on top of the caramel, in a single tight-fitting layer, to cover the caramel completely. Sprinkle the praline mixture over the peaches and bake in the center of the preheated oven for 30 to 35 minutes, or until nicely browned.

7.   Remove from the oven and let cool until just warm. Serve with a small scoop of ice cream or a dollop of sugared crème fraîche.

Yellow Squash Soup with a Garnish of Summer Vegetables

Sauté of Chicken Bressane

Gratin of Raspberries and Blueberries in Ginger Cream

This light and easily accomplished menu harmoniously blends a diversity of tastes and textures and highlights an array of summer produce within a variety of cuisines.

Pungent Madras curry powder gives a vigorous boost to the mellow summer squash soup. The soup can be made well ahead of time, and it's equally good served hot or slightly chilled, depending upon the weather.

Follow the soup with the gutsy sauté of chicken perfumed with plenty of garlic and fresh rosemary, typical Mediterranean flavorings. Vary the dish by substituting fresh thyme or savory for the rosemary or by using all three herbs. The traditional accompaniment of baked tomatoes topped with homemade bread crumbs, herbs, and a generous drizzle of fruity olive oil is still my first choice. Crusty French bread is all you need to dunk into the delicious pan juices.

For a family dinner, you may want to keep this meal quite simple by ending it with some sliced ripe peaches accompanied by sugared crème fraîche or some honey-flavored mascarpone. For a more elaborate dessert, however, the gratin of summer berries nestled under a blanket of ginger-flavored pastry cream adds an elegant touch to this otherwise homey summer meal.

■ ■

YELLOW SQUASH SOUP WITH A GARNISH OF SUMMER VEGETABLES

serves 4–6

3 tablespoons unsalted butter
1 large Bermuda onion, peeled and finely minced
2 teaspoons curry powder, preferably Madras
1¾ pounds small yellow squash,

trimmed and cubed
4 cups Chicken Stock (page 470)
Salt and freshly ground white pepper
½ cup heavy cream
Pinch of cayenne pepper

Dollops of Crème Fraîche (page 491)
1 large tomato, seeded and diced

1 small zucchini, diced and steamed for
   1 minute
3 tablespoons finely minced fresh chives

1. In a 3-quart casserole melt the butter over low heat. Add the onion and cook until soft but not browned. Stir in the curry powder and cook for 1 minute, tossing with the onions.

2. Add the yellow squash and chicken stock and season with salt and white pepper. Bring to a boil, reduce the heat, and simmer, partially covered, until the squash is very soft, about 25 minutes.

3. Strain the soup into a bowl and puree the vegetables until smooth in the workbowl of a food processor or blender. Add the puree to the strained broth and whisk until smooth. Return the soup to the casserole, add the cream and cayenne, and simmer for an additional 10 minutes.

4. Taste and correct the seasoning. Serve the soup hot in individual bowls with a dollop of crème fraîche in the center and a sprinkling of tomato, zucchini, and fresh chives.

SAUTÉ OF CHICKEN BRESSANE

serves 6

A European chicken, even today, is a completely different animal from the domesticated, flavorless poultry produced in America. Plump, juicy, and not fatty, a European bird requires only the simplest preparation. In this homey dish the key ingredients—garlic, herbs, and a splash of white wine—will enhance the flavor of any chicken you use. The cooked, unpeeled garlic cloves easily slip out of their skins and are delicious spread on crusty bread. Serve with baked tomatoes, or add a different seasonal touch with Potato and Farmstand Vegetable "Paella" (page 208) or Lemon- and Parsley-Scented Braised Bulgur (page 210).

6–8 small whole chicken legs (½ pound
    each, thighs and legs attached)
4 tablespoons unsalted butter
2 teaspoons peanut oil
Coarse salt and freshly ground black
    pepper
16 garlic cloves, unpeeled
¼ cup brandy or cognac

1 large sprig of fresh thyme
1 large sprig of fresh rosemary
½ cup dry white wine
¾ cup Chicken Stock (page 470)
1 teaspoon potato flour mixed with a
    little chicken broth
4 tablespoons unsalted butter, for
    enrichment, optional

1. Dry the chicken legs thoroughly with paper towels. In a heavy 12-inch skillet, heat 2 tablespoons of the butter together with 1 teaspoon of the oil, over medium-high heat. When very hot, add half the chicken legs; do not crowd the pan. Sauté the legs for 5 to 8 minutes, or until nicely browned on all sides.

2. Remove the browned chicken to a side dish and discard the fat if it has burned. Replace with the remaining butter and oil and continue sautéing the remaining chicken until nicely browned. Transfer the chicken to the side dish containing the first batch, season with salt and pepper, and set aside.

3. Discard all but 2 tablespoons of fat from the skillet. Add the garlic cloves and cognac, ignite, and, when the flames subside, return the chicken legs to the skillet. Add the herbs and reduce the heat to low. Add half of the white wine, cover the skillet tightly, and simmer the chicken for 10 minutes.

4. Add the remaining wine and continue braising for another 10 to 15 minutes, or until the chicken is cooked and the juices run pale yellow. Transfer to a warm serving dish with the garlic cloves and reserve.

5. Thoroughly degrease the pan juices and return them to the skillet. Place the skillet over medium heat, add the stock, bring to a boil, and simmer until it is slightly reduced.

6. Whisk in the potato-starch mixture and correct the seasoning. Remove the skillet from the heat. Discard the herbs and whisk in the optional butter for enrichment, 1 tablespoon at a time. Spoon the sauce over the chicken legs and serve hot, accompanied by sautéed broccoli rape (page 429).

GRATIN OF RASPBERRIES AND
BLUEBERRIES IN GINGER CREAM

serves 4–6

CREAM
1 (1-inch) piece of fresh ginger, peeled
   and thinly sliced
1½ cups light cream or ¾ cup heavy
   cream and ¾ cup whole milk
5 large egg yolks
½ cup granulated sugar
1 teaspoon vanilla extract

BERRIES
1 pint fresh raspberries
1 pint fresh blueberries

TOPPING
2–3 tablespoons dark-brown sugar

1. Bring water to a boil in a small saucepan, add the ginger slices, and blanch for 1 minute. Drain and finely mince. Combine the ginger with the light cream in a medium saucepan and simmer over low heat for 15 to 20 minutes, or until fragrant. Reserve.

2. Combine the yolks and sugar in the top of a double boiler and whisk until fluffy and pale yellow. Add the hot ginger cream and whisk until well blended.

3. Place the mixture over simmering water and whisk constantly until the sauce heavily coats a spoon. Be careful not to curdle the yolks.

4. Immediately transfer the ginger cream to a mixing bowl, whisk in the vanilla, and set aside. At this point the cream can be made up to 3 days ahead of time and kept covered in the refrigerator.

5. Preheat the broiler.

6. Place about 9 raspberries in the center of 4 to 6 individual round au gratin dishes with a 6-inch diameter. Surround with some of the blueberries and spoon about 3 to 4 tablespoons of the ginger cream over the berries, covering them completely.

7. Sprinkle the custard with brown sugar and place the dishes under the broiler, 6 inches from the source of heat, until the brown sugar is caramelized and bubbly. Remove from the broiler and serve at once.

Note: This dessert can be made in a single round baking dish such as a quiche pan, rather than in individual dishes. You may use just raspberries or blueberries or substitute blackberries for the blueberries.

Risotto of Leek and Tomatoes Cipriani

Lilac-Smoked Cornish Hens with Grilled Eggplant Aioli

Tricolor of Melons and Kiwifruit in Lime Caramel

Risotto is a perfect dish to show off the great flavors of summer. Here this wonderful rice takes center stage with those perennial summer favorites: basil, tomatoes, and leeks. The risotto can also be served as a main course for lunch or for a casual summer supper preceded by chilled beet soup or seafood salad.

I find that the lightly smoked Cornish hens are an ideal choice. These can be roasted in advance and served at room temperature. Be sure to use the hot coals to roast the eggplants and possibly some green bell peppers as well. The grilled peppers, peeled and served with coarse salt and a sprinkling of virgin olive oil, make an interesting and quick side dish. A salad of cucumbers in a yogurt dressing with a touch of mint will complete this main course.

Summer desserts usually revolve around fresh fruit simply prepared, and melons are unquestionably one of the key warm-weather fruit. Unfortunately, the wonderful variety of ripe melons that you find in California, Arizona, and Texas are rarely available in the Northeast. Still, if you are on the lookout for good ripe melons and create your menus accordingly, you will come across good honeydews, cantaloupe, casabas, and Cranshaws. Whatever your selection, you will find that you can make a quick and dazzling fruit potpourri, to which kiwis will add color and interest. I usually make a large quantity of lime caramel at a time and infuse fresh peppermint from my garden in it for two or three days. Just before serving, you may want to add an additional julienne of mint leaves or tiny whole spearmint leaves to further enhance the flavor.

■ ■

RISOTTO OF LEEK AND TOMATOES CIPRIANI

serves 4–6

1 cup tightly packed fresh basil leaves, stemmed and washed

1 cup heavy cream

5 tablespoons unsalted butter

1 large leek, trimmed of all but 2 inches of greens, thinly sliced and well rinsed

2 large ripe tomatoes, peeled, seeded, and chopped

1 cup Italian rice, preferably Arborio, well rinsed

3–4 cups hot Chicken Stock (page 470)

Salt and freshly ground black pepper

⅓ cup freshly grated Parmesan cheese

Arborio Rice

When everything Italian became the rage in this country, everyone expected that Arborio rice—the key ingredient in the north Italian classic risotto—would soon become the hot new grain. So far this has not happened, although both home cooks and restaurant chefs have come to recognize the enormous versatility of this unique grain. Arborio rice is a roundish, short-grained kernel that is harvested primarily in the Po Valley in Italy. The grain has been grown in Italy since the Renaissance and became the basis for the most famous rice dish in Italy—Risotto alla Milanese—to which saffron adds its golden color. Many Italians claim that it is in fact this dish that has inspired Spanish cooks to create their famous paella—another rice dish that pairs saffron and Arborio.

Arborio rice usually comes packed in small canvas sacks. It is available in gourmet and specialty stores, although some supermarkets are starting to carry the rice in boxes (my favorite is Tesori). The best-quality Arborio is marked *superfino;* here the kernels are larger and rounder but lesser quality Arborio marked *fino* can also be used with success. Because of its high starch content, it can absorb up to 4 times its original weight in liquid. Arborio should not be limited to the making of risotto. It makes the best rice puddings, tastes great in rice salads, and works perfectly for rice stuffings with vegetables or fruit, such as baked apples and pears. I prefer Arborio to any other rice because the grain has great flavor and its unique texture adds a special character to any dish.

■ ■ ■

1. In the workbowl of a food processor combine the basil leaves with 3 tablespoons of the heavy cream and process until a smooth paste is formed. Add the remaining cream, pulse to combine, and set aside.

2. Melt 3 tablespoons of the butter in a heavy 3-quart saucepan over low heat. Add the leek and cook for 3 to 4 minutes, or until soft. Add the tomatoes and cook or until the tomato liquid has evaporated. Add the rice and stir until it is well blended with the leek-tomato mixture.

3. Add ½ cup of the stock and cook over medium-low heat, stirring constantly, until all the stock is absorbed by the rice. Regulate the heat so that the rice absorbs the liquid but does not dry out too quickly. Continue adding ½ cup of stock at a time as the liquid is absorbed.

4. After about 20 minutes, add the stock ¼ cup at a time. If the rice is still very chewy, partially cover the saucepan and cook the rice a little more slowly, letting it absorb the liquid and stirring it every 3 minutes. Keep an eye on the risotto at all times; the result must be a creamy rice with a kernel that is soft on the outside but still slightly chewy on the inside.

5. When the rice is done, season it carefully with salt and pepper. Add the

basil cream and remaining butter and stir well. Cook for another 2 to 3 minutes, or until the cream is absorbed by the rice.

6. Add the Parmesan and stir until well blended. Transfer to a hot serving dish and serve immediately.

LILAC-SMOKED CORNISH HENS WITH GRILLED EGGPLANT AIOLI

serves 4

4 Cornish hens, about 1–1¼ pounds each
Juice of 1 lemon
Juice of 1 orange
3 tablespoons finely minced fresh Italian parsley
2 tablespoons finely minced fresh thyme
3 large cloves garlic, peeled and finely minced
⅓ cup extra-virgin olive oil

1 teaspoon whole freeze-dried green peppercorns, crushed
Coarse salt and freshly ground black pepper
8 pieces of fresh, green lilac wood

GARNISH
Sprigs of fresh thyme

1 recipe Grilled Eggplant Aioli (recipe follows)

1. Dry the Cornish hens thoroughly with paper towels and set aside.

2. Combine the lemon and orange juices, parsley, thyme, garlic, oil, and green peppercorns in a mixing bowl and whisk until well blended. Place the hens in a large, heavy, zip-lock plastic bag, pour the marinade over them, and seal the bag. Place the bag in a shallow dish and refrigerate for 6 hours or overnight, turning the bag several times.

3. Prepare the charcoal grill for grilling over indirect heat (see Spice-Charred Grilled Rib Eye with Black Radish and Mustard Sauce, page 155).

4. Remove the hens from the marinade, truss, and season on all sides with coarse salt and black pepper. Reserve the marinade. Place the hens on the cooking grill over the drip pan, add 4 pieces of the lilac wood to the pile of burning coals, cover the kettle, and cook for 40 to 45 minutes without turning,

or until a meat thermometer inserted into the thigh area registers 180° F. The juices should run pale yellow. Baste often with the marinade and, after 20 minutes, add 8 briquettes and the remaining lilac wood to the pile of burning coals through the openings near the grill handle.

5. Transfer the hens to a cutting board, remove the trussing string, and cut each hen in half. Place the hens on a warm platter, garnish with sprigs of fresh thyme, and serve with the eggplant aioli on the side.

GRILLED EGGPLANT AIOLI

serves 4

1 large eggplant, about 1½ pounds
2 large egg yolks
2 large garlic cloves, peeled and
  mashed
2 teaspoons Dijon mustard

1 tablespoon sherry vinegar
½–⅔ cup extra-virgin olive oil
Salt and freshly ground black pepper
2 tablespoons finely minced chives or
  scallions

1. Prepare the charcoal grill.

2. Place the eggplant directly on the hot coals and grill until the skin is charred and the eggplant is tender, turning the eggplant to char evenly. Transfer to a cutting board and cool. When the eggplant is cool enough to handle, remove the stem and cut open lengthwise. With a spoon, scoop out the flesh, being careful not to include any of the charred skin, and place in the workbowl of a food processor or a blender. Puree the eggplant until smooth and reserve.

3. In a small, heavy bowl, combine the yolks, garlic, mustard, and vinegar. Whisk the mixture until smooth and start adding the oil drop by drop, whisking constantly as you would for mayonnaise. The mixture should thicken and have the consistency of a loose mayonnaise. Season with salt and pepper and add the eggplant puree. Blend thoroughly, taste, and adjust the seasoning. Add the minced chives and serve warm or at room temperature.

Note: If the eggplant has a lot of seeds, pass it through a fine sieve after it has been pureed. You can also make the sauce in a blender, but not the food processor. When using the blender use 2 whole eggs instead of just the yolks; you will probably need as much as ¾ cup of oil. Do not add the garlic until the sauce is done. For a lighter flavor use half peanut oil and half extra-virgin olive oil.

TRICOLOR OF MELONS AND KIWIFRUIT
IN LIME CARAMEL

serves 4–6

3 ripe kiwifruit, peeled
1 ripe cantaloupe, peeled and seeded
1 ripe honeydew melon, peeled and seeded
1 small ripe Cranshaw melon, peeled and seeded
1 ripe mango, peeled and cubed, optional
¾ cup granulated sugar

2 tablespoons plus ¼ cup hot water
Juice of 2 limes
¼ cup dark rum
1 cup tiny fresh peppermint leaves

GARNISH
Zest of 1 lime, cut into fine julienne
Sprigs of fresh mint

1. Cut the kiwi in half lengthwise and then crosswise into ½-inch slices. Reserve.

2. Cut the melons into rounds with a medium-sized melon-ball cutter. You will need about 6 cups of melon balls in all.

3. Place the melon balls in a shallow, round glass serving bowl together with the kiwis and optional mango and set aside.

4. In a small saucepan, combine the sugar and 2 tablespoons of hot water, bring to a boil over high heat, and stir once to dissolve the sugar. Continue to boil, without stirring, until the mixture turns a hazelnut brown. Remove the saucepan from the heat and immediately add the ¼ cup hot water, averting your face, as the mixture will bubble violently. Return to low heat and stir the caramel until smooth. Remove from the heat, add the lime juice, rum, and mint leaves, cover, and chill.

5. Strain the caramel syrup and spoon over the fruit. Garnish with the lime zest and sprigs of fresh mint and serve chilled in individual bowls.

Sicilian Braised Vegetables in Sweet Balsamic Dressing

Spice-Charred Grilled Rib Eye with Black Radish and Mustard Sauce

Red-Potato Salad in Sour Cream Vinaigrette

Drunken Blueberry Fool with Cassis

Beef seems to be regaining much of its popularity, especially in the summer when a grilled steak is still very much a favorite. Planning a meal around beef can be a challenge if you want to break out of the traditional mold into a more contemporary main course. Instead of standard side dishes, such as potatoes and a green salad, you can go with a variety of seasonal salads and vegetable dishes that give the whole menu a certain theme.

A well-seasoned new-potato salad is almost essential for this kind of outdoor summer menu. Again, it is at its best when prepared at least one day ahead of time. Any leftover salad can become part of an antipasto or served as an accompaniment to cold roast chicken.

If you prefer not to make the Sicilian braised vegetables, you might want to take advantage of the hot coals before the steaks are cooked and grill some colorful peppers, thickly sliced Bermuda onions, and sliced eggplant. Drizzle the grilled vegetables with fruity olive oil, coarse salt, and a splash of balsamic vinegar. To complete this colorful buffet, make a cool salad of cherry tomatoes, cut in half and tossed in a basil vinaigrette. Don't forget to serve crusty bread, a bowl of oil-cured olives, and some fresh, crisp summer radishes.

For those who want a light and simple dessert to offset the heat of summer, drunken blueberry fool will quickly become a favorite. It takes only minutes to prepare, and any leftovers can be stored in the freezer and served as a frozen mousse sprinkled with additional fresh blueberries.

SICILIAN BRAISED VEGETABLES IN SWEET BALSAMIC DRESSING

serves 6

8 tablespoons olive oil

2 large red onions, peeled and thinly sliced

2 large cloves garlic, peeled and finely minced

1 medium eggplant, about 1½ pounds, unpeeled and cut into ¾-inch cubes

Coarse salt

1 red bell pepper, cored, seeded, and cut into ¾-inch cubes

1 yellow bell pepper, cored, seeded, and cut into ¾-inch cubes

Freshly ground black pepper

3 small zucchini, trimmed and cut into ¾-inch cubes

2 large ripe tomatoes, peeled, seeded, and diced, optional

3 tablespoons balsamic vinegar

1½ tablespoons granulated sugar

½ cup golden raisins, plumped in hot water and drained

2–3 tablespoons pine nuts, oven toasted or sautéed in olive oil

2 tablespoons finely minced fresh Italian parsley

1. In a large, heavy cast-iron skillet heat 3 tablespoons of the oil over high heat. Add the onions and garlic and sauté for 2 to 3 minutes. Reduce heat and simmer, partially covered, for 30 to 35 minutes, stirring often, until very soft.

2. While the onions are braising, place the cubed eggplant in a bowl of ice water, add 1 teaspoon coarse salt, and set aside for 30 minutes. Drain the eggplant. Dry thoroughly with paper towels and set aside.

3. Add the red and yellow peppers to the onion mixture and continue to simmer until the peppers are tender. Transfer the mixture to a side bowl and reserve.

4. Add 3 tablespoons of oil to the skillet and, when hot, add the drained eggplant. Season with salt and pepper and sauté over medium heat until nicely browned. You may need to add a little more oil to the skillet. Transfer the eggplant to the bowl with the onion mixture and set aside.

5. Add 2 tablespoons of oil to the skillet and, when hot, add the zucchini. Sauté over medium heat for 3 to 5 minutes, or until nicely browned. Season with salt and pepper and return the reserved vegetables to the skillet together with the optional tomatoes. Simmer the mixture for another 5 minutes.

6. Combine the vinegar and sugar in a small bowl and blend thoroughly. Add to the vegetable mixture together with the plumped raisins and cook for 5 minutes longer. Taste and correct the seasoning.

7. Fold the pine nuts and parsley into the vegetables and let cool.

8. Serve at room temperature as an accompaniment to roasted or grilled meats or poultry.

SPICE-CHARRED GRILLED RIB EYE WITH BLACK RADISH AND MUSTARD SAUCE

serves 4–6

The rib eye is one of the most flavorful cuts of beef, suitable for both oven-roasting and grill. It holds up well to robust seasonings, as here, with this dry marinade and pungent horseradish and mustard sauce. The sauce, which is traditionally served with boiled beef in Austria, seems to work very well with grilled rib eye. The grated black radish—quite common in central European cooking—intensifies the flavor and texture of the sauce. If it's not available, white radishes are a good substitute, but you may need more horseradish.

RIB EYE
2 tablespoons crushed black peppercorns
1 tablespoon coarse salt
2 tablespoons imported medium paprika
1 tablespoon cayenne pepper
2 teaspoons celery seed
1 tablespoon dried thyme
1 tablespoon dried oregano
1 (4½-pound) rib-eye roast, bone in
2 tablespoons Dijon mustard
2 large cloves garlic, peeled and mashed

2 tablespoons grated raw onion

SAUCE
2 tablespoons sour cream
1 tablespoon Dijon mustard
3 tablespoons prepared horseradish
1 teaspoon granulated sugar
½ cup freshly grated black radish
Salt and freshly ground white pepper
½ cup heavy cream, whipped

GARNISH
Sprigs of fresh thyme and oregano

1. Start by marinating the rib eye: In a small bowl combine the peppercorns, salt, paprika, cayenne, celery seed, thyme, and oregano and mix well. Reserve.

2. Dry the rib eye thoroughly with paper towels and set aside.

3. In another small bowl combine the mustard, garlic, and onion. Mix well and spread over the entire surface of the rib eye. Sprinkle all of the dry spice mixture over the mustard to create a thick layer of spices. Let the rib eye marinate at room temperature for 1 hour.

4. While the roast is marinating prepare the sauce: In a medium bowl combine the sour cream, mustard, horseradish, sugar, and black radish. Mix well

and season with salt and white pepper. Fold in the whipped cream gently but thoroughly. Cover and refrigerate until serving.

5. Prepare the charcoal grill. Use a round medium-sized charcoal, not gas, grill, about 22½ inches in diameter, preferably Weber. Open all vents.

6. You will be grilling over indirect heat. Place 26 charcoal briquettes on the lower grill. Using an electric or cylindrical starter, ignite the charcoal. When the coals are white-hot, carefully push them to one side of the grill. Set a rectangular aluminum drip pan on the other side of the charcoal. Position the cooking grill in the kettle with one handle directly over the pile of coals. This will allow you to add charcoal through the opening by the grill handle during grilling.

7. Place the rib eye on the cooking grill over the drip pan. Cover the kettle and cook the rib eye, without turning, about 12 minutes per pound for medium-rare, or until a meat thermometer inserted in the center registers 125° F. Every 20 minutes, add 8 briquettes to the pile of burning coals through the opening by the grill handle.

8. When the rib eye is done, transfer to a carving board and let sit for 10 minutes before carving.

9. To carve, cut the meat off the bone in one piece, keeping the knife close to the bone while cutting. Slice the meat into 4 to 6 slices. Serve at once with the black radish and mustard sauce and garnish with sprigs of fresh thyme and oregano.

Note: If using a gas grill, it is best to heat only one side of the grill. Keep the rib eye on the cooking grill over the side not heated, cover, and open all vents. If this is not possible, grill on medium heat, covered, turning often, and use an instant thermometer to check roasting time frequently.

RED-POTATO SALAD IN SOUR CREAM VINAIGRETTE

serves 6

2½ pounds small new red potatoes
Salt
3 tablespoons red wine vinegar
6 tablespoons olive oil
½ cup sour cream

½ cup mayonnaise, preferably
  homemade (page 477)
1 tablespoon Dijon mustard
1 large clove garlic, peeled and mashed
Freshly ground black pepper

1 green bell pepper, cored, seeded, and finely diced
4 tablespoons finely minced scallions

3 small dill gherkins, finely diced
½ cup finely diced pimientos

1. Cook the potatoes in boiling salted water about 20 to 25 minutes or until just tender. Drain and, when cool enough to handle, peel and cut the potatoes in half if large. Place them in a salad bowl. Sprinkle with 2 tablespoons of the vinegar and 4 tablespoons of oil, toss lightly, and set aside.

2. Combine the remaining vinegar, oil, sour cream, and mayonnaise in a bowl, add the Dijon mustard and garlic, and whisk until smooth. Season the dressing with salt and pepper and set aside.

3. Add the green pepper, scallions, gherkins, pimientos, and dressing to the potatoes. Toss gently and refrigerate for 2 to 4 hours before serving. To serve, return to room temperature and correct the seasoning.

Variation: Add 2 tablespoons finely minced anchovies and 2 tablespoons finely minced fresh tarragon to the dressing. Add ¼ cup of tiny well-drained capers to the salad.

DRUNKEN BLUEBERRY FOOL WITH CASSIS

serves 6

1 pint fresh blueberries
¼ cup granulated sugar
2–3 tablespoons cassis
½ cup Crème Fraîche (page 491)
½ cup superfine sugar

½ cup heavy cream, whipped
2 teaspoons finely grated lemon rind

GARNISH
Blueberries
Sprigs of fresh mint

1. In a medium saucepan combine half of the blueberries and the granulated sugar. Place over medium heat and, when the blueberries begin to exude their juices, reduce the heat and simmer for 10 minutes, stirring often.

2. Transfer the blueberries to the workbowl of a food processor and process until smooth. Pass the puree through a fine sieve into a bowl and stir in the cassis. Set aside to cool completely.

3. In a large mixing bowl, combine the crème fraîche, superfine sugar, and blueberry puree and whisk until well blended. Fold in the whipped cream, grated lemon rind, and remaining uncooked blueberries. Cover and refrigerate 2 hours before serving.

4. To serve, spoon the fool into parfait glasses. Garnish each with a few fresh blueberries and sprig of mint. Serve chilled with ladyfingers.

CHICKEN SALAD IN SWEET MUSTARD VINAIGRETTE WITH BEETS AND CUCUMBERS

serves 4–6

1 large seedless or gourmet cucumber,
    peeled, cut in half lengthwise, and
    seeded
Juice of 1 large lemon
2 teaspoons granulated sugar
1 tablespoon Dijon mustard
3 tablespoons finely minced fresh dill
3 tablespoons finely minced scallions

5 tablespoons olive oil
Salt and freshly ground black pepper
1 whole skinless chicken breast, about
    10 ounces, cooked (page 158) and
    cut into ¾-inch cubes
4–5 medium beets (about 1 pound),
    cooked (page 98), peeled, and cut
    into ½-inch dice

1. Cut the cucumber crosswise into thin slices (half moons) and set aside.

2. Combine the lemon juice, sugar, mustard, dill, and scallions in a salad bowl. Add the olive oil in a slow stream and whisk the dressing until well blended and creamy. Season with salt and pepper. Add the chicken, beets, and cucumber and toss gently. Chill for 4 to 6 hours before serving. Serve slightly chilled, accompanied by thinly sliced, buttered pumpernickel bread.

Poached Chicken Breast

There are innumerable uses for poached chicken breasts, from salads to casseroles to filling for pasta and crepes. Since most American chicken tastes somewhat bland, I like to give the breasts a flavor boost by adding a few aromatic vegetables to the poaching medium.

1 quart Chicken Stock (page 470)
1 small carrot, trimmed and cut into ½-inch pieces
1 leek, trimmed of all but 3 inches of greens and washed well
1 stalk celery, trimmed and cut into 1-inch pieces
1 (10-ounce) chicken breast, skin removed

Pour the stock into a 3-quart saucepan, add the carrot, leek, and celery, and bring to a boil. Add the chicken, reduce the heat, and simmer, uncovered, for 5 to 7 minutes, or until the juices run pale yellow. Remove the chicken from the stock, cool, and use as directed. Strain the stock and reserve for another recipe.

■ ■ ■

HERBED ZUCCHINI AND POTATO FRITTATA

serves 4–5

Potato frittata is probably the most popular dish in Catalonia, where it is called *tortilla de patatas*. You can sample a wedge of it, served at room temperature, at almost any tapas bar. Countless cooks make it at home, where it is usually the basic starter to the midday meal, and one can see schoolchildren nibbling on enormous sandwiches filled with sliced tomatoes and this hearty omelet. Until recently I made potato frittata as it is most often served in Catalonia, but I find that more and more I tend to look for simple but interesting additions to the basic recipe. Braised red onions, roasted peppers, a julienne of zucchini, and a variety of herbs are all good additions.

Warm or at room temperature, frittata is excellent for a summer brunch, lunch, or picnic. It is so undemanding a dish that it can be reheated easily in a low oven. You can serve it as a light main course accompanied by homemade tomato sauce and grilled or sautéed sausages or some rashers of bacon. With a crusty French bread and one or two interesting cheeses, you'll have a simple and satisfying summer supper.

4 tablespoons unsalted butter
1 tablespoon extra-virgin olive oil
1 medium zucchini, trimmed and thinly sliced
Salt and freshly ground black pepper
2 medium new potatoes, peeled and thinly sliced
8 extra-large eggs

3 tablespoons heavy cream
4 tablespoons finely minced fresh Italian parsley
4 tablespoons finely minced scallions
1 tablespoon finely minced fresh thyme
½ cup crumbled feta cheese or mild goat cheese

1. Preheat the oven to 375° F.

2. In a 10-inch ovenproof nonstick skillet heat half the butter and oil over medium-high heat, add the zucchini, and sauté until browned on all sides. Season with salt and pepper. Remove from the skillet and reserve.

3. Add the remaining butter and oil to the skillet and add the potatoes in 2 slightly overlapping layers, seasoning each layer with salt and pepper. Cook for 2 minutes over high heat, or until the potatoes start to brown. Lower the heat and cover the skillet. Cook the potatoes until tender, about 10 minutes, turning once if necessary. Remove the skillet from the heat and arrange the zucchini over the potatoes. Set aside.

4. In a bowl combine the eggs, cream, parsley, scallions, thyme, and cheese.

Season the mixture with salt and pepper and whisk until well blended. Pour the egg mixture over the potatoes, lifting them gently so that the eggs are well incorporated into the vegetables.

5. Return the skillet to medium-low heat and cook for 2 minutes, or until the eggs start to set.

6. Place the skillet in the center of the preheated oven and bake the frittata until the custard is set and the top is lightly browned, about 15 minutes. Slide the frittata onto a serving platter and serve with sliced tomatoes or a tossed green salad.

MOZZARELLA TOASTS IN TOMATO VINAIGRETTE

serves 6

Sliced tomatoes interlaced with fresh mozzarella and showered with a julienne of basil has become a summer standby. Although it is a delicious dish when all the ingredients are fresh, it can also become rather boring. Still, I like to combine the two main ingredients in a variety of preparations, such as this one. Although the roasted pepper garnish adds a zesty touch, it is not essential, especially if you are short of time. A drizzle of extra-virgin olive oil, a grinding of black pepper, a thin slice of red onion, and some tiny whole basil leaves work very well indeed.

VINAIGRETTE

3 large ripe tomatoes, peeled, seeded, and chopped
1 tablespoon finely minced red onion
2 tablespoons sherry vinegar
1/2 cup extra-virgin olive oil
Salt and freshly ground black pepper

TOASTS

12 slices of French bread, cut 1/4 inch thick
2 cloves garlic, peeled and slightly mashed
Extra-virgin olive oil

PEPPERS

2 large red bell peppers, roasted, peeled (page 483), and cut into 1/4-inch julienne strips
2 large yellow bell peppers, roasted, peeled (page 483), and cut into 1/4-inch julienne strips
1 small red onion, peeled and thinly sliced
2 teaspoons sherry vinegar
4 tablespoons extra-virgin olive oil
12 thin slices fresh mozzarella

GARNISH

Tiny black oil-cured olives
Tiny fresh basil or Italian parsley leaves

1. Start by making the vinaigrette: In the container of a blender or food processor, combine the tomatoes, onion, vinegar, and olive oil and process until

smooth. Season with salt and pepper, pass through a fine sieve into a bowl, and set aside.

2. Preheat the broiler.

3. To make the toasts: Rub each side of the French bread with the mashed garlic and drizzle with a little oil. Place on a baking sheet and run under the broiler until each side is golden brown. Do not let the toasts burn. Reserve.

4. To make the peppers: In a small bowl combine the peppers, onion, vinegar, and oil. Toss gently and set aside.

5. To serve, place 2 to 3 tablespoons of tomato vinaigrette on each of 6 individual serving plates. Top with 2 toasts and place a slice of mozzarella on each. Top with some of the peppers and season with a large grinding of black pepper. Garnish with black olives and tiny leaves of basil or parsley.

Variation: Use smoked mozzarella or a creamy goat cheese mixed with some fresh thyme.

EGGPLANT SANDWICHES À L'ANTIBOISE
serves 6–8

Eggplant sandwiches are extremely popular both in the Middle East and in Mediterranean Europe, where they are often used for flavorful leftovers. Ratatouille, an onion jam, smoked or fresh mozzarella, or a julienne of roasted peppers are just some of the fillings that come to mind. I usually make whatever filling I've chosen a day or two in advance and sauté the eggplant several hours ahead of time. Once they're filled, the sandwiches can be set aside and reheated thirty minutes before serving. The entire dish can be done in advance, and it is good hot, at room temperature, or lightly chilled. Instead of the tomato sauce, you can simply drizzle the sandwiches with extra-virgin olive oil and a touch of sherry vinegar.

3 small eggplants, unpeeled (¾–
   1 pound each)
Coarse salt
All-purpose flour for dredging
8–10 tablespoons olive oil
1 cup tightly packed fresh basil leaves
2 tablespoons finely minced fresh thyme
2 tablespoons freshly grated Parmesan
   cheese
1 small clove garlic, peeled and mashed

6 ounces mild goat cheese, crumbled
Freshly ground black pepper
12–16 small black oil-cured olives,
   pitted and cut in half
2 cups Fresh Tomato Sauce (page 476)
   or Quick Tomato Sauce (page 474)

GARNISH
Small fresh basil leaves
Tiny black oil-cured olives

1. Cut the eggplant crosswise into even ½-inch slices, about 12 to 16 slices total (enough for 6 to 8 sandwiches). Place the eggplant slices in a bowl, add ice water to cover and 1 teaspoon of coarse salt, and let soak for 30 minutes.

2. Drain the eggplant and dry thoroughly with paper towels. Dredge lightly in flour, shaking off the excess.

3. Heat 3 tablespoons of oil in a large skillet over medium-high heat. Add the eggplant slices a few at a time and sauté until nicely browned on both sides. Transfer with a slotted spatula to paper towels and set aside. Continue to sauté the remaining eggplant slices, adding more oil to the skillet as needed.

4. Combine the basil leaves with the thyme, Parmesan, and garlic in the workbowl of a food processor or blender. Add 2 to 3 tablespoons of olive oil and process until smooth. Add the goat cheese and a large grinding of black pepper and process again until smooth. Taste and correct the seasoning. Transfer the goat cheese paste to a bowl, cover, and refrigerate for 30 minutes before using.

5. With a small knife, spread a layer of the goat cheese paste on each of 6 to 8 eggplant slices. Place 4 pitted olive halves on each, then top with the remaining eggplant slices to complete the sandwiches.

6. Place the sandwiches in a rectangular well-oiled au gratin dish. Spoon a little tomato sauce over each, cover the dish tightly with foil, and set aside. This can be done several hours ahead of time or the day before.

7. Preheat the oven to 350° F.

8. Thirty minutes before serving, place the baking dish in the center of the preheated oven and bake until the sandwiches are just heated through.

9. Serve as an appetizer on individual serving plates and garnish with leaves of fresh basil and black olives. Serve warm, accompanied by crusty French bread.

NIÇOISE FUSILLI SALAD

serves 6

In 1979 I included a pasta salad in *Market to Kitchen,* and at the time it seemed new and unusual. Since then, pasta salads have become as popular as cole slaw, and unfortunately you find them made with a combination of the most unlikely ingredients. Yet, a good pasta salad can still be memorable, and I often find myself craving a well-seasoned version of this summer favorite.

2 tablespoons olive oil
½ pound imported fusilli, preferably De Cecco
8 tablespoons Basil Vinaigrette (page 215)
Salt and freshly ground black pepper
1 small red onion, peeled and finely minced
1 small zucchini, trimmed and thinly sliced
1 red bell pepper, cored, seeded, and thinly sliced

1 yellow bell pepper, cored, seeded, and thinly sliced
¼ pound green beans, cooked and cut into 1-inch pieces
1 (7½-ounce) can tuna, packed in oil, drained and flaked
18 Niçoise olives or small black oil-cured olives
12 ripe cherry tomatoes, cut in half

1. In a large casserole, bring plenty of salted water to a boil with the olive oil. Add the pasta and cook for 8 to 10 minutes, or until al dente. Immediately add 2 cups cold water to stop further cooking. Drain well and transfer to a large serving bowl.

2. Toss the warm pasta with the basil vinaigrette, season with salt and pepper, and set aside to cool.

3. Add the red onion, zucchini, red and yellow peppers, green beans, tuna, and olives and toss the salad gently. Cover and refrigerate overnight.

4. Remove the salad from the refrigerator an hour before serving. Add the tomatoes, toss the salad, and taste and correct the seasoning. Serve at room temperature.

Variation: You can use other shaped pasta such as penne or ziti. All summer squash, both green and yellow, work well in the salad.

MULTICOLORED PEPPER, ONION, AND
TWO-CHEESE PIZZA

serves 2

For a casual lunch or supper, a summer pizza—hot from the oven and bursting with seasonal color—is a favorite appetizer. Cooked toppings can be made a day or two ahead, leaving you time to choose whatever combination of cheeses and herbs suits your mood. My own preference is for cheeses with some character—Gorgonzola, smoked mozzarella, and well-seasoned goat cheeses—combined with fresh herbs such as thyme, oregano, and marjoram.

With some garlicky salami, black oil-cured olives, and a platter of roasted peppers any one of the three pizzas that follow will make a casual, satisfying main course.

3 ounces Gorgonzola cheese
4 ounces Fontina cheese
4 tablespoons plus 1 teaspoon extra-virgin olive oil
1 large green bell pepper, cored, seeded, and thinly sliced
1 large red bell pepper, cored, seeded, and thinly sliced
1 large yellow bell pepper, cored, seeded, and thinly sliced
1 large red onion, peeled, quartered, and thinly sliced

2 large cloves garlic, peeled and thinly sliced
Salt and freshly ground black pepper
½ recipe Pizza Dough (page 487)
All-purpose flour for dusting surface
2 teaspoons fresh thyme leaves
2 teaspoons fresh oregano leaves
12 small black oil-cured olives, pitted and cut in half

1. Place the Gorgonzola and Fontina in the freezer for about 15 minutes, or until quite firm.

2. While the cheeses are in the freezer prepare the pizza topping: In a large, heavy skillet heat 3 tablespoons of the olive oil over medium heat. Add the peppers, onion, and garlic and cook, stirring often, for 15 minutes, or until crisp-tender and lightly browned. Season with salt and pepper and set aside.

3. When both cheeses are firm, grate the Gorgonzola and Fontina. Place in the refrigerator and reserve.

4. Preheat the oven to 425° F. Brush a 12-inch black pizza pan with 1 teaspoon of the olive oil and set aside.

5. Roll out the pizza dough on a lightly floured surface into a 9-inch circle.

Transfer the dough to the prepared pizza pan and, using the tips of your fingers, stretch the dough gently from the center outward to the edge of the pan. If the dough becomes too elastic, let rest for 5 minutes and begin again.

6. Spread the pepper and onion mixture evenly over the dough, leaving a ½-inch border around the edge. Sprinkle the top with the shredded Gorgonzola and then with the Fontina. Sprinkle with the thyme and oregano and drizzle with the remaining olive oil.

7. Place in the center of the preheated oven and bake for 18 to 20 minutes, or until the top of the pizza is bubbly and the crust is crisp and nicely browned.

8. Remove from the oven, sprinkle with the black olives, and cut into wedges. Serve at once.

Variation: Substitute a mild goat cheese for the Gorgonzola, or add grated smoked mozzarella to the other cheeses.

STIR-FRIED SWISS CHARD, BACON, AND PINE-NUT PIZZA

serves 2

1½ pounds Swiss chard
3 tablespoons plus 1 teaspoon olive oil
4 medium cloves garlic, peeled and
   thinly sliced
Salt and freshly ground black pepper
½ recipe Pizza Dough (page 487)
2 slices meaty bacon, about 2 ounces,
   cut crosswise into ½-inch pieces

1 teaspoon *herbes de Provence*
1 cup coarsely grated smoked
   mozzarella, optional
2 tablespoons pine nuts, sautéed in 1
   tablespoon olive oil
3 tablespoons freshly grated Parmesan
   cheese

1. Start by preparing the Swiss chard: Remove the leaves from the Swiss chard, rinse thoroughly, and tear into 1½-inch pieces. Set aside. You should have about 1 pound of leaves; reserve the stalks for another preparation.

2. Bring water to a boil in a vegetable steamer. Add the Swiss chard leaves, cover, and steam until just wilted, about 3 to 4 minutes. Remove from the steamer and, when cool enough to handle, gently squeeze to remove the excess moisture and set aside.

3. In a large skillet heat 3 tablespoons of the oil over medium heat. Add the garlic and Swiss chard, and sauté for 2 to 3 minutes or until the remaining moisture has evaporated. Season with salt and pepper and reserve.

4. Preheat the oven to 425° F. Brush a 12-inch black pizza pan lightly with the remaining oil and set aside.

5. Roll out the pizza dough on a lightly floured surface into a 9-inch circle. Transfer the dough to the prepared pizza pan and, with your fingertips, stretch the dough gently from the center outward to the edge of the pan. If the dough becomes too elastic, let rest for 5 minutes and begin again.

6. Spread the Swiss chard mixture evenly over the surface of the dough, leaving a ½-inch border around the edge. Top with the bacon and sprinkle with the *herbes de Provence*, a good grinding of black pepper, and the optional smoked mozzarella.

7. Place in the center of the preheated oven and bake for 15 to 20 minutes, or until the top of the pizza is bubbly and the crust is crisp and nicely browned. Remove from the oven, sprinkle with the pine nuts and Parmesan, cut into wedges, and serve at once.

Variation: Instead of the Parmesan, use ½ cup crumbled goat cheese and sprinkle it over the pizza before baking. In this case, serve on the side a concassé of raw tomatoes: 2 large ripe tomatoes unpeeled, seeded, and diced, mixed with 2 tablespoons finely minced shallots, 2 teaspoons sherry vinegar, and 2 to 3 tablespoons extra-virgin olive oil. Season the concassé with salt and a large grinding of black pepper.

PESTO, TOMATO, AND
SUMMER-SQUASH PIZZA

serves 2

1 teaspoon plus 1 tablespoon extra-virgin olive oil

Coarse yellow cornmeal for sprinkling on pizza pan

½ recipe Pizza Dough (page 487)

3 tablespoons Basil Paste (page 170)

1 medium clove garlic, mashed

3–4 ripe plum tomatoes, cut crosswise into ¼-inch slices

1 small yellow squash, cut crosswise into ⅛-inch slices

1 small zucchini, trimmed and cut crosswise into ⅛-inch slices

1 teaspoon fresh thyme leaves

1 teaspoon fresh oregano leaves

Freshly ground black pepper

Coarse salt

GARNISH

12 small black oil-cured olives

3 tablespoons freshly grated Parmesan cheese

1. Preheat the oven to 425° F. Brush a 12-inch black pizza pan lightly with 1 teaspoon of the oil, then sprinkle with cornmeal and set aside.

2. Roll out the pizza dough on a lightly floured surface into a 9-inch circle. Transfer the dough to the prepared pizza pan and, with your fingertips, stretch the dough gently from the center outward to the edge of the pan. If the dough becomes too elastic, let rest for 5 minutes and begin again.

3. Spread the basil paste and mashed garlic over the surface of the dough, leaving a ½-inch border around the edge. Top with alternating slices of tomato, yellow squash, and zucchini in a decorative pattern to cover the basil paste completely. Sprinkle with thyme, oregano, a good grinding of black pepper, coarse salt, and the remaining olive oil.

4. Place the pizza in the center of the preheated oven and bake for 15 to 20 minutes, or until the bottom and edges of the dough are nicely browned and the squash is tender.

5. Remove from the oven, sprinkle with the black olives and Parmesan, cut into wedges, and serve at once.

SPICY CORN AND ZUCCHINI SOUP

serves 6

This is one of my favorite end-of-season soups. Hearty and delicious, it can easily be served as a main course on those very first cool days in early fall. Of course, in warmer regions such as California and Florida this is a perfect soup for both winter and spring.

4 ounces lean slab bacon, cut into ¼-inch dice
2 tablespoons extra-virgin olive oil
1 large Spanish onion, peeled, quartered, and thinly sliced
2 teaspoons finely minced jalapeño pepper
2 tablespoons all-purpose flour
4 large ripe tomatoes, peeled, seeded, and chopped
4 cups Chicken Stock (page 470) or bouillon

Salt and freshly ground black pepper
1 tablespoon minced fresh thyme
Pinch of cayenne
1½ cups cooked fresh corn (about 2 large ears)
1 large zucchini, trimmed and finely diced
½ cup heavy cream

GARNISH
Finely minced fresh parsley or basil

1. In a medium saucepan, bring water to a boil, add the bacon, and blanch for 1 to 2 minutes. Drain and dry thoroughly on paper towels. Reserve.

2. In a large casserole heat the oil over medium heat, add the bacon, and sauté until almost crisp. Remove with a slotted spoon to a side dish and reserve. Discard all but 2 tablespoons of fat from the casserole. Add the onion and jalapeño pepper, and cook until soft but not browned, about 3 to 4 minutes.

3. Sprinkle the flour over the onion mixture and stir with a wooden spoon until well incorporated. Add the tomatoes and stock and bring to a boil. Reduce the heat, season with salt, pepper, thyme, and a pinch of cayenne, and simmer, partially covered, for 25 to 30 minutes.

4. Add the corn and zucchini and continue to simmer the soup until the zucchini is tender, about 2 or 3 minutes longer.

5. Return the bacon to the casserole together with the cream and heat through. Taste the soup and correct the seasoning. Serve the soup hot or at almost room temperature garnished with minced parsley or sprigs of fresh basil and crusty French bread.

Note: This soup is equally delicious at room temperature or even lightly chilled. If you plan to serve the soup this way, omit the flour, particularly if you are going to use homemade stock, which is quite gelatinous. You may also vary the soup by adding a creamy basil paste to it. Puree 1 cup of basil leaves with the cream and add to the soup. One-half cup each of diced red and yellow pepper will give the soup additional color and texture.

GALICIAN SUMMER VEGETABLE SOUP

serves 5–6

Every Mediterranean country has its own version of minestrone, each set apart from the others by regional seasonings and herbs and, sometimes, the addition of local greens. The French pistou is rich with the fragrant flavor of basil, an herb that is practically unknown in Catalan cooking. Other countries, especially those of North Africa, use spices such as cinnamon and cumin in their vegetable soups. I grew up on this simple version which we always made when plum tomatoes were ripe and bursting with the taste of summer.

3 tablespoons olive oil
2 cloves garlic, peeled and finely minced
1 small dry red chili pepper, crumbled
2 leeks, trimmed of all but 2 inches of greens, diced and well washed
2 medium red potatoes, peeled and diced
2 medium carrots, peeled and diced
Salt and freshly ground black pepper
6–8 ripe Italian plum tomatoes, peeled, seeded, and coarsely chopped

4–5 cups Chicken Stock (page 476) or bouillon
1 cup diced yellow squash
1 cup diced zucchini
1 large red bell pepper, cored, seeded, and diced

GARNISH
A bowl of yogurt mixed with a little garlic

1. In a large, heavy casserole heat the oil over medium heat, add the garlic and chili pepper, and cook for 30 seconds without letting the garlic brown. Add the leeks, potatoes, and carrots and season the mixture with salt and pepper.

2. Add the tomatoes and stock. Bring to a boil, reduce the heat, and simmer the soup for 30 minutes.

3. Add the squash, zucchini, and red pepper and continue to cook until all the vegetables are tender but not falling apart, about 5 to 10 minutes longer. Taste and correct the seasoning. Serve hot or slightly chilled accompanied by garlic yogurt.

Note: I usually add whatever other seasonal vegetables I have in the garden at the moment. A few diced yellow and green beans or a handful of corn kernels will give additional texture and flavor to the soup.

PISTOU OF RED BEANS, TOMATO, AND SUMMER SQUASH

serves 6

I had not realized that red kidney beans are used extensively in Italian cooking until I was served this substantial peasant soup in the enchanting town of Todi. What a delicious surprise! The soup came to the table lukewarm with a bottle of virgin olive oil and a bowl of fragrant green pesto on the side. I have since varied the soup using either white or red kidney beans and adding various seasonal vegetables such as diced green peppers and substituting cooked fresh corn for the pasta. This version is as close to the original as I can remember and has become one of my favorite summer supper dishes.

SOUP
3 tablespoons olive oil
1 large onion, peeled and finely minced
2 large cloves garlic, peeled and finely minced
2 tablespoons fresh oregano leaves or 1 teaspoon dried
2 tablespoons fresh thyme leaves or 1 teaspoon dried
6 large ripe tomatoes, peeled, seeded, and chopped
3 cups bean poaching liquid or Chicken Stock (page 470)
Salt and freshly ground black pepper
3 cups cooked red kidney beans (page 484)
1 small yellow squash, trimmed and finely diced

1 small zucchini, trimmed and finely diced
¼ cup tiny pasta or orzo

BASIL PASTE
¼ cup fresh parsley leaves
2 cups fresh basil leaves
1 large clove garlic, peeled and mashed
¼ cup extra-virgin olive oil
3 tablespoons freshly grated Parmesan cheese

GARNISH
Tiny fresh parsley leaves
Bowl of freshly grated Parmesan cheese

1. Start by making the soup: In a large casserole heat the oil over low heat. Add the onion, garlic, oregano, and thyme and cook for 10 minutes, or until the onion is soft but not browned.

2. Add the tomatoes and bean liquid or stock, and season with salt and pepper. Bring to a boil, reduce the heat, and simmer, partially covered, for 20 minutes.

3. Add the cooked beans, yellow squash, zucchini, and pasta to the soup and cook 15 minutes longer.

4. While the soup is cooking prepare the basil paste: Combine the parsley, basil, and garlic in the workbowl of a food processor and process until finely minced. Add the oil and process until a smooth paste is formed. Add the Parmesan and blend well. Transfer to a bowl, cover, and reserve.

5. Just before serving add the basil paste to the soup and mix well. Correct the seasoning and just heat through. Serve at once, garnished with tiny leaves of parsley and a sprinkling of fresh Parmesan.

CREAM OF TOMATO AND TANGY LEMON SOUP

serves 4–5

Despite years of lamenting the lack of flavor in commercially available tomatoes, not much seems to change. Those of us who can grow them do; others buy them in season at farmstands and farmers markets. But the season is all too short no matter where we live, and the quest for the truly vine-ripened tomato continues. It is because of the brevity of the tomato season that the simplest of preparations, such as this robustly flavored tomato soup, is always appreciated. I have found that the addition of lemon juice and zest heightens the taste of even less-than-perfect tomatoes.

4 tablespoons unsalted butter
2 medium onions, peeled and finely minced
½ teaspoon ground coriander
2 tablespoons all-purpose flour
8 large ripe tomatoes, peeled, seeded, and chopped
4 cups Chicken Stock (page 470) or bouillon

Salt and freshly ground black pepper
¾ cup sour cream
2 tablespoons fresh lemon juice
1 teaspoon grated lemon rind

GARNISH
Tiny fresh cilantro or Italian parsley leaves

1. In a heavy 3-quart casserole, melt the butter over low heat. Add the onions and cook for 6 to 8 minutes, or until soft but not browned.

2. Add the coriander and flour and blend well into the onions with a wooden spoon. Add the tomatoes and stock and season with salt and pepper. Bring the soup to a boil, reduce the heat, and simmer for 30 minutes.

3. Cool the soup slightly and puree in batches in a food processor or blender until the soup is very smooth.

4. Return the soup to the casserole, whisk in the sour cream and lemon juice, and correct the seasoning. Keep warm.

5. Add the lemon rind just before serving. Ladle the soup into individual soup bowls and garnish with small leaves of cilantro or Italian parsley.

Variation: For a Middle Eastern flavor, add ¼ teaspoon each of ground cumin, cinnamon, and cardamom to the soup base along with the ground coriander. I find this variation particularly refreshing toward the end of summer, when tomatoes are plentiful.

CREAMY GAZPACHO WITH
LIME-MARINATED BAY SCALLOPS

serves 5–6

For someone like me, who grew up in Spain, gazpacho is the ever-present soup, served almost daily during the long Mediterranean summer season. No matter how ingenious were the variations offered by home cooks and restaurant chefs, by the end of summer I would tire of it. But that was until I came to America and sampled the imitations that were labeled as gazpacho. That was when I began to miss the real thing.

Recently, while spending the summer in San Miguel de Allende in Mexico, I was served a gazpacho at the home of a friend whose cook combined this traditional Spanish soup with a quintessentially Mexican scallop seviche. I found the result both whimsical and delicious and have been serving it ever since.

SCALLOPS

1 pound small bay scallops
Juice of 2 limes
1/3 cup extra-virgin olive oil
3 tablespoons finely minced jalapeño
   pepper
2 tablespoons finely minced fresh
   cilantro leaves
2 tablespoons finely minced fresh basil
   leaves
Pinch of coarse salt and freshly ground
   black pepper

SOUP

8 large ripe tomatoes, peeled, seeded,
   and chopped

2 pickling cucumbers, peeled and
   coarsely chopped
1 small red bell pepper, cored, seeded,
   and cubed
1–2 cups tomato juice
1/2 cup diced red onion
Coarse salt and freshly ground black
   pepper
Dash of cayenne
1 large clove garlic, peeled and mashed
2 1/2 tablespoons sherry vinegar
1/3 cup extra-virgin olive oil

GARNISH

Tiny fresh basil leaves

1.   Start by marinating the scallops: In a mixing bowl combine the scallops with the lime juice, olive oil, jalapeño pepper, cilantro, and basil. Add a pinch of salt and a grinding of pepper, cover with foil, and let the scallops marinate for 3 to 4 hours in the refrigerator.

2.   While the scallops are marinating, prepare the soup: In the container of a blender or food processor combine half the tomatoes, cucumbers, and red bell pepper, adding a little tomato juice if necessary, and process until smooth. Transfer to a bowl and add the remaining tomatoes, cucumbers, and red pepper to the processor. Again process until smooth.

3.   Add the onion and pulse quickly. Add to the bowl containing the first batch and season the gazpacho with salt, pepper, cayenne, and the crushed garlic. Add the vinegar and oil and blend thoroughly. Cover the soup and chill until 20 minutes before serving.

4.   Just before serving drain the scallops and place a few in each individual serving bowl, divide the soup among the bowls, and garnish each with a tiny basil leaf. Serve chilled but not cold.

Note: Traditionally, a gazpacho is made by soaking some day-old peasant bread in the sherry vinegar and olive oil, which thickens the soup and gives it its characteristic consistency. For this version, you may finely slice day-old French bread and rub the slices with the cut side of a garlic clove, brush with fruity olive oil, and broil on both sides. Serve these croutons on the side or place 1 in each bowl and then top with the scallops. You may also omit the scallops and serve the soup only with the croutons.

AVOCADO SOUP À LA MEXICAINE WITH TRADITIONAL SALSA TOPPING

serves 4–6

Here is a soup that I fell in love with while visiting American friends in Cuernavaca. Their Mexican cook had learned a variety of French dishes while working for a French family and, with imagination and a great deal of flair, had created some of her own interpretations combining French techniques with Mexican ingredients. I was especially taken with this soup, which follows the classic technique of a traditional velouté. The richness of the mild but distinctive avocado offset by the assertive Mexican salsa was a lovely idea.

Unfortunately, the taste of the true Mexican lime cannot be duplicated with our limes. Still, the piquant marriage of flavors and textures has made this soup one of my preferred summer starters.

You may vary the recipe by adding to each serving three or four raw or steamed marinated bay scallops or some diced cooked marinated shrimp. Since cilantro seems to be an herb that is either loved or hated, you can substitute a fine julienne of basil in the salsa for a less authentic but just as interesting variation.

SALSA

2 large ripe tomatoes, unpeeled, seeded, and diced
2 tablespoons finely minced red onion
2 medium cloves garlic, peeled and finely minced
2 large Anaheim peppers, roasted, peeled (page 483), and finely diced
2 tablespoons finely minced fresh cilantro
Juice of 1 lime
1 small avocado, peeled, pitted, and diced
4–5 tablespoons olive oil
Salt and freshly ground black pepper

SOUP

4 tablespoons unsalted butter
1 large onion, peeled and finely minced
2 large cloves garlic, peeled and finely minced
1–2 teaspoons finely minced jalapeño pepper
2 tablespoons all-purpose flour
5 cups Chicken Stock (page 470)
2 medium avocados, peeled, pitted, and diced
Juice of ½ lemon
½ cup sour cream
Salt and freshly ground white pepper

GARNISH

Tiny fresh cilantro leaves

1. Start by making the salsa: In a medium bowl combine all of the ingredients. Season with salt and pepper and set aside.

2. To make the soup: Melt the butter over medium heat in a heavy 4-quart casserole. Add the onion, garlic, and jalapeño pepper and cook until soft but not brown. Add the flour and cook for 1 minute, stirring constantly. Add the chicken stock all at once and whisk until it comes to a boil. Reduce the heat, partially cover, and simmer for 15 minutes, or until the onion and pepper are very tender.

3. Strain the soup into a bowl and set the broth aside. Transfer the onion-pepper mixture to the workbowl of a food processor and process until very smooth. Add the avocados, lemon juice, and sour cream and again process until smooth.

4. Return the broth to the casserole. Bring to a simmer, whisk in the avocado puree, and just heat through. Do not let come to a boil. Taste and correct the seasoning, adding a dash more lemon juice if necessary and a large grinding of white pepper.

5. Serve the soup at room temperature in individual bowls and garnish each with a dollop of salsa and a few tiny cilantro leaves.

BROCHETTE OF SWORDFISH WITH LEEK AND TOMATO COMPOTE AND BALSAMIC VINEGAR SAUCE

serves 5–6

SWORDFISH
½ cup extra-virgin olive oil
Juice of 1 large lemon
3 tablespoons light soy sauce
2½ pounds swordfish steaks, skin
   removed and cut into 1½-inch cubes

COMPOTE
2 large leeks, all but 2 inches of greens
   removed
3 tablespoons olive oil
3 large ripe tomatoes, peeled, seeded,
   and finely chopped
1 teaspoon fennel seed
Salt and freshly ground black pepper

SAUCE
¼ cup excellent-quality balsamic
   vinegar
½ cup extra-virgin olive oil
Salt and freshly ground black pepper
Drops of lemon juice

25–30 bay leaves
Coarse salt and freshly ground black
   pepper

GARNISH
Finely minced fresh Italian parsley or
   julienned fresh basil leaves

1. Place 5 or 6 wooden skewers in a pan of water to cover and let soak for 1 hour.

2. Start by marinating the swordfish: In a shallow bowl, combine the olive oil, lemon juice, and soy sauce and whisk until well blended. Add the swordfish cubes and toss gently to coat with the marinade. Cover with foil and refrigerate for 1 to 2 hours.

3. While the swordfish is marinating prepare the compote: Dice the leeks, run under warm water to remove all sand and grit, and drain well. In a 10-inch cast-iron skillet, heat the olive oil over low heat, add the leeks, and cook for 1 to 2 minutes, or until the leeks start to soften. Add the tomatoes and fennel seed, season with salt and pepper, and simmer, partially covered, for 25 to 30 minutes, or until all the tomato liquid has evaporated and the mixture is thick. Taste and correct the seasoning. Set aside. The compote can be made a day or two ahead of time and reheated in the microwave or served at room temperature.

4. To prepare the balsamic vinegar sauce: In a small bowl, combine the vinegar and olive oil, and season with salt and pepper. Whisk the mixture until it has emulsified. Add drops of lemon juice to taste and reserve.

5. Prepare the charcoal grill.

6. Remove the swordfish from the marinade and thread on the wooden skewers, alternating each cube with a bay leaf, using 4 to 5 cubes of swordfish for each skewer.

7. When the coals are red-hot, sprinkle the brochettes with coarse salt and black pepper and place over the hot coals. Grill for only 4 to 6 minutes total or until the swordfish is nicely browned but still a bit rare in the center. Turn once during grilling and do not overcook. Remove the brochettes from the grill.

8. To serve, spread 2 to 3 tablespoons of the leek and tomato compote in a 3-inch circle in the center of each individual serving plate. Top each with 1 brochette and drizzle the plate with a thin ribbon of the balsamic vinegar sauce. Garnish with minced parsley or basil and serve at once.

Note: Excellent-quality balsamic vinegar is a must, since its sweet, pungent flavor is the key to the success of this dish. You may also skillet-grill the swordfish instead of outdoor grilling.

This leek and tomato compote is remarkably versatile. A half cup of it can be used as the base for a risotto. Fold 1 cup of the mixture into the pan juices of a simple veal roast, or serve it at room temperature drizzled with extra-virgin olive oil and sprinkled with cracked black pepper.

PENNE WITH PAN-SEARED PEPPERED TUNA, TOMATOES, AND GREMOLATA

serves 6

The combination of a variety of grilled foods with pasta is an appealing Cal-Italian concept. The technique is particularly suitable for restaurant cooking, where the grill is in constant use, but it's far less practical for most home cooks to heat up the grill just for a quick pasta dish.

Fortunately, this dish tastes just as good made with pan-seared tuna or swordfish. Be sure to use the best extra-virgin olive oil and cook the fish only until medium-rare. It's best to cut the fish into bite-size pieces at the last minute so it doesn't dry out. Add a generous sprinkling of minced Italian parsley and some briny, meaty capers to give the dish a sprightly finishing touch.

SAUCE
3 tablespoons extra-virgin olive oil
2 large shallots, peeled and finely minced
6 large ripe tomatoes, peeled, seeded, and chopped
1 tablespoon fresh thyme leaves
2 tablespoons finely minced fresh basil
1 tablespoon fresh oregano or 1 teaspoon dried
Salt and freshly ground black pepper

GREMOLATA
1 large lemon
½ cup finely minced fresh parsley
2 tablespoons large capers, drained

1 can flat anchovies, drained and finely minced

TUNA
2 fresh tuna steaks, about 8–10 ounces each
2 teaspoons crushed black peppercorns
3 tablespoons extra-virgin olive oil
Coarse salt

1 pound imported penne, preferably De Cecco

GARNISH
Tiny fresh basil leaves or miniature basil

1. Start by making the tomato sauce: Heat the oil in a medium casserole over medium-low heat. Add the shallots and cook for 1 minute without browning. Add the tomatoes, thyme, basil, and oregano and cook for 25 to 30 minutes, or until most of the tomato liquid has evaporated. Season with salt and pepper.

2. Transfer the mixture to the workbowl of a food processor and process until smooth. Set aside.

3. To prepare the gremolata: Finely grate the rind of the lemon, avoiding the white pith. Reserve. Peel the lemon, removing all of the remaining rind and white pith with a very sharp knife. Cut the lemon crosswise into thin slices and then quarter each slice. Set aside.

4. In a small bowl combine the grated lemon rind, parsley, capers, and anchovies and mix well. Reserve the gremolata.

5. To prepare the tuna: Sprinkle each side of the tuna steaks heavily with crushed black pepper and pat firmly. Place in a flat dish, cover, and chill for 30 minutes.

6. Heat the oil in a large nonstick skillet over high heat. When the oil is smoking hot, add the tuna steaks and sear for 2 minutes on each side, or until nicely browned on the outside but still rare in the center. Remove from the skillet, season with coarse salt, and set aside.

7. Bring plenty of salted water to a boil in a large casserole, add the pasta, and cook until al dente. Immediately add 2 cups cold water to the pot to stop further cooking. Drain well and return the pasta to the casserole. Add the reserved tomato sauce, the lemon quarters, and the gremolata. Toss gently and keep warm.

8. Slice the tuna crosswise into ⅓-inch slices and break each slice into bite-size pieces. Add to the pasta and again toss gently. Correct the seasoning, adding a large grinding of black pepper, and serve at once garnished with tiny leaves of basil.

CEDAR-ROASTED SALMON WITH
CHIVE-BUTTER SAUCE

serves 4

The first time I saw someone roasting salmon on a cedar plank was in a Seattle restaurant several years ago. I was totally fascinated by this cooking method, as were other professional chefs across the country, and cedar- or oak-plank roasting now appears on the menus of an increasing number of restaurants.

What makes this cooking method a natural for the home cook is its simplicity and the special advantage it offers in cooking the fish virtually without fat. The cedar plank, which is first heated in the oven for about an hour, imparts a slight but very pleasant smoky taste to the salmon that I find most appealing. Other fish that lend themselves very well to this technique are tuna, kingfish, or mackerel. The plank itself can be bought at any lumbermill in the country. Simply rinsed, it can be used over and over again.

The richly flavored chive-butter sauce that accompanies the salmon in this recipe can be replaced by a straightforward mustard-flavored mayonnaise and sour cream mixture enlivened by fresh herbs. Choose a seasonal vegetable, such as the glazed cucumbers or a sauté of peas and asparagus. Tiny boiled and parsleyed new potatoes will complete this main course nicely.

1 untreated hardwood plank, preferably
   cedar or oak, measuring 18 × 9 ×
   ¾-inch thick
3 tablespoons olive oil
1½ pounds salmon fillets, skinned
½ teaspoon coarse salt
¼ teaspoon freshly ground black
   pepper

SAUCE
¼ cup Noilly Prat vermouth
¼ cup dry white wine

2 tablespoons finely minced fresh
   shallots
¼ cup heavy cream
½ pound unsalted butter, slightly chilled,
   cut into 16 pieces
2–3 tablespoons finely minced fresh
   chives
Salt and freshly ground white pepper
Drops of lemon juice to taste

GARNISH
Fresh chives, cut into 3-inch matchsticks

1. Preheat the oven to 250° F.

2. Brush the smooth side of the hardwood plank with 1 tablespoon of the oil.

Place in the center of the preheated oven, oiled side up, and bake until it starts to brown, about 30 to 40 minutes.

3. Brush both sides of the salmon with the remaining oil and sprinkle with the salt and pepper. Set aside.

4. Remove the plank from the oven and raise the oven temperature to 375° F. Place the salmon on the plank with what would have been the skin side down. Place the plank in the center of the preheated oven and roast the salmon for 16 to 20 minutes, or until cooked through but still very moist.

5. While the salmon is roasting, prepare the butter sauce: In a heavy-bottomed 2½-quart saucepan, preferably copper, combine the vermouth, white wine, and shallots. Cook over high heat until reduced to a glaze. Add the cream, bring to a boil, and reduce by half.

6. Reduce the heat to very low. Add the butter, 2 tablespoons at a time, whisking constantly and making sure that each portion of butter is absorbed before adding the next. The sauce should be smooth and creamy. Add the minced chives. Season with salt, pepper, and drops of lemon juice to taste. Cover and keep over pan of lukewarm water until the salmon is done.

7. When the salmon is done, remove from the oven and cut into serving pieces. Place on individual plates, spoon some sauce over each portion, and sprinkle with the chive matchsticks. You may also serve the salmon right on the plank for a more informal dinner or Sunday brunch, with the sauce on the side.

How to Tell When Fish Is Done

The old-fashioned method of testing the doneness of fish by determining if it flakes easily is no longer valid. Fish that flakes easily is overcooked. We have now learned to recognize and appreciate seafood that is properly cooked—that is, slightly underdone. The new test is to pierce the fish—be it a steak, scallop, or fillet—with the tip of a sharp knife for 5 to 10 seconds, then test the temperature of the knife tip against the inside of your wrist. If the knife is cold, the fish is not done; if the knife is warm, the fish is just cooked; if the knife is hot, the fish is overdone. Use this test more than once as the fish cooks.

■ ■ ■

SALMON AND LOBSTER FRITTERS IN SORREL CREAM

serves 4–6

If you're looking for an elegant summer main course, this would be a good choice. Although it may sound rather expensive, it is not. Small lobsters are often offered by supermarkets as the weekly special, and salmon has become a basic in every supermarket or fish store in the country. Although the sorrel cream is a perfect accompaniment, you can opt for a less rich sauce made with simply fresh diced tomatoes tossed in virgin olive oil, some minced chives, and a touch of vinegar, folded into a little homemade crème fraîche.

CREAM
2 cups tightly packed fresh sorrel leaves, stemmed and washed
1/3 cup Fish Stock (page 472) or bouillon
1 1/2 cups Crème Fraîche (page 491)
Salt and freshly ground white pepper

FRITTERS
2 whole lobsters, about 1 1/4 pounds each
1 1/2 tablespoons unsalted butter
2 tablespoons finely minced fresh shallots
1 large clove garlic, peeled and finely minced
1/4 cup cooked fresh corn kernels, finely minced
2 tablespoons finely minced scallions

1/2 pound poached fresh salmon fillet, preferably from the center cut, flaked (page 70)
2 large eggs, lightly beaten
1/4 cup heavy cream
1/2 cup fine homemade bread crumbs
Coarse salt
1/4 teaspoon cracked black pepper
Large pinch of cayenne pepper
1/4 teaspoon paprika
All-purpose flour for dredging
12 tablespoons Clarified Butter (page 478)

GARNISH
Tiny fresh cilantro or Italian parsley leaves
Steamed snow peas

1. Start by making the sorrel cream: Place the sorrel leaves in a small saucepan with the water that clings to its leaves. Place over medium-low heat and cook, covered, until just wilted, about 2 to 3 minutes.

2. Drain the sorrel in a colander and press lightly to extract all of the excess liquid. Place the sorrel in the workbowl of a food processor together with the fish stock or bouillon and process until smooth. Transfer the puree to a medium saucepan and place over low heat. Whisk in the crème fraîche, season with salt and pepper, and reduce slightly. Cover and keep warm.

3. Preheat the oven to 200° F.

4. To prepare the fritters: Bring a large pot of salted water to a boil. Add the lobsters, return to a boil, and simmer for 5 minutes. Drain and set the lobsters aside to cool.

5. Remove all the meat from the lobsters, which will be mainly from the tail and claws, finely dice, and set aside.

6. Melt the butter in a small skillet over low heat. Add the shallots and garlic and cook for 1 minute without browning. Transfer to a large mixing bowl and add the diced lobster, corn, scallions, salmon, eggs, cream, and bread crumbs. Fold gently until thoroughly combined. Season the mixture carefully with salt and add the cracked black pepper, pinch of cayenne, and paprika.

7. With floured hands, shape the lobster-salmon mixture into round, thick fritters using about ¼ cup of the mixture for each fritter. Dredge lightly in flour and set aside.

8. In a 10-inch nonstick skillet heat about 3 to 4 tablespoons of the clarified butter over medium-low heat. Add 4 fritters, without crowding the skillet, and cook for 3 minutes on each side, or until nicely browned. Transfer with a spatula to a plate, cover, and keep warm in the preheated oven. Continue sautéing the remaining fritters, adding more butter to the skillet for each batch.

9. Place 2 to 3 fritters on each individual serving plate and spoon a little of the sorrel cream over each portion. Garnish with leaves of cilantro and a few steamed snow peas and serve at once.

Note: Tiny boiled and buttered new potatoes, a roasted medley of summer vegetables, braised summer squash, and wilted summer spinach are all delicious accompaniments.

PERNOD-SCENTED MUSSELS IN TOMATO, MINT, AND SAFFRON SAUCE

serves 4–6

It's unfortunate that mussels are not more widely available. To me, they are the most interesting of all shellfish. Their delicious broth doubles as a fish stock, and the plump sweet mussel is a delicacy much appreciated in all countries bordering the Atlantic as well as the Mediterranean. Mussels are now farmed in Maine and shipped all over the country. Although they tend to lack the depth of flavor of the wild mollusk, they are still very good indeed. The great advantage of farmed mussels is that they are usually quite clean and barely need scrubbing.

Here mussels are prepared in the style of the Camargue, an area of France near the town of Sète, close to the Spanish border where some of the best mussels come from. Saffron and Pernod are key ingredients to this preparation, lending a unique taste to the sauce.

9 tablespoons unsalted butter
2 large shallots, peeled and finely minced
½ cup dry white wine
¼ cup Pernod
1 tablespoon fresh thyme leaves or 1 teaspoon dried
1 teaspoon saffron threads
1 teaspoon whole black peppercorns
4 pounds fresh small mussels, cleaned and scrubbed well

3 large ripe tomatoes, peeled, seeded, and pureed
½ cup Crème Fraîche (page 491)
1 Beurre Manie (page 478)
2 tablespoons finely minced fresh mint
4 tablespoons fresh basil leaves, cut into fine julienne

GARNISH
Tiny fresh mint leaves
Tiny cooked red potatoes

1. In a large casserole, melt 3 tablespoons of the butter over medium-high heat. Add the shallots and cook for 1 minute without browning. Add the white wine, Pernod, thyme, saffron, and black peppercorns. Bring to a boil, add the mussels, and steam, covered, until all the mussels open. Discard any that do not. Transfer the mussels to a bowl with a slotted spoon, cover, and reserve.

2. Add the tomatoes and crème fraîche to the casserole. Bring to a simmer and reduce by half. Strain the sauce through a fine sieve and return to the casserole. Place over medium heat and whisk in bits of beurre manie until the sauce lightly coats a spoon. Reduce the heat and whisk in the remaining butter, 1 tablespoon at a time, until the sauce is smooth and creamy. Add the mint and basil, taste and correct the seasoning, and keep warm.

3. Remove and discard the empty half shell from each mussel. Place 12 to 14 of the mussels on the half shell in each individual soup bowl. Spoon the sauce over and garnish with tiny leaves of mint. Serve hot accompanied by tiny cooked red potatoes and lots of crusty French bread.

Variation: Remove the mussels from their broth and reserve. Mix the mussels and sauce into a well-seasoned pilaf of rice, or serve over linguine tossed in virgin olive oil, minced garlic, and fresh parsley.

FUSILLI WITH SAUTÉED SHRIMP, TOMATO, AND GINGER FONDUE

serves 4–5

½ pound small shrimp
8 tablespoons olive oil
1 small dry red chili pepper, broken
2 large cloves garlic, peeled and finely minced
1½ teaspoons finely minced fresh ginger
6 small shallots, peeled and thinly sliced
4 large ripe tomatoes, peeled, seeded, and chopped
2 teaspoons fresh oregano leaves

2 teaspoons fresh thyme leaves
Salt and freshly ground black pepper
2 small zucchini, trimmed, quartered lengthwise, and thinly sliced
½ pound imported fusilli, preferably De Cecco
4 tablespoons finely minced fresh basil
2 tablespoons finely minced fresh Italian parsley
¾ cup mild goat cheese, crumbled

1. Peel the shrimp, reserving both shrimp and shells separately.

2. In a large, heavy skillet heat 2 tablespoons of olive oil over medium-high heat. When the oil is hot, add the chili pepper and shrimp shells and sauté quickly until the shells turn bright pink. Remove the shells and chili pepper with a slotted spoon and discard.

3. Add 1 tablespoon of oil to the skillet and, when the oil is very hot, add the shrimp without crowding the pan. Sauté the shrimp until bright pink and lightly browned. Immediately remove from the skillet with a slotted spoon and reserve.

4. Reduce the heat, add 2 tablespoons of oil to the skillet, and add the garlic, ginger, and shallots. Cook until the shallots are soft but not browned. Add the tomatoes, oregano, thyme, and salt and pepper, reduce the heat, and simmer, covered, until the ginger is very soft.

5. In another heavy skillet, heat the remaining oil over medium-high heat. Add the zucchini and sauté quickly until lightly browned. Season with salt and pepper, add to the tomato sauce, and simmer for 5 minutes. Keep the sauce warm.

6. Dice the shrimp and set aside.

7. Bring plenty of salted water to a boil in a large pot. Add the fusilli and cook

until al dente. Immediately add 2 cups of cold water to the pot to stop further cooking.

8. Add the diced shrimp to the tomato-zucchini sauce and remove from the heat.

9. Drain the fusilli well, return it to the pot, and toss with the shrimp, zucchini, and ginger sauce. Add the basil, minced parsley, and goat cheese. Taste and correct the seasoning, adding a large grinding of black pepper. Serve at once with a crusty loaf of French bread.

SAUTÉ OF SHRIMP WITH TOMATO FONDUE AND FRAGRANT INDIAN SPICES

serves 6

Sautéed shrimp is a good quick main course at any time of the year. Add a well-seasoned tomato fondue and some diced summer vegetables and you have a light but full-flavored seasonal dish. For a variation, I like adding a touch of Indian spices and hot chili peppers, which make the whole dish come alive. I often grill the shrimp and then let them braise for another minute or two in the tomato and vegetable mixture. Serve the shrimp either hot or at room temperature accompanied by braised bulgur.

FONDUE
2 tablespoons extra-virgin olive oil
2 tablespoons finely minced shallots
4 large very ripe tomatoes, peeled, seeded, and diced
2 teaspoons tomato paste
5 large cloves garlic, peeled and finely minced
1½ tablespoons finely minced fresh ginger
1 teaspoon Garam Masala (page 485) or curry powder
Salt and freshly ground black pepper

SHRIMP
1½ pounds medium shrimp, peeled
3–4 tablespoons extra-virgin olive oil
2 small dry red chili peppers, broken
Freshly ground black pepper
1 medium red bell pepper, cut in half, seeded, and thinly sliced
1 medium green bell pepper, cut in half, seeded, and thinly sliced
1 small zucchini, cut into ¼-inch julienne strips
1 tablespoon fresh lemon juice
½ cup finely minced fresh cilantro or parsley

1. Start by preparing the fondue: In a heavy-bottomed 2-quart saucepan, heat the oil over medium heat. Add the shallots and cook for 1 minute, or until soft but not browned. Add the diced tomatoes and tomato paste. Reduce the

heat to low and simmer, partially covered, stirring often, for 30 to 35 minutes or until all the tomato liquid has evaporated and a thick puree is formed. Set aside.

2. Transfer the fondue to the workbowl of a food processor and add the minced garlic, ginger, and garam masala. Process for 20 seconds, or until smooth. Season with salt and pepper and set aside.

3. Dry the shrimp thoroughly on paper towels and set aside.

4. In a large, heavy skillet, heat the oil over medium heat and add the chili peppers. Cook until they turn dark, about 30 seconds, then remove from the skillet and discard. Raise the heat to high and, when the oil is almost smoking, add the shrimp and cook for 1 to 2 minutes, or until they turn bright pink and are lightly browned. Season with a good grinding of pepper, remove the shrimp with a slotted spoon to a dish, and set aside.

5. Reduce the heat to medium-low. Add the reserved tomato fondue and cook for 1 minute to heat through. Add the red and green peppers, zucchini, lemon juice, and cilantro and cook for 2 minutes, or until the vegetables are crisp tender. Return the shrimp to the skillet and just heat through. The shrimp should be well coated with the sauce. Taste and correct the seasoning. Serve at once on individual plates, accompanied by a crusty loaf of French bread.

A SOUTHERN STEW OF CHICKEN, OKRA, AND TOMATOES

serves 4

Unless you have grown up with okra, it is not a vegetable that you'll fall in love with at first bite. In fact, it was not until I started to grow okra in my garden that I began to appreciate this interesting summer vegetable. True Southerners will find that this stew is not authentic, and it is not meant to be. It is, however, a delicious and full-flavored one-dish meal particularly suitable for weekend entertaining since it can be prepared in advance and reheated either in the regular oven or in the microwave. It is also delicious served at room temperature drizzled with a touch of virgin olive oil and a large grinding of black pepper. For a more substantial flavor, add ½ pound of chorizo or andouille sausage to the stew. Be sure to omit one of the chili peppers if the sausage is spicy.

2 tablespoons corn oil
¼ pound slab bacon, cut into ½-inch cubes and blanched
12 chicken wings, cut in half at joint and tips removed and discarded
Salt and freshly ground black pepper
2 small dry red chili peppers, broken
3 cups finely diced yellow onions
1 green bell pepper, cored, seeded, and diced
2 medium cloves garlic, peeled and finely minced
6 medium ripe tomatoes, peeled, seeded, and chopped

¾ pound okra, cut crosswise into ¼-inch slices
2 cups Chicken Stock (page 470)
½ pound smoked shoulder pork butt, cut into 1-inch cubes
2 tablespoons finely minced fresh oregano

GARNISH
Finely minced fresh parsley or cilantro

1. In a 4-quart casserole, heat the oil over medium heat, add the bacon, and cook until almost crisp. Remove with a slotted spoon to a side dish and reserve.

2. Add the chicken wings to the casserole and sauté until nicely browned on all sides. Transfer to a side dish, season with salt and pepper, and reserve.

3. Discard all but 2 tablespoons of fat from the casserole. Add the chili peppers and cook until dark. Add the onions, green pepper, and garlic and cook the mixture for 10 minutes, or until soft and lightly browned.

4. Add the tomatoes, okra, and chicken stock to the casserole. Bring to a boil, add the pork butt and return the chicken wings and bacon to the casserole, and season with salt, pepper, and oregano. Reduce the heat, partially cover, and simmer the stew for 1½ hours.

5. Let the stew cool completely, and carefully skim the fat off the top. Reheat slowly on top of the stove until hot and spoon into deep soup bowls preferably over a well-seasoned parsleyed pilaf of rice. Sprinkle with parsley or cilantro and serve hot.

Note: If you can find canned smoked chipotle peppers, add 2 or 3 of them to the stew; they will impart an interesting smoky flavor.

SAUTÉED CHICKEN WITH ROASTED
GARLIC, ROSEMARY, AND PEPPERS

serves 4–6

1 large head garlic (12–16 cloves), unpeeled, plus 2 large cloves, peeled and sliced
4 tablespoons extra-virgin olive oil
1 large red bell pepper, cored, seeded, and cut into 1½-inch cubes
1 yellow bell pepper, cored, seeded, and cut into 1½-inch cubes
Coarse salt and freshly ground black pepper
1 tablespoon finely minced fresh thyme
1 tablespoon unsalted butter
1 (3-pound) chicken, back removed and cut into eighths

2 small chicken legs separated into thighs and drumsticks
1 large sprig of fresh rosemary
⅓ cup dry white wine
2 tablespoons sherry vinegar
1–1½ cups Chicken Stock (page 470) or bouillon
1–1½ teaspoons Beurre Manie (page 478)
2 tablespoons finely minced fresh parsley

1. Preheat the oven to 350° F.

2. Break the head of garlic into cloves. Place the cloves on a sheet of heavy-duty aluminum foil, sprinkle with 1 tablespoon olive oil, close the foil tightly, and bake for 30 minutes, or until tender when pierced with the tip of a sharp knife.

3. Place the cubed red and yellow peppers on another sheet of aluminum foil, season with a touch of salt and pepper, and sprinkle with fresh thyme and 1 tablespoon olive oil. Close the foil tightly and place in the oven. Roast the peppers for 5 to 7 minutes, or until just tender. Remove from the oven and set aside.

4. When the garlic is tender, remove from the oven and let cool. Peel the cloves and set aside, together with the peppers.

5. In a heavy 12-inch skillet, heat 2 tablespoons of the olive oil and the butter over high heat. Add the chicken pieces and sauté until nicely browned on all sides. You may need to do this in 2 batches. When all the chicken has been sautéed, transfer it to a platter and season with salt and pepper.

6. Remove all but 2 tablespoons of fat from the skillet, add the sliced garlic cloves, the rosemary sprig, and the white wine, and cook for about 5 minutes, or until the wine is almost completely evaporated. Add the sherry vinegar and cook down to 1 tablespoon, about 1 minute.

7. Return the chicken to the skillet and add ½ cup of the stock. Lower the

heat and simmer the chicken, tightly covered, for 25 minutes, adding a little stock every few minutes and shaking the pan to make sure the chicken and garlic do not burn.

8. When the chicken is cooked, transfer it to a side dish and reserve. Thoroughly degrease the pan juices, discard the sliced garlic and rosemary, and add the remaining stock, the roasted garlic cloves, and the roasted peppers and just heat through. If the sauce is too thin whisk in bits of beurre manie until the sauce lightly coats a spoon. Taste the sauce and correct the seasoning, spoon over the chicken, and garnish with the parsley. Serve accompanied by buttery polenta, roasted new potatoes, or just crusty French bread.

SAUTÉED CHICKEN BREASTS WITH TOMATOES, BASIL, AND A CONFIT OF GARLIC

serves 6

CONFIT

2 heads garlic, separated into cloves
   and peeled
¾ cup olive oil

CHICKEN

2½ tablespoons olive oil
1 large shallot, peeled and finely minced
2 large ripe tomatoes, peeled, seeded,
   and chopped
2 tablespoons finely minced fresh basil
Salt and freshly ground black pepper
3 whole chicken breasts, skinned, cut in
   half, and all fat removed

1½ tablespoons unsalted butter
½ cup Chicken Stock (page 470)
⅓ cup Crème Fraîche (page 491)
1 Beurre Manie (page 478)

GARNISH

1 large ripe tomato, peeled, seeded,
   and cut into ½-inch dice
1 tablespoon finely minced fresh Italian
   parsley
2 tablespoons julienne of fresh basil

1. Start by making the confit: In a small saucepan, combine the garlic cloves and olive oil. Place over very low heat and stew for 1½ hours, or until the garlic is pale beige and very soft. Remove from the heat and let the garlic cool in the oil.

2. To prepare the chicken: In a 10-inch nonstick skillet, heat 2 tablespoons of the olive oil over medium-low heat, add the shallot, and cook for 1 minute without browning. Add the tomatoes and basil, season with salt and pepper, and cook the mixture until all the tomato water has evaporated and the mixture is thick, about 10 to 15 minutes. Reserve.

3. Dry the chicken breasts thoroughly on paper towels. Melt the butter together with the remaining olive oil in a large, heavy skillet over high heat. Add the chicken breasts and sauté quickly until nicely browned on both sides. Season with salt and pepper. Add ⅓ cup of the chicken stock to the skillet, reduce the heat, and simmer, partially covered, for 5 minutes. Transfer the chicken breasts to a side dish and reserve.

4. Add the remaining stock, the reserved tomato mixture, and the crème fraîche. Bring to a simmer and whisk in bits of beurre manie until the sauce lightly coats a spoon.

5. Remove the garlic cloves from the oil with a slotted spoon and add to the skillet together with the chicken breasts and just heat through. Taste and correct the seasoning. Transfer the chicken breasts to a serving platter and keep warm.

6. Add the diced tomato, parsley, and basil garnish to the sauce and just heat through. Spoon the sauce over the chicken and serve immediately.

GRILLED QUAIL WITH RED ONION–CANDIED GINGER–THYME JAM

serves 4

QUAIL
½ cup olive oil
Juice of 4 limes
4 tablespoons light soy sauce
2 large cloves garlic, peeled and mashed
1 (1-inch) piece of fresh ginger, peeled and pressed through a garlic press
3 tablespoons finely minced scallions
1 teaspoon fresh rosemary leaves
2 teaspoons fresh thyme leaves
¾–1 teaspoon cracked black pepper
8 fresh whole quail, about 5 ounces each, cut through the backbone

GINGER
¼ cup granulated sugar
¼ cup water

1 tablespoon very fine julienne of fresh ginger

ONIONS
3 tablespoons olive oil
1½ tablespoons unsalted butter
6 medium onions, peeled, halved, and thinly sliced
2 teaspoons fresh thyme leaves
Salt and freshly ground black pepper
⅓ cup dry white wine
¾ cup heavy cream
1–2 teaspoons fresh lemon juice

Coarse salt

GARNISH
Sprigs of fresh thyme and rosemary

1. Start by marinating the quail: Combine the oil, lime juice, soy sauce, garlic, ginger, scallions, rosemary, thyme, and cracked pepper in a small bowl and whisk until well blended. Place the quail in a plastic zip-lock bag, add the marinade, and seal the bag. Place the bag in a shallow dish and turn to coat the birds well with the marinade. Refrigerate for 3 to 4 hours.

2. While the quail are marinating, prepare the candied ginger: In a small saucepan, combine the sugar and water, place over medium heat, and simmer for 3 minutes, or until the sugar has completely dissolved. Add the ginger and simmer until it is just tender, about 5 minutes. Remove from the heat and set aside.

3. To prepare the onions: In a large, heavy skillet heat the oil and the butter over high heat. Add the onions and cook for 5 minutes, stirring constantly, until they start to brown. Reduce the heat and cook, partially covered, until they are soft and nicely browned, about 30 minutes.

4. Add the thyme to the onions and season with salt and pepper. Raise the heat, add the white wine, and cook until the wine is reduced to a glaze. Add the heavy cream and reduce slightly.

5. Add the candied ginger and its cooking syrup to the onions, then add the lemon juice, and correct the seasoning. Cover and keep warm.

6. Prepare the charcoal grill.

7. Remove the quail from the marinade and break the breastbone so that they will lie flat on the grill. Sprinkle with coarse salt and, when the coals are red hot and fiery, place the quail skin side down on the grid. Grill for 3 to 4 minutes on each side, or until medium to medium-rare and nicely browned. Do not overcook or the quail will have a very strong and gamy taste.

8. Remove the quail from the grill, garnish with thyme and rosemary, and serve at once with the jam.

GRILLED TURKEY BREAST WITH SPICY YOGURT PASTE AND TWO-BEAN FRICASSEE

serves 6–8

1 whole turkey breast, bone in, about 5¾ pounds

MARINADE

2 teaspoons cracked black pepper
1½ teaspoons coarse salt
½ teaspoon cayenne pepper
1 teaspoon celery seed
1½ teaspoons dried thyme
1½ teaspoons dried oregano
2 teaspoons imported hot paprika
1 tablespoon Dijon mustard
2 large cloves garlic, peeled and mashed
2 tablespoons grated onion
⅓ cup plain yogurt

BEANS

¾ pound green beans, trimmed and cut into 1½-inch pieces
4 tablespoons extra-virgin olive oil

1 cup smoked pork shoulder butt, diced
1 large Bermuda onion, peeled, quartered, and thinly sliced
2 large cloves garlic, peeled and thinly sliced
2 large ripe tomatoes, peeled, seeded, and chopped
1 large sprig of fresh thyme or 1 teaspoon dried
1 large sprig of fresh oregano or 1 teaspoon dried
Salt and freshly ground black pepper
1½ cups dried white beans (cannellini), cooked (page 484), and ½ cup bean cooking liquid reserved

GARNISH

2 tablespoons finely minced fresh parsley
1 recipe Aioli (page 477)

1. Start by marinating the turkey breast: With a sharp knife, remove each half of the breast from the bone in one piece. Dry with paper towels and reserve.

2. In a small bowl combine all of the marinade ingredients and whisk until a smooth paste is formed. Spread each of the breast halves generously with the yogurt paste, cover, and refrigerate for 2 to 4 hours.

3. While the turkey is marinating, prepare the beans: Bring water to a boil in a vegetable steamer, add the green beans, and steam, covered, for 5 minutes or until just tender. Remove and set aside.

4. In a 10-inch skillet heat 2 tablespoons of the olive oil over medium heat, add the pork butt, and cook for 2 minutes. Remove with a slotted spoon to a side dish and reserve.

5. Add the remaining oil to the skillet and when hot, add the onion and garlic and cook, stirring often, for 5 minutes or until the onion begins to brown.

Reduce the heat and cook for 15 minutes longer, or until soft and lightly browned.

6. Add the tomatoes, thyme, and oregano and season with salt and pepper. Cook for 10 minutes, or until most of the tomato juices have evaporated. Reduce the heat to low. Add the cooked white beans and ⅓ cup of the bean liquid, season with salt and pepper, and simmer, covered, for 20 minutes. If the juices run dry, add the remaining bean liquid. Add the reserved green beans and simmer for 5 minutes longer. Set aside.

7. Prepare the charcoal grill for grilling over indirect heat (see Apple-Smoked Turkey with Onion and Chestnut Jam, page 247).

8. Remove the turkey-breast halves from the refrigerator and grill over indirect heat, covered, without turning (see the recipe cited above) for 45 minutes, or until the internal temperature registers 165° F on a meat thermometer and the juices run pale yellow.

9. While the turkey is on the grill, slowly reheat the bean fricassee, cover, and keep warm.

10. When the turkey breasts are done, transfer to a cutting board and let rest for 5 minutes. Cut the breasts crosswise into ¼-inch slices. Serve the turkey warm or at room temperature with the warm bean fricassee, garnished with minced parsley and a drizzle of aioli.

GRILLED FLANK STEAKS IN SPICY CHILI MARINADE

serves 6

Flank steak is a flavorful and relatively inexpensive cut of meat which, I find, is at its best when marinated and quickly grilled. Many marinades work well, particularly those based on a variety of spices, including ginger and soy sauce. Leftover steak is excellent in sandwiches the next day or thinly sliced and served with a well-seasoned green salad. Be sure to use a very hot fire, and grill the meat quickly to ensure that it cooks to medium-rare or else it will toughen and dry out.

MARINADE

2 large cloves garlic, peeled and finely minced

1 (1-inch) piece of fresh ginger, peeled and finely minced

1 medium onion, peeled and thinly sliced

2 tablespoons finely minced fresh parsley

1 tablespoon cumin seeds, toasted and crushed

1 tablespoon pure chili powder

1 teaspoon turmeric

1 teaspoon dried oregano

½ cup olive oil

Juice of 1 large lemon

2 tablespoons soy sauce

2 teaspoons dry sherry

1 teaspoon coarsely ground black pepper

2 flank steaks, each weighing 1½–2 pounds

Coarse salt

Vegetable oil for brushing on grill

Sprigs of fresh cilantro

Salsa Topping (page 174)

1. In a large mixing bowl combine all the ingredients for the marinade and set aside. Place the flank steaks in a large zip-lock plastic bag, pour the marinade over them, and close the bag. Place the bag in a shallow dish and refrigerate for 24 hours, turning the steaks in the marinade several times.

2. The next day, prepare the charcoal grill.

3. Thirty minutes before grilling the steaks, remove them from the refrigerator. Remove from the marinade and set the steaks and marinade aside separately.

4. When the coals are red-hot, sprinkle coarse salt on each side of the steaks. Brush the grill with vegetable oil and place the steaks over the hot coals. Grill 5 to 6 minutes on one side, or until nicely browned and the steaks begin to tighten up and slightly curl, brushing often with the marinade. Turn over and

grill for another 3 to 4 minutes for medium-rare, or until a meat thermometer registers an internal temperature of 130° to 135° F.

5. Transfer the steaks to a cutting board and let rest for 3 or 4 minutes before carving. Cut the meat across the grain on the bias into thin slices. Garnish with sprigs of cilantro and serve warm or at room temperature with the vinaigrette.

Note: The steaks may also be broiled instead of grilled. Place over a pan on the broiler rack and broil 6 to 8 inches from the source of heat. The broiling time will be about the same as for outdoor grilling, but it is best to check the doneness with a meat thermometer.

CHARCOAL-GRILLED TARTARE OF BEEF WITH CAPERS AND CHIVE ESSENCE

serves 4

TARTARE
1½ tablespoons prepared chili sauce
¾ teaspoon Worcestershire sauce
1½ tablespoons Dijon mustard
1 teaspoon dry mustard
Large pinch of cayenne pepper
¼ teaspoon cracked black pepper
4 flat anchovy fillets, drained and finely minced
2 tablespoons finely minced shallots
1¾ pounds freshly ground lean sirloin
Coarse salt

SAUCE
2 tablespoons Chicken Stock (page 470) or bouillon

2 tablespoons minced fresh chives
2 tablespoons fresh parsley leaves
1 small clove garlic, peeled
1 small shallot, peeled and cut in half
8 tablespoons unsalted butter
Salt and freshly ground black pepper

TOASTS
4 slices good-quality white bread
2 to 3 tablespoons fruity olive oil

GARNISH
1½ tablespoons tiny capers, drained
2 tablespoons finely minced fresh chives

1. Start by preparing the tartare: In a large mixing bowl, combine the chili and Worcestershire sauce, mustards, cayenne, and black pepper, anchovies, and shallots and whisk until well blended. Add the ground sirloin and mix well. Season with coarse salt to taste, cover, and refrigerate for 1 hour.

2. While the sirloin is marinating, prepare the sauce: Combine the stock, chives, parsley, garlic, and shallot in a mini-blender and process until very smooth and bright green.

3. Melt the butter over medium-low heat and add the chive puree, whisking until smooth. Season with salt and pepper and set aside, covered, over a pan of lukewarm water.

4. To prepare the toasts: With a sharp knife, remove the crusts from the bread and cut each slice in half diagonally to create triangles. Brush both sides with olive oil and reserve.

5. Prepare the charcoal grill.

6. Remove the tartare from the refrigerator and form into 4 thick oval patties. When the coals are red-hot, season the patties with coarse salt and black pepper and place directly above the coals. Grill very quickly, about 2 minutes per side, or until very brown but still rare in the center. Remove the patties from the grill and reserve.

7. Place the bread triangles above the coals and grill until golden brown on both sides.

8. Transfer 2 toasts to each of 4 individual serving plates, top each with a grilled beef tartare, and drizzle with some of the warm chive sauce. Sprinkle each with some capers and minced chives and serve at once.

GRILLED VEAL ROAST À LA FLORENTINE

serves 6–8

ROAST
1 cup tightly packed fresh spinach
2½ pounds boneless shoulder of veal, butterflied
Coarse salt and freshly ground black pepper
1 tablespoon finely minced fresh thyme
1 tablespoon finely minced fresh rosemary
1 large clove garlic, peeled and finely minced
3 thin slices prosciutto
¼ cup extra-virgin olive oil
Juice of 1 large lemon
1 (3½-ounce) can light tuna, packed in oil and drained

2–3 flat anchovy fillets, drained
3 tablespoons Chicken Stock (page 470) or bouillon
½ teaspoon egg yolk
1½–2 tablespoons fresh lemon juice
Salt and freshly ground black pepper
¾ cup olive oil
2 tablespoons finely minced fresh chives

1 teaspoon *herbes de Provence*
½ teaspoon cracked black pepper
4–5 sprigs of fresh rosemary

GARNISH
Sprigs of fresh rosemary, thyme, and/or chives

1. Start by preparing the veal roast: Carefully remove and discard all stems from the spinach, wash thoroughly, and drain very well with paper towels. Set aside.

2. Lay the veal roast flat with what would have been the skin side down. Sprinkle with coarse salt, black pepper, the thyme, rosemary, and garlic. Place the prosciutto slices over the veal to cover it completely and top with the spinach leaves. Carefully roll the veal up from the short side, truss, and place in a shallow dish. Reserve.

3. Whisk together the olive oil and lemon juice in a small bowl, pour the mixture over the veal, and marinate, covered with foil, at room temperature for 1 hour.

4. While the veal is marinating prepare the sauce: In the workbowl of a food processor, combine the tuna, anchovies, and chicken stock and process until very smooth. Set aside.

5. In a medium bowl, combine the egg yolk and lemon juice, season with salt and pepper, and whisk until well blended. Add the olive oil slowly, drop by drop, whisking constantly, until the mixture has emulsified and all of the oil has been added. Whisk in the reserved tuna mixture and minced chives and correct the seasoning. Cover and chill.

6. Prepare the charcoal grill for grilling over indirect heat (see Apple-Smoked Turkey with Onion and Chestnut Jam, page 247).

7. Remove the veal roast from the marinade, sprinkle with coarse salt, the *herbes de Provence,* and cracked pepper and tuck the rosemary sprigs between the trussing string. Grill over indirect heat, covered, without turning for 1 hour or until the internal temperature reaches 155° to 160° F on a meat thermometer. Baste often with the marinade.

8. When the veal is done, transfer to a cutting board, remove the trussing string and sprigs of rosemary, and let rest for 10 minutes before carving.

9. Cut the veal crosswise into ¼- to ½-inch-thick slices. Place a few slices on each individual serving plate, spoon some of the tonnato over each portion, and garnish with sprigs of fresh herbs.

CORN, POLENTA, AND JALAPEÑO PEPPER FRITTERS

serves 6–8

These spicy corn fritters are an excellent accompaniment to all grilled meats and seafood, but they work equally well as an appetizer drizzled with a garlic-, anchovy-, or basil-flavored butter or topped with a julienne of roasted peppers dressed in a sherry vinegar vinaigrette. The fritter batter will keep for two or three days in the refrigerator, so make a double batch. You can serve the fritters as an easy starter one day and as a vegetable the next.

1½ cups whole milk
1 cup stone-ground cornmeal
3 tablespoons unsalted butter, softened
1 teaspoon baking powder
1 teaspoon salt
2 large eggs, lightly beaten
¼ cup very finely diced red bell pepper
½ cup very finely minced scallions
1–2 teaspoons very finely minced
    jalapeño pepper

2 tablespoons finely minced fresh basil
1 cup coarsely chopped cooked fresh
    corn, about 2 ears
¼ cup all-purpose flour
5–6 tablespoons Clarified Butter (page
    478)

GARNISH
Sour cream
Finely minced fresh chives

1. In a small bowl combine ½ cup of the milk and the cup of cornmeal and stir until well blended and set aside.

2. Bring the remaining milk to a simmer in a medium saucepan. Add the cornmeal mixture to the simmering milk and whisk until the mixture becomes very thick, about 2 to 3 minutes. Remove the saucepan from the heat, transfer the polenta to a large mixing bowl, and let cool slightly.

3. Add the softened butter, baking powder, salt, eggs, red pepper, scallions, jalapeño pepper, basil, corn, and flour to the polenta and stir until well blended.

4. Melt 1 tablespoon of clarified butter in a 10-inch nonstick skillet over medium heat. Drop the batter by heaping tablespoonfuls into the hot skillet, about 1 to 2 inches apart. You will have to cook the fritters in batches of 4 or 5. Cook until nicely browned and bubbles appear around the edges and in the

center, about 2 minutes. Carefully turn the fritters over and cook until the other side is nicely browned as well, about 1 to 2 minutes longer. Transfer the fritters to a double layer of paper towels. Continue making fritters, adding 1 tablespoon of clarified butter to the skillet for each batch.

5. Serve warm with dollops of sour cream and a sprinkling of fresh chives.

Variation: This batter also can be used to make a light cornbread. Prepare the batter as directed above, but instead of adding whole beaten eggs, separate the eggs and whisk the yolks into the batter. Beat the whites with a pinch of salt and fold gently into the batter. Pour the batter into a hot, greased 10-inch cast-iron skillet and bake in a preheated 400° F oven for 20 minutes. Serve cut into wedges.

BRAISED OKRA IN SPICY MIDDLE EASTERN TOMATO SAUCE

serves 6

Okra is usually thought of as a Southern vegetable, and as such it is rarely part of the seasonal repertoire of cooks in the Northeast or Midwest. Although there is a definite regional characteristic to this vegetable, it is now available fresh in many markets throughout the country and is well worth experimenting with. Unfortunately, most okra arrives at the market in a tired state, with pods that lack freshness and eye appeal. This is quickly changing, now that the demand for the vegetable is growing. Personally, I love to grow my own okra, and I use it often. I especially like to combine it with a zesty, somewhat spicy tomato fondue and serve it at room temperature as an accompaniment to grilled or roasted poultry, seafood, and meats. You can easily turn the ragout into a one-dish meal by adding a few cooked and diced shrimp or some sautéed sliced Italian sausage.

6 large ripe tomatoes, peeled, seeded, and chopped
Coarse salt
4 tablespoons olive oil
2 small dry red chili peppers, crushed
2 large onions, peeled, quartered, and thinly sliced
5 medium cloves garlic, peeled and finely minced

1 tablespoon finely minced fresh ginger
1 teaspoon ground cumin
¼ teaspoon ground cardamom
⅛ teaspoon freshly grated nutmeg
1 teaspoon ground coriander
1 pound okra, cut crosswise into ¼-inch slices
Salt and freshly ground black pepper

**3 tablespoons finely minced fresh
   cilantro or parsley**

1. Sprinkle the tomatoes with coarse salt, place in a colander over a bowl, and let drain for 30 minutes. Reserve the tomatoes and their juices separately.

2. In a deep 10-inch skillet heat the oil over medium heat, add the chili peppers, and cook until very dark. Remove with a slotted spoon and discard. Add the onions, garlic, and ginger, reduce the heat, and cook the mixture for 20 minutes, or until very soft and nicely browned.

3. Add the cumin, cardamom, nutmeg, and coriander and blend into the onion mixture. Add the tomatoes and okra and season with salt and pepper. Simmer, partially covered, for 25 to 30 minutes or until the okra is tender. As the okra cooks, it tends to thicken the sauce; if the sauce is too thick, thin with a little of the reserved tomato juices. Taste and correct the seasoning.

4. Sprinkle with the optional cilantro or parsley and serve the okra hot as an accompaniment to barbecued steak or spare ribs or grilled chicken.

Note: This dish can be made several days ahead of time and reheated, covered, in a low (250° F) oven. It is also delicious as an accompaniment to fried eggs, and I often use the leftovers as a filling for omelets.

Hot Peppers

In most parts of the country there are now a bewildering number of hot peppers to choose from, and trying to figure out which is which can be overwhelming. I usually cook with 2 or 3 types that are available fresh year-round.

The Anaheim, which is quite mild, is my favorite since it lends itself to grilling, and the large ones are good for stuffing. The jalapeño is smaller than the Anaheim, and the serrano is even smaller and hotter, but both are pretty much interchangeable. I like to grow the cayenne, an elongated red pepper that usually is available in late summer at produce markets and farm-stands. It can be substituted for any other hot pepper.

The flesh of the hot pepper is not spicy. Instead, the heat comes from the seeds and the ribs that hold the seeds. If you cut a pepper in half lengthwise and remove the seeds and ribs, the intensity of the pepper will be reduced to a minimum.

You can substitute dry hot chili peppers in many cooked preparations calling for fresh hot peppers. You can control the spiciness of the dish by not breaking the pepper in half and by removing it from the skillet once it has darkened.

■■■

CREAMY SAUTÉ OF CORN AND
YELLOW SQUASH

serves 4–6

What I like best about this simple sauté is that it is all one glowing shade of yellow, a soothing change from the otherwise brilliantly colorful summer kitchen. The combination of these two seasonal favorites is both lovely and flavorful. If you are tempted to garnish the dish with a touch of green, use tiny cilantro leaves, or spice it up with minced jalapeño peppers. Serve the sauté with grilled or Cedar-Roasted Salmon (page 180), a Brochette of Swordfish (page 176), or as part of a luncheon or dinner buffet table.

2 large ears fresh corn, cooked
3 tablespoons unsalted butter
2 teaspoons peanut oil
2 tablespoons finely minced Anaheim or jalapeño pepper
3–4 cups thinly sliced small yellow squash

Salt and freshly ground black pepper
½ cup heavy cream
1 tablespoon finely minced fresh chives
1 tablespoon finely minced fresh cilantro

1. With a sharp knife remove the corn kernels from the cob and reserve.

2. Heat the butter and oil in a large, deep skillet over medium heat. When hot, add the minced hot pepper and cook without browning for 1 minute.

3. Add the yellow squash, season with salt and pepper, and reduce the heat. Simmer the squash, covered, for 3 to 4 minutes, or until just tender

4. Add the corn and cream and toss gently with the squash. Raise the heat and simmer the mixture, uncovered, until the cream is almost evaporated and has coated the vegetables well. Add the chives and cilantro, correct the seasoning, and serve hot as an accompaniment to roasted, braised, or grilled poultry or sautéed or grilled seafood.

Note: You may substitute yogurt for the cream in this dish. Be sure to use squash that are no more than 4 to 5 inches long and cook them over low heat for only a short time; otherwise, the dish will become mushy and uninteresting. You may substitute yellow zucchini for the squash; they are slightly firmer and have less water content, which gives them a better overall texture.

SAUTÉED YOUNG GREEN BEANS WITH
GINGER, GARLIC, AND PLUM TOMATOES

serves 4–6

Freshly picked garden string beans need little in the way of adornment. When young and tender, they are at their best simply cooked and tossed with a little browned butter. But, inevitably at some point in the season, I find myself with larger, stringier specimens that, while still tender, require some creative thinking. Here is a vegetable preparation that makes excellent use of the larger green beans commonly found at produce stands and supermarkets. A further advantage is that the dish can be made in advance and reheated either on top of the stove or in the microwave.

Salt
1¼ pounds young string beans, trimmed and cut on the diagonal into 1½-inch pieces
4 tablespoons extra-virgin olive oil
1 large red bell pepper, cored, seeded, and cut
2 large shallots, peeled and thinly sliced

2 large cloves garlic, peeled and finely minced
1 tablespoon finely minced ginger
8 firm but ripe Italian plum tomatoes, unpeeled and quartered, or cut into eighths, if large
Freshly ground black pepper
3 tablespoons finely minced fresh cilantro or Italian parsley

1. In a large saucepan, bring lightly salted water to a boil, add the green beans, and cook for 3 to 5 minutes or until just tender. Run under cold water to stop further cooking, drain, and reserve.

2. In a heavy 12-inch skillet heat the oil over medium heat. Add the pepper strips and sauté for 1 to 2 minutes, or until just soft.

3. Add the shallots, garlic, and ginger and cook for 1 minute without browning. Immediately add the tomatoes and green beans and sauté over medium-high heat until just heated through. Season with salt and pepper, add the cilantro or parsley, and transfer to a serving dish. Serve hot or at room temperature.

GRILLED STUFFED PEPPERS
À LA BASQUAISE

serves 6–8

Stuffed peppers are a Mediterranean and central European classic that deserve to be used more creatively by the everyday cook. Whether filled with braised bulgur, a simple pilaf of rice enhanced by a sprinkling of fresh herbs, or some leftover vegetable paella, stuffed peppers can easily become an inexpensive, undemanding, and versatile summer main course. In this recipe, the peppers are first grilled, which gives them an additional smoky taste. When using young, freshly picked green peppers, it is best not to grill them since their delicate flesh will not hold up.

6 small or medium bell peppers, a
   mixture of green, red, and yellow

FILLING
6 tablespoons olive oil
1 small dry red chili pepper, broken
1 small onion, peeled and finely minced
2 medium cloves garlic, peeled and
   finely minced
2 tablespoons finely minced fresh
   parsley
2 large ripe tomatoes, peeled and finely
   chopped
Salt and freshly ground black pepper
¼ cup finely minced fresh basil

2 hard-boiled eggs, peeled and finely
   minced
¼ cup thinly sliced pitted green olives
2 tablespoons tiny capers, drained
⅓ cup homemade bread crumbs
1 (3½-ounce) can light tuna, packed in
   oil and drained

GARNISH
1 large clove garlic, peeled and finely
   minced
2 tablespoons finely minced fresh
   parsley
12 flat anchovy fillets
12 small black oil-cured olives

1. Prepare the charcoal grill.

2. When the coals are extremely hot and glowing red, place the whole peppers directly on the coals and char quickly, turning often, until the skins have blackened evenly on all sides.

3. Remove the peppers from the grill and immediately cut out and discard the stems. This will allow steam to escape so that the peppers, when cool, will stay as crisp as possible. Set the peppers aside to cool completely.

4. When the peppers are cool, carefully rub off all of the blackened skin. If necessary, wipe with damp paper towels to remove excess pieces of skin. Do not run under water. Cut the peppers in half lengthwise, remove all seeds and membranes, and set aside.

5.   Preheat the oven to 375° F.

6.   To prepare the filling: In a heavy medium skillet heat 3 tablespoons of oil over medium heat. Add the chili pepper and cook until dark. Remove with a slotted spoon and discard. Add the onion to the skillet and cook until soft but not browned. Add the garlic, parsley, tomatoes, salt, and pepper and cook for about 15 minutes or until the tomato juices have evaporated.

7.   Add the basil, hard-boiled eggs, olives, capers, bread crumbs, and tuna. Mix well and correct the seasoning.

8.   Place the grilled pepper halves, cut side up, in a large, heavy baking dish. Fill each with some of the tomato-tuna mixture. Drizzle with the remaining oil and place in the center of the preheated oven. Bake for 15 minutes or until heated through and lightly browned.

9.   Remove from the oven and let cool. Sprinkle with minced garlic and parsley and garnish each pepper half with an anchovy fillet and a black olive. Serve at room temperature as part of an hors d'oeuvre table or as an accompaniment to grilled meats.

Note: It is very important to grill the peppers quickly over very hot coals so that they will retain their shape and stay somewhat crisp when peeled. If you char the peppers over a very hot gas flame or directly on the hot coils of an electric stove, the result will be equally delicious but quite different in terms of shape and texture.

MOROCCAN STUFFED BAKED TOMATOES

serves 8

Too little has been said in praise of stuffed vegetables. Although herbed baked tomatoes have always been part of my cooking repertoire, I had never really spent much time experimenting with filling vegetables until I came back from a recent trip to Turkey. There, the extraordinary variety of peppers, eggplants, tomatoes, and zucchini, which are stuffed with savory rice, bulgur, and meats, is truly overwhelming. Now I look forward to the abundance of summer vegetables, especially the medium-sized tomatoes, peppers, and zucchini that are perfect for stuffing. Remember that many grains make delicious stuffings, as do sausage meats and pureed vegetables. The key is to braise the stuffed vegetables slowly so that they do not burst. Always serve them at room temperature with a drizzle of virgin olive oil and plenty of cracked black pepper.

8 ripe medium tomatoes
¾ cup Chicken Stock (page 470) or bouillon
½ cup instant couscous
5 tablespoons unsalted butter
2 tablespoons very finely diced red bell pepper
2 tablespoons very finely diced zucchini, avoiding the seedy center

1 small leek, trimmed of all but 2 inches of greens, very finely diced and washed well
2 tablespoons water
Salt and freshly ground black pepper
2–3 tablespoons finely minced fresh parsley
2 tablespoons extra-virgin olive oil

1. Preheat the oven to 350° F.

2. Cut a ⅓-inch slice off each tomato on the side opposite the stem end and set these caps aside. Hollow out the tomatoes with a spoon, removing and discarding the seeds, and leaving a ⅓-inch-thick shell. Reserve.

3. Bring the stock or bouillon to a boil in a small saucepan. Stir in the couscous, return to a boil, cover, and remove from the heat. Set aside for 5 minutes, then add 2 tablespoons of the butter while still warm and mix well. Reserve.

4. In a small skillet melt the remaining butter over medium heat. Add the red pepper, zucchini, leek, and water and season with salt and pepper. Reduce the heat, cover, and cook the mixture for 5 minutes, or until just tender.

5. Add the vegetable mixture to the couscous, together with the parsley, mix well, and correct the seasoning.

6. Fill each of the tomato cavities with some of the couscous mixture, mounding slightly. Top each tomato with 1 of the reserved tomato caps and place the tomatoes, cap side up, in a heavy baking dish. Drizzle with olive oil and set the dish in the center of the preheated oven. Bake for 15 to 20 minutes or until the tomatoes are tender, basting often with the pan juices.

7. Remove from the oven and serve as a vegetable accompaniment to roast or grilled veal or as part of an hors d'oeuvre table.

BRAISED NEW POTATOES WITH CUCUMBERS, TOMATOES, AND SUMMER HERBS

serves 4–6

This combination of summer vegetables is very popular in Austrian wine bistros. There, the warm braised potatoes are tossed with cider vinegar, and the whole dish is often topped with a dollop of sour cream. After we moved to Spain, my mother adapted the dish to local ingredients, using fruity olive oil and herbs and omitting the sour cream, which is virtually unknown in Spain. I much prefer this lighter version, but if you like, you may vary it by tossing the cucumber and tomato mixture in cider vinegar and sour cream.

CUCUMBERS AND TOMATOES
2 pickling cucumbers, peeled, seeded, and cut into ¼-inch dice
2 ripe medium tomatoes, unpeeled, seeded, and cut into ¼-inch dice
3 tablespoons extra-virgin olive oil
2 tablespoons finely minced fresh parsley
6 fresh basil leaves, cut into fine julienne
Salt and freshly ground black pepper

POTATOES
3 tablespoons extra-virgin olive oil
1½ pounds small new potatoes, unpeeled and cut into ¼-inch slices
2 large cloves garlic, peeled and thinly sliced
1 tablespoon fresh thyme leaves
Salt and freshly ground black pepper
⅓–½ cup Chicken Stock (page 470) or bouillon

1. Start by marinating the cucumbers and tomatoes: In a medium bowl combine the cucumbers and tomatoes. Toss with the olive oil, parsley, and basil. Season with salt and pepper, cover, and set aside to marinate at room temperature for 1 to 2 hours

2. To prepare the potatoes: In a heavy 10-inch skillet heat the olive oil over medium-low heat. Add the sliced potatoes and garlic in layers, seasoning each layer with thyme, salt, and pepper. Add ⅓ cup chicken stock, cover, and simmer for 30 to 35 minutes, or until tender, turning the potatoes twice during cooking. If the liquid evaporates during cooking, add the remaining stock.

3. When the potatoes are done, remove the cover and sprinkle the top with the marinated cucumbers and tomatoes. Serve hot, directly from the skillet.

Note: The potatoes are equally delicious served at room temperature.

POTATO AND FARMSTAND
VEGETABLE "PAELLA"

serves 6

This dish has little in common with a traditional seafood paella, or even the Catalan vegetable paella, both of which are made with rice. Here, cubed red potatoes are substituted for the rice, creating more of a fricassee than a paella. This homey dish is often made in Spain with potatoes that are first lightly poached in a saffron-infused broth that renders the potato cubes a bright yellow color. They are then added to the vegetable mixture and left to braise until they are done. Simple as it is, it is one of my favorite summer vegetable dishes, full of lively flavor and color. It can be served with just about anything: poultry, grilled seafood, sausages, and shish kebobs. The paella can be made either spicy or benign and topped with a garlic-flavored yogurt.

4 tablespoons extra-virgin olive oil
2 medium red onions, peeled, cut in half, and thinly sliced
1 teaspoon finely minced garlic
1 tablespoon finely minced fresh thyme
1 tablespoon finely minced fresh oregano
Salt and freshly ground black pepper
2 medium zucchini, trimmed and cut into ¾-inch cubes
3 medium red new potatoes, peeled and cut into ¾-inch cubes

1 large green bell pepper, seeded, cored, and thinly sliced
4 large ripe tomatoes, peeled, seeded, and chopped
1 large red bell pepper, seeded, cored, and cut into ¾-inch cubes

GARNISH
2 tablespoons finely minced fresh Italian parsley

1. In a large, heavy skillet heat the oil over medium heat, add the onions, garlic, and herbs, season with salt and pepper, and cook the mixture for 10 minutes or until soft and lightly browned.

2. Add the zucchini, potatoes, green pepper, and tomatoes and season with salt and pepper. Cover the skillet and simmer over low heat for 25 minutes, or until the vegetables are tender but not falling apart.

3. Add the red pepper and continue to simmer for 3 to 4 minutes or until the pepper is just tender.

4. With a slotted spoon, transfer the vegetables to a side dish. Raise the heat and cook the tomato mixture for 2 or 3 minutes, or until it thickens. Return the vegetables to the skillet and correct the seasoning. Sprinkle with parsley and serve hot as an accompaniment to meats, fish, and poultry.

THYME-BRAISED ZUCCHINI
IN SOUR CREAM

serves 4

After years of sautéing zucchini, I now find that braising over low heat brings out their subtle, delicate flavor. You may braise the zucchini well in advance, but be sure not to overcook them. Add the sour cream at the last minute when reheating the dish. If you can get yellow zucchini in the summer, use them for this dish. Since the water content is much lower in this new variety of zucchini (not to be confused with yellow squash), their texture will remain firmer and the finished dish will be more interesting.

4 tablespoons unsalted butter
4 medium zucchini, about 1 1/4 pounds, trimmed and thinly sliced
Coarse salt and freshly ground black pepper

Juice of 1/2 lemon
2 tablespoons finely minced fresh thyme
3/4 cup sour cream or Crème Fraîche (page 491)

1. In a large, heavy skillet, melt the butter over low heat. Add the zucchini and season with salt, pepper, lemon juice, and thyme. Cover the skillet and braise the zucchini over low heat for 6 to 8 minutes, or until tender. Shake the skillet back and forth several times to make sure that the zucchini cook evenly.

2. Uncover the skillet. Add the sour cream, fold gently into the mixture, and just heat through. Taste and correct the seasoning and serve at once.

LEMON- AND PARSLEY-SCENTED
BRAISED BULGUR

serves 4–6

5 tablespoons unsalted butter
½ cup finely minced scallions
½ cup finely diced green bell pepper
3 tablespoons water
1 cup quick-cooking bulgur
Salt and freshly ground black pepper

2⅓ cups Chicken Stock (page 470) or bouillon
Juice of ½ lemon
Grated rind of ½ lemon
2 tablespoons finely minced fresh Italian parsley

1. In a heavy saucepan melt 3 tablespoons of the butter over medium-low heat. Add the scallions, green pepper, and water and cover and simmer until the pepper is tender, about 5 minutes. Uncover the saucepan, raise the heat, and cook until all the water has evaporated.

2. Add the bulgur to the vegetable mixture and toss gently. Season with salt and pepper and stir in the stock. Bring to a boil, reduce the heat to low, and simmer, tightly covered, for 20 to 25 minutes, or until the bulgur is tender and all the stock has been absorbed.

3. Add the lemon juice, lemon rind, parsley, and the remaining butter and fold gently into the bulgur. Taste and correct the seasoning and serve hot as an accompaniment to grilled poultry, seafood, or lamb.

SWEET STEAMED SCALLOP SALAD IN TOMATO AND BASIL VINAIGRETTE

serves 4–5

Summer is not considered the best scallop season since this most delicious of shellfish is at its most abundant from fall through spring. Still, small Maine sea scallops are widely available fresh in the summer, and while they are not quite as sweet as the winter varieties, they are excellent used in this zesty salad. It is one of my favorite additions to an hors d'oeuvre table, and it also doubles as an appetizer or even as a light-lunch main course when placed in the center of a plate of baby lettuces or used as a filling for ripe avocados.

1 pound bay scallops
½ red bell pepper, cored, seeded, and cut into fine julienne
½ yellow bell pepper, cored, seeded, and cut into fine julienne
1 small red onion, peeled, cut in half, and thinly sliced
Salt and freshly ground black pepper
1 large ripe tomato, peeled and seeded
2 tablespoons sherry vinegar
7 tablespoons olive oil

1 large clove garlic, peeled and mashed in a garlic press
10–12 large fresh basil leaves, cut into fine julienne
1 large avocado, peeled and pitted
Juice of 1 lime
1 tablespoon extra-virgin olive oil

GARNISH
Fresh cilantro leaves or Italian parsley leaves

1. Bring water to a boil in a large steamer, add the scallops, cover, and steam for 3 to 5 minutes, or just until they turn opaque. Do not overcook. Transfer the scallops to a bowl, add the peppers and onion, and toss gently. Season with salt and pepper. Reserve.

2. In the workbowl of a food processor combine the tomato, sherry vinegar, and olive oil and process until smooth. Season with salt and pepper, add the crushed garlic, and pour the dressing over the salad. Add the basil leaves, toss gently, and chill the salad until serving.

3. Cut the avocado lengthwise into ½-inch slices. Arrange 4 avocado slices on each individual serving plate, forming an open square. Sprinkle the avocado slices with lime juice and drops of extra-virgin olive oil and season lightly with salt and pepper. Just before serving, fill the avocado squares with the scallop

salad and garnish with fresh cilantro leaves or Italian parsley. Serve accompanied by crusty French bread.

Note: To save time, you need not peel the tomato, just dice it and remove the seeds.

If you can find small ripe avocados, halve and pit them, season the cavities, and fill with the scallop salad. The salad will keep well if made a day ahead of time. Add the julienne of basil leaves at the last moment.

Instead of the avocado garnish, arrange the scallop salad on Boston or Bibb lettuce leaves lightly seasoned with drops of lime juice and extra-virgin olive oil.

CORSICAN EGGPLANT SALAD WITH ONIONS, OLIVES, AND CAPERS

serves 6–8

A marinated vegetable salad is a delicious and inexpensive way to take advantage of the bounty of summer produce—from either the farmstand or your own garden. I usually make one whenever I find myself with the usual garden surplus of peppers, zucchini, tomatoes, and eggplant. Once cooked and marinated, the vegetable mixture will keep for up to two weeks. Include this salad as part of an antipasto table with some sliced prosciutto, oil-cured olives, and, if you like, some fresh mozzarella dressed in virgin olive oil, sherry vinegar, and plenty of cracked black pepper. The salad is also wonderful served with a juicy hamburger or your favorite summer sandwich.

3 medium eggplants, about 3 pounds in all, unpeeled and cut into ¾-inch cubes
Salt
½–¾ cup olive oil
2 medium red onions, peeled, quartered, and thinly sliced
2 tablespoons tiny capers, drained
½ cup small black oil-cured olives, pitted and cut in half

3 tablespoons finely minced fresh Italian parsley
12–18 whole cloves garlic confit (page 94)
5 tablespoons Port and Shallot Vinaigrette (page 481)
Freshly ground black pepper

1. Place the eggplant cubes in a large colander. Sprinkle with salt and let drain for 1 hour.

2. Pat the eggplant dry with plenty of paper towels and set aside.

3. In a heavy skillet heat 3 tablespoons of the olive oil over medium-high heat. Add one third of the eggplant and sauté until lightly browned on all sides.

Transfer to a double layer of paper towels and set aside.

4. Continue to sauté the remaining eggplant, adding oil to the skillet for each batch, and transfer the cooked eggplant to double layers of paper towels to drain.

5. Add 3 tablespoons of oil to the skillet, add the onions, and cook over medium-low heat for 10 minutes.

6. Transfer the sautéed eggplant to a large serving bowl. Add the cooked onion mixture, together with the capers, olives, parsley, and garlic confit. Pour the vinaigrette over the vegetables, season with salt and pepper, and toss gently. Cover and refrigerate overnight.

7. The next day, bring the salad back to room temperature and serve as part of an hors d'oeuvre table.

RED AND WHITE RADISH SALAD WITH ARUGULA AND YOGURT VINAIGRETTE

serves 4

½ cup plain yogurt
2 tablespoons red wine vinegar
3 tablespoons extra-virgin olive oil
1 large clove garlic, peeled and mashed
¼ cup finely minced red onion
2 tablespoons finely minced fresh mint
Salt and freshly ground black pepper
1 cup daikon radish, cut into small

matchsticks (¼ × ¼ × 1 inch)
10–12 red radishes, trimmed and thinly sliced
8–10 white icicle radishes, trimmed and thinly sliced
3 cups arugula leaves, washed and dried

1. In a large serving bowl, combine the yogurt, vinegar, oil, and garlic and whisk until well blended. Add the onion and fresh mint and season with salt and pepper.

2. Add the daikon and red and white radishes to the bowl and toss gently to coat with the yogurt dressing. Cover and refrigerate 4 to 6 hours, or overnight.

3. Remove the salad from the refrigerator 20 minutes before serving.

4. Just before serving, add the arugula leaves and toss gently with the radish salad. Taste and correct the seasoning. Serve at once with thinly sliced buttered black bread.

VINE-RIPENED TOMATO SALAD IN
BASIL-YOGURT DRESSING

serves 4–6

Here's a good addition to your summer repertoire. When the weather is very hot, I arrange the salad on a bed of lettuce and use it as the centerpiece for the meal with, perhaps, some sliced mozzarella and crusty semolina bread as accompaniments.

Tomatoes like the company of their garden friends, so if you use this as one of several buffet salads, you might add a cucumber-and-radish salad and some cool green beans in a dill and mustard vinaigrette. For a more substantial meal, the Grilled Flank Steaks (page 195) or Lilac-Smoked Cornish Hens (page 150) would complete this lighthearted warm-weather menu.

DRESSING
1 medium clove garlic, peeled and finely
   minced
2 teaspoons red wine vinegar
2 tablespoons extra-virgin olive oil
3 tablespoons yogurt
1 tablespoon Basil Paste (page 170)
Salt and freshly ground black pepper

SALAD
3 tablespoons finely minced red onion

1 medium red bell pepper, cored,
   seeded, and finely diced
8–10 small ripe tomatoes (about 1–1½
   inches in diameter), cut into sixths
¼ cup fresh cilantro leaves

GARNISH
3–4 tablespoons crumbled feta cheese
Leaves of garden lettuces

1. First prepare the dressing: In the workbowl of a food processor or blender, combine the garlic, vinegar, oil, yogurt, and basil paste and process until smooth. Season with salt and pepper to taste and set aside.

2. To make the salad: Combine the onion, pepper, tomatoes, and cilantro leaves in a serving bowl, pour the dressing over the vegetables, and toss gently. Correct the seasoning.

3. Garnish with the crumbled feta cheese and serve on leaves of garden lettuces.

FARMSTAND ZUCCHINI, CHERRY TOMATO, AND TUNA SALAD IN A BASIL VINAIGRETTE

serves 6

VINAIGRETTE
1 cup loosely packed fresh basil leaves, stems removed
½ cup extra-virgin olive oil
2–3 tablespoons red wine vinegar
2 medium cloves garlic, peeled and mashed
Salt and freshly ground black pepper

8 small zucchini, about 1¾ pounds total

1 small red onion, peeled, quartered, and thinly sliced
18 ripe cherry tomatoes, cut in half

GARNISH
10–12 small black oil-cured olives
1 (3½-ounce) can light tuna, packed in oil, drained and flaked
Sprigs of fresh basil

1. In the container of a blender or food processor, combine the basil, oil, 2 tablespoons of the vinegar, and the garlic and blend until quite smooth. Season with salt and pepper and set the vinaigrette aside for 30 minutes.

2. Bring lightly salted water to a boil in a large saucepan, add the whole zucchini, and blanch for 3 minutes. Remove and run under cold water to stop further cooking. Drain well on paper towels. Trim the zucchini, cut in half lengthwise, and then crosswise into ½-inch slices.

3. Combine the zucchini, red onion, and cherry tomatoes in a serving bowl. Add the basil vinaigrette, season with salt and pepper, and toss gently. Set the salad aside for 2 to 3 hours to develop flavor.

4. Just before serving, taste and correct the seasoning, adding a large grinding of black pepper and the remaining vinegar if necessary. If the salad was refrigerated, remove from the refrigerator 20 minutes before serving. Garnish with the black olives, tuna, and sprigs of fresh basil and serve with crusty Italian bread.

Zucchini

Unless you are using tender summer zucchini, which have very few seeds, you should discard the seedy inside of the vegetable since it is watery and tasteless. To do so, using a sharp knife and cutting lengthwise, remove the skin and ¼ inch of the pulp. You will be left with the seedy center, which should be discarded. This step is most important when julienning or cubing the vegetable for use in pasta or rice preparations, but not for sautéing or braising as a vegetable by itself.

■ ■ ■

TWO-TONE ZUCCHINI AND
PEPPER SALAD PARMA

serves 4–6

The immense popularity of zucchini is due not only to their easy cultivation and overall availability, but also to their amazing versatility. Somehow, when I think that I have literally tried every preparation possible for this lovely vegetable, I come across an interesting variation such as this, which was part of a simple antipasto table at Marino's, a popular trattoria in Parma.

3 medium zucchini, about 1 pound, trimmed
3 medium yellow squash, about 1 pound, trimmed
6 tablespoons extra-virgin olive oil
1 tablespoon sherry vinegar
Juice of 1 lemon, or more to taste
2 large cloves garlic, peeled and mashed

1 medium red bell pepper, cored, seeded, and thinly sliced
1 medium yellow bell pepper, cored, seeded, and thinly sliced
⅓ cup fine julienne of fresh basil
¼ cup finely minced fresh parsley
Salt and freshly ground black pepper

1. Cut the zucchini and yellow squash in half lengthwise and then crosswise into ¼-inch slices. Reserve.

2. Bring water to a boil in a vegetable steamer, add the zucchini and yellow squash, and steam, covered, for 3 minutes, or until barely tender. Remove from the steamer and set aside.

3. Combine the oil, vinegar, lemon juice, and garlic in a serving bowl and whisk until well blended. Add the steamed zucchini and squash, the red and yellow peppers, basil, and parsley. Toss gently in the vinaigrette. Season with salt and pepper, cover, and chill for 1 to 2 hours.

4. Thirty minutes before serving, bring the salad back to room temperature. Taste and correct the seasoning, adding more lemon juice if necessary, and serve as an appetizer accompanied by thinly sliced prosciutto or hard salami.

ZUCCHINI SALAD WITH PINK-LEMON
MAYONNAISE

serves 4

If you happen to grow your own zucchini, you'll want to come up with some fresh ways of preparing this prolific vegetable. Even if you don't grow them, zucchini and summer squash play such a major role in the summer kitchen that creativity in using them is a must. This salad, which is at its best made a day ahead, doubles as an appetizer and is also good with simply grilled or sautéed fish steaks. For a variation, and color contrast, use both zucchini and yellow squash. If you are a gardener be sure to add one or even two plants of yellow zucchini to your vegetable patch. These are far less watery than their green relatives and their taste and texture are wonderful in both sautéed and steamed dishes.

⅓–½ cup Mayonnaise (page 477)
2 tablespoons diced pimientos
1 teaspoon finely grated lemon rind
Salt and freshly ground black pepper
3 tablespoons olive oil
Juice of 1 lemon
6–8 small zucchini, about 1 pound, trimmed and cut crosswise into ¼-inch slices

3 tablespoons finely minced fresh parsley

GARNISH
1 tablespoon finely minced fresh parsley
1 tablespoon finely minced fresh chives
4–6 radishes, thinly sliced

1. Combine the mayonnaise and pimientos in the workbowl of a food processor or blender and process until the pimientos are very finely minced but not smooth. Add the lemon rind and season with salt and pepper. Transfer the pink mayonnaise to a small bowl, cover, and chill.

2. In a medium serving bowl combine the olive oil and lemon juice and season with salt and pepper. Whisk until well blended and set aside.

3. Bring salted water to a boil in a vegetable steamer. Add the zucchini, cover, and steam for 4 to 6 minutes, or until just tender. Transfer the zucchini to the serving bowl, toss to coat evenly with the vinaigrette, and set aside to cool completely.

4. Drain the zucchini well and return to the bowl. Add the minced parsley together with ⅓ cup of the pink mayonnaise and toss the salad gently, being careful not to break the zucchini slices. Taste and correct the seasoning, adding a large grinding of pepper, cover, and chill for 2 to 4 hours before serving.

5. Thirty minutes before serving, remove the salad from the refrigerator and bring back to room temperature. Taste and correct the seasoning; if the salad

seems a bit dry, fold in the remaining mayonnaise. Sprinkle with finely minced parsley and chives, and garnish with sliced radishes.

Variation: The salad may also be served on individual serving plates as part of a light luncheon. Do not combine the mayonnaise with the zucchini; instead, marinate the zucchini in the lemon vinaigrette and use the mayonnaise as part of the garnish. Mound the salad in the center of salad plates. Spoon the mayonnaise around the salad and decorate with anchovy fillets and a sprinkling of minced parsley. Serve at once.

COOL RICE SALAD WITH FRAGRANT INDIAN SPICES

serves 6–8

2 tablespoons unsalted butter
1 teaspoon plus 5–7 tablespoons extra-virgin olive oil
1 medium onion, peeled and finely minced
1–2 teaspoons finely minced fresh jalapeño pepper
1 teaspoon tomato paste
1½ teaspoons Madras curry powder
½ teaspoon ground cumin
1½ cups Italian rice, preferably Arborio
3–4 cups Chicken Stock (page 470)

2 large cloves garlic, peeled and finely minced
½ cup finely minced fresh parsley
3 medium scallions, finely minced
4 small dill gherkins, finely diced
1 tablespoon tiny capers, drained
½ cup finely diced green bell pepper
½ cup finely diced red bell pepper
3 tablespoons sherry vinegar
Salt and freshly ground black pepper
3–4 tablespoons finely minced fresh cilantro, optional

1. Heat the butter and 1 teaspoon of the olive oil in a 3-quart saucepan over medium heat. Add the onion and cook until soft but not browned. Add the jalapeño pepper and cook until soft.

2. Add the tomato paste, curry powder, and cumin and stir the mixture for 1 minute. Add the rice and stir to coat completely with the cumin mixture, then add 3 cups of the stock. Bring to a boil, reduce the heat, and simmer the rice mixture, covered, for 25 to 30 minutes, or until the rice is tender. If the liquid evaporates before the rice is done, add the remaining stock. Remove the rice to a serving bowl and set aside to cool.

3. Add 5 tablespoons of the remaining olive oil, the garlic, parsley, scallions, gherkins, capers, green and red peppers, and vinegar to the rice. Toss gently, but thoroughly, and season with salt and pepper. Cover and set aside at room temperature for 1 to 2 hours.

4. Just before serving, taste and correct the seasoning. Add the remaining oil if necessary, the optional cilantro, and a large grinding of black pepper. Serve with grilled meats.

BANANA-CARAMEL SORBET WITH
CHOCOLATE SAUCE

serves 6

¾ cup granulated sugar
2 tablespoons water plus ½ cup hot
  water
¼ cup hot heavy cream
3 large very ripe bananas, peeled and
  cut into pieces

1 tablespoon fresh lemon juice or more
  to taste
1 recipe warm Basic Chocolate Sauce
  (page 489)

1. In a 2-quart saucepan, combine the sugar and the 2 tablespoons of water and place the pan over high heat. Bring to a boil, stir once to dissolve the sugar, and continue to boil, without stirring, until the mixture is hazelnut brown. Be careful not to let the caramel become too dark. Remove the pan from the heat and immediately add the ½ cup hot water and the hot cream, averting your face and stirring constantly until the foam subsides. Return the pan to the heat and stir until the cream and water are completely incorporated. Set aside to cool completely.

2. Combine the bananas and lemon juice in the workbowl of a food processor and process until smooth. Add the cooled caramel cream and process until completely incorporated.

3. Freeze the mixture in an electric ice cream machine, following the manufacturer's instructions. Serve very cold with warm chocolate sauce.

COEUR À LA CRÈME WITH SUMMER BERRIES

serves 8

Some desserts never seem to go out of style, and some actually become better as time goes on—better because they epitomize the good things of earlier times, when such basics as butter, cream, and eggs were fresh and full of flavor. This is true of coeur à la crème, an old-fashioned country dessert that calls for three simple ingredients: heavy cream, sugar, and cream cheese. The coeur is usually served with a bowl of fresh strawberries, but I prefer the somewhat more complex layering of textures in this recipe: a strawberry or raspberry sauce and freshly sliced fruit. You may add the blueberries and blackberries to the sauce or use them as a garnish.

¼ pound mascarpone
¼ pound cream cheese, softened
½ cup confectioners' sugar
2 cups heavy cream (preferably not ultrapasteurized), lightly whipped

SAUCE
2 (10-ounce) packages frozen raspberries in syrup, thawed and drained, reserving the syrup from 1 package
¾ cup good-quality red currant jelly

2–4 tablespoons kirsch or Grand Marnier
2 cups sliced fresh strawberries
1 tablespoon fresh lemon juice

GARNISH
2 cups mixed berries, such as raspberries, blueberries, and/or blackberries
1 cup tiny balls of cantaloupe melon
Tiny fresh mint leaves

1. Line 8 individual coeur à la crème molds with a double layer of cheesecloth, so that the cheesecloth extends 1 inch over each side. Set aside.

2. In a mixing bowl combine the mascarpone and cream cheese with the sugar and whisk until quite smooth. Add the heavy cream and fold gently but thoroughly into the cream cheese mixture. Place 2 tablespoons of the mixture into each of the cheesecloth-lined molds. Place the molds on a jelly-roll pan and refrigerate overnight.

3. To prepare the sauce: Place the thawed raspberries in the workbowl of a food processor together with the reserved raspberry syrup and process until smooth. Pass the puree through a fine sieve into a bowl and set aside.

4. Heat the jelly in a small saucepan over low heat. Remove from the heat, add the kirsch, and whisk the jelly into the puree until well blended. Fold in the sliced strawberries together with the lemon juice and chill until serving.

5. Unmold the coeurs à la crème onto a serving plate, peel off and discard the cheesecloth, and spoon some of the sauce around them. Garnish the sauce with fresh mixed berries, tiny balls of melon, and mint leaves. Serve at once.

BLUEBERRY-GINGER CUSTARD TART

serves 6

I can't think of a single Asian ingredient that has been adopted with more enthusiasm by American cooks than fresh ginger. It has become a kitchen basic, finding its way into marinades, vinaigrettes, both savory and sweet sauces, and even custards. Ginger has a natural affinity for many vegetables, but it is also wonderful with a variety of fruits, especially pears and blueberries.

In this tart the custard is subtly flavored with fresh ginger. If you like your ginger more assertive, add a tablespoon of very finely minced preserved stem ginger, which lends a sweeter, more assertive taste and additional texture to the custard.

If you'd rather not make a tart shell, you can bake the custard in shallow flan dishes, then top them with the blueberry mixture.

CUSTARD
3 large eggs
½ cup granulated sugar
½ teaspoon vanilla extract
1 teaspoon finely grated lemon rind
1 (½-inch) piece of fresh ginger, peeled and mashed through a garlic press
1 cup hot heavy cream
1 prebaked 9-inch Basic Tart Shell (page 492), in a pan with a removable bottom

BLUEBERRIES
1 pint fresh blueberries, rinsed
2 teaspoons finely grated lemon zest
⅓ cup granulated sugar
1–2 tablespoons cassis, optional

GARNISH
1 cup fresh blueberries, rinsed
Tiny fresh mint leaves

1. Preheat the oven to 400° F.

2. In a large mixing bowl combine the eggs, sugar, vanilla, lemon rind, and ginger and whisk until well blended. Add the hot cream and again whisk until well blended.

3. Transfer the mixture to a heavy-bottomed 2-quart saucepan and place over medium-low heat. Whisk constantly for 2 minutes, or until the mixture is just hot.

4. Immediately pour the custard into the prepared tart shell and bake in the center of the preheated oven for 10 minutes, or until the custard is barely set. Remove from the oven and let cool completely.

5. While the tart is cooling, prepare the blueberries: Combine the blueberries, lemon zest, and sugar in another heavy 2-quart saucepan. Bring to a boil, reduce the heat, and simmer, stirring often, for 10 to 15 minutes, or until the juices have reduced and the mixture has thickened. Remove the blueberry mixture from the heat, stir in the optional cassis, and let cool slightly.

6. Pour the blueberry mixture over the top of the custard and spread to completely cover the entire surface. Set aside to cool.

7. Just before serving, sprinkle the tart with the fresh blueberries and garnish with tiny mint leaves. Serve at room temperature, cut into wedges.

Variation: At the end of the summer, instead of blueberries, use Bartlett pears, cut into eighths and sautéed in butter and sugar.

BING CHERRY TART WITH
HAZELNUT-BUTTER CUSTARD

serves 6

After a seemingly endless stream of fresh-fruit medleys and compotes, I find that a summer fruit tart is not only a welcome relief but an especially appropriate dessert for a formal dinner. Whenever I have a free moment—and if it happens to be a cool day, so much the better—I roll out the pastry crust, which can be frozen right in the tart pan. I then decide which fruit to use, depending on availability and ripeness. Poached cherries are at the top of my list of summer favorites, and I like to use them in this buttery tart. Brown butter gives the custard a slight taste of hazelnuts that sets off the flavor of the cherries. Other fruits—fresh raspberries or wine-poached nectarines—can be substituted for the cherries with equally interesting results.

CHERRIES
3 cups water
1 cup granulated sugar
1 (3-inch) piece of cinnamon stick

3 whole cloves
1 pound Bing cherries, pitted

6 tablespoons unsalted butter
1 (3-inch) piece of vanilla bean, split
3 large eggs
⅔ cup granulated sugar
¼ cup all-purpose flour
¾ cup heavy cream
2 tablespoons finely grated orange rind
½ teaspoon vanilla extract

1 partially baked 9-inch Basic Tart Shell (page 492), in a pan with a removable bottom
3 tablespoons dark-brown sugar combined with a pinch of ground cinnamon

GARNISH
Bowl of lightly sweetened Crème Fraîche (page 491)

1. Start by poaching the cherries: In a large saucepan combine the water, sugar, cinnamon stick, and cloves. Bring to a boil, reduce the heat, and simmer for 5 minutes. Add the cherries and simmer for 5 minutes longer or until just tender. Remove the pan from the heat and let the cherries cool completely in their poaching liquid.

2. Transfer the cooled cherries to a colander and let drain for 30 minutes to remove all the liquid. Set aside.

3. Preheat the oven to 375° F.

4. To prepare the custard: Melt the butter in a medium saucepan. Add the vanilla bean and cook over medium heat, stirring constantly, until the butter turns a golden brown. Immediately pour the brown vanilla butter into a bowl, remove and discard the vanilla bean, and set the butter aside to cool slightly.

5. In a large mixing bowl combine the eggs, sugar, and flour and whisk until well blended. Add the cream, orange rind, vanilla extract, and vanilla butter and whisk until smooth.

6. Arrange the drained cherries evenly over the bottom of the tart shell and pour the custard over them. Place in the center of the preheated oven and bake for 35 to 40 minutes, or until the custard is set and nicely browned. A knife when inserted should come out clean. Remove the tart from the oven and let cool completely.

7. Just before serving, preheat the broiler.

8. Sprinkle the entire surface of the tart with the mixture of brown sugar and cinnamon. Make a rim of foil and place it on the edge of the tart to protect the crust. Place the tart in the broiler, 6 inches from the source of heat, until the sugar melts and is caramelized. Do not burn.

9. Remove the tart from the broiler and cut into wedges. Serve at once, accompanied by a bowl of lightly sweetened crème fraîche.

KIR-POACHED CHERRIES WITH
PRALINE MERINGUE

serves 6

Cherries and grapes are considered finger-fruits, and indeed a bowl of lightly chilled fresh cherries is a simple and welcome dessert during their relatively short season. But cherries are actually more interesting when poached in a red– or white wine–cinnamon–flavored sugar syrup. What's more, the poached cherries will keep for several months stored in glass jars in the refrigerator. They can be used for many exciting desserts such as bread pudding, clafoutis, cherry tart, and in this handsome presentation where they are tucked under a swirl of praline meringue.

CHERRIES
2⅔ cups dry white wine
⅓ cup cassis
1 (3-inch) piece of cinnamon stick
3 whole cloves
1 cup granulated sugar
1 pound Bing cherries, pitted

MERINGUE
4 large egg whites
½ cup superfine sugar
½ teaspoon ground cinnamon
⅓ cup Praline Powder (page 489)

1. Start by poaching the cherries: In a large saucepan combine the wine, cassis, cinnamon, cloves, and sugar. Bring to a boil, reduce heat, and simmer 5 minutes. Add the cherries and simmer for 5 to 7 minutes, or until just tender.

2. Remove the cherries with a slotted spoon and divide between 6 porcelain or ovenproof bowls and set aside.

3. Bring the poaching liquid to a boil and reduce by half. Pour about 2 to 3 tablespoons of syrup into each of the bowls and chill for 2 to 4 hours.

4. Preheat the oven to 400° F.

5. To make the meringue: In a large bowl whisk the egg whites until soft peaks form. Gradually add the sugar and continue whisking until the meringue is shiny and firm. Gently fold in the cinnamon and praline powder.

6. Spoon the meringue into a pastry bag fitted with a plain-tipped nozzle and pipe the meringue in a large circular mound over the cherries to completely cover each surface. The cherries must not be exposed. If you do not feel comfortable using a pastry bag, you may spoon the meringue onto the cherries instead. Set in the upper third of the preheated oven and bake for 2 minutes or until the tops are nicely browned. Serve immediately.

POACHED FIGS IN PORT AND
RED WINE SYRUP

serves 4–6

Figs of many varieties abound in every southern European and North African country, and this most sensual of fruits is synonymous with a Mediterranean summer. The fig tree in our garden in Barcelona bore fruit in such profusion that during the season we ate figs at almost every meal. Unfortunately, in the Northeast, figs are always expensive and often not as ripe as they should be. Even so, they are always delicious when poached, a welcome change from the usual selection of summer fruits. Served with crème fraîche, this is a delicious dessert; you can also use it as an accompaniment for grilled quail, squab, or duck breasts.

2½ cups dry red wine
1 cup tawny port
¾ cup granulated sugar
6 whole black peppercorns
4 coriander seeds
1 (3-inch) piece of cinnamon stick
2 whole cloves

2 whole allspice berries
20–24 fresh, small whole purple figs

GARNISH
Tiny fresh mint leaves
Crème Fraîche (page 491)

1. Combine all the ingredients except the figs in a large saucepan. Bring to a boil, reduce the heat, and simmer for 10 to 15 minutes.

2. Add the figs and continue to simmer for 10 minutes longer or until the figs are tender but not falling apart. Remove from the heat and let the figs cool completely in the poaching liquid. Carefully remove the figs with a slotted spoon to a shallow serving bowl and set aside.

3. Strain the poaching liquid and return to the saucepan. Bring to a boil and reduce until syrupy. Spoon the syrup over the figs and toss gently to glaze the figs. Let cool completely.

4. To serve, place some figs in glass compote dishes. Garnish with tiny mint leaves and serve at room temperature accompanied by a bowl of crème fraîche.

MANGO AND LEMON BAVARIANS

serves 8

3 large ripe mangoes
1 envelope unflavored gelatin
3 tablespoons water
4 large egg yolks

½ cup granulated sugar
1¼ cups hot whole milk
1 teaspoon vanilla extract
1 cup heavy cream, whipped

1. Peel the mangoes with a sharp knife and cut the pulp away from the pit. Place the pulp in the workbowl of a food processor and process until smooth. You should have about 1½ cups mango puree. Reserve.

2. In a small saucepan combine the gelatin and water and set aside for 5 minutes to soften. Place the saucepan over low heat and stir until the gelatin is completely dissolved. Keep warm.

3. Combine the egg yolks and sugar in a medium mixing bowl and whisk until fluffy and pale yellow. Add the hot milk, whisk until well blended, and transfer to the top of a double boiler. Set over simmering water and whisk constantly until the mixture thickens and heavily coats a spoon.

4. Immediately transfer the custard to a large mixing bowl. Add the dissolved gelatin, mango puree, and vanilla extract and whisk until well blended. Chill the mixture until it begins to set, whisking it every few minutes. Do not let set completely.

5. Fold in the whipped heavy cream, gently but thoroughly, and spoon the mixture into 8 (4-ounce) porcelain ramekins. Chill until completely set. Serve chilled with Crisp Brown Sugar and Almond Cookies (page 450).

COMPOTE OF PEACHES AND BLUEBERRIES

serves 8

Whenever I think of poached fresh-fruit compotes, I always remember the beautiful and elegant cart at the Connaught Grill in London. Wheeled to your table, the cart is filled with a seemingly endless array of the most glorious compotes displayed in oversized silver bowls, with a huge pitcher of Devonshire cream ready to be poured over your choice of fruits. Personally, I find summer fruit compotes to be the season's most satisfying and welcome desserts. Unfortunately, a truly delicious compote is usually made with a large amount of sugar, something many of us are trying to avoid. This compote, therefore, is even more welcome since it requires a small amount of sugar, due to the natural sweetness of the ripe blueberries.

With a "u-pick" blueberry patch near my house, I am especially lucky and can start making this refreshing compote as soon as the ripe berries are ready to be picked. Store-bought blueberries and peaches can also be extremely sweet, and I suggest that you make this compote whenever the ripe fruit is available. Besides peaches, you can add nectarines, plums, and fresh figs to the compote. Served lightly chilled with sugared crème fraîche and an additional sprinkling of fresh raspberries, it becomes a showy and most elegant dessert.

6 large freestone peaches
2 pints fresh blueberries, rinsed
¾–1 cup granulated sugar
12 small fresh whole black figs, optional
12 purple plums, pitted and quartered,
   optional

GARNISH
Dollops of sweetened Crème Fraîche
   (page 491)
Tiny fresh mint leaves

1. Bring plenty of water to a boil in a large saucepan. Drop the peaches into the water and blanch for 30 seconds. Immediately remove the peaches and plunge into a bowl of ice water. When cool enough to handle, peel the peaches with a sharp knife. Cut in half, remove the pit, and cut each half into 3 wedges. Reserve.

2. In a very large casserole combine the peaches, blueberries, and ¾ cup sugar or more, depending upon the sweetness of the fruit. Bring to a simmer, stirring often, and when the berries just begin to exude some of their juices,

add the optional figs and plums. Simmer the mixture for 10 to 12 minutes, or until the peaches, figs, and plums are just tender. Do not cook too long or the compote will turn into jelly.

3. Remove the compote from the heat and cool completely. Serve slightly chilled in individual dessert bowls with a dollop of sweetened crème fraîche and a few leaves of fresh mint.

Fruit Compotes

Almost any fruit can be turned into a delicious and refreshing fruit compote. Generations of cooks have used this method as a way to make use of either overripe or underripe fruit. Therefore it is almost impossible to give exact proportions of sugar for any given compote because so much depends on the type of fruit used and its ripeness. As a rule of thumb, I usually use 1 cup of sugar to 3 cups of either water or wine and like to add 2 to 3 thin slices of lemon peel, a cinnamon stick, 3 cloves, and a 3-inch piece of fresh vanilla bean for additional flavor and interest to about 2 to 3 pounds of fruit.

In recent years I have started to add less sugar and instead often combine berries such as blueberries, strawberries, or raspberries which exude a natural sweet-and-sour juice that has a purity of taste that is quite refreshing.

When making a fruit compote out of underripe fruit, you will need more sugar. Poach the fruit well ahead of time, letting it absorb sweetness in the syrup for several days before serving.

PECAN-ROASTED PEACHES IN SAUTERNES

serves 6

Even with a glut of wonderful peaches that I can get from two nearby peach farms, the season for this spectacular fruit seems to be here one day and gone the next. True, you can buy good peaches at supermarkets and farmstands, but it is not quite the same experience as walking up to a tree and filling your basket with the sun-ripened fruit.

My favorite way of cooking peaches is to poach them in either a sugar syrup or ginger- or peppercorn-flavored red wine. For a homey dessert try warm baked peaches filled with nuts and a brown-sugar mixture and accompanied by a freshly made vanilla ice cream or, even better, your own peach sorbet.

6 large semi-ripe freestone peaches
½ cup Sauternes or cream sherry
1 cup pecan halves, finely minced
4 tablespoons dark-brown sugar

5 tablespoons unsalted butter, softened
2 tablespoons all-purpose flour
1 cup Crème Fraîche (page 491)
3 tablespoons granulated sugar

1. Preheat the oven to 350° F.

2. Cut the peaches in half with a sharp knife and remove the pits. Arrange the peaches in a heavy baking dish, cavity side up. Spoon the Sauternes or sherry into the dish and reserve.

3. In a bowl, combine the pecans, brown sugar, butter, and flour and blend thoroughly. Fill the cavity of each peach half with some of the pecan mixture.

4. Place the peaches in the center of the preheated oven and bake for 20 minutes. If the topping still has not browned, run the baking dish quickly under the broiler.

5. Cool the peaches to room temperature, then transfer them to individual dessert plates. Combine the crème fraîche and granulated sugar in a small bowl, stir the mixture into the pan juices, and spoon the sauce over the peaches. Serve warm.

Note: Nectarines are a good substitute for peaches in the early part of the season. The pecan mixture can be used as a topping for early Bartlett pears with delicious results.

KIRSCH-SOAKED PINEAPPLE IN
PINEAPPLE ZABAGLIONE

serves 4

PINEAPPLE
1 large ripe pineapple, trimmed
2 tablespoons good-quality kirsch
2 tablespoons granulated sugar

ZABAGLIONE
4 large egg yolks
½ cup granulated sugar

4 ounces frozen pineapple concentrate,
  thawed
2 tablespoons good-quality kirsch
2 tablespoons Sauternes
1 cup heavy cream, whipped

GARNISH
Tiny fresh mint leaves

1.   Start by preparing the pineapple: With a sharp knife, remove and discard the peel and core from the pineapple and cut the pulp into 1-inch cubes. Place the cubes in a bowl, add the kirsch and sugar, and toss gently but thoroughly. Set aside to macerate for 1 hour.

2.   To prepare the zabaglione: In a medium bowl, combine the egg yolks and sugar and whisk until fluffy and pale yellow. Add the pineapple concentrate, kirsch, and Sauternes and whisk until well blended. Transfer the mixture to a very heavy 2-quart saucepan, preferably copper, and place over low heat. Whisk constantly until the mixture thickens and heavily coats a spoon; do not overcook or the yolks will curdle. Immediately remove from the heat and transfer the mixture to a large bowl. Refrigerate until completely cool. Fold the whipped cream into the zabaglione, gently but thoroughly.

3.   To serve, place some of the macerated pineapple in individual shallow serving dishes, spoon some of the zabaglione over each portion, and garnish with tiny mint leaves.

FIELD STRAWBERRIES IN CRÈME ROUGE

When someone asks me for a quick summer dessert, I immediately think of this one. Nothing could be simpler or more refreshing than truly ripe strawberries perfumed with orange liqueur and topped with a gorgeous pink cloud of pureed berries and cream.

CRÈME ROUGE
2 (10-ounce) packages frozen
    strawberries in syrup (preferably Bird's
    Eye), defrosted and well drained
½ cup superfine sugar
1½ cups heavy cream, whipped
¾ cup sour cream

STRAWBERRIES
3 pints medium fresh strawberries,
    washed and hulled

4–5 tablespoons superfine sugar
3 tablespoons Grand Marnier,
    Cointreau, or Triple Sec

GARNISH
Sprigs of fresh mint

1. Start by making the crème rouge: In the workbowl of a food processor or blender, combine the well-drained strawberries and sugar and process until smooth. Pass the puree through a fine sieve into a bowl and set aside.

2. Whisk together the whipped cream and sour cream in a large bowl. Fold in the pureed strawberries, cover, and refrigerate for 2 to 4 hours.

3. To prepare the strawberries: In another large bowl, combine the fresh strawberries with 3 tablespoons of sugar and the liqueur. Let marinate for 2 hours. Taste and if the strawberries are not sweet enough, add the remaining sugar.

4. To serve, place the fresh strawberries with their juices in individual dessert bowls. Top with some crème rouge and garnish with sprigs of fresh mint. Serve at once.

Note: Any leftover crème rouge can be combined with fresh blueberries and served as a dessert on its own with sponge cake or ladyfingers.

CHOCOLATE MOUSSE ICE CREAM

makes about 5 cups

Although a fruit dessert is thought of as the ideal ending for a summer meal, that doesn't mean one's craving for chocolate automatically disappears with the first warm breezes of June. Quite the contrary, I've found, since this rich, homemade chocolate ice cream is one of the most popular summer desserts I teach. I serve it either by itself or with Honey Ice Cream (page 233) or Banana-Caramel Sorbet (page 219). If you let the ice cream melt to a somewhat saucelike consistency, it will be delicious served with the first late-summer poached pears or over sautéed bananas.

6 ounces bittersweet chocolate, preferably Tobler or Lindt, broken into small pieces

2 tablespoons strong prepared coffee or 1 teaspoon instant espresso mixed with 2 tablespoons water

2 tablespoons unsweetened cocoa, preferably Droste's

2½ cups heavy cream, preferably not ultrapasteurized

1 cup whole milk

6 large egg yolks

⅔ cup granulated sugar

1. Combine the broken chocolate and coffee in the top of a double boiler, place over simmering water, and stir until the chocolate has melted and the mixture is quite smooth. Add the cocoa and stir gently until thoroughly incorporated. Keep warm over a pan of warm water.

2. In a small saucepan combine 2 cups of the heavy cream and the milk, place over medium heat, and stir until just hot; do not boil. Set aside.

3. Beat the egg yolks and sugar in a large mixing bowl until fluffy and pale yellow. Whisk in the hot cream-milk mixture and transfer to a medium saucepan. Place over medium heat and whisk constantly until thick and smooth, adjusting the heat, if necessary, so as not to curdle the custard.

4. Remove the custard from the heat, add the warm chocolate-coffee mixture, and whisk until smooth. Let the mousse cool to room temperature, cover, and refrigerate until thoroughly chilled. This is best done the day before.

5. Place the chocolate mousse in an electric ice cream machine and freeze, following the manufacturer's instructions, for 15 minutes or until thick but still soft.

6. Meanwhile beat the remaining ½ cup of cream and add to the ice cream

machine while still running. Freeze the mixture completely. Immediately remove the ice cream from the machine and serve at once.

HONEY CUSTARD ICE CREAM

makes 1 quart

Summer is unimaginable without ice cream, and a good homemade ice cream or sorbet provides an instant pleasure that cannot be duplicated by even the best commercial preparation. Of course, you can create intriguing flavors and textures with spur-of-the-moment additions, but, best of all, you'll find that the consistency of freshly made ice cream is hard to beat. This one adds a magical touch to the Pecan-Roasted Peaches in Sauternes (page 229) or the Compote of Peaches and Blueberries (page 227), but it's just as appealing served with bananas tossed in a rum caramel.

2 cups whole milk
1 cup heavy cream, preferably not
  ultrapasteurized
5 large egg yolks

8 ounces honey, preferably clover,
  Swedish heather, or other soft-spread
  honey
1 teaspoon vanilla extract

1. Combine the milk and cream in a medium saucepan, bring to a simmer, and keep warm.

2. In a large bowl combine the egg yolks and honey and whisk until fluffy and pale yellow. Pour in the warm milk-cream mixture and whisk until well blended. Transfer to the top of a double boiler, place over simmering water, and whisk constantly until the mixture is thick and heavily coats a spoon.

3. Remove the custard from the heat and whisk in the vanilla extract. Let the custard cool completely. Cover and refrigerate until quite cold. For best results, make the custard the day before and refrigerate overnight.

4. Pour the custard into an electric ice cream machine and freeze, following the manufacturer's instructions. Serve cold.

THE

FALL

KITCHEN

Returning to Barcelona recently in late October, I was moved to visit the old market in the neighborhood where we lived when I was a child and where we shopped almost daily. The market was strangely empty, and the bustle and liveliness I remembered were gone. But I found my mother's egg lady in her usual place, and when I asked her why there were so many closed stands, she told me that the market would soon be torn down to make room for a large parking lot.

The impact her words had on me was devastating. A part of my life that I had expected to remain forever would soon disappear. Overcome by nostalgia, I started to meander through the half-deserted market. Soon, a familiar aroma drew me to the place where, as always, the chestnut man was roasting fragrant, delicious nuts over his wood fire in an ancient black steel skillet. Schoolchildren lined up for their afternoon snack clutching their pennies in their hands and eagerly watching the vendor as he expertly rolled little paper packets in his hand, filling them with hot chestnuts.

I felt better; it was reassuring to see that this, at least, had not changed. Fall was here, and the damp, chilly air brought with it the smell of a burning fire which mingled with the intoxicating scent of ripe pears and wild Boletus mushrooms, which were at the height of their season. My spirits lifted; this, after all, is what made the change of seasons so exciting—the expectation and then the fulfillment of tasting again, after a long year, these wonderful seasonal foods, the joy in taking advantage of the new season while the old one is still fresh in your mind.

When I was a student, summers were never long enough. I wanted them to last forever. The grill, cool soups, and fresh fruit were an uncomplicated and most satisfying way to eat. In those days, fall always came too soon, bringing with it the responsibilities of studies and city life. But later, when cooking turned into an avocation, and I started to appreciate every season for what it had to offer, autumn came to mean a time of seasonal abundance, one of great variety and vigor.

Now it is my season of choice. The selection of produce is so different and exciting. The brisk cool air is exhilarating and so perfectly suited for heartier meals. And, the best news is that my old market in Barcelona was saved from destruction and has regained the vitality of yesteryear.

1

Wild Mushroom and Scallion Soup
Apple-Smoked Turkey with Onion and Chestnut Jam
Chocolate-Walnut Cake with Candied Fruit

2

Roasted Red Pepper Fritters with Caper, Anchovy,
and Garlic Butter
Old-Fashioned Blanquette of Veal with Wild Mushrooms
and Pearl Onions
Caramelized Banana Tatin

3

Spiced Double Winter Squash Soup
Penne with Cabbage and Sausage in Pink Tomato Sauce
Caramelized Apple Mousse

4

Angel-Hair Pasta with Chanterelles and Tarragon Cream
Roast Ducks in Acacia Honey and Cider Vinegar Sauce
Hazelnut-Chocolate Soufflés with White Chocolate–Mocha Sauce

5

Leek and Stilton Cheese Tart
Roast Pork with Purple Figs and Port Sauce
Caramelized Rice Pudding

FALL APPETIZERS

Sautéed Scallops with a Marmalade of Belgian Endives Flamande

Leek, Chèvre, and Roast-Pepper Tart

Fettuccine with Cremini Mushrooms and Smoked Trout

Fusilli Umbriana with Radicchio and Bacon in a
Creamy Tomato Sauce

Penne with Plum Tomatoes, Broccoli Rape, Garlic, and Anchovies

Crisp Potato, Onion, and Bacon Pizza

Roasted Peppers, Melted Onions, and Anchovy Pizza

Fennel-Infused Saffron Risotto with Fall Vegetables

Cazuela of White Beans and Sausage Asturiana

Spiced Sweet Potato and Jalapeño Pepper Soup

Farm-Style Soup of Cabbage and Root Vegetables

Grilled Eggplant Soup with Roasted Garlic

Curried Carrot and Turnip Soup

Middle Eastern Black Bean Soup with Concassé of Tomatoes

Basque Plum Tomato and Red Onion Soup

FALL MAIN COURSES

Sautéed Medallions of Monkfish à la Valenciana

Lemon- and Herb-Marinated Red Snapper à la Catalan

Navarin of Scallops à la Fermière

Lemon-Herb Grilled Chicken "Under a Brick"

Sauté of Chicken Breasts with a Fricassee of Mushrooms and Dill

Roast Duck with Cabernet-Poached Pears Bordalou

Roast Duck with Glazed Butternut Squash and Lingonberries

Fillets of Beef in Red Wine, Shallot, and Honey Essence

Oxtail Ragout in Cabbage Leaves

Braised Veal with Glazed Onions and Chestnuts

Veal Medallions with Chanterelles and Apples

Ragout of Lamb Shanks Niçoise

Grilled Leg of Lamb in Mustard and Herb Marinade

Pan-Seared Lamb Chops with Cabernet and Heather
Honey—Onion Jam

Citrus-Marinated Roast Loin of Pork

Curried Ragout of Pork with Butternut Squash

Fricassee of Snails à l'Esperance

Ziti with Sautéed Eggplant, Roasted Peppers,
and Onion Marmalade

FALL VEGETABLES

Ragout of Red Cabbage with Red Wine and Grenadine

Steamed Cauliflower in Piquant Caper and Anchovy Sauce

Deep-Fried Cauliflower with Chili Mayonnaise

Braised Endive and Apple Compote

Sweet-and-Sour Ragout of Fennel and Pearl Onions

Sauté of Cinnamon, Leeks, and Cremini Mushrooms

Wild Mushroom–Rice Timbales

Old-Fashioned Roasted Onions

Mediterranean Roasted Pepper and Eggplant Fricassee

Parsnip and Potato Pancakes

Gratin of Polenta with Goat Cheese, Leeks, and Ham

Belgian Potato Crepes

Cocido of Potatoes and Leeks with Crème Fraîche

Spaghetti Squash with Ginger and Brown-Sugar Cream

Sautéed Green Tomatoes Alfredo

Marinated and Grilled Green Tomatoes

Oven-Roasted Fall Vegetables with Roasted Garlic Vinaigrette

Cabbage and Carrot Slaw with Roquefort Dressing

Mixed Garden Salad with Lemon and Mint Vinaigrette

Sicilian Roasted Pepper Marmalade

Broccoli, Cauliflower, Snow Pea, and Carrot Salad

FALL DESSERTS

Brown Sugar–Glazed Bananas with Mascarpone and Honey

Sauté of Apples in Caramel Cream

Bittersweet Chocolate Soufflés in White Chocolate–Ginger Sauce

Chocolate and Pear Tart

Chocolate Crepes with Grand Marnier Cream and Orange Sauce

Purple Fig and Almond Tart

Gratin of Poached Pears with Red Wine Sabayon

Wild Mushroom and Scallion Soup
■
Apple-Smoked Turkey with Onion and Chestnut Jam
■
Chocolate-Walnut Cake with Candied Fruit

I have been using this menu for years, and I am always reluctant to change it since it works so well. I may add a new vegetable or a different accompaniment to the cake, but I love this particular combination of tastes and so it has become one of my fall classics.

Anyone who has ever eaten at the Villa Cipriani in Asolo, a small town one hour north of Venice, is bound to remember two of the restaurant's signature dishes: the incredible spaghettini carbonara and the spectacular porcini soup. At the restaurant, this soup is made with both fresh and dried porcini, which would make it prohibitively expensive to duplicate. Instead, I use the imported Chilean mushrooms which are inexpensive and have an intense smoky taste, perfect for this soup.

The same earthiness is the theme for the menu's main course, juicy fresh turkey, which takes on a different but subtle flavor when roasted over fresh apple wood. Picking apples at a local orchard is always an early fall weekend highlight, and while you're about it, why not pick up a few twigs for your grill. The green apple wood, used as a light smoking agent, imparts a rich yet subtle taste to the turkey, and it is also a wonderful choice for grilling other poultry and pork. If peeling fresh chestnuts is too much of a chore, vacuum-packed whole French chestnuts work beautifully in this recipe. You may also combine them with tiny white pearl onions for an even more interesting garnish. If you really want to go all out, the braised spaghetti squash (page 333) will add texture and interest to what could easily be served as a holiday menu.

Be sure to leave room for the rich chocolate cake, which is one of my mother's treasured recipes. I say treasured because my mother insists that it can be made only with the candied fruit she brings from Barcelona each year. I used to serve it Viennese-style, with a bowl of whipped cream on the side, but I now prefer a cognac-flavored crème anglaise, to which I add diced candied orange peel. But, in fact, this cake needs no embellishments. It is rich, delicious, and perfect for this fall menu.

■ ■

WILD MUSHROOM AND SCALLION SOUP

serves 6

2 ounces dried Chilean mushrooms
6 cups beef bouillon
4 tablespoons unsalted butter
3 bunches scallions, trimmed of 2 inches of greens and finely minced
2 medium cloves garlic, peeled and mashed
Salt and freshly ground black pepper

3 tablespoons all-purpose flour
1 cup heavy cream

GARNISH
6 fresh mushrooms, stemmed and thinly sliced
3–4 tablespoons finely minced fresh parsley

1. Rinse the dried mushrooms thoroughly under warm running water to remove all traces of sand and grit. Combine the mushrooms in a medium saucepan with the bouillon and bring to a boil. Reduce the heat and simmer, covered, for 30 minutes, or until the mushrooms are tender. Drain and finely mince. Reserve the broth and minced mushrooms separately.

2. In a heavy-bottomed 3½- to 4-quart casserole, melt the butter over low heat. Add the scallions, garlic, and 2 to 3 tablespoons of the reserved mushroom broth. Season with salt and pepper, cover, and braise the scallions for 3 minutes, or until very soft.

3. Add the flour and stir until well blended. Add the remaining reserved broth and whisk until well blended. Add the reserved mushrooms, bring to a boil, reduce the heat, and simmer, covered, for 20 minutes.

4. Add the cream and just heat through. Taste and correct the seasoning and serve hot in individual soup bowls, garnished with a few thinly sliced raw mushrooms and a sprinkling of parsley.

APPLE-SMOKED TURKEY WITH ONION
AND CHESTNUT JAM

serves 6

1 whole fresh turkey, about 7 pounds
Peanut oil for brushing on turkey
Coarse salt and freshly ground black
   pepper

JAM
2 teaspoons olive oil
3 tablespoons unsalted butter
4 large onions peeled, quartered, and
   thinly sliced

Pinch of granulated sugar
Salt and freshly ground black pepper
½ cup Chicken Stock (page 470) or
   bouillon
2 cups roasted and peeled fresh
   chestnuts
2 tablespoons dark-brown sugar

GARNISH
Sprigs of fresh watercress

1. Dry the turkey thoroughly on paper towels and truss. Brush the entire surface with peanut oil and sprinkle with coarse salt and pepper inside and out. Set aside.

2. To prepare the grill, use a round charcoal grill about 22½ inches in diameter. Open all vents and place 30 briquettes on the lower grill. When the coals are white-hot, carefully push an equal amount to each side of the lower grill. Place a rectangular drip pan in the center of the lower grill between the coals. Position the cooking grill in the kettle with the handles directly over each pile of coals. This will allow you to add briquettes through the openings near the grill handles during smoking.

3. Place the trussed and seasoned turkey on the center of the cooking grill, breast side up, directly over the roasting pan. Add a few twigs of apple wood to each pile of coals. Place the cover on the kettle and cook the turkey about 12 minutes per pound or to an internal temperature of 180° F. The turkey will need no basting or turning, but every 20 minutes add 6 briquettes and a few twigs of apple wood to each pile of burning coals through the openings at each side of the cooking grill.

4. While the turkey is roasting, prepare the jam: In a large skillet, heat the olive oil and butter over medium-high heat. Add the onions, sprinkle with a pinch of sugar, and cook for 5 minutes until the onions begin to brown; be careful not to burn them. Season with salt and pepper, reduce the heat to low, and cook, partially covered, until the onions are soft and nicely browned, 45 to 50 minutes.

5. Bring the stock to a boil in a small saucepan. Add the chestnuts, reduce the heat, and simmer until tender, 8 to 10 minutes. Drain and set aside.

6. When the onions are done, add the brown sugar and stir until the sugar has melted. Taste and correct the seasoning. Add the chestnuts and keep warm.

7. When the turkey is done, transfer to a carving board and remove the trussing string. Let sit for 15 minutes before carving.

8. If there are any accumulated juices in the roasting pan, degrease them and add to the jam. Carve the turkey, place on a large serving platter, and garnish with the watercress. Serve with the jam on the side.

Note: Other green woods, especially alder, cherry, and lilac, are equally good used with poultry. If they are unavailable, use store-bought mesquite chips and soak them for 20 minutes. Do not use hickory, which is too strong and will overwhelm the taste of the bird.

CHOCOLATE-WALNUT CAKE WITH CANDIED FRUIT

serves 6

CAKE
4½ ounces bittersweet chocolate, broken into pieces
1 tablespoon strong prepared coffee
6 extra-large eggs, separated
¾ cup plus 2 teaspoons granulated sugar
¾ cup dark raisins, plumped in hot water and drained
⅓ cup excellent-quality candied fruit, finely minced
1 tablespoon orange marmalade, finely minced
½ teaspoon finely grated lemon rind, optional
½ cup corn oil

1 tablespoon all-purpose flour
1 tablespoon fine bread crumbs
6 ounces walnuts, finely ground
Pinch of salt

GLAZE
3½ ounces bittersweet chocolate, broken into pieces
2 tablespoons confectioners' sugar
2 tablespoons strong prepared coffee
1 tablespoon unsalted butter
3 drops lemon juice

GARNISH
Whipped cream

1. Preheat the oven to 350° F.

2. Butter the bottom and sides of a 9-inch cake pan with a removable bottom or springform pan. Sprinkle lightly with flour, shaking out the excess, and set aside.

3. To make the cake: Combine the chocolate and coffee in a small bowl and place in a low oven for 3 or 4 minutes, or until the chocolate is melted.

4. Combine the yolks and ¾ cup of sugar in a large bowl and beat with an electric hand beater until the mixture is pale yellow and forms a ribbon. Add the melted chocolate and whisk until well blended. Reserve.

5. In a small bowl combine the raisins, candied fruit, marmalade, and optional lemon rind, add the oil, and blend well. Add the flour and bread crumbs and stir until well blended. Add the fruit mixture to the chocolate mixture and fold gently. Add the ground nuts and fold gently but thoroughly and reserve.

6. In a large bowl beat the whites with the remaining sugar and a touch of salt until firm peaks form. Fold the whites into the yolk mixture gently but thoroughly.

7. Pour the batter into the prepared pan, place in the center of the preheated oven, and bake for 50 minutes to 1 hour, or until a toothpick when inserted into the center comes out clean. Do not overbake; the cake should be moist.

8. While the cake is baking, prepare the glaze: Combine the chocolate, sugar, and coffee in a small saucepan, place over the lowest heat possible, and stir until the chocolate is just melted and smooth. Remove from the heat and let cool slightly. Add the butter and drops of lemon juice and stir until well blended and smooth. Set aside at room temperature until the cake is ready to be glazed. Do not refrigerate.

9. When the cake is done, remove from the oven and place on a wire rack to cool completely.

10. Run a knife around the edge of the pan and remove the ring from the cake. Invert the cake onto a serving platter and remove the bottom of the pan carefully by running a long, sharp knife between the cake and the pan. Spread evenly with the glaze and let the cake sit for at least 2 to 4 hours, or overnight.

11. Serve cut into wedges with a dollop of whipped cream.

Note: If you let the cake sit for several hours, its consistency will greatly improve to a dense, mousselike texture. The cake will also be easier to slice. This dessert freezes beautifully for several weeks.

Roasted Red Pepper Fritters with Caper, Anchovy,
and Garlic Butter
■
Old-Fashioned Blanquette of Veal with Wild Mushrooms
and Pearl Onions
■
Caramelized Banana Tatin

I still vividly remember choosing the menus for *The Seasonal Kitchen*. Fall meant red peppers, leeks, apples, pears, squashes, and some wonderful late-harvest peaches. Now red peppers are a year-round staple, and for those who do not garden, they are only dimly associated with the fall. But when I choose a menu, I like to base it on what nature had in mind when giving us seasonal produce, and a fall menu would not be complete without red peppers.

Here, the peppers are roasted, peeled, and pureed and then combined in a light fritter batter, which lends a brilliant color and a smoky taste to these light pancakes. Instead of the caper, anchovy, and garlic butter, the herbed sour cream and smoked-salmon sauce (page 29) is also a good choice.

I often opt for a complete change of pace when putting together a menu, and after the somewhat nouvelle American starter, a hearty ragout is a nice old-fashioned main course.

This one is an adaptation of a traditional blanquette of veal, a dish that had virtually disappeared for many years and is now making a well-deserved come-back. But where once the sauce was thick and floury, it is now based on a reduction of the poaching liquid and thickened only slightly. Earthy fall mushrooms and piquant lemon juice further enhance the ragout, giving it depth of taste and interesting texture. It is one of those undemanding dishes that can be made successfully a day or two ahead of time and reheated without changes in flavor.

One of the most appealing accompaniments to this creamy full-bodied ragout is fresh fettuccine tossed in a vibrant saffron butter and laced with braised leeks and a julienne of zucchini.

Here again, I follow the theme with an updated version of a classic dessert which, with crème brûlée, has become synonymous with new bistro cooking. The creative possibilities of the tarte tatin seem unlimited. At one time it was prepared only with apples, but now you can sample versions made with pears and quince. Personally, I find the best choices are fruits that are not too juicy, such as figs and bananas. In fact, aside from apples, bananas are my favorite choice since the fruit does not need preliminary cooking, and there is no danger of a soggy tart. Serve the tart with Honey Custard Ice Cream (page 233) or a caramel cream which turns this dessert into a truly memorable ending.

ROASTED RED PEPPER FRITTERS WITH CAPER, ANCHOVY, AND GARLIC BUTTER

serves 6

SAUCE
8 tablespoons unsalted butter
4–6 flat anchovy fillets, finely minced
2 medium cloves garlic, peeled and finely minced
1 tablespoon tiny capers, drained
2 tablespoons finely minced fresh parsley
Freshly ground black pepper

FRITTERS
2 large red bell peppers, roasted, peeled, seeded (page 483), and chopped

3 large egg yolks
1½ cups whole milk
1 tablespoon grated onion
1½ cups all-purpose flour
½ teaspoon baking powder
3 tablespoons unsalted butter, melted
Salt and freshly ground white pepper
Clarified Butter (page 478) for sautéing

GARNISH
Fresh Italian parsley leaves

1. Start by making the sauce: In a small, heavy saucepan, melt the butter over low heat, add the anchovies, garlic, capers, and parsley, and season with a large grinding of black pepper. Stir until the anchovies have completely melted, cover, and keep warm.

2. To prepare the fritters: In the workbowl of a food processor combine the roasted peppers and egg yolks and process until smooth. Add the milk and grated onion and again process until smooth.

3. Combine the flour and baking powder in a mixing bowl. Add the pureed pepper mixture and melted butter and whisk until well blended; do not over-

mix. Season the mixture with salt and pepper and let the batter rest for 1 to 2 hours.

4. Preheat the oven to 200° F.

5. To cook the fritters, heat a large nonstick skillet over medium heat and brush the surface lightly with clarified butter. Add the batter to the skillet by the tablespoonful without crowding. You will have to cook the fritters in batches of 4 or 5. Cook until lightly browned on both sides. Transfer to a serving platter, cover, and keep warm in the oven. Continue making the fritters until all the batter is used, brushing the skillet with a little clarified butter for each batch.

6. To serve the fritters, spoon a little of the sauce over them and garnish with leaves of Italian parsley. Serve at once.

Note: For a more interesting presentation, place 3 to 4 fritters on individual serving plates. Drizzle with a little of the brown butter and dollops of smoked-salmon sauce (page 29). Garnish each plate with sprigs of fresh dill and steamed sprigs of asparagus or thin strips of roasted red and yellow peppers.

OLD-FASHIONED BLANQUETTE OF VEAL WITH WILD MUSHROOMS AND PEARL ONIONS

serves 6

3 pounds boneless shoulder of veal, cut into 1-inch cubes
2–3 pounds meaty veal bones
2 large carrots, peeled and cut in half
2 large leeks, trimmed of all but 2 inches of greens, rinsed
2 small whole onions, peeled and stuck with a whole clove
1 parsley root, optional
2–3 stalks celery with leaves, cut in half
6 white peppercorns
Salt
1 pint small pearl onions, peeled

¾ cup heavy cream
2 large egg yolks
Juice of 1 large lemon
7 tablespoons unsalted butter
3 tablespoons all-purpose flour
Freshly ground black pepper
1 pound Cremini mushrooms, wiped, stemmed, and cubed
½ pound shiitake mushrooms, stemmed and quartered
⅓ cup finely minced fresh dill or 3 tablespoons finely minced fresh tarragon, optional

1.   In a large casserole, combine the veal, veal bones, carrots, leeks, whole onions, optional parsley root, celery, and peppercorns. Bring to a boil, reduce the heat, and season lightly with salt. Simmer the veal for 1 hour to 1 hour and 15 minutes, or until tender but not falling apart. From time to time skim the surface of all fat.

2.   When the veal is done, strain and discard the vegetables. Remove any meat from the bones and discard the bones. Place all the veal in a bowl, cover with a little broth, and reserve.

3.   Combine 5 cups of the remaining broth and the pearl onions in a saucepan. Bring to a boil, reduce the heat, and simmer until the onions are tender. Strain and reserve the onions and the broth.

4.   Return the broth to the saucepan and reduce to 2½ cups. Taste several times and, if the broth becomes too salty, stop the reduction.

5.   While the broth is reducing, combine the cream, yolks, and lemon juice in a small bowl, whisk until smooth, and set aside.

6.   In a heavy, medium-sized casserole, melt 3 tablespoons of butter over medium heat. Add the flour and cook for 1 minute without browning. Add the reduced broth all at once and whisk constantly until the mixture comes to a boil. Reduce the heat, add the cream-yolk mixture, and just heat through. Taste and correct the seasoning, add the reserved pearl onions, and set aside.

7.   In a large skillet melt 2 tablespoons butter over medium heat, add the Cremini mushrooms, and sauté until nicely browned. Season with salt and pepper and add to the sauce.

8.   Add the remaining 2 tablespoons butter to the skillet. When hot add the shiitake mushrooms and sauté for 2 to 3 minutes, or until lightly browned. Season with salt and pepper and add to the sauce.

9.   Strain the veal and add to the casserole together with the optional dill or tarragon. Place over low heat and just heat through. Taste and correct the seasoning. Serve hot in individual dishes.

CARAMELIZED BANANA TATIN

serves 6–8

½ cup granulated sugar
2 tablespoons water
4 large semi-ripe bananas
2 tablespoons unsalted butter

½ recipe Basic Tart Shell dough (page 492)
1 recipe Honey Custard Ice Cream (page 233)

1. Preheat the oven to 375° F.

2. In a small saucepan combine the sugar and water. Bring to a boil and stir once to dissolve the sugar. Continue to boil, without stirring, until the mixture turns a hazelnut brown.

3. Immediately remove the saucepan from the heat and pour the caramel into the bottom of a 9-inch round, nonstick cake pan, tilting the pan to coat the bottom as evenly as possible. As you are pouring, turn the cake pan to create an even layer of caramel. Set aside to cool completely.

4. Peel and cut the bananas into ¼-inch diagonal slices and reserve.

5. Dot the caramel in the cake pan with the butter and arrange the bananas in a single layer, tightly overlapping them to create a circular pattern. Set aside.

6. Roll out the dough on a lightly floured surface into a circle about 9½ inches in diameter and about ¼ inch thick.

7. Place the dough over the bananas, covering them completely, and tuck the edges into the pan. Prick with a fork, place in the center of the preheated oven, and bake for 30 to 35 minutes, or until the dough is crisp and nicely browned.

8. Remove the pan from the oven, invert the banana tatin carefully onto a serving platter, and cool.

9. Serve the tart cut into wedges, drizzled with a bit of the banana pan juices and topped with a small scoop of ice cream.

Note: If the bananas are very ripe, they may exude quite a bit of juice. Remove the excess juice with a bubble baster and reserve. Once the tart is unmolded, drizzle with the reserved juice.

Spiced Double Winter Squash Soup

Penne with Cabbage and Sausage in Pink Tomato Sauce

Caramelized Apple Mousse

This menu clearly demonstrates how many new and innovative recipes have been created by home cooks and professional chefs on both sides of the Atlantic in the past ten years by rediscovering the use of traditional vegetables in new preparations. Of all the winter squashes, I far prefer the acorn, and during one recent fall weekend I experimented with this version of a curried acorn squash soup that has since become a favorite for my family and friends.

What really sets this soup apart is the seasoning, which breaks it away from the standard pumpkin-pie spices so commonly used with all winter squash. Instead, the delicate taste of the squash is enhanced by hot peppers, chili, ginger, and cumin. Acorn squash is a perfect carrier for these intense flavors, and the dish is an example of how these flavors complement each other. Finally, the addition of spaghetti squash and a few cooked red kidney beans adds to the visual and textural excitement of the soup. The soup improves when allowed to develop flavor, so make it a day or two ahead of time.

Another old-timer, curly cabbage, particularly the crinkly Savoy, has also been gaining in status and its use greatly updated. Of course, there is nothing new about this wonderful vegetable. In peasant cooking of all European cuisines, cabbage has always played a key role, but teaming it with pasta is a relatively new way of serving it. Here the still-crunchy cabbage is tossed with full-flavored sausage, a creamy tomato sauce, and tubular pasta, such as penne. You may vary the dish by using diced smoked ham instead of prosciutto and adding a large grating of smoked mozzarella just before serving. I also like the rather more unusual combination of crisp bacon and some small pan-seared bay scallops, which seem to hold their own in this earthy pasta dish. Follow this hearty main course with a well-seasoned green salad and possibly one or two well-selected Italian cheeses, such as a Tallegio and a sweet Gorgonzola.

As for dessert, I suggest returning to one that is quite traditional and very rustic. This creamy apple mousse calls for one of the most common fall apples, the McIntosh, but for anyone who has access to a farmstand or orchard, there are a variety of more interesting apples to choose from, especially at the beginning of the season. Good ripe pears and quince make a superb mousse as well. So keep these fruits in mind when the season begins. The key to the success of any truly seasonal menu is to highlight its star fruits and vegetables at their peak of ripeness and this menu offers you just that opportunity.

SPICED DOUBLE WINTER SQUASH SOUP

serves 6

SOUP
2 whole acorn squash, about 1¼
  pounds each
2 tablespoons unsalted butter
1 large onion, peeled and finely minced
2 teaspoons finely minced fresh green
  chili pepper
2 teaspoons finely minced fresh ginger
2 cloves garlic, peeled and finely minced
1 teaspoon pure ground chili powder
Large pinch of ground cumin
4 cups Chicken Stock (page 470)

Salt and freshly ground white pepper
1 cup heavy cream

SQUASH
2 tablespoons unsalted butter
2½–3 cups cooked spaghetti squash
  (page 333)
2 tablespoons dark-brown sugar

GARNISH
Finely minced fresh cilantro
1 cup cooked red kidney beans (see
  Cooked White Beans, page 484)

1. Preheat the oven to 350° F.

2. Start by making the soup: Place the whole acorn squash on a baking sheet, set in the preheated oven, and roast until tender when pierced with a sharp knife, about 35 to 45 minutes, turning once during roasting. Remove from the oven and let cool.

3. Cut each acorn squash in half. Scoop out and discard the seeds. Scoop out the pulp and set aside.

4. In a heavy 4-quart casserole, melt the butter over medium heat. Add the onion, chili pepper, ginger, and garlic and cook for about 3 minutes or until soft but not browned. Add the chili powder and cumin and cook for 1 minute longer. Add the chicken stock and acorn squash and season with salt and white pepper. Bring the soup to a boil, reduce the heat, and simmer, partially covered, for 25 minutes.

5. Strain the soup into a bowl and set the broth aside. Transfer the vegetables to the workbowl of a food processor and process until smooth. Whisk the vegetable puree into the reserved broth and return to the casserole.

6. Place the soup over low heat. Add the heavy cream and simmer for 10 minutes longer. Taste and correct the seasoning and keep warm.

7. To prepare the spaghetti squash: In a medium skillet melt the butter over

medium-low heat. Add the spaghetti squash and the brown sugar and toss gently until the sugar has melted and the squash is heated through.

8. To serve, ladle the soup into warm individual shallow soup bowls. Place a spoonful of spaghetti squash in the center of each portion and garnish with cilantro and a sprinkling of red kidney beans.

PENNE WITH CABBAGE AND SAUSAGE IN PINK TOMATO SAUCE

serves 6

Salt
1 pound wedge of green cabbage, with core
2 medium leeks, all but 2 inches of greens removed
3 tablespoons unsalted butter
½ cup julienne of prosciutto
⅓ cup Chicken Stock (page 470)
Freshly ground black pepper

1½ cups Quick Tomato Sauce (page 474)
⅓ cup heavy cream
1 tablespoon olive oil
¾ pound sweet Italian sausage
¾ pound imported penne
¾ cup finely diced or grated smoked mozzarella
⅓ cup freshly grated Parmesan cheese

1. Bring salted water to a boil in a large saucepan, add the cabbage, and cook for 5 to 7 minutes or until just tender. Drain, and when cool enough to handle, squeeze out all the excess moisture with your hands. Cut out and discard the core. Slice the cabbage crosswise into ½-inch julienne and set aside.

2. Cut the leeks crosswise into thin slices. Place in a colander, run under warm water to remove all sand and grit, drain, and reserve.

3. In a large, heavy skillet, melt 1 tablespoon of the butter over medium-low heat, add the prosciutto, and cook for 15 seconds. Remove with a slotted spoon to a side dish and set aside.

4. Add the remaining butter to the skillet. When melted, add the sliced leeks together with a little stock and cook, covered, until just soft. Add the cabbage,

season with salt and pepper, add the remaining stock, and cook for 5 minutes longer. Add the tomato sauce and cream, reduce slightly, and set aside.

5. Heat the olive oil in a medium skillet over medium heat. Add the sausage and cook quickly until nicely browned on all sides. Do not overcook. Transfer to a cutting board and cut crosswise into ¼-inch slices. Add to the tomato-cabbage mixture and keep warm.

6. In a large casserole, bring plenty of salted water to a boil. Add the penne and cook until al dente. Immediately add 2 cups of cold water to the pot to stop further cooking, drain well, and return the penne to the casserole.

7. Add the cabbage mixture together with the prosciutto to the casserole and toss to coat the penne. Add the smoked mozzarella and half of the Parmesan, toss gently, and correct the seasoning. Immediately transfer to a serving dish, sprinkle with the remaining Parmesan, and serve at once.

Note: The mozzarella must not in fact melt but should just heat through and retain its shape. For best results, you may place the piece of mozzarella, whole, in the freezer for 20 to 30 minutes before using. Grate it cold into the pasta, toss gently, and serve at once.

CARAMELIZED APPLE MOUSSE

serves 6–8

APPLES
2 tablespoons unsalted butter
4–5 medium McIntosh apples, peeled, cored, and quartered
½ teaspoon ground cinnamon
1 tablespoon plus ½ cup granulated sugar
Pinch of freshly grated nutmeg
2 tablespoons water

CUSTARD
4 large egg yolks
1 teaspoon cornstarch

⅓ cup granulated sugar
1½ cups warm whole milk
1 teaspoon vanilla extract
1 envelope unflavored gelatin
1 cup heavy cream, whipped
Flavorless oil for brushing mold

SAUCE
1 cup apricot preserves
2 tablespoons fresh lemon juice
1 teaspoon finely grated lemon zest
2 tablespoons apricot brandy
2 tablespoons confectioners' sugar

1. Start by cooking the apples: In a heavy skillet, melt the butter over low heat. Add the apples, cinnamon, 1 tablespoon of sugar, and a pinch of nutmeg and cook for 8 to 10 minutes or until the apples are very soft and can be mashed easily with a fork. Stir the apples several times to keep them from scorching.

2. While the apples are cooking make the caramel: Combine the remaining sugar with 2 tablespoons of water in a small, heavy saucepan and place over high heat. Bring to a boil and stir only once to dissolve the sugar. Boil until the mixture turns a hazelnut brown. Remove from the heat, add to the cooked apples, and cook the apples with the caramel for 2 minutes. Transfer to the workbowl of a food processor and process until smooth. Set aside to cool.

3. To prepare the custard: In the top of a double boiler, combine the yolks, cornstarch, and sugar and whisk the mixture until smooth. Add the warm milk and whisk until well blended. Set the mixture over simmering water and whisk constantly until it heavily coats a spoon. Add the vanilla and gelatin and whisk for 1 or 2 minutes or until the gelatin is dissolved. (You may also dissolve the gelatin separately in a little water.) Transfer the mixture to a bowl and chill until it starts to set. Whisk in the apple puree and whipped cream. Taste the mousse, adding more cinnamon or nutmeg to taste.

4. Lightly brush the inside of a 6-cup ring mold with flavorless oil. Pour in the mousse mixture and chill for 4 to 6 hours.

5. Now prepare the sauce: In a small saucepan, combine the apricot preserves, lemon juice, lemon zest, apricot brandy, and confectioners' sugar. Heat the mixture until the preserves melt and the sugar has completely dissolved; add more apricot brandy if needed. Pass the sauce through a sieve and reserve.

6. Unmold the mousse onto a platter and spoon some sauce around it. Pass the rest of the sauce in a sauceboat.

Angel-Hair Pasta with Chanterelles and Tarragon Cream

Roast Ducks in Acacia Honey and Cider Vinegar Sauce

Hazelnut-Chocolate Soufflés with White Chocolate—Mocha Sauce

I still remember a time, not so long ago, when fresh chanterelles were a delicacy seldom found in American markets. Now, thanks to our appetite for fresh produce, various kinds of mushrooms are available throughout the country. Fresh chanterelles are picked throughout the fall in the Northwest and are in good supply in produce markets during the season. This menu highlights these delicate mushrooms in a simple fresh pasta dish that leaves room for a substantial and assertive main course.

I seem to fall back on roast duck quite often when I entertain in cool weather. Duck is such a "good-natured" bird that you can roast it hours ahead of time and then reheat it, without drying, just before serving. The sweet, lightly gamy meat is a wonderful foil for many fresh and dried fruits and condiments. The honey and cider sauce in this recipe offers a complexity of flavors that is equally good with rabbit or roast pork, which also have an affinity for full-bodied sweet sauces.

A light chocolate dessert makes a nice ending to this rather elegant menu. I particularly like these traditional and very Viennese soufflés. What is wonderful about these little nutty confections is that they can be prepared in advance since they can also be served as fallen soufflés and will pop right out of their molds. You may serve them with soft vanilla ice cream folded into some whipped cream, or as I suggest here, with a white chocolate sauce. If all of this sounds too rich, a perfectly ripe pear accompanied by thinly sliced, well-aged Parmesan is an equally good dessert choice for this time of year.

ANGEL-HAIR PASTA WITH CHANTERELLES
AND TARRAGON CREAM

serves 4–6

For years American cooks had to contend with dried imported chanterelles without ever enjoying the superb delicate taste of this exquisite fall mushroom. Now, thanks to modern transportation, during the season fresh chanterelles are shipped daily from the Pacific Northwest to cities everywhere and can be found in abundance in fine produce markets and some supermarkets across the country. While they are expensive, they are well worth splurging on during their relatively brief season.

1 pound fresh chanterelles
8 tablespoons unsalted butter
2 tablespoons finely minced shallots
Salt and freshly ground white pepper
1¼ cups heavy cream
2 tablespoons finely minced fresh
   tarragon

1 tablespoon chicken glaze: 1 cup of
   Chicken Stock (page 470) reduced to
   1 tablespoon, optional
1 pound fresh angel-hair pasta

1. Trim the chanterelles, removing all but ½ inch of their stems. Rinse carefully under cold water. Slice the chanterelles into fine julienne strips or, if they are small, dice them and set aside.

2. In a large, heavy skillet melt half the butter over medium heat, add the chanterelles, and sauté for 2 minutes. Add the shallots and continue sautéing until the chanterelles are cooked through and all the juices have evaporated. Season with salt and white pepper and set aside.

3. In another deep, heavy skillet, heat the remaining butter over medium heat. When the butter starts to brown, add the cream and tarragon. Season lightly with salt and pepper and simmer until the cream is reduced by one third. Add the optional chicken glaze and the chanterelles. Cover and keep warm.

4. In a large casserole bring salted water to a boil. Add the pasta and cook for 3 minutes, or until al dente; do not overcook. Drain immediately and return to the casserole. Add the cream and mushroom mixture. Toss lightly and serve immediately.

Note: If you cannot get fresh chanterelles, use ½ ounce of dried ones. Reconstitute the dried chanterelles in well-flavored chicken stock, cut into fine julienne, and reduce the poaching liquid to 2 tablespoons. Use this reduction to flavor the cream. You may also add a julienne of 6 to 8 all-purpose mushrooms quickly sautéed in a little butter for additional texture. Do not attempt to make this dish with dried tarragon.

ROAST DUCKS IN ACACIA HONEY AND CIDER VINEGAR SAUCE

serves 4

1 teaspoon whole coriander seeds
4 tablespoons light soy sauce
3 tablespoons honey, preferably Acacia
½ teaspoon ground coriander
2 ducks, about 4–4½ pounds each
Salt and freshly ground white pepper
2 teaspoons dried thyme
4 tablespoons unsalted butter
1 teaspoon corn oil
1½ cups chicken bouillon
2 large shallots, peeled and finely
  minced

1 tablespoon tomato paste
¼ cup cider vinegar
¾ cup Brown Chicken Stock (page 467)
1 Beurre Manie (page 478)
2–3 tablespoons unsalted butter, for
  enrichment, optional

GARNISH
Sprigs of fresh watercress
Tiny whole glazed carrots

1. Several hours in advance of roasting the duck, or the day before if possible, prepare the soy and honey mixture. In a small skillet, toast the coriander seeds until fragrant. Crush in a mortar and pestle and combine with the soy sauce, honey, and ground coriander in a small bowl, blend well, and set aside.

2. Preheat the oven to 350° F.

3. Dry the ducks thoroughly with paper towels. Season with salt, white pepper, and thyme, and truss the ducks.

4. Heat 2 tablespoons butter and the oil in a large baking dish. Add the ducks and brown on all sides. Add a little chicken bouillon and roast the ducks in the center of the oven for 2 hours and 45 minutes, basting every 15 minutes with a little bouillon and removing some of the fat at the same time. When the ducks are done, remove to a baking sheet and set aside.

5. Strain the pan juices into a bowl and thoroughly degrease. Set the juices aside.

6. Add the remaining butter to the pan over medium heat, add the shallots, and sauté until soft and lightly browned. Add the tomato paste and cider vinegar, bring to a boil, and cook until the vinegar has almost evaporated.

7. Add the honey and soy mixture, bring to a boil, and add the brown chicken stock. Lower the heat and simmer the sauce until somewhat reduced, tasting it to see that it does not become too salty.

8. Whisk in bits of the beurre manie and cook until the sauce lightly coats a spoon. Taste and correct the seasoning. Strain the sauce into a saucepan, whisk in the optional butter for enrichment, and keep the sauce warm.

9. Preheat the broiler. Place the ducks in the broiler, 6 to 8 inches from the source of heat, and broil until the skin on all sides is nicely browned and crisp.

10. Transfer the ducks to a cutting board, carve, and place on a warm serving platter. Garnish with watercress and tiny whole glazed carrots. Serve at once with the sauce on the side.

HAZELNUT-CHOCOLATE SOUFFLÉS WITH
WHITE CHOCOLATE–MOCHA SAUCE

serves 6

5 tablespoons unsalted butter, softened
⅓ cup granulated sugar
3 extra-large egg yolks
5 ounces extra-bittersweet chocolate
½ cup ground hazelnuts
4 extra-large egg whites
Pinch of salt

GARNISH
1 recipe White Chocolate–Mocha Sauce
  (page 488)
1½ cups heavy cream, whipped

1. Preheat the oven to 375° F. Generously butter the bottom and sides of 6 (4-ounce) porcelain ramekins. Sprinkle with sugar, shaking out the excess, and set aside.

2. In a large mixing bowl combine the butter and sugar and beat until fluffy and light. Add the yolks and beat until thoroughly incorporated.

3. Place the chocolate in the top of a double boiler and set over simmering water until the chocolate has melted. Stir until well blended, remove from the heat, and cool slightly. Add the chocolate and ground hazelnuts to the yolk mixture and fold gently but thoroughly.

4. In a large bowl, beat the whites with a pinch of salt until stiff but not dry. Gently fold the whites into the yolk-chocolate mixture to incorporate thoroughly. Fill the prepared ramekins with some of the soufflé mixture and reserve.

5. Place the ramekins in a heavy baking dish, add boiling water enough to come halfway up the sides of the ramekins, and bake the soufflés in the preheated oven for 20 minutes or until a knife when inserted in the center comes out almost clean; the soufflés should still be somewhat soft. Remove from the oven and let the soufflés cool completely. Unmold onto individual plates and surround with white chocolate–mocha sauce and a large dollop of whipped cream.

Leek and Stilton Cheese Tart

Roast Pork with Purple Figs and Port Sauce

Caramelized Rice Pudding

This elegant and quite traditional fall menu highlights some of the best of the season's produce—leeks and fresh figs. The menu also provides a great deal of room for changes and creative interpretations.

I first sampled the leek and Stilton-cheese tart at Gidleigh Park, a beautiful American-owned country inn at the edge of the Cornish moors. Much of the food was delicious and beautifully prepared, but this tart was outstanding and a real find. The marriage of buttery braised leeks with the delicate but sharp-edged taste of Stilton cheese is truly inspired. I have since used the idea in a soup combining the cheese with the mellow flavor of celery, but the tart has become one of my star fall appetizers. It is best to make the tart shell in advance and freeze it, then bake the entire tart hours ahead of time. Just before serving, reheat it in a low oven. Any leftovers make a light supper delicious when served with a green salad that has been tossed with a warm shallot dressing and sautéed cubed bacon.

If you plan to serve the roast pork with fresh figs, you will have to plan ahead since the figs are at their best when marinated in port for at least twenty-four hours. It is best to make this dish only when you see fresh, truly ripe figs at your market, because unripe figs are simply not worth the effort. I only make this dish spontaneously, when good figs appear in the fall, and I am usually asked by my students whether dried figs are an acceptable substitute. They are, but do not expect the same results. In fact, dried figs are almost a completely different fruit; they will produce a grainier texture and result in a much sweeter sauce. Serve a turnip and potato puree with the pork or a pilaf of rice enhanced by diced dried porcini using the intensely flavored porcini poaching broth (page 324).

Following this highly seasonal and richly flavored main course, I would choose something creamy like this updated version of rice pudding, a dessert I have loved since childhood. A dollop of the lemony pudding is nestled in individual ramekins or in a large, shallow porcelain quiche pan, spread with rich custard and baked. The pudding is then topped with brown sugar that is caramelized under the broiler. The dessert is similar in appearance to crème brûlée, but its texture is quite different and unique. You may vary the flavoring of both the custard and the rice pudding, adding cranberries or currants to the pudding and flavoring the custard with orange zest or grated fresh ginger. Whatever you choose should be seasonal. I prefer clear flavors such as lemon, ginger, and orange, which seem to enhance both the pudding and the custard.

■■■

LEEK AND STILTON CHEESE TART

serves 6

2 tablespoons unsalted butter
4 medium leeks, trimmed of all but 2 inches of greens, finely sliced and washed
Salt and freshly ground white pepper
5 ounces Stilton cheese, crust trimmed and crumbled
1½ cups Crème Fraîche (page 491)

1 large egg
3 large egg yolks
2 tablespoons finely minced fresh chives, optional
4 tablespoons freshly grated Parmesan cheese
1 prebaked 10-inch Basic Tart Shell (page 492), in a porcelain quiche pan

1. Preheat the oven to 350° F.

2. In a 10-inch skillet, melt the butter over medium heat. Add the leeks and 2 tablespoons water and season with salt and pepper. Reduce the heat and cook, covered, for 8 minutes or until just tender. Set aside.

3. Combine the Stilton cheese, crème fraîche, egg, and yolks in the workbowl of a food processor and process until smooth. Transfer to a large mixing bowl. Add the sautéed leeks, optional chives, and 2 tablespoons of the Parmesan and whisk until well blended. Season with salt and pepper.

4. Pour the mixture into the prebaked tart shell and spread the leeks evenly in the custard. Sprinkle the top with the remaining Parmesan and bake in the center of the preheated oven for 35 to 45 minutes, or until the custard is set; a knife when inserted into the center should come out clean.

5. Remove from the oven and let sit for 15 minutes before cutting and serving. Serve warm, cut into wedges, with a well-seasoned green salad.

ROAST PORK WITH PURPLE FIGS
AND PORT SAUCE

serves 6

16–18 small fresh whole purple figs
1¼ cups tawny port
2½–3 pounds boneless pork shoulder butt, rolled and tied
Salt and freshly ground black pepper
1½ tablespoons dried thyme
4 tablespoons unsalted butter
1 teaspoon peanut oil
1–1¼ cups Brown Chicken Stock (page 467) or beef bouillon

2 tablespoons finely minced shallots
2 large ripe tomatoes, peeled, seeded, and chopped
1 tablespoon excellent-quality pure uncooked honey
1 Beurre Manie (page 478)
6 tablespoons unsalted butter, for enrichment

1. Rinse the figs in a colander and drain; do not remove the stems. Combine the figs and port in a medium bowl and marinate overnight.

2. The next day, preheat the oven to 375° F.

3. Season the pork with salt, pepper, and thyme. In a large, heavy flame-proof baking dish or black cast-iron skillet, melt 2 tablespoons of butter with the oil over high heat. Add the pork and brown nicely on all sides. Add a little stock to the dish, place in the center of the preheated oven, and roast for 1 hour and 30 minutes, basting with a little brown stock every 10 to 15 minutes until the internal temperature on a meat thermometer registers 165° F.

4. While the pork is roasting, prepare the sauce: In a small, heavy saucepan melt the remaining butter over medium heat. Add the shallots and sauté until soft but not browned. Add the tomatoes and cook for about 10 minutes or until the tomato liquid has evaporated and the mixture is quite thick.

5. Transfer the tomato mixture to the workbowl of a food processor and process until smooth. Return the tomato puree to the saucepan and add ⅓ cup of the stock. Drain the figs and reserve the fruit and port separately. Add ½ cup of the port and the honey to the saucepan. Bring to a boil, reduce the heat, and simmer until reduced by one third. Set aside.

6. Bring the remaining port to a boil in a large, shallow saucepan and reduce to a thick syrup. Add the figs and roll in the warm syrup until they are well heated throughout. The figs should be glossy and must not fall apart. Reserve.

7.   When the roast is done, transfer it to a carving board, remove and discard the string, and set aside.

8.   Degrease the pan juices and add to the saucepan of tomato sauce. Bring to a boil and whisk in bits of beurre manie until the sauce lightly coats a spoon. Whisk in the butter for enrichment, 1 tablespoon at a time, and correct the seasoning. Keep warm.

9.   Cut the pork roast into thin slices and place in a decorative pattern on a serving platter. Surround with the figs and spoon the sauce over the pork. Serve immediately, accompanied by a carrot and sweet potato puree or a potato and celery-root puree.

Note: This dish can also be made with dried figs, which must be cooked in the port until plumped and tender. The result will be a far sweeter sauce, and you may have to omit the honey.

CARAMELIZED RICE PUDDING

serves 6–8

PUDDING
1½ cups whole milk
¼ cup converted rice
¼ cup granulated sugar
1 (3-inch) piece of vanilla bean, split
2 (3-inch) strips of lemon zest
¼ cup good-quality candied fruit, finely
   diced, or ¼ cup golden raisins,
   plumped

CUSTARD
6 large egg yolks

¼ cup granulated sugar
2 cups hot heavy cream
1 teaspoon grated lemon rind
1 teaspoon vanilla extract

TOPPING
12–16 teaspoons granulated brown
   sugar

1.   Start by making the pudding: Bring the milk to a simmer in a heavy medium saucepan, add the rice, sugar, vanilla bean, and lemon zest and simmer, covered, over low heat until the rice is very soft, about 1 hour. Stir from

time to time to prevent the rice from scorching. Ten minutes before the rice is done, add the candied fruit or raisins.

2.   When the rice is done, remove from the heat and divide between 6 to 8 (5-ounce) porcelain ramekins and set aside.

3.   Preheat the oven to 375° F.

4.   To prepare the custard: In a medium bowl combine the egg yolks and sugar and whisk until fluffy and pale yellow. Add the hot heavy cream, lemon rind, and vanilla and whisk until well blended. Pour the custard equally into each of the ramekins.

5.   Set the ramekins in a large, shallow baking dish. Fill the dish with boiling water to come halfway up the sides of the ramekins. Place the baking dish in the center of the preheated oven and bake for 12 to 15 minutes, or until the custards are barely set. Do not overcook.

6.   When the custards are done, remove from the water bath and place on wire racks to cool slightly.

7.   Preheat the broiler.

8.   Sprinkle the top of each custard with 2 teaspoons of granulated brown sugar. Place under the broiler 6 inches from the source of heat and broil until the sugar is melted and caramelized. Serve warm.

SAUTÉED SCALLOPS WITH A MARMALADE
OF BELGIAN ENDIVES FLAMANDE

serves 6

The Belgian endive, once reserved only for the salad bowl, has stepped into the limelight which it shares with many other unusual cooked greens, such as radicchio and arugula. During my college years in Geneva, my roommates and I ate boiled endive almost daily. Sometimes topped with grated cheese, other times with brown butter, it was one of the cheapest vegetables around. For years afterward, I couldn't bear to eat another endive, but now I love it, especially when sautéed and cooked down into a somewhat sweet and tangy "marmalade."

ENDIVES
12 large Belgian endives
2 tablespoons unsalted butter
1 teaspoon granulated sugar
Salt and freshly ground white pepper
¼ cup water

SCALLOPS
1¼–1½ pounds small sea scallops
Coarse salt and freshly ground white pepper
1 tablespoon unsalted butter
2 teaspoons olive oil

SAUCE
1 large shallot, peeled and finely minced
¼ cup fresh lime juice
¼ cup Noilly Prat vermouth
10 tablespoons unsalted butter, softened
¼ cup Crème Fraîche (page 491)
Salt and freshly ground black pepper

GARNISH
Finely minced fresh parsley

1. Start by braising the endives: Remove any wilted outer leaves and discard. Core and slice the endives into a long, fine julienne. Set aside.

2. In a large skillet, melt the butter over medium-high heat. Add the endives and sauté until lightly browned, stirring constantly so as not to burn them. When evenly browned, add the sugar and season with salt and pepper. Add the water, reduce the heat, and simmer, covered, until the endives are tender, about 10 to 12 minutes. Reserve.

3. To make the sauce: In a heavy saucepan, combine the shallot, lime juice, and vermouth, bring to a boil, and cook until reduced to 2 tablespoons.

4. Remove the saucepan from the heat and, when the mixture is slightly cooled, whisk in 2 to 3 tablespoons of butter. Return the saucepan to low heat and continue adding butter, whisking constantly, until the sauce is thick and emulsified. Season with salt and pepper.

5. Pass the sauce through a fine sieve and return to the saucepan. Add the crème fraîche and whisk until well blended. Taste and correct the seasoning and place the saucepan in a larger saucepan or a deep baking dish filled with hot water. Cover and keep warm.

6. To prepare the scallops: Cut in half horizontally and dry thoroughly with paper towels. Season with coarse salt and freshly ground white pepper. In a large skillet, melt the butter, together with the oil, over high heat. When very hot, add the scallops without crowding the pan. Sauté for 1 to 2 minutes on each side or until nicely browned. Do not overcook. Remove the scallops to a side dish and reserve.

7. Reheat the endives over low heat.

8. Place 1 large spoonful of braised endives on individual appetizer plates. Top with 6 scallop halves and spoon a little sauce over each portion. Sprinkle with minced parsley and serve immediately.

Note: The endives can be prepared up to a day ahead of time and reheated. The sauce can be made 2 to 3 hours ahead of time and kept warm in a water bath. Be sure to keep the water warm but not simmering or the sauce will start to curdle. You may omit the sauce completely for a simpler version of this dish and serve the scallops over the endives with some melted hazelnut butter.

LEEK, CHÈVRE, AND ROAST-PEPPER TART

serves 6

I think that what I miss most about Europe, especially France, Italy, and Spain, is the wonderful sidewalk food. In France there is always a pastry shop or charcuterie around the corner, where you can pick up a slice of either a eweet or savory tart and nibble at it while you walk down the street. At Saint-Jean-du-Luz, I stopped to buy a slice of this brilliantly red tart and sat down at a sidewalk café to watch the world go by. By the time I left Saint-Jean, I had eaten six slices of tart and couldn't wait to get home to test the recipe to see if I could duplicate it. I think I have it right. Anyway, it is delicious.

4 medium leeks, trimmed of all but 2 inches of greens
2 tablespoons unsalted butter
2 tablespoons water
Salt and freshly ground white pepper
2 medium red bell peppers, roasted, peeled (page 483), and finely diced
2 large eggs

2 extra-large egg yolks
1 cup heavy cream
1 partially baked 10-inch Basic Tart Shell (page 492), in a pan with a removable bottom
4 ounces mild goat cheese, crumbled
1 tablespoon fresh thyme leaves

1. Preheat the oven to 350° F.

2. Cut the leeks in half lengthwise and then crosswise into thin slices. Place in a colander and run under warm water to remove all sand. Drain and set aside.

3. In a medium skillet, melt the butter over medium heat. Add the leeks and water and season with salt and pepper. Reduce the heat, cover, and braise the leeks 8 to 10 minutes, or until tender and all the water has evaporated. Reserve.

4. Be sure to remove all traces of blackened skin from the red peppers and combine them in the container of a blender with the eggs and yolks. Blend until quite smooth. Add the heavy cream and blend until smooth. Season with salt and pepper and set aside.

5. Place the cooked leeks on the bottom of the partially baked tart shell and pour the red-pepper mixture over the leeks. Place in the center of the preheated oven and bake for 30 to 35 minutes, or until the custard is set. When the tart is almost set, sprinkle it with the crumbled goat cheese and thyme leaves and continue to bake until completely set. A knife inserted into the center should come out clean.

6. Remove the tart from the oven and let rest for 5 minutes before serving. Serve hot, cut into wedges, as part of a light lunch.

FETTUCCINI WITH CREMINI MUSHROOMS AND SMOKED TROUT

serves 4–5

There was a time when you wouldn't dream of combining non-Italian ingredients in a pasta dish. But those days are virtually gone, and pasta of all kinds has become an American and international food. As long as a harmony is maintained between the ingredients, many non-Italian pasta dishes, such as this one, can be extremely pleasing. If you can't find good smoked trout, you may use smoked haddock blanched in milk instead. Smoked salmon in combination with vodka is now a very popular addition to fresh pasta in Italy and it can also be a good choice. But be careful that the heat of the sauce does not toughen the salmon.

1 pound Cremini (brown) mushrooms
8 tablespoons unsalted butter
Salt and freshly ground black pepper
1 cup Crème Fraîche (page 491) or heavy cream
3 tablespoons finely minced fresh dill
4 tablespoons freshly grated Parmesan cheese

1 pound fettuccini, preferably homemade (page 486)
2 small smoked trout, boned and cut into fine slivers

GARNISH
Small sprigs of fresh dill

1. Rinse the mushrooms, remove the stems, and cut the caps into ½-inch cubes.

2. In a large, heavy skillet, melt 3 tablespoons of butter over medium heat. Add the mushrooms and sauté until nicely browned. Season lightly with salt and pepper and set aside.

3. In a shallow skillet, melt the remaining butter over medium heat. Whisk constantly until the butter starts to brown and becomes hazelnut in color. Add the crème fraîche or heavy cream, bring to a simmer and reduce slightly. Season with salt and pepper and add the dill, half the Parmesan, and the sautéed mushrooms. Keep warm.

4. In a large pot, bring plenty of salted water to a boil, add the fettuccine, and cook for 3 to 4 minutes, or until al dente; do not overcook.

5. Drain the fettuccine and combine with the sauce in the skillet. Toss gently and simmer for another minute or two. Taste and correct the seasoning, adding a generous grinding of black pepper.

6. Transfer the fettuccine to individual serving plates. Top with the slivers of smoked trout and sprigs of fresh dill, and serve immediately.

Note: A combination of three herbs, such as tarragon, chives, and parsley, is an interesting substitute for the dill.

FUSILLI UMBRIANA WITH RADICCHIO AND BACON IN A CREAMY TOMATO SAUCE

serves 4

4 tablespoons unsalted butter
1 cup finely diced slab bacon, blanched
2 teaspoons olive oil
1 large head radicchio, separated and torn into 1½-inch pieces
1 cup Quick Tomato Sauce (page 474)

½ cup heavy cream
Salt and freshly ground black pepper
½ pound imported fusilli, preferably De Cecco
⅓ cup freshly grated Parmesan cheese

1. In a 10-inch, black cast-iron skillet melt 2 tablespoons of the butter over medium heat, add the bacon, and cook until lightly browned. Remove with a slotted spoon to a side dish and reserve.

2. Remove and discard all but 2 tablespoons of the fat from the skillet. Add the olive oil and the radicchio, and sauté quickly, stirring constantly, until just wilted. Transfer to a side dish and reserve.

3. Combine the tomato sauce and cream in a small saucepan, season with salt and pepper, and just heat through. Keep warm.

4. Bring plenty of salted water to a boil in a large casserole. Add the fusilli and cook until al dente. Add 2 cups of cold water to the pot to stop further cooking and drain the fusilli well in a colander.

5. Return the pasta to the casserole. Add the bacon, radicchio, creamy tomato sauce, and remaining butter. Sprinkle with the Parmesan and toss gently. Taste and correct the seasoning and serve at once with a crusty loaf of French bread and extra Parmesan on the side.

PENNE WITH PLUM TOMATOES, BROCCOLI RAPE, GARLIC, AND ANCHOVIES

serves 4

TOMATOES
3 tablespoons extra-virgin olive oil
1 small dry red chili pepper, broken
2 medium cloves garlic, peeled and
finely minced
¼ cup finely minced fresh parsley
1 pound ripe Italian plum tomatos,
peeled and cut crosswise into ¼-inch
slices
Salt and freshly ground black pepper

BROCCOLI RAPE
Salt
1¼ pounds broccoli rape, trimmed of all
but 2 inches of stalks

3 tablespoons extra-virgin olive oil
2 medium cloves garlic, peeled and
sliced
4–6 flat anchovy fillets, drained and
finely minced

PASTA
1 tablespoon olive oil
½ pound imported penne, preferably
De Cecco
½ cup freshly grated Parmesan cheese

1. Start by preparing the tomatoes: In a heavy 10-inch skillet, heat the olive oil over medium heat, add the chili pepper, and cook about 30 seconds or until dark. Remove with a slotted spoon and discard. Add the minced garlic, parsley, and sliced tomatoes. Cook for 3 to 5 minutes. The tomatoes should be soft but still retain their shape, and the mixture should be quite juicy. Season with salt and pepper and keep warm.

2. To prepare the broccoli rape: In a large casserole, bring plenty of salted water to a boil, add the broccoli rape, and blanch for 1 minute. Drain well and set aside.

3. In a large skillet, heat the olive oil over medium-low heat, add the garlic and anchovy fillets, and cook for 30 seconds. Add the broccoli rape, toss gently with the garlic mixture, and just heat through. Season with salt and pepper and keep warm.

4. To make the pasta: In a large casserole bring plenty of salted water to a boil. Add the olive oil and pasta and cook until al dente. Add 2 cups of cold water to the pot to stop further cooking. Drain well and return to the casserole.

5. Place the casserole over very low heat and add the reserved stewed to-matoes together with the broccoli rape and half of the Parmesan. Toss gently

and just heat through. Correct the seasoning and transfer to a serving bowl.

6. Serve at once with the remaining Parmesan on the side and a good crusty loaf of Italian bread.

CRISP POTATO, ONION, AND BACON PIZZA

serves 2

4 tablespoons plus 1 teaspoon olive oil
2 large onions, preferably Bermuda, peeled and thinly sliced
Salt and freshly ground black pepper
All-purpose flour for dusting the surface
½ recipe Pizza Dough (page 487)
½ large baking potato, peeled and cut into paper-thin slices

3 ounces slab bacon, finely diced and blanched
1 teaspoon herbes de Provence
½ teaspoon fresh oregano leaves
1 large clove garlic, peeled and thinly sliced
3 tablespoons freshly grated Parmesan cheese

1. In a large, heavy skillet heat 3 tablespoons of oil over high heat. Add the onions and cook for 3 to 4 minutes, or until the onions begin to brown. Reduce the heat to low and cook for 35 to 40 minutes, partially covered, until the onions are soft and caramelized. Season with salt and pepper and set aside.

2. Preheat the oven to 450° F. Brush a 12-inch black pizza pan lightly with 1 teaspoon of oil and reserve.

3. On a lightly floured surface, roll out the pizza dough into a 9-inch circle. Transfer the dough to the prepared pan and stretch gently from the center outward to the edge of the pan. If the dough becomes too elastic, let it rest for 5 minutes and begin again.

4. Spread the caramelized onions over the surface of the dough, leaving a ½-inch border at the edge of the dough.

5. Place the potato slices over the onions in a thin layer, not overlapping. Sprinkle the potatoes with the bacon, herbes de Provence, oregano, garlic, salt, and a large grinding of pepper. Drizzle with the remaining olive oil and place in the center of the preheated oven. Bake for 15 to 18 minutes, or until the potatoes are tender and lightly browned.

6. Remove the pizza from the oven and sprinkle with the Parmesan. Cut into wedges and serve at once.

ROASTED PEPPERS, MELTED ONIONS, AND ANCHOVY PIZZA

serves 2

5 tablespoons plus 1 teaspoon olive oil
4 large red onions, peeled and thinly
  sliced
3 teaspoons fresh thyme leaves
Salt and freshly ground black pepper
5 large bell peppers, a mixture of green,
  red, and yellow, roasted, peeled
  (page 483), and very thinly sliced

2 large cloves garlic, peeled and finely
  minced
1 can anchovy fillets, drained and finely
  minced
All-purpose flour for dusting the surface
½ recipe Pizza Dough (page 487)

1. In a large, heavy skillet heat 3 tablespoons of oil over high heat. Add the onions and cook for 3 to 4 minutes, stirring constantly, until the onions begin to brown. Reduce the heat to low, add 2 teaspoons of the thyme, and season with salt and pepper. Continue to cook, partially covered, until the onions are soft and caramelized, about 40 to 45 minutes. Reserve.

2. Preheat the oven to 425° F. Brush a 12-inch black pizza pan with 1 teaspoon of oil and set aside.

3. Heat the remaining oil in a medium saucepan over medium-low heat. Add the peppers, garlic, and anchovies and cook, stirring constantly, until the anchovies have completely melted. Season with black pepper and reserve.

4. On a lightly floured surface, roll out the pizza dough into a 9-inch circle. Transfer the dough to the prepared pan and stretch gently from the center outward to the edge of the pan. If the dough becomes too elastic, let it rest for 5 minutes and begin again.

5. Spread the onions on the surface of the dough, leaving a ½-inch border at the edge of the dough.

6. Top with the roasted pepper mixture and sprinkle with the remaining thyme. Place the pizza in the center of the preheated oven and bake for 15 to 20 minutes, or until the crust is nicely browned and the top is bubbly.

7. Remove from the oven and serve hot, cut into wedges.

Variation: Top the pizza with 5 ounces thinly sliced smoked mozzarella and bake for 15 to 20 minutes, until the mozzarella is melted and the crust is nicely browned.

FENNEL-INFUSED SAFFRON RISOTTO WITH FALL VEGETABLES

serves 4–5

While many people get a craving for pasta, I get a craving for rice. And when when it comes to rice, I usually think of a risotto, which is my very favorite rice dish. Here is a takeoff on the traditional risotto Milanese, which is always flavored with saffron and served as an accompaniment to osso buco. I often add a dozen or more small poached mussels to this version and use the mussel broth as part of the stock. Other shellfish, such as small clams and bay scallops, are also good additions, especially in the fall.

2 tablespoons unsalted butter
1 cup finely minced leeks, rinsed under warm water to remove all sand
1 small red pepper, cored, seeded, and finely diced
1 cup finely diced fennel
4–5 cups Chicken Stock (page 470) or bouillon, heated with 1 teaspoon saffron threads
1½ cups Italian rice, preferably Arborio

Salt and freshly ground black pepper
½ cup heavy cream
⅓ cup freshly grated Parmesan cheese
2–3 tablespoons unsalted butter, for enrichment
3 tablespoons finely minced fresh Italian parsley

GARNISH
Small fresh basil leaves

1.   In a heavy 3-quart saucepan melt the butter over medium to low heat, add the leeks, red pepper, and fennel and 2 tablespoons of stock. Cover the saucepan and cook until the vegetables are soft.

2.   Add the rice and 2½ cups of stock, stir gently, and season with salt and pepper. Cover the saucepan and simmer the rice over low heat for 15 minutes.

3.   Uncover the rice. Start adding more stock, ¼ cup at a time, stirring the rice constantly, for 10 to 15 minutes, or until the rice is tender on the outside but still chewy on the inside. Do not let the rice dry out—it should have a creamy consistency—and do not drown the rice with too much stock. When the risotto is done (you may not need to add all of the stock), add the cream and heat through.

4.   Add the Parmesan, correct the seasoning, and stir in the additional butter and the parsley. Garnish with basil and serve immediately.

Note: The precooking of the rice can be done hours ahead of time. If you plan to do that, cook the rice without the vegetables for 15 minutes and then transfer it to a colander and set aside covered with a towel. When ready to serve, finish cooking and add the vegetables.

CAZUELA OF WHITE BEANS AND
SAUSAGE ASTURIANA

serves 6

2 cups white kidney beans
8–10 cups light beef broth or bouillon
1 medium onion, unpeeled
¾ pound smoked pork shoulder butt
6 whole black peppercorns
Several sprigs of fresh parsley
1 bay leaf
2 tablespoons unsalted butter plus 2
   teaspoons olive oil

1 small jalapeño or Anaheim pepper,
   finely minced
1 large onion, peeled and finely minced
1 large red pepper, cored, ribs
   removed, and finely cubed
1½ teaspoons paprika
½ pound chorizo or cooked sweet
   Italian sausage

1.   Soak the beans in water to cover overnight.

2.   Preheat the oven to 350° F.

3.   The next day, drain the beans and transfer to a casserole. Cover with beef
broth or bouillon, add the unpeeled onion, pork butt, peppercorns, parsley
sprigs, and bay leaf, and bring to a boil on top of the stove. Cover tightly and
place in the center of the preheated oven. Cook the beans for 1 hour, or until
tender but not falling apart. Remove the casserole from the oven and drain the
beans, discarding the bay leaf and parsley, but reserving the pork, cooking
liquid, and beans. Puree 1½ cups of beans in the food processor and set aside.

4.   In a large, heavy saucepan, heat the butter and oil over medium heat. Add
the chili pepper and cook for 1 minute. Add the minced onion and continue
sautéing for 5 minutes or until the onion is soft and lightly browned. Add the
red pepper and paprika and continue to cook for 2 more minutes, blending the
paprika well into the onion mixture.

5.   Add the bean cooking liquid, the beans, and the bean puree. Bring the
soup to a simmer and cook for another 25 minutes.

6.   While the soup is simmering dice the pork butt and slice the chorizo or

Italian sausage. Add to the soup and taste and correct the seasoning.

7. Serve the soup hot, accompanied by crusty French bread.

Note: If you are using fresh Italian sausage it is best to sauté it in a little olive oil until almost done. Then slice it and add to the soup and simmer until it is just heated through.

SPICED SWEET POTATO AND JALAPEÑO PEPPER SOUP

serves 6

Yams and sweet potatoes, together with winter squash, have been stereo-typed for years. They start appearing almost magically around Thanks-giving and are a traditional side dish for holiday dinners. These beautiful vegetables seem to be treated in the same two or three ways throughout the fall and winter months—often in combination with cinnamon, brown sugar, and nutmeg. But sweet potatoes and yams are much more versa-tile, and I especially like them teamed with Indian spices and hot peppers, as they are in this cold-weather soup.

2 tablespoons unsalted butter
1 large onion, peeled and finely minced
3 medium sweet potatoes, peeled and diced
6 cups Chicken Stock (page 470)
Salt and freshly ground white pepper
1 cup sour cream

Juice of ½ lime
1 teaspoon freshly grated lemon rind
½ cup heavy cream
1 jalapeño pepper, thinly sliced
1½ cups cooked fresh corn kernels
2–3 tablespoons tiny fresh cilantro leaves

1. In a 4-quart casserole melt the butter over medium heat. Add the onion and cook for 3 to 4 minutes or until the onion begins to brown. Reduce the heat, cover, and cook until the onion is soft and nicely browned. Add the sweet potatoes and chicken stock and season with salt and pepper. Bring to a boil, reduce the heat, and simmer, partially covered, for 25 minutes, or until the potatoes are quite soft.

2. While the soup is cooking, prepare the garnish: In a small bowl, combine

the sour cream, lime juice, and lemon rind and whisk until well blended. Reserve.

3. When the potatoes are tender, strain the soup into a bowl, place the vegetables in the workbowl of a food processor, and process until smooth. Whisk the vegetable puree back into the broth and return to the casserole.

4. Whisk in the heavy cream and add the sliced pepper and corn kernels. Cook for 5 minutes more, or until the pepper is crisp-tender. Taste and correct the seasoning.

5. Serve the soup hot in individual bowls garnished with a dollop of the sour cream mixture and sprinkled with tiny cilantro leaves.

Variation: Add 2 cups of diced carrots to the soup base with several large sprigs of mint. Garnish the soup with mint instead of cilantro.

FARM-STYLE SOUP OF CABBAGE AND ROOT VEGETABLES

serves 6

Every country has at least one soup that combines a number of seasonal ingredients in a full-flavored broth. Some, such as the Italian minestrone and the Russian borscht, have become classics. Others, such as this typically Austrian soup, deserve to be. What makes this one especially interesting to the busy cook is that, unlike most soups, it can be made successfully with a good-quality commercial broth. Be sure to let the soup develop flavor for a day or two and serve it piping hot with black bread and some sweet butter.

2 large leeks, all but 2 inches of greens
  removed
3 tablespoons unsalted butter
3 carrots, peeled and cut into ½-inch
  cubes
3 medium turnips, peeled and cut into
  ½-inch cubes
2 large red potatoes, peeled and cut
  into ½-inch cubes
1 pound smoked pork shoulder butt, cut
  into 1-inch cubes
7–8 cups beef bouillon or White Stock
  (page 471)

CROUTONS
2–3 tablespoons Clarified Butter (page
  478)
6 slices of French bread
4 tablespoons sweet Gorgonzola (dolce)

4 cups tightly packed, shredded Savoy
  cabbage
Salt and freshly ground black pepper
1½–2 tablespoons red wine vinegar
1 teaspoon granulated sugar
¾ cup heavy cream

1. Cut the leeks into ½-inch pieces. Place in a colander, rinse thoroughly under warm water to remove all sand and grit, drain, and reserve.

2. In a heavy 4-quart casserole, melt the butter over medium heat. Add the leeks, carrots, turnips, and potatoes and toss well in the butter. Add the pork butt and bouillon or stock. Bring to a boil, reduce the heat, and simmer, partially covered, for 25 to 30 minutes.

3. While the soup is simmering, prepare the croutons: Heat the clarified butter in a heavy skillet over medium heat. Add the bread slices and brown nicely on both sides. Do not burn. Transfer to a double layer of paper towels to drain. Spread 2 teaspoons of Gorgonzola on each crouton and set aside.

4. Add the cabbage to the soup, season with a pinch of salt and a large grinding of pepper, and continue to simmer until the cabbage is tender.

5. In a small bowl, combine the vinegar, sugar, and cream and mix thoroughly. Add to the soup and simmer for a few minutes longer. Taste and correct the seasoning; the soup should have a sweet-and-sour taste. Keep warm.

6. Preheat the broiler.

7. Place the croutons under the broiler, cheese side up and 6 inches from the source of heat, and broil until the cheese has just melted.

8. Ladle the soup into deep individual serving bowls and top each with a crouton and a large grinding of black pepper. Serve at once.

GRILLED EGGPLANT SOUP WITH
ROASTED GARLIC

serves 6

I first sampled a version of this soup in a Balkan restaurant in Vienna. I was totally intrigued by its smoky flavor, which reminded me of the eggplant "caviar" I have been making for years. After several tests, I feel that with the addition of roasted or poached garlic cloves, I have come close to the taste I remember. Be sure to roast the eggplant on a grill, directly on hot coals or over a gas flame. Oven-roasting or broiling simply will not give you the same results. The soup is equally good served at room temperature, topped with a dollop of garlic-flavored sour cream and a sprig of fresh cilantro.

CROUTONS
12 slices of French bread, about ⅓ inch thick
2 large cloves garlic, peeled and cut in half
2 tablespoons olive oil

GARLIC CREAM
12 large cloves garlic, roasted and peeled (page 482)
1¼ cups heavy cream

SOUP
2 tablespoons unsalted butter
1 teaspoon corn oil

2 medium onions, peeled and coarsely chopped
2 medium eggplants, about 1–1½ pounds each, roasted, peeled, and cubed (page 138)
5 cups Chicken Stock (page 470)
Salt and freshly ground white pepper

GARNISH
Dollops of sour cream
Sprigs of fresh Italian parsley

1. Preheat the broiler.

2. Start by making the croutons: Rub each side of the French bread with garlic. Brush each side with olive oil and place on a cookie sheet. Place in the broiler about 6 inches from the source of heat and broil until lightly browned on both sides. Remove and set aside.

3. To prepare the garlic cream: Combine the roasted garlic and heavy cream in the workbowl of a food processor and process until smooth. Reserve.

4. To make the soup: In a heavy 4-quart casserole, melt the butter together with the oil over high heat. Add the onions and cook, stirring constantly, until the onions begin to brown, about 5 minutes. Reduce the heat to low, cover, and

cook, stirring often, until the onions are soft and nicely browned, about 30 to 35 minutes.

5. Add the eggplant and chicken stock and season with salt and pepper. Bring the soup to a boil, reduce the heat, and simmer, partially covered, for 20 minutes.

6. Transfer the soup to the workbowl of a food processor or blender, process until smooth, and pass through a fine sieve. Return the strained soup to the casserole, bring to a boil, and whisk in the reserved garlic cream. Taste and correct the seasoning and simmer, uncovered, for 10 minutes, whisking from time to time. Keep warm.

7. To serve, place 2 croutons in each of 6 individual serving bowls. Pour some of the soup over the croutons and garnish each with a dollop of sour cream and a parsley sprig. Serve at once.

CURRIED CARROT AND TURNIP SOUP

serves 6

6 cups plus 2 tablespoons Chicken Stock
  (page 470)
3 medium carrots, peeled and cut into
  ½-inch cubes
3 small turnips, peeled and cut into
  ½-inch cubes
3 tablespoons unsalted butter
1 large onion, peeled and finely minced
1 tablespoon dark-brown sugar
1 teaspoon imported curry powder,
  preferably Madras

⅛ teaspoon ground coriander
Large grating of nutmeg
2 tablespoons all-purpose flour
½ cup heavy cream
Salt and freshly ground black pepper

GARNISH

2 ripe medium tomatoes, unpeeled,
  seeded, and finely diced, optional
2 tablespoons fresh cilantro leaves

1. In a large saucepan, combine 6 cups of the chicken stock, the carrots, and turnips. Bring to a boil, reduce the heat, and simmer, partially covered, for 30 to 35 minutes, or until the vegetables are very tender.

2. Cool the soup slightly. Transfer to the workbowl of a food processor or blender and process until very smooth. Set aside.

3. In a heavy 3-quart casserole, melt the butter over low heat. Add the onion and cook until soft but not brown. Add the brown sugar, curry powder, ground

coriander, nutmeg, and flour. Add the 2 tablespoons of stock and cook, stirring constantly, for 2 minutes. Be careful not to burn the sugar.

4. Add the pureed soup, reduce the heat, and whisk in the cream. Season with salt and pepper to taste and just heat through.

5. To serve, ladle the soup into individual soup bowls and garnish each with 1 tablespoon of the optional diced raw tomato and a sprinkling of fresh cilantro leaves.

Note: The soup can be prepared several days ahead of time and reheated. It can also be served chilled.

MIDDLE EASTERN BLACK BEAN SOUP WITH CONCASSÉ OF TOMATOES

serves 6

3 tablespoons olive oil
1 small dry red chili pepper, broken
2 cups finely diced onion
2 large cloves garlic, peeled and finely minced
1¼ teaspoon ground cumin
1 tablespoon pure chili powder
1 teaspoon Spanish paprika
4–6 large ripe tomatoes, peeled, seeded, and chopped or 1 (2-pound) can Italian plum tomatoes, drained and chopped
Salt and freshly ground black pepper
3 cups Chicken Stock (page 470)

1 pound dry black turtle beans, cooked and drained, plus 3 cups bean cooking liquid (page 484)

GARNISH
1¼ cups sour cream
1 ripe medium tomato, seeded and diced
2 tablespoons extra-virgin olive oil
2 tablespoons fresh cilantro leaves
2 tablespoons finely minced red onion or chives
2 teaspoons finely grated lemon rind

1. In a heavy 4-quart casserole heat the oil over low heat. Add the chili pepper, onion, and garlic and cook the mixture for about 6 to 8 minutes, or until the onion is very soft and lightly browned. Add the cumin, chili powder, and paprika and stir well into the onion mixture.

2. Add the tomatoes, season with salt and pepper, and add the stock. Bring to a boil, reduce the heat, and simmer, covered, for 20 minutes.

3. In the workbowl of a food processor, process ½ cup of the cooked black beans until smooth. Add the puree, the remaining beans, and the bean cooking liquid to the casserole. Simmer the soup for an additional 10 minutes.

4. While the soup is simmering, prepare the garnish: In a small bowl combine all the garnish ingredients, mix well, and reserve.

5. When the soup is done, correct the seasoning. If the soup is too thick, thin it with additional stock. Serve hot in individual soup bowls, garnished with a dollop of the sour cream mixture and accompanied by a crusty loaf of bread.

Note: This soup is best made 2 to 3 days ahead of time, and it freezes well, too.

In late summer I often substitute 1 cup of freshly cooked corn kernels for 1 cup of the beans. The corn adds a lovely color and texture to the soup.

BASQUE PLUM TOMATO AND RED ONION SOUP

serves 4

No matter how you look at it, the fresh tomato season is too short wherever you live. Soups such as this, however, can be made with excellent results using canned tomatoes. Zesty, slightly spicy and full of flavor, the soup can be varied greatly, and I usually make a double batch, changing it according to the produce available at the local farmstand. For a Provençal touch, I add 1 teaspoon of saffron threads to the chicken broth and serve the soup with a bowl of garlic-infused yogurt. The soup is delicious eaten at room temperature on those early fall days that are so welcome after a hot and humid summer.

3 tablespoons olive oil
2 medium red onions, peeled and finely diced
1 large clove garlic, peeled and finely minced
1–2 teaspoons finely minced jalapeño pepper
1 green bell pepper, cored, seeded, and diced
1 teaspoon Spanish paprika
1 tablespoon fresh thyme or 1 teaspoon dried

1 tablespoon fresh oregano or 1 teaspoon dried
4–5 large ripe tomatoes, peeled, seeded, and chopped (in early fall) or 1 (2-pound) can Italian plum tomatoes, drained and coarsely chopped
3 cups Chicken Stock (page 470) or bouillon
Salt and freshly ground black pepper
½–¾ cup sour cream or Crème Fraîche (page 491)

1 large ripe tomato, seeded and diced
1 tablespoon extra-virgin olive oil

1 teaspoon sherry vinegar
2 tablespoons finely minced fresh Italian
parsley

1. In a 3-quart casserole, heat the oil over low heat. Add the onions and garlic and cook the mixture for about 5 minutes or until just soft. Add the peppers and cook for another 2 to 3 minutes without browning.

2. Add the paprika, thyme, oregano, and tomatoes. Add the stock and season with salt and pepper. Bring to a boil, reduce the heat, and simmer, partially covered, for 45 minutes.

3. While the soup is simmering, prepare the garnish: In a small bowl combine the tomato, oil, and vinegar, toss gently, and set aside.

4. Cool the soup slightly, transfer to the workbowl of a food processor, and process until smooth. Return the soup to the casserole, whisk in the sour cream, and just heat through. Taste and correct the seasoning.

5. Serve the soup hot, garnished with the marinated tomato and minced Italian parsley.

Note: Taste the jalapeño pepper before using it. If it is very hot, you may want to use only half rather than all of it.

SAUTÉED MEDALLIONS OF MONKFISH À LA VALENCIANA

serves 4–5

Monkfish, a popular and inexpensive fish, is well worth seeking out. Here it is prepared in a way fairly common in and around Valencia—the home of the classic paella, which is also usually prepared with monkfish.

Be sure to look for monkfish that has been properly handled: All of the bluish white membrane should be removed. The fresh fish should have an almost white, translucent flesh. Stay away from monkfish that is grayish and opaque. If you can't get it in perfect condition, use fresh shrimp instead, which will do very nicely in this dish.

3 tablespoons olive oil
2 large cloves garlic, peeled and thinly sliced
1 large sprig of fresh thyme
1 medium red onion, peeled, quartered, and thinly sliced
2 large ripe tomatoes, peeled, seeded, and chopped
1 medium zucchini, quartered and finely cubed
Salt and freshly ground black pepper

1 red bell pepper, roasted and peeled (page 483), and cut into fine julienne
3 cups Cooked White Beans (page 484)
1½ pounds fresh well-trimmed monkfish
All-purpose flour for dredging
2 tablespoons unsalted butter

GARNISH
Fresh lemon juice
Extra-virgin olive oil
Finely minced fresh Italian parsley

1. In a heavy 10-inch skillet heat 2 tablespoons of olive oil over low heat. Add the sliced garlic, thyme, and onion and sauté the mixture until the onion is soft but not browned. Add the tomatoes and zucchini and season with salt and pepper. Simmer over medium heat until the zucchini is tender and most of the tomato liquid has evaporated.

2. Add the julienne of roasted pepper and the beans, cover the skillet, and simmer the mixture for 5 minutes. Taste and correct the seasoning and keep warm.

3. Cut the monkfish crosswise into ½-inch slices, or medallions. Dry the medallions on paper towels, season with salt and pepper, and dredge lightly in flour, shaking off the excess.

4. In a large nonstick skillet heat the remaining olive oil together with the

butter over medium-high heat. Add the monkfish, without crowding the skillet, and sauté for 2 minutes on each side or until nicely browned. Do not overcook.

5. Place the vegetable mixture on a large oval serving platter. Top with the monkfish medallions and sprinkle with lemon juice and extra-virgin olive oil. Garnish with the parsley and serve immediately.

Note: The vegetable mixture can be prepared several days ahead of time. You may also use a combination of scallops and shrimp and fold the seafood mixture gently into the vegetable stew as it is often prepared in Spain.

LEMON- AND HERB-MARINATED RED SNAPPER À LA CATALAN

serves 6

Although I grew up in Barcelona, a city with an abundance of great markets and an immense variety of fresh fish, my mother would only serve a whole roasted fish on special occasions. But when she did, she did it well—always perfectly cooked and beautifully garnished. The memory has stayed with me, and now I too like to serve a whole roasted fish, with its head still attached and surrounded by glistening shellfish, as the centerpiece for a dinner party. Just remember to choose a truly fresh fish. Look at the gills to be sure that they are red, glossy, and fresh-looking, then have the fishmonger remove them for you, leaving the head on.

1 whole red snapper (3½–4 pounds), cleaned, with head left on

MARINADE
2 tablespoons olive oil
Juice of 1 lemon
1 teaspoon whole coriander seeds
Coarse salt

FILLING
3 tablespoons soft bread crumbs
3 tablespoons finely minced fresh parsley
3 medium cloves garlic, peeled and finely minced
2 tablespoons olive oil
Salt and freshly ground white pepper

SAUCE
24 small mussels, well scrubbed and debearded
¼ cup dry white wine
1½ tablespoons olive oil, preferably extra-virgin
12–14 medium shrimp, peeled
2 tablespoons finely minced shallots
2 tablespoons finely minced fresh parsley
2 large cloves garlic, peeled and finely minced
6 large ripe tomatoes, peeled, seeded, and chopped
1 teaspoon fresh thyme
Salt and freshly ground white pepper
1 cup reserved mussel stock
1 lemon, unpeeled and thinly sliced

1.   Place the fish in a large baking dish, sprinkle with the olive oil, lemon juice, coriander seeds, and coarse salt. Cover the fish and marinate in the refrigerator for 1 to 2 hours. (This is particularly important to do if you have bought the fish a day in advance.)

2.   To prepare the filling: In a small bowl combine the bread crumbs, parsley, minced garlic, and olive oil. Season the mixture with salt and pepper and mix well.

3.   Drain the fish and fill the cavity with the bread crumb mixture. Sew up the opening or fasten with toothpicks. Season the fish with a little more salt and pepper and return to the baking dish.

4.   Preheat the oven to 350° F.

5.   To make the sauce: In an enameled saucepan, combine the mussels with the wine, bring to a boil, and steam the mussels, partially covered, for about 5 minutes, until they open. With a slotted spoon remove the mussels and shell all but 8 of them. Reserve both the shelled and unshelled mussels. Strain the stock through a double layer of cheesecloth or paper towel and reserve separately.

6.   In a large, heavy skillet, heat the olive oil. Add the shrimp and cook over fairly high heat until they turn a bright pink. Remove with a slotted spoon to a side dish and reserve. Add the shallots, parsley, and garlic to the skillet and cook for 1 minute. Then add the tomatoes, thyme, salt, and white pepper. Cook the mixture until most of the tomato liquid has evaporated, add the reserved mussel stock, and pour over the fish in the baking dish.

7.   Place the sliced lemon over the fish, cover with buttered parchment paper, and place in the center of the preheated oven. Bake the fish for 30 to 35 minutes.

8.   Add the reserved shrimp and shelled mussels to the baking dish and continue to bake for another 10 minutes.

9.   Raise the heat to broil. Uncover the fish, place the whole unshelled mussels in the baking dish, and run the fish under the broiler for 2 to 3 minutes. Serve right out of the baking dish accompanied by French bread or parsleyed new potatoes.

Note: The fish pan juices will be quite thin. If you want a heavier sauce, remove the fish, mussels, and shrimp to a serving platter. Place the baking dish with its pan juices over direct heat and cook until the sauce reaches the desired consistency, then spoon the sauce around the fish.

NAVARIN OF SCALLOPS À LA FERMIÈRE

serves 6

Navarin of lamb is a traditional bistro and home-cooked dish that combines tender baby lamb with young root vegetables in a hearty but delicate stew. Even before the recent burst of trendy bistro fare, chefs such as Pierre Gagnaire, Jacques Lameloise, and others sought to incorporate the concept of the traditional navarin into lighter and more unusual combinations, using a variety of seafood.

Here is my interpretation of Jacques Lameloise's navarin of scallops, one of the signature dishes of the lovely three-star restaurant bearing his name. To me, the creamy, sweet Lameloise sauce is the perfect finishing touch, but you may lighten the dish by making a vegetable stock or light chicken stock, reducing it to ½ cup, and enriching it with a little unsalted butter and a touch of crème fraîche.

1¾ cups Lameloise Sauce (page 479)

VEGETABLE GARNISH
1¼ pounds red new potatoes, peeled and cut into ¾-inch cubes
1 pound turnips, peeled and cut into ¾-inch cubes
2 medium carrots, peeled and cut into 1-inch matchsticks
1 cup snow peas, tips and strings removed

SCALLOPS
1½ pounds small sea scallops
Salt and freshly ground black pepper
All-purpose flour for dredging
4 tablespoons unsalted butter
2 teaspoons virgin olive oil

GARNISH
Finely minced fresh Italian parsley
Tomato Concassé (page 89)

1. In the top of a double boiler, heat the Lameloise sauce over simmering water, whisking constantly until hot. Reduce the heat, cover, and keep warm, whisking often to keep the sauce smooth.

2. To prepare the vegetable garnish: In a large saucepan, bring salted water to a boil, add the potatoes, and cook for 12 to 15 minutes, or until just tender. Remove with a slotted spoon to a side dish and reserve. Add the turnips and carrots and cook for about 5 minutes or until just tender. Drain and add to the potatoes.

3. Bring water to a boil in a vegetable steamer, add the snow peas, set over simmering water, and steam, covered, until just tender. Add to the remaining vegetables, cover, and keep the vegetables warm.

4. To prepare the scallops: Season them with salt and pepper and dredge

lightly with flour, shaking off the excess. In a 10-inch nonstick skillet, heat 2 tablespoons of butter with the oil over medium-high heat. When very hot, add the scallops and sauté until nicely browned on both sides. Do not overcook; the scallops should still feel spongy. Remove the scallops with a slotted spoon to individual shallow soup bowls.

5.   Discard the fat from the skillet and melt the remaining 2 tablespoons of butter over medium heat. Add the reserved vegetables and toss in the butter for 1 to 2 minutes. Divide the vegetables among the plates and spoon the sauce over each portion. Garnish with minced parsley and tomato concassé.

Note: The sauce can be made 2 to 3 days ahead of time and reheated. The vegetable garnish can be cooked several hours ahead of time and reheated in the butter. Do not refrigerate.

You can cut the potato and turnips into olive shapes instead of cubes, using an oval melon-ball cutter, which can be purchased in any local cookware shop.

Do not attempt to make this dish with either frozen scallops or the tiny bay scallops from the Carolinas, both of which are quite bland and become rubbery when poached; they will not stand up in this dish.

Scallops

Scallops are my favorite shellfish because of their enormous versatility. This delicious sweet shellfish has been plentiful in America for a long time but their quality varies widely depending on where they come from. Scallop connoisseurs will agree that the small bay scallop from the inlets of Long Island and northern New England are among the sweetest and most delicate. The larger ones that come from the deep cold waters of the Atlantic are not as good. Both are at their best from October through April. They should be plump and glossy. When buying scallops, avoid those that sit in a pool of liquid since this is a sure sign that they were frozen and lack freshness. Also avoid scallops that look dry or seem slightly grayish. Always buy scallops at a reputable fish market and, if possible, sniff them; they should smell sweet and not at all fishy. I always choose scallops that are uniform in size and prefer them on the larger side. Don't cube a large scallop because the pieces will dry out in cooking. However, you can slice them crosswise into fine slices which, when very fresh, can be sprinkled with lime juice and some fruity olive oil and enjoyed raw.

Most people are not aware that the northeast Atlantic scallop season is actually short since small scallops from the Carolinas and Florida are available for much of the year. Although these Calico scallops are also labled "bay" they have really nothing in common—other than their size—with the Atlantic bay scallops and are tasteless and watery and turn rubbery within a minute of cooking. Several varieties of sea scallops are also imported from Mexico, Japan, Chile, and New Zealand, and as long as they are fresh they can be quite tasty.

After buying scallops be sure to transfer them immediately to a glass or stainless-steel bowl and cover lightly with foil. Use them within a day or two of purchase.

■ ■ ■

LEMON-HERB GRILLED CHICKEN "UNDER A BRICK"

serves 4

1 small chicken, about 2½–3 pounds

MARINADE
2 teaspoons dark soy sauce
Juice of 1 large lemon
2 teaspoons fresh thyme leaves
1 teaspoon *herbes de Provence*
3 large cloves garlic, peeled and
    crushed with the back of a knife

Large grinding of black pepper
⅓ cup olive oil

"UNDER A BRICK"
2 sprigs of fresh thyme
2 sprigs of fresh rosemary
4 thin slices of garlic
½ cup Chicken Stock (page 470)
Coarse salt and freshly ground black
    pepper

1.   Start by preparing the chicken: With sharp kitchen scissors, remove the backbone of the chicken and discard. Split the chicken in half and place each half between 2 sheets of wax paper. With a mallet, pound the chicken all over to flatten it. Set aside.

2.   To prepare the marinade: In a small bowl combine all the ingredients for the marinade and whisk until well blended. Place the chicken halves in a large zip-lock bag, pour the marinade over the chicken, and seal the bag. Place the bag in a large, shallow dish and refrigerate overnight, turning the bag several times.

3.   The next day, remove the chicken from the marinade. Carefully life the skin over each breast and thigh portion and tuck 1 sprig each of thyme and rosemary and 2 slices of garlic between the skin and flesh of each half. Reserve the chicken and marinade separately. Wrap 2 bricks completely with foil and set aside.

4.   Preheat the oven to 500° F.

5.   Place a dry 12-inch cast-iron skillet over medium-high heat. When it begins to smoke, add the marinade to the skillet and immediately add the chicken halves, skin side down. Place a brick on each half and cook for 3 to 4 minutes, or until nicely browned.

6.   Remove the bricks, turn the chicken over, and return the bricks. Add ¼ cup of chicken broth to the skillet, place the chicken in the center of the preheated oven, and bake for 18 minutes, adding a couple of tablespoons of stock every 8 minutes.

7.   Again remove the bricks, turn the chicken over so that it is again skin side

down, and bake for 5 minutes longer or until the chicken is done. The juices should run pale yellow.

8. Remove the chicken from the oven and transfer to a serving platter. Spoon the pan juices over the chicken, season with coarse salt and black pepper, and serve at once.

SAUTÉ OF CHICKEN BREASTS WITH A FRICASSEE OF MUSHROOMS AND DILL

serves 6

1 ounce dried porcini mushrooms
1½ cups Chicken Stock (page 470)
6 tablespoons unsalted butter
10 ounces cultivated mushrooms,
   stemmed, wiped, and thinly sliced
Salt and freshly ground black pepper
3 whole chicken breasts, boned, skin
   removed, cut in half
All-purpose flour for dredging

2 teaspoons peanut oil
3 tablespoons finely minced shallots
½ cup heavy cream, optional
1 Beurre Manie (page 478)
3 tablespoons finely minced fresh dill

GARNISH
Sprigs of fresh dill

1. Place the porcini in a strainer and run under warm water to remove all sand and grit.

2. In a small saucepan, combine the porcini and stock and bring to a simmer. Cover and continue to simmer for 25 to 30 minutes or until the mushrooms are tender. Remove the porcini with a slotted spoon to a cutting board, cut into a fine julienne, and reserve. Strain the stock through a double layer of cheesecloth and set aside.

3. In another large, heavy skillet, melt 3 tablespoons of the butter over medium-high heat. Add the cultivated mushrooms and sauté quickly, until nicely browned. Add the porcini and cook for 1 minute longer. Season with salt and pepper and reserve.

4. Dry the chicken breasts thoroughly on paper towels. Season both sides with salt and pepper and dredge lightly in flour, shaking off excess. Reserve.

5. In a large, heavy skillet, melt the remaining butter together with the oil over medium-high heat. Add chicken breasts, without crowding the skillet and

sauté until nicely browned on both sides. Transfer the chicken breasts to a side dish.

6. Add the shallots to the skillet and cook for 1 minute or until soft and lightly browned. Add ¼ cup of the reserved porcini stock to the skillet. Return the chicken breasts to the skillet, reduce the heat, and simmer, covered, for 5 minutes, or until chicken is done.

7. Transfer the chicken breasts to a warm serving platter, cover, and keep warm.

8. Add the mushroom mixture, remaining porcini stock, and optional heavy cream to the skillet and reduce by half. Whisk in bits of beurre manie until the sauce lightly coats a spoon. Add the minced dill and correct the seasoning. Spoon the sauce over the chicken breasts, garnish with sprigs of fresh dill, and serve at once.

ROAST DUCK WITH CABERNET-POACHED PEARS BORDALOU

serves 4–6

I first had this dish in Valence, a small town halfway between Lyon and the Riviera. Valence is not known for its regional food, yet one of the most prestigious three-star restaurants in France, Pic, has been a favorite of my parents' for the past thirty years. I knew it well as a child when my father always had us stop there for lunch on our way to Geneva where I went to school.

Fall is a particularly exciting time in that region, and at Pic's all kinds of game are available on the menu during the hunting season. One of the house specialties is wild duck, quickly roasted and served medium-rare with wine-poached pears. The superb, light gamy flavor of a wild duck is in perfect harmony with the pears. Even when made with our domestic ducks, it is still a wonderful fall and winter dish, well worth adding to your repertoire. For a beautifully balanced main course, serve the ducks accompanied by a puree of celery root and apples and Wild Mushroom–Rice Timbales (page 324).

Pears

Like apples, pears are often taken for granted and considered practically a year-round fruit. Nothing could be further from the truth. Good pears are synonymous with fall and in fact, more often than not, they are very hard to find unless you live in California or the Pacific Northwest, or have access to a farm or local produce stand in the Northeast. Also, like apples, not all pears are interchangeable in cooking so knowing which type of pear works in a recipe is often crucial to the dish's success.

The best pear season is of course the late summer and early fall when both golden and red Bartletts come to market. Besides being a great eating pear, Bartletts work well poached, baked, or sautéed. A truly ripe Bartlett served with an excellent mellow piece of Parmesan cheese is a simple yet memorable ending to a meal.

Another good and sometimes fantastic eating pear is the Comice. Superbly juicy, this pear has a far longer season than Bartletts and is available off and on during the cold-weather months. However it is strictly for eating and does not take well to baking or sautéing.

The two most available pears are the Bosc and the Anjou. The Boscs—with a brown skin that turns somewhat russet in color when ripe—is probably the most shapely of the pear varieties, and can be quite sweet. I like the fact that it remains crunchy even when fully ripe. It is a good eating pear and lends itself well to sautéing or poaching in red wine. As for the Anjou, the best thing going for it is a relatively modest price and the fact that it is available from October through May. Unfortunately, although a ripe Anjou can be good, it is almost impossible to find. I only use this pear for oven baking because with the addition of sugar, butter, and some maple syrup or cream the flavor can be vastly improved. In the late spring you may come upon some ripe Anjous and these are good for sautéing with a touch of butter and sugar.

The two most popular miniature pears are the Seckel and the Forelle. The Seckel (which is grown both on the east and west coasts) can taste quite woody and uninteresting but is lovely as a colorful centerpiece for a dinner table. The Western Seckel is only available in the fall. A superb eating pear, it is also delicious when baked with a little white wine, butter, and sugar.

Personally I love the rather bland sweet flavor of the Forelle pear; it is crunchy and a nice size pear to nibble. It is not interesting for cooking and its season is often so short that one can easily miss it, but if you see this charming pear in the market, don't pass it up; it makes a welcome addition to a fresh fruit dessert basket.

■■■

PEARS
1¼ cups granulated sugar
1 (3-inch) piece of cinnamon stick
10 whole black peppercorns
2 whole cloves
4 cups dry red wine, preferably
    Cabernet Sauvignon
4 Bosc pears, peeled

2 ducks, each weighing about 4½
    pounds

Salt and freshly ground black pepper
1 teaspoon dried thyme
3 tablespoons unsalted butter
2 teaspoons peanut oil
3 small onions, peeled and cut in half
    crosswise
1–1½ cups hot chicken bouillon
⅓ cup finely minced shallots
½ cup ruby port
1 cup Brown Chicken Stock (page 467)
Drops of lemon juice to taste

1. Start by poaching the pears: In a large casserole combine the sugar, cinnamon stick, peppercorns, cloves, and wine. Bring the mixture to a simmer and cook until the sugar has dissolved. Add the peeled pears and bring to a boil. Reduce the heat and simmer the pears until just tender; do not overcook. Remove the pears to a side dish with a slotted spoon. Strain the poaching liquid into a saucepan, discarding the spices, and reduce the syrup over medium-high heat to 1½ cups. Reserve.

2. Preheat the oven to 350° F.

3. Dry the ducks thoroughly with paper towels, season with salt and pepper and thyme, and truss. In a large flameproof rectangular baking pan, melt the butter and oil over medium heat. Add the ducks and brown on all sides. Discard all but 3 tablespoons of fat from the pan, add the onions, cut side down, to the pan, and place in the center of the preheated oven. Roast the ducks for 2 hours and 45 minutes, removing the fat every 10 minutes and basting the ducks with a little of the hot bouillon. When they are done, transfer the ducks to a broiling pan and reserve.

4. Thoroughly degrease the pan juices. Return the juices to the pan, place over high heat, and reduce to a glaze. Add the shallots and port and reduce again to a glaze. Add the brown chicken stock and the reserved pear syrup. Continue cooking until the sauce is reduced to 1 or 1¼ cups. Strain, return to the saucepan, and correct the seasoning, adding drops of lemon juice to taste. Keep warm.

5. Preheat the broiler.

6. Run the ducks under the broiler until crisp, carve, and place on a serving platter. Cut the pears into eighths and heat briefly in the sauce. Place the pears around the ducks and spoon the sauce over them. Serve at once.

Variation: In late summer and early fall, use wine-poached freestone peaches instead of the pears. In late fall and early winter, substitute prunes or figs for the pears.

ROAST DUCK WITH GLAZED BUTTERNUT SQUASH AND LINGONBERRIES

serves 4–6

This recipe really began on a certain Thanksgiving, when I tried to come up with an imaginative side dish for a roast turkey. The combination of a sweet onion jam, butternut squash, and lingonberries all seemed to fit into the holiday menu. But since turkey seems to lose its charm soon after Christmas, I have varied the recipe and now serve the onion and squash mixture with roast duck. If you are a devotee of turkey or baked ham, by all means try them with this combination. It might become a favorite addition to your Thanksgiving and Christmas menus. For an additional side dish, I would choose a gratin of potatoes or a pilaf of wild mushrooms.

DUCKS
2 fresh ducks, each weighing about 4½ pounds
Salt and freshly ground black pepper
1 teaspoon dried thyme
3 tablespoons unsalted butter
2 teaspoons peanut oil
3 small onions, peeled and cut in half through stem end
2 cups hot Brown Chicken Stock (page 467)

VEGETABLES AND SAUCE
5 tablespoons unsalted butter
2 large onions, peeled, quartered, and thinly sliced

Salt and freshly ground black pepper
Pinch of granulated sugar
2 cups peeled and cubed butternut squash
2 tablespoons dark-brown sugar
2 tablespoons lingonberries in syrup, drained
Drops of lemon juice to taste
1 tablespoon arrowroot mixed with a little cold stock, optional

GARNISH
Sprigs of fresh watercress

1. Preheat the oven to 400° F.

2. Dry the ducks thoroughly with paper towels. Season with salt, pepper, and thyme and truss.

3. Melt the butter and oil in a large rectangular flameproof baking pan, add the ducks, and brown quickly on all sides. Discard all but 3 tablespoons of fat, add the onions to the pan cut side down, and roast the ducks in the center of the preheated oven for 15 minutes. Reduce the oven temperature to 350° F and continue to roast for 2 hours and 30 minutes, turning the ducks once during roasting. Remove the fat every 15 to 20 minutes and baste with 2 to 3 table-

spoons of stock. The ducks are done when the juices run pale yellow when pierced with the tip of a sharp knife.

4. While the ducks are roasting, prepare the vegetables: In a heavy 10-inch skillet, melt 3 tablespoons of butter over medium-high heat, add the sliced onions, and cook for 5 minutes, stirring often, until the onions begin to brown. Reduce the heat to low and season with salt and pepper and a pinch of sugar. Cover and cook for 20 to 25 minutes, stirring often, until the onions are very soft and nicely browned. Set aside.

5. In a small saucepan, bring salted water to a boil, add the squash, and cook for 1 to 2 minutes, or until barely tender. Drain well.

6. In another heavy 10-inch skillet, melt the remaining butter over medium-high heat. Add the drained squash, sprinkle with the brown sugar, and sauté quickly until lightly caramelized, shaking the pan back and forth to brown the squash evenly. Season with salt and pepper and set aside.

7. When the ducks are done, transfer them to a baking sheet, remove and discard the trussing string, and keep the ducks warm.

8. Discard the onions from the roasting pan, thoroughly degrease the pan juices, and place the roasting pan over medium heat. Return the pan juices together with the remaining stock and bring to a boil. Reduce the heat, add the reserved caramelized onions and butternut squash, and simmer until just heated through. Add the lingonberries and drops of lemon juice, and correct the seasoning. The sauce should have a sweet yet somewhat tart flavor. If the sauce seems too thin, whisk in enough of the arrowroot mixture until the sauce lightly coats a spoon. Keep the sauce warm.

9. Preheat the broiler.

10. Place the ducks in the broiler 6 inches from the source of heat and broil until crisp. Carve the ducks, place on a serving platter, and garnish with sprigs of watercress. Serve the lingonberry-squash sauce on the side.

FILLETS OF BEEF IN RED WINE, SHALLOT, AND HONEY ESSENCE

serves 4

For some mysterious reason, the beef fillet is considered the most elegant and finest cut of meat. Personally, I find it rather boring and not as tasty as other cuts of beef. But I must admit that some new and interesting preparations have come my way and now I find myself seeking this cut whenever I need a quick, flavorful main course.

2 tablespoons finely minced shallots
1 cup dry red wine, preferably Bordeaux
1 teaspoon freeze-dried green peppercorns, crushed
1 cup Brown Chicken Stock (page 467)
2–4 tablespoons unsalted butter
1 teaspoon peanut oil

4 beef fillet steaks, cut 1 inch thick
Coarse salt and freshly ground black pepper
1–1½ teaspoons Swedish-style honey
1 Beurre Manie (page 478)
3–4 tablespoons unsalted butter, for enrichment, optional

1. In a small, heavy saucepan, combine the shallots, wine, and green peppercorns, bring to a boil, and cook until reduced to 2 tablespoons. Add the stock, bring back to a boil, and reduce by one third. Reserve.

2. In a heavy 10-inch skillet, melt 2 tablespoons of butter together with the oil over high heat. When almost smoking, add the fillets and sauté for 3 minutes on each side. Season with salt and pepper and transfer to a side dish. Reserve.

3. If the fat in the pan has burned, discard it and add 2 tablespoons fresh butter to the skillet. Add the wine and stock mixture, bring to a boil, and whisk in the honey. Season carefully with salt and pepper. Whisk in bits of beurre manie and simmer the sauce until it coats a spoon. Remove from the heat and whisk in the optional butter for enrichment. Taste the sauce and correct the seasoning.

4. Return the beef to the skillet along with any juices that may have accumulated on the plate. Coat the fillets thoroughly with sauce, transfer to individual serving plates, and serve immediately.

OXTAIL RAGOUT IN CABBAGE LEAVES

serves 6

4 pounds oxtails, trimmed and cut
    crosswise into 2-inch pieces
Salt and freshly ground black pepper
Flour for dredging
2–3 tablespoons peanut oil
2 tablespoons unsalted butter
2 stalks celery, diced
2 carrots, peeled and diced
2 medium onions, peeled and diced

2 large cloves garlic, peeled and finely
    minced
1 (2-pound) can Italian plum tomatoes,
    drained and chopped
1 cup dry red wine
3–4 cups beef bouillon
1 Bouquet Garni (page 485)
1 head Savoy cabbage

1. Preheat the oven to 350° F.

2. Season the oxtails with salt and pepper and dredge lightly in flour, shaking off the excess. Set aside.

3. In a large flameproof casserole, heat half the oil and butter over medium-high heat. Add the oxtail pieces and brown on all sides. Remove with a slotted spoon to a side dish and reserve.

4. Discard the oil and replace with the remaining oil and butter. Reduce the heat and add the celery, carrots, onions, and garlic. Cook the vegetable mixture for 5 to 7 minutes until soft and lightly browned.

5. Add the tomatoes, wine, and 3 cups of stock and bring to a boil. Return the oxtails to the casserole. If the liquid does not cover the oxtails, add the remaining stock and bury the bouquet garni among the oxtail pieces. Cover the casserole tightly and set in the center of the oven. Braise for 3 hours, or until the meat is very tender and almost falling off the bones.

6. When done remove the casserole from the oven, let the stew cool completely, and then refrigerate for at least several hours and preferably overnight.

7. The next day carefully skim all the fat from the surface of the pan juices. Bring to a boil. Strain the sauce through a fine sieve into a smaller saucepan, pressing well on the vegetables to extract all their juices; discard the vegetables. Place the pan juices over very low heat and cook until well reduced and slightly syrupy. Taste the sauce and correct the seasoning.

8. While the sauce is reducing, remove the meat from the bones and shred it, being careful not to include any of the remaining fat. Set the meat aside.

9. In a large casserole bring salted water to a boil. Add the cabbage and cook for 5 minutes. Drain the cabbage and run under cold water to stop it from further cooking. Dry the cabbage on a clean kitchen towel and carefully detach 12 to 14 of the largest leaves. Place the core end of the leaves facing you and with a sharp knife cut out the center rib.

10. Fill each cabbage leaf with some of the oxtail meat and roll the cabbage leaves into cylinders, folding in the sides, encasing the meat completely.

11. Butter a rectangular flameproof baking dish large enough to hold the cabbage rolls in one layer. Place the rolls seam side down in the baking dish and pour the strained, reduced sauce over them. Cover tightly with aluminum foil and reserve.

12. Preheat the oven to 350° F.

13. Place the baking dish in the center of the oven and braise for 45 minutes to an hour, or until the cabbage is very tender and nicely glazed. Serve directly from the baking dish accompanied by parsleyed new potatoes and crusty French bread to sop up the sauce.

Note: This dish is really ideal for leftovers and I usually plan 2 menus whenever I make it. The first is the simply braised oxtails accompanied by white beans or served over saffron-flavored fettuccine.

I then make the stuffed cabbage rolls with leftover oxtails. If you plan to make the oxtails just for the stuffed cabbage, be sure to braise them at least a day ahead of time so that the pan juices can be properly degreased. Be sure to use a beef bouillon that is not salty and be restrained when seasoning the oxtails since the sauce will get quite salty as it reduces. Once the cabbage rolls are in the oven, the sauce will reduce even more and the result is an absolutely delicious full-flavored, gutsy, fall or winter main course.

BRAISED VEAL WITH GLAZED ONIONS AND CHESTNUTS

A ragout of veal is an enormously pleasing cool-weather main course. Gently braised with lots of garlic cloves, it becomes intensely flavored but not at all garlicky. Instead, the garlic turns quite sweet and, when pureed with the pan juices, slightly thickens the sauce, giving it more body and a lovely texture.

Although fresh chestnuts appear in many supermarkets and grocery stores during the fall, it is not easy to determine their freshness. Look for large, firm chestnuts with one side completely flat. If you cannot find good, fresh nuts, it is best to use the French vacuum-packed peeled chestnuts that come in a jar, usually under the name of Minerve, which, I find, give excellent results. Do not use canned chestnuts, as they are too soft and will fall apart as soon as they are added to the ragout.

3 pounds boneless shoulder of veal, cut into 1½-inch cubes
2 to 4 tablespoons unsalted butter
3 to 4 tablespoons peanut oil
Salt and freshly ground white pepper
2 teaspoons all-purpose flour
16 cloves garlic, peeled
2¼ cups Brown Chicken Stock (page 467)
1 large sprig of fresh thyme or 1 teaspoon dried

2 small yellow onions, peeled and cut in half
16–18 small white onions, peeled
1 tablespoon dark-brown sugar
16–18 chestnuts, roasted and peeled
½ cup heavy cream, optional
1 Beurre Manie (page 478)

GARNISH
Finely minced fresh parsley

1. Preheat the oven to 350° F. Dry the veal well on paper towels and set aside.

2. In a large, heavy flameproof casserole, heat 2 tablespoons of butter and 1 tablespoon of oil over fairly high heat. Add the veal cubes a few at a time—do not crowd the pan—and sauté until nicely browned on all sides. Season with salt and pepper and remove to a side dish. Continue sautéing until all the veal has been browned, adding additional butter and oil if necessary.

3. Return all the veal to the casserole, sprinkle with flour, and toss the meat over medium-high heat until it is covered with flour and lightly glazed. Add the garlic cloves, 2 cups stock, thyme, and yellow onions and bring to a boil. Cover

tightly, place in the center of the preheated oven, and braise for 1 hour and 30 minutes.

4. While the meat is braising, caramelize the white onions: In a heavy skillet, heat 2 tablespoons of oil over medium heat. Add the onions and shake the skillet back and forth until the onions are nicely browned. Add the brown sugar and shake the pan until the onions are well coated with the sugar. Add the remaining ¼ cup of stock and the chestnuts, partially cover, and simmer until tender. Reserve the onions and chestnuts with their pan juices.

5. When the meat is tender, when pierced with a fork, remove it to a side dish. Transfer the pan juices with the garlic cloves to the workbowl of a food processor and puree until smooth. Return to the casserole.

6. Add the optional heavy cream to the casserole, bring to a boil, and add bits of beurre manie, whisking constantly until the sauce coats a spoon. Return the meat, onions, chestnuts, and the juices to the casserole and just heat through. Spoon the ragout into a serving dish, sprinkle with parsley, and serve immediately.

Chestnuts

The appearance of chestnuts in the market is a true sign of fall, and one of the most intoxicating aromas is that of a roasted chestnut. Unfortunately, the only chestnuts available in the United States are imported, and their quality varies greatly. The best chestnuts come from Italy, where they have been cultivated since Roman times. The very large Spanish chestnut is famous for its flavor and size; it is not often available outside the large metropolitan areas.

Look for large chestnuts with one definite flat side; these will be easier to peel and will not disintegrate when cooked. Be sure to buy chestnuts in a market with a large turnover, because contrary to common belief, this nut has a limited shelf life. A recent boon to cooks are the frozen peeled chestnuts that are imported from Italy and available in fine gourmet and specialty food stores.

Do not use canned chestnuts. However, the vacuum-packed chestnuts from France, sold under the Minerve label, are excellent and quite widely available.

The 2 basic methods for peeling chestnuts are roasting and boiling. In both cases, be sure to cut a slit in the flat side of the nut with a sharp knife and remove the outer and inner skin after blanching. It's much easier to peel chestnuts in the microwave than by using conventional methods. Again, cut a slit in the flat part of each nut, place the chestnuts on a microwavable baking sheet, and cook on medium-high heat for 8 to 10 minutes per pound. The shell and inner skin can be peeled right off, and the nut will be ready for poaching or braising.

■ ■ ■

VEAL MEDALLIONS WITH CHANTERELLES AND APPLES

serves 4–5

APPLES
3 tablespoons unsalted butter
2 Golden Delicious apples, peeled, cored, and cut into eighths
1 teaspoon granulated sugar
Salt and freshly ground white pepper

CHANTERELLES
2 tablespoons unsalted butter
½ pound fresh chanterelles, stemmed and cut into ½-inch pieces
Salt and freshly ground black pepper

VEAL
4 tablespoons unsalted butter
1 teaspoon peanut oil

1 tenderloin of veal, about 1¼ pounds, cut into ½-inch slices (medallions), or 4 loin veal chops, boned and cut crosswise into ½-inch medallions
Salt and freshly ground black pepper
1 medium shallot, peeled and finely minced
½ cup Noilly Prat vermouth
2 large ripe tomatoes, peeled, seeded, and chopped
1 cup Brown Chicken Stock (page 467)
½ cup heavy cream
1 Beurre Manie (page 478)

1. Preheat the oven to 200° F.

2. Start by preparing the apples: In a heavy 10-inch skillet, melt the butter over medium heat. Add the apple wedges and a large pinch of sugar, season lightly with salt and pepper, and sauté until the apples are nicely browned and glazed, about 5 to 6 minutes. Cover with foil and place in the preheated oven and keep warm.

3. To prepare the chanterelles: In a large skillet, melt the butter over medium heat. Add the chanterelles, season with salt and pepper, and sauté for 2 to 3 minutes, or until all the liquid has evaporated. Set the chanterelles aside.

4. To prepare the veal: In another large skillet, melt 2 tablespoons of butter together with the oil over high heat, add the veal medallions, and sauté until nicely browned on both sides. Season with salt and pepper, transfer the veal medallions to a side dish, and reserve.

5. Add the remaining butter to the skillet and, when melted, add the shallot and cook over medium heat for 1 to 2 minutes, scraping the browned particles from the bottom of the skillet. Cook the shallot until soft and lightly browned. Add the vermouth, bring to a boil, and reduce to a glaze.

6. Add the tomatoes and stock, bring to a boil, and cook the sauce until it is reduced by half. Strain the sauce through a fine sieve and return to the skillet. Add the heavy cream and reserved chanterelles and simmer the sauce for another 2 to 3 minutes.

7. Whisk in just enough of the beurre manie until the sauce lightly coats a spoon. Taste and correct the seasoning. Return the veal medallions along with any accumulated pan juices to the skillet. Heat through, gently, over low heat. Do not overcook the veal.

8. Place the veal medallions on a deep serving platter. Spoon the sauce over them and garnish with the sautéed apples. Serve accompanied by roasted potatoes or potato rosti (page 382).

Note: The apples and chanterelles can be sautéed several hours ahead of time and reheated right in the skillet or in the microwave. When fresh chanterelles are not available, you can substitute 1 ounce of dried chanterelles reconstituted in ½ cup brown chicken stock.

Fresh Mushrooms—Cultivated and Wild

The familiar all-purpose mushroom is moving over to make room for cultivated exotic and wild mushrooms. As the popularity of mushrooms increases so does the number of types available in markets throughout the country. Mushroom farms—which just a short time ago only cultivated the white button mushroom—are now growing shiitake, Enoki, Cremini (also called brown mushrooms), and oyster mushrooms. As with all-purpose mushrooms these new varieties are available year-round and although often labeled "wild" they are indeed cultivated and therefore lack the unique intense flavor of a wild mushroom. Some, like the Cremini and the shiitake, can be quite flavorful when used properly and they do add interest and texture to a variety of dishes. The Enoki and oyster mushrooms are rather bland and are best used only as a garnish.

The two most available wild mushrooms here are chanterelles in the fall and fresh morels in the spring. Both have a unique flavor and texture that needs little adornment. They work wonderfully with pasta, veal, poultry, and eggs. Fresh morels in the United States seem to often lack the depth of flavor they have in Europe and are better used dried.

Fresh cultivated mushrooms should never be soaked in water. Simply brush them gently with the edge of a clean towel or a damp paper towel. Wild mushrooms need more thorough cleaning and may need to be rinsed in plenty of water and their stems trimmed. All mushrooms should be used within a day or two of purchase and cleaned just before cooking.

■ ■ ■

RAGOUT OF LAMB SHANKS NIÇOISE

serves 6

Traditionally, lamb stews are synonymous with spring, when young baby lamb is at its best. To this day I remember my grandmother's ragout, to which she added fava beans, baby artichokes, tiny turnips, and sweet carrots. Although the vegetables far outweighed the few pieces of tender lamb, the result was a flavor-packed dish that I have never forgotten. With American lamb, it is not necessary to wait for spring since its quality does not change very much year-round. Now I find that a good lamb stew is welcome as soon as the cool weather sets in. This preparation, typical of the Nice region of France, makes use of the season's last ripe tomatoes and garden fresh basil. The intense flavors blend and improve when allowed to develop over a day or even two. Roasted peppers and sautéed cubed eggplant can be added to the ragout for additional texture and interest.

SHANKS

6 whole lamb shanks
2 tablespoons unsalted butter
2 tablespoons olive oil
Salt and freshly ground black pepper
3 cups finely minced onions
3 large cloves garlic, peeled and finely minced
⅓ cup dry white wine
4 large ripe tomatoes, peeled, seeded, and chopped, or 1 (2-pound) can Italian plum tomatoes, drained and chopped
2 tablespoons fresh thyme leaves
1 sprig of fresh rosemary
2 tablespoons fresh oregano leaves or 1½ teaspoons dried
2 cups Brown Chicken Stock (page 467) or beef bouillon

VEGETABLES

3 tablespoons unsalted butter
1 large leek, trimmed of all but 2 inches of greens, cut into ½-inch cubes, and washed
1 large zucchini, trimmed and diced, discarding the seedy center (see page 215)
1 large red bell pepper, cored, seeded, and cut into ½-inch cubes
1 large yellow bell pepper, cored, seeded, and cut into ½-inch cubes
Salt and freshly ground black pepper
2 tablespoons water
1 tablespoon finely grated lemon zest
Juice of 1 lemon
4 tablespoons finely minced fresh basil
4 anchovy fillets, drained and finely minced, optional

2 teaspoons cornstarch, mixed with a little cold stock or bouillon

GARNISH

3 tablespoons finely minced fresh parsley mixed with 2 medium cloves garlic, peeled and finely minced

1.   Preheat the oven to 350° F.

2.   Start by braising the lamb shanks: Dry the lamb shanks thoroughly with paper towels.

3.   Melt the butter with 1 tablespoon of the oil in a 12-inch cast-iron skillet over medium-high heat, add the shanks, and brown nicely on all sides. Transfer to an oval casserole or large, deep cast-iron skillet, season with salt and pepper, and reserve.

4.   Reduce the heat to medium and add the remaining oil to the skillet. Add the onions and garlic and sauté for 5 to 7 minutes, or until nicely browned. Add the wine, bring to a boil, and reduce to a glaze. Add the tomatoes, thyme, and rosemary, bring to a boil, and transfer to the casserole containing the lamb. Add the stock, cover tightly, and place in the center of the preheated oven. Braise the lamb for 1½ to 1¾ hours, or until tender when pierced with a fork.

5.   While the lamb is braising, prepare the vegetables: In a large skillet melt the butter over medium-low heat. Add the leek, zucchini, and peppers, season with salt and pepper, and add the water. Reduce the heat and simmer, covered, for 8 to 10 minutes, or until tender.

6.   Add the lemon zest, lemon juice, basil, and optional anchovies to the vegetables and set aside.

7.   When the shanks are done, transfer them with a slotted spoon to a side dish. Degrease the pan juices and strain them through a fine sieve back into the casserole. Place over high heat and reduce by half. Whisk in enough of the cornstarch mixture to thicken the sauce so that it lightly coats a spoon. Taste and correct the seasoning.

8.   Return the shanks and reserved vegetables to the casserole and just heat through. Garnish with the parsley and garlic mixture and serve hot directly from the casserole with plenty of crusty French bread.

GRILLED LEG OF LAMB IN MUSTARD AND HERB MARINADE

serves 6

Grilled butterflied leg of lamb is a summer favorite, but it need not be restricted to that season. In fact, because of lamb's natural affinity for legumes, which are more of a hearty cool-weather food, this dish is wonderful and feasible in most parts of the country all the way into the late fall. I like to serve the lamb accompanied by the oven-roasted vegetables (page 336), which can also be cooked right on the grill, and the grilled eggplant fritters (page 138), which take advantage of the hot coals as well.

MARINADE
1 teaspoon dry mustard
1 tablespoon Dijon mustard
1 teaspoon granulated sugar
3 tablespoons water
3 cloves garlic, peeled and mashed
3 tablespoons plus ½ cup olive oil
2 small dry red chili peppers, broken
⅓ cup fresh lemon juice
2 teaspoons coarsely ground black
   pepper
1 teaspoon coarse salt
2 teaspoons dried oregano
4 tablespoons finely minced fresh
   rosemary

3 tablespoons red wine vinegar

1 leg of lamb (about 6–6½ pounds),
   boned and butterflied
2 red bell peppers
2 yellow bell peppers
Vegetable oil for brushing on grill
Coarse salt and coarsely ground black
   pepper

GARNISH
Sprigs of fresh Italian parsley

1. Start by preparing the marinade: In a small bowl, combine the dry mustard, Dijon mustard, sugar, water, and garlic and mix well. Let stand for 30 minutes to develop flavor.

2. In a small skillet, heat the 3 tablespoons of olive oil over medium heat, add the chili peppers, and cook until just dark. Do not burn. Set aside to cool.

3. Combine the mustard mixture, oil–chili pepper mixture, remaining olive oil, lemon juice, black pepper, salt, oregano, rosemary, and vinegar in a medium mixing bowl and mix well. Transfer the marinade to a shallow baking dish, add the leg of lamb, and turn to coat evenly with the marinade. Cover and marinate 2 to 4 hours at room temperature or in the refrigerator overnight.

4. Prepare the charcoal grill.

5. When the coals are very hot, place the red and yellow peppers directly on top of them. Char evenly on all sides, being careful not to burn past the skin. Wrap the peppers in damp paper towels and set aside to cool completely.

6. Brush the grill with vegetable oil. Remove the lamb from the marinade and pat dry. Sprinkle each side with coarse salt and pepper. Place the lamb over the coals and grill for 10 minutes, brushing with the reserved marinade. Turn and grill another 15 minutes for medium-rare. The lamb will register an internal temperature of 135° to 140° F (in the thickest part) on a meat thermometer for medium-rare.

7. While the lamb is grilling, peel the peppers and trim them. Slice into a fine julienne and set aside.

8. When the lamb is done, transfer to a carving board. Cut crosswise on the bias into thin slices. Serve hot, garnished with the julienned peppers, Italian parsley, and accompanied by braised white beans.

Note: The lamb can also be grilled on a gas grill set at medium-high heat.

PAN-SEARED LAMB CHOPS WITH CABERNET AND HEATHER HONEY—ONION JAM

serves 4

Come early fall, I yearn for this dish, which combines a buttery tomato sauce with the irresistible flavor of a honey- and Cabernet-sweetened onion marmalade. Nothing is simpler than pan-seared lamb chops cooked on the grill, and here their full-bodied flavor greatly heightens the satisfying taste of this main course.

JAM
3 tablespoons unsalted butter
1 tablespoon olive oil
4 large onions or 2 Bermuda onions, peeled, quartered, and thinly sliced
Salt and freshly ground black pepper
⅓ cup sherry vinegar
2 cups Cabernet Sauvignon or dry red wine
2 tablespoons Swedish heather honey,

clover honey, or other soft-spread honey
1 tablespoon dark-brown sugar

SAUCE
6 tablespoons unsalted butter
1 tablespoon extra-virgin olive oil
2 medium shallots, peeled and finely minced

(cont.)

1 large clove garlic, peeled and finely
   minced
1 (2-pound) can Italian plum tomatoes,
   drained and coarsely chopped, juice
   reserved
Salt and freshly ground black pepper
1 tablespoon fresh thyme leaves or 1
   teaspoon dried
1 tablespoon fresh oregano leaves or 1
   teaspoon dried

LAMB CHOPS
8 rib lamb chops, each about ½ inch
   thick

1 large clove garlic, peeled and cut in
   half
Coarse salt and freshly ground black
   pepper
1 tablespoon fresh rosemary leaves or 1
   teaspoon dried
2 tablespoons olive oil
1 tablespoon unsalted butter

GARNISH
Sprigs of fresh rosemary

1.   Start by making the jam: In a deep 10-inch skillet, heat the butter and oil over medium heat. Add the onions, season with salt and pepper, and cook for 20 to 25 minutes, stirring often, until soft and lightly browned. Be careful not to burn.

2.   Add the vinegar and cook until completely evaporated. Reduce the heat, add the wine, and cook until it is completely evaporated and the onions are very soft, about 30 to 35 minutes.

3.   Add the honey and brown sugar and cook 5 minutes longer. Taste and correct the seasoning; the onions should have a sweet-and-sour taste. Set aside.

4.   To prepare the sauce: In a 2-quart saucepan, heat 2 tablespoons of the butter and the olive oil over medium heat. Add the shallots and garlic and cook for 1 minute without browning. Add the tomatoes, salt, pepper, thyme, and oregano. Reduce the heat, cover, and simmer the sauce for 1 hour. If the sauce is too thick, thin with a little of the reserved tomato juice.

5.   Transfer the sauce to the workbowl of a food processor and process until smooth. Return the sauce to the saucepan, place over medium-low heat, and whisk in the remaining butter, 1 tablespoon at a time. Taste and correct the seasoning. Keep the sauce warm, but do not let it come to a boil.

6.   To prepare the lamb chops: Rub both sides of the chops with the cut garlic clove. Season with salt, pepper, and rosemary.

7.   Heat 1 tablespoon of the oil and ½ tablespoon of butter in each of 2 cast-iron skillets over high heat. When the oil is almost smoking, add 4 chops

to each and sauté for 3 minutes. Turn the chops and sauté for another 3 to 4 minutes, or until done. Do not overcook; the chops should be medium-rare.

8. To serve, place 2 tablespoons of onion jam in the center of each individual serving plate. Arrange 2 lamb chops over each portion and spoon some tomato sauce in a circle around the chops. Garnish with a sprig of rosemary and serve immediately.

Note: Both the onion jam and tomato sauce, to the point of adding the remaining 8 tablespoons butter, can be prepared days ahead of time and reheated. Add the butter to the tomato sauce just before serving.

CITRUS-MARINATED ROAST LOIN OF PORK

serves 4–5

Marinated meats, poultry, and seafood have become so popular that almost every home cook has at least one or two such recipes. But marinades can be overpowering and must not be used with too heavy a hand. They should complement the food, not dominate it. Here is an assertive marinade that is particularly suitable for pork or venison. I have also used it with great success for squabs and quail, but the marinating period must not exceed two hours.

2¾ pounds pork shoulder butt or loin of pork, boned, rolled, and tied

MARINADE
2 medium onions, peeled, cut in half, and thinly sliced
1 tablespoon finely minced fresh ginger
½ teaspoon ground cardamom
1 teaspoon ground cinnamon
½ teaspoon ground coriander
1½ teaspoons dried oregano
Salt and freshly ground black pepper

1½ teaspoons finely minced garlic
Juice of 1 orange
Juice of ½ lemon
¼ cup olive oil

SAUCE
2 tablespoons unsalted butter
2 teaspoons peanut oil
2 cups Brown Chicken Stock (page 467)
½ cup good-quality red currant jelly
1 Beurre Manie (page 478)
Juice of 1 lemon, optional

1. Place the pork butt or loin in a deep bowl just large enough to hold the meat.

2. In a small bowl combine all the ingredients for the marinade, pour over

the pork, cover with foil, and refrigerate for 24 hours, turning the pork frequently in the marinade.

3. The next day, remove the pork from the marinade and dry thoroughly on paper towels. Strain the marinade and reserve.

4. Preheat the oven to 375° F.

5. In a large, heavy cast-iron skillet, heat the butter and oil over medium-high heat. Add the pork butt or loin and brown nicely on all sides. Place the skillet in the center of the preheated oven and roast the pork for 1 hour and 15 minutes to 1 hour and 30 minutes, or until it registers 160° F on a meat thermometer. Baste every 10 minutes with a little stock, using ½ cup in all. Fifteen minutes before the pork is done, baste with some of the strained marinade. When the pork is done, remove it to a cutting surface.

6. Degrease the pan juices carefully and return them to the skillet with the remaining stock, currant jelly, and marinade. Bring to a boil and whisk in bits of beurre manie until the sauce is syrupy and lightly coats a spoon. Taste and correct the seasoning, adding a large grinding of black pepper. If the sauce is too sweet, add optional lemon juice to taste. Keep warm.

7. Cut the pork crosswise into ⅛-inch slices and place 3 to 4 slices on each individual serving plate. Spoon some sauce around the pork slices and serve accompanied by wild-rice timbales.

Note: The marinated pork is even better grilled. Grill, covered over indirect heat (page 247), and baste with the marinade every 10 to 15 minutes. Since in this case you will not have a sauce, be sure to add 1 or 2 interesting side vegetable dishes, such as a red onion jam and sautéed caramelized apples.

CURRIED RAGOUT OF PORK WITH BUTTERNUT SQUASH

serves 6

To me, ragouts are the backbone of the late fall and winter kitchen. Hearty, full of flavor, and easy to prepare, they fill the need for homey, gutsy food that does not let you down. I grew up with ragouts of all kinds: lamb, veal, sometimes beef, but mostly pork. The best cut is the shoulder butt, which is tender and extremely flavorful. Aromatic spices and the all-American butternut squash lift this ragout out of the ordinary and highlight its seasonality. Since butternut squash is rather bland, be sure to make this dish a day ahead to give the squash plenty of time to absorb the flavors of the ragout.

3 pounds boneless pork butt, trimmed and cut into 1½-inch cubes
4 tablespoons unsalted butter
1 tablespoon peanut oil
Salt and freshly ground black pepper
2 large onions, peeled and finely chopped
2 large cloves garlic, peeled and finely minced
1 tablespoon Madras curry powder
2 tablespoons tomato paste
⅛ teaspoon ground cloves
2 small dry red chili peppers, broken

¼ teaspoon ground coriander
¼ teaspoon ground cinnamon
2 cups Brown Chicken Stock (page 467)
1 butternut squash (about 2 pounds), peeled, seeded, and cut into 1-inch cubes
2 tablespoons dark-brown sugar
1 tablespoon arrowroot mixed with 2 tablespoons stock

GARNISH
2 tablespoons finely minced fresh cilantro

1. Preheat the oven to 350° F.

2. Dry the pork cubes thoroughly with paper towels. In a heavy 12-inch cast-iron skillet, heat 2 tablespoons of the butter and the oil over medium-high heat. Add the pork cubes a few at a time and sauté until nicely browned on all sides, about 5 minutes per batch. Do not crowd the pan. Transfer each batch of pork as it is browned to a heavy 4-quart casserole, season with salt and pepper, and set aside.

3. Discard all but 2 tablespoons of fat from the pan. Add the onions and garlic and cook for 2 or 3 minutes or until soft but not browned. Add the curry powder, tomato paste, cloves, chili peppers, ground coriander, and cinnamon and cook for 1 minute longer. Add the brown chicken stock and stir until well blended.

4. Transfer the onion-stock mixture to the casserole. Bring to a boil, cover

the casserole tightly, and place in the center of the preheated oven. Braise for 1 hour and 30 minutes, or until the meat is tender when pierced with a fork.

5. While the pork is braising, steam the butternut squash in a vegetable steamer for 3 to 4 minutes. It should still be quite crisp. Remove and reserve. Melt the remaining butter in a 10-inch skillet, add the brown sugar and the butternut squash, and sauté for 2 to 3 minutes, or until the squash is almost tender and nicely caramelized. Reserve.

6. Fifteen minutes before the pork is done, add the caramelized squash cubes to the ragout, replace the cover, and continue cooking.

7. When the meat is done, remove it carefully with the squash to a side dish. Carefully degrease the pan juices. Return the pan juices to the casserole, bring to a boil, and whisk in just enough of the arrowroot mixture until the sauce lightly coats a spoon. Return the meat and squash to the casserole and just heat through. Taste and correct the seasoning. Transfer the ragout to a serving dish, sprinkle with the fresh cilantro, and serve hot.

Note: You may add cubed carrots, turnips, or small pearl onions to the butternut squash. Other winter squashes such as acorn or spaghetti squash cannot be used as a substitute, but would make good side dishes for the ragout.

FRICASSEE OF SNAILS À L'ESPERANCE

serves 4

I first met Marc Meneau in the early seventies at the Mondavi winery in the Napa Valley. I am not sure, but I think that at the time his restaurant in the picture-perfect village of Vezelay, two hours southeast of Paris, had one star in the Michelin guide. Since then, I have spent several wonderful days with Marc in his kitchen and have seen the restaurant receive its second and third stars. It is now one of France's great country inns, with a sublime cuisine that equals some of the most creative food anywhere. Here is a simple dish that Marc likes to serve on small circles of buttery, light puff pastry. But it is equally good when presented in individual au gratin dishes with plenty of crusty French bread to sop up the delicious sauce.

SNAILS
3 cups Chicken Stock (page 470)
2 dozen snails (if large, cut in half), drained and rinsed

SAUCE
16 cloves Double-Poached Garlic (page 482)
2 tablespoons finely minced shallots

2 small cloves raw garlic, peeled and
　finely minced
¼ cup sherry vinegar
⅓ cup Crème Fraîche (page 491) or
　heavy cream
12 tablespoons unsalted butter, cut into
　12 pieces and slightly chilled
Salt and freshly ground white pepper

VEGETABLE GARNISH
1 large leek, trimmed of all but 2 inches
　of greens, well rinsed and cut into

⅓-inch dice, about 1½ cups
3 medium carrots, peeled and cut into
　⅓-inch dice, about 1½ cups
2 tablespoons unsalted butter
3 tablespoons finely minced fresh
　parsley

GARNISH
Tomato Concassé (page 89)
2–3 fresh basil leaves, cut into fine
　julienne, optional

1.　Start by poaching the snails: Bring the stock to a boil in a medium sauce-pan, reduce the heat, and add the snails. Simmer for 4 minutes. Set the snails aside in the stock until needed.

2.　To prepare the sauce: In the workbowl of a food processor puree the double-poached garlic until smooth. Set aside.

3.　Combine the shallots, minced garlic, and sherry vinegar in a 2½-quart copper saucepan or heavy-bottomed saucepan. Bring to a boil and cook until the vinegar is reduced to a glaze. Add the garlic puree and crème fraîche, reduce the heat to low, and start adding the butter, 2 tablespoons at a time, whisking constantly, until the butter has been absorbed by the crème fraîche and the sauce is thick and velvety smooth. You must regulate the heat so that the sauce becomes creamy and does not liquefy at any given time. If this happens add an ice cube and whisk vigorously. When the sauce becomes creamy again, remove any remaining ice cube and continue adding the remaining butter.

4.　When all the butter has been added, season the sauce with salt and white pepper and place over a pan of lukewarm water, cover, and set aside.

5.　To prepare the vegetable garnish: Bring salted water to a boil in a medium saucepan. Add the leek and carrots and cook for 2 to 3 minutes, or until just tender. Drain and set aside.

6.　In a large nonstick skillet, melt the butter over medium heat and add the cooked leek and carrots. Remove the snails from the poaching liquid, add to the skillet together with the parsley, and season with salt and pepper. Simmer for 1 to 2 minutes, or until just heated through.

7.　Remove the skillet from the heat and add the sauce. Taste and correct the seasoning and serve the fricassee at once in individual au gratin dishes, garnished with the concassé of tomatoes and a sprinkling of the optional basil.

ZITI WITH SAUTÉED EGGPLANT, ROASTED PEPPERS, AND ONION MARMALADE

serves 4

1 pound Japanese eggplants or small
    eggplants, unpeeled and cut into
    ¾-inch cubes
Coarse salt
7 tablespoons olive oil
2 large cloves garlic, peeled and thinly
    sliced
3 medium red onions, peeled, quartered,
    and thinly sliced

Freshly ground black pepper
2 medium red bell peppers, roasted,
    peeled (page 483), and thinly sliced
1 teaspoon dried oregano
½ pound imported ziti, preferably De
    Cecco
½ cup freshly grated Parmesan cheese

1. Place the eggplant cubes in a colander, sprinkle with coarse salt, and let drain for 30 minutes to 1 hour.

2. While the eggplants are draining, prepare the onions: In a heavy skillet, heat 3 tablespoons of the olive oil over medium-high heat. Add the garlic and onions and cook for 2 minutes, stirring constantly, until the onions begin to brown. Reduce the heat, cover, and cook until the onions are soft and nicely browned, about 35 to 40 minutes. Season with salt and pepper, remove from the skillet, and reserve.

3. Pat the eggplant dry on plenty of paper towels. Add 2 more tablespoons of oil to the skillet, add half the eggplant, and cook over medium-high heat, stirring often, until nicely browned on all sides. Transfer the eggplant to a colander and set aside. Add the remaining oil to the skillet and sauté the remaining eggplant. Transfer to the colander and let drain.

4. Combine the cooked onions, eggplant, roasted peppers, and oregano in a large skillet. Place over medium-low heat and simmer for 10 minutes. Taste and correct the seasoning and keep warm.

5. Bring plenty of salted water to a boil in a large casserole. Add the ziti and cook until al dente. Add 2 cups of cold water to the casserole to stop further cooking and drain the pasta well.

6. Return the pasta to the casserole. Add the onion and eggplant mixture and place over low heat just long enough to heat through. Correct the seasoning and add ¼ cup of the Parmesan. Toss well and serve at once with the remaining Parmesan on the side.

RAGOUT OF RED CABBAGE WITH RED WINE AND GRENADINE

serves 6

Pierre Gagnaire is among the most creative chefs working in France to-day. In his sun-swept, California-modern restaurant in the small town of Saint-Étienne, he creates extraordinary dishes with everyday ingredients. Pierre is often criticized by other chefs for being "too creative" and for taking too many chances. Personally, he has inspired me to do likewise. Here is one of his signature dishes, a unique combination of red cabbage and grenadine. With a deliciously mellow sweet-and-sour flavor, it makes a superb accompaniment to a roast duck, grilled or roast squab, or smoked pork roast. The cabbage can be prepared as much as a week ahead of time and reheated either in a low oven or in the microwave.

2 tablespoons unsalted butter
2 large onions, peeled, quartered, and thinly sliced
3 tablespoons dark-brown sugar
2 cups dry red wine
1 cup grenadine
1 cup red wine vinegar

3 pounds red cabbage, cored and thinly sliced
Salt and freshly ground black pepper
Zest of 1 large juice orange, cut into fine julienne
2 tablespoons peanut oil
2 tablespoons all-purpose flour

1. Preheat the oven to 350° F.

2. In a heavy 3-quart flameproof casserole melt the butter over medium-high heat, add the onions, and cook, stirring often, for 5 minutes, or until they begin to brown. Reduce the heat, partially cover, and cook until the onions are soft and nicely browned, about 30 minutes.

3. Add the brown sugar to the onions and, when melted, add the wine, grenadine, and vinegar. Add the red cabbage, season with salt and pepper, and bring to a boil. Cover tightly, place in the center of the preheated oven, and braise for 3 hours.

4. While the cabbage is braising, prepare the orange zest: In a small sauce-

pan bring water to a boil, add the orange zest, and simmer for 5 minutes. Drain and add to the casserole 30 minutes before the cabbage is done.

5. When the cabbage is done, remove from the oven and set aside.

6. Heat the peanut oil in a small skillet over medium heat. Whisk in the flour and cook until the mixture turns hazelnut brown; be careful not to burn. Add to the cabbage, place the casserole over low heat, and cook until the sauce is thickened. Taste and correct the seasoning and serve hot as an accompaniment to roast duck.

STEAMED CAULIFLOWER IN PIQUANT CAPER AND ANCHOVY SAUCE

serves 4–6

To people who do not garden, cauliflower does not seem particularly seasonal. Yet, it is at its best in the early fall. In the Northeast and Northwest, wonderful white heads tucked into deep layers of green leaves can be found at most farmstands. That is when I like to prepare this delicious vegetable in many ways. There are few vegetables that are more disappointing than a less-than-fresh cauliflower. Later in the season, when cauliflower is only available cellophane-wrapped, be sure to check it carefully and buy only fresh-looking white heads with much of the greens attached. Look over the greens to be sure that they are not starting to yellow.

SAUCE
1 large egg yolk
1 tablespoon lemon juice
6 flat anchovy fillets, drained
1 large clove garlic, peeled and mashed
1½ teaspoons Dijon mustard
½ cup olive oil
Dash of Worcestershire sauce
Drops of Tabasco sauce
2 tablespoons finely minced chives or scallions
1 tablespoon tiny capers, drained
Salt and freshly ground black pepper

CAULIFLOWER
Salt
1 cup whole milk
1 large head cauliflower, trimmed of all leaves

GARNISH
Tiny fresh Italian parsley leaves
Tiny black Niçoise olives, optional
2 hard-boiled eggs, peeled and cut in half, optional

1.   Start by making the sauce: In the workbowl of a food processor combine the yolk, lemon juice, anchovies, garlic, and mustard and process until smooth. Slowly add the oil in droplets and continue processing the sauce until it is thick and smooth. Add the Worcestershire sauce and a dash of Tabasco and process again.

2.   Transfer the sauce to a small bowl and fold in the chives or scallions and capers. Taste the sauce and correct the seasoning. Chill until 15 minutes before serving.

3.   In a large casserole, bring plenty of salted water to a boil. Add the milk and the whole cauliflower. Cook until you can pierce the florets easily with the tip of a sharp knife. Do not overcook. A truly fresh head will need no more than about 5 to 7 minutes of cooking.

4.   Remove the cauliflower, drain, and transfer to a round serving dish. Cool completely and spoon the sauce over it. Garnish the cauliflower with tiny leaves of fresh Italian parsley and the optional tiny black Niçoise olives and hard-boiled eggs.

Note: You may separate the cauliflower into florets and steam them instead of poaching them. You can also combine broccoli and cauliflower florets for an interesting combination of color and taste. You can then toss both vegetables in the sauce and marinate the mixture overnight. Bring the vegetables to room temperature before serving.

You can also serve the sauce as a dip, as a topping to hard-boiled eggs, and with other steamed vegetables such as broccoli or asparagus.

DEEP-FRIED CAULIFLOWER WITH CHILI MAYONNAISE

serves 6

Cauliflower fritters used to be quite fashionable in British cooking of the forties and fifties. I still remember as a child when my parents took me to London for the first time. We were at the Savoy Grill, and my father ordered a side dish of deep-fried cauliflower. The crisp florets arrived, nestled in an enormous white linen napkin garnished with tiny sprigs of parsley. The waiter held out a platter of what looked like golden nuggets while we helped ourselves. It was so impressive that I felt I was eating a rare delicacy. I don't think that cauliflower, fried or not, has ever tasted that good again.

4 large dried red ancho chili pods
1 recipe Mayonnaise (page 477)
1 large clove garlic, peeled and mashed
    through a garlic press
Salt and freshly ground black pepper
2 teaspoons red wine vinegar, optional

BATTER
1 cup all-purpose flour
½ teaspoon coarse salt

3 tablespoons olive oil
2 large eggs, separated
1 cup flat beer

1 medium head cauliflower, about 1½
    pounds, trimmed of all leaves and
    separated into small florets
4 cups peanut or corn oil for deep
    frying
Coarse salt

1. Place the chili pods in a large mixing bowl, cover completely with cold water, and let sit overnight. The next day drain the chili pods and remove and discard the stems. Carefully peel the pods, removing all the tough, leathery skin, without removing too much of the flesh.

2. Combine the peeled chili pods, the mayonnaise, and garlic in a food processor or blender and process until smooth. Season with salt and pepper to taste and add the optional vinegar. Cover and chill until serving time.

3. To prepare the batter: In the workbowl of a food processor or blender combine the flour, salt, oil, egg yolks, and beer and process until smooth. Transfer to a large mixing bowl and let the batter rest unrefrigerated for 2 hours.

4. In a mixing bowl, beat the egg whites until firm. Fold the whites gently but thoroughly into the batter and set aside.

5. While the batter is resting, prepare the cauliflower: Bring salted water to a boil in a vegetable steamer. Add the cauliflower florets, cover, and steam for 3 to 4 minutes, or until tender. Remove and let cool completely.

6. In a large saucepan, heat the peanut or corn oil to 375° F. Dip the cauliflower florets into the batter, letting the excess drip off. Add the florets, a few at a time, to the hot oil and fry until golden brown on all sides, about 3 to 4 minutes. Remove with a slotted spoon to a double layer of paper towels to drain. Sprinkle with coarse salt and serve at once with the chili mayonnaise.

Belgian Endives

Belgian endives belong to the chicory family, which also includes escarole and the various Italian chicories such as radicchio. To buy them in the right condition choose those that are still in the box protected from the light by the wrapping paper. Belgian endives that are exposed to the light will turn pale green in a matter of a day or two and become quite bitter. Many supermarkets package endives 2 to a package, and in many cities it is, unfortunately, the only way to purchase them. In this case choose only endives that are pale yellow, uniform in size, and without traces of brown spots.

To clean endives, rinse them quickly under cold running water, never letting them soak in the water since this makes them bitter. Remove any wilted outer leaves and with the tip of a sharp knife cut out the core. Then cut the vegetable into fine julienne or crosswise into rounds.

Whole endives can be blanched or steamed. I usually add the heel of a stale rye bread to the water, which removes any of the bitterness. When tender, drain the endives and braise in a little butter and lemon juice with a touch of sugar either in a skillet on top of the stove or in the oven.

You may also sauté endives and caramelize them much as you would onions, with a touch of butter and sugar. Cooked, they make a delicious vegetable that works well alongside veal, poultry, and all seafood.

■ ■ ■

BRAISED ENDIVE AND APPLE COMPOTE

serves 4–5

2 pounds Belgian endives, cored
3 Golden Delicious apples
4–6 tablespoons unsalted butter
2 teaspoons granulated sugar

Salt and freshly ground white pepper
3–5 tablespoons Chicken Stock (page 470) or bouillon

1. Rinse the endives quickly under cold water, remove any outer leaves that are wilted or discolored, and cut lengthwise into ½-inch julienne strips. Set aside.

2. Choose semi-ripe apples; they should still be green in color. Peel and core the apples. Cut crosswise into ¼-inch slices, and then into julienne strips.

3. In a large, heavy cast-iron skillet, melt 2 tablespoons of butter over medium heat. Add the julienne of apples and 1 teaspoon sugar and sauté until the apples are nicely browned on all sides. Lower the heat, cover the skillet, and cook the apples until tender, or for about 5 minutes. Transfer the apples to a side dish and reserve.

4. Add the remaining butter to the skillet and, when very hot, add the julienne of endives, season with salt, pepper, and the remaining sugar, and sauté over fairly high heat until the endives are nicely browned and caramelized.

5. Add 3 tablespoons of stock, lower the heat, and braise the endives, covered, until tender, about 10 minutes, adding a little more stock if the mixture seems too dry.

6. Add the cooked apples, toss into the endives, and just heat through. Taste and correct the seasoning.

7. Serve the compote hot as an accompaniment to sautéed pork tenderloins, veal scaloppine, or sautéed calves' liver.

SWEET-AND-SOUR RAGOUT OF FENNEL AND PEARL ONIONS

serves 4

3 tablespoons unsalted butter
1 tablespoon olive oil
1 pint pearl onions, peeled
Salt and freshly ground black pepper
2 tablespoons dark-brown sugar
1 tablespoon tomato paste
2 tablespoons sherry vinegar

1 cup Chicken Stock (page 470) or bouillon
1 large sprig of fresh thyme
1 large clove garlic, peeled and crushed
2 large fennel bulbs, trimmed of outer leaves and finely sliced
1 teaspoon granulated sugar

1. Melt 2 tablespoons of butter and half the olive oil in a heavy 10-inch skillet over medium heat. Add the onions and sauté until they are brown on all sides. (The onions will not brown evenly.) Season with salt and pepper.

2. Add the brown sugar and tomato paste and toss the onions until they are well coated with the sugar. Add the sherry vinegar and ½ cup of chicken stock and bring to a boil. Reduce the heat and bury the thyme and garlic clove among the onions. Simmer, tightly covered, for 15 to 20 minutes, or until the onions

are tender. You may need to add a little more broth during cooking. Set the onions aside.

3. In another large, heavy skillet, heat the remaining butter and oil over medium heat. When very hot, add the sliced fennel and season with the granulated sugar, salt, and pepper and sauté until nicely browned on both sides. Add ⅓ cup of stock, cover the skillet, and continue braising the fennel for 10 minutes, or until tender.

4. Combine the onions and fennel in the same skillet and simmer gently for another few minutes. Do not overcook. Taste and correct the seasoning. Serve hot as an accompaniment to roasted or sautéed meats or seafood.

Note: The ragout is also delicious served at room temperature as an accompaniment to a pâté or to cold, roasted, or grilled chicken. The onions can be made several days ahead of time. I usually double or triple the recipe and serve them with roast pork or turkey.

Fennel

Fennel is a Mediterranean favorite, especially appreciated by Italian cooks, which may be the reason it is often referred to as Florence fennel.

While Italians consider fennel mostly as a vegetable, in France it is used primarily as an herb and as a flavoring agent to delicate fish. The dried fennel stalks found in markets throughout Provence are used as a bed for, or added to, a small fire, and their smoke imparts a delicious flavor to striped or sea bass. Fennel seeds are an important addition to a court bouillon and are called for in many pickling recipes.

I love the taste of braised or roasted fennel. Simply cooked it can be pureed and added to a potato puree. To braise fennel, cut the bulbs into quarters or eighths. Blanch them for 3 minutes, then braise in a heavy cast-iron skillet with a touch of butter and olive oil until nicely browned and slightly caramelized. Other vegetables that work well with fennel are carrots, pearl onions, and turnips. To roast fennel, quarter the bulbs, toss them in virgin olive oil, and place in a heavy baking dish. Season with a touch of fresh thyme and roast at 375° F, or until the fennel is tender and nicely browned. Serve roasted fennel with a sprinkling of virgin olive oil and a touch of balsamic vinegar.

After buying fennel, remove all the feathery tops and store the bulbs in plastic bags in the vegetable bin. You may have to remove the outer leaves of the bulbs if they are bruised or brown. Fennel will keep easily for a week to 10 days. To heighten the taste of anise in any fennel preparation, add 1 teaspoon of Pernod or dried fennel seeds to the dish.

■ ■ ■

SAUTÉ OF CINNAMON, LEEKS, AND CREMINI MUSHROOMS

serves 4

The idea could not be simpler, yet it never would have occurred to me to combine these two vegetables in a quick sauté, much less to flavor the dish with cinnamon. Pierre Gagnaire came up with this delicious sauté during one of my recent visits to his kitchen in Saint-Étienne. It is typical of his cooking and his overall philosophy of trying everything spontaneously. His advice, "Take the simplest and most straightforward ingredients and with a touch of the unexpected, make them memorable." Serve the sauté as a side dish to a roast veal, simply grilled chops, sautéed pork tenderloins, or a roast duck.

3 medium leeks, trimmed of all but 2 inches of greens
6 medium Cremini mushrooms, wiped

3–4 tablespoons unsalted butter
Salt and freshly ground black pepper
Large pinch of ground cinnamon

1. Cut the leeks in half lengthwise and then crosswise into ¾-inch pieces. Rinse under warm water to remove all sand. Drain.

2. Place the leeks in a vegetable steamer and set over simmering water. Cover and steam for 3 to 5 minutes, or until just tender. Remove and set aside.

3. Cut off and discard the mushroom stems. Cut the caps into ¼-inch slices. In a 10-inch nonstick skillet, melt the butter over medium-high heat. Add the mushrooms and sauté quickly for 3 minutes or until nicely browned. Add the leeks, toss with the mushrooms, and season with salt, pepper, and a large pinch of ground cinnamon. Just heat through. Serve at once as an accompaniment to grilled salmon.

WILD MUSHROOM–RICE TIMBALES

serves 6–8

A rice timbale is a dream come true for every cook who wants to entertain in style with as much of the meal as possible prepared in advance. The secret of a perfect rice timbale is to use Italian rice or a medium-grain Carolina rice. The starchiness of the rice allows it to be pressed into any timbale mold and retain that particular shape once cooked. I usually place the timbale molds in a pan of hot (not boiling) water and unmold them onto a platter or into individual serving plates just before serving. The porcini, of course, add a wonderful flavor to the rice which cannot be achieved with any other kind of mushroom, but you can vary the flavors in the rice timbales according to the seasons.

1 ounce dried porcini mushrooms
2½ cups Chicken Stock (page 470)
2 tablespoons unsalted butter
1 small onion, peeled and finely minced

1 cup Italian rice, preferably Arborio, rinsed and drained
Salt and freshly ground black pepper
3 tablespoons finely minced fresh parsley

1. Place the mushrooms in a strainer under cold water to remove all sand.

2. In a 2-quart saucepan, combine the washed mushrooms and chicken stock and place over medium-high heat. Bring to a boil, reduce the heat to low, and simmer the mushrooms, covered, for 25 to 30 minutes, or until tender.

3. Remove the mushrooms from the stock with a slotted spoon. Taste a piece of 1 mushroom and if still gritty, rinse quickly under cold water. Drain them well. Transfer to a cutting board, finely dice, and set the mushrooms aside. Strain the stock through a double layer of cheesecloth and reserve separately.

4. In a heavy 3-quart saucepan, melt the butter over medium heat. Add the onion and cook for 1 to 2 minutes, or until soft but not browned. Add the rice, season with salt and pepper, and stir to coat well with the butter. Cook for 1 minute, or until opaque. Add the reserved mushroom poaching liquid, reduce the heat to medium-low, and simmer, covered, for 25 to 30 minutes, or until the rice is tender.

5. Fold in the parsley together with the reserved diced mushrooms and correct the seasoning, adding a large grinding of pepper. Set aside.

6. Lightly oil 6 to 8 French timbale molds (5 ounces each), pack each tightly with some of the pilaf, and unmold onto individual plates or a platter. Serve with roast turkey or veal.

Note: The timbale molds can be packed, covered, and kept warm in a hot-water bath and then unmolded just before serving. You may also serve this rice as a pilaf instead of molding it.

Variation: Use 1 large red pepper, peeled, cored, and finely diced, instead of the mushrooms. Herbs such as tarragon, parsley, chives, and dill can be added to or substituted for the mushrooms. Spices such as curry, cumin, garam masala, or a combination of all three are excellent flavoring for the timbales.

OLD-FASHIONED ROASTED ONIONS

serves 4–6

Roasted onions are a classic of the Provençal hors d'oeuvre table. Although onions are an all-season vegetable, this preparation is usually served during the fall season. For the best flavor, roast the onions a day ahead. I like serving them in the early fall along with some sausages or a country pâté, as part of a buffet table, or as an accompaniment to grilled tuna steaks.

8 tablespoons extra-virgin olive oil
4 large Bermuda onions, peeled and cut crosswise into thick slices
1 large clove garlic, peeled and finely minced
1½ teaspoons balsamic vinegar

1½ teaspoons sherry vinegar
2 tablespoons finely minced fresh parsley
2 teaspoons fresh thyme leaves
Salt and freshly ground black pepper

1. Preheat the oven to 350° F.

2. Brush the bottom of a very heavy baking pan with 2 tablespoons of olive oil. Place the onion slices in the pan in a single layer, slightly overlapping if necessary, and drizzle with 3 tablespoons oil. Bake in the center of the preheated oven for 1 hour and 15 minutes to 1 hour and 30 minutes, or until tender and golden brown. Watch carefully to see that the onions do not burn. Turn the slices once or twice with a spatula during roasting. When the onions are done, transfer to a shallow serving platter and set aside.

3. In a small bowl combine the remaining oil, garlic, vinegars, parsley, and thyme. Whisk until well blended and season with salt and pepper. Pour the vinaigrette over the onions, toss gently, and let marinate at room temperature for 30 minutes to 1 hour. Serve at room temperature as part of an hors d'oeuvre table.

MEDITERRANEAN ROASTED PEPPER AND
EGGPLANT FRICASSEE

serves 5–6

In its ingredients and adaptability, this fricassee is similar to the famous Sicilian dish, caponata. Usually served as an hors d'oeuvre or vegetable course, caponata is a medley of braised vegetables in a sweet-and-sour vinaigrette. Here, however, some of the vegetables are grilled, and they are not bound with tomatoes. Versions of this vegetable medley are also popular throughout northern Spain and Catalonia, where the combination of sweet-and-sour flavors has been used in vegetable preparations for several centuries.

1 small eggplant, about 1 pound
8–9 tablespoons olive oil
2 large red peppers, roasted, peeled, and seeded (page 483), cut into fine julienne strips
2 large yellow peppers, roasted, peeled, and seeded (page 483), cut into fine julienne strips

⅓ cup golden raisins
1 large shallot, peeled and finely minced
1 tablespoon sherry vinegar
1½ tablespoons balsamic vinegar
1 medium clove garlic, peeled and crushed
Salt and freshly ground black pepper

1. Cut the unpeeled eggplant crosswise into ½-inch slices, stack the slices, and cut into thick julienne strips. Place the eggplant in a bowl, cover with ice water, and let soak for 1 hour. Drain the eggplant and dry thoroughly with paper towels.

2. In a large cast-iron skillet heat 2 to 3 tablespoons of the olive oil, add the eggplant strips, and sauté until nicely browned on all sides. Transfer to a double layer of paper towels to drain and then combine in a bowl with the roasted peppers and set aside.

3. Place the raisins in a small saucepan with water to cover. Bring to a boil, reduce the heat, and simmer for 3 minutes. Drain the raisins thoroughly and add to the eggplant-pepper mixture.

4. Combine the shallot, sherry vinegar, balsamic vinegar, and 6 tablespoons olive oil and garlic in a small bowl and whisk until smooth. Season with salt and pepper and spoon the dressing over the eggplant mixture. Toss lightly, cover, and marinate at room temperature for at least 4 to 6 hours before serving.

Note: This dish is at its best when prepared several days ahead of time and brought back to room temperature 30 minutes before serving. In Spain, where this kind of vegetable medley is very popular, quartered small onions and whole wild mushrooms are often added. All the vegetables are usually grilled over a wood fire, which gives the whole dish a very appealing smoky flavor. I have had good results using both a gas or Weber grill.

You can add 1 or 2 yellow squash or golden zucchini to the medley. Cube the zucchini and sauté in a little olive oil. Drain in a colander and add to the pepper and eggplant fricassee.

PARSNIP AND POTATO PANCAKES

makes about 24–28

The old-fashioned parsnip, once the most treasured winter root vegetable, has been added to the list of packaged mass-merchandised vegetables that, like carrots and cauliflower, are available almost everywhere practically year-round. If you are a gardener, the parsnip is well worth planting and digging up when the ground is almost frozen. What you will find is a sugary-sweet vegetable that is wonderful in soups and fritters, simply braised or steamed, and used in quick breads or even elegant cakes. Most of the time, however, we have to fall back on the packaged parsnip. While it is a far cry from the home garden variety, it is still worth experimenting with and lends itself to many interesting preparations. These light and flavorful pancakes make an interesting side dish to pan-seared salmon fillets, sautéed sea scallops, medallions of veal, or a roast chicken or turkey.

2 large parsnips, peeled and finely grated, about 2 cups
2 large baking potatoes, peeled and finely grated, about 2½ cups
2 large eggs, lightly beaten

2 tablespoons all-purpose flour
Salt and freshly ground black pepper
4 to 5 tablespoons Clarified Butter (page 478)

1. Squeeze the excess moisture from the grated parsnips with your hands and transfer to a large mixing bowl. Add the grated potatoes, eggs, and flour and mix until well blended. Season with salt and pepper.

2. In a 10-inch nonstick skillet melt 2 teaspoons of butter over medium heat. Cook 4 pancakes at a time by adding about 2 tablespoons of batter into the skillet for each pancake; do not cover the skillet. Cook for 2 to 3 minutes or until nicely browned on each side. Transfer to a platter and keep warm. Continue to make about 4 pancakes at a time, using 2 teaspoons of butter for each batch.

3. Serve the pancakes hot as an accompaniment to grilled quail or other grilled meats.

GRATIN OF POLENTA WITH GOAT CHEESE, LEEKS, AND HAM

serves 4–5

Polenta preparations of all kinds have taken the country by storm. The Italian technique of boiling fine yellow cornmeal in either milk or water, then spreading the thick mass onto an oiled baking sheet, has innumerable creative possibilities. In *The Peasant Kitchen*, I wrote about a polenta gnocchi that is still one of my all-time favorite recipes. In that preparation as well as here, small round disks are cut out of the polenta sheet, then placed in a well-buttered baking dish, and topped with a basil and anchovy butter. The finished dish is an intoxicating, delicious appetizer that also doubles as a simple main course. Here, the polenta disks form little round sandwiches, trapping the delicate goat cheese, leeks, and ham, and the whole dish is baked until the flavors blend and marry. It is not an authentic Italian recipe, but rather a combination of fresh ingredients that complement one another beautifully. Serve the gratin as an appetizer or a side dish to a simple roast chicken, sautéed veal scaloppine, or roast leg of lamb.

POLENTA
4½ cups water
1 teaspoon salt
1½ cups yellow cornmeal
Freshly ground black pepper

GRATIN
11 ounces Montrachet goat cheese or other tubular, mild goat cheese
3 medium leeks, trimmed of all but 2 inches of greens

7 tablespoons unsalted butter
1 cup very finely diced smoked ham
⅓ cup Chicken Stock (page 470) or water
Salt and freshly ground white pepper
3 tablespoons freshly grated Parmesan cheese

GARNISH
Sprigs of fresh Italian parsley

1. Start by preparing the polenta: In a heavy 2½-quart saucepan, combine the water and salt and bring to a boil. Sprinkle in the cornmeal very slowly, stirring constantly, until all has been added. Reduce the heat and simmer, covered, for 15 minutes, stirring often. Season with pepper.

2. Sprinkle a jelly-roll pan (10½ × 15½ inches), preferably nonstick, lightly with water and pour in the polenta. Spread evenly with a wet spatula into a thin layer to cover the bottom of the pan completely. Place the pan in the refrigerator for 1 hour, or until the polenta is cold and very firm.

3. To prepare the gratin: Place the Montrachet in the freezer in 1 piece for at least 1 hour, or until it is firm and can be sliced easily without falling apart.

4. Remove the cheese from the freezer and cut crosswise into ¼-inch-thick slices with a very sharp knife; you will need 20 to 21 slices, in all. Place the slices on a plate and refrigerate until needed.

5. Quarter the leeks lengthwise, then cut crosswise into thin slices. Place the slices in a colander, run under warm water to remove all sand, and drain thoroughly. Set aside.

6. In a 10-inch nonstick skillet, melt 1 tablespoon of the butter over medium-low heat. Add the diced ham and sauté for 1 to 2 minutes. Remove with a slotted spoon to a side dish and reserve.

7. Add 2 more tablespoons of butter to the skillet. Add the leeks and stock and season with salt and pepper. Cover and simmer for 10 minutes, or until the leeks are tender. Remove the cover, raise the heat, and cook, stirring constantly, until all the stock has evaporated. Add the reserved ham, correct the seasoning, and set aside.

8. Preheat the oven to 375° F.

9. Remove the polenta from the refrigerator and, using a 2-inch plain round cookie cutter, cut it into disks. You will have 40 to 42 disks, enough to make 20 or 21 polenta sandwiches.

10. Butter the inside of a heavy rectangular baking dish with 1 tablespoon of the butter. Place half of the disks of polenta on the bottom of the dish in a single layer, slightly overlapping if necessary. Top each disk with a spoonful of the leek-ham mixture and 1 slice of the goat cheese. Top each with another disk of polenta to form sandwiches.

11. Melt the remaining butter in a small saucepan and drizzle it over the sandwiches. Sprinkle with the Parmesan, place the dish in the center of the preheated oven, and bake for 15 to 20 minutes or until piping hot. Remove from the oven, garnish with sprigs of parsley, and serve at once.

BELGIAN POTATO CREPES

serves 4–6

The key to these light, buttery potato crepes is the starch content of the potatoes, which must result in a creamy and somewhat starchy puree. If the potatoes are not the right texture, the small crepe will not hold together. Look for all-purpose russet potatoes. If you cannot find them, use good-quality baking potatoes which are also starchy. These crepes are well worth experimenting with. When successful, there are few potato preparations that are as delicious or satisfying. I often serve the crepes as an appetizer, much like a blini with a side dish of dill-flavored sour cream, finely sliced smoked salmon, or just melted brown butter.

1¼ pounds all-purpose potatoes, peeled
¼ cup whole milk
3 tablespoons all-purpose flour
3 large eggs
4 large egg whites

4 tablespoons Crème Fraîche (page 491)
Salt and freshly ground black pepper
4–6 tablespoons Clarified Butter (page 478)

1. Bring salted water to a boil in a medium saucepan. Add the potatoes and cook until very tender. Drain the potatoes and pass through a foodmill or ricer. Do not use a food processor. Transfer the potatoes to a large bowl and combine with the milk. Chill for 15 minutes.

2. In a medium bowl combine the flour and whole eggs and whisk until well blended. Add the egg whites and crème fraîche and whisk until smooth. Fold in the mashed potatoes and season with salt and pepper. The mixture should have the consistency of a thick crepe batter.

3. Heat 2 teaspoons of the clarified butter in a 10-inch nonstick skillet over medium heat. Drop the potato batter by the tablespoonful into the hot butter without crowding the skillet. Cook for 2 to 3 minutes or until golden brown on both sides, turning once with a spatula. You will have to cook the crepes in batches of 4 to 5. Transfer to a platter and keep warm. Continue making the crepes, adding 2 teaspoons of the butter to the skillet for each batch. Serve hot as an accompaniment to roasted or grilled meats and sautéed fish dishes.

Note: Ideally the crepes should be the size of silver-dollar pancakes—about 2 inches in diameter—and similar in texture.

Variation: Add 1 or 2 tablespoons of finely minced dill, parsley, or chives to the potato batter. If the batter seems too thin and the crepes seem too runny, you may have to increase the amount of flour by 1 tablespoon to make sure the batter holds together.

COCIDO OF POTATOES AND LEEKS WITH CRÈME FRAÎCHE

serves 4

Don't expect this dish to dazzle. It is homey and flavorful, typical of my mother's cooking. As I was growing up, it was everyday fare but I loved it, and now so does my family. Literally, a cocido means a stew, and that is exactly what this dish reminds me of. For best results, use truly fresh leeks and excellent potatoes, which should retain some of their shape after braising. You may substitute heavy cream for the crème fraîche, or omit it entirely and drizzle a little extra-virgin olive oil over the stew just before serving.

4 medium leeks, trimmed of all but 3 inches of greens
4 tablespoons unsalted butter
6–8 medium russet potatoes, peeled and cubed
½–¾ cup Chicken Stock (page 470), bouillon, or water

Salt and freshly ground white pepper
3 tablespoons Crème Fraîche (page 491) or heavy cream
2 tablespoons finely minced fresh parsley, optional

Leeks

The humble leek is emerging as one of our most important and interesting vegetables. For years it was considered a gourmet item, but now you can find truly fresh leeks in good supply practically year-round. Leeks are a must in stocks and many stews and soups, and are a versatile and delicious vegetable by themselves.

Leeks have an additional advantage: When properly stored, they will keep for 2 to 3 weeks in the refrigerator. Look for leeks that are about ½ inch in diameter with deep green, crisp tops.

To store leeks, cut off all but 3 inches of their tops and place them, unwashed, in plastic bags in the vegetable bin. Before using them, slit the green part with a sharp knife and rinse thoroughly in warm, not cold, water.

To julienne leeks evenly, cut them crosswise into 1½-inch pieces. Peel off 2 layers at a time, place the layers flat on a cutting board, and cut into fine strips. Reserve the cores for soups or for other leek preparations in which they are diced or sliced crosswise.

In the cool climate of the Northwest, you can find pencil-thin leeks sold in bunches. These are now being shipped across the country, and I always keep an eye out for them. Poach these tender leeks in a deep skillet with water to cover and 2 tablespoons of butter. Serve at room temperature or slightly chilled with a mustardy vinaigrette. These young leeks can also be steamed and then grilled and served with a piquant mayonnaise.

■ ■ ■

---

1. Slice the leeks lengthwise and then crosswise into ½-inch slices. Place in a colander and rinse thoroughly under warm water. Set aside.

2. In a heavy 10-inch skillet, melt the butter over medium heat, add the leeks, potatoes, and stock, and season with salt and pepper. Cover and simmer until the potatoes are very tender. If the juices run dry, add the remaining stock.

3. Uncover the skillet and raise the heat. Continue to cook until the stock has evaporated. Add the crème fraîche and optional parsley and fold gently. Taste and correct the seasoning. Serve hot as an accompaniment to roast or braised chicken, veal scaloppine, roast pork, or sautéed calves' liver.

Variation: Large green beans cut into 2-inch pieces and steamed for 3 to 4 minutes can be substituted for the leeks. Allow the beans to overcook a little; that is part of the charm of this dish.

SPAGHETTI SQUASH WITH GINGER AND BROWN-SUGAR CREAM

serves 4

Spaghetti squash is relatively new to the vegetable scene. Although by now we have become familiar with the light yellow melon-shaped squash, it is still rare to find it prepared creatively by either the restaurant or the home cook. With a name like spaghetti squash, this rather mild vegetable was immediately associated with tomato-based sauces and other pasta preparations. In fact, it has no affinity for tomatoes. Instead, it is delicious seasoned with Indian or Middle Eastern spices or teamed with certain fruits, such as apples and pears, which enhance this rather bland but texturally interesting vegetable.

1 medium spaghetti squash
1 (½-inch) slice of fresh ginger, peeled
¾ cup heavy cream
4 tablespoons unsalted butter

2 semi-tart apples, peeled, cored, and cut into ¼-inch julienne
2 tablespoons dark-brown sugar
Salt and freshly ground black pepper
Large pinch of freshly grated nutmeg

1. Preheat the oven to 350° F.

2. Place the whole spaghetti squash on a baking sheet and bake in the center of the preheated oven for 45 minutes to 1 hour, or until the squash can be pierced easily with the tines of a fork.

3. While the squash is baking prepare the ginger cream: In a small saucepan combine the ginger and heavy cream and bring just to a simmer. Reduce the heat to low and steep the ginger in the cream for 20 minutes. Remove from the heat, strain the cream into a bowl, and set aside. Discard the ginger.

4. Remove the squash from the oven and let it cool completely. Cut the squash in half lengthwise and remove and discard the seeds.

5. With a large fork, carefully scoop out the strands of spaghetti squash and reserve 4 cups. Save the remaining squash for another preparation.

6. In a 10-inch skillet melt the butter over medium-low heat. Add the apples and cook for about 5 minutes or until just tender. Add the brown sugar and cook until the sugar has melted and the apples are glazed.

7. Add the reserved squash to the skillet and toss gently with the glazed apples. Season with salt, pepper, and a large pinch of nutmeg. Add the reserved ginger cream and cook gently for 4 to 5 minutes or until the cream has been absorbed by the squash.

8. Serve hot as an accompaniment to roast duck, the Braised Veal Glazed with Onions and Chestnuts (page 302), or the Sauté of Pork Tenderloins with Prunes and Sweet-and-Sour Caramel (page 423).

Note: For a quick preparation, toss the warm spaghetti squash with minced green onions, a touch of sesame oil, and toasted sesame seeds and serve as an accompaniment to marinated and grilled chicken or pork.

SAUTÉED GREEN TOMATOES ALFREDO

serves 4–6

4 medium green tomatoes, about 1 pound, cut crosswise into ⅓-inch slices
All-purpose flour for dredging
6 tablespoons olive oil
Salt and freshly ground black pepper
3 tablespoons unsalted butter

½ cup heavy cream
4 tablespoons freshly grated Parmesan cheese

GARNISH
Freshly grated Parmesan cheese

1. Pat the green tomato slices dry on paper towels. Dredge lightly in flour, shaking off the excess.

2. In a large, heavy skillet heat 3 tablespoons oil over medium heat. Add half of the tomato slices, without crowding the pan, and sauté for 2 minutes or until tender and nicely browned on both sides. Transfer with a slotted spatula to a double layer of paper towels. Add the remaining oil to the skillet and continue sautéing until all the tomatoes are done. Season with salt and pepper and set aside.

3. Wipe the skillet clean, add the butter, and whisk over medium heat until lightly browned. Do not burn. Immediately add the cream, bring to a boil, and reduce slightly. Add the Parmesan, season with salt and pepper, and, when the cheese has just melted, return the tomato slices to the skillet and just heat through. Serve hot directly from the skillet, with extra Parmesan and crusty Italian bread.

MARINATED AND GRILLED GREEN TOMATOES

serves 4

4 medium green tomatoes, about 1
  pound, cut crosswise into ⅓-inch slices
3 tablespoons extra-virgin olive oil
Coarse salt and freshly ground black
  pepper
2 tablespoons red wine vinegar

1 large clove garlic, peeled and thinly
  sliced
1 tablespoon tiny fresh basil leaves
2 tablespoons thinly sliced red onion,
  optional

1. Prepare the charcoal grill.

2. Brush the tomato slices with olive oil, using 1 tablespoon in all. Sprinkle with coarse salt and pepper. Grill the slices for 2 minutes on each side or until nicely browned.

3. Place the grilled tomatoes in a shallow serving dish. Sprinkle with the vinegar, garlic, basil leaves, optional red onion, and remaining olive oil. Let marinate for 6 hours or overnight.

4. Taste and correct the seasoning and serve with a crusty loaf of French bread as part of an hors d'oeuvre table.

OVEN-ROASTED FALL VEGETABLES WITH ROASTED GARLIC VINAIGRETTE

serves 6

Oven-roasted vegetables have always been popular in Mediterranean Europe and the Middle East. In the market in Cannes and throughout the Nice region, you can buy an amazing array of vegetables already roasted and ready to serve. Colorful peppers, thick slices of onions, zucchini, and rich purple eggplants are set out in huge baking dishes, glistening with olive oil, roasted garlic cloves, and sprigs of thyme.

Re-creating the vegetable mixture at home is quick and simple. The results are far better when using a gas or convection oven, but you will get good results in an electric oven as well. If you plan to serve the vegetables as a salad course or an appetizer, the garlic vinaigrette is an excellent choice, and it can be made a day or two in advance. If you are pressed for time, just drizzle the vegetables with an excellent-quality extra-virgin olive oil and a touch of either sherry or balsamic vinegar.

VINAIGRETTE
8 large cloves Roasted Garlic (page 482), peeled
1 tablespoon balsamic vinegar
1 teaspoon Dijon mustard
6 tablespoons olive oil
Coarse salt and freshly ground black pepper

VEGETABLES
3 tablespoons extra-virgin olive oil
3 large cloves garlic, peeled and thinly sliced
2 large red bell peppers, cored, seeded, and cut into eighths

2 large yellow bell peppers, cored, seeded, and cut into eighths
2 medium zucchini, trimmed and cut into ¼-inch slices
6 small Italian eggplants or 2 small eggplants (about 2½ pounds total), quartered lengthwise and then cut in half crosswise
Coarse salt and freshly ground black pepper

GARNISH
Sprigs of fresh parsley

1. Start by making the vinaigrette: Combine all the ingredients in the workbowl of a food processor and process until very smooth. Season with salt and pepper and set aside.

2. Preheat the oven to 400° F.

3. To roast the vegetables: Heat the oil in a large, heavy flameproof baking

dish over low heat. Add the garlic and cook for 30 seconds. Place the vegetables in the baking dish in a single layer, overlapping slightly. Set the baking dish in the center of the preheated oven and roast the vegetables for 45 minutes to 1 hour, turning them carefully with a spatula once or twice during roasting, until they are tender and nicely browned. Season with coarse salt and pepper and let cool slightly.

4. While the vegetables are still warm, drizzle with the garlic vinaigrette and let sit for 30 minutes to 1 hour.

5. Serve at room temperature as part of an hors d'oeuvre table or as an accompaniment to grilled meats.

CABBAGE AND CARROT SLAW WITH ROQUEFORT DRESSING

serves 4–6

1¼ pounds green cabbage, trimmed and finely shredded, about 5 cups
Salt
2 ounces Roquefort cheese or other mild blue cheese, crumbled
5 tablespoons olive oil
1–1½ tablespoons red wine vinegar

2 anchovy fillets, drained and finely minced
1 small clove garlic, peeled and mashed
2 medium scallions, trimmed and finely minced
3 tablespoons sour cream
Freshly ground black pepper
1 cup finely shredded carrots

1. Place the cabbage in a colander, sprinkle with 1 teaspoon of salt, and let drain for 2 to 3 hours. Squeeze out the excess moisture from the cabbage with your hands. Transfer the cabbage to a serving bowl and set aside.

2. In the workbowl of a food processor or blender combine the Roquefort, olive oil, 1 tablespoon of the vinegar, the anchovies, and garlic and process until smooth. Transfer to a small mixing bowl, add the scallions and sour cream, and whisk until well blended. Season with salt and pepper.

3. Pour the dressing over the cabbage, add the carrots, and toss gently. Cover and refrigerate overnight.

4. The next day, bring the salad back to room temperature. Taste and correct the seasoning, adding more vinegar, if necessary, and a large grinding of black pepper. Serve chilled as an accompaniment to a frittata or grilled hamburgers.

MIXED GARDEN SALAD WITH LEMON AND MINT VINAIGRETTE

serves 6

3 cups tightly packed finely shredded
  green cabbage
Salt
¼ cup olive oil
2–3 tablespoons fresh lemon juice
1 large clove garlic, peeled and mashed
2 tablespoons finely minced fresh mint
2 tablespoons finely minced fresh
  parsley

Freshly ground black pepper
1 small green bell pepper, cored,
  seeded, and thinly sliced
2 pickling cucumbers, peeled, seeded,
  and cut into fine julienne
8 radishes, trimmed and cut into julienne
½ small red onion, peeled and thinly
  sliced
10 ripe cherry tomatoes, cut in half

1. Place the cabbage in a colander, sprinkle with 1 teaspoon of salt, and let drain for 2 to 3 hours. Squeeze out the excess moisture from the cabbage with your hands, place the cabbage in a serving bowl, and set aside.

2. In a small bowl combine the olive oil, lemon juice, garlic, mint, and parsley, whisk until well blended, and season with salt and pepper. Pour the dressing over the cabbage. Add the green pepper, cucumbers, radishes, and onion. Toss gently, cover, and refrigerate overnight.

3. The next day, bring the salad back to room temperature. Add the cherry tomatoes and correct the seasoning. Serve as part of an hors d'oeuvre table.

SICILIAN ROASTED PEPPER MARMALADE

serves 6

7 tablespoons olive oil
⅓ cup pine nuts
3 tablespoons balsamic vinegar
2 small cloves garlic, peeled and
  mashed
3 large red bell peppers, roasted,

peeled (page 483), and cut into fine
  julienne
3 large yellow bell peppers, roasted,
  peeled (page 483), and cut into fine
  julienne
½ cup golden raisins, plumped
Salt and freshly ground black pepper

1.   Heat 1 tablespoon of oil in a small skillet over medium heat. Add the pine nuts and stir constantly until lightly browned. Immediately remove from the heat, transfer to a small bowl, and set aside.

2.   In a medium mixing bowl, combine the vinegar, the remaining oil, and the garlic and whisk until well blended. Add the peppers and raisins, toss gently, and season with salt and pepper. Serve at room temperature as an accompaniment to grilled or roasted meats.

BROCCOLI, CAULIFLOWER, SNOW PEA, AND CARROT SALAD

serves 4–6

VINAIGRETTE
6 tablespoons olive oil, preferably extra-virgin
2–3 tablespoons red wine vinegar
2 teaspoons Dijon mustard
2 tablespoons finely minced shallots
1 small clove garlic, peeled and mashed
Salt and freshly ground black pepper

SALAD
3 cups cauliflower florets
3 cups broccoli florets
2 ounces snow peas, tips and strings removed and cut in half lengthwise on the diagonal
1 large carrot, peeled and cut into thin matchsticks

1.   Start by making the vinaigrette: Combine the oil, 2 tablespoons vinegar, mustard, shallots, and garlic in a small jar, cover tightly, and shake the jar until the dressing is well blended. Season with salt and pepper and reserve.

2.   Bring lightly salted water to a boil in a vegetable steamer, add the cauliflower florets, and steam, covered, for 3 to 4 minutes, or until tender. Run under cold water to stop further cooking and drain on paper towels. Set aside.

3.   Add the broccoli florets to the steamer, cover, and steam for 4 to 5 minutes, or until tender. Run under cold water, drain on paper towels, and set aside.

4.   Add the snow peas and carrot, cover, and steam for 2 minutes. Run under cold water and drain on paper towels.

5.   Combine the cauliflower, broccoli, snow peas, and carrot in a large serving bowl, toss gently with the vinaigrette, cover, and chill overnight.

6.   The next day bring the salad back to room temperature. Taste and correct the seasoning, adding a large grinding of pepper and more vinegar if necessary. Serve as part of an hors d'oeuvre table or with grilled meats.

BROWN SUGAR–GLAZED BANANAS WITH MASCARPONE AND HONEY

serves 6

Mascarpone cheese is relatively new to the American gourmet scene but is now available in many upscale supermarkets and specialty cheese shops. The sweetened cheese adds a rich and delicious finishing touch to a great variety of fruits in every season, but I particularly like it here with pan-sautéed bananas glazed with brown sugar and orange juice. Of course, if you cannot get mascarpone, a good-quality vanilla ice cream is an excellent substitute. You may also serve caramelized apples and pears in this way, but I find that using a year-round fruit, such as bananas, adds interest and variety to my fall dessert repertoire.

1 cup mascarpone
1½ tablespoons pure uncooked honey
4 tablespoons unsalted butter
3 tablespoons dark-brown sugar
½ teaspoon vanilla extract

½ teaspoon ground cinnamon
½ cup freshly squeezed orange juice
Grated zest of 1 orange
Grated zest of 1 lemon
6 semi-ripe bananas, peeled

1. In a small bowl combine the mascarpone and honey and whisk until well blended. Set aside.

2. In a large, heavy skillet melt the butter over low heat. Add the brown sugar and stir for 1 to 2 minutes, or until it melts. Add the vanilla, cinnamon, orange juice, and orange and lemon zest.

3. Cut a ¼-inch slice lengthwise off each banana. Place the bananas in the skillet, cut side down, and heat for 2 to 3 minutes, spooning the pan juices over them. The bananas should be just warmed through, but not soft. Transfer the bananas to a serving platter and reserve.

4. Raise the heat, bring the pan juices to a boil, and reduce until they are syrupy and heavily coat a spoon.

5. Pour the sauce over the bananas and serve immediately, accompanied by the mascarpone and honey sauce.

SAUTÉ OF APPLES IN CARAMEL CREAM

serves 4–5

Some simple, homey desserts are so good that it is hard to improve on them. That is certainly true of this flavorful sauté of apples. All you need are crisp Golden Delicious, Granny Smith, or any other apple that holds its shape in cooking, good butter, sugar, and fresh non-ultrapasteurized heavy cream. A typical Normandy country dessert, it is also popular, with some variations, in Austria, where caramelized nuts, raisins, and apple brandy are often added to the sauté. Although you may choose to try some of these variations, I personally like it best in its simplest form, which brings into focus the clear, delicious, and uncomplicated flavors of the basic ingredients.

5 large Granny Smith apples
3–4 tablespoons unsalted butter
½ cup plus 1 tablespoon granulated
  sugar

¾ cup heavy cream
2 tablespoons water

GARNISH
Bowl of good-quality vanilla ice cream

1. Peel the apples, cut in half, remove the core using an apple corer, and cut into ⅛-inch slices. In a 10-inch cast-iron skillet melt the butter over fairly high heat. Add the apples and sauté until lightly browned. Add 1 tablespoon of the sugar and continue sautéing the apples until they are soft and caramelized, about 4 to 5 minutes.

2. While the apples are sautéing, heat the cream in a small saucepan and keep hot.

3. In a heavy 2½ quart saucepan combine the remaining sugar and water. Bring to a boil over high heat and stir once to dissolve the sugar. Continue to boil until the mixture turns a deep hazelnut brown. Immediately remove from the heat and add the hot cream, averting your face; the mixture will bubble violently, but will settle back quickly.

4. Return the saucepan to low heat and whisk until the cream and caramel are well blended and the sauce is smooth.

5. Add the sauce to the skillet of sautéed apples and fold gently. Transfer the apples to individual serving plates and serve with a side dish of vanilla ice cream.

Note: Bartlett pears or a combination of raw pears and poached quince can be used successfully instead of apples.

BITTERSWEET CHOCOLATE SOUFFLÉS IN WHITE CHOCOLATE–GINGER SAUCE

serves 12

6 ounces bittersweet chocolate,
  preferably Lindt, broken into pieces
4 tablespoons whole milk
5 tablespoons granulated sugar
4 large egg yolks
6 large egg whites

1 recipe White Chocolate–Ginger Sauce
(page 488)

GARNISH
½ pint fresh raspberries
Tiny fresh mint leaves

1. Preheat the oven to 450° F. Lightly butter 12 (2-ounce) porcelain ramekins and reserve.

2. Combine the chocolate, milk, and 4 tablespoons sugar in a medium saucepan, place over low heat, and stir until the chocolate has melted and the mixture is smooth. Remove from the heat, add the egg yolks, and whisk again until smooth. Set aside.

3. In a large mixing bowl combine the egg whites with the remaining sugar and beat until the peaks are just firm.

4. Whisk one fifth of the beaten whites into the chocolate mixture to lighten it. Then fold the lightened chocolate mixture gently but thoroughly into the remaining whites.

5. Spoon the mixture into the prepared ramekins and place them in a large baking dish. Add enough boiling water to the baking dish to come halfway up the sides of the ramekins. Place the baking dish in the center of the preheated oven and bake for 5 to 7 minutes. The soufflés will rise and become spongy to the touch, but they should still be a bit runny in the center. Do not overcook.

6. Remove the soufflés from the oven and let cool slightly, until they are just warm. Run a knife around the sides of the soufflés. Unmold and invert onto individual serving plates. Surround the warm soufflés with the white chocolate–ginger sauce and garnish the sauce with raspberries and mint leaves.

CHOCOLATE AND PEAR TART

serves 6

Although I love chocolate, I have never really liked the way it tastes when teamed with most fruits. But the combination of pears and chocolate is one that I have loved since my school years in Geneva, when I was introduced to pears belle Hélenè, a classic Swiss confection in which vanilla ice cream is topped with a poached pear and hot, buttery chocolate sauce.

In this rather homey tart, roughly grated chocolate is tucked under poached pears, covered with a light flan mixture, and baked with a sprinkling of slivered almonds. Since the chocolate is not visible, it comes as a pleasant taste surprise. For a truly impressive presentation, serve with Chocolate Mousse Ice Cream (page 232). Although this is one of the few tarts that can be made a day ahead of time, I still prefer putting it together a few hours before I serve it.

2 tablespoons unsalted butter
3 large slightly underripe pears, preferably Comice, peeled, cored, and cut into ½-inch wedges
5 tablespoons granulated sugar
½ cup heavy cream
1 large egg
1 partially baked 10-inch Basic Tart Shell (page 492), in a pan with a removable bottom

5 ounces bittersweet chocolate, preferably Lindt or Tobler, coarsely chopped
½ cup sliced blanched almonds

GARNISH
Bowl of lightly sweetened Crème Fraîche (page 491)

1. Preheat the oven to 375° F.

2. In a large skillet melt the butter over medium-high heat. Add the pears, sprinkle with 2 tablespoons sugar, and sauté quickly until nicely browned. Reduce the heat and cook until tender, about 15 minutes. Drain the pears in a colander and set aside.

3. Combine the heavy cream, egg, and 1 tablespoon sugar in a mixing bowl and whisk until smooth. Reserve.

4. Sprinkle the bottom of the partially baked tart shell with the chopped chocolate. Arrange the pears in a decorative pattern over the chocolate and drizzle with the cream-egg mixture. Sprinkle the top with the sliced almonds and remaining sugar, place in the center of the preheated oven, and bake for 30 to 35 minutes, or until the custard is set and the almonds are nicely browned. Remove from the oven and let cool.

5. Serve the tart cut into wedges with a bowl of lightly sweetened crème fraîche.

CHOCOLATE CREPES WITH GRAND MARNIER CREAM AND ORANGE SAUCE

serves 6

It is truly sad when great foods fall out of flavor. The crepe was a victim of a 1970s fad during which everything and anything was used as a filling for crepes, which themselves were tasteless and dry. Then the response came: Until quite recently, scarcely a restaurant or home cook served crepes either as a starter or closer to a meal.

But now a crepe revival is under way, and crepes of all kinds are reappearing on the menus of top restaurants. Maison Blanche in Paris serves this buttery chocolate crepe in a pool of fresh orange coulis with a bittersweet chocolate sorbet on the side. Even without the sorbet, it is a winner and well worth the effort.

CREPES
4 large eggs
½ cup unsweetened cocoa
½ cup all-purpose flour
¼ cup granulated sugar
1 cup hot milk
3 tablespoons unsalted butter, melted and cooled

SAUCE
6 tablespoons unsalted butter
8 tablespoons granulated sugar
½ cup freshly squeezed orange juice
1 teaspoon finely grated orange rind
⅓ cup Grand Marnier
1 recipe Crème Chiboust (page 118)

1. Start by making the crepes: In the workbowl of a food processor combine all of the ingredients for the crepes except for the melted butter and process until smooth. Transfer to a large mixing bowl, whisk in the melted butter, and let stand at room temperature for 1 hour.

2. To make the crepes, follow the procedure given for Dessert Crepes (page 493).

3. To prepare the orange sauce: In a heavy saucepan melt the butter with the sugar and cook the mixture for 3 to 4 minutes, or until the butter turns a light brown and the sugar starts to caramelize. Immediately add the orange juice, bring to a boil, and whisk until the sugar is melted and the sauce is smooth. Add the orange rind and Grand Marnier and whisk until well blended. If the sauce is not sweet enough, add more sugar to taste. Keep the sauce warm.

4. Preheat the oven to 200° F.

5. Fifteen minutes before serving, wrap the crepes in aluminum foil, place in the center of the preheated oven, and keep warm.

6. Remove the crepes from the oven, spread about 2 tablespoons of the Crème Chiboust on each, and fold in half and then in half again (into triangles).

7. Place 2 crepes on each individual serving plate and spoon a little of the warm orange sauce over each portion. Serve at once.

Variations: Instead of the Crème Chiboust, fill the crepes with Apples in Caramel Cream (page 341), using the pan juices as a sauce. Or fill the crepes with sautéed or poached sliced pears and top with a good vanilla ice cream or a buttery chocolate sauce (see the Basics section).

The crepes can be flavored with coffee extract instead of chocolate and filled with a light mocha mousse or even a coffee-flavored whipped cream and finely diced poached pears.

PURPLE FIG AND ALMOND TART

serves 6–8

¾ cup slivered blanched almonds, about 3 ounces
½ cup granulated sugar
6 tablespoons unsalted butter, softened
2 large eggs
1 teaspoon vanilla extract
1 partially baked 9-inch Basic Tart Shell

(page 492), in a pan with a removable bottom
1 pound small, fresh whole purple figs
¼ cup sliced blanched almonds

GARNISH
Dusting of confectioners' sugar
Crème Fraîche (page 491)

1. Preheat the oven to 375° F.

2. In the workbowl of a food processor combine the slivered almonds with ¼ cup of the sugar and process until the almonds are finely ground. Be careful not to overprocess or the mixture will turn into a paste. Reserve.

3. Combine the remaining sugar and softened butter in the bowl of an electric mixer and beat on medium speed until fluffy. Add the eggs, reserved almond mixture, and vanilla and beat until smooth. Pour the mixture into the prepared tart shell, smooth evenly with a spatula, and set aside.

4. Cut off and discard the stem of each fig, cut the figs lengthwise into ¼-inch slices, and arrange on top of the almond mixture in concentric circles, slightly overlapping the slices.

5. Sprinkle the top with the sliced almonds, place the tart in the center of the preheated oven, and bake for 35 to 40 minutes, or until the mixture is set and nicely browned.

6. Remove from the oven. Serve warm, sprinkled with confectioners' sugar and cut into wedges, with a bowl of crème fraîche.

GRATIN OF POACHED PEARS WITH RED WINE SABAYON

serves 8

PEARS
2 cups dry red wine
1 cup granulated sugar
1 (3-inch) piece of cinnamon stick
1 (2-inch) piece of vanilla bean
2 whole cloves
4 slightly underripe Bartlett pears, peeled

SABAYON
5 large egg yolks
½ cup granulated sugar

⅓ cup pear poaching liquid
2 tablespoons Cointreau, or other orange liqueur
1 cup heavy cream, whipped
Confectioners' sugar

GARNISH
Tiny fresh mint leaves

1. Start by poaching the pears: Combine the wine, sugar, cinnamon stick, vanilla bean, and cloves in a 3-quart saucepan. Bring to a boil, reduce the heat, and simmer for 5 minutes. Add the pears and return the syrup to a boil. Reduce the heat and simmer the pears until tender, about 20 to 30 minutes, turning from time to time in the poaching liquid. Remove the saucepan from the heat and set the pears aside to cool in the poaching liquid.

2. Remove the pears from the poaching liquid, reserving ⅓ cup of the liquid, and drain the pears. Cut the pears in half and remove the cores. Place each pear half, cut side down, on a cutting board. Beginning at the stem end, but without cutting through the stem itself, cut each pear lengthwise into ¼-inch slices using a sharp knife. The slices should still be attached at the stem end.

3. Transfer the pear halves to a large au gratin dish, about 8 × 12 inches, and press down lightly on each pear half to fan out the slices. Reserve.

4. To prepare the sabayon: In the top of a double boiler, combine the egg yolks and granulated sugar and beat until fluffy and pale yellow. Add the reserved pear poaching liquid and Cointreau and whisk until well blended.

5. Set over simmering water and whisk constantly until the mixture is thick and heavily coats a spoon. Be careful not to overcook or the yolks will curdle. Remove the custard from the heat and transfer to a large mixing bowl. Let cool completely.

6. Preheat the broiler.

7. Fold the whipped cream gently but thoroughly into the sabayon and spoon the mixture over the pears Sprinkle heavily with confectioners' sugar and place the dish in the broiler, 6 inches from the source of heat. Broil until the sabayon is nicely browned and the sugar has caramelized, which happens quite fast, so be careful not to burn the sugar.

8. Immediately remove the pears from the broiler, garnish with the mint, and serve at once.

THE

WINTER

KITCHEN

I never truly appreciated the winter kitchen until we bought a house in New England. For me, growing up in a Mediterranean country, winter was never much of a season, and before you knew it spring would appear with an intensity the winter never had. Now, after the abundance and energy of fall, winter, too, is a much appreciated season. Somehow, without the diversion of outdoor activities, the pace gets slower and there's time to experiment, to savor more complex dishes.

With temperatures often hovering around the teens in the Northeast, and the countryside covered with snow, I like to spend more time cooking and baking. Making yeast breads, savory tarts, and puff pastry from scratch seems to come more naturally when there's plenty of time. Cozy, comforting food like a full-flavored pot-au-feu, stuffed cabbage, and an innumerable variety of hearty soups laced with root vegetables, beans, or lentils are more appreciated now than ever before. In fact, winter food is enjoying an unprecedented popularity, possibly because so many of the dishes that take time to prepare and develop flavor are synonymous with the winter kitchen and our yearning for the real taste of yesterday's foods.

I often hear people say that winter produce is limited. Personally, I find that these seasonal limitations bring out my creativity as a cook. This may sound like a cliché in a time of almost year-round availability of nonseasonal fruits and vegetables, but I am more interested in creating a robust onion and potato dish, a hearty ragout, or a custardy bread pudding than in celebrating the globality of fresh foods by serving asparagus or raspberries in winter. Instead, I turn toward such vintage vegetables as carrots, onions, potatoes, and celery. These, together with legumes, are the vegetables that have been the mainstay of generations of cooks everywhere and the key components of my favorite soups, hearty stews, and flavorful sauces.

Happily, over the past few years, the winter kitchen has become increasingly varied due to the growing selection of cool-weather vegetables—celery root, broccoli rape, fennel, Belgian endive, and other winter greens. These additions are helping to make the season as interesting as spring, summer, or fall.

For me, the winter kitchen offers some very special moments. I often start the morning by combining the various elements for a hearty winter soup or a richly flavored stock and placing the pot on the back burner while I take my morning walk. By the time I return, I can see that the kitchen windows have fogged up, and I can smell the fragrant aroma of the simmering broth. With a freshly brewed cup of coffee, I like to sit down and plan my day, often reaching for a spoon and tasting the broth as it begins to develop flavor. There is something about these moments that brings back the past even as it celebrates the present. And then, during the slow passage of the season, I begin to lighten my meals and their preparation in anticipation of the upcoming spring.

1

Cream of Roasted Pepper Soup with Crisp Croutons and Aioli
Rigatoni with a Ragout of Veal and Mushrooms
Espresso Sabayon Gritti

2

Skillet-Braised Spinach and Smoked-Salmon Toasts
Ragout of Beef with Wild Mushrooms
Soufflé Glacé au Grand Marnier

3

Spaghetti Squash and Scallion Fritters
Leg of Lamb and Curried Pinto Beans with Middle Eastern Spices
Bittersweet Chocolate Fondant with Rum Raisins

4

Warm Goat Cheese Timbales on a Bed of Greens with
Honey-Cumin Vinaigrette
Marinated Roast Pork Tenderloins with White Bean and Garlic
Puree
Caramelized Pear and Prune Flans

5

Roasted Potatoes with Gruyère on a Bed of Greens
Cabbage Packages with Spiced Lamb and Rice Stuffing
Baked Apples in Brown Sugar and Cinnamon Wine

WINTER APPETIZERS

Rillettes of Salmon with Potato Rosti

Sauté of Calamari with Caramelized Onions

Ragout of Lobster and Scallops with Broccoli and Roasted Shiitake
Mushrooms

Scallop Flans à l'Arlesienne in a Roasted Pepper and Shallot
Butter Sauce

Old-Fashioned Burgundian Escargot Tart

Spaghettini with Clams in Braised Fennel and Saffron Cream

Orecchiette with Kale, Sausage, and Garlic Cream

Risotto with Cremini Mushrooms and Sweet Gorgonzola

Risotto Veneziana

Goat Cheese Gnocchi with Pink Alfredo Sauce

WINTER SOUPS

Devon Velouté of Celery and Stilton Cheese

Catalan Kale, Potato, and Garlic Soup

Cream of Leek, Potato, Lemon, and Dill Soup

Cream of Lentil Soup with Garlic, Bacon, and Smoked Trout

A Spicy Middle Eastern Soup of Lamb Shanks and Lentils

Lemon–Split Pea Soup

Curried Lentil Soup with Root Vegetables

WINTER MAIN COURSES

Monkfish Scaloppine with Warm Cannelini Bean and Garlic Puree

Fillets of Salmon with Lentils à la Bourgeoise

Pot-au-Feu of Red Snapper with Winter Vegetables

Sauté of Cornish Hens Forestière

Boneless Chicken with Morels au Farci

Roast Poussins à la Provençale with Lemon and Cracked Pepper

Roast Duck in Raisin Sauce

Ragout of Veal with Sherry Vinegar and Shallot Essence

Fricassee of Veal Shanks with Glazed Carrots in Tarragon Sauce

Pan-Seared Sauté of Calves' Liver in Mustard and Green
Peppercorn Sauce

Roast Loin of Lamb with Red Wine–Shallot Sauce

Sauté of Pork Tenderloins with Prunes and Sweet-and-Sour
Caramel

Penne with Three-Mushroom Ragout, Ham, and Parmesan Cream

Catalan White Bean Ragout la Bodega

WINTER VEGETABLES

Fricassee of Brussels Sprouts, Peppers, and Onions

Sauté of Broccoli Rape with Prosciutto and Pine Nuts

Cabbage and Leek Flan à la Vaudoise

Skillet-Glazed Cucumbers

Marmalade of Belgian Endives

Braised Winter Lentils with Thyme and Garlic

Fennel Fricassee with Carrots, Bacon, and Pearl Onions

Potato and Leek Rosti

Shallot Puree with Garlic and Thyme

"Lasagna" of Sweet Potatoes, Turnips, and Tart Apples

Spinach and Belgian Endive Salad with Fried Goat Cheese

Belgian Endive and Stilton Cheese Salad in Anchovy and Shallot Vinaigrette

Fennel, Radish, and Cucumber Salad with Sherry Vinegar and Lemon Vinaigrette

Winter Salad with a Julienne of Omelet Crepes

Roasted Pear Salad with Walnut and Shallot Vinaigrette

WINTER DESSERTS

Viennese Chocolate Soufflé Cake

Two-Tone Chocolate Pâté with Coffee Crème Anglaise

Double Espresso and Caramel Custards

Crisp Brown Sugar and Almond Cookies

Crepes with Caramelized Pear and Raisin Jam

Mocha Crepes with Coffee Sabayon

Cranberry, Apple, and Pear Mousse

Meringue-Topped Lemon-Butter Tart

Orange and Grapefruit Salad in Pineapple and Black Pepper Syrup

Potpourri of Winter Fruit in Vanilla-Rum Syrup

Roasted Pears with Brown Sugar and Ginger Crisp

Cream of Roasted Pepper Soup with Crisp Croutons and Aioli

Rigatoni with a Ragout of Veal and Mushrooms

Espresso Sabayon Gritti

This is a homey, everyday kind of menu. It is good for a family dinner, but I often serve it to friends for an informal supper.

The soup does take some advance planning so that it has time to develop full flavor, but is also delicious when done at the last moment. It is an easy soup to assemble, particularly if you are one of those cooks who roasts a large batch of red peppers at one time and stores them in the refrigerator topped with a layer of olive oil. Since the imported red peppers are extremely expensive, I have also experimented using a good brand of commercially packed roasted peppers with excellent results.

In midwinter, rigatoni in an earthy veal ragout fills the craving for a hearty pasta dish. Rather than the standard Bolognese sauce, which is made with ground meat, I prefer this chunky, intensely flavored ragout. You may add a variety of sautéed mushrooms to the ragout, and if you splurge on an ounce of imported porcini, the result will be even more aromatic. Another terrific addition toward spring would be two cooked artichoke hearts, either cubed or finely sliced, which will lend their own nutty taste and special texture to the dish.

After this filling main course, I would follow with a palate-cleansing, well-seasoned salad and then finish with the espresso sabayons. These are a light and richly textured dessert that can be made a day in advance. For a more elaborate presentation you may place a few rum-soaked raisins at the bottom of each ramekin. It makes for a nice little discovery and provides a charming finishing touch for the meal.

CREAM OF ROASTED PEPPER SOUP WITH CRISP CROUTONS AND AIOLI

serves 6

2 tablespoons unsalted butter
1 tablespoon corn oil
1 medium onion, peeled and finely
   minced
½ cup finely diced carrot
½ cup finely diced celery
2½ tablespoons all-purpose flour
6–8 large red bell peppers roasted,
   peeled (page 483), and diced

5–6 cups Chicken Stock (page 470)
1 cup heavy cream
Salt and freshly ground white pepper

GARNISH
Aioli (page 477)
Garlic Croutons (page 282)

1. In a large flameproof casserole, heat the butter and oil over low heat. Add the onion, carrot, and celery and cook the mixture for 4 minutes or until soft but not browned. Add the flour and stir well into the vegetable mixture. Add the roasted peppers and stock, bring to a boil, reduce the heat, and simmer for 25 minutes.

2. Remove the soup from the heat and cool slightly. Transfer the soup to the workbowl of a food processor and process until smooth. Return the pureed soup to the casserole, add the cream, and season with salt and pepper. Just heat through. Serve with a dollop of aioli and sprinkled with garlic croutons.

Note: When peeling the roasted peppers make sure to remove all traces of the charred skins.

RIGATONI WITH A RAGOUT OF VEAL AND MUSHROOMS

serves 6

Rigatoni is a particularly sturdy pasta, and its texture lends itself well to hearty preparations. I first had this dish in a Milanese restaurant where the cook used leftover osso buco, which made great sense since the combination of the juicy veal with pasta resulted in a full-flavored one-dish meal. Veal knucklebones or boned shoulder of veal are suitable for this ragout, and it's easy to double the recipe so that you have another main course on hand. By adding a seasonal vegetable such as pearl onions, green or red pepper, or cubed butternut squash, you can transform the leftover ragout into another delicious main course.

1 ounce dried porcini mushrooms, rinsed
1½ cups beef broth
2½ pounds veal shoulder, cut into
  ¾-inch cubes
3–5 tablespoons unsalted butter
2–3 tablespoons olive oil
Salt and freshly ground black pepper
¾ pound Cremini mushrooms, stems
  removed and cubed
2 large onions, peeled and finely
  minced

2 large cloves garlic, peeled and finely
  minced
¾ cup dry white wine
1 (35-ounce) can Italian plum tomatoes,
  drained and chopped
1 Bouquet Garni (page 485)
¾ pound imported rigatoni, preferably
  De Cecco
½ cup freshly grated Parmesan cheese
1 teaspoon finely grated lemon rind,
  optional

1. In a small saucepan combine the porcini with the beef broth. Bring to a boil, reduce the heat, and simmer, covered, for 20 minutes. Drain, reserving the poaching liquid. Finely julienne the porcini and set aside.

2. Preheat the oven to 350° F.

3. Dry the veal thoroughly on paper towels. In a large, heavy skillet heat 2 tablespoons of the butter and 1 tablespoon of oil over medium heat. Cook the veal cubes a few at a time without crowding the pan, and sauté until nicely browned on all sides. With a slotted spoon, transfer the veal to a heavy casserole and continue browning the remaining veal, adding more butter and oil when necessary. Transfer all the veal to the casserole. Season with salt and pepper and reserve.

4. Add a little more butter and oil to the skillet, add the Cremini mushrooms, and sauté until nicely browned. Transfer to the casserole.

5. Add a little butter and 2 teaspoons of oil to the skillet and add the onions and garlic. Cook the mixture, scraping the bottom of the pan, until the onion and garlic mixture is nicely browned. Add the wine, bring to a boil, and reduce to 2 tablespoons. Add the tomatoes and season with salt and pepper. Bring the sauce to a boil and add to the veal. Bury the bouquet garni in the casserole and add the reserved mushroom bouillon.

6. Cover the casserole tightly and place in the center of the preheated oven. Braise the veal for 1 hour and 15 minutes to 1 hour and 30 minutes, or until the meat is very tender but not falling apart.

7. When the veal is done remove from the oven and degrease the pan juices carefully. Add the reserved julienne of porcini, cover, and keep warm.

8. Bring plenty of salted water to a boil in a large pot. Add the rigatoni and cook until al dente. Drain the pasta thoroughly and toss gently into the veal ragout with half the Parmesan cheese. Taste and correct the seasoning. Transfer to a warm, deep serving bowl and sprinkle with the remaining Parmesan and optional lemon rind. Serve immediately.

Note: Any leftover rigatoni can be reheated successfully in the microwave oven. You can further perk up these delicious leftovers by adding a red or green pepper (cored and finely sliced), which is sautéed in olive oil, to the pasta prior to reheating it.

ESPRESSO SABAYON GRITTI

serves 6–8

5 large egg yolks
⅓ cup granulated sugar
½ cup strong prepared coffee, preferably espresso
1 teaspoon vanilla extract

½ cup heavy cream, whipped

GARNISH
Cocoa powder
Grated bittersweet chocolate

1. In a large mixing bowl, combine the egg yolks and sugar. Whisk until the mixture is pale yellow and fluffy. Whisk in the prepared coffee and transfer to a heavy 2-quart saucepan. Place over medium-low heat and whisk constantly until the mixture has doubled in volume and is thick and creamy and heavily coats a spoon. Do not let the sabayon come to a boil.

2. Immediately transfer the sabayon to a large mixing bowl, add the vanilla extract, and chill for 15 minutes, or until just cool.

3. Fold in the whipped cream gently but thoroughly and spoon the mixture into 6 to 8 (4-ounce) ramekins. Cover and chill for 2 to 4 hours before serving.

4. Just before serving, heavily dust the top of each sabayon with cocoa powder and sprinkle with grated chocolate. Serve at once.

Note: The sabayon can be made a day ahead and garnished just before serving.

Skillet-Braised Spinach and Smoked-Salmon Toasts
Ragout of Beef with Wild Mushrooms
Soufflé Glacé au Grand Marnier

Here is one of my favorite winter menus. It has a nice balance and works beautifully for a casual dinner party since everything but sautéing the spinach can be done well in advance.

Good-quality smoked salmon is very important in the appetizer. Taste before buying it to make sure that it's not too salty. If you do not have a good source for the salmon, you can make a quick topping with a mixture of sour cream, finely minced dill, and red salmon caviar. A light fish mousse made with smoked haddock or blue fish would also be a delicious topping for the spinach.

Follow this appetizer with the earthy full-bodied ragout, a bracing cold-weather main course that truly delivers. Unlike a stew, in which all the elements are cooked together (often beyond recognition), in a ragout vegetables are cooked separately and added at the last moment, so that each component retains its own individual texture and flavor. I first sampled this rich mahogany-colored ragout many years ago at a country inn in the Aosta Valley in the Piedmont region of Italy. I still remember the intense taste of that dish to which the porcini mushrooms had lent their unique flavor. As an accompaniment, I like to serve one or two purees, such as a sweet potato and carrot puree and a celery root, apple, and potato puree. Steamed broccoli florets or snow peas would add a nice touch of color to the dinner plate. Before moving on to dessert, serve a well-seasoned salad, perhaps accompanied by two or three interesting cheeses.

The soufflé glacé may sound ambitious, but in fact is one of the easiest desserts I prepare. You have to plan in advance, since the soufflé must be frozen. The best accompaniment is a fresh orange salad but a raspberry or strawberry sauce made with frozen berries will taste just as good. Any leftover soufflé can be refrozen and served several weeks later.

SKILLET-BRAISED SPINACH AND
SMOKED-SALMON TOASTS

serves 6–8

1 cup sour cream
2 teaspoons fresh lemon juice
2 teaspoons finely grated lime zest
2 pounds fresh spinach, stemmed
6 tablespoons unsalted butter
Salt and freshly ground black pepper
6 tablespoons Clarified Butter (page 478)

8 slices of day-old dense white bread, about ⅓ inch thick, crusts removed
½ pound smoked salmon, finely minced
¼ cup finely minced fresh dill

GARNISH
Sprigs of fresh dill

1. In a small bowl combine the sour cream, lemon juice, and lime zest, whisk until well blended, and set aside.

2. Wash the spinach thoroughly and place in a large saucepan with the water that clings to its leaves. Cover and cook over low heat until just wilted. Drain thoroughly, pressing down lightly on the leaves, and reserve.

3. Melt the butter in a 10-inch skillet over medium-low heat. Add the spinach and sauté for 2 to 3 minutes, stirring often. Season with salt and a large grinding of pepper. Keep warm.

4. Melt ¼ cup of the clarified butter in a 10-inch skillet over medium heat. Add 4 slices of bread and sauté until crisp and lightly browned on both sides. Transfer to a side dish and reserve. Repeat with the remaining clarified butter and bread.

5. Place the sautéed bread on a serving platter, top each piece with 2 tablespoons of spinach, and spread evenly over each piece of toast.

6. Combine the salmon and dill in a medium bowl. Form the mixture into 8 (1-inch) balls and place a ball on top of each of the spinach toasts. Garnish with sprigs of fresh dill and serve at once with a dollop of sour cream mixture.

RAGOUT OF BEEF WITH WILD MUSHROOMS

serves 6–8

Beef ragouts deserve to be more popular. Come winter, few dishes are as satisfying as this flavor-packed one-dish meal in which beef is gently braised for a long time with two types of mushrooms that deepen its taste and impart a wonderful aroma. Serve the ragout accompanied by a puree of sweet potatoes and carrots.

4 pounds boneless beef chuck, cut into
   1-inch cubes
7 tablespoons unsalted butter
3–4 teaspoons peanut oil
Salt and freshly ground black pepper
3 cups finely minced onions
1 tablespoon finely minced garlic
1 teaspoon imported Hungarian paprika
1 tablespoon tomato paste
2 cups beef bouillon

1½ pounds Cremini (brown)
   mushrooms, stemmed and cut into
   ¾-inch cubes
1 ounce dried porcini or Chilean
   mushrooms, thoroughly rinsed under
   warm water to remove all sand and
   drained
1 Beurre Manie (page 478)

GARNISH
Finely minced fresh Italian parsley

1. Preheat the oven to 350° F.

2. Dry the beef thoroughly on paper towels.

3. In a 12-inch black cast-iron skillet, melt 2 tablespoons of the butter with 2 tablespoons of the oil over medium-high heat. Add the beef cubes a few at a time and sauté until well browned on all sides. Transfer to a heavy 4-quart casserole and reserve. Continue to sauté the beef until all the meat has been browned, adding more butter and oil to the skillet as necessary. Season the beef with salt and freshly ground pepper and transfer to the casserole.

4. If the fat in the skillet has burned, discard it and replace with 2 tablespoons of butter and a teaspoon of oil. Reduce the heat, add the onions and garlic, and cook over low heat until the onions become a rich brown color. Do not burn. Add the paprika and tomato paste and mix well into the onion mixture. Add 1 cup beef bouillon, bring to a boil, and remove from the heat. Add the onion mixture to the beef and set aside.

5. In a medium skillet, heat the remaining butter and 1 teaspoon oil over medium heat. Add the Cremini mushrooms and sauté until lightly browned. Transfer to the casserole and reserve.

6. In a small saucepan, combine the porcini mushrooms with 1 cup of bouillon. Bring to a boil, reduce the heat, and simmer, covered, for 15 minutes, or until tender. Strain the broth through cheesecloth into the casserole of beef and set the mushrooms aside separately.

7. Place the casserole over high heat and bring to a boil. Cover the casserole, place in the center of the preheated oven, and braise the beef for 1 hour and 30 minutes, or until tender.

8. While the beef is braising, cut the porcini mushrooms into fine julienne and reserve.

9. When the beef is done, remove it with a slotted spoon to a side dish. If the pan juices are too thin, bring to a boil on top of the stove and reduce slightly. Whisk in bits of beurre manie until the sauce heavily coats a spoon. Taste and correct the seasoning.

10. Return the beef to the casserole, add the julienne of porcini, and just heat through.

11. Transfer the ragout to a serving dish, sprinkle with minced Italian parsley, and serve accompanied by well-seasoned polenta or buttery mashed potatoes and follow with a well-seasoned salad.

Note: This ragout needs time to develop flavor. It is best when made at least a day ahead of time. It also freezes well.

If you cannot get Cremini mushrooms, you may use 2 pounds of all-purpose mushrooms instead.

SOUFFLÉ GLACÉ AU GRAND MARNIER

serves 6–8

4 large eggs, separated
5 tablespoons plus ½ cup granulated
  sugar
⅓ cup Grand Marnier
3 tablespoons water
2 cups heavy cream, whipped

GARNISH
2 large oranges, peeled and sliced, or
  Spiced Oranges in Red Wine and
  Cinnamon Syrup (page 114)
Tiny fresh mint leaves

1. Lightly oil the inside of a 1½-quart soufflé dish that is 3½ inches high. Wrap a foil collar around the dish to extend 4 inches above the top of the dish and secure with kitchen string. Lightly oil the inside of the foil and set the dish aside.

2. In the bowl of an electric mixer combine the egg yolks, 5 tablespoons of sugar, and the Grand Marnier and beat on medium-high speed until fluffy, pale yellow, and double in volume. Transfer to a medium bowl and set aside.

3. Rinse the mixer bowl and beaters so that they are very clean and add the egg whites. Beat on medium speed until just soft peaks form. Do not overbeat. Reserve in the mixer bowl.

4. In a small saucepan, combine the remaining ½ cup of sugar and water. Place over high heat and bring to a boil. Stir once to dissolve the sugar and continue to boil without stirring until the syrup reaches 234° F (soft-ball stage) on a candy thermometer.

5. With the electric mixer running, immediately pour the sugar syrup slowly, drop by drop, into the whites, beating constantly, until all the syrup has been absorbed by the whites. Continue beating until the egg-white mixture cools completely and the whites are stiff and glossy. This is called an Italian meringue.

6. Fold the reserved egg-yolk mixture into the meringue, gently but thoroughly. Fold in the whipped cream, again gently but thoroughly, and spoon the mixture into the prepared soufflé dish. Tap once on the counter to remove large air bubbles. Smooth the top with a spatula, cover, and place in the freezer overnight.

7. The next day carefully peel off the foil collar. Serve large scoops of soufflé in individual dessert dishes and surround with sliced oranges or Spiced Oranges in Red Wine and Cinnamon Syrup and garnish with tiny mint leaves.

Spaghetti Squash and Scallion Fritters

Leg of Lamb and Curried Pinto Beans with Middle Eastern Spices

Bittersweet Chocolate Fondant with Rum Raisins

For years spaghetti squash has suffered the fate of a poorly named vegetable. Although, admittedly, the strands of this rather bland winter squash, when baked, remind one of spaghetti, it is here that the similarity ends. Rather than teaming it with pasta sauces, I prefer to season it with Oriental or Middle Eastern spices or to use it solely for its texture, as in these light, savory fritters. The fritter recipe evolved one day when I found myself with some leftover squash, and they have since become a seasonal favorite. You may serve the fritters simply with a brown Noisette butter or some sour cream flavored with lemon juice and lime zest. The fritters hold up nicely and can be made about an hour ahead of time and kept warm in a low oven.

In winter, a leg of lamb is usually larger and somewhat gamier than in spring, and it demands a more assertive side dish. Although all legumes have an affinity to lamb, the curried pinto beans are one of my preferred choices. The spiciness of the beans is mellowed by the lamb pan juices. I always make some extra beans since they reheat well and are delicious served with pan-seared sausages or sautéed lamb chops.

For a total change of pace I like to serve the elegant chocolate fondant for dessert. Again, it is at its best when made one or two days ahead of time, and it will actually keep for two to three weeks in the refrigerator. As with most chocolate desserts, the fondant is best when served with a light custard sauce, either coffee- or vanilla-flavored. The rum raisins give the dessert a luxurious finishing touch.

SPAGHETTI SQUASH AND SCALLION FRITTERS

makes 24–28

1 medium spaghetti squash, about 1½ pounds
3 medium scallions, including greens, finely minced
½ cup all-purpose unbleached flour
1 teaspoon baking powder
2 large eggs, lightly beaten
Salt and freshly ground black pepper

6–8 tablespoons Clarified Butter (page 478)
Coarse salt

GARNISH
Sour cream
Tiny fresh cilantro leaves

1. Preheat the oven to 375° F.

2. Place the whole spaghetti squash on a baking sheet and bake in the center of the preheated oven for 45 to 50 minutes, or until easily pierced with a fork, turning once during roasting. Remove from the oven and let cool.

3. Cut the squash in half lengthwise. With the tines of a fork, scoop out the flesh, which will separate into spaghettilike strands. Measure 4 cups of the strands and place them in a large mixing bowl. Add the scallions, flour, baking powder, and eggs and mix gently until thoroughly combined. Season the batter with salt and pepper. Reserve the remaining squash for another preparation.

4. Heat 1½ tablespoons of clarified butter in a 10-inch nonstick skillet over medium heat. Drop heaping tablespoonfuls of the batter into the hot butter about 1 inch apart, without crowding the skillet. Flatten each fritter slightly to form 2-inch circles and sauté for 2½ to 3 minutes, or until nicely browned. Turn the fritters over and cook until nicely browned, about 1 to 2 minutes. Transfer to a serving platter and keep warm.

5. Continue making fritters, adding a little clarified butter to the skillet for each batch.

6. Sprinkle the fritters with coarse salt. Serve warm, garnished with a dollop of sour cream and tiny cilantro leaves.

LEG OF LAMB AND CURRIED PINTO BEANS
WITH MIDDLE EASTERN SPICES

serves 6

1 leg of lamb, about 6–6½ pounds
Salt and freshly ground black pepper
1 teaspoon imported paprika
2 teaspoons Dijon mustard
2 large cloves garlic, peeled and
  mashed
1 tablespoon fresh rosemary leaves
3 anchovy fillets, drained and finely
  minced
2 tablespoons all-purpose flour
2 tablespoons unsalted butter

3 small onions, peeled and cut in half
3 cups hot Brown Chicken Stock (page
  467)
2 heads garlic, separated into cloves
  and peeled
1 Beurre Manie (page 478)
Curried Pinto Beans (recipe follows)

GARNISH
Sprigs of fresh watercress

1.   Rub the lamb with the salt, pepper, paprika, and Dijon mustard. Make tiny slits along the bottom of the roast with the tip of a sharp knife and insert bits of mashed garlic, rosemary, and minced anchovies. Sprinkle the roast evenly with the flour and let stand at room temperature for 1 hour.

2.   Preheat the oven to 350° F.

3.   In a large, heavy-bottomed baking dish, melt the butter over medium-high heat. Place the roast in the pan and add the halved onions, cut side down. Sear the roast quickly on all sides until nicely browned.

4.   Add 2 tablespoons of stock to the baking dish, place the lamb in the center of the preheated oven, and roast for 1 hour and 20 minutes, or until the internal temperature registers 145° F on a meat thermometer; about 12 minutes per pound for medium-rare. Baste every 15 minutes with a little hot stock, reserving 1 cup of stock for the garlic.

5.   While the lamb is roasting, prepare the garlic. In a 1-quart saucepan, combine the peeled garlic cloves and 1 cup of stock. Bring to a boil, reduce the heat, and simmer for 50 minutes to 1 hour, or until the garlic is very tender and cooked almost to a puree. Transfer to the workbowl of a food processor and process until smooth. Set aside.

6.   When the lamb is done, transfer to a serving platter. Degrease the pan juices carefully and discard the onions. Return the juices to the baking dish and add the garlic puree and remaining stock. Place over high heat, bring to a boil, and whisk in bits of beurre manie until the sauce lightly coats a spoon. Serve the lamb thinly sliced surrounded by the curried beans and garnish with fresh watercress. Serve the sauce on the side.

CURRIED PINTO BEANS

3 tablespoons olive oil
2 large Bermuda onions, peeled and finely sliced
2 small dry red chili peppers, broken
3 large cloves garlic, peeled and finely minced
1 tablespoon tomato paste
4 large ripe tomatoes, peeled, seeded, and chopped, or 1 (2-pound) can Italian plum tomatoes, drained and chopped
1 tablespoon imported curry powder, preferably Madras

½ teaspoon Garam Masala (page 485)
¼ teaspoon powdered ginger
Salt and freshly ground black pepper
4 cups cooked Pinto Beans (page 484)
2 red ball peppers, roasted, peeled (page 483), and thinly sliced, about 1 cup

GARNISH
3 tablespoons finely minced fresh parsley

1. In a large, heavy skillet heat the oil over medium-high heat. Add the onions and chili peppers and cook, stirring constantly, until the onions start to brown, about 10 minutes. Reduce the heat, partially cover the skillet, and continue cooking the onions for 30 to 40 minutes or until they are soft and nicely browned.

2. Add the garlic, tomato paste, tomatoes, curry powder, garam masala, and ginger, bring the mixture to a boil, and season with salt and pepper. Reduce the heat and simmer the onion and tomato mixture, uncovered, for 30 minutes, or until all the tomato juices have evaporated and cooked to a thick puree.

3. Add the cooked pinto beans and roasted peppers and continue simmering for another 10 minutes, or until the flavors are well blended. Taste and correct the seasoning. Garnish with minced parsley and serve with roast leg of lamb or grilled meats.

BITTERSWEET CHOCOLATE FONDANT
WITH RUM RAISINS

serves 8

RAISINS
1 (1-pound) box dark raisins, plumped
½ cup dark rum
1½ cups water

FONDANT
10½ ounces bittersweet chocolate, preferably Lindt or Tobler, broken into pieces

10 tablespoons unsalted butter
1 cup heavy cream, lightly whipped

Diced poached pears
Tiny fresh mint leaves

OPTIONAL GARNISH
White Chocolate–Mocha Sauce (page 488)

1. Start by macerating the raisins: Combine the raisins, rum, and water in a jar. Cover tightly and let marinate for at least 4 hours. Reserve.

2. To prepare the fondant: Line the bottom and long sides of a 3½-cup loaf pan with a single piece of parchment paper large enough to extend over each of the long sides by 3 inches. Lightly oil the parchment and set the pan aside.

3. Combine the chocolate and butter in a very heavy saucepan, place over low heat, and stir until the chocolate has completely melted and the mixture is smooth. Transfer to a medium mixing bowl and refrigerate for about 20 minutes or until the mixture just begins to thicken. Watch carefully and do not let the mixture set completely.

4. Add the whipped cream to the chocolate mixture and whisk quickly until just combined. Do not overbeat. Transfer the fondant to the prepared loaf pan, tap gently on a flat surface to remove air bubbles, and smooth the top evenly with a spatula. Cover and refrigerate for 6 hours or overnight.

5. Run a hot, wet knife between the fondant and pan, along each of the short sides, then carefully lift the fondant out of the mold with the aid of the parchment paper. Invert onto a serving platter and peel off the parchment. Chill until ready to serve.

6. Drain the raisins and place in a serving bowl.

7. To serve, cut the fondant into ⅓-inch slices using a hot, wet knife and place a slice in the center of each individual serving plate. Sprinkle with some rum raisins and surround with the optional white chocolate sauce and a few optional diced pears. Garnish with tiny mint leaves and serve at once.

Note: The rum raisins can be stored indefinitely in a tightly covered jar. Use them as a topping for ice cream, and in tarts, rice puddings, and apple crisps. You may also serve the fondant surrounded by a Coffee Crème Anglaise (page 490).

Warm Goat Cheese Timbales on a Bed of Greens with Honey-Cumin Vinaigrette

■

Marinated Roast Pork Tenderloins with White Bean and Garlic Puree

■

Caramelized Pear and Prune Flans

In winter I enjoy starting a meal with a salad that combines cool greens with a warm garnish, and now that interesting greens are available at any time of year, it's possible to do so. Here, the addition of a warm custard timbale adds a comforting note. The light flavor of goat cheese harmonizes well with the more assertive honey-cumin vinaigrette. Other soft cheeses such as Gorgonzola can be used instead of the goat cheese, and a julienne of prosciutto or sautéed, finely diced bacon would add further interest to this starter.

The marinated pork tenderloins are a quick and extremely versatile main course, allowing for a variety of accompaniments, but the white bean puree is by far one of my favorite side dishes. It is a forthright peasant dish that needs the intense, yet sweet taste of roasted garlic. It's likely that originally the puree was an inspired way to use leftover beans. Now the dish appears on the menus of chic restaurants, which are turning toward such old-fashioned dishes and finding them as good as ever. For a touch of color, I would add some sautéed broccoli rape, braised broccoli florets, or steamed snow peas tossed in a little lemon and anchovy butter.

This main course has the advantage of being rather light, so you'll have plenty of room left to enjoy an interesting cheeseboard and something sweet as well. The caramelized pear flans taste just as good when made with apples. A dollop of sweetened crème fraîche would further enhance this simple, yet flavorful dessert.

■ ■

WARM GOAT CHEESE TIMBALES ON A BED OF GREENS WITH HONEY-CUMIN VINAIGRETTE

serves 6

SALAD
2 tablespoons red wine vinegar
6–8 tablespoons olive oil
2 teaspoons pure uncooked honey
½ teaspoon ground cumin
Salt and freshly ground black pepper
2 cups each: radicchio, endive, Bibb lettuce, arugula, and green-leaf lettuce, leaves separated, washed, and dried

TIMBALES
8 ounces good-quality goat cheese, such as a Cornilly, young Crottin, or

Coach Farm
6 large eggs
4 tablespoons unsalted butter, softened
2 tablespoons freshly grated Parmesan cheese
2 teaspoons fresh thyme leaves, minced
Salt and freshly ground white pepper

GARNISH
Fresh thyme leaves
Tiny black oil-cured olives

1. Start by making the salad: In a large mixing bowl, combine the vinegar, oil, honey, and cumin. Whisk until well blended and season with salt and pepper. Top with the radicchio, endive, Bibb lettuce, arugula and green-leaf lettuce. Do not toss. Cover the salad and refrigerate until needed.

2. Preheat the oven to 350° F. Butter 6 to 8 (4-ounce) porcelain ramekins. Then line the bottoms with wax paper and butter the paper. Set aside.

3. To prepare the timbales: In the workbowl of a food processor, combine the goat cheese, eggs, and Parmesan and process until smooth. Add the thyme and season with salt and pepper to taste.

4. Pour the timbale mixture into the buttered ramekins, place them in a large baking pan, and pour enough boiling water into the pan to reach halfway up the sides of the ramekins. Place the pan in the center of the preheated oven and cover it with a sheet of buttered wax paper. Bake for 20 to 25 minutes, or until the tops are lightly browned and a knife, when inserted, comes out clean. When the timbales are done, remove from the oven and set aside until they are just warm.

5. Toss the greens with the vinaigrette at the bottom of the bowl and place some salad on each of 6 to 8 individual serving plates. Unmold a warm timbale in the center of each portion of salad. Garnish with the thyme and olives and serve at once.

MARINATED ROAST PORK TENDERLOINS
WITH WHITE BEAN AND GARLIC PUREE

serves 6

PORK

2 large cloves garlic, peeled and sliced
5 tablespoons olive oil
Juice of 1 large lemon
1 ½ tablespoons fresh thyme leaves
2 teaspoons cracked black pepper
2 whole pork tenderloins, trimmed,
    about 1 pound each
4 tablespoons unsalted butter
Coarse salt
1 ¼ cups Brown Chicken Stock (page
    467)

PUREE

2 cups Cooked White Beans plus ½ cup
    Bean Cooking Liquid (page 484)
5–6 tablespoons Aioli (page 477)
Salt and freshly ground black pepper

1 Beurre Manie (page 478)

GARNISH

1 ½ cups Cooked White Beans (page
    484) tossed with the juice of ½ lemon
    and 2 tablespoons finely minced fresh
    parsley

1. Start by marinating the pork: Combine the garlic, 4 tablespoons of oil, lemon juice, thyme, and cracked pepper in a small bowl and mix well. Place the pork tenderloins in a heavy zip-lock bag, pour the marinade over the pork, and seal the bag. Place the bag in a shallow dish and let stand at room temperature for 2 hours, turning often in the marinade.

2. Preheat the oven to 425° F.

3. Remove the tenderloins from the marinade, dry thoroughly with paper towels, and set aside.

4. Heat the remaining 1 tablespoon oil together with 1 tablespoon of butter in a large cast-iron skillet or baking dish over high heat. Add the tenderloins to the skillet and brown nicely on all sides. Season with coarse salt and add ¼ cup of the stock to the skillet. Place in the center of the preheated oven and roast for 16 to 20 minutes or until the juices run clear, adding a little stock and basting often with the pan juices. Do not overcook.

5. While the pork is in the oven, make the puree: In the workbowl of a food processor, combine the beans and bean cooking liquid and process until quite smooth. Add the aioli and process again until thoroughly combined. Season with salt and pepper and transfer to the top of a double boiler. Place over simmering water, cover, and keep warm.

6. When the tenderloins are done, remove from the oven and transfer to a carving board. Set aside.

7. Degrease the pan juices thoroughly and return them to the skillet together with the remaining stock. Bring to a boil, reduce slightly, and whisk in bits of beurre manie until the sauce lightly coats a spoon. Remove from the heat, add the remaining butter, and whisk until smooth. Taste and correct the seasoning.

8. Cut the pork crosswise into thin slices. Place some of the bean puree in the center of 6 warm individual serving plates. Top each with about 4 to 5 slices of pork. Spoon some of the sauce over each portion and garnish with some of the cooked beans. Serve at once.

CARAMELIZED PEAR AND PRUNE FLANS

serves 6

2 ripe pears, about 1 pound, preferably
　Anjou, peeled, quartered, and cored
2 tablespoons unsalted butter
3½ tablespoons granulated sugar
2 large eggs

1 large egg yolk
½ cup heavy cream
10 pitted prunes, diced
Large grating of fresh nutmeg

1. Preheat the oven to 350° F.

2. Cut each pear quarter crosswise into ¼-inch slices and set aside.

3. In a 10-inch cast-iron skillet, melt the butter over medium-high heat. Add the pear slices, sprinkle with 1½ tablespoons of the sugar, and cook for about 5 minutes without turning, or until the pears begin to brown. Reduce the heat to medium and cook until the pears are soft and nicely caramelized. Turn the pears often to avoid burning.

4. Combine the eggs, egg yolk, and remaining sugar and whisk until pale yellow. Add the heavy cream and whisk until well blended. Fold in the pears and prunes and season with the fresh nutmeg.

5. Pour the mixture evenly into 6 (4-ounce) ramekins. Place the ramekins in a large, shallow baking dish, fill the baking dish with enough boiling water to come halfway up the sides of the ramekins, and set the dish in the center of the preheated oven. Bake for 25 minutes, or until the flans are set. A knife inserted into the center of a flan should come out clean.

6. Remove the flans from the oven and serve warm.

Roasted Potatoes with Gruyère on a Bed of Greens

Cabbage Packages with Spiced Lamb and Rice Stuffing

Baked Apples in Brown Sugar and Cinnamon Wine

I've come to realize that a menu can dazzle by its simplicity. Instead of trying to duplicate restaurant food, which involves a lot of presentation and assembly, it often makes more sense for the home cook to choose from the wealth of comforting everyday food that is so satisfying, especially on a chilly winter evening.

Begin with a typically Swiss country salad topped with crisp roasted potatoes and dusted with a shower of nutty Gruyère. Serve it with crusty French bread and sweet butter and you are off to a good start.

The cabbage packages are a sophisticated, earthy main course, at their best when made a day or two ahead of time to allow the dish to develop its full flavor. A touch of Middle Eastern spices added to the sweet, piquant sauce imparts a resonance to the lamb stuffing. Do not attempt to make this dish unless you can get Savoy cabbage; white cabbage will not give you the same results. You may substitute a mixture of pork and veal for the ground lamb or make a savory stuffing of cooked rice, spinach, pine nuts, and raisins.

When making this dish, plan ahead so that you will have some leftovers. They taste wonderful and make a memorable appetizer served at room temperature with a light drizzle of virgin olive oil and a sprinkling of fresh lemon juice.

As an accompaniment to the cabbage, I usually opt simply for crusty bread or a red-pepper-and-parsley-flecked couscous, which is light, colorful, and in keeping with the rest of the meal.

I love the taste of a well-reduced spiced wine and serve it year-round with a variety of cooked and raw seasonal fruits. In fall or winter, it lends its rich color and spice to buttery baked apples. Be sure to use Rome Beauties or another good baking apple that will keep its shape and not burst during baking. You can vary this homey dessert by filling the cavities of the apples with diced prunes or macerated raisins.

ROASTED POTATOES WITH GRUYÈRE ON A BED OF GREENS

serves 4

SALAD

1 recipe Sherry and Lemon Vinaigrette
 (page 480)
¼ cup thinly sliced red onion
6 cups loosely packed mixed salad
 greens, including endives, radicchio,
 Bibb lettuce, and arugula, all leaves
 separated, washed, and dried

POTATOES

8 small red potatoes, quartered
2 tablespoons unsalted butter, melted
1 tablespoon olive oil
Coarse salt and freshly ground black
 pepper
2 teaspoons fresh thyme leaves, minced

½ cup finely diced slab bacon, blanched
½ cup finely shredded Gruyère cheese

1. Start by making the salad: Place the vinaigrette at the bottom of a large salad bowl, add the red onion, and top with the endives, radicchio, Bibb lettuce, and arugula, in that order, but do not toss. Cover and chill until serving.

2. Preheat the oven to 400° F.

3. To prepare the potatoes: Place the potatoes on a heavy baking sheet. Drizzle with 1 tablespoon of the melted butter and the olive oil and sprinkle with coarse salt, pepper, and thyme. Roast in the center of the preheated oven for 40 minutes, or until the potatoes are tender and nicely browned, turning once or twice during roasting.

4. While the potatoes are roasting, heat the remaining butter in a small skillet over medium heat. Add the bacon and cook until almost crisp. Remove with a slotted spoon to a side dish and reserve.

5. When the potatoes are done, remove from the oven and preheat the broiler.

6. Toss the salad gently and divide between 4 individual serving plates. Sprinkle with the reserved bacon and set aside.

7. Divide the potatoes into 4 equal portions on the baking sheet, placing the pieces of roasted potatoes quite close to one another within each portion. Sprinkle each portion with 2 tablespoons of the Gruyère, place in the broiler, 6 inches from the source of heat, and broil until the cheese has just melted.

8. Remove the potatoes from the broiler. Carefully remove the potatoes from the baking sheet, 1 portion at a time, using a large, flat spatula, and place on top of the salads. Serve at once.

CABBAGE PACKAGES WITH SPICED LAMB AND RICE STUFFING

serves 6

CABBAGE ROLLS
1/4 cup Italian rice, preferably Arborio
1 1/4 pounds ground lamb
1 teaspoon ground cumin
1/8 teaspoon cayenne
1 teaspoon dried oregano
1/8 teaspoon ground coriander
2 tablespoons minced fresh Italian
  parsley
1 large clove garlic, peeled and mashed
1 tablespoon grated raw onion
Salt and freshly ground black pepper
2 tablespoons water
12 large Savoy cabbage leaves,
  blanched

SAUCE
4 tablespoons olive oil

1 small dry red chili pepper
2 large Bermuda onions, peeled,
  quartered, and thinly sliced
5 large cloves garlic, peeled and finely
  minced
1/2 teaspoon ground cinnamon
1/2 teaspoon ground ginger
1 teaspoon ground cumin
1 teaspoon dried oregano
2 tablespoons dark-brown sugar
1 (2-pound) can Italian plum tomatoes,
  pureed but not strained
Salt and freshly ground black pepper

GARNISH
Braised bulgur or pilaf of rice

1. Start by filling the cabbage leaves: In a small saucepan bring 3/4 cup of salted water to a boil. Add the rice, reduce the heat, and simmer the rice, covered, for 20 minutes, or until tender and all the water has been absorbed.

2. In a large bowl combine the lamb, cooked rice, cumin, cayenne, oregano, coriander, parsley, garlic, and onion. Season the mixture with salt and pepper, add the water, and mix well. Reserve.

3. Place the cabbage leaves on a large wooden board and with a small, sharp knife, remove the center ribs. Place a heaping tablespoonful of stuffing at the bottom of each leaf and roll up, tucking in the sides and enclosing the filling completely. Reserve.

4. Preheat the oven to 350° F.

5. To prepare the sauce: In a large, deep skillet heat the oil over high heat, add the chili pepper and onions, and cook the mixture for 5 minutes, or until the onions start to brown. Add the garlic, lower the heat, and continue cooking for about 30 minutes, or until the onions are nicely browned and very soft. Add the cinnamon, ginger, cumin, oregano, and brown sugar and cook for another 2 to 3 minutes. Add the tomato puree and season with salt and pepper. Bring to a boil, reduce the heat, and simmer, uncovered, for 15 minutes. Remove and discard the chili pepper.

6. Place the rolls, seam side down, in the sauce, cover tightly, and reserve.

7. Place in the center of the oven and bake for 1 hour and 30 minutes, basting several times with the sauce. Uncover and continue baking until the pan juices are somewhat reduced, about 15 minutes. Transfer to a serving platter, spoon the sauce over them, and surround with braised bulgur.

BAKED APPLES IN BROWN SUGAR AND CINNAMON WINE

serves 4

4 large Rome Beauty apples	¾ cup granulated sugar
4 tablespoons unsalted butter	1 cinnamon stick, broken in half
2 tablespoons dark-brown sugar	2 whole cloves
2 teaspoons all-purpose flour	1 (2-inch) strip of lemon zest
1 cup water	4–5 whole black peppercorns
2½ cups dry red wine	Crème Fraîche (page 491)

1. Preheat the oven to 350° F.

2. Core the apples with an apple corer. With a sharp paring knife, cut ½ inch of peel off the top of each apple and set the apples aside.

3. In a large bowl, combine the butter, brown sugar, and flour and mix with your hands until well blended but still a bit coarse. Fill the cavity of each cored apple with some of this mixture and set aside.

4. Butter a rectangular or oval baking dish in which the apples will fit snugly. Add the stuffed apples and pour ½ cup water into the baking dish. Loosely cover the apples with foil and bake in the center of the preheated oven for 1 hour and 30 minutes, basting every 10 to 15 minutes with the pan juices.

5. While the apples are baking, prepare the spiced wine: Combine the remaining water, wine, granulated sugar, cinnamon stick, cloves, lemon zest, and peppercorns in a medium saucepan. Bring to a boil, reduce the heat, and simmer the wine until it is reduced by half and becomes syrupy. Discard the cloves, cinnamon stick, and lemon zest and set the spiced wine aside.

6. The apples are done when they can be pierced easily with the tip of a paring knife. Spoon the spiced wine over them, return to the oven, and bake, covered, another 10 minutes.

7. Transfer the apples to individual serving dishes and spoon some warm wine sauce over them. Serve warm with lightly sweetened crème fraîche.

RILLETTES OF SALMON WITH POTATO ROSTI

serves 8

Similar to the English potted salmon, this simple but delicious hors d'oeuvre can be turned into an impressive, festive starter to a meal. Serve the rillettes with Roasted Red Pepper Fritters (page 251), three fritters per person, drizzled with brown butter and each one topped with a teaspoon of the salmon rillettes.

RILLETTES
4 ounces fresh salmon fillet, preferably the center cut
Salt and freshly ground white pepper
¼ cup dry white wine
1 tablespoon finely minced shallots
4 tablespoons unsalted butter, softened
4 ounces smoked salmon
Juice of ½ lemon
Finely grated zest of ½ lemon
2–3 tablespoons finely minced fresh dill
1 tablespoon finely minced fresh chives
2 tablespoons Mayonnaise (page 477)

4 ounces salmon caviar, optional
4–6 tablespoons cool Clarified Butter (page 478)

ROSTI
4 medium baking potatoes, about 2 pounds, peeled and shredded
Salt and freshly ground black pepper
8 tablespoons unsalted butter
8 teaspoons olive oil

GARNISH
Sprigs of fresh dill
2 tablespoons finely minced fresh chives

1. Season the fresh salmon with salt and pepper.

2. In a small saucepan, combine the wine and shallots and bring to a boil. Add the seasoned salmon, reduce the heat, and simmer gently until the salmon is just opaque. Remove the saucepan from the heat and let the salmon cool completely in its poaching liquid. Remove the salmon from the poaching liquid, flake, and set aside.

3. In the workbowl of a food processor, combine the butter and smoked salmon and process until smooth. Transfer the mixture to a large mixing bowl and fold in the reserved flaked fresh salmon, the lemon juice, lemon zest, dill, chives, mayonnaise, and optional caviar. Taste and correct the seasoning.

4. Pack the rillettes into a crock, tap twice on the counter to remove any air

bubbles, and smooth the top evenly with a spatula. Pour the cool clarified butter over the surface to cover completely, and refrigerate the rillettes.

5. About 30 minutes before serving, remove the rillettes from the refrigerator and prepare the rosti: Place the shredded potatoes in a mixing bowl and season with salt and pepper.

6. It is preferable to cook the rosti in 2 (8-inch) nonstick skillets simultaneously. Melt 2 tablespoons of the butter with 2 teaspoons of oil in each of the skillets over medium heat. Place half of the shredded seasoned potatoes in each skillet, pressing them down lightly into an even layer and covering the bottom of each skillet completely. Cover and cook for 7 to 10 minutes, or until nicely browned. Regulate the heat so that the potatoes do not burn. Invert each onto a plate with the browned side faceup.

7. Melt another 2 tablespoons butter with 2 teaspoons oil in each of the skillets. Carefully slide the rosti back into each skillet, cover, and cook for 10 to 15 minutes longer, or until nicely browned on the outside and tender on the inside.

8. Carefully slide each rosti onto a cutting surface and cut each into 4 wedges. Place each wedge on an individual serving plate, top with a dollop of the rillettes, and garnish with sprigs of fresh dill and a sprinkling of fresh chives.

Note: You may also serve the rillettes, lightly chilled, right from the crock as a pâté with thinly sliced French bread.

Variation: Smoked haddock or smoked bluefish (a total of about 8 ounces) can be substituted for both the cooked and raw salmon in this preparation. Be sure to season the smoked fish pâté with plenty of fresh lemon juice, and add finely diced raw red onion instead of shallots.

SAUTÉ OF CALAMARI WITH CARAMELIZED ONIONS

serves 4

In Spain, very tiny calamari the size of fingernails are called cipirones, and their incredibly succulent taste is impossible to describe. This is a delicacy that can run as high as $25 a pound, but for aficionados the price is not a deterrent. Whenever cipirones are in season, I make a beeline for my favorite Catalan bistro where cipirones in onion marmalade are the specialty of the house. I take my time, savoring every bite and mopping up the sauce with crusty bread. If someone were to ask me to name my very favorite food, the cipirones in onion sauce would head my list.

Although there is no way to duplicate the taste of these choice Mediterranean morsels, I often make this dish with tiny fresh calamari with excellent results. Still, if you're ever in Barcelona, head for the Hostal San Giordi and sample the original. I think you will understand why so many Catalans are passionate about this simple dish.

4½ tablespoons extra-virgin olive oil
1 small dry red chili pepper, broken
3 large onions, peeled, quartered, and
 thinly sliced
Salt and freshly ground black pepper
1¼ pounds fresh calamari, including

tentacles, cleaned and cut crosswise
into ¼-inch slices

GARNISH
3 tablespoons finely minced fresh
 parsley

1. In a heavy 10-inch skillet, heat 3 tablespoons of the olive oil over medium-high heat. Add the chili pepper and cook until dark, about 30 seconds. Remove with a slotted spoon and discard.

2. Add the onions to the hot chili oil and cook for 5 minutes, stirring often, until the onions begin to brown. Reduce the heat, cover, and cook 30 to 40 minutes, or until the onions are nicely browned and caramelized, stirring from time to time. Season with salt and pepper and set aside.

3. In another 10-inch skillet, heat the remaining olive oil over high heat. Add the sliced calamari and sauté quickly, stirring constantly, for 30 seconds to 1 minute, or until the calamari just turn opaque.

4. Add the onions to the skillet of calamari, season with salt and pepper, and toss the mixture together. Set aside in the skillet for at least 30 minutes before serving.

5. Just before serving return the calamari and onions to low heat and just heat through. Taste and correct the seasoning. Sprinkle with minced parsley and serve hot with crusty French bread.

RAGOUT OF LOBSTER AND SCALLOPS WITH BROCCOLI AND ROASTED SHIITAKE MUSHROOMS

serves 6

At first glance this dish may seem expensive, but small lobsters are usually quite reasonable, and this ragout is well worth the expense. Serve the ragout as an appetizer or main course for a Sunday lunch. When dealing with something as delicate as scallops, it is best to serve them at the start of the meal or at an informal dinner, when you can take a few minutes out to do some last-minute cooking. Vary the ragout with garnishes such as fresh peas and pencil-thin asparagus spears in the spring, a tomato concassé in the summer, and a julienne of braised leeks or Belgian endives in the fall.

VEGETABLES
18 medium fresh shiitake mushrooms, stemmed
1 tablespoon olive oil
Coarse salt and freshly ground white pepper
2 cups fresh broccoli florets

SAUCE
6 tablespoons finely minced shallots
¼ cup plus 2 teaspoons Noilly Prat vermouth
¼ cup plus 2 teaspoons Chablis
2 cups heavy cream

LOBSTER AND SCALLOPS
2 live lobsters, about 1½ pounds each
Salt and freshly ground white pepper
2 tablespoons unsalted butter
1 teaspoon olive oil
1 pound fresh bay or small sea scallops

8 tablespoons unsalted butter, lightly chilled

GARNISH
Tiny fresh chervil or cilantro leaves
18 snow peas, steamed

1. Preheat the oven to 400° F.

2. Start by roasting the shiitakes: Place the mushrooms, gill side down, on a small, heavy baking sheet. Drizzle with the oil, season with coarse salt and pepper, and bake in the center of the preheated oven for 15 minutes. Remove the mushrooms from the oven, and lower the oven heat to 350° F. Transfer the mushrooms to a cutting board, quarter, and set aside.

3. Bring water to a boil in a vegetable steamer, add the broccoli, and steam, covered, for about 4 minutes, or until just tender. Remove from the steamer and reserve.

4. To prepare the sauce: In a large saucepan, combine the shallots, ¼ cup

of the vermouth, and ¼ cup of the Chablis. Place over high heat, bring to a boil, and reduce to a glaze. Add the heavy cream, reduce the heat, and simmer slowly until reduced by one third. Strain the sauce into a bowl and return it to the saucepan. Set aside.

5. To prepare the lobster: Bring plenty of salted water to a boil in a large pot. Plunge the lobsters into the water and cook for 3 minutes. Drain well and transfer to a cutting board. Cut the lobsters in half lengthwise, and sprinkle with salt and pepper on the cut side (the side exposing the flesh).

6. Melt the 2 tablespoons butter together with the oil in a large, heavy skillet or baking dish. Remove from the heat, add the lobster halves, cut side down, and place in the center of the preheated oven. Roast for 5 minutes, turning often, until the lobster is barely done. (The lobster should still be a bit under-cooked in the center, since it will be reheated in the sauce.)

7. Remove the lobster from the oven, remove and discard the shells, and cut the meat into ¾-inch pieces. Reserve.

8. Add the reserved cream mixture to the skillet or baking dish, place over medium heat, and cook for 1 minute, scraping the bottom of the pan well. Transfer to a saucepan and return to medium heat. Add the scallops and cook for 30 seconds. Remove with a slotted spoon and set aside.

9. Reduce the heat to low and add the remaining vermouth and Chablis. Whisk in the 8 tablespoons of lightly chilled butter, 1 tablespoon at a time, until thoroughly incorporated and the sauce is smooth.

10. Add the broccoli florets and reserved lobster and scallops, fold gently, and just heat through. Taste and correct the seasoning. Serve immediately in shallow soup bowls, sprinkled with chervil or cilantro and garnished with a few steamed snow peas.

SCALLOP FLANS À L'ARLESIENNE IN A ROASTED PEPPER AND SHALLOT BUTTER SAUCE

serves 8

It's almost a cliché to say that before the advent of the food processor, certain categories of recipes required so much skill and dedication that they were only rarely attempted by the home cook. Among these is seafood mousse, which took a lot of technique and was a formidable recipe. Now it takes only a few minutes to make in the processor. Unfortunately, a food processor is just a tool, and it is up to the cook to have an understanding of the ingredients and to know the art of proper seasoning.

These light, airy timbales are studded with tiny cubes of colorful vegetables and look especially appealing surrounded by a vivid red pepper sauce. You can also bake the mousse in a loaf pan and serve it as a hot or room-temperature pâté, accompanied by a stir-fry of spinach or escarole, or a light roasted pepper vinaigrette.

FLANS
3 tablespoons unsalted butter
¼ cup finely diced carrots
¼ cup finely diced zucchini, skin only
¼ cup finely diced leeks, well rinsed
⅓ cup water
Salt and freshly ground white pepper
1 cup sea scallops
3 large eggs
1¾ cups cold heavy cream
Pinch of cayenne pepper

SAUCE
12 tablespoons unsalted butter, very cold

2 tablespoons finely minced shallots
¼ cup sherry vinegar or white wine vinegar
2 tablespoons cold heavy cream
2 medium red bell peppers, roasted, peeled (page 483), and coarsely chopped
2 tablespoons finely minced fresh chives
Salt and freshly ground white pepper

GARNISH
Finely minced fresh parsley and/or fresh chives

1. Preheat the oven to 350° F.

2. Butter 8 (4-ounce) porcelain ramekins or French timbale molds. Line the bottoms with parchment paper, then butter the paper. Set aside.

3. To prepare the flans: In a medium skillet melt the butter over medium heat. Add the carrots, zucchini, leeks, and water. Season with salt and pepper, cover, and simmer until tender. Drain the vegetables and set aside.

4.   In the workbowl of a food processor, combine the scallops and eggs and process until very smooth. With the machine running, add the cream in a slow, steady stream until it is absorbed by the scallop mixture. Transfer to a large bowl, season with salt, pepper, and a pinch of cayenne, and fold in the cooked vegetable mixture. Pour the mixture into the prepared molds. Place the molds in a shallow baking dish and add enough boiling water to the baking dish to come halfway up the sides of the molds. Place the baking dish in the center of the preheated oven and bake for 30 to 40 minutes or until the flans are set.

5.   While the flans are baking, prepare the sauce: Remove the butter from the refrigerator, cut into ½-inch pieces, and let stand for 20 minutes.

6.   In a heavy-bottomed saucepan, preferably copper or enameled cast-iron, combine the minced shallots and vinegar. Bring to a boil over high heat and continue to boil, whisking often, until the liquid has reduced to about 1 teaspoon.

7.   Reduce the heat, add the heavy cream, and reduce slightly. Add the butter, 1 piece at a time, whisking constantly until the sauce is smooth and very creamy and lightly coats a spoon. Remove from the heat and set aside.

8.   In the workbowl of a food processor add the chopped roasted peppers and one third of the butter sauce. Process until smooth. Whisk the red pepper puree and minced chives into the remaining butter sauce. Season with salt and pepper and transfer to the top of a double boiler. Set over lukewarm water, cover, and reserve.

9.   When the flans are done, invert them onto a platter or individual serving plates, peel off the parchment, and discard. Surround the flans with some of the red pepper butter sauce, garnish with a sprinkling of finely minced parsley and/or chives, and serve at once.

OLD-FASHIONED BURGUNDIAN ESCARGOT TART

serves 6

Burgundy is known for its snails. The famous escargots à la bourgui-gnonne, in which the snails are poached and packed back into their shells with a flavorful herb and garlic butter, has become a classic preparation featured on restaurant menus throughout the world. Snails have now been liberated, and many interesting appetizers have been created, com-bining them with wild mushrooms, intense red wine–butter sauces, and even using them as fillings for the chic "beggars purses." This tart, how-ever, is a traditional farmhouse recipe that I seek out in local pastry shops whenever I am in the region. It is extremely easy to make and well worth trying at home. Serve the tart as an appetizer followed by a light main course such as veal scaloppine, sautéed calves' liver, or roast squabs.

1 can (1 dozen) large escargots, drained and quartered, about 4½ ounces
2 cups Chicken Stock (page 470) or bouillon
1 Bouquet Garni (page 485)
4 tablespoons unsalted butter
2 tablespoons finely minced shallots
3 large cloves garlic, peeled and finely minced

2 tablespoons finely minced fresh parsley
1½ cups heavy cream
2 large eggs
2 large egg yolks
Salt and freshly ground black pepper
1 partially baked 10-inch Basic Tart Shell (page 492) in a porcelain quiche pan
2 tablespoons freshly grated Parmesan cheese

1. Preheat the oven to 350° F.

2. Place the drained escargots in a small saucepan together with the chicken stock and bouquet garni. Bring to a boil, reduce the heat, and simmer for 20 minutes. Drain and reserve the escargots.

3. In a small, heavy skillet, melt the butter over low heat. Add the shallots and garlic, and cook for about 2 minutes, or until soft but not browned. Add the reserved escargots and parsley and toss with the shallots and garlic for 2 minutes. Set aside.

4. Combine the heavy cream, eggs, and egg yolks in a large mixing bowl. Season with salt and pepper and whisk until well blended. Add the escargot mixture, including the butter and herbs, and spoon into the partially baked tart

shell, dispersing the escargots evenly in the custard. Sprinkle with the Parmesan and bake in the center of the preheated oven for 35 minutes, or until the top is nicely browned and the custard is set.

5. Remove from the oven and let cool for 5 minutes. Serve hot, cut into wedges.

SPAGHETTINI WITH CLAMS IN BRAISED FENNEL AND SAFFRON CREAM

serves 5–6

Saffron, which was once limited to three classic preparations—the Spanish paella, the French bouillabaisse, and the Italian saffron risotto that traditionally accompanies osso buco—has emerged as one of the key spices in the last decade. It has a particular affinity for seafood and rice, but it is also wonderful in sauces that accompany vegetables. Here, the intensely flavored spice is mellowed by heavy cream and, combined with reduced clam broth, it creates a rich and flavorful sauce that is equally good folded into a risotto or spooned around grilled or sautéed scallops, shrimp, or salmon fillets.

1¼ cups heavy cream
½ teaspoon saffron threads
8 tablespoons unsalted butter
1 large fennel bulb, trimmed of all greens and stalks and cut into fine julienne
Salt and freshly ground black pepper
2 tablespoons water
1 Bouquet Garni (page 485)
1 teaspoon fennel seeds
6 whole black peppercorns
¼ cup dry white wine

¼ cup Noilly Prat vermouth
2 large cloves garlic, peeled and crushed
2 tablespoons finely minced shallots
20 littleneck clams, well scrubbed and rinsed
1 pound imported spaghettini, preferably De Cecco
Drops of Pernod, optional

GARNISH
Finely minced fennel greens or fresh Italian parsley

1. In a medium saucepan, combine the heavy cream and saffron, bring to a boil, and immediately reduce the heat. Simmer for 10 to 15 minutes, or until reduced to 1 cup. Strain into a bowl and set aside.

2. Melt 2 tablespoons of the butter in a 10-inch skillet over medium heat.

Add the fennel and toss gently in the butter. Season with salt and pepper and add the water. Bring to a boil, reduce the heat, and braise, covered, for 6 to 8 minutes, or until just tender. If all the water has not evaporated, remove the cover, raise the heat, and cook, stirring often, until the juices have evaporated. Reserve.

3.   Place the bouquet garni, fennel seeds, and peppercorns in the center of an 8-inch square of cheesecloth. Gather up the sides to enclose the herbs and spices completely. Tie with a piece of kitchen string and set the sachet aside.

4.   In a large, heavy casserole, combine the white wine, vermouth, garlic, shallots, sachet, and clams. Bring to a boil, reduce the heat, and cook, covered, shaking the pan back and forth and transferring the clams with a slotted spoon to a bowl as they open. Continue to cook until all the clams have opened. Discard those that do not.

5.   Remove all but 6 clams from their shells. Discard the shells and dice the clams. Reserve the diced and unshelled clams in a medium bowl with a little of their cooking liquid.

6.   Strain the clam cooking liquid through a double layer of cheesecloth into a large, heavy skillet. Place over high heat, bring to a boil, and reduce slightly. Add the reserved saffron cream, bring back to a boil, and again reduce until slightly thickened. Remove from the heat and add the remaining butter, 2 tablespoons at a time, whisking until the sauce is smooth and velvety. Taste and correct the seasoning, adding a large grinding of pepper, and return to very low heat. Keep warm.

7.   Bring plenty of salted water to a boil in a large pot, add the spaghettini, and cook for 3 to 4 minutes, or until al dente. Immediately add 2 cups cold water to the pot to stop further cooking and drain well.

8.   Add the spaghettini to the skillet of warm sauce together with the cooked fennel julienne and diced clams. Toss gently to just heat through. Taste and correct the seasoning. If you want a more pronounced fennel taste, add a few optional drops of Pernod.

9.   Serve at once in individual shallow soup bowls. Sprinkle each portion with some minced fennel greens or parsley and garnish each portion with 1 whole clam in its shell.

Note: This dish is even better when made with mussels. Cultivated Maine mussels are now widely available and are far less expensive than clams. If you plan to use only the sauce (and not the mussels) with grilled or sautéed seafood, store the shelled mussels, covered, with some of their broth in the refrigerator and serve with a quick tomato sauce folded into spaghettini the next day.

ORECCHIETTE WITH KALE, SAUSAGE, AND GARLIC CREAM

serves 4–5

Here is a hearty pasta dish that can easily double as a light main course. If you plan to serve it as an appetizer, follow with a light main course such as a pan-seared salmon fillet, sautéed shrimp simply tossed with garlic and parsley, or pan-grilled chicken breasts. The charming shape of the orecchiette makes for a nice change of pace, but if you cannot get them, penne and fusilli both work well in this dish. Don't shy away from kale. Just be sure that you buy this delicious winter green when it is truly fresh and use it the same day since it tends to lose flavor quickly.

1 pound fresh kale
Salt
6 tablespoons unsalted butter
1½ cups heavy cream
Freshly ground black pepper
2 tablespoons olive oil

1 pound sweet Italian fennel sausage
¾ pound orecchiette pasta, preferably imported
2 large cloves garlic, peeled and finely minced
½ cup freshly grated Parmesan cheese

1. Remove and discard the stems and large ribs from the kale. Tear the leaves into 1½-inch pieces. Bring plenty of salted water to a boil in a large pot, add the kale leaves, and cook for 2 minutes. Drain thoroughly and press down lightly with the back of a spoon to remove all the moisture. Reserve.

2. In a large, heavy skillet melt 4 tablespoons of butter over medium heat and whisk until a light hazelnut brown. Whisk in the heavy cream, immediately reduce the heat to low, and simmer until the mixture is reduced by a third. Season with salt and pepper and keep warm.

3. Heat 1 tablespoon of olive oil in a medium skillet over medium-high heat. Add the sausage and cook for about 5 minutes, or until nicely browned on all sides. Remove from the skillet and drain on paper towels. Cut the sausage crosswise into ⅓-inch slices and set aside.

4. Bring plenty of salted water to a boil in a large pot. Add the remaining olive oil and orecchiette and cook for 10 to 12 minutes, or until al dente. Immediately add 2 cups of cold water to stop further cooking and drain the pasta thoroughly.

5. While the pasta is cooking, melt the remaining butter in a large, heavy skillet. Add the garlic and kale and cook for 2 minutes, or until just heated through. Add the warm cream sauce, sliced sausage, and cooked pasta. Toss

gently and cook 2 minutes, or until the sauce lightly coats the pasta. Add ⅓ cup Parmesan. Taste and correct the seasoning, adding a large grinding of pepper.

6. Sprinkle with the remaining Parmesan and serve hot, accompanied by a crusty loaf of Italian bread.

RISOTTO WITH CREMINI MUSHROOMS AND SWEET GORGONZOLA

serves 4

No matter how creative you want to be, there are certain magic combinations that are hard to beat. A risotto with porcini mushrooms and a touch of cream is simply heavenly, and so is the currently popular one that combines the rice with four cheeses. Here is a version that is a lovely addition to your risotto repertoire, one that combines the good flavors of both risotti. For a more intense flavor, poach ½ ounce of dark imported Chilean mushrooms in 1 cup of the chicken broth and use both the diced reconstituted mushrooms and the intense broth to make the risotto.

5 tablespoons unsalted butter
1 medium shallot, peeled and finely minced
½ pound Cremini mushrooms, stemmed and diced
Salt and freshly ground white pepper
1 cup Italian rice, preferably Arborio

3–3½ cups Chicken Stock (page 470) or bouillon
½ cup heavy cream
3 tablespoons mild Gorgonzola
⅓ cup freshly grated Parmesan cheese
2 tablespoons finely minced fresh Italian parsley
½ cup cooked fresh peas, optional

1. In a 3-quart saucepan melt 3 tablespoons of the butter over low heat, add the shallot, and sauté for about 2 minutes, or until just soft. Add the diced mushrooms and continue to sauté for 5 minutes, or until the mushrooms are tender. Season with salt and pepper. Add the rice and toss well in the mushroom mixture. Add 2 cups of stock and bring to a boil. Reduce the heat and simmer, covered, for 15 minutes.

2. While the rice is cooking combine the heavy cream and Gorgonzola in the workbowl of a food processor and process until smooth. Reserve.

3. After 15 minutes, uncover the rice and raise the heat slightly. Start adding the remaining stock, ¼ cup at a time, stirring constantly, and continue to cook the rice for about 10 minutes, or until soft but still slightly chewy.

4. Add the Gorgonzola cream and 2 tablespoons of Parmesan. Season the risotto with salt and pepper, and add the remaining butter, parsley, and the optional peas. Taste and correct the seasoning.

5. Serve immediately in individual shallow soup bowls sprinkled with the remaining Parmesan.

Note: A good-quality Gorgonzola is of great importance in this dish. All supermarket blue cheese will be too salty, as would most Gorgonzolas. However, if you can find Bleu de Bresse or another mild, creamy blue cheese, reduce the amount used to 2 tablespoons and add 2 tablespoons of butter for enrichment.

Cooking with Blue Cheeses

A small amount of blue cheese can turn a risotto, soup, or salad into a memorable dish. But buying the right blue is not always easy. Supermarket Roquefort, for instance, whether domestic or imported, is very often too salty, but because it is sold in packaged, prewrapped pieces, there is no way to tell until you get it home. I prefer to buy blue cheeses at a reputable cheese store, one with a large turnover, where the cheese is handled properly and where you can ask for a taste.

My favorite blues for cooking are Gorgonzola Dolce or Gorgonzola Naturale among the Italian cheeses; English Stilton; and Fourme d'Ambert, Bleu d'Auvergne, and Bleu de Bresse among the French. I rarely find a Roquefort that is mild enough for cooking, but it does work well when combined with other mild cheeses such as Fontina and cream chese.

■ ■ ■

RISOTTO VENEZIANA

serves 4–6

Seafood risotto is the signature dish of Venice. There, it comes in many variations, some with shrimp, others with mussels, clams, or lobster or a combination of seafood. A full-bodied fish broth will, of course, turn a good risotto into a great one, but you can achieve excellent results with packaged fish bouillon, in which you have poached the shrimp shells with a carrot, a celery stalk, and an herb bouquet consisting of some fresh thyme, a few parsley sprigs, and six to eight black peppercorns.

¼ cup finely minced shallots
½ cup dry white wine
12–16 small littleneck clams, scrubbed well and rinsed
½–¾ pound small fresh shrimp, unpeeled
2 tablespoons extra-virgin olive oil
1½ cups Italian rice, preferably Arborio

Salt and freshly ground white pepper
4 cups Fish Stock (page 472), fish bouillon, or Chicken Stock (page 470)
½ cup heavy cream
2 large egg yolks
Juice of ½ lemon

GARNISH
½ cup finely minced fresh parsley

1. In a medium saucepan, combine 2 tablespoons of the shallots, the wine, and the clams. Cover and steam over medium heat for 4 or 5 minutes, or until all the clams open. Discard any clams that do not. Remove the clams with a slotted spoon and reserve.

2. Add the shrimp to the saucepan and cook until just pink. Transfer the shrimp with a slotted spoon to a side dish. Remove and discard the shells of both the clams and the shrimp, and dice both shellfish. Set aside. Strain the broth through a double layer of cheesecloth into a small bowl and keep warm.

3. In a heavy 3-quart casserole, heat the oil over medium heat. Add the remaining shallots and cook for 1 to 2 minutes, or until just soft. Add the rice, season lightly with salt and pepper, and toss to coat with the oil and shallots. Add the clam-shrimp broth and stir constantly until all the liquid has been absorbed by the rice.

4. Add ½ cup of stock or bouillon and continue stirring and cooking until it has been absorbed by the rice. Continue adding ½ cup of stock at a time until the rice is tender on the outside but still chewy on the inside, about 30 minutes.

5. Combine the heavy cream, egg yolks, and lemon juice in a small bowl and whisk until well blended.

6. Add the diced shellfish and lemon-cream mixture to the rice and fold

gently. Just heat through, but do not allow the mixture to come to a boil. Taste and correct the seasoning, adding a large grinding of pepper, and serve at once garnished with minced parsley.

Note: Chances are the clam broth will be quite salty, so do not salt the risotto until the very end. If you are going to use a fish bouillon cube, dilute it with more water than is called for in the directions so that the broth is not too salty.

GOAT CHEESE GNOCCHI WITH PINK ALFREDO SAUCE

serves 6

I first sampled these delicate goat cheese gnocchi at Alain Ducasse's restaurant at the Hôtel de Paris in Monte Carlo. There the gnocchi were served as a garnish to a cool cream of broccoli soup. I found that the broccoli somewhat overpowered these delicate dumplings and have since served them as an appetizer accompanied by either a fresh basil-flavored tomato sauce or a pink Alfredo sauce. You may add finely minced fresh thyme, basil, or a mixture of herbs to the sauce.

GNOCCHI

8 ounces goat cheese, preferably Coach Farm; do not use Montrachet
1 large egg
1 large egg yolk
⅓ cup all-purpose flour
1 tablespoon freshly grated Parmesan cheese

SAUCE

5 tablespoons unsalted butter
1 cup heavy cream
Salt and freshly ground white pepper
3 tablespoons Quick Tomato Sauce (page 474)
3–4 tablespoons freshly grated Parmesan cheese
2–3 tablespoons finely minced fresh thyme, optional

1. Start by making the gnocchi: In a large bowl combine the goat cheese, egg, yolk, flour, and Parmesan and blend together with a wooden spoon until the mixture has the consistency of cream cheese. Cover and chill for at least 2 to 3 hours, or until very cold.

2. While the goat cheese mixture is chilling, prepare the sauce: In a large, heavy skillet melt the butter over medium heat and whisk constantly until the butter becomes a hazelnut brown. Do not burn. Immediately add the heavy cream, whisking constantly until the mixture comes to a boil. Reduce by one quarter. Season with salt and pepper. Add the tomato sauce and Parmesan and whisk until well blended. Set aside.

3. Place a large bowl of ice water near the work surface. Reserve.

4. In a large, shallow saucepan, bring salted water to a boil and reduce to a simmer. With a tablespoon in one hand, take some of the very cold goat cheese mixture and press it into the spoon, using the side of the bowl. With a teaspoon in your other hand, remove a small amount of the batter from the tablespoon, sliding it off the tablespoon into the simmering water. When the gnocchi float to the surface, remove them immediately with a slotted spoon and plunge them into the bowl of ice water. Continue making the gnocchi with the remaining goat cheese mixture, cooking them in batches, if necessary.

5. When all the gnocchi have been cooked, carefully drain, place on a large side dish, and reserve.

6. When ready to serve, reheat the sauce in a large skillet over low heat. Slide the gnocchi into the sauce and just heat through. Taste and correct the seasoning. Transfer to warm individual serving dishes, garnish with optional thyme, and serve at once.

DEVON VELOUTÉ OF CELERY AND STILTON CHEESE

serves 6–8

I had heard of Gidleigh Park through a friend who raved about an American-owned inn at the edge of the Moors in England. The area had just suffered a major storm, and most of the electricity had been knocked out, so I did not expect much in the way of food. I was happily surprised by the fresh and creative dishes the kitchen turned out. I spent the next two days in the rather rudimentary kitchen of this lovely inn and several of their dishes have since become my favorites.

8 ounces Stilton cheese
4 tablespoons unsalted butter
6 large stalks celery, diced (about 3 cups)
1 large onion, peeled and finely minced
Salt and freshly ground white pepper
7 cups Chicken Stock (page 470)
1½ tablespoons all-purpose flour

1¼ cups heavy cream

GARNISH
1 medium leek, trimmed of all but 1 inch of greens, cut into fine julienne and well washed
1 medium carrot, peeled, trimmed, and cut into fine julienne
1 large stalk celery, cut into fine julienne

1. Trim the Stilton of its outer yellow-brown crust and use only the center. Crumble the cheese and set aside.

2. In a large, heavy casserole, melt 2 tablespoons butter over medium heat, add the diced celery and onion, and season with salt and pepper. Partially cover and cook for about 5 minutes, or until the onion is just soft. Add the stock and bring to a boil. Reduce the heat and simmer, covered, for 25 minutes, or until the vegetables are very tender.

3. Strain the soup through a colander set over a large bowl and reserve the vegetables and stock separately.

4. Transfer the cooked vegetables to the workbowl of a food processor, add ½ cup of the reserved stock, and process until very smooth. Whisk the pureed vegetables into the bowl containing the remaining reserved stock and set aside.

5. Return the casserole to medium heat. Melt the remaining butter, add the flour, and whisk until well blended. Cook for 1 to 2 minutes without browning. Add the pureed soup and whisk until well blended. Bring to a boil, reduce the heat, and simmer for 15 minutes.

6. In the workbowl of a food processor, combine the heavy cream and Stilton and process until smooth. Whisk the cream-cheese mixture into the soup, stirring until well blended, and just heat through. Do not bring to a boil.

7. Add the julienne of leek, carrot, and celery and just heat through. Taste and correct the seasoning, adding a large grinding of pepper. Serve hot.

Note: Stilton is a wonderful cheese that can be added to other vegetable soups, in particular those made with celery root, leeks, and potatoes or turnips and carrots. For a more intense celery taste, add a medium peeled and diced celery root to this soup.

CATALAN KALE, POTATO, AND GARLIC SOUP

serves 6

16 large cloves garlic, peeled
3 tablespoons unsalted butter
2 teaspoons olive oil
1 large Spanish onion, peeled and finely minced
6–8 cups Chicken Stock (page 470)

1½ pounds kale, stemmed, washed, and torn into 1-inch pieces
2 medium all-purpose potatoes, peeled and finely diced
Salt and freshly ground white pepper
½ cup heavy cream, optional

1. Bring water to a boil in a small saucepan, add the garlic cloves, and blanch for 2 minutes. Drain and set aside.

2. In a 4-quart casserole melt the butter with the oil over medium heat. Add the onion and cook for 3 or 4 minutes, or until soft but not brown. Add the stock, kale, diced potatoes, and blanched garlic cloves and season with salt and pepper. Bring to a boil, reduce the heat, partially cover, and simmer for 25 to 30 minutes, or until the vegetables are tender.

3. With a slotted spoon, remove about 2 cups of vegetables from the broth, excluding the garlic cloves, and reserve. Strain the soup into a large bowl. Transfer the drained vegetables to the workbowl of a food processor or blender and proccess until smooth.

4. Whisk the pureed vegetables into the strained broth and return to the casserole. Add the reserved 2 cups of cooked vegetables and the optional heavy cream. Bring to a boil, reduce the heat, and simmer for 10 minutes. Taste and correct the seasoning and serve hot, accompanied by a crusty loaf of French bread.

CREAM OF LEEK, POTATO, LEMON, AND DILL SOUP

serves 6

3 tablespoons unsalted butter
4 large leeks, trimmed of all but 2 inches of greens, thinly sliced and well rinsed
3 medium all-purpose potatoes, peeled and cubed
Salt and freshly ground white pepper
6 cups Chicken Stock (page 470)
½–¾ cup heavy cream
Juice of 1 large lemon

GARNISH
2 tablespoons unsalted butter
2 large leeks, trimmed of all but 2 inches of greens, cut into fine julienne and well rinsed
Salt and freshly ground white pepper
2–3 tablespoons water

3–4 tablespoons finely minced fresh dill
Tiny sprigs of fresh dill

1. Melt the butter in a large casserole over medium heat. Add the leeks and potatoes. Season with salt and pepper and toss the vegetables in the butter for 2 to 3 minutes.

2. Add the stock to the casserole. Bring to a boil, reduce the heat, and simmer the soup, covered, for 45 minutes, or until the vegetables are very tender.

3. In a small bowl, combine the heavy cream and lemon juice. Whisk until well blended and set aside.

4. While the soup is simmering, prepare the garnish: In a medium skillet melt the butter over medium-low heat, add the julienne of leeks, and season with salt and pepper. Toss the leeks in the butter for 1 minute. Add 2 tablespoons of the water, cover, and braise the leeks until tender, about 3 to 5 minutes. If all the water has evaporated before the leeks are tender, add the remaining water. If all the water has not evaporated and the leeks are tender, uncover and cook until the water has evaporated. Reserve.

5. Strain the soup through a colander set over a bowl. Transfer the vegetables to the workbowl of a food processor and process until smooth. Add the pureed vegetables to the reserved broth and whisk until smooth. Return the soup to the casserole, add the reserved lemon-cream mixture and the minced dill, and just heat through. Taste and correct the seasoning and serve hot in individual soup bowls, garnished with the reserved julienne of leeks and tiny sprigs of fresh dill.

CREAM OF LENTIL SOUP WITH GARLIC, BACON, AND SMOKED TROUT

serves 5–6

I don't know of a food that has so consistently been taken for granted as the lentil. It is also amazing that this inexpensive staple is used in almost every part of the world and yet, over the years, I have only come across a few memorable preparations. At home, lentils were part of our everyday kitchen since they can be bought already cooked in every Spanish market. So it wasn't unusual for my mother to add one or two cups to a variety of soups. When I first came across this unusual soup in a restaurant in southern Holland, I wondered why I had not thought about this natural combination of ingredients and flavors before.

3 tablespoons unsalted butter
2 medium cloves garlic, peeled and finely minced
2 medium onions, peeled and finely minced
1½ cups dried lentils
1 (8-ounce) piece of slab bacon, cut in half and blanched

7–8 cups Chicken Stock (page 470)
1 large sprig of fresh thyme or 1 teaspoon dried
Salt and freshly ground white pepper
1 cup heavy cream

GARNISH
6 ounces smoked trout, shredded
Tiny fresh Italian parsley leaves

1. In a 4-quart casserole, melt the butter over medium-low heat. Add the garlic and onions and cook for about 5 minutes, or until soft but not browned. Add the lentils to the casserole together with the bacon, stock, and thyme and season with salt and pepper. Bring to a boil, reduce the heat, and simmer,

covered, for 1 hour and 15 minutes to 1 hour and 30 minutes, or until the lentils are very tender.

2. Pour the lentils into a colander set over a large bowl and reserve the lentils and the cooking liquid separately. Remove the bacon, dice it, and set aside.

3. Transfer the lentils to the workbowl of a food processor and puree until very smooth. Return the pureed lentils to the casserole, add the reserved cooking liquid, bacon, and heavy cream, and place over medium-low heat. Simmer for 10 minutes and correct the seasoning.

4. Serve the soup hot, in individual bowls, each sprinkled with some smoked trout and tiny parsley leaves.

Note: Dress the lentils in a vinaigrette or serve them as a warm puree accompanying grilled sausages. The bacon can be omitted from the soup completely, or you may use smoked pork shoulder butt instead.

A SPICY MIDDLE EASTERN SOUP OF LAMB SHANKS AND LENTILS

serves 6

A full-bodied lamb stock is easy to make and it opens the way to a large variety of interesting and intensely flavorful soups and sauces. The stock is made with lamb shanks and is prepared much like a chicken stock. Once cooked, the shanks can be boned, diced, and added to the soup or combined with rice and seasonings into a tasty stuffing for cabbage leaves. The well-degreased broth adds a depth of flavor to the pan juices of a roast leg of lamb or simply pan-sautéed lamb chops. Leftover broth can be frozen and used in many soups such as this one or any winter legume soup.

Other legumes, particularly pinto beans, black beans, red kidney beans, and white Great Northern beans, can be substituted for the lentils. Remember that all legumes other than lentils and split peas must be pre-soaked overnight. Almost as thick as a stew, this hearty soup doubles as a one-dish meal when served with a loaf of good bread. Start with a well-seasoned salad and finish with a lemon rice pudding or Baked Apples in Brown Sugar and Cinnamon Wine (page 381).

4 whole lamb shanks (about 4 pounds,
   in all), cut into 2-inch pieces
2 large carrots, peeled and cut in half
2 large stalks celery, including leaves,
   cut in half
2 medium leeks, including all but 2
   inches of greens, well rinsed
1 bay leaf
1 parsley root, including greens, or 2
   large sprigs of Italian parsley, optional
6 whole black peppercorns
1 teaspoon dried thyme
Large pinch of salt
10–12 cups water

3 tablespoons olive oil
2 small dry red chili peppers, broken
1 large Spanish onion, peeled and finely
   minced
2 large cloves garlic, peeled and finely
   minced
1 large stalk celery, diced
1 large carrot, peeled and diced
1 medium turnip, peeled and diced
1 (2-pound) can Italian plum tomatoes,
   drained and chopped
1 teaspoon dried oregano
Salt and freshly ground black pepper
¼ cup tiny pasta, preferably imported
   tubettini
2 cups cooked lentils (page 433)

1. Start by making the stock: Combine all the ingredients in a large stockpot and cover with the water. Bring to a boil, reduce the heat, and simmer, partially covered, for 1½ to 2 hours, or until the lamb is tender but not falling apart. Skim the stock carefully from time to time. When the stock is done, strain it into a large bowl and discard the vegetables. Cool completely. You should have about 9 to 10 cups.

2. Remove meat from the shanks in bite-size pieces; cover and refrigerate.

3. Refrigerate the stock overnight. The next day degrease the stock thoroughly, bring to a boil, and set aside.

4. To prepare the soup: In a large casserole heat the olive oil over medium-low heat. Add the chili peppers and cook for about 1 minute, or until dark. Remove and discard. Add the minced onion and garlic and cook until soft but not brown. Add the diced celery, carrot, turnip, tomatoes, and oregano, season with salt and pepper, and simmer for 10 minutes, stirring often.

5. Add 6 cups of the reserved stock. Bring to a boil, reduce the heat, and simmer, partially covered, for 30 to 35 minutes, or until the vegetables are tender. Ten minutes before the vegetables are done, add the pasta and continue simmering.

6. Add the lentils and reserved lamb to the soup and just heat through. Taste and correct the seasoning. Serve the soup hot.

Note: Any leftover stock can be frozen and used instead of a brown stock in all lamb preparations.

LEMON–SPLIT PEA SOUP

serves 6–8

Although certain legumes are interchangeable in many preparations, split peas have a distinctive personality that allows little room in terms of variation or experimentation. The classic Dutch split pea soup, made with smoky pigs' knuckles or bacon, is really hard to improve upon, and I never seem to come across other ways of preparing this delicious legume. But one day I found myself without either bacon or smoked pork butt and through improvisation came up with this soup, which I find somewhat lighter and just as satisfying as the Dutch classic.

2 tablespoons unsalted butter
1 tablespoon olive oil
2 large onions, peeled and finely diced
2 teaspoons finely minced jalapeño or Anaheim pepper
1 large carrot, peeled and finely diced
2 large stalks celery, scraped and finely diced
1 parsley root, scraped and finely diced, optional
2 teaspoons ground cumin
1½ cups split green peas, well rinsed

8–10 cups Chicken Stock (page 470) or bouillon
6 whole black peppercorns
Salt and freshly ground white pepper
½ cup heavy cream

GARNISH
1 cup sour cream
3 tablespoons finely minced fresh cilantro or dill
2 teaspoons finely grated lemon zest
1 cup snow peas, cut into very fine julienne and steamed, optional

1. In a large, heavy casserole, heat the butter and oil over medium-low heat. Add the onions and hot pepper and cook the mixture until the onions are soft and start to brown, about 7 minutes. Add the carrot, celery, and optional parsley root and continue to cook until the vegetables are soft, about 3 to 4 minutes.

2. Add the cumin and stir until incorporated into the onion mixture. Add the split peas, toss them well in the vegetable mixture, then add 8 cups of the stock and the peppercorns. Season with salt and pepper. Bring to a boil, reduce the heat, and simmer, covered, for 45 minutes, or until the peas are very soft and almost falling apart.

3. Transfer the soup to the workbowl of a food processor and process in batches until smooth. Return to the casserole and if the soup is too thick, thin it with some additional stock. Add the heavy cream, heat through, and correct the seasoning. Set aside.

4. In a bowl, combine the sour cream, cilantro, and 2 teaspoons of lemon zest.

5. Just before serving, heat the soup through. Serve in individual shallow soup bowls with a dollop of the sour cream mixture and the optional snow peas.

Note: Snow peas cut into fine, long julienne are an interesting and easy garnish. Steam them for 2 minutes well ahead of time and keep them in a small bowl of cold water. When ready to serve, drain the snow peas and either sauté quickly in a touch of butter to heat through or heat in the microwave.

Other herbs such as dill or tarragon can be substituted for the cilantro.

CURRIED LENTIL SOUP WITH ROOT VEGETABLES

serves 6

2 tablespoons unsalted butter
1 teaspoon olive oil
1 medium onion, peeled and finely minced
1 large clove garlic, peeled and finely minced
1½ teaspoons Madras curry powder
1 teaspoon ground cumin
1 large leek, trimmed of all but 2 inches of greens, diced and well washed
2 medium carrots, peeled and diced

1 medium parsnip, peeled and diced
1 cup dried lentils
6—8 cups Chicken Stock (page 470) or bouillon
Salt and freshly ground black pepper
½ cup Crème Fraîche (page 491) or sour cream
Pinch of cayenne pepper

GARNISH
Tiny fresh cilantro leaves
Bowl of sour cream

1. In a heavy 3-quart casserole, melt the butter together with the oil over low heat. Add the onion and garlic and cook for 3 or 4 minutes, or until the onion is soft but not browned. Add the curry and cumin and blend well into the onion mixture. Add the leek, carrots, parsnip, lentils, and 6 cups of stock or bouillon and season with salt and pepper. Bring to a boil, reduce the heat, and simmer the soup, covered, for 1 hour to 1 hour and 15 minutes, or until the lentils are very tender. If the soup gets too thick, add the remaining stock.

2. Whisk in the crème fraîche or sour cream. Taste the soup and correct the seasoning, adding a pinch of cayenne. Serve the soup in individual bowls, accompanied by crusty French bread and garnished with tiny cilantro leaves. Serve a bowl of sour cream on the side.

MONKFISH SCALOPPINE WITH WARM CANNELINI BEAN AND GARLIC PUREE

serves 6

Here is an easygoing bistro supper dish that can be kept quite simple or embellished and made more elegant with a variety of interesting little garnishes. Monkfish, with its lobsterlike texture, works well with the hearty puree of white beans. A simple garnish of minced parsley and a drizzle of fresh lemon juice is all it needs, but I sometimes use a more colorful garnish, such as a julienne of mixed vegetables. The combination of leeks, carrots, and mushrooms, slowly braised in a touch of butter and chicken broth, complements the fish nicely.

For a more showy presentation, I might fry a fine julienne of leeks and top each serving with a mound of them. Winter greens such as spinach and celery leaves can also be deep-fried and used instead of the leeks. Other fish scaloppine, especially swordfish steaks cut ½ inch thick, are a good substitute for the monkfish.

PUREE
1 cup cooked cannelini or Great
   Northern beans (page 484)
½–¾ cup bean poaching liquid
1 large clove garlic, peeled and crushed
   in a garlic press
2 tablespoons unsalted butter, softened
2 tablespoons extra-virgin olive oil
Salt and freshly ground black pepper

MONKFISH
2 pounds trimmed monkfish, cut
   crosswise on the bias into ½-inch
   slices (scaloppine)
1½ tablespoons fresh thyme leaves
Coarsely cracked black pepper

2 tablespoons olive oil
4 tablespoons unsalted butter
2 medium cloves garlic, peeled and
   thickly sliced
2 sprigs of fresh thyme
Coarse salt
Juice of 1 large lemon
½ cup bean poaching liquid

GARNISH
2–3 tablespoons finely minced fresh
   parsley
Thinly sliced roasted red pepper (page
   483), optional

1. Start by making the bean puree: In the workbowl of a food processor combine the cooked beans, ½ cup bean liquid, garlic, butter, and olive oil and process until smooth. Season with salt and pepper to taste. You may need to add the remaining bean liquid if the puree is too thick. Transfer the puree to the top of a double boiler. Place over simmering water, cover, and keep warm.

2. To prepare the monkfish: Dry the monkfish scaloppine well on paper towels. Season with thyme and lots of cracked pepper. Set aside.

3. Sauté the monkfish in 2 batches: In a large nonstick skillet heat 2 tablespoons butter with 2 teaspoons oil over high heat. When almost smoking, add half of the monkfish scaloppine, without crowding the pan, half of the garlic, and 1 branch of thyme. Cook the monkfish for 1½ minutes on each side, or until nicely browned and just cooked through. Remove the scaloppine to a side dish, season with coarse salt, and reserve.

4. Discard the garlic and thyme. Add 2 more tablespoons butter together with 2 teaspoons oil to the skillet. Add the remaining scaloppine, garlic, and thyme. When done, remove from the skillet, season, and set aside.

5. Discard the remaining garlic and thyme and the fat, if it has burned. Add the lemon juice and bean liquid, reduce the heat, and simmer for 1 minute. Taste and correct the seasoning. Keep warm.

6. To serve, place some of the bean puree in the center of each of 6 individual serving plates. Top each with some of the scaloppine in a slightly overlapping pattern and spoon some of the sauce over each portion. Garnish with minced parsley and optional slivers of roasted pepper. Serve at once.

Note: If you do not have time to roast a pepper, you may dice 1 large red pepper and braise it in a little butter and water until tender. Sprinkle over the finished dish.

For maximum flavor be sure to cook the beans 1 or 2 days ahead of time and leave them in their cooking liquid. This is also a good way to avoid last-minute preparation and allows you to serve the beans simply warmed up and tossed in a little extra-virgin olive oil and lemon juice, if you like.

FILLETS OF SALMON WITH LENTILS À LA BOURGEOISE

serves 4

For many years salmon was a spring fish, and it was always paired with such delicate seasonal accompaniments as cucumbers, spinach, asparagus, and peas. Now that farmed salmon can be bought in any season, chefs and home cooks are looking for creative new ways to treat it.

I first sampled this combination of salmon with lentils at Alain Passard's restaurant in Paris in the late eighties, and I thought the idea was brilliant. Finally, salmon was being treated as an earthy fish, and Passard's full-bodied red wine sauce enhanced it even further. Now versions of this dish can be found on many restaurant menus, but as often happens with a newly popular recipe, most of them are a far cry from the original. Salmon with lentils has become one of my very favorite cool-weather main courses, and it's well on its way to becoming a home classic.

LENTILS
1 cup dried Spanish or small French green lentils (see Note below)
6 ounces smoked pork shoulder butt, cut into large cubes
1 small onion, peeled and stuck with 1 whole clove
1 small carrot, peeled and cut in half
1 Bouquet Garni (page 485)

VEGETABLES
1 large carrot, peeled and finely diced
2 small turnips, peeled and finely diced
1 tablespoon unsalted butter
½ cup finely diced slab bacon, blanched
1 tablespoon finely minced shallots
2 teaspoons finely minced garlic
¾ cup heavy cream or reserved lentil broth

Salt and freshly ground black pepper

SAUCE
2 tablespoons finely minced shallots
1½ cups Red Wine Fish Stock (page 473)
2–3 tablespoons Crème Fraîche (page 491)
10 tablespoons unsalted butter

SALMON
4 salmon fillets, about 6–7 ounces each, with skin on and preferably center cut
Coarse salt and freshly ground black pepper
4 tablespoons Clarified Butter (page 478)

1. Start by cooking the lentils: In a large saucepan combine the lentils, pork butt, onion, carrot, and bouquet garni and cover with 2 inches of water. Bring to a boil, reduce the heat, and simmer, covered, until tender, about 20 minutes. Drain the lentils, discard the vegetables and pork, and set aside. If making the vegetables without cream, reserve ¾ cup lentil broth.

2. To prepare the vegetables: Place the carrot and turnips in a vegetable steamer, set over simmering water, and steam, covered, until the vegetables are just tender. Remove from the steamer and reserve.

3. Melt the butter in a 10-inch skillet over medium heat. Add the bacon and cook until lightly browned. Remove with a slotted spoon to a side dish and reserve. Add the shallots and garlic to the skillet and cook for 1 minute without browning. Add the cooked lentils, steamed vegetables, bacon, and heavy cream and simmer until the cream has been absorbed. Season with salt and pepper and keep warm.

4. To make the sauce: In a heavy 2½-quart saucepan, preferably copper, combine the shallots and red wine fish stock, bring to a boil, and reduce to ½ cup. Add the crème fraîche, reduce the heat to low, and whisk in the butter a few tablespoons at a time until the sauce is smooth and velvety. Taste and correct the seasoning and keep warm over a pan of lukewarm water.

5. To prepare the salmon: Dry the salmon fillets thoroughly on paper towels. Season with coarse salt and pepper. Heat the clarified butter in 2 (10-inch) nonstick skillets over high heat. Place 2 salmon fillets in each skillet, skin side down, and cover loosely with foil. Cook for 2 minutes, lower the heat slightly, and cook for an additional 4 to 5 minutes without turning. (You may turn the fillets once for about 1 minute if you want them to brown.)

6. Place 1 fillet on each of 4 warm individual serving plates. Surround with lentils and spoon the red wine butter sauce over them. Serve at once.

Note: Tiny French green lentils from Puy are increasingly available in specialty food shops, the most notable being Dean & Deluca in New York City. Use all-purpose lentils if you can't find those from Puy.

POT-AU-FEU OF RED SNAPPER WITH
WINTER VEGETABLES

serves 4–6

12–16 pearl onions, peeled
1 cup Chicken Stock (page 470) or
    bouillon
6 tablespoons unsalted butter
3 medium turnips, peeled and cut into
    ½-inch cubes
3 medium carrots, peeled and cut into
    ½-inch cubes
Salt and freshly ground black pepper
Pinch of granulated sugar
2 medium shallots, peeled and thinly
    sliced

1 cup Fish Stock (page 472)
1 cup heavy cream
1 Beurre Manie (page 478)
4–6 red snapper fillets, about 6–7
    ounces each

GARNISH
Tiny fresh Italian parsley leaves
12 small new potatoes, unpeeled, cut in
    half and cooked until just tender

1. In a small saucepan combine the onions and chicken stock and bring to a boil. Reduce the heat and cook for 7 to 10 minutes, or until the onions are just tender. Strain the stock into a bowl and set the onions aside.

2. In a heavy 10-inch cast-iron skillet melt 2 tablespoons of butter over medium heat. Add the turnips and carrots, season lightly with salt and pepper, and add a sprinkling of sugar. Sauté for 2 minutes, or until the vegetables start to glaze. Add the reserved chicken broth, cover the skillet tightly, and simmer until the vegetables are tender, about 5 minutes.

3. Add the reserved pearl onions and continue to simmer until all the broth has evaporated and the vegetables are nicely glazed. Set aside.

4. In a heavy 2-quart saucepan melt 2 more tablespoons of butter, then add the shallots, fish stock, and cream, bring to a boil, and cook until reduced by one third. Whisk in a little of the beurre manie, just enough to slightly thicken the sauce. Remove the sauce from the heat and whisk in the remaining butter. Taste and correct the seasoning. Keep the sauce warm in the top of a double boiler or in a pan of hot water. Reserve.

5. Preheat the oven to 400° F.

6. Butter a large baking dish, season the fish fillets with salt and pepper, and place in the dish. Cover with buttered parchment paper and bake the fillets for 3 minutes on each side, or until they just turn opaque. Leave in the oven for 2 or 3 minutes longer if necessary. Remove the fillets from the oven and lift off

the skin. With a large spatula, transfer the fish to a large serving platter and cover loosely with foil.

7. Quickly reheat the vegetables, adding whatever juices have accumulated around the fillets. Place the vegetables all around the snapper fillets and spoon the sauce over them. Garnish with whole Italian parsley leaves and serve with tiny boiled new potatoes in their skin. Serve immediately.

Note: The vegetable garnish and the sauce can be made several hours ahead of time and reheated.

You can use other fish in this dish, such as sea bass, striped bass, or white snapper. Be sure to buy the whole fish so that you have the bones for making a full-bodied stock, which is most important for the success of the sauce.

SAUTÉ OF CORNISH HENS FORESTIÈRE

serves 4

The days of frozen Cornish hens are over. In most parts of the country fresh Cornish hens have become a supermarket staple. But every once in a while I will find myself in a city or town where the only hens available are frozen. If that is the case, use small fresh chickens, about 2½ pounds each for this recipe, and cut them into eighths. This dish is actually at its best made with fresh rabbit. But because that idea still brings tears to the eyes of many, who immediately visualize their pet bunny rabbit being devoured, it is worth compromising, since this is an easily made, intensely flavored dish that is perfect for a winter dinner. Good accompaniments are caramelized sweet potatoes and oven-roasted fennel.

4 Cornish hens, about 1¼ pounds each, quartered
2 large shallots, peeled and finely minced
2 large cloves garlic, peeled and finely minced, plus 1 large clove, peeled and crushed
3 tablespoons finely minced fresh thyme leaves
1 cup white wine, preferably Riesling
Freshly ground black pepper
3 tablespoons unsalted butter
1 cup finely diced slab bacon, blanched

Coarse salt
24 pearl onions, peeled
2 cups Brown Chicken Stock (page 467)
3 tablespoons olive oil
1 pound fresh mushrooms (preferably Cremini), stemmed, wiped, and cubed
6 fresh shiitake mushrooms, stemmed and quartered, optional
1 Beurre Manie (page 478)

GARNISH
2–3 tablespoons finely minced fresh Italian parsley
Braised Savoy cabbage

1. In a porcelain or glass dish, combine the hens, shallots, minced garlic, and thyme. Add the wine and season the hens with ground pepper. Cover the dish, refrigerate, and marinate for 12 to 14 hours, turning the hens once or twice.

2. Drain the hens and dry thoroughly with paper towels.

3. In a 12-inch cast-iron skillet, melt 2 tablespoons of the butter over medium heat. Add the bacon and sauté until almost crisp. Remove the bacon with a slotted spoon to a side dish and reserve.

4. Raise the heat and add the remaining butter to the skillet. When hot, add the Cornish hens, skin side down, and sauté until nicely browned. Remove the hens from the skillet, season with coarse salt and pepper, and reserve.

5. Add the onions to the skillet and sauté for 3 or 4 minutes, or until lightly browned, shaking the pan back and forth. The onions will not brown evenly. Add the crushed garlic to the onions together with the reserved bacon. Return the Cornish hens to the skillet and add 1 cup of stock. Bring to a boil, reduce the heat, and simmer, tightly covered, for 15 minutes.

6. While the hens are braising, heat the olive oil in a 10-inch skillet over medium-high heat. Add the cubed mushrooms and optional shiitakes and sauté until nicely browned. Season lightly with salt and pepper and set aside.

7. After the hens have braised for 15 minutes, add the mushrooms and ½ cup of stock to the skillet. Cover and continue braising the hens for another 10 minutes, or until they are just done and their juices run pale yellow. Transfer the hens and vegetables to a serving dish and keep warm.

8. Degrease the pan juices thoroughly and return to the skillet with the remaining ½ cup stock. Bring to a boil and whisk in just enough beurre manie until the sauce lightly coats a spoon. Taste and correct the seasoning and spoon the sauce over the hens. Garnish with parsley and serve immediately over a bed of braised cabbage.

BONELESS CHICKEN WITH MORELS AU FARCI

serves 6

A boned, stuffed chicken leg served in a creamy sauce and enhanced by morel mushrooms is an impressive dish suitable for elegant entertaining. I like the idea of elevating the humble chicken by using it in new and unusual ways. Stuffing a whole chicken, and the legs, is a popular way to stretch the number of servings. Chicken is one bird that has its bones in all the right places, and even someone who has never boned one before will have no trouble. Once the legs are stuffed, they are gently braised and enriched with morels. Morels have a unique texture and flavor, but good-quality porcini taste equally delicious (and are equally expensive) and are more readily available. Serve the chicken accompanied by fresh fettucine tossed in a saffron-flavored butter or parsleyed rice timbales. A simple vegetable such as buttered spinach or the Marmalade of Belgian Endives (page 433) will be just the right finishing touch.

MUSHROOMS
1½ cups Brown Chicken Stock (page 467) or bouillon
¾ cup whole dried morels, stemmed and thoroughly rinsed of all grit

FILLING
¼ pound ground pork
¼ pound ground veal
1 tablespoon finely minced fresh shallots
1 medium clove garlic, peeled and mashed
1 large egg
1 tablespoon finely minced fresh thyme leaves
1 tablespoon finely minced fresh tarragon leaves

Salt and freshly ground white pepper
⅓ cup very cold heavy cream

CHICKEN
6–8 medium chicken legs (from 3-pound chickens), boned in one piece
Salt and freshly ground white pepper
All-purpose flour for dredging
2 tablespoons unsalted butter
1 teaspoon peanut oil
¼ cup dry white wine
1 cup heavy cream
1 Beurre Manie (page 478)

GARNISH
Finely minced fresh parsley

1. Start by preparing the mushrooms: In a 1-quart saucepan, combine the stock and morels. Bring to a boil, reduce the heat, and simmer, covered, for 25 to 30 minutes, or until tender. Strain into a bowl, reserving both the stock and morels separately.

2. To prepare the filling: In the workbowl of a food processor, combine the pork, veal, shallots, garlic, egg, thyme, tarragon, salt, and pepper and process

until the mixture is smooth. With the machine running, add the cold heavy cream very slowly until it is thoroughly incorporated. Season highly with salt and pepper. Transfer the mixture to a mixing bowl, cover, and refrigerate for 30 minutes to 1 hour.

3. To prepare the chicken: Dry the boned chicken legs thoroughly on paper towels and season both sides with salt and pepper. Place the boned legs, skin side down, opened flat and place a heaping tablespoonful of the filling in the center of each leg. Roll up the chicken from the short side, tucking in the long sides to completely enclose the filling, and truss each with kitchen string. Season with salt and pepper and dredge lightly in flour, shaking off the excess.

4. In a heavy 10-inch skillet, heat the butter and oil over medium heat. When the butter and oil are hot, add the stuffed legs, without crowding the pan, and brown nicely on all sides. Add the wine and reduce to 1 tablespoon. Add 1 cup of the reserved mushroom cooking stock. Bring to a boil, reduce the heat, and simmer, covered, for 35 minutes, or until the chicken juices run clear. Remove the chicken legs to a plate, cover, and set aside.

5. Degrease the pan juices and return to the skillet. Bring to a boil and reduce by half. Add the heavy cream and reduce by a third. Add the morels and whisk in bits of beurre manie, until the sauce lightly coats a spoon. Taste and correct the seasoning. Return the chicken legs to the skillet and just heat through.

6. Transfer the legs to a serving platter. Spoon the sauce over and garnish with the minced parsley. Serve accompanied by buttered fettuccine and a julienne of sautéed zucchini.

Note: You may substitute ½ cup fresh spinach leaves, cooked, squeezed, and chopped, for the thyme and tarragon.

ROAST POUSSINS À LA PROVENÇALE WITH LEMON AND CRACKED PEPPER

serves 6

A poussin is a small free-range chicken that, while more expensive and harder to find than commercial chickens, truly tastes like the bird of yesteryear. As the quest for real taste continues to grow, poussins are becoming increasingly available, and to me they are well worth the high price tag, especially when it comes to simple and straightforward preparations such as this, in which there is little to mask the basic taste of the bird. If you cannot get poussins, you may use fresh Cornish hens or even 2½-pound broilers or fryers. In any case, why not investigate the possibility that there is a nearby farm or butcher who can provide free-range chickens. The difference in taste is so amazing that it is bound to change the way you feel about cooking and eating chicken.

6 fresh poussins, each weighing about 1 pound
Juice of 2 lemons
3 tablespoons finely minced fresh thyme
3 tablespoons fresh rosemary leaves
1½ teaspoons freeze-dried green peppercorns, crushed
8 tablespoons extra-virgin olive oil

1 tablespoon unsalted butter
Coarse salt and cracked black pepper
1–1½ cups Chicken Stock (page 470)
1 Beurre Manie (page 478)
4 tablespoons unsalted butter, for enrichment, optional

GARNISH
Sprigs of fresh thyme

1. Dry the poussins thoroughly on paper towels.

2. In a small bowl combine the lemon juice, thyme, rosemary, peppercorns, and 6 tablespoons of the olive oil and whisk until well blended. Place the poussins in a heavy zip-lock bag, pour the marinade over them, and seal the bag. Place the bag in a shallow dish and refrigerate for 4 to 6 hours, turning the bag several times during marinating.

3. Preheat the oven to 425° F.

4. Remove the poussins from the marinade and dry with paper towels. Heat the remaining olive oil together with the butter in a large, heavy baking dish over medium-high heat. Add the poussins and brown nicely on all sides. You may have to brown the birds in 2 batches. Season the poussins with coarse salt and cracked pepper and add a little stock to the baking dish. Place the dish in the center of the preheated oven and roast the poussins for 25 to 30 minutes,

or until the juices run pale yellow, adding a little stock to the pan every 5 to 10 minutes. Start basting after the first 15 minutes of roasting.

5. Remove the poussins from the oven, transfer to a broiler pan, and set aside.

6. Preheat the broiler.

7. Place the baking dish containing the pan juices over medium heat. Add the remaining stock, bring to a boil, and reduce by one quarter. Add bits of the beurre manie until the sauce lightly coats a spoon. Whisk in the optional 4 tablespoons butter until the sauce is velvety. Taste and correct the seasoning and keep warm.

8. Place the poussins in the broiler, about 6 inches from the source of heat, and broil until the skin is nicely browned and crisp. Serve on individual plates with some of the sauce spooned on the side and garnished with sprigs of fresh thyme.

Note: Serve the poussins accompanied by the Fennel Fricassee (page 434) or a creamy gratin of potatoes and skillet-braised broccoli.

ROAST DUCK IN RAISIN SAUCE

serves 4–6

About three years ago someone mentioned to me that the new kid on the block in Paris was Alain Passard and that his restaurant, called Arpège, was worth trying. I got in touch with Passard, who immediately let me spend several days in his tiny kitchen. What sets this young and ambitious chef apart is the utter simplicity and straightforwardness of his ingredients and the unadulterated flavors he achieves with such ease. This exquisite raisin sauce, which at Arpège is served with roast squab and fresh figs, is a true example of Passard's creativity.

DUCKS

2 ducks, about 4½–5 pounds each
Coarse salt and freshly ground black
  pepper
2 teaspoons dried thyme
2 tablespoons unsalted butter or
  rendered duck fat
2 teaspoons peanut oil

2 cups Brown Chicken Stock (page 467)
  or bouillon

SAUCE

1 pound dark raisins
6–8 tablespoons unsalted butter
Salt and freshly ground black pepper
Juice of ½ lemon

1. Preheat the oven to 350° F.

2. Dry the ducks thoroughly on paper towels and season with salt, pepper, and thyme. In a large cast-iron skillet heat the butter or fat with the oil over medium heat. Add 1 of the ducks and brown nicely on all sides. Transfer the browned duck to a large roasting pan. Place the second duck in the skillet, brown it nicely on all sides, and add to the roasting pan. Roast the ducks in the center of the preheated oven for 2½ hours, removing all of the accumulated fat from the pan and basting the ducks with a little stock every 15 to 20 minutes.

3. While the ducks are roasting, prepare the sauce: Combine the raisins with water to cover in a medium saucepan. Bring to a boil, reduce the heat, and simmer for 20 minutes or until the poaching liquid is reduced to about ½ cup. Remove the pan from the heat and add the butter, 2 tablespoons at a time, to the raisins, incorporating well without letting the butter liquefy.

4. Transfer the sauce to the workbowl of a food processor or blender and process until smooth. Strain the sauce back into the saucepan and season with salt, pepper, and lemon juice. Keep warm.

5. When the ducks are done, remove from the oven and transfer to a carving board. Let stand for 15 minutes, then carve the ducks, removing the backbone, quarter, and place them on a serving platter. Keep warm.

6. Degrease the pan juices thoroughly and add to the raisin sauce. Taste and correct the seasoning and spoon the sauce over the ducks. Serve at once.

RAGOUT OF VEAL WITH SHERRY VINEGAR AND SHALLOT ESSENCE

serves 4–5

3 pounds boneless veal shoulder, cut
   into 1¼-inch cubes
2–3 tablespoons unsalted butter
1 tablespoon peanut oil
Salt and freshly ground black pepper
1 teaspoon granulated sugar
1 tablespoon all-purpose flour
¼ cup cognac
½ cup dry white wine

1 cup finely minced shallots
4 large cloves garlic, peeled and
   mashed
1 teaspoon dried thyme
½ cup sherry vinegar
1–1½ cups Brown Chicken Stock (page
   467)
2 tablespoons dark-brown sugar
1 Beurre Manie (page 478), optional

1. Preheat the oven to 375° F.

2. Dry the veal cubes thoroughly on paper towels. In a heavy skillet, heat 2 tablespoons of the butter and the oil over medium-high heat. Add the veal, a few pieces at a time, and brown well on all sides. Transfer the veal to a side dish and continue browning the remaining pieces, adding a little more butter to the skillet when necessary.

3. When all the veal has been browned, return to the skillet, season well with salt and pepper, sprinkle with the granulated sugar and flour, and toss until the veal is glazed. Transfer to a heavy 3-quart casserole.

4. Add the cognac to the skillet and reduce to a glaze. Add the wine and cook until reduced by half. Add the shallots to the skillet and cook for 1 minute, scraping the bottom of the pan well. When the shallots are soft, add the garlic, thyme, and vinegar, bring to a boil, and reduce by half.

5. Pour the vinegar mixture over the veal and add 1 cup of the stock. The stock should not cover the veal completely. Cover the casserole tightly and braise the ragout in the center of the preheated oven for 1 hour and 30 minutes, or until the meat is fork tender, adding the remaining stock halfway through the cooking. When the veal is done transfer the pieces to a side dish with a slotted spoon and reserve.

6. Bring the pan juices to a boil in the casserole and add the brown sugar. Taste and correct the seasoning and whisk in bits of beurre manie, if necessary, until the sauce lightly coats a spoon. Return the veal to the casserole and toss gently to coat it well with the sauce and just heat through. Serve hot accompanied by Marmalade of Belgian Endives (page 433).

Note: The ragout can be prepared a day or two in advance and reheated. Other braised vegetables such as carrots and turnips are excellent accompaniments.

FRICASSEE OF VEAL SHANKS WITH GLAZED CARROTS IN TARRAGON SAUCE

serves 4–5

Veal shanks are traditionally used in a familiar Italian dish, osso buco. However, veal knucklebones are extremely versatile, and I like to use them in many of the classic preparations that usually call for chicken. Here they are poached and combined with tender young carrots and a velvety sauce. Other garden vegetables such as peas, new potatoes, and beans may be added for a variation.

VEAL

4 pounds veal shanks, cut into ¾-inch slices
2 medium carrots, peeled and cut in half
1 large leek, trimmed of all but 3 inches of greens and well rinsed
2 medium stalks celery, trimmed and cut in half
4 whole black peppercorns
Pinch of salt

CARROTS

2 tablespoons unsalted butter
1½ pounds carrots, peeled and cut into ¾-inch cubes

1 teaspoon granulated sugar

SAUCE

½ cup heavy cream
2 large egg yolks
2 tablespoons finely minced fresh tarragon
3 tablespoons unsalted butter
1½ tablespoons all-purpose flour
2 cups reserved broth
Salt and freshly ground black pepper

GARNISH

12–15 tiny cooked new potatoes, tossed in a little butter and minced parsley

1. Start by preparing the veal: Combine the veal shanks, carrots, leek, celery, peppercorns, and pinch of salt in a large pot and add water to cover. Bring to a boil, reduce the heat, and simmer for 1 hour and 15 minutes to 1 hour and 30 minutes, or until the veal is tender but not falling off the bones. Skim the top carefully from time to time.

2. While the veal is poaching, prepare the carrots: Melt the butter in a heavy 10-inch skillet over medium-low heat, add the carrot cubes, and sprinkle with sugar. Simmer the carrots, covered, for about 8 minutes, or until just tender. Remove and set aside.

3. When the veal is done, carefully remove all the fat from the surface of the veal broth. Remove 2 cups broth and reserve. Set the veal aside in the remaining broth until needed. Keep warm.

4. To prepare the sauce: In a small mixing bowl combine the heavy cream, egg yolks, and tarragon. Reserve. Melt the 4 tablespoons of butter in a heavy

saucepan over medium heat. Add the flour and cook, stirring constantly, for 1 to 2 minutes without browning. Whisk in the 2 cups of reserved veal broth all at once and simmer for 5 minutes. Whisk in the egg yolk–cream mixture and just heat through. Do not let the sauce come to a boil or the eggs will curdle. Season with salt and pepper.

5. To serve, remove the veal shanks from the broth with a slotted spoon and transfer to a shallow serving dish. Top with the glazed carrots and spoon the sauce over them. Garnish with the parsleyed new potatoes and serve at once.

PAN-SEARED SAUTÉ OF CALVES' LIVER IN MUSTARD AND GREEN PEPPERCORN SAUCE

serves 4–6

This dish belongs in the "fast and easy" category. When choosing a main course that takes you away from the dining table in the middle of the meal, be sure that the remaining components of the menu have been completed in advance and that the only thing left to do is sauté the liver, which will take all of two to three minutes. Be sure to choose rosy, pink liver and buy it only at a reputable butcher. Don't buy presliced liver; it will have lost much of its natural juices. Have the butcher slice it while you wait and use it the same day. Now that there are so many interesting mustards around, I would choose the Maille herb mustard, which has a brilliant green color and, when combined with green peppercorns, makes a wonderful quick sauce. A leek and potato rosti is the best accompaniment to this main course.

1 teaspoon dry mustard
1 tablespoon water
7 tablespoons unsalted butter, 4 of them softened
1 teaspoon Dijon mustard
1 tablespoon peanut oil
4 cups thinly sliced yellow onions
Salt and freshly ground white pepper
1½ pounds calves' liver, thinly sliced

All-purpose flour for dredging
½ cup dry white wine
¾ cup Brown Chicken Stock (page 467)
¾ cup heavy cream
1–2 teaspoons green peppercorns, packed in brine, drained and crushed

GARNISH
3 tablespoons finely minced fresh parsley

1. In a small bowl dissolve the dry mustard in the water and let stand for 10 minutes to develop flavor.

2. In a small mixing bowl combine the 3 tablespoons of softened butter with the Dijon mustard and the dry mustard mixture. Blend thoroughly with a wooden spoon and reserve.

3. In a large, heavy skillet melt 2 tablespoons of the butter together with the oil over high heat. Add the onions, season with salt and pepper, and cook for 10 minutes, or until lightly browned, stirring often. Lower the heat, partially cover the pan, and continue cooking the onions for 25 to 30 minutes longer, or until they are very soft and nicely browned. Transfer the onions to a side dish and set aside.

4. Slice the liver crosswise into ½-inch strips. Season with salt and pepper and dredge lightly in flour, shaking off the excess. Add the remaining 2 tablespoons of butter to the skillet and melt over high heat. When the butter is very hot, add the liver and sauté for 2 to 3 minutes, tossing the strips in the butter. The liver should still be quite pink in the center. With a slotted spoon, remove the liver to a side dish and reserve.

5. Add the wine to the skillet and cook until it is reduced to a glaze. Add the stock, bring to a boil, and cook until reduced to 3 tablespoons. Add the heavy cream and green peppercorns and reduce slightly.

6. Return the liver and onions to the skillet and cook the mixture for 1 minute or until just heated through. Do not overcook. Remove the skillet from the heat, add the mustard-butter, and stir until it is well incorporated. Taste and correct the seasoning. Transfer to a serving platter, sprinkle with minced parsley, and serve at once, accompanied by crusty French bread.

ROAST LOIN OF LAMB WITH RED
WINE–SHALLOT SAUCE

serves 4–5

SAUCE
6 tablespoons unsalted butter
5 ounces shallots, peeled and finely
  minced
1 large clove garlic, peeled and crushed
1 small bay leaf
1 large sprig of fresh thyme
½ cup ruby port
2½ cups dry red wine
6 whole white peppercorns
1 cup Brown Chicken Stock (page 467)
¼ cup heavy cream
1 large sprig of fresh tarragon
1 Beurre Manie (page 478)
Salt and freshly ground black pepper

LAMB
2 loins of lamb, about 2–2½ pounds
  total

2 medium cloves garlic, peeled and
  mashed
2 teaspoons dried thyme
1 teaspoon dried rosemary
Coarse salt and freshly ground black
  pepper
2 tablespoons unsalted butter
2 teaspoons olive oil
½ cup Brown Chicken Stock (page 467)

GARNISH
1 recipe warm Shallot Puree with Garlic
  and Thyme (page 437)
8–10 Roasted Shallots (page 483)
Sprigs of fresh watercress

1. Start by making the sauce: Melt 3 tablespoons of the butter in a 4-quart casserole over medium heat. Add the minced shallots and garlic and cook for 2 minutes without browning. Add the bay leaf, thyme, port, red wine, and peppercorns and slowly reduce by two thirds, skimming the top often.

2. Add the stock, heavy cream, and sprig of fresh tarragon and reduce to 1½ cups. Strain the sauce through a fine sieve, pressing down on the shallots to extract all of their juices; you should have about 1 cup of sauce, in all.

3. Transfer the strained sauce to a small saucepan, bring to a boil, and whisk in bits of beurre manie, until the sauce lightly coats a spoon. Reduce the heat, add the remaining butter, and whisk until well blended. Correct the seasoning and keep warm.

4. Preheat the oven to 375° F.

5. Dry the loins of lamb thoroughly on paper towels. Make tiny slits on the underside of the lamb and insert bits of mashed garlic. Season with thyme and rosemary. Sprinkle with coarse salt and pepper and set aside.

6. In a large, heavy skillet, melt the butter with the oil over medium-high heat. Add the lamb and brown nicely on all sides. Add a little of the stock and place in the center of the preheated oven. Roast to an internal temperature of 135° F for medium-rare, adding a little stock to the pan when necessary.

7. When the lamb is done, transfer to a cutting board. Degrease the pan juices carefully and add them to the red wine sauce.

8. To serve, cut the lamb crosswise into ½-inch slices and place 4 slices on each of 4 to 5 individual serving plates. Spoon some of the sauce around the lamb and place a mound of shallot puree in the pool of sauce. Garnish each plate with 2 roasted shallots and sprigs of fresh watercress. Serve at once.

SAUTÉ OF PORK TENDERLOINS WITH PRUNES AND SWEET-AND-SOUR CARAMEL

serves 6

Until I traveled through southwest France, I never realized that a prune could be truly great. But in the small town of Agen—the prune capital of France—I learned to appreciate this winter fruit, and now a familiar run-of-the-mill prune will never fill my needs in quite the same way.

Agen has a bustling indoor market where several stands carry the best of the local crop: prunes, pitted and stuffed with a prune jam that is laced with rum, Cointreau, mint, or cognac, and basic unpitted prunes. A friendly shopkeeper sold me 8 kilograms (16 pounds) of various prunes and recommended that I stop at a local bistro whose specialty was rabbit stewed in a prune and orange sauce. I loved it. I have since tried the sauce with various other meats and liked it best with pork tenderloins. But if you are a fan of rabbit, substitute it for the tenderloins and braise it with the prunes and the orange sauce over low heat for 25 to 30 minutes.

24 large unpitted prunes
1¼ cups water
¼ cup granulated sugar
¼ cup red wine vinegar
1 cup hot Brown Chicken Stock (page 467)
2 whole pork tenderloins, trimmed of all fat, about 1 pound each
All-purpose flour for dredging

6 tablespoons unsalted butter
1 teaspoon peanut oil
Salt and cracked black pepper
½ cup fresh orange juice
1 teaspoon cornstarch mixed with 1 tablespoon cold stock or water

GARNISH
Sprigs of fresh watercress

1. Place the prunes and water in a medium saucepan. Bring to a boil, reduce the heat, and simmer for about 15 minutes, or until the prunes are soft and the liquid has reduced to ½ cup. Drain and reserve the prunes and prune cooking liquid separately.

2. In a medium saucepan combine the sugar and vinegar. Bring to a boil over high heat and continue to boil until the mixture turns a hazelnut brown. Be careful not to burn. Remove from the heat and immediately add the hot stock, averting your face. The mixture will bubble violently, but will settle in a few minutes. Return the saucepan to low heat and stir until the caramel is dissolved. Reserve.

3. Dry the tenderloins thoroughly on paper towels. Cut crosswise into ½-inch slices (medallions). Dredge lightly in flour, shaking off the excess. In a large, heavy skillet melt 2 tablespoons of the butter together with the peanut oil over medium-high heat. Add the pork medallions, without crowding the skillet, and sauté until nicely browned, about 2 minutes per side. Season with salt and pepper and transfer to a side dish. Reserve.

4. Add the reserved prune cooking liquid to the skillet, bring to a boil, and reduce to 2 tablespoons, whisking often. Add the orange juice, return to a boil, and reduce to 2 tablespoons.

5. Add the caramelized stock, again bring back to a boil, and whisk in just enough of the cornstarch mixture until the sauce lightly coats a spoon. Add the prunes, pork medallions, and any juices that may have accumulated and just heat through. Do not overcook.

6. Remove the medallions and prunes from the sauce. Place 4 or 5 medallions and 4 prunes in a decorative pattern on each of 6 warm individual serving plates. Whisk the remaining butter into the sauce, taste and correct the seasoning, and spoon a little over each portion. Garnish with sprigs of watercress and serve immediately accompanied by a sweet potato puree.

PENNE WITH THREE-MUSHROOM RAGOUT, HAM, AND PARMESAN CREAM

serves 4

1 ounce dried porcini mushrooms
1 cup chicken bouillon
4 tablespoons unsalted butter
1 cup finely cubed smoked ham
1 teaspoon finely minced fresh garlic
2 tablespoons finely minced fresh parsley
8–10 medium Cremini mushrooms, stemmed and cut into ½-inch cubes
Freshly ground black pepper
2–3 tablespoons olive oil

8 medium fresh shiitake mushrooms, stemmed and thinly sliced
¾ cup heavy cream
Salt
½ pound imported penne, preferably De Cecco
⅓ cup freshly grated Parmesan cheese

GARNISH
2–3 tablespoons finely minced fresh parsley
Bowl of freshly grated Parmesan cheese

1. Place the porcini in a small strainer, rinse well under warm water to remove all sand, and drain thoroughly.

2. Combine the porcini with the bouillon in a small saucepan. Bring to a boil, reduce the heat, and simmer for 20 minutes. Remove the porcini with a slotted spoon, cut into a fine julienne, and set aside. Reserve the broth separately.

3. In a heavy 10-inch skillet melt 2 tablespoons of the butter over medium heat. Add the ham and sauté for 2 minutes. Remove to a side dish with a slotted spoon and set aside. Add the remaining butter to the skillet and, when hot, add the garlic and parsley and cook for 1 minute. Add the Cremini mushrooms and sauté quickly until nicely browned. Season with a large grinding of pepper. Add the reserved porcini broth, and cook until it is reduced to 2 tablespoons. Add the julienne of porcini and the ham, and cook for 2 minutes longer.

4. In another skillet, heat the olive oil over medium-high heat, add the shiitake mushrooms, and sauté quickly until lightly browned. Add to the Cremini mushroom mixture together with the heavy cream and just heat through. Cover, and keep warm.

5. Bring plenty of salted water to a boil in a large pot, add the penne, and cook until al dente. Immediately add 2 cups cold water to the pot to stop further cooking, and drain the pasta well.

6. Return the penne to the casserole and add the mushroom mixture and the Parmesan cheese and toss well. Taste and correct the seasoning and transfer to a serving dish. Sprinkle with the minced parsley and serve at once with a side bowl of Parmesan.

CATALAN WHITE BEAN RAGOUT LA BODEGA

serves 4–6

When I first served this dish for a Sunday dinner, my family was somewhat surprised. "Is that all there is?" was the unasked question. But for me, this gutsy one-dish meal holds its own against the most elegant roast or ragout. The key to its success is of course the freshness of the dried beans, which, when cooked, should be tender yet not split or mushy. Other beans, especially pintos, can be substituted for the Great Northern beans in this dish.

2 cups dried white beans, preferably Great Northern
1 large carrot, peeled and cut in half
1 large stalk celery, cut in half
1 medium onion, peeled and stuck with 1 whole clove
1 Bouquet Garni (page 485)
3 garlic sausages or knockwursts
1 tablespoon unsalted butter
1½ cups slab bacon, cubed, blanched, and drained
1 small dry red chili pepper, broken
1 Bermuda onion, peeled, quartered, and thinly sliced

2 large cloves garlic, peeled and finely minced
2 tablespoons finely minced fresh parsley
1½ cups canned whole Italian plum tomatoes, drained and chopped
½ teaspoon dried marjoram
1 teaspoon dried thyme
Salt and freshly ground black pepper

GARNISH
2 tablespoons finely minced fresh parsley
2 tablespoons thinly sliced pimientos

1. Soak the beans in cold water overnight. The next day drain, rinse, and set aside.

2. Preheat the oven to 350° F.

3. Bring 3 quarts of water to a boil in a large, heavy flameproof casserole. Add the beans, carrot, celery, onion stuck with a clove, and the bouquet garni. Return to a boil, cover, and place in the center of the preheated oven. Braise the beans for 2 hours, or until tender. Twenty minutes before the beans are done, add the sausages to the casserole.

4. While the beans are braising, melt the butter in a large skillet over medium heat. Add the bacon and cook until lightly browned. Remove with a slotted spoon to a side dish and reserve.

5. Discard all but 2 tablespoons fat from the skillet. Add the chili pepper, cook until dark, about 30 seconds, and discard. Add the sliced onion and cook until soft and lightly browned. Add the garlic, parsley, tomatoes, marjoram, and thyme. Cook over medium-high heat until the tomato juices have evaporated and the mixture is thick. Season with salt and pepper, add the reserved bacon to the skillet, and set aside.

6. When the beans are done, pour them into a colander set over a large bowl. Discard the vegetables and bouquet garni. Slice the sausages and add them to the tomato mixture, together with the cooked beans. Add 1 cup of bean cooking liquid, and place the skillet in the center of the oven. Bake the ragout for 30 minutes, or until most of the bean water has evaporated; it must not be dry.

7. Remove from the oven, sprinkle with the minced parsley and sliced pimientos, and serve right out of the skillet, accompanied by a well-seasoned salad and French bread.

Variation: Add 2 small zucchini, cubed and sautéed in olive oil, to the finished ragout. Also you may add 1 red bell pepper, cored, quartered, and finely julienned, to the onion and tomato mixture.

Canned Tomatoes

Canned tomatoes fill a need for at least 6 months of the year. In fact they are a pantry staple that I keep in plentiful supply year-round. The best canned tomatoes are imported plum tomatoes from the San Marzano region of Italy and are labled as such. They are springy, with a full sweet flavor, and come packed in tomato juice, not water. They are also quite expensive. Other excellent imported tomatoes come from the Naples area and are somewhat more reasonably priced. As for domestic brands, there are 3 or 4 excellent ones, but unfortunately they are not easy to find; they may be available in your local supermarket one week and not the next. Personally, I do not recommend the Vitello or Progresso brands since they are watery and, once you have drained them, you are left with very little tomato pulp. Always buy whole canned plum tomatoes, never those that have been crushed or sliced. Drain tomatoes by leaving them in a colander for several hours at room temperature.

For the best flavor, braise the drained tomatoes gently in extra-virgin olive oil with minced shallots, a touch of sugar, and a sprinkling of fresh herbs for 1 to 1½ hours. Once stewed, the tomatoes can be covered with additional oil and refrigerated in a jar. They will keep for up to 10 days. For a quick sauce, simply puree the tomatoes in the food processor, adding a little chicken stock if the mixture is too thick.

For those who garden or have access to a good farmstand, freezing whole tomatoes is well worthwhile. I follow the Italian method of freezing the Roma plum tomatoes whole, without peeling them. Although they get mushy, they are flavorful and excellent for soups and sauces.

■ ■ ■

FRICASSEE OF BRUSSELS SPROUTS, PEPPERS, AND ONIONS

serves 4–5

As every Mediterranean cook knows, Brussels sprouts must be blanched for five to eight minutes before they can be used in any preparation. This process removes their cabbagy aroma, makes them far more digestible, and readies them for a variety of tasty and easy recipes. A cool-weather sauté of red onions with colorful two-tone peppers becomes a perfect backdrop to the brilliant green sprouts. Large sprouts will look more appealing if cut in half. Serve the fricassee as a side dish to a roast leg of lamb, pan-seared lamb chops, sautéed pork tenderloins, or the Roast Cornish Hens in Garlic Cream with Pan-Sautéed Escarole (page 35).

Salt
1 pound small to medium Brussels
  sprouts, trimmed and rinsed
5 tablespoons extra-virgin olive oil
3 medium red onions, peeled, quartered,
  and thinly sliced

2 large cloves garlic, peeled and thinly
  sliced
1 medium red bell pepper, cored,
  seeded, and cut into fine julienne
1 medium yellow bell pepper, cored,
  seeded, and cut into fine julienne
Freshly ground black pepper

1. In a large saucepan bring salted water to a boil. Add the Brussels sprouts, return to a boil, reduce the heat, and simmer for 7 minutes. Drain the sprouts thoroughly and cut in half through the root end. Set aside.

2. Heat 3 tablespoons of olive oil in a heavy 10-inch skillet over medium-high heat. Add the onions and garlic and cook for 5 minutes, stirring often, until the onions begin to brown. Add the red and yellow peppers. Season with salt and black pepper, reduce the heat, and cook, covered, until the onions are quite soft and the peppers are tender, about 20 minutes.

3. Preheat the oven to 350° F.

4. Add the Brussels sprouts to the skillet, toss with the onion and pepper mixture, and drizzle with the remaining olive oil. Cover the skillet tightly and place in the center of the preheated oven. Bake for 20 minutes or until the

vegetables are bubbling hot. Serve directly out of the skillet as an accompaniment to a roast loin of lamb or the Ragout of Veal with Sherry Vinegar and Shallot Essence (page 418).

Variation: Add ½ cup cubed blanched bacon to the fricassee. Sauté the bacon in a little olive oil until almost crisp, remove it from the skillet with a slotted spoon, and use the remaining bacon fat to cook the onions and peppers. Add 12 double-poached garlic cloves (page 482) to the fricassee, then sauté them in a little olive oil with a touch of sugar and add to the Brussels sprouts.

SAUTÉ OF BROCCOLI RAPE WITH PROSCIUTTO AND PINE NUTS

serves 4

2 pounds broccoli rape
Salt
4 tablespoons olive oil
2 tablespoons pine nuts

½ cup finely diced prosciutto
2 large cloves garlic, peeled and thinly sliced
Freshly ground black pepper

1. Remove all but 2 inches of the stems from the broccoli rape and most of the large leaves. Rinse under warm water and drain. Bring plenty of salted water to a boil in a large pot, add the broccoli rape, and cook for 1 minute. Immediately transfer to a colander and run under cold water to stop further cooking. Drain thoroughly and set aside.

2. In a small skillet, heat 1 tablespoon of oil over low heat, add the pine nuts, and sauté until lightly browned, shaking the skillet to prevent them from burning. Remove from the skillet with a slotted spoon and reserve.

3. Heat the remaining oil in a large skillet over medium-low heat, add the prosciutto, and sauté for 30 seconds. Remove with a slotted spoon to a side dish and reserve.

4. Add the garlic to the skillet and cook for 30 seconds. Add the broccoli rape, season with salt and pepper, and cook for 2 to 3 minutes, tossing lightly in the oil until just heated through. Add the reserved pine nuts and prosciutto and serve hot as an accompaniment to sautéed chicken, pork, or shrimp.

CABBAGE AND LEEK FLAN À LA VAUDOISE

serves 6–8

The traditional Spanish flan is simply an unmolded crème caramel. The word *flan,* however, is used nowadays for savory preparations as well, especially in France where each region may have at least two or three versions. Any number of ingredients are suitable for a savory flan, but the key is to work with those that will not interfere with the custard's delicate consistency and will, at the same time, produce a tasty dish. Zucchini, summer squash, and tomatoes are either too watery or bland, or both, to use in a savory flan.

I find that winter greens and vegetables make some of the very best flans, and the combination of leeks, onions, and cabbage is one of my favorites. Not only do these vegetables enjoy each other's company, they also work well with several cheeses such as goat cheese, Stilton, Fontina, smoked mozzarella, or a combination of two or three of these.

Serve the flan either as a starter to the meal, possibly with a dollop of quick Tomato Fondue (page 474) on the side, or as a side dish to any roast, spooning some of the pan juices over each serving.

½ cup freshly grated Parmesan cheese
2 tablespoons dried bread crumbs
Salt
1 very small head green cabbage, trimmed and quartered
6 large leeks, trimmed of all but 2 inches of greens
4 tablespoons unsalted butter

¼ cup water
Freshly ground white pepper
1 cup diced slab bacon, blanched and drained
1 large onion, peeled and finely minced
¾ cup coarsely grated Gruyère cheese
7 large eggs
1 cup heavy cream

1. Preheat the oven to 350° F. Generously butter a 10-inch porcelain quiche pan, sprinkle with 2 tablespoons of Parmesan and the bread crumbs, and set aside.

2. Bring plenty of salted water to a boil in a large pot, add the cabbage, and cook for 4 minutes. Drain the cabbage and dry thoroughly in a kitchen towel. Remove the core and finely slice the cabbage; you will need only 3 cups for this recipe. Reserve the remaining cabbage for another use.

3. Finely dice the leeks and rinse thoroughly under warm water to remove all sand.

4. In a 10-inch nonstick skillet, melt 2 tablespoons of the butter over medium heat. Add the leeks and water, season with salt and pepper, cover, and

braise for about 5 minutes, or until the leeks are tender and all the water has evaporated. Transfer to a large mixing bowl and set aside.

5. Melt the remaining butter in the skillet, add the bacon, and cook until lightly browned. Remove with a slotted spoon to the bowl containing the leeks. Discard all but 3 tablespoons of fat from the skillet, add the onion, and cook for 2 or 3 minutes, or until soft but not browned. Add the reserved sliced cabbage, season with salt and pepper, and braise, partially covered, for 5 minutes.

6. Add the cabbage mixture and the Gruyère cheese to the leeks and bacon, season with salt and pepper, and set aside.

7. In a medium bowl combine the eggs, heavy cream, and 3 tablespoons of the Parmesan. Season with salt and pepper and whisk until well blended. Pour over the leek and cabbage mixture and fold gently but thoroughly. Correct the seasoning.

8. Spoon the mixture into the prepared pan, sprinkle with the remaining Parmesan, and bake in the center of the preheated oven for 1 hour, or until the top is nicely browned and the tip of a knife inserted in the center comes out clean. Serve warm directly from the pan as a simple main course accompanied by a well-seasoned green salad or as a brunch dish.

Note: The flan can be baked in advance and reheated, covered, in a 250° F oven. Any leftovers can be refrigerated and reheated successfully the next day.

SKILLET-GLAZED CUCUMBERS

serves 6

Sautéed cucumbers have been part of my cooking repertoire for years. I use them mostly in the winter and early spring when seedless cucumbers are in good supply. Their light texture and delicate flavor make a welcome change after weeks of root vegetables, legumes, and cabbages.

A few years ago I found a great gadget at Dehillerin, a wonderful restaurant-supply house in Paris. It was an olive-shaped cutter for which I have found no other use than to turn cucumbers and zucchini into "olives." With the skin left on, the result is a lovely glazed vegetable that surprises you with its rather sweet, assertive flavor and pretty shape. Of course, you can use a melon-ball cutter instead, or simply cut the cucumbers into battonets, which are 1¼-inch sticks about ¼ inch thick. When properly sautéed and glazed, the taste will be the same.

Serve the cucumbers as a side dish to pan-seared salmon, tuna, or scallops. They are also delicious with all dishes using chicken breasts or Cornish hens.

2 seedless (gourmet) cucumbers, unpeeled
Coarse salt
3 tablespoons unsalted butter

2 teaspoons granulated sugar
Salt and freshly ground black pepper
½ cup water

1. Cut the cucumbers lengthwise into quarters. If the cucumbers seem seedy, remove the seeds with a sharp paring knife. Cut into 1¼-inch matchsticks and place in a colander. Sprinkle with coarse salt and let drain for 1 hour.

2. Heat the butter in a 12-inch skillet over medium heat. Sauté the cucumbers for 2 or 3 minutes, add the sugar, and season with salt and pepper. Add the water, bring to a boil, reduce the heat, and simmer the cucumbers until they are crisp-tender and all the water has been absorbed. Taste and correct the seasoning. Serve hot.

Note: Cucumbers have a great affinity to herbs such as dill, mint, chervil, chives, and tarragon. Add the fresh herb just before serving. In warm regions like Florida, where good garden vine-ripe tomatoes are available as early as January and February, add 2 tomatoes, peeled, seeded, and finely chopped, to the cucumbers and braise the 2 vegetables together, adding a touch of fresh dill just before serving. This makes a wonderful side dish to braised red snapper, sautéed grouper, or pompano.

Variation: To make Cucumbers au Fines Herbes: When the cucumbers are almost done, add ½ cup of heavy cream or crème fraîche (page 491), 3 tablespoons finely minced dill, 2 tablespoons minced chives, and/or 2 tablespoons finely sliced mint leaves to the skillet. Simmer the cucumbers for another 3 to 4 minutes, or until they have absorbed the cream.

MARMALADE OF BELGIAN ENDIVES

serves 4–6

4 tablespoons unsalted butter
2 teaspoons peanut oil
2 pounds Belgian endives, cored and cut
  lengthwise into ½-inch julienne

Salt and freshly ground white pepper
Large pinch of granulated sugar

1. In a heavy 12-inch skillet melt the butter with the peanut oil over medium-high heat, add the endives, and do not stir until the endives begin to brown.

2. Season with salt, pepper, and a pinch of sugar. Toss the endives and continue to sauté for 5 to 8 minutes or until they have become nicely browned and caramelized.

3. Serve the endives with a ragout, such as the Ragout of Veal with Sherry Vinegar and Shallot Essence (page 418) or as an accompaniment to sautéed veal or beef.

Variation: Add 1½ tablespoons sherry vinegar to the caramelized endives and cook until reduced to a glaze. Add 1½ tablespoons pure honey, reduce the heat, and braise for another 3 to 5 minutes or until the endives are nicely glazed. Taste and correct the seasoning, adding drops of lemon juice to taste.

BRAISED WINTER LENTILS WITH THYME AND GARLIC

serves 6

2 cups dried lentils
1 medium carrot, peeled and cut in half
1 stalk celery, cut in half
1 medium onion, peeled and stuck with
  a whole clove
Salt
6 cloves Double-Poached Garlic (page
  482)

1 cup heavy cream
3 ounces mild goat cheese, crumbled
3 tablespoons unsalted butter
1 cup finely diced slab bacon, blanched
  and drained
1 tablespoon fresh thyme leaves
Freshly ground black pepper

1. In a 4-quart casserole combine the lentils, carrot, celery, onion, and a

pinch of salt. Add water to cover by 1 inch. Bring to a boil, reduce the heat to low, and simmer, covered, for 20 to 25 minutes, or until the lentils are just tender. Drain, discard the vegetables, and set the lentils aside.

2. In the workbowl of a food processor combine the garlic, heavy cream, and goat cheese and process until smooth. Reserve.

3. Melt the butter in a large, heavy skillet. Add the bacon and cook until lightly browned. Remove all but 2 tablespoons of fat from the skillet and add the lentils, garlic-cream, and thyme. Simmer over low heat until the cream has been completely absorbed by the lentils. Taste and correct the seasoning and serve at once as an accompaniment to roast pork, duck, or grilled salmon.

FENNEL FRICASSEE WITH CARROTS, BACON, AND PEARL ONIONS

serves 4–6

This hearty winter fricassee is more than a vegetable dish or an accompaniment to a main course, it is also one of my favorite starters to a weekday meal. I even serve it as the main course, along with a gutsy cool-weather soup, for a casual Sunday supper. An earthy combination of bacon, onions, and fennel is gently braised to create a deeply flavored and satisfying dish. Other vegetables, such as cubed turnips or diced celery root, can be added with excellent results. The dish need not be prepared at the last minute and, in fact, tastes better when made in advance and reheated in a low oven before serving. Leftovers can also be added to the pan juices of a roasting chicken to make a delicious one-dish meal.

4 small fennel bulbs, trimmed of all but 1 inch of stalks
Salt
2 large carrots, peeled and cubed
1 pint pearl onions
1 tablespoon unsalted butter
1 teaspoon olive oil
1 cup finely cubed slab bacon, blanched and drained

1 teaspoon granulated sugar
Freshly ground black pepper
½–¾ cup Brown Chicken Stock (page 467)
2 tablespoons finely minced fresh parsley

1. Quarter the fennel bulbs, but do not remove the core. In a large saucepan bring salted water to a boil, add the fennel, and cook for 5 minutes. Remove with a slotted spoon to paper towels and reserve. Add the carrots to the saucepan and cook for 3 to 4 minutes, or until just tender. Drain and reserve.

2. Bring water to a boil in a medium saucepan, add the onions, and cook for 1 minute. Drain and, when cool enough to handle, peel and cut a cross in the root to prevent the onions from bursting when cooking. Reserve.

3. In a large, heavy skillet, heat the butter and oil over medium-high heat, add the bacon cubes, and sauté until just crisp. Remove the cubes with a slotted spoon and reserve.

4. Add the onions to the skillet. Sauté for 2 to 3 minutes, or until lightly browned, shaking the skillet back and forth. Add the sugar and continue sautéing for 2 to 3 minutes longer, or until the onions are brown and caramelized. Add the fennel, carrots, and bacon and season with salt and pepper. Add ½ cup of the stock, reduce the heat, and simmer, covered, for 30 to 35 minutes, adding a little stock every 10 minutes. When the vegetables are tender, transfer with a slotted spoon to a serving dish and sprinkle with parsley.

5. If the pan juices have not reduced, bring them to a boil and cook until they are quite syrupy. Spoon over the vegetables and serve as a separate dish accompanied by crusty French bread.

POTATO AND LEEK ROSTI

serves 4–6

Rosti is the national potato dish of Switzerland. The potatoes are cooked in their jackets, cooled, then coarsely grated and fried in butter until crisp on both sides.

Rosti can be either great or barely edible. At the restaurant Kronenhalle in Zurich, one of the best and oldest establishments in Switzerland, the rosti is always fabulous! Crisp, crunchy, and buttery, it is one of those dishes you feel must be easy to duplicate at home, but, as with all such deceptively simple recipes, you can easily miss.

On a recent visit to the Kronenhalle, I had a long discussion about rosti with our waitress: the type of potatoes to use, whether to cook them in butter or oil, the right kind of skillet—nonstick versus old-fashioned cast iron—and, finally, variations on the basic recipe. Our waitress told me that her secret was to add a julienne of melted leeks or onions to the potato mixture and that a combination of oil and butter is better than all butter. I am not quite sure how acceptable the added leeks are to a traditionalist, but I like the combination, especially when serving rosti with seafood and poultry.

5 large red potatoes

2 large leeks, trimmed of all but 2 inches of greens

6 tablespoons unsalted butter

Salt and freshly ground white pepper

2–3 tablespoons water

1 tablespoon peanut oil

1. Bring plenty of water to boil in a large saucepan. Add the potatoes, reduce the heat, and simmer until just tender. Do not overcook. Drain the potatoes and refrigerate for 2 to 4 hours or overnight.

2. Cut the leeks crosswise into 1-inch pieces and then into fine julienne. Rinse thoroughly under warm water, drain, and reserve.

3. In a heavy 10-inch skillet, melt 2 tablespoons of the butter over medium-low heat. Add the leeks and season with salt and pepper. Add the water, cover the skillet tightly, and braise until tender, about 5 minutes. Uncover the skillet, raise the heat, and let all the water evaporate. Transfer the leeks to a colander and set aside.

4. Peel the potatoes and grate through the large holes of a 4-sided grater. Transfer to a large mixing bowl, add the reserved leeks, and toss gently. Season lightly with salt and pepper.

5. Melt 2 tablespoons of butter and half the oil in a 10-inch nonstick skillet over medium heat. Add the potato mixture and press lightly with a spatula into an even layer that covers the bottom of the skillet completely. Cook for 3 to 4 minutes, or until the bottom is nicely browned and crisp. Invert the rosti onto a large plate with the browned side faceup.

6. Add the remaining butter and oil to the skillet and, when hot, slide the rosti back into the skillet. Cook for another 3 minutes, or until the other side is nicely browned.

7. Carefully slide the rosti onto a round serving platter and serve hot, cut into wedges like a pie.

Note: The rosti can be made an hour or two ahead of time and kept warm in a 200° F oven. Leftovers can be wrapped in foil and reheated in a low oven or placed on a plate and briefly reheated in the microwave.

I like to serve a wedge of crisp rosti as a topping for a mixed salad as an appetizer or a light lunch.

You can also add a shower of shaved Gruyère and run the rosti under the broiler.

SHALLOT PUREE WITH GARLIC AND THYME

se·rves 4–5

7 tablespoons unsalted butter
16 large shallots, about ¾ pound,
   peeled and cut in half
1 large clove garlic, peeled and crushed
½ teaspoon fresh thyme leaves

Large pinch of granulated sugar
Salt and freshly ground white pepper
¼–½ cup water
2–3 tablespoons heavy cream
2 teaspoons red wine vinegar

1. In a heavy 3-quart saucepan, melt 4 tablespoons of the butter over medium-low heat. Add the shallots, garlic, and thyme. Sprinkle with the sugar and season with salt and pepper. Add ¼ cup of water, cover, and simmer until very tender, about 20 to 25 minutes, adding a little more water to the pan if the juices run dry.

2. When the shallots are very tender, transfer to the workbowl of a food processor, add the heavy cream and remaining butter, and process until smooth. Add the vinegar and correct the seasoning. Serve warm as an accompaniment to lamb or grilled meats.

Shallots

Most chefs will agree that the fresh shallot with its rich complex flavor is one of the most important flavorants for the preparation of hot and cold sauces and mandatory in innumerable dishes.

Unlike onions, they have the ability to melt down when heated and will actually sweeten sauces such as fish sauce, beurre blanc, tomato fondue, or vinaigrette. Fresh shallots keep better in the refrigerator than in an onion bin unless stored in a very cool, dry place, and minced shallots will keep for weeks in a jar covered with white wine in the refrigerator.

Never mince shallots in a food processor. Try to mince them neatly with a sharp paring knife and avoid going over them again with a large chef's knife. Overwork makes them exude their onion juice and lose their sweetness. Just like garlic cloves, shallots can be stewed in olive oil into a confit or roasted, unpeeled, in the oven until tender. They will slip right out of their skins and are a wonderful accompaniment to many grilled or roasted meats.

■ ■ ■

"LASAGNA" OF SWEET POTATOES, TURNIPS, AND TART APPLES

serves 6

Because of their natural sweetness, yams work well with a variety of fruits as well as root vegetables that are themselves naturally sweet, such as carrots, turnips, and parsnips. This is one of my basic recipes, and I like to improvise on it by adding one or two finely sliced medium carrots and some diced smoked ham or bacon.

This dessertlike gratin is best served with fowl, such as quail, duck, or squab, which harmonize well with sweet side dishes, but it is also delicious as an accompaniment to a fiery chicken or lamb curry. When the weather permits I often serve the gratin with a roast turkey smoked in the Weber kettle over apple or lilac wood (page 247).

3 tablespoons unsalted butter

4 medium turnips (about 1 pound), peeled and thinly sliced

2 medium sweet potatoes (about 1 pound), peeled and thinly sliced

2 tart apples, peeled, cored, and thinly sliced

Salt and freshly ground white pepper

1⅓ cups hot heavy cream

1. Preheat the oven to 350° F. Butter a 2-quart oval au gratin dish with 1 tablespoon of the butter and set aside.

2. Bring water to a boil in a vegetable steamer, add the turnips, and steam, covered, for 4 to 6 minutes or until barely tender. Remove from the steamer and transfer to a large mixing bowl.

3. Add the potatoes to the steamer, cover, and steam for 6 minutes or until barely tender. Remove from the steamer and add to the bowl containing the turnips with the sliced apples. Season with salt and pepper and toss gently.

4. Place the vegetable mixture in the prepared gratin dish, pressing down lightly with a spatula. Pour the hot cream over the vegetables and dot with the remaining butter. Place the dish in the center of the preheated oven, cover with foil, and bake for 1 hour, or until the vegetables are tender when pierced with the tip of a sharp knife and all of the cream has been absorbed.

5. Remove the "lasagna" from the oven, let it cool slightly, and serve, cut into squares.

Note: If you are going to add carrots to the gratin, be sure to peel and quarter them, and then slice them slightly on the diagonal lengthwise. This procedure is best done on a mandoline, but if you don't have one, cut the carrots slightly on the diagonal crosswise, making the slices as large (but not thick) as possible.

SPINACH AND BELGIAN ENDIVE SALAD WITH FRIED GOAT CHEESE

serves 4–6

The success of a spinach salad depends greatly on the type of spinach you use. I find that the most suitable spinach for salads is the flat-leafed variety called New Zealand spinach. It is common in California and other warm-weather regions, but is available in the Northeast only from early spring through the summer. When it comes to cooking spinach, the cool-weather curly variety is far more flavorful, but it lacks the delicate texture necessary for most salads. New Zealand spinach is usually sold in bunches rather than bags and, in fact, looks more like sorrel than true spinach. Because of its rather assertive taste, a salad made with spinach and other character greens, such as Belgian endives and arugula, can be dressed with a full-flavored, zesty vinaigrette, becoming an interesting accompaniment to fried goat cheese or other cheeses such as fried smoked mozzarella.

Since fried goat cheese is almost a staple in my cooking repertoire, I often serve it in a pool of homemade tomato sauce, surrounded by roasted vegetables, or as an accompaniment to roasted peppers which have been cut into thick julienne and dressed in sherry vinegar and virgin olive oil.

1 cylinder (11 ounces) goat cheese, preferably Montrachet
2 tablespoons balsamic vinegar
6–8 tablespoons extra-virgin olive oil
2 teaspoons Dijon mustard
1 large clove garlic, peeled and mashed
8–10 medium radishes, trimmed and thinly sliced
1 small red onion, peeled, cut in half, and thinly sliced

Salt and freshly ground black pepper
5–6 cups lightly packed fresh spinach, stems removed, washed, and dried
4 medium Belgian endives, cored and cut into fine julienne
2 large eggs
2 tablespoons whole milk
All-purpose flour for dredging
1½ cups homemade bread crumbs
1 cup peanut or corn oil

1. Place the Montrachet in the freezer while preparing the salad.

2. In a small jar, combine the vinegar, olive oil, mustard, and garlic. Close tightly and shake until the dressing is well blended.

3. Pour the dressing into a large salad bowl. Add the radishes and onion and toss to coat well. Sprinkle with salt and a large grinding of pepper and marinate for 30 minutes. Top with the spinach and endives, but do not toss. Cover and chill until serving time.

4. In a shallow bowl, combine the eggs and milk and lightly beat. Place the flour and bread crumbs on individual plates or wax paper. With a sharp knife cut the goat cheese crosswise into ¼-inch slices. Dip each slice lightly in the flour, then into the egg-milk mixture, and then in the bread crumbs. Place the breaded slices on a cookie sheet and chill for 30 minutes.

5. In a heavy 10-inch skillet, heat the peanut oil over medium-high heat. When very hot, add the cheese slices and sauté quickly until nicely browned on both sides. You may have to sauté in 2 batches. Transfer the cheese slices to a double layer of paper towels and reserve while you finish the salad.

6. Toss the salad and correct the seasoning. Place on individual salad plates and top each portion with 2 or 3 of the fried cheese slices. Serve immediately.

Note: The entire salad can be prepared a day ahead. Do not toss the salad until serving. Prepare the fried goat cheese disks at any time and freeze them. Just before serving, drop them into the hot oil, still frozen. They are done when a bit of melted cheese appears on the surface.

BELGIAN ENDIVE AND STILTON CHEESE SALAD IN ANCHOVY AND SHALLOT VINAIGRETTE

serves 4

SALAD
2 flat anchovy fillets, drained and finely minced
2 tablespoons finely minced shallots
1½ tablespoons sherry vinegar
5 tablespoons walnut oil
Salt and freshly ground black pepper

6 large endives, cored and cut into fine julienne

GARNISH
3 ounces Stilton cheese, crumbled
2 tablespoons finely diced radish
2 tablespoons finely diced cucumber

1. In a medium bowl combine the anchovies, shallots, vinegar, and oil,

whisk until well blended, and season with salt and pepper. Top with the endives, but do not toss. Cover and refrigerate until serving.

2. Twenty minutes before serving, toss the salad gently and divide between 4 individual serving plates. Sprinkle each portion with some of the cheese, garnish with radish and cucumber, and serve at once.

FENNEL, RADISH, AND CUCUMBER SALAD WITH SHERRY VINEGAR AND LEMON VINAIGRETTE

2 large fennel bulbs, trimmed and cored
1 small seedless (gourmet) cucumber, peeled
Salt
2 cups thinly sliced radishes
2 tablespoons sherry vinegar

2 teaspoons lemon juice
2 tablespoons finely minced shallots
6 tablespoons extra-virgin olive oil
Freshly ground black pepper

GARNISH
12 tiny black oil-cured olives

1. Cut the fennel bulbs in half and then crosswise into thin slices. Place the slices in a salad bowl and reserve.

2. Cut the cucumber in half lengthwise and remove all the seeds with a grapefruit spoon. Cut the cucumber crosswise into thin slices. Place the slices in a colander, sprinkle with salt, and let drain for 1 hour. Dry the cucumber slices on paper towels and add them to the fennel together with the sliced radishes.

3. In a small bowl combine the sherry vinegar, lemon juice, shallots, and olive oil. Whisk the dressing until smooth and creamy.

4. Pour the dressing over the salad, season with salt and pepper, and toss gently. Cover and chill until serving. Just before serving, garnish the salad with black olives.

WINTER SALAD WITH A JULIENNE OF OMELET CREPES

serves 6

In a country where pancakes and fritters appear on the menus of diners and four-star restaurants alike, I am always surprised that the omelet crepe has never caught on. It is much simpler to make than either an omelet or a crepe and is one of those versatile foods that deserves to be more appreciated. Omelet crepes are pretty standard fare in Austrian cooking, where they are almost always cut into julienne and added to chicken or beef consommé. They are wonderful filled with leftover ratatouille, a creamy goat cheese, and thyme, or braised spinach with a few cubes of diced bacon. A tossed well-seasoned salad topped with a julienne of warm omelet crepes doubles as a light lunch. For a variation, add a julienne of smoked ham and a fine julienne of Gruyère or Comte and serve with crusty bread and some sweet butter.

SALAD
8 cups mixed salad greens, including
   Bibb, radicchio, romaine, and frisee
1–1½ recipes Lemon and Sherry
   Vinaigrette (page 480)

CREPES
4 large eggs
2 tablespoons sifted all-purpose flour
2 tablespoons heavy cream

¼ teaspoon salt
Freshly ground white pepper
3 tablespoons finely minced scallions
2 tablespoons unsalted butter

Salt and freshly ground black pepper

GARNISH
Fillets of flat anchovies, drained
Tiny black oil-cured olives

1. Start by making the salad: Separate the leaves of the various greens, rinse well, and dry thoroughly.

2. Place the Vinaigrette in the bottom of a large salad bowl and top with the greens, but do not toss. Cover and refrigerate until serving.

3. To prepare the crepes: In a medium mixing bowl combine the eggs, flour, heavy cream, and salt. Whisk together until smooth and well blended. Season with pepper and fold in the scallions. (The batter will yield 6 crepes.)

4. Lightly butter a 6-inch crepe pan and place over medium heat. Add one sixth of the egg mixture and cook until the omelet is set, tilting the pan back and forth while loosening the omelet with the tip of a sharp knife and lifting it to let any extra batter seep underneath. The omelet will be quite thick. When

lightly browned, carefully turn the omelet over with the aid of a spatula and cook for 1 minute more, or until lightly browned.

5. Slide the omelet onto a cutting surface and make 5 more omelets with the remaining batter, always brushing the pan with butter before adding the batter. When all the omelets are done, cut them crosswise into thin strips and set aside.

6. Just before serving, add the omelet strips to the salad bowl and toss the salad gently but thoroughly. Taste and correct the seasoning. Garnish with anchovies and black olives and serve lightly chilled.

ROASTED PEAR SALAD WITH WALNUT AND SHALLOT VINAIGRETTE

serves 6

I was once told that spontaneity is the hallmark of a good cook. This simple salad is just the kind of dish to make when you see the right pears in the market. Choose semi- or almost ripe pears. In fall, you will find Bartletts in good supply; in winter, check out the Anjou pears, which are often quite green at this time of year; and later, in the spring, you may come upon a good batch of Boscs. All three are delicious in this salad. Serve the pears lukewarm with well-seasoned greens, some creamy Bleu de Bresse, or just coarsely chopped walnuts. Be sure to choose buttery greens and mix them with some Belgian endives, which will not conflict with the mild taste of the pears.

3 ripe Bartlett or Anjou pears
1 tablespoon granulated sugar
⅓ cup Sauternes or Riesling
Large grinding of black pepper
Large grating of freshly grated nutmeg
3 tablespoons unsalted butter
2 tablespoons minced shallots
2 tablespoons sherry vinegar
6–8 tablespoons walnut oil

Salt and freshly ground black pepper
6 cups assorted young greens, including frisee, Bibb, Belgian endive, arugula, oak leaf, and corn lettuce

GARNISH
6 ounces Gorgonzola Dolce or other mild, creamy blue cheese, shaped into 6 small balls

1. Preheat the oven to 375° F.

2. Peel the pears, cut in half, and core. Place the pears, cut side down, in a well-buttered baking dish. Sprinkle with sugar. Add the wine, season the pears with pepper and nutmeg, and dot with the butter. Bake for 25 minutes, or until the pears are quite tender but not falling apart. Transfer the pears to a side dish and reserve.

3. Place the baking pan over direct heat and reduce the pan juices to 2 tablespoons. Transfer the pan juices to a bowl, add the shallots and vinegar, and whisk in the walnut oil slowly; the dressing should be well emulsified. Taste and correct the seasoning.

4. Transfer the dressing to the bottom of a large salad bowl. Add the greens and toss lightly. Divide the greens between 6 serving plates, place a ball of Gorgonzola in the center of each plate, and top with half a roasted pear. Serve immediately, accompanied by crusty French bread.

VIENNESE CHOCOLATE SOUFFLÉ CAKE

serves 10–12

I adore all chocolate cakes, and especially the dense, fallen-soufflé types, where even one bite is so satisfying at the end of a meal. Here is a variation of such a cake that will give you plenty of room for variation by adding diced candied fruit to the mousse. Flavorings such as rum, cognac, or Grand Marnier can also further enhance this cake. I always make the cake the day before, cover it tightly, and serve it the next day, garnished with dollops of whipped cream and showered with powdered sugar. For a dressier presentation, serve the cake in individual portions on a pool of white chocolate sauce (page 488) and garnished with caramelized sautéed pears.

9 ounces bittersweet chocolate, preferably Lindt or Tobler, broken into pieces

2 teaspoons instant coffee dissolved in 2 tablespoons water

9 large eggs, separated

3 tablespoons plus 1⅓ cups granulated sugar

9 ounces unsalted butter, softened

3 tablespoons unsweetened cocoa

4 ounces shelled pecans, coarsely chopped, optional

Confectioners' sugar for dusting

1. Preheat the oven to 200° F. Generously butter the bottom and sides of 2 (9-inch) round cake pans with removable bottoms or 2 (9-inch) springform pans. Sprinkle each with flour, shaking out the excess, and set aside.

2. Combine the chocolate and dissolved instant coffee in a heavy-bottomed 2-quart saucepan and place in the center of the preheated oven until the chocolate has just melted, about 5 minutes. Remove from the oven, stir until smooth, and reserve.

3. Raise the oven temperature to 350° F.

4. In the bowl of an electric mixer, combine the egg whites and 3 tablespoons of sugar and beat on medium speed for 3 minutes, or until the whites are shiny and hold firm peaks. Transfer to a large mixing bowl and set aside.

5. Add the remaining 1⅓ cups sugar and the softened butter to the bowl of the mixer and beat on medium speed until creamy. Add the egg yolks and beat

until fluffy. Reduce the speed to very low, add the cocoa and melted chocolate mixture, and beat until thoroughly combined.

6. Transfer the chocolate-butter-egg mixture to a large stainless-steel mixing bowl. Fold in one third of the beaten egg whites to lighten the mixture, then fold the lightened chocolate mixture into the remaining beaten egg whites, gently but thoroughly. Fold in the optional pecans, again gently but thoroughly.

7. Measure 1 cup of the chocolate mixture and set aside. Divide the remaining chocolate mixture evenly between the 2 prepared pans and spread evenly with a spatula. Place in the center of the preheated oven and bake for 35 minutes. The cakes will still be slightly moist when a toothpick is inserted.

8. When the cakes are done, remove from the oven and let cool for 20 minutes on wire racks. Carefully remove the sides of the pans and invert 1 cake onto a round cake platter. Carefully remove the bottom of the cake pan and spread the reserved chocolate mixture in an even layer over the warm cake. Invert the second cake on top of the first and carefully remove the bottom of its cake pan. Cool completely, cover, and refrigerate overnight.

9. The next day remove the cake from the refrigerator 30 minutes before serving. Dust heavily with confectioners' sugar and serve, cut into wedges.

Note: After the cake has cooled completely, you may, instead of refrigerating, serve the cake at room temperature the same day it is baked.

TWO-TONE CHOCOLATE PÂTÉ WITH COFFEE CRÈME ANGLAISE

It is amazing how much food terminology, once reserved for very specific preparations, has been so generally applied that almost nothing really means what it used to. The flan is a good example, and so is the pavé. Now you will find many dishes called pavés and pâtés, both savory and sweet, made with leeks, asparagus, seafood, or whatever the creative cook may come up with. A chocolate pâté is therefore a pâté only in the sense that it is molded into a long and narrow rectangular loaf pan, then served in thin slices much like a pâté.

This pâté is a wonderfully rich dessert, the perfect ending to a cool-weather meal. It can be served in a number of creative ways: sitting in a pool of white chocolate sauce, over a light fresh purée of poached pears, a mocha crème anglaise, or simply accompanied by a bowl of whipped cream; all work well with this dense yet creamy truffle-textured chocolate dessert.

3 ounces hazelnuts, with skins
2 tablespoons superfine sugar
Flavorless oil
3 ounces white chocolate, preferably Lindt, broken into small pieces
6 ounces bittersweet chocolate, preferably Lindt, broken into small pieces

2 tablespoons strong prepared coffee
9 ounces unsalted butter, softened
2 large eggs, separated
¼ cup dried currants, soaked in 2 tablespoons scotch whiskey, optional
Pinch of salt

GARNISH
Coffee Crème Anglaise (page 490)

1. Preheat the oven to 300° F.

2. Place the hazelnuts on a baking sheet and bake in the center of the preheated oven for 8 minutes or until lightly toasted and aromatic. Remove from the oven and, while still warm, place the nuts in a kitchen towel and rub together to remove most of the skins.

3. Transfer the hazelnuts to the workbowl of a food processor, add the sugar, and pulse on and off until the nuts are finely ground. Reserve.

4. Lightly oil a narrow 4-cup French terrine with flavorless oil. Line the

terrine with a sheet of parchment paper that extends 2 inches over each long side. Set aside.

5. Place the white chocolate in a medium saucepan, and the dark chocolate and coffee in another medium saucepan. Place both pans in the center of the oven and set aside until both chocolates have just melted. Remove from the oven, stir until smooth, and transfer each batch of chocolate to a separate stainless-steel mixing bowl. Set both chocolates aside to cool slightly; they should still be lukewarm to the touch.

6. While the chocolates are cooling, combine the butter and egg yolks in an electric mixer and beat until smooth and fluffy.

7. Whisk one third of the butter–egg yolk mixture into the white chocolate until smooth and creamy. If the chocolate begins to look curdled, do not worry; keep whisking, and after about 3 seconds it will become quite smooth. Set aside.

8. Add the remaining butter–egg yolk mixture to the dark chocolate, whisking until smooth. Fold in the ground hazelnuts and currants, and set aside.

9. In a large mixing bowl combine the egg whites with a pinch of salt and beat until soft peaks form. Fold one third of the whites into the white chocolate gently but thoroughly. Fold the remaining whites into the dark-chocolate mixture.

10. Spoon half of the dark-chocolate mixture into the prepared terrine. Spread evenly with a spatula and place in the freezer for 10 minutes.

11. Spoon the white-chocolate mixture into the terrine and smooth evenly. Place in the freezer for 10 minutes. Repeat the procedure with the remaining dark-chocolate mixture, cover, and refrigerate overnight.

12. The next day run a knife along the short sides of the terrine. Unmold onto a narrow rectangular serving platter. Peel off the parchment paper and serve, cut into ½-inch slices, accompanied by the coffee crème anglaise.

DOUBLE ESPRESSO AND CARAMEL CUSTARDS

serves 8

A simple pot de crème flavored with vanilla, chocolate, or coffee is pretty much of a classic. When served in elegant Limoges ramekins, it becomes a most impressive dessert. Although this old-fashioned custard has practically disappeared from the menus of chic restaurants, you can still find wonderful versions at many bistros. I continue to make it, since I find it to be a perfect backdrop to seasonal embellishments. Topped with raspberries, blackberries, stewed blueberries, or a dollop of fresh strawberry coulis, it is both refreshing and eye-appealing. In cool weather, I prefer flavorings such as coffee, chocolate, and citrus.

I think this recipe is probably the best of all: caramelized cream enhanced by strong coffee, a combination that is both rich and pleasing.

1 teaspoon instant espresso coffee
¼ cup strong brewed coffee
2 cups heavy cream

1 cup granulated sugar
½ cup water
8 large egg yolks

1. Preheat the oven to 350° F.

2. Combine the instant espresso, strong coffee, and heavy cream in a medium saucepan, bring to a simmer, and keep warm.

3. In a large saucepan, combine the sugar and water and bring to a boil, stirring once to dissolve the sugar. Continue to boil, without stirring, until the mixture turns a hazelnut brown. Remove the saucepan from the heat, and very slowly add the warm cream mixture, averting your face and whisking constantly. Return the saucepan to low heat and whisk until the caramel is completely dissolved.

4. Transfer the caramel cream to a stainless-steel mixing bowl and cool slightly. Add the egg yolks and whisk until well blended. Pour through a strainer lined with a double layer of cheesecloth and placed over a bowl. Divide the mixture between 8 (5-ounce) porcelain ramekins.

5. Place the ramekins in a large shallow baking dish and place the dish in the center of the preheated oven. Fill the baking dish with boiling water to come halfway up the sides of the ramekins and bake for 15 to 18 minutes, or until the custards are just set.

6. Remove the ramekins from the oven and let cool until just warm. Serve warm or at room temperature.

CRISP BROWN SUGAR AND ALMOND COOKIES

makes 4 1/2–5 dozen

These lacy, buttery cookies are the house specials of Comme Chez Soi, the three-star restaurant in Brussels. There they are made fresh twice a day as part of the platter of petits fours and truffles that is served with coffee after dessert.

In the week I spent in their kitchen, I managed to consume a large quantity of these delicious morsels, realizing, a little too late, that cookie crumbs are as fattening as whole cookies.

For a non-cookie baker such as myself, these fragile, lacy disks have become a year-round favorite. Although the dough can be made at any time and frozen for later use, they do not hold up well in humid weather, and I therefore reserve them for the cooler months.

It is best to be spontaneous when it comes to making these cookies. Whip up a batch or two of the dough, wrap it tightly, and freeze. When you feel like it, bake a few cookies and serve them with good homemade ice cream, a chocolate mousse, or a selection of two or three homemade sorbets.

¾ cup slivered almonds, blanched
1⅓ cups dark-brown sugar, tightly packed

8 tablespoons unsalted butter, cut into 8 pieces
1½ teaspoons water
1 cup all-purpose flour

1. Preheat the oven to 350° F.

2. Place the almonds on a baking sheet and bake in the center of the preheated oven for 5 minutes, or until lightly toasted. Remove from the oven and reserve.

3. Raise the oven temperature to 400° F and generously butter 4 or 5 baking sheets. Set aside.

4. In the workbowl of a food processor, combine the toasted almonds and brown sugar and process until the nuts are very finely ground. Add the butter, water, and flour and process until thoroughly combined.

5. Divide the cookie dough in half and, on a lightly floured surface, roll each portion into a cylinder, 1½ inches in diameter. Wrap each cylinder in plastic wrap and refrigerate for 45 minutes to 1 hour, or until firm.

6. With a sharp knife, cut each cylinder crosswise into ¼-inch slices and place the slices on the prepared cookie sheets about 2 inches apart. Bake in the center of the preheated oven for 4 to 5 minutes, or until lightly browned.

7. Remove the cookies from the oven and let rest for 5 minutes on the baking sheets. With a metal spatula, carefully transfer the cookies to wire racks and cool completely. The cookies should be thin, crisp, and lacy.

Note: After the dough is shaped into cylinders and wrapped, it may be frozen and saved for a later use. When ready to use, thaw in the refrigerator until the cylinders can be cut easily with a knife.

CREPES WITH CARAMELIZED PEAR AND RAISIN JAM

serves 6–8

I first had a version of this dish at one of those freestanding fast-food creperies in Normandy. The crepe was dry, as they usually are when bought at one of these stands, but the caramelized-pear *tatin*like filling was delicious. When asked about it, the young woman making the crepes told me that her father owned a bakery nearby and that the filling was indeed the same they used for a pear tatin tart. I re-created the recipe using buttery homemade crepes, and it's a winner, particularly if you like pears as much as I do.

½ cup golden raisins, plumped
¼ cup pear brandy
10 tablespoons unsalted butter
1 cup granulated sugar
2 teaspoons fresh lemon juice
½ teaspoon baking powder
6–8 slightly underripe pears (3 pounds)

peeled, cored, and cut into ½-inch dice
⅓ cup Grand Marnier or other orange liqueur
1–2 navel oranges, zest removed and cut into fine julienne
12–16 Dessert Crepes (page 493)
Good-quality vanilla ice cream

1. In a small mixing bowl combine the raisins and pear brandy. Set aside.

2. In a heavy 3-quart saucepan, melt the butter over medium-low heat. Add the sugar and lemon juice. Raise the heat to high, stirring once to dissolve the sugar, and boil the mixture until it starts to turn a hazelnut brown. Immediately add the baking powder and stir until dissolved. Add the diced pears and stir until the foaming subsides. Cook the pears for 7 to 10 minutes, or until they are tender and golden brown. Remove the pan from the heat, transfer the pears to a bowl, using a slotted spoon, and set aside. Reserve the sauce.

3. Preheat the oven to 350° F.

4. Add the raisins, together with the Grand Marnier and orange julienne, to the sauce. Return the saucepan to the heat and simmer over medium-low heat for 1 to 2 minutes, or until the orange is tender. Set the sauce aside.

5. Place 2 tablespoons of the cooked pears toward the bottom of each crepe. Roll up the crepes to enclose the filling and place them, seam side down and side by side, in a shallow baking dish. Pour the reserved sauce over the crepes, cover with foil, and bake for 20 to 25 minutes, or until very hot.

6. Place 2 crepes on each individual serving plate. Spoon a little of the pan juices over the crepes and place a small scoop of vanilla ice cream beside them. Serve at once.

MOCHA CREPES WITH COFFEE SABAYON

serves 6

Adding a touch of coffee to a basic dessert crepe batter may turn the crepes a toasty hazelnut color, but it does not impart an assertive coffee taste to them. Filling the crepes with a coffee sabayon does the trick, highlighting the intense coffee taste and providing a wonderful soufflélike texture to this dessert.

CREPES
1½ cups all-purpose flour, sifted
1 tablespoon granulated sugar
3 large eggs
1½ cups whole milk
1 tablespoon strong prepared coffee
½ teaspoon vanilla extract
Pinch of salt
¼ cup unsalted butter, melted

SAUCE
½ cup granulated sugar
3 tablespoons plus ⅓ cup hot water
6 tablespoons unsalted butter, softened

FILLING
1 recipe Espresso Sabayon Gritti (page 363)

GARNISH
Heavy dusting of confectioners' sugar

1. Start by making the crepes: In the workbowl of a food processor combine all the ingredients except for the melted butter and process until smooth. Transfer to a mixing bowl, whisk in the melted butter, and let stand at room temperature for 1 hour. Make the crepes following the directions for Dessert Crepes on page 493.

2. Preheat the oven to 200° F.

3. Wrap the crepes in aluminum foil and place in the center of the preheated oven to keep warm while making the sauce.

4. Next, prepare the caramel-butter sauce: In a medium saucepan, combine the sugar and the 3 tablespoons of hot water. Bring to a boil over high heat, stir once to dissolve the sugar, and continue to boil without stirring until the mixture turns a hazelnut brown. Remove from the heat and immediately add the remaining ⅓ cup hot water, averting your face. When the bubbling has subsided, return the pan to low heat and stir until smooth. Add the soft butter and again whisk until the sauce is smooth. Keep warm.

5. To serve, remove the crepes from the oven, place a spoonful of the coffee sabayon in the center of each warm crepe, and fold in quarters. Place 2 to 3 crepes on each individual serving plate and drizzle with some of the caramel butter sauce. Dust with confectioners' sugar and serve at once.

Note: Dessert crepes can be prepared 2 to 3 days ahead of time. Personally, I prefer not to freeze them and usually refrigerate them on a large round plate with a sheet of wax paper between each crepe. It is a good idea to double this batter, flavoring one part with coffee and the other with an orange liqueur and some vanilla extract. If using orange-flavored liqueur, substitute for coffee sabayon a filling made with either Cointreau or Grand Marnier. Top these crepes with the orange-butter sauce on page 112.

CRANBERRY, APPLE, AND PEAR MOUSSE

serves 8

My earliest taste of American cranberries was in Paris in the early sixties. There I was invited to my very first American Thanksgiving dinner, and the hostess introduced us to Ocean Spray cranberry sauce, which she was able to find at Fauchon, the elegant gourmet shop on place Madeleine. I was immediately intrigued by the texture and slightly sour taste, especially after I bought my own can of the sauce and doused it liberally with sugared crème fraîche. Little did I know at the time how much better a homemade cranberry sauce could be. Now, as soon as I see fresh cranberries in the market, I make batches of my own sauce and relish, which I serve with pancakes at breakfast and as a side dish to roast pork, duck, and turkey. But I still believe that cranberries deserve to be featured more prominently at the dessert table, and this mousse evolved spontaneously one day out of my desire to establish this unique fruit as the star ending for a cool-weather menu.

Flavorless oil

JAM
2½ cups (10 ounces) fresh cranberries
1 ripe Golden Delicious apple, peeled, cored, and very finely diced
1 ripe Bartlett pear, peeled, cored, and very finely diced
⅔ cup granulated sugar
⅓ cup golden raisins
⅓ cup fresh orange juice
1 teaspoon ground cinnamon
1 teaspoon finely grated orange zest

MOUSSE
4 large egg yolks
1 teaspoon cornstarch
⅔ cup granulated sugar
1½ cups warm milk
1 tablespoon unflavored powdered gelatin
1 teaspoon vanilla extract
1 cup heavy cream, whipped

GARNISH
Sprigs of fresh mint
Whipped cream

1. Lightly coat the inside of a 6-cup ring mold with a flavorless oil such as almond or cold-pressed peanut oil.

2. Start by making the cranberry jam: Combine all the cranberry ingredients in a heavy 3-quart saucepan. Bring to a boil, reduce the heat, and simmer, uncovered, for 30 to 35 minutes, or until the mixture is thick, stirring often. Transfer to the workbowl of a food processor and process until smooth.

3. To prepare the mousse: In a large mixing bowl, combine the egg yolks, cornstarch, and sugar and whisk until pale yellow. Add the warm milk and whisk until well blended. Transfer the mixture to the top of a double boiler,

place over simmering water, and whisk constantly until the mixture heavily coats a spoon, about 6 to 8 minutes. Be careful not to curdle the egg yolks. Add the powdered gelatin and vanilla and whisk until the gelatin is dissolved. Immediately transfer the custard to a large mixing bowl and refrigerate until it just begins to set. Do not let the custard set fully.

4. When the custard begins to set, fold in the reserved cranberry puree. Add the whipped cream and fold gently but thoroughly. Transfer the mixture to the prepared mold, smooth evenly with a spatula, cover, and chill until set, about 4 to 6 hours.

5. To unmold the mousse, dip the pan briefly into warm water and invert onto a serving platter. Garnish with sprigs of mint and serve the whipped cream on the side.

Note: You may also serve the cranberry jam as a side dish. In that case, cut the apple and pear a little larger and do not puree in the food processor. You may double or triple the cranberry jam (this recipe yields 3 cups) and store in jars for a later use.

MERINGUE-TOPPED LEMON-BUTTER TART
serves 6–8

I always like either a pear or lemon dessert after a hearty winter main course. Even though this tart is rather rich, the clean lemon taste makes it a light, refreshing, and welcome dessert. You can garnish each serving plate with a dollop of cranberry jam (page 454) or a small pool of either strawberry or raspberry coulis made with the frozen sugarless fruit. I prefer to reserve these garnishes for the spring or early summer, when the fresh fruit is available, but these sauces add a nice and colorful finishing touch to the dessert plate, whenever they're made.

CANDIED LEMON
1/3 cup granulated sugar
2 tablespoons water
Zest of 1 lemon, cut into fine julienne, blanched and drained

LEMON BUTTER
2 large eggs
2 large egg yolks
1/2 cup granulated sugar
1/2 cup fresh lemon juice
Zest of 1 lemon, finely minced

10 tablespoons unsalted butter, cut into 10 pieces and chilled

1 prebaked 9-inch Basic Tart Shell (page 492), in a pan with a removable bottom

MERINGUE
4 large egg whites
Pinch of salt
1/2 cup granulated sugar
Confectioners' sugar

1. Start by making the candied lemon: In a heavy 2-quart saucepan, combine the sugar and water, bring to a boil, and continue to boil for 3 minutes. Reduce the heat, add the blanched julienne of lemon zest, and cook for 2 minutes longer. With a slotted spoon, remove the zest to a wire cake rack and reserve.

2. To prepare the lemon butter: In a medium stainless-steel mixing bowl, combine the eggs and egg yolks, add the sugar, lemon juice, and lemon zest and whisk until well blended. Transfer the mixture to another 2-quart saucepan, preferably copper. Place over medium-low heat, add the butter, 2 pieces at a time, and whisk until the butter is completely incorporated and the mixture is thick, about 3 to 4 minutes. Adjust the heat so that the mixture does not come to a boil or the eggs will curdle.

3. Immediately transfer the lemon-butter to a stainless-steel mixing bowl, place a piece of plastic wrap directly on the surface of the lemon-butter, and chill for 6 hours or overnight.

4. Preheat the broiler.

5. Spoon the chilled lemon-butter into the prebaked tart shell and spread evenly with a spatula. Set aside.

6. To make the meringue: In a large mixing bowl combine the egg whites and pinch of salt. Beat with an electric mixer at high speed until the whites hold soft peaks. Add the granulated sugar, 2 to 3 tablespoons at a time, beating constantly, until the egg whites are firm and shiny.

7. Spread two thirds of the meringue in a smooth, even layer over the lemon filling to cover it completely. Spoon the remaining meringue into a pastry bag fitted with a ½-inch star tip, and pipe the meringue into a crisscross pattern over the top, then decorate with a scalloped border. Sprinkle heavily with confectioners' sugar and place in the broiler, 6 inches from the source of heat. Broil until nicely browned, 1 to 2 minutes. Be careful not to burn.

8. Sprinkle with the candied lemon peel, cool, and cut into wedges. Serve within 1 hour.

ORANGE AND GRAPEFRUIT SALAD IN
PINEAPPLE AND BLACK PEPPER SYRUP

serves 6

Can a fruit compote be as memorable as a chocolate soufflé? Well, almost. When I first tasted this medley of perfectly sliced fresh citrus in a pool of spicy black pepper–pineapple syrup, I thought it was one of the most interesting compotes I had ever tasted. It was served as part of the spectacular breakfast tray at Jean Bardet Restaurant in Tours, and since my own breakfasts tend to be much more modest in scale, this refreshing combination of oranges, grapefruit, and pineapple quickly joined the ranks of my cool-weather dessert repertoire.

Be sure to slice the fruit no more than three to four hours ahead of time. Leftovers will not look as attractive, but will taste even better.

6 large navel oranges
2 small ruby-red Indian River grapefruits
Zest of 1 large orange, cut into fine
  julienne
2 tablespoons grenadine
1½ cups rosé wine

1½ cups unsweetened pineapple juice
1¼ cups granulated sugar
8 whole black peppercorns
1 (2-inch) piece of vanilla bean
¼ cup tiny fresh mint leaves

1. With a very sharp paring knife, peel the oranges and grapefruits, removing all the rind and white pith. Cut the flesh between the membranes into segments and transfer the segments to a mixing bowl. Set aside.

2. Bring water to a boil in a small saucepan, add the julienne of orange zest, and cook for 5 minutes. Drain, place in a small bowl, add the grenadine, and reserve.

3. In a heavy 2-quart saucepan combine the wine, pineapple juice, sugar, peppercorns, and vanilla bean. Bring to a boil, reduce the heat, and simmer until the mixture is reduced to 1½ cups. Remove from the heat and let cool completely.

4. Drain the orange and grapefruit segments and reserve ¼ cup of their juices. Combine the segments, the reserved ¼ cup juice, and the pineapple-wine syrup in a mixing bowl. Toss gently and let the fruit macerate for 30 minutes.

5. Just before serving, drain the julienne of orange zest and add it with the mint to the fruit salad. Stir gently and serve chilled in individual glass bowls or wine goblets.

POTPOURRI OF WINTER FRUIT IN
VANILLA-RUM SYRUP

serves 6–8

SYRUP
½ cup granulated sugar
3 tablespoons water
1 (3-inch) piece of vanilla bean, split
  lengthwise
½ cup dark rum
3 tablespoons Grand Marnier or other
  orange liqueur
¾ cup golden raisins

FRUIT
2 large navel oranges
2 cups fresh pineapple, cored and cut
  into ½-inch cubes
1 large Golden Delicious apple, peeled,
  cored, and thinly sliced
2 large bananas, peeled and sliced
⅓ cup lightly toasted pine nuts, optional

1. Start by making the syrup: In a small saucepan, combine the sugar, water, and vanilla bean. Bring to a boil, reduce the heat, and simmer for 2 minutes, or until the sugar is completely dissolved. Transfer to a large bowl and remove the vanilla bean. Scrape the seeds from the vanilla bean and return them to the syrup together with the rum, Grand Marnier, and raisins. Let macerate at room temperature for 48 hours.

2. To prepare the fruit: Peel the oranges with a very sharp knife, removing the zest and all of the white pith. Cut the flesh between the membranes into segments, and cut each segment in half crosswise.

3. Add the orange segments, pineapple, apple, and bananas to the vanilla-rum syrup. Toss gently, cover, and refrigerate until just cold, about 20 minutes.

4. Serve chilled, sprinkled with the optional pine nuts.

Macerated Raisins

For a wonderful spur-of-the-moment garnish for all kinds of desserts there are few better choices than a spoonful of macerated raisins.

1 cup golden or dark raisins
1/3 cup granulated sugar
1/3 cup water

3/4 cup rum, Cointreau, Grand Marnier, cognac, or a combination

1. In a small saucepan combine the raisins with water to cover. Bring to a boil, reduce the heat, and simmer for 3 minutes. Drain.
2. In another saucepan combine the sugar and 1/3 cup of water, bring to a boil, and cook for 3 minutes, or until the sugar is completely dissolved. Remove from the heat, add 3/4 cup of rum or the liqueur of your choice and the raisins, and let cool. Transfer to a jar, seal, and let the raisins macerate for a week before using. They will keep, at room temperature or in the refrigerator, for up to 3 months.

■■■

ROASTED PEARS WITH BROWN SUGAR AND GINGER CRISP

serves 6

6 ripe medium pears, preferably Bartlett
1/4–1/3 cup granulated sugar
1 teaspoon ground ginger
1 tablespoon finely grated orange zest

3/4 cup dark-brown sugar, tightly packed
3/4 cup all-purpose flour
8 tablespoons unsalted butter, 6 tablespoons cut into 1/2-inch pieces

1. Preheat the oven to 375° F. Peel the pears and cut into quarters or if the pears are large, cut into eighths. Place in a bowl and sprinkle with the granulated sugar, ginger, and grated orange zest. Set aside.

2. In a small bowl combine the brown sugar, flour, and the 6 tablespoons of butter cut into 1/2-inch pieces. Work the mixture with your hands until it resembles cornmeal. Reserve.

3. In a large oval baking dish heat the remaining butter, add the pears in 1 layer, and sprinkle with the brown-sugar topping. Place the dish in the center of the oven and bake for 30 minutes, or until golden and crisp.

4. Cool for 30 minutes before serving. Serve warm with vanilla ice cream or lightly sweetened whipped cream flavored with 2 tablespoons sour cream.

Note: If the pears are underripe, you may need as much as 1/2 cup of granulated sugar. You may substitute Golden Delicious apples, peeled and cored, for the pears.

B A S I C S

BASIC RECIPES

Brown Chicken Stock

Brown Lamb Stock

Chicken Stock

White Stock

Fish Stock

Red Wine Fish Stock

Tomato Fondue

Quick Tomato Sauce

Fresh Tomato Sauce

Mayonnaise

Aioli

Clarified Butter

Beurre Manie

Lameloise Sauce

Provençal Vinaigrette

Sherry and Lemon Vinaigrette

Honey-Cumin Vinaigrette

Port and Shallot Vinaigrette

Emerald Vinaigrette

Double-Poached Garlic

Roasted Garlic

Roasted Red or Green Bell Peppers

Roasted Shallots

Cooked White Beans

Bouquet Garni

Garam Masala

Fresh Pasta

Pizza Dough

White Chocolate and Mint Sauce

Basic Chocolate Sauce

Praline Powder

Crème Anglaise

Crème Fraîche

Marinated Orange Julienne in Grenadine

Basic Tart Shell

Dessert Crepes

Many of the recipes in this book are quite simple, and the method of preparation follows a few basic techniques with which you will soon be familiar. But I don't believe that any step in cooking should be followed blindly; if you understand the overall concept of a particular dish, its techniques can easily be applied to many recipes. If you know that all custards should be baked in a water bath to prevent them from curdling, you will be able to make various timbales, flans, and molded purees without having to look them up in a cookbook. Or, if you know how to properly deglaze a pan, your sauces are bound to have better texture and flavor.

To me the secret of success of the simplest dish lies in careful advance preparation. Be it mincing a shallot, julienning a leek, or sautéing a chicken breast, every step can make a difference in both taste and texture. Browning the meat properly for a ragout is essential; reducing a sauce to the right consistency is a must. Steps such as skimming a stock, emulsifying a vinaigrette, or even properly seeding a tomato may seem unnecessary chores, but they are as important to the result as careful seasoning.

I never cease to be amazed at the intricacy of the *mise-en-place*, which translates to "put into place," of even the simplest of French restaurants, and after years of spending time in many of Europe's and America's best, as well as simpler, restaurants, I am more convinced than ever that the initial preparation, the touches, and the garnishes are what make the difference between a good and a great dish.

It is essential for a cook to understand that the techniques in cooking are the grammar of the language of food, and that with this understanding any cook can be inventive and give his or her own individual touch to any dish.

Once you have mastered the basic recipes that follow, you will be able to cook with ease and more spontaneity. With a basic stock, a quick tomato sauce, or your favorite vinaigrette, you can create a variety of soups, main courses, and salads on the spur of the moment. When it comes to more complex dishes, such as those calling for homemade stocks, try to make them from scratch. Their depth of flavor cannot be duplicated by commercial bouillons, and while I am a great believer in shortcuts whenever possible, I don't believe that it is worth going to the trouble of trying to reproduce a wonderful dish only to find that the flavor is disappointing because the sauce lacks body. A homemade stock will greatly enhance any soup while canned broth will never give you the same results no matter how much it is doctored up. This is also true for a homemade vinaigrette that is properly emulsified and in perfect proportion with your greens. It too will be memorable and is well worth the extra five minutes it may take to make. As for slicing, chopping, mincing, and julienning, you cannot beat the very old-fashioned method of using a good sharp knife. Don't be tempted to use the food processor for everything; take the time to learn to mince a shallot or an onion by hand, and the result will not let you down.

Ultimately it is small techniques such as these that can make the difference, because after you've learned how to choose quality ingredients, understanding the fundamental techniques of cooking is the most important step in creating good food.

BROWN CHICKEN STOCK

makes 1½–2 quarts

Brown chicken stock is by far the most important stock in the everyday kitchen. Although something of a chore to make, this stock repays your effort in allowing you to create more complex and sophisticated sauces. Add brown stock to simple pan juices and you'll have a sauce with depth of taste and texture.

Brown chicken stock is, in a sense, a much simplified brown stock—a classic preparation which calls for meaty veal bones and a very long simmering. Since veal bones are sometimes hard to come by, I make the stock with chicken wings, which give it the gelatinous quality, and with beef, which adds richness of color and flavor.

3 tablespoons corn or peanut oil
2 large onions, peeled and diced, plus 2
  medium onions, unpeeled and cut in
  half crosswise
20–24 chicken wings
3–4 pounds short ribs of beef, beef
  shanks, or meaty beef neck bones
3 large carrots, peeled and diced
3 large stalks celery, diced

1 tablespoon tomato paste, optional
1 parsley root with greens attached,
  peeled, optional
1 teaspoon dried thyme
1 bay leaf, crumbled
6 whole black peppercorns
Large pinch of salt
3 large sprigs of fresh parsley

1. Preheat the broiler.

2. In a heavy 10-inch skillet, preferably cast-iron, heat the oil over medium-high heat. Add the peeled and diced onions and cook, stirring constantly, for 2 to 3 minutes, or until they begin to brown. Reduce the heat to low, partially cover, and cook for 25 to 30 minutes, or until the onions are soft and nicely caramelized. Do not let them burn.

3. Meanwhile, prepare the chicken and beef: Cut each chicken wing at its joints into 3 pieces. Place the pieces in a single flat layer in a roasting pan and broil 4 to 6 inches from the source of heat until nicely browned on both sides, turning them once during broiling. Be careful not to burn. Remove the wings from the roasting pan and transfer to an 8- to 10-quart stockpot. Set aside.

4. Place the beef ribs or bones in a single layer in the roasting pan and broil until nicely browned on both sides, again being careful not to burn. Transfer

the beef to the stockpot containing the wings. When the onions are done, transfer them also to the stockpot and reserve.

5.   Transfer the pan juices to a large skillet, raise the heat, and when hot, add the unpeeled onions, cut side down. Cook for 5 to 8 minutes, or until very brown, and add the onions to the stockpot.

6.   Add the carrots and celery to the skillet and cook, stirring often, for 10 to 15 minutes or until nicely browned. (The pieces will not brown evenly.) Add the optional tomato paste and cook for 1 minute longer, or until the vegetables are well coated. Add to the stockpot together with the optional parsley root, thyme, bay leaf, peppercorns, salt, and parsley sprigs. Add enough water to cover the ingredients by 2 inches and bring to a boil. Reduce the heat and simmer the stock, partially covered, for 3 to 4 hours, skimming often, until the liquid is considerably reduced and has become deep brown in color.

7.   When the stock is done, strain it through a cheesecloth-lined colander placed over a large bowl and let cool completely. Refrigerate, uncovered, overnight.

8.   The next day, thoroughly degrease the stock, return it to a large pot, and bring to a boil. Remove from the heat and again let cool completely. Transfer to containers or jars, cover tightly, and refrigerate or freeze.

Note: All stocks must be cooled uncovered to prevent them from turning sour. If the stock is stored in the refrigerator, bring it back to a boil every 3 to 4 days, simmering it for 3 to 4 minutes. Transfer to a jar and cool again, uncovered, then chill. Stocks can be frozen successfully for 6 to 8 weeks.

BROWN LAMB STOCK

makes 1½–2 quarts

5 pounds meaty lamb bones
2 meaty lamb shanks
1 tablespoon peanut oil
2 large onions, peeled and diced
2 large celery stalks, diced

2 carrots, peeled and diced
1 Bouquet Garni (page 485)
1 teaspoon whole black peppercorns
Large pinch of salt

1. Preheat the broiler.

2. Place the lamb bones and shanks in a single layer in a large roasting pan and broil 6 inches from the source of heat until the bones are nicely browned on all sides. Turn the bones once to broil evenly and be sure not to burn them. Transfer the browned bones to a large stockpot and set aside.

3. Transfer 2 tablespoons of the lamb drippings to a 10-inch cast-iron skillet. Add the oil and heat over moderately high heat. Add the diced onions, celery, and carrots and sauté until nicely browned. Do not let the onions burn. Add the browned vegetables, bouquet garni, peppercorns, and salt to the stockpot. Add enough water to cover by 2 inches, then bring to a boil. Reduce the heat and simmer the stock, partially covered, for 2 to 3 hours. When the stock is done, strain through a cheesecloth-lined colander placed over a bowl and cool completely. Chill the stock overnight.

4. The next day, degrease the stock thoroughly. Pour it into a pot, return to a boil, and cool completely. Transfer to storage containers, cover, and refrigerate or freeze (see Note on page 468).

CHICKEN STOCK

makes 2–2½ quarts

Nothing can be simpler to make than chicken stock. The ingredients can be assembled in a matter of minutes and, since almost all of them are basics that keep well for weeks in the refrigerator, you can make the stock on the spur of the moment. I usually make the stock using the wings only or sometimes the whole chicken, but never the backs. While backs are inexpensive and plentiful they are also fatty, and if the stock is allowed to boil as it cooks instead of simmering gently, the fat will work itself back into the stock, and it will be almost impossible to degrease it properly.

All soups benefit from a homemade well-reduced chicken stock. Vegetables such as carrots, leeks, kohlrabi, cabbage, and turnips are wonderful when poached in chicken stock, and the pan juices of poultry basted with chicken stock are incomparably better than those in which canned bouillon was used. Chicken stock can be frozen successfully for two months, so it is well worth having it on hand at all times.

1 (3-pound) chicken, quartered
10–14 chicken wings, or 2 pounds
  chicken necks and gizzards
2 large carrots, peeled and cut in half
2 large stalks celery, with tips cut in half
1 parsley root, scraped (optional), or 1
  small bunch of fresh parsley

1 bay leaf
1 large leek, well rinsed, with 3 inches
  of greens included, or 1 large onion,
  peeled and stuck with a clove
Large pinch of salt
6–8 whole black peppercorns

1. Combine all the ingredients in a large stockpot. Add 12 to 14 cups cold water, or enough to cover the ingredients by 2 inches. Bring to a boil slowly on top of the stove, carefully skimming the scum that rises to the top. Partially cover the pot and simmer the stock for 1½ to 2 hours. When done, cool the stock, uncovered.

2. Strain the stock through a cheesecloth-lined colander placed over a bowl and, when completely cool, refrigerate it, uncovered, overnight.

3. The following day, thoroughly degrease the stock. Transfer to a casserole, bring to a boil, and pour into 1-quart containers. Again, cool the stock, uncovered, and, when completely cool, cover and refrigerate or freeze (see Note on page 468).

WHITE STOCK

makes 2 quarts

4–5 pounds meaty beef bones,
   preferably shin bones
2 large carrots, peeled and cut in half
2–3 stalks celery, with leaves, cut in half
2 leeks, trimmed and washed
2 medium onions, peeled

6 whole black peppercorns
1 parsley root, optional
1 Bouquet Garni (page 485)
1½ teaspoons salt
12 cups water

1. Combine all the ingredients in a large, tall, but narrow stockpot. Bring to a boil, reduce the heat, and simmer, partially covered, for 3 to 4 hours, skimming the surface several times during cooking.

2. Strain the stock through a cheesecloth-lined colander placed over a large bowl, and set aside at room temperature, uncovered, until completely cool. Place in the refrigerator overnight.

3. The next day, remove and discard all the fat from the surface and transfer the stock to a large saucepan. Bring to a boil, remove from the heat, and let cool completely. Refrigerate for up to 1 week, returning the stock to a boil every few days, or freeze in covered containers for up to 1 month.

FISH STOCK

Fish stock is one of the quickest stocks to make; it can also be one of the least expensive if the fish market gives you the trimmings. The most flavorful trimmings (which should include the heads) are those of red or white snapper or sea bass. After a short cooking time of twenty to thirty minutes, the stock is ready for use. Although this fish stock freezes very well, I rarely make it in great quantity unless I know I'll be using it soon, because the flavor becomes fishier and more acidic after six to eight weeks. The following recipe calls for wine, but you can make a neutral fish stock by omitting the wine. A neutral stock can be kept frozen for a longer time than one made with wine.

6 pounds fish trimmings, preferably snapper heads and bones
4 tablespoons unsalted butter
2 large leeks, trimmed, well washed and finely sliced
2 medium carrots, peeled and finely sliced
2 medium stalks celery, finely diced

1½ cups excellent-quality dry white wine, preferably Blanc de Blanc, Mâcon Blanc, or Côtes du Rhône
6 cups water
1 Bouquet Garni (page 485)
10 whole black peppercorns
1 teaspoon salt

1. Rinse the fish trimmings thoroughly. Transfer the trimmings to a large bowl and place under the cold-water tap. Let cold water run over the trimmings for 2 hours, or until they are completely white and free of blood. Remove and discard the gills and set the trimmings aside.

2. Melt the butter in a large enamel casserole over low heat. Add the leeks, carrots, and celery and cook for 2 minutes. Place the fish trimmings on top of the vegetables, cover, and simmer for 10 minutes. Add the wine, water, bouquet garni, pepper corns, and salt and bring to a boil. Reduce the heat and simmer the stock, partially covered, for 30 to 40 minutes, skimming carefully every 10 minutes.

3. Strain the stock through a cheesecloth-lined colander placed over a large bowl. If the flavor of the stock is not intense enough, return to the casserole and reduce by one third.

4. Let the stock cool completely, uncovered. Transfer to 1-quart jars, cover tightly, and refrigerate or freeze.

5. Refrigerate the stock for up to 1 week, returning it to a boil every 2 days, or freeze it for up to 6 weeks.

RED WINE FISH STOCK

makes about 3 cups

4 pounds fish trimmings, preferably red snapper heads and bones

3 tablespoons unsalted butter, or a mixture of olive oil and butter

1 large carrot, peeled and diced

2 medium leeks, finely diced and rinsed well under warm water

4 shallots, peeled and thinly sliced

2 large stalks celery, minced

2 large cloves garlic, peeled and finely minced

3 cups good-quality dry red wine, preferably Bordeaux

1½ cups Chicken Stock (page 460) or bouillon

2 sprigs of fresh thyme

1 large sprig of fresh Italian parsley

6 whole black peppercorns

Pinch of salt

1. Rinse the fish bones under plenty of cold water. Transfer the trimmings to a large bowl and place under the cold-water tap. Let cold water run over the trimmings for 2 hours, until all visible blood has been removed. Remove the gills and set the trimmings aside.

2. In a large casserole, melt the butter over medium heat. Add the carrot, leeks, shallots, celery, and garlic and sauté for about 2 minutes, or until just tender. Add the fish bones, cover, and braise for 5 minutes. Add the wine, stock, thyme, parsley, and peppercorns and season with a pinch of salt. Bring to a boil, reduce the heat, and simmer the stock, skimming often, for 25 minutes. Strain into a large stainless-steel bowl through a colander lined with a double layer of cheesecloth, and set aside to cool completely. Refrigerate the stock overnight.

3. The next day, remove and discard all the fat from the surface of the stock and bring back to a boil. Remove from the heat, let cool, and either freeze or keep covered in the refrigerator for up to 1 week, bringing the stock back to a boil every couple of days.

TOMATO FONDUE

makes about 3 cups

2 (2-pound) cans Italian plum
   tomatoes
⅓ cup extra-virgin olive oil
½ cup finely minced shallots

3–4 large cloves garlic, peeled and
   finely minced
1 teaspoon dried thyme
1 teaspoon dried oregano
Salt and freshly ground black pepper

1.   Place the tomatoes in a large strainer or colander and let stand at room temperature to drain for 4 hours. Coarsely chop the tomatoes and set aside.

2.   In a large, heavy-bottomed saucepan, heat the oil over medium heat. Add the shallots and garlic and cook the mixture for 1 to 2 minutes, or until soft but not browned. Add the drained tomatoes, thyme, and oregano and season with salt and pepper. Raise the heat to medium-high and cook for 15 minutes, stirring often to prevent the mixture from scorching. Reduce the heat, partially cover, and simmer the sauce for 40 to 45 minutes, or until very thick, stirring often. Taste and correct the seasoning.

3.   Transfer the fondue to a jar, cool thoroughly, and cover tightly. Store in the refrigerator for up to 3 weeks. The fondue can also be frozen for 2 to 3 months.

QUICK TOMATO SAUCE

makes 4 cups

3 tablespoons extra-virgin olive oil
2 medium onions, peeled and finely
   minced
2 large cloves garlic, peeled and finely
   minced

2 (2-pound) cans Italian plum tomatoes,
   drained and chopped, juice reserved
1 teaspoon granulated sugar
1½ teaspoons dried thyme
1½ teaspoons dried oregano
Salt and freshly ground black pepper

1.   In a large, heavy casserole, heat the olive oil over medium-low heat. Add the onions and garlic and cook for about 5 minutes, or until soft and lightly browned. Add the tomatoes, sugar, thyme, and oregano and season with salt and pepper. Reduce the heat and simmer, uncovered, for 45 minutes, stirring often. If the sauce seems too thick, add a bit of the reserved tomato juice.

2.   Transfer the sauce to the workbowl of a food processor and puree until smooth. Taste and correct the seasoning. Let the sauce cool completely, cover, and chill. The sauce will keep 3 to 4 days in the refrigerator, or up to 2 weeks in the freezer.

Olive Oil

So much has been said and written about olive oil that by now both cooks and noncooks know a fair amount about this Mediterranean staple. But even if you know enough to distinguish among extra-virgin, virgin, and pure olive oils, buying can still be a confusing experience. The best oils, labeled "cold-pressed extra-virgin oil," are made from hand-picked ripe olives that are pressed in large circular stone presses without the use of heat. The result is a deep green, superbly fruity and assertive oil with a rather short shelf life. It is becoming increasingly rare. Subsequent pressings of the olives are labeled "virgin" and "pure," the last of which is made by using heat and certain chemical solvents. These oils lack the depth of flavor of an extra-virgin oil but they are less expensive and perfectly adequate for cooking and for salad dressings that include herbs, garlic, or mustard.

The major dilemma facing the consumer is how to choose the right olive oil from the large assortment available in a variety of price ranges. Personally I prefer the "boutique" oils produced in small quantity in the south of France, Italy, and Spain. These oils, which are pressed and bottled on the premises, are made with locally grown olives and are far more flavorful than most supermarket varieties and therefore more expensive. My favorite among these are Extra-Virgin James Plagniol, Louis de Regis, Hilarie Fabré, Old Monk, in cans, and Extra-Virgin Sasso.

Extra-virgin, virgin, and pure olive oils that are marked imported from Italy are not actually produced in Italy but rather are processed and bottled there. These oils are usually a mixture of Moroccan, Spanish, Turkish, Greek, Lebanese, and Lybian oils, all of which are shipped to cooperatives in Lucca where they are refined and bottled. These oils, while less expensive and perfectly acceptable for many preparations, lack the depth of flavor and individuality of the great regional olive oils.

A top-quality extra-virgin olive oil should not be used in cooking but instead reserved for salads or used in finished pasta dishes and as a final flavorful touch to grilled vegetables and pizzas. For sautéing and simple marinades I generally opt for a good brand of pure olive oil such as Sasso, which is light and fruity, or Colavita, which is heavier and more peppery. I believe in testing many olive oils and never buying them blindly. To judge the quality of an olive oil, simply dip a piece of French bread into the oil and taste it. Once you've used this simple test on a variety of oils you will be able to tell which you prefer and which one seems to enhance a dish rather than detract from it.

■ ■ ■

FRESH TOMATO SAUCE

makes about 4–5 cups

3 tablespoons extra-virgin olive oil
1 large onion, peeled and finely minced
1 medium stalk celery, finely minced
1 medium carrot, peeled and finely diced
3 tablespoons finely minced fresh parsley

3 large cloves garlic, peeled and finely minced
¼ cup fresh basil leaves, stemmed
1 large sprig of fresh oregano
4–5 pounds ripe tomatoes, peeled, seeded, and coarsely chopped
1 tablespoon tomato paste, optional
Salt and freshly ground black pepper

1. In a large, heavy casserole, heat the oil over low heat. Add the onion, celery, carrot, parsley, garlic, basil, and oregano and cook the mixture for about 5 minutes over low heat, partially covered, or until soft but not browned.

2. Add the tomatoes and the optional tomato paste and season with salt and pepper. Bring to a boil, reduce the heat, and simmer the sauce, covered, for 1½ to 2 hours, stirring several times to prevent the sauce from scorching.

3. Strain the sauce through a fine sieve, or pass it through a food mill, and correct the seasoning. Let cool, then refrigerate in covered jars. You may freeze the sauce for up to 2 months.

Note: The tomato paste is optional depending upon the ripeness of the tomatoes. If they are very ripe and sweet, you will not need the paste.

If the sauce seems too thin, uncover and let it reduce to the desired consistency.

If you prefer a heartier, chunkier sauce, do not strain it, simply process quickly in the food processor or blender.

MAYONNAISE

makes 1–1¼ cups

1 large egg
1 large egg yolk
1½ tablespoons red wine vinegar or
  lemon juice

2 teaspoons Dijon mustard
Salt and freshly ground black pepper
¾–1 cup peanut oil

1. In the container of a blender or food processor, combine the egg, egg yolk, vinegar, mustard, salt, and pepper. Process until smooth.

2. With the motor running, add the oil, drop by drop, until the mixture starts to thicken. Add the remaining oil in a slow, steady stream until all of it has been incorporated into the egg mixture and the mayonnaise is thick. Transfer to a bowl or jar, cover, and refrigerate until ready to use.

Note: A mayonnaise made in the blender is practically foolproof, and it will also absorb far less oil than one made in the food processor. For truly great texture, you should try to make the mayonnaise by hand. A chef's trick for a mayonnaise made by hand is to prepare it in a small, heavy saucepan, which won't move around while you're adding the oil with one hand and whisking with the other.

AIOLI

makes about 1 cup

2 teaspoons red wine vinegar
2 large whole eggs
2 large cloves garlic, peeled and
  crushed

Salt and freshly ground black pepper
½–¾ cup fruity olive oil

1. In the container of a blender, combine the vinegar, eggs, and garlic. Season with salt and pepper and process for 1 minute. With the machine running, add the oil, drop by drop, until the mixture begins to emulsify. Add the remaining oil in a slow, steady stream until a loose, saucelike mayonnaise is formed.

2. Taste and correct the seasoning. Transfer the aioli to a bowl or jar, cover tightly, and keep in the refrigerator until needed. The sauce will keep for up to 1 week.

Note: Instead of the raw garlic, you can use 6 to 8 cloves of Roasted Garlic (page 482) for a mild, yet interesting flavor.

CLARIFIED BUTTER

The process of clarifying butter removes the milky residue and impurities that make butter burn quickly. Clarifying is a simple technique that works best when done with at least eight ounces of butter. Once clarified, the butter can be refrigerated in a tightly sealed jar for at least two weeks, and it can be frozen for several months. Clarified butter is used for sautéing delicate foods, such as calves' liver, fish fillets, and anything that is breaded. It is also a must for frying croutons. Of course, you can use clarified butter in any recipe that calls for regular butter for sautéing.

Unsalted butter

Melt any desired quantity of unsalted butter in a heavy saucepan over low heat. As soon as the butter is melted and very foamy, remove the pan from the heat and, with a spoon, carefully skim off the foam and discard. Strain the clear yellow liquid through a fine sieve, taking care to leave all of the milky residue at the bottom of the pan. The strained clear yellow liquid is the clarified butter, which can be stored in a covered jar in the refrigerator for 2 to 3 weeks.

BEURRE MANIE

makes 8

A beurre manie is used in both classic and peasant cooking to thicken sauces. It is a flour and butter paste that can be formed into a ball and kept successfully for several weeks. You will usually only need one to two teaspoons of beurre manie for any recipe in this book, since it is preferable to reduce the sauce naturally before thickening it so as to intensify the flavor.

8 tablespoons unsalted butter, slightly softened   8 tablespoons all-purpose flour

1. Combine the butter and flour in the workbowl of a food processor and process until smooth.

2. Refrigerate the mixture until it is firm enough to shape into balls with your hands. Divide the mixture into 8 equal parts and shape each into a ball, using your hands. Place the balls in a tightly covered jar and refrigerate until needed. The beurre manie can also be frozen for 2 to 3 months.

LAMELOISE SAUCE

makes 4 cups

1 cup finely minced shallots
1 cup dry white wine
1½ cups dry Noilly Prat vermouth

2 quarts non-ultrapasteurized heavy
 cream only
Salt and freshly ground white pepper

1. In a heavy 5-quart saucepan, combine the shallots, white wine, and vermouth. Place the pan over high heat, bring the mixture to a boil, and continue to boil until reduced to a glaze. When the liquid has reduced to ½ cup, stir often so that the shallots do not brown.

2. Whisk in the heavy cream and bring to a boil, then immediately reduce the heat so that the cream is at a simmer. Continue to simmer, uncovered, for 1½ to 2 hours, or until the cream is reduced by half, stirring often to prevent the shallots from sticking to the bottom of the pan.

3. Remove the pan from the heat and strain the mixture through a fine sieve placed over a stainless-steel bowl, pressing down on the shallots to extract all of their juices. Season lightly with salt and white pepper. The sauce will be quite thick.

4. Place a piece of plastic wrap directly on top of the sauce to prevent a skin from forming. Let cool completely, then refrigerate until needed. If you are not using the sauce right away, transfer it to a tightly covered jar. Store the sauce for up to 2 weeks in the refrigerator, or up to 2 months in the freezer.

Note: The cream must reduce slowly to prevent it from browning around the edges of the pan, which will impart a burnt flavor to the sauce.

PROVENÇAL VINAIGRETTE

6–8 tablespoons olive oil
2 tablespoons red wine vinegar or
 sherry vinegar
2 teaspoons Dijon mustard

1 medium clove garlic, peeled and
 mashed
Salt and freshly ground black pepper

Combine the oil, vinegar, mustard, and garlic in a small jar, cover, and shake until the dressing is well blended and smooth. Season with salt and pepper and chill. Bring back to room temperature 30 minutes before making the salad and shake again until smooth.

SHERRY AND LEMON VINAIGRETTE

2 tablespoons sherry vinegar
1 tablespoon fresh lemon juice
2 tablespoons finely minced shallots

8 tablespoons extra-virgin olive oil
Salt and freshly ground black pepper

Combine the vinegar, lemon juice, and shallots in a medium mixing bowl and whisk until well blended. Add the oil, slowly whisking constantly until the mixture is smooth and creamy. Season with salt and pepper.

HONEY-CUMIN VINAIGRETTE

6–8 tablespoons olive oil
2 tablespoons red wine vinegar
2 teaspoons pure honey

½ teaspoon ground cumin
Salt and freshly ground black pepper

Combine the oil, vinegar, honey, and cumin in a small jar, cover, and shake until well mixed. Season with salt and pepper and chill. Return to room temperature before serving with grilled shrimp, mixed greens, or as an accompaniment to grilled fish.

PORT AND SHALLOT VINAIGRETTE

makes about 1 cup

¼ cup tawny port
¼ cup red wine vinegar
1 large shallot, peeled and finely minced

⅔ cup extra-virgin olive oil
Salt and freshly ground black pepper

1. Combine the port, vinegar, shallot, and oil in a small bowl. Whisk until smooth and creamy, and season with salt and pepper to taste. Let the dressing stand at room temperature for 1 hour to develop flavor.

2. Just before serving, whisk the dressing again until creamy. Serve with cold poached leeks or asparagus.

Note: The dressing may be made ahead and refrigerated. Bring back to room temperature and whisk until creamy just before serving.

EMERALD VINAIGRETTE

makes about 1⅓ cups

½ cup fresh parsley leaves
½ cup fresh chervil leaves
⅓ cup Chicken Stock (page 470)
½ large egg yolk
1 tablespoon sherry vinegar

2–3 teaspoons Dijon mustard with green herbs, preferably Maille
⅓ cup first-press or cold-pressed peanut oil, preferably French
⅓ cup mild olive oil
Salt and freshly ground white pepper

1. In the container of a blender combine the parsley, chervil, and stock. Blend at top speed until very smooth.

2. Strain the herb mixture through a fine sieve into a medium mixing bowl, pressing down on the herbs to extract all their juices. Add the egg yolk, vinegar, and mustard and whisk until well blended.

3. Combine both oils in a liquid measure and add to the green mixture, a drop at a time, whisking constantly, until all the oil has been added. The vinaigrette will be somewhat loose but quite creamy. Season with salt and pepper and serve with sautéed salmon fillets and steamed vegetables.

DOUBLE-POACHED GARLIC

2–3 heads garlic, separated into cloves
  and peeled

1. Bring plenty of water to a boil in a medium saucepan. Add the garlic cloves and bring back to a boil. Cook for 1 minute.

2. Drain and discard the poaching liquid. Return the blanched garlic to the saucepan, cover with fresh water, and again bring to a boil. Reduce the heat and simmer for 15 to 20 minutes, or until the garlic is tender when pierced with the tip of a sharp knife. Drain again. The garlic is now ready to be used.

Note: The garlic may be double-poached days in advance. Simply place the poached garlic in a jar, cover with olive oil, seal tightly, and store in the refrigerator until needed. The garlic will keep for up to 3 months.

ROASTED GARLIC

2 heads garlic, unpeeled
2 teaspoons extra-virgin olive oil
Coarse salt and freshly ground black
  pepper

Large pinch of granulated sugar
Olive oil

1. Preheat the oven to 375° F.

2. Separate the garlic into cloves; do not peel. Place all the garlic cloves in the center of a large piece of aluminum foil. Drizzle with the 2 teaspoons of extra-virgin olive oil and sprinkle with coarse salt and black pepper. Bring the ends of the foil up over the garlic and crimp to seal in the cloves completely.

3. Place the foil packet in the center of the preheated oven and roast for 45 minutes or until the garlic is soft to the touch. Open the foil packet, sprinkle the cloves with a large pinch of sugar, and continue to roast the garlic, uncovered, for 15 minutes or until golden brown.

4. Remove the garlic from the oven and, when cool enough to handle, peel the cloves carefully so that they remain whole. Place the garlic in a small container and pour in enough olive oil to cover the cloves completely. Cover the container and store in the refrigerator for 2 to 3 months.

ROASTED RED OR GREEN BELL PEPPERS

Red or green bell peppers

1.   Outdoor grilling: Prepare the grill.

2.   Place the peppers on the grill directly on top of very hot coals and grill, turning as necessary, until the skins of the peppers are blackened and somewhat charred on all sides. Remove from the grill and wrap each pepper in a damp paper towel. Set aside to cool completely.

3.   When cool enough to handle, rub off the charred skin with your fingers or do it under slowly running water, core, and remove all the seeds. The peppers are now ready to be used in the appropriate recipe.

4.   Indoor grilling: You may also roast the peppers indoors on either an electric or a gas stove. For an electric stove, place the peppers directly on the coils of the burner over medium-high heat to char on all sides. For a gas stove, pierce the peppers onto the tines of a long fork through the stem end. Hold over a medium-high flame and char on all sides. In both cases remove from the burner, wrap in damp paper towels, cool, and peel.

Note: Roasted peppers will keep for several days in the refrigerator, covered with olive oil, in a tightly covered jar. You may also roast the variety peppers (jalapeños, serranos, yellow, and orange) and fresh tomatoes in the same manner. Tomatoes need not be wrapped in damp towels before they are peeled.

ROASTED SHALLOTS

makes 12

12 large shallots, unpeeled
2 teaspoons olive oil
Coarse salt and freshly ground black
  pepper

Large pinch of granulated sugar
Dash of water

1.   Preheat the oven to 375° F

2.   Place the shallots in the center of a large piece of foil. Drizzle with the olive oil and sprinkle with coarse salt, pepper, and a large pinch of granulated sugar. Add a bit of water, gather up the foil, and crimp tightly to enclose the shallots completely. Place the packet in the center of the preheated oven and roast the shallots until tender, 30 to 40 minutes. During the last 10 minutes, open up the foil in order to caramelize the shallots.

3.   Remove the shallots from the oven and cool. Peel and store them in the refrigerator in a tightly covered jar for up to 1 week.

COOKED WHITE BEANS

makes about 6 cups

1 pound dried white beans, preferably
  Great Northern
Water
1 large stalk celery, including leaves, cut
  in half
1 large carrot, peeled and cut in half

1 small onion, peeled and stuck with a
  clove
1 Bouquet Garni (page 485)
6 whole black peppercorns
½ pound smoked pork shoulder butt,
  quartered
Salt

1. Place the dried beans in a large bowl with plenty of water to cover. Let soak overnight at room temperature.

2. The next day, preheat the oven to 325° F.

3. Drain the beans and place in a large, heavy flameproof casserole with a tight-fitting lid. Cover with water by 2 inches and add the celery, carrot, onion, bouquet garni, peppercorns, and pork butt. Bring to a boil on top of the stove. Immediately cover tightly, place the casserole in the center of the preheated oven, and cook the beans for 1 hour and 15 minutes, or until just tender. Do not add salt until the beans are almost done. Remove the casserole from the oven and let the beans cool in their cooking liquid.

4. Discard the vegetables and bouquet garni and drain the beans. They are now ready to use.

Note: Use the same method for cooking dry navy pea beans, flageolets, kidney beans, black turtle beans, cannelini beans, and pinto beans. The cooking time will vary with each bean from about 45 minutes to 1 hour and 15 minutes.
  The cooked pork butt can be diced and used in combination with the beans as well.

BOUQUET GARNI

makes 1 bouquet

A bouquet garni is a seasoning bag composed essentially of three aromatic herbs. It is used in French and Mediterranean cooking for flavoring many soups, ragouts, and stocks.

3–4 large sprigs of fresh parsley, preferably Italian
1 bay leaf

1 large sprig of fresh thyme or 1 teaspoon dried
1 stalk celery, optional
6–8 whole black peppercorns, optional

1. If using fresh herbs, tie them together with a small piece of kitchen string.

2. If using dried thyme, place the entire bouquet in a piece of cheesecloth and tie it with a kitchen string.

GARAM MASALA

makes about ¼ cup

1 tablespoon whole cumin seeds
1 tablespoon whole coriander seeds
8 whole cloves
2 teaspoons whole cardamom pods

1½ teaspoons whole black peppercorns
1 (2-inch) piece of cinnamon stick
1 teaspoon turmeric

1. Preheat the oven to 250° F.

2. Place the cumin, coriander, cloves, cardamom, peppercorns, and cinnamon stick on a cookie sheet and bake until the spices are lightly browned and fragrant, about 15 to 20 minutes. Do not let brown.

3. Transfer the spices to a blender or spice mill, and process until pulverized. Add the turmeric and process for 5 seconds, or until completely incorporated. Store in an airtight container in the refrigerator.

FRESH PASTA

makes 1 pound

2¼ cups very fine semolina flour
3 extra-large eggs
Large pinch of salt

1 teaspoon olive oil
All-purpose flour for dusting

1. In the workbowl of a food processor, combine the semolina flour, eggs, salt, and oil and process until thoroughly combined.

2. Dust a wooden surface with flour and transfer the dough to it. Knead the dough with your hands for 2 to 3 minutes, or until it forms a perfectly smooth ball. If the dough seems a bit dry or very hard when kneading, wet your hands with a little water and continue kneading. If the dough is a little sticky, add 1 to 2 tablespoons of flour and continue kneading. The moisture content of the dough will depend upon the size of the eggs used. When the dough is smooth and elastic, divide it into 4 or 6 pieces, flatten each into a small rectangle, and cover with a linen towel.

3. Place the kneading attachment in an electric or a hand-crank pasta machine, and turn the knob until the rollers are at their largest opening. Dust each piece of dough lightly with flour. Working with 1 piece of dough at a time, run each through the largest opening several times.

4. When the dough seems very smooth, turn the knob to the next notch, which will be a smaller opening, and run each piece of dough through the machine. Keep turning the knob, narrowing the opening of the rollers a little more, each time you have run the sheets of dough through the machine. When the pasta is at the desired thickness, place the pasta sheets on well-floured cloth towels and let them dry for about 10 to 15 minutes. Be careful not to let them dry out too much, otherwise they will become brittle and will break during cutting.

5. Replace the kneading attachment with the cutting attachment (broad one for fettuccine and thin for linguine), and run each sheet of dough through the machine. Spread the pasta on the floured towels, turning from time to time, until completely dry.

6. Place the dry pasta in plastic bags, twist tie, and keep in a very dry spot. The dry pasta will keep for months.

Variation: For green, or spinach, fettuccine, steam 2 to 3 ounces of fresh spinach, and drain and squeeze out all excess moisture. Add to the food processor along with the semolina, eggs, salt, and oil and process until smooth. You might need to add 2 to 3 extra tablespoons of semolina flour, if the mixture is too sticky.

Note: Fine semolina flour is available by mail order through various food catalogs, such as Williams-Sonoma. You can use all-purpose flour for making pasta, but the texture of the dough will be somewhat softer and you may need up to 1 cup more flour than is called for in the recipe.

PIZZA DOUGH

makes 2 (12-inch) pizzas

¼ cup warm water (110° F)
1 teaspoon active dry yeast
½ teaspoon granulated sugar
¾ cup cold water

3 tablespoons olive oil
2¾ cups all-purpose flour
1½ teaspoons salt

1. Start by proofing the yeast: In a small bowl combine the warm water, yeast, and sugar. Mix until the yeast is dissolved and set aside for 10 minutes, or until foamy.

2. Combine the cold water and olive oil in a 1-cup liquid measure. Reserve.

3. In the workbowl of a food processor combine the flour and salt. Stir down the yeast mixture and, with the motor running, add it to the processor through the feed tube. Add the cold water–oil mixture in a steady stream until it has been absorbed by the flour. Process for 1 minute, or until a smooth dough is formed. The dough will be very soft and a bit sticky. If the dough is extremely sticky, add 1 to 2 tablespoons of flour and process for 15 seconds.

4. With lightly floured hands, remove the dough carefully from the processor and knead for 30 seconds. Divide in half and place each piece in a lightly floured plastic bag. Tie and refrigerate for 30 minutes to relax the dough. At this point the dough will keep refrigerated overnight, but not longer. After the dough has relaxed, it is ready to be used in the appropriate recipe.

WHITE CHOCOLATE AND MINT SAUCE

makes 2½–3 cups

12 ounces white chocolate, broken into pieces
1½ cups heavy cream

6 tablespoons finely minced fresh mint
2 cups whipped heavy cream

1. Place the chocolate and heavy cream in a medium saucepan over low heat and whisk until the chocolate has melted. Add the mint and steep for 5 minutes, stirring often.

2. Strain the sauce into a small bowl and let cool to room temperature, whisking from time to time. Fold in the whipped cream, gently but thoroughly, and serve with a chocolate pâté or chocolate soufflés.

Note: The sauce will keep for up to 1 week in the refrigerator, in a tightly covered jar. Whisk again just before serving.

Variation: *White Chocolate–Mocha Sauce*: Omit the mint and substitute 2 teaspoons instant espresso. Combine the espresso with the white chocolate and heavy cream in a medium saucepan placed over low heat, and whisk until the chocolate has melted. Steep for 5 minutes and proceed with step 2 of the original recipe.

White Chocolate–Ginger Sauce: Omit the mint, substitute 1 tablespoon of fresh ginger mashed through a garlic press, and combine with the white chocolate and heavy cream in a medium saucepan over low heat. Whisk until the chocolate has melted. Steep for 5 minutes and proceed with step 2 the original recipe.

BASIC CHOCOLATE SAUCE

makes 2 cups

8 ounces bittersweet chocolate,
  preferably Lindt, broken into pieces
½ cup brewed espresso or strong coffee
2–3 tablespoons granulated sugar
⅓ cup heavy cream
8 tablespoons unsalted butter, cut into
  pieces

OPTIONAL
2 tablespoons dark rum, brandy, or
  orange-flavored liqueur, or 2
  tablespoons currants soaked in 2
  tablespoons scotch

In a heavy 2-quart saucepan combine the chocolate, coffee, and sugar. Place over low heat and stir constantly until the chocolate is completely melted. Add the heavy cream and heat through. Add the butter, 1 piece at a time, incorporating each piece before the next is added. The sauce should be very smooth. Stir in any of the optional flavorings, transfer to the top of a double boiler, and place over lukewarm water. Cover the sauce and set aside until serving.

PRALINE POWDER

makes 1 cup

2 cups sliced almonds, lightly toasted
1 cup granulated sugar

¼ cup water

1. Line a baking sheet with aluminum foil and spread the sliced almonds on it in a single layer.

2. In a heavy-bottomed saucepan, combine the sugar and water, bring to a boil, and stir once to dissolve the sugar. Continue to boil, without stirring, until the mixture turns a light hazelnut brown. Do not let the caramel become too dark or it will taste burned.

3. Immediately pour the hot caramel over the almonds and set aside to cool completely. In about 10 minutes the caramel will harden. Break the mixture into small pieces and pulverize in a blender or food processor. Transfer to a jar with a tight cover. The praline will keep for up to 4 months in a cool, dry place.

Note: You can use praline powder in a variety of preparations: as a topping for ice cream; sprinkled on crepes which are then gratinéed, or add a couple of tablespoons to your favorite dessert recipe.
   Praline powder can also be made with walnuts, pecans, or hazelnuts to give each dish a unique flavor.

CRÈME ANGLAISE

serves 6–8

¾ cup whole milk
¾ cup heavy cream
6 large egg yolks

½ cup granulated sugar
1 teaspoon vanilla extract
1 cup heavy cream, whipped

1. Combine the milk and heavy cream in a small saucepan, bring to a simmer over low heat, and keep warm.

2. In the top of a double boiler, combine the egg yolks and sugar and beat until thick and pale yellow. Gradually add the hot milk mixture and whisk until smooth and well blended. Place over simmering water and whisk constantly until the custard is thick and heavily coats a spoon. Do not let the custard come to a boil or the yolks will curdle.

3. Remove from the heat, whisk in the vanilla, and immediately transfer to a mixing bowl. Cool completely, fold in the whipped cream gently but thoroughly, and chill until serving time.

Variations: *Coffee Crème Anglaise:* Dissolve 2 teaspoons instant espresso powder in the warm milk-and-cream mixture. Proceed with original recipe.

Brandy Crème Anglaise: Add 3 tablespoons cognac with the vanilla and immediately transfer to a mixing bowl. Proceed with the original recipe.

For a lighter custard, use milk instead of heavy cream. Do not attempt to use low-fat milk.

CRÈME FRAÎCHE

makes 2 cups

4 tablespoons buttermilk

2 cups non-ultrapasteurized heavy cream

Combine the buttermilk and heavy cream in a glass jar and whisk until well blended. Cover the jar loosely and set aside in a warm place until the cream becomes very thick. Stir thoroughly, cover, and refrigerate. The crème fraîche will keep from 10 days to 2 weeks; it cannot be frozen.

Note: Crème fraîche can take from 10 to 24 hours to sour and thicken—it depends mainly on the temperature of your kitchen. Be sure to keep the mixture in a warm, draft-free place.

MARINATED ORANGE JULIENNE IN GRENADINE

makes about 1/3 cup

Fine julienne of orange zest macerated in grenadine adds interest and color to many dishes. The grenadine turns the zest a vivid red, making it a lovely garnish for poached or baked pears, an orange salad, fruit compotes, and for roast duck, turkey, or pork.

Zest of 2 oranges
1/4 cup granulated sugar

1/2 cup water
3/4 cup grenadine

With a sharp paring knife, cut the zest into fine julienne, taking care not to include any part of the white pith. In a small, heavy saucepan, combine the sugar and water. Place over high heat, bring to a boil, and cook for 3 minutes. Reduce the heat, add the orange zest, and cook until just tender. Drain and place the zest in a small bowl. Add the grenadine and marinate the zest for 2 to 3 days before using.

Note: The zest will keep 3 to 4 months in the liqueur, in a tightly covered jar, refrigerated.

BASIC TART SHELL

makes 1 (10-inch) shell

1½ cups all-purpose unbleached flour
¼–½ teaspoon salt

9 tablespoons unsalted butter, cut into 9
    pieces and chilled
3–4 tablespoons ice water

1. In the workbowl of a food processor, combine the flour, salt, and butter. Pulse quickly on and off until the mixture resembles oatmeal. Add 3 tablespoons of the ice water and pulse quickly until the mixture begins to come together. Do not let it form a ball.

2. Transfer the mixture to a large mixing bowl and gather into a ball. If the mixture is too crumbly, add the remaining 1 tablespoon of ice water and work the dough quickly into a ball.

3. Flatten the ball into a disk, wrap in foil, and refrigerate for at least 30 minutes, or until firm enough to roll.

4. On a lightly floured surface, roll out the dough into a circle about 12 inches in diameter and ⅛ inch thick. Loosely roll the circle around the rolling pin, then unroll it over a 10-inch porcelain quiche pan. Press it into the bottom and sides of the pan, being careful not to stretch the dough. Fold the excess overhang into the pan and press it firmly against the sides to create a double layer of dough which will reinforce the sides of the shell. Prick the bottom of the shell all over with a fork, cover with aluminum foil, and place in the freezer for at least 30 minutes before baking. At this point, the dough can be covered and kept frozen for up to 2 weeks.

5. Preheat the oven to 425° F.

6. To partially bake: Line the frozen shell with parchment paper, fill with large dried beans, and place on a cookie sheet. Place in the center of the preheated oven and bake for 12 minutes, or until the sides are set.

7. Remove the paper and beans and continue to bake for 4 to 6 minutes longer, until the dough is set but not browned. Remove from the oven and place the pan on a wire rack to cool until needed. The tart shell is now partially baked.

8. To prebake: When the shell is returned to the oven after the beans have been removed, bake for 10 to 12 minutes longer, or until golden brown.

9. For a dessert tart: Use only a pinch of salt instead of ½ teaspoon. Also add 2 tablespoons confectioners' sugar to the food processor along with the flour, salt, and butter and proceed with the recipe.

DESSERT CREPES

makes 16–18 crepes

1½ cups sifted all-purpose flour
¾ cup whole milk
¾ cup water
3 large eggs
1 tablespoon granulated sugar

1 teaspoon vanilla extract
2 tablespoons cognac or Cointreau,
   optional
6 tablespoons unsalted butter, melted
   and cooled

1. In the container of a blender or food processor combine the flour, milk, water, eggs, sugar, vanilla, and optional cognac and blend at top speed or process for 1 minute. Transfer to a mixing bowl and whisk in ¼ cup of the melted butter. Let stand at room temperature for 1 hour.

2. Heat a 6-inch crepe pan over medium heat and brush it very lightly with some of the remaining melted butter. When the pan is very hot, remove it from the heat and add 2 tablespoons crepe batter. Quickly turn the pan in all directions to coat its entire surface, then immediately pour any excess batter back into the mixing bowl. Return the pan to the heat and cook the crepe for 1 or 2 minutes, or until it is lightly browned, regulating the heat as necessary. Loosen the crepe with the tip of a sharp knife, lift the edge, and then quickly, with the tips of your fingers, turn the crepe. Brown the crepe lightly on the other side and slide it onto a plate. Continue making crepes, brushing the pan lightly with melted butter after every 3 crepes.

3. If the crepes are to be used within a day or two, they can simply be stacked on a plate, covered, and refrigerated. Bring the crepes back to room temperature before filling. If they are to be frozen, place a sheet of wax paper between each crepe and wrap them in heavy-duty foil. Defrost the crepes at room temperature before using, and place in a 250° F oven until they are lightly heated through.

Vegetables

The proper storage of fresh produce is as important as its purchase. In my travels as a cooking teacher, I have seen such poor handling of greens, vegetables, and fruit that it defeats the purpose of choosing the best fresh stuff in the first place.

All **root vegetables** such as carrots, turnips, kohlrabi, and beets should be bought with their tops—an indication of freshness—but remove the tops before storing the vegetables in plastic bags, or they will wilt within a day or two.

Trimming off the tops is also a must for **leeks** and **scallions.** Remove 1 inch from the scallion tops and one-third of the leek tops, then store in plastic bags, unwashed. They will keep for up to 3 weeks.

Radishes are another vegetable that should be bought only with their crisp tops, and here again these should be removed and the radishes stored in a bag. If you will not be using all the radishes at once, check the bag every 2 or 3 days, rinse off any mold that may have formed because of leftover tops, and store again. Radishes will keep well for 2 weeks.

All **greens,** both hardy winter greens and delicate salad greens, will last longer when placed in perforated plastic bags. Never wash greens until you are ready to use them. Aside from iceberg lettuce, romaine has the longest refrigerator shelf life, and it can even be extended a few days by slicing off 1 inch of the tops of the leaves before storing the romaine in a plastic bag. After a day or two, remove the outer leaves, which may have turned brown, and the lettuce will still be good for another 2 or 3 days.

Spinach is the only green that can be stored in the refrigerator for a longer period of time. Loose spinach should be stemmed and kept unwashed in a plastic bag. Bagged spinach should be opened and checked carefully. Discard any moldy leaves, and place the unwashed greens in a plastic bag. Still better, I steam spinach, and this way it can be kept for another 3 to 4 days in a covered bowl.

The **artichoke** may seem rather sturdy, but in fact it does not take well to long storage. I find that it does best when left unwrapped in the vegetable bin and cooked within 2 or 3 days. Even by that time the outer leaves may have some brown spots on them, but they can be removed and half an inch cut off the tops of each artichoke. Be sure to rub the cut part with lemon juice to prevent the artichoke from darkening. Cooked artichokes will keep for another 2 or 3 days.

There are various opinions as to the right way to handle **asparagus.** Some cooks believe that standing the spear upright in a jar of water is the best way. I personally believe that method is impractical and instead find that asparagus keep well for several days when I cut an inch off the stems, wrap the bases in damp paper towels, then store the asparagus in a plastic bag. Either way, this vegetable should be cooked within a day or two of purchase.

Beans such as fava, lima, and other shell beans have an extremely short shelf life and should be cooked within a day or two. Shelling them prolongs the shelf life for another 1 or 2 days, but just like peas these vegetables will become starchy when stored for a length of time. String beans, on the other hand, can be stored for up to 5 days in a perforated plastic bag.

Other summer classic vegetables such as zucchini, peppers, and cucumbers are highly perishable when bought garden fresh at farmers markets or produce stands during the summer season. But, when waxed, both cucumbers and peppers can be stored for a week or more. Be sure to leave them loose in the vegetable bin.

Although it is not necessary to refrigerate shallots, garlic, and onions, I find that storing them loose in the refrigerator bin keeps them fresh longer.

Potatoes are a definite dilemma since they turn sweet when refrigerated, but unless you can store them in a cool spot, they tend to get soft and will start to "root" in a few days, so it is best to buy potatoes fresh and use them soon.

The fact that both corn and peas turn to starch almost as soon as they are picked makes them among the most perishable vegetables, so it does not really matter how you store them. It is best to buy both these vegetables on the day that you plan to serve them.

Snow and sugar peas, on the other hand, can be stored in a plastic bag for several days with excellent results.

If possible, do not refrigerate tomatoes at all. But if you must, then it is best to place them in an uncovered bowl so that they do not get bruised.

Mushrooms are another delicate, rather capricious vegetable. All-purpose mushrooms are best kept in a brown paper bag, while wild mushrooms can be stored in plastic bags for a day or two.

Broccoli is one of the most good-natured vegetables when it comes to storage. I usually cut off the stems, scrape them, and cut them in half lengthwise, then keep the florets and stems separately in plastic bags. The broccoli will keep well for up to 5 days.

Cauliflower, a member of the broccoli family, seems to be suitable for longer storage, but its taste changes drastically after a few days and becomes quite cabbagy. For best results, remove all the outer green leaves, break the head into florets, and store in a plastic bag for no more than 2 days. Always be sure to trim away any brown spots before cooking.

Basic vegetables such as celery, carrots, and onions are easily stored for 2 to 3 weeks or even longer. Celery and carrots should be put into plastic bags and may need trimming every few days, while onions can be stored in a cool place or loose in the vegetable bin.

Other good-natured vegetables are **fennel** and **celery root**. Both will keep for 2 weeks stored in the fridge. Be sure to remove the fennel tops as soon as you buy it, and you may also have to remove the outer layer of the vegetable before using it. Celery root should be stored in a plastic bag, but be sure that it does not get wet since it can easily become moldy.

When all is said and done the key to tasty vegetables is freshness, and while proper storage does increase the shelf life of many of them, you should still be ready to cook or eat them within 2 to 3 days.

Fruit

The art of buying and storing fruit is very similar to that of vegetables, just as challenging and, unfortunately, even more disappointing. Your eye tends to steer you in the direction of the perfect-looking apple or peach, yet it does not take long to realize that perfect looks do not guarantee perfect taste.

Because ripe fruit spoils and bruises easily, most fruit that you buy at green grocers and supermarkets is picked green, which means that you have to allow 2 to 5 days from the time you buy the fruit to the time you serve it. Unfortunately, fruit never ripens completely once it is off the tree, and is never as sweet as the tree-ripened fruit that you can buy during the season at roadside stands. Only bananas ripen completely off the tree, so they are a fruit you can buy green and expect good results from.

All other fruit should be left to ripen in a bowl and, once ripe, should be refrigerated. Remove all berries from their containers, place in a bowl, and discard any bruised or moldy berries. Do not wash either fruit or berries until you are ready to serve them.

Liquid and Dry Measure Equivalencies

CUSTOMARY	METRIC
¼ teaspoon	1.25 milliliters
½ teaspoon	2.5 milliliters
1 teaspoon	5 milliliters
1 tablespoon	15 milliliters
1 fluid ounce	30 milliliters
¼ cup	60 milliliters
⅓ cup	80 milliliters
½ cup	120 milliliters
1 cup	240 milliliters
1 pint (2 cups)	480 milliliters
1 quart (4 cups, 32 ounces)	960 milliliters (.96 liter)
1 gallon (4 quarts)	3.84 liters
1 ounce (by weight)	28 grams
¼ pound (4 ounces)	114 grams
1 pound (16 ounces)	454 grams
2.2 pounds	1000 grams (1 kilogram)

Oven Temperature Equivalencies

DESCRIPTION	°FAHRENHEIT	°CELSIUS
Cool	200	90
Very slow	250	120
Slow	300–325	150–160
Moderately slow	325–350	160–180
Moderate	350–375	180–190
Moderately hot	375–400	190–200
Hot	400–450	200–230
Very hot	450–500	230–260

fettuccine (*cont.*)
in saffron cream with
sauté of shrimp and
scallops, 86–87
field strawberries in crème
rouge, 231
fig(s):
poached, in port and red
wine syrup, 225
purple, and almond tart,
345–46
purple, roast pork with
port sauce and,
266–67
first courses, *see appetizer
and soup headings*
fish:
brochette of swordfish
with leek and tomato
compote and balsamic
vinegar sauce, 176–
177
fillets of sea bass with
vegetable pilaf and
chive-butter sauce,
76–77
fried sardines with
wilted spinach and
lemon vinaigrette, 51
oven-braised shad with
new potatoes, 74–75
penne with pan-seared
peppered tuna, toma-
toes, and gremolata,
178–79
pickled halibut with
spicy peppers and cit-
rus zest, 50
sauté of swordfish with
pepper and anchovy
marmalade, 141–42
stock, 472
stock, red wine, 473
testing doneness of, 181
see also anchovy(ies);
monkfish; mussels;
red snapper; salmon;
scallop(s); shellfish;
shrimp
flank steaks, grilled, in spicy
chili marinade,
195–96
flans:
cabbage and leek, à la
Vaudoise, 430–31

caramelized pear and
prune, 377
scallop, à l'Arlesienne in
roasted pepper and
shallot butter sauce,
387–88
fondant, bittersweet chocolate, with rum raisins,
372–73
fondue, tomato, 94, 95, 186–
187, 474
Fontina, in multicolored pepper, onion, and two-
cheese pizza, 164–65
fool, drunken blueberry, with
cassis, 157
fricassees:
of Brussels sprouts, pep-
pers, and onions,
428–29
fennel, with carrots, ba-
con, and pearl onions,
434–35
Mediterranean roasted
pepper and eggplant,
327–28
of mushrooms and dill,
293–94
of pearl onions and fresh
garden peas, 90–91
of salmon with baby
vegetables Lameloise,
71–72
of snails à l'esperance,
314–15
two-bean, 193–94
of veal shanks with
glazed carrots in tarra-
gon sauce, 419–20
fried sardines with wilted
spinach and lemon
vinaigrette, 51
frittata, herbed zucchini and
potato, 159–60
fritters:
charcoal-roasted egg-
plant, with yogurt and
mint sauce, 138
corn, polenta, and jala-
peño, 199–200
roasted red pepper, with
caper, anchovy, and
garlic butter, 251–52
salmon and lobster, in
sorrel cream, 182–83

spaghetti squash and
scallion, 370
spinach, with herbed
sour cream and
smoked-salmon sauce,
29–30
fruit:
candied, chocolate-
walnut cake with,
248–49
compotes, 228
storage of, 497
see also specific fruits
fusilli:
salad, Niçoise, 163
with sautéed shrimp,
tomato, and ginger
fondue, 185–86
Umbriana with radicchio
and bacon in creamy
tomato sauce, 273

Gagnaire, Pierre, 290, 317,
324
Galician summer vegetable
soup, 168
garam masala, 485
garlic:
aioli, 477
caper, and anchovy but-
ter, 251
cloves, glazed, butter-
braised spinach with,
104–5
confit, 94, 95, 190
cream, 35
double-poached, 482
kale, and potato soup,
Catalan, 399–
400
roasted, 189, 482
roasted, vinaigrette, 336,
337
storage of, 496
and white bean puree,
376
gazpacho, creamy, with lime-
marinated bay scal-
lops, 172–73
ginger:
blueberry custard tart,
221–22
candied, red onion–
thyme jam, 191, 192
cream, 147

---

mango:
  and lemon bavarians, 226
  sweet spring, soufflés, 41
marinades:
  citrus, 311–12
  lemon-coriander, 288, 289
  lemon-herb, 292
  lime-herb, 136–37
  mustard-herb, 308
  spicy chili, 195
  tangerine-spice, 80
marinated:
  and grilled green tomatoes, 335
  orange julienne in grenadine, 491
  roast pork tenderloins with white bean and garlic puree, 376–77
  white mushroom salad in crème fraîche and chive vinaigrette, 109
Marino's (Parma), 216
marmalades:
  of Belgian endives, 433
  of Belgian endives flamande, 269
  onion, 316
  pepper and anchovy, 141–42
  Sicilian roasted pepper, 338–39
mascarpone:
  brown sugar–glazed bananas with honey and, 340
  and strawberry mousse, sugar-laced crepes with, 112–13
mayonnaise, 477
  aioli, 477
  bright-green Provençal, 34
  chili, 320
  pink-lemon, 217
medallions of Maryland crab and shrimp mousse in curry-lemon sauce, 81–82
Mediterranean roasted pepper and eggplant fricassee, 327–28

melons, tricolor of kiwifruit and, in lime caramel, 152
Meneau, Marc, 314
meringue:
  custards, very lemony, 49
  praline, Kir-poached cherries with, 224
  puffs, caramelized, in strawberry sauce, 120–21
  -topped lemon-butter tart, 455–56
Mexican avocado soup with traditional salsa topping, 174–75
Middle Eastern:
  black bean soup with concassé of tomatoes, 284–85
  spices, leg of lamb and curried pinto beans with, 371
  spicy soup of lamb shanks and lentils, 402–3
  spicy tomato sauce, braised okra in, 200–201
mint:
  and lemon vinaigrette, 338
  pineapple coulis, 118
  sauté of cucumbers with peas and, 99
  and white chocolate sauce, 488
  and yogurt sauce, 138
mixed garden salad with lemon and mint vinaigrette, 338
mocha crepes with coffee sabayon, 452–53
monkfish:
  grilled brochettes of, in tangerine and spice marinade, 80–81
  sautéed medallions of, à la Valenciana, 287–288
  scaloppine with warm cannelini bean and garlic puree, 406–7
morels, 305

boneless chicken au farci with, 413–14
Moroccan stuffed baked tomatoes, 205–6
mousse:
  caramelized apple, 258–59
  chocolate, ice cream, 232–33
  cranberry, apple, and pear, 454–55
  Maryland crab and shrimp, medallions in curry-lemon sauce, 81–82
  strawberry and mascarpone, sugar-laced crepes with, 112–13
mozzarella toasts in tomato vinaigrette, 160–61
multicolored pepper, onion, and two- cheese pizza, 164–65
mushroom(s):
  braised veal shanks with peas and, primavera, 30–31
  chanterelles, angel-hair pasta with tarragon cream and, 260–61
  chanterelles, veal medallions with apples and, 304–5
  concassé of crème fraîche, dill and, 135
  Cremini, 31
  Cremini, fettuccini with smoked trout and, 272–73
  Cremini, risotto with sweet Gorgonzola and, 393–94
  Cremini, sauté of cinnamon, leeks and, 324
  fresh, 305
  fricassee of dill and, 293–94
  marinated white, salad in crème fraîche and chive vinaigrette, 109
  morels, boneless chicken au farci with, 413–14
  rigatoni with ragout of veal and, 361–63
  roasted shiitake, ragout

I N D E X

rice (*cont.*)
  salad with fragrant In-
    dian spices, cool, 218
  and spiced lamb stuff-
    ing, cabbage packages
    with, 380–81
  wild mushroom tim-
    bales, 324–25
  see also risotto
ricotta and orange soufflés,
  fallen, 45
rigatoni with ragout of veal
  and mushrooms,
  361–63
rillettes of salmon with potato rosti, 382–83
risotto:
  with Cremini mush-
    rooms and sweet Gor-
    gonzola, 393–94
  fennel-infused saffron,
    with fall vegetables,
    277
  of leek and tomatoes
    Cipriani, 148–50
  with shrimp and radic-
    chio Treviso, 39
  Veneziana, 395–96
roast:
  chicken in vinegar and
    garlic sauce, 88–89
  Cornish hens in garlic
    cream with pan-
    sautéed escarole,
    35–36
  duck in raisin sauce,
    417
  ducks in acacia honey
    and cider vinegar
    sauce, 262–63
  duck with Cabernet-
    poached pears
    Bordalou, 294–96
  duck with glazed butter-
    nut squash and lin-
    gonberries, 297–98
  leg of lamb with garlic
    confit in tomato fon-
    due, 94–95
  loin of lamb with red
    wine–shallot sauce,
    422–23
  loin of pork, citrus-
    marinated, 311–12
  marinated pork tender-

loins with white bean
  and garlic puree,
  376–77
pork with purple figs
  and port sauce,
  266–67
poussins à la Provençale
  with lemon and
  cracked pepper,
  415–16
squabs with fricassee
  of pearl onions and
  fresh garden peas,
  90–91
veal with glazed fennel
  and tomato jam,
  96–97
roasted:
  garlic, 189, 482
  onions, old-fashioned,
    326
  pear salad with walnut
    and shallot vinaigrette,
    443–44
  pears with brown sugar
    and ginger crisp, 459
  potatoes with Gruyère
    on bed of greens,
    379
  shallots, 483
  see also pepper(s), bell,
    roasted
root vegetables:
  curried lentil soup with,
    405
  farm-style soup of cab-
    bage and, 280–81
  storage of, 495
  *see also specific root veg-
    etables*
Roquefort, 394
  dressing, 337
rosti, potato and leek,
  435–36
rum:
  raisins, 372, 373
  vanilla syrup, 458

sabayon:
  coffee, mocha crepes
    with, 452–53
  espresso, Gritti, 363
  red wine, 346, 347
  white-wine, 139
  see also zabaglione

saffron, 87
  cream, 390
  cream, fettuccine in,
    with sauté of shrimp
    and scallops, 86–87
  risotto with fall vegeta-
    bles, fennel-infused,
    277
salad dressings, *see* dressings; vinaigrettes
salads, fall, 336–39
  broccoli, cauliflower,
    snow pea, and carrot,
    339
  cabbage and carrot slaw
    with Roquefort dress-
    ing, 337
  mixed garden, with
    lemon and mint vinai-
    grette, 338
  oven-roasted fall vegeta-
    bles with roasted gar-
    lic vinaigrette, 336–37
  Sicilian roasted pepper
    marmalade, 338–39
salads, spring, 106–10
  of carrots in ginger,
    mustard, and cilantro
    vinaigrette, 107
  Catalan broccoli, 106
  endive, radish, and wa-
    tercress, in double-
    mustard dressing, 108
  marinated white mush-
    room, in crème fraîche
    and chive vinaigrette,
    109
  pan-grilled shrimp, in
    sesame-scallion vinai-
    grette, 55–56
  peasant, with Provençal
    vinaigrette, 110
  warm shrimp, with crisp
    vegetables Niçoise,
    54
salads, summer, 211–18
  chicken, in sweet mus-
    tard vinaigrette with
    beets and cucumbers,
    158
  cool rice, with fragrant
    Indian spices, 218
  Corsican eggplant, with
    onions, olives, and ca-
    pers, 212–13

white, sabayon, 139
see also red wine
winter salad with julienne of omelet crepes, 442–43

yellow squash:
    creamy sauté of corn and, 202
    pesto, tomato, and summer-squash pizza, 167
    pistou of red beans, tomato, and summer squash, 170–71
    soup with garnish of summer vegetables, 144–45
yogurt:
    basil dressing, 214

and mint sauce, 138
paste, spicy, 193
vinaigrette, 213

zabaglione:
    pineapple, Kirsch-soaked pineapple in, 230
    see also sabayon
ziti:
    with asparagus, peas, and lemon cream, 63–64
    with sautéed eggplant, roasted peppers, and onion marmalade, 316
zucchini, 215
    and corn soup, spicy, 169–70
    farmstand, cherry to-

mato, and tuna salad in basil vinaigrette, 215
and pepper salad Parma, two-tone, 216
pesto, tomato, and summer-squash pizza, 167
pistou of red beans, tomato, and summer squash, 170–71
and potato herbed frittata, 159–60
salad with pink-lemon mayonnaise, 217–18
storage of, 496
thyme-braised, in sour cream, 209